Springer Texts in Statistics

Advisors:
George Casella Stephen Fienberg Ingram Olkin

Springer
New York
Berlin
Heidelberg
Barcelona
Hong Kong
London
Milan
Paris
Singapore
Tokyo

Springer Texts in Statistics

Alfred: Elements of Statistics for the Life and Social Sciences
Berger: An Introduction to Probability and Stochastic Processes
Bilodeau and Brenner: Theory of Multivariate Statistics
Blom: Probability and Statistics: Theory and Applications
Brockwell and Davis: Introduction to Times Series and Forecasting, Second Edition
Chow and Teicher: Probability Theory: Independence, Interchangeability, Martingales, Third Edition
Christensen: Advanced Linear Modeling: Multivariate, Time Series, and Spatial Data—Nonparametric Regression and Response Surface Maximization, Second Edition
Christensen: Log-Linear Models and Logistic Regression, Second Edition
Christensen: Plane Answers to Complex Questions: The Theory of Linear Models, Third Edition
Creighton: A First Course in Probability Models and Statistical Inference
Davis: Statistical Methods for the Analysis of Repeated Measurements
Dean and Voss: Design and Analysis of Experiments
du Toit, Steyn, and Stumpf: Graphical Exploratory Data Analysis
Durrett: Essentials of Stochastic Processes
Edwards: Introduction to Graphical Modelling, Second Edition
Finkelstein and Levin: Statistics for Lawyers
Flury: A First Course in Multivariate Statistics
Jobson: Applied Multivariate Data Analysis, Volume I: Regression and Experimental Design
Jobson: Applied Multivariate Data Analysis, Volume II: Categorical and Multivariate Methods
Kalbfleisch: Probability and Statistical Inference, Volume I: Probability, Second Edition
Kalbfleisch: Probability and Statistical Inference, Volume II: Statistical Inference, Second Edition
Karr: Probability
Keyfitz: Applied Mathematical Demography, Second Edition
Kiefer: Introduction to Statistical Inference
Kokoska and Nevison: Statistical Tables and Formulae
Kulkarni: Modeling, Analysis, Design, and Control of Stochastic Systems
Lehmann: Elements of Large-Sample Theory
Lehmann: Testing Statistical Hypotheses, Second Edition
Lehmann and Casella: Theory of Point Estimation, Second Edition
Lindman: Analysis of Variance in Experimental Design
Lindsey: Applying Generalized Linear Models
Madansky: Prescriptions for Working Statisticians

(continued after index)

Liu
Best of Luck

Neil H. Timm

Applied Multivariate Analysis

With 42 Figures

Springer

Neil H. Timm
Department of Education in Psychology
School of Education
University of Pittsburgh
Pittsburgh, PA 15260
timm@pitt.edu

Library of Congress Cataloging-in-Publication Data
Timm, Neil H.
Applied multivariate analysis / Neil H. Timm.
p. cm. — (Springer texts in statistics)
Includes bibliographical references and index.
ISBN 0-387-95347-7 (alk. paper)
1. Multivariate analysis. I. Title. II. Series.
QA278 .T53 2002
519.5'35–dc21 2001049267

ISBN 0-387-95347-7

3 2 1

www.springer-ny.com

Springer-Verlag New York Berlin Heidelberg
A member of BertelsmannSpringer Science+Business Media GmbH

To my wife
Verena

Preface

Univariate statistical analysis is concerned with techniques for the analysis of a single random variable. This book is about applied multivariate analysis. It was written to provide students and researchers with an introduction to statistical techniques for the analysis of continuous quantitative measurements on several random variables simultaneously. While quantitative measurements may be obtained from any population, the material in this text is primarily concerned with techniques useful for the analysis of continuous observations from multivariate normal populations with linear structure. While several multivariate methods are extensions of univariate procedures, a unique feature of multivariate data analysis techniques is their ability to control experimental error at an exact nominal level and to provide information on the covariance structure of the data. These features tend to enhance statistical inference, making multivariate data analysis superior to univariate analysis.

While in a previous edition of my textbook on multivariate analysis, I tried to precede a multivariate method with a corresponding univariate procedure when applicable, I have not taken this approach here. Instead, it is assumed that the reader has taken basic courses in multiple linear regression, analysis of variance, and experimental design. While students may be familiar with vector spaces and matrices, important results essential to multivariate analysis are reviewed in Chapter 2. I have avoided the use of calculus in this text. Emphasis is on applications to provide students in the behavioral, biological, physical, and social sciences with a broad range of linear multivariate models for statistical estimation and inference, and exploratory data analysis procedures useful for investigating relationships among a set of structured variables. Examples have been selected to outline the process one employs in data analysis for checking model assumptions and model development, and for exploring patterns that may exist in one or more dimensions of a data set.

To successfully apply methods of multivariate analysis, a comprehensive understanding of the theory and how it relates to a flexible statistical package used for the analysis has become critical. When statistical routines were being developed for multivariate data analysis over twenty years ago, developing a text using a single comprehensive statistical package was risky. Now, companies and software packages have stabilized, thus reducing the risk. I have made extensive use of the Statistical Analysis System (SAS) in this text. All examples have been prepared using Version 8 for Windows. Standard SAS procedures have been used whenever possible to illustrate basic multivariate methodologies; however, a few illustrations depend on the Interactive Matrix Language (IML) procedure. All routines and data sets used in the text are contained on the Springer-Verlag Web site, http://www.springer-ny.com/detail.tpl?ISBN=0387953477 and the author's University of Pittsburgh Web site, http://www.pitt.edu/~timm.

Acknowledgments

The preparation of this text has evolved from teaching courses and seminars in applied multivariate statistics at the University of Pittsburgh. I am grateful to the University of Pittsburgh for giving me the opportunity to complete this work. I would like to express my thanks to the many students who have read, criticized, and corrected various versions of early drafts of my notes and lectures on the topics included in this text. I am indebted to them for their critical readings and their thoughtful suggestions. My deepest appreciation and thanks are extended to my former student Dr. Tammy A. Mieczkowski who read the entire manuscript and offered many suggestions for improving the presentation. I also wish to thank the anonymous reviewers who provided detail comments on early drafts of the manuscript which helped to improve the presentation. However, I am responsible for any errors or omissions of the material included in this text. I also want to express special thanks to John Kimmel at Springer-Verlag. Without his encouragement and support, this book would not have been written.

This book was typed using Scientific WorkPlace Version 3.0. I wish to thank Dr. Melissa Harrison, Ph.D., of Far Field Associates who helped with the LaTeX commands used to format the book and with the development of the author and subject indexes. This book has taken several years to develop and during its development it went through several revisions. The preparation of the entire manuscript and every revision was performed with great care and patience by Mrs. Roberta S. Allan, to whom I am most grateful. I am also especially grateful to the SAS Institute for permission to use the Statistical Analysis System (SAS) in this text. Many of the large data sets analyzed in this book were obtained from the Data and Story Library (DASL) sponsored by Cornell University and hosted by the Department of Statistics at Carnegie Mellon University (http://lib.stat.cmu.edu/DASL/). I wish to extend my thanks and appreciation to these institutions for making available these data sets for statistical analysis. I would also like to thank the authors and publishers of copyrighted material for making available the statistical tables and many of the data sets used in this book.

Finally, I extend my love, gratitude, and appreciation to my wife Verena for her patience, love, support, and continued encouragement throughout this project.

Neil H. Timm, Professor
University of Pittsburgh

Contents

List of Tables

List of Figures

1
Introduction

1.1 Overview

In this book we present applied multivariate data analysis methods for making inferences regarding the mean and covariance structure of several variables, for modeling relationships among variables, and for exploring data patterns that may exist in one or more dimensions of the data. The methods presented in the book usually involve analysis of data consisting of n observations on p variables and one or more groups. As with univariate data analysis, we assume that the data are a random sample from the population of interest and we usually assume that the underlying probability distribution of the population is the multivariate normal (MVN) distribution. The purpose of this book is to provide students with a broad overview of methods useful in applied multivariate analysis. The presentation integrates theory and practice covering both formal linear multivariate models and exploratory data analysis techniques.

While there are numerous commercial software packages available for descriptive and inferential analysis of multivariate data such as SPSS™, S-Plus™, Minitab™, and SYSTAT™, among others, we have chosen to make exclusive use of SAS™, Version 8 for Windows.

1.2 Multivariate Models and Methods

Multivariate analysis techniques are useful when observations are obtained for each of a number of subjects on a set of variables of interest, the dependent variables, and one wants to relate these variables to another set of variables, the independent variables. The data collected are usually displayed in a matrix where the rows represent the observations and the columns the variables. The $n \times p$ data matrix $\mathbf{Y}$ usually represents the dependent variables and the $n \times q$ matrix $\mathbf{X}$ the independent variables.

When the multivariate responses are samples from one or more populations, one often first makes an assumption that the sample is from a multivariate probability distribution. In this text, the multivariate probability distribution is most often assumed to be the multivariate normal (MVN) distribution. Simple models usually have one or more means $\boldsymbol{\mu}_i$ and covariance matrices Σ_i.

One goal of model formulation is to estimate the model parameters and to test hypotheses regarding their equality. Assuming the covariance matrices are unstructured and unknown one may develop methods to test hypotheses regarding fixed means. Unlike univariate analysis, if one finds that the means are unequal one does not know whether the differences are in one dimension, two dimensions, or a higher dimension. The process of locating the dimension of maximal separation is called discriminant function analysis. In models to evaluate the equality of mean vectors, the independent variables merely indicate group membership, and are categorical in nature. They are also considered to be fixed and nonrandom. To expand this model to more complex models, one may formulate a linear model allowing the independent variables to be nonrandom and contain either continuous or categorical variables. The general class of multivariate techniques used in this case are called linear multivariate regression (MR) models. Special cases of the MR model include multivariate analysis of variance (MANOVA) models and multivariate analysis of covariance (MANCOVA) models.

In MR models, the same set of independent variables, $\mathbf{X}$, is used to model the set of dependent variables, $\mathbf{Y}$. Models which allow one to fit each dependent variable with a different set of independent variables are called seemingly unrelated regression (SUR) models. Modeling several sets of dependent variables with different sets of independent variables involve multivariate seemingly unrelated regression (MSUR) models. Oftentimes, a model is overspecified in that not all linear combinations of the independent set are needed to "explain" the variation in the dependent set. These models are called linear multivariate reduced rank regression (MRR) models. One may also extend MRR models to seemingly unrelated regression models with reduced rank (RRSUR) models. Another name often associated with the SUR model is the completely general MANOVA (CGMANOVA) model since growth curve models (GMANOVA) and more general growth curve (MGGC) models are special cases of the SUR model. In all these models, the covariance structure of $\mathbf{Y}$ is unconstrained and unstructured.

In formulating MR models, the dependent variables are represented as a linear structure of both fixed parameters and fixed independent variables. Allowing the variables to remain fixed and the parameters to be a function of both random and fixed parameters leads to classes of linear multivariate mixed models (MMM). These models impose a structure on Σ so that both the means and the variance and covariance components of Σ are estimated. Models included in this general class are random coefficient models, multilevel models, variance component models, panel analysis models and models used to analyze covariance structures. Thus, in these models, one is usually interested in estimating both the mean and the covariance structure of a model simultaneously.

A general class of models that define the dependent and independent variables as random, but relate the variables using fixed parameters are the class of linear structure relation (LISREL) models or structural equation models (SEM). In these models, the variables may be both observed and latent. Included in this class of models are path analysis, factor analysis, simultaneous equation models, simplex models, circumplex models, and numerous test theory models. These models are used primarily to estimate the covariance structure in the data. The mean structure is often assumed to be zero.

Other general classes of multivariate models that rely on multivariate normal theory include multivariate time series models, nonlinear multivariate models, and others. When the dependent variables are categorical rather than continuous, one can consider using multinomial logit or probit models or latent class models. When the data matrix contains n subjects (examinees) and p variables (test items), the modeling of test results for a group of examines is called item response modeling.

Sometimes with multivariate data one is interested in trying to uncover the structure or data patterns that may exist. One may wish to uncover dependencies both within a set of variables and uncover dependencies with other variables. One may also utilize graphical methods to represent the data relationships. The most basic displays are scatter plots or a scatter plot matrix involving two or three variables simultaneously. Profile plots, star plots, glyph plots, biplots, sunburst plots, contour plots, Chernoff faces, and Andrews' Fourier plots can also be utilized to display multivariate data.

Because it is very difficult to detect and describe relationships among variables in large dimensional spaces, several multivariate techniques have been designed to reduce the dimensionality of the data. Two commonly used data reduction techniques include principal component analysis and canonical correlation analysis. When one has a set of dissimilarity or similarity measures to describe relationships, multidimensional scaling techniques are frequently utilized. When the data are categorical, the methods of correspondence analysis, multiple correspondence analysis, and joint correspondence analysis are used to geometrically interpret and visualize categorical data.

Another problem frequently encountered in multivariate data analysis is to categorize objects into clusters. Multivariate techniques that are used to classify or cluster objects into categories include cluster analysis, classification and regression trees (CART), classification analysis and neural networks, among others.

1.3 Scope of the Book

In reviewing applied multivariate methodologies, one observes that several procedures are model oriented and have the assumption of an underlying probability distribution. Other methodologies are exploratory and are designed to investigate relationships among the "multivariables" in order to visualize, describe, classify, or reduce the information under analysis. In this text, we have tried to address both aspects of applied multivariate analysis. While Chapter 2 reviews basic vector and matrix algebra critical to the manipulation of multivariate data, Chapter 3 reviews the theory of linear models, and Chapters 4–6 and

10 address standard multivariate model based methods. Chapters 7-9 include several frequently used exploratory multivariate methodologies.

The material contained in this text may be used for either a one-semester course in applied multivariate analysis for nonstatistics majors or as a two-semester course on multivariate analysis with applications for majors in applied statistics or research methodology. The material contained in the book has been used at the University of Pittsburgh with both formats. For the two-semester course, the material contained in Chapters 1–4, selections from Chapters 5 and 6, and Chapters 7–9 are covered. For the one-semester course, Chapters 1–3 are covered; however, the remaining topics covered in the course are selected from the text based on the interests of the students for the given semester. Sequences have included the addition of Chapters 4–6, or the addition of Chapters 7–10, while others have included selected topics from Chapters 4–10. Other designs using the text are also possible. No text on applied multivariate analysis can discuss all of the multivariate methodologies available to researchers and applied statisticians. The field has made tremendous advances in recent years. However, we feel that the topics discussed here will help applied professionals and academic researchers enhance their understanding of several topics useful in applied multivariate data analysis using the Statistical Analysis System (SAS), Version 8 for Windows.

All examples in the text are illustrated using procedures in base SAS, SAS/STAT, and SAS/ETS. In addition, features in SAS/INSIGHT, SAS/IML, and SAS/GRAPH are utilized. All programs and data sets used in the examples may be downloaded from the Springer-Verlag Web site, http://www.springer.com/editorial/authors.html. The programs and data sets are also available at the author's University of Pittsburgh Web site, http://www.pitt.edu/~timm. A list of the SAS programs, with the implied extension .sas, discussed in the text follow.

Chapter 3	Chapter 4	Chapter 5	Chapter 6	Chapter 7
Multinorm	m4_3_1	m5_3_1	m6_4_1	m7_3_1
Norm	MulSubSel	m5_5_1	m6_4_2	m7_3_2
m3_7_1	m4_5_1	m5_5_2	m6_4_3	m7_5_1
m3_7_2	m4_5_1a	m5_7_1	m6_6_1	
Box-Cox	m4_5_2	m5_7_2	m6_6_2	
Ramus	m4_7_1	m5_9_1	m6_8_1	
Unorm	m4_9_1	m5_9_2	m6_8_2	
m3_8_1	m4_9_2	m5_13_1		
m3_8_7	m4_11_1	m5_14_1		
m3_9a	m4_13_1a			
m3_9d	m4_13_1b			
m3_9e	m4_15_1			
m3_9f	Power			
m3_10a	m4_17_1			
m3_10b				
m3_11_1				

Chapter 8	Chapter 9	Chapter 10	Other
m8_2_1	m9_4_1	m10_4_1	Xmacro
m8_2_2	m9_4_2	m10_4_2	Distnew
m8_3_1	m9_4_3	m10_6_1	
m8_3_2	m9_4_3a	m10_6_2	
m8_3_3	m9_4_4	m10_8_1	
m8_6_1	m9_4_4		
m8_8_1	m9_6_1		
m8_8_2	m9_6_2		
m8_10_1	m9_6_3		
m8_10_2			
m8_10_3			

Also included on the Web site is the Fortran program Fit.For and the associated manual: Fit-Manual.ps, a postscript file. All data sets used in the examples and some of the exercises are also included on the Web site; they are denoted with the extension .dat. Other data sets used in some of the exercises are available from the Data and Story Library (DASL) Web site, http://lib.stat.cmu.dat/DASL/. The library is hosted by the Department of Statistics at Carnegie Mellon University, Pittsburgh, Pennsylvania.

2

Vectors and Matrices

2.1 Introduction

In this chapter, we review the fundamental operations of vectors and matrices useful in statistics. The purpose of the chapter is to introduce basic concepts and formulas essential to the understanding of data representation, data manipulation, model building, and model evaluation in applied multivariate analysis. The field of mathematics that deals with vectors and matrices is called linear algebra; numerous texts have been written about the applications of linear algebra and calculus in statistics. In particular, books by Carroll and Green (1997), Dhrymes (2000), Graybill (1983), Harville (1997), Khuri (1993), Magnus and Neudecker (1999), Schott (1997), and Searle (1982) show how vectors and matrices are useful in applied statistics. Because the results in this chapter are to provide the reader with the basic knowledge of vector spaces and matrix algebra, results are presented without proof.

2.2 Vectors, Vector Spaces, and Vector Subspaces

a. Vectors

Fundamental to multivariate analysis is the collection of observations for d variables. The d values of the observations are organized into a meaningful arrangement of d real[1] numbers, called a vector (also called, a d-variate response or a multivariate vector valued observa-

[1] All vectors in this text are assumed to be real valued.

tion). Letting y_i denote the ith observation where i goes from 1 to d, the $d \times 1$ vector $\mathbf{y}$ is represented as

$$\mathbf{y} = \begin{bmatrix} y_1 \\ y_2 \\ \vdots \\ y_d \end{bmatrix} \tag{2.2.1}$$

This representation of $\mathbf{y}$ is called a column vector of order d, with d rows and 1 column. Alternatively, a vector may be represented as a $1 \times d$ vector with 1 row and d columns. Then, we denote $\mathbf{y}$ as $\mathbf{y}'$ and call it a row vector. Hence,

$$\mathbf{y}' = [y_1,\ y_2,\ \ldots,\ y_d] \tag{2.2.2}$$

Using this notation, $\mathbf{y}$ is a column vector and $\mathbf{y}'$, the transpose of $\mathbf{y}$, is a row vector. The dimension or order of the vector $\mathbf{y}$ is d where the index d represents the number of variables, elements or components in $\mathbf{y}$. To emphasize the dimension of $\mathbf{y}$, the subscript notation $\mathbf{y}_{d\times 1}$ or simply $\mathbf{y}_d$ is used.

The vector $\mathbf{y}$ with d elements represents, geometrically, a point in a d-dimensional Euclidean space. The elements of $\mathbf{y}$ are called the coordinates of the vector. The null vector $\mathbf{0}_{d\times 1}$ denotes the origin of the space; the vector $\mathbf{y}$ may be visualized as a line segment from the origin to the point $\mathbf{y}$. The line segment is called a position vector. A vector $\mathbf{y}$ with n variables, $\mathbf{y}_n$, is a position vector in an n-dimensional Euclidean space. Since the vector $\mathbf{y}$ is defined over the set of real numbers R, the n-dimensional Euclidean space is represented as R^n or in this text as V_n.

Definition 2.2.1 *A vector $\mathbf{y}_{n\times 1}$ is an ordered set of n real numbers representing a position in an n-dimensional Euclidean space V_n.*

b. Vector Spaces

The collection of $n \times 1$ vectors in V_n that are closed under the two operations of vector addition and scalar multiplication is called a (real) vector space.

Definition 2.2.2 *An n-dimensional vector space is the collection of vectors in V_n that satisfy the following two conditions*

1. *If $\mathbf{x} \in V_n$ and $\mathbf{y} \in V_n$, then $\mathbf{z} = \mathbf{x} + \mathbf{y} \in V_n$*

2. *If $\alpha \in R$ and $\mathbf{y} \in V_n$, then $\mathbf{z} = \alpha\mathbf{y} \in V_n$*

(The notation $\in$ is set notation for "is an element of.")

For vector addition to be defined, $\mathbf{x}$ and $\mathbf{y}$ must have the same number of elements n. Then, all elements z_i in $\mathbf{z} = \mathbf{x} + \mathbf{y}$ are defined as $z_i = x_i + y_i$ for $i = 1, 2, \ldots, n$. Similarly, scalar multiplication of a vector $\mathbf{y}$ by a scaler $\alpha \in R$ is defined as $z_i = \alpha y_i$.

c. *Vector Subspaces*

Definition 2.2.3 *A subset, S, of V_n is called a subspace of V_n if S is itself a vector space. The vector subspace S of V_n is represented as $S \subseteq V_n$.*

Choosing $\alpha = 0$ in Definition 2.2.2, we see that $\mathbf{0} \in V_n$ so that every vector space contains the origin $\mathbf{0}$. Indeed, $S = \{\mathbf{0}\}$ is a subspace of V_n called the null subspace. Now, if α and β are elements of R and $\mathbf{x}$ and $\mathbf{y}$ are elements of V_n, then all linear combinations $\alpha\mathbf{x} + \beta\mathbf{y}$, are in V_n. This subset of vectors is called V_k, where $V_k \subseteq V_n$. The subspace V_k is called a subspace, linear manifold or linear subspace of V_n. Any subspace V_k, where $0 < k < n$, is called a proper subspace. The subset of vectors containing only the zero vector and the subset containing the whole space are extreme examples of vector spaces called improper subspaces.

Example 2.2.1 *Let*

$$\mathbf{x} = \begin{bmatrix} 1 \\ 0 \\ 0 \end{bmatrix} \quad \text{and} \quad \mathbf{y} = \begin{bmatrix} 0 \\ 1 \\ 0 \end{bmatrix}$$

The set of all vectors S of the form $\mathbf{z} = \alpha\mathbf{x}+\beta\mathbf{y}$ represents a plane (two-dimensional space) in the three-dimensional space V_3. Any vector in this two-dimensional subspace, $S = V_2$, can be represented as a linear combination of the vectors $\mathbf{x}$ and $\mathbf{y}$. The subspace V_2 is called a proper subspace of V_3 so that $V_2 \subseteq V_3$.

Extending the operations of addition and scalar multiplication to k vectors, a linear combination of vectors $\mathbf{y}_i$ is defined as

$$\mathbf{v} = \sum_{i=1}^{k} \alpha_i \mathbf{y}_i \in V \tag{2.2.3}$$

where $\mathbf{y}_i \in V$ and $\alpha_i \in R$. The set of vectors $\mathbf{y}_1, \mathbf{y}_2, \ldots, \mathbf{y}_k$ are said to span (or generate) V, if

$$V = \{\mathbf{v} \mid \mathbf{v} = \sum_{i=1}^{k} \alpha_i \mathbf{y}_i\} \tag{2.2.4}$$

The vectors in V satisfy Definition 2.2.2 so that V is a vector space.

Theorem 2.2.1 *Let $\{\mathbf{y}_1, \mathbf{y}_2, \ldots, \mathbf{y}_k\}$ be the subset of k, $n \times 1$ vectors in V_n. If every vector in V is a linear combination of $\mathbf{y}_1, \mathbf{y}_2, \ldots, \mathbf{y}_k$ then V is a vector subspace of V_n.*

Definition 2.2.4 *The set of $n \times 1$ vectors $\{\mathbf{y}_1, \mathbf{y}_2, \ldots, \mathbf{y}_k\}$ are linearly dependent if there exists real numbers $\alpha_1, \alpha_2, \ldots, \alpha_k$ not all zero such that*

$$\sum_{i=1}^{k} \alpha_i \mathbf{y}_i = \mathbf{0}$$

Otherwise, the set of vectors are linearly independent.

For a linearly independent set, the only solution to the equation in Definition 2.2.4 is given by $\alpha_1 = \alpha_2 = \cdots = \alpha_k = 0$. To determine whether a set of vectors are linearly independent or linearly dependent, Definition 2.2.4 is employed as shown in the following examples.

Example 2.2.2 *Let*

$$\mathbf{y}_1 = \begin{bmatrix} 1 \\ 1 \\ 1 \end{bmatrix}, \quad \mathbf{y}_2 = \begin{bmatrix} 0 \\ 1 \\ -1 \end{bmatrix}, \quad \text{and} \quad \mathbf{y}_3 = \begin{bmatrix} 1 \\ 4 \\ -2 \end{bmatrix}$$

To determine whether the vectors $\mathbf{y}_1, \mathbf{y}_2$, and $\mathbf{y}_3$ are linearly dependent or linearly independent, the equation

$$\alpha_1\mathbf{y}_1 + \alpha_2\mathbf{y}_2 + \alpha_3\mathbf{y}_3 = \mathbf{0}$$

is solved for α_1, α_2, and α_3. From Definition 2.2.4,

$$\alpha_1 \begin{bmatrix} 1 \\ 1 \\ 1 \end{bmatrix} + \alpha_2 \begin{bmatrix} 0 \\ 1 \\ -1 \end{bmatrix} + \alpha_3 \begin{bmatrix} 1 \\ 4 \\ -2 \end{bmatrix} = \begin{bmatrix} 0 \\ 0 \\ 0 \end{bmatrix}$$

$$\begin{bmatrix} \alpha_1 \\ \alpha_1 \\ \alpha_1 \end{bmatrix} + \begin{bmatrix} 0 \\ \alpha_2 \\ -\alpha_2 \end{bmatrix} + \begin{bmatrix} \alpha_3 \\ 4\alpha_3 \\ -2\alpha_3 \end{bmatrix} = \begin{bmatrix} 0 \\ 0 \\ 0 \end{bmatrix}$$

This is a system of three equations in three unknowns

$$\begin{aligned}
(1)\quad & \alpha_1 + \alpha_3 = 0 \\
(2)\quad & \alpha_1 + \alpha_2 + 4\alpha_3 = 0 \\
(3)\quad & \alpha_1 - \alpha_2 - 2\alpha_3 = 0
\end{aligned}$$

From equation (1), $\alpha_1 = -\alpha_3$. Substituting α_1 into equation (2), $\alpha_2 = -3\alpha_3$. If α_1 and α_2 are defined in terms of α_3, equation (3) is satisfied. If $\alpha_3 \neq 0$, there exist real numbers α_1, α_2, and α_3, not all zero such that

$$\sum_{i=1}^{3} \alpha_i = 0$$

Thus, $\mathbf{y}_1$, $\mathbf{y}_2$, and $\mathbf{y}_3$ are linearly dependent. For example, $\mathbf{y}_1 + 3\mathbf{y}_2 - \mathbf{y}_3 = \mathbf{0}$.

Example 2.2.3 *As an example of a set of linearly independent vectors, let*

$$\mathbf{y}_1 = \begin{bmatrix} 0 \\ 1 \\ 1 \end{bmatrix}, \quad \mathbf{y}_2 = \begin{bmatrix} 1 \\ 1 \\ -2 \end{bmatrix}, \quad \textit{and} \quad \mathbf{y}_3 = \begin{bmatrix} 3 \\ 4 \\ 1 \end{bmatrix}$$

Using Definition 2.2.4,

$$\alpha_1 \begin{bmatrix} 0 \\ 1 \\ 1 \end{bmatrix} + \alpha_2 \begin{bmatrix} 1 \\ 1 \\ -2 \end{bmatrix} + \alpha_3 \begin{bmatrix} 3 \\ 4 \\ 1 \end{bmatrix} = \begin{bmatrix} 0 \\ 0 \\ 0 \end{bmatrix}$$

is a system of simultaneous equations

$$(1) \qquad \alpha_2 + 3\alpha_3 = 0$$

$$(2) \qquad \alpha_1 + \alpha_2 + 4\alpha_3 = 0$$

$$(3) \qquad \alpha_1 - 2\alpha_2 + \alpha_3 = 0$$

From equation (1), $\alpha_2 = -3\alpha_3$. Substituting $-3\alpha_3$ for α_2 into equation (2), $\alpha_1 = -\alpha_3$; by substituting for α_1 and α_2 into equation (3), $\alpha_3 = 0$. Thus, the only solution is $\alpha_1 = \alpha_2 = \alpha_3 = 0$, or $\{\mathbf{y}_1, \mathbf{y}_2, \mathbf{y}_3\}$ is a linearly independent set of vectors.

Linearly independent and linearly dependent vectors are fundamental to the study of applied multivariate analysis. For example, suppose a test is administered to n students where scores on k subtests are recorded. If the vectors $\mathbf{y}_1, \mathbf{y}_2, \ldots, \mathbf{y}_k$ are linearly independent, each of the k subtests are important to the overall evaluation of the n students. If for some subtest the scores can be expressed as a linear combination of the other subtests

$$\mathbf{y}_k = \sum_{i=1}^{k-1} \alpha_i \mathbf{y}_i$$

the vectors are linearly dependent and there is redundancy in the test scores. It is often important to determine whether or not a set of observation vectors is linearly independent; when the vectors are not linearly independent, the analysis of the data may need to be restricted to a subspace of the original space.

Exercises 2.2

1. For the vectors

$$\mathbf{y}_1 = \begin{bmatrix} 1 \\ 1 \\ 1 \end{bmatrix} \quad \text{and} \quad \mathbf{y}_2 = \begin{bmatrix} 2 \\ 0 \\ -1 \end{bmatrix}$$

find the vectors

(a) $2\mathbf{y}_1 + 3\mathbf{y}_2$

(b) $\alpha\mathbf{y}_1 + \beta\mathbf{y}_2$

(c) $\mathbf{y}_3$ such that $3\mathbf{y}_1 - 2\mathbf{y}_2 + 4\mathbf{y}_3 = \mathbf{0}$

2. For the vectors and scalars defined in Example 2.2.1, draw a picture of the space S generated by the two vectors.

3. Show that the four vectors given below are linearly dependent.

$$\mathbf{y}_1 = \begin{bmatrix} 1 \\ 0 \\ 0 \end{bmatrix}, \quad \mathbf{y}_2 = \begin{bmatrix} 2 \\ 3 \\ 5 \end{bmatrix}, \quad \mathbf{y}_3 = \begin{bmatrix} 1 \\ 0 \\ 1 \end{bmatrix}, \quad \text{and} \quad \mathbf{y}_4 = \begin{bmatrix} 0 \\ 4 \\ 6 \end{bmatrix}$$

4. Are the following vectors linearly dependent or linearly independent?

$$\mathbf{y}_1 = \begin{bmatrix} 1 \\ 1 \\ 1 \end{bmatrix}, \quad \mathbf{y}_2 = \begin{bmatrix} 1 \\ 2 \\ 3 \end{bmatrix}, \quad \mathbf{y}_3 = \begin{bmatrix} 2 \\ 2 \\ 3 \end{bmatrix}$$

5. Do the vectors

$$\mathbf{y}_1 = \begin{bmatrix} 2 \\ 4 \\ 2 \end{bmatrix}, \quad \mathbf{y}_2 = \begin{bmatrix} 1 \\ 2 \\ 3 \end{bmatrix}, \quad \text{and} \quad \mathbf{y}_3 = \begin{bmatrix} 6 \\ 12 \\ 10 \end{bmatrix}$$

span the same space as the vectors

$$\mathbf{x}_1 = \begin{bmatrix} 0 \\ 0 \\ 2 \end{bmatrix} \quad \text{and} \quad \mathbf{x}_2 = \begin{bmatrix} 2 \\ 4 \\ 10 \end{bmatrix}$$

6. Prove the following laws for vector addition and scalar multiplication.

   (a) $\mathbf{x} + \mathbf{y} = \mathbf{y} + \mathbf{x}$ (commutative law)
   (b) $(\mathbf{x} + \mathbf{y}) + \mathbf{z} = \mathbf{x} + (\mathbf{y} + \mathbf{z})$ (associative law)
   (c) $\alpha(\beta\mathbf{y}) = (\alpha\beta)\mathbf{y} = (\beta\alpha)\mathbf{y} = \alpha(\beta\mathbf{y})$ (associative law for scalars)
   (d) $\alpha\,(\mathbf{x} + \mathbf{y}) = \alpha\mathbf{x} + \alpha\mathbf{y}$ (distributive law for vectors)
   (e) $(\alpha + \beta)\mathbf{y} = \alpha\mathbf{y} + \beta\mathbf{y}$ (distributive law for scalars)

7. Prove each of the following statements.

   (a) Any set of vectors containing the zero vector is linearly dependent.
   (b) Any subset of a linearly independent set is also linearly independent.
   (c) In a linearly dependent set of vectors, at least one of the vectors is a linear combination of the remaining vectors.

2.3 Bases, Vector Norms, and the Algebra of Vector Spaces

The concept of dimensionality is a familiar one from geometry. In Example 2.2.1, the subspace S represented a plane of dimension two, a subspace of the three-dimensional space V_3. Also important is the minimal number of vectors required to span S.

a. Bases

Definition 2.3.1 *Let* $\{\mathbf{y}_1, \mathbf{y}_2, \ldots, \mathbf{y}_k\}$ *be a subset of k vectors where* $\mathbf{y}_i \in V_n$. *The set of k vectors is called a basis of* V_k *if the vectors in the set span* V_k *and are linearly independent. The number k is called the dimension or rank of the vector space.*

Thus, in Example 2.2.1 $S \equiv V_2 \subseteq V_3$ and the subscript 2 is the dimension or rank of the vector space. It should be clear from the context whether the subscript on V represents the dimension of the vector space or the dimension of the vector in the vector space. Every vector space, except the vector space $\{\mathbf{0}\}$, has a basis. Although a basis set is not unique, the number of vectors in a basis is unique. The following theorem summarizes the existence and uniqueness of a basis for a vector space.

Theorem 2.3.1 *Existence and Uniqueness*

1. *Every vector space has a basis.*
2. *Every vector in a vector space has a unique representation as a linear combination of a basis.*
3. *Any two bases for a vector space have the same number of vectors.*

b. Lengths, Distances, and Angles

Knowledge of vector lengths, distances and angles between vectors helps one to understand relationships among multivariate vector observations. However, prior to discussing these concepts, the inner (scalar or dot) product of two vectors needs to be defined.

Definition 2.3.2 *The inner product of two vectors* $\mathbf{x}$ *and* $\mathbf{y}$, *each with n elements, is the scalar quantity*

$$\mathbf{x}'\mathbf{y} = \sum_{i=1}^{n} x_i y_i$$

In textbooks on linear algebra, the inner product may be represented as $(\mathbf{x}, \mathbf{y})$ or $\mathbf{x} \cdot \mathbf{y}$. Given Definition 2.3.2, inner products have several properties as summarized in the following theorem.

Theorem 2.3.2 *For any conformable vectors* $\mathbf{x}$, $\mathbf{y}$, $\mathbf{z}$, *and* $\mathbf{w}$ *in a vector space V and any real numbers* α *and* β, *the inner product satisfies the following relationships*

1. $\mathbf{x}'\mathbf{y} = \mathbf{y}'\mathbf{x}$
2. $\mathbf{x}'\mathbf{x} \geq 0$ *with equality if and only if* $\mathbf{x} = \mathbf{0}$
3. $(\alpha\mathbf{x})'(\beta\mathbf{y}) = \alpha\beta(\mathbf{x}'\mathbf{y})$
4. $(\mathbf{x} + \mathbf{y})'\mathbf{z} = \mathbf{x}'\mathbf{z} + \mathbf{y}'\mathbf{z}$
5. $(\mathbf{x} + \mathbf{y})'(\mathbf{w} + \mathbf{z}) = \mathbf{x}'(\mathbf{w} + \mathbf{z}) + \mathbf{y}'(\mathbf{w} + \mathbf{z})$

If $\mathbf{x} = \mathbf{y}$ in Definition 2.3.2, then $\mathbf{x}'\mathbf{x} = \sum_{i=1}^{n} x_i^2$. The quantity $(\mathbf{x}'\mathbf{x})^{1/2}$ is called the Euclidean vector norm or length of $\mathbf{x}$ and is represented as $\|\mathbf{x}\|$. Thus, the norm of $\mathbf{x}$ is the positive square root of the inner product of a vector with itself. The norm squared of $\mathbf{x}$ is represented as $\|\mathbf{x}\|^2$. The Euclidean distance or length between two vectors $\mathbf{x}$ and $\mathbf{y}$ in V_n is $\|\mathbf{x} - \mathbf{y}\| = [(\mathbf{x} - \mathbf{y})'(\mathbf{x} - \mathbf{y})]^{1/2}$. The cosine of the angle between two vectors by the law of cosines is

$$\cos\theta = \mathbf{x}'\mathbf{y}/\ \|\mathbf{x}\|\ \|\mathbf{y}\| \quad 0^\circ \leq \theta \leq 180^\circ \tag{2.3.1}$$

Another important geometric vector concept is the notion of orthogonal (perpendicular) vectors.

Definition 2.3.3 *Two vectors* $\mathbf{x}$ *and* $\mathbf{y}$ *in* V_n *are orthogonal if their inner product is zero.*

Thus, if the angle between $\mathbf{x}$ and $\mathbf{y}$ is 90°, then $\cos\theta = 0$ and $\mathbf{x}$ is perpendicular to $\mathbf{y}$, written as $\mathbf{x} \perp \mathbf{y}$.

Example 2.3.1 *Let*

$$\mathbf{x} = \begin{bmatrix} -1 \\ 1 \\ 2 \end{bmatrix} \quad \text{and} \quad \mathbf{y} = \begin{bmatrix} 1 \\ 0 \\ -1 \end{bmatrix}$$

The distance between $\mathbf{x}$ *and* $\mathbf{y}$ *is then* $\|\mathbf{x} - \mathbf{y}\| = [(\mathbf{x} - \mathbf{y})'(\mathbf{x} - \mathbf{y})]^{1/2} = \sqrt{14}$ *and the cosine of the angle between* $\mathbf{x}$ *and* $\mathbf{y}$ *is*

$$\cos\theta = \mathbf{x}'\mathbf{y}/\ \|\mathbf{x}\|\ \|\mathbf{y}\| = -3/\sqrt{6}\sqrt{2} = -\sqrt{3}/2$$

so that the angle between $\mathbf{x}$ *and* $\mathbf{y}$ *is* $\theta = \cos^{-1}(-\sqrt{3}/2) = 150°$.

If the vectors in our example have unit length, so that $\|\mathbf{x}\| = \|\mathbf{y}\| = 1$, then the $\cos\theta$ is just the inner product of $\mathbf{x}$ and $\mathbf{y}$. To create unit vectors, also called normalizing the vectors, one proceeds as follows

$$\mathbf{u}_x = \mathbf{x}/\ \|\mathbf{x}\| = \begin{bmatrix} -1/\sqrt{6} \\ 1/\sqrt{6} \\ 2/\sqrt{6} \end{bmatrix} \quad \text{and} \quad \mathbf{u}_y = \mathbf{y}/\ \|\mathbf{y}\| = \begin{bmatrix} 1/\sqrt{2} \\ 0/\sqrt{2} \\ -1/\sqrt{2} \end{bmatrix}$$

and the $\cos\theta = \mathbf{u}_x'\mathbf{u}_y = -\sqrt{3}/2$, the inner product of the normalized vectors. The normalized orthogonal vectors $\mathbf{u}_x$ and $\mathbf{u}_y$ are called orthonormal vectors.

Example 2.3.2 *Let*

$$\mathbf{x} = \begin{bmatrix} -1 \\ 2 \\ -4 \end{bmatrix} \quad \text{and} \quad \mathbf{y} = \begin{bmatrix} -4 \\ 0 \\ 1 \end{bmatrix}$$

Then $\mathbf{x}'\mathbf{y} = 0$; *however, these vectors are not of unit length.*

Definition 2.3.4 *A basis for a vector space is called an orthogonal basis if every pair of vectors in the set is pairwise orthogonal; it is called an orthonormal basis if each vector additionally has unit length.*

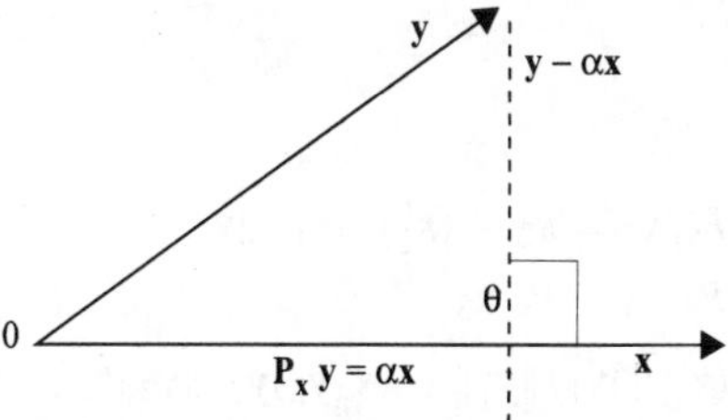

FIGURE 2.3.1. Orthogonal Projection of $\mathbf{y}$ on $\mathbf{x}$, $\mathbf{P_x y} = \alpha\mathbf{x}$

The standard orthonormal basis for V_n is $\{\mathbf{e}_1, \mathbf{e}_2, \ldots, \mathbf{e}_n\}$ where $\mathbf{e}_i$ is a vector of all zeros with the number one in the ith position. Clearly the $\|\mathbf{e}_i\| = 1$ and $\mathbf{e}_i \perp \mathbf{e}_j$; for all pairs i and j. Hence, $\{\mathbf{e}_1, \mathbf{e}_2, \ldots, \mathbf{e}_n\}$ is an orthonormal basis for V_n and it has dimension (or rank) n. The basis for V_n is not unique. Given any basis for $V_k \subseteq V_n$ we can create an orthonormal basis for V_k. The process is called the Gram-Schmidt orthogonalization process.

c. *Gram-Schmidt Orthogonalization Process*

Fundamental to the Gram-Schmidt process is the concept of an orthogonal projection. In a two-dimensional space, consider the vectors $\mathbf{x}$ and $\mathbf{y}$ given in Figure 2.3.1. The orthogonal projection of $\mathbf{y}$ on $\mathbf{x}$, $P_\mathbf{x}\mathbf{y}$, is some constant multiple, $\alpha\mathbf{x}$ of $\mathbf{x}$, such that $P_\mathbf{x}\mathbf{y} \perp (\mathbf{y} - P_\mathbf{x}\mathbf{y})$.

Since the cos θ =cos 90° $= 0$, we set $(\mathbf{y} - \alpha\mathbf{x})'\alpha\mathbf{x}$ equal to 0 and we solve for α to find $\alpha = (\mathbf{y}'\mathbf{x}) / \|\mathbf{x}\|^2$. Thus, the projection of $\mathbf{y}$ on $\mathbf{x}$ becomes

$$P_\mathbf{x}\mathbf{y} = \alpha\mathbf{x} = (\mathbf{y}'\mathbf{x})\mathbf{x} / \|\mathbf{x}\|^2$$

Example 2.3.3 *Let*

$$\mathbf{x} = \begin{bmatrix} 1 \\ 1 \\ 1 \end{bmatrix} \quad \text{and} \quad \mathbf{y} = \begin{bmatrix} 1 \\ 4 \\ 2 \end{bmatrix}$$

Then, the

$$P_\mathbf{x}\mathbf{y} = \frac{(\mathbf{y}'\mathbf{x})\mathbf{x}}{\|\mathbf{x}\|^2} = \frac{7}{3}\begin{bmatrix} 1 \\ 1 \\ 1 \end{bmatrix}$$

Observe that the coefficient α in this example is no more than the average of the elements of $\mathbf{y}$. This is always the case when projection an observation onto a vector of 1s (the equiangular or unit vector), represented as $\mathbf{1}_n$ or simply $\mathbf{1}$. $P_\mathbf{1}\mathbf{y} = \overline{y}\mathbf{1}$ for any multivariate observation vector $\mathbf{y}$.

To obtain an orthogonal basis $\{\mathbf{y}_1, \ldots, \mathbf{y}_r\}$ for any subspace V of V_n, spanned by any set of vectors $\{\mathbf{x}_1, \mathbf{x}_2, \ldots, \mathbf{x}_k\}$, the preceding projection process is employed sequentially as follows

$$\begin{aligned}
\mathbf{y}_1 &= \mathbf{x}_1 \\
\mathbf{y}_2 &= \mathbf{x}_2 - P_{\mathbf{y}_1}\mathbf{x}_2 = \mathbf{x}_2 - (\mathbf{x}_2'\mathbf{y}_1)\mathbf{y}_1/\,\|\mathbf{y}_1\|^2 \qquad && \mathbf{y}_2 \perp \mathbf{y}_1 \\
\mathbf{y}_3 &= \mathbf{x}_3 - P_{\mathbf{y}_1}\mathbf{x}_3 - P_{\mathbf{y}_2}\mathbf{x}_3 \\
&= \mathbf{x}_3 - (\mathbf{x}_3'\mathbf{y}_1)\mathbf{y}_1/\|\mathbf{y}_1^2\| - (\mathbf{x}_3'\mathbf{y}_2)\mathbf{y}_2/\|\mathbf{y}_2\|^2 \qquad && \mathbf{y}_3 \perp \mathbf{y}_2 \perp \mathbf{y}_1
\end{aligned}$$

or, more generally

$$\mathbf{y}_i = \mathbf{x}_i - \sum_{j=1}^{i-1} c_{ij}\mathbf{y}_j \quad \text{where} \quad c_{ij} = (\mathbf{x}_i'\mathbf{y}_j)/\,\|\mathbf{y}_j\|^2$$

deleting those vectors $\mathbf{y}_i$ for which $\mathbf{y}_i = \mathbf{0}$. The number of nonzero vectors in the set is the rank or dimension of the subspace V and is represented as V_r, $r \leq k$. To find an orthonormal basis, the orthogonal basis must be normalized.

Theorem 2.3.3 *(Gram-Schmidt) Every r-dimensional vector space, except the zero-dimensional space, has an orthonormal basis.*

Example 2.3.4 *Let V be spanned by*

$$\mathbf{x}_1 = \begin{bmatrix} 1 \\ -1 \\ 1 \\ 0 \\ 1 \end{bmatrix}, \quad \mathbf{x}_2 = \begin{bmatrix} 2 \\ 0 \\ 4 \\ 1 \\ 2 \end{bmatrix}, \quad \mathbf{x}_3 = \begin{bmatrix} 1 \\ 1 \\ 3 \\ 1 \\ 1 \end{bmatrix}, \quad \text{and} \quad \mathbf{x}_4 = \begin{bmatrix} 6 \\ 2 \\ 3 \\ -1 \\ 1 \end{bmatrix}$$

To find an orthonormal basis, the Gram-Schmidt process is used. Set

$$\mathbf{y}_1 = \mathbf{x}_1 = \begin{bmatrix} 1 \\ -1 \\ 1 \\ 0 \\ 1 \end{bmatrix}$$

$$\mathbf{y}_2 = \mathbf{x}_2 - (\mathbf{x}_2'\mathbf{y}_1)\mathbf{y}_1/\,\|\mathbf{y}_1\|^2$$

$$= \begin{bmatrix} 2 \\ 0 \\ 4 \\ 1 \\ 2 \end{bmatrix} - \frac{8}{4}\begin{bmatrix} 1 \\ -1 \\ 1 \\ 0 \\ 1 \end{bmatrix} = \begin{bmatrix} 0 \\ 2 \\ 2 \\ 1 \\ 0 \end{bmatrix}$$

$$\mathbf{y}_3 = \mathbf{x}_3 - (\mathbf{x}_3'\mathbf{y}_1)\mathbf{y}_1/\,\|\,\mathbf{y}_1\,\|^2 - (\mathbf{x}_3'\mathbf{y}_2)\mathbf{y}_2/\,\|\,\mathbf{y}_2\,\|^2 = \mathbf{0}$$

so delete $\mathbf{y}_3$;

$$\mathbf{y}_4 = \begin{bmatrix} 6 \\ 2 \\ 3 \\ -1 \\ 1 \end{bmatrix} - (\mathbf{x}_4'\mathbf{y}_1)\mathbf{y}_1/\ \|\mathbf{y}_1\|^2 - (\mathbf{x}_4'\mathbf{y}_2)\mathbf{y}_2/\ \|\mathbf{y}_2\|^2$$

$$= \begin{bmatrix} 6 \\ 2 \\ 3 \\ -1 \\ 1 \end{bmatrix} - \frac{8}{4}\begin{bmatrix} 1 \\ -1 \\ 1 \\ 0 \\ 1 \end{bmatrix} - \frac{9}{9}\begin{bmatrix} 0 \\ 2 \\ 2 \\ 1 \\ 0 \end{bmatrix} = \begin{bmatrix} 4 \\ 2 \\ -1 \\ -2 \\ -1 \end{bmatrix}$$

Thus, an orthogonal basis for V *is* $\{\mathbf{y}_1, \mathbf{y}_2, \mathbf{y}_4\}$. *The vectors must be normalized to obtain an orthonormal basis; an orthonormal basis is* $\mathbf{u}_1 = \mathbf{y}_1/\sqrt{4}, \mathbf{u}_2 = \mathbf{y}_2/3$, *and* $\mathbf{u}_3 = \mathbf{y}_4/\sqrt{26}$.

d. Orthogonal Spaces

Definition 2.3.5 *Let* $V_r = \{\mathbf{x}_1, \ldots, \mathbf{x}_r\} \subseteq V_n$. *The orthocomplement subspace of* V_r *in* V_n, *represented by* $V^\perp$, *is a vector subspace of* V_n *which consists of all vectors* $\mathbf{y} \in V_n$ *such that* $\mathbf{x}_i'\mathbf{y} = 0$ *and we write* $V_n = V_r \oplus V^\perp$.

The vector space V_n is the direct sum of the subspaces V_n and $V^\perp$. The intersection of the two spaces only contain the null space. The dimension of V_n, $\dim V_n$, is equal to the $\dim V_r + \dim V^\perp$ so that the $\dim V^\perp = n - r$. More generally, we have the following result.

Definition 2.3.6 *Let* $S_1, S_2, \ldots, S_k$ *denote vector subspaces of* V_n. *The direct sum of these vector spaces, represented as* $\bigoplus_{i=1}^k S_i$, *consists of all unique vectors* $\mathbf{v} = \sum_{i=1}^k \alpha_i \mathbf{s}_i$ *where* $\mathbf{s}_i \in S_i, i = 1, \ldots, k$ *and the coefficients* $\alpha_i \in R$.

Theorem 2.3.4 *Let* $S_1, S_2, \ldots, S_k$ *represent vector subspaces of* V_n. *Then,*

1. $V = \bigoplus_{i=1}^k S_i$ *is a vector subspace of* V_n, $V \subseteq V_n$.
2. *The intersection of* S_i *is the null space* $\{\mathbf{0}\}$.
3. *The intersection of* V *and* $V^\perp$ *is the null space.*
4. *The* dim $V = n - k$ *so that* $\dim V \oplus V^\perp = n$.

Example 2.3.5 *Let*

$$V = \left\{ \begin{bmatrix} 1 \\ 0 \\ 1 \end{bmatrix}, \begin{bmatrix} 0 \\ 1 \\ -1 \end{bmatrix} \right\} = \{\mathbf{x}_1, \mathbf{x}_2\} \text{ and } \mathbf{y} \in V_3$$

We find $V^\perp$ using Definition 2.3.5 as follows

$$\begin{aligned} V^\perp &= \{\mathbf{y} \in V_3 \mid (\mathbf{y}'\mathbf{x}) = 0 \textit{ for any } \mathbf{x} \in V\} \\ &= \{\mathbf{y} \in V_3 \mid (\mathbf{y} \perp V\} \\ &= \{\mathbf{y} \in V_3 \mid (\mathbf{y} \perp \mathbf{x}_i\} \qquad (i = 1, 2) \end{aligned}$$

A vector $\mathbf{y}' = [y_1, y_2, y_3]$ must be found such that $\mathbf{y} \perp \mathbf{x}_1$ and $\mathbf{y} \perp \mathbf{x}_2$. This implies that $y_1 - y_3 = 0$, or $y_1 = y_3$, and $y_2 = y_3$, or $y_1 = y_2 = y_3$. Letting $y_i = 1$,

$$V^\perp = \begin{bmatrix} 1 \\ 1 \\ 1 \end{bmatrix} = \mathbf{1} \quad \text{and} \quad V_3 = V^\perp \oplus V$$

Furthermore, the

$$P_1\mathbf{y} = \begin{bmatrix} \bar{y} \\ \bar{y} \\ \bar{y} \end{bmatrix} \quad \text{and} \quad P_V\mathbf{y} = \mathbf{y} - P_1\mathbf{y} = \begin{bmatrix} y_1 - \bar{y} \\ y_2 - \bar{y} \\ y_3 - \bar{y} \end{bmatrix}$$

Alternatively, from Definition 2.3.6, an orthogonal basis for V is

$$V = \left\{ \begin{bmatrix} 1 \\ 0 \\ -1 \end{bmatrix}, \begin{bmatrix} -1/2 \\ 1 \\ -1/2 \end{bmatrix} \right\} = \{\mathbf{v}_1, \mathbf{v}_2\} = S_1 \oplus S_2$$

and the $P_V\mathbf{y}$ becomes

$$P_{\mathbf{v}_1}\mathbf{y} + P_{\mathbf{v}_2}\mathbf{y} = \begin{bmatrix} y_1 - \bar{y} \\ y_2 - \bar{y} \\ y_3 - \bar{y} \end{bmatrix}$$

Hence, a unique representation for $\mathbf{y}$ is $\mathbf{y} = P_1\mathbf{y} + P_V\mathbf{y}$ as stated in Theorem 2.3.4. The dim $V_3 = \dim \mathbf{1} + \dim V^\perp$.

In Example 2.3.5, $V^\perp$ is the orthocomplement of V relative to the whole space. Often $S \subseteq V \subseteq V_n$ and we desire the orthocomplement of S relative to V instead of V_n. This space is represented as V/S and $V = (V/S) \oplus S = S_1 \oplus S_2$. Furthermore, $V_n = V^\perp \oplus (V/S) \oplus S = V^\perp \oplus S_1 \oplus S_2$. If the dimension of V is k and the dimension of S is r, then the dimension of $V^\perp$ is $n - k$ and the dim V/S is $k - r$, so that $(n - k) + (k - r) + r = n$ or the dim $V_n = \dim V^\perp + \dim(V/S) + \dim S$ as stated in Theorem 2.3.4. In Figure 2.3.2, the geometry of subspaces is illustrated with $V_n = S \oplus (V/S) \oplus V^\perp$.

$$y_{ij} = \mu + \alpha_i + e_{ij} \qquad i = 1, 2 \quad \text{and} \quad j = 1, 2$$

The algebra of vector spaces has an important representation for the analysis of variance (ANOVA) linear model. To illustrate, consider the two group ANOVA model

Thus, we have two groups indexed by i and two observations indexed by j. Representing the observations as a vector,

$$\mathbf{y}' = [y_{11}, y_{12}, y_{21}, y_{22}]$$

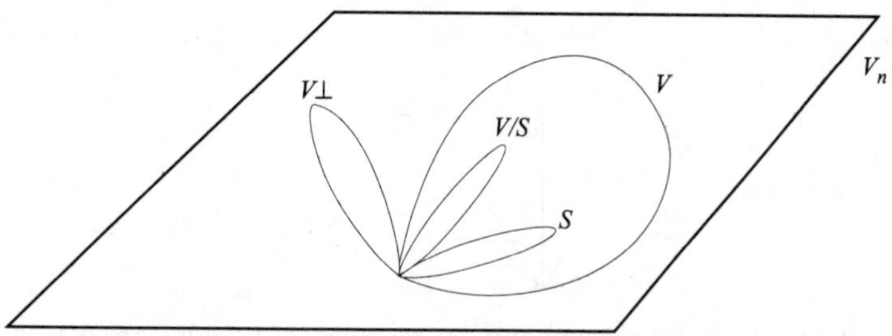

FIGURE 2.3.2. The orthocomplement of S relative to V, V/S

and formulating the observation vector as a linear model,

$$\mathbf{y} = \begin{bmatrix} y_{11} \\ y_{12} \\ y_{21} \\ y_{22} \end{bmatrix} = \begin{bmatrix} 1 \\ 1 \\ 1 \\ 1 \end{bmatrix} \mu + \begin{bmatrix} 1 \\ 1 \\ 0 \\ 0 \end{bmatrix} \alpha_1 + \begin{bmatrix} 0 \\ 0 \\ 1 \\ 1 \end{bmatrix} \alpha_2 + \begin{bmatrix} e_{11} \\ e_{12} \\ e_{21} \\ e_{22} \end{bmatrix}$$

The vectors associated with the model parameters span a vector space V often called the design space. Thus,

$$V = \left\{ \begin{bmatrix} 1 \\ 1 \\ 1 \\ 1 \end{bmatrix} \begin{bmatrix} 1 \\ 1 \\ 0 \\ 0 \end{bmatrix} \begin{bmatrix} 0 \\ 0 \\ 1 \\ 1 \end{bmatrix} \right\} = \{\mathbf{1}, \mathbf{a}_1, \mathbf{a}_2\}$$

where $\mathbf{1}$, $\mathbf{a}_1$, and $\mathbf{a}_2$ are elements of V_4. The vectors in the design space V are linearly dependent. Let $A = \{\mathbf{a}_1, \mathbf{a}_2\}$ denote a basis for V. Since $\mathbf{1} \subseteq A$, the orthocomplement of the subspace $\{\mathbf{1}\} \equiv \mathbf{1}$ relative to A, denoted by $A/\mathbf{1}$ is given by

$$\begin{aligned} A/\mathbf{1} &= \{\mathbf{a}_1 - P_{\mathbf{1}}\mathbf{a}_1, \mathbf{a}_2 - P_{\mathbf{1}}\mathbf{a}_2\} \\ &= \left\{ \begin{bmatrix} 1/2 \\ 1/2 \\ -1/2 \\ -1/2 \end{bmatrix} \begin{bmatrix} -1/2 \\ -1/2 \\ 1/2 \\ 1/2 \end{bmatrix} \right\} \end{aligned}$$

The vectors in $A/\mathbf{1}$ span the space; however, a basis for $A/\mathbf{1}$ is given by

$$A/\mathbf{1} = \begin{bmatrix} 1 \\ 1 \\ -1 \\ -1 \end{bmatrix}$$

where $(A/\mathbf{1}) \oplus \mathbf{1} = A$ and $A \subseteq V_4$. Thus, $(A/\mathbf{1}) \oplus \mathbf{1} \oplus A^\perp = V_4$. Geometrically, as shown in Figure 2.3.3, the design space $V \equiv A$ has been partitioned into two orthogonal subspaces $\mathbf{1}$ and $A/\mathbf{1}$ such that $A = \mathbf{1} \oplus (A/\mathbf{1})$, where $A/\mathbf{1}$ is the orthocomplement of $\mathbf{1}$ relative to A, and $A \oplus A^\perp = V_4$.

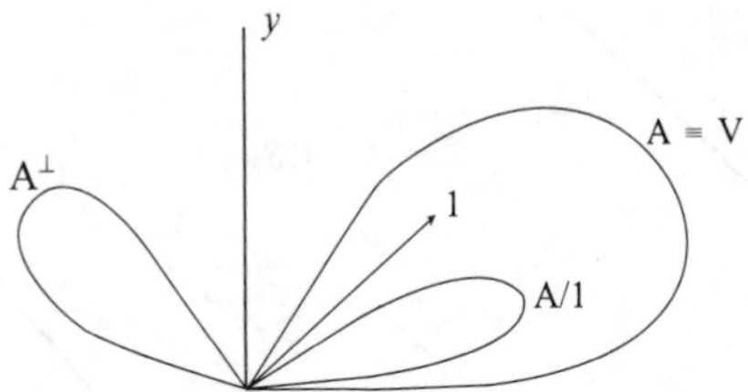

FIGURE 2.3.3. The orthogonal decomposition of V for the ANOVA

The observation vector $\mathbf{y} \in V_4$ may be thought of as a vector with components in various orthogonal subspaces. By projecting $\mathbf{y}$ onto the orthogonal subspaces in the design space A, we may obtain estimates of the model parameters. To see this, we evaluate $P_A\mathbf{y} = P_{\mathbf{1}}\mathbf{y} + P_{A/\mathbf{1}}\mathbf{y}$.

$$P_{\mathbf{1}}\mathbf{y} = \overline{y}\begin{bmatrix} 1 \\ 1 \\ 1 \\ 1 \end{bmatrix} = \widehat{\mu}\begin{bmatrix} 1 \\ 1 \\ 1 \\ 1 \end{bmatrix}$$

$$\begin{aligned} P_{A/\mathbf{1}}\mathbf{y} &= P_A\mathbf{y} - P_{\mathbf{1}}\mathbf{y} \\ &= \frac{(\mathbf{y}'\mathbf{a}_1)\mathbf{a}_1}{\|\mathbf{a}_1\|^2} + \frac{(\mathbf{y}'\mathbf{a}_2)\mathbf{a}_2}{\|\mathbf{a}_2\|^2} - \frac{(\mathbf{y}'\mathbf{1})\mathbf{1}}{\|\mathbf{1}\|^2} \\ &= \sum_{i=1}^{2}\left[\frac{(\mathbf{y}'\mathbf{a}_i)}{\|\mathbf{a}_i\|^2} - \frac{(\mathbf{y}'\mathbf{1})}{\|\mathbf{1}\|^2}\right]\mathbf{a}_i \\ &= \sum_{i=1}^{2}(\overline{y}_i - \overline{y})\mathbf{a}_i = \sum_{i=1}^{2}\widehat{\alpha}_i\mathbf{a}_i \end{aligned}$$

since $(A/\mathbf{1})\perp\mathbf{1}$ and $\mathbf{1} = \mathbf{a}_1 + \mathbf{a}_2$. As an exercise, find the projection of $\mathbf{y}$ onto $A^\perp$ and the $\|P_{A/\mathbf{1}}\mathbf{y}\|^2$.

From the analysis of variance, the coefficients of the basis vectors for $\mathbf{1}$ and $A/\mathbf{1}$ yield the estimators for the overall effect μ and the treatment effects α_i for the two-group ANOVA model employing the restriction on the parameters that $\alpha_1 + \alpha_2 = 0$. Indeed, the restriction creates a basis for $A/\mathbf{1}$. Furthermore, the total sum of squares, $\|\mathbf{y}\|^2$, is the sum of squared lengths of the projections of $\mathbf{y}$ onto each subspace, $\|\mathbf{y}\|^2 = \|P_{\mathbf{1}}\mathbf{y}\|^2 + \|P_{A/\mathbf{1}}\mathbf{y}\|^2 + \|P_{A^\perp}\mathbf{y}\|^2$. The dimensions of the subspaces for I groups, corresponding to the decomposition of $\|\mathbf{y}\|^2$, satisfy the relationship that $n = 1 + (I-1) + (n-I)$ where the dim $A = I$ and $\mathbf{y} \in V_n$. Hence, the degrees of freedom of the subspaces are the dimensions of the orthogonal vector spaces $\{\mathbf{1}\}$, $\{A/\mathbf{1}\}$ and $\{A^\perp\}$for the design space A. Finally, the $\|P_{A/\mathbf{1}}\mathbf{y}\|^2$ is the hypothesis sum of squares and the $\|P_{A^\perp}\mathbf{y}\|^2$ is the error sum of squares. Additional relationships between linear algebra and linear models using ANOVA and regression models are contained in the exercises for this section. We conclude this section with some inequalities useful in statistics and generalize the concepts of distance and vector norms.

e. *Vector Inequalities, Vector Norms, and Statistical Distance*

In a Euclidean vector space, two important inequalities regarding inner products are the Cauchy-Schwarz inequality and the triangular inequality.

Theorem 2.3.5 *If* **x** *and* **y** *are vectors in a Euclidean space* V*, then*

1. $(\mathbf{x}'\mathbf{y})^2 \leq \|\mathbf{x}\|^2 \, \|\mathbf{y}\|^2$ *(Cauchy-Schwarz inequality)*
2. $\|\mathbf{x}+\mathbf{y}\| \leq \|\mathbf{x}\| + \|\mathbf{y}\|$ *(Triangular inequality)*

In terms of the elements of **x** and **y**, (1) becomes

$$\left(\sum_i x_i y_i\right)^2 \leq \left(\sum_i x_i^2\right)\left(\sum_i y_i^2\right) \tag{2.3.2}$$

which may be used to show that the zero-order Pearson product-moment correlation coefficient is bounded by ± 1. Result (2) is a generalization of the familiar relationship for triangles in two-dimensional geometry.

The Euclidean norm is really a member of Minkowski's family of norms (L_p-norms)

$$\|\mathbf{x}\|_p = \left\{\sum_{i=1}^{n} |x_i|^p\right\}^{1/p} \tag{2.3.3}$$

where $1 \leq p < \infty$ and **x** is an element of a normed vector space V. For $p = 2$, we have the Euclidean norm. When $p = 1$, we have the minimum norm, $\|\mathbf{x}\|_1$. For $p = \infty$, Minkowski's norm is not defined, instead we define the maximum or infinity norm of **x** as

$$\|\mathbf{x}\|_\infty = \max_{1 \leq i \leq n} |x_i| \tag{2.3.4}$$

Definition 2.3.7 *A vector norm is a function defined on a vector space that maps a vector into a scalar value such that*

1. $\|\mathbf{x}\|_p \geq 0$, and $\|\mathbf{x}\|_p = 0$ *if and only if* $\mathbf{x} = \mathbf{0}$,
2. $\|\alpha\mathbf{x}\|_p = |\alpha| \, \|\mathbf{x}\|_p$ *for* $\alpha \in R$,
3. $\|\mathbf{x}+\mathbf{y}\|_p \leq \|\mathbf{x}\|_p + \|\mathbf{y}\|_p$,

for all vectors **x** *and* **y**.

Clearly the $\|\mathbf{x}\|_2 = (\mathbf{x}'\mathbf{x})^{1/2}$ satisfies Definition 2.3.7. This is also the case for the maximum norm of **x**. In this text, the Euclidean norm (L_2-norm) is assumed unless noted otherwise. Note that $(||\mathbf{x}||_2)^2 = ||\mathbf{x}||^2 = \mathbf{x}'\mathbf{x}$ is the Euclidean norm squared of **x**.

While Euclidean distances and norms are useful concepts in statistics since they help to visualize statistical sums of squares, non-Euclidean distance and non-Euclidean norms are often useful in multivariate analysis. We have seen that the Euclidean norm generalizes to a more general function that maps a vector to a scalar. In a similar manner, we may generalize the concept of distance. A non-Euclidean distance important in multivariate analysis is the statistical or Mahalanobis distance.

To motivate the definition, consider a normal random variable X with mean zero and variance one, $X \sim N(0, 1)$. An observation $\mathbf{x}_o$ that is two standard deviations from the mean lies a distance of two units from the origin since the $\|\mathbf{x}_o\| = (0^2 + 2^2)^{1/2} = 2$ and the probability that $0 \leq x \leq 2$ is 0.4772. Alternatively, suppose $Y \sim N(0, 4)$ where the distance from the origin for $\mathbf{y}_o = \mathbf{x}_o$ is still 2. However, the probability that $0 \leq y \leq 2$ becomes 0.3413 so that y is closer to the origin than x. To compare the distances, we must take into account the variance of the random variables. Thus, the squared distance between x_i and x_j is defined as

$$D_{ij}^2 = (x_i - x_j)^2/\sigma^2 = (x_i - x_j)(\sigma^2)^{-1}(x_i - x_j) \tag{2.3.5}$$

where σ^2 is the population variance. For our example, the point $\mathbf{x}_o$ has a squared statistical distance $D_{ij}^2 = 4$ while the point $\mathbf{y}_o = 2$ has a value of $D_{ij}^2 = 1$ which maintains the inequality in probabilities in that Y is "closer" to zero statistically than X. D_{ij} is the distance between x_i and x_j, in the metric of σ^2 called the Mahalanobis distance between x_i and x_j. When $\sigma^2 = 1$, Mahalanobis' distance reduces to the Euclidean distance.

Exercises 2.3

1. For the vectors

$$\mathbf{x} = \begin{bmatrix} -1 \\ 3 \\ 2 \end{bmatrix}, \quad \mathbf{y} = \begin{bmatrix} 1 \\ 2 \\ 0 \end{bmatrix}, \quad \text{and} \quad \mathbf{z} = \begin{bmatrix} 1 \\ 1 \\ 2 \end{bmatrix}$$

and scalars $\alpha = 2$ and $\beta = 3$, verify the properties given in Theorem 2.3.2.

2. Using the law of cosines

$$\|\mathbf{y} - \mathbf{x}\|^2 = \|\mathbf{x}\|^2 + \|\mathbf{y}\|^2 - 2\,\|\mathbf{x}\|\,\|\mathbf{y}\| \cos\theta$$

derive equation (2.3.1).

3. For the vectors

$$\mathbf{y}_1 = \begin{bmatrix} 2 \\ -2 \\ 1 \end{bmatrix} \quad \text{and} \quad \mathbf{y}_2 = \begin{bmatrix} 3 \\ 0 \\ -1 \end{bmatrix}$$

(a) Find their lengths, and the distance and angle between them.

(b) Find a vector of length 3 with direction cosines

$$\cos\alpha_1 = y_1/\,\|\mathbf{y}\| = 1/\sqrt{2} \quad \text{and} \quad \cos\alpha_2 = y_2/\,\|\mathbf{y}\| = -1/\sqrt{2}$$

where α_1 and α_2 are the cosines of the angles between $\mathbf{y}$ and each of its references axes $\mathbf{e}_1 = \begin{bmatrix} 1 \\ 0 \end{bmatrix}$, and $\mathbf{e}_2 = \begin{bmatrix} 0 \\ 1 \end{bmatrix}$.

(c) Verify that $\cos^2\alpha_1 + \cos^2\alpha_2 = 1$.

4. For

$$\mathbf{y} = \begin{bmatrix} 1 \\ 9 \\ -7 \end{bmatrix} \quad \text{and} \quad V = \left\{ \mathbf{v}_1 = \begin{bmatrix} 2 \\ 3 \\ 1 \end{bmatrix}, \mathbf{v}_2 = \begin{bmatrix} 5 \\ 0 \\ 4 \end{bmatrix} \right\}$$

   (a) Find the projection of $\mathbf{y}$ onto V and interpret your result.

   (b) In general, if $\mathbf{y} \perp V$, can you find the $P_V\mathbf{y}$?

5. Use the Gram-Schmidt process to find an orthonormal basis for the vectors in Exercise 2.2, Problem 4.

6. The vectors

$$\mathbf{v}_1 = \begin{bmatrix} 1 \\ 2 \\ -1 \end{bmatrix} \quad \text{and} \quad \mathbf{v}_2 = \begin{bmatrix} 2 \\ 3 \\ 0 \end{bmatrix}$$

   span a plane in Euclidean space.

   (a) Find an orthogonal basis for the plane.

   (b) Find the orthocomplement of the plane in V_3.

   (c) From (a) and (b), obtain an orthonormal basis for V_3.

7. Find an orthonormal basis for V_3 that includes the vector $\mathbf{y}' = [-1/\sqrt{3}, 1/\sqrt{3}, -1/\sqrt{3}]$.

8. Do the following.

   (a) Find the orthocomplement of the space spanned by $\mathbf{v}' = [4, 2, 1]$ relative to Euclidean three dimensional space, V_3.

   (b) Find the orthocomplement of $\mathbf{v}' = [4, 2, 1]$ relative to the space spanned by $\mathbf{v}_1' = [1, 1, 1]$ and $\mathbf{v}_1' = [2, 0, -1]$.

   (c) Find the orthocomplement of the space spanned by $\mathbf{v}_1' = [1, 1, 1]$ and $\mathbf{v}_2' = [2, 0, -1]$ relative to V_3.

   (d) Write the Euclidean three-dimensional space as the direct sum of the relative spaces in (a), (b), and (c) in all possible ways.

9. Let V be spanned by the orthonormal basis

$$\mathbf{v}_1 = \begin{bmatrix} 1/\sqrt{2} \\ 0 \\ 1/\sqrt{2} \\ 0 \end{bmatrix} \quad \text{and} \quad \mathbf{v}_2 = \begin{bmatrix} 0 \\ -1/\sqrt{2} \\ 0 \\ -1/\sqrt{2} \end{bmatrix}$$

   (a) Express $\mathbf{x}' = [0, 1, 1, 1]$ as $\mathbf{x} = \mathbf{x}_1 + \mathbf{x}_2$, where $\mathbf{x}_1 \in V$ and $\mathbf{x}_2 \in V^{\perp}$.

   (b) Verify that the $\|P_V\mathbf{x}\|^2 = \|P_{\mathbf{v}_1}\mathbf{x}\|^2 + \|P_{\mathbf{v}_2}\mathbf{x}\|^2$.

   (c) Which vector $\mathbf{y} \in V$ is closest to $\mathbf{x}$? Calculate the minimum distance.

10. Find the dimension of the space spanned by

$$\begin{array}{ccccc} \mathbf{v}_1 & \mathbf{v}_2 & \mathbf{v}_3 & \mathbf{v}_4 & \mathbf{v}_5 \\ \begin{bmatrix} 1 \\ 1 \\ 1 \\ 1 \end{bmatrix} & \begin{bmatrix} 1 \\ 1 \\ 0 \\ 0 \end{bmatrix} & \begin{bmatrix} 0 \\ 0 \\ 1 \\ 1 \end{bmatrix} & \begin{bmatrix} 1 \\ 0 \\ 1 \\ 0 \end{bmatrix} & \begin{bmatrix} 0 \\ 1 \\ 0 \\ 1 \end{bmatrix} \end{array}$$

11. Let $\mathbf{y}_n \in V_n$, and $V = \{\mathbf{1}\}$.

   (a) Find the projection of $\mathbf{y}$ onto $V^\perp$, the orthocomplement of V relative to V_n.

   (b) Represent $\mathbf{y}$ as $\mathbf{y} = \mathbf{x}_1 + \mathbf{x}_2$, where $\mathbf{x}_1 \in V$ and $\mathbf{x}_2 \in V^\perp$. What are the dimensions of V and $V^\perp$?

   (c) Since $\|\mathbf{y}\|^2 = \|\mathbf{x}_1\|^2 + \|\mathbf{x}_2\|^2 = \|\mathbf{P}_V\mathbf{y}\|^2 + \left\|\mathbf{P}_{V^\perp}\mathbf{y}\right\|^2$, determine a general form for each of the components of $\|\mathbf{y}\|^2$. Divide $\left\| P_{V^\perp}\mathbf{y}\right\|^2$ by the dimension of $V^\perp$. What do you observe about the ratio $\left\| P_{V^\perp}\mathbf{y}\right\|^2 / \dim V^\perp$?

12. Let $\mathbf{y}_n \in V_n$ be a vector of observations, $\mathbf{y}' = [y_1, y_2, \ldots, y_n]$ and let $V = \{\mathbf{1}, \mathbf{x}\}$ where $\mathbf{x}' = [x_1, x_2, \ldots, x_n]$.

   (a) Find the orthocomplement of $\mathbf{1}$ relative to V (that is, $V/\mathbf{1}$) so that $\mathbf{1} \oplus (V/\mathbf{1}) = V$. What is the dimension of $V/\mathbf{1}$?

   (b) Find the projection of $\mathbf{y}$ onto $\mathbf{1}$ and also onto $V/\mathbf{1}$. Interpret the coefficients of the projections assuming each component of $\mathbf{y}$ satisfies the simple linear relationship $y_i = \alpha + \beta(x_i - \overline{x})$.

   (c) Find $\mathbf{y} - P_V\mathbf{y}$ and $\|\mathbf{y} - P_V\mathbf{y}\|^2$. How are these quantities related to the simple linear regression model?

13. For the I Group ANOVA model $y_{ij} = \mu + \alpha_i + e_{ij}$ where $i = 1, 2, \ldots, I$ and $j = 1, 2, \ldots, n$ observations per group, evaluate the square lengths $\|\mathbf{P_1 y}\|^2$, $\left\|\mathbf{P}_{A/1}\mathbf{y}\right\|^2$, and $\left\|\mathbf{P}_{A^\perp}\mathbf{y}\right\|^2$ for $V = \{\mathbf{1}, \mathbf{a}_1, \ldots, \mathbf{a}_I\}$. Use Figure 2.3.3 to relate these quantities geometrically.

14. Let the vector space V be spanned by

$$\left\{ \begin{array}{c} \mathbf{v}_1 \\ \begin{bmatrix} 1\\1\\1\\1\\1\\1\\1\\1 \end{bmatrix} \\ 1 \end{array} \; \begin{array}{cc} \mathbf{v}_2 & \mathbf{v}_3 \\ \begin{bmatrix} 1\\1\\1\\1\\0\\0\\0\\0 \end{bmatrix} & \begin{bmatrix} 0\\0\\0\\0\\1\\1\\1\\1 \end{bmatrix} \\ \multicolumn{2}{c}{A,} \end{array} \; \begin{array}{cc} \mathbf{v}_4 & \mathbf{v}_5 \\ \begin{bmatrix} 1\\1\\0\\0\\1\\1\\0\\0 \end{bmatrix} & \begin{bmatrix} 0\\0\\1\\1\\0\\0\\1\\1 \end{bmatrix} \\ \multicolumn{2}{c}{B,} \end{array} \; \begin{array}{cccc} \mathbf{v}_6 & \mathbf{v}_7 & \mathbf{v}_8 & \mathbf{v}_9 \\ \begin{bmatrix} 1\\1\\0\\0\\0\\0\\0\\0 \end{bmatrix} & \begin{bmatrix} 0\\0\\1\\1\\0\\0\\0\\0 \end{bmatrix} & \begin{bmatrix} 0\\0\\0\\0\\1\\1\\0\\0 \end{bmatrix} & \begin{bmatrix} 0\\0\\0\\0\\0\\0\\1\\1 \end{bmatrix} \\ \multicolumn{4}{c}{AB} \end{array} \right\}$$

(a) Find the space $A+B = \mathbf{1} \oplus (A/\mathbf{1}) \oplus (B/\mathbf{1})$ and the space $AB/(A+B)$ so that $V = \mathbf{1} \oplus (A/\mathbf{1}) \oplus (B/\mathbf{1}) + [AB/(A+B)]$. What is the dimension of each of the subspaces?

(b) Find the projection of the observation vector $\mathbf{y} = [y_{111}, y_{112}, y_{211}, y_{212}, y_{311}, y_{312}, y_{411}, y_{412}]$ in V_8 onto each subspace in the orthogonal decomposition of V in (a). Represent these quantities geometrically and find their squared lengths.

(c) Summarize your findings.

15. Prove Theorem 2.3.4.

16. Show that Minkowski's norm for $p = 2$ satisfies Definition 2.3.7.

17. For the vectors $\mathbf{y}' = [y_1, \ldots, y_n]$ and $\mathbf{x}' = [x_1, \ldots, x_n]$ with elements that have a mean of zero,

(a) Show that $s_y^2 = \|\mathbf{y}\|^2 /(n-1)$ and $s_x^2 = \|\mathbf{x}\|^2 / (n-1)$.

(b) Show that the sample Pearson product moment correlation between two observations $\mathbf{x}$ and $\mathbf{y}$ is $r = \mathbf{x}'\mathbf{y} / \|\mathbf{x}\| \|\mathbf{y}\|$.

2.4 Basic Matrix Operations

The organization of real numbers into a rectangular or square array consisting of n rows and d columns is called a matrix of order n by d and written as $n \times d$.

Definition 2.4.1 *A matrix* $\mathbf{Y}$ *of order* $n \times d$ *is an array of scalars given as*

$$\mathbf{Y}_{n\times d} = \begin{bmatrix} y_{11} & y_{12} & \cdots & y_{1d} \\ y_{21} & y_{22} & \cdots & y_{2d} \\ \vdots & \vdots & & \vdots \\ y_{n1} & y_{n2} & \cdots & y_{nd} \end{bmatrix}$$

The entries y_{ij} of $\mathbf{Y}$ are called the elements of $\mathbf{Y}$ so that $\mathbf{Y}$ may be represented as $\mathbf{Y} = [y_{ij}]$. Alternatively, a matrix may be represented in terms of its column or row vectors as

$$\mathbf{Y}_{n\times d} = [\mathbf{v}_1, \mathbf{v}_2, \ldots, \mathbf{v}_d] \quad \text{and} \quad \mathbf{v}_j \in V_n \tag{2.4.1}$$

or

$$\mathbf{Y}_{n\times d} = \begin{bmatrix} \mathbf{y}_1' \\ \mathbf{y}_2' \\ \vdots \\ \mathbf{y}_n' \end{bmatrix} \quad \text{and} \quad \mathbf{y}_i' \in V_d$$

Because the rows of $\mathbf{Y}$ are usually associated with subjects or individuals each $\mathbf{y}_i'$ is a member of the person space while the columns $\mathbf{v}_j$ of $\mathbf{Y}$ are associated with the variable space. If $n = d$, the matrix $\mathbf{Y}$ is square.

a. *Equality, Addition, and Multiplication of Matrices*

Matrices like vectors may be combined using the operations of addition and scalar multiplication. For two matrices **A** and **B** of the same order, matrix addition is defined as

$$\mathbf{A}+\mathbf{B}=\mathbf{C} \quad \text{if and only if} \quad \mathbf{C}=\left[c_{ij}\right]=\left[a_{ij}+b_{ij}\right] \tag{2.4.2}$$

The matrices are conformable for matrix addition only if both matrices are of the same order and have the same number of row and columns.

The product of a matrix **A** by a scalar α is

$$\alpha\mathbf{A}=\mathbf{A}\alpha=[\alpha a_{ij}] \tag{2.4.3}$$

Two matrices **A** and **B** are equal if and only if $[a_{ij}]=[b_{ij}]$. To extend the concept of an inner product of two vectors to two matrices, the matrix product $\mathbf{AB}=\mathbf{C}$ is defined if and only if the number of columns in **A** is equal to the number of rows in **B**. For two matrices $\mathbf{A}_{n\times d}$ and $\mathbf{B}_{d\times m}$, the matrix (inner) product is the matrix $\mathbf{C}_{n\times m}$ such that

$$\mathbf{AB}=\mathbf{C}=[c_{ij}] \quad \text{for} \quad c_{ij}=\sum_{k=1}^{d} a_{ik}b_{kj} \tag{2.4.4}$$

From (2.4.4), we see that **C** is obtained by multiplying each row of **A** by each column of **B**. The matrix product is conformable if the number of columns in the matrix **A** is equal to the number of rows in the matrix **B**. The column order is equal to the row order for matrix multiplication to be defined. In general, $\mathbf{AB}\neq\mathbf{BA}$. If $\mathbf{A}=\mathbf{B}$ and **A** is square, then $\mathbf{AA}=\mathbf{A}^2$. When $\mathbf{A}^2=\mathbf{A}$, the matrix **A** is said to be idempotent.

From the definitions and properties of real numbers, we have the following theorem for matrix addition and matrix multiplication.

Theorem 2.4.1 *For matrices* **A**, **B**, **C**, *and* **D** *and scalars* α *and* β, *the following properties hold for matrix addition and matrix multiplication.*

1. $\mathbf{A}+\mathbf{B}=\mathbf{B}+\mathbf{A}$
2. $(\mathbf{A}+\mathbf{B})+\mathbf{C}=\mathbf{A}+(\mathbf{B}+\mathbf{C})$
3. $\alpha(\mathbf{A}+\mathbf{B})=\alpha\mathbf{A}+\beta\mathbf{B}$
4. $(\alpha+\beta)\mathbf{A}=\alpha\mathbf{A}+\beta\mathbf{A}$
5. $(\mathbf{AB})\mathbf{C}=\mathbf{A}(\mathbf{BC})$
6. $\mathbf{A}(\mathbf{B}+\mathbf{C})=\mathbf{AB}+\mathbf{AC}$
7. $(\mathbf{A}+\mathbf{B})\mathbf{C}=\mathbf{AC}+\mathbf{BC}$
8. $\mathbf{A}+(-\mathbf{A})=\mathbf{0}$
9. $\mathbf{A}+\mathbf{0}=\mathbf{A}$
10. $(\mathbf{A}+\mathbf{B})(\mathbf{C}+\mathbf{D})=\mathbf{A}(\mathbf{C}+\mathbf{D})+\mathbf{B}(\mathbf{C}+\mathbf{D})=\mathbf{AC}+\mathbf{AD}+\mathbf{BC}+\mathbf{BD}$

Example 2.4.1 *Let*

$$\mathbf{A} = \begin{bmatrix} 1 & 2 \\ 3 & 7 \\ -4 & 8 \end{bmatrix} \quad \text{and} \quad \mathbf{B} = \begin{bmatrix} 2 & 2 \\ 7 & 5 \\ 3 & 1 \end{bmatrix}$$

Then

$$\mathbf{A} + \mathbf{B} = \begin{bmatrix} 3 & 4 \\ 10 & 12 \\ -1 & 9 \end{bmatrix} \quad \text{and} \quad 5(\mathbf{A} + \mathbf{B}) = \begin{bmatrix} 15 & 20 \\ 50 & 60 \\ -5 & 45 \end{bmatrix}$$

For our example, **AB** *and* **BA** *are not defined. Thus, the matrices are said to not be conformable for matrix multiplication. The following is an example of matrices that are conformable for matrix multiplication.*

Example 2.4.2 *Let*

$$\mathbf{A} = \begin{bmatrix} -1 & 2 & 3 \\ 5 & 1 & 0 \end{bmatrix} \quad \text{and} \quad \mathbf{B} = \begin{bmatrix} 1 & 2 & 1 \\ 1 & 2 & 0 \\ 1 & 2 & -1 \end{bmatrix}$$

Then

$$\mathbf{AB} = \begin{bmatrix} (-1)(1) + 2(1) + 3(1) & -1(2) + 2(2) + 3(2) & -1(1) + 2(0) + 3(-1) \\ \\ 5(1) + 1(1) + 0(1) & 5(2) + 1(2) + 0(2) & 5(1) + 1(0) + 0(-1) \end{bmatrix}$$

$$= \begin{bmatrix} 4 & 8 & -4 \\ \\ 6 & 12 & 5 \end{bmatrix}$$

Alternatively, if we represent **A** and **B** as

$$\mathbf{A} = [\mathbf{a}_1, \mathbf{a}_2, \ldots, \mathbf{a}_d] \quad \text{and} \quad \mathbf{B} = \begin{bmatrix} \mathbf{b}_1' \\ \mathbf{b}_2' \\ \vdots \\ \mathbf{b}_n' \end{bmatrix}$$

Then the matrix product is defined as an "outer" product

$$\mathbf{AB} = \sum_{k=1}^{d} \mathbf{a}_k \mathbf{b}_k'$$

where each $\mathbf{C}_k = \mathbf{a}_k \mathbf{b}_k'$ is a square matrix, the number of rows is equal to the number of columns. For the example, letting

$$\mathbf{a}_1 = \begin{bmatrix} -1 \\ 5 \end{bmatrix}, \quad \mathbf{a}_2 = \begin{bmatrix} 2 \\ 1 \end{bmatrix}, \quad \mathbf{a}_3 = \begin{bmatrix} 3 \\ 0 \end{bmatrix}$$

$$\mathbf{b}_1' = [1, 2, 1], \quad \mathbf{b}_2' = [1, 2, 0], \quad \mathbf{b}_3' = [1, 2, -1]$$

Then

$$\sum_{k=1}^{3} \mathbf{a}_k \mathbf{b}_k' = \mathbf{C}_1 + \mathbf{C}_2 + \mathbf{C}_3$$

$$= \begin{bmatrix} -1 & -2 & -1 \\ 5 & 10 & 5 \end{bmatrix} + \begin{bmatrix} 2 & 4 & 0 \\ 1 & 2 & 0 \end{bmatrix} + \begin{bmatrix} 3 & 6 & -3 \\ 0 & 0 & 0 \end{bmatrix}$$

$$= \begin{bmatrix} 4 & 8 & -4 \\ 6 & 12 & 5 \end{bmatrix} = \mathbf{AB}$$

Thus, the inner and outer product definitions of matrix multiplication are equivalent.

b. *Matrix Transposition*

In Example 2.4.2, we defined **B** in terms of row vectors and **A** in terms of column vectors. More generally, we can form the transpose of a matrix. The transpose of a matrix $\mathbf{A}_{n\times d}$ is the matrix $\mathbf{A}'_{d\times n}$ obtained from $\mathbf{A} = [a_{ij}]$ by interchanging rows and columns of **A**. Thus,

$$\mathbf{A}'_{d\times n} = \begin{bmatrix} a_{11} & a_{21} & \cdots & a_{n1} \\ a_{12} & a_{22} & \cdots & a_{n2} \\ \vdots & \vdots & & \vdots \\ a_{1d} & a_{2d} & \cdots & a_{nd} \end{bmatrix} \tag{2.4.5}$$

Alternatively, if $\mathbf{A} = [a_{ij}]$ then $\mathbf{A}' = [a_{ji}]$. A square matrix **A** is said to be symmetric if and only if $\mathbf{A} = \mathbf{A}'$ or $[a_{ij}] = [a_{ji}]$. A matrix **A** is said to be skew-symmetric if $\mathbf{A} = -\mathbf{A}'$. Properties of matrix transposition follow.

Theorem 2.4.2 *For matrices* **A**, **B**, *and* **C** *and scalars* α *and* β, *the following properties hold for matrix transposition.*

1. $(\mathbf{AB})' = \mathbf{B}'\mathbf{A}'$
2. $(\mathbf{A}+\mathbf{B})' = \mathbf{A}' + \mathbf{B}'$
3. $(\mathbf{A}')' = \mathbf{A}$
4. $(\mathbf{ABC})' = \mathbf{C}'\mathbf{B}'\mathbf{A}'$
5. $(\alpha\mathbf{A})' = \alpha\mathbf{A}'$
6. $(\alpha\mathbf{A}+\beta\mathbf{B})' = \alpha\mathbf{A}' + \beta\mathbf{B}'$

Example 2.4.3 *Let*

$$\mathbf{A} = \begin{bmatrix} 1 & 3 \\ -1 & 4 \end{bmatrix} \quad \text{and} \quad \mathbf{B} = \begin{bmatrix} 2 & 1 \\ 1 & 1 \end{bmatrix}$$

Then

$$\mathbf{A}' = \begin{bmatrix} 1 & -1 \\ 3 & 4 \end{bmatrix} \quad \text{and} \quad \mathbf{B}' = \begin{bmatrix} 2 & 1 \\ 1 & 1 \end{bmatrix}$$

$$\mathbf{AB} = \begin{bmatrix} 5 & 4 \\ 2 & 3 \end{bmatrix} \quad \text{and} \quad (\mathbf{AB})' = \begin{bmatrix} 5 & 2 \\ 4 & 3 \end{bmatrix} = \mathbf{B}'\mathbf{A}'$$

$$(\mathbf{A} + \mathbf{B})' = \begin{bmatrix} 3 & 0 \\ 4 & 5 \end{bmatrix} = \mathbf{A}' + \mathbf{B}'$$

The transpose operation is used to construct symmetric matrices. Given a data matrix $\mathbf{Y}_{n\times d}$, the matrix $\mathbf{Y}'\mathbf{Y}$ is symmetric, as is the matrix $\mathbf{YY}'$. However, $\mathbf{Y}'\mathbf{Y} \neq \mathbf{YY}'$ since the former is of order $d \times d$ where the latter is an $n \times n$ matrix.

c. *Some Special Matrices*

Any square matrix whose off-diagonal elements are 0s is called a diagonal matrix. A diagonal matrix $\mathbf{A}_{n\times n}$ is represented as $\mathbf{A} = \text{diag}[a_{11}, a_{22}, \ldots, a_{nn}]$ or $\mathbf{A} = \text{diag}[a_{ii}]$ and is clearly symmetric. If the diagonal elements, $a_{ii} = 1$ for all i, then the diagonal matrix $\mathbf{A}$ is called the identity matrix and is written as $\mathbf{A} = \mathbf{I}_n$ or simply $\mathbf{I}$. Clearly, $\mathbf{IA} = \mathbf{AI} = \mathbf{A}$ so that the identity matrix behaves like the number 1 for real numbers. Premultiplication of a matrix $\mathbf{B}_{n\times d}$ by a diagonal matrix $\mathbf{R}_{n\times n} = \text{diag}[r_{ii}]$ multiplies each element in the ith row of $\mathbf{B}_{n\times d}$ by r_{ii}; postmultiplication of $\mathbf{B}_{n\times d}$ by a diagonal matrix $\mathbf{C}_{d\times d} = \text{diag}[c_{jj}]$ multiplies each element in the jth column of $\mathbf{B}$ by c_{jj}. A matrix $\mathbf{0}$ with all zeros is called the null matrix.

A square matrix whose elements above (or below) the diagonal are 0s is called a lower (or upper) triangular matrix. If the elements on the diagonal are 1s, the matrix is called a unit lower (or unit upper) triangular matrix.

Another important matrix used in matrix manipulation is a permutation matrix. An elementary permutation matrix is obtained from an identity matrix by interchanging two rows (or columns) of $\mathbf{I}$. Thus, an elementary permutation matrix is represented as $\mathbf{I}_{i,i'}$. Premultiplication of a matrix $\mathbf{A}$ by $\mathbf{I}_{i,i'}$, creates a new matrix with interchanged rows of $\mathbf{A}$ while postmultiplication by $\mathbf{I}_{i,i'}$, creates a new matrix with interchanged columns.

Example 2.4.4 *Let*

$$\mathbf{X} = \begin{bmatrix} 1 & 1 & 0 \\ 1 & 1 & 0 \\ 1 & 0 & 1 \\ 1 & 0 & 1 \end{bmatrix} \quad \text{and} \quad \mathbf{I}_{1,2} = \begin{bmatrix} 0 & 1 & 0 \\ 1 & 0 & 0 \\ 0 & 0 & 1 \end{bmatrix}$$

Then

$$\mathbf{A} = \mathbf{X}'\mathbf{X} = \begin{bmatrix} 4 & 2 & 2 \\ 2 & 2 & 0 \\ 2 & 0 & 2 \end{bmatrix} \quad \textit{is symmetric}$$

$$\mathbf{I}_{1,2}\mathbf{A} = \begin{bmatrix} 2 & 2 & 0 \\ 4 & 2 & 2 \\ 2 & 0 & 2 \end{bmatrix} \quad \textit{interchanges rows 1 and 2 of } \mathbf{A}$$

$$\mathbf{A}\mathbf{I}_{1,2} = \begin{bmatrix} 2 & 4 & 2 \\ 2 & 2 & 0 \\ 0 & 2 & 2 \end{bmatrix} \quad \textit{interchanges columns 1 and 2 of } \mathbf{A}$$

More generally, an $n \times n$ permutation matrix is any matrix that is constructed from $\mathbf{I}_n$ by permuting its columns. We may represent the matrix as $\mathbf{I}_{n,\,n}$ since there are $n!$ different permutation matrices of order n.

Finally, observe that $\mathbf{I}_n\mathbf{I}_n = \mathbf{I}_n^2 = \mathbf{I}_n$ so that $\mathbf{I}_n$ is an idempotent matrix. Letting $\mathbf{J}_n = \mathbf{1}_n\mathbf{1}'_n$, the matrix $\mathbf{J}_n$ is a symmetric matrix of ones. Multiplying $\mathbf{J}_n$ by itself, observe that $\mathbf{J}_n^2 = n\mathbf{J}_n$ so that $\mathbf{J}_n$ is not idempotent. However, $n^{-1}\mathbf{J}_n$ and $\mathbf{I}_n - n^{-1}\mathbf{J}_n$ are idempotent matrices. If $\mathbf{A}^2_{n\times n} = \mathbf{0}$, the matrix $\mathbf{A}$ is said to be nilpotent. For $\mathbf{A}^3 = \mathbf{0}$, the matrix is tripotent and if $\mathbf{A}^k = \mathbf{0}$ for some finite $k > 0$, it is k – potent. In multivariate analysis and linear models, symmetric idempotent matrices occur in the context of quadratic forms, Section 2.6, and in partitioning sums of squares, Chapter 3.

d. Trace and the Euclidean Matrix Norm

An important operation for square matrices is the trace operator. For a square matrix $\mathbf{A}_{n\times n} = [a_{ij}]$, the trace of $\mathbf{A}$, represented as tr($\mathbf{A}$), is the sum of the diagonal elements of $\mathbf{A}$. Hence,

$$\text{tr}\,(\mathbf{A}) = \sum_{i=1}^{n} a_{ii} \tag{2.4.6}$$

Theorem 2.4.3 *For square matrices* $\mathbf{A}$ *and* $\mathbf{B}$ *and scalars* α *and* β, *the following properties hold for the trace of a matrix.*

1. $\text{tr}(\alpha\mathbf{A}+\beta\mathbf{B}) = \alpha\,\text{tr}(\mathbf{A}) + \beta\,\text{tr}(\mathbf{B})$
2. $\text{tr}(\mathbf{AB}) = \text{tr}\,(\mathbf{BA})$
3. $\text{tr}(\mathbf{A}') = \text{tr}(\mathbf{A})$
4. $\text{tr}(\mathbf{A}'\mathbf{A}) = \text{tr}(\mathbf{A}\mathbf{A}') = \sum_{i,j} a_{ij}^2$ *and equals* 0, *if and only if* $\mathbf{A} = \mathbf{0}$.

Property (4) is an important property for matrices since it generalizes the Euclidean vector norm squared to matrices. The Euclidean norm squared of $\mathbf{A}$ is defined as

$$\|\mathbf{A}\|^2 = \sum_i \sum_j a_{ij}^2 = \text{tr}(\mathbf{A}'\mathbf{A}) = \text{tr}(\mathbf{A}\mathbf{A}')$$

The Euclidean matrix norm is defined as

$$\|\mathbf{A}\| = \left\{\operatorname{tr}\left(\mathbf{A}'\mathbf{A}\right)\right\}^{1/2} = \left\{\operatorname{tr}\left(\mathbf{A}\mathbf{A}'\right)\right\}^{1/2} \tag{2.4.7}$$

and is zero only if $\mathbf{A} = \mathbf{0}$. To see that this is merely a Euclidean vector norm, we introduce the vec $(\cdot)$ operator.

Definition 2.4.2 *The vec operator for a matrix* $\mathbf{A}_{n\times d}$ *stacks the columns of* $\mathbf{A}_{n\times d} = [\mathbf{a}_1, \mathbf{a}_2, \ldots, \mathbf{a}_d]$ *sequentially, one upon another, to form a* $nd \times 1$ *vector* $\mathbf{a}$

$$\mathbf{a} = \operatorname{vec}(\mathbf{A}) = \begin{bmatrix} \mathbf{a}_1 \\ \mathbf{a}_2 \\ \vdots \\ \mathbf{a}_d \end{bmatrix}$$

Using the vec operator, we have that the

$$\begin{aligned} \operatorname{tr}\left(\mathbf{A}'\mathbf{A}\right) &= \sum_{i=1}^{d} \mathbf{a}_i'\mathbf{a}_i = [(\operatorname{vec}\mathbf{A})'][\operatorname{vec}(\mathbf{A})] = \mathbf{a}'\mathbf{a} \\ &= \|\mathbf{a}\|^2 \end{aligned}$$

so that $\|\mathbf{a}\| = \left(\mathbf{a}'\mathbf{a}\right)^{1/2}$, the Euclidean vector norm of $\mathbf{a}$. Clearly $\|\mathbf{a}\|^2 = 0$ if and only if all elements of $\mathbf{a}$ are zero. For two matrices $\mathbf{A}$ and $\mathbf{B}$, the Euclidean matrix norm squared of the matrix difference $\mathbf{A} - \mathbf{B}$ is

$$\|\mathbf{A} - \mathbf{B}\|^2 = \operatorname{tr}\left[(\mathbf{A} - \mathbf{B})(\mathbf{A} - \mathbf{B})'\right] = \sum_{i,j} \left(a_{ij} - b_{ij}\right)^2$$

which may be used to evaluate the "closeness" of $\mathbf{A}$ to $\mathbf{B}$. More generally, we have the following definition of a matrix norm represented as $\|\mathbf{A}\|$.

Definition 2.4.3 *The matrix norm of* $\mathbf{A}_{n\times d}$ *is any real-valued function represented as* $\|\mathbf{A}\|$ *which satisfies the following properties.*

1. $\|\mathbf{A}\| \geq 0$, *and* $\|\mathbf{A}\| = 0$ *if and only if* $\mathbf{A} = \mathbf{0}$.
2. $\|\alpha\mathbf{A}\| = |\alpha|\,\|\mathbf{A}\|$ for $\alpha \in R$
3. $\|\mathbf{A} + \mathbf{B}\| \leq \|\mathbf{A}\| + \|\mathbf{B}\|$ *(Triangular inequality)*
4. $\|\mathbf{A}\mathbf{B}\| \leq \|\mathbf{A}\|\,\|\mathbf{B}\|$ *(Cauchy-Schwarz inequality)*

Example 2.4.5 *Let*

$$\mathbf{X} = \begin{bmatrix} 1 & 1 & 0 \\ 1 & 1 & 0 \\ 1 & 0 & 1 \\ 1 & 0 & 1 \end{bmatrix} = [\mathbf{x}_1, \mathbf{x}_2, \mathbf{x}_3]$$

Then

$$\mathbf{x} = \operatorname{vec} \mathbf{X} = \begin{bmatrix} \mathbf{x}_1 \\ \mathbf{x}_2 \\ \mathbf{x}_3 \end{bmatrix}$$

$$\operatorname{tr}\left(\mathbf{X}'\mathbf{X}\right) = 8 = (\operatorname{vec} \mathbf{X})' \operatorname{vec}(\mathbf{X})$$

$$\|\mathbf{x}\| = \sqrt{8}$$

More will be said about matrix norms in Section 2.6.

e. *Kronecker and Hadamard Products*

We next consider two more definitions of matrix multiplication called the direct or Kronecker product and the dot or Hadamard product of two matrices. To define these products, we first define a partitioned matrix.

Definition 2.4.4 *A partitioned matrix is obtained from a* $n \times m$ *matrix* $\mathbf{A}$ *by forming submatrices* $\mathbf{A}_{ij}$ *of order* $n_i \times m_j$ *such that the* $\sum\limits_i n_i = n$ *and* $\sum\limits_j m_j = m$. *Thus,*

$$\mathbf{A} = [\mathbf{A}_{ij}]$$

The elements of a partitioned matrix are the submatrices $\mathbf{A}_{ij}$. A matrix with matrices $\mathbf{A}_{ii}$ as diagonal elements and zero otherwise is denoted as diag $[\mathbf{A}_{ii}]$ and is called a block diagonal matrix.

Example 2.4.6 *Let*

$$\mathbf{A} = \begin{bmatrix} 1 & 2 & \vdots & 0 & 1 \\ \cdots & \cdots & \cdots & \cdots & \cdots \\ 1 & -1 & \vdots & 3 & 1 \\ 2 & 3 & \vdots & 2 & -1 \end{bmatrix} = \begin{bmatrix} \mathbf{A}_{11} & \mathbf{A}_{12} \\ \mathbf{A}_{21} & \mathbf{A}_{22} \end{bmatrix}$$

$$\mathbf{B} = \begin{bmatrix} 1 & \vdots & 1 \\ 1 & \vdots & -1 \\ \cdots & \cdots & \cdots \\ 2 & \vdots & 0 \\ 0 & \vdots & 5 \end{bmatrix} = \begin{bmatrix} \mathbf{B}_{11} & \mathbf{B}_{12} \\ \mathbf{B}_{21} & \mathbf{B}_{22} \end{bmatrix}$$

Then

$$\mathbf{AB} = \sum_{k=1}^{2} \mathbf{A}_{ik}\mathbf{B}_{kj} = \begin{bmatrix} 3 & \vdots & 4 \\ \cdots & \cdots & \cdots \\ 6 & \vdots & 7 \\ 9 & \vdots & 6 \end{bmatrix}$$

The matrix product is defined only if the elements of the partitioned matrices are conformable for matrix multiplication. The sum

$$\mathbf{A} + \mathbf{B} = \left[\mathbf{A}_{ij} + \mathbf{B}_{ij}\right]$$

is not defined for this example since the submatrices are not conformable for matrix addition.

The direct or Kronecker product of two matrices $\mathbf{A}_{n\times m}$ and $\mathbf{B}_{p\times q}$ is defined as the partitioned matrix

$$\mathbf{A} \otimes \mathbf{B} = \begin{bmatrix} a_{11}\mathbf{B} & a_{12}\mathbf{B} & \cdots & a_{1m}\mathbf{B} \\ a_{21}\mathbf{B} & a_{22}\mathbf{B} & \cdots & a_{2m}\mathbf{B} \\ \vdots & \vdots & & \vdots \\ a_{n1}\mathbf{B} & a_{n2}\mathbf{B} & \cdots & a_{nm}\mathbf{B} \end{bmatrix} \tag{2.4.8}$$

of order $np \times mq$. This definition of multiplication does not depend on matrix conformability and is always defined.

Kronecker or direct products have numerous properties. For a comprehensive discussion of the properties summarized in Theorem 2.4.4 (see, for example Harville, 1997, Chapter 16).

Theorem 2.4.4 *Let* **A**, **B**, **C**, *and* **D** *be matrices,* **x** *and* **y** *vectors, and* α *and* β *scalars. Then*

1. $\mathbf{x}' \otimes \mathbf{y} = \mathbf{yx}' = \mathbf{y} \otimes \mathbf{x}'$
2. $\alpha\mathbf{A} \otimes \beta\mathbf{B} = \alpha\beta(\mathbf{A} \otimes \mathbf{B})$
3. $(\mathbf{A} \otimes \mathbf{B}) \otimes \mathbf{C} = \mathbf{A} \otimes (\mathbf{B} \otimes \mathbf{C})$
4. $(\mathbf{A} + \mathbf{B}) \otimes \mathbf{C} = (\mathbf{A} \otimes \mathbf{C}) + (\mathbf{B} \otimes \mathbf{C})$
5. $\mathbf{A} \otimes (\mathbf{B} + \mathbf{C}) = (\mathbf{A} \otimes \mathbf{B}) + \mathbf{A} \otimes \mathbf{C}$
6. $(\mathbf{A} \otimes \mathbf{B})(\mathbf{C} \otimes \mathbf{D}) = (\mathbf{AC} \otimes \mathbf{BD})$
7. $(\mathbf{A} \otimes \mathbf{B})' = \mathbf{A}' \otimes \mathbf{B}'$
8. $\text{tr}(\mathbf{A} \otimes \mathbf{B}) = \text{tr}\,(\mathbf{A})\ \text{tr}\,(\mathbf{B})$
9. $[\mathbf{A}_1,\ \mathbf{A}_2] \otimes \mathbf{B} = [\mathbf{A}_1 \otimes \mathbf{B},\ \mathbf{A}_2 \otimes \mathbf{B}]$ *for a partitioned matrix* $\mathbf{A} = [\mathbf{A}_1,\ \mathbf{A}_2]$

10. $\mathbf{I} \otimes \mathbf{A} = \begin{bmatrix} \mathbf{A} & \mathbf{0} & \cdots & \mathbf{0} \\ \mathbf{0} & \mathbf{A} & \cdots & \mathbf{0} \\ \vdots & \vdots & & \vdots \\ \mathbf{0} & \mathbf{0} & \cdots & \mathbf{A} \end{bmatrix} = \text{diag}\,[\mathbf{A}]$, *a block diagonal matrix.*

11. $(\mathbf{I} \otimes \mathbf{x})\mathbf{A}(\mathbf{I} \otimes \mathbf{x}') = \mathbf{A} \otimes \mathbf{x}\mathbf{x}'$

12. *In general,* $\mathbf{A} \otimes \mathbf{B} \neq \mathbf{B} \otimes \mathbf{A}$

Another matrix product that is useful in multivariate analysis is the dot matrix product or the Hadamard product. For this product to be defined, the matrices **A** and **B** must be of the same order, say $n \times m$. Then, the dot product or Hadamard product is the element by element product defined as

$$\mathbf{A} \odot \mathbf{B} = [a_{ij} b_{ij}] \tag{2.4.9}$$

For a discussion of Hadamard products useful in multivariate analysis see Styan (1973). Some useful properties of Hadamard products are summarized in Theorem 2.4.5 (see, for example, Schott, 1997, p. 266).

Theorem 2.4.5 *Let* **A**, **B**, *and* **C** *be* $n \times m$ *matrices, and* $\mathbf{x}_n$ *and* $\mathbf{y}_m$ *any vectors. Then*

1. $\mathbf{A} \odot \mathbf{B} = \mathbf{B} \odot \mathbf{A}$
2. $(\mathbf{A} \odot \mathbf{B})' = \mathbf{A}' \odot \mathbf{B}'$
3. $(\mathbf{A} \odot \mathbf{B}) \odot \mathbf{C} = \mathbf{A} \odot (\mathbf{B} \odot \mathbf{C})$
4. $(\mathbf{A} + \mathbf{B}) \odot \mathbf{C} = (\mathbf{A} \odot \mathbf{C}) + (\mathbf{B} \odot \mathbf{C})$
5. *For* $\mathbf{J} = \mathbf{1}_n \mathbf{1}_n'$, *a matrix of all* **1***s,* $\mathbf{A} \odot \mathbf{J} = \mathbf{A}$
6. $\mathbf{A} \odot \mathbf{0} = \mathbf{0}$
7. *For* $n = m, \mathbf{I} \odot \mathbf{A} = \text{diag}[a_{11}, a_{22}, \ldots, a_{nn}]$
8. $\mathbf{1}_n'(\mathbf{A} \odot \mathbf{B})\mathbf{1}_m = \text{tr}(\mathbf{A}\mathbf{B}')$
9. *Since* $\mathbf{x} = \text{diag}\,[\mathbf{x}]\,\mathbf{1}_n$ and $\mathbf{y} = \text{diag}\,[\mathbf{y}]\,\mathbf{1}_m$, $\mathbf{x}'\,(\mathbf{A} \odot \mathbf{B})\,\mathbf{y} = \text{tr}\left(\text{diag}\,[\mathbf{x}]\,\mathbf{A}\,\text{diag}\,[\mathbf{y}]\,\mathbf{B}'\right)$ *where* diag [**x**] *or* diag [**y**] *refers to the construction of a diagonal matrix by placing the elements of the vector* **x** *(or* **y***) along the diagonal and* 0*s elsewhere.*
10. $\text{tr}\{(\mathbf{A}' \odot \mathbf{B}')\mathbf{C}\} = \text{tr}\{\mathbf{A}'(\mathbf{B}' \odot \mathbf{C})\}$

Example 2.4.7 *Let*

$$\mathbf{A} = \begin{bmatrix} 1 & 2 \\ 3 & 4 \end{bmatrix} \quad \text{and} \quad \mathbf{B} = \begin{bmatrix} 1 & 2 \\ 0 & 3 \end{bmatrix}$$

Then

$$\mathbf{A} \otimes \mathbf{B} = \begin{bmatrix} 1\mathbf{B} & 2\mathbf{B} \\ 3\mathbf{B} & 4\mathbf{B} \end{bmatrix} = \begin{bmatrix} 1 & 2 & 2 & 4 \\ 0 & 3 & 0 & 6 \\ 3 & 6 & 4 & 8 \\ 0 & 9 & 0 & 12 \end{bmatrix}$$

and

$$\mathbf{A} \odot \mathbf{B} = \begin{bmatrix} 1 & 4 \\ 0 & 12 \end{bmatrix}$$

In Example 2.4.7, observe that $\mathbf{A} \odot \mathbf{B}$ is a submatrix of $\mathbf{A} \otimes \mathbf{B}$. Schott (1997) discusses numerous relationships between Kronecker and Hadamard products.

f. Direct Sums

The Kronecker product is an extension of a matrix product which resulted in a partitioned matrix. Another operation of matrices that also results in a partitioned matrix is called the direct sum. The direct sum of two matrices **A** and **B** is defined as

$$\mathbf{A} \oplus \mathbf{B} = \begin{bmatrix} \mathbf{A} & \mathbf{0} \\ \mathbf{0} & \mathbf{B} \end{bmatrix}$$

More generally, for k matrices $\mathbf{A}_{11}\ \mathbf{A}_{22}, \ldots, \mathbf{A}_{kk}$ the direct sum is defined as

$$\bigoplus_{i=1}^{k} \mathbf{A}_{ii} = \operatorname{diag}\left[\mathbf{A}_{ii}\right] \tag{2.4.10}$$

The direct sum is a block diagonal matrix with matrices $\mathbf{A}_{ii}$ as the ith diagonal element. Some properties of direct sums are summarized in the following theorem.

Theorem 2.4.6 *Properties of direct sums.*

1. $(\mathbf{A} \oplus \mathbf{B}) + (\mathbf{C} \oplus \mathbf{D}) = (\mathbf{A} + \mathbf{C}) \oplus (\mathbf{B} + \mathbf{D})$
2. $(\mathbf{A} \oplus \mathbf{B})\,(\mathbf{C} \oplus \mathbf{D}) = (\mathbf{AC}) \oplus (\mathbf{BD})$
3. $\operatorname{tr}\left[\bigoplus_{i=1}^{k} \mathbf{A}_i\right] = \sum_i \operatorname{tr}(\mathbf{A}_i)$

Observe that for all $\mathbf{A}_{ii} = \mathbf{A}$, that direct sum $\bigoplus_i \mathbf{A}_{ii} = \mathbf{I} \otimes \mathbf{A} = \operatorname{diag}[\mathbf{A}]$, property (10) in Theorem 2.4.4.

g. The Vec(·) and Vech(·) Operators

The vec operator was defined in Definition 2.4.2 and using the vec(·) operator, we showed how to extend a Euclidean vector norm to a Euclidean matrix norm. Converting a matrix to a vector has many applications in multivariate analysis. It is most useful when working with random matrices since it is mathematically more convenient to evaluate the distribution of a vector. To manipulate matrices using the vec(·) operator requires some "vec" algebra. Theorem 2.4.7 summarizes some useful results.

Theorem 2.4.7 *Properties of the* vec(·) *operator.*

1. $\text{vec}(\mathbf{y}) = \text{vec}(\mathbf{y}') = \mathbf{y}$

2. $\text{vec}(\mathbf{yx}') = \mathbf{x} \otimes \mathbf{y}$

3. $\text{vec}(\mathbf{A} \otimes \mathbf{x}) = \text{vec}(\mathbf{A}) \otimes \mathbf{x}$

4. $\text{vec}(\alpha\mathbf{A}+\beta\mathbf{B}) = \alpha\, \text{vec}(\mathbf{A}) + \beta\, \text{vec}(\mathbf{B})$

5. $\text{vec}(\mathbf{ABC}) = (\mathbf{C}' \otimes \mathbf{A})\, \text{vec}(\mathbf{B})$

6. $\text{vec}(\mathbf{AB}) = (\mathbf{I} \otimes \mathbf{A})\, \text{vec}(\mathbf{B}) = (\mathbf{B}' \otimes \mathbf{I})\, \text{vec}(\mathbf{B}' \otimes \mathbf{I})\, \text{vec}(\mathbf{A})$

7. $\text{tr}(\mathbf{A}'\mathbf{B}) = (\text{vec}\ \mathbf{A})'\, \text{vec}(\mathbf{B})$

8. $\text{tr}(\mathbf{ABC}) = \text{vec}(\mathbf{A}')(\mathbf{I} \otimes \mathbf{B})\, \text{vec}(\mathbf{C})$

9. $\text{tr}(\mathbf{ABCD}) = \left(\text{vec}(\mathbf{A}')\right)' (\mathbf{D}' \otimes \mathbf{B})\, \text{vec}(\mathbf{C})$

10. $\text{tr}(\mathbf{AX}'\mathbf{BXC}) = (\text{vec}(\mathbf{X}))' (\mathbf{CA} \otimes \mathbf{B}')\, \text{vec}(\mathbf{X})$

Again, all matrices in Theorem 2.4.7 are assumed to be conformable for the stated operations.

The vectors vec **A** and vec **A**′ contain the same elements, but in a different order. To relate vec **A** to vec **A**′, a vec- permutation matrix may be used. To illustrate, consider the matrix

$$\mathbf{A}_{n\times m} = \begin{bmatrix} a_{11} & a_{12} \\ a_{21} & a_{22} \\ a_{31} & a_{32} \end{bmatrix} \quad \text{where} \quad \mathbf{A}' = \begin{bmatrix} a_{11} & a_{21} & a_{31} \\ a_{12} & a_{22} & a_{32} \end{bmatrix}$$

Then

$$\text{vec}\,\mathbf{A} = \begin{bmatrix} a_{11} \\ a_{21} \\ a_{31} \\ a_{12} \\ a_{22} \\ a_{32} \end{bmatrix} \quad \text{and} \quad \text{vec}\,\mathbf{A}' = \begin{bmatrix} a_{11} \\ a_{12} \\ a_{21} \\ a_{22} \\ a_{31} \\ a_{32} \end{bmatrix}$$

To create vec **A**′ from vec **A**, observe that

$$\begin{bmatrix} 1 & 0 & 0 & 0 & 0 & 0 \\ 0 & 0 & 0 & 1 & 0 & 0 \\ 0 & 1 & 0 & 0 & 0 & 0 \\ 0 & 0 & 0 & 0 & 1 & 0 \\ 0 & 0 & 1 & 0 & 0 & 0 \\ 0 & 0 & 0 & 0 & 0 & 1 \end{bmatrix} \text{vec}\,\mathbf{A} = \text{vec}\,\mathbf{A}'$$

and that

$$\text{vec}\,\mathbf{A} = \begin{bmatrix} 1 & 0 & 0 & 0 & 0 & 0 \\ 0 & 0 & 1 & 0 & 0 & 0 \\ 0 & 0 & 0 & 0 & 1 & 0 \\ 0 & 1 & 0 & 0 & 1 & 0 \\ 0 & 0 & 0 & 1 & 0 & 0 \\ 0 & 0 & 0 & 0 & 0 & 1 \end{bmatrix} \text{vec}\,\mathbf{A}'$$

Letting $\mathbf{I}_{nm}$ vec $\mathbf{A} =$ vec $\mathbf{A}'$, the vec-permutation matrix $\mathbf{I}_{nm}$ of order $nm \times nm$ converts vec $\mathbf{A}$ to vec $\mathbf{A}'$. And, letting $\mathbf{I}_{mn}$ be the vec-permutation matrix that converts vec $\mathbf{A}'$ to vec $\mathbf{A}$, observe that $\mathbf{I}'_{nm} = \mathbf{I}_{mn}$.

Example 2.4.8 *Let*

$$\underset{3\times 2}{\mathbf{A}} = \begin{bmatrix} a_{11} & a_{12} \\ a_{21} & a_{22} \\ a_{31} & a_{32} \end{bmatrix} \quad \text{and} \quad \underset{2\times 1}{\mathbf{y}} = \begin{bmatrix} y_1 \\ y_2 \end{bmatrix}$$

Then

$$\mathbf{A} \otimes \mathbf{y} = \begin{bmatrix} a_{11}\mathbf{y} & a_{12}\mathbf{y} \\ a_{21}\mathbf{y} & a_{22}\mathbf{y} \\ a_{31}\mathbf{y} & a_{32}\mathbf{y} \end{bmatrix} = \begin{bmatrix} a_{11}y_1 & a_{12}y_1 \\ a_{11}y_2 & a_{12}y_2 \\ a_{21}y_1 & a_{22}y_1 \\ a_{21}y_2 & a_{22}y_2 \\ a_{31}y_1 & a_{32}y_1 \\ a_{31}y_2 & a_{32}y_2 \end{bmatrix}$$

$$\mathbf{y} \otimes \mathbf{A} = \begin{bmatrix} y_1\mathbf{A} \\ y_2\mathbf{A} \end{bmatrix} = \begin{bmatrix} y_1a_{11} & y_1a_{12} \\ y_1a_{21} & y_1a_{22} \\ y_1a_{31} & y_1a_{32} \\ y_2a_{11} & y_2a_{12} \\ y_2a_{21} & y_2a_{22} \\ y_2a_{31} & y_2a_{32} \end{bmatrix}$$

and

$$\begin{bmatrix} 1 & 0 & 0 & 0 & 0 & 0 \\ 0 & 0 & 1 & 0 & 0 & 0 \\ 0 & 0 & 0 & 0 & 1 & 0 \\ 0 & 1 & 0 & 0 & 0 & 0 \\ 0 & 0 & 0 & 1 & 0 & 0 \\ 0 & 0 & 0 & 0 & 0 & 1 \end{bmatrix} (\mathbf{A} \otimes \mathbf{y}) = (\mathbf{y} \otimes \mathbf{A})$$

$$\mathbf{I}_{np}(\mathbf{A} \otimes \mathbf{y}) = \mathbf{y} \otimes \mathbf{A}$$

or

$$\begin{aligned} \mathbf{A} \otimes \mathbf{y} &= \mathbf{I}'_{np}(\mathbf{y} \otimes \mathbf{A}) \\ &= \mathbf{I}_{pn}(\mathbf{y} \otimes \mathbf{A}) \end{aligned}$$

From Example 2.4.8, we see that the vec-permutation matrix allows the Kronecker product to commute. For this reason, it is also called a commutation matrix; see Magnus and Neudecker (1979).

Definition 2.4.5 *A vec-permutation (commutation) matrix of order $nm \times nm$ is a permutation matrix $\mathbf{I}_{nm}$ obtained from the identity matrix of order $nm \times nm$ by permuting its columns such that $\mathbf{I}_{nm}$ vec $\mathbf{A} =$ vec $\mathbf{A}'$.*

A history of the operator is given in Henderson and Searle (1981). An elementary overview is provided by Schott (1997) and Harville (1997).

Another operation that is used in many multivariate applications is the vech($\cdot$) operator defined for square matrices that are symmetric. The vech($\cdot$) operator is similar to the vec($\cdot$) operator, except only the elements in the matrix on or below the diagonal of the symmetric matrix are included in vech($\mathbf{A}$).

Example 2.4.9 *Let*

$$\mathbf{A} = \mathbf{X}'\mathbf{X} = \begin{bmatrix} 1 & 2 & 3 \\ 2 & 5 & 6 \\ 3 & 6 & 8 \end{bmatrix}_{n \times n}$$

Then

$$\text{vech}\,\mathbf{A} = \begin{bmatrix} 1 \\ 2 \\ 3 \\ 5 \\ 6 \\ 8 \end{bmatrix}_{n(n+1)/2\times 1} \quad \text{and} \quad \text{vec}\,\mathbf{A} = \begin{bmatrix} 1 \\ 2 \\ 3 \\ 2 \\ 5 \\ 6 \\ 2 \\ 6 \\ 8 \end{bmatrix}_{n^2\times 1}$$

Also, observe that the relationships between vech($\mathbf{A}$) and vec($\mathbf{A}$) is as follows:

$$\begin{bmatrix} 1 & 0 & 0 & 0 & 0 & 0 \\ 0 & 1 & 0 & 0 & 0 & 0 \\ 0 & 0 & 1 & 0 & 0 & 0 \\ 0 & 1 & 0 & 0 & 0 & 0 \\ 0 & 0 & 0 & 1 & 0 & 0 \\ 0 & 0 & 0 & 0 & 1 & 0 \\ 0 & 0 & 1 & 0 & 0 & 0 \\ 0 & 0 & 0 & 0 & 1 & 0 \\ 0 & 0 & 0 & 0 & 0 & 1 \end{bmatrix}_{n^2\times n(n+1)2} \text{vech}\,\mathbf{A} = \text{vec}\,\mathbf{A}$$

Example 2.4.9. leads to the following theorem.

Theorem 2.4.8 *Given a symmetric matrix $\mathbf{A}_{n\times n}$ there exist unique matrices $\mathbf{D}_n$ of order $n^2 \times n(n+1)/2$ and $\mathbf{D}_n^+$ of order $n\,(n+1)\,/2 \times n^2$(its Moore-Penrose inverse) such that*

$$\text{vec}\,\mathbf{A} = \mathbf{D}_n\,\text{vech}\,\mathbf{A} \quad \text{and} \quad \mathbf{D}_n^+\,\text{vec}\,\mathbf{A} = \text{vech}\,\mathbf{A}$$

The definition of the matrix $\mathbf{D}_n^+$ is reviewed in Section 2.5. For a discussion of vec(·) and vech (·) operators, the reader is referred to Henderson and Searle (1979), Harville (1997), and Schott (1997). Magnus and Neudecker (1999, p. 49) call the matrix $\mathbf{D}_n$ a duplication matrix and $\mathbf{D}_n^+$ an elimination matrix, Magnus and Neudecker (1980). The vech (·) operator is most often used when evaluating the distribution of symmetric matrices which occur in multivariate analysis; see McCulloch (1982), Fuller (1987), and Bilodeau and Brenner (1999).

Exercises 2.4

1. Given

$$\mathbf{A} = \begin{bmatrix} 1 & 2 \\ 0 & -1 \\ 4 & 5 \end{bmatrix}, \mathbf{B} = \begin{bmatrix} 3 & 0 \\ -1 & 1 \\ 2 & 7 \end{bmatrix}, \mathbf{C} = \begin{bmatrix} 1 & 2 \\ -3 & 5 \\ 0 & -1 \end{bmatrix}, \text{ and } \mathbf{D} = \begin{bmatrix} 1 & 1 \\ -1 & 2 \\ 6 & 0 \end{bmatrix}$$

and $\alpha = 2$, and $\beta = 3$, verify the properties in Theorem 2.4.1.

2. For

$$\mathbf{A} = \begin{bmatrix} 1 & -2 & 3 \\ 0 & 4 & 2 \\ 1 & 2 & 1 \end{bmatrix} \quad \text{and} \quad \mathbf{B} = \begin{bmatrix} 1 & 1 & 2 \\ 0 & 0 & 4 \\ 2 & -1 & 3 \end{bmatrix}$$

(a) Show $\mathbf{AB} \neq \mathbf{BA}$. The matrices do not commute.

(b) Find $\mathbf{A}'\mathbf{A}$ and $\mathbf{AA}'$.

(c) Are either $\mathbf{A}$ or $\mathbf{B}$ idempotent?

(d) Find two matrices $\mathbf{A}$ and $\mathbf{B}$ not equal to zero such that $\mathbf{AB} = \mathbf{0}$, but neither $\mathbf{A}$ or $\mathbf{B}$ is the zero matrix.

3. If $\mathbf{X} = \left[\mathbf{1}, \mathbf{x}_1, \mathbf{x}_2, \ldots, \mathbf{x}_p\right]$ and $\mathbf{x}_i$ and $\mathbf{e}$ are $n \times 1$ vectors while $\boldsymbol{\beta}$ is a $k \times 1$ vector where $k = p + 1$, show that $\mathbf{y} = \mathbf{1} + \sum_{i=1}^{p} \beta_i \mathbf{x}_i + \mathbf{e}$ may be written as $\mathbf{y} = \mathbf{X}\boldsymbol{\beta} + \mathbf{e}$.

4. For $\alpha = 2$ and $\beta = 3$, and $\mathbf{A}$ and $\mathbf{B}$ given in Problem 2, verify Theorem 2.4.2.

5. Verify the relationships denoted in (a) to (e) and prove (f).

(a) $\mathbf{1}_n'\mathbf{1}_n = n$ and $\mathbf{1}_n\mathbf{1}_n' = \mathbf{J}_n$ (a matrix of 1's)

(b) $(\mathbf{J}_n\mathbf{J}_n) = \mathbf{J}_n^2 = n\mathbf{J}_n$

(c) $\mathbf{1}_n'\left(\mathbf{I}_n - n^{-1}\mathbf{J}_n\right) = \mathbf{0}_n'$

(d) $\mathbf{J}_n'\left(\mathbf{I}_n - n^{-1}\mathbf{J}_n\right) = \mathbf{0}_{n\times n}$

(e) $\left(\mathbf{I}_n - n^{-1}\mathbf{J}_n\right)^2 = \mathbf{I}_n - n^{-1}\mathbf{J}_n$

(f) What can you say about $\mathbf{I} - \mathbf{A}$ if $\mathbf{A}^2 = \mathbf{A}$?

6. Suppose $\mathbf{Y}_{n\times d}$ is a data matrix. Interpret the following quantities statistically.
   (a) $\mathbf{1}'\mathbf{Y}/n$
   (b) $\mathbf{Y}_c = \mathbf{Y} - \mathbf{1}(\mathbf{1}'\mathbf{Y}/n)$
   (c) $\mathbf{Y}_c'\mathbf{Y}_c/(n-1) = \mathbf{Y}'\left(\mathbf{I}_n - n^{-1}\mathbf{J}_n\right)\mathbf{Y}/\left(n-1\right)$
   (d) For $\mathbf{D} = \left[\sigma_{ii}^2\right]$ and $\mathbf{Y}_z = \mathbf{Y}_c\mathbf{D}^{-1/2}$, what is $\mathbf{Y}_z'\mathbf{Y}_z/(n-1)$.

7. Given

$$\mathbf{A} = \begin{bmatrix} \sigma_1^2 & \sigma_{12} \\ \sigma_{21} & \sigma_2^2 \end{bmatrix} \quad \text{and} \quad \mathbf{B} = \begin{bmatrix} 1/\sigma_1 & 0 \\ 0 & 1/\sigma_2 \end{bmatrix}$$

form the product $\mathbf{B}'\mathbf{A}\mathbf{B}$ and interpret the result statistically.

8. Verify Definition 2.4.2 using matrices **A** and **B** in Problem 2.

9. Prove Theorems 2.4.4 through 2.4.7 and represent the following ANOVA design results and models using Kronecker product notation.
   (a) In Exercise 2.3, Problem 13, we expressed the ANOVA design geometrically. Using matrix algebra verify that
      i. $\|\mathbf{P}_1\mathbf{y}\|^2 = \mathbf{y}'\left(a^{-1}\mathbf{J}_a \otimes n^{-1}\mathbf{J}_n\right)\mathbf{y}$
      ii. $\left\|\mathbf{P}_{A/1}\mathbf{y}\right\|^2 = \mathbf{y}'\left[\left(\mathbf{I}_a - a^{-1}\mathbf{J}_a\right) \otimes n^{-1}\mathbf{J}_n\right]\mathbf{y}$
      iii. $\left\|\mathbf{P}_{A\perp}\mathbf{y}\right\|^2 = \mathbf{y}'\left[\mathbf{I}_a \otimes \left(\mathbf{I}_n - n^{-1}\mathbf{J}_n\right)\right]\mathbf{y}$

      for $\mathbf{y}' = [y_{11}, y_{12}, \ldots, y_{1n}, \ldots, y_{a1}, \ldots, y_{an}]$
   (b) For $i = 2$ and $j = 2$, verify that the ANOVA model has the structure $\mathbf{y} = (\mathbf{1}_2 \otimes \mathbf{1}_2)\,\boldsymbol{\mu} + (\mathbf{I}_2 \otimes \mathbf{1}_2)\,\boldsymbol{\alpha} + \mathbf{e}$.
   (c) For $\mathbf{X} \equiv V$ in Exercise 23, Problem 14, show that
      i. $\mathbf{X} = [\mathbf{1}_2 \otimes \mathbf{1}_2 \otimes \mathbf{1}_2, \mathbf{I}_2 \otimes \mathbf{1}_2 \otimes \mathbf{1}_2, \mathbf{1}_2 \otimes \mathbf{I}_2 \otimes \mathbf{1}_2, \mathbf{I}_2 \otimes \mathbf{I}_2 \otimes \mathbf{I}_2]$
      ii. $AB = [\mathbf{v}_2 \odot \mathbf{v}_4, \mathbf{v}_2 \odot \mathbf{v}_5, \mathbf{v}_3 \odot \mathbf{v}_4, \mathbf{v}_3 \odot \mathbf{v}_5]$

10. For

$$\mathbf{A} = \begin{bmatrix} 1 & -2 \\ 2 & 1 \end{bmatrix}, \quad \mathbf{B} = \begin{bmatrix} 1 & 2 \\ 5 & 3 \end{bmatrix}, \quad \mathbf{C} = \begin{bmatrix} 2 & 6 \\ 0 & 1 \end{bmatrix}, \quad \text{and} \quad \mathbf{D} = \begin{bmatrix} 0 & 4 \\ 1 & 1 \end{bmatrix}$$

and scalars $\alpha = \beta = 2$, verify Theorem 2.4.2, 2.4.3, and 2.4.4.

11. Letting $\underset{n\times d}{\mathbf{Y}} = \underset{n\times k}{\mathbf{X}}\,\underset{k\times d}{\mathbf{B}} + \underset{n\times d}{\mathbf{U}}$ where $\mathbf{Y} = [\mathbf{v}_1, \mathbf{v}_2, \ldots, \mathbf{v}_d]$, $\mathbf{B} = \left[\boldsymbol{\beta}_1, \boldsymbol{\beta}_2, \ldots, \boldsymbol{\beta}_d\right]$, and $\mathbf{U} = [\mathbf{u}_1, \mathbf{u}_2, \ldots, \mathbf{u}_d]$, show that $\text{vec}(\mathbf{Y}) = (\mathbf{I}_d \otimes \mathbf{X})\,\text{vec}(\mathbf{B}) + \text{vec}(\mathbf{U})$ is equivalent to $\mathbf{Y} = \mathbf{X}\mathbf{B} + \mathbf{U}$.

12. Show that the covariances of the elements of $\mathbf{u} = \text{vec}(\mathbf{U})$ has the structure $\Sigma \otimes \mathbf{I}$ while the structure of the covariance of $\text{vec}\left(\mathbf{U}'\right)$ is $\mathbf{I} \otimes \Sigma$.

13. Find a vec-permutation matrix so that we may write

$$\mathbf{B} \otimes \mathbf{A} = \mathbf{I}_{np} (\mathbf{A} \otimes \mathbf{B}) \mathbf{I}_{mq}$$

for any matrices $\mathbf{A}_{n \times m}$ and $\mathbf{B}_{p \times q}$.

14. Find a matrix $\mathbf{M}$ such that $\mathbf{M}\ \text{vec}(\mathbf{A}) = \text{vec}(\mathbf{A} + \mathbf{A}')/\mathbf{2}$ for any matrix $\mathbf{A}$.

15. If $\mathbf{e}_i$ is the ith column of $\mathbf{I}_n$ verify that

$$\text{vec}(\mathbf{I}_n) = \sum_{i=1}^{n} (\mathbf{e}_i \otimes \mathbf{e}_i)$$

16. Let Δ_{ij} represent an $n \times m$ indicator matrix that has zeros for all elements except for element $\delta_{ij} = 1$. Show that the commutation matrix has the structure.

$$\mathbf{I}_{nm} = \sum_{i=1}^{n} \sum_{j=1}^{m} (\Delta_{ij} \otimes \Delta'_{ij}) = \sum_{i=1}^{n} \sum_{j=1}^{m} (\Delta'_{ij} \otimes \Delta_{ij})' = \mathbf{I}_{mn}$$

17. For any matrices $\mathbf{A}_{n \times m}$ and $\mathbf{B}_{p \times q}$, verify that

$$\text{vec}(\mathbf{A} \otimes \mathbf{B}) = (\mathbf{I}_m \otimes \mathbf{I}_{qn} \otimes \mathbf{I}_p)(\text{vec}\,\mathbf{A} \otimes \text{vec}\,\mathbf{B})$$

18. Prove that $\sum_{i=1}^{k} \mathbf{A}_i = tr\,(\mathbf{I} \otimes \mathbf{A})$, if $\mathbf{A}_1 = \mathbf{A}_2 = \cdots = \mathbf{A}_k = \mathbf{A}$.

19. Let $\mathbf{A}_{n \times n}$ be any square matrix where the $n^2 \times n^2$ matrix $\mathbf{I}_{nn}$ is its vec-permutation (commutation) matrix, and suppose we define the $n^2 \times n^2$ symmetric and idempotent matrix $\mathbf{P} = \left(\mathbf{I}_{n^2} + \mathbf{I}_{nn}\right)/2$. Show that

    (a) $\mathbf{P}\ \text{vec}\,\mathbf{A} = \text{vec}\left(\mathbf{A} + \mathbf{A}'\right)/2$

    (b) $\mathbf{P}\,(\mathbf{A} \otimes \mathbf{A}) = \mathbf{P}\,(\mathbf{A} \otimes \mathbf{A})\,\mathbf{P}$

20. For square matrices $\mathbf{A}$ and $\mathbf{B}$ of order $n \times n$, show that $\mathbf{P}\,(\mathbf{A} \otimes \mathbf{B})\,\mathbf{P} = \mathbf{P}\,(\mathbf{B} \otimes \mathbf{A})\,\mathbf{P}$ for $\mathbf{P}$ defined in Problem 19.

2.5 Rank, Inverse, and Determinant

a. *Rank and Inverse*

Using (2.4.1), a matrix $\mathbf{A}_{n \times m}$ may be represented as a partitioned row or column matrix. The m column n-vectors span the column space of $\mathbf{A}$, and the n row m-vectors generate the row space of $\mathbf{A}$.

Definition 2.5.1 *The rank of a matrix* $\mathbf{A}_{n \times m}$ *is the number of linearly independent rows (or columns) of* $\mathbf{A}$.

The rank of $\mathbf{A}$ is denoted as rank($\mathbf{A}$) or simply $r(\mathbf{A})$ is the dimension of the space spanned by the rows (or columns) of $\mathbf{A}$. Clearly, $0 \leq r(\mathbf{A}) \leq \min(n, m)$. For $\mathbf{A} = \mathbf{0}$, the $r(\mathbf{A}) = 0$. If $m \leq n$, the $r(\mathbf{A})$ cannot exceed m, and if the $r(\mathbf{A}) = r = m$, the matrix $\mathbf{A}$ is said to have full column rank. If $\mathbf{A}$ is not of full column rank, then there are $m - r$ dependent column vectors in $\mathbf{A}$. Conversely, if $n \leq m$, there are $n - r$ dependent row vectors in $\mathbf{A}$. If the $r(\mathbf{A}) = n$, $\mathbf{A}$ is said to have full row rank.

To find the rank of a matrix $\mathbf{A}$, the matrix is reduced to an equivalent matrix which has the same rank as $\mathbf{A}$ by premultiplying $\mathbf{A}$ by a matrix $\mathbf{P}_{n\times n}$ that preserves the row rank of $\mathbf{A}$ and by postmultiplying $\mathbf{A}$ by a matrix $\mathbf{Q}_{m\times m}$ that preserves the column rank of $\mathbf{A}$, thus reducing $\mathbf{A}$ to a matrix whose rank can be obtained by inspection. That is,

$$\mathbf{PAQ} = \begin{bmatrix} \mathbf{I}_r & \mathbf{0} \\ \mathbf{0} & \mathbf{0} \end{bmatrix} = \mathbf{C}_{n\times m} \tag{2.5.1}$$

where the $r(\mathbf{PAQ}) = r(\mathbf{C}) = r$. Using $\mathbf{P}$ and $\mathbf{Q}$, the matrix $\mathbf{C}$ in (2.5.1) is called the canonical form of $\mathbf{A}$. Alternatively, $\mathbf{A}$ is often reduced to diagonal form. The diagonal form of $\mathbf{A}$ is represented as

$$\mathbf{P}^*\mathbf{AQ}^* = \begin{bmatrix} \mathbf{D}_r & \mathbf{0} \\ \mathbf{0} & \mathbf{0} \end{bmatrix} = \Lambda \tag{2.5.2}$$

for some sequence of row and column operations.

If we could find a matrix $\mathbf{P}^{-1}_{n\times n}$ such that $\mathbf{P}^{-1}\mathbf{P} = \mathbf{I}_n$ and a matrix $\mathbf{Q}^{-1}_{m\times m}$ such that $\mathbf{QQ}^{-1} = \mathbf{I}_m$, observe that

$$\begin{aligned} \mathbf{A} &= \mathbf{P}^{-1}\begin{bmatrix} \mathbf{I}_r & \mathbf{0} \\ \mathbf{0} & \mathbf{0} \end{bmatrix}\mathbf{Q}^{-1} \\ &= \mathbf{P}_1\mathbf{Q}_1 \end{aligned} \tag{2.5.3}$$

where $\mathbf{P}_1$ and $\mathbf{Q}_1$ are $n \times r$ and $r \times m$ matrices of rank r. Thus, we have factored the matrix $\mathbf{A}$ into a product of two matrices $\mathbf{P}_1\,\mathbf{Q}_1$ where $\mathbf{P}_1$ has column rank r and $\mathbf{Q}_1$ has row rank r.

The inverse of a matrix is closely associated with the rank of a matrix. The inverse of a square matrix $\mathbf{A}_{n\times n}$ is the unique matrix $\mathbf{A}^{-1}$ that satisfies the condition that

$$\mathbf{A}^{-1}\mathbf{A} = \mathbf{I}_n = \mathbf{AA}^{-1} \tag{2.5.4}$$

A square matrix $\mathbf{A}$ is said to be nonsingular if an inverse exists for $\mathbf{A}$; otherwise, the matrix $\mathbf{A}$ is singular. A matrix of full rank always has a unique inverse. Thus, in (2.5.3) if the $r(\mathbf{P}) = n$ and the $r(\mathbf{Q}) = m$ and matrices $\mathbf{P}$ and $\mathbf{Q}$ can be found, the inverses $\mathbf{P}^{-1}$ and $\mathbf{Q}^{-1}$ are unique.

In (2.5.4), suppose $\mathbf{A}^{-1} = \mathbf{A}'$, then the matrix $\mathbf{A}$ said to be an orthogonal matrix since $\mathbf{A}'\mathbf{A} = \mathbf{I} = \mathbf{AA}'$. Motivation for this definition follows from the fact that the columns of $\mathbf{A}$ form an orthonormal basis for $\mathbf{V}_n$. An elementary permutation matrix $\mathbf{I}_{n,m}$ is orthogonal. More generally, a vec-permutation (commutation) matrix is orthogonal since $\mathbf{I}'_{nm} = \mathbf{I}_{nm}$ and $\mathbf{I}^{-1}_{nm} = \mathbf{I}_{mn}$.

Finding the rank and inverse of a matrix is complicated and tedious, and usually performed on a computer. To determine the rank of a matrix, three basic operations called elementary operations are used to construct the matrices **P** and **Q** in (2.5.1). The three basic elementary operations are as follows.

(a) Any two rows (or columns) of **A** are interchanged.

(b) Any row of **A** is multiplied by a nonzero scalar α.

(c) Any row (or column) of **A** is replaced by adding to the replaced row (or column) a nonzero scalar multiple of another row (or column) of **A**.

In (a), the elementary matrix is no more than a permutation matrix. In (b), the matrix is a diagonal matrix which is obtained from **I** by replacing the (i, i) element by $\alpha_{ii} > 0$. Finally, in (c) the matrix is obtained from **I** by replacing one zero element with $\alpha_{ij} \neq 0$.

Example 2.5.1 *Let*

$$\mathbf{A} = \begin{bmatrix} a_{11} & a_{12} \\ a_{21} & a_{22} \end{bmatrix}$$

Then

$$\mathbf{I}_{1,2} = \begin{bmatrix} 0 & 1 \\ 1 & 0 \end{bmatrix} \quad \textit{and } \mathbf{I}_{1,2}\mathbf{A} \textit{ interchanges rows 1 and 2 in } \mathbf{A}$$

$$\mathbf{D}_{1,1}(\alpha) = \begin{bmatrix} \alpha & 0 \\ 0 & 1 \end{bmatrix} \quad \textit{and } \mathbf{D}_{1,1}(\alpha)\mathbf{A} \textit{ multiplies row 1 in } \mathbf{A} \textit{ by } \alpha$$

$$\mathbf{E}_{2,1}(\alpha) = \begin{bmatrix} 1 & 0 \\ \alpha & 1 \end{bmatrix} \quad \textit{and } \mathbf{E}_{2,1}(\alpha)\mathbf{A} \textit{ replaces row 2 in } \mathbf{A} \textit{ by adding to it } \alpha \textit{ times row 1 of } \mathbf{A}$$

Furthermore, the elementary matrices in Example 2.5.1 are nonsingular since the unique inverse matrices are

$$\mathbf{E}_{1,2}^{-1}(\alpha) = \begin{bmatrix} 1 & 0 \\ \alpha & 0 \end{bmatrix}, \quad \mathbf{D}_{1,1}^{-1}(\alpha) = \begin{bmatrix} \alpha^{-1} & 0 \\ 0 & 1 \end{bmatrix}, \quad \mathbf{I}_{1,2}^{-1} = \mathbf{I}_{1,2}'$$

To see how to construct **P** and **Q** to find the rank of **A**, we consider an example.

Example 2.5.2 *Let*

$$\mathbf{A} = \begin{bmatrix} 1 & 2 \\ 3 & 9 \\ 5 & 6 \end{bmatrix}, \qquad \mathbf{E}_{2,1}(-3) = \begin{bmatrix} 1 & 0 & 0 \\ -3 & 1 & 0 \\ 0 & 0 & 1 \end{bmatrix},$$

$$\mathbf{E}_{3,1}(-5) = \begin{bmatrix} 1 & 0 & 0 \\ 0 & 1 & 0 \\ -5 & 0 & 1 \end{bmatrix}, \qquad \mathbf{E}_{3,2}(4/3) = \begin{bmatrix} 1 & 0 & 0 \\ 0 & 1 & 0 \\ 0 & 4/3 & 1 \end{bmatrix},$$

$$\mathbf{D}_{2,2}(1/3) = \begin{bmatrix} 1 & 0 & 0 \\ 0 & 1/3 & 0 \\ 0 & 0 & 1 \end{bmatrix}, \qquad \mathbf{E}_{1,2}(-2) = \begin{bmatrix} 1 & -2 \\ 0 & 1 \end{bmatrix}$$

Then

$$\underbrace{\mathbf{D}_{2,2}(1/3)\mathbf{E}_{3,2}(4/3)\,\mathbf{E}_{3,1}(-5)\mathbf{E}_{2,1}(-3)}_{\mathbf{P}}\ \ \mathbf{A}\ \ \underbrace{\mathbf{E}_{1,2}(-2)}_{\mathbf{Q}} = \begin{bmatrix} 1 & 0 \\ 0 & 1 \\ 0 & 0 \end{bmatrix} = \begin{bmatrix} \mathbf{I}_2 & \mathbf{0} \\ 0 & 0 \end{bmatrix}$$

$$\begin{bmatrix} 1 & 0 & 0 \\ -1 & 1/3 & 0 \\ -1 & 4/3 & 1 \end{bmatrix} \mathbf{A} \begin{bmatrix} 1 & -2 \\ 0 & 1 \end{bmatrix} = \begin{bmatrix} 1 & 0 \\ 0 & 1 \\ 0 & 0 \end{bmatrix}$$

so that the $r(\mathbf{A}) = 2$. *Alternatively, the diagonal form is obtained by not using the matrix* $\mathbf{D}_{2,2}(1/3)$

$$\begin{bmatrix} 1 & 0 & 0 \\ -3 & 1 & 0 \\ -9 & 4/3 & 1 \end{bmatrix} \mathbf{A} \begin{bmatrix} 1 & -2 \\ 0 & 1 \end{bmatrix} = \begin{bmatrix} 1 & 0 \\ 0 & 3 \\ 0 & 0 \end{bmatrix}$$

From Example 2.5.2, the following theorem regarding the factorization of $\mathbf{A}_{n\times m}$ is evident.

Theorem 2.5.1 *For any matrix* $\mathbf{A}_{n\times m}$ *of rank r, there exist square nonsingular matrices* $\mathbf{P}_{n\times n}$ *and* $\mathbf{Q}_{m\times m}$ *such that*

$$\mathbf{PAQ} = \begin{bmatrix} \mathbf{I}_r & \mathbf{0} \\ \mathbf{0} & \mathbf{0} \end{bmatrix}$$

or

$$\mathbf{A} = \mathbf{P}^{-1} \begin{bmatrix} \mathbf{I}_r & \mathbf{0} \\ \mathbf{0} & \mathbf{0} \end{bmatrix} \mathbf{Q}^{-1} = \mathbf{P}_1\mathbf{Q}_1$$

where $\mathbf{P}_1$ *and* $\mathbf{Q}_1$ *are* $n \times r$ *and* $r \times m$ *matrices of rank r. Furthermore, if* $\mathbf{A}$ *is square and symmetric there exists a matrix* $\mathbf{P}$ *such that*

$$\mathbf{PAP}' = \begin{bmatrix} \mathbf{D}_r & \mathbf{0} \\ \mathbf{0} & \mathbf{0} \end{bmatrix} = \Lambda$$

and if the $r(\mathbf{A}) = n$, *then* $\mathbf{PAP}' = \mathbf{D}_n = \Lambda$ *and* $\mathbf{A} = \mathbf{P}^{-1}\Lambda(\mathbf{P}')^{-1}$.

Given any square nonsingular matrix $\mathbf{A}_{n\times n}$, elementary row operations when applied to $\mathbf{I}_n$ transforms $\mathbf{I}_n$ into $\mathbf{A}^{-1}$. To see this, observe that $\mathbf{PA} = \mathbf{U}_n$ where $\mathbf{U}_n$ is a unit upper triangular matrix and only $n(n-1)/2$ row operations $\mathbf{P}^*$ are needed to reduce $\mathbf{U}_n$ to $\mathbf{I}_n$; or $\mathbf{P}^*\mathbf{PA} = \mathbf{I}_n$; hence $\mathbf{A}^{-1} = \mathbf{P}^*\mathbf{PI}_n$ by definition. This shows that by operating on $\mathbf{A}$ and $\mathbf{I}_n$ with $\mathbf{P}^*\mathbf{P}$ simultaneously, $\mathbf{P}^*\mathbf{PA}$ becomes $\mathbf{I}_n$ and that $\mathbf{P}^*\mathbf{P}\,\mathbf{I}_n$ becomes $\mathbf{A}^{-1}$.

Example 2.5.3 *Let*

$$\mathbf{A} = \begin{bmatrix} 2 & 3 & 1 \\ 1 & 2 & 3 \\ 3 & 1 & 2 \end{bmatrix}$$

To find $\mathbf{A}^{-1}$*, write*

$$(\mathbf{A}\,|\mathbf{I}\|\text{ row totals}) = \left[\begin{array}{ccc|ccc||c} 2 & 3 & 1 & 1 & 0 & 0 & 7 \\ 1 & 2 & 3 & 0 & 1 & 0 & 7 \\ 3 & 1 & 2 & 0 & 0 & 1 & 7 \end{array}\right]$$

Multiply row one by $1/2$*, and subtract row one from row two. Multiply row three by* $1/3$*, and subtract row one from three.*

$$\left[\begin{array}{ccc|ccc||c} 1 & 3/2 & 1/2 & 1/2 & 0 & 0 & 7/2 \\ 0 & 1/2 & 5/2 & -1/2 & 1 & 0 & 7/2 \\ 0 & -7/6 & 1/6 & -1/2 & 0 & 1/3 & -7/6 \end{array}\right]$$

Multiply row two by 2 *and row three by* $-6/7$*. Then subtract row three from row two. Multiple row three by* $-7/36$*.*

$$\left[\begin{array}{ccc|ccc||c} 1 & 3/2 & 0 & 23/26 & -7/36 & -1/36 & 105/36 \\ 0 & 1 & 0 & 7/18 & 1/18 & -5/18 & 7/6 \\ 0 & 0 & 1 & -5/18 & 7/18 & 1/18 & 7/6 \end{array}\right]$$

Multiply row two by $-3/2$*, and add to row one.*

$$\left[\begin{array}{ccc|ccc||c} 1 & 0 & 0 & 1/18 & -5/18 & 7/18 & 7/6 \\ 0 & 1 & 0 & 7/18 & 1/18 & -5/18 & 7/6 \\ 0 & 0 & 1 & -5/18 & 7/18 & 1/18 & -7/6 \end{array}\right] = \left(\mathbf{I}\,\middle|\mathbf{A}^{-1}\middle\|\text{ row totals}\right)$$

Then

$$\mathbf{A}^{-1} = (1/18)\begin{bmatrix} 1 & -5 & 7 \\ 7 & 1 & -5 \\ -5 & 7 & 1 \end{bmatrix}$$

This inversion process is called Gauss' matrix inversion technique. The totals are included to systematically check calculations at each stage of the process. The sum of the elements in each row of the two partitions must equal the total when the elementary operations are applied simultaneous to $\mathbf{I}_n$, $\mathbf{A}$ and the column vector of totals.

When working with ranks and inverses of matrices, there are numerous properties that are commonly used. Some of the more important ones are summarized in Theorem 2.5.2 and Theorem 2.5.3. Again, all operations are assumed to be defined.

Theorem 2.5.2 *For any matrices* $\mathbf{A}_{n\times m}$, $\mathbf{B}_{m\times p}$, *and* $\mathbf{C}_{p\times q}$, *some properties of the matrix rank follow.*

1. $r(\mathbf{A}) = r(\mathbf{A}')$
2. $r(\mathbf{A}) = r(\mathbf{A}'\mathbf{A}) = r(\mathbf{A}\mathbf{A}')$
3. $r(\mathbf{A}) + r(\mathbf{B}) - n \le r(\mathbf{AB}) \le \min[r(\mathbf{A}), r(\mathbf{B})]$ *(Sylvester's law)*
4. $r(\mathbf{AB}) + r(\mathbf{BC}) \le r(\mathbf{B}) + r(\mathbf{ABC})$

5. *If* $r(\mathbf{A}) = m$ *and the* $r(\mathbf{B}) = p$, *then the* $r(\mathbf{AB}) \leq p$
6. $r(\mathbf{A} \otimes \mathbf{B}) = r(\mathbf{A})r(\mathbf{B})$
7. $r(\mathbf{A} \odot \mathbf{B}) \leq r(\mathbf{A})r(\mathbf{B})$
8. $r(\mathbf{A}) = \sum_{i=1}^{k} r(\mathbf{A}_i)$ for $\mathbf{A} = \bigoplus_{i=1}^{k} \mathbf{A}_i$
9. *For a partitioned matrix* $\mathbf{A} = [\mathbf{A}_1, \mathbf{A}_2, \ldots, \mathbf{A}_k]$, *the* $r\left(\sum_{i=1}^{k} \mathbf{A}_i\right) \leq r(\mathbf{A}) \leq \sum_{i=1}^{k} r(\mathbf{A}_i)$
10. *For any square, idempotent matrix* $\mathbf{A}_{n\times n}$ $(\mathbf{A}^2 = \mathbf{A})$, *of rank* $r < n$
    (a) $\operatorname{tr}(\mathbf{A}) = r(\mathbf{A}) = r$
    (b) $r(\mathbf{A}) + r(\mathbf{I} - \mathbf{A}) = n$

The inverse of a matrix, like the rank of a matrix, has a number of useful properties as summarized in Theorem 2.5.3.

Theorem 2.5.3 *Properties of matrix inversion.*

1. $(\mathbf{AB})^{-1} = \mathbf{B}^{-1}\mathbf{A}^{-1}$
2. $(\mathbf{A}')^{-1} = (\mathbf{A}^{-1})'$, *the inverse of a symmetric matrix is symmetric.*
3. $(\mathbf{A}^{-1})^{-1} = \mathbf{A}$
4. $(\mathbf{A} \otimes \mathbf{B})^{-1} = \mathbf{A}^{-1} \otimes \mathbf{B}^{-1}$ {*compare this with (1)*}
5. $(\mathbf{I} + \mathbf{A})^{-1} = \mathbf{A}(\mathbf{A} + \mathbf{I})^{-1}$
6. $(\mathbf{A} + \mathbf{B})^{-1} = \mathbf{A}^{-1} - \mathbf{A}^{-1}\mathbf{B}(\mathbf{A} + \mathbf{B})^{-1} = \mathbf{A}^{-1} - \mathbf{A}^{-1}(\mathbf{A}^{-1} + \mathbf{B}^{-1})^{-1}\mathbf{A}^{-1}$ *so that* $\mathbf{B}(\mathbf{A} + \mathbf{B})^{-1}\mathbf{A} = (\mathbf{A}^{-1} + \mathbf{B}^{-1})^{-1}$
7. $(\mathbf{A}^{-1} + \mathbf{B}^{-1})^{-1} = (\mathbf{I} + \mathbf{AB}^{-1})^{-1}$
8. $(\mathbf{A} + \mathbf{CBD})^{-1} = \mathbf{A}^{-1} - \mathbf{A}^{-1}\mathbf{C}(\mathbf{B}^{-1} + \mathbf{DA}^{-1}\mathbf{C})^{-1}\mathbf{DA}^{-1}$ *so that for* $\mathbf{C} = \mathbf{Z}$ *and* $\mathbf{D} = \mathbf{Z}'$ *we have that* $(\mathbf{A} + \mathbf{ZBZ}')^{-1} = \mathbf{A}^{-1} - \mathbf{A}^{-1}\mathbf{Z}(\mathbf{B}^{-1} + \mathbf{Z}'\mathbf{AZ})^{-1}\mathbf{Z}'\mathbf{A}$.
9. *For a partitioned matrix*

$$\mathbf{A} = \begin{bmatrix} \mathbf{A}_{11} & \mathbf{A}_{12} \\ \mathbf{A}_{21} & \mathbf{A}_{22} \end{bmatrix}, \mathbf{A}^{-1} = \begin{bmatrix} \mathbf{B}_{11} & \mathbf{B}_{12} \\ \mathbf{B}_{21} & \mathbf{B}_{22} \end{bmatrix}$$

where

$$\begin{aligned} \mathbf{B}_{11} &= (\mathbf{A}_{11} - \mathbf{A}_{12}\mathbf{A}_{22}^{-1}\mathbf{A}_{21})^{-1} \\ \mathbf{B}_{12} &= -\mathbf{B}_{11}\mathbf{B}_{12}\mathbf{A}_{22}^{-1} \\ \mathbf{B}_{21} &= \mathbf{A}_{22}^{-1}\mathbf{A}_{21}\mathbf{B}_{11} \\ \mathbf{B}_{22} &= \mathbf{A}_{22}^{-1} + \mathbf{A}_{22}^{-1}\mathbf{A}_{21}\mathbf{B}_{11}\mathbf{A}_{12}\mathbf{A}_{22}^{-1} \end{aligned}$$

provided all inverses exist.

b. *Generalized Inverses*

For an inverse of a matrix to be defined, the matrix $\mathbf{A}$ must be square and nonsingular. Suppose $\mathbf{A}_{n \times m}$ is rectangular and has full column rank m; then the $r(\mathbf{A}'\mathbf{A}) = m$ and the inverse of $\mathbf{A}'\mathbf{A}$ exists. Thus, $\left[(\mathbf{A}'\mathbf{A})^{-1}\mathbf{A}'\right]\mathbf{A} = \mathbf{I}_m$. However, $\mathbf{A}[(\mathbf{A}'\mathbf{A})^{-1}\mathbf{A}'] \neq \mathbf{I}_n$ so $\mathbf{A}$ has a left inverse, but not a right inverse. Alternatively, if the $r(\mathbf{A}) = n$, then the $r(\mathbf{A}\mathbf{A}') = n$ and $\mathbf{A}\mathbf{A}'(\mathbf{A}\mathbf{A}')^{-1} = \mathbf{I}_n$ so that $\mathbf{A}$ has a right inverse. Multiplying $\mathbf{I}_m$ by $\mathbf{A}$, we see that $\mathbf{A}(\mathbf{A}'\mathbf{A})^{-1}\mathbf{A}'\mathbf{A} = \mathbf{A}$ and multiplying $\mathbf{I}_n$ by $\mathbf{A}$, we also have that $\mathbf{A}\mathbf{A}'(\mathbf{A}\mathbf{A}')^{-1}\mathbf{A} = \mathbf{A}$. This leads to the definition of a generalized or g-inverse of $\mathbf{A}$.

Definition 2.5.2 *A generalized inverse of any matrix $\mathbf{A}_{n \times m}$, denoted by $\mathbf{A}^-$, is any matrix of order $m \times n$ that satisfies the condition*

$$\mathbf{A}\mathbf{A}^-\mathbf{A} = \mathbf{A}$$

Clearly, the matrix $\mathbf{A}^-$ is not unique. To make the g-inverse unique, additional conditions must be satisfied. This leads to the Moore-Penrose inverse $\mathbf{A}^+$. A Moore-Penrose inverse for any matrix $\mathbf{A}_{n \times m}$ is the unique matrix $\mathbf{A}^+$ that satisfies the four conditions

$$\begin{array}{llll} (1) & \mathbf{A}\mathbf{A}^+\mathbf{A} = \mathbf{A} & (3) & (\mathbf{A}\mathbf{A}^+)' = \mathbf{A}\mathbf{A}^+ \\ (2) & \mathbf{A}^+\mathbf{A}\mathbf{A}^+ = \mathbf{A} & (4) & \left(\mathbf{A}^+\mathbf{A}\right)' = \mathbf{A}^+\mathbf{A} \end{array} \tag{2.5.5}$$

To prove that the matrix $\mathbf{A}^+$ is unique, let $\mathbf{B}$ and $\mathbf{C}$ be two Moore-Penrose inverse matrices. Using properties (1) to (4) in (2.5.5, observe that the matrix $\mathbf{B} = \mathbf{C}$ since

$$\begin{aligned} &\mathbf{B} = \mathbf{BAB} = \mathbf{B}(\mathbf{AB})' = \mathbf{BB}'\mathbf{A}' = \mathbf{BB}'(\mathbf{ACA})' = \mathbf{BB}'\mathbf{A}'\mathbf{C}'\mathbf{A}' = \\ &\mathbf{B}(\mathbf{AB})'(\mathbf{AC})' = \mathbf{BABAC} = \mathbf{BAC} = \mathbf{BACAC} = (\mathbf{BA})'(\mathbf{CA})'\mathbf{C} = \\ &\mathbf{A}'\mathbf{B}'\mathbf{A}'\mathbf{C}'\mathbf{C} = (\mathbf{ABA})'\mathbf{CC} = \mathbf{A}'\mathbf{C}'\mathbf{C} = (\mathbf{CA})'\mathbf{C} = \mathbf{CAC} = \mathbf{C} \end{aligned}$$

This shows that the Moore-Penrose inverse is unique. The proof of existence is left as an exercise. From (2.5.5), $\mathbf{A}^-$ only satisfies conditions (1). Further, observe that if $\mathbf{A}$ has full column rank, the matrix

$$\mathbf{A}^+ = (\mathbf{A}'\mathbf{A})^{-1}\mathbf{A}' \tag{2.5.6}$$

satisfies conditions (1)–(4), above. If a square matrix $\mathbf{A}_{n \times n}$ has full rank, then $\mathbf{A}^{-1} = \mathbf{A}^- = \mathbf{A}^+$. If the $r(\mathbf{A}) = n$, then $\mathbf{A}^+ = \mathbf{A}'(\mathbf{A}\mathbf{A}')^{-1}$. If the columns of $\mathbf{A}$ are orthogonal, then $\mathbf{A}^+ = \mathbf{A}'$. If $\mathbf{A}_{n \times n}$ is idempotent, then $\mathbf{A}^+ = \mathbf{A}$. Finally, if $\mathbf{A} = \mathbf{A}'$, then $\mathbf{A}^+ = (\mathbf{A}')^+ = (\mathbf{A}^+)'$so $\mathbf{A}^+$ is symmetric. Other properties of $\mathbf{A}^+$ are summarized in Theorem 2.5.4.

Theorem 2.5.4 *For any matrix $\mathbf{A}_{n \times m}$, the following hold.*

1. $(\mathbf{A}^+)^+ = \mathbf{A}$
2. $\left(\mathbf{A}^+\right)' = (\mathbf{A}')^+$
3. $\mathbf{A}^+ = (\mathbf{A}'\mathbf{A})^+\mathbf{A}' = \mathbf{A}'(\mathbf{A}\mathbf{A}')^+$
4. *For* $\mathbf{A} = \mathbf{P}_1\mathbf{Q}_1$, $\mathbf{A}^+ = \mathbf{Q}_1'(\mathbf{P}_1'\mathbf{A}\mathbf{Q}_1')^{-1}\mathbf{P}_1'$ *where the* $r(\mathbf{P}_1) = r(\mathbf{Q}_1) = r$.
5. $(\mathbf{A}'\mathbf{A}^+) = \mathbf{A}^+(\mathbf{A}')^+ = \mathbf{A}^+(\mathbf{A}^+)'$

6. $(\mathbf{AA}^+)^+ = \mathbf{AA}^+$

7. $r(\mathbf{A}) = \mathbf{r}(\mathbf{A}^+) = r(\mathbf{AA}^+) = r(\mathbf{A}^+\mathbf{A})$

8. *For any matrix* $\mathbf{B}_{m\times p}$, $(\mathbf{AB})^+ \neq \mathbf{B}^+\mathbf{A}^+$.

9. *If* $\mathbf{B}$ *has full row rank,* $(\mathbf{AB})(\mathbf{AB})^+ = \mathbf{AA}^+$.

10. *For* $\mathbf{A} = \bigoplus_{i=1}^{k} \mathbf{A}_{ii}$, $\mathbf{A}^+ = \bigoplus_{i=1}^{k} \mathbf{A}_{ii}^+$.

While (2.5.6) yields a convenient Moore-Penrose inverse for $\mathbf{A}_{n\times m}$ when the $r(\mathbf{A}) = m$, we may use Theorem 2.5.1 to create $\mathbf{A}^-$ when the $r(\mathbf{A}) = r < m \leq n$. We have that

$$\mathbf{PAQ} = \Lambda = \begin{bmatrix} \mathbf{D}_r & \mathbf{0} \\ \mathbf{0} & \mathbf{0} \end{bmatrix}$$

Letting

$$\Lambda^- = \begin{bmatrix} \mathbf{D}_r^{-1} & \mathbf{0} \\ \mathbf{0} & \mathbf{0} \end{bmatrix}$$

$\Lambda\Lambda^-\Lambda = \Lambda$, and a g-inverse of $\mathbf{A}$ is

$$\mathbf{A}^- = \mathbf{Q}\Lambda^-\mathbf{P} \tag{2.5.7}$$

Example 2.5.4 *Let*

$$\mathbf{A} = \begin{bmatrix} 2 & 4 \\ 2 & 2 \\ -2 & 0 \end{bmatrix}$$

Then with

$$\mathbf{P} = \begin{bmatrix} 1 & 0 & 0 \\ -1 & 1 & 0 \\ -1 & 2 & 1 \end{bmatrix} \quad \text{and} \quad \mathbf{Q} = \begin{bmatrix} 1 & -2 \\ 0 & 1 \end{bmatrix}$$

$$\mathbf{PAQ} = \begin{bmatrix} 2 & 0 \\ 0 & -2 \\ 0 & 0 \end{bmatrix}$$

Thus

$$\Lambda^- = \begin{bmatrix} 1/2 & 0 & 0 \\ 0 & -1/2 & 0 \end{bmatrix} \quad \text{and} \quad \mathbf{A}^- = \mathbf{Q}\Lambda^-\mathbf{P} = \begin{bmatrix} -1/2 & 0 & 0 \\ 1/2 & -1/2 & 0 \end{bmatrix}$$

Since $r(\mathbf{A}) = 2 = n$, *we have that*

$$\mathbf{A}^+ = (\mathbf{A}'\mathbf{A})^{-1}\mathbf{A}' = (1/96)\begin{bmatrix} 20 & -12 \\ -12 & 12 \end{bmatrix}\mathbf{A}'$$
$$= (1/96)\begin{bmatrix} -8 & 16 & -40 \\ 24 & 0 & 24 \end{bmatrix}$$

Example 2.5.5 *Let*

$$\mathbf{A} = \begin{bmatrix} 4 & 2 & 2 \\ 2 & 2 & 0 \\ 2 & 0 & 2 \end{bmatrix}$$

Choose

$$\mathbf{P} = \begin{bmatrix} 1/4 & 0 & 0 \\ -1/2 & 1 & 0 \\ -1 & 1 & 1 \end{bmatrix}, \mathbf{Q} = \begin{bmatrix} 1/4 & -1/2 & -1 \\ 0 & 1 & 1 \\ 0 & 0 & 1 \end{bmatrix}$$

and

$$\Lambda^- = \begin{bmatrix} 4 & 0 & 0 \\ 0 & 1 & 0 \\ 0 & 0 & 0 \end{bmatrix}$$

Then

$$\mathbf{A}^- = \begin{bmatrix} 1/2 & -1/2 & 0 \\ -1/2 & 1 & 0 \\ 0 & 0 & 0 \end{bmatrix}$$

Theorem 2.5.5 summarizes some important properties of the generalized inverse matrix $\mathbf{A}^-$.

Theorem 2.5.5 *For any matrix $\mathbf{A}_{n \times m}$, the following hold.*

1. $(\mathbf{A}')^- = (\mathbf{A}^-)'$
2. *If* $\mathbf{A} = \mathbf{P}^{-1}\mathbf{A}^-\mathbf{Q}^{-1}$, *then* $(\mathbf{PAQ})^- = \mathbf{QA}^-\mathbf{P}$.
3. $r(\mathbf{A}) = r(\mathbf{AA}^-) = r(\mathbf{A}^-\mathbf{A}) \le r(\mathbf{A}^-)$
4. *If* $(\mathbf{A}'\mathbf{A})^-$ *is a g-inverse of* $\mathbf{A}$, *then* $\mathbf{A}^- = (\mathbf{A}'\mathbf{A})^-\mathbf{A}'$, $\mathbf{A}^+ = \mathbf{A}'(\mathbf{AA}')^-\mathbf{A}(\mathbf{A}'\mathbf{A})^-\mathbf{A}'$ *and* $\mathbf{A}(\mathbf{A}'\mathbf{A})^-\mathbf{A}'$ *is unique and symmetric and called an orthogonal projection matrix.*
5. *For* $\mathbf{A} = \begin{bmatrix} \mathbf{A}_{11} & \mathbf{A}_{12} \\ \mathbf{A}_{21} & \mathbf{A}_{22} \end{bmatrix}$ and $\mathbf{A}^- = \begin{bmatrix} \mathbf{A}_{11}^{-1} & \mathbf{0} \\ \mathbf{0} & \mathbf{0} \end{bmatrix}$ *for some nonsingular matrix* $\mathbf{A}_{11}$ *of* $\mathbf{A}$, *then*

$$\mathbf{A}^- = \begin{bmatrix} \mathbf{A}_{11}^{-1} - \mathbf{A}_{11}^{-1}\mathbf{A}_{12}\mathbf{A}_{11}^{-1} & \mathbf{0} \\ \mathbf{0} & \mathbf{0} \end{bmatrix}$$

6. *For*

$$\mathbf{M} = \begin{bmatrix} \mathbf{A} & \mathbf{B} \\ \mathbf{B}' & \mathbf{C} \end{bmatrix}$$

$$\mathbf{M}^- = \begin{bmatrix} \mathbf{A}^- + \mathbf{A}^-\mathbf{BF}^-\mathbf{B}'\mathbf{A}' & -\mathbf{A}^-\mathbf{BF}^- \\ -\mathbf{F}^-\mathbf{B}'\mathbf{A}' & \mathbf{F}^- \end{bmatrix}$$

$$= \begin{bmatrix} \mathbf{A}^- & \mathbf{0} \\ \mathbf{0} & \mathbf{0} \end{bmatrix} + [-\mathbf{A}^-, \mathbf{B}]\,\mathbf{F}^-\,[-\mathbf{B}'\mathbf{A}^-, \mathbf{I}]$$

where $\mathbf{F} = \mathbf{C} - \mathbf{B}'\mathbf{A}^-\mathbf{B}$.

The Moore-Penrose inverse and g-inverse of a matrix are used to solve systems of linear equations discussed in Section 2.6. We close this section with some operators that map a matrix to a scalar value. For a further discussion of generalized inverses, see Boullion and Odell (1971), Rao and Mitra (1971), Rao (1973a), and Harville (1997).

c. *Determinants*

Associated with any $n \times n$ square matrix $\mathbf{A}$ is a unique scalar function of the elements of $\mathbf{A}$ called the determinant of $\mathbf{A}$, written $|\mathbf{A}|$. The determinant, like the inverse and rank, of a matrix is difficult to compute. Formally, the determinant of a square matrix $\mathbf{A}$ is a real-valued function defined by

$$|\mathbf{A}| = \sum^{n!} (-1)^k a_{1i_1}, a_{2i_2}, \ldots, a_{ni_n} \tag{2.5.8}$$

where the summation is taken over all $n!$ permutations of the elements of $\mathbf{A}$ such that each product contains only one element from each row and each column of $\mathbf{A}$. The first subscript is always in its natural order and the second subscripts are $1, 2, \ldots, n$ taken in some order. The exponent k represents the necessary number of interchanges of successive elements in a sequence so that the second subscripts are placed in their natural order $1, 2, \ldots, n$.

Example 2.5.6 *Let*

$$\mathbf{A} = \begin{bmatrix} a_{11} & a_{12} \\ a_{21} & a_{22} \end{bmatrix}$$

Then

$$|\mathbf{A}| = (-1)^k a_{11}a_{22} + (-1)^k a_{12}a_{21}$$

$$|\mathbf{A}| = a_{11}a_{22} - a_{12}a_{21}$$

Let

$$\mathbf{A} = \begin{bmatrix} a_{11} & a_{12} & a_{13} \\ a_{21} & a_{22} & a_{23} \\ a_{31} & a_{32} & a_{33} \end{bmatrix}$$

Then

$$\begin{aligned} |\mathbf{A}| &= (-1)^k a_{11}a_{22}a_{33} + (-1)^k a_{12}a_{23}a_{31} + (-1)^k a_{13}a_{21}a_{32} + (-1)^k a_{11}a_{23}a_{32} \\ &\quad + (-1)^k a_{12}a_{21}a_{33} + (-1)^k a_{13}a_{22}a_{31} \\ &= a_{11}a_{22}a_{33} + a_{12}a_{23}a_{31} + a_{13}a_{21}a_{32} - a_{11}a_{23}a_{32} - a_{12}a_{21}a_{33} - a_{13}a_{22}a_{31} \end{aligned}$$

Representing $\mathbf{A}$ in Example 2.5.6 as

$$[\mathbf{A} \mid \mathbf{B}] = \left[\begin{array}{ccc|cc} & & (-) & (-) & (-) \\ a_{11} & a_{12} & a_{13} & a_{11} & a_{12} \\ a_{21} & a_{22} & a_{23} & a_{21} & a_{22} \\ a_{31} & a_{32} & a_{33} & a_{31} & a_{32} \\ & & (+) & (+) & (+) \end{array}\right]$$

where **B** is the first two columns of **A**, observe that the $|\mathbf{A}|$ may be calculated by evaluating the diagonal products on the matrix $[\mathbf{A} \mid \mathbf{B}]$, similar to the 2×2 case where $(+)$ signs represent plus "diagonal" product terms and $(-)$ signs represent minus "diagonal" product terms in the array in the evaluation of the determinant.

Expression (2.5.8) does not provide for a systematic procedure for finding the determinant. An alternative expression for the $|\mathbf{A}|$ is provided using the cofactors of a matrix **A**. By deleting the ith row and j^{th} column of **A** and forming the determinant of the resulting sub-matrix, one creates the minor of the element which is represented as $|\mathbf{M}_{ij}|$. The cofactor of a_{ij} is $C_{ij} = (-1)^{i+j}\,|\mathbf{M}|_{ij}$, and is termed the signed minor of the element. The $|\mathbf{A}|$ defined in terms of cofactors is

$$|\mathbf{A}| = \sum_{j=1}^{n} a_{ij} C_{ij} \qquad \text{for any } i \tag{2.5.9}$$

$$|\mathbf{A}| = \sum_{i=1}^{n} a_{ij} C_{ij} \qquad \text{for any } j \tag{2.5.10}$$

These expressions are called the row and column expansion by cofactors, respectively, for finding the $|\mathbf{A}|$.

Example 2.5.7 *Let*

$$\mathbf{A} = \begin{bmatrix} 6 & 1 & 0 \\ 3 & -1 & 2 \\ 4 & 0 & -1 \end{bmatrix}$$

Then

$$\begin{aligned} |\mathbf{A}| &= (6)\,(-1)^{1+1} \begin{vmatrix} -1 & 2 \\ 0 & -1 \end{vmatrix} + (1)\,1^{(1+2)} \begin{vmatrix} 3 & 2 \\ 4 & -1 \end{vmatrix} + (0)\,1^{(1+3)} \begin{vmatrix} 3 & -1 \\ 4 & 0 \end{vmatrix} \\ &= 6\,(1-0) + (-1)\,(-3-8) \\ &= 17 \end{aligned}$$

Associated with a square matrix is the adjoint (or adjugate) matrix of **A**. If C_{ij} is the cofactor of an element a_{ij} in the matrix **A**, the adjoint of **A** is the transpose of the cofactors of **A**

$$\operatorname{adj} \mathbf{A} = [C_{ij}]' = [C_{ji}] \tag{2.5.11}$$

Example 2.5.8 *For* **A** *in Example 2.5.7, the*

$$\operatorname{adj} \mathbf{A} = \begin{bmatrix} 1 & 11 & 4 \\ 1 & -6 & 4 \\ 2 & -12 & -9 \end{bmatrix}' = \begin{bmatrix} 1 & 1 & 2 \\ 11 & -6 & -12 \\ 4 & 4 & -9 \end{bmatrix}$$

and the

$$(\operatorname{adj} \mathbf{A})\mathbf{A} = \begin{bmatrix} 17 & 0 & 0 \\ 0 & 17 & 0 \\ 0 & 0 & 17 \end{bmatrix} = \begin{bmatrix} |\mathbf{A}| & 0 & 0 \\ 0 & |\mathbf{A}| & 0 \\ 0 & 0 & |\mathbf{A}| \end{bmatrix}$$

Example 2.5.8 motivates another method for finding $\mathbf{A}^{-1}$. In general,

$$\mathbf{A}^{-1} = \frac{\text{adj}\,\mathbf{A}}{|\mathbf{A}|} \tag{2.5.12}$$

where if the $|\mathbf{A}| \neq 0$, $\mathbf{A}$ is nonsingular. In addition, $|\mathbf{A}|^{-1} = 1/\,|\mathbf{A}|$. Other properties of the determinant are summarized in Theorem 2.5.6.

Theorem 2.5.6 *Properties of determinants.*

1. $|\mathbf{A}| = |\mathbf{A}'|$

2. $|\mathbf{AB}| = |\mathbf{BA}|$

3. *For an orthogonal matrix,* $|\mathbf{A}| = \pm 1$.

4. *If* $\mathbf{A}^2 = \mathbf{A}$, ($\mathbf{A}$ is idempotent), *then the* $|\mathbf{A}| = 0$ *or* 1.

5. *For* $\mathbf{A}_{n\times n}$ *and* $\mathbf{B}_{m\times m}$
$$|\mathbf{A} \otimes \mathbf{B}| = |\mathbf{A}|^m \, |\mathbf{B}|^n$$

6. *For* $\mathbf{A} = \begin{bmatrix} \mathbf{A}_{11} & \mathbf{A}_{12} \\ \mathbf{A}_{21} & \mathbf{A}_{22} \end{bmatrix}$, *then*
$$|\mathbf{A}| = |\mathbf{A}_{11}| \left|\mathbf{A}_{22} - \mathbf{A}_{21}\mathbf{A}_{11}^{-1}\mathbf{A}_{12}\right| = |\mathbf{A}_{22}| \left|\mathbf{A}_{11} - \mathbf{A}_{12}\mathbf{A}_{22}^{-1}\mathbf{A}_{21}\right|,$$
provided $\mathbf{A}_{11}$ *and* $\mathbf{A}_{22}$ *are nonsingular.*

7. *For* $\mathbf{A} = \bigoplus_{i=1}^{k} \mathbf{A}_{ii}$, $|\mathbf{A}| = \prod_{i=1}^{k} |\mathbf{A}_{ii}|$.

8. $|\alpha \mathbf{A}| = \alpha^n \, |\mathbf{A}|$

Exercises 2.5

1. For
$$\mathbf{A} = \begin{bmatrix} 1 & 0 & 2 \\ 3 & 1 & 5 \\ 5 & 2 & 8 \\ 0 & 0 & 1 \end{bmatrix}$$
Use Theorem 2.5.1 to factor $\mathbf{A}$ into the product
$$\underset{4\times 3}{\mathbf{A}} = \underset{4\times r}{\mathbf{P}_1} \; \underset{r\times 3}{\mathbf{Q}_1}$$
where $r = R(\mathbf{A})$.

2. For

$$\mathbf{A} = \begin{bmatrix} 2 & 1 & 2 \\ 1 & 0 & 4 \\ 2 & 4 & -16 \end{bmatrix}$$

   (a) Find $\mathbf{P}$ and $\mathbf{P}'$ such that $\mathbf{P}'\mathbf{A}\mathbf{P} = \Lambda$.

   (b) Find a factorization for $\mathbf{A}$.

   (c) If $\mathbf{A}^{1/2}\mathbf{A}^{1/2} = \mathbf{A}$, define $\mathbf{A}^{1/2}$.

3. Find two matrices $\mathbf{A}$ and $\mathbf{B}$ such that the $r = r(\mathbf{AB}) \leq \min[r(\mathbf{A}), r(\mathbf{B})]$.

4. Prove that $\mathbf{AB}$ is nonsingular if $\mathbf{A}$ has full row rank and $\mathbf{B}$ has full column rank.

5. Verify property (6) in Theorem 2.5.2.

6. For

$$\mathbf{A} = \begin{bmatrix} 1 & 0 & 0 & 1 \\ 3 & 5 & 1 & 2 \\ 1 & -1 & 2 & -1 \\ 0 & 3 & 4 & 1 \end{bmatrix}$$

   (a) Find $\mathbf{A}^{-1}$ using Gauss' matrix inversion method.

   (b) Find $\mathbf{A}^{-1}$ using formula (2.5.12).

   (c) Find $\mathbf{A}^{-1}$ by partitioning $\mathbf{A}$ and applying property (9) in Theorem 2.5.3.

7. For

$$\mathbf{A} = \begin{bmatrix} 2 & 1 & -1 \\ 0 & 2 & 3 \\ 1 & 1 & 1 \end{bmatrix}, \quad \mathbf{B} = \begin{bmatrix} 1 & 2 & 3 \\ 1 & 0 & 0 \\ 2 & 1 & 1 \end{bmatrix}, \quad \text{and} \quad \mathbf{C} = \begin{bmatrix} 1 & 0 & 1 \\ 0 & 2 & 0 \\ 3 & 0 & 2 \end{bmatrix}$$

   Verify Theorem 2.5.3.

8. For

$$\mathbf{A} = \begin{bmatrix} 1 & 0 \\ 2 & 5 \end{bmatrix} \quad \text{and} \quad \mathbf{B} = \begin{bmatrix} 1 & 0 & 1 \\ 0 & 2 & 0 \\ 3 & 0 & 2 \end{bmatrix}$$

   Find the $r(\mathbf{A} \otimes \mathbf{B})$ and the $|\mathbf{A} \otimes \mathbf{B}|$.

9. For $\mathbf{I}_n$ and $\mathbf{J}_n = \mathbf{1}_n\mathbf{1}_n'$, verify

$$(\alpha\mathbf{I}_n + \beta\mathbf{J}_n)^{-1} = \left[\mathbf{I}_n - \frac{\alpha}{\alpha + n\beta}\mathbf{J}_n\right] / \alpha$$

   for $\alpha + n\beta \neq 0$.

10. Prove that $(\mathbf{I} + \mathbf{A})^{-1}$ is $\mathbf{A}\,(\mathbf{A} + \mathbf{I})^{-1}$.

11. Prove that $|\mathbf{A}|\,|\mathbf{B}| \leq |\mathbf{A} \odot \mathbf{B}|$

12. For a square matrix $\mathbf{A}_{n\times n}$ that is idempotent where the $r(\mathbf{A}) = r$, prove

    (a) $\text{tr}(\mathbf{A}) = r(\mathbf{A}) = r$;

    (b) $(\mathbf{I} - \mathbf{A})$ is idempotent;

    (c) $r(\mathbf{A}) + r(\mathbf{I} - \mathbf{A}) = n$.

13. For the Toeplitz matrix

$$\mathbf{A} = \begin{bmatrix} 1 & \beta & \beta^2 \\ \alpha & 1 & \beta \\ \alpha^2 & \alpha & 1 \end{bmatrix}$$

Find $\mathbf{A}^{-1}$ for $\alpha\beta \neq 1$.

14. For the Hadamard matrices

$$\mathbf{H}_{4\times 4} = \begin{bmatrix} 1 & 1 & 1 & 1 \\ -1 & -1 & 1 & 1 \\ 1 & -1 & 1 & -1 \\ 1 & -1 & -1 & 1 \end{bmatrix}$$

$$\mathbf{H}_{2\times 2} = \begin{bmatrix} 1 & 1 \\ 1 & -1 \end{bmatrix}$$

Verify that

    (a) $|\mathbf{H}_{n\times n}| = \pm\, n^{n/2}$;

    (b) $n^{-1/2}\mathbf{H}_{n\times n}$;

    (c) $\mathbf{H}'\mathbf{H} = \mathbf{H}\mathbf{H}' = n\mathbf{I}_n$ for $\mathbf{H}_{n\times n}$.

15. Prove that for $\mathbf{A}_{n\times m}$ and $\mathbf{B}_{m\times p}$, $|\mathbf{I}_n + \mathbf{AB}| = |\mathbf{I}_m + \mathbf{BA}|$.

16. Find the Moore-Penrose and a g-inverse for the matrices

$$(a)\ \begin{bmatrix} 1 \\ 2 \\ 0 \end{bmatrix}, \quad (b)\ [0, 1, 2,], \quad \text{and} \quad (c)\ \begin{bmatrix} 1 & 2 \\ 0 & -1 \\ 1 & 0 \end{bmatrix}$$

17. Find g-inverses for

$$\mathbf{A} = \begin{bmatrix} 8 & 4 & 4 \\ 4 & 4 & 0 \\ 4 & 0 & 4 \end{bmatrix} \quad \text{and} \quad \mathbf{B} = \begin{bmatrix} 0 & 0 & 0 & 2 \\ 0 & 1 & 2 & 3 \\ 0 & 4 & 5 & 6 \\ 0 & 7 & 8 & 9 \end{bmatrix}$$

18. For

$$\mathbf{X} = \begin{bmatrix} 1 & 1 & 0 \\ 1 & 1 & 0 \\ 1 & 0 & 1 \\ 1 & 0 & 1 \end{bmatrix} \quad \text{and} \quad \mathbf{A} = \mathbf{X}'\mathbf{X}$$

Find the $r(\mathbf{A})$, a g-inverse $\mathbf{A}^-$ and the matrix $\mathbf{A}^+$.

19. Verify that each of the matrices $\mathbf{P}_1 = \mathbf{A}'\left(\mathbf{A}'\mathbf{A}\right)^{-1}\mathbf{A}$, $\mathbf{P}_2 = \mathbf{A}'\left(\mathbf{A}'\mathbf{A}\right)^{-}\mathbf{A}$, and $\mathbf{P}_3 = \mathbf{A}'\left(\mathbf{A}'\mathbf{A}\right)^{+}\mathbf{A}$ are symmetric, idempotent, and unique. The matrices $\mathbf{P}_i$ are called projection matrices. What can you say about $\mathbf{I} - \mathbf{P}_i$?

20. Verify that

$$\mathbf{A}^- + \mathbf{Z} - \mathbf{A}^-\mathbf{A}\mathbf{X}\mathbf{A}\mathbf{A}^-$$

is a g-inverse of $\mathbf{A}$.

21. Verify that $\mathbf{B}^-\mathbf{A}^-$ is a g-inverse of $\mathbf{AB}$ if and only if $\mathbf{A}^-\mathbf{A}\mathbf{B}\mathbf{B}^-$ is idempotent.

22. Prove that if the Moore-Penrose inverse $\mathbf{A}^+$, which satisfies (2.5.5), always exists.

23. Show that for any $\mathbf{A}^-$ of a symmetric matrix $\left(\mathbf{A}^-\right)^2$ is a g-inverse of $\mathbf{A}^2$ if $\mathbf{A}^-\mathbf{A}$ is symmetric.

24. Show that $\left(\mathbf{A}'\mathbf{A}\right)^-\mathbf{A}'$ is a g-inverse of $\mathbf{A}$, given $\left(\mathbf{A}'\mathbf{A}\right)^-$ is a g-inverse of $\left(\mathbf{A}'\mathbf{A}\right)$.

25. Show that $\mathbf{A}\mathbf{A}^+ = \mathbf{A}\left(\mathbf{A}'\mathbf{A}\right)^-\mathbf{A}$.

26. Let $\mathbf{D}_n$ be a duplication matrix of order $n^2 \times n\,(n+1)/2$ in that $\mathbf{D}_n \operatorname{vech} \mathbf{A} = \operatorname{vec} \mathbf{A}$ for any symmetric matrix $\mathbf{A}$. Show that

    (a) $\operatorname{vech} \mathbf{A} = \mathbf{D}_n^- \operatorname{vec} \mathbf{A}$;

    (b) $\mathbf{D}_n\mathbf{D}_n^+ \operatorname{vec} \mathbf{A} = \mathbf{P} \operatorname{vec} \mathbf{A}$ where $\mathbf{P}$ is a projection matrix, a symmetric, and idempotent matrix;

    (c) $\left[\mathbf{D}_n'\left(\mathbf{A} \otimes \mathbf{A}\right)\mathbf{D}_n\right]^{-1} = \mathbf{D}_n^+\left(\mathbf{A}^{-1} \otimes \mathbf{A}^{-1}\right)\mathbf{D}_n^{+'}$.

2.6 Systems of Equations, Transformations, and Quadratic Forms

a. *Systems of Equations*

Generalized inverses are used to solve systems of equations of the form

$$\mathbf{A}_{n\times m}\mathbf{x}_{m\times 1} = \mathbf{y}_{n\times 1} \tag{2.6.1}$$

where $\mathbf{A}$ and $\mathbf{y}$ are known. If $\mathbf{A}$ is square and nonsingular, the solution is $\widehat{\mathbf{x}} = \mathbf{A}^{-1}\mathbf{y}$. If $\mathbf{A}$ has full column rank, then $\mathbf{A}^+ = (\mathbf{A}'\mathbf{A})^{-1}\mathbf{A}'$ so that $\widehat{\mathbf{x}} = \mathbf{A}^+\mathbf{y} = (\mathbf{A}'\mathbf{A})^{-1}\mathbf{A}'\mathbf{y}$ provides the unique solution since $(\mathbf{A}'\mathbf{A})^{-1}\mathbf{A}'\mathbf{A} = \mathbf{A}^{-1}$. When the system of equations in (2.6.1) has a unique solution, the system is consistent. However, a unique solution does not have to exist for the system to be consistent (have a solution). If the $r(\mathbf{A}) = r < m \le n$, the system of equations in (2.6.1) will have a solution $\widehat{\mathbf{x}} = \mathbf{A}^{-}\mathbf{y}$ if and only if the system of equations is consistent.

Theorem 2.6.1 *The system of equations* $\mathbf{Ax} = \mathbf{y}$ *is consistent if and only if* $\mathbf{AA}^{-}\mathbf{y} = \mathbf{y}$, *and the general solution is* $\widehat{\mathbf{x}} = \mathbf{A}^{-}\mathbf{y} + (\mathbf{I}_m - \mathbf{A}^{-}\mathbf{A})\mathbf{z}$ *for arbitrary vectors* $\mathbf{z}$*; every solution has this form.*

Since Theorem 2.6.1 is true for any g-inverse of $\mathbf{A}$, it must be true for $\mathbf{A}^{+}$ so that $\widehat{\mathbf{x}} = \mathbf{A}^{+}\mathbf{y} + (\mathbf{I}_m - \mathbf{A}^{+}\mathbf{A})\mathbf{z}$. For a homogeneous system where $\mathbf{y} = \mathbf{0}$ or $\mathbf{Ax} = \mathbf{0}$, a general solution becomes $\widehat{\mathbf{x}} = (\mathbf{I}_m - \mathbf{A}^{-}\mathbf{A})\mathbf{z}$. When $\mathbf{y} \neq \mathbf{0}$, (2.6.1) is called a nonhomogeneous system of equations.

To solve a consistent system of equations, called the normal equations in many statistical applications, three general approaches are utilized when the $r(\mathbf{A}) = r < m \le n$. These approaches include (1) restricting the number of unknowns, (2) reparameterization, and (3) generalized inverses.

The method of adding restrictions to solve (2.6.1) involves augmenting the matrix $\mathbf{A}$ of rank r by a matrix $\mathbf{R}$ of rank $m - r$ such that the $r(\mathbf{A}'\mathbf{R}') = r + (m - r) = m$, a matrix of full rank. The augmented system with side conditions $\mathbf{Rx} = \boldsymbol{\theta}$ becomes

$$\begin{bmatrix} \mathbf{A} \\ \mathbf{R} \end{bmatrix} \mathbf{x} = \begin{bmatrix} \mathbf{y} \\ \boldsymbol{\theta} \end{bmatrix} \tag{2.6.2}$$

The unique solution to (2.6.2) is

$$\widehat{\mathbf{x}} = \left(\mathbf{A}'\mathbf{A} + \mathbf{R}'\mathbf{R}\right)^{-1}\left(\mathbf{A}'\mathbf{y} + \mathbf{R}'\boldsymbol{\theta}\right) \tag{2.6.3}$$

For $\boldsymbol{\theta} = \mathbf{0}, \widehat{\mathbf{x}} = (\mathbf{A}'\mathbf{A} + \mathbf{R}'\mathbf{R})^{-1}\mathbf{A}'\mathbf{y}$ so that $\mathbf{A}^{+} = (\mathbf{A}'\mathbf{A} + \mathbf{R}'\mathbf{R})^{-1}\mathbf{A}'$ is a Moore-Penrose inverse.

The second approach to solving (2.6.1) when $r(\mathbf{A}) = r < m \le n$ is called the reparameterization method. Using this method we can solve the system for r linear combinations of the unknowns by factoring $\mathbf{A}$ as a product where one matrix is known. Factoring $\mathbf{A}$ as

$$\underset{n \times m}{\mathbf{A}} = \underset{n \times r}{\mathbf{B}} \; \underset{r \times m}{\mathbf{C}}$$

and substituting $\mathbf{A} = \mathbf{BC}$ into (2.6.1),

$$\begin{aligned} \mathbf{Ax} &= \mathbf{y} \\ \mathbf{BCx} &= \mathbf{y} \\ (\mathbf{B}'\mathbf{B})\mathbf{Cx} &= \mathbf{B}'\mathbf{y} \\ \mathbf{Cx} &= (\mathbf{B}'\mathbf{B})^{-1}\mathbf{B}'\mathbf{y} \\ \widehat{\mathbf{x}}^{*} &= (\mathbf{B}'\mathbf{B})^{-1}\mathbf{B}'\mathbf{y} \end{aligned} \tag{2.6.4}$$

a unique solution for the reparameterized vector $\mathbf{x}^* = \mathbf{Cx}$ is realized. Here, $\mathbf{B}^+ = (\mathbf{B}'\mathbf{B})^{-1}\mathbf{B}'$ is a Moore-Penrose inverse.

Because $\mathbf{A} = \mathbf{BC}$, $\mathbf{C}$ must be selected so that the rows of $\mathbf{C}$ are in the row space of $\mathbf{A}$. Hence, the

$$r\begin{bmatrix} \mathbf{A} \\ \mathbf{C} \end{bmatrix} = r(\mathbf{A}) = r(\mathbf{C}) = r$$
$$\mathbf{B} = \mathbf{B}(\mathbf{CC}')(\mathbf{CC}')^{-1} = \mathbf{AC}'(\mathbf{CC}')^{-1} \tag{2.6.5}$$

so that $\mathbf{B}$ is easily determined given the matrix $\mathbf{C}$. In many statistical applications, $\mathbf{C}$ is a contrast matrix.

To illustrate these two methods, we again consider the two group ANOVA model

$$y_{ij} = \mu + \alpha_i + e_{ij} \qquad i = 1, 2 \text{ and } j = 1, 2$$

Then using matrix notation

$$\begin{bmatrix} y_{11} \\ y_{12} \\ y_{21} \\ y_{22} \end{bmatrix} = \begin{bmatrix} 1 & 1 & 0 \\ 1 & 1 & 0 \\ 1 & 0 & 1 \\ 1 & 0 & 1 \end{bmatrix} \begin{bmatrix} \mu \\ \alpha_1 \\ \alpha_2 \end{bmatrix} + \begin{bmatrix} e_{11} \\ e_{12} \\ e_{21} \\ e_{22} \end{bmatrix} \tag{2.6.6}$$

For the moment, assume $\mathbf{e}' = [e_{11}, e_{12}, e_{21}, e_{22}] = \mathbf{0}'$ so that the system becomes

$$\begin{array}{ccc} \mathbf{A} & \mathbf{x} = & \mathbf{y} \\ \begin{bmatrix} 1 & 1 & 0 \\ 1 & 1 & 0 \\ 1 & 0 & 1 \\ 1 & 0 & 1 \end{bmatrix} & \begin{bmatrix} \mu \\ \alpha_1 \\ \alpha_2 \end{bmatrix} = & \begin{bmatrix} y_{11} \\ y_{12} \\ y_{21} \\ y_{22} \end{bmatrix} \end{array} \tag{2.6.7}$$

To solve this system, recall that the $r(\mathbf{A}) = r(\mathbf{A}'\mathbf{A})$ and from Example 2.5.5, the $r(\mathbf{A}) = 2$. Thus, $\mathbf{A}$ is not of full column rank.

To solve (2.6.7) using the restriction method, we add the restriction that $\alpha_1 + \alpha_2 = 0$. Then, $\mathbf{R} = \begin{bmatrix} 0 & 1 & 1 \end{bmatrix}$, $\boldsymbol{\theta} = 0$, and the $r(\mathbf{A}'\mathbf{R}') = 3$. Using (2.6.3),

$$\begin{bmatrix} \widehat{\mu} \\ \widehat{\alpha}_1 \\ \widehat{\alpha}_2 \end{bmatrix} = \begin{bmatrix} 4 & 2 & 2 \\ 2 & 3 & 1 \\ 2 & 1 & 3 \end{bmatrix}^{-1} \begin{bmatrix} 4y_{..} \\ 2y_{1.} \\ 2y_{2.} \end{bmatrix}$$

where

$$y_{..} = \sum_i^I \sum_j^J y_{ij}/IJ = \sum_i^2 \sum_j^2 y_{ij}/(2)(2)$$
$$y_{i.} = \sum_j^J y_{ij}/J = \sum_j^2 y_{ij}/2$$

for $I = J = 2$. Hence,

$$\begin{bmatrix} \widehat{\mu} \\ \widehat{\alpha}_1 \\ \widehat{\alpha}_2 \end{bmatrix} = \frac{1}{16}\begin{bmatrix} 8 & -4 & -4 \\ -4 & 8 & 0 \\ -4 & 0 & 8 \end{bmatrix}^{-1}\begin{bmatrix} 4y_{..} \\ 2y_{1.} \\ 2y_{2.} \end{bmatrix}$$

$$= \begin{bmatrix} 2y_{..} - (y_{1.}/2 - (y_{2.}/2) \\ y_{1.} - y_{..} \\ y_{2.} - y_{..} \end{bmatrix} = \begin{bmatrix} y_{..} \\ y_{1.} - y_{..} \\ y_{2.} - y_{..} \end{bmatrix}$$

is a unique solution with the restriction $\alpha_1 + \alpha_2 = 0$.

Using the reparameterization method to solve (2.6.7), suppose we associate with $\mu + \alpha_i$ the parameter μ_i. Then, $(\mu_1 + \mu_2)/2 = \mu + (\alpha_1 + \alpha_2)/2$. Thus, under this reparameterization the average of μ_i is the same as the average of $\mu + \alpha_i$. Also observe that $\mu_1 - \mu_2 = \alpha_1 - \alpha_2$ under the reparameterization. Letting

$$\mathbf{C} = \begin{bmatrix} 1 & 1/2 & 1/2 \\ 0 & 1 & -1 \end{bmatrix}$$

be the reparameterization matrix, the matrix

$$\mathbf{C}\begin{bmatrix} \mu \\ \alpha_1 \\ \alpha_2 \end{bmatrix} = \begin{bmatrix} 1 & 1/2 & 1/2 \\ 0 & 1 & -1 \end{bmatrix}\begin{bmatrix} \mu \\ \alpha_1 \\ \alpha_2 \end{bmatrix} = \begin{bmatrix} \mu + (\alpha_1 + \alpha_2)/2 \\ \alpha_1 - \alpha_2 \end{bmatrix}$$

has a natural interpretation in terms of the original model parameters. In addition, the $r(\mathbf{C}) = r(\mathbf{A}'\mathbf{C}') = 2$ so that $\mathbf{C}$ is in the row space of $\mathbf{A}$. Using (2.6.4),

$$(\mathbf{CC}') = \begin{bmatrix} 3/2 & 0 \\ 0 & 2 \end{bmatrix}, \qquad (\mathbf{CC}')^{-1} = \begin{bmatrix} 2/3 & 0 \\ 0 & 1/2 \end{bmatrix}$$

$$\mathbf{B} = \mathbf{AC}'(\mathbf{CC}')^{-1} = \begin{bmatrix} 1 & 1/2 \\ 1 & 1/2 \\ 1 & -1/2 \\ 1 & -1/2 \end{bmatrix}$$

so

$$\widehat{\mathbf{x}}^* = \mathbf{Cx} = (\mathbf{BB}')^{-1}\mathbf{B}'\mathbf{y}$$

$$\begin{bmatrix} \widehat{x}_1^* \\ \widehat{x}_2^* \end{bmatrix} = \begin{bmatrix} \mu + (\alpha_1 + \alpha_2)/2 \\ \alpha_1 - \alpha_2 \end{bmatrix} = \begin{bmatrix} 4 & 0 \\ 0 & 1 \end{bmatrix}^{-1}\begin{bmatrix} 4y_{..} \\ y_{1..} - y_{..} \end{bmatrix}$$

$$= \begin{bmatrix} y_{..} \\ y_{1.} - y_{..} \end{bmatrix}$$

For the parametric function $\psi = \alpha_1 - \alpha_2$, $\widehat{\psi} = \widehat{x}_2^* = y_{1.} - y_{2.}$ which is identical to the restriction result since $\widehat{\alpha}_1 = y_{1.} - y_{..}$ and $\widehat{\alpha}_2 = y_{2.} - y_{..}$. Hence, the estimated contrast is $\widehat{\psi} = \widehat{\alpha}_1 - \widehat{\alpha}_2 = y_{1.} - y_{2.}$. However, the solution under reparameterization is only the same as the restriction method if we know that $\alpha_1 + \alpha_2 = 0$. Then, $\widehat{x}_1^* = \widehat{\mu} = y_{..}$. When this is not the case, $\widehat{x}_1^*$ is estimating $\mu + (\alpha_1 + \alpha_2)/2$. If $\alpha_1 = \alpha_2 = 0$ we also have that $\widehat{\mu} = y_{..}$ for both procedures.

To solve the system using a g-inverse, recall from Theorem 2.5.5, property (4), that $(\mathbf{A}'\mathbf{A})^-\mathbf{A}'$ is a g-inverse of $\mathbf{A}$ if $(\mathbf{A}'\mathbf{A})^-$ is a g-inverse of $\mathbf{A}'\mathbf{A}$. From Example 2.5.5,

$$(\mathbf{A}'\mathbf{A}) = \begin{bmatrix} 4 & 2 & 2 \\ 2 & 2 & 0 \\ 2 & 0 & 2 \end{bmatrix} \quad \text{and} \quad (\mathbf{A}'\mathbf{A})^- = \begin{bmatrix} 1/2 & -1/2 & 0 \\ -1/2 & 1 & 0 \\ 0 & 0 & 0 \end{bmatrix}$$

so that

$$\mathbf{A}^-\mathbf{y} = (\mathbf{A}'\mathbf{A})^-\mathbf{A}'\mathbf{y} = \begin{bmatrix} y_{.} \\ y_{1.} - y_{2.} \\ 0 \end{bmatrix}$$

Since

$$\mathbf{A}^-\mathbf{A} = (\mathbf{A}'\mathbf{A})^-(\mathbf{A}'\mathbf{A}) = \begin{bmatrix} 1 & 0 & 1 \\ 0 & 1 & -1 \\ 0 & 0 & 0 \end{bmatrix}$$

$$\mathbf{I} - \mathbf{A}^-\mathbf{A} = \begin{bmatrix} 0 & 0 & -1 \\ 0 & 0 & 1 \\ 0 & 0 & 1 \end{bmatrix}$$

A general solution to the system is

$$\begin{bmatrix} \widehat{\mu} \\ \widehat{\mu}_1 \\ \widehat{\mu}_2 \end{bmatrix} = \begin{bmatrix} y_{2.} \\ y_{1.} - y_{2.} \\ 0 \end{bmatrix} + (\mathbf{I} - \mathbf{A}^-\mathbf{A})\,\mathbf{z}$$

$$= \begin{bmatrix} y_{2.} \\ y_{1.} - y_{2.} \\ 0 \end{bmatrix} + \begin{bmatrix} -z_3 \\ z_3 \\ z_3 \end{bmatrix}$$

Choosing $z_3 = y_{2.} - y_{..}$, a solution is

$$\begin{bmatrix} \mu \\ \widehat{\alpha}_1 \\ \widehat{\alpha}_2 \end{bmatrix} = \begin{bmatrix} y_{..} \\ y_{1.} - y_{..} \\ y_{2.} - y_{..} \end{bmatrix}$$

which is consistent with the restriction method. Selecting $z_3 = y_{2.}$; $\widehat{\mu} = 0, \widehat{\alpha}_1 = y_{..}$ and $\widehat{\alpha}_2 = y_{2.}$ is another solution. The solution is not unique. However, for either selection of $\mathbf{z}$, $\widehat{\psi} = \widehat{\alpha}_1 - \widehat{\alpha}_2$ is unique. Theorem 2.6.1 only determines the general form for solutions of $\mathbf{Ax} = \mathbf{y}$. Rao (1973a) established the following result to prove that certain linear combinations of the unknowns in a consistent system are unique, independent of the g-inverse $\mathbf{A}^-$.

Theorem 2.6.2 *The linear combination $\psi = \mathbf{a}'\mathbf{x}$ of the unknowns, called parametric functions of the unknowns, for the consistent system $\mathbf{Ax} = \mathbf{y}$ has a unique solution $\widehat{\mathbf{x}}$ if and only if $\mathbf{a}'(\mathbf{A}^-\mathbf{A}) = \mathbf{a}'$. Furthermore, the solutions are given by $\mathbf{a}'\widehat{\mathbf{x}} = \mathbf{t}'(\mathbf{A}^-\mathbf{A})\mathbf{A}^-\mathbf{y}$ for $r = r(\mathbf{A})$ linearly independent vectors $\mathbf{a}' = \mathbf{t}'\mathbf{A}^-\mathbf{A}$ for arbitrary vectors $\mathbf{t}'$.*

Continuing with our illustration, we apply Theorem 2.6.2 to the system defined in (2.6.7) to determine if unique solutions for the linear combinations of the unknowns $\alpha_1 - \alpha_2$, $\mu + (\alpha_1 + \alpha_2)$, and μ can be found. To check that a unique solution exists, we have to verify that $\mathbf{a}'\left(\mathbf{A}^{-}\mathbf{A}\right) = \mathbf{a}'$.

For $\alpha_1 - \alpha_2$, $\mathbf{a}'(\mathbf{A}^{-}\mathbf{A}) = [0, 1, -1]\begin{bmatrix} 1 & 0 & 1 \\ 0 & 1 & -1 \\ 0 & 0 & 0 \end{bmatrix} = [0, 1, -1] = \mathbf{a}'$

For $\mu + (\alpha_1 - \alpha_2)/2$, $\mathbf{a}' = [1, 1/2, 1/2]$ and

$$\mathbf{a}'(\mathbf{A}^{-}\mathbf{A}) = [1, 1/2, 1/2]\begin{bmatrix} 1 & 0 & 1 \\ 0 & 1 & -1 \\ 0 & 0 & 0 \end{bmatrix} = [1, 1/2, 1/2] = \mathbf{a}'.$$

Thus both $\alpha_1 - \alpha_2$ and $\mu + (\alpha_1 - \alpha_2)/2$ have unique solutions and are said to be estimable.

For μ, $\mathbf{a}' = \begin{bmatrix} 1, & 0, & 0 \end{bmatrix}$ and $\mathbf{a}'(\mathbf{A}'\mathbf{A}) = \begin{bmatrix} 1, & 0, & 1 \end{bmatrix} \neq \mathbf{a}'$. Hence no unique solution exists for μ so the parameter μ is not estimable. Instead of checking each linear combination, we find a general expression for linear combinations of the parameters given an arbitrary vectors $\mathbf{t}$. The linear parametric function

$$\begin{aligned} \mathbf{a}'\mathbf{x} = \mathbf{t}'\left(\mathbf{A}'\mathbf{A}\right)\mathbf{x} &= [t_0, t_1, t_2]\begin{bmatrix} 1 & 0 & 0 \\ 0 & 1 & -1 \\ 0 & 0 & 0 \end{bmatrix}\begin{bmatrix} \mu \\ \alpha_1 \\ \alpha_2 \end{bmatrix} \\ &= t_0(\mu + \alpha_1) + t_1(\alpha_1 - \alpha_2) \end{aligned} \tag{2.6.8}$$

is a general expression for all linear combinations of $\mathbf{x}$ for arbitrary $\mathbf{t}$. Furthermore, the general solution is

$$\begin{aligned} \mathbf{a}'\widehat{\mathbf{x}} = \mathbf{t}'(\mathbf{A}'\mathbf{A})^{-}\mathbf{A}'\mathbf{y} &= \mathbf{t}'\left[(\mathbf{A}'\mathbf{A})^{-}(\mathbf{A}'\mathbf{A})\right](\mathbf{A}'\mathbf{A})^{-}\mathbf{A}'\mathbf{y} \\ &= \mathbf{t}'\begin{bmatrix} 1 & 0 & 1 \\ 0 & 1 & -1 \\ 0 & 0 & 0 \end{bmatrix}\begin{bmatrix} y_{2.} \\ y_{1.} - y_{2.} \\ 0 \end{bmatrix} \\ &= t_0 y_{2.} + t_1(y_{1.} - y_{2.}) \end{aligned} \tag{2.6.9}$$

By selecting t_0, t_1, and t_2 and substituting their values into (2.6.8) and (2.6.9), one determines whether a linear combination of the unknowns exists, is estimable. Setting $t_0 = 0$ and $t_1 = 1$, $\psi_1 = \mathbf{a}'\mathbf{x} = \alpha_1 - \alpha$ and $\mathbf{a}'\widehat{\mathbf{x}} = \widehat{\boldsymbol{\psi}}_1 = y_{1.} - y_{2.}$ has a unique solution. Setting $t_0 = 1$ and $t_1 = 1/2$,

$$\begin{aligned} \boldsymbol{\psi}_2 = \mathbf{a}'\mathbf{x} &= 1(\mu + \alpha_2) + (\alpha_1 - \alpha_2)/2 \\ &= \mu + (\alpha_1 + \alpha_2)/2 \end{aligned}$$

and

$$\widehat{\boldsymbol{\psi}}_2 = \mathbf{a}'\widehat{\mathbf{x}} = \frac{y_{1.} + y_{2.}}{2} = y_{..}$$

which shows that ψ_2 is estimated by $y_{..}$. No elements of t_0, t_1, and t_2 may be chosen to estimate μ; hence, $\widehat{\mu}$ has no unique solution so that μ is not estimable. To make μ estimable, we must add the restriction $\alpha_1 + \alpha_2 = 0$. Thus, restrictions add "meaning" to nonestimable linear combinations of the unknowns, in order to make them estimable. In addition, the restrictions become part of the model specification. Without the restriction the parameter μ has no meaning since it is not estimable. In Section 2.3, the restriction on the sum of the parameters α_i orthogonalized A into the subspaces **1** and $A/\mathbf{1}$.

b. Linear Transformations

The system of equations $\mathbf{Ax} = \mathbf{y}$ is typically viewed as a linear transformation. The $m \times 1$ vector $\mathbf{x}$ is operated upon by the matrix $\mathbf{A}$ to obtain the $n \times 1$ image vector $\mathbf{y}$.

Definition 2.6.1 *A transformation is linear if, in carrying* $\mathbf{x}_1$ *into* $\mathbf{y}_1$ *and* $\mathbf{x}_2$ *into* $\mathbf{y}_2$*, the transformation maps the vector* $\alpha_1\mathbf{x}_1 + \alpha_2\mathbf{x}_2$ *into* $\alpha_1\mathbf{y}_1 + \alpha_2\mathbf{y}_2$ *for every pair of scalars* α_1 *and* α_2.

Thus, if $\mathbf{x}$ is an element of a vector space U and $\mathbf{y}$ is an element of a vector space V, a linear transformation is a function $T : U \longrightarrow V$ such that $T(\alpha_1\mathbf{x}_1 + \alpha_2\mathbf{x}_2) = \alpha_1 T(\mathbf{x}_1) + \alpha_2 T(\mathbf{x}_2) = \alpha_1\mathbf{y}_1 + \alpha_2\mathbf{y}_2$. The null or kernel subspace of the matrix A, denoted by $N(\mathbf{A})$ or K_A is the set of all vectors satisfying the homogeneous transformation $\mathbf{Ax} = \mathbf{0}$. That is, the null or kernel of $\mathbf{A}$ is the linear subspace of V_n such that $N(\mathbf{A}) \equiv K_A = \{\mathbf{x} \mid \mathbf{Ax} = \mathbf{0}\}$. The dimension of the kernel subspace is $\dim\{K_A\} = m - r(\mathbf{A})$. The transformation is one-to-one if the dimension of the kernel space is zero; then, $r(\mathbf{A}) = m$. The complement subspace of K_A is the subspace $K_{A'} = \{\mathbf{y} \mid \mathbf{A}'\mathbf{y} = \mathbf{0}\}$.

Of particular interest in statistical applications are linear transformations that map vectors of a space onto vectors of the same space. The matrix $\mathbf{A}$ for this transformation is now of order n so that $\mathbf{A}_{n\times n}\mathbf{x}_{n\times 1} = \mathbf{y}_{n\times 1}$. The linear transformation is nonsingular if and only if the $|\mathbf{A}| = 0$. Then, $\mathbf{x} = \mathbf{A}^{-1}\mathbf{y}$ and the transformation is one-to-one since $N(\mathbf{A}) = \{\mathbf{0}\}$. If $\mathbf{A}$ is less than full rank, the transformation is singular and many to one. Such transformations map vectors into subspaces.

Example 2.6.1 *As a simple example of a nonsingular linear transformation in Euclidean two-dimensional space, consider the square formed by the vectors* $\mathbf{x}_1' = [1, 0]$, $\mathbf{x}_2' = [0, 1]$, $\mathbf{x}_1 + \mathbf{x}_2 = [1, 1]$ *under the transformation*

$$\mathbf{A} = \begin{bmatrix} 1 & 4 \\ 0 & 1 \end{bmatrix}$$

Then

$$\mathbf{Ax}_1 = \mathbf{y}_1 = \begin{bmatrix} 1 \\ 0 \end{bmatrix}$$

$$\mathbf{Ax}_2 = \mathbf{y}_2 = \begin{bmatrix} 4 \\ 1 \end{bmatrix}$$

$$\mathbf{A}(\mathbf{x}_1 + \mathbf{x}_2) = \mathbf{y}_1 + \mathbf{y}_2 = \begin{bmatrix} 5 \\ 1 \end{bmatrix}$$

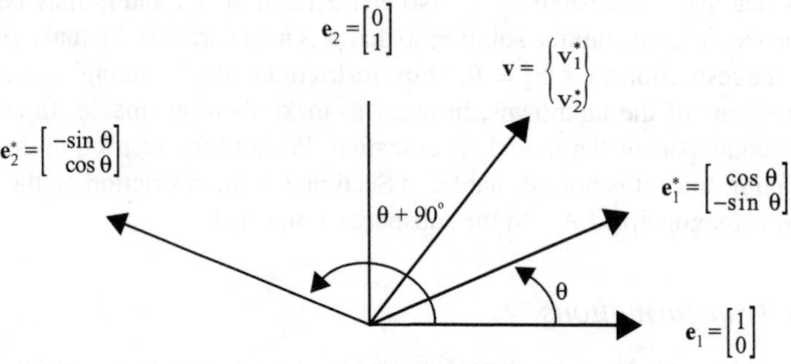

FIGURE 2.6.1. Fixed-Vector Transformation

Geometrically, observe that the parallel line segments {[0, 1], [1, 1]} *and* {[0, 0], [1, 0]} *are transformed into parallel line segments* {[4, 1], [5, 1]} *and* {[0, 0], [1, 0]} *as are other sides of the square. However, some lengths, angles, and hence distances of the original figure are changed under the transformation.*

Definition 2.6.2 *A nonsingular linear transformaiion* $\mathbf{Tx} = \mathbf{y}$ *that preserves lengths, distances and angles is called an orthogonal transformation and satisfies the condition that* $\mathbf{TT}' = \mathbf{I} = \mathbf{T}'\mathbf{T}$ *so that* $\mathbf{T}$ *is an orthogonal matrix.*

Theorem 2.6.3 *For an orthogonal transformation matrix* $\mathbf{T}$

1. $|\mathbf{T}'\mathbf{AT}| = |\mathbf{A}|$
2. *The product of a finite number of orthogonal matrices is itself orthogonal.*

Recall that if $\mathbf{T}$ is orthogonal that the $|\mathbf{T}| = \pm 1$. If the $|\mathbf{T}| = 1$, the orthogonal matrix transformation may be interpreted geometrically as a rigid rotation of coordinate axes. If the $|\mathbf{T}| = -1$ the transformation is a rotation, followed by a reflection.

For a fixed angle θ, let $\mathbf{T} = \begin{bmatrix} \cos\theta & \sin\theta \\ -\sin\theta & \cos\theta \end{bmatrix}$ and consider the point $\mathbf{v} = [v_1, v_2] = [v_1^*, v_2^*]$ relative to the old coordinates $\mathbf{e}_1 = \begin{bmatrix} 1 \\ 0 \end{bmatrix}$, $\mathbf{e}_2 = \begin{bmatrix} 0 \\ 1 \end{bmatrix}$ and the new coordinates $\mathbf{e}_1^*$and $\mathbf{e}_2^*$. In Figure 2.6.1, we consider the point $\mathbf{v}$ relative to the two coordinate systems $\{\mathbf{e}_1, \mathbf{e}_2\}$ and $\{\mathbf{e}_1^*, \mathbf{e}_2^*\}$.

Clearly, relative to $\{\mathbf{e}_1, \mathbf{e}_2\}$,

$$\mathbf{v} = v_1\mathbf{e}_1 + v_2\mathbf{e}_2$$

However, rotating $\mathbf{e}_1 \longrightarrow \mathbf{e}_1^*$ and $\mathbf{e}_2 \longrightarrow \mathbf{e}_2^*$, the projection of $\mathbf{e}_1$ onto $\mathbf{e}_1^*$ is $\|\mathbf{e}_1\| \cos\theta = \cos\theta$ and the projection of $\mathbf{e}_1$onto $\mathbf{e}_2^*$ is $\cos(\theta + 90°) = -\sin\theta$ or,

$$\mathbf{e}_1 = (\cos\theta)\,\mathbf{e}_1^* + (-\sin\theta)\,\mathbf{e}_2^*$$

Similarly,

$$\mathbf{e}_2 = (\cos\theta)\,\mathbf{e}_1^* + (-\sin\theta)\,\mathbf{e}_2^*$$

Thus,

$$\mathbf{v} = (\mathrm{v}_1 \cos\theta + \mathrm{v}_2 \sin\theta)\, \mathbf{e}_1^* + (\mathrm{v}_1 (-\sin\theta) + \mathrm{v}_2 \cos\theta)\, \mathbf{e}_2^*$$

or

$$\begin{bmatrix} \mathrm{v}_1^* \\ \mathrm{v}_2^* \end{bmatrix} = \begin{bmatrix} \cos\theta & \sin\theta \\ -\sin\theta & \cos\theta \end{bmatrix} \begin{bmatrix} \mathrm{v}_1 \\ \mathrm{v}_2 \end{bmatrix}$$
$$\mathbf{v}^* = \mathbf{T}\mathbf{v}$$

is a linear transformation of the old coordinates to the new coordinates. Let θ_{ij} be the angle of the ith old axes and the jth new axes: $\theta_{11} = \theta, \theta_{22} = \theta, \theta_{21} = \theta - 90°$ and $\theta_{12} = \theta + 90$. Using trigonometric identities

$$\begin{aligned} \cos\theta_{21} &= \cos\theta\ \cos(-90°) - \sin\theta \sin(-90°) \\ &= \cos\theta\ (0) - \sin\theta\,(-1) \\ &= \sin\theta \end{aligned}$$

and

$$\begin{aligned} \cos\theta_{12} &= \cos\theta\ \cos -90° - \sin\theta \sin -90° \\ &= \cos\theta\ (0) - \sin\theta\,(1) \\ &= -\sin\theta \end{aligned}$$

we observe that the transformation becomes

$$\mathbf{v}^* = \begin{bmatrix} \cos\theta_{11} & \cos\theta_{21} \\ \cos\theta_{12} & \cos\theta_{22} \end{bmatrix} \mathbf{v}$$

For three dimensions, the orthogonal transformation is

$$\begin{bmatrix} \mathrm{v}_1^* \\ \mathrm{v}_2^* \\ \mathrm{v}_3^* \end{bmatrix} = \begin{bmatrix} \cos\theta_{11} & \cos\theta_{21} & \cos\theta_{31} \\ \cos\theta_{12} & \cos\theta_{22} & \cos\theta_{32} \\ \cos\theta_{13} & \cos\theta_{23} & \cos\theta_{33} \end{bmatrix} \begin{bmatrix} \mathrm{v}_1 \\ \mathrm{v}_2 \\ \mathrm{v}_3 \end{bmatrix}$$

Extending the result to n-dimensions easily follows. A transformation that transforms an orthogonal system to a nonorthogonal system is called an oblique transformation. The basis vectors are called an oblique basis. In an oblique system, the axes are no longer at right angles. This situation arises in factor analysis discussed in Chapter 8.

c. *Projection Transformations*

A linear transformation that maps vectors of a given vector space onto a vector subspace is called a projection. For a subspace $V_r \subseteq V_n$, we saw in Section 2.3 how for any $\mathbf{y} \in V_n$ that the vector $\mathbf{y}$ may be decomposed into orthogonal components

$$\mathbf{y} = P_{V_r}\mathbf{y} + P_{V_{n-r}^{\perp}}\mathbf{y}$$

such that $\mathbf{y}$ is in an r-dimensional space and the residual is in an $(n-r)$-dimensional space. We now discuss projection matrices which make the geometry of projections algebraic.

Definition 2.6.3 *Let $\mathbf{P}_V$ represent a transformation matrix that maps a vector $\mathbf{y}$ onto a subspace V. The matrix $\mathbf{P}_V$ is an orthogonal projection matrix if and only if $\mathbf{P}_V$ is symmetric and idempotent.*

Thus, an orthogonal projection matrix $\mathbf{P}_V$ is a matrix such that $\mathbf{P}_V = \mathbf{P}'_V$ and $\mathbf{P}^2_V = \mathbf{P}_V$. Note that $\mathbf{I} - \mathbf{P}_V$ is also an orthogonal projection matrix. The projection transformation $\mathbf{I} - \mathbf{P}_V$ projects $\mathbf{y}$ onto $V^\perp$, the orthocomplement of V relative to V_n. Since $\mathbf{I} - \mathbf{P}_V$ projects $\mathbf{y}$ onto $V^\perp$, and $V_n = V_r \oplus V_{n-r}$ where $V_{n-r} = V^\perp$,we see that the rank of a projection matrix is equal to the dimension of the space that is being projected onto.

Theorem 2.6.4 *For any orthogonal projection matrix $\mathbf{P}_V$, the*

$$\begin{aligned} r\left(\mathbf{P}_V\right) &= \dim V = r \\ r\left(\mathbf{I} - \mathbf{P}_V\right) &= \dim V^\perp = n - r \end{aligned}$$

for $V_n = V_r \oplus V^\perp_{n-r}$.

The subscript V on the matrix $\mathbf{P}_V$ is used to remind us that $\mathbf{P}_V$ projects vectors in V_n onto a subspace $V_r \subseteq V_n$. We now remove the subscript to simplify the notation. To construct a projection matrix, let $\mathbf{A}$ be an $n \times r$ matrix where the $r(\mathbf{A}) = r$ so that the columns span the r-dimensional subspace $V_r \subseteq V_n$. Consider the matrix

$$\mathbf{P} = \mathbf{A}\left(\mathbf{A}'\mathbf{A}\right)^{-1}\mathbf{A}' \tag{2.6.10}$$

The matrix is a projection matrix since $\mathbf{P} = \mathbf{P}'$, $\mathbf{P}^2 = \mathbf{P}$ and the $r\left(\mathbf{P}\right) = r$. Using $\mathbf{P}$ defined in (2.6.10), observe that

$$\begin{aligned} \mathbf{y} &= \mathbf{P}_{V_r}\mathbf{y} + \mathbf{P}_{V^\perp}\mathbf{y} \\ &= \mathbf{P}\mathbf{y} + \left(\mathbf{I} - \mathbf{P}\right)\mathbf{y} \\ &= \mathbf{A}\left(\mathbf{A}'\mathbf{A}\right)^{-1}\mathbf{A}'\mathbf{y} + [\mathbf{I} - \mathbf{A}\left(\mathbf{A}'\mathbf{A}\right)^{-1}\mathbf{A}']\mathbf{y} \end{aligned}$$

Furthermore, the norm squared of $\mathbf{y}$ is

$$\begin{aligned} \|\mathbf{y}\|^2 &= \|\mathbf{P}\mathbf{y}\|^2 + \|\left(\mathbf{I} - \mathbf{P}\right)\mathbf{y}\|^2 \\ &= \mathbf{y}'\mathbf{P}\mathbf{y} + \mathbf{y}'\left(\mathbf{I} - \mathbf{P}\right)\mathbf{y} \end{aligned} \tag{2.6.11}$$

Suppose $\mathbf{A}_{n\times m}$ is not of full column rank, $r\left(\mathbf{A}\right) = r < m \leq n$, Then

$$\mathbf{P} = \mathbf{A}\left(\mathbf{A}'\mathbf{A}\right)^{-}\mathbf{A}'$$

is a unique orthogonal projection matrix. Thus, $\mathbf{P}$ is the same for any g-inverse $\left(\mathbf{A}'\mathbf{A}\right)^{-}$. Alternatively, one may use a Moore-Penrose inverse for $\left(\mathbf{A}'\mathbf{A}\right)$. Then, $\mathbf{P} = \mathbf{A}\left(\mathbf{A}'\mathbf{A}\right)^{+}\mathbf{A}'$.

Example 2.6.2 *Let*

$$\mathbf{V} = \begin{bmatrix} 1 & 1 & 0 \\ 1 & 1 & 0 \\ 1 & 0 & 1 \\ 1 & 0 & 1 \end{bmatrix} \quad \text{and} \quad \mathbf{y} = \begin{bmatrix} y_{11} \\ y_{12} \\ y_{21} \\ y_{22} \end{bmatrix}$$

where

$$(\mathbf{V}'\mathbf{V})^{-} \begin{bmatrix} 0 & 0 & 0 \\ 0 & 1/2 & 0 \\ 0 & 0 & 1/2 \end{bmatrix}$$

Using the methods of Section 2.3, we can obtain the $\mathbf{P}_V\mathbf{y}$ *by forming an orthogonal basis for the column space of* $\mathbf{V}$. *Instead, we form the projection matrix*

$$\mathbf{P} = \mathbf{V}\left(\mathbf{V}'\mathbf{V}\right)^{-}\mathbf{V}' = \begin{bmatrix} 1/2 & 1/2 & 0 & 0 \\ 1/2 & 1/2 & 0 & 0 \\ 0 & 0 & 1/2 & 1/2 \\ 0 & 0 & 1/2 & 1/2 \end{bmatrix}$$

Then the $\mathbf{P}_V\mathbf{y}$ *is*

$$\mathbf{x} = \mathbf{V}\left(\mathbf{V}'\mathbf{V}\right)^{-}\mathbf{V}'\mathbf{y} = \begin{bmatrix} y_{1.} \\ y_{1.} \\ y_{2.} \\ y_{2.} \end{bmatrix}$$

Letting $\mathbf{V} \equiv A$ *as in Figure 2.3.3,*

$$\begin{aligned} \mathbf{x} &= \mathbf{P}_1\mathbf{y} + \mathbf{P}_{A/1}\mathbf{y} \\ &= \bar{y}\begin{bmatrix} 1 \\ 1 \\ 1 \\ 1 \end{bmatrix} + (y_{1.} - \bar{y})\begin{bmatrix} 1 \\ 1 \\ -1 \\ -1 \end{bmatrix} \\ &= \begin{bmatrix} y_{1.} \\ y_{1.} \\ y_{2.} \\ y_{2.} \end{bmatrix} \end{aligned}$$

leads to the same result.

To obtain the $\mathbf{P}_{V^\perp}\mathbf{y}$, *the matrix* $\mathbf{I} - \mathbf{P}$ *is constructed. For our example,*

$$\mathbf{I} - \mathbf{P} = \mathbf{I} - \mathbf{V}\left(\mathbf{V}'\mathbf{V}\right)^{-}\mathbf{V}' = \begin{bmatrix} 1/2 & 1/2 & 0 & 0 \\ 1/2 & 1/2 & 0 & 0 \\ 0 & 0 & 1/2 & 1/2 \\ 0 & 0 & -1/2 & 1/2 \end{bmatrix}$$

so that

$$\mathbf{e} = (\mathbf{I} - \mathbf{P})\,\mathbf{y} = \begin{bmatrix} (y_{11} - y_{12})/2 \\ (y_{12} - y_{11})/2 \\ (y_{21} - y_{22})/2 \\ (y_{22} - y_{21})/2 \end{bmatrix} = \begin{bmatrix} y_{11} - y_{1.} \\ y_{21} - y_{1.} \\ y_{21} - y_{2.} \\ y_{22} - y_{2.} \end{bmatrix}$$

is the projection of $\mathbf{y}$ *onto* $V^{\perp}$. *Alternatively,* $\mathbf{e} = \mathbf{y} - \mathbf{x}$.

In Figure 2.3.3, the vector space $V \equiv A = (A/\mathbf{1}) \oplus \mathbf{1}$. *To create matrices that project* $\mathbf{y}$ *onto each subspace, let* $\mathbf{V} \equiv \{\mathbf{1}, \mathbf{A}\}$ *where*

$$\mathbf{1} = \begin{bmatrix} 1 \\ 1 \\ 1 \\ 1 \end{bmatrix} \quad \text{and} \quad \mathbf{A} = \begin{bmatrix} 1 & 0 \\ 1 & 0 \\ 0 & 1 \\ 0 & 1 \end{bmatrix}$$

Next, define $\mathbf{P}_1, \mathbf{P}_2$, and $\mathbf{P}_3$ *as follows*

$$\begin{aligned} \mathbf{P}_1 &= \mathbf{1}\left(\mathbf{1}'\mathbf{1}\right)^{-}\mathbf{1}' \\ \mathbf{P}_2 &= \mathbf{V}\left(\mathbf{V}'\mathbf{V}\right)^{-}\mathbf{V}' - \mathbf{1}\left(\mathbf{1}'\mathbf{1}\right)^{-}\mathbf{1}' \\ \mathbf{P}_3 &= \mathbf{I} - \mathbf{V}\left(\mathbf{V}'\mathbf{V}\right)^{-}\mathbf{V}' \end{aligned}$$

so that

$$\mathbf{I} = \mathbf{P}_1 + \mathbf{P}_2 + \mathbf{P}_3$$

Then, the quantities $\mathbf{P}_i\mathbf{y}$

$$\begin{aligned} \mathbf{P}_1\mathbf{y} &= \mathbf{P}_1\mathbf{y} \\ \mathbf{P}_{A/1}\mathbf{y} &= \mathbf{P}_2\mathbf{y} \\ \mathbf{P}_{A^{\perp}}\mathbf{y} &= \mathbf{P}_3\mathbf{y} \end{aligned}$$

are the projections of $\mathbf{y}$ *onto the orthogonal subspaces.*

One may also represent V using Kronecker products. Observe that the two group ANOVA model has the form

$$\begin{bmatrix} y_{11} \\ y_{12} \\ y_{21} \\ y_{22} \end{bmatrix} = (\mathbf{1}_2 \otimes \mathbf{1}_2)\,\mu + (\mathbf{I}_2 \otimes \mathbf{1}_2)\begin{bmatrix} \alpha_1 \\ \alpha_2 \end{bmatrix} + \begin{bmatrix} e_{11} \\ e_{12} \\ e_{21} \\ e_{22} \end{bmatrix}$$

Then, it is also easily established that

$$\begin{aligned} \mathbf{P}_1 &= \mathbf{1}_2\left(\mathbf{1}_2'\mathbf{1}_2\right)^{-1}\mathbf{1}_2' = (\mathbf{J}_2 \otimes \mathbf{J}_2)/4 = \mathbf{J}_4/4 \\ \mathbf{P}_2 &= \frac{1}{2}(\mathbf{I}_2 \otimes \mathbf{1}_2)(\mathbf{I}_2 \otimes \mathbf{1}_2)' - \mathbf{J}_4/4 = \frac{1}{2}(\mathbf{I}_2 \otimes \mathbf{J}_2) - \mathbf{J}_4/4 \\ \mathbf{P}_3 &= (\mathbf{I}_2 \otimes \mathbf{I}_2) - (\mathbf{I}_2 \otimes \mathbf{J}_2)/4 \end{aligned}$$

so that $\mathbf{P}_i$ *may be calculated from the model.*

By employing projection matrices, we have illustrated how one may easily project an observation vector onto orthogonal subspaces. In statistics, this is equivalent to partitioning a sum of squares into orthogonal components.

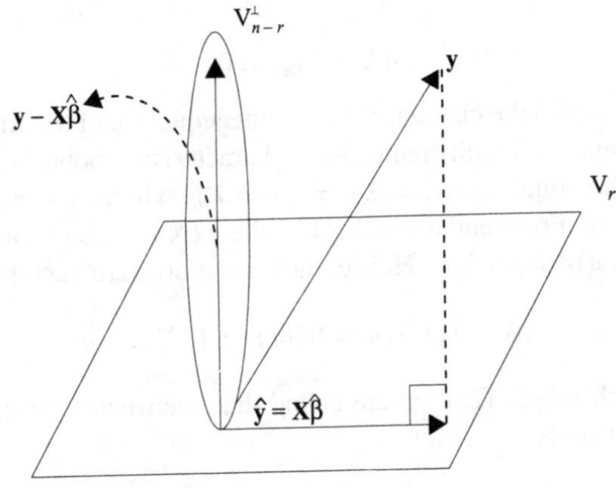

FIGURE 2.6.2. $\|\mathbf{y}\|^2 = \|\mathbf{P}_{V_r}\mathbf{y}\|^2 + \|\mathbf{P}_{V_{n-r}}\mathbf{y}\|^2$.

Example 2.6.3 *Consider a model that relates one dependent variable* y *to* $x_1, x_2, \ldots, x_k$ *linearly independent variables by the linear relationship*

$$y = \beta_0 + \beta_1 x_1 + \beta_2 x_2 + \cdots + \beta_k x_k + e$$

where e is a random error. This model is the multiple linear regression model, which, using matrix notation, may be written as

$$\underset{n\times 1}{\mathbf{y}} = \underset{n\times(k+1)}{\mathbf{X}} \underset{(k+1)\times 1}{\boldsymbol{\beta}} + \underset{n\times 1}{\mathbf{e}}$$

Letting $\mathbf{X}$ *represent the space spanned by the columns of* $\mathbf{X}$*, the projection of* $\mathbf{y}$ *onto* $\mathbf{X}$ *is*

$$\widehat{\mathbf{y}} = \mathbf{X}\left(\mathbf{X}'\mathbf{X}\right)^{-1}\mathbf{X}'\mathbf{y}$$

Assuming $\mathbf{e} = \mathbf{0}$*, the system of equations is solved to obtain the best estimate of* $\boldsymbol{\beta}$*. Then, the best estimate of* $\mathbf{y}$ *using the linear model is* $\mathbf{X}\widehat{\boldsymbol{\beta}}$ *where* $\widehat{\boldsymbol{\beta}}$ *is the solution to the system* $\mathbf{y} = \mathbf{X}\boldsymbol{\beta}$ *for unknown* $\boldsymbol{\beta}$*. The least squares estimate* $\widehat{\boldsymbol{\beta}} = \left(\mathbf{X}'\mathbf{X}\right)^{-1}\mathbf{X}'\mathbf{y}$ *minimizes the sum of squared errors for the fitted model* $\widehat{\mathbf{y}} = \mathbf{X}\widehat{\boldsymbol{\beta}}$*. Furthermore,*

$$\begin{aligned}
\|(\mathbf{y} - \widehat{\mathbf{y}})\|^2 &= (\mathbf{y} - \mathbf{X}\widehat{\boldsymbol{\beta}})'(\mathbf{y} - \mathbf{X}\widehat{\boldsymbol{\beta}}) \\
&= \mathbf{y}'(\mathbf{I}_n - \mathbf{X}\left(\mathbf{X}'\mathbf{X}\right)^{-1}\mathbf{X}')\mathbf{y} \\
&= \left\|\mathbf{P}_{V^{\perp}}\mathbf{y}\right\|^2
\end{aligned}$$

is the squared distance of the projection of $\mathbf{y}$ *onto the orthocomplement of* $V_r \subseteq V_n$.

Figure 2.6.2 represents the squared lengths geometrically.

d. *Eigenvalues and Eigenvectors*

For a square matrix $\mathbf{A}$ of order n the scalar λ is said to be the eigenvalue (or characteristic root or simply a root) of $\mathbf{A}$ if $\mathbf{A} - \lambda\mathbf{I}_n$ is singular. Hence the determinant of $\mathbf{A} - \lambda\mathbf{I}_n$ must equal zero

$$|\mathbf{A} - \lambda \mathbf{I}_n| = 0 \tag{2.6.12}$$

Equation (2.6.12) is called the characteristic or eigenequation of the square matrix **A** and is an n-degree polynomial in λ with eigenvalues (characteristic roots) $\lambda_1, \lambda_2, \ldots, \lambda_n$. If some subset of the roots are equal, say $\lambda_1 = \lambda_2 = \cdots = \lambda_m$, where $m < n$, then the root is said to have multiplicity m. From equation (2.6.12), the $r\,(\mathbf{A} - \lambda_k \mathbf{I}_n) < n$ so that the columns of $\mathbf{A} - \lambda_k \mathbf{I}_n$ are linearly dependent. Hence, there exist nonzero vectors $\mathbf{p}_i$ such that

$$(\mathbf{A} - \lambda_k \mathbf{I}_n)\,\mathbf{p}_i = \mathbf{0} \text{ for } i = 1, 2, \ldots, n \tag{2.6.13}$$

The vectors $\mathbf{p}_i$ which satisfy (2.6.13) are called the eigenvectors or characteristic vectors of the eigenvalues or roots λ_i.

Example 2.6.4 *Let*

$$\mathbf{A} = \begin{bmatrix} 1 & 1/2 \\ 1/2 & 1 \end{bmatrix}$$

Then

$$\begin{aligned} |\mathbf{A} - \lambda \mathbf{I}_2| = \left| \begin{bmatrix} 1-\lambda & 1/2 \\ 1/2 & 1-\lambda \end{bmatrix} \right| &= 0 \\ (1-\lambda)^2 - 1/4 &= 0 \\ \lambda^2 - 2\lambda + 3/4 &= 0 \\ (\lambda - 3/2)\,(\lambda - 1/2) &= 0 \end{aligned}$$

Or, $\lambda_1 = 3/2$ *and* $\lambda_2 = 1/2$. *To find* $\mathbf{p}_1$ *and* $\mathbf{p}_2$, *we employ Theorem 2.6.1. For* λ_1,

$$\begin{aligned} \widehat{\mathbf{x}} &= \left[\mathbf{I} - (\mathbf{A} - \lambda_1 \mathbf{I})^- (\mathbf{A} - \lambda_1 \mathbf{I})\right] \mathbf{z} \\ &= \left\{ \begin{bmatrix} 1 & 0 \\ 0 & 1 \end{bmatrix} - \begin{bmatrix} 0 & 0 \\ 0 & -2 \end{bmatrix} \begin{bmatrix} -1/2 & 1/2 \\ 1/2 & -1/2 \end{bmatrix} \right\} \mathbf{z} \\ &= \begin{bmatrix} 1 & 0 \\ 1 & 0 \end{bmatrix} \mathbf{z} \\ &= \begin{bmatrix} z_1 \\ z_1 \end{bmatrix} \end{aligned}$$

Letting $z_1 = 1, \widehat{\mathbf{x}}_1' = [1, 1]$. *In a similar manner, with* $\lambda_2 = 1/2, \widehat{\mathbf{x}}_2' = [z_2, -\ z_2]$. *Setting* $z_2 = 1, \widehat{\mathbf{x}}_2' = [1, -1]$, *and the matrix* $\mathbf{P}_0$ *is formed:*

$$\mathbf{P}_0 = [\widehat{\mathbf{x}}_1, \widehat{\mathbf{x}}_2] = \begin{bmatrix} 1 & 1 \\ 1 & -1 \end{bmatrix}$$

Normalizing the columns of $\mathbf{P}_0$, *the orthogonal matrix becomes*

$$\mathbf{P}_1 = \begin{bmatrix} 1/\sqrt{2} & 1/\sqrt{2} \\ 1/\sqrt{2} & -1/\sqrt{2} \end{bmatrix}$$

However, the $|\mathbf{P}_1| = -1$ *so that* $\mathbf{P}_1$ *is not a pure rotation. However, by changing the signs of the second column of* $\mathbf{P}_0$ *and selecting* $z_1 = -1$, *the orthogonal matrix* $\mathbf{P}$ *where the* $|\mathbf{P}| = 1$ *is*

$$\mathbf{P} = \begin{bmatrix} 1/\sqrt{2} & -1/\sqrt{2} \\ 1/\sqrt{2} & 1/\sqrt{2} \end{bmatrix} \quad \text{and} \quad \mathbf{P}' = \begin{bmatrix} 1/\sqrt{2} & 1/\sqrt{2} \\ -1/\sqrt{2} & 1/\sqrt{2} \end{bmatrix}$$

Thus, $\mathbf{P}' = \mathbf{T}$ *is a rotation of the axes* $\mathbf{e}_1, \mathbf{e}_2$ *to* $\mathbf{e}_1^*, \mathbf{e}_2^*$ *where* $\theta_{ij} = 45°$ *in Figure 2.6.1.*

Our example leads to the spectral decomposition theorem for a symmetric matrix $\mathbf{A}$.

Theorem 2.6.5 *(Spectral Decomposition) For a (real) symmetric matrix* $\mathbf{A}_{n\times n}$, *there exists an orthogonal matrix* $\mathbf{P}_{n\times n}$ *with columns* $\mathbf{p}_i$ *such that*

$$\mathbf{P}'\mathbf{A}\mathbf{P} = \Lambda,\ \mathbf{A}\mathbf{P} = \mathbf{P}\Lambda,\ \mathbf{P}\mathbf{P}' = \mathbf{I} = \sum_i \mathbf{p}_i\mathbf{p}_i' \quad \text{and} \quad \mathbf{A} = \mathbf{P}\Lambda\mathbf{P}' = \sum_i \lambda_i \mathbf{p}_i\mathbf{p}_i'$$

where Λ *is a diagonal matrix with diagonal elements* $\lambda_1 \geq \lambda_2 \geq \cdots \geq \lambda_n$.

If the $r(\mathbf{A}) = r \leq n$, then there are r nonzero elements on the diagonal of Λ . A symmetric matrix for which all $\lambda_i > 0$ is said to be positive definite (*p.d.*) and positive semidefinite (*p.s.d.*) if some $\lambda_i > 0$ and at least one is equal to zero. The class of matrices taken together are called non-negative definite (*n.n.d*) or Gramian. If at least one $\lambda_i = 0$, $\mathbf{A}$ is clearly singular.

Using Theorem 2.6.5, one may create the square root of a square symmetric matrix. By Theorem 2.6.5, $\mathbf{A} = \mathbf{P}\Lambda^{1/2}\Lambda^{1/2}\mathbf{P}'$ and $\mathbf{A}^{-1} = \mathbf{P}\Lambda^{-1/2}\Lambda^{-1/2}\mathbf{P}'$. The matrix $\mathbf{A}^{1/2} = \mathbf{P}\Lambda^{1/2}$ is called the square root matrix of the square symmetric matrix $\mathbf{A}$ and the matrix $\mathbf{A}^{-1/2} = \mathbf{P}\Lambda^{-1/2}$ is called the square root matrix of $\mathbf{A}^{-1}$ since $\mathbf{A}^{1/2}\mathbf{A}^{1/2} = \mathbf{A}$ and $(\mathbf{A}^{1/2})^{-1} = \mathbf{A}^{-1/2}$. Clearly, the factorization of the symmetric matrix $\mathbf{A}$ is not unique. Another common factorization method employed in statistical applications is called the Cholesky or square root factorization of a matrix. For this procedure, one creates a unique lower triangular matrix $\mathbf{L}$ such that $\mathbf{L}\mathbf{L}' = \mathbf{A}$ The lower triangular matrix $\mathbf{L}$ is called the Cholesky square root factor of the symmetric matrix $\mathbf{A}$. The matrix $\mathbf{L}'$ in the matrix product is an upper triangular matrix. By partitioning the lower triangular matrix in a Cholesky factorization into a product of a unit lower triangular matrix time a diagonal matrix, one obtains the LDU decomposition of the matrix where U is a unit upper triangular matrix.

In Theorem 2.6.5, we assumed that the matrix $\mathbf{A}$ is symmetric. When $\mathbf{A}_{n\times m}$ is not symmetric, the singular-value decomposition (SVD) theorem is used to reduce $\mathbf{A}_{n\times m}$ to a diagonal matrix; the result readily follows from Theorem 2.5.1 by orthonormalizing the matrices $\mathbf{P}$ and $\mathbf{Q}$. Assuming that n is larger than m, the matrix $\mathbf{A}$ may be written as $\mathbf{A} = \mathbf{P}\mathbf{D}_r\mathbf{Q}'$ where $\mathbf{P}'\mathbf{P} = \mathbf{Q}'\mathbf{Q} = \mathbf{Q}\mathbf{Q}' = \mathbf{I}_m$. The matrix $\mathbf{P}$ contains the orthonormal eigenvectors of the matrix $\mathbf{A}\mathbf{A}'$, and the matrix $\mathbf{Q}$ contains the orthonormal eigenvectors of $\mathbf{A}'\mathbf{A}$. The diagonal elements of $\mathbf{D}_r$ contain the positive square root of the eigenvalues of $\mathbf{A}\mathbf{A}'$ or $\mathbf{A}'\mathbf{A}$, called the singular values of $\mathbf{A}_{n\times m}$. If $\mathbf{A}$ is symmetric, then $\mathbf{A}\mathbf{A}' = \mathbf{A}'\mathbf{A} = \mathbf{A}^2$ so that the singular values are the eigenvalues of $\mathbf{A}$. Because most matrices $\mathbf{A}$ are usually symmetric in statistical applications, Theorem 2.6.5 will usually suffice for the study of multivariate analysis. For symmetric matrices **A**, some useful results of the eigenvalues of (2.6.12) follow.

Theorem 2.6.6 *For a square symmetric matrix* $\mathbf{A}_{n\times n}$, *the following results hold.*

1. $\text{tr}(\mathbf{A}) = \sum_i \lambda_i$
2. $|\mathbf{A}| = \prod_i \lambda_i$
3. $r\,(\mathbf{A})$ *equals the number of nonzero* λ_i.
4. *The eigenvalues of* $\mathbf{A}^{-1}$ *are* $1/\lambda_i$ *if* $r\,(\mathbf{A}) = n$.
5. *The matrix* **A** *is idempotent if and only if each eigenvalue of* **A** *is* 0 *or* 1.
6. *The matrix* **A** *is singular if and only if one eigenvalue of* **A** *is zero.*
7. *Each of the eigenvalues of the matrix* **A** *is either* $+1$ *or* -1, *if* **A** *is an orthogonal matrix.*

In (5), if **A** is only idempotent and not symmetric each eigenvalue of **A** is also 0 or 1; however, now the converse is not true.

It is also possible to generalize the eigenequation (2.6.12) for an arbitrary matrix **B** where **B** is a (real) symmetric *p.d.* matrix **B** and **A** is a symmetric matrix of order *n*

$$|\mathbf{A} - \lambda\mathbf{B}| = 0 \tag{2.6.14}$$

The homogeneous system of equations

$$|\mathbf{A} - \lambda_i\mathbf{B}|\,\mathbf{q}_i = \mathbf{0} \qquad (i = 1, 2, \ldots, n) \tag{2.6.15}$$

has a nontrivial solution if and only if (2.6.14) is satisfied. The quantities λ_i and $\mathbf{q}_i$ are the eigenvalues and eigenvectors of **A** in the metric of **B**. A generalization for Theorem 2.6.5 follows.

Theorem 2.6.7 *For (real) symmetric matrices* $\mathbf{A}_{n\times n}$ *and* $\mathbf{B}_{n\times n}$ *where* **B** *is p.d., there exists a nonsingular matrix* $\mathbf{Q}_{n\times n}$ *with columns* $\mathbf{q}_i$ *such that*

$$\mathbf{Q}'\mathbf{A}\mathbf{Q} = \Lambda \quad \text{and} \quad \mathbf{Q}'\mathbf{B}\mathbf{Q} = \mathbf{I}$$

$$\mathbf{A} = (\mathbf{Q}')^{-1}\,\Lambda\mathbf{Q}^{-1} \quad \text{and} \quad (\mathbf{Q}')^{-1}\mathbf{Q}^{-1} = \mathbf{B}$$

$$\mathbf{A} = \sum_i \lambda_i\mathbf{x}_i\mathbf{x}_i' \quad \text{and} \quad \mathbf{B} = \sum_i \mathbf{x}_i\mathbf{x}_i'$$

where $\mathbf{x}_i$ *is the* i^{th} *column of* $(\mathbf{Q}')^{-1}$ *and* Λ *is a diagonal matrix with eigenvalues* $\lambda_1 \geq \lambda_2 \geq \cdots \geq \lambda_n$ *for the equation* $|\mathbf{A} - \lambda\mathbf{B}| = 0$.

Thus, the matrix $\mathbf{Q}$ provides a simultaneous diagonalization of $\mathbf{A}$ and $\mathbf{B}$. The solution to $|\mathbf{A} - \lambda\mathbf{B}| = 0$ is obtained by factoring $\mathbf{B}$ using Theorem 2.6.5: $\mathbf{P}'\mathbf{B}\mathbf{P} = \Lambda$ or $\mathbf{B} = \mathbf{P}\Lambda\mathbf{P}' = (\mathbf{P}\Lambda^{1/2})(\Lambda^{1/2}\mathbf{P}') = \mathbf{P}_1\mathbf{P}_1'$ so that $\mathbf{P}_1^{-1}\mathbf{B}(\mathbf{P}_1') = \mathbf{I}$. Using this result, and the transformation $\mathbf{q}_i = (\mathbf{P}_1')^{-1}\mathbf{x}_i$, (2.6.15) becomes

$$\left[\mathbf{P}_1^{-1}\mathbf{A}(\mathbf{P}_1')^{-1} - \lambda_i\mathbf{I}\right]\mathbf{x}_i = \mathbf{0} \tag{2.6.16}$$

so that we have reduced (2.6.14) to solving $|\mathbf{P}_1^{-1}\mathbf{A}(\mathbf{P}_1')^{-1} - \lambda\mathbf{I}| = 0$ where $\mathbf{P}_1^{-1}\mathbf{A}(\mathbf{P}_1')^{-1}$ is symmetrical. Thus, roots of (2.6.16) are the same as the roots of (2.6.14) and the vectors are related by $\mathbf{q}_i = (\mathbf{P}_1')^{-1}\mathbf{x}_i$.

Alternatively, the transformation $\mathbf{q}_i = \mathbf{B}^{-1}\mathbf{x}_i$ could be used. Then $(\mathbf{A}\mathbf{B}^{-1} - \lambda_i\mathbf{I})\mathbf{x}_i = \mathbf{0}$; however, the matrix $\mathbf{A}\mathbf{B}^{-1}$ is not necessarily symmetric. In this case, special iterative methods must be used to find the roots and vectors, Wilkinson (1965).

The eigenvalues $\lambda_1, \lambda_2, \ldots, \lambda_n$ of $|\mathbf{A} - \lambda\mathbf{B}| = \mathbf{0}$ are fundamental to the study of applied multivariate analysis. Theorem 2.6.8 relates the roots of the various characteristic equations where the matrix $\mathbf{A}$ is associated with an hypothesis test matrix $\mathbf{H}$ and the matrix $\mathbf{B}$ is associated with an error matrix $\mathbf{E}$.

Theorem 2.6.8 *Properties of the roots of* $|\mathbf{H} - \lambda\mathbf{E}| = 0$.

1. *The roots of* $|\mathbf{E} - v(\mathbf{H} + \mathbf{E})| = 0$ *are related to the roots of* $|\mathbf{H} - \lambda\mathbf{E}| = 0$*:*

$$\lambda_i = \frac{1 - v_i}{v_i} \quad \text{or} \quad v_i = \frac{1}{1 + \lambda_i}$$

2. *The roots of* $|\mathbf{H} - \theta(\mathbf{H} + \mathbf{E})| = 0$ *are related to the roots of* $|\mathbf{H} - \lambda\mathbf{E}| = 0$*:*

$$\lambda_i = \frac{\theta_i}{1 - \theta_i} \quad \text{or} \quad \theta_i = \frac{\lambda_i}{1 + \lambda_i}$$

3. *The roots of* $|\mathbf{E} - v(\mathbf{H} + \mathbf{E})| = 0$ *are*

$$v_i = (1 - \theta_i)$$

Theorem 2.6.9 *If* $\alpha_1, \ldots, \alpha_n$ *are the eigenvalues of* $\mathbf{A}$ *and* $\beta_1, \beta_2, \ldots, \beta_m$ *are the eigenvalues of* $\mathbf{B}$*. Then*

1. The eigenvalues of $\mathbf{A} \otimes \mathbf{B}$ are $\alpha_i\beta_j$ $(i = 1, \ldots, n;\ j = 1, \ldots, m)$.
2. The eigenvalues of $\mathbf{A} \oplus \mathbf{B}$ are $\alpha_1, \ldots, \alpha_n, \beta_1, \beta_2, \ldots, \beta_m$.

e. *Matrix Norms*

In Section 2.4, the Euclidean norm of a matrix $\mathbf{A}_{n \times m}$ was defined as the $\{\mathrm{tr}(\mathbf{A}'\mathbf{A})\}^{1/2}$. Solving the characteristic equation $|\mathbf{A}'\mathbf{A} - \lambda\mathbf{I}| = 0$, the Euclidean norm becomes $\|\mathbf{A}\|_2 =$

$\{\sum_i \lambda_i\}^{1/2}$ where λ_i is a root of $\mathbf{A}'\mathbf{A}$. The spectral norm is the square root of the maximum root of $\mathbf{A}'\mathbf{A}$. Thus, $\|\mathbf{A}\|_s = \max \sqrt{\lambda_i}$. Extending the Minkowski vector norm to a matrix, a general matrix (L_p norm) norm is $\|\mathbf{A}\|_p = \left\{\sum_i \lambda_i^{p/2}\right\}^{1/p}$ where λ_i are the roots of $\mathbf{A}'\mathbf{A}$, also called the von Neumann norm. For $p = 2$, it reduces to the Euclidean norm. The von Neumann norm satisfies Definition 2.4.2.

f. Quadratic Forms and Extrema

In our discussion of projection transformations, the norm squared of $\mathbf{y}$ in (2.6.11) was constructed as the sum of two products of the form $\mathbf{y}'\mathbf{A}\mathbf{y} = Q$ for a symmetric matrix $\mathbf{A}$. The quantity Q defined by

$$f(\mathbf{y}) = \sum_{i=1}^{n}\sum_{j=1}^{n} a_{ij} y_i y_j = Q \tag{2.6.17}$$

is called a quadratic form of $\mathbf{y}_{n\times 1}$ for any symmetric matrix $\mathbf{A}_{n\times n}$. Following the definition for matrices, a quadratic form $\mathbf{y}'\mathbf{A}\mathbf{y}$ is said to be

1. Positive definite *(p.d.)* if $\mathbf{y}'\mathbf{A}\mathbf{y} > \mathbf{0}$ for all $\mathbf{y} \neq \mathbf{0}$ and is zero only if $\mathbf{y} = \mathbf{0}$.
2. Positive semidefinite (p.s.d.) if $\mathbf{y}'\mathbf{A}\mathbf{y} > \mathbf{0}$ for all $\mathbf{y}$ and equal zero for at least one nonzero value of $\mathbf{y}$.
3. Non-negative definite *(n.n.d.)* or Gramian if $\mathbf{A}$ is *p.d.* or *p.s.d.*

Using Theorem 2.6.5, every quadratic form can be reduced to a weighted sum of squares using the transformation $\mathbf{y} = \mathbf{P}\mathbf{x}$ as follows

$$\mathbf{y}'\mathbf{A}\mathbf{y} = \sum_i \lambda_i x_i^2$$

where the λ_i are the roots of $|\mathbf{A} - \lambda\mathbf{I}| = 0$ since $\mathbf{P}'\mathbf{A}\mathbf{P} = \Lambda$.

Quadratic forms arise naturally in multivariate analysis since geometrically they represent an ellipsoid in an n-dimensional space with center at the origin and $Q > 0$. When $\mathbf{A} = \mathbf{I}$, the ellipsoid becomes spherical. Clearly the quadratic form $\mathbf{y}'A\mathbf{y}$ is a function of $\mathbf{y}$ and for $\mathbf{y} = \alpha\mathbf{y}$ $(\alpha > 0)$, it may be made arbitrarily large or small. To remove the scale changes in $\mathbf{y}$, the general quotient $\mathbf{y}'\mathbf{A}\mathbf{y}/\mathbf{y}'\mathbf{B}\mathbf{y}$ is studied.

Theorem 2.6.10 *Let* $\mathbf{A}$ *be a symmetric matrix of order n and* $\mathbf{B}$ *a p.d. matrix where* $\lambda_1 \geq \lambda_2 \geq \cdots \geq \lambda_n$ *are the roots of* $|\mathbf{A} - \lambda\mathbf{B}| = 0$. *Then for any* $\mathbf{y} \neq \mathbf{0}$,

$$\lambda_n \leq \mathbf{y}'\mathbf{A}\mathbf{y}/\mathbf{y}'\mathbf{B}\mathbf{y} \leq \lambda_1$$

and

$$\min_{\mathbf{y}\neq 0} \left(\mathbf{y}'\mathbf{A}\mathbf{y}/\mathbf{y}'\mathbf{B}\mathbf{y}\right) = \lambda_n$$

$$\max_{\mathbf{y}\neq 0} \left(\mathbf{y}'\mathbf{A}\mathbf{y}/\mathbf{y}'\mathbf{B}\mathbf{y}\right) = \lambda_1$$

For $\mathbf{B} = \mathbf{I}$, the quantity $\mathbf{y}'\mathbf{A}\mathbf{y}/\mathbf{y}'\mathbf{y}$ is known as the Rayleigh quotient.

g. *Generalized Projectors*

We defined a vector $\mathbf{y}$ to be orthogonal to $\mathbf{x}$ if the inner product $\mathbf{y}'\mathbf{x} = 0$. That is $\mathbf{y} \perp \mathbf{x}$ in the metric of $\mathbf{I}$ since $\mathbf{y}'\mathbf{I}\mathbf{x} = 0$. We also found the eigenvalues of $\mathbf{A}$ in the metric of $\mathbf{I}$ when we solved the eigenequation $|\mathbf{A} - \lambda\mathbf{I}| = 0$. More generally, for a $p.d.$ matrix $\mathbf{B}$, we found the eigenvalues of $\mathbf{A}$ in the metric of $\mathbf{B}$ by solving $|\mathbf{A} - \lambda\mathbf{B}| = 0$. Thus, we say that $\mathbf{y}$ is B-constrained orthogonal to $\mathbf{x}$ if $\mathbf{y}'\mathbf{B}\mathbf{x} = \mathbf{0}$ or $\mathbf{y}$ is orthogonal to $\mathbf{x}$ in the metric of $\mathbf{B}$ and since $\mathbf{B}$ is $p.d.$, $\mathbf{y}'\mathbf{B}\mathbf{y} > 0$. We also saw that an orthogonal projection matrix $\mathbf{P}$ is symmetric $(\mathbf{P} = \mathbf{P}')$ and idempotent $(\mathbf{P}^2 = \mathbf{P})$ and has the general structure $\mathbf{P} = \mathbf{X}\left(\mathbf{X}'\mathbf{X}\right)^{-}\mathbf{X}'$. Inserting a symmetric matrix $\mathbf{B}$ between $\mathbf{X}'\mathbf{X}$ and postmultiplying $\mathbf{P}$ by $\mathbf{B}$, the matrix $P_{\mathbf{X}/\mathbf{B}} = \mathbf{X}\left(\mathbf{X}'\mathbf{B}\mathbf{X}\right)^{-}\mathbf{X}'\mathbf{B}$ is constructed. This leads one to the general definition of an "affine" projector.

Definition 2.6.4 *The affine projection in the metric of a symmetric matrix* $\mathbf{B}$ *is the matrix* $P_{\mathbf{X}/\mathbf{B}} = \mathbf{X}\left(\mathbf{X}'\mathbf{B}\mathbf{X}\right)^{-}\mathbf{X}'\mathbf{B}$

Observe that the matrix $P_{\mathbf{X}/\mathbf{B}}$ is not symmetric, but that it is idempotent. Hence, the eigenvalues of $P_{\mathbf{X}/\mathbf{B}}$ are either 0 or 1. In addition, $\mathbf{B}$ need not be $p.d.$ If we let V represent the space associated with $\mathbf{X}$ and $V_{\mathbf{x}\perp}$ the space associated with $\mathbf{B}$ where $V_n = V_{\mathbf{x}} \oplus V_{\mathbf{x}\perp}$ so that the two spaces are disjoint, then $P_{\mathbf{X}/\mathbf{B}}$ is the projector onto $V_{\mathbf{x}\perp}$. Or, $P_{\mathbf{X}/\mathbf{B}}$ is the projector onto $\mathbf{X}$ along the kernel $\mathbf{X}'\mathbf{B}$ and we observe that $\mathbf{X}[(\mathbf{X}'\mathbf{B}\mathbf{X})^{-}\mathbf{X}'\mathbf{B}]\mathbf{X} = \mathbf{X}$ and $\mathbf{X}\left(\mathbf{X}'\mathbf{B}\mathbf{X}\right)^{-}\mathbf{X}'\mathbf{B} = \mathbf{0}$.

To see how we may use the affine projector, we return to Example 2.6.4 and allow the variance of $\mathbf{e}$ to be equal to $\sigma^2\mathbf{V}$ where $\mathbf{V}$ is known. Now, the projection of $\mathbf{y}$ onto $\mathbf{X}$ is

$$\widehat{\mathbf{y}} = \mathbf{X}[\left(\mathbf{X}'\mathbf{V}^{-1}\mathbf{X}\right)^{-}\mathbf{X}'\mathbf{V}^{-1}\mathbf{y}] = \mathbf{X}\widehat{\boldsymbol{\beta}} \tag{2.6.18}$$

along the kernel $\mathbf{X}'\mathbf{V}^{-1}$. The estimate $\widehat{\boldsymbol{\beta}}$ is the generalized least squares estimator of $\widehat{\boldsymbol{\beta}}$. Also,

$$\|\mathbf{y} - \widehat{\mathbf{y}}\|^2 = (\mathbf{y} - \mathbf{X}\widehat{\boldsymbol{\beta}})'\mathbf{V}^{-1}(\mathbf{y} - \mathbf{X}\widehat{\boldsymbol{\beta}}) \tag{2.6.19}$$

is minimal in the metric of $\mathbf{V}^{-1}$. A more general definition of a projector is given by Rao and Yanai (1979).

Exercises 2.6

1. Using Theorem 2.6.1, determine which of the following systems are consistent, and if consistent whether the solution is unique. Find a solution for the consistent systems

(a) $$\begin{bmatrix} 2 & -3 & 1 \\ 6 & -9 & 3 \end{bmatrix} \begin{bmatrix} x \\ y \\ z \end{bmatrix} = \begin{bmatrix} 5 \\ 10 \end{bmatrix}$$

(b) $$\begin{bmatrix} 1 & 1 & 0 \\ 1 & 0 & -1 \\ 0 & 1 & 1 \\ 1 & 0 & 1 \end{bmatrix} \begin{bmatrix} x \\ y \\ z \end{bmatrix} = \begin{bmatrix} 6 \\ -2 \\ 8 \\ 0 \end{bmatrix}$$

(c) $\begin{bmatrix} 1 & 1 \\ 2 & -3 \end{bmatrix}\begin{bmatrix} x \\ y \end{bmatrix} = \begin{bmatrix} 0 \\ 0 \end{bmatrix}$

(d) $\begin{bmatrix} 1 & 1 & -1 \\ 2 & -1 & 1 \\ 1 & 4 & -4 \end{bmatrix}\begin{bmatrix} x \\ y \\ z \end{bmatrix} = \begin{bmatrix} 0 \\ 0 \\ 0 \end{bmatrix}$

(e) $\begin{bmatrix} 1 & 1 & 1 \\ 1 & -1 & 1 \\ 1 & 2 & 1 \\ 3 & -1 & 3 \end{bmatrix}\begin{bmatrix} x \\ y \\ z \end{bmatrix}\begin{bmatrix} 2 \\ 6 \\ 0 \\ 14 \end{bmatrix}$

2. For the two-group ANOVA model where

$$\begin{bmatrix} y_{11} \\ y_{12} \\ y_{21} \\ y_{22} \end{bmatrix} = \begin{bmatrix} 1 & 1 & 0 \\ 1 & 1 & 0 \\ 1 & 0 & 1 \\ 1 & 0 & 1 \end{bmatrix}\begin{bmatrix} \mu \\ \alpha_1 \\ \alpha_2 \end{bmatrix}$$

solve the system using the restrictions

$$(1) \qquad \alpha_2 = 0$$

$$(2) \quad \alpha_1 - \quad \alpha_2 = 0$$

3. Solve Problem 2 using the reparameterization method for the set of new variables

$$(1)\ \mu + \alpha_1 \text{ and } \mu + \alpha_2 \qquad (2)\ \mu \text{ and } \alpha_1 + \alpha_2$$

$$(3)\ \mu + \alpha_1 \text{ and } \alpha_1 - \alpha_2 \qquad (4)\ \mu + \alpha_1 \text{ and } \alpha_2$$

4. In Problem 2, determine whether unique solutions exist for the following linear combinations of the unknowns and, if they do, find them.

$$(1)\mu + \alpha_1 \qquad (2)\mu + \alpha_1 + \alpha_2$$

$$(3)\alpha_1 - \alpha_2/2 \qquad (4)\alpha_1$$

5. Solve the following system of equations, using the g-inverse approach.

$$\begin{bmatrix} 1 & 1 & 0 & 1 & 0 \\ 1 & 1 & 0 & 1 & 0 \\ 1 & 1 & 0 & 0 & 1 \\ 1 & 1 & 0 & 0 & 1 \\ 1 & 0 & 1 & 1 & 0 \\ 1 & 0 & 1 & 1 & 0 \\ 1 & 0 & 1 & 0 & 1 \\ 1 & 0 & 1 & 0 & 1 \end{bmatrix}\begin{bmatrix} \mu \\ \alpha_1 \\ \alpha_2 \\ \beta_1 \\ \beta_2 \end{bmatrix} = \begin{bmatrix} y_{111} \\ y_{112} \\ y_{121} \\ y_{122} \\ y_{211} \\ y_{212} \\ y_{221} \\ y_{222} \end{bmatrix} = \mathbf{y}$$

For what linear combination of the parameter vector do unique solutions exist? What is the general form of the unique solutions?

6. For Problem 5 consider the vector spaces

$$\mathbf{1} = \left\{\begin{matrix}1\\1\\1\\1\\1\\1\\1\\1\end{matrix}\right\}, \mathbf{A} = \left\{\begin{matrix}1 & 0\\1 & 0\\1 & 0\\1 & 0\\0 & 1\\0 & 1\\0 & 1\\0 & 1\end{matrix}\right\}, \mathbf{B} = \left\{\begin{matrix}1 & 0\\1 & 0\\0 & 1\\0 & 1\\1 & 0\\1 & 0\\0 & 1\\0 & 1\end{matrix}\right\}, \mathbf{X} = \left\{\begin{matrix}1 & 1 & 0 & 1 & 0\\1 & 1 & 0 & 1 & 0\\1 & 1 & 0 & 0 & 1\\1 & 1 & 0 & 0 & 1\\1 & 0 & 1 & 1 & 0\\1 & 0 & 1 & 1 & 0\\1 & 0 & 1 & 0 & 1\\1 & 0 & 1 & 0 & 1\end{matrix}\right\}$$

   (a) Find projection matrices for the projection of $\mathbf{y}$ onto $\mathbf{1}$, $\mathbf{A}/\mathbf{1}$ and $\mathbf{B}/\mathbf{1}$.
   (b) Interpret your findings.
   (c) Determine the length squares of the projections and decompose $||\mathbf{y}||^2$ into a sum of quadratic forms.

7. For each of the symmetric matrices

$$\mathbf{A} = \begin{bmatrix} 2 & 1 \\ 1 & 2 \end{bmatrix} \quad \text{and} \quad \mathbf{B} = \begin{bmatrix} 1 & 1 & 0 \\ 1 & 5 & -2 \\ 0 & -2 & 1 \end{bmatrix}$$

   (a) Find their eigenvalues
   (b) Find n mutually orthogonal eigenvectors and write each matrix as $\mathbf{P}\Lambda\mathbf{P}'$.

8. Determine the eigenvalues and eigenvectors for the $n \times n$ matrix $\mathbf{R} = [r_{ij}]$ where $r_{ij} = 1$ for $i = j$ and $r_{ij} = r \neq 0$ for $i \neq j$.

9. For $\mathbf{A}$ and $\mathbf{B}$ defined by

$$\mathbf{A} = \begin{bmatrix} 498.807 & 426.757 \\ 426.757 & 374.657 \end{bmatrix} \text{ and } \mathbf{B} = \begin{bmatrix} 1838.5 & -334.750 \\ -334.750 & 12,555.25 \end{bmatrix}$$

   solve $|\mathbf{A} - \lambda\mathbf{B}| = 0$ for λ_i and $\mathbf{q}_i$.

10. Given the quadratic forms

$$3y_1^2 + y_2^2 + 2y_3^2 + y_1y_3 \quad \text{and} \quad y_1^2 + 5y_2^2 + y_3^2 + 2y_1y_2 - 4y_2y_3$$

   (a) Find the matrices associated with each form.
   (b) Transform both to each the form $\sum_i \lambda_i x_i^2$.
   (c) Determine whether the forms are *p.d.*, *p.s.d.*, or neither, and find their ranks.
   (d) What is the maximum and minimum value of each quadratic form?

11. Use the Cauchy-Schwarz inequality (Theorem 2.3.5) to show that

$$(\mathbf{a}'\mathbf{b})^2 \leq (\mathbf{a}'\mathbf{G}\mathbf{a})\,(\mathbf{b}'\mathbf{G}^{-1}\mathbf{b})$$

for a *p.d.* matrix $\mathbf{G}$.

2.7 Limits and Asymptotics

We conclude this chapter with some general comments regarding the distribution and convergence of random vectors. Because the distribution theory of a random vectors depend on the calculus of probability that involves multivariable integration theory and differential calculus, which we do not assume in this text, we must be brief. For an overview of the statistical theory for multivariate analysis, one may start with Rao (1973a), or at a more elementary level the text by Casella and Berger (1990) may be consulted.

Univariate data analysis is concerned with the study of a single random variable Y characterized by a cumulative distribution function $F_Y(y) = P[Y \leq y]$ which assigns a probability that Y is less than or equal to a specific real number $y < \infty$, for all $y \in R$. Multivariate data analysis is concerned with the study of the simultaneous variation of several random variables $Y_1, Y_2, \ldots, Y_d$ or a random vector of d-observations, $\mathbf{Y}' = [Y_1, Y_2, \ldots, Y_d]$.

Definition 2.7.1 *A random vector $\mathbf{Y}_{d\times 1}$ is characterized by a joint cumulative distribution function $F_{\mathbf{Y}}(\mathbf{y})$ where*

$$F_{\mathbf{Y}}(\mathbf{y}) = P[\mathbf{Y} \leq \mathbf{y}] = P[Y_1 \leq y_1, Y_2 \leq y_2 \leq \cdots \leq Y_d \leq y_d]$$

assigns a probability to any real vector $\mathbf{y}' = [y_1, y_2, \ldots, y_d]$, $\mathbf{y} \varepsilon V_d$. The vector $\mathbf{Y}' = [Y_1, Y_2, \ldots, Y_d]$ is said to have a multivariate distribution.

For a random vector $\mathbf{Y}$, the cumulative distribution function always exists whether all the elements of the random vector are discrete or (absolutely) continuous or mixed. Using the fundamental theorem of calculus, when it applies, one may obtain from the cumulative distribution function the probability density function for the random vector $\mathbf{Y}$ which we shall represent as $f_{\mathbf{Y}}(\mathbf{y})$. In this text, we shall always assume that the density function exists for a random vector. And, we will say that the random variables $Y_i \in \mathbf{Y}$ are (statistically) independent if the density function for $\mathbf{Y}$ factors into a product of marginal probability density functions; that is,

$$f_{\mathbf{y}}(\mathbf{y}) = \prod_{i=1}^{d} f_{Y_i}(y_i)$$

for all $\mathbf{y}$. Because many multivariate distributions are difficult to characterize, some basic notions of limits and asymptotic theory will facilitate the understanding of multivariate estimation theory and hypothesis testing.

Letting $\{\mathbf{y}_n\}$ represent a sequence of real vectors $\mathbf{y}_1, \mathbf{y}_2, \ldots$, for $n = 1, 2, \ldots$, and $\{c_n\}$ a sequence of positive real numbers, we say that $\mathbf{y}_n$ tends to zero more rapidly than the sequence c_n as $n \longrightarrow \infty$ if the

$$\lim_{n \longrightarrow \infty} \frac{\|\mathbf{y}_n\|}{c_n} = 0 \tag{2.7.1}$$

Using small *oh* notation, we write that

$$\mathbf{y}_n = o(c_n) \text{ as } n \longrightarrow \infty \tag{2.7.2}$$

which shows that $\mathbf{y}_n$ converges more rapidly to zero than c_n as $n \longrightarrow \infty$. Alternatively, suppose the $\|\mathbf{y}_n\|$ is bounded, there exist real numbers K for all n, then, we write that

$\|\mathbf{y}_n\| \leq Kc_n$ for some K. Using big Oh notation

$$\mathbf{y}_n = O(c_n) \tag{2.7.3}$$

These concepts of order are generalized to random vectors by defining convergence in probability.

Definition 2.7.2 *A random vector* $\mathbf{Y}_n$ *converges in probability to a random vector* $\mathbf{Y}$ *written as* $\mathbf{Y}_n \xrightarrow{p} \mathbf{Y}$, *if for all* ϵ *and* $\delta > 0$ *there is an* N *such that for all* $n > N$, $P(\|\mathbf{Y}_n - \mathbf{Y}\| > \epsilon) < \delta$. *Or,* $\lim_{n\to\infty} \{\|\mathbf{Y}_n - \mathbf{Y}\| > 0\} = 0$ *and written as* $\text{plim}\{\mathbf{Y}_n\} = \mathbf{Y}$. *Thus, for the elements in the vectors* $\mathbf{Y}_n - \mathbf{Y}$, $\{\mathbf{Y}_n - \mathbf{Y}\}$, $n = 1, 2, \ldots$ *converge in probability to zero.*

Employing order of convergence notation, we write that

$$\mathbf{Y}_n = o_p(\mathbf{c}_n) \tag{2.7.4}$$

when the sequence plim $\frac{\|\mathbf{Y}_n\|}{c_n} = 0$. Furthermore, if the $\|\mathbf{Y}_n\|$ is bounded in probability by the elements in $\mathbf{c}_n$, we write

$$\mathbf{Y}_n = O(\mathbf{c}_n) \tag{2.7.5}$$

if for $\varepsilon > 0$ the $P\{\|\mathbf{Y}_n\| \leq c_n K\} \leq \varepsilon$ for all n; see Ferguson (1996).

Associating with each random vector a cumulative distribution function, convergence in law or distribution is defined.

Definition 2.7.3 $\mathbf{Y}_n$ *converges in law or distribution to* $\mathbf{Y}$ *written as* $\mathbf{Y}_n \xrightarrow{d} \mathbf{Y}$, *if the limit* $\lim_{n\to\infty} = \mathbf{F}_Y(y)$ *for all points* $\mathbf{y}$ *at which* $F_\mathbf{Y}(\mathbf{y})$ *is continuous.*

Thus, if a parameter estimator $\widehat{\boldsymbol{\beta}}_n$ converges in distribution to a random vector $\boldsymbol{\beta}$, then $\widehat{\boldsymbol{\beta}}_n = O_p(\mathbf{1})$. Furthermore, if $\widehat{\boldsymbol{\beta}}_n - \boldsymbol{\beta} = O_p(\mathbf{c}_n)$ and if $\mathbf{c}_n = o_p(\mathbf{1})$, then the plim $\{\widehat{\boldsymbol{\beta}}_n\} = \boldsymbol{\beta}$ or $\widehat{\boldsymbol{\beta}}_n$ is a consistent estimator of $\boldsymbol{\beta}$. To illustrate this result for a single random variable, we know that $\sqrt{n}\left(\overline{\mathrm{X}}_n - \mu\right) \xrightarrow{d} N\left(0, \sigma^2\right)$. Hence, $\left(\overline{\mathrm{X}}_n - \mu\right) = O_p\left(1/\sqrt{n}\right) = o_p(1)$ so that $\overline{\mathrm{X}}_n$ converges in probability to μ, plim $\{\overline{\mathrm{X}}_n\} = \mu$. The asymptotic distribution of $\overline{\mathrm{X}}_n$ is the normal distribution with mean μ and asymptotic variance σ^2/n as $n \longrightarrow \infty$. If we assume that this result holds for finite n, we say the estimator is asymptotically efficient if the variance of any other consistent, asymptotically normally distributed estimator exceeds σ^2/n. Since the median converges in distribution to a normal distribution, $\sqrt{2_n/\pi}\,(M_n - \mu) \xrightarrow{d} N\left(0, \sigma^2\right)$, the median is a consistent estimator of μ; however, the mean is more efficient by a factor of $\pi/2$.

Another important asymptotic result for random vectors in Slutsky's Theorem.

Theorem 2.7.1 *If* $\overline{\mathbf{X}}_n \xrightarrow{d} \mathbf{X}$ *and* plim $\{\mathbf{Y}_n\} = \mathbf{c}$. *Then*

1. $\begin{bmatrix} \mathbf{X}_n \\ \mathbf{Y}_n \end{bmatrix} \xrightarrow{d} \begin{bmatrix} \mathbf{x} \\ \mathbf{c} \end{bmatrix}$

2. $\mathbf{Y}_n\mathbf{X}_n \xrightarrow{d} \mathbf{cX}$

Since $\mathbf{X}_n \xrightarrow{d} \mathbf{x}$ implies plim $\{\mathbf{X}_n\} = \mathbf{X}$, convergence in distribution may be replaced with convergence in probability. Slutsky's result also holds for random matrices. Thus if $\mathbf{Y}_n$ and $\mathbf{X}_n$ are random matrices such that if plim $\{\mathbf{Y}_n\} = \mathbf{A}$ and plim $\{\mathbf{X}_n\} = \mathbf{B}$, then plim $\{\mathbf{Y}_n\mathbf{X}_n^{-1}\} = \mathbf{AB}^{-1}$.

Exercises 2.7

1. For a sequence of positive real numbers $\{c_n\}$, show that

   (a) $\mathrm{O}(c_n) = \mathrm{O}_p(c_n) = c_n \mathrm{O}(1)$

   (b) $o(c_n) = o_p(c_n) = c_n o(1)$

2. For a real number $\alpha > 0$, Y_n converges to the αth mean of Y if the expectation of $|Y_n - Y|^\alpha \longrightarrow 0$, written $E\,|Y_n - Y|^\alpha \longrightarrow 0$ as $n \longrightarrow \infty$, this is written as $Y_n \xrightarrow{gm} Y_n$ convergence in quadratic mean. Show that if $Y_n \xrightarrow{\alpha} Y$ for some α,that $Y_n \xrightarrow{p} Y$.

   (a) Hint: Use Chebyshev's Inequality.

3. Suppose $X_n \xrightarrow{d} N(0,\ 1)$. What is the distribution of X_n^2.

4. Suppose $X_n - E(X_n)/\sqrt{\operatorname{var} X_n} \xrightarrow{d} X$ and $E(X_n - Y_n)^2 / \operatorname{var} X_n \longrightarrow 0$. What is the distribution of $Y_n - E(Y_n)/\sqrt{\operatorname{var} Y_n}$?

5. Asymptotic normality of t. If $X_1, X_2, \ldots,$ is a sample from $N(\mu,\ \sigma^2)$, then $\overline{X}_n \xrightarrow{d} \mu$ and $\Sigma X_j^2/n \xrightarrow{d} E(X^2)$ so $s_n^2 = \Sigma X_j^2/n - \overline{X}_n^2 \xrightarrow{d} E(X^2) - \mu^2 = \sigma_1^2$. Show that $\sqrt{n-1}\left(\overline{X}_n - \mu\right)/s_n \xrightarrow{d} N(0,\ 1)$.

3

Multivariate Distributions and the Linear Model

3.1 Introduction

In this chapter, the multivariate normal distribution, the estimation of its parameters, and the algebra of expectations for vector- and matrix-valued random variables are reviewed. Distributions commonly encountered in multivariate data analysis, the linear model, and the evaluation of multivariate normality and covariance matrices are also reviewed. Finally, tests of locations for one and two groups are discussed. The purpose of this chapter is to familiarize students with multivariate sampling theory, evaluating model assumptions, and analyzing multivariate data for one- and two-group inference problems.

The results in this chapter will again be presented without proof. Numerous texts at varying levels of difficulty have been written that discuss the theory of multivariate data analysis. In particular, books by Anderson (1984), Bilodeau and Brenner (1999), Jobson (1991, 1992), Muirhead (1982), Seber (1984), Srivastava and Khatri (1979), Rencher (1995) and Rencher (1998) may be consulted, among others.

3.2 Random Vectors and Matrices

Multivariate data analysis is concerned with the systematic study of p random variables $\mathbf{Y}' = [Y_1, Y_2, ..., Y_p]$. The expected value of the random $p \times 1$ vector is defined as the vector of expectations

$$E(\mathbf{Y}) = \begin{bmatrix} E(Y_1) \\ E(Y_2) \\ \vdots \\ E(Y_p) \end{bmatrix}$$

More generally, if $\mathbf{Y}_{n\times p} = \left[Y_{ij}\right]$ is a matrix of random variables, then the $E(\mathbf{Y})$ is the matrix of expectations with elements $[E(Y_{ij})]$. For constant matrices $\mathbf{A}$, $\mathbf{B}$, and $\mathbf{C}$, the following operation for expectations of matrices is true

$$E\left(\mathbf{AY}_{n\times p}\mathbf{B}+\mathbf{C}\right) = \mathbf{A}E\left(\mathbf{Y}_{n\times p}\right)\mathbf{B}+\mathbf{C} \tag{3.2.1}$$

For a random vector $\mathbf{Y}' = [Y_1, Y_2, \ldots, Y_p]$, the mean vector is

$$\boldsymbol{\mu} = E(\mathbf{Y}) = \begin{bmatrix} \mu_1 \\ \mu_2 \\ \vdots \\ \mu_p \end{bmatrix}$$

The covariance matrix of a random vector $\mathbf{Y}$ is defined as the $p \times p$ matrix

$$\begin{aligned}
\operatorname{cov}(\mathbf{Y}) &= E\left\{[\mathbf{Y}-E(\mathbf{Y})][\mathbf{Y}-E(\mathbf{Y})]'\right\} \\
&= E\left\{[\mathbf{Y}-\boldsymbol{\mu}][\mathbf{Y}-\boldsymbol{\mu}]'\right\} \\
&= \begin{bmatrix} \sigma_{11} & \sigma_{12} & \cdots & \sigma_{1p} \\ \sigma_{21} & \sigma_{22} & \cdots & \sigma_{2p} \\ \vdots & \vdots & & \vdots \\ \sigma_{p1} & \sigma_{p2} & \cdots & \sigma_{pp} \end{bmatrix} = \boldsymbol{\Sigma}
\end{aligned}$$

where

$$\sigma_{ij} = \operatorname{cov}\left(Y_i, Y_j\right) = E\left\{\left[Y_i - \mu_i\right]\left[Y_j - \mu_j\right]\right\}$$

and $\sigma_{ii} = \sigma_i^2 = E[(Y_i - \mu_i)^2] = \operatorname{var} Y_i$. Hence, the diagonal elements of Σ must be nonnegative. Furthermore, Σ is symmetric so that covariance matrices are nonnegative definite matrices. If the covariance matrix of a random vector $\mathbf{Y}$ is not positive definite, the components Y_i of $\mathbf{Y}$ are linearly related and the $|\Sigma| = 0$. The multivariate analogue of the variance σ^2 is the covariance matrix Σ. Wilks (1932) called the determinant of the covariance matrix, $|\Sigma|$, the generalized variance of a multivariate normal distribution. Because the determinant of the covariance matrix is related to the product of the roots of the characteristic equation $|\Sigma - \lambda\mathbf{I}|$, even though the elements of the covariance matrix may be large, the generalized variance may be close to zero. Just let the covariance matrix be a diagonal matrix where all diagonal elements are large and one variance is nearly zero. Thus, a small value for the generalized variance does not necessary imply that all the elements in the covariance matrix are small. Dividing the determinant of Σ by the product of the variances for each of the p variables, we have the bounded measure $0 \leq \left(|\Sigma| / \prod_{i=1}^{p} \sigma_{ii}\right)^2 \leq 1$.

Theorem 3.2.1 *A $p \times p$ matrix Σ is a covariance matrix if and only if it is nonnegative definite (n.n.d.).*

Multiplying a random vector $\mathbf{Y}$ by a constant matrix $\mathbf{A}$ and adding a constant vector $\mathbf{c}$, the covariance of the linear transformation $\mathbf{z} = \mathbf{AY} + \mathbf{c}$ is seen to be

$$\operatorname{cov}(\mathbf{z}) = \mathbf{A}\boldsymbol{\Sigma}\mathbf{A}' \tag{3.2.2}$$

since the cov $(\mathbf{c}) = \mathbf{0}$. For the linear combination $\mathbf{z} = \mathbf{a}'\mathbf{Y}$ and a constant vector $\mathbf{a}$, the cov $\left(\mathbf{a}'\mathbf{Y}\right) = \mathbf{a}'\Sigma\mathbf{a}$.

Extending (3.2.2) to two random vectors $\mathbf{X}$ and $\mathbf{Y}$, the

$$\text{cov}(\mathbf{X}, \mathbf{Y}) = E\{[\mathbf{Y} - \boldsymbol{\mu}_Y][\mathbf{X} - \boldsymbol{\mu}_{\mathbf{X}}]'\} = \Sigma_{\mathbf{XY}} \tag{3.2.3}$$

Properties of the cov $(\cdot)$ operator are given in Theorem 3.2.2.

Theorem 3.2.2 *For random vectors* $\mathbf{X}$ *and* $\mathbf{Y}$*, scalar matrices* $\mathbf{A}$ *and* $\mathbf{B}$*, and scalar vectors* $\mathbf{a}$ *and* $\mathbf{b}$*, the*

1. cov $\left(\mathbf{a}'\mathbf{X}, \mathbf{b}'\mathbf{Y}\right) = \mathbf{a}'\Sigma_{XY}\mathbf{b}$
2. cov $(\mathbf{X}, \mathbf{Y}) =$ cov $(\mathbf{Y}, \mathbf{X})$
3. $\text{cov}(\mathbf{a} + \mathbf{AX}, \mathbf{b} + \mathbf{BY}) = \mathbf{A}\,\text{cov}(\mathbf{X}, \mathbf{Y})\mathbf{B}'$

The zero-order Pearson correlation between two random variables Y_i and Y_j is given by

$$\rho_{ij} = \frac{\sigma_{ij}}{\sigma_i \sigma_j} = \frac{\text{cov}\left(Y_i, Y_j\right)}{\sqrt{\text{var}\,(Y_i)\,\text{var}\left(Y_j\right)}} \quad \text{where} \quad -1 \leq \rho_{ij} \leq 1$$

The correlation matrix for the random p-vector $\mathbf{Y}$ is

$$\mathbf{P} = \left[\rho_{ij}\right] \tag{3.2.4}$$

Letting $(\text{diag}\,\Sigma)^{-1/2}$ represent the diagonal matrix with diagonal elements equal to the square root of the diagonal elements of Σ, the relationship between $\mathbf{P}$ and Σ is established

$$\mathbf{P} = (\text{diag}\,\Sigma)^{-1/2}\,\Sigma\,(\text{diag}\,\Sigma)^{-1/2}$$
$$\Sigma = (\text{diag}\,\Sigma)^{1/2}\,\mathbf{P}\,(\text{diag}\,\Sigma)^{1/2}$$

Because the correlation matrix does not depend on the scale of the random variables, it is used to express relationships among random variables measured on different scales. Furthermore, since the $|\Sigma| = |\,(\text{diag}\,\Sigma)^{1/2}\,\mathbf{P}\,(\text{diag}\,\Sigma)^{1/2}\,|$ we have that $0 \leq |\mathbf{P}|^2 \leq 1$. Takeuchi, et al. (1982, p. 246) call the $|\mathbf{P}|$ the generalized alienation coefficient. If the elements of $\mathbf{Y}$ are independent its value is one and if elements are dependent it value is zero. Thus, the determinant of the correlation matrix may be interpreted as an overall measure of association or nonassociation.

Partitioning a random p-vector into two subvectors: $\mathbf{Y} = [\mathbf{Y}_1', \mathbf{Y}_2']'$, the covariance matrix of the partitioned vector is

$$\text{cov}\,(\mathbf{Y}) = \begin{bmatrix} \text{cov}\,(\mathbf{Y}_1, \mathbf{Y}_1) & \text{cov}\,(\mathbf{Y}_1, \mathbf{Y}_2) \\ \text{cov}\,(\mathbf{Y}_2, \mathbf{Y}_1) & \text{cov}\,(\mathbf{Y}_2, \mathbf{Y}_2) \end{bmatrix} = \begin{bmatrix} \Sigma_{11} & \Sigma_{12} \\ \Sigma_{21} & \Sigma_{22} \end{bmatrix}$$

where $\Sigma_{ij} = \text{cov}(\mathbf{Y}_i, \mathbf{Y}_j)$. To evaluate whether $\mathbf{Y}_1$ and $\mathbf{Y}_2$ are uncorrelated, the following theorem is used.

Theorem 3.2.3 *The random vectors* $\mathbf{Y}_1$ *and* $\mathbf{Y}_2$ *are uncorrelated if and only if* $\Sigma_{12} = \mathbf{0}$.

The individual components of $\mathbf{Y}_i$ are uncorrelated if and only if Σ_{ii} is a diagonal matrix. If $\mathbf{Y}_i$ has cumulative distribution function $(c.d.f)$, $F_{\mathbf{Y}_i}(\mathbf{y}_i)$, with mean $\boldsymbol{\mu}_i$ and covariance matrix Σ_{ii}, we write $\mathbf{Y}_i \sim (\boldsymbol{\mu}_i, \Sigma_{ii})$.

Definition 3.2.1 *Two (absolutely) continuous random vectors* $\mathbf{Y}_1$ *and* $\mathbf{Y}_2$ *are (statistically) independent if the probability density function of* $\mathbf{Y} = [\mathbf{Y}_1', \mathbf{Y}_2']'$ *is obtained from the product of the marginal densities of* $\mathbf{Y}_1$ *and* $\mathbf{Y}_2$*:*

$$f_{\mathbf{Y}}(\mathbf{y}) = f_{\mathbf{Y}_1}(\mathbf{y}_1)\, f_{\mathbf{Y}_2}(\mathbf{y}_2)$$

The probability density function or the joint density of $\mathbf{Y}$ is obtained from $F_{\mathbf{Y}}(\mathbf{y})$ using the fundamental theorem of calculus. If $\mathbf{Y}_1$ and $\mathbf{Y}_2$ are independent, then the $\text{cov}(\mathbf{Y}_1, \mathbf{Y}_2) = \mathbf{0}$. However, the converse is not in general true.

In Chapter 2, we defined the Mahalanobis distance for a random variable. It was an "adjusted" Euclidean distance which represented statistical closeness in the metric of $1/\sigma^2$ or $(\sigma^2)^{-1}$. With the first two moments of a random vector defined, suppose we want to calculate the distance between $\mathbf{Y}$ and $\boldsymbol{\mu}$. Generalizing (2.3.5), the Mahalanobis distance between $\mathbf{Y}$ and $\boldsymbol{\mu}$ in the metric of Σ is

$$D_{\Sigma}(\mathbf{Y}, \boldsymbol{\mu}) = [(\mathbf{Y} - \boldsymbol{\mu})'\Sigma^{-1}(\mathbf{Y} - \boldsymbol{\mu})]^{1/2} \tag{3.2.5}$$

If $\mathbf{Y} \sim (\boldsymbol{\mu}_1, \Sigma)$ and $\mathbf{X} \sim (\boldsymbol{\mu}_2, \Sigma)$, then the Mahalanobis distance between $\mathbf{Y}$ and $\mathbf{X}$, in the metric of Σ, is the square root of

$$D^2_{\Sigma}(\mathbf{X}, \mathbf{Y}) = (\mathbf{X} - \mathbf{Y})'\Sigma^{-1}(\mathbf{X} - \mathbf{Y})$$

which is invariant under linear transformations $\mathbf{z}_{\mathbf{X}} = \mathbf{AX} + \mathbf{a}$ and $\mathbf{z}_{\mathbf{Y}} = \mathbf{AY} + \mathbf{b}$. The covariance matrix Σ of $\mathbf{X}$ and $\mathbf{Y}$ becomes $\Omega = \mathbf{A}\Sigma\mathbf{A}'$ under the transformations so that $D^2_{\Sigma}(\mathbf{X}, \mathbf{Y}) = \mathbf{z}_{\mathbf{X}}'\Omega\mathbf{z}_{\mathbf{Y}} = D^2_{\Omega}(\mathbf{z}_{\mathbf{X}}, \mathbf{z}_{\mathbf{Y}})$.

The Mahalanobis distances, D, arise in a natural manner when investigating the separation between two or more multivariate populations, the topic of discriminant analysis discussed in Chapter 7. It is also used to assess multivariate normality.

Having defined the mean and covariance matrix for a random vector $\mathbf{Y}_{p\times 1}$ and the first two moments of a random vector, we extend the classical measures of skewness and kurtosis, $E[(Y - \mu)^3]/\sigma^3 = \mu_3'/\sigma^3$ and $E[(Y - \mu)^4]/\sigma^4 = \mu_4'/\sigma^4$ of a univariate variable Y, respectively, to the multivariate case. Following Mardia (1970), multivariate skewness and kurtosis measures for a random p-variate vector $\mathbf{Y}_p \sim (\boldsymbol{\mu}, \Sigma)$ are, respectively, defined as

$$\beta_{1,p} = E\{(\mathbf{Y} - \boldsymbol{\mu})'\,\Sigma^{-1}\,(\mathbf{X} - \boldsymbol{\mu})\}^3 \tag{3.2.6}$$

$$\beta_{2,p} = E\{(\mathbf{Y} - \boldsymbol{\mu})'\,\Sigma^{-1}\,(\mathbf{Y} - \boldsymbol{\mu})\}^2 \tag{3.2.7}$$

where $\mathbf{Y}_p$ and $\mathbf{X}_p$ are identically and independent identically distributed $(i.i.d.)$. Because $\beta_{1,p}$ and $\beta_{2,p}$ have the same form as Mahalanobis' distance, they are also seen to be invariant under linear transformations.

The multivariate measures of skewness and kurtosis are natural generalizations of the univariate measures

$$\sqrt{\beta_1} = \sqrt{\beta_{1,1}} = \mu_3'/\sigma^3 \tag{3.2.8}$$

and

$$\beta_2 = \beta_{2,1} = \mu_4'/\sigma^4 \tag{3.2.9}$$

For a univariate normal random variable, $\gamma_1 = \sqrt{\beta_1} = 0$ and $\gamma_2 = \beta_2 - 3 = 0$.

Exercises 3.2

1. Prove Theorems 3.2.2 and 3.2.3.

2. For $\mathbf{Y} \sim N(\boldsymbol{\mu}, \Sigma)$ and constant matrices $\mathbf{A}$ and $\mathbf{B}$, prove the following results for quadratic forms.

   (a) $E(\mathbf{Y}'\mathbf{A}\mathbf{Y}) = \text{tr}(\mathbf{A}\Sigma) + \boldsymbol{\mu}'\mathbf{A}\boldsymbol{\mu}$
   (b) $\text{cov}\left(\mathbf{Y}'\mathbf{A}\mathbf{Y}\right) = \mathbf{3}\,\text{tr}\,(\mathbf{A}\Sigma)^2 + 4\boldsymbol{\mu}'\mathbf{A}\Sigma\mathbf{A}\boldsymbol{\mu}$
   (c) $\text{cov}\left(\mathbf{Y}'\mathbf{A}\mathbf{Y}, \mathbf{Y}'\mathbf{B}\mathbf{Y}\right) = 2\,\text{tr}\,(\mathbf{A}\Sigma\mathbf{B}\Sigma) + 4\boldsymbol{\mu}'\mathbf{A}\Sigma\mathbf{B}\boldsymbol{\mu}$
   $\left[\text{Hint}: E\left(\mathbf{Y}\mathbf{Y}'\right) = \Sigma + \boldsymbol{\mu}\boldsymbol{\mu}', \text{and the tr}\left(\mathbf{A}\mathbf{Y}\mathbf{Y}'\right) = \mathbf{Y}'\mathbf{A}\mathbf{Y}.\right]$

3. For $\mathbf{X} \sim (\boldsymbol{\mu}_1, \Sigma)$ and $\mathbf{Y} \sim (\boldsymbol{\mu}_2, \Sigma)$, where $\boldsymbol{\mu}_1' = [1, 1]$, $\boldsymbol{\mu}_2' = [0, 0]$ and $\Sigma = \begin{bmatrix} 2 & 1 \\ 1 & 2 \end{bmatrix}$, find $D^2(\mathbf{X}, \mathbf{Y})$.

4. Graph contours for ellipsoids of the form $(\mathbf{Y} - \boldsymbol{\mu})'\Sigma^{-1}(\mathbf{Y} - \boldsymbol{\mu}) = c^2$ where $\boldsymbol{\mu}' = [2, 2]$ and $\Sigma = \begin{bmatrix} 2 & 1 \\ 1 & 2 \end{bmatrix}$.

5. For the equicorrelation matrix $\mathbf{P} = (1 - \rho)\mathbf{I} + \rho\mathbf{1}\mathbf{1}'$ and $-(p-1)^{-1} < \rho < 1$ for $p \geq 2$, show that the Mahalanobis squared distance between $\boldsymbol{\mu}_1' = [\alpha, \mathbf{0}']$ and $\boldsymbol{\mu}_2' = \mathbf{0}$ in the metric of $\mathbf{P}$ is

$$D^2(\boldsymbol{\mu}_1, \boldsymbol{\mu}_2) = \alpha\left\{\frac{1 + (p-2)\rho}{(1-\rho)\left[1 + (p-1)\rho\right]}\right\}$$

   $\left[\text{Hint: } \mathbf{P}^{-1} = (1-\rho)^{-1}\left\{\mathbf{I} - \rho\left[1 + (p-1)\rho\right]^{-1}\mathbf{1}\mathbf{1}'\right\}.\right]$

6. Show that $\beta_{2,p}$ may be written as $\beta_{2,p} = \text{tr}[\{\mathbf{D}_p'(\Sigma^{-1} \otimes \Sigma^{-1})\mathbf{D}_p\}\Gamma] + p$ where $\mathbf{D}_p$ is a duplication matrix in that $\mathbf{D}_p\,\text{vech}(\mathbf{A}) = \text{vec}(\mathbf{A})$ and $\Gamma = \text{cov}[\text{vech}\{(\mathbf{Y} - \boldsymbol{\mu})(\mathbf{Y} - \boldsymbol{\mu})'\}]$.

7. We noted that the $|\mathbf{P}|^2$ may be used as an overall measure of multivariate association, construct a measure of overall multivariate association using the functions $||\mathbf{P}||^2$, and the $\text{tr}(\mathbf{P})$?

3.3 The Multivariate Normal (MVN) Distribution

Derivation of the joint density function for the multivariate normal is complex since it involves calculus and moment-generating functions or a knowledge of characteristic functions which are beyond the scope of this text. To motivate its derivation, recall that a random variable Y_i has a normal distribution with mean μ_i and variance σ^2, written $Y_i \sim N(\mu_i, \sigma^2)$, if the density function of Y_i has the form

$$f_{Y_i}(y_i) = \frac{1}{\sigma\sqrt{2\pi}} \exp\{-(y_i - \mu_i)^2/2\sigma^2\} \qquad -\infty < y_i < \infty \tag{3.3.1}$$

Letting $\mathbf{Y}' = [Y_1, Y_2, \ldots, Y_p]$ where each Y_i is independent normal with mean μ_i and variance σ^2, we have from Definition 3.2.1, that the joint density of $\mathbf{Y}$ is

$$\begin{aligned}
f_{\mathbf{Y}}(\mathbf{y}) &= \prod_{i=1}^{p} f_{Y_i}(y_i) \\
&= \prod_{i=1}^{p} \frac{1}{\sigma\sqrt{2\pi}} \exp\{-(y_i - \mu_i)^2/2\sigma^2\} \\
&= (2\pi)^{-p/2}\left(\frac{1}{\sigma^p}\right)\exp\{-\sum_{i=1}^{p}(y_i - \mu_i)^2/2\sigma^2\} \\
&= (2\pi)^{-p/2}\left|\left(\sigma^2\mathbf{I}_p\right)\right|^{-1/2}\exp\{-(\mathbf{y}-\boldsymbol{\mu})'\left(\sigma^2\mathbf{I}_p\right)^{-1}(\mathbf{y}-\boldsymbol{\mu})/2\}
\end{aligned}$$

This is the joint density function of an independent multivariate normal distribution, written as $\mathbf{Y} \sim N_p(\boldsymbol{\mu}, \sigma^2\mathbf{I})$, where the mean vector and covariance matrix are

$$E(\mathbf{Y}) = \boldsymbol{\mu} = \begin{bmatrix} \mu_1 \\ \mu_2 \\ \vdots \\ \mu_p \end{bmatrix}, \quad \text{and} \quad \operatorname{cov}(\mathbf{Y}) = \begin{bmatrix} \sigma^2 & 0 & \cdots & 0 \\ 0 & \sigma^2 & \cdots & 0 \\ \vdots & \vdots & & \vdots \\ 0 & 0 & \cdots & \sigma^2 \end{bmatrix} = \sigma^2\mathbf{I}_p,$$

respectively.

More generally, replacing $\sigma^2\mathbf{I}_p$ with a positive definite covariance matrix Σ, a generalization of the independent multivariate normal density to the multivariate normal (MVN) distribution is established

$$f(\mathbf{y}) = (2\pi)^{-p/2}|\Sigma|^{-1/2}\exp\{-(\mathbf{y}-\boldsymbol{\mu})'\Sigma^{-1}(\mathbf{y}-\boldsymbol{\mu})/2\} \qquad -\infty < y_i < \infty \tag{3.3.2}$$

This leads to the following theorem.

Theorem 3.3.1 *A random p-vector* $\mathbf{Y}$ *is said to have a p-variate normal or multivariate normal (MVN) distribution with mean* $\boldsymbol{\mu}$ *and p.d. covariance matrix* Σ *written* $\mathbf{Y} \sim N_p(\boldsymbol{\mu}, \Sigma)$, *if it has the joint density function given in (3.3.2). If* Σ *is not p.d., the density function of* $\mathbf{Y}$ *does not exist and* $\mathbf{Y}$ *is said to have a singular multivariate normal distribution.*

If $\mathbf{Y} \sim N_p(\boldsymbol{\mu}, \Sigma)$ independent of $\mathbf{X} \sim N_p(\boldsymbol{\mu}, \Sigma)$ then multivariate skewness and kurtosis become $\beta_{1,p} = 0$ and $\beta_{2,p} = p(p+2)$. Multivariate kurtosis is sometimes defined as $\gamma = \beta_{2,p} - p(p+2)$ to also make its value zero. When comparing a general spherical symmetrical distribution to a MVN distribution, the multivariate kurtosis index is defined as $\xi = \beta_{2,p}/p(p+2)$. The class of distributions that maintain spherical symmetry are called elliptical distributions. An overview of these distributions may be found in Bilodeau and Brenner (1999, Chapter 13).

Observe that the joint density of the MVN distribution is constant whenever the quadratic form in the exponent is constant. The constant density ellipsoid $(\mathbf{Y} - \boldsymbol{\mu})' \Sigma^{-1} (\mathbf{Y} - \boldsymbol{\mu}) = c$ has center at $\boldsymbol{\mu}$ while Σ determines its shape and orientation. In the bivariate case,

$$\mathbf{Y} = \begin{bmatrix} Y_1 \\ Y_2 \end{bmatrix}, \boldsymbol{\mu} = \begin{bmatrix} \mu_1 \\ \mu_2 \end{bmatrix}, \Sigma = \begin{bmatrix} \sigma_{11} & \sigma_{12} \\ \sigma_{21} & \sigma_{22} \end{bmatrix} = \begin{bmatrix} \sigma_1^2 & \rho\sigma_1\sigma_2 \\ \rho\sigma_1\sigma_2 & \sigma_2^2 \end{bmatrix}$$

For the MVN to be nonsingular, we need $\sigma_1^2 > 0$, $\sigma_2^2 > 0$ and the $|\Sigma| = \sigma_1^2\sigma_2^2(1-\rho^2) > 0$ so that $-1 < \rho < 1$. Then

$$\Sigma^{-1} = \frac{1}{1-\rho^2} \begin{bmatrix} \frac{1}{\sigma_1^2} & \frac{-\rho}{\sigma_1\sigma_2} \\ \frac{-\rho}{\sigma_1\sigma_2} & \frac{1}{\sigma_2^2} \end{bmatrix}$$

and the joint probability density of $\mathbf{Y}$ yields the bivariate normal density

$$f(\mathbf{y}) = \frac{\exp\left\{\frac{-1}{2(1-\rho^2)}\left[\left(\frac{y_1-\mu_1}{\sigma_1}\right)^2 - 2\rho\left(\frac{y_1-\mu_1}{\sigma_1}\right)\left(\frac{y_2-\mu_2}{\sigma_2}\right) + \left(\frac{y_2-\mu_2}{\sigma_2}\right)^2\right]\right\}}{2\pi\sigma_1\sigma_2\left(1-\rho^2\right)^{1/2}}$$

Letting $Z_i = \left(Y_i - \mu_i\right)/\sigma_i (i = 1, 2)$, the joint bivariate normal becomes the standard bivariate normal

$$f(\mathbf{z}) = \frac{\exp\frac{-1}{2(1-\rho^2)}\left(z_1^2 - 2\rho z_1 z_2 + z_2^2\right)}{2\pi\left(1-\rho^2\right)^{1/2}} \qquad -\infty < z_i < \infty$$

The exponent in the standard bivariate normal distribution is a quadratic form

$$Q = [z_1, z_2]' \Sigma^{-1} \begin{bmatrix} z_1 \\ z_2 \end{bmatrix} = \frac{z_1^2 - 2\rho z_1 z_2 + z_2^2}{1-\rho^2} > 0$$

where

$$\Sigma^{-1} = \frac{1}{1-\rho^2}\begin{bmatrix} 1 & -\rho \\ -\rho & 1 \end{bmatrix} \quad \text{and} \quad \Sigma = \begin{bmatrix} 1 & \rho \\ \rho & 1 \end{bmatrix}$$

which generates concentric ellipses about the origin. Setting $\rho = 1/2$, the ellipses have the form $Q = z_1^2 - z_1 z_2 + z_2^2$ for $Q > 0$. Graphing this function in the plane with axes z_1 and z_2 for $Q = 1$ yields the constant density ellipse with semi-major axis a and semi-minor axis b, Figure 3.3.1.

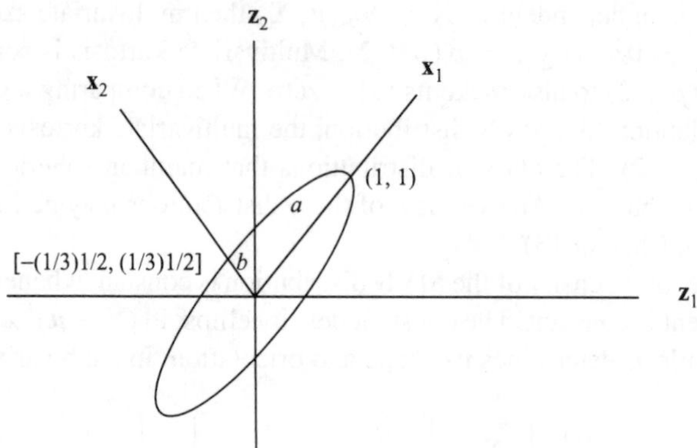

FIGURE 3.3.1. $\mathbf{z}'\Sigma^{-1}\mathbf{z} = z_1^2 - z_1 z_2 + z_2^2 = 1$

Performing an orthogonal rotation of $\mathbf{x} = \mathbf{P}'\mathbf{z}$, the quadratic form for the exponent of the standard MVN becomes

$$\mathbf{z}'\Sigma^{-1}\mathbf{z} = \lambda_1^* x_1^2 + \lambda_2^* x_2^2 = \frac{1}{\lambda_2} x_1^2 + \frac{1}{\lambda_1} x_2^2$$

where $\lambda_1 = 1 + \rho = 3/2$ and $\lambda_2 = 1 - \rho = 1/2$ are the roots of $|\Sigma - \lambda \mathbf{I}| = 0$ and $\lambda_2^* = 1/\lambda_1$ and $\lambda_1^* = 1/\lambda_2$ are the roots of $|\ \Sigma^{-1} - \lambda^* \mathbf{I} = 0$. From analytic geometry, the equation of an ellipse, for $Q = 1$, is given by

$$\left(\frac{1}{b}\right)^2 x_1^2 + \left(\frac{1}{a}\right)^2 x_2^2 = 1$$

Hence $a^2 = \lambda_1$ and $b^2 = \lambda_2$ so that each half-axis is proportional to the inverse of the squared lengths of the eigenvalues of Σ. As Q varies, concentric ellipsoids are generated so that $a = \sqrt{Q\lambda_1}$ and $b = \sqrt{Q\lambda_2}$.

a. *Properties of the Multivariate Normal Distribution*

The multivariate normal distribution is important in the study of multivariate analysis because numerous population phenomena may be approximated by the distribution and the distribution has very nice properties. In large samples the distributions of multivariate parameter estimators tend to multivariate normality.

Some important properties of a random vector $\mathbf{Y}$ having a MVN distribution follow.

Theorem 3.3.2 *Properties of normally distributed random variables.*

1. *Linear combinations of the elements of* $\mathbf{Y} \sim \mathbf{N}[\boldsymbol{\mu}, \Sigma]$ *are normally distributed. For a constant vector* $\mathbf{a} \neq \mathbf{0}$ *and* $\mathbf{X} = \mathbf{a}'\mathbf{Y}$, *then* $\mathbf{X} \sim N_1\left(\mathbf{a}'\boldsymbol{\mu}, \mathbf{a}'\Sigma\mathbf{a}\right)$.
2. *The normal distribution of* $\mathbf{Y}_p \sim N_p[\boldsymbol{\mu}, \Sigma]$ *is invariant to linear transformations. For a constant matrix* $\mathbf{A}_{q\times p}$ *and vector* $\mathbf{b}_{q\times 1}$, $\mathbf{X} = \mathbf{A}\mathbf{Y}_p + \mathbf{b} \sim N_q(\mathbf{A}\boldsymbol{\mu} + \mathbf{b}, \mathbf{A}\Sigma\mathbf{A}')$.

3. *Partitioning* $\mathbf{Y} = [\mathbf{Y}_1', \mathbf{Y}_2']'$, $\boldsymbol{\mu} = \begin{bmatrix} \mu_1 \\ \mu_2 \end{bmatrix}$ and $\Sigma = \begin{bmatrix} \Sigma_{11} & \Sigma_{12} \\ \Sigma_{21} & \Sigma_{22} \end{bmatrix}$, *the subvectors of* $\mathbf{Y}$ *are normally distributed.* $\mathbf{Y}_1 \sim N_{p_1}\left[\boldsymbol{\mu}_1, \Sigma_{11}\right]$ *and* $\mathbf{Y}_2 \sim N_{p_2}\left[\boldsymbol{\mu}_2, \Sigma_{22}\right]$ *where* $p_1 + p_2 = p$*. More generally, all marginal distributions for any subset of random variables are normally distributed. However, the converse is not true, marginal normality does not imply multivariate normality.*

4. *The random subvectors* $\mathbf{Y}_1$ *and* $\mathbf{Y}_2$ *of* $\mathbf{Y} = [\mathbf{Y}_1', \mathbf{Y}_2']'$ *are independent if and only if* $\Sigma = \text{diag}[\Sigma_{11}, \Sigma_{22}]$*. Thus, uncorrelated normal subvectors are independent under multivariate normality.*

5. *The conditional distribution of* $\mathbf{Y}_1 \mid \mathbf{Y}_2$ *is normally distributed,*

$$\mathbf{Y}_1 \mid \mathbf{Y}_2 \sim N_{p_1}\left[\boldsymbol{\mu}_1 + \Sigma_{12}\Sigma_{22}^{-1}(\mathbf{y}_2 - \boldsymbol{\mu}_1), \Sigma_{11} - \Sigma_{12}\Sigma_{22}^{-1}\Sigma_{21}\right]$$

Writing the mean of the conditional normal distribution as

$$\begin{aligned} \boldsymbol{\mu} &= (\boldsymbol{\mu}_1 - \Sigma_{12}\Sigma_{22}^{-1}\boldsymbol{\mu}_2) + \Sigma_{12}\Sigma_{22}^{-1}\mathbf{y}_2 \\ &= \boldsymbol{\mu}_0 + \mathbf{B}_1'\mathbf{y}_2 \end{aligned}$$

$\boldsymbol{\mu}$ *is called the regression function of* $\mathbf{Y}_1$ *on* $\mathbf{Y}_2 = \mathbf{y}_2$ *with regression coefficients* $\mathbf{B}_1'$*. The matrix* $\Sigma_{11.2} = \Sigma_{11} - \Sigma_{12}\Sigma_{22}^{-1}\Sigma_{21}$ *is called the partial covariance matrix with elements* $\sigma_{ij.\,p_1+1,\ldots,\,p_1+p_2}$*. A similar result holds for* $\mathbf{Y}_2 \mid \mathbf{Y}_1$.

6. *Letting* $\mathbf{Y}_1 = Y$*, a single random variable and letting the random vector* $\mathbf{Y}_2 = \mathbf{X}$*, a random vector of independent variables, the population coefficient of determination or population squared multiple correlation coefficient is defined as the maximum correlation between* $\mathbf{Y}$ *and linear functions* $\boldsymbol{\beta}'\mathbf{X}$*. The population coefficient of determination or the squared population multiple correlation coefficient is*

$$\rho_{Y\mathbf{X}}^2 = \boldsymbol{\sigma}_{Y\mathbf{X}}'\Sigma_{\mathbf{XX}}^{-1}\boldsymbol{\sigma}_{\mathbf{X}Y}/\sigma_{YY}$$

If the random vector $\mathbf{Z} = (Y, \mathbf{X}')'$ *follows a multivariate normal distribution, the population coefficient of determination is the square of the zero-order correlation between the random variable* Y *and the population predicted value of* Y *which we see from (5) has the form* $\widehat{Y} = \mu_Y + \boldsymbol{\sigma}_{Y\mathbf{X}}'\Sigma_{\mathbf{XX}}^{-1}(\mathbf{x} - \boldsymbol{\mu}_{\mathbf{X}})$.

7. *For* $\mathbf{X} = \Sigma^{-1/2}(\mathbf{Y} - \boldsymbol{\mu})$ *where* $\Sigma^{-1/2}$*is the symmetric positive definite square root of* Σ^{-1} *then* $\mathbf{X} \sim N_p(\mathbf{0}, \mathbf{I})$ or $X_i \sim IN(0, 1)$.

8. *If* $\mathbf{Y}_1$ and $\mathbf{Y}_2$ *are independent multivariate normal random vectors, then the sum* $\mathbf{Y}_1 + \mathbf{Y}_2 \sim N(\boldsymbol{\mu}_1 + \boldsymbol{\mu}_2, \Sigma_{11} + \Sigma_{22})$*. More generally, if* $\mathbf{Y}_i \sim IN_p(\boldsymbol{\mu}_i, \Sigma_i)$ and $a_1, a_2, \ldots, a_n$ *are fixed constants, then the sum of n p-variate vectors*

$$\sum_{i=1}^{n} a_i\mathbf{Y}_i \sim N_p\left[\sum_{i=1}^{n} a_i\boldsymbol{\mu}_i, \sum_{i=1}^{n} a_i^2\Sigma_i\right]$$

From property (7), we have the following theorem.

Theorem 3.3.3 *If $\mathbf{Y}_1, \mathbf{Y}_2, \ldots, \mathbf{Y}_n$ are independent MVN random vectors with common mean $\boldsymbol{\mu}$ and covariance matrix Σ*, then $\overline{\mathbf{Y}} = \sum_{i=1}^{n} \mathbf{Y}_i/n$ *is MVN with mean $\boldsymbol{\mu}$ and covariance matrix Σ/n, $\overline{\mathbf{Y}} \sim N_p(\boldsymbol{\mu}, \Sigma/n)$.*

b. *Estimating $\boldsymbol{\mu}$ and Σ*

From Theorem 3.3.3, observe that for a random sample from a normal population that $\overline{\mathbf{Y}}$ is an unbiased and consistent estimator of $\boldsymbol{\mu}$, written as $\widehat{\boldsymbol{\mu}} = \overline{\mathbf{Y}}$. Having estimated $\boldsymbol{\mu}$, the $p \times p$ sample covariance matrix is

$$
\begin{aligned}
\mathbf{S} &= \sum_{i=1}^{n} (\mathbf{y}_i - \overline{\mathbf{y}})(\mathbf{y}_i - \overline{\mathbf{y}})' / (n-1) \\
&= \sum_{i=1}^{n} [(\mathbf{y}_i - \boldsymbol{\mu}) - (\overline{\mathbf{y}} - \boldsymbol{\mu})][(\mathbf{y}_i - \boldsymbol{\mu}) - (\overline{\mathbf{y}} - \boldsymbol{\mu})]' / (n-1) \\
&= \left[\sum_{i=1}^{n} (\mathbf{y}_i - \boldsymbol{\mu})(\mathbf{y}_i - \boldsymbol{\mu})' + n\left(\overline{\mathbf{y}}_i - \boldsymbol{\mu}\right)\left(\overline{\mathbf{y}}_i - \boldsymbol{\mu}\right)' \right] / (n-1)
\end{aligned} \tag{3.3.3}
$$

where $E(\mathbf{S}) = \Sigma$ so that $\mathbf{S}$ is an unbiased estimator of Σ. Representing the sample as a matrix $\mathbf{Y}_{n \times p}$ so that

$$
\mathbf{Y} = \begin{bmatrix} \mathbf{y}_1' \\ \mathbf{y}_2' \\ \vdots \\ \mathbf{y}_n' \end{bmatrix}
$$

$\mathbf{S}$ may be written as

$$
\begin{aligned}
(n-1)\,\mathbf{S} &= \mathbf{Y}'\left[\mathbf{I}_n - \mathbf{1}_n\left(\mathbf{1}_n'\mathbf{1}_n\right)^{-1}\mathbf{1}_n'\right]\mathbf{Y} \\
&= \mathbf{Y}'\mathbf{Y} - n\overline{\mathbf{y}}\overline{\mathbf{y}}'
\end{aligned} \tag{3.3.4}
$$

where $\mathbf{I}_n$ is the identity matrix and $\mathbf{1}_n$ is a vector of n 1s. While the matrix $\mathbf{S}$ is an unbiased estimator, a biased estimator, called the maximum likelihood estimator under normality is $\widehat{\Sigma} = \frac{(n-1)}{n}\mathbf{S} = \sum_{i=1}^{n}(\mathbf{y}_i - \overline{\mathbf{y}})(\mathbf{y}_i - \overline{\mathbf{y}})'/n = \mathbf{E}/n$. The matrix $\mathbf{E}$ is called the sum of squares and cross-products matrix, SSCP and the $|\mathbf{S}|$ is the sample estimate of the generalized variance.

In Theorem 3.3.3, we assumed that the observations $\mathbf{Y}_i$ represent a sample from a normal distribution. More generally, suppose $\mathbf{Y}_i \sim (\boldsymbol{\mu}, \Sigma)$ is an independent sample from any distribution with mean $\boldsymbol{\mu}$ and covariance matrix Σ. Theorem 3.3.4 is a multivariate version of the Central Limit Theorem (CLT).

Theorem 3.3.4 *Let* $\{\mathbf{Y}_i\}_{i=1}^{\infty}$ *be a sequence of random p-vectors with finite mean* $\boldsymbol{\mu}$ *and covariance matrix* Σ. *Then*

$$n^{1/2}(\overline{\mathbf{Y}} - \boldsymbol{\mu}) = n^{-1/2}\sum_{i=1}^{n}(\mathbf{Y}_i - \boldsymbol{\mu}) \xrightarrow{d} N_p(\mathbf{0}, \Sigma)$$

Theorem 3.3.4 is used to show that $\mathbf{S}$ is a consistent estimator of Σ. To obtain the distribution of a random matrix $\mathbf{Y}_{n\times p}$, the vec $(\cdot)$ operator is used. Assuming a random sample of n p-vectors $\mathbf{Y}_i \sim (\boldsymbol{\mu}, \Sigma)$, consider the random matrix $\mathbf{X}_i = (\mathbf{Y}_i - \boldsymbol{\mu})(\mathbf{Y}_i - \boldsymbol{\mu})'$. By Theorem 3.3.4,

$$n^{-1/2}\sum_{i=1}^{n}[\text{vec}(\mathbf{X}_i) - \text{vec}\,(\Sigma)] \xrightarrow{d} N_{p^2}\,(\mathbf{0}, \Omega)$$

where

$$\Omega = \text{cov}[\text{vec}(\mathbf{X}_i)]$$

and

$$n^{-1/2}\,(\overline{\mathbf{y}} - \boldsymbol{\mu}) \xrightarrow{d} N_p\,(\mathbf{0}, \Sigma)$$

so that

$$n^{-1/2}[\text{vec}(\mathbf{E}) - n\ \text{vec}(\Sigma)] \xrightarrow{d} N_{p^2}(\mathbf{0}, \Omega).$$

Because $\mathbf{S} = (n-1)^{-1}\mathbf{E}$ and the replacement of n by $n-1$ does not effect the limiting distribution, we have the following theorem.

Theorem 3.3.5 *Let* $\{\mathbf{Y}_i\}_{i=1}^{\infty}$ *be a sequence of independent and identically distributed* $p \times 1$ *vectors with finite fourth moments and mean* $\boldsymbol{\mu}$ *and covariance matrix* Σ. *Then*

$$(n-1)^{-1/2}\,\text{vec}(\mathbf{S} - \Sigma) \xrightarrow{d} N_{p^2}(\mathbf{0}, \Omega)$$

Theorem 3.3.5 can be used to show that $\mathbf{S}$ is a consistent estimate of Σ, $\mathbf{S} \xrightarrow{p} \Sigma$ since $\mathbf{S} - \Sigma = O_p(n-1)^{1/2} = o_p(1)$. The asymptotic normal distribution in Theorem 3.3.5 is singular because $\Omega_{p^2\times p^2}$ is singular. To illustrate the structure of Ω under normality, we consider the bivariate case.

Example 3.3.1 *Let* $\mathbf{Y} \sim N_2(\boldsymbol{\mu}, \Sigma)$. *Then*

$$\Omega = \left[\begin{pmatrix} 1 & 0 & 0 & 0 \\ 0 & 1 & 0 & 0 \\ 0 & 0 & 1 & 0 \\ 0 & 0 & 0 & 1 \end{pmatrix} + \begin{pmatrix} 1 & 0 & 0 & 0 \\ 0 & 0 & 1 & 0 \\ 0 & 1 & 0 & 0 \\ 0 & 0 & 0 & 1 \end{pmatrix} \right] (\Sigma \otimes \Sigma)$$

$$= \begin{bmatrix} 2 & 0 & 0 & 0 \\ 0 & 1 & 1 & 0 \\ 0 & 1 & 1 & 0 \\ 0 & 0 & 0 & 2 \end{bmatrix} \begin{bmatrix} \sigma_{11}\sigma_{11} & \sigma_{11}\sigma_{12} & \sigma_{12}\sigma_{11} & \sigma_{12}\sigma_{12} \\ \sigma_{11}\sigma_{21} & \sigma_{11}\sigma_{22} & \sigma_{12}\sigma_{21} & \sigma_{12}\sigma_{22} \\ \sigma_{21}\sigma_{11} & \sigma_{21}\sigma_{12} & \sigma_{22}\sigma_{11} & \sigma_{22}\sigma_{12} \\ \sigma_{21}\sigma_{21} & \sigma_{21}\sigma_{22} & \sigma_{22}\sigma_{21} & \sigma_{22}\sigma_{22} \end{bmatrix}$$

$$= \begin{bmatrix} \sigma_{11}\sigma_{11} + \sigma_{11}\sigma_{11} & \cdots & \sigma_{12}\sigma_{12} + \sigma_{12}\sigma_{12} \\ \sigma_{11}\sigma_{21} + \sigma_{21}\sigma_{11} & \cdots & \sigma_{12}\sigma_{22} + \sigma_{22}\sigma_{12} \\ \sigma_{11}\sigma_{21} + \sigma_{21}\sigma_{11} & \cdots & \sigma_{12}\sigma_{22} + \sigma_{22}\sigma_{12} \\ \sigma_{21}\sigma_{21} + \sigma_{21}\sigma_{21} & \cdots & \sigma_{22}\sigma_{22} + \sigma_{22}\sigma_{22} \end{bmatrix}$$

$$= \left[\sigma_{ik}\sigma_{jm} + \sigma_{im}\sigma_{jk} \right] \tag{3.3.5}$$

See Magnus and Neudecker (1979, 1999) or Muirhead (1982, p. 90).

Because the elements of $\mathbf{S}$ are duplicative, the asymptotic distribution of the elements of $\mathbf{s} = \text{vech}(\mathbf{S})$ are also MVN. Indeed,

$$\sqrt{(n-1)}\,(\mathbf{s} - \boldsymbol{\sigma}) \xrightarrow{d} N\,(\mathbf{0}, \Gamma)$$

where $\boldsymbol{\sigma} = \text{vech}(\Sigma)$ and $\Gamma = \text{cov}\left[\text{vech}\,(\mathbf{Y} - \boldsymbol{\mu})\,(\mathbf{Y} - \boldsymbol{\mu})'\right]$. Or,

$$\sqrt{(n-1)}\Gamma^{-1/2}\,(\mathbf{s} - \boldsymbol{\sigma}) \xrightarrow{d} N\left(\mathbf{0}, \mathbf{I}_{p^*}\right)$$

where $p^* = p(p+1)/2$. While the matrix Ω in (3.3.5) is not of full rank, the matrix Γ is of full rank. Using the duplication matrix in Exercise 3.2, problem 6, the general form of Γ under multivariate normality is $\Gamma = 2\mathbf{D}_p^+(\Sigma \otimes \Sigma)\mathbf{D}_p^+$ for $\mathbf{D}_p^+ = (\mathbf{D}_p'\mathbf{D}_p)^{-1}\mathbf{D}_p'$, Schott (1997, p. 285, Th. 7.38).

c. *The Matrix Normal Distribution*

If we write that a random matrix $\mathbf{Y}_{n \times p}$ is normally distributed, we say $\mathbf{Y}$ has a matrix normal distribution written as $\mathbf{Y} \sim N_{n,\,p}(\mathbf{M}, \mathbf{V} \otimes \mathbf{W})$ where $\mathbf{V}_{p \times p}$ and $\mathbf{W}_{n \times n}$ are positive definite matrices, $E(\mathbf{Y}) = \mathbf{M}$ and the $\text{cov}(\mathbf{Y}) = \mathbf{V} \otimes \mathbf{W}$. To illustrate, suppose $\mathbf{y} = \text{vec}(\mathbf{Y}) \sim N_{np}(\boldsymbol{\beta}, \Omega = \Sigma \otimes \mathbf{I}_n)$, then the density of $\mathbf{y}$ is

$$(2\pi)^{np/2}\,|\Omega|^{-1/2} \exp\left\{-\frac{1}{2}(\mathbf{y} - \boldsymbol{\beta})'\Omega^{-1}(\mathbf{y} - \boldsymbol{\beta})\right\}$$

However, $|\Omega|^{-1/2} = \left|\Sigma_p \otimes \mathbf{I}_n\right|^{-1/2} = |\Sigma|^{-n/2}$ using the identity $|\mathbf{A}_m \otimes \mathbf{B}_n| = |\mathbf{A}|^n\,|\mathbf{B}|^m$. Next, recall that $\text{vec}(\mathbf{ABC}) = (\mathbf{C}' \otimes \mathbf{A})\,\text{vec}(\mathbf{B})$ and that the $\text{tr}(\mathbf{A}'\mathbf{B}) = (\text{vec}\,\mathbf{A})'\,\text{vec}\,\mathbf{B}$. For $\boldsymbol{\beta} = \text{vec}\,(\mathbf{M})$, we have that

$$\begin{aligned}(\mathbf{y}-\boldsymbol{\beta})'\,\Omega^{-1}\,(\mathbf{y}-\boldsymbol{\beta}) &= [\text{vec}\,(\mathbf{Y}-\mathbf{M})]'\left[(\Sigma \otimes \mathbf{I}_n)^{-1}\right]\text{vec}\,(\mathbf{Y}-\mathbf{M}) \\ &= \text{tr}\left[\Sigma^{-1}\,(\mathbf{Y}-\mathbf{M})'\,(\mathbf{Y}-\mathbf{M})\right]\end{aligned}$$

This motivates the following definition for the distribution of a random normal matrix $\mathbf{Y}$ where Σ is the covariance matrix among the columns of $\mathbf{Y}$ and $\mathbf{W}$ is the covariance matrix among the rows of $\mathbf{Y}$.

Definition 3.3.1 *The data matrix $\mathbf{Y}_{n\times p}$ has a matrix normal distribution with parameters $\mathbf{M}$ and covariance matrix $\Sigma \otimes \mathbf{W}$. The multivariate density of $\mathbf{Y}$ is*

$$(2\pi)^{-np/2}\,|\Sigma|^{-n/2}\,|\mathbf{W}|^{-p/2}\,\text{etr}\left[-\frac{1}{2}\Sigma^{-1}\,(\mathbf{Y}-\mathbf{M})'\,\mathbf{W}^{-1}\,(\mathbf{Y}-\mathbf{M})\right]$$

$\mathbf{Y} \sim N_{n,\,p}\,(\mathbf{M},\,\Sigma \otimes \mathbf{W})$ *or* $\text{vec}\,(\mathbf{Y}) \sim N_{np}\,(\text{vec}\,\mathbf{M},\,\Sigma \otimes \mathbf{W})$.

As a simple illustration of Definition 3.3.1, consider $\mathbf{y} = \text{vec}(\mathbf{Y}')$. Then the distribution of $\mathbf{y}$ is

$$(2\pi)^{-np/2}\left|\mathbf{I}_n \otimes \Sigma_p\right|^{-1/2}\exp\left[-\frac{1}{2}\,(\mathbf{y}-\mathbf{m})'\left|\mathbf{I}_n \otimes \Sigma_p\right|^{-1}(\mathbf{y}-\mathbf{m})\right]$$

$$(2\pi)^{-np/2}\,|\Sigma|^{-n/2}\,\text{etr}\left[-\frac{1}{2}\Sigma^{-1}\,(\mathbf{Y}-\mathbf{M})'\,(\mathbf{Y}-\mathbf{M})\right]$$

where $E(\mathbf{Y}) = \mathbf{M}$ and $\mathbf{m} = \text{vec}(\mathbf{M}')$. Thus $\mathbf{Y}' \sim N_{n,\,p}(\mathbf{M}', \mathbf{I}_n \otimes \Sigma)$. Letting $\mathbf{Y}_1, \mathbf{Y}_2, \ldots, \mathbf{Y}_n \sim IN_p\,(\boldsymbol{\mu},\,\Sigma)$,

$$E\,(\mathbf{Y}) = \begin{bmatrix}\boldsymbol{\mu}' \\ \boldsymbol{\mu}' \\ \vdots \\ \boldsymbol{\mu}'\end{bmatrix} = \mathbf{1}_n\,\boldsymbol{\mu}' = \mathbf{M}$$

so that the density of $\mathbf{Y}'$ is

$$(2\pi)^{-np/2}\,|\Sigma|^{-n/2}\,\text{etr}\left[-\frac{1}{2}\Sigma^{-1}\left(\mathbf{Y}-\mathbf{1}\boldsymbol{\mu}'\right)'\left(\mathbf{Y}-\mathbf{1}\boldsymbol{\mu}'\right)\right]$$

More generally, suppose

$$E(\mathbf{Y}) = \mathbf{XB}$$

where $\mathbf{X}_{n\times q}$ is a known "design" matrix and $\mathbf{B}_{q\times p}$ is an unknown matrix of parameters. The matrix normal distribution of $\mathbf{Y}'$ is

$$(2\pi)^{-np/2}\,|\Sigma|^{-n/2}\,\text{etr}\left[-\frac{1}{2}\Sigma^{-1}\,(\mathbf{Y}-\mathbf{XB})'\,(\mathbf{Y}-\mathbf{XB})\right]$$

or

$$(2\pi)^{-np/2} |\Sigma|^{-n/2} \text{etr}\left[-\frac{1}{2}(\mathbf{Y}-\mathbf{XB})\,\Sigma^{-1}\,(\mathbf{Y}-\mathbf{XB})'\right]$$

The expression for the covariance structure of $\mathbf{Y}_{n\times p}$ depends on whether one is considering the structure of $\mathbf{y} = \text{vec}\,(\mathbf{Y})$ or $\mathbf{y}^* = \text{vec}\left(\mathbf{Y}'\right)$. Under independence and identically distributed ($i.i.d.$) observations, the cov $(\mathbf{y}) = \Sigma_p \otimes \mathbf{I}_n$ and the cov $(\mathbf{y}^*) = \mathbf{I}_n \otimes \Sigma_p$. In the literature, the definition of a matrix normal distribution may differ depending on the "orientation" of $\mathbf{Y}$. If the cov $(\mathbf{y}) = \Sigma \otimes \mathbf{W}$ or the cov $(\mathbf{y}^*) = \mathbf{W} \otimes \Sigma$ the data has a dependency structure where $\mathbf{W}$ is a structure among the rows of $\mathbf{Y}$ and Σ is the structure of the columns of $\mathbf{Y}$.

Exercises 3.3

1. Suppose $\mathbf{Y} \sim N_4(\boldsymbol{\mu}, \Sigma)$, where

$$\boldsymbol{\mu} = \begin{bmatrix} 1 \\ 2 \\ 3 \\ 4 \end{bmatrix} \quad \text{and} \quad \Sigma = \begin{bmatrix} 3 & 1 & 0 & 0 \\ 1 & 4 & 0 & 0 \\ 0 & 0 & 1 & 4 \\ 0 & 0 & 2 & 0 \end{bmatrix}$$

   (a) Find the joint distribution of Y_1 and Y_2 and of Y_3 and Y_4.

   (b) Determine ρ_{12} and ρ_{24}.

   (c) Find the length of the semimajor axis of the ellipse association with this MVN variable $\mathbf{Y}$ and a construct $Q = 100$.

2. Determine the MVN density associated with the quadratic form

$$Q = 2y_1^2 + y_2^2 + 3y_3^2 + 2y_1y_2 + 2y_1y_3$$

3. For the bivariate normal distribution, graph the ellipse of the exponent for $\mu_1 = \mu_2 = 0$, $\sigma_1^2 + \sigma_2^2 = 1$, and $Q = 2$ and $\rho = 0, .5,$ and $.9$.

4. The matrix of partial correlations has as elements

$$\rho_{ij.p+1,\ldots,p+q} = \frac{\sigma_{ij.p+1,\ldots,p+q}}{\sqrt{\sigma_{ii.p+1,\ldots,p+q}}\sqrt{\sigma_{jj.p+1,\ldots,p+q}}}$$

   (a) For $\mathbf{Y} = \begin{bmatrix} Y_1 \\ Y_2 \end{bmatrix}$ and $Y_3 = y_3$, find $\rho_{12.3}$.

   (b) For $Y_1 = Y_1$ and $\mathbf{Y}_2 = \begin{bmatrix} Y_2 \\ Y_3 \end{bmatrix}$ show that $\sigma_{1.2}^2 = \sigma_1^2 - \sigma_{12}'\Sigma_{22}^{-1}\sigma_{21} = |\Sigma| / |\Sigma_{22}|$.

   (c) The maximum correlation between $Y_1 \equiv Y$ and the linear combination $\beta_1 Y_2 + \beta_2 Y_3 \equiv \beta'\mathbf{X}$ is called the multiple correlation coefficient and represented as $\rho_{0(12)}$. Show that $\sigma_{1.2}^2 = \sigma_1^2(1 - \rho_{0(12)}^2)$ and assuming that the variables are jointly multivariate normal, derive the expression for $\rho_{0(12)}^2 = \rho_{Y\mathbf{X}}^2$.

5. For the $p^2 \times p^2$ commutation matrix $\mathbf{K} = \sum_{ij} \Delta_{ij} \otimes \Delta'_{ij}$ where Δ_{ij} is a $p \times p$ matrix of zeros with only $\delta_{ij} = 1$ and $\mathbf{Y} \sim N_p(\boldsymbol{\mu}, \Sigma)$, show that

$$\begin{aligned} \operatorname{cov}\left\{\operatorname{vec}(\mathbf{Y}-\boldsymbol{\mu})(\mathbf{Y}-\boldsymbol{\mu})'\right\} &= \left(I_{p^2}+\mathbf{K}\right)(\Sigma \otimes \Sigma)\left(I_{p^2}+\mathbf{K}\right) \\ &= \left(I_{p^2}+\mathbf{K}\right)(\Sigma \otimes \Sigma) \end{aligned}$$

since $\left(I_{p^2}+\mathbf{K}\right)$ is idempotent.

6. If $\mathbf{Y} \sim N_{n,p}[\mathbf{XB}, \Sigma \otimes \mathbf{I}_n]$, $\widehat{\mathbf{B}} = \left(\mathbf{X}'\mathbf{X}\right)^{-1}\mathbf{X}'\mathbf{Y}$, and $\widehat{\Sigma} = (\mathbf{Y}-\mathbf{X}\widehat{\mathbf{B}})'(\mathbf{Y}-\mathbf{X}\widehat{\mathbf{B}})/n$. Find the distribution of $\widehat{\mathbf{B}}$.

7. If $\mathbf{Y}_p \sim N_p[\boldsymbol{\mu}, \Sigma]$ and one obtains a Cholesky factorization of $\Sigma = \mathbf{LL}'$, what is the distribution of $\mathbf{X} = \mathbf{LY}$?

3.4 The Chi-Square and Wishart Distributions

The chi-square distribution is obtained from a sum of squares of independent normal zero-one, $N(0, 1)$, random variables and is fundamental to the study of the analysis of variance methods. In this section, we review the chi-square distribution and generalize several results, in an intuitive manner, to its multivariate analogue known as the Wishart distribution.

a. *Chi-Square Distribution*

Recall that if $Y_1, Y_2, \ldots, Y_n$ are independent normal random variables with mean $\mu_i = 0$ and variance $\sigma^2 = 1$, $Y_i \sim IN(0, 1)$, or, employing vector notation $\mathbf{Y} \sim N_n(\mathbf{0}, \mathbf{I})$, then

$$Q = \mathbf{Y}'\mathbf{Y} = \textstyle\sum_{i=1}^n Y_i^2 \sim \chi^2(n) \qquad 0 < Q < \infty$$

$Q = \mathbf{Y}'\mathbf{Y}$ has a central χ^2 distribution with n degrees of freedom. Letting $Y_i \sim IN(\mu_i, \sigma^2)$, results in the noncentral chi-square distribution.

Definition 3.4.1 *If the random n-vector* $\mathbf{Y} \sim N_n(\boldsymbol{\mu}, \sigma^2\mathbf{I})$, *then* $\mathbf{Y}'\mathbf{Y}/\sigma^2$ *has a noncentral* χ^2 *distribution with n degrees of freedom and noncentrality parameter* $\gamma = \boldsymbol{\mu}'\boldsymbol{\mu}/\sigma^2$.

For $\boldsymbol{\mu} = \mathbf{0}$, the noncentral chi-square distribution reduces to a central chi-square distribution. For $\mathbf{Y} \sim N_n(\boldsymbol{\mu}, \mathbf{I})$, then $\mathbf{Y}'\mathbf{Y} \sim \chi^2(n, \gamma)$ with $\gamma = \boldsymbol{\mu}'\boldsymbol{\mu}$ so that $\gamma = \|\boldsymbol{\mu}\|^2$ is a norm squared. The further $\boldsymbol{\mu}$ is from zero, the larger the noncentrality parameter γ or the norm squared of $\boldsymbol{\mu}$. Because $\mathbf{Y}'\mathbf{Y}$ in Definition 3.4.1 is a special case of the quadratic form $\mathbf{Y}'\mathbf{AY}$, with $\mathbf{A} = \mathbf{I}$ and since $\mathbf{I}^2 = \mathbf{I}$, we have the following more general result.

Theorem 3.4.1 *Let* $\mathbf{Y} \sim N_n\left(\boldsymbol{\mu}, \sigma^2\mathbf{I}\right)$ *and* $\mathbf{A}$ *be a symmetric matrix of rank r. Then we have* $\mathbf{Y}'\mathbf{AY}/\sigma^2 \sim \chi^2(r, \gamma)$, *where* $\gamma = \boldsymbol{\mu}'\mathbf{A}\boldsymbol{\mu}/\sigma^2$, *if and only if* $\mathbf{A} = \mathbf{A}^2$.

Example 3.4.1 *As an example of Theorem 3.4.1, suppose* $\mathbf{Y} \sim N_n(\boldsymbol{\mu}, \sigma^2\mathbf{I}_n)$. *Then*

$$\frac{(n-1)s^2}{\sigma^2} = \frac{\sum_{i=1}^n \left(Y_i - \overline{Y}\right)^2}{\sigma^2} = \frac{\mathbf{Y}'[\mathbf{I} - \mathbf{1}\left(\mathbf{1}'\mathbf{1}\right)^{-1}\mathbf{1}]\mathbf{Y}}{\sigma^2} = \frac{\mathbf{Y}'\mathbf{AY}}{\sigma^2}$$

However, $\mathbf{A}' = \mathbf{A}$ *and* $\mathbf{A}^2 = \mathbf{A}$ *since* $\mathbf{A}$ *is a projection matrix and the* $r(\mathbf{A}) = \text{tr}(\mathbf{A}) = n-1$. *Hence*

$$\frac{(n-1)\,s^2}{\sigma^2} \sim \chi^2\,(n-1,\ \gamma = 0)$$

since $\gamma = E\left(\mathbf{Y}'\right)\mathbf{A}E(\mathbf{Y})/\sigma^2 = 0$. *Thus,* $(n-1)\,s^2 \sim \sigma^2\chi^2(n-1)$.

Theorem 3.4.2 generalizes Theorem 3.4.1 to a vector of dependent variables in a natural manner by setting $\mathbf{Y} = \mathbf{FX}$ and $\mathbf{FF}' = \Sigma$.

Theorem 3.4.2 *If* $\mathbf{Y} \sim N_p\,(\boldsymbol{\mu}, \Sigma)$. *Then the quadratic form* $\mathbf{Y}'\mathbf{AY} \sim \chi^2(r, \gamma)$, *where* $\gamma = \left(\boldsymbol{\mu}'\mathbf{A}\boldsymbol{\mu}\right)$ *and the* $r(\mathbf{A}) = r$, *if and only if* $\mathbf{A}\Sigma\mathbf{A} = \mathbf{A}$ *or* $\mathbf{A}\Sigma$ *is idempotent.*

Example 3.4.2 *An important application of Theorem 3.4.2 follows:*

Let $\mathbf{Y}_1, \mathbf{Y}_2, \ldots, \mathbf{Y}_n$ *be n independent p-vectors from any distribution with mean* $\boldsymbol{\mu}$ *and nonsingular covariance matrix* Σ. *Then by the CLT,* $\sqrt{n}(\overline{\mathbf{Y}} - \boldsymbol{\mu}) \xrightarrow{d} N_p(\mathbf{0}, \Sigma)$. *By Theorem 3.4.2,* $T^2 = n(\overline{\mathbf{Y}} - \boldsymbol{\mu})'\Sigma^{-1}\left(\overline{\mathbf{Y}} - \boldsymbol{\mu}\right) = nD^2 \xrightarrow{d} \chi^2\,(p)$ *for* $n - p$ *large since* $\Sigma^{-1}\Sigma\Sigma^{-1} = \Sigma$. *The distribution is exactly* $\chi^2\,(p)$ *if the sample is from a multivariate normal distribution.*

Thus, comparing nD^2 with a χ^2 critical value may be used to evaluate multivariate normality. Furthermore, nD^2 for known Σ may be used to test $H_0 : \boldsymbol{\mu} = \boldsymbol{\mu}_0$ vs. $H_1 : \boldsymbol{\mu} \neq \boldsymbol{\mu}_0$. The critical value of the test with significance level α is represented as

$$\Pr[nD^2 \geq \chi^2_{1-\alpha}\,(p) \mid H_0] = \alpha$$

where $\chi^2_{1-\alpha}$ is the upper $1 - \alpha$ chi-square critical value. For $\boldsymbol{\mu} \neq \boldsymbol{\mu}_0$, the noncentrality parameter is

$$\gamma = n\left(\boldsymbol{\mu} - \boldsymbol{\mu}_0\right)'\,\Sigma^{-1}\left(\boldsymbol{\mu} - \boldsymbol{\mu}_0\right)$$

The above result is for a single quadratic form. More generally we have Cochran's Theorem.

Theorem 3.4.3 *If* $\mathbf{Y} \sim N_n(\boldsymbol{\mu}, \sigma^2\mathbf{I}_n)$ *and* $\mathbf{Y}'\mathbf{Y}/\sigma^2 = \sum_{i=1}^n \mathbf{Y}'\mathbf{A}_i\mathbf{Y}$ *where* $r(\mathbf{A}_i) = r$ *and* $\sum_{i=1}^n \mathbf{A}_i = \mathbf{I}_n$, *then the quadratic forms* $\mathbf{Y}'\mathbf{A}_i\mathbf{Y}/\sigma^2 \sim \chi^2(r_i, \gamma_i)$, *where* $\gamma_i = \boldsymbol{\mu}'\mathbf{A}_i\boldsymbol{\mu}/\sigma^2$ *are statistically independent for all i if and only if* $\sum_{i=1}^n r_i = n$ and $\sum_i r\,(\mathbf{A}_i) = r(\sum_i \mathbf{A}_i)$.

Cochran's Theorem is used to establish the independence of quadratic forms. The ratios of independent quadratic forms normalized by their degrees of freedom are used to test hypotheses regarding means. To illustrate Theorem 3.4.3, we show that $\overline{Y}$ and s^2 are statistically independent. Let $\mathbf{Y} \sim N_n\left(\mu\mathbf{1}, \sigma^2\mathbf{I}\right)$ and let $\mathbf{P} = \mathbf{1}(\mathbf{1}'\mathbf{1})^{-1}\mathbf{1}'$ be the averaging projection matrix. Then

$$\frac{\mathbf{Y}'\mathbf{IY}}{\sigma^2} = \frac{\mathbf{Y}'\mathbf{PY}}{\sigma^2} + \frac{\mathbf{Y}'\,(\mathbf{I} - \mathbf{P})\,\mathbf{Y}}{\sigma^2}$$

$$\frac{\sum_{i=1}^n Y_i^2}{\sigma^2} = \frac{n\overline{Y}^2}{\sigma^2} + \frac{(n-1)\,s^2}{\sigma^2}$$

Since the $r(\mathbf{I}) = n = r(\mathbf{P}) + r(\mathbf{I} - \mathbf{P}) = 1 + (n-1)$, the quadratic forms are independent by Theorem 3.4.3, or $\overline{Y}$ is independent of s^2.

Example 3.4.3 *Let* $Y \sim N_4(\boldsymbol{\mu}, \sigma^2 I)$

$$\mathbf{A} = \begin{bmatrix} 1 & 1 & 0 \\ 1 & 1 & 0 \\ 1 & 0 & 1 \\ 1 & 0 & 1 \end{bmatrix} = [\mathbf{A}_1\, \mathbf{A}_2] \quad \text{and} \quad \mathbf{y} = \begin{bmatrix} y_{11} \\ y_{12} \\ y_{21} \\ y_{22} \end{bmatrix}$$

where

$$\mathbf{A}_1 = \begin{bmatrix} 1 \\ 1 \\ 1 \\ 1 \end{bmatrix} \quad \text{and} \quad \mathbf{A}_2 \begin{bmatrix} 1 & 0 \\ 1 & 0 \\ 0 & 1 \\ 0 & 1 \end{bmatrix}$$

In Example 2.6.2, projection matrices of the form

$$\begin{aligned} \mathbf{P}_1 &= \mathbf{A}_1 \left(\mathbf{A}_1'\mathbf{A}_1\right)^{-} \mathbf{A}_1' \\ \mathbf{P}_2 &= \mathbf{A} \left(\mathbf{A}'\mathbf{A}\right)^{-} \mathbf{A}' - \mathbf{A}_1 \left(\mathbf{A}_1'\mathbf{A}_1\right)^{-} \mathbf{A}_1' \\ \mathbf{P}_3 &= \mathbf{I} - \mathbf{A} \left(\mathbf{A}'\mathbf{A}\right)^{-} \mathbf{A}' \end{aligned}$$

were constructed to project the observation vector $\mathbf{y}$ *onto orthogonal subspaces. The projection matrices were constructed such that* $\mathbf{I} = \mathbf{P}_1 + \mathbf{P}_2 + \mathbf{P}_3$ *where* $\mathbf{P}_i\mathbf{P}_j = \mathbf{0}$ *for* $i \neq j$ *and each* $\mathbf{P}_i$ *is symmetric and idempotent so that the* $r(\mathbf{I}) = \sum_i r(\mathbf{P}_i)$. *Forming an equation of quadratic forms, we have that*

$$\mathbf{y}'\mathbf{y} = \sum_{i=1}^{3} \mathbf{y}'\mathbf{P}_i\mathbf{y}$$

or

$$\|\mathbf{y}\|^2 = \sum_{i=1}^{3} \|\mathbf{P}_i\mathbf{y}\|^2$$

For $\mathbf{P}_1$, $\mathbf{P}_2$, *and* $\mathbf{P}_3$ *given in Example 2.6.3, it is easily verified that*

$$\begin{aligned} \|\mathbf{P}_1\mathbf{y}\|^2 &= \mathbf{y}'\mathbf{P}_1\mathbf{y} = 4^2 y_{..} \\ \|\mathbf{P}_2\mathbf{y}\|^2 &= \mathbf{y}'\mathbf{P}_2\mathbf{y} = \sum_i 2\left(y_{i.} - y_{..}\right)^2 \\ \|\mathbf{P}_3\mathbf{y}\|^2 &= \mathbf{y}'\mathbf{P}_3\mathbf{y} = \sum_i \sum_j 2\left(y_{ij} - y_{i.}\right)^2 \end{aligned}$$

Hence, the total sum of squares has the form

$$\mathbf{y}'\mathbf{I}\mathbf{y} = \sum_i \sum_j y_{ij}^2 = 4y_{..}^2 + \sum_i 2(y_{i.} - y_{..})^2 + \sum_i \sum_j \left(y_{ij} - y_{i.}\right)^2$$

or

$$\sum_i \sum_j (y_{ij} - y_{..})^2 = \sum_i 2(y_{i.} - y_{..})^2 + \sum_i \sum_j (y_{ij} - y_{i.})^2$$
$$\text{“Total about the Mean” } SS = \text{Between } SS + \text{Within } SS$$

where the degrees of freedom are the ranks of $r(\mathbf{I} - \mathbf{P}_1) = n - 1$, $r(\mathbf{P}_2) = I - 1$, *and* $r(\mathbf{P}_3) = n - I$ *for* $n = 4$ *and* $I = 2$. *By Theorem 3.4.3, the sum of squares* (SS) *are independent and may be used to test hypotheses in analysis of variance, by forming ratios of independent chi-square statistics.*

b. *The Wishart Distribution*

We saw that the asymptotic distribution of $\mathbf{S}$ is MVN. To derive the distribution of $\mathbf{S}$ in small samples, suppose $\mathbf{Y}_i \sim IN_p(\mathbf{0}, \Sigma)$. Then $\mathbf{y}_i = \text{vec}(\mathbf{Y}_{n\times p}) \sim N_{np}(\mathbf{0}, \Sigma \otimes \mathbf{I}_n)$. Let $\mathbf{Q} = \mathbf{Y}'\mathbf{Y} = \sum_{i=1}^{n} \mathbf{Y}_i\mathbf{Y}_i'$ represent the SSCP matrix.

Definition 3.4.2 *If* $\mathbf{Q} = \mathbf{Y}'\mathbf{Y}$ *and the matrix* $\mathbf{Y} \sim N_{n,\,p}(\mathbf{0}, \Sigma \otimes \mathbf{I}_n)$. *Then* $\mathbf{Q}$ *has a central p-dimensional Wishart distribution with n degrees of freedom and covariance matrix* Σ, *written as* $\mathbf{Q} \sim W_p(n, \Sigma)$.

For $E(\mathbf{Y}) = \mathbf{M}$ and $\mathbf{M} \neq \mathbf{0}, \mathbf{Q}$ has a noncentral Wishart distribution with noncentrality parameter $\Gamma = \mathbf{M}'\mathbf{M}\Sigma^{-1}$, written as $\mathbf{Q} \sim W_p(n, \Sigma, \Gamma)$. More formally, $\mathbf{Q} \sim W_p(n, \Sigma, \Gamma = \mathbf{M}'\mathbf{M}\Sigma^{-1})$ if and only if $\mathbf{a}'\mathbf{Q}\,\mathbf{a}/\mathbf{a}'\Sigma\mathbf{a} \sim \chi^2(n, \mathbf{a}'\mathbf{M}'\mathbf{M}\mathbf{a}\,/\mathbf{a}'\mathbf{M}\mathbf{a})$ for all nonnull vectors $\mathbf{a}$. In addition $E(\mathbf{Q}) = n\Sigma + \Sigma\Gamma = n\Sigma + \mathbf{M}'\mathbf{M}$ and $E(\mathbf{Y}'\mathbf{A}\mathbf{Y}) = \text{tr}(\mathbf{A})\Sigma + \mathbf{M}'\mathbf{A}\mathbf{M}$ for a symmetric matrix $\mathbf{A}_{n\times n}$. For a comprehensive treatment of the noncentral Wishart distribution, see Muirhead (1982).

If $\mathbf{Q} \sim W_p(n, \Sigma)$, then the distribution of $\mathbf{Q}^{-1}$ is called an inverted Wishart distribution. That is $\mathbf{Q}^{-1} \sim W_p^{-1}(n + p + 1,\ \Sigma^{-1})$ and

$$E(\mathbf{Q}^{-1}) = \Sigma^{-1}/(n - p - 1)$$

for $n - p - 1 > 0$. Or, if $\mathbf{P} \sim W_p^{-1}(n^*, \mathbf{V}^{-1})$ then $E(\mathbf{P}) = \mathbf{V}^{-1}/(n^* - 2p - 2)$.

The Wishart distribution is a multivariate extension of the chi-square distribution and arises in the derivation of the distribution of the sample covariance matrix $\mathbf{S}$. For a random sample of n p-vectors, $\mathbf{Y}_i \sim N_p(\boldsymbol{\mu}, \Sigma)$ for $i = 1, \ldots, n$ and $n \geq p$,

$$(n - 1)\mathbf{S} = \sum_{i=1}^{n} (\mathbf{Y}_i' - \overline{\mathbf{Y}})(\mathbf{Y}_i - \overline{\mathbf{Y}})' \sim W_p(n - 1,\ \Sigma) \tag{3.4.1}$$

or

$$\mathbf{S} \sim W_p[n - 1, \Sigma/(n - 1)]$$

so that $\mathbf{S}$ has a central Wishart distribution. Result (3.4.1) follows from the multivariate extension of Theorem 3.4.1. Furthermore, if $\mathbf{A}_{q\times p}$ is a matrix of constants where the $r(\mathbf{A}) = r \geq p$, then$(n-1)\mathbf{A}\mathbf{S}\mathbf{A}' \sim W_q(n-1,\ \mathbf{A}\Sigma\mathbf{A}')$. If $\mathbf{F}\mathbf{F}' = \Sigma$ so that $\mathbf{I} = \mathbf{F}^{-1}\Sigma(\mathbf{F}')^{-1}$

then $\mathbf{I} = \mathbf{F}'\Sigma\mathbf{F}$. Hence, letting $\mathbf{A} = \mathbf{F}'$ we have that $(n-1)\mathbf{F}'\mathbf{S}\mathbf{F} \sim W_p(n-1, \mathbf{I})$. Partitioning the matrix $\mathbf{Q} \sim W_p(n, \Sigma)$ where

$$\mathbf{Q} = \begin{bmatrix} \mathbf{Q}_{11} & \mathbf{Q}_{12} \\ \mathbf{Q}_{21} & \mathbf{Q}_{22} \end{bmatrix}, (n-1)\,\mathbf{Q} = \mathbf{S} = \begin{bmatrix} \mathbf{S}_{11} & \mathbf{S}_{12} \\ \mathbf{S}_{21} & \mathbf{S}_{22} \end{bmatrix}, \text{ and } \Sigma = \begin{bmatrix} \Sigma_{11} & \Sigma_{12} \\ \Sigma_{21} & \Sigma_{22} \end{bmatrix}$$

we have the following result.

Theorem 3.4.4 *For a $p_1 \times p_1$ matrix $\mathbf{Q}_{11}$ and a $p_2 \times p_2$ matrix $\mathbf{Q}_{22}$ where $p_1 + p_2 = p$,*

1. $\mathbf{Q}_{11} \sim W_{p_1}(n, \Sigma_{11})$ *or* $(n-1)\,\mathbf{S}_{11} \sim W_{p_1}[(n-1), \Sigma_{11}]$
2. $\mathbf{Q}_{22} \sim W_{p_2}(n, \Sigma_{22})$ *or* $(n-1)\,\mathbf{S}_{22} \sim W_{p_2}[(n-1), \Sigma_{22}]$
3. *If $\Sigma_{12} = 0$, then $\mathbf{Q}_{11}$ and $\mathbf{Q}_{22}$ are independent, or $\mathbf{S}_{11}$ and $\mathbf{S}_{22}$ are independent.*
4. $\mathbf{Q}_{11.2} = \mathbf{Q}_{11} - \mathbf{Q}_{12}\mathbf{Q}_{22}^{-1}\mathbf{Q}_{21} \sim W_{p}, [n - p_2 \sim \Sigma_{11.2}]$ *where* $\Sigma_{11.2} = \Sigma_{11} - \Sigma_{12}\Sigma_{22}^{-1}\Sigma_{21}$ *or* $(n-1)\,\mathbf{S}_{11.2} \sim W_{p_1}[n - p_2, \Sigma_{11.2}]$ *and $\mathbf{Q}_{11.2}$ is independent of $\mathbf{Q}_{22}$ or $\mathbf{S}_{11.2}$ and $\mathbf{S}_{22}$ are independently distributed. Similar results hold for $\mathbf{Q}_{22.1}$ and $\mathbf{S}_{22.1}$.*
5. *The conditional distribution of $\mathbf{Q}_{12}$ given $\mathbf{Q}_{22}$ follows a matrix multivariate normal*

$$\mathbf{Q}_{12} \mid \mathbf{Q}_{22} \sim N_{p_1,\, p-p_2}\left(\mathbf{Q}_{12}\mathbf{Q}_{22}^{-}\mathbf{Q}_{21},\ \Sigma_{11.2} \otimes \mathbf{Q}_{22}\right)$$

In multivariate analysis, the sum of independent Wishart distributions follows the same rules as in the univariate case. Matrix quadratic forms are often used in multivariate mixed models. Also important in multivariate analysis are the ratios of independently distributed Wishart matrices or, more specifically, the determinant and trace of matrix products or ratios which are functions of the eigenvalues of matrices. To construct distributions of roots of Wishart matrices, independence needs to be established. The multivariate extension for Cochran's Theorem follows.

Theorem 3.4.5 *If $\mathbf{Y}_i \sim IN_p(\boldsymbol{\mu}, \Sigma)$ for $i = 1, \ldots, n$ and $\mathbf{Y}'\mathbf{Y} = \sum_{i=1}^k \mathbf{Y}'\mathbf{P}_i\mathbf{Y}$ where $\sum_{i=1}^k \mathbf{P}_i = \mathbf{I}_n$, the forms $\mathbf{Y}'\mathbf{P}_i\mathbf{Y} \sim W_p(r_i, \Sigma, \Gamma_i)$ are statistically independent for all i if and only if $\sum_{i=1}^k r_i = n$. If $r_i < p$, the Wishart density does not exist.*

Example 3.4.4 *Suppose $\mathbf{Y}_i \sim IN_p(\boldsymbol{\mu}, \Sigma)$. Then $\mathbf{Y}'[\mathbf{I} - \mathbf{1}(\mathbf{1}'\mathbf{1})^{-1}\mathbf{1}']\mathbf{Y} \sim W_p(n-1, \Sigma, \Gamma = \mathbf{0})$ and $\mathbf{Y}'[\mathbf{1}(\mathbf{1}'\mathbf{1})^{-1}\mathbf{1}']\mathbf{Y} \sim W_p(1, \Sigma, \Gamma_2)$ are independent since*

$$\begin{aligned} \mathbf{Y}'\mathbf{Y} &= \mathbf{Y}'[\mathbf{I} - \mathbf{1}(\mathbf{1}'\mathbf{1})^{-1}\mathbf{1}']\mathbf{Y} + \mathbf{Y}'[\mathbf{1}(\mathbf{1}'\mathbf{1})^{-1}\mathbf{1}']\mathbf{Y} \\ \mathbf{Y}'\mathbf{Y} &= \mathbf{Y}'\mathbf{P}_1\mathbf{Y} + \mathbf{Y}'\mathbf{P}_2\mathbf{Y} \\ \mathbf{Y}'\mathbf{Y} &= (n-1)\mathbf{S} + n\overline{\mathbf{Y}}\overline{\mathbf{Y}}' \end{aligned}$$

or

$$W_p(n, \Sigma, \Gamma) = W_p(n-1, \Sigma, \Gamma_1 = \mathbf{0}) + W_p(1, \Sigma, \Gamma_2)$$

where $\Gamma = \Gamma_1 + \Gamma_2$ so that variance covariances matrix $\mathbf{S}$ and the vector of means $\overline{\mathbf{Y}}$ are independent. The matrices $\mathbf{P}_i$ are projection matrices.

In Theorem 3.4.5, each row of $\mathbf{Y}_{n \times p}$ is assumed to be independent. More generally, assume that $\mathbf{y}^* = \text{vec}\left(\mathbf{Y}'\right)$ has structure $\text{cov}\left(\mathbf{y}^*\right) = \mathbf{W} \neq \mathbf{I}$ so that the observations are no longer independent. Wong et al. (1991) provide necessary and sufficient conditions to ensure that $\mathbf{Y}'\mathbf{P}_i\mathbf{Y}$ still follow a Wishart distribution. Necessary and sufficient conditions for independence of the Wishart matrices is more complicated; see Young et al. (1999).

The $|\mathbf{S}|$, the sample generalized variance of a normal random sample, is distributed as quantity $|\Sigma|/(n-1)^p$ times a product of independent chi-square variates

$$|\mathbf{S}| \sim \frac{|\Sigma|}{(n-1)^p} \prod_{i=1}^{p} \chi^2 (n-i) \tag{3.4.2}$$

as shown by Muirhead (1982, p. 100). The sample mean and variance of the generalized variance are

$$E\left(|\mathbf{S}|\right) = |\Sigma| \prod_{i=1}^{p} [1 - (i-1)/(n-1)] \tag{3.4.3}$$

$$\begin{aligned} &\text{var}|\mathbf{S}| \\ &= |\Sigma|^2 \prod_{i=1}^{p} [1 - (i-1)/(n-1)]\left\{\prod_{j=1}^{p} [1 - (j-3)/(n-1)] - \prod_{j=1}^{p} [1 - (j-1)/(n-1)]\right\} \end{aligned} \tag{3.4.4}$$

so that the $E\left(|\mathbf{S}|\right) < |\Sigma|$ for $p > 1$. Thus, the determinant of the sample covariance matrix underestimates the determinant of the population covariance matrix. The asymptotic distribution of the sample generalized variance is given by Anderson (1984, p. 262). The distribution of the quantity $\sqrt{(n-1)}(|\mathbf{S}|/|\Sigma| - 1)$ is asymptotically normally distributed with mean zero and variance $2p$. Distributions of the ratio of determinants of some matrices are reviewed briefly in the next section.

In Example 3.3.1 we illustrated the form of the matrix $\text{cov}\left\{\text{vec}\left(\mathbf{S}\right)\right\} = \Omega$ for the bivariate case. More generally, the structure of Ω found in more advanced statistical texts is provided in Theorem 3.4.6.

Theorem 3.4.6 *If* $\mathbf{Y}_i \sim IN_p\left(\boldsymbol{\mu}, \Sigma\right)$ *for* $i = 1, 2, \ldots, n$ *so that* $(n-1)\,\mathbf{S} \sim W_p\left(n-1, \Sigma\right)$. *Then*

$$\Omega = \text{cov}\left(\text{vec}\, \mathbf{S}\right) = 2\mathbf{P}\left(\Sigma \otimes \Sigma\right)\mathbf{P}/\left(n-1\right)$$

where $\mathbf{P} = \left(\mathbf{I}_{p^2} + \mathbf{K}\right)/2$ *and* $\mathbf{K}$ *is a commutation matrix.*

Exercises 3.4

1. If $\mathbf{Y} \sim N_p(\boldsymbol{\mu}, \Sigma)$, prove that $(\mathbf{Y} - \boldsymbol{\mu})'\Sigma^{-1}(\mathbf{Y} - \boldsymbol{\mu}) \sim \chi^2(p)$.

2. If $\mathbf{Y} \sim N_p(\mathbf{0}, \Sigma)$, show that $\mathbf{Y}'\mathbf{A}\mathbf{Y} = \sum_{j=1}^{p} \lambda z_j^2$ where the λ_j are the roots of $|\Sigma^{1/2}\mathbf{A}\Sigma^{1/2} - \lambda\mathbf{I}| = 0$, $\mathbf{A} = \mathbf{A}'$ and $Z_i \sim N\left(0, 1\right)$.

3. If $\mathbf{Y} \sim N_p(\mathbf{0}, \mathbf{P})$ where $\mathbf{P}$ is a projection matrix, show that the $\|\mathbf{Y}\|^2 \sim \chi^2(p)$.
4. Prove property (4) in Theorem 3.4.4.
5. Prove that $E(\mathbf{S}) = \Sigma$ and that the cov $\{\text{vec}(\mathbf{S})\} = 2(\mathbf{I}_{p^2} + \mathbf{K})(\Sigma \otimes \Sigma)/(n-1)$ where $\mathbf{K}$ is a commutation matrix defined in Exercises 3.3, Problem 5.
6. What is the distribution of $\mathbf{S}^{-1}$? Show that $E(\mathbf{S}^{-1}) = \Sigma^{-1}(n-1)/(n-p-2)$ and that $E(\widehat{\Sigma}^{-1}) = n\Sigma^{-1}/(n-p-1)$.
7. What is the mean and variance of the $tr(\mathbf{S})$ under normality?

3.5 Other Multivariate Distributions

a. *The Univariate t and F Distributions*

When testing hypotheses, two distributions employed in univariate analysis are the t and F distributions.

Definition 3.5.1 *Let X and Y be independent random variables such that $X \sim N(\mu, \sigma^2)$ and $Y \sim \chi^2(n, \gamma)$. Then $t = X/\sqrt{Y/n} \sim t(n, \gamma), -\infty < t < \infty$.*

The statistic t has a noncentral t distribution with n degrees of freedom and noncentrality parameter $\gamma = \mu/\sigma$. If $\mu = 0$, the noncentral t distribution reduces to the central t distribution known as Student's t-distribution.

A distribution closely associated with the t distribution is R.A. Fisher's F distribution.

Definition 3.5.2 *Let H and E be independent random variables such that $H \sim \chi^2(v_h, \gamma)$ and $E = \chi^2(v_e, \gamma = 0)$. Then the noncentral F distribution with v_h and v_e degrees of freedom, and noncentrality parameter γ is the ratio*

$$F = \frac{H/v_h}{E/v_e} \sim F(v_h, v_e, \gamma) 0 \leq F \leq \infty$$

b. *Hotelling's T^2 Distribution*

A multivariate extension of Student's t distribution is Hotelling's T^2 distribution.

Definition 3.5.3 *Let $\mathbf{Y}$ and $\mathbf{Q}$ be independent random variables where $\mathbf{Y} \sim N_p(\boldsymbol{\mu}, \Sigma)$ and $\mathbf{Q} \sim W_p(n, \Sigma)$, and $n > p$. Then Hotelling's T^2 (1931) statistic*

$$T^2 = n\mathbf{Y}'\mathbf{Q}^{-1}\mathbf{Y}$$

has a distribution proportional to a noncentral F distribution

$$\frac{n-p+1}{p}\frac{T^2}{n} \sim F(p, n-p+1, \gamma)$$

where $\gamma = \boldsymbol{\mu}'\Sigma^{-1}\boldsymbol{\mu}$.

The T^2 statistic occurs when testing hypotheses regarding means in one- and two-sample multivariate normal populations discussed in Section 3.9.

Example 3.5.1 *Let $\mathbf{Y}_1, \mathbf{Y}_2, \ldots, \mathbf{Y}_n$ be a random sample from a MVN population, $\mathbf{Y}_i \sim IN_p(\boldsymbol{\mu}, \Sigma)$. Then $\overline{\mathbf{Y}} \sim N_p(\boldsymbol{\mu}, \Sigma/n)$ and $(n-1)\mathbf{S} \sim W_p(n-1, \Sigma)$, and $\overline{\mathbf{Y}}$ and $\mathbf{S}$ are independent. Hence, for testing $H_0 : \boldsymbol{\mu} = \boldsymbol{\mu}_0$ vs. $H_1 : \boldsymbol{\mu} \neq \boldsymbol{\mu}_0$, $T^2 = n(\overline{\mathbf{Y}} - \boldsymbol{\mu}_0)'\mathbf{S}^{-1}(\overline{\mathbf{Y}} - \boldsymbol{\mu}_0)$ or*

$$\frac{n-p}{p}\frac{T^2}{n-1} = \frac{n(n-p)}{p(n-1)}\left(\overline{\mathbf{Y}} - \boldsymbol{\mu}_0\right)'\mathbf{S}^{-1}\left(\overline{\mathbf{Y}} - \boldsymbol{\mu}_0\right) \sim F(p, n-p, \gamma)$$

where

$$\gamma = n(\boldsymbol{\mu} - \boldsymbol{\mu}_0)'\Sigma^{-1}(\boldsymbol{\mu} - \boldsymbol{\mu}_0)$$

is the noncentrality parameter. When H_0 is true, the noncentrality parameter is zero and T^2 has a central F distribution.

Example 3.5.2 *Let $\mathbf{Y}_1, \mathbf{Y}_2, \ldots, \mathbf{Y}_{n_1} \sim IN(\boldsymbol{\mu}_1, \Sigma)$ and $\mathbf{X}_1, \mathbf{X}_2, \ldots, \mathbf{X}_{n_2} \sim IN_p(\boldsymbol{\mu}_2, \Sigma)$ where $\overline{\mathbf{Y}} = \sum_{i=1}^{n_1} \mathbf{Y}_i/n_1$ and $\overline{\mathbf{X}} = \sum_{i=1}^{n_2} \mathbf{X}_i/n_2$. An unbiased estimator of Σ in the pooled covariance matrix*

$$\mathbf{S} = \frac{1}{n_1+n_2-2}\left[\sum_{i=1}^{n_1}(\mathbf{Y}_i - \overline{\mathbf{Y}})(\mathbf{Y}_i - \overline{\mathbf{Y}})' + \sum_{i=1}^{n_2}(\mathbf{X}_i - \overline{\mathbf{X}})(\mathbf{X}_i - \overline{\mathbf{X}})'\right]$$

Furthermore, $\overline{\mathbf{X}}$, $\overline{\mathbf{Y}}$, and $\mathbf{S}$ are independent, and

$$\left(\frac{n_1 n_2}{n_1+n_2}\right)^{1/2}(\overline{\mathbf{Y}} - \overline{\mathbf{X}}) \sim N_p\left[\left(\frac{n_1 n_2}{n_1+n_2}\right)^{1/2}(\boldsymbol{\mu}_1 - \boldsymbol{\mu}_2), \Sigma\right]$$

and

$$(n_1+n_2-2)\mathbf{S} \sim W_p(n_1+n_2-2, \Sigma)$$

Hence, to test $H_0 : \boldsymbol{\mu}_1 = \boldsymbol{\mu}_2$ vs. $H_1 : \boldsymbol{\mu}_1 \neq \boldsymbol{\mu}_2$, the test statistic is

$$\begin{aligned} T^2 &= \left(\frac{n_1 n_2}{n_1+n_2}\right)(\overline{\mathbf{Y}} - \overline{\mathbf{X}})'\mathbf{S}^{-1}(\overline{\mathbf{Y}} - \overline{\mathbf{X}}) \\ &= \left(\frac{n_1 n_2}{n_1+n_2}\right)D^2 \end{aligned}$$

By Definition 3.5.3,

$$\frac{n_1+n_2-p-1}{p}\frac{T^2}{n_1+n_2-2} \sim F(p, n_1+n_2-p-1, \gamma)$$

where the noncentrality parameter is

$$\gamma = \left(\frac{n_1 n_2}{n_1+n_2}\right)(\boldsymbol{\mu}_1 - \boldsymbol{\mu}_2)'\Sigma^{-1}(\boldsymbol{\mu}_1 - \boldsymbol{\mu}_2)$$

Example 3.5.3 *Replacing* **Q** *by* **S** *in Definition 3.5.3, Hotelling's* T^2 *statistic follows an* F *distribution*

$$(n-p)\,T^2/\,(n-1)\,p \sim F\,(p, n-p, \gamma)$$

For $\gamma = 0$,

$$E(T^2) = (n-1)\,p/\,(n-p-2)$$

$$\text{var}(T^2) = \frac{2p\,(n-1)^2\,(n-2)}{(n-p-2)^2\,(n-p-4)}$$

By Theorem 3.4.2, $T^2 \xrightarrow{d} \chi^2\,(p)$ *as* $n \longrightarrow \infty$. *However, for small values of* n, *the distribution of* T^2 *is far from chi-square. If* $X^2 \sim \chi^2\,(p)$, *then* $E\left(X^2\right) = p$ *and the* var $\left(X^2\right) = 2p$. *Thus, if one has a statistic* $T^2 \xrightarrow{d} \chi^2\,(p)$, *a better approximation for small to moderate sample sizes is the statistic*

$$\frac{(n-p)\,T^2}{(n-1)\,p} \mathrel{\dot\sim} F\,(p, n-p, \gamma)$$

c. *The Beta Distribution*

A distribution closely associated with the F distribution is the beta distribution.

Definition 3.5.4 *Let* H *and* E *be independent random variables such that* $H \sim \chi^2(v_h, \gamma)$ *and* $E \sim \chi^2(v_e, \gamma = 0)$. *Then*

$$B = \frac{H}{H+E} \sim \text{beta}\,(v_h/2, v_e/2, \gamma)$$

has a noncentral (Type I) beta distribution and

$$V = H/E \sim \textit{Inverted beta}\ (v_h/2, v_e/2, \gamma)$$

has a (Type II) beta or inverted noncentral beta distribution.

From Definition 3.5.4,

$$B = H/(H+E) = \frac{v_h F/v_e}{1 + v_h F/v_e}$$
$$= \frac{H/E}{1+H/E} = V/(1+V)$$

where $v_e V/v_h \sim F(v_h, v_e, \gamma)$. Furthermore, $B = 1 + (1+V)^{-1}$ so that the percentage points of the beta distribution can be related to a monotonic decreasing function of F

$$1 - B(a, b) = B'(b, a) = (1 + 2aF(2a, 2b)/2b)^{-1}$$

Thus, if t is a random variable such that $t \sim t(v_e)$, then

$$1 - B(1, v_e) = B'(v_e, 1) = \frac{1}{1 + t^2/v_e}$$

so that large values of t^2 correspond to small values of B'.

To extend the beta distribution in the central multivariate case, we let $\mathbf{H} \sim W_p(v_h, \Sigma)$ and $\mathbf{E} \sim W_p(v_e, \Sigma)$. Following the univariate example, we set

$$\begin{aligned} \mathbf{B} &= (\mathbf{E}+\mathbf{H})^{-1/2}\,\mathbf{H}\,(\mathbf{E}+\mathbf{H})^{-1/2} \\ \mathbf{V} &= \mathbf{E}^{-1/2}\mathbf{H}\mathbf{E}^{-1/2} \end{aligned} \tag{3.5.1}$$

where $\mathbf{E}^{-1/2}$ and $(\mathbf{E}+\mathbf{H})^{-1/2}$are the symmetric square root matrices of $\mathbf{E}^{-1}$ and $(\mathbf{E}+\mathbf{H})^{-1}$ in that $\mathbf{E}^{-1/2}\,\mathbf{E}^{-1/2} = \mathbf{E}^{-1}$ and $(\mathbf{E}+\mathbf{H})^{-1/2}(\mathbf{E}+\mathbf{H})^{-1/2} = (\mathbf{E}+\mathbf{H})^{-1}$.

Definition 3.5.5 *Let $\mathbf{H} \sim W_p(v_h, \Sigma)$ and $\mathbf{E} \sim W_p(v_e, \Sigma)$ be independent Wishart distributions where $v_h \geq p$ and $v_e \geq p$. Then $\mathbf{B}$ in (3.5.1) follows a central p-variate multivariate (Type I) beta distribution with $v_h/2$ and $v_e/2$ degrees of freedom, written as $\mathbf{B} \sim B_p(v_h/2, v_e/2)$. The matrix $\mathbf{V}$ in (3.5.1) follows a central p-variate multivariate (Type II) beta or inverted beta distribution with $v_h/2$ and $v_e/2$ degrees of freedom, sometimes called a matrix F density.*

Again $\mathbf{I}_p - \mathbf{B} \sim B_p(v_e/2, v_h/2)$ as in the univariate case. An important function of $\mathbf{B}$ in multivariate data analysis is $|\mathbf{I}_p - \mathbf{B}|$ due to Wilks (1932). The statistic

$$\Lambda = |\mathbf{I}_p - \mathbf{B}| = \frac{|\mathbf{E}|}{|\mathbf{E}+\mathbf{H}|} \sim U(p, v_h, v_e) \quad 0 \leq \Lambda \leq 1$$

is distributed as a product of independent beta random variables on $(v_e - i + 1)/2$ and $v_h/2$ degrees of freedom for $i = 1, \ldots, p$.

Because Λ is a ratio of determinants, by Theorem 2.6.8 we can relate Λ to the product of roots

$$\Lambda = \prod_{i=1}^{s}(1-\theta_i) = \prod_{i=1}^{s}(1+\lambda_i)^{-1} = \prod_{i=1}^{s} \mathrm{v}_i \tag{3.5.2}$$

for $i = 1, 2, \ldots, s = \min(v_h, p)$ where θ_i, λ_i and v_i are the roots of $|\mathbf{H} - \theta(\mathbf{E}+\mathbf{H})| = 0$, $|\mathbf{H} - \lambda\mathbf{E}| = 0$, and $|\mathbf{E} - \mathrm{v}(\mathbf{H}+\mathbf{E})| = 0$, respectively.

One of the first approximations to the distribution of Wilks' likelihood ratio criterion Λ was developed by Bartlett (1947). Letting $X_B^2 = -[v_e - (p - v_h + 1)/2] \log \Lambda$, Bartlett showed that the statistic X^2 converges to a chi-square distribution with degrees of freedom $v = pv_h$. Wall (1968) developed tables for the exact distribution of Wilks' likelihood ratio criterion using an infinite series approximation. Coelho (1998) obtained a closed form solution. One of the most widely used approximations to the Λ criterion was developed by Rao (1951, 1973a, p. 556). Rao approximated the distribution of Λ with an F distribution as follows.

$$\frac{1-\Lambda^{1/d}}{\Lambda^{1/d}} \frac{fd - 2\lambda}{pv_h} \sim F(pv_h, fd - 2\lambda) \tag{3.5.3}$$

$$\begin{aligned} f &= v_e - (p - v_h + 1)/2 \\ d^2 &= \frac{p^2 v_h^2 - 4}{p^2 + v_h^2 - 5} \quad \text{for} \quad p^2 + v_h^2 - 5 > 0 \text{ or } d = 1 \\ \lambda &= (pv_h - 2)/4 \end{aligned}$$

The approximation is exact for p or v_h equal to 1 or 2 and accurate to three decimal places if $p^2 + v_h^2 \leq f/3$; see Anderson (1984, p. 318).

Given (3.5.2) and Theorem 2.6.8, other multivariate test statistics are related to the distribution of the roots of the $|\mathbf{B}|$ or $|\mathbf{H} - \theta(\mathbf{E} + \mathbf{H})| = 0$. In particular, the Bartlett (1939), Lawley (1938), and Hotelling (1947, 1951) (BLH) trace criterion is

$$\begin{aligned} T_o^2 &= v_e \operatorname{tr}(\mathbf{HE}^{-1}) \\ &= v_e \sum_{i=1}^{s} \theta_i / (1 - \theta_i) \\ &= v_e \sum_{i=1}^{s} \lambda_i \\ &= v_e \sum_{i=1}^{s} (1 - \mathrm{v}_i) / \mathrm{v}_i \end{aligned} \tag{3.5.4}$$

The Bartlett (1939), Nanda (1950), Pillai (1955) (BNP) trace criterion is

$$\begin{aligned} V^{(s)} &= \operatorname{tr}\left[\mathbf{H}(\mathbf{E} + \mathbf{H})^{-1}\right] \\ &= \sum_{i=1}^{s} \theta_i = \sum_{i=1}^{s} \left(\frac{\lambda_i}{1 + \lambda_i}\right) = \sum_{i=1}^{s} (1 - \mathrm{v}_i) \end{aligned} \tag{3.5.5}$$

The Roy (1953) maximum root criterion is

$$\theta_1 = \frac{\lambda_1}{1 + \lambda_1} = 1 - \mathrm{v}_1 \tag{3.5.6}$$

Tables for these statistics were developed by Pillai (1960) and are reproduced in Kres (1983). Relating the eigenvalues of the criteria to an asymptotic chi-square distribution, Berndt and Savin (1977) established a hierarchical inequality among the test criteria developed by Bartlett-Nanda-Pillai (BNP), Wilks (W), and Bartlett-Lawley-Hotelling (BLH). The inequality states that the BLH criterion has the largest value, followed by the W criterion and the BNP criterion. The larger the roots, the larger the difference among the criteria. Depending on the criterion selected, one may obtain conflicting results when testing linear hypotheses. No criterion is uniformly best, most powerful against all alternatives. However, the critical region for the statistic $V^{(s)}$ in (3.5.5) is locally best invariant. All the criteria may be adequately approximated using the F distribution; see Pillai (1954, 1956), Roy (1957), and Muirhead (1982, Th. 10.6.10).

Theorem 3.5.1 *Let* $\mathbf{H} \sim W_p(v_h, \Sigma)$ *and* $\mathbf{E} \sim W_p(v_e, \Sigma)$ *be independent Wishart distributions under the null linear hypothesis with* v_h *degrees of freedom for the hypothesis test matrix and* v_e *degrees of freedom for error for the error test matrix on p normal random variables where* $v_e \geq p$, $s = \min(v_h, p)$, $M = (|v_h - p| - 1)/2$ *and* $N = (v_e - p - 1)/2$. *Then*

$$\left(\frac{2N + s + 1}{2M + s + 1}\right)\left(\frac{V}{s - V}\right) \xrightarrow{d} F(v_1, v_2)$$

where $v_1 = s\,(2M + s + 1)$ *and* $v_2 = s(2N + s + 1)$

$$\frac{2\,(sN + 1)}{s^2\,(2M + s + 1)}\,\frac{T_o^2}{v_e} \xrightarrow{d} F\,(v_1, v_2)$$

where $v_1 = s\,(2M + s + 1)$ *and* $v_2 = 2\,(sN + 1)$. *Finally,*

$$v_2\lambda_1/v_1 \doteq F^{\max}\,(v_1, v_2) \leq F$$

where $v_1 = \max\,(v_h, p)$ *and* $v_2 = v_e - v_1 + v_h$. *For* $v_h = 1$, $(v_e - p + 1)\,\lambda_1/p = F$ *exactly with* $v_1 = p$ *and* $v_2 = v_e - p + 1$ *degrees of freedom.*

When $s \neq 1$, the statistic $F^{\max}$ for Roy's criterion does not follow an F distribution. It provides an upper bound on the F statistic and hence results in a lower bound on the level of significance or p-value. Thus, in using the approximation for Roy's criterion one can be sure that the null hypothesis is true if the hypothesis is accepted. However, when the null hypothesis is rejected this may not be the case since $F^{\max} \leq F$, the true value. Muller et al. (1992) develop F approximations for the Bartlett-Lawley-Hotelling, Wilks and Bartlett-Nanda-Pillai criteria that depend on measures of multivariate association.

d. *Multivariate t, F, and* χ^2 *Distributions*

In univariate and multivariate data analysis, one is often interested in testing a finite number of hypotheses regarding univariate- and vector-valued population parameters simultaneously or sequentially, in some planned order. Such procedures are called simultaneous test procedures (STP) and often involve contrasts in means. While the matrix $\mathbf{V}$ in (3.5.1) follows a matrix variate F distribution, we are often interested in the joint distribution of F statistics when performing an analysis of univariate and multivariate data using STP methods. In this section, we define some multivariate distributions which arise in STP.

Definition 3.5.6 *Let* $\mathbf{Y} \sim N_p(\mu, \Sigma = \sigma^2\mathbf{P})$ *where* $\mathbf{P} = \left[\rho_{ij}\right]$ *is a correlation matrix and* $s^2/\sigma^2 \sim \chi^2(n, \gamma = 0)$ *independent of* $\mathbf{Y}$. *Setting* $T_i = Y_i\sqrt{n}/s$ *for* $i = 1, \ldots, p$. *Then the joint distribution of* $\mathbf{T}' = [T_1, T_2, \ldots, T_p]$ *is a central or noncentral multivariate t distribution with n degrees of freedom.*

The matrix $\mathbf{P} = \left[\rho_{ij}\right]$ is called the correlation matrix of the accompanying MVN distribution. The distribution is central or noncentral depending on whether $\boldsymbol{\mu} = \mathbf{0}$ or $\boldsymbol{\mu} \neq \mathbf{0}$, respectively. When $\rho_{ij} = \rho\,(i \neq j)$,the structure of $\mathbf{P}$ is said to be equicorrelated. The multivariate t distribution is a joint distribution of correlated t statistics which is clearly not the same as Hotelling's T^2 distribution which involves the distribution of a quadratic form. Using this approach, we generalize the chi-square distribution to a multivariate chi-square distribution which is a joint distribution of p correlated chi-square random variables.

Definition 3.5.7 *Let* $\mathbf{Y}_i$ *be m independent MVN random p-vectors with mean* $\boldsymbol{\mu}$ *and covariance matrix* Σ, $\mathbf{Y}_i \sim IN_p\,(\boldsymbol{\mu}, \Sigma)$. *Define* $X_j = \sum_{i=1}^{m} Y_{ij}^2$ *for* $j = 1, 2, \ldots, p$. *Then the joint distribution of* $\mathbf{X}' = \left[X_1, X_2, \ldots, X_p\right]$ *is a central or noncentral multivariate chi-square distribution with m degrees of freedom.*

Observe that X_j is the sum of m independent normal random variables with mean μ_j and variance $\sigma_j^2 = \sigma_{jj}$. For $m = 1$, $\mathbf{X}$ has a multivariate chi-square distribution with 1 degree of freedom. The distribution is central or noncentral if $\boldsymbol{\mu} = \mathbf{0}$ or $\boldsymbol{\mu} \neq \mathbf{0}$, respectively. In many applications, $m = 1$ so that $\mathbf{Y} \sim N_p(\boldsymbol{\mu}, \Sigma)$ and $\mathbf{X} \sim \chi_1^2(p, \gamma)$, a multivariate chi-square with one degree of freedom.

Having defined a multivariate chi-square distribution, we define a multivariate F-distribution with (m, n) degrees of freedom.

Definition 3.5.8 *Let* $\mathbf{X} \sim \chi_m^2(p, \gamma)$ *and* $\mathbf{Y}_i \sim IN_p(\boldsymbol{\mu}, \Sigma)$ *for* $i = 1, 2, \ldots, m$. *Define* $F_j = nX_j\sigma_{00}/mX_0\sigma_{jj}$ *for* $j = 1, \ldots, p$ *and* $X_0/\sigma_{00} \sim \chi^2(n)$ *independent of* $\mathbf{X}' = [X_1, X_2, \ldots, X_p]$. *Then the joint distribution of* $\mathbf{F}' = [F_1, F_2, \ldots, F_p]$ *is a multivariate F with* (m, n) *degrees of freedom.*

For $m = 1$, the multivariate F distribution is equivalent to a multivariate t^2 distribution or for $T_i = \sqrt{F_i}$, the distribution of $\mathbf{T}' = [T_1, T_2, \ldots, T_p]$ is multivariate t, also known as the Studentized Maximum Modulus distribution used in numerous univariate STP; see Hochberg and Tamhane (1987) and Nelson (1993). We will use the distribution to test multivariate hypotheses involving means using the finite intersection test (FIT) principle; see Timm (1995).

Exercises 3.5

1. Use Definition 3.5.1 to find the distribution of $\sqrt{n}\left(\overline{y} - \mu_0\right)/s$ if $Y_i \sim N\left(\mu, \sigma^2\right)$ and $\mu \neq \mu_0$.

2. For $\mu = \mu_0$ in Problem 1, what is the distribution of $F/[(n-1)+F]$?

3. Verify that for large values of v_e, $X_B^2 = -\left[v_e - (p - v_h + 1)/2\right] \ln \Lambda \mathrel{\dot\sim} \chi^2(pv_h)$ by comparing the chi-square critical value and the critical value of an F-distribution with degrees of freedom pv_h and v_e.

4. For $\mathbf{Y}_i \sim IN_p(\boldsymbol{\mu}, \Sigma)$ for $i = 1, 2, \ldots, n$, verify that $\Lambda = 1/\left[1 + T^2/(n-1)\right]$.

5. Let $Y_{ij} \sim IN\left(\mu_i, \sigma^2\right)$ for $i = 1, \ldots, k$, $j = 1, \ldots, n$. Show that

$$T_i = \left(y_{i.} - \overline{y}_{..}\right)/s\sqrt{(k-1)/nk} \sim t\left[k(n-1)\right]$$

if $\mu_1 = \mu_2 = \cdots = \mu_k = \mu$ and that $\mathbf{T}' = [T_1, T_2, \ldots, T_k]$ and have a central multivariate t-distribution with $v = n(k-1)$ degree of freedom and equicorrelation structure $\mathbf{P} = \left[\rho_{ij} = \rho\right]$ for $i \neq j$ where $\rho = -1/(k-1)$; see Timm (1995).

6. Let $Y_{ij} \sim IN\left(\mu_i, \sigma^2\right)$ $i = 1, \ldots, k$, $j = 1, \ldots, n_i$ and σ^2 known. For $\psi_g = \sum_{i=1}^k c_{ig}\mu_i$ and $\widehat{\psi}_g = \sum_{i=1}^k c_{ig}\widehat{\mu}_i$ where $E\left(\widehat{\mu}_i\right) = \mu_i$ define $X_g = \widehat{\psi}_g/\sqrt{d_g\sigma}$ where $d_g = \sum_{i=1}^k \left(c_{ig}/n_i\right)$. Show that $X_1, X_2, \ldots, X_q$ for $g = 1, 2, \ldots, q$ is multivariate chi-square with one degree of freedom.

3.6 The General Linear Model

The linear model is fundamental to the analysis of both univariate and multivariate data. When formulating a linear model, one observes a phenomenon represented by an observed data vector (matrix) and relates the observed data to a set of linearly independent fixed variables. The relationship between the random dependent set and the linearly independent set is examined using a linear or nonlinear relationship in the vector (matrix) of parameters. The parameter vector (matrix) may be assumed either as fixed or random. The fixed or random set of parameters are usually considered to be independent of a vector (matrix) of errors. One also assumes that the covariance matrix of the random parameters of the model has some unknown structure. The goals of the data analysis are usually to estimate the fixed and random model parameters, evaluate the fit of the model to the data, and to test hypotheses regarding model parameters. Model development occurs with a calibration sample. Another goal of model development is to predict future observations. To validate the model developed using a calibration sample, one often obtains a validation sample.

To construct a general linear model for a random set of correlated observations, an observation vector $\mathbf{y}_{N\times K}$ is related to a vector of K parameters represented by a vector $\boldsymbol{\beta}_{K\times 1}$ through a known nonrandom design matrix $\mathbf{X}_{N\times K}$ plus a random vector of errors $\mathbf{e}_{N\times 1}$ with mean zero, $E(\mathbf{e}) = \mathbf{0}$,and covariance matrix $\Omega = \text{cov}(\mathbf{e})$. The representation for the linear model is

$$\begin{aligned} \mathbf{y}_{N\times 1} &= \mathbf{X}_{N\times K}\,\boldsymbol{\beta}_{K\times 1} + \mathbf{e}_{N\times 1} \\ E(\mathbf{e}) &= \mathbf{0} \quad \text{and} \quad \text{cov}(\mathbf{e}) = \Omega \end{aligned} \tag{3.6.1}$$

We shall always assume that $E(\mathbf{e}) = \mathbf{0}$ when writing a linear model. Model (3.6.1) is called the general linear model (GLM) or the Gauss-Markov setup. The model is linear since the ith element of $\mathbf{y}$ is related to the ith row of $\mathbf{X}$ as $y_i = \mathbf{x}_i'\boldsymbol{\beta}$; y_i is modeled by a linear function of the parameters. We only consider linear models in this text; for a discussion of nonlinear models see Davidian and Giltinan (1995) and Vonesh and Chinchilli (1997). The procedure NLMIXED in SAS may be used to analyze these models. The general nonlinear model used to analyze non-normal data is called the generalized linear model. These models are discussed by McCullagh and Nelder (1989) and McCulloch and Searle (2001).

In (3.6.1), the elements of $\boldsymbol{\beta}$ can be fixed, random, or both (mixed) and $\boldsymbol{\beta}$ can be either unrestricted or restricted. The structure of $\mathbf{X}$ and Ω may vary and, depending on the form and structure of $\mathbf{X}$, Ω, and $\boldsymbol{\beta}$, the GLM is known by many names. Depending on the structure of the model, different approaches to parameter estimation and hypothesis testing are required. In particular, one may estimate $\boldsymbol{\beta}$ and Ω making no assumptions regarding the distribution of $\mathbf{y}$. In this case, generalized least squares (GLS) theory and minimum quadratic norm unbiased estimation (MINQUE) theory is used to estimate the model parameters; see Rao (1973a) and Kariya (1985). In this text, we will usually assume that the vector $\mathbf{y}$ in (3.6.1) has a multivariate normal distribution; hence, maximum likelihood (ML) theory will be used to estimate model parameters and to test hypotheses using the likelihood ratio (LR) principle. When the small sample distribution is unknown, large sample tests may be developed. In general, these will depend on the Wald principle developed by Wald (1943), large sample distributions of LR statistics, and Rao's Score principle developed by Rao

(1947) or equivalently Silvey's (1959) Lagrange multiplier principle. An introduction to the basic principles may be found in Engle (1984) and Mittelhammer et al. (2000), while a more advanced discussion is given by Dufour and Dagenais (1992). We now review several special cases of (3.6.1).

a. *Regression, ANOVA, and ANCOVA Models*

Suppose each element y_i in the vector $\mathbf{y}_N$ is related to k linearly independent predictor variables

$$y_i = \beta_0 + x_{i1}\beta_1 + x_{i2}\beta_2 + \cdots + x_{ik}\beta_k + e_i \tag{3.6.2}$$

For $i = 1, 2, \ldots, n$, the relationship between the dependent variable Y and the k independent variables $x_1, x_2, \ldots, x_k$ is linear in the parameters. Furthermore, assume that the parameters $\beta_0, \beta_1, \ldots, \beta_k$ are free to vary over the entire parameter space so that there is no restriction on $\boldsymbol{\beta}'_q = \left[\beta_0, \beta_1, \ldots, \beta_k\right]$ where $q = k+1$ and that the errors e_i have mean zero and common, unknown variance σ^2. Then using (3.6.1) with $N = n$ and $K = q = k + 1$, the univariate (linear) regression (UR) model is

$$\begin{bmatrix} y_1 \\ y_2 \\ \vdots \\ y_n \end{bmatrix} = \begin{bmatrix} 1 & x_{11} & x_{12} & \cdots & x_{1k} \\ 1 & x_{21} & x_{22} & \cdots & x_{2k} \\ \vdots & \vdots & \vdots & & \vdots \\ 1 & x_{n1} & x_{n1} & \cdots & x_{nk} \end{bmatrix} \begin{bmatrix} \beta_0 \\ \beta_1 \\ \vdots \\ \beta_k \end{bmatrix} + \begin{bmatrix} e_1 \\ e_2 \\ \vdots \\ e_n \end{bmatrix}$$

$$\mathbf{y}_{n\times 1} = \mathbf{X}_{n\times q} \boldsymbol{\beta}_{q\times 1} + \mathbf{e}_{n\times 1}$$

$$\operatorname{cov}(\mathbf{y}) = \sigma^2 \mathbf{I}_n \tag{3.6.3}$$

where the design matrix $\mathbf{X}$ has full column rank, $r(\mathbf{X}) = q$. If the $r(\mathbf{X}) < q$ so that $\mathbf{X}$ is not of full column rank and $\mathbf{X}$ contains indicator variables, we obtain the analysis of variance (ANOVA) model.

Often the design matrix $\mathbf{X}$ in (3.6.3) is partitioned into two sets of independent variables, a matrix $\mathbf{A}_{n\times q_1}$ that is not of full rank and a matrix $\mathbf{Z}_{n\times q_2}$ that is of full rank so that $\mathbf{X} = [\mathbf{A}\ \mathbf{Z}]$ where $q = q_1 + q_2$. The matrix $\mathbf{A}$ is the ANOVA design matrix and the matrix $\mathbf{Z}$ is the regression design matrix, also called the matrix of covariates. For $N = n$ and $\mathbf{X} = [\mathbf{A}\ \mathbf{Z}]$, model (3.6.3) is called the ANCOVA model. Letting $\boldsymbol{\beta}' = \left[\boldsymbol{\alpha}'\ \boldsymbol{\gamma}'\right]$, the analysis of covariance ANCOVA model has the general linear model form

$$\mathbf{y} = [\mathbf{A}\ \mathbf{Z}] \begin{bmatrix} \boldsymbol{\alpha} \\ \boldsymbol{\gamma} \end{bmatrix} + \mathbf{e}$$

$$\mathbf{y} = \mathbf{A}\boldsymbol{\alpha} + \mathbf{Z}\boldsymbol{\gamma} + \mathbf{e} \tag{3.6.4}$$

$$\operatorname{cov}(\mathbf{y}) = \sigma^2 \mathbf{I}_n$$

Assuming the observation vector $\mathbf{y}$ has a multivariate normal distribution with mean $\mathbf{X}\boldsymbol{\beta}$ and covariance matrix $\Omega = \sigma^2\mathbf{I}_n$, $\mathbf{y} \sim \mathbf{N}_n\left(\mathbf{X}\boldsymbol{\beta}, \sigma^2\mathbf{I}_n\right)$, the ML estimates of $\boldsymbol{\beta}$ and σ^2 are

$$\begin{aligned}\widehat{\boldsymbol{\beta}} &= \left(\mathbf{X}'\mathbf{X}\right)^{-1}\mathbf{X}'\mathbf{y} \\ \widehat{\sigma}^2 &= (\mathbf{y} - \mathbf{X}\widehat{\boldsymbol{\beta}})'(\mathbf{y} - \mathbf{X}\widehat{\boldsymbol{\beta}})/n \\ &= \mathbf{y}'[\mathbf{I} - \mathbf{X}\left(\mathbf{X}'\mathbf{X}\right)^{-1}\mathbf{X}']\mathbf{y}/n \\ &= E/n \end{aligned} \tag{3.6.5}$$

The estimator $\widehat{\boldsymbol{\beta}}$ is only unique if the rank of the design matrix $r(\mathbf{X}) = q$, $\mathbf{X}$ has full column rank; when the rank of the design matrix is less than full rank, $r(\mathbf{X}) = r < q$, $\widehat{\boldsymbol{\beta}} = \left(\mathbf{X}'\mathbf{X}\right)^{-}\mathbf{X}'\mathbf{y}$. Then, Theorem 2.6.2 is used to find estimable functions of $\boldsymbol{\beta}$. Alternatively, the methods of reparameterization or adding side conditions to the model parameters are used to obtain unique estimates. To obtain an unbiased estimator of σ^2, the restricted maximum likelihood (REML) estimate is $s^2 = E/(n-r)$ where $r = r(\mathbf{X}) \leq q$ is used; see Searle et al. (1992, p. 452).

To test the hypothesis of the form $H_o : \mathbf{C}\boldsymbol{\beta} = \boldsymbol{\xi}$, one uses the likelihood ratio test which has the general form

$$\Lambda = \lambda^{2/n} = E/(E + H) \tag{3.6.6}$$

where E is defined in (3.6.5) and

$$H = (\mathbf{C}\widehat{\boldsymbol{\beta}} - \boldsymbol{\xi})'\left[\mathbf{C}\left(\mathbf{X}'\mathbf{X}\right)^{-1}\mathbf{C}'\right]^{-1}(\mathbf{C}\widehat{\boldsymbol{\beta}} - \boldsymbol{\xi}) \tag{3.6.7}$$

The quantities E and H are independent quadratic forms and by Theorem 3.4.2, $H \sim \sigma^2\chi^2(v_h, \delta)$ and $E \sim \sigma^2\chi^2(v_e = n - r)$. For additional details, see Searle (1971) or Rao (1973a).

The assumption of normality was needed to test the hypothesis H_o. If one only wants to estimate the parameter vector $\boldsymbol{\beta}$, one may estimate the parameter using the least squares criterion. That is, one wants to find an estimate for the parameter $\boldsymbol{\beta}$ that minimizes the error sum of squares, $\mathbf{e}'\mathbf{e} = (\mathbf{y} - \mathbf{X}\boldsymbol{\beta})'(\mathbf{y} - \mathbf{X}\boldsymbol{\beta})$. The estimate for the parameter vector $\boldsymbol{\beta}$ is called the ordinary least squares (OLS) estimate for the parameter β. Using Theorem 2.6.1, the general form of the OLS estimate is $\widehat{\boldsymbol{\beta}}_{OLS} = \left(\mathbf{X}'\mathbf{X}\right)^{-}\mathbf{X}'\mathbf{y} + (\mathbf{I} - \mathbf{H})\mathbf{z}$ where $\mathbf{H} = (\mathbf{X}'\mathbf{X})^{-}(\mathbf{X}'\mathbf{X})$ and $\mathbf{z}$ is an arbitrary vector; see Rao (1973a). The OLS estimate always exists, but need not be unique. When the design matrix $\mathbf{X}$ has full column rank, the ML estimate is equal to the OLS estimator.

The decision rule for the likelihood ratio test is to reject H_o if $\Lambda < c$ where c is determined such that the $P(\Lambda < c|H_o) = \alpha$. From Definition 3.5.4, Λ is related to a noncentral (Type I) beta distribution with degrees of freedom $v_h/2$ and $v_e/2$. Because the percentage points of the beta distribution are easily related to a monotonic function of F as illustrated in Section 3.5, the null hypothesis $H_o : \mathbf{C}\boldsymbol{\beta} = \boldsymbol{\xi}$ is rejected if $F = v_e H / v_e E \geq F^{1-\alpha}(v_h, v_e)$where $F^{1-\alpha}(v_h,\ v_e)$ represents the upper $1 - \alpha$ percentage point of the central F distribution for a test of size alpha.

In the UR model, we assumed that the structure of $\Omega = \sigma^2\mathbf{I}_n$. More generally, suppose $\Omega = \Sigma$ where Σ is a known nonsingular covariance matrix so that $\mathbf{y} \sim N_n(\mathbf{X}\boldsymbol{\beta},\ \Omega = \Sigma)$.

The ML estimate of $\boldsymbol{\beta}$ is

$$\widehat{\boldsymbol{\beta}}_{ML} = \left(\mathbf{X}'\Sigma^{-1}\mathbf{X}\right)^{-1}\mathbf{X}'\Sigma^{-1}\mathbf{y} \tag{3.6.8}$$

To test the hypothesis $H_o : \mathbf{C}\boldsymbol{\beta} = \boldsymbol{\xi}$ for this case, Theorem 3.4.2 is used. The Wald W statistic, Rao's score statistic, and the LR statistic all have the following form

$$X^2 = (\mathbf{C}\widehat{\boldsymbol{\beta}}_{ML} - \boldsymbol{\xi})'[\mathbf{C}\left(\mathbf{X}'\Sigma^{-1}\mathbf{X}\right)^{-1}\mathbf{C}']^{-1}(\mathbf{C}\widehat{\boldsymbol{\beta}}_{ML} - \boldsymbol{\xi}) \tag{3.6.9}$$

and follow a noncentral chi-square distribution; see Breusch (1979). The test of $H_o : \mathbf{C}\boldsymbol{\beta} = \boldsymbol{\xi}$ is to reject H_o if $X^2 \geq \chi^2_{1-\alpha}(v_h)$ where $v_h = r(\mathbf{C})$. For known Ω, model (3.6.1) is also called the weighted least squares or generalized least squares model when one makes no distribution assumptions regarding the observation vector $\mathbf{y}$. The generalized least squares estimate for $\boldsymbol{\beta}$ is obtained by minimizing the error sum of squares in the metric of the inverse of the covariance matrix, $\mathbf{e}'\mathbf{e} = (\mathbf{y} - \mathbf{X}\boldsymbol{\beta})'\Sigma^{-1}(\mathbf{y} - \mathbf{X}\boldsymbol{\beta})$ and is often called Aitken's generalized least squares (GLS) estimator. This method of estimation is only applicable because the covariance matrix is nonsingular. The GLS estimate for $\boldsymbol{\beta}$ is identical to the ML estimate and the GLS estimate for $\boldsymbol{\beta}$ is equal to the OLS estimate if and only if $\Omega\mathbf{X} = \mathbf{X}\mathbf{F}$ for some nonsingular conformable matrix $\mathbf{F}$; see Zyskind (1967). Rao (1973b) discusses a unified theory of least squares for obtaining estimators for the parameter $\boldsymbol{\beta}$ when the covariance structure has the form $\Omega = \sigma^2\mathbf{V}$ when $\mathbf{V}$ is singular and only assumes that $E(\mathbf{y}) = \mathbf{X}\boldsymbol{\beta}$ and that the $\text{cov}(\mathbf{y}) = \Omega = \sigma^2\mathbf{V}$. Rao's approach is to find a matrix $\mathbf{T}$ such that $(\mathbf{y} - \mathbf{X}\boldsymbol{\beta})'\mathbf{T}^-(\mathbf{y} - \mathbf{X}\boldsymbol{\beta})$ is minimized for $\boldsymbol{\beta}$. Rao shows that for $\mathbf{T}^- = \Sigma^-$, a singular matrix, that an estimate of the parameter $\boldsymbol{\beta}$ is $\widehat{\boldsymbol{\beta}}_{GM} = \left(\mathbf{X}'\Sigma^-\mathbf{X}\right)^{-1}\mathbf{X}'\Sigma^-\mathbf{y}$ sometimes called the generalized Gauss-Markov estimator. Rao also shows that the generalized Gauss-Markov estimator reduces to the ordinary least squares estimator if and only if $\mathbf{X}'\Sigma\mathbf{Q} = \mathbf{0}$ where the matrix $\mathbf{Q} = \mathbf{X}^{\perp} = \mathbf{I} - \mathbf{X}\left(\mathbf{X}'\mathbf{X}\right)^-\mathbf{X}'$ is a projection matrix. This extends Zyskind's result to matrices that are nonsingular. In the notation of Rao, Zyskind's result for a nonsingular matrix Σ is equivalent to the condition that $\mathbf{X}'\Sigma^{-1}\mathbf{Q} = \mathbf{0}$.

Because $\mathbf{y} \sim N_n(\mathbf{X}\boldsymbol{\beta}, \Omega)$, the maximum likelihood estimate is normally distributed as follows

$$\widehat{\boldsymbol{\beta}}_{ML} \sim N_n\left[\boldsymbol{\beta}, \left(\mathbf{X}'\Sigma^{-1}\mathbf{X}\right)^{-1}\right] \tag{3.6.10}$$

When Σ is unknown, asymptotic theory is used to test $H_o : \mathbf{C}\boldsymbol{\beta} = \boldsymbol{\xi}$. Given that we can find a consistent estimate $\widehat{\Sigma} \xrightarrow{p} \Sigma$, then

$$\widehat{\widehat{\boldsymbol{\beta}}}_{FGLS} = (\mathbf{X}'\widehat{\Sigma}^{-1}\mathbf{X}')^{-1}\mathbf{X}'\mathbf{y} \xrightarrow{p} \widehat{\boldsymbol{\beta}}_{ML} \tag{3.6.11}$$

$$\text{cov}(\widehat{\widehat{\boldsymbol{\beta}}}_{FGLS}) = \frac{1}{n}\left(\frac{X'\widehat{\Sigma}^{-1}X}{n}\right)^{-1} \xrightarrow{p} \left(\mathbf{X}\Sigma^{-1}\mathbf{X}\right)^{-1}$$

where $\widehat{\widehat{\boldsymbol{\beta}}}_{FGLS}$ is a feasible generalized least squares estimate of $\boldsymbol{\beta}$ and $\mathbf{X}'\Sigma^{-1}\mathbf{X}/n$ is the information matrix of $\boldsymbol{\beta}$. Because Σ is unknown, the standard errors for the parameter vector $\boldsymbol{\beta}$ tend to be underestimated; see Eaton (1985). To test the hypothesis $H_o : \mathbf{C}\boldsymbol{\beta} = \boldsymbol{\xi}$, the statistic

$$W = (\mathbf{C}\widehat{\widehat{\boldsymbol{\beta}}}_{FGLS} - \boldsymbol{\xi})'[\mathbf{C}(\mathbf{X}'\widehat{\Sigma}^{-1}\mathbf{X})\mathbf{C}']^{-1}(\mathbf{C}\widehat{\widehat{\boldsymbol{\beta}}}_{FGLS} - \boldsymbol{\xi}) \xrightarrow{d} \chi^2(v_h) \tag{3.6.12}$$

where $v_h = r(\mathbf{C})$ is used. When n is small, W/v_h may be approximated by an distribution with degrees of freedom $v_h = r(\mathbf{C})$ and $v_e = n - r(\mathbf{X})$; see Zellner (1962).

One can also impose restrictions on the parameter vector $\boldsymbol{\beta}$ of the form $\mathbf{R}\boldsymbol{\beta} = \boldsymbol{\theta}$ and test hypotheses with the restrictions added to model (3.6.3). This linear model is called the restricted GLM. The reader is referred to Timm and Carlson (1975) or Searle (1987) for a discussion of this model. Timm and Mieczkowski (1997) provide numerous examples of the analyses of restricted linear models using SAS software.

One may also formulate models using (3.6.1) which permit the components of $\boldsymbol{\beta}$ to contain only random effects or more generally both random and fixed effects. For example, suppose in (3.6.2) we add a random component so that

$$y_i = \mathbf{x}_i'\boldsymbol{\beta} + \alpha_i + e_i \tag{3.6.13}$$

where $\boldsymbol{\beta}$ is a fixed vector of parameters and α_i and e_i are independent random errors with variances σ_α^2 and σ^2, respectively. Such models involve the estimation of variance components. Searle et al. (1992), McCulloch and Searle (2001), and Khuri et al. (1998) provide an extensive review of these models.

Another univariate extension of (3.6.2) is to assume that y_i has the linear form

$$y_i = \mathbf{x}_i'\boldsymbol{\beta} + \mathbf{z}_i'\boldsymbol{\alpha}_i + e_i \tag{3.6.14}$$

where $\boldsymbol{\alpha}_i$ and e_i are independent and $\mathbf{z}_i'$ is a vector of known covariates. Then Ω has the structure, $\Omega = \Phi + \sigma^2\mathbf{I}$ where Φ is a covariance matrix of random effects. The model is important in the study of growth curves where the random $\boldsymbol{\alpha}_i$ are used to estimate random growth differences among individuals. The model was introduced by Laird and Ware (1982) and is called the general univariate (linear) mixed effect model. A special case of this model is Swamy's (1971) random coefficient regression model. Vonesh and Chinchilli (1997) provide an excellent discussion of both the random coefficient regression and the general univariate mixed effect models. Littell et al. (1996) provide numerous illustrations using SAS software. We discuss this model in Chapter 6. This model is a special case of the general multivariate mixed model. Nonlinear models used to analyze non-normal data with both fixed and random components are called generalized linear mixed models. These models are discussed by Littell et al. (1996, Chapters 11) and McCulloch and Searle (2001), for example.

b. *Multivariate Regression, MANOVA, and MANCOVA Models*

To generalize (3.6.3) to the multivariate (linear) regression model, a model is formulated for each of p correlated dependent, response variables

$$\begin{array}{ccccccccc}
\mathbf{y}_1 & = & \beta_{01}\mathbf{1}_n & + & \beta_{11}\mathbf{x}_1 & + \cdots + & \beta_{k1}\mathbf{x}_k & + & \mathbf{e}_1 \\
\mathbf{y}_2 & = & \beta_{02}\mathbf{1}_n & + & \beta_{12}\mathbf{x}_2 & + \cdots + & \beta_{k2}\mathbf{x}_k & + & \mathbf{e}_2 \\
\vdots & & \vdots & & \vdots & & \vdots & & \vdots \\
\mathbf{y}_p & = & \beta_{0p}\mathbf{1}_n & + & \beta_{1p}\mathbf{x}_p & + \cdots + & \beta_{kp}\mathbf{x}_k & + & \mathbf{e}_p
\end{array} \tag{3.6.15}$$

Each of the vectors $\mathbf{y}_j, \mathbf{x}_j$ and $\mathbf{e}_j$, for $j = 1, 2, \ldots, p$ are $n \times 1$ vectors. Hence, we have n observations for each of p variables. To represent (3.6.15) in matrix form, we construct matrices using each variable as a column vector. That is,

$$\begin{aligned}
\mathbf{Y}_{n\times p} &= \left[\mathbf{y}_1, \mathbf{y}_2, \ldots, \mathbf{y}_p\right] \\
\mathbf{X}_{n\times q} &= [\mathbf{1}_n, \mathbf{x}_1, \mathbf{x}_2, \ldots, \mathbf{x}_k] \\
\mathbf{B}_{q\times p} &= \left[\boldsymbol{\beta}_1, \boldsymbol{\beta}_2, \ldots, \boldsymbol{\beta}_p\right] \\
&= \begin{bmatrix} \beta_{01} & \beta_{01} & \cdots & \beta_{0p} \\ \beta_{11} & \beta_{11} & \cdots & \beta_{1p} \\ \vdots & \vdots & & \vdots \\ \beta_{k1} & \beta_{k2} & \cdots & \beta_{kp} \end{bmatrix} \\
\mathbf{E}_{n\times p} &= \left[\mathbf{e}_1, \mathbf{e}_2, \ldots, \mathbf{e}_p\right]
\end{aligned} \tag{3.6.16}$$

Then for $q = k + 1$, the matrix linear model for (3.6.15) becomes

$$\begin{aligned}
\mathbf{Y}_{n\times p} &= \mathbf{X}_{n\times q}\mathbf{B}_{q\times p} + \mathbf{E}_{n\times p} \\
&= \left[\mathbf{X}\boldsymbol{\beta}_1, \mathbf{X}\boldsymbol{\beta}_2, \ldots, \mathbf{X}\boldsymbol{\beta}_p\right] + \left[\mathbf{e}_1, \mathbf{e}_2, \ldots, \mathbf{e}_p\right]
\end{aligned} \tag{3.6.17}$$

Model (3.6.17) is called the multivariate (linear) regression (MR) model , or MLMR model. If the $r(\mathbf{X}) < q = k+1$, so that the design matrix is not of full rank, the model is called the multivariate analysis of variance (MANOVA) model. Partitioning $\mathbf{X}$ into two matrices as in the univariate regression model, $\mathbf{X} = [\mathbf{A}, \mathbf{Z}]$ and $\mathbf{B}' = [\Theta', \Gamma']$, model (3.6.17) becomes the multivariate analysis of covariance (MANCOVA) model.

To represent the MR model as a GLM, the vec $(\cdot)$ operator is employed. Let $\mathbf{y} = \text{vec}(\mathbf{Y})$, $\boldsymbol{\beta} = \text{vec}(\mathbf{B})$ and $\mathbf{e} = \text{vec}(\mathbf{Y})$. Since the design matrix $\mathbf{X}_{n\times q}$ is the same for each of the p dependent variables, the GLM for the MR model is as follows

$$\begin{bmatrix} \mathbf{y}_1 \\ \mathbf{y}_2 \\ \vdots \\ \mathbf{y}_p \end{bmatrix} = \begin{bmatrix} \mathbf{X} & \mathbf{0} & \cdots & \mathbf{0} \\ \mathbf{0} & \mathbf{0} & \cdots & \mathbf{0} \\ \vdots & \vdots & & \vdots \\ \mathbf{0} & \mathbf{0} & \cdots & \mathbf{X} \end{bmatrix} \begin{bmatrix} \boldsymbol{\beta}_1 \\ \boldsymbol{\beta}_2 \\ \vdots \\ \boldsymbol{\beta}_p \end{bmatrix} + \begin{bmatrix} \mathbf{e}_1 \\ \mathbf{e}_2 \\ \vdots \\ \mathbf{e}_p \end{bmatrix}$$

Or, for $N = np$ and $K = pq = p(k+1)$, we have the vector form of the MLMR model

$$\begin{aligned}
\mathbf{y}_{N\times 1} &= \left(\mathbf{I}_p \otimes \mathbf{X}\right)_{N\times K} \boldsymbol{\beta}_{K\times 1} + \mathbf{e}_{K\times 1} \\
\text{cov}\,(\mathbf{y}) &= \Sigma \otimes \mathbf{I}_n
\end{aligned} \tag{3.6.18}$$

To test hypotheses, we assume that $\mathbf{E}$ in (3.6.17) has a matrix normal distribution, $\mathbf{E} \sim N_{n,\ p}(\mathbf{0}, \Sigma \otimes \mathbf{I}_n)$ or using the row representation that $\mathbf{E}' \sim N_{n,\ p}(\mathbf{0}, \mathbf{I}_n \otimes \Sigma)$. Alternatively, by (3.6.18), $\mathbf{e} \sim N_{np}(\mathbf{0}, \Sigma \otimes \mathbf{I}_n)$. To obtain the ML estimate of $\boldsymbol{\beta}$ given (3.6.18), we associate the covariance structure $\Sigma$with $\Sigma \otimes \mathbf{I}_n$ and apply (3.6.8), even though Σ is unknown. The unknown matrix drops out of the product. To see this, we have by substitution that

$$\begin{aligned}
\widehat{\boldsymbol{\beta}}_{ML} &= \left[\left(\mathbf{I}_p \otimes \mathbf{X}\right)' (\Sigma \otimes \mathbf{I}_n)^{-1} \left(\mathbf{I}_p \otimes \mathbf{X}\right)\right]^{-1} \left(\mathbf{I}_p \otimes \mathbf{X}\right)' (\Sigma \otimes \mathbf{I}_n)^{-1} \mathbf{y} \\
&= \left(\Sigma^{-1} \otimes \mathbf{X}'\mathbf{X}\right)^{-1} \left(\Sigma^{-1} \otimes \mathbf{X}'\right) \mathbf{y}
\end{aligned} \tag{3.6.19}$$

However, by property (5) in Theorem 2.4.7, we have that

$$\widehat{\boldsymbol{\beta}}_{ML} = \text{vec}\left[\left(\mathbf{X}'\mathbf{X}\right)^{-1}\mathbf{X}'\mathbf{Y}\right]$$

by letting $\mathbf{A} = \left(\mathbf{X}'\mathbf{X}\right)^{-1}\mathbf{X}'$ and $\mathbf{C}' = \mathbf{I}_p$. Thus,

$$\widehat{\mathbf{B}}_{ML} = \left(\mathbf{X}'\mathbf{X}\right)^{-1}\mathbf{X}'\mathbf{Y} \tag{3.6.20}$$

using the matrix form of the model. This is also the OLS estimate of the parameter matrix.

Similarly using (3.6.19), the

$$\begin{aligned}\text{cov}\left(\widehat{\boldsymbol{\beta}}_{ML}\right) &= \left[\left(\mathbf{I}_p \otimes \mathbf{X}\right)'\left(\Sigma \otimes \mathbf{I}_n\right)^{-1}\left(\mathbf{I}_p \otimes \mathbf{X}\right)\right]^{-1} \\ &= \Sigma \otimes \left(\mathbf{X}'\mathbf{X}\right)^{-1}\end{aligned}$$

Finally, the ML estimate of Σ is

$$\widehat{\Sigma} = \mathbf{Y}'\left[\mathbf{I}_n - \mathbf{X}\left(\mathbf{X}'\mathbf{X}\right)^{-1}\mathbf{X}'\right]\mathbf{Y}/n \tag{3.6.21}$$

or the restricted maximum likelihood (REML) unbiased estimate is $\mathbf{S} = \mathbf{E}/(n-q)$ where $q = r(\mathbf{X})$. Furthermore $\widehat{\boldsymbol{\beta}}_{ML}$ and $\widehat{\Sigma}$ are independent, and $n\ \widehat{\Sigma} \sim W_p(n-q,\ \Sigma)$. Again, the Wishart density only exists if $n \geq p+q$.

In the above discussion, we have assumed that $\mathbf{X}$ has full column rank q. If the $r(\mathbf{X}) = r < q$, then $\widehat{\mathbf{B}}$ is not unique since $(\mathbf{X}'\mathbf{X})^{-1}$ is replaced with a g-inverse. However, $\widehat{\Sigma}$ is still unique since $\left(\mathbf{I}_n - \mathbf{X}(\mathbf{X}'\mathbf{X})^{-}\mathbf{X}'\right)$ is a unique projection matrix by property (4), Theorem 2.5.5. The lack of a unique inverse only affects which linear parametric functions of the parameters are estimable and hence testable. Theorem 2.7.2 is again used to determine the parametric functions in $\boldsymbol{\beta} = \text{vec}(\mathbf{B})$ that are estimable.

The null hypothesis tested for the matrix form of the MR model takes the general form

$$H : \mathbf{CBM} = \mathbf{0} \tag{3.6.22}$$

where $\mathbf{C}_{g\times q}$ is a known matrix of full row rank g, $g \leq q$ and $\mathbf{M}_{p\times u}$ is a matrix of full column rank $u \leq p$. Hypothesis (3.6.22) is called the standard multivariate hypothesis. To test (3.6.22) using the vector form of the model, observe that $\text{vec}(\mathbf{C}\widehat{\mathbf{B}}\mathbf{M}) = (\mathbf{M}' \otimes \mathbf{C})\,\text{vec}\,\widehat{\mathbf{B}}$ so that (3.6.22) is equivalent to testing $H : \mathbf{L}\boldsymbol{\beta} = \mathbf{0}$ when $\mathbf{L}$ is a matrix of order $gu \times pq$ of rank $v = gu$. Assuming $\Omega = \Sigma \otimes \mathbf{I}_n$ is known,

$$\widehat{\boldsymbol{\beta}}_{ML} \sim N_{gu}(\boldsymbol{\beta}, \mathbf{L}[(\mathbf{I}_n \otimes \mathbf{X})'\Omega^{-1}(\mathbf{I}_p \otimes \mathbf{X})]^{-1}\mathbf{L}') \tag{3.6.23}$$

Simplifying the structure of the covariance matrix,

$$\text{cov}\left(\widehat{\boldsymbol{\beta}}_{ML}\right) = \left(\mathbf{M}'\Sigma\mathbf{M}\right) \otimes \left(\mathbf{C}\left(\mathbf{X}'\mathbf{X}\right)^{-1}\mathbf{C}'\right) \tag{3.6.24}$$

For known Σ, the likelihood ratio test of H is to reject H if $X^2 > c_\alpha$ where c_α is chosen such that the $P(X^2 > c_\alpha \mid H) = \alpha$ and $X^2 = \widehat{\boldsymbol{\beta}}'_{ML}[(\mathbf{M}'\Sigma\mathbf{M}) \otimes (\mathbf{C}(\mathbf{X}'\mathbf{X})^{-1}\mathbf{C}')]^{-1}\widehat{\boldsymbol{\beta}}_{ML}$.

However, we can simplify X^2 since

$$
\begin{aligned}
X^2 &= [\text{vec}(\mathbf{C}\widehat{\mathbf{B}}\mathbf{M})]'\left[\left(\mathbf{M}'\Sigma\mathbf{M}\right)^{-1} \otimes (\mathbf{C}\left(\mathbf{X}'\mathbf{X}\right)^{-1})\right] \text{vec}(\mathbf{C}\widehat{\mathbf{B}}\mathbf{M}) \\
&= [\text{vec}(\mathbf{C}\widehat{\mathbf{B}}\mathbf{M})]' \,\text{vec}[(\mathbf{C}\left(\mathbf{X}'\mathbf{X}\right)^{-1} \mathbf{C}']^{-1}(\mathbf{C}\widehat{\mathbf{B}}\mathbf{M}) \left(\mathbf{M}'\Sigma\mathbf{M}\right)^{-1} \\
&= \text{tr}[(\mathbf{C}\widehat{\mathbf{B}}\mathbf{M})'[\mathbf{C}\left(\mathbf{X}'\mathbf{X}\right)^{-1} \mathbf{C}']^{-1}(\mathbf{C}\widehat{\mathbf{B}}\mathbf{A}) \left(\mathbf{M}'\Sigma\mathbf{M}\right)^{-1}]
\end{aligned}
\tag{3.6.25}
$$

Thus to test $H : \mathbf{L}\boldsymbol{\beta} = \mathbf{0}$, the hypothesis is rejected if X^2 in (3.6.25) is larger than a chi-square critical value with $v = gu$ degrees of freedom. Again, by finding a consistent estimate of Σ, $X^2 \xrightarrow{d} \chi^2(v = gu)$. Thus an approximate test of H is available if Σ is estimated by $\widehat{\Sigma} \xrightarrow{p} \Sigma$.

However, one does not use the approximate chi-square test when Σ is unknown since an exact likelihood ratio test exists for $H : \mathbf{L}\boldsymbol{\beta} = \mathbf{0} \Longleftrightarrow \mathbf{CBM} = \mathbf{0}$. The hypothesis and error SSCP matrices under the MR model are

$$
\begin{aligned}
\mathbf{H} &= (\mathbf{C}\widehat{\mathbf{B}}\mathbf{M})'[\mathbf{C}(\mathbf{X}'\mathbf{X})^{-1}\mathbf{C}']^{-1}(\mathbf{C}\widehat{\mathbf{B}}\mathbf{M}) \\
\mathbf{E} &= \mathbf{M}'\mathbf{Y}'[\mathbf{I}_n - \mathbf{X}(\mathbf{X}'\mathbf{X})^{-1}\mathbf{X}']\mathbf{Y}\mathbf{M} \\
&= (n-q)\mathbf{M}'\mathbf{S}\mathbf{M}
\end{aligned}
\tag{3.6.26}
$$

Using Theorem 3.4.5, it is easily established that $\mathbf{E}$ and $\mathbf{H}$ have independent Wishart distributions

$$
\begin{aligned}
\mathbf{E} &\sim W_u(v_e = n - q, \mathbf{M}'\Sigma\mathbf{M}, \Gamma = \mathbf{0}) \\
\mathbf{H} &\sim W_u\left(v_h = g, \mathbf{M}'\Sigma\mathbf{M}, \left(\mathbf{M}'\Sigma\mathbf{M}\right)^{-1} \Delta = \Gamma\right)
\end{aligned}
$$

where the noncentrality parameter matrix Γ is

$$
\Gamma = (\mathbf{M}'\Sigma\mathbf{M})^{-1} (\mathbf{CBM})' (\mathbf{C}(\mathbf{X}'\mathbf{X})^{-1}\mathbf{C}')^{-1}(\mathbf{CBM}) \tag{3.6.27}
$$

To test $\mathbf{CBM} = \mathbf{0}$, one needs the joint density of the roots of $\mathbf{HE}^{-1}$ which is extremely complicated; see Muirhead (1982, p. 449). In applied problems, the approximations summarized in Theorem 3.5.1 are adequate for any of the four criteria. Exact critical values are required for establishing exact $100\,(1-\alpha)$ simultaneous confidence intervals.

For $\mathbf{H}$ and $\mathbf{E}$ defined in (3.6.26) and $v_e = n - q$, an alternative expression for T_o^2 defined in (3.5.4) is

$$
T_o^2 = \text{tr}[(\mathbf{C}\widehat{\mathbf{B}}\mathbf{M})[\mathbf{C}\left(\mathbf{X}'\mathbf{X}\right)^{-1} \mathbf{C}']^{-1}(\mathbf{C}\widehat{\mathbf{B}}\mathbf{M}) \left(\mathbf{M}'\mathbf{E}\mathbf{M}\right)^{-1} \tag{3.6.28}
$$

Comparing T_o^2 with X^2 in (3.6.25), we see that for known Σ, T_o^2 has a chi-square distribution. Hence, for $v_e\mathbf{E}/n = \widehat{\Sigma}$, T_o^2 has an asymptotic chi-square distribution.

In our development of the multivariate linear model, to test hypotheses of form H_o : $\mathbf{CBM} = \mathbf{0}$, we have assumed that the covariance matrix for the $\mathbf{y}^* = \text{vec}(\mathbf{Y}'_{n\times p})$ has covariance structure $\mathbf{I}_n \otimes \Sigma_p$ so the rows of $\mathbf{Y}_{n\times p}$ are independent identically normally distributed with common covariance matrix Σ_p. This structure is a sufficient, but not necessary condition for the development of exact tests. Young et al. (1999) have developed necessary and sufficient conditions for the matrix $\mathbf{W}$ in the expression cov $(\mathbf{y}^*) = \mathbf{W}_n \otimes \Sigma_p = \Omega$ for tests to remain exact. They refer to such structures of Ω as being independence distribution-preserving (IDP).

Any of the four criteria may be used to establish simultaneous confidence intervals for parametric functions of the form $\psi = \mathbf{a}'\mathbf{B}\mathbf{m}$. Details will be illustrated in Chapter 4 when we discuss applications using the four multivariate criteria, approximate single degree of freedom F tests for C planned comparisons, and stepdown finite intersection tests. More general extended linear hypotheses will also be reviewed. We conclude this section with further generalizations of the GLM also discussed in more detail with illustrations in later chapters.

c. *The Seemingly Unrelated Regression (SUR) Model*

In developing the MR model in (3.6.15), observe that the jth equation for $j = 1, \ldots, p$, has the GLM form $\mathbf{y}_j = \mathbf{X}\boldsymbol{\beta}_j + \mathbf{e}_j$ where $\boldsymbol{\beta}'_j = \left[\beta_{1j}, \ldots, \beta_{kj}\right]$. The covariance structure of the errors $\mathbf{e}_j$ is $\text{cov}(\mathbf{e}_j) = \sigma_{jj}\mathbf{I}_n$ so that each $\boldsymbol{\beta}_j$ can be estimated independently of the others. The dependence is incorporated into the model by the relationship $\text{cov}(\mathbf{y}_i,\ \mathbf{y}_j) = \sigma_{ij}\mathbf{I}_n$. Because the design matrix is the same for each variable and $\mathbf{B} = [\boldsymbol{\beta}_1, \boldsymbol{\beta}_2, \ldots, \boldsymbol{\beta}_p]$ has a simple column form, each $\boldsymbol{\beta}_j$ may be estimated independently as $\widehat{\boldsymbol{\beta}}_j = \left(\mathbf{X}'\mathbf{X}\right)^{-1}\mathbf{X}'\mathbf{y}_j$ for $j = 1, \ldots, p$. The cov $\left(\widehat{\boldsymbol{\beta}}_i, \widehat{\boldsymbol{\beta}}_j\right) = \sigma_{ij}\left(\mathbf{X}'\mathbf{X}\right)^{-1}$.

A simple generalization of (3.6.17) is to replace $\mathbf{X}$ with $\mathbf{X}_j$ so that the regression model (design matrix) may be different for each variable

$$\begin{aligned} E(\mathbf{Y}_{n\times p}) &= [\mathbf{X}_1\boldsymbol{\beta}_1, \mathbf{X}_2\boldsymbol{\beta}_2, \ldots, \mathbf{X}_p\boldsymbol{\beta}_p] \\ \text{cov}[\text{vec}(\mathbf{Y}')] &= \mathbf{I}_n \otimes \Sigma \end{aligned} \tag{3.6.29}$$

Such a model may often be more appropriate since it allows one to fit different models for each variable. When fitting the same model to each variable using the MR model, some variables may be overfit. Model (3.6.29) is called S.N. Srivastava's multiple design multivariate (MDM) model or Zellner's seemingly unrelated regression (SUR) model. The SUR model is usually written as p correlated regression models

$$\begin{aligned} \underset{(n\times 1)}{\mathbf{y}_j} &= \underset{(n\times q_j)}{\mathbf{X}} \ \underset{(q_j\times 1)}{\boldsymbol{\beta}_j} + \underset{(n\times 1)}{\mathbf{e}_j} \\ &\text{cov}\left(\mathbf{y}_i,\ \mathbf{y}_j\right) = \sigma_{ij}\mathbf{I}_n \end{aligned} \tag{3.6.30}$$

for $j = 1, 2, \ldots, p$. Letting $\mathbf{y}' = [\mathbf{y}'_1, \mathbf{y}'_2, \ldots, \mathbf{y}'_p]$ with $\boldsymbol{\beta}'$and $\mathbf{e}'$partitioned similarly and the design matrix defined by $\mathbf{X} = \bigoplus_{j=1}^{p}\mathbf{X}_j$, with $N = np$, $K = \sum_j q_j = \sum_j (k_j + 1)$ and the $r(X_j) = q_j$, model (3.6.30) is again seen to be special case of the GLM (3.6.1). Alternatively, letting

$$\begin{aligned} \mathbf{Y} &= [\mathbf{y}_1, \mathbf{y}_2, \ldots, \mathbf{y}_p] \\ \mathbf{X} &= [\mathbf{x}_1, \mathbf{x}_2, \ldots, \mathbf{x}_q] \end{aligned}$$

$$\widetilde{\mathbf{B}} = \begin{bmatrix} \boldsymbol{\beta}_{11} & \mathbf{0} & \cdots & \mathbf{0} \\ \mathbf{0} & \boldsymbol{\beta}_{22} & \cdots & \mathbf{0} \\ \vdots & \vdots & & \vdots \\ \mathbf{0} & \mathbf{0} & \cdots & \boldsymbol{\beta}_{pp} \end{bmatrix}$$

where $\boldsymbol{\beta}_j{}' = [\beta_{0j}, \beta_{1j}, \ldots, \beta_{kj}]$, the SUR model may be written as

$$\mathbf{Y}_{n \times p} = \mathbf{X}_{n \times q} \widetilde{\mathbf{B}}_{q \times p} + \mathbf{E}_{n \times p}$$

which is a MR model with restrictions.

The matrix version of (3.6.30) is called the multivariate seemingly unrelated regression (MSUR) model. The model is constructed by replacing $\mathbf{y}_j$ and $\boldsymbol{\beta}_j$ in (3.6.30) with matrices. The MSUR model is called the correlated multivariate regression equations (CMRE) model by Kariya et al. (1984). We review the MSUR model in Chapter 5.

d. *The General MANOVA Model (GMANOVA)*

Potthoff and Roy (1964) extended the MR and MANOVA models to the growth curve model (GMANOVA). The model was first introduced to analyze growth in repeated measures data that have the same number of observations per subject with complete data. The model has the general form

$$\begin{aligned} \mathbf{Y}_{n \times p} &= \mathbf{A}_{n \times q} \mathbf{B}_{q \times k} \mathbf{Z}_{k \times p} + \mathbf{E}_{n \times p} \\ \text{vec}\,(\mathbf{E}) &\sim N_{np}\,(\mathbf{0}, \Sigma \otimes \mathbf{I}_n) \end{aligned} \tag{3.6.31}$$

The matrices $\mathbf{A}$ and $\mathbf{Z}$ are assumed known with $n \geq p$ and $k \leq p$. Letting the $r(\mathbf{A}) = q$ and the $r(\mathbf{Z}) = p$, (3.6.31) is again a special case of (3.6.1) if we define $\mathbf{X} = \mathbf{A} \otimes \mathbf{Z}'$. Partitioning $\mathbf{Y}$, $\mathbf{B}$ and $\mathbf{E}$ rowwise,

$$\mathbf{Y} = \begin{bmatrix} \mathbf{y}_1' \\ \mathbf{y}_2' \\ \vdots \\ \mathbf{y}_n' \end{bmatrix}_{n \times p}, \quad \mathbf{B} = \begin{bmatrix} \boldsymbol{\beta}_1' \\ \boldsymbol{\beta}_2' \\ \vdots \\ \boldsymbol{\beta}_q' \end{bmatrix}_{q \times k}, \quad \text{and} \quad \mathbf{E} = \begin{bmatrix} \mathbf{e}_1' \\ \mathbf{e}_2' \\ \vdots \\ \mathbf{e}_n' \end{bmatrix}$$

so that (3.6.31) is equivalent to the GLM

$$\begin{aligned} \mathbf{y}^* = \text{vec}\left(\mathbf{Y}'\right) &= \left(\mathbf{A} \otimes \mathbf{Z}'\right) \text{vec}\left(\mathbf{B}'\right) + \text{vec}\left(\mathbf{E}'\right) \\ \text{cov}\left(\mathbf{Y}'\right) &= \mathbf{I} \otimes \Sigma \end{aligned} \tag{3.6.32}$$

A further generalization of (3.6.31) was introduced by Chinchilli and Elswick (1985) and Srivastava and Carter (1983). The model is called the MANOVA-GMANOVA model and has the following structure

$$\mathbf{Y} = \mathbf{X}_1 \mathbf{B}_1 \mathbf{Z}_1 + \mathbf{X}_2 \mathbf{B}_2 + \mathbf{E} \tag{3.6.33}$$

where the GMANOVA component contains growth curves and the MR or MANOVA component contains covariates associated with baseline data. Chinchilli and Elswick (1985) provide ML estimates and likelihood ratio tests for the model.

Patel (1983, 1986) and von Rosen (1989, 1990, 1993) consider the more general growth curve (MGGC) model, also called the sum-of-profiles model,

$$\mathbf{Y} = \sum_{i=1}^{r} \mathbf{X}_i \mathbf{B}_i \mathbf{Z}_i + \mathbf{E} \tag{3.6.34}$$

by Verbyla and Venables (1988a, b). Using two restrictions on the design matrices

$$r(\mathbf{X}_1) + p \leq n \text{ and } \mathbf{X}_r'\mathbf{X}_r \subseteq \mathbf{X}_{r-1}'\mathbf{X}_{r-1} \subseteq \cdots \subseteq \mathbf{X}_1'\mathbf{X}_1$$

von Rosen was able to obtain closed-form expressions for ML estimates of all model parameters. He did not obtain likelihood ratio tests of hypotheses. A canonical form of the model was also considered by Gleser and Olkin (1970). Srivastava and Khatri (1979, p. 197) expressed the sum-of-profiles model as a nested growth model. They developed their model by nesting the matrices $\mathbf{Z}_i$ in (3.6.34). Details are available in Srivastava (1997).

Without imposing the nested condition on the design matrices, Verbyla and Venables (1988b) obtained generalized least squares estimates of the model parameters for the MGGC model using the MSUR model. Unique estimates are obtained if the

$$r\left[\left(\mathbf{X}_1 \otimes \mathbf{Z}_1'\right), \left(\mathbf{X}_2 \otimes \mathbf{Z}_2'\right), \ldots, \left(\mathbf{X}_r \otimes \mathbf{Z}_2'\right)\right] = q$$

To see this, one merely has to write the MGGC model as a SUR model

$$\text{vec}\left(\mathbf{Y}'\right) = \left[\left(\mathbf{X}_1 \otimes \mathbf{Z}_1'\right), \left(\mathbf{X}_2 \otimes \mathbf{Z}_2'\right), \ldots, \left(\mathbf{X}_r \otimes \mathbf{Z}_r'\right)\right] \begin{bmatrix} \text{vec}\left(\mathbf{B}_1'\right) \\ \text{vec}\left(\mathbf{B}_2'\right) \\ \vdots \\ \text{vec}\left(\mathbf{B}_r'\right) \end{bmatrix} + \text{vec}\left(\mathbf{E}'\right) \tag{3.6.35}$$

Hecker (1987) called this model the completely general MANOVA (CGMANOVA) model. Thus, the GMANOVA and CGMANOVA models are SUR models.

One may add restrictions to the MR, MANOVA, GMANOVA, CGMANOVA, and their extensions. Such models belong to the class of restricted multivariate linear models; see Kariya (1985). In addition, the elements of the parameter matrix may be only random, or mixed containing both fixed and random parameters. This leads to multivariate random effects and multivariate mixed effects models, Khuri et al. (1998). Amemiya (1994) and Thum (1997) consider a general multivariate mixed effect repeated measures model.

To construct a multivariate (linear) mixed model (MMM) from the MGLM, the matrix $\mathbf{E}$ of random errors is modeled. That is, the matrix $\mathbf{E} = \mathbf{Z}\mathbf{U}$ where $\mathbf{Z}$ is a known nonrandom matrix and $\mathbf{U}$ is a matrix of random effects. Hence, the MMM has the general structure

$$\underset{n\times r}{\mathbf{Y}} = \underset{n\times q}{\mathbf{X}} \; \underset{q\times r}{\mathbf{B}} + \underset{n\times h}{\mathbf{Z}} \; \underset{h\times r}{\mathbf{U}} \tag{3.6.36}$$

where $\mathbf{B}$ is the matrix of fixed effects. When $\mathbf{XB}$ does not exist in the model, the model is called the random coefficient model or a random coefficient regression model. The data matrix $\mathbf{Y}$ in (3.6.36) is of order $(n \times r)$ where the rows of $\mathbf{Y}$ are a random sample of n

observations on r responses. The subscript r is used since the r responses may be a vector of p-variables over t occasions (time) so that $r = pt$.

Because model (3.6.36) contains both fixed and random effects, we may separate the model into its random and fixed components as follows

$$\begin{aligned} \mathbf{XB} &= \sum_i \mathbf{K}_i \mathbf{B}_i \\ \mathbf{ZU} &= \sum_j \mathbf{K}_j \mathbf{U}_j \end{aligned} \tag{3.6.37}$$

The matrices $\mathbf{K}_i$ and $\mathbf{K}_j$ are known and of order $(n \times r_i)$ of rank r_i; the matrices $\mathbf{B}_i$ of order $(r_i \times r)$ contain the fixed effects, while the matrices $\mathbf{U}_j$ of rank r_j contain the random effects. The rows of the matrices $\mathbf{U}_j$ are assumed to be independent MVN as $N\left(\mathbf{0}, \Sigma_j\right)$. Writing the model using the rows of $\mathbf{Y}$ or the columns of $\mathbf{Y}'$, $\mathbf{y}^* = \text{cov}\left(\mathbf{Y}'\right)$,

$$\text{cov}\left(\mathbf{y}^*\right) = \sum_j \left(\mathbf{V}_j \otimes \Sigma_j\right) \tag{3.6.38}$$

Model (3.6.36) with structure (3.6.38) is discussed in Chapter 6. There we will review random coefficient models and mixed models. Models with $r = pt$ are of special interest. These models with multiple-response, repeated measures are a p-variate generalization of Scheffé's mixed model, also called double multivariate linear models and treated in some detail by Reinsel (1982, 1984) and Boik (1988, 1991). Khuri et al. (1998) provide an overview of the statistical theory for univariate and multivariate mixed models.

Amemiya (1994) provides a generalization of the model considered by Reinsel and Boik, which permits incomplete data over occasions. The matrix version of Amemiya's general multivariate mixed model is

$$\begin{aligned} \underset{n_i \times p}{\mathbf{Y}_i} &= \underset{n_i \times k}{\mathbf{X}_i} \ \underset{k \times p}{\mathbf{B}} + \underset{n_i \times h}{\mathbf{Z}_i} \ \underset{h \times p}{\mathbf{A}_i} + \underset{n_i \times p}{\mathbf{E}_i} \\ \text{cov}\left(\text{vec}\, \mathbf{Y}_i'\right) &= \left(\mathbf{Z}_i \otimes \mathbf{I}_p\right) \Phi \left(\mathbf{Z}_i \otimes \mathbf{I}_p\right)' + \mathbf{I}_{n_i} \otimes \Sigma_e \end{aligned} \tag{3.6.39}$$

for $i = 1, 2, \ldots, n$ and where $\Phi_{hp \times hp} = \text{cov}\left(\text{vec}\, \mathbf{A}_i'\right)$ and Σ_e is the $p \times p$ covariance matrix of the ith row of $\mathbf{E}_i$. This model is also considered by Thum (1997) and is reviewed in Chapter 6

Exercises 3.6

1. Verify that $\widehat{\boldsymbol{\beta}}$ in (3.6.5) minimizes the error sum of squares

$$\sum_{i=1}^n e_i^2 = (\mathbf{y} - \mathbf{X}\widehat{\boldsymbol{\beta}})'(\mathbf{y} - \mathbf{X}\widehat{\boldsymbol{\beta}})$$

   using projection operators.

2. Prove that H and E in (3.6.6) are independent.

3. Verify that $\widehat{\boldsymbol{\beta}}$ in (3.6.8) minimizes the weighted error sum of squares

$$\sum_{i=1}^{n} e_i^2 = (\mathbf{y} - \mathbf{X}\widehat{\boldsymbol{\beta}})'\Sigma^{-1}(\mathbf{y} - \mathbf{X}\widehat{\boldsymbol{\beta}})$$

4. Prove that **H** and **E** under the MR model are independently distributed.
5. Obtain the result given in (3.6.23).
6. Represent (3.6.39) as a GLM.

3.7 Evaluating Normality

Fundamental to parameter estimation and tests of significance for the models considered in this text is the assumption of multivariate normality. Whenever parameters are estimated, we would like them to have optimal properties and to be insensitive to mild departures from normality, i.e., to be robust to non-normality, and from the effects of outliers. Tests of significance are said to be robust if the size of the test α and the power of the test are only marginally effected by departures from model assumptions such as normality and restrictions placed on the structure of covariance matrices when sampling from one or more populations.

The study of robust estimation for location and dispersion of model parameters, the identification of outliers, the analysis of multivariate residuals, and the assessment of the effects of model assumptions on tests of significance and power are as important in multivariate analysis as they are in univariate analysis. However, the problems are much more complex. In multivariate data analysis there is no natural one-dimensional order to the observations, hence we can no longer just investigate the extremes of the distribution to locate outliers or identify data clusters in only one dimension. Clusters can occur in some subspace and outliers may not be extreme in any one dimension. Outliers in multivariate samples effect not only the location and variance of a variable, but also its orientation in the sample as measured by the covariance or correlation with other variables. Residuals formed from fitting a multivariate model to a data set in the presence of extreme outliers may lead to the identification of spurious outliers. Upon replotting the data, they are often removed. Finally, because non-normality can occur in so many ways robustness studies of Type I errors and power are difficult to design and evaluate.

The two most important problems in multivariate data analysis are the detection of outliers and the evaluation of multivariate normality. The process is complex and first begins with the assessment of marginal normality, a variable at a time; see Looney (1995). The evaluation process usually proceeds as follows.

1. Evaluate univariate normality by performing the Shapiro and Wilk (1965) W test a variable at a time when sample sizes are less than or equal to 50. The test is known to show a reasonable sensitivity to nonnormality; see Shapiro et al. (1968). For $50 < n \leq 2000$, Royston's (1982, 1992) approximation is recommended and is implemented in the SAS procedure UNIVARIATE; see SAS Institute (1990, p. 627).

2. Construct normal probability quantile-vs-quantile (Q-Q) plots a variable at a time which compare the cumulative empirical distribution with the expected order values of a normal density to informally assess the lack of linearity and the presence of extreme values; see Wilk and Gnanadesikan (1968) and Looney and Gulledge (1985).

3. If variables are found to be non-normal, transform them to normality using perhaps a Box and Cox (1964) power transformation or some other transformation such as a logit.

4. Locate and correct outliers using graphical techniques or tests of significance as outlined by Barnett and Lewis (1994).

The goals of steps (1) to (4) are to evaluate marginal normality and to detect outliers. If $r + s$ outliers are identified for variable i, two robust estimators of location, trimmed and Winsorized means, may be calculated as

$$\begin{aligned} y_T^\star(r, s) &= \sum_{i=r+1}^{n-s} y_i / (n - r - s) \\ y_W^\star(r, s) &= \left\{ r y_{r+1} + \sum_{i=r+1}^{n-s} y_i + s y_{n-s} \right\} / n \end{aligned} \tag{3.7.1}$$

respectively, for a sample of size n. If the proportion of observations at each extreme are equal, $r = s$, the estimate $y_W^\star$ is called an α-Winsorized mean. To create an α-trimmed mean, a proportion α of the ordered sample $y_{(i)}$ from the lower and upper extremes of the distribution is discarded. Since the proportion may not be an integer value, we let $\alpha n = r + \mathrm{w}$ where r is an integer and $0 < \mathrm{w} < 1$. Then,

$$y_{T(\alpha)}^\star(r, r) = \left\{ (1 - \mathrm{w}) y_{r+1} + \sum_{i=r+2}^{n-r-1} y_i + (1 - \mathrm{w}) y_{n-r} \right\} / n(1 - 2\alpha) \tag{3.7.2}$$

is an α-trimmed mean; see Gnanadesikan and Kettenring (1992). If r is an integer, then the r-trimmed or α-trimmed mean for $\alpha = r/n$ reduces to formula (3.7.1) with $r = s$ so that

$$y_{T(\alpha)}^\star(r, r) = \sum_{i=r+1}^{n-r} y_i / (n - 2r) \tag{3.7.3}$$

In multivariate analysis, Winsorized data ensures that the number of observations for each of the p variables remains constant over the n observations. Trimmed observations cause complicated missing value problems when not applied to all variables simultaneously. In univariate analysis, trimmed means are often preferred to Winsorized means. Both are special cases of an L-estimator which is any linear combination of the ordered sample. Another class of robust estimators are M-estimators. Huber (1981) provides a comprehensive discussion of such estimators, the M stands for maximum likelihood.

Using some robust estimate of location m^*, a robust estimate of the sample variance (scale) parameter σ^2 is defined as

$$\widetilde{s}_{ii}^2 = \sum_{i=1}^{k} (y_i - m^*)^2 / (k-1) \tag{3.7.4}$$

where $k \leq n$, depending on the "trimming" process. In obtaining an estimate for σ^2, we see an obvious conflict between protecting the estimate from outliers versus using the data in the tails to increase precision. Calculating a trimmed variance from an α-trimmed sample or a Winsorized-trimmed variance from a α-Winsorized sample leads to estimates that are not unbiased, and hence correction factors are required based on the moments of order statistics. However, tables of coefficients are only available for $n \leq 15$ and $r = s$.

To reduce the bias and improve consistency, the Winsorized-trimmed variance suggested by Huber (1970) may be used for an α-trimmed sample. For $\alpha = r/n$ and $r = s$

$$\begin{aligned} \widetilde{s}_{ii}^2\,(H) = \Bigg\{ & (r+1)\left(y_{r+1} - y_{T(\alpha)}^{\star}\right)^2 + \sum_{i=r+2}^{n-r-1} \left(y_i - y_{T(\alpha)}^{\star}\right)^2 \\ & + (r+1)\left(y_{n-r} - y_{T(\alpha)}^{\star}\right)^2 \Bigg\} / [n - 2r - 1] \end{aligned} \tag{3.7.5}$$

which reduces to s^2 if $r = 0$. The numerator in (3.7.5) is a Winsorized sum of squares. The denominator is based on the trimmed mean value $t = n - 2r$ observations and not n which would have treated the Winsorized values as "observed." Alternatively, we may write (3.7.5) as

$$\widetilde{s}_{ii}^2 = \textstyle\sum_{k=1}^{n} \left(\widetilde{y}_{ik} - y_{ik}^{\star}\right)^2 / (n - 2r_i - 1) \qquad i = 1, 2, \ldots, p \tag{3.7.6}$$

where $y_{ik}^{\star}$ is an α-trimmed mean and $\widetilde{y}_{ik}$ is either an observed sample value or a Winsorized value that depends on α for each variable. Thus, the trimming value r_i may be different for each variable.

To estimate the covariance between variables i and j, we may employ the Winsorized sample covariance suggested by Mudholkar and Srivastava (1996). A robust covariance estimate is

$$\widetilde{s}_{ij} = \sum_{k=1}^{n} \left(\widetilde{y}_{ik} - y_{ik}^{\star}\right)\left(\widetilde{y}_{jk} - y_{jk}^{\star}\right) / (n - 2\overline{r} - 1) \tag{3.7.7}$$

for all pairs $i, j = 1, 2, \ldots, p$. The average $\overline{r} = (r_1 + r_2)/2$ is the average number of Winsorized observations in each pairing. The robust estimate of the covariance matrix is

$$\mathbf{S}_{\mathrm{w}}^{\star} = \left[\widetilde{s}_{ij}\right]$$

Depending on the amount of "Winsorizing," the matrix $\mathbf{S}^{\star}$ may not be positive definite. To correct this problem, the covariance matrix is smoothed by solving $\mid \mathbf{S}^{\star} = \lambda \mathbf{I} \mid = 0$. Letting

$$\widehat{\Sigma}^{\star} = \mathbf{P} \Lambda^{\star} \mathbf{P}' \tag{3.7.8}$$

where $\Lambda^\star$ contains only the positive roots of $\mathbf{S}^\star$ and $\mathbf{P}$ is the matrix of eigenvectors; $\widehat{\Sigma}^\star$ is positive definite, Bock and Peterson (1975). Other procedures for finding robust estimates of Σ are examined by Devlin et al. (1975). They use a method of "shrinking" to obtain a positive definite estimate for Σ.

The goals of steps (1) to (4) are to achieve marginal normality in the data. Because marginal normality does not imply multivariate normality, one next analyzes the data for multivariate normality and multivariate outliers. Sometimes the evaluation of multivariate normality is done without investigating univariate normality since a MVN distribution ensures marginal normality.

Romeu and Ozturk (1993) investigated ten tests of goodness-of-fit for multivariate normality. Their simulation study shows that the multivariate tests of skewness and kurtosis proposed by Mardia (1970, 1980) are the most stable and reliable tests for assessing multivariate normality.

Estimating skewness by

$$\widehat{\beta}_{1,\,p} = \sum_{i=1}^{n}\sum_{j=1}^{n}\left[(\mathbf{y}_i - \overline{\mathbf{y}})' \mathbf{S}^{-1} (\mathbf{y}_j - \overline{\mathbf{y}})\right]^3 / n^2 \tag{3.7.9}$$

Mardia showed that the statistic $X^2 = n\widehat{\beta}_{1,\,p}/6 \xrightarrow{d} \chi^2(v)$ where $v = p\,(p+1)\,(p+2)/6$. He also showed that the sample estimate of multivariate kurtosis

$$\widehat{\beta}_{2,\,p} = \sum_{i=1}^{n}\left[(\mathbf{y}_i - \overline{\mathbf{y}})' \mathbf{S}^{-1} (\mathbf{y}_i - \overline{\mathbf{y}})\right]^2 / n \tag{3.7.10}$$

converges in distribution to a $N(\mu, \sigma^2)$ distribution with mean $\mu = p(p+2)$ and variance $\sigma^2 = 8p(p+2)/n$. Thus, subtracting μ from $\widehat{\beta}_{2,\,p}$ and then dividing by σ, $Z = \left(\widehat{\beta}_{2,\,p} - \mu\right)/\sigma \xrightarrow{d} N\,(0,\ 1)$. Rejection of normality using Mardia's tests indicates either the presences of multivariate outliers or that the distribution is significantly different from a MVN distribution. If we fail to reject, the distribution is assumed to be MVN. Small sample empirical critical values for the skewness and kurtosis tests were calculated by Romeu and Ozturk and are provided in Appendix A, Table VI-VIII. If the multivariate tests are rejected, we have to either identify multivariate outliers and/or transform the vector sample data to achieve multivariate normality. While Andrews et al., (1971) have developed a multivariate extension of the Box-Cox power transformation, determining the appropriate transformation is complicated; see Chambers (1977), Velilla and Barrio (1994), and Bilodeau and Brenner (1999, p. 95). An alternative procedure is to perform a data reduction transformation and to analyze the sample using some subset of linear combinations of the original variables such as principal components, discussed in Chapter 8, which may be more nearly normal. Another option is to identify directions of possible nonnormality and then to estimate univariate Box-Cox power transformations of projections of the original variables onto a set of direction vectors to improve multivariate normality; see Gnanadesikan (1997).

Graphical displays of the data are needed to visually identify multivariate outliers in a data set. Seber (1984) provides an overview of multivariate graphical techniques. Many of the procedures are illustrated in Venables and Ripley (1994) using S-plus. SAS/INSIGHT

(1993) provides a comprehensive set of graphical displays for interacting with multivariate data. Following any SAS application on the PC, one may invoke SAS/INSIGHT by using the Tool Bar: and clicking on the option "Solutions." From the new pop-up menu, one selects the option "analysis" from this menu and finally from the last menu one selects the option "Interactive Data Analysis." Clicking on this last option opens the interactive mode of SAS/INSIGHT. The WORK library contains data sets created by the SAS application. By clicking on the WORK library, the names of the data sets created in the SAS procedure are displayed in the window. By clicking on a specific data set, one may display the data created in the application. To analyze the data displayed interactively, one selects from the Tool Bar the option "Analyze." This is illustrated more fully to locate potential outliers in a multivariate data set using plotted displays in Example 3.7.3. Friendly (1991), using SAS procedures and SAS macros, has developed numerous graphs for plotting multivariate data. Other procedures are illustrated in Khattree and Naik (1995) and Timm and Mieczkowski (1997). Residual plots are examined in Chapter 4 when the MR model is discussed. Robustness of multivariate tests is also discussed in Chapter 4. We next discuss the generation of a multivariate normal distribution and review multivariate Q-Q plots to help identify departures from multivariate normality and outliers.

To visually evaluate whether a multivariate distribution has outliers, recall from Theorem 3.4.2 that if $\mathbf{Y}_i \sim N_p(\boldsymbol{\mu}, \Sigma)$ then the quadratic form

$$\Delta_i^2 = (\mathbf{Y}_i - \boldsymbol{\mu})' \Sigma^{-1} (\mathbf{Y}_i - \boldsymbol{\mu}) \sim \chi^2(p)$$

The Mahalanobis distance estimate of Δ_i^2 in the sample is

$$D_i^2 = (\mathbf{y}_i - \overline{\mathbf{y}})' \mathbf{S}^{-1} (\mathbf{y}_i - \overline{\mathbf{y}}) \tag{3.7.11}$$

which converges to a chi-square distribution with p degrees of freedom. Hence, to evaluate multivariate normality one may plot the ordered squared Mahalanobis distances $D_{(i)}^2$ against the expected order statistics of a chi-square distribution with sample quantilies $\chi_p^2[(i-1/2)/n] = q_i$ where q_i $(i = 1, 2, \ldots, n)$ is the $100(i-1/2)/n$ sample quantile of the chi-square distribution with p degrees of freedom. The plotting correction $(i - .375)/(n + .25)$ may also be used. This is the value used in the SAS UNIVARIATE procedure for constructing normal Q-Q plots. For a discussion of plotting corrections, see Looney and Gulledge (1985). If the data are multivariate normal, plotted pairs $(D_{(i)}, q_i)$ should be close to a line. Points far from the line are potential outliers. Clearly a large value of D_i^2 for one value may be a candidate. Formal tests for multivariate outliers are considered by Barnett and Lewis (1994). Given the complex nature of multivariate data these tests have limited value.

The exact distribution of $b_i = nD_i^2/(n-1)^2$ follows a beta $[a = p/2, b = (n - p - 1)/2]$ distribution and not a chi-square distribution; see Gnanadesikan and Kettenring (1972). Small (1978) found that as p gets large ($p > 5\%$ of n) relative to n that the chi-square approximation may not be adequate unless $n \geq 25$ and recommends a beta plot. He suggested using a beta $[\alpha, \beta]$ distribution with $\alpha = (a-1)/2a$ and $\beta = (b-1)/2b$ and the ordered statistics

$$b_{(i)}^{\star} = \text{beta}_{\alpha, \beta}[(i-\alpha)/(n-\alpha-\beta+1)] \tag{3.7.12}$$

Then, the ordered $b_{(i)}$ are plotted against the expected order statistics $b^{\star}_{(i)}$. Gnanadesikan and Kettenring (1972) consider a more general plotting scheme using Γ plots to assess normality. A gamma plot fits a scaled chi-square or gamma distribution to the quantity $(\mathbf{y}_i - \overline{\mathbf{y}})' \Gamma (\mathbf{y}_i - \overline{\mathbf{y}})$, by estimating a shape parameter (η) and scale parameter (λ).

Outliers in a multivariate data set inflate/deflate $\overline{\mathbf{y}}$ and $\mathbf{S}$, and sample correlations. This tends to reduce the size of $D^2_{(i)}$. Hence, robust estimates of $\boldsymbol{\mu}$ and Σ in plots may help to identify outliers. Thus, the "robustified" ordered distances

$$D^2_{(i)} = (\mathbf{y}_i - \mathbf{m}^*)' (\mathbf{S}^{\star})^{-1} (\mathbf{y}_i - \mathbf{m}^*)$$

may be plotted to locate extreme outliers. The parameter $\mathbf{m}^*$ and $\mathbf{S}^{\star}$ are robust estimates of $\boldsymbol{\mu}$ and Σ.

Singh (1993) recommends using robust M-estimators derived by Maronna (1976) to robustify plots. However, we recommend using estimates obtained using the multivariate trimming (MVT) procedure of Gnanadesikan and Kettenring (1972) since Devlin et al. (1981) showed that the procedure is less sensitive to the number of extreme outliers, called the breakdown point. For M-estimators the breakdown value is $\leq (1/p)$ regardless of the proportion of multivariate outliers. The S estimator of Davies (1987) also tends to have high breakdowns in any dimension; see Lopuaä and Rousseeuw (1991). For the MVT procedure the value is equal to α, the fraction of multivariate observations excluded from the sample. To obtain the robust estimates, one proceeds as follows.

1. Because the MVT procedure is sensitive to starting values, use the Winsorized sample covariance matrix $\mathbf{S}^{\star}_{\text{w}}$ using (3.7.7) to calculate its elements and the α-trimmed mean vector calculated for each variable. Then, calculate Mahalanobis (Mhd) distances

$$D^2_{(i)} = \left(\mathbf{y}_i - \mathbf{y}^{\star}_{T(\alpha)}\right)' \left(\mathbf{S}^{\star}_{\text{w}}\right)^{-1} \left(\mathbf{y}_i - \mathbf{y}^{\star}_{T(\alpha)}\right)$$

2. Set aside a proportion α_1 of the n vector observations based on the largest $D^2_{(i)}$ values.

3. Calculate the trimmed multivariate mean vector over the retained vectors and the sample covariance matrix

$$\mathbf{S}^{\star}_{\alpha_1} = \sum_{n-r} \left(\mathbf{y}_i - \mathbf{y}^{\star}_{T(\alpha_1)}\right) \left(\mathbf{y}_i - \mathbf{y}^{\star}_{T(\alpha_1)}\right)' / (n - r - 1)$$

for $\alpha_1 = r/n$. Smooth $\mathbf{S}^{\star}_{\alpha_1}$ to ensure that the matrix is positive definite.

4. Calculate the $D^2_{(i)}$ values using the α_1 robust estimates

$$D^2_{(i)} = \left(\mathbf{y}_i - \mathbf{y}^{\star}_{T(\alpha_1)}\right)' \left(\mathbf{S}^{\star}_{\alpha_1}\right)^{-1} \left(\mathbf{y}_i - \mathbf{y}^{\star}_{T(\alpha_1)}\right)$$

and order the $D^2_{(i)}$ to find another subset of vectors α_2 and repeat step 3.

The process continues until the trimmed mean vector $\mathbf{y}^{\star}_T$ and robust covariance matrix $\mathbf{S}^{\star}_{\alpha_i}$ converges to $\mathbf{S}^{\star}$. Using the robust estimates, the raw data are replotted. After making appropriate data adjustments for outliers and lack of multivariate normality using some data transformations, Mardia's test for skewness and kurtosis may be recalculated to affirm multivariate normality of the data set under study.

Example 3.7.1 (Generating MVN Distributions) *To illustrate the analysis of multivariate data, several multivariate normal distributions are generated. The data generated are used to demonstrate several of the procedures for evaluating multivariate normality and testing hypotheses about means and covariance matrices.*

By using the properties of the MVN distribution, recall that if $\mathbf{z} \sim N_p(\mathbf{0}, \mathbf{I}_p)$*, then* $\mathbf{y} = \mathbf{zA} + \boldsymbol{\mu} \sim N_p(\boldsymbol{\mu}, \Sigma = \mathbf{A}'\mathbf{A})$ *. Hence, to generate a MVN distribution with mean* $\boldsymbol{\mu}$ *and covariance matrix* Σ*, one proceeds as follows.*

1. Specify $\boldsymbol{\mu}$ and Σ.
2. Obtain a Cholesky decomposition for Σ; call it $\mathbf{A}$.
3. Generate a $n \times p$ matrix of $N(0, 1)$ random variables named $\mathbf{Z}$.
4. Transform $\mathbf{Z}$ to $\mathbf{Y}$ using the expression $\mathbf{Y} = \mathbf{ZA} + \mathbf{U}$ where $\mathbf{U}$ is created by repeating the row vector $\mathbf{u}'$ n times producing an $n \times p$ matrix.

In program m3_7_1.sas three data sets are generated, each consisting of two independent groups and $p = 3$ variables. Data set A is generated from normally distributed populations with the two groups having equal covariance matrices. Data set B is also generated from normally distributed populations, but this time the two groups do not have equal covariance matrices. Data set C consists of data generated from a non-normal distribution.

Example 3.7.2 (Evaluating Multivariate Normality) *Methods for evaluating multivariate normality include, among other procedures, evaluating univariate normality using the Shapiro-Wilk tests a variable at a time, Mardia's test of multivariate skewness and kurtosis, and multivariate chi-square and beta Q-Q plots. Except for the beta Q-Q plots, there exists a SAS Institute (1998) macro % MULTINORM that performs these above mentioned tests and plots. The SAS code in program m3_7_2.sas demonstrates the use of the macro to evaluate normality using data sets generated in program m3_7_1.sas. Program m3_7_2.sas also includes SAS PROC IML code to produce both chi-square Q-Q and beta Q-Q plots.*

The full instructions for using the MULTINORM macro are included with the macro program. Briefly, the data = statement is where the data file to be analyzed is specified, the var = statement is where the variable names are specified, and then in the plot = statement one can specify whether to produce the multivariate chi-square plot.

Using the data we generated from a multivariate normally distributed population (data set A, group 1 from program m3_7_1.sas), program m3_7_2.sas produces the output in Table 3.7.1 to evaluate normality.

For this data, generated from a multivariate normal distribution with equal covariance matrices, we see that for each of the three variables individually we do not reject the null hypothesis of univariate normality based on the Shapiro-Wilk tests. We also do not reject the null hypothesis of multivariate normality based on Mardia's tests of multivariate skewness and kurtosis. It is important to note that p-values for Mardia's test of skewness and kurtosis are large sample values. Table VI-VIII in Appendix A must be used with small sample sizes.

When $n < 25$*, one should construct beta Q-Q plots, and not chi-square Q-Q plots. Program m3_7_2.sas produces both plots. The outputs are shown in Figures 3.7.1 and 3.7.2. As expected, the plots display a linear trend.*

TABLE 3.7.1. Univariate and Multivariate Normality Tests, Normal Data–Data Set A, Group 1

Variable	N	Test	Multivariate Skewness & Kurtosis	Test Statistic Value	p-value
COL 1	25	Shapiro-Wilk	.	0.96660	0.56055
COL 2	25	Shapiro-Wilk	.	0.93899	0.14030
COL 3	25	Shapiro-Wilk	.	0.99013	0.99592
	25	Mardia Skewness	0.6756	3.34560	0.97208
	25	Mardia Kurtosis	12.9383	−0.94105	0.34668

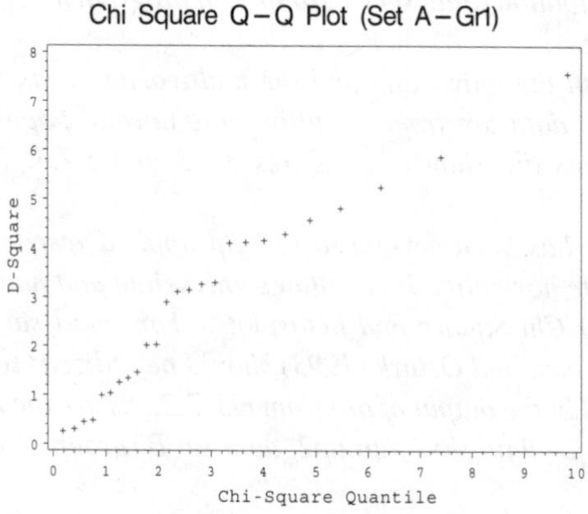

FIGURE 3.7.1. Chi-Square Plot of Normal Data in Set A, Group 1.

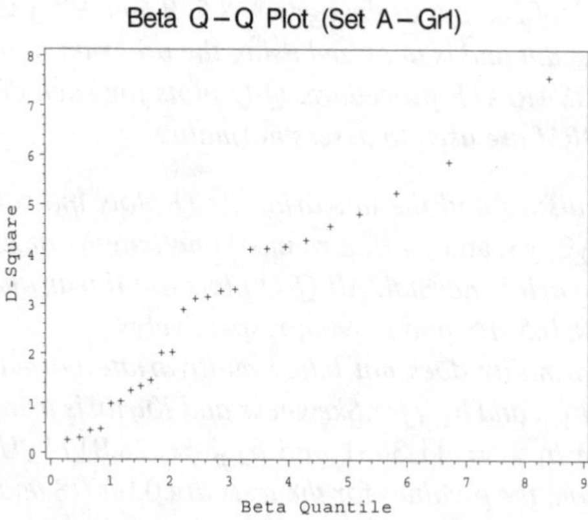

FIGURE 3.7.2. Beta Plot of Normal Data in Data Set A, Group 1

TABLE 3.7.2. Univariate and Multivariate Normality Tests Non-normal Data, Data Set C, Group 1

Variable	N	Test	Multivariate Skewness and Kurtosis	Test Statistic Value	Non-Normal Data p-value
COL 1	25	Shapiro-Wilk	.	0.8257	0.000630989
COL 2	25	Shapiro-Wilk	.	0.5387	0.000000092
COL 3	25	Shapiro-Wilk	.	0.8025	0.000250094
	25	Mardia Skewness	14.6079	72.3441	0.000000000
	25	Mardia Kurtosis	31.4360	7.5020	0.000000000

We next evaluate the data that we generated not from a multivariate normal distribution but from a Cauchy distribution (data set C, group 1, in program m3_7_1.sas); the test results are given in Table 3.7.2.

We can see from both the univariate and the multivariate tests that we reject the null hypothesis and that the data are from a multivariate normal population. The chi-square Q-Q and beta Q-Q plots are shown in Figures 3.7.3 and 3.7.4. They clearly display a nonlinear pattern.

Program m3_7_2.sas has been developed to help applied researchers evaluate the assumption of multivariate normality. It calculates univariate and multivariate test statistics and provides both Q-Q Chi-Square and beta plots. For small sample sizes, the critical values developed by Romeu and Ozturk (1993) should be utilized; see Table VI-VIII in Appendix A. Also included in the output of program m3_7_2.sas are the tests for evaluating the multivariate normality for data set A, group2, data set B (groups 1 and 2) and data set C, group2.

Example 3.7.3 (Normality and Outliers) *To illustrate the evaluation of normality and the identification of potential outliers, the ramus bone data from Elston and Grizzle (1962) displayed in Table 3.7.3 are utilized. The dependent variables represent the measurements of the ramus bone length of 20 boys at the ages 8, 8.5, 9, and 9.5 years of age. The data set is found in the file ramus.dat and is analyzed using the program ramus.sas. Using program ramus.sas, the SAS UNIVARIATE procedure, Q-Q plots for each dependent variable, and the macro %MULTINORM are used to assess normality.*

The Shapiro-Wilk statistics and the univariate Q-Q plots indicate that each of the dependent variables $y1$, $y2$, $y3$, and $y4$ *(the ramus bone lengths at ages 8, 8.5, 9, and 9.5) individually appear univariate normal. All Q-Q plots are linear and the W statistics have p-values 0.3360, 0.6020, 0.5016, and 0.0905, respectively.*

Because marginal normality does not imply multivariate normality, we also calculate Mardia's test statistics $b_{1,p}$ *and* $b_{2,p}$ *for Skewness and Kurtosis using the macro %MULTINORM. The values are* $b_{1,p} = 11.3431$ *and* $b_{2,p} = 28.9174$*. Using the large sample chi-square approximation, the p-values for the tests are* 0.00078 *and* 0.11249*, respectively. Because n is small, tables in Appendix A yield a more accurate test. For* $\alpha = 0.05$*, we again conclude that the data appear skewed.*

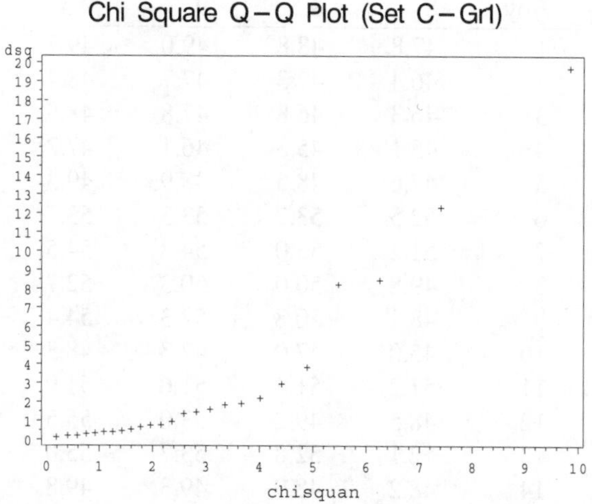

FIGURE 3.7.3. Chi-Square Plot of Non-normal Data in Data Set C, Group 2.

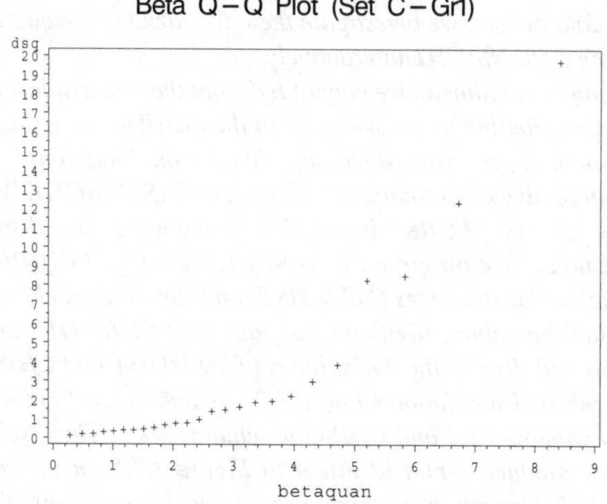

FIGURE 3.7.4. Beta Plot of Non-normal Data in Data Set C, Group 2.

TABLE 3.7.3. Ramus Bone Length Data

Boy	Age in Years			
	8	8.5	9	9.5
1	47.8	48.8	49.0	49.7
2	46.4	47.3	47.7	48.4
3	46.3	46.8	47.8	48.5
4	45.1	45.3	46.1	47.2
5	47.6	48.5	48.9	49.3
6	52.5	53.2	53.3	53.7
7	51.2	53.0	54.3	54.5
8	49.8	50.0	50.3	52.7
9	48.1	50.8	52.3	54.4
10	45.0	47.0	47.3	48.3
11	51.2	51.4	51.6	51.9
12	48.5	49.2	53.0	55.5
13	52.1	52.8	53.7	55.0
14	48.2	48.9	49.3	49.8
15	49.6	50.4	51.2	51.8
16	50.7	51.7	52.7	53.3
17	47.2	47.7	48.4	49.5
18	53.3	54.6	55.1	55.3
19	46.2	47.5	48.1	48.4
20	46.3	47.6	51.3	51.8

To evaluate the data further, we investigate the multivariate chi-square Q-Q plot shown in Figure 3.7.5 using SAS/INSIGHT interactively.

While the plot appears nonlinear, we cannot tell from the plot displayed which of the observations may be contributing to the skewness of the distribution. Using the Tool Bar following the execution of the program ramus.sas, we click on "Solutions," select "Analysis," and then select "Interactive Data Analysis." This opens SAS/INSIGHT. With SAS/INSIGHT open, we select the Library "WORK" by clicking on the word. This displays the data sets used in the application of the program ramus.sas. The data set "CHIPLOT" contains the square of the Mahalanobis distances (MANDIST) and the ordered chi-square Q-Q values (CHISQ). To display the values, highlight the data set "CHIPLOT" and select "Open" from the menu. This will display the coordinates of MAHDIST and CHISQ. From the Tool Bar select "Analyze" and the option "Fit(YX)." Clicking on "Fit(YX)," move variable MAHDIST to window "Y" and CHISQ to window "X". Then, select "Apply" from the menu. This will produce a plot identical to Figure 3.7.5 on the screen. By holding the "Ctrl" key and clicking on the extreme upper most observations, the numbers 9 *and* 12 *will appear on your screen. These observations have large Mahalanobis squared distances:* 11.1433 *and* 8.4963 *(the same values calculated and displayed in the output for the example). None of the distances exceed the chi-square critical value of* 11.07 *for alpha*

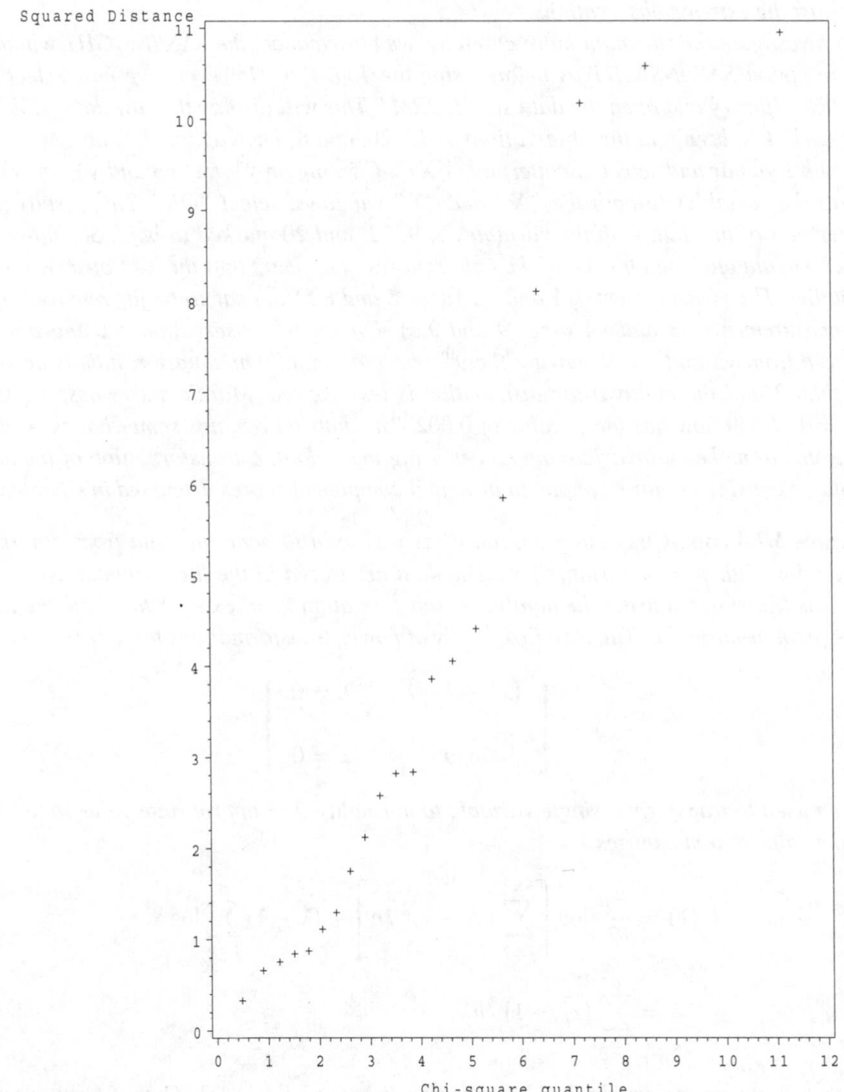

FIGURE 3.7.5. Ramus Data Chi-square Plot

$= 0.05$ *for evaluating a single outlier. By double clicking on an extreme observation, the window "Examine Observations" appears. Selecting each of the extreme observations* 9, 12, 20, *and* 8, *the chi-square residual values are* -1.7029, 1.1893, 2.3651, *and* 1.3783, *respectively. While the* 9th *observation has the largest distance value, the imbedded* 20th *observation has the largest residual. This often happens with multivariate data. One must look past the extreme observations.*

To investigate the raw data more carefully, we close/cancel the SAS/INSIGHT windows and re-option SAS/INSIGHT as before using the Tool Bar. However, we now select the "WORK" library and open the data set "NORM." This will display the raw data. Holding the "Ctrl" key, highlight the observations 9, 12, 20, *and* 8. *Then again click on "Analyze" from the Tool bar and select "Scatterplot* (YX)*". Clicking on* $y1$, $y2$, $y3$ *and* $y4$, *and moving all the variables to both the "X" and "Y" windows, select "OK." This results in a scatter plot of the data with the variables* 8, 9, 12, *and* 20 *marked in bold. Scanning the plots by again clicking on each of the bold squares, it appears that the* 20th *observation is an outlier. The measurements* $y1$ *and* $y2$ *(ages* 8 *and* 8.5) *appear to be far removed from the measurements* $y3$ *and* $y4$ *(ages* 9 *and* 9.5). *For the* 9th *observation,* $y1$ *appears far removed from* $y3$ *and* $y4$. *Removing the* 9th *observation, all chi-square residuals become less than* 2 *and the multivariate distribution is less skewed. Mardia's skewness statistic* $b_{1,p} = 11.0359$ *now has the p-value of* 0.002. *The data set remains somewhat skewed. If one wants to make multivariate inferences using these data, a transformation of the data should be considered, for example, to principal component scores discussed in Chapter* 8.

Example 3.7.4 (Box-Cox) *Program norm.sas was used to generate data from a normal distribution with* $p = 4$ *variables,* y_i. *The data are stored in the file norm.dat. Next, the data was transformed using the nonlinear transformation* $x_i = \exp(y_i)$ *to create the data in the file non-norm.dat. The Box-Cox family of power transformations for* $x > 0$

$$y = \left\{ \begin{array}{ll} \left(x^{\lambda} - 1\right)/\lambda & \lambda \neq 0 \\ \\ \log x & \lambda = 0 \end{array} \right\}$$

is often used to transform a single variable to normality. The appropriate value to use for λ *is the value that maximizes*

$$L(\lambda) = \frac{-n}{2} \log \left[\sum_{i=1}^{n} (y_i - \overline{y})^2 / n \right] + (\lambda - 1) \sum_{i-1}^{n} \log x_i$$

$$\overline{y} = \sum_{i-1}^{n} \left(x_i^{\lambda} - 1\right) / n\lambda$$

Program Box-Cox.sas graphs $L(\lambda)$ *for values of* $\lambda : -1.0\,(.1)\,1.3$. *Output from executing the program indicates that the parameter* $\lambda \simeq 0$ *for the Box-Cox transformation for the graph of* $L(\lambda)$ *to be a maximum. Thus, one would use the logarithm transformation to achieve normality for the transformed variable. After making the transformation, one should always verify that the transformed variable does follow a normal distribution. One may also use the macro ADXTRANS available in SAS/AC software to estimate the optimal*

Box-Cox transformation within the class of power transformations of the form $y = x^{\lambda}$. Using the normal likelihood, the value of λ is estimated and an associated 95% confidence interval is created for the parameter lambda. The SAS macro is illustrated in the program unorm.sas. Again, we observe that the Box-Cox parameter $\lambda \simeq 0$.

Exercises 3.7

1. Use program m3_7_1.sas to generate a multivariate normal data set of $n_1 = n_2 = 100$ observations with mean structure

$$\boldsymbol{\mu}_1' = \begin{bmatrix} 42.0 & 28.4 & 41.2 & 31.2 & 33.4 \end{bmatrix}$$
$$\boldsymbol{\mu}_2' = \begin{bmatrix} 50.9 & 35.0 & 49.6 & 37.9 & 44.9 \end{bmatrix}$$

and covariance matrix

$$S = \begin{bmatrix} 141.49 & & & & (\text{Sym}) \\ 33.17 & 53.36 & & & \\ 52.59 & 31.62 & 122.44 & & \\ 14.33 & 8.62 & 31.12 & 64.69 & \\ 21.44 & 16.63 & 33.22 & 31.83 & 49.96 \end{bmatrix}$$

where the seed is 101999.

2. Using the data in Problem 1, evaluate the univariate and multivariate normality of the data using program m3_7_2.sas.

3. After matching subjects according to age, education, former language training, intelligence and language aptitude, Postovsky (1970) investigated the effects of delay in oral practice at the beginning of second-language learning. The data are provided in Timm (1975, p. 228). Using an experimental condition with a 4-week delay in oral practice and a control condition with no delay, evaluation was carried out for language skills: listening (L), speaking (S), reading (R), and writing (W). The data for a comprehensive examination given at the end of the first 6 weeks follow in Table 3.7.4.

   (a) For the data in Table 3.7.4, determine whether the data for each group, Experimental and Control, are multivariate normal. If either group is nonnormal, find an appropriate transformation to ensure normality.

   (b) Construct plots to determine whether there are outliers in the transformed data. For the groups with outliers, create robust estimates for the joint covariance matrix.

4. For the Reading Comprehension data found on the Internet link at http://lib.stat.cmu.edu/DASL/Datafiles/ReadingTestScore.html from a study of the effects of instruction on reading comprehension in 66 children, determine if the observations follow a multivariate normal distribution and if there are outliers in the data. Remove the outliers, and recalculate the mean and covariance matrix. Discuss your findings.

TABLE 3.7.4. Effects of Delay on Oral Practice.

	Experimental Group				Control Group			
Subject	L	S	R	W	L	S	R	W
1	34	66	39	97	33	56	36	81
2	35	60	39	95	21	39	33	74
3	32	57	39	94	29	47	35	89
4	29	53	39	97	22	42	34	85
5	37	58	40	96	39	61	40	97
6	35	57	34	90	34	58	38	94
7	34	51	37	84	29	38	34	76
8	25	42	37	80	31	42	38	83
9	29	52	37	85	18	35	28	58
10	25	47	37	94	36	51	36	83
11	34	55	35	88	25	45	36	67
12	24	42	35	88	33	43	36	86
13	25	59	32	82	29	50	37	94
14	34	57	35	89	30	50	34	84
15	35	57	39	97	34	49	38	94
16	29	41	36	82	30	42	34	77
17	25	44	30	65	25	47	36	66
18	28	51	39	96	32	37	38	88
19	25	42	38	86	22	44	22	85
20	30	43	38	91	30	35	35	77
21	27	50	39	96	34	45	38	95
22	25	46	38	85	31	50	37	96
23	22	33	27	72	21	36	19	43
24	19	30	35	77	26	42	33	73
25	26	45	37	90	30	49	36	88
26	27	38	33	77	23	37	36	82
27	30	36	22	62	21	43	30	85
28	36	50	39	92	30	45	34	70

5. Use PROC UNIVARIATE to verify that each variable in file non-norm.dat is nonnormal. Use the macro %MULTINORM to create a chi-square Q-Q plot for the four variables. Use programs Box-Cox.sas and norm.sas to estimate the parameter λ for a Box-Cox transformation of each of the other variables in the file non-norm.dat. Verify that all the variables are multivariate normal, after an appropriate transformation.

3.8 Tests of Covariance Matrices

a. *Tests of Covariance Matrices*

In multivariate analysis, as in univariate analysis, when testing hypotheses about means, three assumptions are essential for valid tests

1. independence
2. multivariate normality, and
3. equality of covariance matrices for several populations or that a covariance matrix has a specific pattern for one or more populations.

In Section 3.7 we discussed evaluation of the multivariate normal assumption. We now assume data are normally distributed and investigate some common likelihood ratio tests of covariance matrices for one or more populations. The tests are developed using the likelihood ratio principle which compares the likelihood function under the null hypothesis to the likelihood function over the entire parameter space (the alternative hypothesis) assuming multivariate normality. The ratio is often represented by the statistic λ. Because the exact distribution of the lambda statistic is often unknown, large sample results are used to obtain tests. For large samples and under very general conditions, Wald (1943) showed that $-2 \log \lambda$ converges in distribution to a chi-square distribution under the null hypothesis where the degrees of freedom are f. The degrees of freedom is obtained by subtracting the number of independent parameters estimated for the entire parameter space minus the number of independent parameters estimated under the null hypothesis. Because tests of covariance matrices involves variances and covariances, and not means, the tests are generally very sensitive to lack of multivariate normality.

b. *Equality of Covariance Matrices*

In testing hypotheses regarding means in k independent populations, we often require that the independent covariance matrices $\Sigma_1, \Sigma_2, \ldots, \Sigma_k$ be equal. To test the hypothesis

$$H : \Sigma_1 = \Sigma_2 = \cdots = \Sigma_k \tag{3.8.1}$$

we construct a modified likelihood ratio statistic; see Box (1949). Let $\mathbf{S}_i$ denote the unbiased estimate of Σ_i for the ith population, with n_i independent p-vector valued observations ($n_i \geq p$) from the MVN distribution with mean $\boldsymbol{\mu}_i$ and covariance matrix Σ_i. Setting $n = \sum_{i=1}^{k} n_i$ and $v_i = n_i - 1$, the pooled estimate of the covariance matrix under H is

$$\mathbf{S} = \sum_{i=1}^{k} v_i \mathbf{S}_i / (n - k) = \mathbf{E}/v_e \tag{3.8.2}$$

where $v_e = n - k$. To test (3.8.1), the statistic

$$W = v_e \log |\mathbf{S}| - \sum_{i=1}^{k} v_i \log |\mathbf{S}_i| \tag{3.8.3}$$

is formed. Box (1949, 1950) developed approximations to W using either a χ^2 or an F approximation. Details are included in Anderson (1984). The test is commonly called Box's M test where M is the likelihood ratio statistic. Multiplying W by $\rho = 1 - C$ where

$$C = \frac{2p^2 + 3p - 1}{6\,(p+1)\,(k-1)} \left[\sum_{i=1}^{k} \frac{1}{v_i} - \frac{1}{v_e} \right] \tag{3.8.4}$$

the quantity

$$X^2 = (1 - C)W = -2\rho \log M \xrightarrow{d} \chi^2(f) \tag{3.8.5}$$

where $f = p\,(p+1)\,(k-1)/2$. Thus, to test H in (3.8.1) the hypothesis is rejected if $X^2 > \chi^2_{1-\alpha}(f)$ for a test of size α. This approximation is reasonable provided $n_i > 20$ and both p and k are less than 6. When this is not the case, an F approximation is used.

To employ the F approximation, one calculates

$$C_0 = \frac{(p-1)\,(p+2)}{6\,(k-1)} \left[\sum_{i=1}^{k} \frac{1}{v_i^2} - \frac{1}{v_e^2} \right] \tag{3.8.6}$$

$$f_0 = (f+2) \,/\, \left| C_0 - C^2 \right|$$

For $C_0 - C^2 > 0$, the statistic

$$F = W \,/\, a \xrightarrow{d} F(f,\, f_0) \tag{3.8.7}$$

is calculated where $a = f/[1 - C - (f/\,f_0)]$. If $C_0 - C^2 < 0$, then

$$F = f_0 W \,/\, f\,(b - W) \xrightarrow{d} F(f,\, f_0) \tag{3.8.8}$$

where $b = f_0/\,(1 - C + 2/f_0)$. The hypothesis of equal covariances is rejected if $F > F^{1-\alpha}_{(f,\, f_0)}$ for a test of size α; see Krishnaiah and Lee (1980).

Both the χ^2 and F approximations are rough approximations. Using Box's asymptotic expansion for X^2 in (3.8.5), as discussed in Anderson (1984, p. 420), the p-value of the test is estimated as

$$\begin{aligned} \alpha_p = P\left(X^2 \geq X_0^2\right) &= P(\chi_f^2 \geq X_0^2) \\ &\quad + \omega \left[P(\chi^2_{f+4} \geq X_0^2) - P\left(\chi_f^2 \geq X_0^2\right) \right] + O\left(v_e^{-3}\right) \end{aligned}$$

where X_0^2 is the calculated value of the test statistic in (3.8.5) and

$$\omega = \frac{p\,(p+1) \left[(p-1)\,(p+2) \left[\sum_{i=1}^{k} \frac{1}{v_i} - \frac{1}{v_e} \right] - 6\,(k-1)\,(1-\rho)^2 \right]}{48\rho^2} \tag{3.8.9}$$

For equal v_i, Lee et al. (1977) developed exact values of the likelihood ratio test for $v_i = (p+1)\,(1)\,20\,(5)\,30$, $p = 2\,(1)\,6$, $k = 2\,(1)\,10$ and $\alpha = 0.01$, 0.05 and 0.10.

TABLE 3.8.1. Box's Test of $\Sigma_1 = \Sigma_2 \chi^2$ Approximation.

XB	V1	PROB XB
1.1704013	6	0.9783214

TABLE 3.8.2. Box's Test of $\Sigma_1 = \Sigma_2$ F Approximation.

FB	V1	PROB FB
0.1949917	16693.132	0.9783382

Layard (1974) investigated the robustness of Box's M test. He states that it is so severely affected by departures from normality as to make it useless; and that under nonnormality and homogeneity of covariance matrices, the M test is a test of multivariate normality. Layard (1974) proposed several robust tests of (3.8.1).

Example 3.8.1 (Testing the Equality of Covariance Matrices) *As an example of testing for the equality of covariance matrices, we utilize the data generated from multivariate normal distributions with equal covariance matrices, data set A generated by program m3_7_1.sas. We generated 25 observations from a normal distribution with* $\boldsymbol{\mu}_1' = [6, 12, 30]$ *and 25 observations with* $\boldsymbol{\mu}_2' = [4, 9, 20]$*; both groups have covariance structure*

$$\Sigma = \begin{bmatrix} 7 & 2 & 0 \\ 2 & 6 & 0 \\ 0 & 3 & 5 \end{bmatrix}$$

Program m3_8_1.sas was written to test $\Sigma_1 = \Sigma_2$ *for data set A. Output for the chi-square and F tests calculated by the program are shown in Tables 3.8.1 and 3.8.2. The chi-square approximation works well when* $n_i > 20$, $p < 6$, *and* $k < 6$. *The F approximation can be used for small* n_i *and* p, *and for* $k > 6$. *By both the chi-square approximation and the F approximation, we fail to reject the null hypothesis of the equality of the covariance matrices of the two groups. This is as we expected since the data were generated from populations having equal covariance matrices.*

The results of Box's M test for equal covariance matrices for data set B, which was generated from multivariate normal populations with unequal covariance matrices, are provided in Table 3.8.3. As expected, we reject the null hypothesis that the covariance matrices are equal.

As yet a third example, Box's M test was performed on data set C which is generated from non-normal populations with equal covariance matrices. The results are shown in Table 3.8.4. Notice that we reject the null hypothesis that the covariance matrices are equal; we however know that the two populations have equal covariance matrices. This illustrates the effect of departures from normality on Box's test; erroneous results can be obtained if data are non-normal.

To obtain the results in Table 3.8.3 and Table 3.8.4, program m3_8_1.sas is executed two times by using the data sets exampl.m371b and exampl.m371c.

TABLE 3.8.3. Box's Test of $\Sigma_1 = \Sigma_2 \chi^2$ Data Set B.

χ^2 Approximation			F Approximation		
XB	VI	PROB XB	FB	VI	PROB FB
43.736477	6	8.337E-8	7.2866025	16693.12	8.6028E-8

TABLE 3.8.4. Box's Test of $\Sigma_1 = \Sigma_2 \chi^2$ Data Set C.

χ^2 Approximation			F Approximation		
XB	VI	PROB XB	FB	VI	PROB FB
19.620669	6	0.0032343	3.2688507	16693.132	0.0032564

In addition to testing for the equality of covariance matrices, a common problem in multivariate analysis is testing that a covariance matrix has a specific form or linear structure. Some examples include the following.

1. Specified Value

$$H : \Sigma = \Sigma_o \; (\Sigma_o \text{ is known})$$

2. Compound Symmetry

$$H : \Sigma = \sigma^2 \begin{bmatrix} 1 & \rho & \rho & \cdots & \rho \\ \rho & 1 & \rho & \cdots & \rho \\ \vdots & \vdots & \vdots & & \vdots \\ \rho & \rho & \rho & \cdots & 1 \end{bmatrix} = \sigma^2 \left[(1-\rho)\mathbf{I} + \rho \mathbf{J} \right]$$

   where $\mathbf{J}$ is a square matrix of 1s, ρ is the intraclass correlation, and σ^2 is the common variance. Both σ^2 and ρ are unknown.

3. Sphericity

$$H : \Sigma = \sigma^2 \mathbf{I} \; (\sigma \text{ unknown})$$

4. Independence, for $\Sigma = (\Sigma_{ij})$

$$H : \Sigma_{ij} = \mathbf{0} \text{ for } i \neq j$$

5. Linear Structure

$$H : \Sigma = \sum_{i=1}^{k} \mathbf{G}_i \otimes \Sigma_i$$

   where $\mathbf{G}_1, \mathbf{G}_2, \ldots, \mathbf{G}_k$ are known $t \times t$ matrices, and $\Sigma_1, \Sigma_2, \ldots, \Sigma_k$ are unknown matrices of order $p \times p$.

Tests of the covariance structures considered in (1)-(5) above have been discussed by Krishnaiah and Lee (1980). This section follows their presentation.

c. *Testing for a Specific Covariance Matrix*

For multivariate data sets that have a large number of observations in which data are studied over time or several treatment conditions, one may want to test that a covariance matrix is equal to a specified value. The null hypothesis is

$$H : \Sigma = \Sigma_o \qquad \text{(known)} \tag{3.8.10}$$

For one population, we let $v_e = n - 1$ and for k populations, $v_e = \sum_i (n_i - 1) = n - k$. Assuming that the n p-vector valued observations are sampled from a MVN distribution with mean $\boldsymbol{\mu}$ and covariance matrix Σ, the test statistic to test (3.8.10) is

$$W = -2 \log \lambda = v_e \left[\log |\Sigma_o| - \log |\mathbf{S}| + \operatorname{tr}\left(\mathbf{S}\Sigma_o^{-1}\right) - p \right]$$

where $\mathbf{S} = \mathbf{E}/v_e$ is an unbiased estimate of Σ. The parameter λ is the standard likelihood ratio criterion. Korin (1968) developed approximations to W using both a χ^2 and an F approximation. Multiplying W by $\rho = 1 - C$ where

$$C = \left(2p^2 + 3p - 1\right) / 6v_e(p + 1) \tag{3.8.11}$$

the quantity

$$X^2 = (1 - C)W = -2\rho \log \lambda \xrightarrow{d} \chi^2(f)$$

where $f = p(p + 1)/2$. Alternatively, the F statistic is

$$F = W / a \xrightarrow{d} F(f,\, f_0) \tag{3.8.12}$$

where

$$\begin{aligned} f_0 &= (f + 2)/\left|C_0 - C^2\right| \\ C_0 &= (p - 1)(p + 2)/6v_e \\ a &= f \,/\, [1 - C - (f \,/\, f_0)] \end{aligned}$$

Again, $H : \Sigma = \Sigma_o$ is rejected if the test statistic is large. A special case of H is to set $\Sigma_o = \mathbf{I}$, a test that the variables are independent and have equal unit variances.

Using Box's asymptotic expansion, Anderson (1984, p. 438), the p-value of the test is estimated as

$$\begin{aligned} \alpha_p &= P(-2\rho \log \lambda \geq X_0^2) \\ &= P(\chi_f^2 \geq X_0^2) + \omega[P(\chi_{f+4}^2 \geq X_0^2) - P(\chi_f^2 \geq X_0^2)] / \rho^2 + O(v_e^{-3}) \end{aligned} \tag{3.8.13}$$

for

$$\omega = p(2p^4 + 6p^3 + p^2 - 12p - 13) \,/\, 288(v_e^2)(p + 1)$$

For $p = 4(1)10$ and small values of v_e, Nagarsenker and Pillai (1973a) have developed exact critical values for W for the significant levels $\alpha = 0.01$ and 0.05.

TABLE 3.8.5. Test of Specific Covariance Matrix Chi-Square Approximation.

S			EO		
7.0874498	3.0051207	0.1585046	6	0	0
3.0051207	5.3689862	3.5164255	0	6	0
0.1585046	3.5164235	5.528464	0		6
	X_SC	DFX_SC			
	48.905088	6	PROB XSC		
			7.7893E-9		

Example 3.8.2 (Testing $\Sigma = \Sigma_o$) *Again we use the first data set generated by program m3_7_1.sas which is from a multivariate normal distribution. We test the null hypothesis that the pooled covariance matrix for the two groups is equal to*

$$\Sigma_o = \begin{bmatrix} 6 & 0 & 0 \\ 0 & 6 & 0 \\ 0 & 0 & 6 \end{bmatrix}$$

The SAS PROC IML code is included in program m3_8_1.sas. The results of the test are given in Table 3.8.5. The results show that we reject the null hypothesis that $\Sigma = \Sigma_o$.

d. Testing for Compound Symmetry

In repeated measurement designs, one often assumes that the covariance matrix Σ has compound symmetry structure. To test

$$H : \Sigma = \sigma^2 [(1-\rho)\mathbf{I} + \rho\mathbf{J}] \tag{3.8.14}$$

we again assume that we have a random sample of vectors from a MVN distribution with mean $\boldsymbol{\mu}$ and covariance matrix Σ. Letting $\mathbf{S}$ be an unbiased estimate of Σ based on v_e degrees of freedom, the modified likelihood ratio statistic is formed

$$M_x = -v_e \log \left\{ |\mathbf{S}| \,/\, \left(s^2\right)^p (1-r)^{p-1} [1 + (p-1)\,r] \right\} \tag{3.8.15}$$

where $\mathbf{S} = [s_{ij}]$ and estimates of σ^2 and $\sigma^2\rho$ are

$$s^2 = \sum_{i=1}^{p} s_{ii}/p \quad \text{and} \quad s^2 r = \sum_{i \neq j} s_{ij} \,/\, p(p-1) \tag{3.8.16}$$

The denominator of M_x is

$$|\mathbf{S}_o| = \left| \begin{bmatrix} s^2 & s^2 r & \cdots & s^2 r \\ s^2 r & s^2 & \cdots & s^2 r \\ \vdots & \vdots & & \vdots \\ s^2 r & s^2 r & \cdots & s^2 \end{bmatrix} \right|$$

TABLE 3.8.6. Test of Comparing Symmetry χ^2 Approximation.

CHIMX	DEMX	PRBCHIMX
31.116647	1	2.4298E-8

so that $M_x = -v_e \log \{|\mathbf{S}| / |\mathbf{S}_o|\}$ where $s^2 r$ is the average of the nondiagonal elements of **S**.

Multiplying M_x by $(1 - C_x)$ for

$$C_x = p(p+1)^2(2p-3)/6(p-1)(p^2+p-4)v_e \tag{3.8.17}$$

Box (1949) showed that

$$X^2 = (1 - C_x)M_x \xrightarrow{d} \chi^2(f) \tag{3.8.18}$$

for $f = (p^2 + p - 4)/2$, provided $n_i > 20$ for each group and $p < 6$. When this is not the case, the F approximation is used. Letting

$$C_{ox} = \frac{p\left(p^2-1\right)(p+2)}{6(p^2+p-4)v_e^2}$$

$$f_{ox} = (f+2) \,/\, \left|C_{ox} - C_x^2\right|$$

the F statistic is

$$F = (1 - C_x - f)M_x/f_{ox} \xrightarrow{d} F(f, f_{ox}) \tag{3.8.19}$$

Again, H in (3.8.16) is rejected for large values of X^2 or F.

The exact critical values for the likelihood ratio test statistic for $p = 4(1)10$ and small values of v_e were calculated by Nagarsenker (1975).

Example 3.8.3 (Testing Compound Symmetry) *To test for compound symmetry we again use data set A, and the sample estimate of* **S** *pooled across the two groups. Thus,* $v_e = n - r$ *where* $r = 2$ *for two groups. The SAS PROC IML code is again provided in program m3_8_1.sas. The output is shown in Table 3.8.6. Thus, we reject the null hypothesis of compound symmetry.*

e. Tests of Sphericity

For the general linear model, we assume a random sample of n p-vector valued observations from a MVN distribution with mean $\boldsymbol{\mu}$ and covariance matrix $\Sigma = \sigma^2\mathbf{I}$. Then, the p variables in each observation vector are independent with common variance σ^2. To test for sphericity or independence given a MVN sample, the hypothesis is

$$H : \Sigma = \sigma^2\mathbf{I} \tag{3.8.20}$$

The hypothesis H also arises in repeated measurement designs. For such designs, the observations are transformed by an orthogonal matrix $\mathbf{M}_{p\times(p-1)}$ of rank $(p-1)$ so the $\mathbf{M}'\mathbf{M} = \mathbf{I}_{(p-1)}$. Then, we are interested in testing

$$H : \mathbf{M}'\Sigma\mathbf{M} = \sigma^2\mathbf{I} \tag{3.8.21}$$

where again σ^2 is unknown. For these designs, the test is sometimes called the test of circularity. The test of (3.8.21) is performed in the SAS procedure GLM by using the REPEATED statement. The test is labeled the "Test of Sphericity Applied to Orthogonal Components." This test is due to Mauchly (1940) and employs Box's (1949) correction for a chi-square distribution, as discussed below. PROC GLM may not be used to test (3.8.20). While it does produce another test of "Sphericity," this is a test of sphericity for the original variables transformed by the nonorthogonal matrix $\mathbf{M}'$. Thus, it is testing the sphericity of the $p-1$ variables in $\mathbf{y}^* = \mathbf{M}'\mathbf{y}$, or that the cov $(\mathbf{y}^*) = \mathbf{M}'\boldsymbol{\Sigma}\mathbf{M} = \sigma^2\mathbf{I}$.

The likelihood ratio statistic for testing sphericity is

$$\lambda_s = \left\{|\mathbf{S}| \,/\, [\operatorname{tr}\mathbf{S} \,/\, p]^p\right\}^{n/2} \tag{3.8.22}$$

or equivalently

$$\Lambda = (\lambda_s)^{2/n} = |\mathbf{S}| \,/[\operatorname{tr}\mathbf{S} \,/\, p]^p \tag{3.8.23}$$

where $\mathbf{S}$ is an unbiased estimate of $\boldsymbol{\Sigma}$ based on $v_e = n-1$ degrees of freedom; see Mauchly (1940). Replacing n by v_e,

$$W = -v_e \log \Lambda \xrightarrow{d} \chi^2(f) \tag{3.8.24}$$

with degrees of freedom $f = (p-1)(p+2)/2$. To improve convergence, Box (1949) showed that for

$$C = (2p^2 + p + 2)/6pv_e$$

that

$$\begin{aligned} X^2 &= -v_e(1-C)\log\Lambda \\ &= -\left(v_e - \frac{2p^2+p+2}{6p}\right)\log\Lambda \xrightarrow{d} \chi^2(f) \end{aligned} \tag{3.8.25}$$

converges more rapidly than W. The hypotheses is rejected for large values of X^2 and works well for $n > 20$ and $p < 6$. To perform the test of circulariy, one replaces $\mathbf{S}$ with $\mathbf{M}'\mathbf{S}\mathbf{M}$ and p with $p-1$ in the test for sphericity. For small samples sizes and large values of p, Box (1949) developed an improved F approximation for the test.

Using Box's asymptotic expansion, the p-value for the test is more accurately estimated using the expression

$$\begin{aligned} \alpha_p &= P(-v_e\rho\log\lambda_2 \geq X_0^2) = P(X^2 \geq X_0^2) \\ &= P\left(\chi_f^2 \geq X_0^2\right) + \omega\left[P\left(\chi_{f+4}^2 \geq X_0^2\right) - P(\chi_f^2 \geq X_0^2)\right] + O(v_e^{-3}) \end{aligned} \tag{3.8.26}$$

for $\rho = 1 - C$ and

$$\omega = \frac{(p+2)(p-1)(2p^3+6p+3p+2)}{288p^2v_e^2\rho^2}$$

For small values of n, $p = 4(1)10$ and $\alpha = 0.05$, Nagarsenker and Pillai (1973) published exact critical values for Λ.

An alternative expression for Λ is found by solving the characteristic equation $|\Sigma - \lambda \mathbf{I}| = 0$ with eigenvalues $\lambda_1, \lambda_2, \ldots, \lambda_p$. Using $\mathbf{S}$ to estimate Σ,

$$\Lambda = \prod_{i=I}^{p} \widehat{\lambda}_i \Big/ \left[\sum_i \widehat{\lambda}_i / p \right]^p \tag{3.8.27}$$

where $\widehat{\lambda}_i$ are the eigenvalues of $\mathbf{S}$. Thus, testing $H : \Sigma = \sigma^2 \mathbf{I}$ is equivalent to testing that the eigenvalues of Σ are equal, $\lambda_1 = \lambda_2 = \cdots = \lambda_p$. Bartlett (1954) developed a test of equal λ_i that is equal to the statistic X^2 proposed by Box (1949). We discuss this test in Chapter 8.

Given the importance of the test of independence with homogeneous variance, numerous tests have been proposed to test $H : \Sigma = \sigma^2 \mathbf{I}$. Because the test is equivalent to an investigation of the eigenvalues of $|\Sigma - \lambda \mathbf{I}| = 0$, there is no uniformly best test of sphericity. However, John (1971) and Sugiura (1972) showed that a locally best invariant test depends on the trace criterion, T, where

$$T = \operatorname{tr}(\mathbf{S}^2) / [\operatorname{tr} \mathbf{S}]^2 \tag{3.8.28}$$

To improve convergence, Sugiura showed that

$$W = \frac{v_e p}{2} \left[\frac{p \operatorname{tr}(\mathbf{S}^2)}{(\operatorname{tr} \mathbf{S})^2} - 1 \right] \xrightarrow{d} \chi^2(f)$$

where $f = (p-1)(p+2)/2 = \frac{1}{2} p(p+1) - 1$. Carter and Srivastava (1983) showed that under a broad class of alternatives that both tests have the same power up to $O(v_e^{-3/2})$.

Cornell et al. (1992) compared the two criteria and numerous other proposed statistics that depend on the roots $\widehat{\lambda}_i$ of $\mathbf{S}$. They concluded that the locally best invariant test was more powerful than any of the others considered, regardless of p and $n \geq p$.

Example 3.8.4 (Test of Sphericity) *In this example we perform Mauchly's test of sphericity for the pooled covariance matrix for data set A. Thus,* $k = 2$. *To test a single group, we would use* $k = 1$. *Implicit in the test is that* $\Sigma_1 = \Sigma_2 = \Sigma$ *and we are testing that* $\Sigma = \sigma^2 \mathbf{I}$. *We also include a test of "pooled" circularity. That* $\mathbf{M}'\Sigma\mathbf{M} = \sigma^2 \mathbf{I}$ *for* $\mathbf{M}'\mathbf{M} = \mathbf{I}_{(p-1)}$. *The results are given in Table 3.8.7. Thus, we reject the null hypothesis that the pooled covariance matrix has spherical or circular structure.*

To test for sphericity in k populations, one may first test for equality of the covariance matrices using the nominal level $\alpha/2$. Given homogeneity, one next tests for sphericity using $\alpha/2$ so that the two tests control the joint test near some nominal level α. Alternatively, the joint hypothesis

$$H : \Sigma_1 = \Sigma_2 = \cdots = \Sigma_k = \sigma^2 \mathbf{I} \tag{3.8.29}$$

may be tested using either a likelihood ratio test or Rao's score test, also called the Lagrange multiplier test. Mendoza (1980) showed that the modified likelihood ratio statistic for testing (3.8.29) is

$$W = -2 \log M = p v_e \log [\operatorname{tr}(\mathbf{A}) / v_e p] - \sum_{i=1}^{k} v_i \log |\mathbf{S}_i|$$

TABLE 3.8.7. Test of Sphericity and Circularity χ^2 Approximation.

	Sphericity (*p*-value)	df	Circularity (*p*-value)	df
Mauchly's test	48.702332 (2.5529E-9)	5	28.285484 (7.2092E-7)	2
Sugiura test	29.82531 (0.000016)	5	21.050999 (0.0000268)	2

where M is the likelihood ratio test statistic of H,

$$n = \sum_{i=1}^{k} n_i,\ v_i = n_i - 1, \quad v_e = n - k, \quad \text{and} \quad \mathbf{A} = \sum_{i=1}^{k} v_i \mathbf{S}_i$$

Letting $\rho = 1 - C$ where

$$C = \frac{\left\{\left[v_e p^2 (p+1)(2p+1) - 2v_e p^2\right]\left[\sum_{i=1}^{k} 1/v_i\right] - 4\right\}}{6v_e p\,[kp\,(p+1) - 2]}$$

Mendoza showed that

$$\chi^2 = (1 - C)\, W = -2\rho \log M \xrightarrow{d} \chi^2\,(f) \tag{3.8.30}$$

where $f = [kp(p+1)/2] - 1$.

An asymptotically equivalent test of sphericity in k populations is Rao's (1947) score test which uses the first derivative of the log likelihood called the vector of efficient scores; see Harris (1984). Silvey (1959) independently developed the test and called it the Lagrange Multiplier Test. Harris (1984) showed that

$$W = \frac{v_e p}{2}\left\{\frac{v_e p\left[\sum_{i=1}^{k} v_i \operatorname{tr}(\mathbf{S}_i)^2\right]}{\left[\sum_{i=1}^{k} v_i \operatorname{tr}(\mathbf{S}_i)\right]^{-1}}\right\} \xrightarrow{d} \chi^2\,(f) \tag{3.8.31}$$

where $f = (kp(p+1)/2) - 1$. When $k = 1$, the score test reduces to the locally best invariant test of sphericity. When $k > 2$, it is not known which test is optimal. Observe that the likelihood ratio test does not exist if $p > n_i$ for some group since the $|\mathbf{S}_i| = 0$. This is not the case for the Rao's score test since the test criterion involves calculating the trace of a matrix.

Example 3.8.5 (Sphericity in k Populations) *To test for sphericity in k populations, we use the test statistic developed by Harris (1984) given in (3.8.31). For the example, we use data set A for $k = 2$ groups. Thus, we are testing that $\Sigma_1 = \Sigma_2 = \sigma^2\mathbf{I}$. Replacing $\mathbf{S}_i$ by $\mathbf{C}'\mathbf{S}_i\mathbf{C}$ where $\mathbf{C}'\mathbf{C} = \mathbf{I}_{(p-1)}$is normalized, we also test that $\mathbf{C}'\Sigma_1\mathbf{C} = \mathbf{C}'\Sigma_2\mathbf{C} = \sigma^2\mathbf{I}$, the test of circularity. Again, program m3_8_1.sas is used. The results are given in Table 3.8.8. Both hypotheses are rejected.*

TABLE 3.8.8. Test of Sphericity and Circularity in k Populations.

	χ^2 Approximation		
	W	DFKPOP	PROB K POP
Sphericity	31.800318	11	0.0008211
Circularity	346.1505	5	< 0.0001

f. Tests of Independence

A problem encountered in multivariate data analysis is the determination of the independence of several groups of normally distributed variables. For two groups of variables, let $\mathbf{Y}_{p\times 1}$ and $\mathbf{X}_{q\times 1}$ represent the two subsets with covariance matrix

$$\Sigma = \begin{bmatrix} \Sigma_{\mathbf{YY}} & \Sigma_{\mathbf{YX}} \\ \Sigma_{\mathbf{XY}} & \Sigma_{\mathbf{XX}} \end{bmatrix}$$

The two sets of variables are independent under joint normality if $\Sigma_{\mathbf{XY}} = \mathbf{0}$. The hypothesis of independence is $H : \Sigma_{\mathbf{XY}} = \mathbf{0}$. This test is related to canonical correlation analysis discussed in Chapter 8.

In this section we review the modified likelihood ratio test of independence developed by Box (1949). The test allows one to test for the independence of k groups with p_i variables per group.

Let $\mathbf{Y}_j \sim IN_p(\boldsymbol{\mu}, \Sigma)$, for $j = 1, 2, \ldots, n$ where $p = \sum_{i=1}^{k} p_i$

$$\boldsymbol{\mu} = \begin{bmatrix} \mu_1 \\ \mu_2 \\ \vdots \\ \mu_k \end{bmatrix} \quad \text{and} \quad \Sigma = \begin{bmatrix} \Sigma_{11} & \Sigma_{12} & \cdots & \Sigma_{1k} \\ \Sigma_{21} & \Sigma_{22} & \cdots & \Sigma_{2k} \\ \vdots & \vdots & & \vdots \\ \Sigma_{k1} & \Sigma_{k2} & \cdots & \Sigma_{kk} \end{bmatrix}$$

then the test of independence is

$$H : \Sigma_{ij} = \mathbf{0} \text{ for } i \neq j = 1, 2, \ldots, k \tag{3.8.32}$$

Letting

$$W = \frac{|\mathbf{S}|}{|\mathbf{S}_{11}| \cdots |\mathbf{S}_{kk}|} = \frac{|\mathbf{R}|}{|\mathbf{R}_{11}| \cdots |\mathbf{R}_{kk}|}$$

where $\mathbf{S}$ is an unbiased estimate of Σ based on v_e degrees of freedom, and $\mathbf{R}$ is the sample correlation matrix, the test statistic is

$$X^2 = (1 - C)v_e \log W \xrightarrow{d} \chi^2(f) \tag{3.8.33}$$

where

$$G_s = \left[\sum_{i=1}^{k} p_i\right]^s - \sum_{i=1}^{k} p_i^s \qquad \text{for } s = 2, 3, 4$$
$$C = (2G_3 + 3G_2)/12 f v_e$$
$$f = G_2/2$$

The hypothesis of independence is rejected for large values of X^2. When p is large, Box's F approximation is used. Calculating

$$f_0 = (f+2)/\left|C_0 - C^2\right|$$
$$C_0 = (G_4 + 2G_3 - G_2)/12 f v_e^2$$
$$V = -v_e \log W$$

for $C_0 - C^2 > 0$, the statistic

$$F = V/a \xrightarrow{d} F(f, f_0) \tag{3.8.34}$$

where $a = f/[1 - C - (f/f_0)]$. If $C_0 - C^2 < 0$ then

$$F = f_0 V/f(b - V) \xrightarrow{d} F(f, f_0) \tag{3.8.35}$$

where $b = f_0/(1 - C + 2/f_0)$. Again, H is rejected for large values of F.

To estimate the p-value for the test, Box's asymptotic approximation is used, Anderson (1984, p. 386). The p-value of the test is estimated as

$$\begin{aligned}\alpha_p &= P\left(-m \log W \geq X_0^2\right)\\ &= P\left(\chi_f^2 \geq X_0^2\right) + \frac{\omega}{m^2}\left[P\left(\chi_{f+4}^2 \geq X_0^2\right) - P\left(\chi_f^2 \geq X_0^2\right)\right] + O\left(m^{-3}\right)\end{aligned} \tag{3.8.36}$$

where

$$m = v_e - \frac{3}{2} - \frac{G_3}{3G_2}$$
$$\omega = G_4/48 - 5G_2/96 - G_3/72G_2$$

A special case of the test of independence occurs when all $p_i = 1$. Then H becomes

$$H : \Sigma = \begin{bmatrix} \sigma_{11} & 0 & \cdots & 0 \\ 0 & \sigma_{22} & \cdots & 0 \\ \vdots & \vdots & & \vdots \\ 0 & 0 & \cdots & \sigma_{pp} \end{bmatrix}$$

which is equivalent to the hypothesis $H : \mathbf{P} = \mathbf{I}$ where $\mathbf{P}$ is the population correlation matrix. For this test,

$$W = \frac{|\mathbf{S}|}{\prod_{i=1}^{p} s_{ii}} = |\mathbf{R}|$$

and X^2 becomes

$$X^2 = [v_e - (2p+5)/6] \log W \xrightarrow{d} \chi^2(f) \tag{3.8.37}$$

where $f = p(p-1)/2$, developed independently by Bartlett (1950, 1954).

Example 3.8.6 (Independence) *Using the pooled within covariance matrix* **S** *based on* $v_e = n_1 + n_2 - 2 = 46$ *degrees of freedom for data set A, we test that the first set of two variables is independent of the third. Program m3_8_1.sas contains the SAS IML code to perform the test. The results are shown in Table 3.8.9. Thus, we reject the null hypothesis that the first two variables are independent of the third variable for the pooled data.*

TABLE 3.8.9. Test of Independence χ^2 Approximation.

INDCH1	INDF	INDPROB
34.386392	2	3.4126E-8

g. *Tests for Linear Structure*

When analyzing general linear mixed models in ANOVA designs, often called components of variance models, the covariance matrix for the observation vectors $\mathbf{y}_n$ has the general structure $\Omega = \sum_{i=1}^{k} \sigma_i^2 \mathbf{Z}_i \mathbf{Z}_i' + \sigma_e^2 \mathbf{I}_n$. Associating Σ with Ω and $\mathbf{G}_i$ with the known matrices $\mathbf{Z}_i \mathbf{Z}_i'$ and $\mathbf{I}_n$, the general structure of Σ is linear where σ_i^2 are the components of variance

$$\Sigma = \sigma_1^2 \mathbf{G}_1 + \sigma_2^2 \mathbf{G}_2 + \cdots + \sigma_k^2 \mathbf{G}_k \tag{3.8.38}$$

Thus, we may want to test for linear structure.

In multivariate repeated measurement designs where vector-valued observations are obtained at each time point, the structure of the covariance matrix for normally distributed observations may have the general form

$$\Sigma = \mathbf{G}_1 \otimes \Sigma_1 + \mathbf{G}_2 \otimes \Sigma_2 + \cdots + \mathbf{G}_k \otimes \Sigma_k \tag{3.8.39}$$

where the $\mathbf{G}_i$ are known commutative matrices and the Σ_i matrices are unknown. More generally, if the $\mathbf{G}_i$ do not commute we may still want to test that Σ has linear structure; see Krishnaiah and Lee (1976).

To illustrate, suppose a repeated measurement design has t time periods and at each time period a vector of p dependent variables are measured. Then for $i = 1, 2, \ldots, n$ subjects an observation vector has the general form $\mathbf{y}' = (\mathbf{y}_1', \mathbf{y}_2', \ldots, \mathbf{y}_t')$ where each $\mathbf{y}_i$ is a $p \times 1$ vector of responses. Assume $\mathbf{y}'$ follows a MVN distribution with mean $\boldsymbol{\mu}$ and covariance matrix

$$\Sigma = \begin{bmatrix} \Sigma_{11} & \Sigma_{12} & \cdots & \Sigma_{1t} \\ \Sigma_{21} & \Sigma_{22} & \cdots & \Sigma_{2t} \\ \vdots & \vdots & & \vdots \\ \Sigma_{t1} & \Sigma_{t2} & \cdots & \Sigma_{tt} \end{bmatrix} \tag{3.8.40}$$

Furthermore, assume there exists an orthogonal matrix $\mathbf{M}_{t \times q}$ of rank $q = t - 1$ such that $(\mathbf{M}' \otimes \mathbf{I}_p)\mathbf{y} = \mathbf{y}^*$ where $\mathbf{M}'\mathbf{M} = \mathbf{I}_q$. Then the covariance structure for $\mathbf{y}^*$ is

$$\Sigma^*_{pq \times pq} = \left(\mathbf{M} \otimes \mathbf{I}_p\right)' \Sigma \left(\mathbf{M} \otimes \mathbf{I}_p\right) \tag{3.8.41}$$

The matrix Σ^* has multivariate sphericity (or circularity) structure if

$$\Sigma^* = \mathbf{I}_q \otimes \Sigma_e \tag{3.8.42}$$

where Σ_e is the covariance matrix for $\mathbf{y}_i$.

Alternatively, suppose Σ has the structure given in (3.8.40) and suppose $\Sigma_{ii} = \Sigma_e + \Sigma_\lambda$ for $i = j$ and $\Sigma_{ij} = \Sigma_\lambda$ for $i \neq j$, then Σ has multivariate compound symmetry structure

$$\Sigma = \mathbf{I}_t \otimes \Sigma_e + \mathbf{J}_{t \times t} \otimes \Sigma_\lambda \tag{3.8.43}$$

where $\mathbf{J}$ is a matrix of 1s. Reinsel (1982) considers multivariate random effect models with this structure. Letting $\Sigma_{ij} = \Sigma_1$ for $i = j$ and $\Sigma_{ij} = \Sigma_\lambda$ for $i \neq j$, (3.8.43) has the form

$$\Sigma = \mathbf{I}_t \otimes \Sigma_1 + (\mathbf{J}_{t \times t} - \mathbf{I}_t)\, \Sigma_2$$

Krishnaiah and Lee (1976, 1980) call this the block version intraclass correlation matrix. The matrix has multivariate compound symmetry structure. These structures are all special cases of (3.8.39).

To test the hypothesis

$$H : \Sigma = \sum_{i=1}^{k} \mathbf{G}_i \otimes \Sigma_i \tag{3.8.44}$$

where the $\mathbf{G}_i$ are known $q \times q$ matrices and Σ_i is an unknown matrix of order $p \times p$, assume we have a random sample of n vectors $\mathbf{y}' = (\mathbf{x}_1', \mathbf{x}_2', \ldots, \mathbf{x}_q')$ from a MVN distribution where the subvectors $\mathbf{x}_i$ are $p \times 1$ vectors. Then the cov $(\mathbf{y}) = \Sigma = \left[\Sigma_{ij}\right]$ where Σ_{ij} are unknown covariance matrices of order $p \times p$, or $\mathbf{y} \sim N_{pq}(\boldsymbol{\mu}, \Sigma)$. The likelihood ratio statistic for testing H in (3.8.44) is

$$\lambda = \left|\widehat{\Sigma}\right|^{n/2} / \left|\sum_{i=1}^{k} \mathbf{G}_i \otimes \widehat{\Sigma}_i\right|^{n/2} \tag{3.8.45}$$

where

$$n\Sigma = \sum_{i=1}^{n} (\mathbf{y}_i - \overline{\mathbf{y}})\,(\mathbf{y}_i - \overline{\mathbf{y}})'$$

and $\widehat{\Sigma}_i$ is the maximum likelihood estimate of Σ_i which is usually obtained using an iterative algorithm, except for some special cases. Then,

$$-2 \log \lambda \xrightarrow{d} \chi^2(f) \tag{3.8.46}$$

As a special case of (3.8.44), we consider testing that Σ^* has multivariate sphericity structure given in (3.8.42), discussed by Thomas (1983) and Boik (1988). Here $k = 1$ and

$\mathbf{I}_q = \mathbf{G}_1$ Assuming $\Sigma_{11}^* = \Sigma_{22}^* = \cdots = \Sigma_{qq}^* = \Sigma_e$, the likelihood ratio statistic for multivariate sphericity is

$$\lambda = \frac{|\widehat{\Sigma}|^{n/2}}{\left|\mathbf{I}_q \otimes \widehat{\Sigma}_e\right|^{n/2}} = \frac{|\widehat{\Sigma}|^{n/2}}{\left|\widehat{\Sigma}_e\right|^{nq/2}} \tag{3.8.47}$$

with $f = [pq(pq+1) - p(p+1)]/2 = p(q-1)(pq+p+1)/2$ and $-2\log\lambda \xrightarrow{d} \chi^2(f)$.

To estimate $\widehat{\Sigma}$, we construct the error sum of square and cross products matrix

$$\mathbf{E} = \left(\mathbf{M} \otimes \mathbf{I}_p\right)' \left[\sum_{i=1}^{n} (\mathbf{y}_i - \overline{\mathbf{y}})(\mathbf{y}_i - \overline{\mathbf{y}})'\right] \left(\mathbf{M} \otimes \mathbf{I}_p\right)$$

Then, $n\widehat{\Sigma} = \mathbf{E}$. Partitioning $\mathbf{E}$ into $p \times p$ submatrices, $\mathbf{E} = \left[\mathbf{E}_{ij}\right]$ for $i, j = 1, 2, \ldots, q = t-1$, $n\widehat{\Sigma}_e = \sum_{i=1}^{q} \mathbf{E}_{ii}/q$. Substituting the estimates into (3.8.47), the likelihood ratio statistic becomes

$$\lambda = \mathbf{E}^{n/2}/|q^{-1}\sum_{i=1}^{q} \mathbf{E}_{ii}|^{nq/2} \tag{3.8.48}$$

as developed by Thomas (1983). If we let α_i $(i = 1, \ldots, pq)$ be the eigenvalues of $\mathbf{E}$, and β_i $(i = 1, \ldots, p)$ the eigenvalues of $\sum_{i=1}^{q} \mathbf{E}_{ii}$, a simple form of (3.8.48) is

$$U = -2\log\lambda = n[q\sum_{i=1}^{p} \log(\beta_i) - \sum_{i=1}^{pq} \log(\alpha_i)] \tag{3.8.49}$$

When p or q are large relative to n, the asymptotic approximation U may be poor. To correct for this, Boik (1988) using Box's correction factor for the distribution of U showed that the

$$\begin{aligned} P(U \le U_o) &= P\left(\rho^* U \le \rho^* U_o\right) \\ &\quad \times (1-\omega) P\left(X_f^2 \le \rho^* U_o\right) + \omega P\left(X_{f+4}^2 \le \rho^* U_o\right) + O\left(v_e^{-3}\right) \end{aligned} \tag{3.8.50}$$

where $f = p(q+1)(pq+p+1)$, and

$$\begin{aligned} \rho &= 1 - p[2p^2\left(q^4-1\right) + 3p\left(q^3-1\right) - \left(q^2-1\right)]/12qf v_e \\ \rho^* &= \rho v_e/n \\ \omega &= \left(2\rho^2\right)^{-1}\left\{\left[\frac{(pq-1)\,pq\,(pq+1)(pq+2)}{24v_e}\right]\right. \\ &\quad \left. - \left[\frac{(p-1)\,p\,(p+1)(p+2)}{24q^2v_e^2}\right]\left[\frac{(f-\rho)^2}{2}\right]\right\} \end{aligned} \tag{3.8.51}$$

and $v_e = n - R(\mathbf{X})$. Hence, the p-value for the test of multivariate sphericity using Box's correction becomes

$$P\left(\rho^* U \ge U_o\right) = (1-\omega) P\left(X_f^2 \ge U_o\right) + \omega P\left(X_{f+4}^2 \ge U_o\right) + O\left(v_e^{-3}\right) \tag{3.8.52}$$

TABLE 3.8.10. Test of Multivariate Sphericity Using Chi-Square and Adjusted Chi-Square Statistics

CHI_2	DF	PVALUE
74.367228	15	7.365E-10
	RHO	
	0.828125	
	OMEGA	
	0.0342649	
RO_CHI_2		CPVALUE
54.742543		2.7772E-6

Example 3.8.7 (Test of Circularity) *For the data from Timm (1980, Table 7.2), used to illustrate a multivariate mixed model (MMM) and a doubly multivariate model (DMM), discussed in Chapter 6, Section 6.9, and illustrated by Boik (1988), we test the hypothesis that Σ^* has the multivariate structure given by (3.8.41). Using (3.8.49), the output for the test using program m3_8_7.sas is provided in Table 3.8.10.*

Since $-2 \log \lambda = 74.367228$ with $df = 15$ with a p-value for the test equal to 7.365×10^{-10} *or using Box's correction,* $\rho^*\mathbf{U} = 54.742543$ *with the p-value* $= 2.7772 \times 10^{-6}$, *we reject the null hypothesis of multivariate sphericity.*

In the case of multivariate sphericity, the matrix $\Sigma^* = \mathbf{I}_q \otimes \Sigma_e$. More generally, suppose Σ^* has Kronecker structure, $\Sigma^* = \Sigma_q \otimes \Sigma_e$ where both matrices are unknown. For this structure, the covariance matrix for the $q = t-1$ contrasts in time is not the identity matrix. Models that permit the analysis of data with a general Kronecker structure are discussed in Chapter 6.

Estimation and tests of covariance matrix structure is a field in statistics called structural equation modeling. While we will review this topic in Chapter 10, the texts by Bollen (1989) and Kaplan (2000) provide a comprehensive treatment of the topic.

Exercises 3.8

1. Generate a multivariate normal distribution with mean structure and covariance structure given in Exercises 3.7.1 for $n_1 = n_2 = 100$ and seed 1056799.

   (a) Test that $\Sigma_1 = \Sigma_2$.

   (b) Test that the pooled $\Sigma = \sigma^2\mathbf{I}$ and that $\Sigma = \sigma^2 \left[(1-\rho)\, I + \rho\mathbf{J}\right]$.

   (c) Test that $\Sigma_1 = \Sigma_2 = \sigma^2\mathbf{I}$ and that $\mathbf{C}'\Sigma_1\mathbf{C} = \mathbf{C}'\Sigma_2\mathbf{C} = \sigma^2\mathbf{I}$.

2. For the data in Table 3.7.3, determine whether the data satisfy the compound symmetry structure or more generally has circularity structure.

3. For the data in Table 3.7.3, determine whether the measurements at age 8 and 8.5 are independent of the measurements at ages 9 and 9.5.

4. Assume the data in Table 3.7.3 represent two variables at time one, the early years (ages 8 and 8.5), and two variables at the time two, the later years (ages 9 and 9.5). Test the hypothesis that the matrix has multivariate sphericity (or circularity) structure.

5. For the data in Table 3.7.4, test that the data follow a MVN distribution and that $\Sigma_1 = \Sigma_2$.

3.9 Tests of Location

A frequently asked question in studies involving multivariate data is whether there is a group difference in mean performance for p variables. A special case of this general problem is whether two groups are different on p variables where one group is the experimental treatment group and the other is a control group. In practice, it is most often the case that the sample sizes of the groups are not equivalent possibly due to several factors including study dropout.

a. *Two-Sample Case,* $\Sigma_1 = \Sigma_2 = \Sigma$

The null hypothesis for the analysis is whether the group means are equal for all variables

$$H : \begin{bmatrix} \mu_{11} \\ \mu_{12} \\ \vdots \\ \mu_{1p} \end{bmatrix} = \begin{bmatrix} \mu_{21} \\ \mu_{22} \\ \vdots \\ \mu_{2p} \end{bmatrix} \text{ or } \boldsymbol{\mu}_1 = \boldsymbol{\mu}_2 \tag{3.9.1}$$

The alternative hypothesis is $A : \boldsymbol{\mu}_1 \neq \boldsymbol{\mu}_2$. The subjects in the control group $i = 1, 2, \ldots, n_1$ are assumed to be a random sample from a multivariate normal distribution, $\mathbf{Y}_i \sim IN_p(\boldsymbol{\mu}_1, \Sigma)$. The subjects in the experimental group, $i = n_1 + 1, \ldots, n_2$ are assumed independent of the control group and multivariate normally distributed: $\mathbf{X}_i \sim IN_p(\boldsymbol{\mu}_2, \Sigma)$. The observation vectors have the general form

$$\begin{aligned} \mathbf{y}_i' &= [y_{i1}, y_{i2}, \ldots, y_{ip}] \\ \mathbf{x}_i' &= [x_{i1}, x_{i2}, \ldots, x_{ip}] \end{aligned} \tag{3.9.2}$$

where $\overline{\mathbf{y}} = \sum_{i=1}^{n_1} \mathbf{y}_i / n_1$ and $\overline{\mathbf{x}} = \sum_{i=n_1+1}^{n_2} \mathbf{x}_i / n_2$. Because $\Sigma_1 = \Sigma_2 = \Sigma$, an unbiased estimate of the common covariance matrix Σ is the pooled covariance matrix $\mathbf{S} = [(n_1 - 1)\,\mathbf{E}_1 + (n_2 - 1)\,\mathbf{E}_2]/(n_1 + n_2 - 2)$ where $\mathbf{E}_i$ is the sum of squares and cross products (SSCP) matrix for the ith group computed as

$$\begin{aligned} \mathbf{E}_1 &= \sum_{i=1}^{n_1} (\mathbf{y}_i - \overline{\mathbf{y}})\,(\mathbf{y}_i - \overline{\mathbf{y}})' \\ \mathbf{E}_2 &= \sum_{i=n_1+1}^{n_2} (\mathbf{x}_i - \overline{\mathbf{x}})\,(\mathbf{x}_i - \overline{\mathbf{x}})' \end{aligned} \tag{3.9.3}$$

To test H in (3.9.1), Hotelling's T^2 statistic derived in Example 3.5.2 is used. The statistic is

$$\begin{aligned} T^2 &= \left(\frac{n_1 n_2}{n_1 + n_2}\right) (\overline{\mathbf{y}} - \overline{\mathbf{x}})' \mathbf{S}^{-1} (\overline{\mathbf{y}} - \overline{\mathbf{x}}) \\ &= \left(\frac{n_1 n_2}{n_1 + n_2}\right) D^2 \end{aligned} \tag{3.9.4}$$

Following the test, one is usually interested in trying to determine which linear combination of the difference in mean vectors led to significance. To determine the significant linear combinations, contrasts of the form $\psi = \mathbf{a}' (\boldsymbol{\mu}_1 - \boldsymbol{\mu}_2) = \mathbf{a}' \Delta$ are constructed where the vector $\mathbf{a}$ is any vector of real numbers. The $100 (1 - \alpha)$ % simultaneous confidence interval has the general structure

$$\widehat{\psi} - c_\alpha \widehat{\sigma}_{\widehat{\psi}} \leq \psi \leq \widehat{\psi} + c_\alpha \widehat{\sigma}_{\widehat{\psi}} \tag{3.9.5}$$

where $\widehat{\psi}$ is an unbiased estimate of ψ, $\widehat{\sigma}_{\widehat{\psi}}$ is the estimated standard deviation of $\widehat{\psi}$, and c_α is the critical value for a size α test. For the two-group problem,

$$\begin{aligned} \widehat{\psi} &= \mathbf{a}' (\overline{\mathbf{y}} - \overline{\mathbf{x}}) \\ \widehat{\sigma}^2_{\widehat{\psi}} &= \left(\frac{n_1 + n_2}{n_1 n_2}\right) \mathbf{a}'\mathbf{S}\mathbf{a} \\ c^2_\alpha &= \frac{p v_e}{v_e - p + 1} F^{1-\alpha} (p,\ v_e - p + 1) \end{aligned} \tag{3.9.6}$$

where $v_e = n_1 + n_2 - 2$.

With the rejection of H, one first investigates contrasts a variable at a time by selecting $\mathbf{a}'_i = (0, 0, \ldots, 0, 1_i, 0, \ldots, 0)$ for $1, 2, \ldots, p$ where the value one is in the ith position. Although the contrasts using these $\mathbf{a}_i$ are easy to interpret, none may be significant. However, when H is rejected there exists at least one vector of coefficients that is significantly different from zero, in that $| \widehat{\psi} | > c_\alpha \widehat{\sigma}_{\widehat{\psi}}$, so that the confidence set does not cover zero.

To locate the maximum contrast, observe that

$$\begin{aligned} T^2 &= \left(\frac{n_1 n_2}{n_1 + n_2}\right) (\overline{\mathbf{y}} - \overline{\mathbf{x}})' \mathbf{S}^{-1} (\overline{\mathbf{y}} - \overline{\mathbf{x}}) \\ &= v_e \left(\frac{1}{n_1} + \frac{1}{n_2}\right)^{-1} (\overline{\mathbf{y}} - \overline{\mathbf{x}})' \mathbf{E}^{-1} (\overline{\mathbf{y}} - \overline{\mathbf{x}}) \\ &= v_e \operatorname{tr} \left[(\overline{\mathbf{y}} - \overline{\mathbf{x}}) \left(\frac{1}{n_1} + \frac{1}{n_2}\right)^{-1} (\overline{\mathbf{y}} - \overline{\mathbf{x}})' \mathbf{E}^{-1} \right] \\ &= v_e \operatorname{tr} \left(\mathbf{H}\mathbf{E}^{-1} \right) \end{aligned} \tag{3.9.7}$$

where $\mathbf{E} = v_e \mathbf{S}$ and $v_e = n_1 + n_2 - 2$ so that $T^2 = v_e \lambda_1$ where λ_1 is the root of $|\mathbf{H} - \lambda \mathbf{E}| = 0$. By Theorem 2.6.10,

$$\lambda_1 = \max_{\mathbf{a}} \left(\mathbf{a}'\mathbf{H}\mathbf{a} / \mathbf{a}'\mathbf{E}\mathbf{a} \right)$$

so that

$$T^2 = v_e \max_{\mathbf{a}} \left(\mathbf{a}'\mathbf{Ha}/\mathbf{a}'\mathbf{Ea}\right) \tag{3.9.8}$$

where $\mathbf{a}$ is the eigenvector of $|\mathbf{H} - \lambda\mathbf{E}| = 0$ associated with the root λ_1. To find a solution, observe that

$$(\mathbf{H} - \lambda\mathbf{E})\,\mathbf{a}_w = \mathbf{0}$$

$$\left[\left(\frac{n_1 n_2}{n_1 + n_2}\right)(\overline{\mathbf{y}} - \overline{\mathbf{x}})\,(\mathbf{y} - \overline{\mathbf{x}})' - \lambda\mathbf{E}\right]\mathbf{a}_w = \mathbf{0}$$

$$\frac{1}{\lambda}\left(\frac{n_1 n_2}{n_1 + n_2}\right)\mathbf{E}^{-1}\,(\overline{\mathbf{y}} - \overline{\mathbf{x}})\,(\overline{\mathbf{y}} - \overline{\mathbf{x}})'\,\mathbf{a}_w = \mathbf{a}_w$$

$$\left[\frac{1}{\lambda}\left(\frac{n_1 n_2}{n_1 + n_2}\right)(\overline{\mathbf{y}} - \overline{\mathbf{x}})'\,\mathbf{a}_w\right]\mathbf{E}^{-1}\,(\overline{\mathbf{y}} - \overline{\mathbf{x}}) = \mathbf{a}_w$$

$$(\text{constant})\,\mathbf{E}^{-1}\,(\overline{\mathbf{y}} - \overline{\mathbf{x}}) = \mathbf{a}_w$$

so that

$$\mathbf{a}_w = \mathbf{E}^{-1}\,(\overline{\mathbf{y}} - \overline{\mathbf{x}}) \tag{3.9.9}$$

is an eigenvector associated with λ_1. Because the solution is not unique, an alternative solution is

$$\mathbf{a}_s = \mathbf{S}^{-1}\,(\overline{\mathbf{y}} - \overline{\mathbf{x}}) \tag{3.9.10}$$

The elements of the weight vector $\mathbf{a}$ are called discriminant weights (coefficients) since any contrast proportional to the weights provide for maximum separation between the two centroids of the experimental and control groups. When the observations are transformed by $\mathbf{a}_s$ they are called discriminant scores. The linear function used in the transformation is called the Fisher's linear discriminant function. If one lets $L_E = \mathbf{a}_s'\mathbf{y}$ represents the observations in the experimental group and $L_C = \mathbf{a}_s'\mathbf{x}$ the corresponding observations in the control group where L_{iE} and L_{iC} are the observations in each group, the multivariate observations are transformed to a univariate problem involving discriminant scores. In this new, transformed, problem we may evaluate the difference between the two groups by using a t statistic that is created from the discriminant scores. The square of the t statistic is exactly Hotelling's T^2 statistic. In addition, the square of Mahalanobis' distance is equal to the mean difference in the sample mean discriminant scores, $D^2 = \overline{L}_E - \overline{L}_C$, when the weights $\mathbf{a}_s = \mathbf{S}^{-1}(\overline{\mathbf{y}} - \overline{\mathbf{x}})$ are used in the linear discriminant function. (Discriminant analysis is discussed in Chapter 7.)

Returning to our two-group inference problem, we can create the linear combination of the mean difference that led to the rejection of the null hypothesis. However, because the linear combination is not unique, it is convenient to scale the vector of coefficients $\mathbf{a}_s$ or $\mathbf{a}_w$ so that the within-group variance of the discriminant scores are unity, then

$$\mathbf{a}_{ws} = \frac{\mathbf{a}_w}{\sqrt{\mathbf{a}_w'\mathbf{S}\mathbf{a}_w}} = \frac{\mathbf{a}_s}{\sqrt{\mathbf{a}_s'\mathbf{S}\mathbf{a}_s}} \tag{3.9.11}$$

This coefficient vector is called the normalized discriminant coefficient vector. Because it is an eigenvector, it is only unique up to a change in sign so that one may use $\mathbf{a}_{ws}$ or $-\mathbf{a}_{ws}$.

Using these coefficients to construct a contrast in the mean difference, the difference in the mean vectors weighted by $\mathbf{a}_{ws}$ yields D, the number of within-group standard deviations separating the mean discriminant scores for the two groups. That is

$$\psi_{ws} = \mathbf{a}'_{ws}\,(\overline{\mathbf{y}} - \overline{\mathbf{x}}) = D \tag{3.9.12}$$

To verify this, observe that

$$\begin{aligned}
\psi_{ws} &= \mathbf{a}'_{ws}\,(\overline{\mathbf{y}} - \overline{\mathbf{x}}) \\
&= \frac{\mathbf{a}'_{w}}{\sqrt{\mathbf{a}'_{w}\mathbf{S}\mathbf{a}_{w}}}\,(\overline{\mathbf{y}} - \overline{\mathbf{x}}) \\
&= \frac{(\overline{\mathbf{y}} - \overline{\mathbf{x}})'\,\mathbf{E}^{-1}\,(\overline{\mathbf{y}} - \mathbf{x})}{\sqrt{(\overline{\mathbf{y}} - \overline{\mathbf{x}})'\,\mathbf{E}^{-1}\mathbf{S}\mathbf{E}^{-1}\,(\overline{\mathbf{y}} - \mathbf{x})}} \\
&= v_e D^2 / \sqrt{v_e^2 D^2} \\
&= D
\end{aligned}$$

Alternatively, using the contrast $\psi_s = \mathbf{a}'_s\,(\overline{\mathbf{y}} - \overline{\mathbf{x}}) = D^2$ and $\psi_{\max} = \frac{n_1 n_2}{n_1 + n_2}\psi_s$. In practice, these contrasts may be difficult to interpret and thus one may want to locate a weight vector $\mathbf{a}$ that only contains 1s and 0s. In this way the parametric function may be more interpretable. To locate the variables that may contribute most to group separation, one creates a scale free vector of weights $\mathbf{a}_{wsa} = (\operatorname{diag}\mathbf{S})^{1/2}\,\mathbf{a}_{ws}$ called the vector of standardized coefficients. The absolute value of the scale-free standardized coefficients may be used to rank order the variables that contributed most to group separation. The standardized coefficients represent the influence of each variable to group separation given the inclusion of the other variables in the study. Because the variables are correlated, the size of the coefficient may change with the deletion or addition of variables in the study.

An alternative method to locate significant variables and to construct contrasts is to study the correlation between the discriminant function $L = \mathbf{a}'\mathbf{y}$ and each variable, ρ_i. The vector of correlations is

$$\boldsymbol{\rho} = \frac{(\operatorname{diag}\Sigma)^{-1/2}\,\Sigma\mathbf{a}}{\sqrt{\mathbf{a}'\Sigma\mathbf{a}}} \tag{3.9.13}$$

Replacing Σ with $\mathbf{S}$, an estimate of $\boldsymbol{\rho}$ is

$$\widehat{\boldsymbol{\rho}} = \frac{(\operatorname{diag}\mathbf{S})^{-1/2}\,\mathbf{S}\mathbf{a}}{\sqrt{\mathbf{a}'\mathbf{S}\mathbf{a}}} \tag{3.9.14}$$

Letting $\mathbf{a} = \mathbf{a}_{ws}$

$$\begin{aligned}
\widehat{\boldsymbol{\rho}} &= (\operatorname{diag}\mathbf{S})^{-1/2}\,\mathbf{S}\mathbf{a}_{ws} \\
&= (\operatorname{diag}\mathbf{S})^{-1/2}\,\mathbf{S}\,(\operatorname{diag}\mathbf{S})^{-1/2}\,(\operatorname{diag}\mathbf{S})^{1/2}\,\mathbf{a}_{ws} \\
&= \mathbf{R}_e\,(\operatorname{diag}\mathbf{S})^{1/2}\,\mathbf{a}_{ws} \\
&= \mathbf{R}_e\mathbf{a}_{wsa}
\end{aligned} \tag{3.9.15}$$

where $\mathbf{a}_{wsa}$ is the within standardized adjusted vector of standardized weights. Investigating $\widehat{\rho}$, the variables associated with low correlations contribute least to the separation of the centroids. Contrasts are constructed by excluding variables with low correlations from the contrast and setting coefficients to one for high correlations. This process often leads to a contrast in the means that is significant and meaningful involving several individual variables; see Bargman (1970). Rencher (1988) shows that this procedure isolates variables that contribute to group separation, ignoring the other variable in the study. This is not the case for standardized coefficients. One may use both procedures to help to formulate meaningful contrasts when a study involves many variables.

Using (3.9.5) to obtain simultaneous confidence intervals for any number of comparisons involving parametric functions of the mean difference ψ as defined in (3.9.6), we know the interval has probability greater than $1 - \alpha$ of including the true population value. If one is only interested in a few comparisons, say p, one for each variable, the probability is considerably larger then $1 - \alpha$. Based on studies by Hummel and Sligo (1971), Carmer and Swanson (1972), and Rencher and Scott (1990), one may also calculate univariate t-tests using the upper $(1 - \alpha)/2$ critical value for each test to facilitate locating significant differences in the means for each variable when the overall multivariate test is rejected. These tests are called protected t-tests, a concept originally suggested by R.A. Fisher. While this procedure will generally control the overall Type I error at the nominal α level for all comparisons identified as significant at the nominal level α, the univariate critical values used for each test may not be used to construct simultaneous confidence intervals for the comparisons The intervals are too narrow to provide an overall confidence level of $100(1 - \alpha)\%$. One must adjust the value of alpha for each comparison to maintain a level not less than $1 - \alpha$ as in planned comparisons, which we discuss next. When investigating planned comparisons, one need not perform the overall test.

In our discussion of the hypothesis $H : \boldsymbol{\mu}_1 = \boldsymbol{\mu}_2$, we have assumed that the investigator was interested in all contrasts $\psi = \mathbf{a}' (\boldsymbol{\mu}_1 - \boldsymbol{\mu}_2)$. Often this is not the case and one is only interested in the p planned comparisons $\psi_i = \mu_{1i} - \mu_{2i}$ for $i = 1, 2, \ldots, p$. In these situations, it is not recommended that one perform the overall T^2 test, but instead one should utilize a simultaneous test procedure (STP). The null hypothesis in this case is

$$H = \bigcap_{i=1}^{p} H_i : \psi_i = 0 \tag{3.9.16}$$

versus the alternative that at least one ψ_i differs from zero. To test this hypothesis, one needs an estimate of each ψ_i and the joint distribution of the vector $\widehat{\theta}' = (\widehat{\psi}_1, \widehat{\psi}_2, \ldots, \widehat{\psi}_p)$. Dividing each element $\widehat{\psi}_i$ by $\widehat{\sigma}_{\widehat{\psi}_i}$, we have a vector of correlated t statistics, or by squaring each ratio, F-tests, $F_i = \widehat{\psi}_i^2 / \widehat{\sigma}_{\widehat{\psi}_i}^2$. However, the joint distribution of the F_i is not multivariate F since the standard errors $\widehat{\sigma}_{\widehat{\psi}_i}^2$ do not depend on a common unknown variance. To construct approximate simultaneous confidence intervals for each of the p contrasts simultaneously near the overall level $1 - \alpha$, we use Šidák's inequality and the multivariate t distribution with a correlation matrix of the accompanying MVN distribution, $\mathbf{P} = \mathbf{I}$, also called the Studentized Maximum Modulus distribution, discussed by Fuchs and Sampson (1987). The approximate Šidák multivariate t, $100\,(1 - \alpha)\,\%$ simultaneous confidence in-

TABLE 3.9.1. MANOVA Test Criteria for Testing $\mu_1 = \mu_2$.

	$s = 1$	$M = 0.5$	$N = 22$		
Statistics	Value	F	NumDF	DenDF	Pr > F
Wilks' lambda	0.12733854	105.0806	3	46	0.0001
Pillai's trace	0.87266146	105.0806	3	46	0.0001
Hotelling-Lawley trace	6.85308175	105.0806	3	46	0.0001
Roy's greatest root	6.85308175	105.0806	3	46	0.0001

tervals have the simple form

$$\widehat{\psi}_i - c_\alpha \widehat{\sigma}_{\widehat{\psi}_i} \leq \psi \leq \widehat{\psi}_i + c_\alpha \widehat{\sigma}_{\widehat{\psi}_i} \tag{3.9.17}$$

where

$$\widehat{\sigma}^2_{\widehat{\psi}} = (n_1 + n_2)\, s^2_{ii} / n_1 n_2 \tag{3.9.18}$$

and s^2_{ii} is the ith diagonal element of $\mathbf{S}$ and c_α is the upper α critical value of the Studentized Maximum Modulus distribution with degrees of freedom $v_e = n_1 + n_2 - 2$ and $p = C$, comparisons. The critical values for c_α for $p = 2\,(16)\,,\ 18\,(2)\,20$, and $\alpha = 0.05$ are given in the Appendix, Table V. As noted by Fuchs and Sampson (1987), the intervals obtained using the multivariate t are always shorter that the corresponding Bonferroni-Dunn or Dunn-Šidák (independent t) intervals that use the Student t distribution to control the overall Type I error near the nominal level α.

If we can, a priori, place an order of importance on the variables in a study, a stepdown procedure is recommended. While one may use Roy's stepdown F statistics, the finite intersection test procedure proposed by Krishnaiah (1979) and reviewed by Timm (1995) is optimal in the Neyman sense, i.e., yielding the smallest confidence intervals. We discuss this method in Chapter 4 for the $k > 2$ groups.

Example 3.9.1 (Testing $\boldsymbol{\mu}_1 = \boldsymbol{\mu}_2$, Given $\Sigma_1 = \Sigma_2$) *We illustrate the test of the hypothesis $H_o : \boldsymbol{\mu}_1 = \boldsymbol{\mu}_2$ using the data set A generated in program m3_7_1.sas. There are three dependent variables and two groups with 25 observations per group. To test that the mean vectors are equivalent, the SAS program m_3_9a.sas is used using the SAS procedure GLM. Because this program is using the MR model to test for differences in the means, the matrices* **H** *and* **E** *are calculated. Hotelling's (1931) T^2 statistic is related to an F distribution using Definition 3.5.3. And, from (3.5.4) $T^2 = v_e \widehat{\lambda}_1$ when $s = 1$ where $\widehat{\lambda}_1$ is the largest root of $|\,\mathbf{H} - \boldsymbol{\lambda}\mathbf{E}\,| = \mathbf{0}$. A portion of the output is provided in Table 3.9.1.*

Thus, we reject the null hypothesis that $\boldsymbol{\mu}_1 = \boldsymbol{\mu}_2$. To relate T^2 to the F distribution, we have from Definition 3.5.3 and (3.5.4) that

$$\begin{aligned} F &= (v_e + p + 1)\, T^2 / p v_e \\ &= (v_e - p + 1)\, v_e \widehat{\lambda}_1 / p v_e \\ &= (v_e - p + 1)\, \widehat{\lambda}_1 / p \\ &= (46)\,(6.85308)\,/3 = 105.0806 \end{aligned}$$

TABLE 3.9.2. Discriminant Structure Vectors, $H : \mu_1 = \mu_2$.

Within Structure $\widehat{\boldsymbol{\rho}}$	Standarized Vector $\mathbf{a}_{wsa}$	Raw Vector $\mathbf{a}_{ws}$
0.1441	0.6189	0.219779947
0.2205	−1.1186	−0.422444930
0.7990	3.2494	0.6024449655

as shown in Table 3.9.1. Rejection of the null hypothesis does not tell us which mean difference led to the significant difference. To isolate where to begin looking, the standardized discriminant coefficient vector and the correlation structure of the discriminate function with each variable is studied.

To calculate the coefficient vectors and correlations using SAS, the /CANONICAL option is used in the MANOVA statement for PROC GLM. SAS labels the vector $\widehat{\boldsymbol{\rho}}$ in (3.9.15) the within canonical structure vector. The vector $\mathbf{a}_{wsa}$ in (3.9.15) is labeled the Standardized Canonical Coefficients and the discriminant weights $\mathbf{a}_{ws}$ in (3.9.11) are labeled as Raw Canonical Coefficients. The results are summarized in Table 3.9.2.

From the entries in Table 3.9.2, we see that we should investigate the significance of the third variable using (3.9.5) and (3.9.6). For $\alpha = 0.05$,

$$c_\alpha^2 = (3)\,(48)\,F^{0.95}\,(3, 46)\,/46 = 144\,(2.807)\,/46 = 8.79$$

so that $c_\alpha = 2.96$. The value of $\widehat{\sigma}_{\widehat{\psi}}$ is obtained from the diagonal of $\mathbf{S}$. Since SAS provides $\mathbf{E}$, we divide the diagonal element by $v_e = n_1 + n_2 - 2 = 48$. The value of $\widehat{\sigma}_{\widehat{\psi}}$ for the third variable is $\sqrt{265.222/48} = 2.35$.

Thus, for $\mathbf{a}' = (0, 0, 1)$, the 95% simultaneous confidence interval for the mean difference in means ψ for variable three, $\widehat{\psi} = 29.76 - 20.13 = 9.63$, is estimated as follows.

$$\begin{array}{ccc} \widehat{\psi} - c_\alpha \widehat{\sigma}_{\widehat{\psi}} & \leq \psi \leq & \widehat{\psi} + c_\alpha \widehat{\sigma}_{\widehat{\psi}} \\ 9.63 - (2.96)\,(2.35) & \leq \psi \leq & 9.63 + (2.96)\,(2.35) \\ 2.67 & \leq \psi \leq & 16.59 \end{array}$$

Since ψ does not include zero, the difference is significant. One may continue to look at any other parametric functions $\psi = \mathbf{a}'(\boldsymbol{\mu}_1 - \boldsymbol{\mu}_2)$ for significance by selecting other variables. While we know that any contrast ψ proportional to $\psi_{ws} = \mathbf{a}'_{ws}(\boldsymbol{\mu}_1 - \boldsymbol{\mu}_2)$ will be significant, the parametric function ψ_{ws} is often difficult to interpret. Hence, one tends to investigate contrasts that involve a single variable or linear combinations of variables having integer coefficients. For this example, the contrast with the largest difference is estimated by $\widehat{\psi}_{ws} = 5.13$.

Since the overall test was rejected, one may also use the protected univariate t-tests to locate significant differences in the means for each variable, but not to construct simultaneous confidence intervals. If only a few comparisons are of interest, adjusted multivariate t critical values may be employed to construct simultaneous confidence intervals for a few comparisons. The critical value for c_α in Table V in the Appendix is less than the multivariate T^2 simultaneous critical value of 2.96 for $C = 10$ planned comparisons using *any of the adjustment methods. As noted previously, the multivariate t (STM) method entry in the table has a smaller critical value than either the Bonferroni-Dunn (BON) or the Dunn-Šidák (SID) methods. If one were only interested in 10 planned comparisons, one would not use the multivariate test for this problem, but instead construct the planned adjusted approximate simultaneous confidence intervals to evaluate significance in the mean vectors.*

b. *Two-Sample Case,* $\Sigma_1 \neq \Sigma_2$

Assuming multivariate normality and $\Sigma_1 = \Sigma_2$, Hotelling's T^2 statistic is used to test $H : \boldsymbol{\mu}_1 = \boldsymbol{\mu}_2$. When $\Sigma_1 \neq \Sigma_2$ we may still want to test for the equality of the mean vectors. This problem is called the multivariate Behrens-Fisher problem. Because $\Sigma_1 \neq \Sigma_2$, we no longer have a pooled estimate for Σ under H. However, an intuitive test statistic for testing $H : \boldsymbol{\mu}_1 = \boldsymbol{\mu}_2$ is

$$X^2 = (\overline{\mathbf{y}} - \overline{\mathbf{x}})' \left(\frac{\mathbf{S}_1}{n_1} + \frac{\mathbf{S}_2}{n_2} \right)^{-1} (\overline{\mathbf{y}} - \overline{\mathbf{x}}) \qquad (3.9.19)$$

where $\mathbf{S}_1 = \mathbf{E}_1/(n_1 - 1)$ and $\mathbf{S}_2 = \mathbf{E}_2/(n_2 - 1)$. $X^2 \xrightarrow{d} \chi^2(p)$ only if we assume that the sample covariance matrices are equal to their population values. In general, X^2 does not converge to either Hotelling's T^2 distribution or to a chi-square distribution. Instead, one must employ an approximation for the distribution of X^2.

James (1954), using an asymptotic expansion for a quadratic form, obtained an approximation to the distribution of X^2 in (3.9.19) as a sum of chi-square distributions. To test $H : \boldsymbol{\mu}_1 = \boldsymbol{\mu}_2$, the null hypothesis is rejected, using James' first-order approximation, if

$$X^2 > \chi^2_{1-\alpha}(p) \left[A + B\chi^2_{1-\alpha}(p) \right]$$

where

$$\mathbf{W}_i = \mathbf{S}_i / n_i \text{ and } \mathbf{W} = \sum_{i=1}^{2} \mathbf{W}_i$$

$$A = 1 + \frac{1}{2p} \sum_{i=1}^{2} \left(\operatorname{tr} \mathbf{W}^{-1}\mathbf{W}_i \right)^2 / (n_i - 1)$$

$$B = \frac{1}{2p(p+2)} \sum_{i=1}^{2} \left[\left(\operatorname{tr} \mathbf{W}^{-1}\mathbf{W}_i \right)^2 + 2 \left(\operatorname{tr} \mathbf{W}^{-1}\mathbf{W}_i \right)^2 \right] / (n_i - 1)$$

and $\chi^2_{1-\alpha}(p)$ is the upper $1 - \alpha$ critical value of a chi-square distribution with p degrees of freedom.

Yao (1965) and Nel and van der Merwe (1986) estimated the distribution of X^2 using Hotelling's T^2 distribution with degrees of freedom p and an approximate degrees of freedom for error. For Yao (1965) the degrees of freedom for error for Hotelling's T^2 statistic is estimated by $\widehat{\nu}$ and for Nel and van der Merwe (1986) the degrees of freedom is estimated by $\widehat{f}$. Nel and van der Merwe (1986) improved upon Yao's result. Both approximations for the error degrees of freedom follow

$$\frac{1}{\widehat{\nu}} = \sum_{i=1}^{2}\left(\frac{1}{n_i - 1}\right)\left\{\frac{(\overline{\mathbf{y}} - \overline{\mathbf{x}})' \mathbf{W}^{-1}\mathbf{W}_i\mathbf{W}^{-1}(\overline{\mathbf{y}} - \overline{\mathbf{x}})}{X^2}\right\}^2$$
$$\widehat{f} = \frac{\text{tr}\,\mathbf{W}^2 + (\text{tr}\,\mathbf{W})^2}{\sum_{i=1}^{2}\left[\text{tr}\,\mathbf{W}_i^2 + (\text{tr}\,\mathbf{W}_i)^2\right] / (n_i - \mathbf{1})} \tag{3.9.20}$$

where the min $(n_1 - 1, n_2 - 1) \leq \widehat{\nu} \leq n_1 + n_2 - 2$. Using the result due to Nel and van der Merwe, the test of $H : \boldsymbol{\mu}_1 = \boldsymbol{\mu}_2$ is rejected if

$$X^2 > T^2_{1-\alpha}(p, \nu) = \frac{p\widehat{f}}{\widehat{f} - p - 1} F^{1-\alpha}\left(p, \widehat{f}\right) \tag{3.9.21}$$

where $F^{1-\alpha}(p, \widehat{f})$ is the upper $1 - \alpha$ critical value of an F distribution. For Yao's test, the estimate for the error degrees of freedom $\widehat{f}$ is replaced by $\widehat{\nu}$ given in (3.9.20).

Kim (1992) obtained an approximate test by solving the eigenequation $|\mathbf{W}_1 - \lambda\mathbf{W}_2| = 0$. For Kim, $H : \boldsymbol{\mu}_1 = \boldsymbol{\mu}_2$ is rejected if

$$F = \frac{\widehat{\nu} - p + 1}{ab\widehat{\nu}}\mathbf{w}'\left[\left(\mathbf{D}^{1/2} + r\mathbf{I}\right)^{-1}\right]^2 \mathbf{w} > F^{1-\alpha}\left(b, \widehat{\nu} - p + 1\right) \tag{3.9.22}$$

where

$$r = \left(\prod_{i=1}^{p}\lambda_i\right)^{1/2p}$$
$$\delta_i = (\lambda_i + 1) \Big/ \left(\lambda_i^{1/2} + r\right)^2$$
$$a = \sum_{i=1}^{p}\delta_i^2 \Big/ \sum_{i=1}^{p}\delta_i$$
$$b = \left(\sum_{i=1}^{p}\delta_i\right)^2 \Big/ \sum_{i=1}^{p}\delta_i^2$$

λ_i and $\mathbf{p}_i$ are the roots and eigenvectors of $|\mathbf{W}_1 - \lambda\mathbf{W}_2| = 0$, $\mathbf{D} = \text{diag}\left[\lambda_1, \lambda_2, \ldots, \lambda_p\right]$, $\mathbf{P} = \left[\mathbf{p}_1, \mathbf{p}_2, \ldots, \mathbf{p}_p\right]$, $\mathbf{w} = \mathbf{P}'(\overline{\mathbf{y}} - \overline{\mathbf{x}})$, and $\widehat{\nu}$ given in (3.9.20) is identical to the approximation provided by Yao (1965).

Johansen (1980), using weighted least squares regression, also approximated the distribution of X^2 by relating it to a scaled F distribution. For Johansen's procedure, H is rejected if

$$X^2 > cF^{1-\alpha}\left(p, f^*\right) \tag{3.9.23}$$

where

$$c = p + 2A - 6A/\left[p(p-1)+2\right]$$

$$A = \sum_{i=1}^{2}\left[\operatorname{tr}\left(\mathbf{I} - \mathbf{W}^{-1}\mathbf{W}_i^{-1}\right)^2 + \left(\operatorname{tr}\left(\mathbf{I} - \mathbf{W}^{-1}\mathbf{W}_i^{-1}\right)\right)^2\right]/2(n_i - 1)$$

$$f^* = p(p+2)/3A$$

Yao (1965) showed that James' procedure led to inflated α levels, her test led to α levels that were less than or equal to the true value α, and that the results were true for equal and unequal sample sizes. Algina and Tang (1988) confirmed Yao's findings and Algina et al. (1991) found that Johansen's solution was equivalent to Yao's test. Kim (1992) showed that his test had a Type I error rate that was always less than Yao's. De la Rey and Nel (1993) showed that Nel and van der Merwe's solution was better than Yao's. Christensen and Rencher (1997) compared the Type I error rates and power for James', Yao's, Johansen's, Nel and van der Merwe's, and Kim's solutions and concluded that Kim's approximation or Nel and van der Merwe's approximation had the highest power for the overall test and always controlled the Type I errors at the level less than or equal to α. While they found James' procedure almost always had the highest power, the Type I error for the tests was almost always slightly larger than the nominal α level. They recommended using Kim's (1992) approximation or the one developed by Nel and van der Merwe (1986). Timm (1999) found James' second-order approximation—James (1954) Equation 6.7 in his paper—to control the overall level at the nominal level when testing the significance of multivariate effect sizes in multiple-endpoint studies. James' second-order approximation may improve the approximation for the two-sample location problem. The procedure should again have higher power and yet also control the overall level of the test nearer to the nominal α level. This needs further investigation.

Myers and Dunlap (2000) recommend extending the simple procedure developed by Alexander and Govern (1994) to the multivariate two group location problem when the covariance matrices are unequal. The method is very simple. To test $H : \boldsymbol{\mu}_1 = \boldsymbol{\mu}_2$, one constructs the weighted centroid

$$\mathbf{c}_p = \left[(\overline{\mathbf{y}} + \overline{\mathbf{x}}) \Big/ \mathrm{w}_i\right] / \left[\sum_{i=1}^{2}(1/\mathrm{w}_i)\right]$$

where the weights w_i are defined using the $1/p^{th}$ root of the covariance matrix for each group

$$\mathrm{w}_i = |\mathbf{S}_i|^{1/p}/n_i$$

Then one calculates Hotelling's statistics T_i^2 for each group as follows

$$T_1^2 = n_1[(\overline{\mathbf{y}} - \mathbf{c}_p)'\,\mathbf{S}_1^{-1}\,(\overline{\mathbf{y}} - \mathbf{c}_p)$$
$$T_2^2 = n_2[(\overline{\mathbf{x}} - \mathbf{c}_p)'\,\mathbf{S}_2^{-1}\,(\overline{\mathbf{x}} - \mathbf{c}_p)$$

or converting each statistic to a corresponding F statistic,

$$F_i = (n_i - p)T_i^2/p(n_i - 1)$$

For each statistic F_i, the p-value ($\widetilde{p}_i$) for the corresponding F distribution with $v_h = p$ and $v_e = (n_i - p - 1)$ degrees of freedom is determined. Because distribution of the sum of two F distributions is unknown, the statistics F_i are combined using additive chi-square statistics. One converts each F_i statistic to a chi-square equivalent statistic using the p-value of the F-statistic. That is, one finds the corresponding chi-square statistic X_i^2 on p degrees of freedom that corresponds to the p-value $1 - \widetilde{p}_i$, the upper tail integral of the chi-square distribution for each statistic F_i. The test statistic A for the two-group location problem is the sum of the chi-square statistics X_i^2 across the two groups

$$A = \sum_{i=1}^{2} X_i^2$$

The statistic A converts the nonadditive T_i^2 statistics to additive chi-square statistics with p-values $\widetilde{p}_i$. The test statistic A is distributed approximately as a chi-square distribution with $v = (g - 1)p$ degrees of freedom where $g = 2$ for the two group location problem. A simulation study performed by Myers and Dunlap (2000) indicates that the procedure maintains the overall Type I error rate for the test of equal mean vectors at the nominal level α and the procedure is easily extended for $g > 2$ groups.

Example 3.9.2 (Testing $\boldsymbol{\mu}_1 = \boldsymbol{\mu}_2$, Given $\Sigma_1 \neq \Sigma_2$) *To illustrate testing $H : \boldsymbol{\mu}_1 = \boldsymbol{\mu}_2$ when $\Sigma_1 \neq \Sigma_2$, we utilize data set B generated in program m3_7_1.sas. There are $p = 3$ variables and $n_1 = n_2 = 25$ observations. Program m3_9a.sas also contains the code for testing $\boldsymbol{\mu}_1 = \boldsymbol{\mu}_2$ using the SAS procedure GLM which assumes $\Sigma_1 = \Sigma_2$. The F statistic calculated by SAS assuming equal covariance matrices is 18.4159 which has a p-value of 5.44E-18. Alternatively, using formula (3.9.19), the X^2 statistic for data set B is $X^2 =$ 57.649696. The critical value for X^2 using formula (3.9.21) is* FVAL = 9.8666146 *where* $\widehat{f} = 33.06309$ *is the approximate degrees of freedom.*

The corresponding p-value for Nel and van der Merwe's test is P-VALF = 0.000854 *which is considerably larger than the p-value for the test generated by SAS assuming $\Sigma_1 = \Sigma_2$, employing the T^2 statistic. When $\Sigma_1 \neq \Sigma_2$ one should not use the T^2 statistic. Approximate* $100\,(1-\alpha)\,\%$ *simultaneous confidence intervals may be again constructed by using (3.9.21) in the formula for c_α^2 given in (3.9.6). Or, one may construct approximate simultaneous confidence intervals by again using the entries in Table V of the Appendix where the degrees of freedom for error is* $\widehat{f} = 33.06309$.

We conclude this example with a nonparametric procedure for nonnormal data based upon ranks. A multivariate extension of the univariate Kruskal-Wallis test procedure for testing the equality of univariate means. While the procedure does not depend on the error structure or whether the data are multivariate normal, it does require continuous data. In addition, the conditional distribution should be symmetrical for each variable if one wants to make inferences regarding the mean vectors instead of the mean rank vectors. Using the nonnormal data in data set C and the incorrect parametric procedure of analysis yields a nonsignificant p-value for the test of equal mean vectors, 0.0165. *Using ranks, the p-value for the test for equal mean rank vectors is* < 0.0001. *To help to locate the variables that led to the significant difference, one may construct protected t-tests or F-tests for each variable using the ranks. The construction of simultaneous confidence intervals is not recommended.*

c. *Two-Sample Case, Nonnormality*

In testing $H : \boldsymbol{\mu}_1 = \boldsymbol{\mu}_2$, we have assumed a MVN distribution with $\Sigma_1 = \Sigma_2$ or $\Sigma_1 \neq \Sigma_2$. When sampling from a nonnormal distribution, Algina et al. (1991) found in comparing the methods of James et al. that in general James' first-order test tended to be outperformed by the other two procedures. For symmetrical distributions and moderate skewness $(-1 < \beta_{1,p} < 1)$ all procedures maintained an α level near the nominal level independent of the ratio of sample sizes and heteroscedasticity.

Using a vector of coordinatewise Winsorized trimmed means and robust estimates $\widetilde{\mathbf{S}}_1^\star$ and $\widetilde{\mathbf{S}}_2^\star$, Mudholkar and Srivastava (1996, 1997) proposed a robust analog of Hotelling's T^2 statistic using a recursive method to estimate the degrees of freedom $\widehat{\nu}$, similar to Yao's procedure. Their statistic maintains a Type I error that is less than or equal to α for a wide variety of nonnormal distributions. Bilodeau and Brenner (1999, p. 226) develop robust Hotelling T^2 statistics for elliptical distributions. One may also use nonparametric procedures that utilize ranks; however, these require the conditional multivariate distributions to be symmetrical in order to make valid inferences about means. The procedure is illustrated in Example 3.9.2. Using PROC RANK, each variable is ranked in ascending order for the two groups. Then, the ranks are processed by the GLM procedure to create the rank test statistic. This is a simple extension of the Kruskal-Wallis test used to test the equality of means in univariate analysis, Neter et al. (1996, p. 777).

d. *Profile Analysis, One Group*

Instead of comparing an experimental group with a control group on p variables, one often obtains experimental data for one group and wants to know whether the group mean for all variables is the same as some standard. In an industrial setting the "standard" is established and the process is in-control (out-of-control) if the group mean is equal (unequal) to the standard. For this situation the variables need not be commensurate. The primary hypothesis is whether the profile for the process is equal to a standard.

Alternatively, the set of variables may be commensurate. In the industrial setting a process may be evaluated over several experimental conditions (treatments). In the social sciences the set of variables may be a test battery that is administered to evaluate psychological traits or vocational skills. In learning theory research, the response variable may be the time required to master a learning task given $i = 1, 2, \ldots, p$ exposures to the learning mechanism. When there is no natural order to the p variables these studies are called profile designs since one wants to investigate the pattern of the means $\mu_1, \mu_2, \ldots, \mu_p$ when they are connected using line segments. This design is similar to repeated measures or growth curve designs where subjects or processes are measured sequentially over p successive time points. Designs in which responses are ordered in time are discussed in Chapters 4 and 6.

In a profile analysis, a random sample of n p-vectors is obtained where $\mathbf{Y}_i \sim IN_p(\boldsymbol{\mu}, \Sigma)$ for $\boldsymbol{\mu}' = [\mu_1, \mu_2, \ldots, \mu_p]$ and $\Sigma = [\sigma_{ij}]$. The observation vectors have the general structure $\mathbf{y}_i' = [y_{i1}, y_{i2}, \ldots, y_{ip}]$ for $i = 1, 2, \ldots, n$. The mean of the n observations is $\overline{\mathbf{y}}$ and Σ is estimated using $\mathbf{S} = \mathbf{E}/(n-1)$. One may be interested in testing that the population mean $\boldsymbol{\mu}$ is equal to some known standard value $\boldsymbol{\mu}_0$; the null hypothesis is

$$H_G : \boldsymbol{\mu} = \boldsymbol{\mu}_0 \tag{3.9.24}$$

If the p responses are commensurate, one may be interested in testing whether the profile over the p responses are equal, i.e., that the profile is level. This hypothesis is written as

$$H_C : \mu_1 = \mu_2 = \cdots = \mu_p \tag{3.9.25}$$

From Example 3.5.1, the test statistic for testing $H_G : \boldsymbol{\mu} = \boldsymbol{\mu}_0$ is Hotelling's T^2 statistic

$$T^2 = n\left(\bar{\mathbf{y}} - \boldsymbol{\mu}_0\right)' \mathbf{S}^{-1}\left(\bar{\mathbf{y}} - \boldsymbol{\mu}_0\right) \tag{3.9.26}$$

The null hypothesis is rejected if, for a test of size α,

$$T^2 > T^2_{1-\alpha}\left(p, n-1\right) = \frac{p\left(n-1\right)}{n-p} F^{1-\alpha}\left(p, n-p\right). \tag{3.9.27}$$

To test H_C, the null hypothesis is transformed to an equivalent hypothesis. For example, by subtracting the pth mean from each variable, the equivalent null hypothesis is

$$H_{C_1^*} : \begin{bmatrix} \mu_1 - \mu_p \\ \mu_2 - \mu_p \\ \vdots \\ \mu_{p-1} - \mu_p \end{bmatrix} = \mathbf{0}$$

This could be accomplished using any variable. Alternatively, we could subtract successive differences in means. Then, H_C is equivalent to testing

$$H_{C_2^*} : \begin{bmatrix} \mu_1 - \mu_2 \\ \mu_2 - \mu_3 \\ \vdots \\ \mu_{p-1} - \mu_p \end{bmatrix} = \mathbf{0}$$

In the above transformations of the hypothesis H_C to H_{C^*}, the mean vector $\boldsymbol{\mu}$ is either postmultiplied by a contrast matrix $\mathbf{M}$ of order $p \times (p-1)$ or premultiplied by a matrix $\mathbf{M}'$ of order $(p-1) \times p$; the columns of $\mathbf{M}$ form contrasts in that the sum of the elements in any column in $\mathbf{M}$ must sum to zero. For C_1^*,

$$\mathbf{M} \equiv \mathbf{M}_1 = \begin{bmatrix} 1 & 0 & 0 & \cdots & 0 \\ 0 & 1 & 0 & \cdots & 0 \\ \vdots & \vdots & \vdots & & \vdots \\ -1 & -1 & -1 & \cdots & -1 \end{bmatrix}.$$

and for C_2^*,

$$\mathbf{M} \equiv \mathbf{M}_2 = \begin{bmatrix} 1 & 0 & \cdots & 0 \\ -1 & 1 & \cdots & 0 \\ 0 & -1 & \cdots & 0 \\ 0 & 0 & \cdots & 0 \\ \vdots & \vdots & & \vdots \\ 0 & 0 & \cdots & -1 \end{bmatrix}$$

Testing H_C is equivalent to testing

$$H_{C^*} : \boldsymbol{\mu}'\mathbf{M} = \mathbf{0}' \tag{3.9.28}$$

or

$$H_{C^*} : \mathbf{M}'\boldsymbol{\mu} = \mathbf{0} \tag{3.9.29}$$

For a random sample of normally distributed observations, to test (3.9.29) each observation is transformed by $\mathbf{M}'$ to create $\mathbf{X}_i = \mathbf{M}'\mathbf{Y}_i$ such that $E(\mathbf{X}_i) = \mathbf{M}'\boldsymbol{\mu}$ and $\text{cov}(\mathbf{X}_i) = \mathbf{M}'\Sigma\mathbf{M}$. By property (2) of Theorem 3.3.2, $\mathbf{X}_i \sim N_{p-1}\left(\mathbf{M}'\boldsymbol{\mu},\ \mathbf{M}'\Sigma\mathbf{M}\right)$, $\overline{\mathbf{X}}_i = \mathbf{M}'\overline{\mathbf{Y}}_i \sim N_{p-1}\left(\mathbf{M}'\boldsymbol{\mu},\ \mathbf{M}'\Sigma\mathbf{M}/n\right)$. Since $(n-1)\mathbf{S}$ has an independent Wishart distribution, following Example 3.5 we have that

$$T^2 = \left(\mathbf{M}'\overline{\mathbf{y}}\right)'\left(\mathbf{M}'\mathbf{S}\mathbf{M}/n\right)^{-1}\left(\mathbf{M}'\overline{\mathbf{y}}\right) = n\left(\mathbf{M}'\overline{\mathbf{y}}\right)'\left(\mathbf{M}'\mathbf{S}\mathbf{M}\right)^{-1}\mathbf{M}'\overline{\mathbf{y}} \tag{3.9.30}$$

has Hotelling's T^2 distribution with degree of freedom $p-1$ and $v_e = n-1$ under the null hypothesis (3.9.29). The null hypothesis H_C, of equal means across the p variables, is rejected if

$$T^2 \geq T^2_{1-\alpha}(p-1, v_e) = \frac{(p-1)(n-1)}{(n-p+1)} F^{1-\alpha}(p-1, n-p+1) \tag{3.9.31}$$

for a test of size α.

When either the test of H_G or H_C is rejected, one may wish to obtain $100(1-\alpha)\%$ simultaneous confidence intervals. For H_G, the intervals have the general form

$$\mathbf{a}'\overline{\mathbf{y}} - c_\alpha\sqrt{\mathbf{a}'\mathbf{S}\mathbf{a}/n} \leq \mathbf{a}'\boldsymbol{\mu} \leq \mathbf{a}'\overline{\mathbf{y}} + c_\alpha\sqrt{\mathbf{a}'\mathbf{S}\mathbf{a}/n} \tag{3.9.32}$$

where $c_\alpha^2 = p(n-1)F^{1-\alpha}(p, n-p)/(n-p)$ for a test of size α and arbitrary vectors $\mathbf{a}$. For the test of H_C, the parametric function $\psi = \mathbf{a}'\mathbf{M}'\boldsymbol{\mu} = \mathbf{c}'\boldsymbol{\mu}$ for $\mathbf{c}' = \mathbf{a}'\mathbf{M}'$. To estimate ψ, $\widehat{\psi} = \mathbf{c}'\overline{\mathbf{y}}$ and the cov $\widehat{\psi} = \mathbf{c}'\mathbf{S}\mathbf{c}/n = \mathbf{a}'\mathbf{M}'\mathbf{S}\mathbf{M}\mathbf{a}/n$. The $100(1-\alpha)\%$ simultaneous confidence interval is

$$\widehat{\psi} - c_\alpha\sqrt{\mathbf{c}'\mathbf{S}\mathbf{c}/n} \leq \psi \leq \widehat{\psi} + c_\alpha\sqrt{\mathbf{c}'\mathbf{S}\mathbf{c}/n} \tag{3.9.33}$$

where $c_\alpha^2 = (p-1)(n-1)F^{1-\alpha}(p-1, n-p-1)/(n-p+1)$ for a test of size α and arbitrary vectors $\mathbf{a}$. If the overall hypothesis is rejected, we know that there exists at least one parametric function that is significant but it may not be a meaningful function of the means. For H_G, it does not include the linear combination $\mathbf{a}'\boldsymbol{\mu}_0$ of the target mean and for H_C, it does not include zero. One may alternatively establish approximate simultaneous confidence sets a-variable-at-a-time using Šidák's inequality and the multivariate t distribution with a correlation matrix of the accompanying MVN distribution $\mathbf{P} = \mathbf{I}$ using the values in the Appendix, Table V.

Example 3.9.3 (Testing H_C: One-Group Profile Analysis) *To illustrate the analysis of a one-group profile analysis, group 1 from data set A (program m3_7_1.sas) is utilized. The data consists of three measures on each of 25 subjects and we want to test $H_C : \mu_1 = \mu_2 = \mu_3$. The observation vectors $\mathbf{Y}_i \sim \text{IN}_3(\boldsymbol{\mu}, \Sigma)$ where $\boldsymbol{\mu}' = [\mu_1, \mu_2, \mu_3]$. While we may test*

TABLE 3.9.3. T^2 Test of $H_C : \mu_1 = \mu_2 = \mu_3$.

	$S = 1$	$M = 0$	$N = 10, 5$		
Statistic	Value	F	Num DF	Den DF	Pr > F
Wilks' lambda	0.01240738	915.37	2	23	0.0001
Pillai's trace	0.98759262	915.37	2	23	0.0001
Hotelling-Lawley trace	79.59717527	915.37	2	23	0.0001
Roy's greatest root	79.59717527	915.37	2	23	0.0001

H_C using the T^2 statistic given in (3.9.30), the SAS procedure GLM employs the matrix $m \equiv \mathbf{M}'$ to test H_C using the MANOVA model program m3_9d.sas illustrates how to test H_C using a model with an intercept, a model with no intercept and contrasts, and the use of the REPEATED statement using PROC GLM. The results are provided in Table 3.9.3.

Because SAS uses the MR model to test H_C, Hotelling's T^2 statistic is not reported. However, relating T^2 to the F distribution and T^2 to T_o^2 we have that

$$\begin{aligned} F &= (n - p + 1)\, T^2 / (p - 1)\,(n - 1) \\ &= (n - p + 1)\, \widehat{\lambda}_1 / (p - 1) \\ &= (23)\,(79.5972)\,/2 = 915.37 \end{aligned}$$

as shown in Table 3.9.3 and H_C is rejected. By using the REPEATED statement, we find that Mauchly's test of circularity is rejected, the chi-square p-value for the test is $p = 0.0007$. Thus, one must use the exact T^2 test and not the mixed model F tests for testing hypotheses. The p-values for the adjusted Geisser-Greenhouse (GG) and Huynh-Feldt (HF) tests are also reported in SAS.

Having rejected H_C, we may use (3.9.32) to investigate contrasts in the transformed variables defined by $\mathbf{M}_1'$. By using the /CANONICAL option on the MODEL statement, we see by using the Standardized and Raw Canonical Coefficient vectors that our investigation should begin with $\psi = \mu_2 - \mu_3$, the second row of $\mathbf{M}_1'$. Using the error matrix

$$\mathbf{M}_1'\mathbf{E}\mathbf{M}_1 = \begin{bmatrix} 154.3152 & \\ 32.6635 & 104.8781 \end{bmatrix}$$

in the SAS output, the sample variance of $\widehat{\psi} = \widehat{\mu}_2 - \widehat{\mu}_3$ is $\widehat{\sigma}_{\widehat{\psi}} = \sqrt{104.8381/24} = 2.09$. For $\alpha = 0.05$,

$$\begin{aligned} c_\alpha^2 &= (p - 1)\,(n - 1)F^{1-\alpha}\,(p - 1, n - p - 1)\,/\,(n - p + 1) \\ &= (2)\,(24)\,(3.42)\,/\,(23) \\ &= 7.14 \end{aligned}$$

so that $c_\alpha = 2.67$. Since $\widehat{\boldsymbol{\mu}}' = [6.1931, 11.4914, 29.7618]$,the contrast $\widehat{\psi} = \widehat{\mu}_2 - \widehat{\mu}_3 = -18.2704$. A confidence interval for ψ is

$$\begin{aligned} -18.2704 - (2.67)\,(2.09) \quad &\le \psi \le \quad -18.2704 + (2.67)\,(2.09) \\ -23.85 \quad &\le \psi \le \quad -12.69 \end{aligned}$$

Since ψ does not include zero, the comparison is significant. The same conclusion is obtained from the one degree of freedom F tests obtained using SAS with the CONTRAST statement as illustrated in the program. When using contrasts in SAS, one may compare the reported p-values to the nominal level of the overall test, only if the overall test is rejected. The F statistic for the comparison $\psi = \mu_2 - \mu_3$ calculated by SAS is $F = 1909.693$ with p-value < 0.0001. The F tests for the comparisons $\psi_1 = \mu_1 - \mu_2$ and $\psi_2 = \mu_1 - \mu_3$ are also significant. Again for problems involving several repeated measures, one may use the discriminant coefficients to locate significant contrasts in the means for a single variable or linear combination of variables.

For our example using the simulated data, we rejected the circularity test so that the most appropriate analysis for the data analysis is to use the exact multivariate T^2 test. When the circularity test is not rejected, the most powerful approach is to employ the univariate mixed model. Code for the mixed univariate model using PROC GLM is included in program m3_9d.sas. Discussion of the SAS code using PROC GLM and PROC MIXED and the associated output is postponed until Section 3.10 where program m3_10a.sas is used for the univariate mixed model analysis. We next review the univariate mixed model for a one-group profile model.

To test H_C we have assumed an arbitrary structure for Σ. When analyzing profiles using univariate ANOVA methods, one formulates the linear model for the elements of $\mathbf{Y}_i$ as

$$\begin{aligned} Y_{ij} &= \mu + s_i + \beta_j + e_{ij} \\ i &= 1, 2, \ldots, n;\ j = 1, 2, \ldots, p \\ e_{ij} &\sim IN\left(0, \sigma_e^2\right) \\ s_i &\sim IN\left(0, \sigma_s^2\right) \end{aligned}$$

where e_{ij} and s_i are jointly independent, commonly known as an unconstrained (unrestricted), randomized block mixed ANOVA model. The subjects form blocks and the within subject treatment conditions are the effects β_j. Assuming the variances of the observations Y_{ij} over the p treatment/condition levels are homogeneous, the covariance structure for the observations is

$$\begin{aligned} \operatorname{var}\left(Y_{ij}\right) &= \sigma_s^2 + \sigma_e^2 \equiv \sigma_Y^2 \\ \operatorname{cov}\left(Y_{ij}, Y_{ij'}\right) &= \sigma_s^2 \\ \rho &= \operatorname{cov}\left(Y_{ij}, Y_{ij'}\right)/\sigma_Y^2 = \sigma_s^2/\left(\sigma_e^2 + \sigma_s^2\right) \end{aligned}$$

so that the covariance structure for $\Sigma_{p\times p}$ is represented as

$$\begin{aligned} \Sigma &= \sigma_s^2\mathbf{J} + \sigma_e^2\mathbf{I} \\ &= \sigma_e^2\left[(1-\rho)\,\mathbf{I} + \rho\mathbf{J}\right] \end{aligned}$$

has compound symmetry structure where $\mathbf{J}$ is a matrix of 1s and ρ is the intraclass correlation coefficient. The compound symmetry structure for Σ is a sufficient condition for an exact univariate F test for evaluating the equality of the treatment effects β_j $(H : \text{all } \beta_j = 0)$ in the mixed ANOVA model; however, it is not a necessary condition.

Huynh and Feldt (1970) showed that only the variances of the differences of all pairs of observations, var $\left(Y_{ij} - Y_{ij'}\right) = \sigma_j^2 + \sigma_{j'}^2 - 2\sigma_{jj'}$ must remain constant for all $j \neq j'$ and $i = 1, 2, \ldots, n$ for exact univariate tests. They termed these covariance matrices "Type H" matrices. Using matrix notation, the necessary and sufficient condition for an exact univariate F test for testing the equality of p correlated treatment differences is that $\mathbf{C}'\Sigma\mathbf{C} = \sigma^2\mathbf{I}$ where $\mathbf{C}'\ (p-1) \times (p-1)$ is an orthogonal matrix calculated from $\mathbf{M}'$ so that $\mathbf{C}'\mathbf{C} = \mathbf{I}_{(p-1)}$; see Rouanet and Lépine (1970). This is the sphericity (circularity) condition given in (3.8.21). When using PROC GLM to analyze a one-group design, the test is obtained by using the REPEATED statement. The test is labeled Mauchly's Criterion Applied to Orthogonal Components.

When the circularity condition is not satisfied, Geisser and Greenhouse (1958) (GG) and Huynh and Feldt (1976) (HF) suggested adjusted conservative univariate F tests for treatment differences. Hotelling's (1931) exact T^2 test of H_C does not impose the restricted structure on Σ; however, since Σ must be positive definite the sample size n must be greater than or equal to p; when this is not the case, one must use the adjusted F tests. Muller et al. (1992) show that the GG test is more powerful than the T^2 test under near circularity; however, the size of the test may be less than α. While the HF adjustment maintains α more near the nominal level it generally has lower power. Based upon simulation results obtained by Boik (1991), we continue to recommend the exact T^2 test when the circularity condition is not met.

e. *Profile Analysis, Two Groups*

One of the more popular designs encountered in the behavioral sciences and other fields is the two independent group profile design. The design is similar to the two-group location design used to compare an experimental and control group except that in a profile analysis p responses are now observed rather than p different variables. For these designs we are not only interested in testing that the means $\boldsymbol{\mu}_1$ and $\boldsymbol{\mu}_2$ are equal, but whether or not the group profiles for the two groups are parallel. To evaluate parallelism of profiles, group means for each variable are plotted to view the mean profiles. Profile analysis is similar to the two-group repeated measures designs where observations are obtained over time; however, in repeated measures designs one is more interested in the growth rate of the profiles. Analysis of repeated measures designs is discussed in Chapters 4 and 6.

For a profile analysis, we let $\mathbf{y}'_{ij} = [y_{ij1}, y_{ij2}, \ldots, y_{ijp}]$ represent the observation vector for the $i = 1, 2,$ groups and the $j = 1, 2, \ldots, n_i$ observations within the ith group as shown in Table 3.9.4. The random observations $\mathbf{y}_{ij} \sim IN_p\left(\boldsymbol{\mu}_i, \Sigma\right)$ where and $\boldsymbol{\mu}_i = \left[\mu_{i1}, \mu_{i2}, \ldots, \mu_{ip}\right]$ and $\Sigma_1 = \Sigma_2 = \Sigma$, a common covariance matrix with an undefined, arbitrary structure.

While one may use Hotelling's T^2 statistic to perform tests, we use this simple design to introduce the multivariate regression (MR) model which is more convenient for extending the analysis to the more general multiple group situation. Using (3.6.17), the MR model for

TABLE 3.9.4. Two-Group Profile Analysis.

Group			Conditions 1	2	$\cdots$	p
	$\mathbf{y}'_{11}$	$=$	y_{111}	y_{112}	$\cdots$	y_{11p}
	$\mathbf{y}'_{12}$	$\cdots$	y_{121}	y_{122}	$\cdots$	y_{12p}
1	$\vdots$	$\vdots$	$\vdots$	$\vdots$	$\vdots$	$\vdots$
	$\mathbf{y}'_{1n_1}$	$=$	y_{1n_11}	y_{1n_12}	$\cdots$	y_{1n_1p}
Mean			$y_{1.1}$	$y_{1.2}$	$\cdots$	$y_{1.p}$
	$\mathbf{y}'_{21}$	$=$	y_{211}	y_{212}	$\cdots$	y_{21p}
	$\mathbf{y}'_{22}$	$\cdots$	y_{221}	y_{222}	$\cdots$	y_{22p}
2	$\vdots$	$\vdots$	$\vdots$	$\vdots$	$\vdots$	$\vdots$
	$\mathbf{y}'_{2n_1}$	$=$	y_{2n_21}	y_{2n_22}	$\cdots$	y_{2n_2p}
Mean			$y_{2.1}$	$y_{2.2}$	$\cdots$	$y_{2.p}$

the design is

$$\underset{n\times p}{\mathbf{Y}} = \underset{n\times 2}{\mathbf{X}} \quad \underset{2\times p}{\mathbf{B}} + \underset{n\times p}{\mathbf{E}}$$

$$\begin{bmatrix} \mathbf{y}'_{11} \\ \mathbf{y}'_{12} \\ \vdots \\ \mathbf{y}'_{1n_1} \\ \mathbf{y}'_{21} \\ \mathbf{y}'_{22} \\ \vdots \\ \mathbf{y}'_{2n_2} \end{bmatrix} = \begin{bmatrix} 1 & 0 \\ 1 & 0 \\ \vdots & \vdots \\ 1 & 0 \\ 0 & 1 \\ 0 & 1 \\ \vdots & \vdots \\ 0 & 0 \end{bmatrix} \begin{bmatrix} \mu_{11}, & \mu_{12}, \dots, & \mu_{1p} \\ \mu_{21}, & \mu_{22}, \dots, & \mu_{2p} \end{bmatrix} + \begin{bmatrix} \mathbf{e}'_{11} \\ \mathbf{e}'_{12} \\ \vdots \\ \mathbf{e}'_{1n_1} \\ \mathbf{e}'_{21} \\ \mathbf{e}'_{22} \\ \vdots \\ \mathbf{e}'_{2n_2} \end{bmatrix}$$

The primary hypotheses of interest in a profile analysis, where the "repeated," commensurate measures have no natural order, are

1. H_P : Are the profiles for the two groups parallel?

2. H_C : Are there differences among conditions?

3. H_G : Are there differences between groups?

The first hypothesis tested in this design is that of parallelism of profiles or the group-by-condition ($G \times C$) interaction hypothesis, H_P. The acceptance or rejection of this hypothesis will effect how H_C and H_G are tested. To aid in determining whether the parallelism hypothesis is satisfied, plots of the sample mean vector profiles for each group should be constructed. Parallelism exists for the two profiles if the slopes of each line segment formed from the $p-1$ slopes are the same for each group. That is, the test of parallelism of profiles in terms of the model parameters is

$$H_P \equiv H_{G\times C}: \begin{bmatrix} \mu_{11}-\mu_{12} \\ \mu_{21}-\mu_{13} \\ \vdots \\ \mu_{1(p-1)}-\mu_{1p} \end{bmatrix} = \begin{bmatrix} \mu_{21}-\mu_{22} \\ \mu_{22}-\mu_{23} \\ \vdots \\ \mu_{2(p-1)}-\mu_{2p} \end{bmatrix} \tag{3.9.34}$$

Using the general linear model form of the hypothesis, $\mathbf{CBM} = \mathbf{0}$, the hypothesis becomes

$$\underset{1\times 2}{\mathbf{C}} \qquad \underset{2\times p}{\mathbf{B}} \qquad \underset{p\times(p-1)}{\mathbf{M}} \qquad = \mathbf{0}$$

$$[1, -1] \quad \begin{bmatrix} \mu_{11} & \mu_{12} & \cdots & \mu_{1p} \\ \mu_{21} & \mu_{22} & \cdots & \mu_{2p} \end{bmatrix} \begin{bmatrix} 1 & 0 & \cdots & 0 & 0 \\ -1 & 1 & \cdots & 0 & 0 \\ 0 & -1 & \cdots & 0 & 0 \\ \vdots & \vdots & & \vdots & \vdots \\ 0 & 0 & \cdots & 1 & 0 \\ 0 & 0 & \cdots & -1 & 1 \\ 0 & 0 & \cdots & 0 & -1 \end{bmatrix} = [0] \tag{3.9.35}$$

Observe that the post matrix $\mathbf{M}$ is a contrast matrix having the same form as the test for differences in conditions for the one-sample profile analysis. Thus, the test of no interaction or parallelism has the equivalent form

$$H_P \equiv H_{G\times C}: \boldsymbol{\mu}_1'\mathbf{M} = \boldsymbol{\mu}_2'\mathbf{M} \tag{3.9.36}$$

or

$$\mathbf{M}'(\boldsymbol{\mu}_1 - \boldsymbol{\mu}_2) = \mathbf{0}$$

The test of parallelism is identical to testing that the transformed means are equal or that their transformed difference is zero. The matrix $\mathbf{C}$ in (3.9.35) is used to obtain the difference while the matrix $\mathbf{M}$ is used to obtain the transformed scores, operating on the "within" conditions dimension.

To test (3.9.36) using T^2, let $\overline{\mathbf{y}}_{i.} = (y_{i.1}, y_{i.2}, \ldots, y_{i.p})$ for $i = 1, 2$. We then have $\mathbf{M}'(\boldsymbol{\mu}_1 - \boldsymbol{\mu}_2) \sim N_{p-1}\left[\mathbf{0}, \mathbf{M}'\boldsymbol{\Sigma}\mathbf{M}/(1/n_1 + 1/n_2)\right]$ so that under the null hypothesis,

$$\begin{aligned} T^2 &= (\mathbf{M}'\overline{\mathbf{y}}_{1.} - \mathbf{M}'\overline{\mathbf{y}}_{2.})' \left[\left(\frac{1}{n_1} + \frac{1}{n_2}\right)\mathbf{M}'\mathbf{S}\mathbf{M}\right]^{-1} (\mathbf{M}'\overline{\mathbf{y}}_{1.} - \mathbf{M}'\overline{\mathbf{y}}_{2.}) \\ &= \left(\frac{n_1 n_2}{n_1 + n_2}\right)(\overline{\mathbf{y}}_{1.} - \overline{\mathbf{y}}_{2.})'\mathbf{M}(\mathbf{M}'\mathbf{S}\mathbf{M})^{-1}\mathbf{M}'(\overline{\mathbf{y}}_{1.} - \overline{\mathbf{y}}_{2.}) \\ &\sim T^2(p-1,\ v_e = n_1 + n_2 - 2) \end{aligned} \tag{3.9.37}$$

where $\mathbf{S} = [(n_1 - 1)\mathbf{E}_1 + (n_2 - 1)\mathbf{E}_2]/(n_1 + n_2 - 2)$; the estimate of Σ obtained for the two-group location problem. $\mathbf{S}$ may be computed as

$$\mathbf{S} = \mathbf{Y}'\left[\mathbf{I} - \mathbf{X}(\mathbf{X}'\mathbf{X})^{-1}\mathbf{X}'\right]\mathbf{Y}/(n_1 + n_2 - 2)$$

The hypothesis of parallelism or no interaction is rejected at the level α if

$$\begin{aligned} T^2 &\geq T^2_{1-\alpha}\left(p-1,\ n_1+n_2-2\right) \\ &= \frac{(n_1+n_2-2)\,(p-1)}{n_1+n_2-p} F^{1-\alpha}\left(p-1, n_1+n_2-p\right) \end{aligned} \tag{3.9.38}$$

using Definition 3.5.3 with $n \equiv v_e = (n_1+n_2-2)$ and $p \equiv p-1$.

Returning to the MR model representation for profile analysis, we have that

$$\begin{aligned} \widehat{\mathbf{B}} &= \left(\mathbf{X}'\mathbf{X}\right)^{-1}\mathbf{X}'\mathbf{Y} = \begin{bmatrix} y_{1.1} & y_{1.2} & \cdots & y_{1.p} \\ y_{2.1} & y_{2.2} & \cdots & y_{2.p} \end{bmatrix} = \begin{bmatrix} \bar{\mathbf{y}}'_{1.} \\ \bar{\mathbf{y}}'_{2.} \end{bmatrix} \\ \left(\mathbf{C}\widehat{\mathbf{B}}\mathbf{M}\right)' &= \mathbf{M}'\left(\bar{\mathbf{y}}_{1.} - \bar{\mathbf{y}}_{2.}\right) \end{aligned}$$

which is identical to (3.9.36). Furthermore,

$$\mathbf{E} = \mathbf{M}'\mathbf{Y}\left(\mathbf{I}_n - \mathbf{X}\left(\mathbf{X}'\mathbf{X}\right)^{-1}\mathbf{X}'\right)\mathbf{Y}\mathbf{M} \tag{3.9.39}$$

for $n = n_1 + n_2$ and $q = r\,(\mathbf{X}) = 2$, $v_e = n_1 + n_2 - 2$. Also

$$\begin{aligned} \mathbf{H} &= (\mathbf{C}\widehat{\mathbf{B}}\mathbf{M})'\left[\mathbf{C}\left(\mathbf{X}'\mathbf{X}\right)^{-1}\mathbf{C}'\right]^{-1}(\mathbf{C}\widehat{\mathbf{B}}\mathbf{M}) \\ &= \left(\frac{n_1 n_2}{n_1+n_2}\right)\mathbf{M}'\left(\bar{\mathbf{y}}_{1.} - \bar{\mathbf{y}}_{2.}\right)\left(\bar{\mathbf{y}}_{1.} - \bar{\mathbf{y}}_{2.}\right)'\mathbf{M} \end{aligned} \tag{3.9.40}$$

Using Wilk's Λ criterion,

$$\Lambda = \frac{|\mathbf{E}|}{|\mathbf{E}+\mathbf{H}|} \sim U\left(p-1, v_h = 1, v_e = n_1+n_2-2\right) \tag{3.9.41}$$

The test of parallelism is rejected at the significance level α if

$$\Lambda < U^{1-\alpha}\left(p-1,\ 1, n_1+n_2-2\right) \tag{3.9.42}$$

or

$$\frac{(n_1+n_2-p)}{(p-1)}\frac{1-\Lambda}{\Lambda} F^{1-\alpha}\left(p-1, n_1+n_2-p\right)$$

Solving the equation $|\mathbf{H} - \boldsymbol{\lambda}\mathbf{E}| = 0$, $\Lambda = (1+\lambda_1)^{-1}$ since $v_h = 1$ and $T^2 = v_e\lambda_1$ so that

$$\begin{aligned} T^2 &= v_e\left(\Lambda^{-1} - 1\right) \\ &= (n_1+n_2-2)\left\{\frac{|\mathbf{E}+\mathbf{H}|}{|\mathbf{E}|} - 1\right\} \\ &= \left(\frac{n_1 n_2}{n_1+n_2}\right)\left(\bar{\mathbf{y}}_{1.} - \bar{\mathbf{y}}_{2.}\right)'\mathbf{M}\left(\mathbf{M}'\mathbf{S}\mathbf{M}\right)^{-1}\mathbf{M}'\left(\bar{\mathbf{y}}_{1.} - \bar{\mathbf{y}}_{2.}\right) \end{aligned}$$

or

$$\Lambda = 1/\left(1 + T^2/v_e\right)$$

Because, $\theta_1 = \lambda_1/(1+\lambda_1)$ one could also use Roy's criterion for tabled values of θ. Or, using Theorem 3.5.1

$$F = \frac{v_e - p + 1}{p}\lambda_1$$

has a central F distribution under the null hypothesis with $v_1 = p$ and $v_2 = v_e - p + 1$ degrees of freedom since $v_h = 1$. For $v_e = n_1 + n_2 - 2$ and $p \equiv p - 1$, $F = (n_1 + n_2 - p)\lambda_1 / (p-1) \sim F(p-1,\ n_1+n_2-p)$. If $v_h \geq 2$ Roy's statistic is approximated using an upper bound on the F statistic which provides a lower bound on the p-value.

With the rejection of parallelism hypothesis, one usually investigates tetrads in the means that have the general structure

$$\begin{aligned}\psi &= \mu_{1j} - \mu_{2j} - \mu_{1j'} + \mu_{2j'} \\ &= \mathbf{c}'(\boldsymbol{\mu}_1 - \boldsymbol{\mu}_2)\,\mathbf{m}\end{aligned} \tag{3.9.43}$$

for $\mathbf{c}' = [1, -1]$ and $\mathbf{m}$ is any column vector of the matrix $\mathbf{M}$. More generally, letting $\mathbf{c}' = \mathbf{a}'\mathbf{M}'$ for arbitrary vectors $\mathbf{a}$, then $\widehat{\psi} = \mathbf{c}'\left(\overline{\mathbf{y}}_{1.} - \overline{\mathbf{y}}_{2.}\right)$and $100(1-\alpha)\%$ simultaneous confidence intervals for the parametric functions ψ have the general form

$$\widehat{\psi} - c_\alpha\widehat{\sigma}_{\widehat{\psi}} \leq \psi \leq \widehat{\psi} + c_\alpha\widehat{\sigma}_{\widehat{\psi}} \tag{3.9.44}$$

where

$$\begin{aligned}\widehat{\sigma}^2_{\widehat{\psi}} &= \frac{n_1+n_2}{n_1 n_2}\mathbf{c}'\mathbf{S}\mathbf{c} \\ c^2_\alpha &= T^2_{1-\alpha}(p-1, n_1+n_2-2)\end{aligned}$$

for a test of size α. Or, c^2_α may be calculated using the F distribution following (3.9.38).

When the test of parallelism is not significant, one averages over the two independent groups to obtain a test for differences in conditions. The tests for no difference in conditions, given parallelism, are

$$\begin{aligned} H_C &: \frac{\mu_{11}+\mu_{21}}{2} = \frac{\mu_{12}+\mu_{22}}{2} = \cdots = \frac{\mu_{1p}+\mu_{2p}}{2} \\ H^W_C &: \frac{n_1\mu_1+n_2\mu_2}{n_1+n_2} = \frac{n_1\mu_1+n_2\mu_2}{n_1+n_2} = \cdots = \frac{n_1\mu_{1p}+n_2\mu_{2p}}{n_1+n_2}\end{aligned}$$

for an unweighted or weighted test of differences in conditions, respectively. The weighted test is only appropriate if the unequal sample sizes result from a loss of subjects that is due to treatment and one would expect a similar loss of subjects upon replication of the study. To formulate the hypothesis using the MR model, the matrix $\mathbf{M}$ is the same as $\mathbf{M}$ in (3.9.35); however, the matrix $\mathbf{C}$ becomes

$$\begin{aligned}\mathbf{C} &= [1/2,\ 1/2] \qquad \text{for } H_C \\ \mathbf{C} &= [n_1/(n_1+n_2),\ n_2/(n_1+n_2)] \qquad \text{for } H^W_C\end{aligned}$$

Using T^2 to test for no difference in conditions given parallel profiles, under H_C

$$\begin{aligned} T^2 &= 4\left(\frac{n_1 n_2}{n_1+n_2}\right)\left(\frac{\overline{\mathbf{y}}_{1.}-\overline{\mathbf{y}}_{2.}}{2}\right)' \mathbf{M}\left(\mathbf{M}'\mathbf{S}\mathbf{M}\right)^{-1}\mathbf{M}'\left(\frac{\overline{\mathbf{y}}_{1.}-\overline{\mathbf{y}}_{2.}}{2}\right) \\ &= 4\left(\frac{n_1 n_2}{n_1+n_2}\right)\mathbf{y}_{..}'\mathbf{M}\left(\mathbf{M}'\mathbf{S}\mathbf{M}\right)^{-1}\mathbf{M}'\mathbf{y}_{..} \\ &\sim T^2_{1-\alpha}\left(p-1,\ n_1+n_2-2\right) \end{aligned} \tag{3.9.45}$$

where $\mathbf{y}_{..}$ is a simple average. Defining the weighed average as $\overline{\mathbf{y}}_{..} = \left(n_1\overline{\mathbf{y}}_{1.}+n_2\overline{\mathbf{y}}_{2.}\right) / \left(n_1+n_2\right)$, the statistic for testing H_C^W is

$$\begin{aligned} T^2 &= \left(n_1+n_2\right)\ \overline{\mathbf{y}}_{..}'\mathbf{M}\left(\mathbf{M}'\mathbf{S}\mathbf{M}\right)^{-1}\mathbf{M}'\overline{\mathbf{y}}_{..} \\ &\sim T^2_{1-\alpha}\left(p-1, n_1+n_2-2\right) \end{aligned} \tag{3.9.46}$$

Simultaneous $100(1-\alpha)\%$ confidence intervals depend on the null hypothesis tested. For H_C and $\mathbf{c}' = \mathbf{a}'\mathbf{M}$, the confidence sets have the general form

$$\mathbf{c}'\mathbf{y}_{..} - c_\alpha\sqrt{\mathbf{c}'\mathbf{S}\mathbf{c}/\left(n_1+n_2\right)} \leq \mathbf{c}'\boldsymbol{\mu} \leq \mathbf{c}'\mathbf{y}_{..} + c_\alpha\sqrt{\mathbf{c}'\mathbf{S}\mathbf{c}/\left(n_1+n_2\right)} \tag{3.9.47}$$

where $c_\alpha^2 = T^2_{1-\alpha}\left(p-1,\ n_1+n_2-2\right)$.

To test for differences in groups, given parallelism, one averages over conditions to test for group differences. The test in terms of the model parameters is

$$\begin{aligned} H_G : \frac{\sum_{j=1}^p \mu_{1j}}{p} &= \frac{\sum_{j=1}^p \mu_{2j}}{p} \\ \mathbf{1}'\boldsymbol{\mu}_1/p &= \mathbf{1}'\boldsymbol{\mu}_2/p \end{aligned} \tag{3.9.48}$$

which is no more than a test of equal population means, a simple t test.

While the tests of H_G and H_C are independent, they both require that the test of the parallelism (interaction) hypothesis be nonsignificant. When this is not the case, the tests for group differences is

$$H_G^* : \boldsymbol{\mu}_1 = \boldsymbol{\mu}_2$$

which is identical to the test for differences in location. The test for differences in conditions when we do not have parallelism is

$$H_C^* : \begin{bmatrix} \mu_{11} \\ \mu_{21} \end{bmatrix} = \begin{bmatrix} \mu_{12} \\ \mu_{22} \end{bmatrix} = \cdots = \begin{bmatrix} \mu_{1p} \\ \mu_{2p} \end{bmatrix} \tag{3.9.49}$$

To test H_C^* using the MR model, the matrices for the hypothesis in the form $\mathbf{CBM} = \mathbf{0}$ are

$$\mathbf{C} = \begin{bmatrix} 1 & 0 \\ 0 & 1 \end{bmatrix} \quad \text{and} \quad \mathbf{M} = \begin{bmatrix} 1 & 0 & \cdots & 0 \\ 0 & 1 & \cdots & 0 \\ \vdots & \vdots & & \vdots \\ 0 & 0 & \cdots & 1 \\ -1 & -1 & \cdots & -1 \end{bmatrix}$$

so that $v_h = r(\mathbf{C}) = 2$. For this test we cannot use T^2 since $v_h \neq 1$; instead, we may use the Bartlett-Lawley-Hotelling trace criterion which from (3.5.4) is

$$T_o^2 = v_e \operatorname{tr}\left(\mathbf{HE}^{-1}\right)$$

for

$$\begin{aligned}\mathbf{H} &= (\mathbf{C}\widehat{\mathbf{B}}\mathbf{M})' \left[\mathbf{C}\left(\mathbf{X}'\mathbf{X}\right)^{-1}\mathbf{C}'\right]^{-1}(\mathbf{C}\widehat{\mathbf{B}}\mathbf{M}) \\ &= \mathbf{M}'\widehat{\mathbf{B}}'\left(\mathbf{X}'\mathbf{X}\right)\widehat{\mathbf{B}}\mathbf{M} \\ \mathbf{E} &= \mathbf{M}'\mathbf{Y}(\mathbf{I}_n - \mathbf{X}\left(\mathbf{X}'\mathbf{X}\right)^{-1}\mathbf{X}')\mathbf{Y}\mathbf{M}\end{aligned} \tag{3.9.50}$$

We can approximate the distribution of T_o^2 using Theorem 3.5.1 with $s = \min(v_h, p - 1 = \min(2, p-1)$, $M = |p-3| - 1$, and $N = (n_1 + n_2 - p - 2)/2$ and relate the statistic to an F distribution with degrees of freedom $v_1 = 2(2M+3)$ and $v_2 = 2(2N+3)$. Alternatively, we may use Wilks' Λ criterion with

$$\Lambda = \frac{|\mathbf{E}|}{|\mathbf{E}+\mathbf{H}|} \sim U(p-1, 2, n_1+n_2-2) \tag{3.9.51}$$

or Roy's test criterion. However, these tests are no longer equivalent. More will be said about these tests in Chapter 4.

Example 3.9.4 (Two-Group Profile Analysis) *To illustrate the multivariate tests of group difference* $\left(H_G^*\right)$*, the test of equal vector profiles across the p conditions* $\left(H_C^*\right)$*, and the test of parallelism of profiles* (H_P)*, we again use data set A generated in program m3_7_1.sas. We may also test* H_C *and* H_G *given parallelism, which assumes that the test of parallelism* (H_P) *is nonsignificant. Again we use data set A and PROC GLM. The code is provided in program m3_9e.sas.*

To interpret how the SAS procedure GLM is used to analyze the profile data, we express the hypotheses using the general matrix product $\mathbf{CBM} = \mathbf{0}$. *For our example,*

$$\mathbf{B} = \begin{bmatrix} \mu_{11} & \mu_{12} & \mu_{13} \\ \mu_{21} & \mu_{22} & \mu_{23} \end{bmatrix}$$

To test $H_G^* : \boldsymbol{\mu}_1 = \boldsymbol{\mu}_2$, *we set* $\mathbf{C} = [1, -1]$ *to obtain the difference in group vectors and* $\mathbf{M} = \mathbf{I}_3$. *The within-matrix* $\mathbf{M}$ *is equal to the identity matrix since we are evaluating the equivalence of the means for each group and p-variables, simultaneously. In PROC GLM, this test is performed with the statement*

```
manova    h = group / printe printh;
```

where the options PRINTE and PRINTH are used to print $\mathbf{H}$ *and* $\mathbf{E}$ *for hypothesis testing. To test* H_C^**, differences among the p conditions (or treatments), the matrices*

$$\mathbf{C} = \mathbf{I}_2 \quad \text{and} \quad \mathbf{M} = \begin{bmatrix} 1 & 0 \\ -1 & 0 \\ 0 & -1 \end{bmatrix}$$

are used. The matrix $\mathbf{M}$ *is used to form differences among conditions (variables/treatments), the within-subject dimension, and the matrix* $\mathbf{C}$ *is set to the identity matrix since we are evaluating p vectors across the two groups, simultaneously. To test this hypothesis using PROC GLM, one uses the CONTRAST statement, the full rank model (NOINT option in the MODEL statement) and the MANOVA statement as follows*

```
contrast 'Mult Cond' group   1   0
group                        0   1;
              (1  -1   0   0   0,
               0   1  -1   0   0,
manova m =
               0   0   0  -1   0,
               0   0   0   1  -1)  prefix = diff / printe printh;
```

where $m = \mathbf{M}'$ *and the group matrix is the identity matrix* $\mathbf{I}_2$*. To test for parallelism of profiles, the matrices*

$$\mathbf{C} = \begin{bmatrix} 1, & -1 \end{bmatrix} \quad \text{and} \quad \mathbf{M} = \begin{bmatrix} 1 & 0 \\ -1 & 1 \\ 0 & -1 \end{bmatrix}$$

are used. The matrix $\mathbf{M}$ *again forms differences across variables (repeated measurements) while* $\mathbf{C}$ *creates the group difference contrast. The matrices* $\mathbf{C}$ *and* $\mathbf{M}$ *are not unique since other differences could be specified; for example,* $\mathbf{C} = [1/2, -1/2]$ *and* $\mathbf{M}' = \begin{bmatrix} 1 & 0 & -1 \\ 0 & 1 & -1 \end{bmatrix}$*. The rank of the matrix is unique. The expression _all_ for h in the SAS code generates the matrix* $\begin{bmatrix} 1 & 1 \\ 1 & -1 \end{bmatrix}$*, the testing for differences in conditions given parallelism; it is included only to obtain the matrix* $\mathbf{H}$*. To test these hypotheses using PROC GLM, the following statements are used.*

```
                        (1  -1   0
manova h = _all_m =
                         0   1  -1)  prefix = diff / printe printh;
```

To test for differences in groups (H_G) *in (3.9.48), given parallelism, we set*

$$\mathbf{C} = \begin{bmatrix} 1, & -1 \end{bmatrix} \quad \text{and} \quad \mathbf{M} = \begin{bmatrix} 1/3 \\ 1/3 \\ 1/3 \end{bmatrix}.$$

To test this hypothesis using PROC GLM, the following statements are used.

```
contrast  'Univ Gr' group   1  -1;
manova    m =  (0.33333  0.33333  0.33333)  prefix = GR/printe printh;
```

To test for conditions given parallelism (H_C) *and parallelism [*$(G \times C)$*, the interaction between groups and conditions], the REPEATED statement is used with the MANOVA statement in SAS.*

As in our discussion of the one group profile example, one may alternatively test H_P*,* H_C*, and* H_G *using an unconstrained univariate mixed ANOVA model. One formulates the* *model as*

$$
\begin{aligned}
Y_{ijk} &= \mu + \alpha_i + \beta_k + (\alpha\beta)_{ik} + s_{(i)j} + e_{ijk} \\
&\quad i = 1, 2;\ j = 1, 2, \ldots, n_i;\ k = 1, 2, \ldots, p \\
s_{(i)j} &\sim IN\left(0, \rho\sigma^2\right) \\
e_{ijk} &\sim IN\left(0, (1-\rho)\,\sigma^2\right)
\end{aligned}
$$

where e_{ijk} and $s_{(i)j}$ are jointly independent, commonly called the unconstrained, split-plot mixed ANOVA design. For each group, Σ_i has compound symmetry structure and $\Sigma_1 = \Sigma_2 = \Sigma$,

$$
\Sigma_1 = \Sigma_2 = \Sigma = \sigma_e^2\left[(1-\rho)\,\mathbf{I} + \rho\mathbf{J}\right] = \sigma_s^2\mathbf{J} + \sigma_e^2\mathbf{I}
$$

where $\rho\sigma^2 = \rho\left(\sigma_s^2 + \sigma_e^2\right)$.

Thus, we have homogeneity of the compound symmetry structures across groups. Again, the compound symmetry assumption is a sufficient condition for split-plot univariate exact univariate F tests of β_k and $(\alpha\beta)_{ik}$. The necessary condition for exact F tests is that Σ_1 and Σ_2 have homogeneous, "Type H" structure; Huynh and Feldt (1970). Thus, we require that

$$
\begin{aligned}
\mathbf{A}'\Sigma_1\mathbf{A} &= \mathbf{A}'\Sigma_2\mathbf{A} \\
&= \mathbf{A}'\Sigma\mathbf{A} \\
&= \lambda\mathbf{I}
\end{aligned}
$$

where $\mathbf{A}'$ is an orthogonalized $(p-1) \times (p-1)$ matrix of $\mathbf{M}'$ used to test H_P and H_C. The whole plot test for the significance of α_i does not depend on the assumption and is always valid.

By using the REPEATED statement in PROC GLM, SAS generates exact univariate F tests for within condition differences (across p variables/treatments) and the group by condition interaction test ($G \times C \equiv P$) given circularity. As shown by Timm (1980), the tests are recovered from the normalized multivariate tests given parallelism. For the one-group example, the SAS procedure performed the test of "orthogonal" sphericity (circularity). For more than one group, the test is not performed. This is because we must test for equality and sphericity. This test was illustrated in Example 3.8.5 using Rao's score test developed by Harris (1984). Finally, PROC GLM calculates the GG and HH adjustments. While these tests may have some power advantage over the multivariate tests under near sphericity, we continue to recommend that one use the exact multivariate test when the circularity condition is not satisfied. In Example 3.8.5 we showed that the tests of circularity is rejected; hence, we must use the multivariate tests for this example. The results are displayed in Table 3.9.5 using Wilk's Λ criterion. The mixed model approach is discussed in Wilks Section 3.10 and in more detail in Chapter 6.

Because the test of parallelism (H_P) is significant for our example, the only valid tests for these data are the test of H_G^* and H_C^*, the multivariate tests for group and condition differences. Observe that the test of H_G^* is no more than the test of location reviewed in

TABLE 3.9.5. MANOVA Table: Two-Group Profile Analysis.

Multivariate Tests				
Test	H Matrix	Λ	F	p-value
H_G^*	$\begin{bmatrix} 48.422 & & \\ 64.469 & 85.834 & \\ 237.035 & 315.586 & 1160.319 \end{bmatrix}$	0.127	105.08	< 0.0001
H_C^*	$\begin{bmatrix} 1241.4135 & \\ 3727.4639 & 11512.792 \end{bmatrix}$	0.0141	174.00	< 0.0001
H_P	$\begin{bmatrix} 5.3178 & \\ 57.1867 & 614.981 \end{bmatrix}$	0.228	79.28	< 0.0001
Multivariate Tests Given Parallelism				
Test	H Matrix	Λ	F	p-value
H_G	280.967	0.3731	80.68	< 0.0001
H_C	$\begin{bmatrix} 1236.11 & \\ 3670.27 & 10897.81 \end{bmatrix}$	0.01666	1387.08	< 0.0001
H_P	$\begin{bmatrix} 5.3178 & \\ 57.1867 & 614.981 \end{bmatrix}$	0.2286	79.28	< 0.0001
Univariate F Tests Given Sphericity (Circularity)				
Test	F-ratios	p-values		
H_G	80.68	< 0.0001		
H_C	1398.37	< 0.0001		
$H_{G\times C}$	59.94	< 0.0001		

Example 3.9.1. We will discuss H_C^* in more detail when we consider a multiple-group example in Chapter 4. The tests of H_G and H_C should not be performed since H_P is rejected. We would consider tests of H_G and H_C only under nonsignificance of H_P since the tests sum over the "between" group and "within" conditions dimensions. Finally, the univariate tests are only exact given homogeneity and circularity across groups.

Having rejected the test H_P of parallelism one may find simultaneous confidence intervals for the tetrads in (3.9.43) by using the **S** *matrix, obtained from* **E** *in SAS. T^2 critical values are related to the F distribution in (3.9.38) and $\widehat{\sigma}_{\widehat{\psi}} = (n_1 + n_2)\,\mathbf{c}'\mathbf{S}\mathbf{c}/n_1 n_2$. Alternatively, one may construct contrasts in SAS by performing single degree of freedom protected F-tests to isolate significance. For $c_1 = [1, -1]$ and $m_1 = [1, -1, 0]$ we* *have* $\psi_1 = \mu_{11} - \mu_{21} - \mu_{12} + \mu_{22}$ *and for* $c_2 = [1, -1]$ *and* $m_2 = [0, 1, -1]$, $\psi_2 = \mu_{12} - \mu_{22} - \mu_{13} + \mu_{23}$. *From the SAS output,* ψ_2 *is clearly significant* $(p-value < 0.0001)$ *while* ψ_1 *is nonsignificant with* $(p - value = 0.3683)$. *To find exact confidence bounds, one must evaluate (3.9.44).*

f. Profile Analysis, $\Sigma_1 \neq \Sigma_2$

In our discussion, we have assumed that samples are from a MVN distribution with homogeneous covariances matrices, $\Sigma_1 = \Sigma_2 = \Sigma$. In addition, we have not restricted the structure of Σ. All elements in Σ have been free to vary. Restrictions on the structure of Σ will be discussed when we analyze repeated measures designs in Chapter 6.

If $\Sigma_1 \neq \Sigma_2$, we may adjust the degrees of freedom for T^2 when testing H_P, H_C, H_C^W, or H_G^*. However, since the test of H_C^* is not related to T^2, we need a more general procedure. This problem was considered by Nel (1997) who developed an approximate degrees of freedom test for hypotheses of the general form

$$H : \underset{g\times q}{\mathbf{C}} \; \underset{q\times p}{\mathbf{B}_1} \; \underset{p\times v}{\mathbf{M}} = \underset{g\times q}{\mathbf{C}} \; \underset{q\times p}{\mathbf{B}_2} \; \underset{p\times v}{\mathbf{M}} \tag{3.9.52}$$

for two independent MR models

$$\underset{n_i\times p}{\mathbf{Y}_i} = \underset{n_i\times q}{\mathbf{X}_i} \; \underset{q\times p}{\mathbf{B}_i} + \underset{n_i\times p}{\mathbf{E}_i} \tag{3.9.53}$$

under multivariate normality and $\Sigma_1 \neq \Sigma_2$.

To test (3.9.52), we first assume $\Sigma_1 = \Sigma_2$. Letting $\widehat{\mathbf{B}}_i = \left(\mathbf{X}_i'\mathbf{X}_i\right)^{-1}\mathbf{X}_i'\mathbf{Y}_i$, $\widehat{\mathbf{B}}_i \sim N_{q,\,p}\left[\mathbf{B}_i, \Omega_i = \Sigma_i \otimes \left(\mathbf{X}_i'\mathbf{X}_i\right)^{-1}\right]$ by Exercise 3.3, Problem 6. Unbiased estimates of Σ_i are obtained using $\mathbf{S}_i = \mathbf{E}_i / (n_i - q)$ where $q = r(\mathbf{X}_i)$ and $\mathbf{E}_i = \mathbf{Y}_i'\left(\mathbf{I}_{n_i} - \mathbf{X}_i\left(\mathbf{X}_i'\mathbf{X}_i\right)^{-1}\mathbf{X}_i'\right)\mathbf{Y}_i$. Finally, we let $v_i = n_i - q$, $v_e = v_1 + v_2$ so that $v_e\mathbf{S} = v_1\mathbf{S}_1 + v_2\mathbf{S}_2$ and $\mathbf{W}_i = \mathbf{C}\left(\mathbf{X}_i'\mathbf{X}_i\right)^{-1}\mathbf{C}'$. Then, the Bartlett-Lawley-Hotelling (BLH) test statistic for testing (3.9.52) with $\Sigma_1 = \Sigma_2$ is

$$\begin{aligned} T_o^2 &= v_e \operatorname{tr}\left(\mathbf{H}\mathbf{E}^{-1}\right) \\ &= \operatorname{tr}\left[\mathbf{M}'(\widehat{\mathbf{B}}_1 - \widehat{\mathbf{B}}_2)'\mathbf{C}'\left(\mathbf{W}_1 + \mathbf{W}_2\right)^{-1}\mathbf{C}(\widehat{\mathbf{B}}_1 - \widehat{\mathbf{B}}_2)\mathbf{M}\left(\mathbf{M}'\mathbf{S}\mathbf{M}\right)^{-1}\right] \end{aligned} \tag{3.9.54}$$

Now assume $\Sigma_1 \neq \Sigma_2$; under H, $\mathbf{C}(\widehat{\mathbf{B}}_1 - \widehat{\mathbf{B}}_2)\mathbf{M} \sim N_{g,\,v}\left[\mathbf{0}, \mathbf{M}'\Sigma_1\mathbf{M} \otimes \mathbf{W}_1 + \mathbf{M}'\Sigma_2\mathbf{M} \otimes \mathbf{W}_2\right]$. The unbiased estimate of the covariance matrix is

$$\mathbf{U} = \mathbf{M}'\mathbf{S}_1\mathbf{M} \otimes \mathbf{W}_1 + \mathbf{M}'\mathbf{S}_2\mathbf{M} \otimes \mathbf{W}_2 \tag{3.9.55}$$

which is distributed independent of $\mathbf{C}\left(\mathbf{B}_1 - \mathbf{B}_2\right)\mathbf{M}$. When $\Sigma_1 \neq \Sigma_2$, the BLH trace statistic can no longer be written as a trace since $\mathbf{U}$ is a sum of Kronecker products. However, using the vec operator it can be written as

$$T_B^2 = \left[\operatorname{vec}\mathbf{C}(\widehat{\mathbf{B}}_1 - \widehat{\mathbf{B}}_2)\mathbf{M}\right]'\mathbf{U}^{-1}\operatorname{vec}\left[\mathbf{C}(\widehat{\mathbf{B}}_1 - \widehat{\mathbf{B}}_2)\mathbf{M}\right] \tag{3.9.56}$$

Defining

$$\mathbf{S}_e = \left[\mathbf{S}_1 \operatorname{tr}\left(\mathbf{W}_1 (\mathbf{W}_1 + \mathbf{W}_2)^{-1}\right) + \mathbf{S}_2 \operatorname{tr}\left(\mathbf{W}_2 (\mathbf{W}_1 + \mathbf{W}_2)^{-1}\right)\right]/g$$
$$\widehat{T}_B^2 = \left[\operatorname{vec}(\mathbf{C}(\widehat{\mathbf{B}}_1 - \widehat{\mathbf{B}}_2)\mathbf{M})\right]' \left[\mathbf{M}'\mathbf{S}_e\mathbf{M} \otimes (\mathbf{W}_1 + \mathbf{W}_2)\right]^{-1} \operatorname{vec}(\mathbf{C}(\widehat{\mathbf{B}}_1 - \widehat{\mathbf{B}}_2)\mathbf{M})$$

Nel (1997), following Nel and van der Merwe (1986), found that $\widehat{T}_B^2$ can be approximated with an F statistic. The hypothesis H in (3.9.52) is rejected for a test of size α if

$$F = \frac{\widehat{f} - \mathrm{v} + 1}{\mathrm{v}} \frac{\widehat{T}_B^2}{\widehat{f}} \geq F^{1-\alpha}\left(\mathrm{v}, \widehat{f} - \mathrm{v} + 1\right) \tag{3.9.57}$$

where $\widehat{f}$ is estimated from the data as

$$\widehat{f} = \frac{\operatorname{tr}\{[\mathbf{D}_\mathrm{v}^+ \otimes \operatorname{vech}(\mathbf{W}_1 + \mathbf{W}_2)]\left(\mathbf{M}'\mathbf{S}_e\mathbf{M} \otimes \mathbf{M}'\mathbf{S}_e\mathbf{M}\right)\left(\mathbf{D}_\mathrm{v} \otimes [\operatorname{vech}(\mathbf{W}_1 + \mathbf{W}_2)]'\right)\}}{\sum_{i=1}^2 \frac{1}{v_i} \operatorname{tr}\{\left(\mathbf{D}_\mathrm{v}^+ \otimes \operatorname{vech} \mathbf{W}_i\right)\left(\mathbf{M}'\mathbf{S}_i\mathbf{M} \otimes \mathbf{M}'\mathbf{S}_i\mathbf{M}\right)\left(\mathbf{D}_\mathrm{v} \otimes [\operatorname{vech}(\mathbf{W}_i)]'\right)\}} \tag{3.9.58}$$

where $\mathbf{D}_\mathrm{v}$ is the unique duplication matrix of order $\mathrm{v}^2 \times \mathrm{v}(\mathrm{v}+1)/2$ defined in Theorem 2.4.8 of the symmetric matrix $\mathbf{A} = \mathbf{M}'\mathbf{S}\mathbf{M}$ where $\mathbf{S}$ is a covariance matrix. That is for a symmetric matrix $\mathbf{A}_{\mathrm{v}\times\mathrm{v}}$, $\operatorname{vec} \mathbf{A} = \mathbf{D}_\mathrm{v} \operatorname{vech} \mathbf{A}$ and the elimination matrix is $\mathbf{D}_\mathrm{v}^+ = (\mathbf{D}_\mathrm{v}' \mathbf{D}_\mathrm{v})^{-1}\mathbf{D}_\mathrm{v}'$, such that $\mathbf{D}_\mathrm{v}^+ \operatorname{vec} \mathbf{A} = \operatorname{vech} \mathbf{A}$. When the $r(\mathbf{C}) = 1$, the approximation in (3.9.57) reduces to Nel and van der Merwe's test for evaluating the equality of mean vectors. For $g = \mathrm{v} = 1$, it reduces to the Welch-Aspin F statistic and if the $r(\mathbf{M}) = 1$ so that $\mathbf{M} = \mathbf{m}$ the statistic simplifies to

$$\widetilde{T}_B^2 = \mathbf{m}'(\widehat{\mathbf{B}}_1 - \widehat{\mathbf{B}}_2)'\mathbf{C}'\left(v_1\mathbf{B}\widehat{\mathbf{G}}_1 + v_2\widehat{\mathbf{G}}_2\right)^{-1}\mathbf{C}(\widehat{\mathbf{B}}_1 - \widehat{\mathbf{B}}_2)\mathbf{m} \tag{3.9.59}$$

where

$$\widehat{\mathbf{G}}_i = (\mathbf{m}'\mathbf{S}_i\mathbf{m})\mathbf{W}_i/v_i$$

Then, $H : \mathbf{CB}_1\mathbf{m} = \mathbf{CB}_2\mathbf{m}$ is rejected if

$$F = \widetilde{T}_B^2/g \geq F^{1-\alpha}\left(g, \widehat{f}\right) \tag{3.9.60}$$

where

$$\widehat{f} = \frac{[\operatorname{vech}(v_1\widehat{\mathbf{G}}_1 + v_2\widehat{\mathbf{G}}_2)]' \operatorname{vech}(v_1\widehat{\mathbf{G}}_1 + v_2\widehat{\mathbf{G}}_2)}{v_1(\operatorname{vech}\widehat{\mathbf{G}}_1)' \operatorname{vech}\widehat{\mathbf{G}}_1 + v_2(\operatorname{vech}\widehat{\mathbf{G}}_2)' \operatorname{vech}\widehat{\mathbf{G}}_2} \tag{3.9.61}$$

Example 3.9.5 (Two-Group Profile Analysis $\Sigma_1 \neq \Sigma_2$) *Expression (3.9.52) may be used to test the multivariate hypotheses of no group difference (H_G^*), equal vector profiles across the p conditions (H_C^*), and the parallelism of profiles (H_P) when the covariance matrices for the two groups are not equal in the population. And, given parallelism, it may be used to test for differences in groups (H_G) or differences in the p conditions (H_C). For the test of conditions given parallelism, we do not have to assume that the covariance matrices have any special structure and for the test for differences in group means we do not require that the population variances be equal. Because the test of parallelism determines how we* *usually proceed with our analysis of profile data, we illustrate how to calculate (3.9.57) to test for parallelism (H_P) when the population covariances are unequal. For this example, the problem solving ability data provided in Table 3.9.9 are used. The data represent the time required to solve four mathematics problems for a new experimental treatment procedure and a control method. The code for the analysis is provided in program m3_9f.sas. Using formula (3.9.57) with Hotelling's approximate T^2 statistic, $\widehat{T}_B^2 = 1.2456693$, the F-statistic is $F = 0.3589843$. The degrees of freedom for the F-statistic for the hypothesis of parallelism are 3 and 12.766423. The degrees of freedom for error is calculated using (3.9.57) and (3.9.58). The p-value for the test of parallelism is 0.7836242. Thus, we do not reject the hypothesis of parallel profiles.*

For this example, the covariance matrices appear to be equal in the population so that we may compare the p-value for the approximate test of parallelism with the p-value for the exact likelihood ratio test. As illustrated in Example 3.9.4, we use PROC GLM to test for parallelism given that the covariance matrices are equal in the population. The exact F-statistic for the test of parallelism is $F = 0.35$ has an associated p-value of 0.7903. Because the Type I error rates for the two procedures are approximately equal, the relative efficiency of the two methods appear to be nearly identical when the covariance matrices are equal. Thus, one would expect to lose little power by using the approximate test procedure when the covariance matrices are equal. Of course, if the covariance matrices are not equal we may not use the exact test. One may modify program m3_9f.sas to test other hypotheses when the covariances are unequal.

Exercises 3.9

1. In a pilot study designed to compare a new training program with the current standard in grammar usage (G), reading skills (R), and spelling (S) to independent groups of students finished the end of the first week of instruction were compared on the three variables. The data are provided in Table 3.9.6

TABLE 3.9.6. Two-Group Instructional Data.

Experimental				Control			
Subject	G	R	S	Subject	G	R	S
1	31	12	24	1	31	50	20
2	52	64	32	2	60	40	15
3	57	42	21	3	65	36	12
4	63	19	54	4	70	29	18
5	42	12	41	5	78	48	24
6	71	79	64	6	90	47	26
7	65	38	52	7	98	18	40
8	60	14	57	8	95	10	10
9	54	75	58				
10	67	22	69				
11	70	34	24				

(a) Test the hypotheses that $\Sigma_1 = \Sigma_2$.

(b) For $\alpha = 0.05$, test the hypothesis than the mean performance on the three dependent variables is the same for both groups; $H_o : \boldsymbol{\mu}_1 = \boldsymbol{\mu}_2$. Perform the test assuming $\Sigma_1 = \Sigma_2$ and $\Sigma_1 \neq \Sigma_2$.

(c) Given that $\Sigma_1 = \Sigma_2$, use the discriminant coefficients to help isolate variables that led to the rejection of H_C.

(d) Find 95% simultaneous confidences for parametric functions that evaluate the mean difference between groups for each variable using (3.9.5). Compare these intervals using the Studentized Maximum Modulus Distribution. The critical values are provided in the Appendix, Table V.

(e) Using all three variables, what is the contrast that led to the rejection of H_o. Can you interpret your finding?

2. Dr. Paul Ammon had subjects listen to tape-recorded sentences. Each sentence was followed by a "probe" taken from one of five positions in the sentence. The subject was to respond with the word that came immediately after the probe word in the sentence and the speed of the reaction time was recorded. The data are given in Table 3.9.7.

Example Statement: The tall man met the young girl who got the new hat.

1 2 3 4 5

Dependent Variable: Speed of response (transformed reaction time).

(a) Does the covariance matrix for this data have Type H structure?

(b) Test the hypothesis that the mean reaction time is the same for the five probe positions.

(c) Construct confidence intervals and summarize your findings.

3. Using that data in Table 3.7.3. Test the hypothesis that the mean length of the ramus bone measurements for the boys in the study are equal. Does this hypothesis make sense? Why or why not? Please discuss your observations.

4. The data in Table 3.9.8 were provided by Dr. Paul Ammon. They were collected as in the one-sample profile analysis example, except that group I data were obtained from subjects with low short-term memory capacity and group II data were obtained from subjects with high short-term memory capacity.

(a) Plot the data.

(b) Are the profiles parallel?

(c) Based on your decision in (b), test for differences among probe positions and differences between groups.

(d) Discuss and summarize your findings.

TABLE 3.9.7. Sample Data: One-Sample Profile Analysis.

	Probe-Word Positions				
Subject	1	2	3	4	5
1	51	36	50	35	42
2	27	20	26	17	27
3	37	22	41	37	30
4	42	36	32	34	27
5	27	18	33	14	29
6	43	32	43	35	40
7	41	22	36	25	38
8	38	21	31	20	16
9	36	23	27	25	28
10	26	31	31	32	36
11	29	20	25	26	25

TABLE 3.9.8. Sample Data: Two-Sample Profile Analysis.

		Probe-Word Positions				
		1	2	3	4	5
Group I	S_1	20	21	42	32	32
	S_2	67	29	56	39	41
	S_3	37	25	28	31	34
	S_4	42	38	36	19	35
	S_5	57	32	21	30	29
	S_6	39	38	54	31	28
	S_7	43	20	46	42	31
	S_8	35	34	43	35	42
	S_9	41	23	51	27	30
	S_{10}	39	24	35	26	32
	Mean	42.0	28.4	41.2	31.2	33.4
Group II	S_1	47	25	36	21	27
	S_2	53	32	48	46	54
	S_3	38	33	42	48	49
	S_4	60	41	67	53	50
	S_5	37	35	45	34	46
	S_6	59	37	52	36	52
	S_7	67	33	61	31	50
	S_8	43	27	36	33	32
	S_9	64	53	62	40	43
	S_{10}	41	34	47	37	46
	Mean	50.9	35.0	49.6	37.9	44.9

TABLE 3.9.9. Problem Solving Ability Data.

		Problems			
	Subject	1	2	3	4
	1	43	90	51	67
	2	87	36	12	14
	3	18	56	22	68
	4	34	73	34	87
C	5	81	55	29	54
	6	45	58	62	44
	7	16	35	71	37
	8	43	47	87	27
	9	22	91	37	78
	1	10	81	43	33
	2	58	84	35	43
	3	26	49	55	84
	4	18	30	49	44
E	5	13	14	25	45
	6	12	8	40	48
	7	9	55	10	30
	8	31	45	9	66

(e) Do these data satisfy the model assumptions of homogeneity and circularity so that one may construct exact univariate F tests?

5. In an experiment designed to investigate problem-solving ability for two groups of subjects, experimental (E) and control (C) subjects were required to solve four different mathematics problems presented in a random order for each subject. The time required to solve each problem was recorded. All problems were thought to be of the same level of difficulty. The data for the experiment are summarized in Table 3.9.9.

   (a) Test that $\Sigma_1 = \Sigma_2$ for these data.

   (b) Can you conclude that the profiles for the two groups are equal? Analyze this question given $\Sigma_1 = \Sigma_2$ and $\Sigma_1 \neq \Sigma_2$.

   (c) In Example 3.9.5, we showed that there is no interaction between groups and conditions. Are there any differences among the four conditions? Test this hypothesis assuming equal and unequal covariance matrices.

   (d) Using simultaneous inference procedures, where are the differences in conditions in (c)?

6. Prove that if a covariance matrix Σ has compound symmetry structure that it is a "Type H" matrix.

3.10 Univariate Profile Analysis

In Section 3.9 we presented the one- and two-group profile analysis models as multivariate models and as univariate mixed models. For the univariate models, we represented the models as unconstrained models in that no restrictions (side conditions) were imposed on the fixed or random parameters. To calculate expected mean squares for balanced/orthogonal mixed models, many students are taught to use rules of thumb. As pointed out by Searle (1971, p. 393), not all rules are the same when applied to mixed models. If you follow Neter et al. (1996, p. 1377) or Kirk (1995, p. 402) certain terms "disappear" from the expressions for expected mean squares (EMS). This is not the case for the rules developed by Searle. The rules provided by Searle are equivalent to obtaining expected mean squares (EMS) using the computer synthesis method developed by Hartley (1967). The synthesis method is discussed in detail by Milliken and Johnson (1992, Chapter 18) and Hocking (1985, p. 336). The synthesis method may be applied to balanced (orthogonal) designs or unbalanced (nonorthogonal) designs. It calculates EMS using an unconstrained model. Applying these rules of thumb to models that include restrictions on fixed and random parameters has caused a controversy among statisticians, Searle (1971, pp. 400-404), Schwarz (1993), Voss (1999), and Hinkelmann et al. (2000).

Because SAS employs the method of synthesis without model constraints, the EMS as calculated in PROC GLM depend on what factors a researcher specifies as random on the RANDOM statement in PROC GLM, in particular, whether interactions between random effects and fixed effects are designated as random or fixed. If any random effect that interacts with a fixed effect or a random effect is designed as random, then the EMS calculated by SAS results in the correct EMS for orthogonal or nonorthogonal models. For any balanced design, the EMS are consistent with EMS obtained using rules of the thumb applied to unconstrained (unrestricted) models as provided by Searle.

If the interaction of random effects with fixed effects is designated as fixed, and excluded from the RANDOM statement in PROC GLM, tests may be constructed assuming one or more of the fixed effects are zero. For balanced designs, this often causes other entries in the EMS table to behave like EMS obtained by rules of thumb for univariate models with restrictions. To ensure correct tests, all random effects that interact with other random effects and fixed effects must be specified on the RANDOM statement in PROC GLM. Then, F or quasi-F tests are created using the RANDOM statement

```
random r r * f / test ;
```

Here, r is a random effect and f is a fixed effect. The MODEL statement is used to specify the model and must include all fixed, random, and nested parameters. When using PROC GLM to analyze mixed models, only the tests obtained from the random statement are valid; see Littell et al. (1996, p. 29).

To analyze mixed models in SAS, one should not use PROC GLM. Instead, PROC MIXED should be used. For balanced designs, the F tests for fixed effects are identical. For nonorthogonal designs they generally do not agree. This is due to the fact that parameter estimates in PROC GLM depend on ordinary least squares theory while PROC MIXED uses generalized least squares theory. An advantage of using PROC MIXED over PROC GLM is that one may estimate variance components in PROC mixed, find confidence intervals for the variance components, estimate contrasts in fixed effects that have correct standard errors, and estimate random effects. In PROC MIXED, the MODEL statement only contains fixed effects while the RANDOM statement contains only random effects. We will discuss PROC MIXED in more detail in Chapter 6; we now turn to the reanalysis of the one-group and two-group profile data.

a. *Univariate One-Group Profile Analysis*

Using program m3_10a.sas to analyze Example 3.9.3 using the unconstrained univariate randomized block mixed model, one must transform the vector observations to elements Y_{ij}. This is accomplished in the data step. Using the RANDOM statement with $subj$, the EMS are calculated and the F test for differences in means among conditions or treatments is $F = 942.9588$. This is the exact value obtained from the univariate test in the MANOVA model. The same value is realized under the Tests of Fixed Effects in PROC MIXED. In addition, PROC MIXED provides point estimates for σ_e^2 and σ_s^2: $\widehat{\sigma}_e^2 = 4.0535$ and $\widehat{\sigma}_s^2 = 1.6042$ with standard errors and upper and lower limits. Tukey-Kramer confidence intervals for simple mean differences are also provided by the software. The F tests are only exact under sphericity of the transformed covariance matrix (circularity).

b. *Univariate Two-Group Profile Analysis*

Assuming homogeneity and circularity, program m3_10b.sas is used to reanalyze the data in Example 3.9.4, assuming a univariate unconstrained split-plot design. Reviewing the tests in PROC GLM, we see that the univariate test of group differences and the test of treatment (condition) differences have a warning that this test assumes one or more other fixed effects are zero. In particular, looking at the table of EMS, the interaction between treatments by groups must be nonsignificant. Or, we need parallel profiles for a valid test.

Because this design is balanced, the Tests of Fixed Effects in PROC MIXED agree with the GLM, F tests. We also have estimates of variance components with confidence intervals. Again, more will be said about these results in Chapter 6. We included a discussion here to show how to perform a correct univariate analysis of these designs when the circularity assumption is satisfied.

3.11 Power Calculations

Because Hotelling's T^2 statistic, $T^2 = n\mathbf{Y}'\mathbf{Q}^{-1}\mathbf{Y}$, is related to an F distribution, by Definition 3.5.3

$$F = \frac{(n-p+1)\,T^2}{pn} \sim F^{1-\alpha}(p, n-p, \gamma) \tag{3.11.1}$$

with noncentrality parameter

$$\gamma = \boldsymbol{\mu}'\Sigma^{-1}\boldsymbol{\mu}$$

one may easily estimate the power of tests that depend on T^2. The power π is the $\Pr[F \geq F^{1-\alpha}(v_h, v_e, \gamma)]$ where $v_h = p$ and $v_e = n - p$. Using the SAS functions FINV and

PROBF, one computes π as follows

$$\begin{aligned} F\_CV &= \text{FINV}\,(1-\alpha, dfh, dfe, \gamma = 0) \\ \pi &= 1 - \text{PROBF}\,(F\_CV, dfh, dfe, \gamma) \end{aligned} \tag{3.11.2}$$

The function FINV returns the critical value for the F distribution and the function PROBF returns the p-value. To calculate the power of the test requires one to know the size of the test α, the sample size n, the number of variables p, and the noncentrality parameter γ which involves both of the unknown population parameters, Σ and $\boldsymbol{\mu}$.

For the two-group location test of $H_o : \boldsymbol{\mu}_1 = \boldsymbol{\mu}_2$, the noncentrality parameter is

$$\gamma = \frac{n_1 n_2}{n_1 + n_2} (\boldsymbol{\mu}_1 - \boldsymbol{\mu}_2)' \, \Sigma^{-1} \, (\boldsymbol{\mu}_1 - \boldsymbol{\mu}_2) \tag{3.11.3}$$

Given n_1, n_2, α, and γ, the power of the test is easily estimated.

Conversely for a given difference $\boldsymbol{\delta} = \boldsymbol{\mu}_1 - \boldsymbol{\mu}_2$ and Σ, one may set $n_1 = n_2 = n_0$ so that

$$\gamma = \frac{n_0^2}{2n_0} \boldsymbol{\delta}' \Sigma^{-1} \boldsymbol{\delta} \tag{3.11.4}$$

By incrementing n_0, the desired power for the test of $H_o : \boldsymbol{\mu}_1 = \boldsymbol{\mu}_2$ may be evaluated to obtain an appropriate sample size for the test.

Example 3.11.1 (Power Calculation) *An experimenter wanted to design a study to evaluate the mean difference in performance between an experimental treatment and a control employing two variables that measured achievement in two related content areas. To test $\boldsymbol{\mu}_E = \boldsymbol{\mu}_C$. Based on a pilot study, the population covariance matrix for the two variables was as follows*

$$\Sigma = \begin{bmatrix} 307 & 280 \\ 280 & 420 \end{bmatrix}$$

The researcher wanted to be able to detect a mean difference in performance of $\boldsymbol{\delta}' = [\boldsymbol{\mu}_1 - \boldsymbol{\mu}_2]' = [1, 5]'$ units. To ensure that the power of the test was at least 0.80, *the researcher wanted to know if five or six subjects per group would be adequate for the study. Using program m3_11_1.sas, the power for $n_0 = 5$ subjects per group or 10 subjects in the study has power, $\pi = 0.467901$. For $n_0 = 6$ subjects per group, the value of $\pi = 0.8028564$. Thus, the study was designed with six subjects per group or* 12 *subjects.*

Power analysis for studies involving multivariate variables is more complicated than univariate power analysis because it involves the prior specification of considerable population structure. Because the power analysis for T^2 tests is a special case of power analysis using the MR model, we will address power analysis more generally for multivariate linear models in Chapter 4.

Exercises 3.11

1. A researcher wants to detect differences of 1, 3, and 5 units on three dependent variables in an experiment comparing two treatments. Randomly assigning an equal

number of subjects to the two treatments, with

$$\Sigma = \begin{bmatrix} 10 & & \\ 5 & 10 & \\ 5 & 5 & 10 \end{bmatrix}$$

and $\alpha = 0.05$, how large a sample size is required to attain the power $\pi = 0.80$ when testing $H : \mu_1 = \mu_2$?

2. Estimate the power of the tests for testing the hypotheses in Exercises 3.7, Problem 4, and Exercises 3.7, Problem 2.

4
Multivariate Regression Models

4.1 Introduction

In Chapter 3, Section 3.6 we introduced the basic theory for estimating the nonrandom, fixed parameter matrix $\mathbf{B}_{q \times p}$ for the multivariate (linear) regression (MR) model $\mathbf{Y}_{n \times p} = \mathbf{X}_{n \times q} \mathbf{B}_{q \times p} + \mathbf{E}_{n \times p}$ and for testing linear hypotheses of the general form $\mathbf{CBM} = \mathbf{0}$. For this model it was assumed that the design matrix $\mathbf{X}$ contains fixed nonrandom variables measured without measurement error, the matrix $\mathbf{Y}_{n \times p}$ contains random variables with or without measurement error, the $E(\mathbf{Y})$ is related to $\mathbf{X}$ by a linear function of the parameters in $\mathbf{B}$, and that each row of $\mathbf{Y}$ has a MVN distribution.

When the design matrix $\mathbf{X}$ contains only indicator variables taking the values of zero or one, the models are called multivariate analysis of variance (MANOVA) models. For MANOVA models, $\mathbf{X}$ is usually not of full rank; however, the model may be reparameterized so that $\mathbf{X}$ is of full rank. When $\mathbf{X}$ contains both quantitative predictor variables also called covariates (or concomitant variables) and indicator variables, the class of regression models is called multivariate analysis of covariance (MANCOVA) models. MANCOVA models are usually analyzed in two steps. First a regression analysis is performed by regressing the dependent variables in $\mathbf{Y}$ on the covariates and then a MANOVA is performed on the residuals. The matrix $\mathbf{X}$ in the multivariate regression model or in MANCOVA models may also be assumed to be random adding an additional level of complexity to the model. In this chapter, we illustrate testing linear hypotheses, the construction of simultaneous confidence intervals and simultaneous test procedures (STP) for the elements of $\mathbf{B}$ for MR, MANOVA and MANCOVA models. Also considered are residual analysis, lack-of-fit tests, the detection of influential observations, model validation and random design matrices. Designs with one, two and higher numbers of factors, with fixed and random covariates, repeated measurement designs and unbalanced data problems are discussed and illustrated. Finally, robustness of test procedures, power calculation issues, and testing means with unequal covariance matrices are reviewed.

4.2 Multivariate Regression

a. *Multiple Linear Regression*

In studies utilizing multiple linear regression one wants to determine the most appropriate linear model to predict only one dependent random variable y from a set of fixed, observed independent variables $x_1, x_2, \ldots, x_k$ measured without error. One can fit a linear model of the form specified in (3.6.3) using the least squares criterion and obtain an unbiased estimate of the unknown common variance σ^2. To test hypotheses, one assumes that $\mathbf{y}$ in (3.6.3) follows a MVN distribution with covariance matrix $\Omega = \sigma^2 \mathbf{I}_n$. Having fit an initial model to the data, model refinement is a necessary process in regression analysis. It involves evaluating the model assumptions of multivariate normality, homogeneity of variance, and independence. Given that the model assumptions are correct, one next obtains a model of best fit. Finally, one may evaluate the model prediction, called model validation. Formal tests and numerous types of plots have been developed to systematically help one evaluate the assumptions of multivariate normality; detect outliers, select independent variables, detect influential observations and detect lack of independence. For a more thorough discussion of the iterative process involved in multiple linear and nonlinear regression analysis, see Neter, Kutner, Nachtsheim and Wasserman (1996).

When the dependent variable $\mathbf{y}$ in a study can be assumed to be independent multivariable normally distributed but the covariance structure cannot be assumed to have the sphericity structure $\Omega = \sigma^2 \mathbf{I}_n$, one may use the generalized least squares analysis. Using generalized least squares, a more general structure for the covariance matrix is assumed. Two common forms for Ω are $\Omega = \sigma^2 \mathbf{V}$ where $\mathbf{V}$ is known and nonsingular called the weighted least squares (WLS) model and $\Omega = \Sigma$ where Σ is known and nonsingular called the generalized least squares (GLS) model. When Σ is unknown, one uses large sample asymptotic normal theory to fit and evaluate models. In the case when Σ is unknown feasible generalized least squares (FGLS) or estimated generalized least squares (EGLS) procedures can be used. For a discussion of these procedures see Goldberger (1991), Neter et al. (1996) and Timm and Mieczkowski (1997, Chapter 4).

When the data contain outliers, or the distribution of $\mathbf{y}$ is nonnormal, but elliptically symmetric, or the structure of $\mathbf{X}$ is unknown, one often uses robust regression, nonparametric regression, smoothing methodologies or bootstrap procedures to fit models to the data vector $\mathbf{y}$, Rousseeuw and Leroy (1987), Buja, Hastie and Tibshirani (1989) and Friedman (1991). When the dependent variable is discrete, generalized linear models introduced by Nelder and Wedderburn (1972) are used to fit models to data. The generalized linear model (GLIM) extends the traditional MVN general linear model theory to models that include the class of distributions known as the exponential family of distributions. Common members of this family are the binomial, Poisson, normal, gamma and inverse gamma distributions. The GLIM combined with quasi-likelihood methods developed by Wedderburn

(1974) allow researchers to fit both linear and nonlinear models to both discrete (e.g., binomial, Poisson) and continuous (e.g., normal, gamma, inverse gamma) random, dependent variables. For a discussion of these models which include logistic regression models, see McCullagh and Nelder (1989), Littell, et al. (1996), and McCulloch and Searle (2001).

b. *Multivariate Regression Estimation and Testing Hypotheses*

In multivariate (linear) regression (MR) models, one is not interested in predicting only one dependent variable but rather several dependent random variables $y_1, y_2, \ldots, y_p$. Two possible extensions with regard to the set of independent variables for MR models are (1) the design matrix $\mathbf{X}$ of independent variables is the same for each dependent variable or (2) each dependent variable is related to a different set of independent variables so that p design matrices are permitted. Clearly, situation (1) is more restrictive than (2) and (1) is a special case of (2). Situation (1) which requires the same design matrix for each dependent variable is considered in this chapter while situation (2) is treated in Chapter 5 where we discuss the seemingly unrelated regression (SUR) model which permits the simultaneous analysis of p multiple regression models.

In MR models, the rows of $\mathbf{Y}$ or $\mathbf{E}$, are assumed to be distributed independent MVN so that vec $(\mathbf{E}) \sim N_{np}\,(\mathbf{0},\,\Sigma \otimes \mathbf{I}_n)$. Fitting a model of the form $E\,(\mathbf{Y}) = \mathbf{XB}$ to the data matrix $\mathbf{Y}$ under MVN, the maximum likelihood (ML) estimate of $\mathbf{B}$ is given in (3.6.20). This ML estimate is identical to the unique best linear unbiased estimator (BLUE) obtained using the multivariate ordinary least squares criterion that the Euclidean matrix norm squared, $\operatorname{tr}\left[(\mathbf{Y}-\mathbf{XB})'\,(\mathbf{Y}-\mathbf{XB})\right] = \|\mathbf{Y}-\mathbf{XB}\|^2$ is minimized over all parameter matrices $\mathbf{B}$ for fixed $\mathbf{X}$, Seber (1984).

For the MR model $\mathbf{Y} = \mathbf{XB} + \mathbf{E}$, the parameter matrix $\mathbf{B}$ is

$$\mathbf{B} = \begin{bmatrix} \boldsymbol{\beta}_0' \\ --- \\ \mathbf{B}_1 \end{bmatrix} = \begin{bmatrix} \beta_{01} & \beta_{02} & \cdots & \beta_{0p} \\ --- & --- & --- & --- \\ \beta_{11} & \beta_{12} & \cdots & \beta_{1p} \\ \beta_{21} & \beta_{22} & \cdots & \beta_{2p} \\ \vdots & \vdots & \vdots & \vdots \\ \beta_{k1} & \beta_{k2} & \cdots & \beta_{kp} \end{bmatrix} \tag{4.2.1}$$

where $q = k + 1$ and is the number of independent variables associated with each dependent variable. The vector $\boldsymbol{\beta}_0'$ contains intercepts while the matrix $\mathbf{B}_1$ contains coefficients associated with independent variables. The matrix $\mathbf{B}$ in (4.2.1) is called the raw score form of the parameter matrix since the elements y_{ij} in $\mathbf{Y}$ have the general form

$$y_{ij} \;=\; \beta_{0j} \;+\; \beta_{1j}\;\; x_{i1} \;\;+\ldots+\;\; \beta_{kj}\;\; x_{ik} \;\;+\;\; e_{ij} \tag{4.2.2}$$

for $i = 1, 2, \ldots, n$ and $j = 1, 2, \ldots, p$.

To obtain the deviation form of the MR model, the means $\overline{x}_j = \sum_{i=1}^n x_{ij}/n,\ j = 1, 2, \ldots, k$ are calculated and the deviation scores $d_{ij} = x_{ij} - \overline{x}_j$, are formed. Then, (4.2.2) becomes

$$y_{ij} = \beta_{0j} + \sum_{h=1}^{k} \beta_{hj}\,\overline{x}_h + \sum_{h=1}^{k} \beta_{hj}\,(x_{ih} - \overline{x}_h) + e_{ij} \tag{4.2.3}$$

Letting

$$\begin{aligned}
\alpha_{0j} &= \beta_{0j} + \sum_{h=1}^{k} \beta_{hj}\overline{x}_h \quad j = 1, 2, \ldots, p \\
\boldsymbol{\alpha}_0' &= \left[\alpha_{01}, \alpha_{02}, \ldots, \alpha_{0p}\right] \\
\mathbf{B}_1 &= \left[\beta_{hj}\right] \quad h = 1, 2, \ldots, k \text{ and } j = 1, 2, \ldots, p \\
\mathbf{X}_d &= \left[d_{ij}\right] \quad i = 1, 2, \ldots, n \text{ and } j = 1, 2, \ldots, p
\end{aligned} \tag{4.2.4}$$

the matrix representation of (4.2.3) is

$$\underset{n \times p}{\mathbf{Y}} = [\mathbf{1}_n\ \mathbf{X}_d] \begin{bmatrix} \boldsymbol{\alpha}_0' \\ \mathbf{B}_1 \end{bmatrix} + \mathbf{E} \tag{4.2.5}$$

where $\mathbf{1}_n$ is a vector of n $1's$. Applying (3.6.21),

$$\widehat{\mathbf{B}} = \begin{bmatrix} \widehat{\boldsymbol{\alpha}}_0' \\ \widehat{\mathbf{B}}_1 \end{bmatrix} = \begin{bmatrix} \overline{\mathbf{y}}' \\ \left(\mathbf{X}_d'\mathbf{X}_d\right)^{-1}\mathbf{X}_d'\mathbf{Y} \end{bmatrix} = \begin{bmatrix} \overline{\mathbf{y}}' \\ \left(\mathbf{X}_d'\mathbf{X}_d\right)^{-1}\mathbf{X}_d\mathbf{Y}_d \end{bmatrix} \tag{4.2.6}$$

where $\mathbf{Y}_d = \left[y_{ij} - \overline{y}_j\right]$, and $\overline{y}_j$ is the mean of the jth dependent variable. The matrix $\mathbf{Y}$ may be replaced by $\mathbf{Y}_d$ since the $\sum_j d_{ij} = 0$ for $i = 1, 2, \ldots, n$. This establishes the equivalence of the raw and deviation forms of the MR model since $\widehat{\beta}_{0j} = \overline{y}_j - \sum_{h=1}^{k} \beta_{hj}\overline{x}_j$. Letting the matrix $\mathbf{S}$ be the partitioned sample covariance matrices for the dependent and independent variables

$$\mathbf{S} = \begin{bmatrix} \mathbf{S}_{yy} & \mathbf{S}_{yx} \\ \mathbf{S}_{xy} & \mathbf{S}_{xx} \end{bmatrix} \tag{4.2.7}$$

and

$$\widehat{\mathbf{B}}_1 = \left(\frac{\mathbf{X}_d'\mathbf{X}_d}{n-1}\right)^{-1} \left(\frac{\mathbf{X}_d'\mathbf{Y}_d}{n-1}\right) = \mathbf{S}_{xx}^{-1}\mathbf{S}_{xy}$$

Because the independent variables are considered to be fixed variates, the matrix $\mathbf{S}_{xx}$ does not provide an estimate of the population covariance matrix. Another form of the MR regression model used in applications is the standard score form of the model. For this form, all dependent and independent variables are standardized to have mean zero and variance one. Replacing the matrix $\mathbf{Y}_d$ with standard scores represented by $\mathbf{Y}_z$ and the matrix $\mathbf{X}_d$ with the standard score matrix $\mathbf{Z}$, the MR model becomes

$$\mathbf{Y}_z = \mathbf{Z}\mathbf{B}_z + \mathbf{E} \tag{4.2.8}$$

and

$$\widehat{\mathbf{B}}_z = \mathbf{R}_{xy}^{-1}\mathbf{R}_{xy} \text{ or } \widehat{\mathbf{B}}_z' = \mathbf{R}_{yx}\mathbf{R}_{xx}^{-1} \tag{4.2.9}$$

where $\mathbf{R}_{xx}$ is a correlation matrix of the fixed $x's$ and $\mathbf{R}_{yx}$ is the sample intercorrelation matrix of the fixed $\mathbf{x}$ and random $\mathbf{y}$ variables. The coefficients in $\widehat{\mathbf{B}}_z$ are called standardized or standard score coefficients. Using the relationships that

$$
\begin{aligned}
\mathbf{R}_{xx} &= (\operatorname{diag}\mathbf{S}_{xx})^{1/2}\,\mathbf{S}_{xx}\,(\operatorname{diag}\mathbf{S}_{xx})^{1/2} \\
\mathbf{R}_{xy} &= \left(\operatorname{diag}\mathbf{S}_{xy}\right)^{1/2}\,\mathbf{S}_{xy}\,\left(\operatorname{diag}\mathbf{S}_{xy}\right)^{1/2}
\end{aligned}
\tag{4.2.10}
$$

$\widehat{\mathbf{B}}_1$ is easily obtained from $\widehat{\mathbf{B}}_z$.

Many regression packages allow the researcher to obtain both raw and standardized coefficients to evaluate the importance of independent variables and their effect on the dependent variables in the model. Because the units of measurement for each independent variable in a MR regression model are often very different, the sheer size of the coefficients may reflect the unit of measurement and not the importance of the variable in the model. The standardized form of the model converts the variables to a scale free metric that often facilitates the direct comparison of the coefficients. As in multiple regression, the magnitude of the coefficients are affected by both the presence of large intercorrelations among the independent variables and the spacing and range of measurements for each of the independent variables. If the spacing is well planned and not arbitrary and the intercorrelations of the independent variables are low so as not to adversely effect the magnitude of the coefficients when variables are added or removed from the model, the standardized coefficients may be used to evaluate the relative simultaneous change in the set Y for a unit change in each X_i when holding the other variables constant.

Having fit a MR model of the form $\mathbf{Y} = \mathbf{XB} + \mathbf{E}$ in (3.6.17), one usually tests hypotheses regarding the elements of $\mathbf{B}$. The most common test is the test of no linear relationship between the two sets of variables or the overall regression test

$$
H_1 : \mathbf{B}_1 = \mathbf{0} \tag{4.2.11}
$$

Selecting $\mathbf{C}_{k\times q} = [\mathbf{0}, \mathbf{I}_k]$ of full row rank k and $\mathbf{M}_{p\times p} = \mathbf{I}_p$, the test that $\mathbf{B}_1 = \mathbf{0}$ is easily derived from the general matrix form of the hypothesis, $\mathbf{CBM} = \mathbf{0}$. Using (3.6.26) and partitioning $\mathbf{X} = [\mathbf{1}\ \mathbf{X}_2]$ where $\mathbf{Q} = \mathbf{I} - \mathbf{1}\left(\mathbf{1}'\mathbf{1}\right)^{-1}\mathbf{1}'$ then

$$
\widehat{\mathbf{B}} = \begin{bmatrix} \widehat{\beta}_0' \\ \widehat{\mathbf{B}}_1 \end{bmatrix} = \begin{bmatrix} \left(\mathbf{1}'\mathbf{1}\right)^{-1}\mathbf{1}'\mathbf{Y} - \left(\mathbf{1}'\mathbf{1}\right)^{-1}\mathbf{1}'\mathbf{X}_2\left(\mathbf{X}_2'\mathbf{Q}\mathbf{X}_2\right)^{-1}\mathbf{X}_2'\mathbf{Q}\mathbf{Y} \\ \left(\mathbf{X}_2'\mathbf{Q}\mathbf{X}_2\right)^{-1}\mathbf{X}_2'\mathbf{Q}\mathbf{Y} \end{bmatrix} \tag{4.2.12}
$$

and $\widehat{\mathbf{B}}_1 = \mathbf{X}_2\left(\mathbf{X}_2'\mathbf{Q}\mathbf{X}_2\right)^{-1}\mathbf{X}_2'\mathbf{Q}\mathbf{Y} = \left(\mathbf{X}_d'\mathbf{X}_d\right)^{-1}\mathbf{X}_d'\mathbf{Y}_d$ since $\mathbf{Q}$ is idempotent and

$$
\begin{aligned}
\mathbf{H} &= \widehat{\mathbf{B}}_1'\left(\mathbf{X}_d'\mathbf{X}_d\right)\widehat{\mathbf{B}}_1 \\
\mathbf{E} &= \mathbf{Y}'\mathbf{Y} - \overline{\mathbf{y}}\,\overline{\mathbf{y}}' - \widehat{\mathbf{B}}_1'\left(\mathbf{X}_d'\mathbf{X}_d\right)\widehat{\mathbf{B}}_1 = \mathbf{Y}_d'\mathbf{Y}_d - \widehat{\mathbf{B}}_1'\left(\mathbf{X}_d'\mathbf{X}_d\right)\widehat{\mathbf{B}}_1
\end{aligned}
\tag{4.2.13}
$$

so that $\mathbf{E} + \mathbf{H} = \mathbf{T} = \mathbf{Y}'\mathbf{Y} - n\overline{\mathbf{y}}\,\overline{\mathbf{y}}' = \mathbf{Y}_d'\mathbf{Y}_d$ is the total sum of squares and cross products matrix, about the mean. The MANOVA table for testing $\mathbf{B}_1 = \mathbf{0}$ is given in Table 4.2.1.

To test $H_1 : \mathbf{B}_1 = \mathbf{0}$, Wilks' Λ criterion from (3.5.2), is

$$
\Lambda = \frac{|\mathbf{E}|}{|\mathbf{E} + \mathbf{H}|} = \prod_{i=1}^{s}\left(1 + \lambda_i\right)^{-1}
$$

TABLE 4.2.1. MANOVA Table for Testing $\mathbf{B}_1 = \mathbf{0}$

Source	df	SSCP	E(MSCP)
$\boldsymbol{\beta}_0$	1	$n\overline{\mathbf{y}}\,\overline{\mathbf{y}}'$	$\Sigma + n\boldsymbol{\beta}_0\boldsymbol{\beta}_0'$
$\mathbf{B}_1 \mid \boldsymbol{\beta}_0$	k	$\mathbf{H} = \widehat{\mathbf{B}}_1'(\mathbf{X}_d'\mathbf{X}_d)\widehat{\mathbf{B}}_1$	$\Sigma + \frac{\mathbf{B}_1'(\mathbf{X}_d'\mathbf{X}_d)\mathbf{B}_1}{k}$
Residual	$n-k-1$	$\mathbf{E} = \mathbf{Y}_d'\mathbf{Y}_d - \widehat{\mathbf{B}}_1'(\mathbf{X}_d'\mathbf{X}_d)\widehat{\mathbf{B}}_1$	Σ
Total	n	$\mathbf{Y}'\mathbf{Y}$	

where λ_i are the roots of $|\mathbf{H} - \lambda\mathbf{E}| = 0$, $s = \min(v_h, p) = \min(k, p)$, $v_h = k$ and $v_e = n - q = n - k - 1$. An alternative form for Λ is to employ sample covariance matrices. Then $\mathbf{H} = \mathbf{S}_{yx}\mathbf{S}_{xx}^{-1}\mathbf{S}_{xy}$ and $\mathbf{E} = \mathbf{S}_{yy} - \mathbf{S}_{yx}\mathbf{S}_{xx}^{-1}\mathbf{S}_{xy}$ so that $|\mathbf{H} - \lambda\mathbf{E}| = 0$ becomes $\left|\mathbf{S}_{yx}\mathbf{S}_{xx}^{-1}\mathbf{S}_{xy} - \lambda(\mathbf{S}_{yy} - \mathbf{S}_{xx}^{-1}\mathbf{S}_{xy})\right| = 0$. From the relationship among the roots in Theorem 2.6.8, $|\mathbf{H} - \theta(\mathbf{H} + \mathbf{E})| = \left|\mathbf{S}_{yx}\mathbf{S}_{xx}^{-1}\mathbf{S}_{xy} - \theta\,\mathbf{S}_{yy}\right| = 0$ so that

$$\Lambda = \prod_{i=1}^{s}(1 + \lambda_i)^{-1} = \prod_{i=1}^{s}(1 - \theta_i) = \frac{\left|\mathbf{S}_{yy} - \mathbf{S}_{yx}\mathbf{S}_{xx}^{-1}\mathbf{S}_{xy}\right|}{\left|\mathbf{S}_{yy}\right|}$$

Finally, letting $\mathbf{S}$ be defined as in (4.2.7) and using Theorem 2.5.6 (6), the Λ criterion for testing $H_1 : \mathbf{B}_1 = \mathbf{0}$ becomes

$$\begin{aligned} \Lambda &= \frac{|\mathbf{E}|}{|\mathbf{E} + \mathbf{H}|} = \prod_{i=1}^{s}(1 + \lambda_i)^{-1} \\ &= \frac{|\mathbf{S}|}{|\mathbf{S}_{xx}|\,\left|\mathbf{S}_{yy}\right|} = \prod_{i=1}^{s}(1 - \theta_i) \end{aligned} \tag{4.2.14}$$

Using (3.5.3), one may relate Λ to an F distribution. Comparing (4.2.14) with the expression for W for testing independence in (3.8.32), we see that testing $H_1 : \mathbf{B}_1 = \mathbf{0}$ is equivalent to testing $\Sigma_{xy} = \mathbf{0}$ or that the set X and Y are independent under joint multivariate normality. We shall see in Chapter 8 that the quantities $r_i^2 = \theta_i = \lambda_i / (1 + \lambda_i)$ are then sample canonical correlations. For the other test criteria, $M = (|p - k| - 1)/2$ and $N = (n - k - p - 2)/2$ in Theorem 3.5.1. To test additional hypotheses regarding the elements of $\mathbf{B}$ other matrices $\mathbf{C}$ and $\mathbf{M}$ are selected. For example, for $\mathbf{C} = \mathbf{I}_q$ and $\mathbf{M} = \mathbf{I}_p$, one may test that all coefficients are zero, $H_o : \mathbf{B} = \mathbf{0}$. To test that any single row of $\mathbf{B}_1$ is zero, a row of $\mathbf{C} = [\mathbf{0}, \mathbf{I}_k]$ would be used with $\mathbf{M} = \mathbf{I}_p$. Failure to reject $H_i : \mathbf{c}_i'\mathbf{B} = \mathbf{0}'$ may suggest removing the variable from the MR model.

A frequently employed test in MR models is to test that some nested subset of the rows of $\mathbf{B}$ are zero, say the last $k - m$ rows. For this situation, the MR model becomes

$$\Omega_o : \mathbf{Y} = [\mathbf{X}_1, \mathbf{X}_2]\begin{bmatrix} \mathbf{B}_1 \\ \mathbf{B}_2 \end{bmatrix} + \mathbf{E} \tag{4.2.15}$$

where the matrix $\mathbf{X}_1$ is associated with $1, x_1, \ldots, x_m$ and $\mathbf{X}_2$ contains the variables x_{m+1}, $\ldots, x_k$ so that $q = k + 1$. With this structure, suppose one is interested in testing H_2 :

$\mathbf{B}_2 = \mathbf{0}$. Then the matrix $\mathbf{C} = [\mathbf{0}_m, \mathbf{I}_{k-m}]$ has the same structure as the test of $\mathbf{B}_1$ with the partition for $\mathbf{X} = [\mathbf{X}_1, \mathbf{X}_2]$ so that $\mathbf{X}_1$ replaces $\mathbf{1}_n$. Now with $\mathbf{Q} = \mathbf{I} - \mathbf{X}_1 \left(\mathbf{X}_1'\mathbf{X}_1\right)^{-1} \mathbf{X}_1'$ and $\widehat{\mathbf{B}}_2 = \left(\mathbf{X}_2'\mathbf{Q}\mathbf{X}_2\right)^{-1} \mathbf{X}_2'\mathbf{Q}\mathbf{Y}$, the hypothesis test matrix becomes

$$\begin{aligned} \mathbf{H} &= \widehat{\mathbf{B}}_2'(\mathbf{X}_2'\mathbf{X}_2 - \mathbf{X}_2'\mathbf{X}_1 \left(\mathbf{X}_1'\mathbf{X}_1\right)^{-1} \mathbf{X}_1'\mathbf{X}_2)\widehat{\mathbf{B}}_2 \\ &= \mathbf{Y}'\mathbf{Q}\mathbf{X}_2 \left(\mathbf{X}_2'\mathbf{Q}\mathbf{X}_2\right)^{-1} \mathbf{X}_2\mathbf{Q}\mathbf{Y} \end{aligned} \tag{4.2.16}$$

Alternatively, one may obtain $\mathbf{H}$ by considering two models: the full model Ω_o in (4.2.15) and the reduced model ω : $\mathbf{Y} = \mathbf{X}_1\mathbf{B}_1 + \mathbf{E}_\omega$ under the hypothesis. Under the reduced model, $\widehat{\mathbf{B}}_1 = \left(\mathbf{X}_1'\mathbf{X}_1\right)^{-1} \mathbf{X}_1'\mathbf{Y}$ and the reduced error matrix $\mathbf{E}_\omega = \mathbf{Y}'\mathbf{Y} - \widehat{\mathbf{B}}_1' \left(\mathbf{X}_1'\mathbf{X}_1\right) \widehat{\mathbf{B}}_1 = \mathbf{Y}'\mathbf{Y} - \widehat{\mathbf{B}}_1'\mathbf{X}_1'\mathbf{Y}$ where $\mathbf{H}_\omega = \widehat{\mathbf{B}}_1' \left(\mathbf{X}_1'\mathbf{X}_1\right) \widehat{\mathbf{B}}_1$ tests $H_\omega : \mathbf{B}_1 = \mathbf{0}$ in the reduced model. Under the full model, $\widehat{\mathbf{B}} = \left(\mathbf{X}'\mathbf{X}\right)^{-1} \mathbf{X}'\mathbf{Y}$ and $\mathbf{E}_{\Omega_o} = \mathbf{Y}'\mathbf{Y} - \widehat{\mathbf{B}}' \left(\mathbf{X}'\mathbf{X}\right) \widehat{\mathbf{B}} = \mathbf{Y}'\mathbf{Y} - \widehat{\mathbf{B}}'\mathbf{X}'\mathbf{Y}$ where $\mathbf{H} = \widehat{\mathbf{B}}' \left(\mathbf{X}'\mathbf{X}\right) \widehat{\mathbf{B}}$ tests $H : \mathbf{B} = \mathbf{0}$ for the full model. Subtracting the two error matrices,

$$\begin{aligned} \mathbf{E}_\omega - \mathbf{E}_{\Omega_o} &= \widehat{\mathbf{B}}' \left(\mathbf{X}'\mathbf{X}\right) \widehat{\mathbf{B}} - \widehat{\mathbf{B}}_1' \left(\mathbf{X}_1'\mathbf{X}_1\right) \widehat{\mathbf{B}}_1 \\ &= \mathbf{Y}'\mathbf{X} \left(\mathbf{X}'\mathbf{X}\right)^{-1} \mathbf{X}'\mathbf{Y} - \widehat{\mathbf{B}}_1' \left(\mathbf{X}_1'\mathbf{X}_1\right) \widehat{\mathbf{B}}_1 \\ &= \mathbf{Y}' \Big[\mathbf{X}_1 \left(\mathbf{X}_1'\mathbf{X}\right)^{-1} \mathbf{X}_1' - \mathbf{X}_1 \left(\mathbf{X}_1'\mathbf{X}_1\right)^{-1} \mathbf{X}_1'\mathbf{X}_2 \left(\mathbf{X}_2'\mathbf{Q}\mathbf{X}_2\right)^{-1} \mathbf{X}_2'\mathbf{Q} \\ &\qquad + \mathbf{X}_2 \left(\mathbf{X}_2'\mathbf{Q}\mathbf{X}_2\right)^{-1} \mathbf{X}_2'\mathbf{Q}\Big] \mathbf{Y} - \widehat{\mathbf{B}}_1' \left(\mathbf{X}_1'\mathbf{X}_1\right) \widehat{\mathbf{B}}_1 \\ &= \mathbf{Y}'\mathbf{X}_2 \left(\mathbf{X}_2'\mathbf{Q}\mathbf{X}_2\right)^{-1} \mathbf{X}_2'\mathbf{Q}\mathbf{Y} - \mathbf{Y}'\mathbf{X}_1 \left(\mathbf{X}_1'\mathbf{X}_1\right)^{-1} \mathbf{X}_1'\mathbf{X}_2 \left(\mathbf{X}_2'\mathbf{Q}\mathbf{X}_2\right)^{-1} \mathbf{X}_2'\mathbf{Q}\mathbf{Y} \\ &= \mathbf{Y}' \Big[\mathbf{I} - \mathbf{X}_1 \left(\mathbf{X}_1'\mathbf{X}_1\right)^{-1} \mathbf{X}_1\Big] \Big[\mathbf{X}_2 \left(\mathbf{X}_2'\mathbf{Q}\mathbf{X}_2\right)^{-1} \mathbf{X}_2'\mathbf{Q}\Big] \mathbf{Y} \\ &= \mathbf{Y}'\mathbf{Q}\mathbf{X}_2 \left(\mathbf{X}_2'\mathbf{Q}\mathbf{X}_2\right)^{-1} \mathbf{X}_2'\mathbf{Q}\mathbf{Y} \\ &= \mathbf{H} \end{aligned}$$

as claimed. Thus, $\mathbf{H}$ is the extra sum of squares and cross products matrix due to $\mathbf{X}_2$ given the variables associated with $\mathbf{X}_1$ are in the model. Finally, to test $H_2 : \mathbf{B}_2 = \mathbf{0}$, Wilks' Λ criterion is

$$\begin{aligned} \Lambda &= \frac{|\mathbf{E}|}{|\mathbf{E} + \mathbf{H}|} \\ &= \frac{\left|\mathbf{E}_{\Omega_o}\right|}{|\mathbf{E}_\omega|} = \prod_{i=1}^{s} (1 + \lambda_i)^{-1} = \prod_{i=1}^{s} (1 - \theta_i) \sim U_{(p, k-m, v_e)} \end{aligned} \tag{4.2.17}$$

where $v_e = n - q = n - k - 1$. For the other criteria, $s = \min(k - m, p)$, $M = (|p - k - m| - 1)/2$ and $N = (n - k - p - 2)/2$.

The test of $H_2 : \mathbf{B}_2 = \mathbf{0}$ is also called the test of additional information since it is being used to evaluate whether the variables $x_{m+1}, \ldots, x_k$ should be in the model given that $x_1, x_2, \ldots, x_m$ are in the model. The tests are being performed in order to uncover and estimate the functional relationship between the set of dependent variables and the set of independent variables. We shall see in Chapter 8 that $\theta_i = \lambda_i / (1 + \lambda_i)$ is the square of a sample partial canonical correlation.

In showing that $\mathbf{H} = \mathbf{E}_\omega - \mathbf{E}_{\Omega_o}$ for the test of H_2, we discuss the test employing the reduction in SSCP terminology. Under ω, recall that $\mathbf{T}_\omega = \mathbf{E}_\omega + \mathbf{H}_\omega$ so that $\mathbf{E}_\omega = \mathbf{T}_\omega - \mathbf{H}_\omega$ is the reduction in the total SSCP matrix due to ω and $\mathbf{E}_{\Omega_o} = \mathbf{T}_{\Omega_o} - \mathbf{H}_{\Omega_o}$ is the reduction in total SSCP matrix due to Ω_o. Thus $\mathbf{E}_\omega - \mathbf{E}_{\Omega_o} = (\mathbf{T}_\omega - \mathbf{H}_\omega) - \left(\mathbf{T}_{\Omega_o} - \mathbf{H}_{\Omega_o}\right) = \mathbf{H}_{\Omega_o} - \mathbf{H}_\omega$ represents the differences in the regression SSCP matrices for fitting $\mathbf{Y} = \mathbf{X}_1\mathbf{B}_1 + \mathbf{X}_2\mathbf{B}_1 + \mathbf{E}$ compared to fitting the model $\mathbf{Y} = \mathbf{X}_1\mathbf{B}_1 + \mathbf{E}$. Letting $R(\mathbf{B}_1, \mathbf{B}_2) = \mathbf{H}_{\Omega_o}$ and $R(\mathbf{B}_1) = \mathbf{H}_\omega$ then $R(\mathbf{B}_1, \mathbf{B}_2) - R(\mathbf{B}_1)$ represents the reduction in the regression SSCP matrix resulting from fitting $\mathbf{B}_2$, having already fit $\mathbf{B}_1$. Hence, the hypothesis SSCP matrix $\mathbf{H}$ is often described at the reduction of fitting $\mathbf{B}_2$, adjusting for $\mathbf{B}_1$. This is written as

$$R(\mathbf{B}_2 \mid \mathbf{B}_1) = R(\mathbf{B}_1, \mathbf{B}_2) - R(\mathbf{B}_1) \tag{4.2.18}$$

The reduction $R(\mathbf{B}_1)$ is also called the reduction of fitting $\mathbf{B}_1$, ignoring $\mathbf{B}_2$. Clearly $R(\mathbf{B}_2 \mid \mathbf{B}_1) \neq R(\mathbf{B}_2)$. However, if $\mathbf{X}_1'\mathbf{X}_2 = \mathbf{0}$, then $R(\mathbf{B}_2) = R(\mathbf{B}_2 \mid \mathbf{B}_1)$ and $\mathbf{B}_1$ is said to be orthogonal to $\mathbf{B}_2$.

One may extend the reduction notation further by letting $\mathbf{B} = (\mathbf{B}_1, \mathbf{B}_2, \mathbf{B}_3)$. Then $R(\mathbf{B}_2 \mid \mathbf{B}_1) = R(\mathbf{B}_1, \mathbf{B}_2) - R(\mathbf{B}_1)$ is not equal to $R(\mathbf{B}_2 \mid \mathbf{B}_1, \mathbf{B}_3) = R(\mathbf{B}_1, \mathbf{B}_2, \mathbf{B}_3) - R(\mathbf{B}_1\mathbf{B}_3)$ unless the design matrix is orthogonal. Hence, the order chosen for fitting variables affects hypothesis SSCP matrices for nonorthogonal designs.

Tests of H_o, H_1, H_2 or H_i are used by the researcher to evaluate whether a set of independent variables should remain in the MR model. If a subset of $\mathbf{B}$ is zero, the independent variables are excluded from the model. Tests of H_o, H_1, H_2 or H_i are performed in SAS using PROC REG and the MTEST statement. For example to test $H_1 : \mathbf{B}_1 = \mathbf{0}$ for k independent variables, the MTEST statement is

```
mtest x1,x2,x3,...,xk / print;
```

where $x1, x2, \ldots, xk$ are names of independent variables separated by commas. The option / PRINT directs SAS to print the hypothesis test matrix. The hypotheses $H_i : [\beta_{i1}, \beta_{i2}, \ldots, \beta_{ip}] = \mathbf{0}'$ are tested using k statements of the form

```
mtest xi /print;
```

for $i = 1, 2, \ldots, k$. For the subtest $H_2 : \mathbf{B}_2 = \mathbf{0}$, the MTEST command is

```
mtest xm,....,xk / print;
```

for a subset of the variable names $xm, \ldots, xk$, again the names are separated by commas. To test that two independent variable coefficients are both equal and equal to zero, the statement

```
mtest x1, x2 / print;
```

is used. To form tests that include the intercept in any of these tests, on must include the variable name intercept in the MTEST statement. The commands will be illustrated with an example.

c. *Multivariate Influence Measures*

Tests of hypotheses are only one aspect of the model refinement process. An important aspect of the process is the systematic analysis of residuals to determine influential observations. The matrix of multivariate residuals is defined as

$$\widehat{\mathbf{E}} = \mathbf{Y} - \mathbf{X}\widehat{\mathbf{B}} = \mathbf{Y} - \widehat{\mathbf{Y}} \tag{4.2.19}$$

where $\widehat{\mathbf{Y}} = \mathbf{X}\widehat{\mathbf{B}}$ is the matrix of fitted values. Letting $\mathbf{P} = \mathbf{X}\left(\mathbf{X}'\mathbf{X}\right)^{-1}\mathbf{X}'$, (4.2.19) is written as $\widehat{\mathbf{E}} = (\mathbf{I} - \mathbf{P})\,\mathbf{Y}$ where $\mathbf{P}$ is the projection matrix. $\mathbf{P}$ is a symmetric idempotent matrix, also called the "hat matrix" since $\mathbf{PY}$ projects $\mathbf{Y}$ into $\widehat{\mathbf{Y}}$. The ML estimate of Σ may be represented as

$$\widehat{\Sigma} = \widehat{\mathbf{E}}'\widehat{\mathbf{E}}/n = \mathbf{Y}'(\mathbf{I} - \mathbf{P})\,\mathbf{Y}/n = \mathbf{E}/n \tag{4.2.20}$$

where $\mathbf{E}$ is the error sum of squares and cross products matrix. Multiplying $\widehat{\Sigma}$ by $n/(n-q)$ where $q = r(\mathbf{X})$, an unbiased estimate of Σ is

$$\mathbf{S} = n\widehat{\Sigma}/(n-q) = \mathbf{E}/(n-q) \tag{4.2.21}$$

The matrix of fitted values may be represented as follows

$$\widehat{\mathbf{Y}} = \begin{bmatrix} \widehat{\mathbf{y}}_1' \\ \widehat{\mathbf{y}}_2' \\ \vdots \\ \widehat{\mathbf{y}}_n' \end{bmatrix} = \mathbf{PY} = \mathbf{P}\begin{bmatrix} \mathbf{y}_1' \\ \mathbf{y}_2' \\ \vdots \\ \mathbf{y}_n' \end{bmatrix}$$

so that

$$\begin{aligned} \widehat{\mathbf{y}}_i' &= \sum_{j=1}^{n} p_{ij}\mathbf{y}_j' \\ &= p_{ii}\mathbf{y}_i' + \sum_{j\neq i}^{n} p_{ij}\mathbf{y}_j' \end{aligned}$$

where $p_{i1}, p_{i2}, \ldots, p_{in}$ are the elements in the ith row of the hat matrix $\mathbf{P}$. The coefficients p_{ii}, the diagonal elements of the hat matrix $\mathbf{P}$, represent the leverage or potential influence an observation $\mathbf{y}_i'$ has in determining the fitted value $\widehat{\mathbf{y}}_i'$. For this reason the matrix $\mathbf{P}$ is also called the leverage matrix. An observation $\mathbf{y}_i'$ with a large leverage value p_{ii} is called a high leverage observation because it has a large influence on the fitted values and regression coefficients in $\mathbf{B}$.

Following standard univariate notation, the subscript '(i)' on the matrix $\mathbf{X}_{(i)}$ is used to indicate that the ith row is deleted from $\mathbf{X}$. Defining $\mathbf{Y}_{(i)}$ similarly, the matrix of residuals with the ith observation deleted is defined as

$$\begin{aligned} \widehat{\mathbf{E}}_{(i)} &= \mathbf{Y}_{(i)} - \mathbf{X}_{(i)}\widehat{\mathbf{B}}_{(i)} \\ &= \mathbf{Y}_{(i)} - \widehat{\mathbf{Y}}_{(i)} \end{aligned} \tag{4.2.22}$$

where $\widehat{\mathbf{B}}_{(i)} = (\mathbf{X}'_{(i)}\mathbf{X}_{(i)})^{-1}\mathbf{X}'_{(i)}\mathbf{Y}_{(i)}$ for $i = 1, 2, \ldots, n$. Furthermore, $\mathbf{S}_{(i)} = \widehat{\mathbf{E}}'_{(i)}\widehat{\mathbf{E}}_{(i)}/(n - q - 1)$. The matrices $\widehat{\mathbf{B}}_i$ and $\mathbf{S}_{(i)}$ are the unbiased estimators of $\mathbf{B}$ and Σ when the ith observation vector $(\mathbf{y}'_i, \mathbf{x}'_i)$ is deleted from both $\mathbf{Y}$ and $\mathbf{X}$.

In multiple linear regression, the residual vector is not distributed $N\left(\mathbf{0}, \sigma^2 (\mathbf{I} - \mathbf{P})\right)$; however, for diagnostic purposes, residuals are "Studentized". The internally Studentized residual is defined as $r_i = \widehat{e}_i / \left[s\,(1 - p_{ii})^{1/2}\right]$ while the externally Studentized residual is defined as $t_i = \widehat{e}_i / \left[s_{(i)}\,(1 - p_{ii})^{1/2}\right]$ where $\widehat{e}_i = y_i - \mathbf{x}'_i\widehat{\boldsymbol{\beta}}_i$. If the $r(\mathbf{X}) = r\left(\mathbf{X}_{(i)}\right) = q$ and $\mathbf{e} \sim N_n\left(\mathbf{0}, \sigma^2\mathbf{I}\right)$, then the r_i are identically distributed as a Beta $(1/2, (n - q - 1)/2)$ distribution and the t_i are identically distributed as a student t distribution; in neither case are the quantities independent, Chatterjee and Hadi (1988, pp. 76–78). The externally Studentized residual is also called the Studentized deleted residual.

Hossain and Naik (1989) and Srivastava and von Rosen (1998) generalize Studentized residuals to the multivariate case by forming statistics that are the squares of r_i and t_i. The internally and externally "Studentized" residuals are defined as

$$r_i^2 = \widehat{\mathbf{e}}'_i\mathbf{S}^{-1}\widehat{\mathbf{e}}_i / (1 - p_{ii}) \quad \text{and} \quad T_i^2 = \widehat{\mathbf{e}}'_i\mathbf{S}_{(i)}^{-1}\widehat{\mathbf{e}}_i / (1 - p_{ii}) \tag{4.2.23}$$

for $i = 1, 2, \ldots, n$ where $\widehat{\mathbf{e}}_i$ is the ith row of $\widehat{\mathbf{E}} = \mathbf{Y} - \mathbf{X}\widehat{\mathbf{B}}$. Because T_i^2 has Hotelling's T^2 distribution and $r_i^2/(n - q) \sim \text{Beta}\,[p/2, (n - q - p)/2]$, assuming no other outliers, an observation $\mathbf{y}'_i$ may be considered an outlier if

$$\frac{(n - q - p)}{p}\frac{T_i^2}{(n - q - 1)} > F^{1-\alpha^*}(p, n - q - 1) \tag{4.2.24}$$

where α^* is selected to control the familywise error rate for n tests at the nominal level α. This is a natural extension of the univariate test procedure for outliers.

In multiple linear regression, Cook's distance measure is defined as

$$\begin{aligned} C_i &= \frac{\left(\widehat{\boldsymbol{\beta}} - \widehat{\boldsymbol{\beta}}_{(i)}\right)' (\mathbf{X}'\mathbf{X}) \left(\widehat{\boldsymbol{\beta}} - \widehat{\boldsymbol{\beta}}_{(i)}\right)}{qs^2} = \frac{\left(\widehat{\mathbf{y}} - \widehat{\mathbf{y}}_{(i)}\right)' \left(\widehat{\mathbf{y}} - \widehat{\mathbf{y}}_{(i)}\right)}{qs^2} \\ &= \frac{1}{q}\frac{p_{ii}}{1 - p_{ii}}r_i^2 \\ &= \frac{1}{q}\frac{p_{ii}}{(1 - p_{ii})^2}\frac{\widehat{e}_i^2}{s} \end{aligned} \tag{4.2.25}$$

where r_i is the internally Studentized residual and is used to evaluate the overall influence of an observation (y_i, x_i) on all n fitted values or all q regression coefficients for $i = 1, 2, \ldots, n$, Cook and Weisberg (1980). That is, it is used to evaluate the overall effect of deleting an observation from the data set. An observation is influential if C_i is larger than the 50th percentile of an F distribution with q and $n - q$ degrees of freedom. Alternatively, to evaluate the effect of the ith observation (y_i, x_i) has on the ith fitted value $\widehat{y}_i$, one may compare the closeness of $\widehat{y}_i$ to $\widehat{y}_{i(i)} = \mathbf{x}'_i\widehat{\boldsymbol{\beta}}_{(i)}$ using the Welsch-Kuh test statistic, Belsley,

Welsch and Kuh (1980), defined as

$$
\begin{aligned}
WK_i &= \frac{\left|\widehat{y}_i - \widehat{y}_{i(i)}\right|}{s_{(i)}\sqrt{p_{ii}}} = \frac{\mathbf{x}_i'\left(\widehat{\boldsymbol{\beta}} - \widehat{\boldsymbol{\beta}}_{(i)}\right)}{s_{(i)}\sqrt{p_{ii}}} \\
&= |t_i|\sqrt{\frac{p_{ii}}{1-p_{ii}}}
\end{aligned}
\tag{4.2.26}
$$

where t_i is an externally Studentized residual. The statistic $WK_i \sim t\sqrt{q/(n-q)}$ for $i = 1, 2, \ldots, n$. The statistic WK_i is also called $(\text{DFFITS})_i$. An observation y_i is considered influential if $WK_i > 2\sqrt{q/(n-q)}$.

To evaluate the influence of the ith observation on the jth coefficient in $\boldsymbol{\beta}$ in multiple (linear) regression, the DFBETA statistics developed by Cook and Weisberg (1980) are calculated as

$$
C_{ij} = \frac{r_i}{(1-p_{ii})^{1/2}} \frac{\mathrm{w}_{ij}}{\left(\mathbf{w}_j'\mathbf{w}_j\right)^{1/2}} \quad i = 1, 2, \ldots, n;\ j = 1, 2, \ldots, q \tag{4.2.27}
$$

where w_{ij} is the ith element of $\mathbf{w}_j = \left(\mathbf{I} - \mathbf{P}_{[j]}\right)\mathbf{x}_j$ and $\mathbf{P}_{[j]}$ is calculated without the jth column of $\mathbf{X}$. Belsley et al. (1980) rescaled C_{ij} to the statistic

$$
D_{ij} = \frac{\widehat{\beta}_j - \widehat{\beta}_{j(i)}}{\sqrt{\operatorname{var}\left(\widehat{\beta}_j\right)}} = \frac{e_i}{\sigma(1-p_{ii})^{1/2}} \frac{\mathrm{w}_{ij}}{\left(\mathbf{w}_j'\mathbf{w}_j\right)^{1/2}} \frac{1}{(1-p_{ii})^{1/2}} \tag{4.2.28}
$$

If σ in (4.2.28) is estimated by $s_{(i)}$, then D_{ij} is called the $(\text{DFBETA})_{ij}$ statistic and

$$
D_{ij} = \frac{t_i}{(1-p_{ii})^{1/2}} \frac{\mathrm{w}_{ij}}{\left(\mathbf{w}_j'\mathbf{w}_j\right)^{1/2}} \tag{4.2.29}
$$

If σ in (4.2.28) is estimated by s, then $D_{ij} = C_{ij}$. An observation y_i is considered inferential on the regression coefficient β_j if the $\left|D_{ij}\right| > 2/\sqrt{n}$.

Generalizing Cook's distance to the multivariate regression model, Cook's distance becomes

$$
\begin{aligned}
C_i &= \left[\operatorname{vec}\left(\mathbf{B} - \mathbf{B}_{(i)}\right)\right]' \left(\mathbf{S} \otimes \mathbf{X}'\mathbf{X}\right)^{-1} \operatorname{vec}\left(\mathbf{B} - \mathbf{B}_{(i)}\right)/q \\
&= \operatorname{tr}\left[\left(\mathbf{B} - \mathbf{B}_{(i)}\right)' \left(\mathbf{X}'\mathbf{X}\right)^{-1} \left(\mathbf{B} - \mathbf{B}_{(i)}\right)\mathbf{S}^{-1}\right]/q \\
&= \operatorname{tr}\left[\left(\widehat{\mathbf{Y}} - \widehat{\mathbf{Y}}_{(i)}\right)' \left(\widehat{\mathbf{Y}} - \widehat{\mathbf{Y}}_{(i)}\right)\right]/q \\
&= \frac{p_{ii}}{1-p_{ii}} r_i^2/q \\
&= \frac{p_{ii}}{(1-p_{ii})^2} \widehat{\mathbf{e}}_i'\mathbf{S}^{-1}\widehat{\mathbf{e}}_i/q
\end{aligned}
\tag{4.2.30}
$$

for $i = 1, 2, \ldots, n$. An observation is influential if C_i is larger than the 50th percentile of a chi square distribution with $v = p(n-q)$ degrees of freedom, Barrett and Ling (1992).

Alternatively, since r_i^2 has a Beta distribution, an observation is influential if $C_i > C_o^* = C_i \times (n-q) \times \text{Beta}^{1-\alpha}(v_1, v_2)$, $v_1 = p/2$ and $v_2 = (n-q-p)/2$. $\text{Beta}^{1-\alpha}(v_1, v_2)$ is the upper critical value of the Beta distribution.

To evaluate the influence of $(\mathbf{y}_i', \mathbf{x}_i')$ on the ith predicted value $\widehat{\mathbf{y}}_i'$ where $\widehat{\mathbf{y}}_i'$ is the ith row of $\widehat{\mathbf{Y}}$, the Welsch-Kuh, DFFITS, type statistic is defined as

$$WK_i = \tfrac{p_{ii}}{1-p_{ii}} T_i^2 \qquad i = 1, 2, \ldots, n \tag{4.2.31}$$

Assuming the rows of $\mathbf{Y}$ follow a MVN distribution and the $r(\mathbf{X}) = r\left(\mathbf{X}_{(i)}\right) = q$, an observation is said to be influential on the ith predicted value $\widehat{\mathbf{y}}_i$ if

$$WK_i > \frac{q}{n-q} \frac{(n-q-1)}{n-q-p} F^{1-\alpha^*}(p, n-q-p) \tag{4.2.32}$$

where α^* is selected to control the familywise error rate for the n tests at some nominal level α. To evaluate the influence of the ith observation $\mathbf{y}_i'$ on the jth row of $\widehat{\mathbf{B}}$, the DFBETA statistics are calculated as

$$D_{ij} = \frac{T_i^2}{1-p_{ii}} \frac{\mathbf{w}_{ij}^2}{\mathbf{w}_j' \mathbf{w}_j} \tag{4.2.33}$$

for $i = 1, 2, \ldots, n$ and $j = 1, 2, \ldots, q$. An observation is considered influential on the coefficient $\widehat{\beta}_{ij}$ of $\widehat{\mathbf{B}}$ if $D_{ij} > 2$ and $n > 30$.

Belsley et al. (1980) use a covariance ratio to evaluate the influence of the ith observation on the $\text{cov}(\widehat{\boldsymbol{\beta}})$ in multiple (linear) regression. The covariance ratio (CVR) for the ith observation is

$$CVR_i = \left[s_{(i)}^2/s^2\right]/(1-p_{ii}) \quad i = 1, 2, \ldots, n \tag{4.2.34}$$

An observation is considered influential if $|\text{CVR}_i - 1| > 3q/n$. For the MR model, Hossain and Naik (1989) use the ratio of determinants of the covariance matrix of $\widehat{\mathbf{B}}$ to evaluate the influence of $\mathbf{y}_i'$ on the covariance matrix of $\widehat{\mathbf{B}}$. For $i = 1, 2, \ldots, n$ the

$$CVR_i = \frac{\left|\text{cov}\left(\text{vec}\,\widehat{\mathbf{B}}_{(i)}\right)\right|}{\left|\text{cov}\left(\text{vec}\,\widehat{\mathbf{B}}\right)\right|} = \left(\frac{1}{1-p_{ii}}\right)^p \left(\frac{|\mathbf{S}_{(i)}|}{\mathbf{S}}\right)^q \tag{4.2.35}$$

If the $\left|\mathbf{S}_{(i)}\right| \approx 0$, then $CVR_i \approx 0$ and if the $|\mathbf{S}| \approx 0$ then $CVR_i \longrightarrow \infty$. Thus, if CVR_i is low or very high, the observation $\mathbf{y}_i'$ is considered influential. To evaluate the influence of $\mathbf{y}_i'$ on the $\text{cov}(\widehat{\mathbf{B}})$, the $|\ \mathbf{S}_i\ |/\mathbf{S} \approx \left[1 + T_i^2/(n-q-1)\right]^{-1} \sim \text{Beta}\left[p/2, (n-q-p)/2\right]$. A CVR_i may be influential if CVR_i is larger than

$$[1/(1-p_{ii})]^p \left[\text{Beta}^{1-\alpha/2}(v_1, v_2)\right]^q$$

or less that the lower value of

$$[1/(1-p_{ii})]^p \left[\text{Beta}^{\alpha/2}(v_1, v_2)\right]^q$$

where $v_1 = p/2$ and $v_2 = [(n - q - p)/2]$ and $\text{Beta}^{1-\alpha/2}$and $\text{Beta}^{\alpha/2}$ are the upper and lower critical values for the Beta distribution. In SAS, one may use the function Betainv $(1 - \alpha, df1, df2)$ to obtain critical values for a Beta distribution.

Finally, we may use the matrix of residuals $\widehat{\mathbf{E}}$ to create chi-square and Beta Q-Q plots, to construct plots of residuals versus predicted values or variables not in the model. These plots are constructed to check MR model assumptions.

d. *Measures of Association, Variable Selection and Lack-of-Fit Tests*

To estimate the coefficient of determination or population squared multiple correlation coefficient ρ^2 in multiple linear regression, the estimator

$$R^2 = \frac{\widehat{\boldsymbol{\beta}}'\mathbf{X}'\mathbf{y} - n\overline{y}^2}{\mathbf{y}'\mathbf{y} - n\overline{y}^2} = \frac{SSR}{SST} = 1 - \frac{SSE}{SST} \tag{4.2.36}$$

is used. It measures the proportional reduction of the total variation in the dependent variable y by using a set of fixed independent variables $x_1, x_2, \ldots, x_k$. While the coefficient of determination in the population is a measure of the strength of a linear relation in the population, the estimator R^2 is only a measure of goodness-of-fit in the sample. Given that the coefficients associated with the independent variables are all zero in the population, $E\left(R^2\right) = k/(n-1)$ so that if $n = k + 1 = q$, $E\left(R^2\right) = 1$. Thus, in small samples the sheer size of R^2 is not the best indication of model fit. In fact Goldberger (1991, p. 177) states: "Nothing in the CR (Classical Regression) model requires R^2 to be high. Hence a high R^2 is not evidence in favor of the model, and a low R^2 is not evidence against it". To reduce the bias for the number of variables in small samples, which discounts the fit when k is large relative to n, R.A. Fisher suggested that the population variances $\sigma^2_{y|x}$ be replaced by its minimum variance unbiased estimate $s^2_{y|x}$ and that the population variance for σ^2_y be replaced by its sample estimator s^2_y, to form an adjusted estimate for the coefficient of determination or population squared multiple correlation coefficient. The adjusted estimate is

$$\begin{aligned} R^2_a &= 1 - \left(\frac{n-1}{n-q}\right)\left(\frac{SSE}{SST}\right) \\ &= 1 - \left(\frac{n-1}{n-q}\right)\left(1 - R^2\right) \\ &= \left(1 - s^2_{y|x} / s^2_y\right) \end{aligned} \tag{4.2.37}$$

and $E\{R^2 - [k(1-R^2)/(n-k-1]\} = E(R^2_a) = 0$, given no linear association between Y and the set of X's. This is the case, since

$$R^2_a = 0 \Longleftrightarrow \sum_i (\widehat{y}_i - \overline{y})^2 = 0 \Longleftrightarrow \widehat{y}_i = \overline{y}_i \text{ for all } i$$

in the sample. The best-fitted model is a horizontal line, and none of the variation in the independent variables is accounted for by the variation in the independent variables. For an overview of procedures for estimating the coefficient of determination for fixed and random independent variables and also the squared cross validity correlation coefficient (ρ_c^2), the population squared correlation between the predicted dependent variable and the dependent variable, the reader may consult Raju et al. (1997).

A natural extension of R^2 in the MR model is to use an extension of Fisher's correlation ratio η^2 suggested by Wilks (1932). In multivariate regression eta squared is called the square of the vector correlation coefficient

$$\eta^2 = 1 - \Lambda = 1 - |\mathbf{E}| / |\mathbf{E} + \mathbf{H}| \tag{4.2.38}$$

when testing $H_1 : \mathbf{B}_1 = \mathbf{0}$, Rozeboom (1965). This measure is biased, thus Jobson (1992, p. 218) suggests the less biased index

$$\eta_a^2 = 1 - n\Lambda / (n - q + \Lambda) \tag{4.2.39}$$

where the $r(\mathbf{X}) = q$. Another measure of association is based on Roy's criterion. It is $\eta_\theta^2 = \lambda_1 / (1 + \lambda_1) = \theta_1 \leq \eta^2$, the square of the largest canonical correlation (discussed in Chapter 8). While other measures of association have been proposed using the other multivariate criteria, there does not appear to be a "best" index since $\mathbf{X}$ is fixed and only $\mathbf{Y}$ varies. More will be said about measures of association when we discuss canonical analysis in Chapter 8.

Given a large number of independent variables in multiple linear regression, to select a subset of independent variables one may investigate all possible regressions and incremental changes in the coefficient of determination $\left(R^2\right)$, the reduction in mean square error (MS_e), models with values of total mean square error $\left(C_p\right)$ near the total number of variables in the model, models with small values of predicted sum of squares (PRESS), and models using the information criteria (AIC, HQIC, BIC and CAIC), McQuarrie and Tsai (1998). To facilitate searching, "best" subset algorithms have been developed to construct models. Search procedures such as forward selection, backward elimination, and stepwise selection methods have also been developed to select subsets of variables. We discuss some extensions of these univariate methods to the MR model.

Before extending R^2 in (4.2.36) to the MR model, we introduce some new notation. When fitting all possible regression models to the $(n \times p)$ matrix $\mathbf{Y}$, we shall denote the pool of possible X variables to be $K = Q - 1$ so that the number of parameters $1 \leq q \leq Q$ and at each step the numbers of X variables is $q - 1 = k$. Then for q parameters or $q - 1 = k$ independent variables in the candidate MR model, the $p \times p$ matrix

$$\mathbf{R}_q^2 = (\widehat{\mathbf{B}}_q' \mathbf{X}_q' \mathbf{Y} - n\overline{\mathbf{y}}\,\overline{\mathbf{y}}') \left(\mathbf{Y}'\mathbf{Y} - n\overline{\mathbf{y}}\,\overline{\mathbf{y}}'\right)^{-1} \tag{4.2.40}$$

is a direct extension of R^2. To convert $\mathbf{R}_q^2$ to a scalar, the determinant or trace functions are used. To ensure that the function of $\mathbf{R}_q^2$ is bounded by 1 and 0, the $\text{tr}(\mathbf{R}_q^2)$ is divided by p. Then $0 < \text{tr}(\mathbf{R}_q^2)/p \leq 1$ attains its maximum when $q = Q$. The goal is to select $q < Q$ or the number of variables $q - 1 = k < K$ and to have the $\text{tr}(\mathbf{R}_q^2)/p$ near one. If the $|\mathbf{R}_q^2|$ is used as a subset selection criterion, one uses the ratio: $|\mathbf{R}_q^2| / |\mathbf{R}_Q^2| \leq 1$ for $q = 1, 2, \ldots, Q$. If the largest eigenvalue is used, it is convenient to normalize $\mathbf{R}_q$ to create a correlation matrix $\mathbf{P}_q$ and to use the measure $\gamma = (\lambda_{\max} - 1) / (q - 1)$ where $\lambda_{\max}$ is the largest root of $\mathbf{P}_q$ for $q = 1, 2, \ldots, Q$.

Another criterion used to evaluate the fit of a subset of variables is the error covariance matrix

$$\begin{aligned}
\mathbf{E}_q &= (n-q)\,\mathbf{S}_q = \mathbf{Y}'\mathbf{Y} - \widehat{\mathbf{B}}_q'\mathbf{X}_q'\mathbf{Y} \\
&= (n-q)\,\mathbf{Y}'\left[\mathbf{I}_n - \mathbf{X}_q'\left(\mathbf{X}_q'\mathbf{X}_q\right)^{-1}\mathbf{X}_q\right]\mathbf{Y} \\
&= (n-q)\,\mathbf{Y}'\left[\mathbf{I}_n - \mathbf{P}_q\right]\mathbf{Y} \\
&= (n-q)\,\widehat{\mathbf{E}}_q'\widehat{\mathbf{E}}_q
\end{aligned} \tag{4.2.41}$$

for $\widehat{\mathbf{E}}_q = \mathbf{Y} - \mathbf{X}_q\widehat{\mathbf{B}}_q = \mathbf{Y} - \widehat{\mathbf{Y}}_q$ for $q = 1, 2, \ldots, Q$. Hence $\mathbf{E}_q$ is a measure of predictive closeness of $\widehat{\mathbf{Y}}$ to $\mathbf{Y}$ for values of q. To reduce $\mathbf{E}_q$ to a scalar, we may use the largest eigenvalue of $\mathbf{E}_q$, the $\operatorname{tr}\left(\mathbf{E}_q\right)$ or the $\left|\mathbf{E}_q\right|$, Sparks, Zucchini and Coutsourides (1985). A value $q < Q$ is selected for the $\operatorname{tr}\left(\mathbf{E}_q\right)$ near the $\operatorname{tr}\left(\mathbf{E}_Q\right)$, for example.

In (4.2.41), we evaluated the overall closeness of $\mathbf{Y}$ to $\widehat{\mathbf{Y}}$ for various values of q. Alternatively, we could estimate each row $\mathbf{y}_i'$ of $\mathbf{Y}$ using $\widehat{\mathbf{y}}_{i(i)}' = \mathbf{x}_i'\widehat{\mathbf{B}}_{q(i)}$ where $\widehat{\mathbf{B}}_{q(i)}$ is estimated by deleting the ith row of $\mathbf{y}$ and $\mathbf{X}$ for various values of q. The quantity $\mathbf{y}_i - \widehat{\mathbf{y}}_{i(i)}$ is called the deleted residual and summing the inner products of these over all observations $i = 1, 2, \ldots, n$ we obtain the multivariate predicted sum of squares (MPRESS) criterion

$$\begin{aligned}
\text{MPRESS}_q &= \sum_{i=1}^{n}\left(\mathbf{y}_i - \widehat{\mathbf{y}}_{i(i)}\right)'\left(\mathbf{y}_i - \widehat{\mathbf{y}}_{i(i)}\right) \\
&= \sum_{i=1}^{n}\widehat{\mathbf{e}}_i'\widehat{\mathbf{e}}_i \,/\, (1 - p_{ii})^2
\end{aligned} \tag{4.2.42}$$

where $\widehat{\mathbf{e}}_i = \mathbf{y}_i - \widehat{\mathbf{y}}_i$ without deleting the ith row of $\mathbf{Y}$ and $\mathbf{X}$, and $p_{ii} = \mathbf{x}_i'\left(\mathbf{X}_q'\mathbf{X}_q\right)^{-1}\mathbf{x}_i$ for the deleted row Chatterjee and Hadi (1988, p. 115). MR models with small MPRESS_q values are considered for selection. Plots of MPRESS_q versus q may facilitate variable selection.

Another criterion used in subset selection is Mallows' (1973) C_q criterion which, instead of using the univariate mean square error,

$$E\left(\widehat{y}_i - \mu_i\right)^2 = \operatorname{var}\left(\widehat{y}_i\right) + \left(E\left(\widehat{y}_i\right) - \mu_i\right)^2,$$

uses the expected mean squares and cross products matrix

$$E\left(\widehat{\mathbf{y}}_i - \boldsymbol{\mu}_i\right)\left(\widehat{\mathbf{y}}_i - \boldsymbol{\mu}_i\right)' = \operatorname{cov}\left(\widehat{\mathbf{y}}_i\right) + \left[E\left(\widehat{\mathbf{y}}_i\right) - \boldsymbol{\mu}_i\right]\left[E\left(\widehat{\mathbf{y}}_i\right) - \boldsymbol{\mu}_i\right]' \tag{4.2.43}$$

where $E\left(\widehat{\mathbf{y}}_i\right) - \boldsymbol{\mu}_i$ is the bias in $\widehat{\mathbf{y}}_i$. However, the $\operatorname{cov}[\operatorname{vec}(\widehat{\mathbf{B}}_q)] = \Sigma \otimes (\mathbf{X}_q'\mathbf{X}_q)^{-1}$ so that the $\operatorname{cov}(\widehat{\mathbf{y}}_i') = \operatorname{cov}(\mathbf{x}_{qi}'\widehat{\mathbf{B}}_q) = (\mathbf{x}_{qi}'(\mathbf{X}_q'\mathbf{X}_q)^{-1}\mathbf{x}_{qi})\Sigma$. Summing over the n observations,

$$\begin{aligned}
\sum_{i=1}^{n}\operatorname{cov}(\widehat{\mathbf{y}}_i') &= [\sum_{i=1}^{n}\mathbf{x}_{qi}'\left(\mathbf{X}_q'\mathbf{X}_q\right)^{-1}\mathbf{x}_{qi}\Big]\Sigma \\
&= \operatorname{tr}[\mathbf{X}_q'(X_q'X)^{-1}\mathbf{X}_q]\Sigma \\
&= \left[\operatorname{tr}\left(\mathbf{X}_q'\mathbf{X}_q\right)^{-1}\left(\mathbf{X}_q'\mathbf{X}_q\right)\right]\Sigma \\
&= q\Sigma
\end{aligned}$$

Furthermore, summing over the bias terms:

$$\sum_{i=1}^{n}\left[\left(E\left(\widehat{\mathbf{y}}_i\right)-\boldsymbol{\mu}_i\right)\right]\left[E\left(\widehat{\mathbf{y}}_i-\boldsymbol{\mu}_i\right)\right]'=(n-q)\,E\left(\mathbf{S}_q-\Sigma\right)$$

where $\mathbf{S}_q = \mathbf{E}_q/(n-q)$. Multiplying both sides of (4.2.43) by Σ^{-1} and summing, the expected mean square error criterion is the matrix

$$\Gamma_q = q\mathbf{I}_p + (n-q)\,\Sigma^{-1}E\left(\mathbf{S}_q-\Sigma\right) \tag{4.2.44}$$

as suggested by Mallows' in univariate regression. To estimate Γ_q, the covariance matrix Σ with Q parameters in the model or $Q-1=K$ variables is $\mathbf{S}_Q = \mathbf{E}_Q/(n-Q)$, so that the sample criterion is

$$\begin{aligned}\mathbf{C}_q &= q\mathbf{I}_p + (n-q)\,\mathbf{S}_Q^{-1}\left(\mathbf{S}_q-\mathbf{S}_Q\right)\\ &= \mathbf{S}_Q^{-1}\mathbf{E}_q + (2q-n)\,\mathbf{I}_p\end{aligned} \tag{4.2.45}$$

When there is no bias in the MR model, $\mathbf{C}_q \approx q\mathbf{I}_p$. Thus, models with values of $\mathbf{C}_q$ near $q\mathbf{I}_p$ are desirable. Using the trace criterion, we desire models in which $\operatorname{tr}\left(\mathbf{C}_q\right)$ is near qp. If the $\mid \mathbf{C}_q \mid$ is used as a criterion, the $\mid \mathbf{C}_q \mid < 0$ if $2q-n<0$. Hence, Sparks, Coutsourides and Troskie (1983) recommend a criterion involving the determinant that is always positive

$$\mid \mathbf{E}_Q^{-1}\mathbf{E}_q \mid \leq \left(\frac{n-q}{n-Q}\right)^p \tag{4.2.46}$$

Using their criterion, we select only subsets among all possible models that meet the criterion as the number of parameters vary in size from $q=1,2,...,Q=K+1$ or as $k=q-1$ variables are included in the model. Among the candidate models, the model with the smallest generalized variance may be the best model. One may also employ the largest root of $\mathbf{C}_q$ as a subset selection criterion. Because the criterion depends on only a single value it has limited value.

Model selection using ad hoc measures of association and distance measures that evaluate the difference between a candidate MR model and the expectation of the true MR model result in matrix measures which must be reduced to a scalar using the determinant, trace or eigenvalue of the matrix measure to assess the "best" subset. The evaluation of the eigenvalues of $\mathbf{R}_q^2$, $\mathbf{E}_q$, MPRESS_q and $\mathbf{C}_q$ involve numerous calculations to obtain the "best" subset using all possible regressions. To reduce the number of calculations involved, algorithms that capitalize on prior calculations have been developed. Barrett and Gray (1994) illustrate the use of the SWEEP operator.

Multivariate extensions of the Akaike Information Criterion (AIC) developed by Akaike (1974) or the corrected AIC (CAIC) measure proposed by Sugiura (1978); Schwartz's (1978) Bayesian Information Criterion (BIC), and the Hannan and Quinn (1979) Information Criterion (HQIC) are information measures that may be extended to the MR model. Recalling that the general AIC measure has the structure

$$-2\,(\text{log - likelihood}) + 2d$$

where d is the number of model parameters estimated; the multivariate AIC criterion is

$$AIC_q = n \log | \widehat{\Sigma}_q | + 2qp + p(p+1) \tag{4.2.47}$$

if maximum likelihood estimates are substituted for $\mathbf{B}$ and Σ in the likelihood assuming multivariate normality, since the constant $np \log (2\pi) + np$ in the log-likelihood does not effect the criterion. The number of parameters in the matrix $\mathbf{B}$ and Σ are qp and $p(p+1)/2$, respectively. The model with the smallest AIC value is said to fit better.

Bedrick and Tsai (1994) proposed a small sample correction to AIC by estimating the Kullback-Leibler discrepancy for the MR model, the log-likelihood difference between the true MR model and a candidate MR motel. Their measure is defined as

$$AIC_{c_q} = (n - q - p - 1) \log | \widehat{\Sigma}_q | + (n+q)\, p$$

Replacing the penalty factor $2d$ in the AIC with $d \log n$ and $2d \log \log n$ where d is the rank of $\mathbf{X}$, the $\mathrm{BIC_q}$ and $\mathrm{HQIC_q}$ criteria are

$$\begin{aligned} BIC_q &= n \log | \widehat{\Sigma}_q | + qp \log n \\ HQIC_q &= n \log | \widehat{\Sigma}_q | + 2qp \log \log n \end{aligned} \tag{4.2.48}$$

One may also calculate the criteria by replacing the penalty factor d with $qp + p(p+1)/2$. Here, small values yield better models. If AIC is defined as the log-likelihood minus d, then models with larger values of AIC are better. When using information criteria in various SAS procedures, one must check the documentation to see how the information criteria are define. Sometimes smallest is best and other times largest is best.

One may also estimate AIC and HQIC using an unbiased estimate for Σ and $\mathbf{B}$ and the small sample correction proposed by Bedrick and Tsai (1994). The estimates of the information criteria are

$$AIC_{u_q} = (n - q - p - 1) \log | \mathbf{S}_q^2 | + (n+q)\, p \tag{4.2.49}$$

$$HQIC_{u_q} = (n - q - p - 1) \log | \mathbf{S}_q^2 | + 2qp \log \log (n) \tag{4.2.50}$$

McQuarrie and Tsai (1998) found that these model selection criteria performed well for real and simulated data whether the true MR model is or is not a member of the class of candidate MR models and generally outperformed the distance measure criterion MPRESS_q, C_q, and R_q^2. Their evaluation involved numerous other criteria.

An alternative to all possible regression procedures in the development of a "best" subset is to employ statistical tests sequentially to obtain the subset of variables. To illustrate, we show how to use Wilks' Λ test of additional information to develop an automatic selection procedure.

To see how we might proceed, we let Λ_F, Λ_R and $\Lambda_{F|R}$ represent the Λ criterion for testing $H_F : \mathbf{B} = \mathbf{0}$, $H_R : \mathbf{B}_1 = \mathbf{0}$, and $H_{F|R} : \mathbf{B}_2 = \mathbf{0}$ where

$$\mathbf{B} = \begin{bmatrix} \mathbf{B}_1 \\ \mathbf{B}_2 \end{bmatrix} \quad \text{and} \quad \mathbf{X} = \begin{bmatrix} \mathbf{X}_1 \\ \mathbf{X}_2 \end{bmatrix}.$$

Then,

$$\Lambda_F = \frac{|\mathbf{E}_F|}{|\mathbf{E}_F + \mathbf{H}_F|} = \frac{|\mathbf{Y}'\mathbf{Y}|}{|\mathbf{Y}'\mathbf{Y} - \widehat{\mathbf{B}}'(\mathbf{X}'\mathbf{X})\widehat{\mathbf{B}}|}$$

$$\Lambda_R = \frac{|\mathbf{E}_R|}{|\mathbf{E}_R + \mathbf{H}_R|} = \frac{|\mathbf{Y}'\mathbf{Y}|}{|\mathbf{Y}'\mathbf{Y} - \widehat{\mathbf{B}}_1'(\mathbf{X}_1'\mathbf{X}_1)\widehat{\mathbf{B}}_1|}$$

$$\Lambda_{F|R} = \frac{|\mathbf{E}_{F|R}|}{|\mathbf{E}_{F|R} + \mathbf{H}_{F|R}|} = \frac{|\mathbf{Y}'\mathbf{Y} - \widehat{\mathbf{B}}_1'(\mathbf{X}_1'\mathbf{X}_1)\widehat{\mathbf{B}}_1|}{|\mathbf{Y}'\mathbf{Y} - \widehat{\mathbf{B}}'(\mathbf{X}'\mathbf{X})\widehat{\mathbf{B}}|}$$

so that

$$\Lambda_F = \Lambda_R \ \Lambda_{F|R}$$

Associating Λ_F with the constant term and the variables $x_1, x_2, \ldots, x_k$ where $q = k + 1$, and Λ_R with the subset of variables $x_1, x_2, \ldots, x_{k-1}$ the significance or nonsignificance of variable x_k is, using (3.5.3), determined by the F statistic

$$\frac{1 - \Lambda_{F|R}}{\Lambda_{F|R}} \frac{v_e - p + 1}{p} \sim F(p, \ v_e - p + 1) \tag{4.2.51}$$

where $v_e = n - q = n - k - 1$. The F statistics in (4.2.51), also called partial F-tests, may be used to develop backward elimination, forward selection, and stepwise procedures to establish a "best" subset of variables for the MR model.

To illustrate, suppose a MR model contains $q = k + 1$ parameters and variables $x_1, x_2, \ldots, x_k$. By the backward elimination procedure, we would calculate F_i in (4.2.51) where the full model contained all the variables and the reduced model contained $k - 1$ variables so that F_i is calculated for each of the k variables. The variable x_i with the smallest $F_i \sim F(p, \ n - k - p)$ would be deleted leaving $k - 1$ variables to be evaluated at the next step. At the second step, the full model would contain $k - 1$ variables and the reduced model $k - 2$ variables. Now, $F_i \sim F(p, \ n - k - p - 1)$. Again, the variable with the smallest F value is deleted. This process continues until F attains a predetermined p-value or exceeds some preselected F critical value.

The forward selection process works in the reverse where variables are entered using the largest calculated F value. However, at the first step we consider only full models where each model contains the constant term and one variable. The one variable model with the smallest $\Lambda \sim U(p, 1, n - 2)$ initiates the process. At the second step, F_i is calculated with the full model containing two variables and the reduced model containing the variable at step one. The model with the largest $F_i \sim F(p, n - p - 1)$, for $k = 2$ is selected. At step k, $F_i \sim F(p, n - k - p)$ and the process stops when the smallest p-value exceeds some preset level or F_i falls below some critical value.

Either the backward elimination or forward selection procedure can be converted to a stepwise process. The stepwise backward process allows each variable excluded to be re-considered for entry. While the stepwise forward regression process allows one to see if a variable already in the model should by dropped using an elimination step. Thus, stepwise procedure require two F criteria or p-values, one to enter variables and one to remove variables.

TABLE 4.2.2. MANOVA Table for Lack of Fit Test

Source	df	SSCP	E(MSCP)
$\mathbf{B}_1$	$m+1$	$\widehat{\mathbf{B}}_1'\mathbf{X}_1'\mathbf{V}^{-1}\mathbf{X}_1\widehat{\mathbf{B}}_1 = \mathbf{H}_1$	$\Sigma + \frac{\mathbf{B}_1'\mathbf{X}_1'\mathbf{V}^{-1}\mathbf{X}_1\mathbf{B}_1}{m+1}$
$\mathbf{B}_2$	$k-m$	$\widehat{\mathbf{B}}_2'\mathbf{X}_2'\mathbf{Q}\mathbf{X}_2\widehat{\mathbf{B}}_2 = \mathbf{H}_2$	$\Sigma + \frac{\mathbf{B}_2'\mathbf{X}_2'\mathbf{Q}\mathbf{X}_2\mathbf{B}_2}{k-m}$
Residual	$c-k-1$	$\mathbf{E}_R = \mathbf{Y}_.'\mathbf{V}^{-1}\mathbf{Y}_. - \mathbf{H}_1 - \mathbf{H}_2$	Σ
Total (Between)	c	$\mathbf{Y}_.'\mathbf{V}^{-1}\mathbf{Y}_.$	
Total Within	$n-c$	$\mathbf{E}_{PE} = \mathbf{Y}'\mathbf{Y} - \mathbf{Y}_.'\mathbf{V}^{-1}\mathbf{Y}_.$	Σ
Total	n	$\mathbf{Y}'\mathbf{Y}$	

For the MR model, we obtained a "best" subset of x variables to simultaneous predict all y variables. If each x variable has a low correlation with a y variable we would want to remove the y variable from the set of y variables. To ensure that all y and x variables should remain in the model, one may reverse the roles of $\mathbf{x}$ and $\mathbf{y}$ and perform a backward elimination procedure on y given the x set to delete y variables.

Having fit a MR model to a data set, one may evaluate the model using a multivariate lack of fit test when replicates or near replicates exist in the data matrix $\mathbf{X}$, Christensen (1989). To develop a lack of fit test with replicates (near replicates) suppose the n rows of $\mathbf{X}$ are grouped into $i = 1, 2, \ldots, c$ groups with n_i rows per group, $1 \le c < n$. Forming replicates of size n_i in the observation vectors $\mathbf{y}_i'$ so that $\mathbf{y}_{i.} = \sum_{i=1}^{n_i} \mathbf{y}_i / n_i$, we have the multivariate weighted least squares (MWLS) model is

$$
\begin{aligned}
\underset{c\times p}{\mathbf{Y}_.} &= \underset{c\times q}{\mathbf{X}}\ \underset{q\times p}{\mathbf{B}} + \underset{c\times p}{\mathbf{E}} \\
\operatorname{cov}\left(\mathbf{y}_{i.}'\right) &= \mathbf{V} \otimes \Sigma, \qquad \mathbf{V} = \operatorname{diag}\left[1/n_i\right] \\
E\left(\mathbf{Y}.\right) &= \mathbf{XB}
\end{aligned}
\tag{4.2.52}
$$

Vectorizing the MWLS model, it is easily shown that the BLUE of $\mathbf{B}$ is

$$
\begin{aligned}
\widehat{\mathbf{B}} &= \left(\mathbf{X}'\mathbf{V}^{-1}\mathbf{X}\right)^{-1}\mathbf{X}'\mathbf{V}^{-1}\mathbf{Y} \\
\operatorname{cov}[\operatorname{vec}(\widehat{\mathbf{B}})] &= \Sigma \otimes \left(\mathbf{X}'\mathbf{V}^{-1}\mathbf{X}\right)
\end{aligned}
\tag{4.2.53}
$$

and that an unbiased estimate of Σ is

$$
\mathbf{S}_R = \mathbf{E}_R / (c-k-1) = \mathbf{Y}_.'\mathbf{V}^{-1}\mathbf{Y}_. - \widehat{\mathbf{B}}'(\mathbf{X}'\mathbf{V}^{-1}\mathbf{X})\widehat{\mathbf{B}} \,/\, (c-k-1)
$$

where $q = k+1$.

Partitioning $\mathbf{X} = [\mathbf{X}_1, \mathbf{X}_2]$ where $\mathbf{X}_1$ contains the variables $x_1, x_2, \ldots, x_m$ included in the model and $\mathbf{X}_2$ the excluded variables, one may test $H_2 : \mathbf{B}_2 = \mathbf{0}$. Letting $\mathbf{Q} = \mathbf{V}^{-1} - \mathbf{V}^{-1}\mathbf{X}_1\left(\mathbf{X}_1'\mathbf{V}^{-1}\mathbf{X}_1\right)^{-1}\mathbf{X}_1'\mathbf{V}^{-1}$, the MANOVA Table 4.2.2 is established for testing H_2 or $H_1 : \mathbf{B}_1 = \mathbf{0}$.

From Table 4.2.2, we see that if $\mathbf{B}_2 = \mathbf{0}$ that the sum of squares and products matrix associated with $\mathbf{B}_2$ may be combined with the residual error matrix to obtain a better estimate of Σ. Adding $\mathbf{H}_2$ to $\mathbf{E}_R$ we obtain the lack of fit error matrix $\mathbf{E}_{LF}$ with degrees of freedom $c - m - 1$. Another estimate of Σ independent of $\mathbf{B}_2 = \mathbf{0}$ is $\mathbf{E}_{PE}/(n - c)$ which is called the pure error matrix. Finally, we can write the pooled error matrix $\mathbf{E}$ as $\mathbf{E} = \mathbf{E}_{PE} + \mathbf{E}_{LF}$ with degrees of freedom $(n - c) + (c - m - 1) = n - m - 1$. The multivariate lack of fit test for the MR model compares the independent matrices $\mathbf{E}_{LF}$ with $\mathbf{E}_{PE}$ by solving the eigenequation

$$|\mathbf{E}_{LF} - \lambda \mathbf{E}_{PE}| = 0 \tag{4.2.54}$$

where $v_h = c - m - 1$ and $v_e = n - c$. We concluded that $\mathbf{B}_2 = \mathbf{0}$ if the lack of fit test is not significant so that the variables in the MR model adequately account for the variables in the matrix $\mathbf{Y}$. Again, one may use any of the criteria to evaluate fit. The parameters for the test criteria are $s = \min(v_h, p)$, $M = [| v_h - p | - 1]/2$ and $N = (v_e - p - 1)/2$ for the other criteria.

e. *Simultaneous Confidence Sets for a New Observation* $\mathbf{y}_{\text{new}}$ *and the Elements of* $\mathbf{B}$

Having fit a MR model to a data set, one often wants to predict the value of a new observation vector $\mathbf{y}'_{\text{new}}$ where $E(\mathbf{y}'_{\text{new}}) = \mathbf{x}'_{\text{new}}\mathbf{B}$. Since $\widehat{\mathbf{y}}'_{\text{new}} = \mathbf{x}'_{\text{new}}\widehat{\mathbf{B}}$ and assuming the $\text{cov}(\mathbf{y}_{\text{new}}) = \Sigma$ where $\mathbf{y}'_{\text{new}}$ is independent of the data matrix $\mathbf{Y}$, one can obtain a prediction interval for $\mathbf{y}'_{\text{new}}$ based on the distribution of $(\mathbf{y}_{\text{new}} - \widehat{\mathbf{y}}_{\text{new}})'$. The

$$\begin{aligned}
\text{cov}(\mathbf{y}_{\text{new}} - \widehat{\mathbf{y}}_{\text{new}})' &= \text{cov}(\mathbf{y}'_{\text{new}}) + \text{cov}(\widehat{\mathbf{y}}'_{\text{new}}) \\
&= \Sigma + \text{cov}(\mathbf{x}'_{\text{new}}\widehat{\mathbf{B}}) \\
&= \Sigma + (\mathbf{x}'_{\text{new}}(\mathbf{X}'\mathbf{X})^{-1}\mathbf{x}_{\text{new}})\Sigma \\
&= (1 + \mathbf{x}'_{\text{new}}(\mathbf{X}'\mathbf{X})^{-1}\mathbf{x}_{\text{new}})\Sigma
\end{aligned} \tag{4.2.55}$$

If $\mathbf{y}'_{\text{new}}$ and the rows of $\mathbf{Y}$ are MVN, then $\mathbf{y}_{\text{new}} - \widehat{\mathbf{y}}_{\text{new}}$ is MVN and independent of $\mathbf{E}$ so that $(1 + \mathbf{x}'_{\text{new}}(\mathbf{X}'\mathbf{X})^{-1}\mathbf{x}_{\text{new}})^{-1}(\mathbf{y}_{\text{new}} - \widehat{\mathbf{y}}_{\text{new}})(\mathbf{y}_{\text{new}} - \widehat{\mathbf{y}}_{\text{new}})' \sim W_p(1, \Sigma, \mathbf{0})$. Using Definition 3.5.3,

$$\frac{(\mathbf{y}_{\text{new}} - \widehat{\mathbf{y}}_{\text{new}})'\mathbf{S}^{-1}(\mathbf{y}_{\text{new}} - \widehat{\mathbf{y}}_{\text{new}})}{(1 + \mathbf{x}'_{\text{new}}(\mathbf{X}'\mathbf{X})^{-1}\mathbf{x}_{\text{new}})} \quad \frac{v_e - p + 1}{p} \sim F(p, v_e - p + 1) \tag{4.2.56}$$

Hence, a $100(1 - \alpha)\%$ prediction ellipsoid for $\mathbf{y}_{\text{new}}$ is all vectors that satisfy the inequality

$$\begin{aligned}
&(\widehat{\mathbf{y}}_{\text{new}} - \mathbf{y}_{\text{new}})'\mathbf{S}^{-1}(\widehat{\mathbf{y}}_{\text{new}} - \mathbf{y}_{\text{new}}) \\
&\qquad \leq \frac{pv_e}{(v_e - p - 1)} F^{1-\alpha}(p, v_e - 1)(1 + \mathbf{x}'_{\text{new}}(\mathbf{X}'\mathbf{X})^{-1}\mathbf{x}_{\text{new}})
\end{aligned}$$

However, the practical usefulness of the ellipsoid is of limited value for $p > 2$. Instead we consider all linear combinations of $\mathbf{a}'\mathbf{y}_{\text{new}}$. Using the Cauchy-Schwarz inequality (Problem 11, Section 2.6), it is easily established that the

$$\max_{\mathbf{a}} \frac{\left[\mathbf{a}'(\widehat{\mathbf{y}}_{\text{new}} - \mathbf{y}_{\text{new}})\right]^2}{\mathbf{a}'\mathbf{S}\mathbf{a}} \leq (\widehat{\mathbf{y}}_{\text{new}} - \mathbf{y}_{\text{new}})'\mathbf{S}^{-1}(\widehat{\mathbf{y}}_{\text{new}} - \mathbf{y}_{\text{new}})$$

Hence, the

$$P\left[\max_{\mathbf{a}} \frac{|\,\mathbf{a}'\left(\widehat{\mathbf{y}}_{\text{new}} - \mathbf{y}_{\text{new}}\right)|}{\sqrt{\mathbf{a}'\mathbf{S}\mathbf{a}}} \leq c_o\right] \geq 1-\alpha$$

for

$$c_\alpha^2 = pv_e F^{1-\alpha}\left(p, v_e - p + 1\right)\left(1 + \mathbf{x}'_{\text{new}}\left(\mathbf{X}'\mathbf{X}\right)^{-1}\mathbf{x}_{\text{new}}\right)/\left(v_e - p + 1\right).$$

Thus, $100\,(1-\alpha)\,\%$ simultaneous confidence intervals for linear combination of $\mathbf{a}'\mathbf{y}_{new}$ for arbitrary $\mathbf{a}$ is

$$\mathbf{a}'\widehat{\mathbf{y}}_{\text{new}} - c_\alpha\sqrt{\mathbf{a}'S\mathbf{a}} \leq \mathbf{a}'\mathbf{y}_{\text{new}} \leq \mathbf{a}'\widehat{\mathbf{y}}_{\text{new}} + c_\alpha\sqrt{\mathbf{a}'S\mathbf{a}} \tag{4.2.57}$$

Selecting $\mathbf{a}' = [0, 1, \ldots, 1_i, 0, \ldots, 0]$, a confidence interval for the ith variable within $\mathbf{y}_{\text{new}}$ is easily obtained. For a few comparisons, the intervals may be considerably larger than $1-\alpha$. Replacing $\widehat{\mathbf{y}}_{\text{new}}$ with $E\left(\widehat{\mathbf{y}}\right)$, $\mathbf{y}_{\text{new}}$ with $E\left(\mathbf{y}\right)$ and $1 + \mathbf{x}'_{\text{new}}\left(\mathbf{X}'\mathbf{X}\right)^{-1}\mathbf{x}_{\text{new}}$ with $\mathbf{x}'\left(\mathbf{X}'\mathbf{X}\right)^{-1}\mathbf{x}$, one may use (4.2.57) to establish simultaneous confidence intervals for the mean response vector.

In addition to establishing confidence intervals for a new observation or the mean response vector, one often needs to establish confidence intervals for the elements in the parameter matrix $\mathbf{B}$ following a test of the form $\mathbf{CBM} = \mathbf{0}$. Roy and Bose (1953) extended Scheffé's result to the MR model. Letting $\mathbf{V} = \text{cov}(\text{vec}\,\widehat{\mathbf{B}}) = \Sigma \otimes \left(\mathbf{X}'\mathbf{X}\right)^{-1}$, they showed using the Cauchy-Schwarz inequality, that the

$$P\{[\text{vec}(\widehat{\mathbf{B}} - \mathbf{B})]'\mathbf{V}^{-1}\,\text{vec}(\widehat{\mathbf{B}} - \mathbf{B})\} \leq \frac{v_e\theta^\alpha}{1-\theta^\alpha} = 1-\alpha \tag{4.2.58}$$

where $\theta^\alpha(s, M, N)$ is the upper α critical value for the Roy's largest root criterion used to reject the null hypotheses. That is, λ_1 is the largest root of $|\mathbf{H} - \lambda\mathbf{E}| = 0$ and $\theta_1 = \lambda_1/\left(1+\lambda_1\right)$ is Roy's largest root criterion for the test $\mathbf{CBM} = 0$. Or one may use the largest root criterion where λ^α is the upper α critical value for λ_1. Then, $100\,(1-\alpha)\,\%$ simultaneous confidence intervals for parametric functions $\psi = \mathbf{c}'\mathbf{Bm}$ have the general structure

$$\mathbf{c}'\widehat{\mathbf{B}}\mathbf{m} - c_\alpha\widehat{\sigma}_{\widehat{\psi}} \leq \psi \leq \mathbf{c}'\widehat{\mathbf{B}}\mathbf{m} + c_\alpha\widehat{\sigma}_{\widehat{\psi}} \tag{4.2.59}$$

where

$$\begin{aligned}
\widehat{\sigma}^2_{\widehat{\psi}} &= \left(\mathbf{m}'\mathbf{S}\mathbf{m}\right)\mathbf{c}'\left(\mathbf{X}'\mathbf{X}\right)^{-1}\mathbf{c} \\
c_\alpha^2 &= v_e\theta^\alpha/\left(1-\theta^\alpha\right) = v_e\lambda^\alpha \\
\mathbf{S} &= \mathbf{E}/v_e
\end{aligned}$$

Letting $U^\alpha, U_o^\alpha = T^2_{o,\,\alpha}/v_e$ and V^α represent the upper α critical values for the other criteria to test $\mathbf{CBM} = \mathbf{0}$, the critical constants in (4.2.59) following Gabriel (1968) are represented as follows

(a) Wilks

$$c_\alpha^2 = v_e[(1-U^\alpha)/U^\alpha] \tag{4.2.60}$$

(b) Bartlett-Nanda-Pillai (BNP)

$$c_\alpha^2 = v_e[V^\alpha/(1 - V^\alpha)]$$

(c) Bartlett-Lawley-Hotelling (BLH)

$$c_\alpha^2 = v_e U_o^\alpha = T_{o,\alpha}^2$$

(d) Roy

$$c_\alpha^2 = v_e[\theta^\alpha/(1 - \theta^\alpha)] = v_e \lambda^\alpha$$

Alternatively, using Theorem 3.5.1, one may use the F distribution to approximate the exact critical values. For Roy's criterion,

$$c_\alpha^2 \approx v_e \left[\frac{v_1}{v_e - v_1 + v_h}\right] F^{1-\alpha}(v_1,\ v_e - v_1 + v_h) \tag{4.2.61}$$

where $v_1 = \max(v_h, p)$. For the Bartlett-Lawley-Hotelling (BLH) criterion,

$$c_\alpha^2 \approx v_e \left[\frac{s v_1}{v_2}\right] F^{1-\alpha}(v_1,\ v_2) \tag{4.2.62}$$

where $v_1 = s(2M + s + 1)$ and $v_2 = 2(sN + 1)$. For the Bartlett-Nanda-Pillai (BNP) criterion, we relate V^α to an F distribution as follows

$$V^\alpha = \left(\frac{s v_1}{v_2} F^{1-\alpha}(v_1,\ v_2)\right) \Big/ \left(1 + \frac{v_1}{v_2} F^{1-\alpha}(v_1,\ v_2)\right)$$

where $v_1 = s(2M + s + 1)$ and $v_2 = s(2N + s + 1)$. Then the critical constant becomes

$$c_\alpha^2 \approx v_e[V^\alpha/(1 - V^\alpha)] \tag{4.2.63}$$

To find the upper critical value for Wilks' test criterion under the null hypothesis, one should use the tables developed by Wall (1968). Or, one may use a chi-square approximation to estimate the upper critical value for U^α. All the criteria are equal when $s = 1$.

The procedure outlined here, as in the test of location, is very conservative for obtaining simultaneous confidence intervals for each of the elements 33 elemts β in the parameter matrix **B**. With the rejection of the overall test, on may again use protected t-tests to evaluate the significance of each element of the matrix **B** and construct approximate $100(1 - \alpha)\%$ simultaneous confidence intervals for each element again using the entries in the Appendix, Table V. If one is only interested in individual elements of **B**, a FIT procedure is preferred, Schmidhammer (1982). The FIT procedure is approximated in SAS using PROC MULTTEST, Westfall and Young (1993).

f. Random X Matrix and Model Validation: Mean Squared Error of Prediction in Multivariate Regression

In our discussion of the multivariate regression model, we have been primarily concerned with the development of a linear model to establish the linear relationship between the matrix of dependent variables **Y** and the matrix of fixed independent variables **X**. The matrix of estimated regression coefficients $\widehat{\mathbf{B}}$ was obtained to estimate the population multivariate linear regression function defined using the matrix of coefficients **B**. The estimation and hypothesis testing process was used to help understand and establish the linear relationship between the random vector variable **Y** and the vector of fixed variables **X** in the population. The modeling process involved finding the population form of the linear relationship. In many multivariate regression applications, as in univariate multiple linear regression, the independent variables are random and not fixed. For this situation, we now assume that the joint distribution of the vector of random variables $\mathbf{Z} = [\mathbf{Y}', \mathbf{X}']' = [Y_1, Y_2, \ldots, Y_p, X_1, X_2, \ldots, X_k]'$ follows a multivariate normal distribution, $\mathbf{Z} \sim \mathbf{N}_{p+k}(\boldsymbol{\mu}_z, \Sigma_z)$ where the mean vector and covariance matrix have the following structure

$$\boldsymbol{\mu}_z = \begin{bmatrix} \boldsymbol{\mu}_y \\ \boldsymbol{\mu}_x \end{bmatrix}, \quad \Sigma = \begin{bmatrix} \Sigma_{yy} & \Sigma_{yx} \\ \Sigma_{xy} & \Sigma_{xx} \end{bmatrix} \tag{4.2.64}$$

The model with random **X** is sometimes called the correlation or structural model. In multiple linear regression and correlation models, interest is centered on estimating the population squared multiple correlation coefficient, ρ^2. The multivariate correlation model is discussed in more detail in Chapter 8 when we discuss canonical correlation analysis. Using Theorem 3.3.2, property (5), the conditional expectation of **Y** given the random vector variable **X** is

$$\begin{aligned} E(\mathbf{Y}|\mathbf{X} = \mathbf{x}) &= \boldsymbol{\mu}_y + \Sigma_{yx}\Sigma_{xx}^{-1}(\mathbf{x} - \boldsymbol{\mu}_x) \\ &= (\boldsymbol{\mu}_y - \Sigma_{yx}\Sigma_{xx}^{-1}\boldsymbol{\mu}_x) + \Sigma_{yx}\Sigma_{xx}^{-1}\mathbf{x} \\ &= \boldsymbol{\beta}_0 + \mathbf{B}_1'\mathbf{x} \end{aligned} \tag{4.2.65}$$

And, the covariance matrix of the random vector **Y** given **X** is

$$\operatorname{cov}(\mathbf{Y}|\mathbf{X} = \mathbf{x}) = \Sigma_{yy} - \Sigma_{yx}\Sigma_{xx}^{-1}\Sigma_{xy} = \Sigma_{y|x} = \Sigma \tag{4.2.66}$$

Under multivariate normality, the maximum likelihood estimators of the population parameters $\boldsymbol{\beta}_0$, $\mathbf{B}_1$, and Σ are

$$\begin{aligned} \widehat{\boldsymbol{\beta}}_0 &= \overline{\mathbf{y}} - \mathbf{S}_{yx}\mathbf{S}_{xx}^{-1}\mathbf{S}_{xx}\overline{\mathbf{x}} \\ \widehat{\mathbf{B}}_1 &= \mathbf{S}_{xx}^{-1}\mathbf{S}_{xy} \\ \widehat{\Sigma} &= (n-1)(\mathbf{S}_{yy} - \mathbf{S}_{yx}\mathbf{S}_{xx}^{-1}\mathbf{S}_{xy})/n \end{aligned} \tag{4.2.67}$$

where the matrices $\mathbf{S}_{ij}$ are formed using deviations about the mean vectors as in (3.3.3). Thus, to obtain the unbiased estimate for the covariance matrix Σ, one may use the matrix $\mathbf{S}_e = n\widehat{\Sigma}/(n-1)$ to correct for the bias. An alternative, minimal variance unbiased REML estimate for the covariance matrix Σ is to use the matrix $\mathbf{S}_{y|x} = \mathbf{E}/(n-q)$ where $q = k+1$ as calculated in the multivariate regression model. From (4.2.67), we see that the ordinary least squares estimate or BLUE of the model parameters are identical to the maximum likelihood (ML) estimate and that an unbiased estimate for the covariance matrix is easily obtained by rescaling the ML estimate for Σ. This result implies that if we assume that the vector $\mathbf{Z}$ follows a multivariate normal distribution, then all estimates and tests for the multivariate regression model conditioned on the independent variables have the same formulation when one considers the matrix of independent variables to be random. However, because the distribution of the columns of the matrix $\widehat{\mathbf{B}}$ do no follow a multivariate normal distribution when the matrix $\mathbf{X}$ is random, power calculations for fixed $\mathbf{X}$ and random $\mathbf{X}$ are not the same. Sampson (1974) discusses this problem in some detail for both the univariate multiple regression model and the multivariate regression model. We discuss power calculations in Section 4.17 for only the fixed $\mathbf{X}$ case. Gatsonis and Sampson (1989) have developed tables for sample size calculations and power for the multiple linear regression model for random independent variables. They show that the difference in power and sample size assuming a fixed variable model when they are really random is very small. They recommend that if one employs the fixed model approach in multiple linear regression when the variables are really random that the sample sizes should be increased by only five observations if the number of independent variables is less than ten; otherwise, the difference can be ignored. Finally, the maximum likelihood estimates for the mean vector $\boldsymbol{\mu}_z$ and covariance matrix Σ_z for the parameters in (4.2.64) follow

$$\widehat{\boldsymbol{\mu}}_z = \begin{bmatrix} \overline{\mathbf{x}} \\ \overline{\mathbf{y}} \end{bmatrix}, \quad \widehat{\Sigma} = \frac{(n-1)}{n} \begin{bmatrix} \mathbf{S}_{yy} & \mathbf{S}_{yx} \\ \mathbf{S}_{xy} & \mathbf{S}_{xx} \end{bmatrix} \tag{4.2.68}$$

Another goal in the development of either a univariate or multivariate regression model is that of model validation for prediction. That is, one is interested in evaluating how well the model developed from the sample, often called the calibration, training, or model-building sample predicts future observations in a new sample called the validation sample. In model validation, one is investigating how well the parameter estimates obtained in the model development phase of the study may be used to predict a set of new observations. Model validation for univariate and multivariate models is a complex process which may involve collecting a new data set, a holdout sample obtained by some a priori data splitting method or by an empirical strategy sometimes referred to as double cross-validation, Lindsay and Ehrenberg (1993). In multiple linear regression, the square of the population multiple correlation coefficient, ρ^2, is used to measure the degree of linear relationship between the dependent variable and the population predicted value of the dependent variables, $\boldsymbol{\beta}'\mathbf{X}$. It represents the square of the maximum correlation between the dependent variable and the population analogue of $\widehat{\mathbf{Y}}$. In some sense, the square of the multiple correlation coefficient is evaluating "model" precision. To evaluate predictive precision, one is interested in how well the parameter estimates developed from the calibration sample predict future observations, usually in a validation sample. One estimate of predictive precision in multiple linear regression is the squared zero-order Pearson product-moment correlation between the fitted values obtained by using the estimates from the calibration sample with the observations in the validation sample, (ρ_c^2), Browne (1975a). The square of the sample coefficient of determination, R_a^2, is an estimate of ρ^2 and not ρ_c^2. Cotter and Raju (1982) show that R_a^2 generally over estimates ρ_c^2. An estimate of ρ_c^2, sometimes called the "shruken" R-squared estimate and denoted by R_c^2 has been developed by Browne (1975a) for the multiple linear regression model with a random matrix of predictors. We discuss precision estimates base upon correlations in Chapter 8. For the multivariate regression model, prediction precision using correlations is more complicated since it involves canonical correlation analysis discussed in Chapter 8. Raju et al. (1997) review many formula developed for evaluating predictive precision for multiple linear regression models.

An alternative, but not equivalent measure of predictive precision is to use the mean squared error of prediction, Stein (1960) and Browne (1975b). In multiple linear regression the mean square error (MSE) of prediction is defined as MSEP $= E[(\mathbf{y}-\widehat{\mathbf{y}}(\mathbf{x}|\widehat{\boldsymbol{\beta}})^2]$, the expected squared difference between the observation vector ("parameter") and its predicted value ("estimator"). To develop a formula for predictive precision for the multivariate regression model, suppose we consider a single future observation $\mathbf{y}_{new}$ and that we are interested determining how well the linear prediction equation $\widehat{\mathbf{y}} = \mathbf{x}'\widehat{\mathbf{B}}$ obtained using the calibration sample predicts the future observation $\mathbf{y}_{new}$ for a new vector of independent variables. Given multivariate normality, the estimators $\widehat{\boldsymbol{\beta}}_0$ and $\widehat{\mathbf{B}}_1$ in (4.2.67) minimize the sample mean square error matrix defined by

$$\sum_{i=1}^{n}(\mathbf{y}_i - \widehat{\boldsymbol{\beta}}_0 - \widehat{\mathbf{B}}_1'\mathbf{x}_i)(\mathbf{y}_i - \widehat{\boldsymbol{\beta}}_0 - \widehat{\mathbf{B}}_1'\mathbf{x}_i)'/n \tag{4.2.69}$$

Furthermore, for $\boldsymbol{\beta}_0 = \boldsymbol{\mu}_y - \Sigma_{yx}\Sigma_{xx}^{-1}\boldsymbol{\mu}_x$ and $\mathbf{B}_1' = \Sigma_{yx}\Sigma_{xx}^{-1}$ in (4.2.65), the expected mean square error matrix $\mathbf{M}$ where

$$\begin{aligned}\mathbf{M} = E(&\mathbf{y} - \boldsymbol{\beta}_0 - \mathbf{B}_1'\mathbf{x})(\mathbf{y}_i - \boldsymbol{\beta}_0 - \mathbf{B}_1'\mathbf{x})' \\ &+ \Sigma_{yy} - \Sigma_{yx}\Sigma_{xx}^{-1}\Sigma_{xy} \\ &+ (\boldsymbol{\beta}_0 - \boldsymbol{\mu}_y + \mathbf{B}_1'\boldsymbol{\mu}_x)(\boldsymbol{\beta}_0 - \boldsymbol{\mu}_y + \mathbf{B}_1'\boldsymbol{\mu}_x)' \\ &+ (\mathbf{B}_1' - \Sigma_{yx}\Sigma_{xx}^{-1})(\Sigma_{xx})(\mathbf{B}_1' - \Sigma_{yx}\Sigma_{xx}^{-1})'\end{aligned} \tag{4.2.70}$$

is minimized. Thus, to evaluate how well a multivariate prediction equation estimates a new observation $\mathbf{y}_{new}$ given a vector of independent variables $\mathbf{x}$, one may use the mean square error matrix $\mathbf{M}$ with the parameters estimated from the calibration sample; this matrix is the mean squared error matrix for prediction $\mathbf{Q}$ defined in (4.2.71) which may be used to evaluate multivariate predictive precision. The mean square error matrix of predictive precision for the multivariate regression model is

$$\begin{aligned}\mathbf{Q} &= E(\mathbf{y} - \widehat{\boldsymbol{\beta}}_0 - \widehat{\mathbf{B}}_1'\mathbf{x})(\mathbf{y} - \widehat{\boldsymbol{\beta}}_0 - \widehat{\mathbf{B}}_1'\mathbf{x})' \\ &= (\Sigma_{yy} - \Sigma_{yx}\Sigma_{xx}^{-1}\Sigma_{xy}) \\ &\quad + (\widehat{\boldsymbol{\beta}}_0 - \boldsymbol{\mu}_y - \widehat{\mathbf{B}}_1'\boldsymbol{\mu}_x)(\widehat{\boldsymbol{\beta}}_0 - \boldsymbol{\mu}_y - \widehat{\mathbf{B}}_1'\boldsymbol{\mu}_x)' \\ &\quad + (\widehat{\mathbf{B}}_1' - \Sigma_{yx}\Sigma_{xx}^{-1})(\Sigma_{xx})(\widehat{\mathbf{B}}_1' - \Sigma_{yx}\Sigma_{xx}^{-1})'\end{aligned} \tag{4.2.71}$$

Following Browne (1975b), one may show that the expected error of prediction is

$$\Delta = \mathbf{E}(\mathbf{Q}) = \Sigma_{y|x}(n+1)(n-2)/n(n-k-2) \tag{4.2.72}$$

where the covariance matrix $\Sigma_{y|x} = \Sigma_{yy} - \Sigma_{yx}\Sigma_{xx}^{-1}\Sigma_{xy}$ is the matrix of partial variances and covariances for the random variable $\mathbf{Y}$ given $\mathbf{X} = \mathbf{x}$. The corresponding value for the expected value of $\mathbf{Q}$, denoted as d^2 by Browne (1975b) for the multiple linear regression model with random $\mathbf{X}$, is $\delta^2 = E(d^2) = \sigma^2(n+1)(n-2)/n(n-k-2)$. Thus Δ is a generalization of δ^2.

In investigating δ^2 for the random multiple linear regression model, Browne (1975b) shows that the value of δ^2 tends to decrease, stabilize, and then increase as the number of predictor variables k increases. Thus, when the calibration sample is small one wants to use a limited number of predictor variables. The situation is more complicated for the random multivariate regression model since we have an expected error of prediction matrix. Recall that if the elements of the determinant of the matrix of partial variances and covariances of the $\Sigma_{y|x}$ are large, one may usually expect that the determinant of the matrix to also be large; however this is not always the case. To obtain a bounded measure of generalized variance, one may divide the determinant of $\Sigma_{y|x}$ by the product of its diagonal elements. Letting σ_{ii} represent the partial variances on the diagonal of the covariance matrix $\Sigma_{y|x}$, the

$$0 \leq |\Sigma_{y|x}| \leq \prod_{i=1}^{p} \sigma_{ii} \tag{4.2.73}$$

and we have that the

$$\frac{|\Sigma_{y|x}|}{\prod_{i=1}^{p} \sigma_{ii}} = |\mathbf{P}_{y|x}| \tag{4.2.74}$$

where $\mathbf{P}_{y|x}$ is the population matrix of partial correlations corresponding to the matrix of partial variances and covariances in $\Sigma_{y|x}$. Using (4.2.73), we have that $0 \leq |\mathbf{P}_{y|x}|^2 \leq 1$.

To estimate Δ, we use the minimum variance unbiased estimator for $\Sigma_{y|x}$ from the calibration sample. Then an unbiased estimator of Δ is

$$\widehat{\Delta}_c = \mathbf{S}_{y|x} \left(\frac{(n+1)(n-2)}{(n-k-1)(n-k-2)} \right) \tag{4.2.75}$$

where $\mathbf{S}_{y|x} = \mathbf{E}/(n-k-1) = \mathbf{E}_q$ is the REML estimate of $\Sigma_{y|x}$ for $q = k-1$ variables. Thus, $\widehat{\Delta}_c$ may also be used to select variables in multivariate regression models. However $\widehat{\Delta}_c$ it is not an unbiased estimate of the matrix $\mathbf{Q}$. Over all calibration samples, one might expect the entire estimation process to be unbiased in that the $E(|\widehat{\Delta}_c| - |\mathbf{Q}|) = 0$. As an exact estimate of the mean square error of prediction using only the calibration sample, one may calculate the determinant of the matrix $\mathbf{S}_{y|x}$ since

$$\frac{|\widehat{\Delta}_c|}{\left(\frac{(n+1)(n-2)}{(n-k-1)(n-k-2)} \right)^p} = |\mathbf{S}_{y|x}| \tag{4.2.76}$$

Using (4.2.74) with population matrices replaced by their corresponding sample estimates, a bounded measure of the mean square error of prediction is $0 \leq |\mathbf{R}_{y|x}|^2 \leq 1$. Using results developed by Ogasawara (1998), one may construct an asymptotic confidence interval for this index of precision or consider other scalar functions of $\widehat{\Delta}_c$. However, the matrix of interest is not $E(\mathbf{Q}) = \Delta$, but $\mathbf{Q}$. Furthermore, the value of the determinant of $\mathbf{R}_{y|x}$ is zero if any eigenvalue of the matrix is near zero so that the determinant may not be a good estimate of the expected mean square error of prediction. To obtain an estimate of the matrix **Q**, a validation sample with $m \doteq n$ observations may be used. Then an unbiased estimate of **Q** is $\mathbf{Q}_*$ where

$$\mathbf{Q}_* = \sum_{i=1}^{m} (\mathbf{y}_i - \widehat{\boldsymbol{\beta}}_0 - \widehat{\mathbf{B}}_1' \mathbf{x}_i)(\mathbf{y}_i - \widehat{\boldsymbol{\beta}}_0 - \widehat{\mathbf{B}}_1' \mathbf{x}_i)' / m \tag{4.2.77}$$

Now, one may compare the $|\Delta_c|$ with the $|\mathbf{Q}_*|$ to evaluate predictive precision. If a validation sample is not available, one might estimate the predictive precision matrix by holding out one of the original observations each time to obtain a MPRESS estimate for $\mathbf{Q}_*$. However, the determinant of the MPRESS estimator is always larger than the determinant of the calibration sample estimate since we are always excluding an observation.

In developing a multivariate linear regression model using a calibration sample and evaluating the predictability of the model using the validation sample, we are evaluating overall predictive "fit". The simple ratio of the squares of the Euclidean norms defined as $1 - ||\mathbf{Q}_*||^2 / ||\widehat{\Delta}_c||^2$ may also be used as a measure of overall multivariate predictive precision. It has the familiar coefficient of determination form. The most appropriate measure of predictive precision using the mean square error criterion for the multivariate regression model requires additional study, Breiman and Friedman (1997).

g. *Exogeniety in Regression*

The concept of exogeniety arises in regression models when both the dependent (endogenous) variables and the independent (exogeneous) variables are jointly defined and random. This occurs in path analysis, simultaneous equation, models discussed in Chapter 10. In regression models, the dependent variable is endogenous since it is determined by the regression function. Whether or not the independent variables are exogeneous depends upon whether or not the variable can be assumed given without loss of information. This depends on the parameters of interest in the system. While joint multivariate normality of the dependent and independent variables is a necessary condition for the independent variable to be exogeneous, the sufficient condition is a concept known as weak exogeniety. Weak exogeniety ensures that estimation and inference for the model parameters (called efficient inference in the econometric literature) may be based upon the conditional density of the dependent variable **Y** given the independent variable $\mathbf{X} = \mathbf{x}$ (rather than the joint density) without loss of information. Engle, Hendry, and Richard (1893) define a set of variables **x** in a model to be weakly exogenous if the full model can be written in terms of a marginal density function for **X** times a conditional density function for $\mathbf{Y}|\mathbf{X} = \mathbf{x}$ such that the estimation of the parameters of the conditional distribution is no less efficient than estimation of the all the parameters in the joint density. This will be the case if none of the parameters in the conditional distribution appears in the marginal distribution for **x**. That is, the parameters in the density function for **X** may be estimated separately, if desired, which implies that the marginal density can be assumed given. More will be said about this in Chapter 10, however, the important thing to notice from this discussion is that merely saying that the variables in a model are exogeneous does not necessary make them exogeneous.

4.3 Multivariate Regression Example

To illustrate the general method of multivariate regression analysis, data provided by Dr. William D. Rohwer of the University of California at Berkeley are analyzed. The data are shown in Table 4.3.1 and contained in the file Rohwer.dat.

The data represent a sample of 32 kindergarten students from an upper-class, white, residential school (Gr). Rohwer was interested in determining how well data from a set of paired-associate (PA), learning-proficiency tests may be used to predict the children's performance on three standardized tests (Peabody Picture Vocabulary Test; PPVT-y_1, Raven Progressive Matrices Test; RPMT-y_2, and a Student Achievement Test, SAT-y_3). The five PA learning proficiency tasks represent the sum of the number of items correct out of 20 (on two exposures). The tasks involved prompts to facilitate learning. The five PA word prompts involved x_1-named (N), x_2-still (S), x_3-named action (NA), x_4-named still (NS) and x_5-sentence still (SS) prompts. The SAS code for the analysis is included in program m4_3_1.sas.

The primary statistical procedure for fitting univariate and multivariate regression models to data in SAS is PROC REG. While the procedure may be used to fit a multivariate model to a data set, it is designed for multiple linear regression. All model selection methods, residual plots, and scatter plots are performed a variable at a time. No provision has yet been made for multivariate selection criteria, multivariate measures of association, multivariate measures of model fit, or multivariate prediction intervals. Researchers must write their own code using PROC IML.

When fitting a multivariate linear regression model, one is usually interested in finding a set of independent variables that jointly predict the independent set. Because some subset of independent variables may predict an independent variable better than others, the MR model may overfit or underfit a given independent variable. To avoid this, one may consider using a SUR model discussed in Chapter 5.

When analyzing a multivariate data set using SAS, one usually begins by fitting the full model and investigates residual plots for each variable, Q-Q plots for each variable, and multivariate Q-Q plots. We included the multinorm.sas macro into the program to produce a multivariate chi-square Q-Q plot of the residuals for the full model. The residuals are also output to an external file (res.dat) so that one may create a Beta Q-Q plot of the residuals. The plots are used to assess normality and whether or not there are outliers in the data set. When fitting the full model, the residuals for y1 $\equiv$ PPVT and y3 $=$ SAT appear normal; however, y2 $=$ RPMT may be skewed right. Even though the second variable is slightly skewed, the chi-square Q-Q plot represents a straight line, thus indicating that the data appear MVN. Mardia's tests of skewness and Kurtosis are also nonsignificant. Finally, the univariate Q-Q plots and residual plots do not indicate the presence of outliers.

Calculating Cook's distance using formula (4.2.30), the largest value, $C_i = 0.85$, does not indicate that the 5th observation is influential. The construction of logarithm leverage plots for evaluating the influence of groups of observation are discussed by Barrett and Ling (1992). To evaluate the influence of a multivariate observation on each row of $\widehat{\mathbf{B}}$ or on the cov(vec $\widehat{\mathbf{B}}$), one may calculate (4.2.33) and (4.2.35) by writing code using PROC IML.

Having determined that the data are well behaved, we next move to the model refinement phase by trying to reduce the set of independent variables needed for prediction. For

TABLE 4.3.1. Rohwer Dataset

PPVT	RPMT	SAT	Gr	N	S	NS	NA	SS
68	15	24	1	0	10	8	21	22
82	11	8	1	7	3	21	28	21
82	13	88	1	7	9	17	31	30
91	18	82	1	6	11	16	27	25
82	13	90	1	20	7	21	28	16
100	15	77	1	4	11	18	32	29
100	13	58	1	6	7	17	26	23
96	12	14	1	5	2	11	22	23
63	10	1	1	3	5	14	24	20
91	18	98	1	16	12	16	27	30
87	10	8	1	5	3	17	25	24
105	21	88	1	2	11	10	26	22
87	14	4	1	1	4	14	25	19
76	16	14	1	11	5	18	27	22
66	14	38	1	0	0	3	16	11
74	15	4	1	5	8	11	12	15
68	13	64	1	1	6	19	28	23
98	16	88	1	1	9	12	30	18
63	15	14	1	0	13	13	19	16
94	16	99	1	4	6	14	27	19
82	18	50	1	4	5	16	21	24
89	15	36	1	1	6	15	23	28
80	19	88	1	5	8	14	25	24
61	11	14	1	4	5	11	16	22
102	20	24	1	5	7	17	26	15
71	12	24	1	9	4	8	16	14
102	16	24	1	4	17	21	27	31
96	13	50	1	5	8	20	28	26
55	16	8	1	4	7	19	20	13
96	18	98	1	4	7	10	23	19
74	15	98	1	2	6	14	25	17
78	19	50	1	5	10	18	27	26

this phase, we depend on univariate selection methods in SAS, e.g. C_q-plots and stepwise methods. We combine univariate methods with multivariate tests of hypotheses regarding the elements of **B** using MTEST statements. The MTEST statements are testing that the regression coefficients associated with of all independent variables are zero for the set of dependent variables simultaneously by separating the independent variables by commas. When a single variable is included in an MTEST statement, the MTEST is used to test that all coefficients for the variable are zero for each dependent variable in the model. We may also test that subsets of the independent variables are zero. To include the intercept in a test, the variable name INTERCEPT must be included in the MTEST statement. Reviewing the multiple regression equations for each variable, the C_q plots, and the backward elimination output one is unsure about which variables jointly prediction the set of dependent variables. Variable NA is significant in predicting PPVT, S is significant in predicting RPMT, and the variables N, NS, and NA are critical in the prediction of SAT. Only the variable SS should be excluded from the model based upon the univariate tests. However, the multivariate tests seem to support retaining only the variables $x2$, $x3$ and $x4$ (S, NS, and NA). The multivariate MTEST with the label N_SS indicates that both independent variable $x_1$and x_5 (N, SS) are zero in the population. Thus, we are led to fit the reduced model which only includes the variables S, NS, and NA.

Fitting the reduced model, the overall measure of association as calculated by η^2 defined in (4.2.39) indicates that 62% of the variation in the dependent set is accounted for by the three independent variables: S, NS and NA. Using the full model, only 70% of the variation is explained. The parameter matrix **B** for the reduced model follows.

$$\mathbf{B} = \begin{bmatrix} 41.695 & 12.357 & -44.093 \\ 0.546 & 0.432 & 2.390 \\ -0.286 & -0.145 & -4.069 \\ 1.7107 & 0.066 & 5.487 \end{bmatrix} \begin{matrix} \text{(Intercept)} \\ \text{(S)} \\ \text{(NS)} \\ \text{(NA)} \end{matrix}$$

Given **B** for the reduced model, one may test the hypothesis $H_o : \mathbf{B} = \mathbf{0}$, $H_1 : \mathbf{B}_1 = \mathbf{0}$, and that a row vector of **B** is simultaneously zero for all dependent variables, $H_i : \boldsymbol{\beta}_i' = \mathbf{0}'$ among others using the MTEST statement in SAS as illustrated in the program m4_3_1.sas. While the tests of H_i are exact, since $s = 1$ for these tests, this is not the case when testing H_o or H_1 since $s > 2$. For these tests $s = 3$. The test that $\mathbf{B}_1 = \mathbf{0}$ is a test of the model or the test of no regression and is labeled $B1$ in the output. Because $s = 3$ for this test, the multivariate criteria are not equivalent and no F approximation is exact. However, all three test criteria indicate that $\mathbf{B}_1 \neq \mathbf{0}$. Following rejection of any null hypothesis regarding the elements of the parameter matrix **B**, one may use (4.2.59) to obtain simultaneous confidence intervals for all parametric functions $\psi = \mathbf{c}'\mathbf{Bm}$. For the test $H_1 : \mathbf{B}_1 = \mathbf{0}$, the parametric functions have the form $\psi = \mathbf{c}'\mathbf{B}_1\mathbf{m}$. There are 9 elements in the parameter matrix $\mathbf{B}_1$. If one is only interested in constructing simultaneous confidence intervals for these elements, formula (4.2.49) tends to generate very wide intervals since it is designed to be used for all bi-linear combinations of the elements of the parameter matrix associated with the overall test and not just a few elements. Because PROC REG in SAS does not generate the confidence sets for parametric functions ψ, PROC IML is used. To illustrate the procedure, a simultaneous confidence interval for β_{42} in the matrix **B** obtained following the test of H_1 by using $\mathbf{c}' = [(0, 0, 0, 1]$, $\mathbf{m}' = [0, 0, 1]$ is illustrated. Using

$\alpha = 0.05$, the approximate critical values for the Roy, BLH, and BNP criteria are 2.97, 4.53, and 5.60, respectively. Intervals for other elements of $\mathbf{B}_1$ may be obtained by selecting other values for **c** and **m**. The approximate simultaneous 95% confidence interval for β_{42} as calculated in the program using the upper bound of the F statistic is (1.5245, 9.4505). While this interval does not include zero, recall that the interval is a lower bound for the true interval and must be used with caution. The intervals using the other criteria are more near their actual values using the F approximations. Using any of the planned comparison procedures for nine intervals, one finds them to be very near Roy's lower bound, for this example. The critical constant for the multivariate t is about 2.98 for the Type I error rate $\alpha = 0.05$, $C = 9$ comparisons and $v_e = 28$.

Continuing with our example, Rohwer's data are reanalyzed using the multivariate forward stepwise selection method and Wilks' Λ criterion, the C_q criterion defined in (4.2.46), the corrected information criteria $AICu_q$ and $HQIC_q$ defined in (4.2.48) and (4.2.49), and the uncorrected criteria: AIC_q, BIC_q, and $HQIC_q$ using program MulSubSel.sas written by Dr. Ali A. Al-Subaihi while he was a doctoral student in the Research Methodology program at the University of Pittsburgh. This program is designed to select the best subset of variables simultaneously for all dependent variables.

The stepwise (STEPWISE), C_q (CP) and $HQICu_q$ (HQ) procedures selected variables $1, 2, 3, 4(N, S, NS, NA)$ while $AICu_q$ $(AICC)$ selected only variables 2 and 4 (S, NS). The uncorrected criteria $AIC_q(AIC)$, $BIC_q(BIC)$, $HQIC_q$ $(HQIC)$ only selected one variable, $4(NA)$. All methods excluded the fifth variable SS. For this example, the number of independent variable is only five and the correlations between the dependent and independent variables are in the moderate range. A Monte Carlo study conducted by Dr. Al-Subaihi indicates that the $HQICu_q$ criterion tends to find the correct multivariate model or to moderately overfit the correct model when the number of variables is not too large and all correlations have moderate values. The $AICu_q$ also frequently finds the correct model, but tends to underfit more often. Because of these problems, he proposed using the reduced rank regression (RRR) model for variable selection. The RRR model is discussed briefly in Chapter 8.

Having found a multivariate regression equation using the calibration sample, as an estimate of the expected mean square error of prediction one may use the determinant of the sample covariance matrix $\mathbf{S}_{y|x}$. While we compute its value for the example, to give meaning to this value, one must obtain a corresponding estimate for a validation sample.

Exercises 4.3

1. Using the data set res.dat for the Rohwer data, create a Beta Q-Q plot for the residuals. Compare the plot obtained with the chi-square Q-Q plot. What do you observe.

2. For the observation $\mathbf{y}'_{\text{new}} = [70, 20, 25]$ find a 95% confidence interval for each element of $\mathbf{y}_{\text{new}}$ using (4.2.57).

3. Use (4.2.58) to obtain simultaneous confidence intervals for the elements in the reduced model parameter matrix $\mathbf{B}_1$.

4. Rohwer collected data identical to the data in Table 4.3.1 for kindergarten students in a low-socioeconomic-status area The data for the $n = 37$ student are provided in table 4.3.2. Does the model developed for the upper-class students adequately predict the performance for the low-socioeconomic-status students? Discuss your findings.

5. For the $n = 37$ students in Table 4.3.2, find the "best" multivariate regression equation and simultaneous confidence intervals for the parameter matrix **B**.

   (a) Verify that the data are approximately multivariate normal.

   (b) Fit a full model to the data.

   (c) Find a best subset of the independent variables.

   (d) Obtain confidence intervals for the elements in **B** for the best subset.

   (e) Calculate η^2 for the final equation.

6. To evaluate the performance of the C_q criterion given in (4.2.46), Sparks et al. (1983) analyzed 25 samples of tobacco leaf for organic and inorganic chemical constituents. The dependent variates considered are defined as follows.

   Y_1: Rate of cigarette burn in inches per 1000 seconds

   Y_2: Percent sugar in the leaf

   Y_3: Percent nicotine

   The fixed independent variates are defined as follows.

   X_1: Percentage of nitrogen

   X_2: Percentage of chlorine

   X_3: Percentage of potassim

   X_4: Percentage of Phosphorus

   X_5: Percentage of calculm

   X_6: Percentage of Magnesium

   The data are given in the file tobacco.sas and organized as $[Y_1, Y_2, Y_3, X_1, X_2, \ldots, X_6]$. Use PROC REG and the program MulSubSel.sas to find the best subset of independent variables. Write up your findings in detail by creating a technical report of your results. Include in your report the evaluation of multivariate normality, evaluation of outliers, model selection criteria, and model validation using data splitting or a holdout procedure.

TABLE 4.3.2. Rohwer Data for Low SES Area

SAT	PPVT	RPMT	N	S	NS	NA	SS
49	48	8	1	2	6	12	16
47	76	13	5	14	14	30	27
11	40	13	0	10	21	16	16
9	52	9	0	2	5	17	8
69	63	15	2	7	11	26	17
35	82	14	2	15	21	34	25
6	71	21	0	1	20	23	18
8	68	8	0	0	10	19	14
49	74	11	9	9	7	16	13
8	70	15	3	2	21	26	25
47	70	15	8	16	15	35	24
6	61	11	5	4	7	15	14
14	54	12	1	12	13	27	21
30	35	13	2	1	12	20	17
4	54	10	1	3	12	26	22
24	40	14	0	2	5	14	8
19	66	13	7	12	21	35	27
45	54	10	0	6	6	14	16
22	64	14	12	8	19	27	26
16	47	16	3	9	15	18	10
32	48	16	0	7	9	14	18
37	52	14	4	6	20	26	26
47	74	19	4	9	14	23	23
5	57	12	0	2	4	11	8
6	57	10	0	1	16	15	17
60	80	11	3	8	18	28	21
58	78	13	1	18	19	34	23
6	70	16	2	11	9	23	11
16	47	14	0	10	7	12	8
45	94	19	8	10	28	32	32
9	63	11	2	12	5	25	14
69	76	16	7	11	18	29	21
35	59	11	2	5	10	23	24
19	55	8	9	1	14	19	12
58	74	14	1	0	10	18	18
58	71	17	6	4	23	31	26
79	54	14	0	6	6	15	14

4.4 One-Way MANOVA and MANCOVA

a. *One-Way MANOVA*

The one-way MANOVA model allows one to compare the means of several independent normally distributed populations. For this design, n_i subjects are randomly assigned to one of k treatments and p dependent response measures are obtained on each subject. The response vectors have the general form

$$\mathbf{y}'_{ij} = \left[y_{ij1}, y_{ij2}, \ldots, y_{ijp}\right] \tag{4.4.1}$$

were $i = 1, 2, \ldots, k$ and $j = 1, 2, \ldots, n_i$. Furthermore, we assume that

$$\mathbf{y}_{ij} \sim IN_p\left(\boldsymbol{\mu}_i, \Sigma\right) \tag{4.4.2}$$

so that the observations are MVN with independent means and common unknown covariance structure Σ.

The linear model for the observation vectors $\mathbf{y}_{ij}$ has two forms, the full rank (FR) or cell means model

$$\mathbf{y}_{ij} = \boldsymbol{\mu}_i + \mathbf{e}_{ij} \tag{4.4.3}$$

and the less than full rank (LFR) overparameterized model

$$\mathbf{y}_{ij} = \boldsymbol{\mu} + \boldsymbol{\alpha}_i + \mathbf{e}_{ij} \tag{4.4.4}$$

For (4.4.3), the parameter matrix for the FR model contains only means

$$\underset{k \times p}{\mathbf{B}} = \begin{bmatrix} \boldsymbol{\mu}'_1 \\ \boldsymbol{\mu}'_2 \\ \vdots \\ \boldsymbol{\mu}'_k \end{bmatrix} = \left[\mu_{ij}\right] \tag{4.4.5}$$

and for the LFR model,

$$\underset{q \times p}{\mathbf{B}} = \begin{bmatrix} \boldsymbol{\mu}' \\ \boldsymbol{\alpha}'_1 \\ \vdots \\ \boldsymbol{\alpha}'_k \end{bmatrix} = \begin{bmatrix} \mu_1 & \mu_2 & \cdots & \mu_p \\ \alpha_{11} & \alpha_{12} & \cdots & \alpha_{1p} \\ \vdots & \vdots & & \vdots \\ \alpha_{k1} & \alpha_{k2} & \cdots & \alpha_{kp} \end{bmatrix} \tag{4.4.6}$$

so that $q = k + 1$ and $\mu_{ij} = \mu_j + \alpha_{ij}$. The matrix $\mathbf{B}$ in the LFR case contain unknown constants μ_j and the treatment effects α_{ij}.

Both models have the GLM form $\mathbf{Y}_{n \times q} = \left[y_{ij}\right] = \mathbf{X}_{n \times q}\mathbf{B}_{q \times p} + \mathbf{E}_{n \times p}$; however, the design matrices of zeros and ones are not the same. For the FR model,

$$\mathbf{X}_{n \times q} = \mathbf{X}_{FR} = \begin{bmatrix} \mathbf{1}_{n_1} & 0 & \cdots & 0 \\ 0 & \mathbf{1}_{n_2} & \cdots & 0 \\ \vdots & \vdots & & \vdots \\ 0 & 0 & \cdots & \mathbf{1}_{n_k} \end{bmatrix} = \bigoplus_{i=1}^{k} \mathbf{1}_{n_i} \tag{4.4.7}$$

and the $r\left(\mathbf{X}_{FR}\right)=k\equiv q-1$. For the LFR model, the design matrix is

$$\mathbf{X}_{n\times q}=\mathbf{X}_{LFR}=\left[\mathbf{1}_n,\mathbf{X}_{FR}\right] \tag{4.4.8}$$

where $n=\sum_i n_i$ is a vector of n 1's and the $r\left(\mathbf{X}_{LFR}\right)=k=q-1<q$ is not of full rank.

When the number of observations in each treatment for the one-way MANOVA model are equal so that $n_1=n_2=\ldots=n_k=r$, the LFR design matrix $\mathbf{X}$ has a balanced structure. Letting $\mathbf{y}_j$ represent the jth column of $\mathbf{Y}$, the linear model for the jth variable becomes

$$\begin{aligned}\mathbf{y}_j&=\mathbf{X}\boldsymbol{\beta}_j+\mathbf{e}_j \qquad j=1,2,\ldots,p\\&=\left(\mathbf{1}_k\otimes\mathbf{1}_r\right)\boldsymbol{\mu}_j+\left(\mathbf{I}_k\otimes\mathbf{1}_r\right)\boldsymbol{\alpha}_j+\mathbf{e}_j\end{aligned} \tag{4.4.9}$$

where $\boldsymbol{\alpha}'_j=\left[\alpha_{1j},\alpha_{2j},\ldots,\alpha_{kj}\right]$ is a vector of k treatment effects, $\boldsymbol{\beta}_j$ is the jth column of $\mathbf{B}$ and $\mathbf{e}_j$ is the jth column of $\mathbf{E}$. Letting $\mathbf{K}_i$ represent the Kronecker or direct product matrices so that $\mathbf{K}_0=\mathbf{1}_k\otimes\mathbf{1}_r$ and $\mathbf{K}_1=\mathbf{I}_k\otimes\mathbf{1}_r$ with $\boldsymbol{\beta}_{0j}=\boldsymbol{\mu}_j$ and $\boldsymbol{\beta}_{1j}=\boldsymbol{\alpha}_j$, an alternative univariate structure for the jth variable in the multivariate one-way model is

$$\mathbf{y}_j=\sum_{i=0}^{2}\mathbf{K}_i\boldsymbol{\beta}_{ij}+\mathbf{e}_j \qquad j=1,2,\ldots,p \tag{4.4.10}$$

This model is a special case of the more general representation for the data matrix $\mathbf{Y}=\left[\mathbf{y}_1,\mathbf{y}_2,\ldots,\mathbf{y}_p\right]$ for balanced designs

$$\underset{n\times p}{\mathbf{Y}}=\mathbf{XB}+\mathbf{E}=\sum_{i=0}^{m}\mathbf{K}_i\mathbf{B}_i+\mathbf{E} \tag{4.4.11}$$

where $\mathbf{K}_i$ are known matrices of order $n\times r_i$ and rank r_i and $\mathbf{B}_i$ are effect matrices of order $r_i\times p$. Form (4.4.11) is used with the analysis of mixed models by Searle, Casella and McCulloch (1992) and Khuri, Mathew and Sinha (1998). We will use this form of the model in Chapter 6.

To estimate the parameter matrix $\mathbf{B}$ for the FR model, with $\mathbf{X}$ defined in (4.4.7), we have

$$\widehat{\mathbf{B}}_{FR}=\left(\mathbf{X}'\mathbf{X}\right)^{-1}\mathbf{X}'\mathbf{Y}=\begin{bmatrix}\mathbf{y}'_{1.}\\\mathbf{y}'_{2.}\\\vdots\\\mathbf{y}'_{k.}\end{bmatrix} \tag{4.4.12}$$

where $\mathbf{y}_{i.}=\sum_{j=1}^{n_i}\mathbf{y}_{ij}/n_i$ is the sample mean for the ith treatment. Hence, $\widehat{\boldsymbol{\mu}}_i=\mathbf{y}_{i.}$ is a vector of sample means. An unbiased estimate of Σ is

$$\begin{aligned}\mathbf{S}&=\mathbf{Y}'\left[\mathbf{I}-\mathbf{X}\left(\mathbf{X}'\mathbf{X}\right)^{-1}\mathbf{X}'\right]\mathbf{Y}/\left(n-k\right)\\&=\sum_{i=1}^{k}\sum_{j=1}^{n_i}\left(\mathbf{y}_{ij}-\mathbf{y}_{i.}\right)\left(\mathbf{y}_{ij}-\mathbf{y}_{i.}\right)'/\left(n-k\right)\end{aligned} \tag{4.4.13}$$

where $v_e=n-k$.

To estimate $\mathbf{B}$ using the LFR model is more complicated since for $\mathbf{X}$ in (4.4.12), $\left(\mathbf{X}'\mathbf{X}\right)^{-1}$ does not exist and thus, the estimate for $\mathbf{B}$ is no longer unique. Using Theorem 2.5.5, a g-inverse for $\mathbf{X}'\mathbf{X}$ is

$$\left(\mathbf{X}'\mathbf{X}\right)^{-} = \begin{bmatrix} 0 & 0 & \cdots & 0 \\ 0 & 1/n_1 & \cdots & 0 \\ \vdots & \vdots & & \vdots \\ 0 & 0 & \cdots & 1/n_k \end{bmatrix} \tag{4.4.14}$$

so that

$$\widehat{\mathbf{B}} = \left(\mathbf{X}'\mathbf{X}\right)^{-} \mathbf{X}'\mathbf{Y} = \begin{bmatrix} \mathbf{0}' \\ \mathbf{y}'_{1.} \\ \vdots \\ \mathbf{y}'_{k.} \end{bmatrix} \tag{4.4.15}$$

which, because of the g-inverse selected, is similar to (4.4.15). Observe that the parameter $\boldsymbol{\mu}$ is not estimable.

Extending Theorem 2.6.2 to the one-way MANOVA model, we consider linear parametric functions $\psi = \mathbf{c}'\mathbf{B}\mathbf{m}$ such that $\mathbf{c}'\mathbf{H} = \mathbf{c}'$ for $\mathbf{H} = \left(\mathbf{X}'\mathbf{X}\right)^{-}\mathbf{X}'\mathbf{X}$ and arbitrary vectors $\mathbf{m}$. Then, estimable functions of ψ have the general form

$$\psi = \mathbf{c}'\mathbf{B}\mathbf{m} = \mathbf{m}'\left(\sum_{i=1}^{k} t_i\,\boldsymbol{\mu} + \sum_{i=1}^{k} t_i\,\boldsymbol{\alpha}_i\right) \tag{4.4.16}$$

and are estimated by

$$\widehat{\psi} = \mathbf{c}'\widehat{\mathbf{B}}\mathbf{m} = \mathbf{m}'\left(\sum_{i=1}^{k} t_i \mathbf{y}_{i.}\right) \tag{4.4.17}$$

for arbitrary vector $\mathbf{t}' = [t_0, t_1, \ldots, t_k]$. By (4.4.16), $\boldsymbol{\mu}$ and the $\boldsymbol{\alpha}_i$ are not estimable; however, all contrasts in the effects vector $\boldsymbol{\alpha}_i$ are estimable. Because $\mathbf{X}\left(\mathbf{X}'\mathbf{X}\right)^{-}\mathbf{X}\mathbf{X}'$ is unique for any g-inverse, the unbiased estimate of Σ for the FR and LFR models are identical.

For the LFR model, the parameter vector $\boldsymbol{\mu}$ has no "natural" interpretation. To give meaning to $\boldsymbol{\mu}$, and to make it estimable, many texts and computer software packages add side conditions or restrictions to the rows of $\mathbf{B}$ in (4.4.6). This converts a LFR model to a model of full rank making all parameters estimable. However, depending on the side conditions chosen, the parameters $\boldsymbol{\mu}$ and $\boldsymbol{\alpha}_i$ have different estimates and hence different interpretations. For example, if one adds the restriction that the $\sum_i \boldsymbol{\alpha}_i = \mathbf{0}$, then $\boldsymbol{\mu}$ is estimated as an unweighted average of the sample mean vectors $\mathbf{y}_{i.}$. If the condition that the $\sum_i n_i\boldsymbol{\alpha}_i = \mathbf{0}$ is selected, then $\boldsymbol{\mu}$ is estimated by a weighted average of the vectors $\mathbf{y}_{i.}$. Representing these two estimates by $\widehat{\boldsymbol{\mu}}_u$ and $\widehat{\boldsymbol{\mu}}_w$, respectively, the parameter estimates for $\boldsymbol{\mu}$ become

$$\widehat{\boldsymbol{\mu}}_u = \sum_{i=1}^{k} \mathbf{y}_i/k = \mathbf{y}_{..} \quad \text{and} \quad \widehat{\boldsymbol{\mu}}_w = \sum_{i=1}^{k} n_i\mathbf{y}_{i.}/k = \overline{\mathbf{y}}_{..} \tag{4.4.18}$$

Now $\boldsymbol{\mu}$ may be interpreted as an overall mean and the effects also become estimable in that

$$\widehat{\boldsymbol{\alpha}}_i = \mathbf{y}_{i.} - \mathbf{y}_{..} \quad \text{or} \quad \widehat{\boldsymbol{\alpha}}_i = \mathbf{y}_{i.} - \overline{\mathbf{y}}_{..} \tag{4.4.19}$$

depending on the weights (restrictions). Observe that one may not interpret $\widehat{\boldsymbol{\alpha}}_i$ unless one knows the "side conditions" used in the estimation process. This ambiguity about the estimates of model parameters is avoided with either the FR cell means model or the overparameterized LFR model. Not knowing the side conditions in more complex designs leads to confusion regarding both parameter estimates and tests of hypotheses.

The SAS procedure GLM allows one to estimate **B** using either the cell means FR model or the LFR model. The default model in SAS is the LFR model; to obtain a FR model the option / NOINT is used on the MODEL statement. To obtain the general form of estimable functions for the LFR solution to the MANOVA design, the option / E is used in the MODEL statement.

The primary hypothesis of interest for the one-way FR MANOVA design is that the k treatment mean vectors, $\boldsymbol{\mu}_i$, are equal

$$H : \boldsymbol{\mu}_1 = \boldsymbol{\mu}_2 = \ldots = \boldsymbol{\mu}_k \tag{4.4.20}$$

For the LFR model, the equivalent hypothesis is the equality of the treatment effects

$$H : \boldsymbol{\alpha}_1 = \boldsymbol{\alpha}_2 = \ldots = \boldsymbol{\alpha}_k \tag{4.4.21}$$

or equivalently that

$$H : \text{all}\, \boldsymbol{\alpha}_i - \boldsymbol{\alpha}_{i'} = \mathbf{0} \tag{4.4.22}$$

for all $i \neq i'$. The hypothesis in (4.4.21) is testable if and only if the contrasts $\boldsymbol{\psi} = \boldsymbol{\alpha}_i - \boldsymbol{\alpha}_{i'}$ are estimable. Using (4.4.21) and (4.4.16), it is easily shown that contrasts in the $\boldsymbol{\alpha}_i$ are estimable so that H in (4.4.21) is testable. This complication is avoided in the FR model since the $\boldsymbol{\mu}_i$ and contrasts of the $\boldsymbol{\mu}_i$ are estimable. In LFR models, individual $\boldsymbol{\alpha}_i$ are not estimable, only contrasts of the $\boldsymbol{\alpha}_i$ are estimable and hence testable. Furthermore, contrasts of $\boldsymbol{\alpha}_i$ do not depend on the g-inverse selected to estimate **B**.

To test either (4.4.20) or (4.4.21), one must again construct matrices **C** and **M** to transform the overall test of the parameters in **B** to the general form $\mathbf{CBM} = \mathbf{0}$. The matrices **H** and **E** have the structure given in (3.6.26). If **X** is not of full rank, $(\mathbf{X'X})^{-1}$ is replaced by any g-inverse $(\mathbf{X'X})^{-}$. To illustrate, we use a simple example for $k = 3$ treatments and $p = 3$ dependent variables. Then the FR and LFR matrices for **B** are

$$\begin{aligned}
\mathbf{B}_{FR} &= \begin{bmatrix} \mu_{11} & \mu_{12} & \mu_{13} \\ \mu_{21} & \mu_{22} & \mu_{23} \\ \mu_{31} & \mu_{32} & \mu_{33} \end{bmatrix} = \begin{bmatrix} \boldsymbol{\mu}'_1 \\ \boldsymbol{\mu}'_2 \\ \boldsymbol{\mu}'_3 \end{bmatrix} \quad \text{or} \\
\mathbf{B}_{LFR} &= \begin{bmatrix} \mu_1 & \mu_2 & \mu_3 \\ \alpha_{11} & \alpha_{12} & \alpha_{13} \\ \alpha_{21} & \alpha_{22} & \alpha_{23} \\ \alpha_{31} & \alpha_{32} & \alpha_{33} \end{bmatrix} = \begin{bmatrix} \boldsymbol{\mu}' \\ \boldsymbol{\alpha}'_1 \\ \boldsymbol{\alpha}'_2 \\ \boldsymbol{\alpha}'_3 \end{bmatrix}
\end{aligned} \tag{4.4.23}$$

To test for differences in treatments, the matrix $\mathbf{M} = \mathbf{I}$ and the contrast matrix $\mathbf{C}$ has the form

$$\mathbf{C}_{FR} = \begin{bmatrix} 1 & 0 & -1 \\ 0 & 1 & -1 \end{bmatrix} \quad \text{or} \quad \mathbf{C}_{LFR} = \begin{bmatrix} 0 & 1 & 0 & -1 \\ 0 & 0 & 1 & -1 \end{bmatrix} \tag{4.4.24}$$

so that in either case, $\mathbf{C}$ has full row rank $v_h = k - 1$. When SAS calculates the hypothesis test matrix $\mathbf{H}$ in PROC GLM, it does not evaluate

$$\mathbf{H} = (\mathbf{C}\widehat{\mathbf{B}}\mathbf{M})' \left(\mathbf{C}\left(\mathbf{X}'\mathbf{X}\right)^{-} \mathbf{C}'\right)^{-1} (\mathbf{C}\widehat{\mathbf{B}}\mathbf{M})$$

directly. Instead, the MANOVA statement can be used with the specification H = TREAT where the treatment factor name is assigned in the CLASS statement and the hypothesis test matrix is constructed by employing the reduction procedure discussed in (4.2.18). To see this, let

$$\mathbf{B}_{LFR} = \begin{bmatrix} \boldsymbol{\mu}' \\ \cdots \\ \boldsymbol{\alpha}_1' \\ \boldsymbol{\alpha}_2' \\ \vdots \\ \boldsymbol{\alpha}_k' \end{bmatrix} = \begin{bmatrix} \mathbf{B}_1 \\ \cdots \\ \mathbf{B}_2 \end{bmatrix}$$

so that the full model Ω_o becomes

$$\begin{aligned} \Omega_o : \mathbf{Y} &= \mathbf{X}\mathbf{B} + \mathbf{E} \\ &= \mathbf{X}_1\mathbf{B}_1 + \mathbf{X}_2\mathbf{B}_2 + \mathbf{E} \end{aligned}$$

To test $\boldsymbol{\alpha}_1 = \boldsymbol{\alpha}_2 = \ldots = \boldsymbol{\alpha}_k$ we set each $\boldsymbol{\alpha}_i$ equal to $\boldsymbol{\alpha}_0$ say so that $\mathbf{y}_{ij} = \boldsymbol{\mu} + \boldsymbol{\alpha}_0 + \mathbf{e}_{ij} = \boldsymbol{\mu}_0 + \mathbf{e}_{ij}$ is the reduced model with design matrix $\mathbf{X}_1$ so that fitting $\mathbf{y}_{ij} = \boldsymbol{\mu}_0 + \boldsymbol{\alpha}_0 + \mathbf{e}_{ij}$ is equivalent to fitting the model $\mathbf{y}_{ij} = \boldsymbol{\mu}_0 + \mathbf{e}_{ij}$. Thus, to obtain the reduced model from the full model we may set all $\boldsymbol{\alpha}_i = \mathbf{0}$. Now if all $\boldsymbol{\alpha}_i = \mathbf{0}$ the reduced model is $\omega : \mathbf{Y} = \mathbf{X}_1\mathbf{B}_1 + \mathbf{E}$ and $R(\mathbf{B}_1) = \mathbf{Y}'\mathbf{X}_1\left(\mathbf{X}_1'\mathbf{X}_1\right)^{-} \mathbf{X}_1'\mathbf{Y} = \widehat{\mathbf{B}}_1'\left(\mathbf{X}_1'\mathbf{X}_1\right)\widehat{\mathbf{B}}_1$. For the full model Ω_o, $R(\mathbf{B}_1, \mathbf{B}_2) = \mathbf{Y}'\mathbf{X}\left(\mathbf{X}'\mathbf{X}\right)^{-}\mathbf{X}'\mathbf{Y} = \widehat{\mathbf{B}}'\left(\mathbf{X}'\mathbf{X}\right)\widehat{\mathbf{B}}$ so that

$$\begin{aligned} \mathbf{H} = \mathbf{H}_\Omega - \mathbf{H}_\omega &= R(\mathbf{B}_2 \mid \mathbf{B}_1) \\ &= R(\mathbf{B}_1, \mathbf{B}_2) - R(\mathbf{B}_1) \\ &= \mathbf{Y}'\mathbf{X}\left(\mathbf{X}'\mathbf{X}\right)^{-} \mathbf{X}'\mathbf{Y} - \mathbf{Y}'\mathbf{X}_1\left(\mathbf{X}_1'\mathbf{X}_1\right)^{-} \mathbf{X}_1'\mathbf{Y} \\ &= \sum_{i=1}^{k} n_i \mathbf{y}_{i.}\mathbf{y}_{i.}' - n\overline{\mathbf{y}}_{..}\overline{\mathbf{y}}_{..}' \\ &= \sum_{i=1}^{l} n_i\left(\mathbf{y}_{i.} - \overline{\mathbf{y}}_{..}\right)\left(\mathbf{y}_{i.} - \overline{\mathbf{y}}_{..}\right)' \end{aligned} \tag{4.4.25}$$

for the one-way MANOVA. The one-way MANOVA table is given in Table 4.4.1.

TABLE 4.4.1. One-Way MANOVA Table

Source	df	$SSCP$	$E\,(SSCP)$
Between	$k-1$	$\mathbf{H}$	$(k-1)\,\Sigma+\Gamma\Sigma$
Within	$n-k$	$\mathbf{E}$	$(n-k)\,\Sigma$
"Total"	$n-1$	$\mathbf{H}+\mathbf{E}$	

The parameter matrix Γ for the FR model is the noncentrality parameter of the Wishart distribution obtained from $\mathbf{H}$ in (4.4.25) by replacing sample estimates with population parameters. That is,

$$\Gamma=\left[\sum_{i=1}^{k}n_i\left(\boldsymbol{\mu}_i-\boldsymbol{\mu}\right)\left(\boldsymbol{\mu}_i-\boldsymbol{\mu}\right)'\right]\Sigma^{-1}.$$

To obtain $\mathbf{H}$ and $\mathbf{E}$ in SAS for the test of no treatment differences, the following commands are used for our example with $p=3$.

FR Model	proc glm;
	class treat;
	model $y1-y3=$ treatment / noint;
	manova $h=$ treat / printe printh;

LFR Model	proc glm; class treat; model $y1-y3=$ treat / e; manova $h=$ treat / printe printh;

In the MODEL statement the variable names for the dependent variables are y_1, y_2, and y_3. The name given the independent classification variable is 'treat'. The PRINTE and PRINTH options on the MANOVA statement directs SAS to print the hypothesis test matrix $\mathbf{H}$ (the hypothesis SSCP matrix) and the error matrix $\mathbf{E}$ (the error SSCP matrix) for the null hypothesis of no treatment effect. With $\mathbf{H}$ and $\mathbf{E}$ calculated, the multivariate criteria again depend on solving $|\mathbf{H} - \lambda \mathbf{E}| = 0$. The parameters for the one-way MANOVA design are

$$\begin{aligned} s &= \min(v_h, u) = \min(k-1, p) \\ M &= (|u - v_e| - 1)/2 = (k - p - 1)/2 \\ N &= (v_e - u - 1)/2 = (n - k - p - 1)/2 \end{aligned} \tag{4.4.26}$$

where $u = r(\mathbf{M}) = p$, $v_h = k - 1$ and $v_e = n - k$.

Because $\boldsymbol{\mu}$ is not estimable in the LFR model, it is not testable. If there were no treatment effect, however, one may fit a mean only model to the data, $\mathbf{y}_{ij} = \boldsymbol{\mu} + \mathbf{e}_{ij}$. Assuming a model with only a mean, we saw that $\boldsymbol{\mu}$ is estimated using unweighted or weighted estimates represented as $\widehat{\boldsymbol{\mu}}_u$ and $\widehat{\boldsymbol{\mu}}_w$. To estimate these parameters in SAS, one would specify Type III estimable functions for unweighted estimates or Type I estimable functions for weighted estimates. While Type I estimates always exist, Type III estimates are only provided with designs that have no empty cells. Corresponding to these estimable functions are **H** matrices and **E** matrices. There are two types of hypotheses for the mean only model; the Type I hypothesis is testing $H_w : \boldsymbol{\mu} = \sum_{i=1}^{n} n_i \boldsymbol{\mu}_i = \mathbf{0}$ and the Type III hypotheses is testing $H_u : \boldsymbol{\mu} = \sum_{i=1}^{n} \boldsymbol{\mu}_i = \mathbf{0}$. To test these in SAS using PROC GLM, one would specify h = INTERCEPT on the MANOVA statement and use the HTYPE $= n$ option where $n = 1$ or 3. Thus, to perform tests on $\boldsymbol{\mu}$ in SAS using PROC GLM for the LFR model, one would have the following statements.

	proc	glm;
	class	treat;
LFR Model	model	$y_1 - y_3 =$ treat / e;
	manova	h = treat / printe printh;
	manova	h = intercept / printe printh htype = 1;
	manova	h = intercept / printe printh htype = 3;

While PROC GLM uses the g-inverse approach to analyze fixed effect MANOVA and MANCOVA designs, it provides for other approaches to the analysis of these designs by the calculation of four types of estimable functions and four types of hypothesis test matrices. We saw the use of the Type I and Type III options in testing the significance of the intercept. SAS also provides Type II and Type IV estimates and tests. Goodnight (1978), Searle (1987), and Littell, Freund and Spector (1991) provide an extensive and detail discussion of the univariate case while Milliken and Johnson (1992) illustrate the procedures using many examples. We will discuss the construction of Type IV estimates and associated tests in Section 4.10 when we discuss nonorthogonal designs.

Analysis of MANOVA designs is usually performed using full rank models with restrictions supplied by the statistical software or input by the user, or by using less than full rank models. No solution to the analysis of MANOVA designs is perfect. Clearly, fixed effect designs with an equal number of observations per cell are ideal and easy to analyze; in the SAS software system PROC ANOVA may be used for such designs. The ANOVA procedure uses unweighted side conditions to perform the analysis. However, in most real world applications one does not have an equal number of observations per cell. For these situations, one has two choices, the FR model or the LFR model. Both approaches have complications that are not easily addressed. The FR model works best in designs that require no restrictions on the population cell means. However, as soon as another factor is introduced into the design restrictions must be added to perform the correct analysis. As designs become more complex so do the restrictions. We have discussed these approaches in Timm and Mieczkowski (1997). In this text we will use either the FR cell means model with no restrictions, or the LFR model.

b. One-Way MANCOVA

Multivariate analysis of covariance (MANCOVA) models are a combination of MANOVA and MR models. Subjects in the one-way MANCOVA design are randomly assigned to k treatments and n_i vectors with p responses are observed. In addition to the vector of dependent variables for each subject, a vector of h fixed or random independent variables, called covariates, are obtained for each subject. These covariates are assumed to be measured without error, and they are believed to be related to the dependent variables and to represent a source of variation that has not been controlled for by the study design to represent a source of variation that has not been controlled in the study. The goal of having covariates in the model is to reduce the determinant of the error covariance matrix and hence increase the precision of the design.

For a fixed set of h covariates, the MANCOVA model may be written as

$$\begin{aligned}
\underset{n\times p}{\mathbf{Y}} &= \underset{n\times q}{\mathbf{X}}\ \underset{q\times p}{\mathbf{B}} + \underset{n\times h}{\mathbf{Z}}\ \underset{h\times p}{\Gamma} + \underset{n\times p}{\mathbf{E}} \\
&= [\mathbf{X}, \mathbf{Z}]\begin{bmatrix} \mathbf{B} \\ \Gamma \end{bmatrix} + \mathbf{E} \\
&= \underset{n\times(q+h)}{\mathbf{A}}\ \underset{(q+h)\times p}{\Theta} + \underset{n\times p}{\mathbf{E}}
\end{aligned} \tag{4.4.27}$$

where $\mathbf{X}$ is the MANOVA design matrix and $\mathbf{Z}$ is the matrix from the MR model containing h covariates. The MANOVA design matrix $\mathbf{X}$ is usually not of full rank, $r(\mathbf{X}) = r < q$, and the matrix $\mathbf{Z}$ of covariates is of full rank h, $r(\mathbf{Z}) = h$.

To find $\widehat{\Theta}$ in (4.4.27), we apply property (6) of Theorem 2.5.5 where

$$\mathbf{A}'\mathbf{A} = \begin{bmatrix} \mathbf{X}'\mathbf{X} & \mathbf{X}'\mathbf{Z} \\ \mathbf{Z}'\mathbf{X} & \mathbf{Z}'\mathbf{Z} \end{bmatrix}$$

Then

$$(\mathbf{A}'\mathbf{A})^{-} = \begin{bmatrix} (\mathbf{X}'\mathbf{X})^{-} & \mathbf{0} \\ \mathbf{0} & \mathbf{0} \end{bmatrix} + \begin{bmatrix} (\mathbf{X}'\mathbf{X})^{-}\mathbf{X}'\mathbf{Z} \\ \mathbf{I} \end{bmatrix} [\ \mathbf{Z}'\mathbf{Q}\mathbf{Z}\ ]^{-} \left[-\mathbf{Z}'\mathbf{X}\left(\mathbf{X}'\mathbf{X}\right)^{-}, \mathbf{I} \right]$$

with $\mathbf{Q}$ defined as $\mathbf{Q} = \mathbf{I} - \mathbf{X}\left(\mathbf{X}'\mathbf{X}\right)^{-}\mathbf{X}'$, we have

$$\widehat{\Theta} = \begin{bmatrix} \widehat{\mathbf{B}} \\ \widehat{\Gamma} \end{bmatrix} = \begin{bmatrix} (\mathbf{X}'\mathbf{X})^{-}\mathbf{X}'(\mathbf{Y} - \mathbf{Z}\widehat{\Gamma}) \\ (\mathbf{Z}'\mathbf{Q}\mathbf{Z})^{-}\mathbf{Z}'\mathbf{Q}\mathbf{Y} \end{bmatrix} \tag{4.4.28}$$

as the least squares estimates of $\widehat{\Theta}$. $\widehat{\Gamma}$ is unique since $\mathbf{Z}$ has full column rank, $\left(\mathbf{Z}'\mathbf{Q}\mathbf{Z}\right)^{-} = \left(\mathbf{Z}'\mathbf{Q}\mathbf{Z}\right)^{-1}$. From (4.4.28) observe that the estimate $\widehat{\mathbf{B}}$ in the MANOVA model is adjusted by the covariates multiplied by $\widehat{\Gamma}$. Thus, in MANCOVA models we are interested in differences in treatment effects adjusted for covariates. Also observe that the matrix Γ is common to all treatments. This implies that $\mathbf{Y}$ and $\mathbf{X}$ are linearly related with a common regression matrix Γ across the k treatments. This is a model assumption that may be tested. In addition, we may test for no association between the dependent variables y and the independent variables z or that $\Gamma = \mathbf{0}$. We can also test for differences in adjusted treatment means.

To estimate Σ given $\widehat{\mathbf{B}}$ and $\widehat{\Gamma}$, we define the error matrix for the combined vector $\binom{\mathbf{y}}{\mathbf{z}}$ as

$$\mathbf{E} = \begin{bmatrix} \mathbf{Y}'\mathbf{Q}\mathbf{Y} & \mathbf{Z}'\mathbf{Q}\mathbf{Y} \\ \mathbf{Y}'\mathbf{Q}\mathbf{Z} & \mathbf{Z}'\mathbf{Q}\mathbf{Z} \end{bmatrix} = \begin{bmatrix} \mathbf{E}_{yy} & \mathbf{E}_{yz} \\ \mathbf{E}_{zy} & \mathbf{E}_{zz} \end{bmatrix} \tag{4.4.29}$$

Then, the error matrix for $\mathbf{Y}$ given $\mathbf{Z}$ is

$$\begin{aligned} \mathbf{E}_{y|z} &= \mathbf{Y}'\left\{ (\mathbf{I} - [\mathbf{X}, \mathbf{Z}]) \begin{bmatrix} \mathbf{X}'\mathbf{X} & \mathbf{X}'\mathbf{Z} \\ \mathbf{Z}'\mathbf{X} & \mathbf{Z}'\mathbf{Z} \end{bmatrix}^{-} \begin{bmatrix} \mathbf{X}' \\ \mathbf{Z}' \end{bmatrix} \right\} \mathbf{Y} \\ &= \mathbf{Y}'\mathbf{Q}\mathbf{Y} - \mathbf{Y}'\mathbf{Q}\mathbf{Z}\ \left(\mathbf{Z}'\mathbf{Q}\mathbf{Z}\right)^{-1}\mathbf{Z}'\mathbf{Q}\mathbf{Y} \\ &= \mathbf{E}_{yy} - \mathbf{E}_{yz}\ \mathbf{E}_{zz}^{-1}\ \mathbf{E}_{zy} \\ &= \mathbf{Y}'\mathbf{Q}\mathbf{Y} - \widehat{\Gamma}'\left(\mathbf{Z}'\mathbf{Q}\mathbf{Z}\right)\widehat{\Gamma} \end{aligned} \tag{4.4.30}$$

To obtain an unbiased estimate of Σ, $\mathbf{E}_{y|z}$ is divided by the $r\,(\mathbf{A}) = n - r - h$

$$\mathbf{S}_{y|z} = \mathbf{E}_{y|z} / \left(n - r - h\right) \tag{4.4.31}$$

The matrix $\widehat{\Gamma}'\left(\mathbf{Z}'\mathbf{Q}\mathbf{Z}\right)\widehat{\Gamma} = \mathbf{E}_{yz}\mathbf{E}_{zz}^{-1}\mathbf{E}_{zy}$ is the regression SSCP matrix for the MR model $\mathbf{Y} = \mathbf{Q}\mathbf{Z}\Gamma + \mathbf{E}$. Thus, to test $H : \Gamma = \mathbf{0}$, or that the covariates have no effect on $\mathbf{Y}$, the hypothesis test matrix is

$$\mathbf{H} = \mathbf{E}_{yz}\mathbf{E}_{zz}^{-1}\mathbf{E}_{zy} = \widehat{\Gamma}'\left(\mathbf{Z}'\mathbf{Q}\mathbf{Z}\right)\widehat{\Gamma} \tag{4.4.32}$$

where the $r(\mathbf{H}) = h$. The error matrix for the test is defined in (4.4.29). The Λ criterion for the test that $\Gamma = \mathbf{0}$ is

$$\Lambda = \frac{\left|\mathbf{E}_{y|z}\right|}{\left|\mathbf{H} + \mathbf{E}_{y|z}\right|} = \frac{\left|\mathbf{E}_{yy} - \mathbf{E}_{yx}\mathbf{E}_{xx}^{-1}\mathbf{E}_{xy}\right|}{\mathbf{E}_{yy}} \tag{4.4.33}$$

where $v_h = h$ and $v_e = n - r - h$. The parameters for the other test criteria are

$$\begin{aligned} s &= \min\left(v_h,\ p\right) = \min\left(h,\ p\right) \\ M &= \left(\left|p - v_h\right| - 1\right)/2 = \left(\left|p - h\right| - 1\right)/2 \\ N &= \left(v_e - p - 1\right)/2 = \left(n - r - h - p - 1\right)/2 \end{aligned} \tag{4.4.34}$$

To test $\Gamma = \mathbf{0}$ using SAS, one must use the MTEST statement in PROC REG. Using SAS Version 8, the test may not currently be tested using PROC GLM.

To find a general expression for testing the hypotheses regarding $\mathbf{B}$ in the matrix Θ is more complicated. By replacing $\mathbf{X}$ in (3.6.26) with the partitioned matrix $[\mathbf{X}, \mathbf{Z}]$ and finding a g-inverse, the general structure of the hypothesis test matrix for hypothesis $\mathbf{CBM} = \mathbf{0}$ is

$$\mathbf{H} = (\mathbf{C}\widehat{\mathbf{B}}\mathbf{M})'\left[\mathbf{C}\left(\mathbf{X}'\mathbf{X}\right)^{-}\mathbf{C}' + \mathbf{C}\left(\mathbf{X}'\mathbf{X}\right)^{-}\mathbf{X}'\mathbf{Z}\left(\mathbf{Z}'\mathbf{Q}\mathbf{Z}\right)^{-1}\mathbf{Z}'\mathbf{X}\left(\mathbf{X}'\mathbf{X}\right)^{-}\mathbf{C}'\right]^{-1}(\mathbf{C}\widehat{\mathbf{B}}\mathbf{M}) \tag{4.4.35}$$

where $v_h = r(\mathbf{C}) = g$. The error matrix is defined in (4.4.29) and $v_e = n - r - h$. An alternative approach for determining $\mathbf{H}$ is to fit a full model (Ω_o) given in (4.4.26) and the reduced model under the hypothesis (ω). Then $\mathbf{H} = \mathbf{E}_\omega - \mathbf{E}_\Omega$. Given a matrix $\mathbf{H}$ and the matrix $\mathbf{E}_{y|z} = \mathbf{E}_{\Omega_o}$, the test criteria depend on the roots of $\left|\mathbf{H} - \lambda\mathbf{E}_{y|z}\right| = 0$. The parameters for testing $H : \mathbf{CBM} = \mathbf{0}$ are

$$\begin{aligned} s &= \min\left(g,\ u\right) \\ M &= \left(\left|u - g\right| - 1\right)/2 \\ N &= \left(n - r - h - g - 1\right)/2 \end{aligned} \tag{4.4.36}$$

where $u = r(\mathbf{M})$. As in the MR model, $\mathbf{Z}$ may be fixed or random.

Critical to the application of the MANCOVA model is the parallelism assumption that $\Gamma_1 = \Gamma_2 = \ldots = \Gamma_I = \Gamma$. To develop a test of parallelism, we consider an I group multivariate intraclass covariance model

$$\begin{aligned} \Omega_o : \quad \begin{bmatrix} \mathbf{Y}_1 \\ \mathbf{Y}_2 \\ \vdots \\ \mathbf{Y}_I \end{bmatrix} &= \begin{bmatrix} \mathbf{X}_1 & \mathbf{Z}_1 & \mathbf{0} & \cdots & \mathbf{0} \\ \mathbf{X}_2 & \mathbf{0} & \mathbf{Z}_2 & \cdots & \mathbf{0} \\ \vdots & \vdots & \vdots & & \vdots \\ \mathbf{X}_I & \mathbf{0} & \mathbf{0} & \cdots & \mathbf{Z}_I \end{bmatrix} \begin{bmatrix} \mathbf{B} \\ \Gamma_1 \\ \Gamma_2 \\ \vdots \\ \Gamma_I \end{bmatrix} + \begin{bmatrix} \mathbf{E}_1 \\ \mathbf{E}_2 \\ \vdots \\ \mathbf{E}_I \end{bmatrix} \\ (n \times p) \qquad & \qquad [n \times Iq^*] \qquad\qquad [Iq^* \times p] \qquad (n \times p) \\ \mathbf{Y} \quad &= \quad [\mathbf{X}, \mathbf{F}] \begin{bmatrix} \mathbf{B} \\ \Delta \end{bmatrix} + \mathbf{E} \end{aligned} \tag{4.4.37}$$

where $q^* = q + h = k^* + 1$ so that the matrices Γ_i vary across the I treatments. If

$$H : \Gamma_1 = \Gamma_2 = \ldots = \Gamma_I = \Gamma \tag{4.4.38}$$

then (4.4.37) reduces to the MANCOVA model

$$\omega : \quad \begin{bmatrix} \mathbf{Y}_1 \\ \mathbf{Y}_2 \\ \vdots \\ \mathbf{Y}_I \end{bmatrix} = \begin{bmatrix} \mathbf{X}_1 & \mathbf{Z}_1 \\ \mathbf{X}_2 & \mathbf{Z}_2 \\ \vdots & \vdots \\ \mathbf{X}_I & \mathbf{Z}_I \end{bmatrix} \begin{bmatrix} \mathbf{B} \\ \Gamma \end{bmatrix} + \begin{bmatrix} \mathbf{E}_1 \\ \mathbf{E}_2 \\ \vdots \\ \mathbf{E}_I \end{bmatrix}$$

$$\mathbf{Y} = [\mathbf{X}, \mathbf{Z}] \begin{bmatrix} \mathbf{B} \\ \Gamma \end{bmatrix} + \mathbf{E}$$

$$\underset{n\times p}{\mathbf{Y}} = \underset{n\times q^*}{\mathbf{A}} \; \underset{q^*\times p}{\Theta} + \underset{n\times p}{\mathbf{E}}$$

To test for parallelism given (4.4.37), we may use (3.6.26). Then we would estimate the error matrix under Ω_o, $\mathbf{E}_{\Omega_o}$, and define $\mathbf{C}$ such that $\mathbf{C}\Delta = \mathbf{0}$. Using this approach,

$$\underset{h(I-1)\times hI}{\mathbf{C}} = \begin{bmatrix} \mathbf{I}_h & \mathbf{0} & \cdots & -\mathbf{I}_h \\ \mathbf{0} & \mathbf{I}_h & \cdots & -\mathbf{I}_h \\ \vdots & \vdots & & \vdots \\ \mathbf{0}_k & \mathbf{0} & \cdots & -\mathbf{I}_h \end{bmatrix}$$

where the $r(\mathbf{C}) = v_h = h(I-1)$. Alternatively, $\mathbf{H}$ may be calculated as in the MR model in (4.2.15) for testing $\mathbf{B}_2 = \mathbf{0}$. Then $\mathbf{H} = \mathbf{E}_\omega - \mathbf{E}_{\Omega_o}$. Under ω, $\mathbf{E}_\omega$ is defined in (4.4.29). Hence, we merely have to find $\mathbf{E}_{\Omega_o}$. To find $\mathbf{E}_{\Omega_o}$, we may again use (4.4.29) with $\mathbf{Z}$ replaced by $\mathbf{F}$ in (4.4.37) and $v_e = n - r(\mathbf{X}, \mathbf{F}) = n - Iq^* = n - r(\mathbf{X}) - Ih$. Alternatively, observe that (4.4.37) represents I independent MANCOVA models so that $\widehat{\Gamma}_i = \left(\mathbf{Z}_i'\mathbf{Q}_i\mathbf{Z}_i\right)^{-1}\mathbf{Z}_i'\mathbf{Q}_i\mathbf{Y}_i$ where $\mathbf{Q}_i = \mathbf{I} - \mathbf{X}_i\left(\mathbf{X}_i'\mathbf{X}_i\right)^{-}\mathbf{X}_i'$. Pooling across the I groups,

$$\begin{aligned} \mathbf{E}_{\Omega_o} &= \sum_{i=1}^{I} (\mathbf{Y}_i'\mathbf{Q}_i\mathbf{Y}_i - \widehat{\Gamma}_i'\left(\mathbf{Z}_i'\mathbf{Q}_i\mathbf{Z}_i\right)\widehat{\Gamma}) \\ &= \mathbf{Y}'\mathbf{Q}\mathbf{Y} \sum_{i=1}^{I} \widehat{\Gamma}_i'\left(\mathbf{Z}_i'\mathbf{Q}_i\mathbf{Z}_i\right)\widehat{\Gamma}_i \end{aligned} \tag{4.4.39}$$

To test for parallelism, or no covariate by treatment interaction, Wilks' Λ criterion is

$$\Lambda = \frac{|\mathbf{E}_{\Omega_o}|}{|\mathbf{E}_\omega|} = \frac{|\mathbf{E}_{yy} - \sum_{i=1}^{I}\widehat{\Gamma}_i'\left(\mathbf{Z}_i'\mathbf{Q}_i\mathbf{Z}_i\right)\widehat{\Gamma}_i|}{|\mathbf{E}_{y|z}|} \tag{4.4.40}$$

with degrees of freedom $v_h = h(I-1)$ and $v_e = n - q - Ih$. Other criteria may also be used.

The one-way MANCOVA model assumes that n_i subjects are assigned to k treatments where the p-variate vector of dependent variables has the FR and LFR linear model structures

$$\mathbf{y}_{ij} = \boldsymbol{\mu}_i + \Gamma'\mathbf{z}_{ij} + \mathbf{e}_{ij} \tag{4.4.41}$$

or the LFR structure

$$\mathbf{y}_{ij} = (\boldsymbol{\mu} + \boldsymbol{\alpha}_i) + \Gamma'\mathbf{z}_{ij} + \mathbf{e}_{ij} \tag{4.4.42}$$

for $i = 1, 2, \ldots, k$ and $j = 1, 2, \ldots, n_i$. The vectors $\mathbf{z}_{ij}$ are h-vectors of fixed covariates, $\Gamma_{h \times p}$ is a matrix of raw regression coefficients and the error vectors $\mathbf{e}_{ij} \sim IN_p(\mathbf{0}, \Sigma)$. As in the MR model, the covariates may also be stochastic or random; estimates and tests remain the same in either case.

Model (4.4.41) and (4.4.42) are the FR and LFR models, for the one-way MANCOVA model. As with the MANOVA design, the structure of the parameter matrix $\mathbf{A} = \left[\begin{smallmatrix}\mathbf{B}\\ \Gamma\end{smallmatrix}\right]$ depends on whether the FR or LFR model is used. The matrix Γ is the raw form of the regression coefficients. Often the covariates are centered by replacing $\mathbf{z}_{ij}$ with overall deviation scores of the form $\mathbf{z}_{ij} - \mathbf{z}_{..}$ where $\mathbf{z}_{..}$ is an unweighted average of the k treatment means $\mathbf{z}_{i.}$. The mean parameter $\boldsymbol{\mu}_i$ or $\boldsymbol{\mu} + \boldsymbol{\alpha}_i$ is estimated by $\mathbf{y}_{i.} - \widehat{\Gamma}'\mathbf{z}_{i.}$. Or, one may use the centered adjusted means

$$\mathbf{y}_{i.}^A = \widehat{\boldsymbol{\mu}}_i + \widehat{\Gamma}'\mathbf{z}_{..} = \mathbf{y}_{i.} - \widehat{\Gamma}'(\mathbf{z}_{i.} - \mathbf{z}_{..}) \tag{4.4.43}$$

These means are called adjusted least squares means (LSMEANS) in SAS. Most software package use the "unweighted" centered adjusted means in that $\mathbf{z}_{..}$ is used in place of $\overline{\mathbf{z}}_{..}$ even with unequal sample sizes.

Given multivariate normality, random assignment of n_i subjects to k treatments, and homogeneity of covariance matrices, one often tests the model assumption that the Γ_i are equal across the k treatments. This test of parallelism is constructed by evaluating whether or not there is a significant covariate by treatment interaction present in the design. If this test is significant, we must use the intraclass multivariate covariance model. For these models, treatment difference may only be evaluated at specified values of the covariate. When the test is not significant, one assumes all $\Gamma_i = \Gamma$ so that the MANCOVA model is most appropriate. Given the MANCOVA model, we first test $H : \Gamma = \mathbf{0}$ using PROC REG. If this test is not significant this means that the covariates do not reduce the determinant of Σ and thus it would be best to analyze the data using a MANOVA model rather than a MANCOVA model.

If $\Gamma \neq \mathbf{0}$, we may test for the significance of treatment difference using PROC GLM. In terms of the model parameters, the test has the same structure as the MANOVA test. The test written using the FR and LFR models follows.

$$\begin{aligned} H &: \boldsymbol{\mu}_1 = \boldsymbol{\mu}_2 = \ldots = \boldsymbol{\mu}_k \quad \text{(FR)} \\ H &: \boldsymbol{\alpha}_1 = \boldsymbol{\alpha}_2 = \ldots = \boldsymbol{\alpha}_k \quad \text{(LFR)} \end{aligned} \tag{4.4.44}$$

The parameter estimates $\widehat{\boldsymbol{\mu}}_i$ or contrasts in the $\boldsymbol{\alpha}_i$ now involve the h covariates and the matrix $\widehat{\Gamma}$. The estimable functions have the form

$$\begin{aligned} \psi &= \mathbf{m}'\left(\sum_{i=1}^{k} t_i\boldsymbol{\mu} + \sum_{i=1}^{k} t_i\boldsymbol{\alpha}_i\right) \\ \widehat{\psi} &= \mathbf{m}'\left\{\sum_{i=1}^{k} t_i[\mathbf{y}_{i.} - \widehat{\Gamma}'(\mathbf{z}_{i.} - \mathbf{z}_{..})]\right\} \end{aligned} \tag{4.4.45}$$

The hypotheses test matrix may be constructed using the reduction procedure.

With the rejection of hypotheses regarding Γ or treatment effects, we may again establish simultaneous confidence sets for estimable functions of **H**. General expressions for the covariance matrices follow

$$\begin{aligned}
\operatorname{cov}(\operatorname{vec}\widehat{\Gamma}) &= \Sigma \otimes \left(\mathbf{Z}'\mathbf{Q}\mathbf{Z}\right)^{-1} \\
\operatorname{var}(\mathbf{c}'\widehat{\Gamma}\mathbf{m}) &= \mathbf{m}'\Sigma\mathbf{m}(\mathbf{c}'\left(\mathbf{Z}'\mathbf{Q}\mathbf{Z}\right)^{-1}\mathbf{c}) \\
\operatorname{var}(\mathbf{c}'\widehat{\mathbf{B}}\mathbf{m}) &= \mathbf{m}'\Sigma\mathbf{m}[\mathbf{c}'\left(\mathbf{X}'\mathbf{X}\right)^{-}\mathbf{c} \\
&\quad + \mathbf{c}'\left(\mathbf{X}'\mathbf{X}\right)^{-}\mathbf{X}'\mathbf{Z}\left(\mathbf{Z}'\mathbf{Q}\mathbf{Z}\right)^{-1}\mathbf{Z}'\mathbf{X}\left(\mathbf{X}'\mathbf{X}\right)^{-}\mathbf{c}] \\
\operatorname{cov}(\mathbf{c}'\widehat{\mathbf{B}}\mathbf{m}, \mathbf{c}\widehat{\Gamma}\mathbf{m}) &= -\mathbf{m}\Sigma\mathbf{m}\left[\mathbf{c}'\left(\mathbf{X}'\mathbf{X}\right)^{-}\mathbf{X}'\mathbf{Z}\left(\mathbf{Z}'\mathbf{Q}\mathbf{Z}\right)^{-1}\mathbf{c}\right]
\end{aligned} \tag{4.4.46}$$

where Σ is estimated by $\mathbf{S}_{y|z}$.

c. *Simultaneous Test Procedures (STP) for One-Way MANOVA / MANCOVA*

With the rejection of the overall null hypothesis of treatment differences in either the MANOVA or MANCOVA designs, one knows there exists at least one parametric function $\psi = \mathbf{c}'\mathbf{B}\mathbf{m}$ that is significantly different from zero for some contrast vector **c** and an arbitrary vector **m**. Following the MR model, the 100 $(1-\alpha)$ % simultaneous confidence intervals have the general structure

$$\widehat{\psi} - c_\alpha\widehat{\sigma}_{\widehat{\psi}} \leq \psi \leq \widehat{\psi} + c_\alpha\widehat{\sigma}_{\widehat{\psi}} \tag{4.4.47}$$

where $\widehat{\psi} = \mathbf{c}'\widehat{\mathbf{B}}\mathbf{m}$ and $\widehat{\sigma}^2_{\widehat{\psi}} = \operatorname{var}(\mathbf{c}'\widehat{\mathbf{B}}\mathbf{m})$ is defined in (4.4.46). The critical constant, c^2_α, depends on the multivariate criterion used for the overall test of no treatment differences. For one-way MANOVA/MANCOVA designs, $\widehat{\psi}$ and $\widehat{\sigma}_{\widehat{\psi}}$ are easy to calculate given the structure of $(\mathbf{X}'\mathbf{X})$. This is not the case for more complicated designs. The ESTIMATE statement in PROC GLM calculates $\widehat{\psi}$ and $\widehat{\sigma}_{\widehat{\psi}}$ for each variable in the model. Currently, SAS does not generate simultaneous confidence intervals for ESTIMATE statements. Instead, a CONTRAST statement may be constructed to test that $H_o : \psi_i = 0$. If the overall test is rejected, one may evaluate each contrast at the nominal level used for the overall test to try to locate significant differences in the group means. SAS approximates the significance of each contrast using the F distribution. As in the test for evaluating the differences in means for the two group location problem, these tests are protected F-tests and may be evaluated using the nominal α level to determine whether any contrast is significant. To construct simultaneous confidence intervals, (4.2.60) must be used or an appropriate F-approximation. To evaluate the significance of a vector contrast $\boldsymbol{\psi} = \mathbf{c}'\mathbf{B}$, one may also use the approximate protected F approximations calculated in SAS. Again, each test is evaluated at the nominal level α when the overall test is rejected.

Instead of performing an overall test of treatment differences and investigating parametric functions of the form $\psi_i = \mathbf{c}_i'\mathbf{B}\mathbf{m}_i$ to locate significant treatment effects, one may, a priori, only want to investigate ψ_i for $i = 1, 2, \ldots, C$ comparisons. Then, the overall test

H : $\mathbf{CBM} = \mathbf{0}$ may be replaced by the null hypothesis $H = \bigcap_{i=1}^{C} H_i : \psi_i = 0$ for $i = 1, 2, \ldots, C$. The hypothesis of overall significance is rejected if at least one H_i is significant. When this is the goal of the study, one may choose from among several single-step STPs to test the null hypothesis; these include the Bonferroni t, Šidák independent t, and the Šidák multivariate t (Studentized maximum modulus procedure). These can be used to construct approximate 100 $(1 - \alpha)$ % simultaneous confidence intervals for the $i = 1, 2 \ldots, C$ contrasts $\psi_i = \mathbf{c}_i' \widehat{\mathbf{B}} \mathbf{m}_i$, Fuchs and Sampson (1987) and Hochberg and Tamhane (1987). Except for the multivariate t intervals, each confidence interval is usually constructed at some level $\alpha^* < \alpha$ to ensure that for all C comparisons the overall level is $\geq 1 - \alpha$. Fuchs and Sampson (1987) show that for $C \leq 30$ the Studentized maximum modulus intervals are "best" in the Neyman sense, the intervals are shortest and have the highest probability of leading to a significant finding that $\psi_i \neq 0$.

A procedure which is superior to any of these methods is the stepdown finite intersection test (FIT) procedure discussed by Krishnaiah (1979) and illustrated in some detail in Schmidhammer (1982) and Timm (1995). A limitation of the FIT procedure is that one must specify both the finite, specific comparisons ψ_i and the rank order of the importance of the dependent variables in $\mathbf{Y}_{n \times p}$ from 1 to p where 1 is the variable of most importance to the study and p is the variable of least importance. To develop the FIT procedure, we use the FR "cell means" MR model so that

$$
\begin{aligned}
\underset{n \times p}{\mathbf{Y}} &= \underset{n \times k}{\mathbf{X}} \underset{k \times p}{\mathbf{B}} + \mathbf{E} \\
\underset{k \times p}{\mathbf{B}} &= \begin{bmatrix} \boldsymbol{\mu}_1' \\ \boldsymbol{\mu}_2' \\ \vdots \\ \boldsymbol{\mu}_k' \end{bmatrix} = \left[\mu_{ij}\right] = \left[\mathbf{u}_1, \mathbf{u}_2, \ldots, \mathbf{u}_p\right]
\end{aligned}
\tag{4.4.48}
$$

where $E(\mathbf{Y}) = \mathbf{XB}$ and each row of $\mathbf{E}$ is MVN with mean $\mathbf{0}$ and covariance matrix Σ. For C specific treatment comparisons, we write the overall hypothesis H as

$$
\begin{aligned}
H &= \bigcap_{i=1}^{C} H_i \quad \text{where} \quad H_i : \psi_i = 0 \\
\psi_i &= \mathbf{c}_i' \mathbf{B} \mathbf{m}_i \qquad\qquad i = 1, 2, \ldots, C
\end{aligned}
\tag{4.4.49}
$$

where $\mathbf{c}_i' = [c_{i1}, c_{i2}, \ldots, c_{ik}]$ is a contrast vector so that the $\sum_{j=1}^{k} c_{ij} = 0$. In many applications, the vectors $\mathbf{m}_i$ are selected to construct contrasts a variable at a time so that $\mathbf{m}_i$ is an indicator vector $\mathbf{m}_j$ (say) that has a one in the jth position and zeros otherwise. For this case, (4.4.49) may be written as $H_{ij} : \theta_{ij} = \mathbf{c}_i' \mathbf{u}_j = 0$. Then, H becomes

$$
H : \bigcap_{i=1}^{C} \bigcap_{j=1}^{p} H_{ij} : \theta_{ij} = 0 \tag{4.4.50}
$$

To test the pC hypotheses H_{ij} simultaneously, the FIT principle is used. That is, F type statistics of the form

$$
F_{ij}^* = \widehat{\theta}_{ij}^2 / \widehat{\sigma}_{\widehat{\theta}_{ij}}
$$

are constructed. The hypothesis H_{ij} is accepted (<) or rejected (>) depending on whether $F_{ij} \lessgtr F_\alpha$ results such that the

$$P\left(F_{ij}^* \leq F_\alpha; i = 1, 2, \ldots, L \text{ and } j = 1, 2, \ldots, p \mid H\right) = 1 - \alpha \tag{4.4.51}$$

The joint distribution of the statistics F_{ij}^* is not multivariate F and involve nuisance parameters, Krishnaiah (1979). To test H_{ij} simultaneously, one could use the Studentized maximum modulus procedure. To remove the nuisance parameters Krishnaiah (1965a, 1965b, 1969) proposed a stepdown FIT procedure that is based on conditional distributions and an assumed decreasing order of importance of the p variables. Using the order of the p variables, let $\mathbf{Y} = [\mathbf{y}_1, \mathbf{y}_2, \ldots \mathbf{y}_p]$, $\mathbf{B} = [\boldsymbol{\beta}_1, \boldsymbol{\beta}_2, \ldots, \boldsymbol{\beta}_p]$, $\mathbf{Y}_j = [\mathbf{y}_1, \mathbf{y}_2, \ldots \mathbf{y}_j]$, $\mathbf{B}_j = [\boldsymbol{\beta}_1, \boldsymbol{\beta}_2, \ldots, \boldsymbol{\beta}_j]$ for $j = 1, 2, \ldots p$ for the model given in (4.4.48). Using property (5) in Theorem 3.3.2 and the realization that the matrix $\Sigma_{1.2} = \Sigma_{11} - \Sigma_{12}\Sigma_{22}^{-1}\Sigma_{21}$ reduces to $\sigma_{1.2}^2 = \sigma_1^2 - \sigma_{12}'\Sigma_{22}^{-1}\sigma_{21} = |\Sigma| / |\Sigma_{22}|$ for one variable, the elements of $\mathbf{y}_{j+1}$ for fixed $\mathbf{Y}_j$ are distributed univariate normal with common variance $\sigma_{j+1}^2 = \left|\Sigma_{j+1}\right| / \left|\Sigma_j\right|$ for $j = 0, 1, 2, \ldots, p-1$ where the $|\Sigma_0| = 1$ and Σ_j is the first principal minor of order j containing the first j rows and j columns of $\Sigma = \left[\sigma_{ij}\right]$. The conditional means are

$$\begin{aligned} E(\mathbf{y}_{j+1}|\mathbf{Y}_j) &= \mathbf{X}\boldsymbol{\eta}_{j+1} + \mathbf{Y}_j\boldsymbol{\gamma}_j \\ &= \left[\mathbf{X}, \mathbf{Y}_j\right]\begin{bmatrix} \boldsymbol{\eta}_{j+1} \\ \boldsymbol{\gamma}_j \end{bmatrix} \end{aligned} \tag{4.4.52}$$

where $\boldsymbol{\eta}_{j+1} = \boldsymbol{\beta}_{j+1} - \mathbf{B}_j\boldsymbol{\gamma}_j$, $\boldsymbol{\gamma}_j' = \left[\sigma_{1,j+1}, \ldots, \sigma_{j,j+1}\right]'\Sigma_j^{-1}$, and $\mathbf{B}_0 = \mathbf{0}$.

With this reparameterization, the hypotheses in (4.4.49) becomes

$$H : \bigcap_{i=1}^{C}\bigcap_{j=1}^{p} H_{ij} : \mathbf{c}_i'\boldsymbol{\eta}_j = 0 \tag{4.4.53}$$

so that the null hypotheses regarding the μ_{ij} are equivalent to testing the null hypotheses regarding the η_{ij} simultaneously or sequentially. Notice that $\boldsymbol{\eta}_{j+1}$ is the mean for variable j adjusting for $j = 0, 1, \ldots, p-1$ covariate where the covariates are a subset of the dependent variables at each step. When a model contains "real" covariates, the dependent variables are sequentially added to the covariates increasing them by one at each step until the final step which would include $h + p - 1$ covariates.

To develop a FIT of (4.4.50) or equivalently (4.4.49), let $\widehat{\xi}_{ij} = \mathbf{c}_i'\widehat{\boldsymbol{\eta}}_j$ where $\widehat{\boldsymbol{\eta}}_j$ is the estimate of the adjusted mean in the MANCOVA model, then for

$$\mathbf{B}_j = \left[\widehat{\boldsymbol{\beta}}_1, \widehat{\boldsymbol{\beta}}_2, \ldots, \widehat{\boldsymbol{\beta}}_j\right] \quad \text{and} \quad \mathbf{S}_j = \mathbf{Y}_j'[\mathbf{I} - \mathbf{X}\left(\mathbf{X}'\mathbf{X}\right)^{-1}\mathbf{X}']\mathbf{Y}_j,$$

the variance of $\mathbf{c}_i'\widehat{\boldsymbol{\eta}}_j = \widehat{\xi}_{ij}$, is

$$\begin{aligned} \sigma_{\xi_{ij}}^2 &= \mathbf{c}_i'[\left(\mathbf{X}'\mathbf{X}\right)^{-1} + \widehat{\mathbf{B}}_j\mathbf{S}_j^{-1}\widehat{\mathbf{B}}_j]\mathbf{c}_i\sigma_{j+1}^2 \\ &= d_{ij}\sigma_{j+1}^2 \end{aligned} \tag{4.4.54}$$

so that an unbiased estimate of $\sigma^2_{\xi_{ij}}$ is $\widehat{\sigma}^2_{\xi_{ij}} = d_{ij}s_j^2/(n-k-j-1)$ where

$$s_j^2/(n-k-j-1)$$

is an unbiased estimate of σ_j^2. Forming the statistics

$$F_{ij} = \frac{(\widehat{\xi}_{ij})^2 (n-k-j+1)}{d_{ij}s_j^2} = \frac{(\widehat{\xi}_{ij})^2 (n-k-j+1)}{\left[\mathbf{c}_i' (\mathbf{X}'\mathbf{X})^{-1} \mathbf{c}_i + \sum_{m=1}^{j-1} \frac{\mathbf{c}_i'\widehat{\eta}_m}{s_m}\right] s_j^2} \tag{4.4.55}$$

where $s_j^2 = \mid \mathbf{S}_j \mid / \mid \mathbf{S}_{j-1} \mid$ and $\mid \mathbf{S}_0 \mid = 1$, the FIT procedure consists of rejecting H if $F_{ij} > f_{j\alpha}$ where the $f_{j\alpha}$ are chosen such that the

$$\begin{aligned}
P\left(F_{ij} \leq f_{j\alpha}; j = 1, 2, \ldots, p \quad \text{and} \quad i = 1, 2, \ldots, C \mid H\right) &\\
= \prod_{j=1}^{p} P\left(F_{ij} \leq f_{j\alpha}; i = 1, 2, \ldots, C \mid H\right) &\\
= \prod_{j=1}^{p} \left(1-\alpha_j\right) = 1 - \alpha. &
\end{aligned}$$

For a given j, the joint distribution of $F_{1j}, F_{2j}, \ldots, F_{Cj}$ is a central C-variate multivariate F distribution with $(1, n-k-j+1)$ degrees of freedom and the statistics F_{ij} in (4.4.55) at each step are independent. When h covariates are in the model, Σ is replaced by $\Sigma_{\mathbf{y}|\mathbf{z}}$ and k is replaced by $h+k$.

Mudholkar and Subbaiah (1980a, b) compared the stepdown FIT of Krishnaiah to Roy's (1958) stepdown F test. They derived approximate 100 $(1-\alpha)$ % level simultaneous confidence intervals for the original population means μ_{ij} and showed that FIT intervals are uniformly shorter than corresponding intervals obtained using Roy's stepdown F tests, if one is only interested in contrasts a variable at a time. For arbitrary contrasts $\psi_{ij} = \mathbf{c}_i'\mathbf{B}\mathbf{m}_j$, the FIT is not uniformly better. In a study by Cox, Krishnaiah, Lee, Reising and Schuurmann (1980) it was shown that the stepdown FIT, is uniformly better in the Neyman sense than Roy's largest root test or Roy's $T^2_{\max}$ test.

The approximate 100 $(1-\alpha)$ % simultaneous confidence intervals for $\theta_{ij} = \mathbf{c}_i'\boldsymbol{\beta}_j$ where $\boldsymbol{\beta}_j$ is the jth column of $\mathbf{B}$, a variable at a time for $i = 1, 2, \ldots, C$ and $j = 1, 2, \ldots, p$ are

$$\begin{aligned}
\widehat{\theta}_{ij} - c_\alpha\sqrt{\mathbf{c}_i' (\mathbf{X}'\mathbf{X})^{-1} \mathbf{c}_i} \quad &\leq \theta_{ij} \leq \quad \widehat{\theta}_{ij} + c_\alpha\sqrt{\mathbf{c}_i' (\mathbf{X}'\mathbf{X})^{-1} \mathbf{c}_i} \\
c_\alpha &= \sum_{q=1}^{j} \mid t_{qj} \mid \sqrt{c_j^*} \\
c_j &= f_{j\alpha}/(n-k-j+1) \\
c_1^* &= c_1, c_j^* = c_j(1 + c_1^* + \ldots + c_{j-1}^*)
\end{aligned} \tag{4.4.56}$$

where t_{qj} are the elements of the upper triangular matrix $\mathbf{T}$ for a Cholesky factorization of

$$\begin{aligned}
\mathbf{E} &= \mathbf{T}'\mathbf{T} \\
&= \mathbf{Y}'[\mathbf{I} - \mathbf{X}(\mathbf{X}'\mathbf{X})^{-1}\mathbf{X}']\mathbf{Y}.
\end{aligned}$$

Replacing θ_{ij} by arbitrary contrasts $\psi_{ij} = \mathbf{c}_i'\mathbf{B}\mathbf{m}_j$ where $\mathbf{h}_j' = \mathbf{T}\mathbf{m}_j$ and $\mathbf{T}'\mathbf{T} = \mathbf{E}$, simultaneous confidence sets for ψ_{ij} become

$$\widehat{\psi}_{ij} - \left(\sum_{j=1}^{p} |h_j| \sqrt{c_j^*}\right)\left(\sqrt{\mathbf{c}_i' (\mathbf{X}'\mathbf{X})^{-1} \mathbf{c}_i}\right) \leq \psi_{ij} \leq \widehat{\psi}_{ij} + \left(\sum_{j=1}^{p} |h_j| \sqrt{c_j^*}\right)\left(\sqrt{\mathbf{c}_i' (\mathbf{X}'\mathbf{X})^{-1} \mathbf{c}_i}\right) \qquad (4.4.57)$$

where c_j^* is defined in (4.4.51). Using the multivariate t distribution, one may also test one-sided hypotheses H_{ij} simultaneously and construct simultaneous confidence sets for directional alternatives.

Currently no SAS procedure has been developed to calculate the F_{ij} statistics in (4.4.45) or to create the approximate simultaneous confidence intervals given in (4.4.57) for the FIT procedure. The problem one encounters is the calculation of the critical values for the multivariate F distribution. The program Fit.For available on the Website performs the necessary calculations for MANOVA designs. However, it only runs on the DEC-Alpha 3000 RISC processor and must be compiled using the older version of the IMSL Library calls. The manual is contained in the postscript file FIT-MANUAL.PS. The program may be run interactively or in batch mode. In batch mode, the interactive commands are placed in an *.com file and the SUBMIT command is used to execute the program. The program offers various methods for approximating the critical values for the multivariate F distribution. One may also approximate the critical values of the multivariate F distribution using a computer intensive bootstrap resampling scheme, Hayter and Tsui (1994). Timm (1996) compared their method with the analytical methods used in the FIT program and found little difference between the two approaches since exact values are difficult to calculate.

4.5 One-Way MANOVA/MANCOVA Examples

a. *MANOVA (Example 4.5.1)*

The data used in the example were taken from a large study by Dr. Stanley Jacobs and Mr. Ronald Hritz at the University of Pittsburgh to investigate risk-taking behavior. Students were randomly assigned to three different direction treatments known as Arnold and Arnold (AA), Coombs (C), and Coombs with no penalty (NC) in the directions. Using the three treatment conditions, students were administrated two parallel forms of a test given under high and low penalty. The data for the study are summarized in Table 4.5.1. The sample sizes for the three treatments are respectively, $n_1 = 30$, $n_2 = 28$, and $n_3 = 29$. The total sample size is $n = 87$, the number of treatments is $k = 3$, and the number of variables is $p = 2$ for the study. The data are provided in the file stan_hz.dat.

TABLE 4.5.1. Sample Data One-Way MANOVA

AA				C				NC			
Low	High	Low	High	Low	High	Low	High	Low	High	Low	High
8	28	31	24	46	13	25	9	50	55	55	43
18	28	11	20	26	10	39	2	57	51	52	49
8	23	17	23	47	22	34	7	62	52	67	62
12	20	14	32	44	14	44	15	56	52	68	61
15	30	15	23	34	4	36	3	59	40	65	58
12	32	8	20	34	4	40	5	61	68	46	53
12	20	17	31	44	7	49	21	66	49	46	49
18	31	7	20	39	5	42	7	57	49	47	40
29	25	12	23	20	0	35	1	62	58	64	22
6	28	15	20	43	11	30	2	47	58	64	54
7	28	12	20	43	25	31	13	53	40	63	64
6	24	21	20	34	2	53	12	60	54	63	56
14	30	27	27	25	10	40	4	55	48	64	44
11	23	18	20	50	9	26	4	56	65	63	40
12	20	25	27					67	56		

The null hypothesis of interest is whether the mean vectors for the two variates are the same across the three treatments. In terms of the effects, the hypothesis may be written as

$$H_o : \begin{bmatrix} \alpha_{11} \\ \alpha_{12} \end{bmatrix} = \begin{bmatrix} \alpha_{21} \\ \alpha_{22} \end{bmatrix} = \begin{bmatrix} \alpha_{31} \\ \alpha_{32} \end{bmatrix} \tag{4.5.1}$$

The code for the analysis of the data in Table 4.5.1 is provided in the programs: m4_5_1.sas and m4_5_1a.sas.

We begin the analysis by fitting a model to the treatment means. Before testing the hypothesis, a chi-square Q-Q plot is generated using the routine multinorm.sas to investigate multivariate normality (program m4_5_1.sas). Using PROC UNIVARIATE, we also generate univariate Q-Q plots using the residuals and investigate plots of residuals versus fitted values. Following Example 3.7.3, the chi-square Q-Q plot for all the data indicate that observation #82 (NC, 64, 22) is an outlier and should be removed from the data set. With the outlier removed (program m4_5_1a.sas), the univariate and multivariate tests, and residual plots indicate that the data are more nearly MVN. The chi-square Q-Q plot is almost linear. Because the data are approximately normal, one may test that the covariance matrices are equal (Exercises 4.5, Problem 1). Using the option HOVTEST = BF on the MEANS statement, the univariate variances appear approximately equal across the three treatment groups.

To test (4.5.1) using PROC GLM, the MANOVA statement is used to create the hypothesis test matrix **H** for the hypothesis of equal means or treatment effects. Solving $|\mathbf{H} - \lambda\mathbf{E}| = 0$, the eigenvalues for the test are $\lambda_1 = 8.8237$ and $\lambda_2 = 4.41650$ since $s = \min(v_h, p) = \min(3, 2) = 2$. For the example, the degrees of freedom for error is $v_e = n - k = 83$. By any of the MANOVA criteria, the equality of group means is rejected using any of the multivariate criteria (p-value < 0.0001).

With the rejection of (4.5.1), using (4.4.47) or (4.2.59) we know there exists at least one contrast $\psi = \mathbf{c}'\mathbf{Bm}$ that is nonzero. Using the one-way MANOVA model, the expression for ψ is

$$\mathbf{c}'\widehat{\mathbf{B}}\mathbf{m} - c_\alpha\sqrt{(\mathbf{m}'\mathbf{S}\mathbf{m})\,\mathbf{c}'\,(\mathbf{X}'\mathbf{X})^-\,\mathbf{c}} \leq \psi \leq \mathbf{c}'\widehat{\mathbf{B}}\mathbf{m} + c_\alpha\sqrt{(\mathbf{m}'\mathbf{S}\mathbf{m})\,\mathbf{c}'\,(\mathbf{X}'\mathbf{X})^-\,\mathbf{c}} \tag{4.5.2}$$

As in the MR model, $\mathbf{m}$ operates on sample covariance matrix $\mathbf{S}$ and the contrast vector $\mathbf{c}$ operates on the matrix $(\mathbf{X}'\mathbf{X})^-$. For a vector $\mathbf{m}$ that has a single element equal to one and all others zero, the product $\mathbf{m}'\mathbf{Sm} = s_i^2$, a diagonal element of $\mathbf{S} = \mathbf{E}/(n - r(\mathbf{X})$. For pairwise comparisons among group mean vectors, the expression

$$\mathbf{c}'\,(\mathbf{X}'\mathbf{X})^-\,\mathbf{c} = \frac{1}{n_i} + \frac{1}{n_j}$$

for any g-inverse. Finally, (4.5.2) involves c_α which depends on the multivariate criterion used for the overall test for treatment differences. The values for c_α were defined in (4.2.60) for the MR model. Because simultaneous confidence intervals allow one to investigate all possible contrast vectors $\mathbf{c}$ and arbitrary vectors $\mathbf{m}$ in the expression $\psi = \mathbf{c}'\mathbf{Bm}$, they generally lead to very wide confidence intervals if one evaluates only a few comparisons. Furthermore, if one locates a significant contrast it may be difficult to interpret when the elements of $\mathbf{c}$ and/or $\mathbf{m}$ are not integer values. Because PROC GLM does not solve (4.5.2) to generate approximate simultaneous confidence intervals, one must again use PROC IML to generate simultaneous confidence intervals for parametric functions of the parameters as illustrated in Section 4.3 for the regression example. In program m4_5_1a.sas we have included IML code to obtain approximate critical values using (4.2.61), (4.2.62) and (4.2.63) [ROY, BLH, and BNP] for the contrast that compares treatment one (AA) versus treatment three (NC) using only the high penalty variable. One may modify the code for other comparisons. PROC TRANSREG is used to generate a full rank design matrix which is input into the PROC IML routine. Contrasts using any of the approximate methods yield intervals that do not include zero for any of the criteria. The length of the intervals depend on the criterion used in the approximation. Roy's approximation yields the shortest interval. The comparison has the approximate simultaneous interval (−31.93, −23.60) for the comparison of group one (AA) with group three (NC) for the variable high penalty. Because these intervals are created from an upper bound statistic, they are most resolute. However, the intervals are created using a crude approximation and must be used with caution. The approximate critical value was calculated as $c_\alpha = 2.49$ while the exact value for the Roy largest root statistic is 3.02. The F approximations for BLH and BNP multivariate criteria are generally closer to their exact values. Hence they may be preferred when creating simultaneous intervals for parametric functions following an overall test. The simultaneous interval for the comparison using the F approximation for the BLH criterion yields the simultaneous interval (−37.19, −18.34) as reported in the output.

To locate significant comparisons in mean differences using PROC GLM, one may combine CONTRAST statements in treatments with MANOVA statements by defining the matrix $\mathbf{M}$. For $\mathbf{M} = \mathbf{I}$, the test is equivalent to using Fisher's LSD method employing Hotelling's T^2 statistics for locating contrasts involving the mean vectors. These protected tests control the per comparison error rate near the nominal level α for the overall test only if the overall test is rejected. However, they may not be used to construct simultaneous confidence intervals. To construct approximate simultaneous confidence intervals for contrast in the mean vectors, one may use

$$c_{\alpha}^{2} = p v_e F_{(p, v_e - p + 1)}^{\alpha^*} / (v_e - p + 1)$$

in (4.5.2) where $\alpha^* = \alpha / C$ is the upper α critical value for the F distribution using, for example, the Bonferroni method where C is the number of mean comparisons. Any number of vectors **m** may be used for each of the Hotelling T^2 tests to investigate contrasts that involve the means of a single variable or to combine means across variables.

Instead of using Hotelling's T^2 statistic to locate significant differences in the means, one may prefer to construct CONTRAST statements that involve the vectors **c** and **m**. To locate significance differences in the means using these contrasts, one may evaluate univariate protected F tests using the nominal level α. Again, with the rejection of the overall test, these protected F tests have an experimental error rate that is near the nominal level α when the overall test is rejected. However, to construct approximate simultaneous confidence intervals for the significant protected F tests, one must again adjust the alpha level for each comparison. Using for example the Bonferroni inequality, one may adjust the overall α level by the number of comparisons, C, so that $\alpha^* = \alpha / C$. If one were interested in all pairwise comparisons for each variable (6 comparisons) and the three comparisons that combine the sum of the low penalty and high penalty variables, then $C = 9$ and $\alpha^* = 0.00556$. Using $\alpha = 0.05$, the p-values for the $C = 9$ comparisons are shown below. They are all significant. The ESTIMATE statement in PROC GLM may be used to produce $\widehat{\psi}$ and $\widehat{\sigma}_{\widehat{\psi}}$ for each contrast specified for each variable. For example, suppose we are interested in all pairwise comparisons $(3 + 3 = 6$ for all variables) and two complex contrasts that compare ave $(1 + 2)$ vs 3 and ave $(2 + 3)$ vs 1 or ten comparisons. To construct approximate simultaneous confidence intervals for 12 the comparisons, the value for c_{α} may be obtained form the Appendix, Table V by interpolation. For $C = 12$ contrasts and degrees of freedom for error equal to 60, the critical values for the BON, SID and STM procedures range between 2.979 and 2.964. Because the Šidák's multivariate t has the smallest value, by interpolation we would use $c_{\alpha} = 2.94$ to construct approximate simultaneous confidence intervals for 12 the comparisons. SAS only produces estimated standard errors, $\widehat{\sigma}_{\widehat{\psi}}$, for contrasts that involve a single variable. The general formula for estimating the standard errors, $\widehat{\sigma}_{\widehat{\psi}} = \sqrt{(\mathbf{m}'\mathbf{S}\mathbf{m})\,\mathbf{c}'\,(\mathbf{X}'\mathbf{X})^{-}\,\mathbf{c}}$, must be used to calculate standard errors for contrasts for arbitrary vectors **m**.

	Variables		
Contrasts	Low	High	Low + High
1 vs 3	.0001	.0001	.0001
2 vs 3	.0001	.0001	.0001
1 vs 2	.0001	.0001	.0209

If one is only interested in all pairwise comparisons. For each variable, one does not need to perform the overall test. Instead, the LSMEANS statement may be used by setting ALPHA equal to $\alpha^* = \alpha / p$ where p is the number of variables and $\alpha = .05$ (say). Then, using the option CL, PDIFF = ALL, and ADJUST = TUKEY, one may directly isolate the planned comparisons that do not include zero. This method again only approximately controls the familywise error rate at the nominal α level since correlations among variables are being ignored. The LSMEANS option only allows one to investigate all pairwise comparisons in unweighted means. The option ADJUST=DUNNETT is used to compare all experimental group means with a control group mean. The confidences intervals for Tukey's method for $\alpha^* = 0.025$ and all pairwise comparisons follow.

	Variables			
Contrasts	Low		High	
1 vs 2	(−28.15	−17.86)	(11.61	20.51)
1 vs 3	(−48.80	−38.50)	(−32.21	−23.31)
2 vs 3	(−25.88	−15.41)	(−48.34	−39.30)

Because the intervals do not include zero, all pairwise comparisons are significant for our example.

Finally, one may use PROC MULTTEST to evaluate the significance of a finite set of arbitrary planned contrasts for all variables simultaneously. By adjusting the p-value for the family of contrasts, the procedure becomes a simultaneous test procedure (STP). For example, using the Šidák method, a hypothesis H_i is rejected if the p-value p_i is less than $1 - (1 - \alpha)^{1/C} = \alpha^*$ where α is the nominal FWE rate for C comparisons. Then the Šidák single-step adjusted p-value is $\widetilde{p}_i = 1 - (1 - p_i)^C$. PROC MULTTEST reports raw p-values p_i and the adjusted values p-values, $\widetilde{p}_i$. One may compare the adjusted $\widetilde{p}_i$ values to the nominal level α to assess significance. For our example, we requested adjusted the p-values using the Bonferroni, Šidák and permutation options. The permutation option resamples vectors without replacement and adjusts p-values empirically. For the finite contrasts used with PROC MULTTEST using the t test option, all comparisons are seen to be significant at the nominal $\alpha = 0.05$ level. Westfall and Young (1993) illustrate PROC MULTTEST in some detail for univariate and multivariate STPs.

When investigating a large number of dependent variables in a MANOVA design, it is often difficult to isolate specific variables that are most important to the significant separation of the centroids. To facilitate the identification of variables, one may use the /CANONICAL option on the MANOVA statement as illustrated in the two group example. For multiple groups, there are $s = \min(v_h, p)$ discriminant functions. For our example, $s = 2$. Reviewing the magnitude of the coefficients for the standardized vectors of canonical variates and the correlations of the within structure canonical variates in each significant dimension often helps in the exploration of significant contrasts. For our example, both discriminant functions are significant with the variable high penalty dominating one dimension and the low penalty variable the other.

One may also use the FIT procedure to analyze differences in mean vectors for the one-way MANOVA design. To implement the method, one must specify all contrasts of interest for each variable, and rank the dependent variables in order of importance from highest to lowest. The Fit.for program generates approximate $100(1 - \alpha)\,\%$ simultaneous confidence intervals for the conditional contrasts involving the η_j and the original means. For the example, we consider five contrasts involving the three treatments as follows.

TABLE 4.5.2. FIT Analysis

Contrast	F_{ij}	Crude Estimate for	of Original	C.Is Means
Variable: Low Penalty				
1	141.679*	−23.0071	(−28.57	−17.45)
2	110.253*	−20.6429	(−26.30	−14.99)
3	509.967*	−43.6500	(−49.21	−38.09)
4	360.501*	−32.1464	(−37.01	−27.28)
5	401.020*	−33.3286	(−38.12	−28.54)
Variable: High Penalty				
1	76.371*	16.0595	(9.67	22.43)
2	237.075*	−43.8214	(−50.32	−37.32)
3	12.681*	−27.7619	(−34.15	−21.37)
4	68.085*	−35.7917	(−41.39	−30.19)
5	1.366	−5.8512	(−11.35	−0.35)

*significant of conditional means for $\alpha = 0.05$

Contrasts	AA	C	NC
1	1	−1	0
2	0	1	−1
3	1	0	−1
4	.5	.5	−1
5	1	−.5	−.5

For $\alpha = 0.05$, the upper critical value for the multivariate F distribution is 8.271. Assuming the order of the variables as Low penalty followed by High penalty, Table 4.5.2 contains the output from the Fit.for program.

Using the FIT procedure, the multivariate overall hypothesis is rejected if any contrast is significant.

b. *MANCOVA (Example 4.5.2)*

To illustrate the one-way MANCOVA design, Rohwer collected data identical to that analyzed in Section 4.3 for $n = 37$ kindergarten students from a residential school in a lower-SES-class area. The data for the second group are given in Table 4.3.2. It is combined with the data in Table 4.3.1 and provided in the file Rohwer2.dat. The data are used to test (4.4.44) for the two independent groups. For the example, we have three dependent variables and five covariates. Program m4_5_2.sas contains the SAS code for the analysis. The code is used to test multivariate normality and to illustrate the test of parallelism

$$H : \Gamma_1 = \Gamma_2 = \Gamma_3 = \Gamma \tag{4.5.3}$$

using both PROC REG and PROC GLM. The MTEST commands in PROC REG allow one to test for parallelism for each covariate and to perform the overall test for all covariates simultaneously. Using PROC GLM, one may not perform the overall simultaneous test. However, by considering interactions between each covariate and the treatment, one may test for parallelism a covariate at a time. Given parallelism, one may test that all covariates are simultaneously zero, $H_o : \Gamma = \mathbf{0}$ or that each covariate is zero using PROC REG. The procedure GLM in SAS may only be used to test that each covariate is zero. It does not allow one to perform the simultaneous test. Given parallelism, one next tests that the group means or effects are equal

$$\begin{aligned} H &: \boldsymbol{\mu}_1 = \boldsymbol{\mu}_2 \text{ (FR)} \\ H &: \boldsymbol{\alpha}_1 = \boldsymbol{\alpha}_2 \text{ (LFR)} \end{aligned} \tag{4.5.4}$$

using PROC GLM.

When using a MANCOVA design to analyze differences in treatments, in addition to the assumptions of multivariate normality and homogeneity of covariance matrices, one must have multivariate parallelism. To test (4.5.3) using PROC REG for our example, the overall test of parallelism is found to be significant at the $\alpha = .05$ level, but not significant for $\alpha = 0.01$. For Wilks' Λ criterion, $\Lambda = 0.62358242$ and the p-value is 0.0277. Reviewing the one degree of freedom tests for each of the covariates N, S, NS, NA, and SS individually, the p-values for the tests are 0.2442, 0.1212, 0.0738, 0.0308, and 0.3509, respectively. These are the tests performed using PROC GLM. Since the test of parallelism is not rejected, we next test $H_o : \Gamma = 0$ using PROC REG. The overall test that all covariates are simultaneously zero is rejected. For Wilks' Λ criterion, $\Lambda = 0.44179289$. All criteria have p-values < 0.0001. However, reviewing the individual tests for each single covariate, constructed by using the MTEST statement in PROC REG or using PROC GLM, we are led to retain only the covariates NA and NS for the study. The p-value for each of the covariates N, S, NS, NA, and SS are : 0.4773, 0.1173, 0.0047, 0.0012, 0.3770. Because only the covariates NA(p-value = 0.0012) and NS (p-value = 0.0047) have p-values less than $\alpha = 0.01$, they are retained. All other covariates are removed from the model. Because the overall test that $\Gamma = 0$ was rejected, these individual tests are again protected F tests. They are used to remove insignificant covariates from the multivariate model.

Next we test (4.4.44) for the revised model. In PROC GLM, the test is performed using the MANOVA statement. Because $s = 1$, all multivariate criteria are equivalent and the test of equal means, adjusted by the two covariates, is significant. The value of the F statistic is 15.47. For the revised model, tests that the coefficient vectors for NA and NS remain significant, however, one may consider removing the covariate NS since the p-value for the test of significance is 0.0257. To obtain the estimate of Γ using PROC GLM, the /SOLUTION option is included on the MODEL statement. The /CANONICAL option performs a discriminant analysis. Again the coefficients may be investigated to form contrasts in treatment effects.

When testing for differences in treatment effects, we may evaluate (4.4.35) with

$$\mathbf{C} = [1, -1, 0, 0] \text{ and } \mathbf{M} = \mathbf{I}$$

This is illustrated in program m4_5_2.sas using PROC IML. The procedure TRANSREG is used to generate a full rank design matrix for the analysis. Observe that the output for

$\mathbf{H}$ and $\mathbf{E}$ agree with that produced by PROC GLM using the MANOVA statement. Also included in the output is the matrix $\mathbf{A}$, where

$$\underset{(4\times 3)}{\mathbf{A}} = \begin{bmatrix} \widehat{\mathbf{B}}_{2\times 3} \\ \widehat{\Gamma}_{2\times 3} \end{bmatrix} = \begin{bmatrix} 51.456 & 11.850 & 8.229 \\ 33.544 & 10.329 & -4.749 \\ \hline 0.117 & 0.104 & -1.937 \\ 1.371 & 0.068 & 2.777 \end{bmatrix}$$

The first two rows of $\mathbf{A}$ are the sample group means adjusted by $\widehat{\Gamma}$ as in (4.4.28). Observe that the rows of $\widehat{\Gamma}$ agree with the 'SOLUTION' output in PROC GLM; however, the matrix $\widehat{\mathbf{B}}$ is not the adjusted means, output by PROC GLM by using the LSMEANS statement. To output the adjusted means in SAS, centered using $\mathbf{Z}_{..}$, one must use the COV and OUT = options on the LSMEANS statement. The matrix of adjusted means is output as follows.

$$\widehat{\mathbf{B}}_{SAS} = \begin{bmatrix} 81.735 & 14.873 & 45.829 \\ 63.824 & 13.353 & 32.851 \end{bmatrix}$$

As with the one-way MANOVA model or any multivariate design analyzed using PROC GLM, the SAS procedure does not generate $100\,(1-\alpha)\,\%$ simultaneous confidence intervals for the matrix $\mathbf{B}$ in the MR model for the MANCOVA design $\mathbf{B}$ is contained in the matrix $\mathbf{A}$. To test hypotheses involving the adjusted means, one may again use CONTRAST statements and define the matrix $\mathbf{M} \equiv \mathbf{m}'$ in SAS with the MANOVA statement to test hypotheses using F statistics by comparing the level of significance with α. These are again protected tests when the overall test is rejected. One may also use the LSMEAN statement. For these comparisons, one usually defines the level of the test at the nominal value of $\alpha^* = \alpha/p$ and uses the ADJUST option to approximate simultaneous confidence intervals For our problem there are three dependent variables simultaneously so we set $\alpha^* = 0.0167$. Confidence sets for all pairwise contrasts in the adjusted means for the TUKEY procedure follow. Also included below are the exact simultaneous confidence intervals for the difference in groups for each variable using the ROY criterion Program m4_5_2.sas contains the difference for $\mathbf{c}' = [1, -1, 0, 0]$ and $\mathbf{m}' = [1, 0, 0]$. By changing $\mathbf{m}$ for each variable, one obtains the other entries in the table. The results follow.

	PPVT	RPMT	SAT
$\widehat{\psi}_{\text{diff}}$	17.912	1.521	−0.546
Lower Limit (Roy)	10.343	−0.546	12.978
Lower Limit (Tukey)	11.534	−0.219	−2.912
Upper Limit (Roy)	25.481	3.587	31.851
Upper Limit (Tukey)	24.285	3.260	28.869

The comparisons indicate that the difference for the variable PPVT is significant since the confidence interval does not include zero. Observe that $\widehat{\psi}_{\text{diff}}$ represents the difference in the rows of $\widehat{\mathbf{B}}$ or $\widehat{\mathbf{B}}_{SAS}$ so that one may use either matrix to form contrasts. Centering does effect the covariance structure of $\widehat{\mathbf{B}}$. In the output from LSMEANS, the columns labeled 'COV' represent the covariance among the elements of $\widehat{\mathbf{B}}_{SAS}$.

A test closely associated with the MANCOVA design is Rao's test for additional information, (Rao, 1973a, p. 551). In many MANOVA or MANCOVA designs, one collects data on p response variables and one is interested in determining whether the additional information provided by the last $(p - s)$ variables, independent of the first s variables, is significant. To develop a test procedure of this hypothesis, we begin with the linear model $\Omega_o : \mathbf{Y} = \mathbf{XB} + \mathbf{U}$ where the usual hypothesis is $H : \mathbf{CB} = \mathbf{0}$. Partitioning the data matrix $\mathbf{Y} = [\mathbf{Y}_1, \mathbf{Y}_2]$ and $\mathbf{B} = [\mathbf{B}_1, \mathbf{B}_2]$, we consider the alternative model

$$\begin{aligned} \Omega_1 &: \mathbf{Y}_1 = \mathbf{XB}_1 + \mathbf{U}_1 \\ H_{01} &: \mathbf{CB}_1 = \mathbf{0} \end{aligned} \tag{4.5.5}$$

where

$$\begin{aligned} E\left(\mathbf{Y}_2 \mid \mathbf{Y}_1\right) &= \mathbf{XB}_2 + \left(\mathbf{Y}_1 - \mathbf{XB}_1\right) \Sigma_{11}^{-1} \Sigma_{12} \\ &= \mathbf{X}\left(\mathbf{B}_2 - \mathbf{B}_1 \Sigma_{11}^{-1} \Sigma_{12}\right) + \mathbf{Y}_1 \Sigma_{11}^{-1} \Sigma_{12} \\ &= \mathbf{X}\Theta + \mathbf{Y}_1 \Gamma \\ \Sigma_{2.1} &= \Sigma_{22} - \Sigma_{21} \Sigma_{11}^{-1} \Sigma_{12} \end{aligned}$$

Thus, the conditional model is

$$\Omega_2 : E\left(\mathbf{Y}_2 \mid \mathbf{Y}_1\right) = \mathbf{X}\Theta + \mathbf{Y}_1 \Gamma \tag{4.5.6}$$

the MANCOVA model. Under Ω_2, testing

$$H_{02} : \mathbf{C}\left(\mathbf{B}_2 - \mathbf{B}_1 \Gamma\right) = \mathbf{0} \tag{4.5.7}$$

corresponds to testing $H_{02} : \mathbf{C}\Theta = \mathbf{0}$. If $\mathbf{C} = \mathbf{I}_p$ and $\Theta = \mathbf{0}$, then the conditional distribution of $\mathbf{Y}_2 \mid \mathbf{Y}_1$ depends only on Γ and does not involve $\mathbf{B}_1$; thus $\mathbf{Y}_2$ provides no additional information on $\mathbf{B}_1$. Because Ω_2 is the standard MANCOVA model with $\mathbf{Y} \equiv \mathbf{Y}_2$ and $\mathbf{Z} \equiv \mathbf{Y}_1$, we may test H_{02} using Wilks' criterion

$$\Lambda_{2.1} = \frac{\left|\mathbf{E}_{22} - \mathbf{E}_{21} \mathbf{E}_{11}^{-1} \mathbf{E}_{12}\right|}{\left|\mathbf{E}_{H22} - \mathbf{E}_{H21} \mathbf{E}_{H11}^{-1} \mathbf{E}_{H12}\right|} \sim U_{p-s,\, v_h,\, v_e} \tag{4.5.8}$$

where $v_e = n - p - s$ and $v_h = r\left(\mathbf{C}\right)$. Because $H\left(\mathbf{CB} = \mathbf{0}\right)$ is true if and only if H_{01} and H_{02} are true, we may partition Λ as $\Lambda = \Lambda_1 \Lambda_{2.1}$ where Λ_1 is from the test of H_{01}; this results in a stepdown test of H (Seber, 1984, p. 472).

Given that we have found a significant difference between groups using the three dependent variables, we might be interested in determining whether variables RPMT and SAT (the variable in set 2) add additional information to the analysis of group differences above that provided by PPVT (the variable in set 1). We calculate $\Lambda_{2.1}$ defined in (4.5.8) using

PROC GLM. Since the p-value for the test is equal to 0.0398, the contribution of set 2 given set 1 is significant at the nominal level $\alpha = 0.05$ and adds additional information in the evaluation of group differences. Hence we should retain the variable in the model.

We have also included in program m4_5_2.sas residual plots and Q-Q plots to evaluate the data set for outliers and multivariate normality. The plots show no outliers and the data appears to be multivariate normal. The FIT procedure may be used with MANCOVA designs by replacing the data matrix $\mathbf{Y}$ with the residual matrix $\mathbf{Y} - \mathbf{Z}\widehat{\Gamma}$.

Exercises 4.5

1. With the outlier removed and $\alpha = 0.05$, test that the covariance matrices are equal for the data in Table 4.5.1 (data set: stan_hz.dat).

2. An experiment was performed to investigate four different methods for teaching school children multiplication (M) and addition (A) of two four-digit numbers. The data for four independent groups of students are summarized in Table 4.5.3.

   (a) Using the data in Table 4.5.3, is there any reason to believe that any one method or set of methods is superior or inferior for teaching skills for multiplication and addition of four-digit numbers?

TABLE 4.5.3. Teaching Methods

Group 1		Group 2		Group 3		Group 4	
A	M	A	M	A	M	A	M
97	66	76	29	66	34	100	79
94	61	60	22	60	32	96	64
96	52	84	18	58	27	90	80
84	55	86	32	52	33	90	90
90	50	70	33	56	34	87	82
88	43	70	32	42	28	83	72
82	46	73	17	55	32	85	67
65	41	85	29	41	28	85	77
95	58	58	21	56	32	78	68
90	56	65	25	55	29	86	70
95	55	89	20	40	33	67	67
84	40	75	16	50	30	57	57
71	46	74	21	42	29	83	79
76	32	84	30	46	33	60	50
90	44	62	32	32	34	89	77
77	39	71	23	30	31	92	81
61	37	71	19	47	27	86	86
91	50	75	18	50	28	47	45
93	64	92	23	35	28	90	85
88	68	70	27	47	27	86	65

(b) What assumptions must you make to answer part a? Are they satisfied?

(c) Are there any significant differences between addition and multiplication skills within the various groups?

3. Smith, Gnanadesikan, and Hughes (1962) investigate differences in the chemical composition of urine samples from men in four weight categories. The eleven variables and two covariates for the study are

y_1 = pH,	y_8 = chloride (mg/ml),
y_2 = modified createnine coefficient,	y_9 = bacon (μ_g/m1),
y_3 = pigment createnine,	y_{10} = choline (μ_g/m1),
y_4 = phosphate (mg/ml),	y_{11} = copper (μ_g/m1),
y_5 = calcium (mg/ml),	x_1 = volume (m1),
y_6 = phosphours (mg/ml),	x_2 = (specific gravity − 1) × 10^3,
y_7 = createnine (mg/ml),	

The data are in the data file SGH.dat.

(a) Evaluate the model assumptions for the one-way MANCOVA design.

(b) Test for the significance of the covariates.

(c) Test for mean differences and construct appropriate confidence sets.

(d) Determine whether variables y_2, y_3, y_4, y_6, y_7, y_{10}, and y_{11} (Set 2) add additional information above those provided by y_1, y_5, y_7, and y_8 (Set 1).

4. Data collected by Tubb et al. (1980) are provided in the data set pottery.dat. The data represent twenty-six samples of Romano-British pottery found a four different sites in Wales, Gwent, and the New Forest. The sites are Llanederyn (L), Caldicot (C), Island Thorns (S), and Ashley Rails (A). The other variables represent the percentage of oxides for the metals: A1, Fe, Mg, Ca, Na measured by atomic absorption spectrophotometry.

(a) Test the hypothesis that the mean percentages are equal for the four groups.

(b) Use the Fit procedure to evaluate whether there are differences between groups. Assume the order A1, Fe, Mg, Ca, and Na.

4.6 MANOVA/MANCOVA with Unequal Σ_i or Nonnormal Data

To test $H : \boldsymbol{\mu}_1 = \boldsymbol{\mu}_2 = \ldots = \boldsymbol{\mu}_k$ in MANOVA/MANCOVA models, both James' (1954) and Johansen's (1980) tests may be extended to the multiple k group case. Letting $\mathbf{S}_i$ be an estimate of the ith group covariance matrix, $\mathbf{W}_i = \mathbf{S}_i / n_i$ and $\mathbf{W} = \sum_{i=1}^{k} \mathbf{W}_i$., we form the statistic

$$X^2 = \sum_{i=1}^{k} (\mathbf{y}_{i.}^A - \overline{\mathbf{y}})' \mathbf{W}_i (\mathbf{y}_{i.}^A - \overline{\mathbf{y}}) \tag{4.6.1}$$

where $\overline{\mathbf{y}} = \mathbf{W}^{-1} \sum_{i=1}^{k} \mathbf{W}_i \mathbf{y}_{i.}^A$, $\widehat{\Gamma} = \mathbf{W}^{-1} \sum_i \mathbf{W}_i \widehat{\Gamma}_i$ is a pooled estimate of Γ, $\widehat{\Gamma}_i = \left(\mathbf{X}_i'\mathbf{X}_i\right)^{-1} \mathbf{X}_i'\mathbf{Y}_i$ and $\mathbf{y}_{i.}^A$ is the adjusted mean for the ith group using $\widehat{\Gamma}$ in (4.4.43). Then, the statistic $X^2 \,\dot{\sim}\, \chi^2 (v_h)$ with degrees of freedom $v_h = p\,(k-1)$. To better approximate the chi-square critical value we may calculate James' first-order or second-order approximations, or use Johansen's F test approximation. Using James' first-order approximation, H is rejected if

$$X^2 > \chi^2_{1-\alpha} (v_h) \left\{ 1 + \frac{1}{2} \left[\frac{k_1}{p} + \frac{k_2 \chi^2_{1-\alpha} (v_h)}{p\,(p+2)} \right] \right\} \tag{4.6.2}$$

where

$$k_1 = \sum_{i=1}^{k} \operatorname{tr} \left(\mathbf{W}^{-1} \mathbf{W}_i \right)^2 / \left(n_i - h - 1 \right)$$

$$k_2 = \sum_{i=1}^{k} \left[\operatorname{tr} \left(\mathbf{W}^{-1} \mathbf{W}_i \right)^2 + 2 \left(\operatorname{tr} \mathbf{W}^{-1} \mathbf{W}_i \right)^2 \right] / \left(n_i - h - 1 \right)$$

and h is the number of covariates. For Johansen's (1980) procedure, the constant A becomes

$$A = \sum_{i=1}^{k} [\operatorname{tr} \left(\mathbf{I} - \mathbf{W}^{-1} \mathbf{W}_i \right)^2 + \left(\operatorname{tr} \mathbf{I} - \mathbf{W}^{-1} \mathbf{W}_i \right)^2] / k \left(n_i - h - 1 \right) \tag{4.6.3}$$

Then (3.9.23) may be used to test for the equality of mean vectors under normality. Finally, one may use the A statistic developed by Myers and Dunlap (2000) as discussed in the two group location problem in Chapter 3, Section 3.9. One would combine the chi-square p-values for the k groups and compare the statistic to the critical value for chi-square distribution with $(k-1)p$ degrees of freedom.

James' and Johansen's procedures both assume MVN samples with unequal Σ_i. Alternatively, suppose that the samples are instead from a nonnormal, symmetric multivariate distribution that have conditional symmetric distributions. Then, Nath and Pavur (1985) show that the multivariate test statistics may still be used if one substitutes ranks from 1 to n for the raw data a variable at a time. This was illustrated for the two group problem.

4.7 One-Way MANOVA with Unequal Σ_i Example

To illustrate the analysis of equal mean vectors given unequal covariance matrices, program m4_7_1.sas is used to reanalyze the data in Table 4.5.1. The code in the program calculates the chi-square statistic adjusted using Johansen's correction. While we still have significance, observe how the correction due to Johansen changes the critical value for the test. Without the adjustment, the critical value is 9.4877. With the adjustment, the value becomes 276.2533. Clearly, one would reject equality of means more often without adjusting the critical value for small sample sizes. For our example, the adjustment has little effect on the conclusion.

Exercises 4.7

1. Modify program m_4_7_1.sas for James' procedure and re-evaluate the data in Table 4.5.1.

2. Analyze the data in Exercises 4.5, problem 2 for unequal covariance matrices.

3. Re-analyze the data in Exercises 4.5, problem 3 given unequal covariance matrices using the chi-square test, the test with James' correction, and the test with Johansen's correction.

4. Analyze the data in Exercises 4.5, problem 3 using the A statistic proposed by Myers and Dunlap discussed in Section3.9 (b) for the two group location problem.

4.8 Two-Way MANOVA/MANCOVA

a. *Two-Way MANOVA with Interaction*

In a one-way MANOVA, one is interested in testing whether treatment differences exist on p variables for one treatment factor. In a two-way MANOVA, $n_o \geq 1$ subjects are randomly assigned to two factors, A and B say, each with levels a and b, respectively, creating a design with ab cells or treatment combinations. Gathering data on p variables for each of the ab treatment combinations, one is interested in testing whether treatment differences exist with regard to the p variables provided there is no interaction between the treatment factors; such designs are called additive models. Alternatively, an interaction may exist for the study, then interest focuses on whether the interaction is significant for some linear combination of variables or for each variable individually. One may formulate the two-way MANOVA with an interaction parameter and test for the presence of interaction. Finding none, an additive model is analyzed. This approach leads to a LFR model. Alternatively, one may formulate the two-way MANOVA as a FR model. Using the FR approach, the interaction effect is not contained in the model equation. The linear model for the two-way

MANOVA design with interaction is

$$\begin{aligned} \mathbf{y}_{ijk} &= \boldsymbol{\mu} + \boldsymbol{\alpha}_i + \boldsymbol{\beta}_j + \boldsymbol{\gamma}_{ij} + \mathbf{e}_{ijk} \quad \text{(LFR)} \\ &= \boldsymbol{\mu}_{ij} + \mathbf{e}_{ijk} \qquad\qquad\qquad \text{(FR)} \end{aligned} \tag{4.8.1}$$

$$\mathbf{e}_{ijk} \sim IN_p(\mathbf{0}, \Sigma) \quad i = 1, 2, \ldots, a;\ j = 1, 2, \ldots, b;\ k = 1, 2, \ldots, n_o > 0.$$

Writing either model in the form $\mathbf{Y}_{n\times p} = \mathbf{X}_{n\times q}\, \mathbf{B}_{q\times p} + \mathbf{E}_{n\times p}$, the $r(\mathbf{X}) = ab < q = 1 + a + b + ab$ for the LFR model and the $r(\mathbf{X}) = ab = q$ for the FR model.

For the FR model, the population cell mean vectors

$$\boldsymbol{\mu}'_{ij} = \left[\mu_{ij1}, \mu_{ij2}, \ldots, \mu_{ijp}\right] \tag{4.8.2}$$

are uniquely estimable and estimated by

$$\mathbf{y}_{ij.} = \sum_{k=1}^{n_o} \mathbf{y}_{ijk}/n_o \tag{4.8.3}$$

Letting

$$\begin{aligned} \mathbf{y}_{i..} &= \sum_{j=1}^{b} \mathbf{y}_{ij.}/b \\ \mathbf{y}_{.j.} &= \sum_{i=1}^{a} \mathbf{y}_{ij.}/a \\ \mathbf{y}_{...} &= \sum_i \sum_j \mathbf{y}_{ij.}/ab \end{aligned} \tag{4.8.4}$$

the marginal means $\boldsymbol{\mu}_{i.} = \sum_j \boldsymbol{\mu}_{ij}/b$ and $\boldsymbol{\mu}_{.j} = \sum_i \boldsymbol{\mu}_{ij}/b$ are uniquely estimable and estimated by $\widehat{\boldsymbol{\mu}}_{i.} = \mathbf{y}_{i..}$ and $\widehat{\boldsymbol{\mu}}_{.j} = \mathbf{y}_{.j.}$, the sample marginal means. Also observe that for any tetrad involving cells (i, j), (i, j'), (i', j) and (i', j') in the ab grid for factors A and B that the tetrad contrasts

$$\begin{aligned} \boldsymbol{\psi}_{i,i',j,j'} &= \boldsymbol{\mu}_{ij} - \boldsymbol{\mu}_{i'j} - \boldsymbol{\mu}_{ij'} + \boldsymbol{\mu}_{i'j'} \\ &= (\boldsymbol{\mu}_{ij} - \boldsymbol{\mu}_{i'j}) - (\boldsymbol{\mu}_{ij'} - \boldsymbol{\mu}_{i'j'}) \end{aligned} \tag{4.8.5}$$

are uniquely estimable and estimated by

$$\widehat{\boldsymbol{\psi}}_{i,i',j,j'} = \mathbf{y}_{ij.} - \mathbf{y}_{i'j.} - \mathbf{y}_{ij'.} + \mathbf{y}_{i'j'.} \tag{4.8.6}$$

These tetrad contrasts represent the difference between the differences of factor A at levels i and i', compared at the levels j and j' of factor B. If these differences are equal for all levels of A and also all levels of B, we say that no interaction exists in the FR design. Thus, the FR model has no interaction effect if and only if all tetrads or any linear combination of the tetrads are simultaneously zero. Using FR model parameters, the test of interaction is

$$H_{AB} : \text{all } \boldsymbol{\mu}_{ij} - \boldsymbol{\mu}_{i'j} - \boldsymbol{\mu}_{ij'} + \boldsymbol{\mu}_{i'j'} = \mathbf{0} \tag{4.8.7}$$

If the H_{AB} hypothesis is not significant, one next tests for significant differences in marginal means for factors A and B, called main effect tests. The tests in terms of FR model parameters are

$$\begin{aligned} H_A &: \text{all } \boldsymbol{\mu}_{i.} \text{ are equal} \\ H_B &: \text{all } \boldsymbol{\mu}_{.j} \text{ are equal} \end{aligned} \tag{4.8.8}$$

This is sometimes called the "no pool" analysis since the interaction SSCP source is ignored when testing (4.8.8).

Alternatively, if the interaction test H_{AB} is not significant one may use this information to modify the FR model. This leads to the additive FR model where the parametric functions (tetrads) in $\boldsymbol{\mu}_{ij}$ are equated to zero and this becomes a restriction on the model. This leads to the restricted MGLM discussed in Timm (1980b). Currently, these designs may not be analyzed using PROG GLM since the SAS procedure does not permit restrictions. Instead the procedure PROG REG may be used, as illustrated in Timm and Mieczkowski (1997). Given the LFR model formulation, PROC GLM may be used to analyze either additive or nonadditive models.

For the LFR model in (4.8.1), the parameters have the following structure

$$\begin{aligned} \boldsymbol{\mu}' &= \left[\mu_1, \mu_2, \ldots, \mu_p\right] \\ \boldsymbol{\alpha}'_i &= \left[\alpha_{i1}, \alpha_{i2}, \ldots, \alpha_{ip}\right] \\ \boldsymbol{\beta}'_j &= \left[\beta_{j1}, \beta_{j2}, \ldots, \beta_{jp}\right] \\ \boldsymbol{\gamma}'_{ij} &= \left[\gamma_{ij1}, \gamma_{ij2}, \ldots, \gamma_{ijp}\right] \end{aligned} \tag{4.8.9}$$

The parameters are called the constant ($\boldsymbol{\mu}$), treatment effects for factor A ($\boldsymbol{\alpha}_i$), treatment effects for factor B $\left(\boldsymbol{\beta}_j\right)$, and interaction effects AB $\left(\boldsymbol{\gamma}_{ij}\right)$. However, because the $r(\mathbf{X}) = ab < q$ is not of full rank and none of the parametric vectors are uniquely estimable. Extending Theorem 2.6.2 to the LFR two-way MANOVA model, the unique BLUEs of the parametric functions $\psi = \mathbf{c}'\mathbf{B}\mathbf{m}$ are

$$\begin{aligned} \psi &= \mathbf{c}'\mathbf{B}\mathbf{m} = \mathbf{m}'\left[\sum_i \sum_j t_{ij}\left(\boldsymbol{\mu} + \boldsymbol{\alpha}_i + \boldsymbol{\beta}_j + \boldsymbol{\gamma}_{ij}\right)\right] \\ \psi &= \mathbf{c}'\widehat{\mathbf{B}}\mathbf{m} = \mathbf{m}'\left[\sum_i \sum_j t_{ij}\mathbf{y}_{ij.}\right] \end{aligned} \tag{4.8.10}$$

where $\mathbf{t}' = \left[t_0, t_1, \ldots, t_a, t'_1, \ldots, t'_b, t_{11}, \ldots, t_{ab}\right]$ is an arbitrary vector such that $\mathbf{c}' = \mathbf{t}'\mathbf{H}$ and $\mathbf{H} = \left(\mathbf{X}'\mathbf{X}\right)^- \left(\mathbf{X}'\mathbf{X}\right)$ for $\left(\mathbf{X}'\mathbf{X}\right)^- = \begin{bmatrix} \mathbf{0} & \mathbf{0} \\ \mathbf{0} & \text{diag}\left[1/n_o\right] \end{bmatrix}$. Thus, while the individual effects are not estimable, weighted functions of the parameters are estimable. The t_{ij} are nonnegative cell weights which are selected by the researcher, Fujikoshi (1993). To illustrate, suppose the elements t_{ij} in the vector $\mathbf{t}$ are selected such that $t_{ij} = t_{i'j'} = 1$, $t_{i'j} = t_{ij'} = -1$ and all other elements are set to zero. Then

$$\boldsymbol{\psi} = \boldsymbol{\psi}_{i,i',j,j'} = \boldsymbol{\gamma}_{ij} - \boldsymbol{\gamma}_{i'j} - \boldsymbol{\gamma}_{ij'} + \boldsymbol{\gamma}_{i'j'} \tag{4.8.11}$$

is estimable, even though the individual $\boldsymbol{\gamma}_{ij}$ are not estimable. They are uniquely estimated by

$$\widehat{\boldsymbol{\psi}} = \widehat{\boldsymbol{\psi}}_{i,i',\,j,\,j'} = \mathbf{y}_{ij.} - \mathbf{y}_{i'j.} - \mathbf{y}_{ij'.} + \mathbf{y}_{i'j'.} \tag{4.8.12}$$

The vector $\mathbf{m}$ is used to combine means across variables. Furthermore, the estimates do not depend on $(\mathbf{X}'\mathbf{X})^{-}$. Thus, an interaction in the LFR model involving the parameters $\boldsymbol{\gamma}_{ij}$ is identical to the formulation of an interaction using the parameters $\boldsymbol{\mu}_{ij}$ in the FR model. As shown by Graybill (1976, p. 560) for one variable, the test of no interaction or additivity has the following four equivalent forms, depending on the model used to represent the two-way MANOVA,

$$\begin{aligned}
&\text{(a)}\quad H_{AB}: \boldsymbol{\mu}_{ij} - \boldsymbol{\mu}_{i'j} - \boldsymbol{\mu}_{ij'} + \boldsymbol{\mu}_{i'j'} = \mathbf{0}\\
&\text{(b)}\quad H_{AB}: \boldsymbol{\gamma}_{ij} - \boldsymbol{\gamma}_{i'j} - \boldsymbol{\gamma}_{ij'} + \boldsymbol{\gamma}_{i'j'} = \mathbf{0}\\
&\text{(c)}\quad H_{AB}: \boldsymbol{\mu}_{ij} = \boldsymbol{\mu} + \boldsymbol{\alpha}_i + \boldsymbol{\beta}_j\\
&\text{(d)}\quad H_{AB}: \boldsymbol{\gamma}_{ij} = \mathbf{0}
\end{aligned} \tag{4.8.13}$$

for all subscripts $i, i',\ j$ and j'. Most readers will recognize (d) which requires adding side conditions to the LFR model to convert the model to full rank. Then, all parameters become estimable

$$\begin{aligned}
\widehat{\boldsymbol{\mu}} &= \mathbf{y}_{...}\\
\widehat{\boldsymbol{\alpha}}_i &= \mathbf{y}_{i..} - \mathbf{y}_{...}\\
\widehat{\boldsymbol{\beta}}_j &= \mathbf{y}_{.j.} - \mathbf{y}_{...}\\
\widehat{\boldsymbol{\gamma}}_{ij} &= \mathbf{y}_{ij.} - \mathbf{y}_{i..} - \mathbf{y}_{.j.} + \mathbf{y}_{...}
\end{aligned}$$

This is the approach used in PROC ANOVA. Models with structure (c) are said to be additive. We discuss the additive model later in this section.

Returning to (4.8.10), suppose the cell weights are chosen such that

$$\sum_j t_{ij} = \sum_j t_{i'j} = 1 \quad \text{for } i \neq i'.$$

Then

$$\boldsymbol{\psi} = \boldsymbol{\alpha}_i - \boldsymbol{\alpha}_{i'} + \sum_{j=1}^{b} t_{ij}(\boldsymbol{\beta}_j + \boldsymbol{\gamma}_{ij}) - \sum_{j=1}^{b} t_{i'j}(\boldsymbol{\beta}_j + \boldsymbol{\gamma}_{i'j})$$

is estimable and estimated by

$$\widehat{\boldsymbol{\psi}} = \sum_j t_{ij}\mathbf{y}_{ij.} - \sum_j t_{i'j}\mathbf{y}_{i'j.}$$

By choosing $t_{ij} = 1/b$, the function

$$\begin{aligned}
\boldsymbol{\psi} &= \boldsymbol{\alpha}_i - \boldsymbol{\alpha}_{i'} + (\boldsymbol{\beta}_. + \boldsymbol{\gamma}_{i.}) - (\boldsymbol{\beta}_. + \boldsymbol{\gamma}_{i'.})\\
&= \boldsymbol{\alpha}_i - \boldsymbol{\alpha}_{i'} + (\boldsymbol{\gamma}_{i.} - \boldsymbol{\gamma}_{i'.})
\end{aligned}$$

is confounded by the parameters $\boldsymbol{\gamma}_{i.}$ and $\boldsymbol{\gamma}_{i'.}$. However, the estimate of $\boldsymbol{\psi}$ is

$$\widehat{\boldsymbol{\psi}} = \mathbf{y}_{i..} - \mathbf{y}_{i'..}$$

This shows that one may not test for differences in the treatment levels of factor A in the presence of interactions. Letting $\boldsymbol{\mu}_{ij} = \boldsymbol{\mu} + \boldsymbol{\alpha}_i + \boldsymbol{\beta}_j + \boldsymbol{\gamma}_{ij}$, $\boldsymbol{\mu}_{i.} = \boldsymbol{\alpha}_i + \boldsymbol{\gamma}_{i.}$ so that $\boldsymbol{\psi} = \boldsymbol{\mu}_{i.} - \boldsymbol{\mu}_{i'.} = \boldsymbol{\alpha}_i - \boldsymbol{\alpha}_{i'} + (\boldsymbol{\gamma}_{i.} - \boldsymbol{\gamma}'_{i.})$. Thus, testing that all $\boldsymbol{\mu}_{i.}$ are equal in the FR model with interaction is identical to testing for treatment effects associated with factor A. The test becomes

$$H_A : \text{all } \boldsymbol{\alpha}_i + \sum_j \boldsymbol{\gamma}_{ij}/b \text{ are equal} \tag{4.8.14}$$

for the LFR model. Similarly,

$$\boldsymbol{\psi} = \boldsymbol{\beta}_j - \boldsymbol{\beta}_{j'} + \sum_i t_{ij}(\boldsymbol{\alpha}_i + \boldsymbol{\gamma}_{ij}) - \sum_i t_{ij'}(\boldsymbol{\alpha}_i + \boldsymbol{\gamma}_{ij'})$$

is estimable and estimated by

$$\widehat{\boldsymbol{\psi}} = \mathbf{y}_{.j.} - \mathbf{y}_{.j'.}$$

provided the cell weights t_{ij} are selected such that the $\sum_i t_{ij} = \sum_i t_{ij'} = 1$ for $j \neq j'$. Letting $t_{ij} = 1/a$, the test of B becomes

$$H_B : \text{all } \boldsymbol{\beta}_j + \sum_i \boldsymbol{\gamma}_{ij} \,/\, a \text{ are equal} \tag{4.8.15}$$

for the LFR model. Using PROC GLM, the tests of H_A, H_B and H_{AB} are called Type III tests and the estimates are based upon LSMEANS.

PROC GLM in SAS employs a different g-inverse for $(\mathbf{X}'\mathbf{X})^-$ so that the general form may not agree with the expression given in (4.8.10). To output the specific structure used in SAS, the / E option on the MODEL statement is used.

The tests H_A, H_B and H_{AB} may also be represented using the general matrix form $\mathbf{CBM} = \mathbf{0}$ where the matrix $\mathbf{C}$ is selected as $\mathbf{C}_A$, $\mathbf{C}_B$ and $\mathbf{C}_{AB}$ for each test and $\mathbf{M} = \mathbf{I}_p$. To illustrate, suppose we consider a 3×2 design where factor A has three levels ($a = 3$) and factor B has two levels ($b = 2$) as shown in Figure 4.8.1

Then forming tetrads $\psi_{i,\, i',\, j,\, j'}$, the test of interaction H_{AB} is

$$H_{AB} : \begin{matrix} \boldsymbol{\gamma}_{11} - \boldsymbol{\gamma}_{21} - \boldsymbol{\gamma}_{12} + \boldsymbol{\gamma}_{22} = \mathbf{0} \\ \\ \boldsymbol{\gamma}_{21} - \boldsymbol{\gamma}_{31} - \boldsymbol{\gamma}_{22} + \boldsymbol{\gamma}_{32} = \mathbf{0} \end{matrix}$$

as illustrated by the arrows in Figure 4.8.1. The matrix $\mathbf{C}_{AB}$ for testing H_{AB} is

$$\underset{2\times 12}{\mathbf{C}_{AB}} = \left[\begin{array}{c} \mathbf{0} \\ 2\times 6 \end{array} \begin{array}{cccccc} \vdots & 1 & -1 & -1 & 1 & 0 & 0 \\ \vdots & 0 & 0 & 1 & -1 & -1 & 1 \end{array} \right]$$

where the $r(\mathbf{C}_{AB}) = v_{AB} = (a-1)(b-1) = 2 = v_h$. To test H_{AB} the hypothesis test matrix H_{AB} and error matrix $\mathbf{E}$ are formed. For the two-way MANOVA design, the error

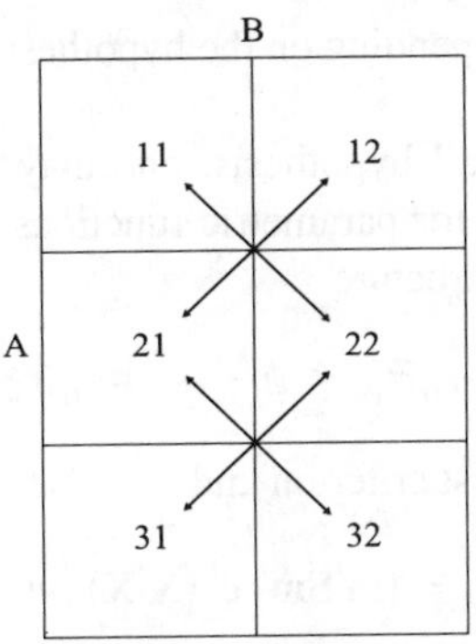

FIGURE 4.8.1. 3 × 2 Design

matrix is

$$
\begin{aligned}
\mathbf{E} &= \mathbf{Y}'\left[\mathbf{I} - \mathbf{X}\left(\mathbf{X}'\mathbf{X}\right)^{-}\mathbf{X}'\right]\mathbf{Y} \\
&= \sum_i \sum_j \sum_k \left(\mathbf{y}_{ijk} - \mathbf{y}_{ij.}\right)\left(\mathbf{y}_{ijk} - \mathbf{y}_{ij.}\right)'
\end{aligned}
\tag{4.8.16}
$$

and $v_e = n - r(\mathbf{X}) = abn_o - ab = ab(n_o - 1)$. SAS automatically creates $\mathbf{C}_{AB}$ so that a computational formula for $\mathbf{H}_{AB}$ is not provided. It is very similar to the univariate ANOVA formula with sum of squares replaced by outer vector products. SAS also creates $\mathbf{C}_A$ and $\mathbf{C}_B$ to test H_A and H_B with $v_a = (a-1)$ and $v_b = (b-1)$ degrees of freedom. Their structure is similar to the one-way MANOVA matrices consisting of $1's$ and $-1's$ to compare the levels of factor A and factor B; however, because main effects are confounded by interactions $\boldsymbol{\gamma}_{ij}$ in the LFR model, there are also $1's$ and $0's$ associated with the $\boldsymbol{\gamma}_{ij}$. For our 3 × 2 design, the hypothesis test matrices for Type III tests are

$$
\mathbf{C}_A = \begin{bmatrix} 0 & \vdots & 1 & 0 & -1 & \vdots & 0 & 0 & \vdots & 1 & 1 & 0 & 0 & -1 & -1 \\ 0 & \vdots & 0 & 1 & -1 & \vdots & 0 & 0 & \vdots & 0 & 0 & 1 & 1 & -1 & -1 \end{bmatrix}
$$

$$
\mathbf{C}_B = \begin{bmatrix} 0 & \vdots & 0 & 0 & 0 & \vdots & 1 & -1 & \vdots & 1 & 1 & 1 & 1 & -1 & -1 \end{bmatrix}
$$

where the $r(\mathbf{C}_A) = v_A = a-1$ and the $r(\mathbf{C}_B) = v_B = (b-1)$. The matrices $\mathbf{C}_A$ and $\mathbf{C}_B$ are partitioned to represent the parameters $\boldsymbol{\mu}$, $\boldsymbol{\alpha}_i$, $\boldsymbol{\beta}_j$ and γ_{ij} in the matrix $\mathbf{B}$. SAS commands to perform the two-way MANOVA analysis using PROC GLM and either the FR or LFR model with interaction are discussed in the example using the reduction procedure.

The parameters s, M and N, required to test the multivariate hypotheses are summarized by

$$
\begin{aligned}
s &= \min(v_h, p) \\
M &= (|v_h - p| - 1)/2 \\
N &= (v_e - p - 1)/2
\end{aligned}
\tag{4.8.17}
$$

where v_h equals v_A, v_B or v_{AB}, depending on the hypothesis of interest, and $v_e = ab(n_o - 1)$.

With the rejection of any overall hypothesis, one may again establish $100\,(1-\alpha)\,\%$ simultaneous confidence intervals for parametric functions $\psi = \mathbf{c}'\mathbf{B}\mathbf{m}$ that are estimable. Confidence sets have the general structure

$$\hat{\psi} - c_\alpha \hat{\sigma}_{\hat{\psi}} \leq \psi \leq \hat{\psi} + c_\alpha \hat{\sigma}_{\hat{\psi}}$$

where c_α depends on the overall test criterion and

$$\hat{\sigma}^2_{\hat{\psi}} = (\mathbf{m}'\mathbf{S}\mathbf{m})\,\mathbf{c}'\left(\mathbf{X}'\mathbf{X}\right)^{-}\mathbf{c} \tag{4.8.18}$$

where $\mathbf{S} = \mathbf{E}/v_e$. For tetrad contrasts

$$\hat{\sigma}^2_{\hat{\psi}} = 4\left(\mathbf{m}'\mathbf{S}\mathbf{m}\right)/\,n_o \tag{4.8.19}$$

Alternatively, if one is only interested in a finite number of parameters μ_{ij} (say), a variable at a time, one may use some approximate method to construct simultaneous intervals or us the stepdown FIT procedure.

b. *Additive Two-Way MANOVA*

If one assumes an additive model for a two-way MANOVA design, which is common in a randomized block design with $n_o = 1$ observation per cell or in factorial designs with $n_o > 1$ observations per cell, one may analyze the design using either a FR or LFR model if all $n_o = 1$; however, if $n_o > 1$ one must use a restricted FR model or a LFR model. Since the LFR model easily solves both situations, we discuss the LFR model. For the additive LFR representation, the model is

$$\begin{aligned} \mathbf{y}_{ijk} &= \boldsymbol{\mu} + \boldsymbol{\alpha}_i + \boldsymbol{\beta}_j + \mathbf{e}_{ijk} \\ \mathbf{e}_{ijk} &\sim IN_p\,(\mathbf{0},\ \Sigma) \\ i &= 1,2,\ldots,a;\ j = 1,2,\ldots,b;\ k = 1,2,\ldots,n_o \geq 1 \end{aligned} \tag{4.8.20}$$

A common situation is to have one observation per cell. Then (4.8.20) becomes

$$\begin{aligned} \mathbf{y}_{ij} &= \boldsymbol{\mu} + \boldsymbol{\alpha}_i + \boldsymbol{\beta}_j + \mathbf{e}_{ij} \\ \mathbf{e}_{ij} &\sim IN_p\,(\mathbf{0},\ \Sigma)\, i = 1,2,\ldots,a;\ j = 1,2,\ldots,b \end{aligned} \tag{4.8.21}$$

We consider the case with $n_o = 1$ in some detail for the 3×2 design given in Figure 4.8.1. The model in matrix form is

$$\begin{bmatrix} \mathbf{y}'_{11} \\ \mathbf{y}'_{12} \\ \mathbf{y}'_{21} \\ \mathbf{y}'_{22} \\ \mathbf{y}'_{31} \\ \mathbf{y}'_{32} \end{bmatrix} = \begin{bmatrix} 1 & 1 & 0 & 0 & 1 & 0 \\ 1 & 1 & 0 & 0 & 0 & 1 \\ 1 & 0 & 1 & 0 & 1 & 0 \\ 1 & 0 & 1 & 0 & 0 & 1 \\ 1 & 0 & 0 & 1 & 1 & 0 \\ 1 & 0 & 0 & 1 & 0 & 1 \end{bmatrix} \begin{bmatrix} \boldsymbol{\mu}' \\ \boldsymbol{\alpha}'_1 \\ \boldsymbol{\alpha}'_2 \\ \boldsymbol{\alpha}'_3 \\ \boldsymbol{\beta}'_1 \\ \boldsymbol{\beta}'_2 \end{bmatrix} + \begin{bmatrix} \mathbf{e}'_{11} \\ \mathbf{e}'_{12} \\ \mathbf{e}'_{21} \\ \mathbf{e}'_{22} \\ \mathbf{e}'_{31} \\ \mathbf{e}'_{32} \end{bmatrix} \tag{4.8.22}$$

where the structure for $\boldsymbol{\mu}, \boldsymbol{\alpha}_i, \boldsymbol{\beta}_j$ follow that given in (4.8.9). Now, $(\mathbf{X}'\mathbf{X})$ is

$$(\mathbf{X}'\mathbf{X}) = \begin{bmatrix} 6 & \vdots & 2 & 2 & 2 & \vdots & 3 & 3 \\ \cdots & \cdots & \cdots & \cdots & \cdots & \cdots & \cdots & \cdots \\ 2 & \vdots & 2 & 0 & 0 & \vdots & 1 & 1 \\ 2 & \vdots & 0 & 2 & 0 & \vdots & 1 & 1 \\ 2 & \vdots & 0 & 0 & 2 & \vdots & 1 & 1 \\ \cdots & \cdots & \cdots & \cdots & \cdots & \cdots & \cdots & \cdots \\ 3 & \vdots & 1 & 1 & 1 & \vdots & 3 & 0 \\ 3 & \vdots & 1 & 1 & 1 & \vdots & 0 & 3 \end{bmatrix} \tag{4.8.23}$$

and a g-inverse is given by

$$(\mathbf{X}'\mathbf{X})^{-} = \begin{bmatrix} -1/6 & 0 & 0 & 0 & 0 & 0 \\ 0 & 1/2 & 0 & 0 & 0 & 0 \\ 0 & 0 & 1/2 & 0 & 0 & 0 \\ 0 & 0 & 0 & 1/2 & 0 & 0 \\ 0 & 0 & 0 & 0 & 1/3 & 0 \\ 0 & 0 & 0 & 0 & 0 & 1/3 \end{bmatrix}$$

so that

$$\mathbf{H} = (\mathbf{X}'\mathbf{X})^{-}(\mathbf{X}'\mathbf{X}) = \begin{bmatrix} -1 & -1/3 & -1/3 & -1/3 & -1/2 & -1/2 \\ 1 & 1 & 0 & 0 & 1/2 & 1/2 \\ 1 & 0 & 1 & 0 & 1/2 & 1/2 \\ 1 & 0 & 0 & 1 & 1/2 & 1/2 \\ 1 & 1/3 & 1/3 & 1/3 & 1 & 0 \\ 1 & 1/3 & 1/3 & 1/3 & 0 & 1 \end{bmatrix}$$

More generally,

$$(\mathbf{X}'\mathbf{X})^{-} = \begin{bmatrix} -1/n & \mathbf{0}' & \mathbf{0}' \\ \mathbf{0} & b^{-1}\mathbf{I}_a & \mathbf{0} \\ \mathbf{0} & \mathbf{0} & a^{-1}\mathbf{I}_b \end{bmatrix}$$

and

$$\mathbf{H} = (\mathbf{X}'\mathbf{X})^{-}(\mathbf{X}'\mathbf{X}) = \begin{bmatrix} -1 & -a^{-1}\mathbf{1}'_a & -b^{-1}\mathbf{1}'_b \\ \mathbf{1}_a & \mathbf{I}_a & b^{-1}\mathbf{J}_{ab} \\ \mathbf{1}_b & b^{-1}\mathbf{J}_{ab} & \mathbf{I}_b \end{bmatrix}$$

Then a solution for $\widehat{\mathbf{B}}$ is

$$\widehat{\mathbf{B}} = (\mathbf{X}'\mathbf{X})^{-} \mathbf{X}'\mathbf{Y} = \begin{bmatrix} -\mathbf{y}'_{..} \\ \cdots \\ \mathbf{y}'_{1.} \\ \vdots \\ \mathbf{y}'_{a.} \\ \cdots \\ \mathbf{y}'_{.1} \\ \vdots \\ \mathbf{y}'_{.b} \end{bmatrix}$$

where

$$\begin{aligned} \mathbf{y}_{i.} &= \sum_{j=1}^{b} \mathbf{y}_{ij}/b \\ \mathbf{y}_{.j} &= \sum_{i=1}^{a} \mathbf{y}_{ij}/a \\ \mathbf{y}_{..} &= \sum_{i}\sum_{j} \mathbf{y}_{ij}/ab \end{aligned}$$

With $\mathbf{c}'\mathbf{H} = \mathbf{c}'$, the BLUE for the estimable functions $\psi = \mathbf{c}'\mathbf{B}\mathbf{m}$ are

$$\begin{aligned} \psi &= \mathbf{c}'\mathbf{B}\mathbf{m} = \mathbf{m}' \left[-t_0 \left(\boldsymbol{\mu} + \boldsymbol{\alpha}_. + \boldsymbol{\beta}_. \right) + \right. \\ & \left. \sum_{i=1}^{a} t_i \left(\boldsymbol{\mu} + \boldsymbol{\alpha}_i + \boldsymbol{\beta}_. \right) + \sum_{j=1}^{b} t'_j \left(\boldsymbol{\mu} + \boldsymbol{\alpha}_i + \boldsymbol{\beta}_. \right) \right] \\ \psi &= \mathbf{c}'\widehat{\mathbf{B}}\mathbf{m} = \mathbf{m}' \left[-t_0 \mathbf{y} + \sum_{i=1}^{a} t_i \mathbf{y}_{i.} + \sum_{j=1}^{b} t'_j \mathbf{y}_{.j} \right] \end{aligned} \tag{4.8.24}$$

where

$$\begin{aligned} \mathbf{t}' &= \left[t_0, t_1, t_2, \ldots, t_a, t'_1, t'_2, \ldots, t'_b \right] \\ \boldsymbol{\alpha}_. &= \sum_i \boldsymbol{\alpha}_i / a \text{ and } \boldsymbol{\beta}_. = \sum_j \boldsymbol{\beta}_j \,/\, b \end{aligned}$$

Applying these results to the 3×2 design, (4.8.24) reduces to

$$\begin{aligned} \boldsymbol{\psi} =& -t_0 \left(\boldsymbol{\mu} + \boldsymbol{\alpha}_. + \boldsymbol{\beta}_. \right) + \left(\boldsymbol{\mu} + \boldsymbol{\alpha}_1 + \boldsymbol{\beta}_. \right) t_1 + \\ & \left(\boldsymbol{\mu} + \boldsymbol{\alpha}_2 + \boldsymbol{\beta}_. \right) t_2 + \left(\boldsymbol{\mu} + \boldsymbol{\alpha}_3 + \boldsymbol{\beta}_. \right) t_3 + \\ & \left(\boldsymbol{\mu} + \boldsymbol{\alpha}_. + \boldsymbol{\beta}_1 \right) t'_1 + \left(\boldsymbol{\mu} + \boldsymbol{\alpha}_. + \beta_2 \right) t'_2 \end{aligned}$$

$$\widehat{\boldsymbol{\psi}} = -t_0 \mathbf{y}_{..} + \sum_{i=1}^{a} t_i \mathbf{y}_{i.} + \sum_{j=1}^{b} t'_i \mathbf{y}_{.j}$$

so that ignoring $\mathbf{m}$, $\boldsymbol{\psi}_1 = \boldsymbol{\beta}_1 - \boldsymbol{\beta}_2$, $\boldsymbol{\psi}_2 = \boldsymbol{\alpha}_i - \boldsymbol{\alpha}_{i'}$ and $\boldsymbol{\psi}_3 = \boldsymbol{\mu} + \boldsymbol{\alpha}_. + \boldsymbol{\beta}_.$ are estimable, and are estimated by $\widehat{\boldsymbol{\psi}}_1 = \mathbf{y}_{.1} - \mathbf{y}_{.2}$, $\widehat{\boldsymbol{\psi}}_2 = \mathbf{y}_{i.} - \mathbf{y}_{i'.}$ and $\widehat{\boldsymbol{\psi}}_3 = \mathbf{y}_{..}$. However, $\boldsymbol{\mu}$ and individual effects $\boldsymbol{\alpha}_i$ and $\boldsymbol{\beta}_j$ are not estimable since for $\mathbf{c}' = [0, 1_i, 0, \ldots, 0]$, $\mathbf{c}'\mathbf{H} \neq \mathbf{c}'$ for any vector $\mathbf{c}$ with a 1 in the ith position. In SAS, the general structure of estimable functions is obtained by using the /E option on the MODEL statement.

For additive models, the primary tests of interest are the main effect tests for differences in effects for factor A or factor B

$$\begin{aligned} H_A &: \text{all } \boldsymbol{\alpha}_i \text{ are equal} \\ H_B &: \text{all } \boldsymbol{\beta}_i \text{ are equal} \end{aligned} \tag{4.8.25}$$

The hypothesis test matrices $\mathbf{C}$ are constructed in a way similar to the one-way MANOVA model. For example, comparing all levels of A (or B) with the last level of A (or B), the matrices become

$$\begin{aligned} \underset{(a-1)\times q}{\mathbf{C}_A} &= \left[\mathbf{0}, \mathbf{I}_{a-1}, -\mathbf{1}, \mathbf{0}_{b\times b}\right] \\ \underset{(b-1)\times q}{\mathbf{C}_B} &= \left[\mathbf{0}, \mathbf{0}_{a\times a}, \mathbf{I}_{b-1}, -\mathbf{1}\right] \end{aligned} \tag{4.8.26}$$

so that $v_A = r(\mathbf{C}_A) = a - 1$ and $v_B = r(\mathbf{C}_B) = b - 1$. Finally, the error matrix $\mathbf{E}$ may be shown to have the following structure

$$\begin{aligned} \mathbf{E} &= \mathbf{Y}'\left[\mathbf{I} - \mathbf{X}\left(\mathbf{X}'\mathbf{X}\right)^{-}\mathbf{X}'\right]\mathbf{Y} \\ &= \sum_{i=1}^{a}\sum_{j=1}^{b}(\mathbf{y}_{ij} - \mathbf{y}_{i.} - \mathbf{y}_{.j} + \mathbf{y}_{..})(\mathbf{y}_{ij} - \mathbf{y}_{i.} - \mathbf{y}_{.j} + \mathbf{y}_{..})' \end{aligned} \tag{4.8.27}$$

with degrees of freedom $v_e = n - r(\mathbf{X}) = n - (a + b - 1) = ab - a - b + 1 = (a-1)(b-1)$.

The parameters s, M, and N for these tests are

$$\begin{array}{ll} \underline{\text{Factor A}} & \underline{\text{Factor B}} \\ s = \min(a-1,\ p) & s = \min(b-1,\ p) \\ M = (|a-p-1|-1)/2 & M = (|b-p-1|-1)/2 \\ N = (n-a-b-p)/2 & N = (n-a-b-p)/2 \end{array} \tag{4.8.28}$$

If the additive model has $n_o > 1$ observations per cell, observe that the degrees of freedom for error becomes $v_e^* = abn_o - (a + b - 1) = ab(n_o - 1) + (a-1)(b-1)$ which is obtained from pooling the interaction degrees of freedom with the within error degrees of freedom in the two-way MANOVA model. Furthermore, the error matrix $\mathbf{E}$ for the design with $n_o > 1$ observations is equal to the sum of the interaction SSCP matrix and the error matrix for the two-way MANOVA design. Thus, one is confronted with the problem of whether to "pool" or "not to pool" when analyzing the two-way MANOVA design. Pooling when the interaction test is not significant, we are saying that there is no interaction so that main effects are not confounded with interaction. Due to lack of power, we could have made a Type II error regarding the interaction term. If the interaction is present, tests of main effects are confounded by interaction. Similarly, we could reject the test of interaction and make a Type I error. This leads one to investigate pairwise comparisons at various levels of the other factor, called simple effects. With a well planned study that has significant power to detect the presences of interaction, we recommend that the "pool" strategy be employed. For further discussion on this controversy see Scheffé (1959, p. 126), Green and Tukey (1960), Hines (1996), Mittelhammer et al. (2000, p. 80) and Janky (2000).

Using (4.8.18), estimated standard errors for pairwise comparisons have a simple structure for treatment differences involving factors A and B follow

$$\begin{aligned} A &: \widehat{\sigma}^2_{\widehat{\psi}} = 2\left(\mathbf{m}'\mathbf{S}\mathbf{m}\right)/bn_o \\ B &: \widehat{\sigma}^2_{\widehat{\psi}} = 2\left(\mathbf{m}'\mathbf{S}\mathbf{m}\right)/an_o \end{aligned} \tag{4.8.29}$$

when $\mathbf{S}$ is the pooled estimate of Σ. Alternatively, one may also use the FIT procedure to evaluate planned comparisons.

c. *Two-Way MANCOVA*

Extending the two-way MANOVA design to include covariates, one may view the two-way classification as a one-way design with ab independent populations. Assuming the matrix of coefficients associated with the vector of covariates is equal over all of the ab populations, the two-way MANCOVA model with interaction is

$$\begin{aligned} \mathbf{y}_{ijk} &= \boldsymbol{\mu} + \boldsymbol{\alpha}_i + \boldsymbol{\beta}_j + \boldsymbol{\gamma}_{ij} + \Gamma'\mathbf{z}_{ijk} + \mathbf{e}_{ijk} \qquad \text{(LFR)} \\ &= \boldsymbol{\mu}_{ij} + \Gamma'\mathbf{z}_{ijk} + \mathbf{e}_{ijk} \qquad \text{(FR)} \end{aligned} \tag{4.8.30}$$

$$\mathbf{e}_{ijk} \sim IN_p\left(\mathbf{0}, \Sigma\right) i = 1, 2, \ldots, a;\ j = 1, 2, \ldots, b;\ k = 1, 2, \ldots, n_o > 0$$

Again estimates and tests are adjusted for covariates. If the ab matrices Γ are not equal, one may consider the multivariate intraclass covariance model for the ab populations.

d. *Tests of Nonadditivity*

When a two-way design has more than one observation per cell, we may test for interaction or nonadditivity. With one observation per cell, we saw that the SSCP matrix becomes the error matrix so that no test of interaction is evident. A test for interaction in the univariate model was first proposed by Tukey (1949) and generalized by Scheffé (1959, p. 144, problem 4.19). Milliken and Graybill (1970, 1971) and Kshirsagar (1993) examine the test using the expanded linear model (ELM) which allows one to include nonlinear terms with conventional linear model theory. Using the MGLM, McDonald (1972) and Kshirsagar (1988) extended the results of Milliken and Graybill to the expanded multiple design multivariate (EMDM) model, the multivariate analogue of the ELM. Because the EMDM is a SUR model, we discuss the test in Chapter 5.

4.9 Two-Way MANOVA/MANCOVA Example

a. *Two-Way MANOVA (Example 4.9.1)*

We begin with the analysis of a two-way MANOVA design. The data for the design are given in file twoa.dat and are shown in Table 4.9.1. The data were obtained from a larger study, by Mr. Joseph Raffaele at the University of Pittsburgh to analyze reading comprehension (C) and reading rate (R). The scores were obtained using subtest scores of the Iowa Test of Basic Skills. After randomly selecting $n = 30$ students for the study and randomly dividing them into six subsamples of size 5, the groups were randomly assigned to two treatment conditions-contract classes and noncontract classes-and three teachers; a total of $n_o = 5$ observations are in each cell. The achievement data for the experiment are conveniently represented by cells in Table 4.9.1.

Calculating the cell means for the study using the MEANS statement, Table 4.9.2 is obtained.

The mathematical model for the example is

$$\begin{aligned} \mathbf{y}_{ijk} &= \boldsymbol{\mu} + \boldsymbol{\alpha}_i + \boldsymbol{\beta}_j + \boldsymbol{\gamma}_{ij} + \boldsymbol{\epsilon}_{ijk} \\ \boldsymbol{\epsilon}_{ijk} &\sim \text{IN}(\mathbf{0}, \Sigma) \qquad i = 1, 2, 3;\ j = 1, 2;\ k = 1, 2, \ldots, n_o = 5 \end{aligned} \tag{4.9.1}$$

In PROC GLM, the MODEL statement is used to define the model in program m4_9_1.sas. To the left of the equal sign one places the names of the dependent variables, Rate and

TABLE 4.9.1. Two-Way MANOVA

		Factor B			
		Contract Class		Noncontract Class	
		R	C	R	C
		10	21	9	14
		12	22	8	15
	Teacher 1	9	19	11	16
		10	21	9	17
		14	23	9	17
		11	23	11	15
		14	27	12	18
Factor A	Teacher 2	14	17	12	18
		15	26	9	17
		14	24	9	18
		8	17	9	22
		7	12	8	18
	Teacher 3	10	18	10	17
		8	17	9	19
		7	19	8	19

TABLE 4.9.2. Cell Means for Example Data

	B_1	B_2	Means
A_1	$\mathbf{y}'_{11.} = [11.00, 21.20]$	$\mathbf{y}'_{12.} = [\ 9.20, 15.80]$	$\mathbf{y}'_{1..} = [10.10, 18.50]$
A_2	$\mathbf{y}'_{21.} = [13.40, 24.80]$	$\mathbf{y}'_{22.} = [10.20, 16.80]$	$\mathbf{y}'_{2.} = [11.80, 20.80]$
A_3	$\mathbf{y}'_{31.} = [\ 8.00, 17.20]$	$\mathbf{y}'_{32.} = [\ 8.80, 19.00]$	$\mathbf{y}'_{3.} = [\ 8.40, 18.10]$
Mean	$\mathbf{y}'_{.1.} = [10.80, 21.07]$	$\mathbf{y}'_{.2.} = [\ 9.40, 17.20]$	$\mathbf{y}'_{...} = [10.10, 19.13]$

Variable 1: reading rate (R)

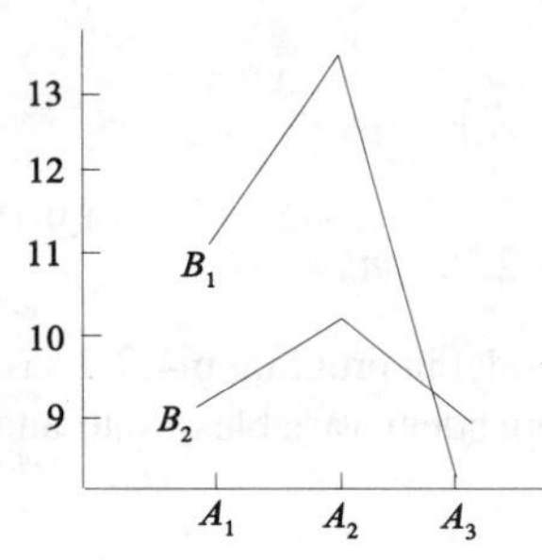

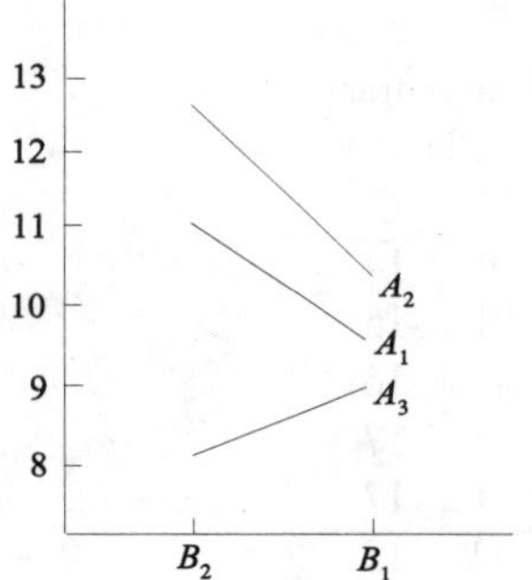

Variable 2: reading comprehension (C)

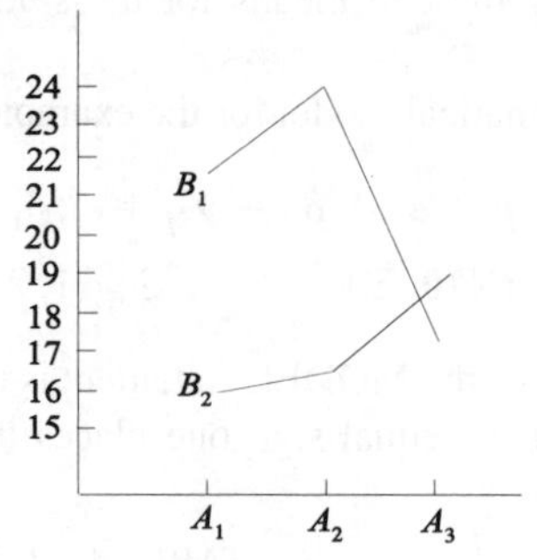

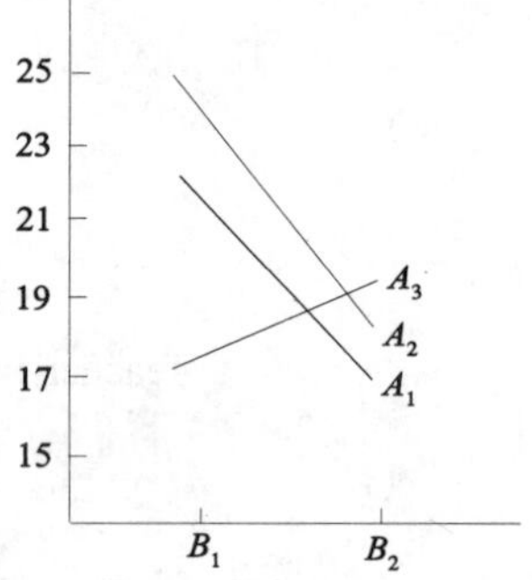

FIGURE 4.9.1. Plots of Cell Means for Two-Way MANOVA

Comp, to the right of the equal sign are the effect names,T, C and T*C. The asterisk between the effect names denotes the interaction term in the model, γ_{ij}.

Before testing the three mean hypotheses of interest for the two-way design, plots of the cell means, a variable at a time, are constructed and shown in Figure 4.9.1. From the plots, it appears that a significant interaction may exist in the data. The hypotheses of interest

TABLE 4.9.3. Two-Way MANOVA Table

Source	df	SSCP
A (Teachers)	2	$\mathbf{H}_A = \begin{bmatrix} 57.8000 & \\ 45.9000 & 42.4667 \end{bmatrix}$
B (Class)	1	$\mathbf{H}_B = \begin{bmatrix} 14.7000 & \\ 40.6000 & 112.1333 \end{bmatrix}$
Interaction AB $(T * C)$	2	$\mathbf{H}_{AB} = \begin{bmatrix} 20.6000 & \\ 51.3000 & 129.8667 \end{bmatrix}$
Within error	24	$\mathbf{E} =$ [given in (4.9.3)]
"Total"	29	$\mathbf{H}+\mathbf{E} \begin{bmatrix} 138.7000 & \\ 157.6000 & 339.4666 \end{bmatrix}$

become

$$
\begin{aligned}
H_{\mathbf{A}} &: \boldsymbol{\alpha}_1 + \frac{\sum_j \boldsymbol{\gamma}_{1j}}{b} = \boldsymbol{\alpha}_2 + \frac{\sum_j \boldsymbol{\gamma}_{2j}}{b} = \boldsymbol{\alpha}_3 + \frac{\sum_j \boldsymbol{\gamma}_{3j}}{b} \\
H_{\mathbf{B}} &: \boldsymbol{\beta}_1 + \frac{\sum_i \boldsymbol{\gamma}_{i1}}{a} = \boldsymbol{\beta}_2 + \frac{\sum_i \boldsymbol{\gamma}_{i2}}{a} \\
H_{\mathbf{AB}} &: \boldsymbol{\gamma}_{11} - \boldsymbol{\gamma}_{31} - \boldsymbol{\gamma}_{12} - \boldsymbol{\gamma}_{32} = \mathbf{0} \\
&\quad \boldsymbol{\gamma}_{21} - \boldsymbol{\gamma}_{31} - \boldsymbol{\gamma}_{22} - \boldsymbol{\gamma}_{32} = \mathbf{0}
\end{aligned}
\tag{4.9.2}
$$

To test any of the hypotheses in (4.9.2), the estimate of **E** is needed. The formula for **E** is

$$
\begin{aligned}
\mathbf{E} &= \mathbf{Y}'\left[\mathbf{I} - \mathbf{X}\left(\mathbf{X}'\mathbf{X}\right)^{-}\mathbf{X}'\right]\mathbf{Y} \\
&= \sum_i \sum_j \sum_k \left(\mathbf{y}_{ijk} - \mathbf{y}_{ij.}\right)\left(\mathbf{y}_{ijk} - \mathbf{y}_{ij.}\right)' \\
&= \begin{bmatrix} 45.6 & \\ 19.8 & 56.0 \end{bmatrix}
\end{aligned}
\tag{4.9.3}
$$

Thus

$$
\mathbf{S} = \frac{\mathbf{E}_e}{v_e} = \begin{bmatrix} 1.9000 & \\ 0.8250 & 2.3333 \end{bmatrix}
$$

where $v_e = n - r\,(\mathbf{X}) = 30 - 6 = 24$.

To find the hypothesis test matrices $\mathbf{H}_A$, $\mathbf{H}_B$ and $\mathbf{H}_{AB}$ using PROC GLM, the MANOVA statement is used. The statement usually contains the model to the right of term $h =$ on the MANOVA statement. This generates the hypothesis test matrices $\mathbf{H}_A$, $\mathbf{H}_B$ and $\mathbf{H}_{AB}$ which we have asked to be printed, along with **E**. From the output, one may construct Table 4.9.3, the MANOVA table for the example.

Using the general formula for $s = \min(v_h, p)$, $M = (\mid v_h - p \mid -1)/2$ and $N = (v_e - p - 1)/2$ with $p = 2$, $v_e = 24$ and v_h defined in Table 4.9.3, the values of s, M, and N for H_A and H_{AB} are $s = 2$, $M = -0.5$, and $N = 10.5$. For $\mathbf{H}_B$, $s = 1$, $M = 0$, and $N = 10.5$. Using $\alpha = 0.05$ for each test, and relating each multivariate criteria to an F statistic, all hypotheses are rejected.

With the rejection of the test of interaction, one does not usually investigate differences in main effects since any difference is confounded by interaction. To investigate interactions in PROC GLM, one may again construct CONTRAST statements which generate one degree of freedom F tests. Because PROC GLM does not add side conditions to the model, individual $\boldsymbol{\gamma}_{ij}$ are not estimable. However, using the cell means one may form tetrads in the $\boldsymbol{\gamma}_{ij}$ that are estimable. The cell mean is defined by the term $T * C$ for our example. The contrasts '11 − 31 − 12 + 32' and '21 − 31 − 22 + 32' are used to estimate

$$\boldsymbol{\psi}_1 = \boldsymbol{\gamma}_{11} - \boldsymbol{\gamma}_{31} - \boldsymbol{\gamma}_{12} - \boldsymbol{\gamma}_{32}$$
$$\boldsymbol{\psi}_2 = \boldsymbol{\gamma}_{21} - \boldsymbol{\gamma}_{31} - \boldsymbol{\gamma}_{22} - \boldsymbol{\gamma}_{32}$$

The estimates from the ESTIMATE statements are

$$\widehat{\boldsymbol{\psi}}_1 = \mathbf{y}_{11.} - \mathbf{y}_{31.} - \mathbf{y}_{12.} + \mathbf{y}_{32.} = \begin{bmatrix} 2.60 \\ 7.20 \end{bmatrix}$$

$$\widehat{\boldsymbol{\psi}}_2 = \mathbf{y}_{21.} - \mathbf{y}_{31.} - \mathbf{y}_{22.} + \mathbf{y}_{32.} = \begin{bmatrix} 4.00 \\ 9.80 \end{bmatrix}$$

The estimate '$c1 - c2$' is estimating $\psi_3 = \boldsymbol{\beta}_1 - \boldsymbol{\beta}_2 + \sum_i \left(\boldsymbol{\gamma}_{i1} - \boldsymbol{\gamma}_{i2}\right)/3$ for each variable. This contrast is confounded by interaction. The estimate for the contrast is

$$\widehat{\boldsymbol{\psi}}_3 = \mathbf{y}_{.1.} - \mathbf{y}_{.2.} = \begin{bmatrix} 1.40 \\ 3.87 \end{bmatrix}$$

The standard error for each variable is labeled 'Std. Error of Estimate' in SAS. Arranging the standard errors as vectors to correspond to the contrasts, the $\widehat{\sigma}_{\widehat{\psi}_i}$ become

$$\widehat{\sigma}_{\widehat{\psi}_2} = \begin{bmatrix} 1.2329 \\ 1.3663 \end{bmatrix} \quad \widehat{\sigma}_{\widehat{\psi}_3} = \begin{bmatrix} 0.5033 \\ 0.5578 \end{bmatrix}$$

To evaluate any of these contrasts using the multivariate criteria, one may estimate

$$\widehat{\psi}_i - c_\alpha \widehat{\sigma}_{\widehat{\psi}_i} \quad \leq \psi_i \leq \quad \widehat{\psi}_i + c_\alpha \widehat{\sigma}_{\widehat{\psi}_i}$$

a variable at a time where c_α is estimated using (4.2.60) exactly or approximately using the F distribution. We use the TRANSREG procedure to generate a FR cell means design matrix and PROC IML to estimate c_α for Roy's criterion to obtain an approximate (lower bound) 95% simultaneous confidence interval for $\theta_{12} = \gamma_{112} - \gamma_{312} - \gamma_{122} + \gamma_{322}$ using the upper bound F statistic. By changing the SAS code from $m = (0\ 1)$ to $m = (1, 0)$ the simultaneous confidence interval for $\theta_{11} = \gamma_{111} - \gamma_{311} - \gamma_{121} + \gamma_{321}$ is obtained. With $c_\alpha = 2.609$, the interaction confidence intervals for each variable follow.

$$-0.616 \leq \theta_{11} \leq 5.816 \quad \text{(Reading Rate)}$$
$$3.636 \leq \theta_{12} \leq 10.764 \quad \text{(Reading Comprehension)}$$

The tetrad is significant for reading comprehension and not the reading rate variable. As noted in the output, the critical constants for the BLH and BNP criteria are again larger, 3.36 and 3.61, respectively. One may alter the contrast vector SAS code for $c = (1\ -1\ 0\ 0\ -1\ 1)$ in program m4_9_1.sas to obtain other tetrads. For example, one may select for example select $c = (0\ 0\ 1\ -1\ -1\ 1)$.

For this MANOVA design, there are only three meaningful tetrads for the study. To generate the protected F tests using SAS, the CONTRAST statement in PROC GLM is used. The p-values for the interactions follow.

Tetrad	Variables	
	R	C
$11 - 31 - 12 + 32$	0.0456	0.0001
$21 - 31 - 22 + 32$	0.0034	0.0001
$11 - 12 - 21 + 22$	0.2674	0.0001

The tests indicate that only the reading comprehension variable appears significant. For $\alpha = 0.05$, $v_e = 24$, and $C = 3$ comparisons, the value for the critical constant in Table V of the Appendix for the multivariate t distribution is $c_\alpha = 2.551$. This value may be used to construct approximate confidence intervals for the interaction tetrads. The standardized canonical variate output for the test of H_{AB} also indicates that these comparisons should be investigated. Using only one discriminant function, the reading comprehension variable has the largest coefficient weight and the highest correlation. Reviewing the univariate and multivariate tests of normality, model assumptions appear tenable.

b. Two-Way MANCOVA (Example 4.9.2)

For our next example, an experiment is designed to study two new reading and mathematics programs in the fourth grade. Using gender as a fixed blocking variable, 15 male and 15 female students are randomly assigned to the current program and to two experimental programs. Before beginning the study, a test was administered to obtain grade-equivalent reading and mathematical levels for the students, labeled ZR and ZM. At the end of the study, 6 months later, similar data (YR and YM) were obtained for each subject. The data for the study are provided in Table 4.9.4.
The mathematical model for the design is

$$\begin{aligned} \mathbf{y}_{ijk} &= \boldsymbol{\mu} + \boldsymbol{\alpha}_i + \boldsymbol{\beta}_j + \boldsymbol{\gamma}_{ij} + \Gamma \mathbf{z}_{ij} + \mathbf{e}_{ijk} \\ i &= 1, 2;\ j = 1, 2, 3;\ k = 1, 2, \ldots, 5 \\ e_{ij} &\sim N_p\left(\mathbf{0}, \Sigma_{y|z}\right) \end{aligned} \tag{4.9.4}$$

The code for the analyses is contained in program m4_9_2.sas.

As with the one-way MANCOVA design, we first evaluate the assumption of parallelism. To test for parallelism, we represent the model as a "one-way" MANCOVA design involving six cells. Following the one-way MANCOVA design, we evaluate parallelism of the regression lines for the six cells by forming the interaction of the factor $T * B$ with each

TABLE 4.9.4. Two-Way MANCOVA

	Control				Experimental 1				Experimental 2			
	YR	YM	ZR	ZM	YR	YM	ZR	ZM	YR	YM	ZR	ZM
	4.1	5.3	3.2	4.7	5.5	6.2	5.1	5.1	6.1	7.1	5.0	5.1
	4.6	5.0	4.2	4.5	5.0	7.1	5.3	5.3	6.3	7.0	5.2	5.2
Males	4.8	6.0	4.5	4.6	6.0	7.0	5.4	5.6	6.5	6.2	5.3	5.6
	5.4	6.2	4.6	4.8	6.2	6.1	5.6	5.7	6.7	6.8	5.4	5.7
	5.2	6.1	4.9	4.9	5.9	6.5	5.7	5.7	7.0	7.1	5.8	5.9
	5.7	5.9	4.8	5.0	5.2	6.8	5.0	5.8	6.5	6.9	4.8	5.1
	6.0	6.0	4.9	5.1	6.4	7.1	6.0	5.9	7.1	6.7	5.9	6.1
Females	5.9	6.1	5.0	6.0	5.4	6.1	5.6	4.9	6.9	7.0	5.0	4.8
	4.6	5.0	4.2	4.5	6.1	6.0	5.5	5.6	6.7	6.9	5.6	5.1
	4.2	5.2	3.3	4.8	5.8	6.4	5.6	5.5	7.2	7.4	5.7	6.0

covariate (ZR and ZM) using PROC GLM.. Both tests are nonsignificant so that we conclude parallelism of regression for the six cells. To perform the simultaneous test that the covariates are both zero, PROC REG may be used.

Given parallelism, one may next test that all the covariates are simultaneously zero. For this test, one must use PROG REG. Using PROC TRANSREG to create a full rank model, and using the MTEST statement the overall test that all covariates are simultaneously zero is rejected for all of the multivariate criteria. Given that the covariates are significant in the analysis, the next test of interest is to determine whether there are significant differences among the treatment conditions. Prior experience indicated that gender should be used as a blocking variable leading to more homogeneous blocks. While we would expect significant differences between blocks (males and females), we do not expect a significant interaction between treatment conditions. We also expect the covariates to be significantly different from zero.

Reviewing the MANCOVA output, the test for block differences (B) and block by treatment interaction ($T * B$) are both nonsignificant while the test for treatment differences is significant (p-value < 0.0001). Reviewing the protected F test for each covariate, we see that only the reading grade-equivalent covariate is significantly different from zero in the study. Thus, one may want to only include a single covariate in future studies. We have again output the adjusted means for the treatment factor. The estimates follow.

Variable	Treatments		
	C	E1	E2
Reading	5.6771	5.4119	6.4110
Mathematics	5.9727	6.3715	6.7758

Using the CONTRAST statement to evaluate significance, we compare each treatment with the control using the protected one degree of freedom tests for each variable and $\alpha = 0.05$. The tests for the reading variable (YR) have p-values of 0.0002 and 0.0001 when one compares the control group with first experimental group (c-e1) and second experimental group (c-e2), respectively. For the mathematics variable (YM), the p-values are 0.0038 and 0.0368. Thus, there appears to be significant differences between the experimental groups and the control group for each of the dependent variables. To form approximate simultaneous confidence intervals for the comparisons, one would have to adjust α. Using the Bonferroni procedure, we may set $\alpha^* = 0.05/4 = 0.0125$. Alternatively, if one is only interested in tests involving the control and each treatment, one might use the DUNNETT option on the LSMEANS statement with $\alpha^* = \alpha/9 = 0.025$ since the study involves two variables. This approach yields significance for both variables for the comparison of e2 with c. The contrast estimates for the contrast ψ of the mean difference with confidence intervals follow.

	$\widehat{\psi}$	C.I. for (e2-c)
Reading	0.7340	(0.2972, 1.1708)
Mathematics	0.8031	(0.1477, 1.4586)

For these data, the Dunnett's intervals are very close to those obtained using Roy's criterion. Again, the TRANSREG procedure is used to generate a FR cell means design matrix and PROC IML is used to generate the approximate simultaneous confidence set. The critical constants for the multivariate Roy, BLH, and BNP criteria are as follows: 2.62, 3.38, and 3.68. Because the largest root criterion is again an upper bound, the intervals reflect lower bounds for the comparisons. For the contrast vector $c_1 = (-.5\ .5\ -.5\ .5\ 0\ 0)$ which compares e2 with c using the FR model, $\widehat{\psi} = 0.8031$ and the interval for Mathematics variable using Roy's criterion is (0.1520, 1.454). To obtain the corresponding interval for Reading, the value of $m = (0\ 1)$ is changed to $m = (1\ 0)$. This yields the interval, (0.3000, 1.1679) again using Roy's criterion.

Exercises 4.9

1. An experiment is conducted to compare two different methods of teaching physics during the morning, afternoon, and evening using the traditional lecture approach and the discovery method. The following table summarizes the test score obtained in the areas of mechanical (M), heat (H), the sound (S) for the 24 students in the study.

	Traditional			Discovery		
	M	H	S	M	H	S
	30	131	34	51	140	36
Morning	26	126	28	44	145	37
8 A.M.	32	134	33	52	141	30
	31	137	31	50	142	33
	41	104	36	57	120	31
Afternoon	44	105	31	68	130	35
2 P.M.	40	102	33	58	125	34
	42	102	27	62	150	39
	30	74	35	52	91	33
Evening	32	71	30	50	89	28
8 P.M.	29	69	27	50	90	28
	28	67	29	53	95	41

(a) Analyze the data, testing for (1) effects of treatments, (2) effects of time of day, and (3) interaction effects. Include in your analysis a test of the equality of the variance-covariance matrices and normality.

(b) In the study does trend analysis make any sense? If so, incorporate it into your analysis.

(c) Summarize the results of this experiment in one paragraph.

2. In an experiment designed to study two new reading and mathematics programs in the fourth grade subjects in the school were randomly assigned to three treatment conditions, one being the old program and two being experimental program. Before beginning the experiment, a test was administered to obtain grade-equivalent reading and mathematics levels for the subjects, labeled R_1 and M_1, respectively, in the table below. At the end of the study, 6 months later, similar data (R_2 and M_2) were obtained for each subject.

Control				Experimental				Experimental			
Y		Z		Y		Z		Y		Z	
R_2	M_2	R_1	M_1	R_2	M_2	R_1	M_1	R_2	M_2	R_1	M_1
4.1	5.3	3.2	4.7	5.5	6.2	5.1	5.1	6.1	7.1	5.0	5.1
4.6	5.0	4.2	4.5	5.0	7.1	5.3	5.3	6.3	7.0	5.2	5.2
4.8	6.0	4.5	4.6	6.0	7.0	5.4	5.6	6.5	6.2	5.3	5.6
5.4	6.2	4.6	4.8	6.2	6.1	5.6	5.7	6.7	6.8	5.4	5.7
5.2	6.1	4.9	4.9	5.9	6.5	5.7	5.7	7.6	7.1	5.8	5.9
5.7	5.9	4.8	5.0	5.2	6.8	5.0	5.8	6.5	6.9	4.8	5.1
6.0	6.0	4.9	5.1	6.4	7.1	6.0	5.9	7.1	6.7	5.9	6.1
5.9	6.1	5.0	6.0	5.4	6.1	5.0	4.9	6.9	7.0	5.0	4.8
4.6	5.0	4.2	4.5	6.1	6.0	5.5	5.6	6.7	6.9	5.6	5.1
4.2	5.2	3.3	4.8	5.8	6.4	5.6	5.5	7.2	7.4	5.7	6.0

(a) Is there any reasons to believe that the programs differ?

(b) Write up your findings in a the report your analysis of all model assumptions.

4.10 Nonorthogonal Two-Way MANOVA Designs

Up to this point in our discussion of the analysis of two-way MANOVA designs, we have assumed an equal number of observations ($n_o \geq 1$) per cell. As we shall discuss in Section 4.16, most two-way and higher order crossed designs are constructed with the power to detect some high level interaction with an equal number of observations per cell. However, in carrying out a study one may find that subjects in a two-way or higher order design drop-out of the study creating a design with empty cells or an unequal and disproportionate number of observations in each cell. This results in a nonorthogonal or unbalanced design. The analysis of two-way and higher order designs with this unbalance require careful consideration since the subspaces associated with the effects are no longer uniquely orthogonal. The order in which effects are entered into the model leads to different decompositions of the test space. In addition to nonorthogonality, an experimenter may find that some observations within a vector are missing. This results in incomplete multivariate data and nonorthogonality. In this section we discuss the analysis of nonorthogonal designs. In Chapter 5 we discuss incomplete data issues where observations are missing within a vector observation.

When confronted with a nonorthogonal design, one must first understand how observations were lost. If observations are lost at random and independent of the treatments one would establish tests of main effects and interactions that do not depend on cell frequencies, this is an unweighted analysis. In this situation, weights are chosen proportional to the reciprocal of the number of levels for a factor (e.g. $1/a$ or $1/b$ for two factors) or the reciprocal of the product of two or more factors (e.g. $1/ab$ for two factor interactions), provided the design has no empty cells. Tests are formed using the Type III option in PROC GLM. If observation loss is associated with the level of treatment and is expected to happen in any replication of the study, tests that depend on cell frequencies are used, this is a weighted analysis. Tests are formed using the Type I option. As stated earlier, the Type II option has in general little value, however, it may be used in designs that are additive. If a design has empty cells, the Type IV option may be appropriate.

a. *Nonorthogonal Two-Way MANOVA Designs with and Without Empty Cells, and Interaction*

The linear model for the two-way MANOVA design with interaction is

$$\begin{aligned} \mathbf{y}_{ijk} &= \boldsymbol{\mu} + \boldsymbol{\alpha}_i + \boldsymbol{\beta}_j + \boldsymbol{\gamma}_{ij} + \mathbf{e}_{ijk} \qquad &\text{(LFR)} \\ &= \boldsymbol{\mu}_{ij} + \mathbf{e}_{ijk} &\text{(FR)} \end{aligned} \tag{4.10.1}$$

$$\mathbf{e}_{ijk} \sim IN_p(\mathbf{0}, \Sigma) \qquad i = 1, 2, \ldots, a;\ j = 1, 2, \ldots, b;\ k = 1, 2, \ldots, n_{ij} \geq 0$$

and the number of observations in a cell may be zero, $n_{ij} = 0$. Estimable functions and tests of hypotheses in two-way MANOVA designs with empty cells depend on the location of the empty cells in the design and the estimability of the population means $\boldsymbol{\mu}_{ij}$. Clearly, the BLUE of $\boldsymbol{\mu}_{ij}$ is again the sample cell mean

$$\widehat{\boldsymbol{\mu}}_{ij} = \mathbf{y}_{ij.} = \textstyle\sum_{k=1}^{n_{ij}} \mathbf{y}_{ijk}/n_{ij}, \qquad n_{ij} > 0 \tag{4.10.2}$$

The parameters $\boldsymbol{\mu}_{ij}$ are not estimable if any $n_{ij} = 0$. Parametric functions of the cell means

$$\boldsymbol{\eta} = \sum_{i,j} K_{ij} \boldsymbol{\mu}_{ij} \tag{4.10.3}$$

are estimable and therefore testable if and only if $K_{ij} = 0$ when $n_{ij} = 0$ and $K_{ij} = 1$ if $n_{ij} \neq 0$. Using (4.10.3), we can immediately define two row (or column) means that are

TABLE 4.10.1. Non-Additive Connected Data Design

	Factor B		
Factor A	μ_{11}	Empty	μ_{13}
	μ_{21}	μ_{22}	μ_{23}
	Empty	μ_{32}	μ_{33}

estimable. The obvious choices are the weighted and unweighted means

$$\overline{\mu}_{i.} = \sum_j n_{ij}\mu_{ij}/n_{i+} \tag{4.10.4}$$

$$\widetilde{\mu}_{i.} = \sum_j K_{ij}\mu_{ij}/K_{i+} \tag{4.10.5}$$

where $n_{i+} = \sum_j n_{ij}$ and $K_{i+} = \sum_j K_{ij}$. If all $K_{ij} \neq 0$, then $\widetilde{\mu}_{i.}$ becomes

$$\mu_{i.} = \sum_j \mu_{ij}/b \tag{4.10.6}$$

the LSMEAN in SAS. The means in (4.10.4) and (4.10.5) depend on the sample cell frequencies and the location of the empty cells, and are not easily interpreted. None of the Type I, Type II, or Type III hypotheses have any reasonable interpretation with empty cells. Furthermore, the tests are again confounded with interaction. PROC GLM does generate some Type IV tests that are interpretable when a design has empty cells. They are balanced simple effect tests that are also confounded by interaction. If a design has no empty cells so that all $n_{ij} > 0$, then one may construct meaningful Type I and Type III tests that compare the equality of weighted and unweighted marginal means. These tests, as in the equal $n_{ij} = n_o$ case, are also confounded by interaction.

Tests of no two-way interaction for designs with all cell frequencies $n_{ij} > 0$ are identical to tests for the case in which all $n_{ij} = n_o$. However, problems occur when empty cells exist in the design since the parameters μ_{ij} are not estimable for the empty cells. Because of the empty cells, the interaction hypothesis for the designs are not identical to the hypothesis for a design with no empty cells. In order to form contrasts for the interaction hypothesis, one must write out a set of linearly independent contrasts as if no empty cells occur in the design and calculate sums and differences of these contrasts in order to eliminate the μ_{ij} that do not exist for the design. The number of degrees of freedom for interaction in any design may be obtained by subtracting the number of degrees of freedom for main effects from the total number of between groups degrees of freedom. Then, $v_{AB} = (f - 1) - (a - 1) - (b - 1) = f - a - b + 1$ where f is the number of nonempty, "filled" cells in the design. To illustrate, we consider the connected data pattern in Table 4.10.1. A design is connected if all nonempty cells may be jointed by row-column paths of filled cells which results in a continuous path that has changes in direction only in filled cells, Weeks and Williams (1964).

Since the number of cells filled in Table 4.10.1 is $f = 7$ and $a = b = 3$, $v_{AB} = f - a - b + 1 = 2$. To find the hypothesis test matrix for testing H_{AB}, we write out the

TABLE 4.10.2. Non-Additive Disconnected Design

	Factor B		
Factor A	μ_{11}	μ_{12}	Empty
	μ_{21}	Empty	μ_{23}
	Empty	μ_{32}	μ_{33}

interactions assuming a complete design

$$\text{a.} \quad \boldsymbol{\mu}_{11} - \underline{\boldsymbol{\mu}_{12}} - \boldsymbol{\mu}_{21} + \boldsymbol{\mu}_{22} = \mathbf{0}$$

$$\text{b.} \quad \boldsymbol{\mu}_{11} - \boldsymbol{\mu}_{13} - \boldsymbol{\mu}_{21} + \boldsymbol{\mu}_{23} = \mathbf{0}$$

$$\text{c.} \quad \boldsymbol{\mu}_{21} - \boldsymbol{\mu}_{22} - \underline{\boldsymbol{\mu}_{31}} + \boldsymbol{\mu}_{32} = \mathbf{0}$$

$$\text{d.} \quad \boldsymbol{\mu}_{21} - \boldsymbol{\mu}_{23} - \underline{\boldsymbol{\mu}_{31}} + \boldsymbol{\mu}_{33} = \mathbf{0}$$

Because contrast (b) contains no underlined missing parameter, we may use it to construct a row of the hypothesis test matrix. Taking the difference between (c) and (d) removes the nonestimable parameter $\boldsymbol{\mu}_{31}$. Hence a matrix with rank 2 to test for no interaction is

$$\mathbf{C}_{AB} = \begin{bmatrix} 1 & -1 & -1 & 0 & 1 & 0 & 0 \\ 0 & 0 & 0 & 1 & -1 & -1 & 1 \end{bmatrix}$$

where the structure of $\mathbf{B}$ using the FR model is

$$\mathbf{B} = \begin{bmatrix} \boldsymbol{\mu}'_{11} \\ \boldsymbol{\mu}'_{13} \\ \boldsymbol{\mu}'_{21} \\ \boldsymbol{\mu}'_{22} \\ \boldsymbol{\mu}'_{23} \\ \boldsymbol{\mu}'_{32} \\ \boldsymbol{\mu}'_{33} \end{bmatrix}$$

An example of a disconnected design pattern is illustrated in Table 4.10.2. For this design, all cells may not be joined by row-column paths with turns in filled cells. The test for interaction now has one degree of freedom since $v_{AB} = f - a - b + 1 = 1$.

Forming the set of independent contracts for the data pattern in Table 4.10.2.

$$\text{a.} \quad \boldsymbol{\mu}_{11} - \boldsymbol{\mu}_{12} - \boldsymbol{\mu}_{21} + \underline{\boldsymbol{\mu}_{22}} = \mathbf{0}$$

$$\text{b.} \quad \boldsymbol{\mu}_{11} - \underline{\boldsymbol{\mu}_{13}} - \boldsymbol{\mu}_{21} + \boldsymbol{\mu}_{23} = \mathbf{0}$$

$$\text{c.} \quad \boldsymbol{\mu}_{21} - \boldsymbol{\mu}_{23} - \underline{\boldsymbol{\mu}_{31}} + \boldsymbol{\mu}_{33} = \mathbf{0}$$

$$\text{d.} \quad \underline{\boldsymbol{\mu}_{22}} - \boldsymbol{\mu}_{23} - \boldsymbol{\mu}_{32} + \boldsymbol{\mu}_{33} = \mathbf{0}$$

TABLE 4.10.3. Type IV Hypotheses for A and B for the Connected Design in Table 4.10.1

Tests of A	Test of B
$\frac{\mu_{11}+\mu_{13}}{2} = \frac{\mu_{21}+\mu_{23}}{2}$	$\frac{\mu_{11}+\mu_{21}}{2} = \frac{\mu_{13}+\mu_{23}}{2}$
$\frac{\mu_{22}+\mu_{23}}{2} = \frac{\mu_{32}+\mu_{33}}{2}$	$\frac{\mu_{22}+\mu_{32}}{2} = \frac{\mu_{33}+\mu_{33}}{2}$
$\mu_{11} = \mu_{21}$	$\mu_{11} = \mu_{13}$
$\mu_{21} = \mu_{32}$	$\mu_{21} = \mu_{22}$
$\mu_{11} = \mu_{23}$	$\mu_{21} = \mu_{23}$
$\mu_{22} = \mu_{33}$	$\mu_{21} = \mu_{23}$
$\mu_{13} = \mu_{33}$	$\mu_{32} = \mu_{33}$

and subtracting (d) from (a), the interaction hypotheses becomes

$$H_{AB} : \mu_{11} - \mu_{12} - \mu_{21} + \mu_{23} + \mu_{32} - \mu_{33} = \mathbf{0}$$

Tests of no interaction for designs with empty cells must be interpreted with caution since the test is not equivalent to the test of no interaction for designs with all cells filled. If H_{AB} is rejected, the interaction for a design with no empty cells would also be rejected. However, if the test is not rejected we cannot be sure that the hypothesis would not be rejected for the complete cell design because nonestimable interactions are excluded from the analysis by the missing data pattern. The excluded interactions may be significant.

For a two-way design with interaction and empty cells, tests of the equality of the means given in (4.10.5) are tested using the Type IV option in SAS. PROC GLM automatically generates Type IV hypothesis; however, to interpret the output one must examine the Type IV estimable functions to determine what hypothesis are being generated and tested. For the data pattern given in Table 4.10.1, all possible Type IV tests that may be generated by PROC GLM are provided in Table 4.10.3 for tests involving means $\widetilde{\mu}_{i.}$.

The tests in Table 4.10.3 are again confounded by interaction since they behave like simple effect tests. When SAS generates Type IV tests, the tests generated may not be the tests of interest for the study. To create your own tests, CONTRAST and ESTIMATE statements can be used. Univariate designs with empty cells are discussed by Milliken and Johnson (1992, Chapter 14) and Searle (1993).

b. *Additive Two-Way MANOVA Designs With Empty Cells*

The LFR linear model for the additive two-way MANOVA design is

$$\begin{aligned} \mathbf{y}_{ijk} &= \boldsymbol{\mu} + \boldsymbol{\alpha}_i + \boldsymbol{\beta}_j + \mathbf{e}_{ijk} \\ \mathbf{e}_{ijk} &\sim IN\left(\mathbf{0}, \Sigma\right) \qquad i = 1, 2, \ldots, a;\ j = 1, 2, \ldots, b;\ k = 1, 2, \ldots, n_{ij} \geq 0 \end{aligned} \tag{4.10.7}$$

where the number of observations per cell n_{ij} is often either zero (empty) or one. Associating $\boldsymbol{\mu}_{ij}$ with $\boldsymbol{\mu} + \boldsymbol{\alpha}_i + \boldsymbol{\beta}_j$ does not reduce (4.10.7) to a full rank cell means model, Timm (1980b). One must include with the cell means model with no interaction a restriction of the form $\boldsymbol{\mu}_{ij} - \boldsymbol{\mu}_{i'j} - \boldsymbol{\mu}_{ij'} + \boldsymbol{\mu}_{i'j'} = \mathbf{0}$ for all cells filled to create a FR model. The restricted MGLM for the additive two-way MANOVA design is then

$$\begin{aligned} \mathbf{y}_{ijk} = &\ \boldsymbol{\mu}_{ij} + \mathbf{e}_{ijk} \qquad i = 1, \ldots, a;\ j = 1, \ldots b; k = 1, \ldots, n_{ij} \geq 0 \\ \boldsymbol{\mu}_{ij} - &\ \boldsymbol{\mu}_{i'j} - \boldsymbol{\mu}_{ij} - \boldsymbol{\mu}_{ij'} + \boldsymbol{\mu}_{i'j'} = \mathbf{0} \\ &\ \mathbf{e}_{ij} \sim IN_p\left(\mathbf{0}, \Sigma\right) \end{aligned} \tag{4.10.8}$$

for a set of $(f - a - b + 1) > 0$ estimable tetrads including sums and differences. Model (4.10.8) may be analyzed using PROG REG while model (4.10.7) is analyzed using PROC GLM.

For the two-way design with interaction, empty cells caused no problem since with $n_{ij} = 0$ the parameter $\boldsymbol{\mu}_{ij}$ was not estimable. For an additive model which contains no interaction, the restrictions on the parameters $\boldsymbol{\mu}_{ij}$ may sometimes be used to estimate the population parameter $\boldsymbol{\mu}_{ij}$ whether or not a cell is empty. To illustrate, suppose the cell with $\boldsymbol{\mu}_{11}$ is empty so that $n_{11} = 0$, but that one has estimates for $\boldsymbol{\mu}_{12}$, $\boldsymbol{\mu}_{21}$, and $\boldsymbol{\mu}_{22}$. Then by using the restriction $\boldsymbol{\mu}_{11} - \boldsymbol{\mu}_{12} - \boldsymbol{\mu}_{21} + \boldsymbol{\mu}_{22} = \mathbf{0}$, an estimate of $\boldsymbol{\mu}_{11}$ is $\widehat{\boldsymbol{\mu}}_{11} = \widehat{\boldsymbol{\mu}}_{12} + \widehat{\boldsymbol{\mu}}_{21} - \widehat{\boldsymbol{\mu}}_{22}$ even though cell (1, 1) is empty. This is not always the case. To see this, consider the data pattern in Table 4.10.2. For this pattern, no interaction restriction would allow one to estimate all the population parameters $\boldsymbol{\mu}_{ij}$ associated with the empty cells. The design is said to be disconnected or disjoint. An additive two-way crossed design is said to be connected if all $\boldsymbol{\mu}_{ij}$ are estimable. This is the case for the data in Table 4.10.1. Thus, given an additive model with empty cells and connected data, all pairwise contrasts of the form

$$\begin{aligned} \boldsymbol{\psi} &= \boldsymbol{\mu}_{ij} - \boldsymbol{\mu}_{i'j} = \boldsymbol{\alpha}_i - \boldsymbol{\alpha}_{i'} &&\text{for all } i,\ i' \\ \boldsymbol{\psi} &= \boldsymbol{\mu}_{ij} - \boldsymbol{\mu}_{ij'} = \boldsymbol{\beta}_j - \boldsymbol{\beta}_{j'} &&\text{for all } j,\ j' \end{aligned} \tag{4.10.9}$$

are estimable as are linear combinations. When a design has all cells filled, the design by default is connected so there is no problem with the analysis. For connected designs, tests for main effects H_A and H_B become, using the restricted full rank MGLM

$$\begin{aligned} H_A &: \boldsymbol{\mu}_{ij} = \boldsymbol{\mu}_{i'j} &&\text{for all } i,\ i' \text{ and } j \\ H_B &: \boldsymbol{\mu}_{ij} = \boldsymbol{\mu}_{ij'} &&\text{for all } i,\ j \text{ and } j' \end{aligned} \tag{4.10.10}$$

Equivalently, using the LFR model, the tests become

$$\begin{aligned} H_A &: \text{all } \boldsymbol{\alpha}_i \text{ are equal} \\ H_B &: \text{all } \boldsymbol{\beta}_j \text{ are equal} \end{aligned} \tag{4.10.11}$$

where contrasts in $\boldsymbol{\alpha}_i$ $(\boldsymbol{\beta}_j)$ involve the LSMEANS $\widehat{\boldsymbol{\mu}}_{i.}$ $(\widehat{\boldsymbol{\mu}}_{.j})$ so that $\boldsymbol{\psi} = \boldsymbol{\alpha}_i - \boldsymbol{\alpha}_{i'}$ is estimated by $\widehat{\boldsymbol{\psi}} = \widehat{\boldsymbol{\mu}}_{i.} - \widehat{\boldsymbol{\mu}}_{i'.}$, for example. Tests of H_A and H_B for connected designs are tested using the Type III option. With unequal n_{ij}, Type I tests may also be constructed. When a design is disconnect, the cell means $\boldsymbol{\mu}_{ij}$ associated with the empty cell are no longer estimable. However, the parametric functions given in Table 10.4.3 remain estimable and

TABLE 4.11.1. Nonorthogonal Design

		Factor B	
		B_1	B_2
Factor A	A_1	[10, 21] [12, 22] $n_{11} = 2$	[9, 17] [8, 13] $n_{12} = 2$
	A_2	[14, 27] [11, 223] $n_{21} = 2$	[12, 18] $n_{22} = 1$
	A_3	[7.151] $n_{31} = 1$	[8, 18] $n_{32} = 1$

are now not confounded by interaction. These may be tested in SAS by using the Type IV option. To know which parametric functions are included in the test, one must again investigate the form of the estimable functions output using the / E option. If the contrast estimated by SAS are not the parametric functions of interest, one must construct CONTRAST statements to form the desired tests.

4.11 Unbalance, Nonorthogonal Designs Example

In our discussion of the MGLM, we showed that in many situations that the normal equations do not have a unique solution. Using any g-inverse, contrasts in the parameters do have a unique solution provided $\mathbf{c}'\mathbf{H} = \mathbf{c}'$ for $\mathbf{H} = (\mathbf{X}'\mathbf{X})^{-} (\mathbf{X}'\mathbf{X})$ for a contrast vector $\mathbf{c}$. This condition, while simple, may be difficult to evaluate for complex nonorthogonal designs. PROC GLM provides users with several options for displaying estimable functions. The option /E on the model statement provides the general form of all estimable functions. The g-inverse of $(\mathbf{X}'\mathbf{X})$ used by SAS to generate the structure of the general form is to set a subset of the parameters to zero, Milliken and Johnson (1992, p. 101). SAS also has four types for sums of squares, Searle (1987, p. 461). Each type (E1, E2, E3, E4) has associated with it estimable functions which may be evaluated to determine testable hypotheses, Searle (1987, p. 465) and Milliken and Johnson (1992, p. 146 and p. 186). By using the XPX and I option on the MODEL statement, PROC GLM will print $(\mathbf{X}'\mathbf{X})$ and $(\mathbf{X}'\mathbf{X})^{-}$ used to obtain $\widehat{\mathbf{B}}$ when requesting the option / SOLUTION. For annotated output produced by PROC GLM when analyzing unbalanced designs, the reader may consult Searle and Yerex (1987).

For our initial application of the analysis of a nonorthogonal design using PROC GLM, the sample data for the two-way MANOVA design given in Table 4.11.1 are utilized using program m4_11_1.sas
The purpose of the example is to illustrate the mechanics of the analysis of a nonorthogonal design using SAS.

When analyzing any nonorthogonal design with no empty cells, one should always use the options / SOLUTION XPX I E E1 E3 SS1 and SS3 on the MODEL statement. Then estimable functions and testable hypotheses using the Type I and Type III sums of squares are usually immediately evident by inspection of the output.

For the additive model and the design pattern given in Table 4.11.1, the general form of the estimable functions follow.

Effect	Coefficients
Intercept	L1
a	L2
a	L3
a	L1−L2−L3
b	L5
b	L1−L5

Setting $L1 = 0$, the tests of main effects always exist and are not confounded by each other. The estimable functions for factor A involve only $L2$ and $L3$ if $L5 = 0$. The estimable functions for factor B are obtained by setting $L2 = 1$, with $L3 = 0$, and by setting $L3 = 1$ and $L2 = 0$. This is exactly the Type III estimable functions. Also observe from the printout that the test of H_A using Type I sums of squares (SS) is confounded by factor B and that the Type I SS for factor B is identical to the Type III SS. To obtain a Type I SS for B, one must reorder the effects in the MODEL statement. In general, only Type III hypotheses are usually most appropriate for nonorthogonal designs whenever the unbalance is not due to treatment.

For the model with interaction, observe that only the test of interaction is not confounded by main effects for either the Type I or Type III hypotheses. The form of the estimable functions are as follows.

Coefficients		Effect
a∗b	11	L7
a∗b	12	−L7
a∗b	21	L9
a∗b	22	−L9
a∗b	31	−L7−L9
a∗b	32	L7+L9

The general form of the interaction contrast involves only coefficients L7 and L9. Setting $L1 = 1$ and all others to zero, the tetrad $'11 - 12 - 31 + 32'$ is realized. Setting $L9 = 1$ and all others to zero, the tetrad $'21-22-31+32'$ is obtained. Summing the two contrasts yields the $'$sum inter$'$ contrast while taking the difference yields the tetrad $'11 - 12 - 21 + 22'$. This demonstrates how one may specify estimable contrasts using SAS. Tests follow those already illustrated for orthogonal designs.

To create a connected two-way design, we delete observation [12, 18] in cell (2, 2). To make the design disconnected, we also delete the observations in cell (2, 1). The statements for the analysis of these two designs are included in program m4_11_1.sas.

When analyzing designs with empty cells, one should always use the / E option to obtain the general form of all estimable parametric functions. One may test Type III or Type IV hypotheses for connected designs (they are equal as seen in the example output); however, only Type IV hypotheses may be useful for disconnected designs. To determine the hypotheses tested, one must investigate the Type IV estimable functions. Whenever a design has empty cells, failure to reject the test of interaction may not imply its nonsignificance since certain cells are being excluded from the analysis. When designs contain empty cells and potential interactions, it is often best to represent the MR model using the NOINT option since only all means are involving in the analysis.

Investigating the test of interaction for the nonorthogonal design with no empty cells, $v_{AB} = (a-1)(b-1) = 2$. For the connected design with an interaction, the cell mean μ_{22} is not estimable. The degrees of freedom for the test of interaction becomes $v_{AB} = f - a - b + 1 = 5 - 3 - 2 + 1 = 1$. While each contrast $\widehat{\psi}_1 = \mu_{11} - \mu_{12} - \mu_{21} + \mu_{22}$ and $\widehat{\psi}_2 = \mu_{21} - \mu_{22} - \mu_{31} + \mu_{32}$ are not estimable, the sum $\psi = \psi_1 + \psi_2$ is estimable. The number of linearly independent contrasts is however one and not two. For the disconnected design, $v_{AB} = f - a - b + 1 = 4 - 3 - 2 + 1 = 1$. For this design, only one tetrad contrast is estimable, for example $\psi = \psi_1 + \psi_2$. Clearly, the test of interaction for the three designs are not equivalent. This is also evident from the calculated p-values for the tests for the three designs. For the design with no empty cells, the p-value is 0.1075 for Wilks' Λ criterion; for the connected design, the p-value is 0.1525; and for the disconnected design, the p-value is 0.2611. Setting the level of the tests at $\alpha = 0.15$, one may erroneously claim nonsignificance for a design with empty cells when if all cells are filled the result would be significant. The analysis of multivariate designs with empty cells is complex and must be analyzed with extreme care.

Exercises 4.11

1. John S. Levine and Leonard Saxe at the University of Pittsburgh obtained data to investigate the effects of social-support characteristics (allies and assessors) on conformity reduction under normative social pressure. The subjects were placed in a situations where three persons gave incorrect answers and a fourth person gave the correct answer. The dependent variables for the study are mean option (O) score and mean visual-perception (V) scores for a nine item test. High scores indicate more conformity. Analyze the following data from the unpublished study (see Table 4.11.2 on page 273) and summarize your findings.

2. For the data in Table 4.9.1, suppose all the observations in cell (2,1) were not collected. The observations for Teacher 2 and in the Contract Class is missing. Then, the design becomes a connected design with an empty cell.

   (a) Analyze the design assuming a model with interaction.

   (b) Analyze the design assuming a model without interaction, an additive model.

TABLE 4.11.2. Data for Exercise 1.

	Assessor			
	Good		Poor	
Ally	O	V	O	V
	2.67	.67	1.44	.11
	1.33	.22	2.78	1.00
	.44	.33	1.00	.11
Good	.89	.11	1.44	.22
	.44	.22	2.22	.11
	1.44	−.22	.89	.11
	.33	.11	2.89	.22
	.78	−.11	.67	.11
			1.00	.67
	1.89	.78	2.22	.11
	1.44	.00	1.89	.33
	1.67	.56	1.67	.33
	1.78	−11	1.89	.78
	1.00	1.11	.78	.22
Poor	.78	.44	.67	.00
	.44	.00	2.89	.67
	.78	.33	2.67	.67
	2.00	.22	2.78	.44
	1.89	.56		
	2.00	.56		
	.67	.56		
	1.44	.22		

(c) Next, suppose that the observations in the cells (1,2) and (3,1) are also missing and that the model is additive. Then the design becomes disconnected. Test for means differences for Factor A and Factor B and interpret your findings.

4.12 Higher Ordered Fixed Effect, Nested and Other Designs

The procedures outlined and illustrated using PROC GLM to analyze two-way crossed MANOVA/MANCOVA designs with fixed effects and random/fixed covariates extend in a natural manner to higher order designs. In all cases there is one within SSCP matrix error matrix **E**. To test hypotheses, one constructs the hypothesis test matrices **H** for main effects or interactions.

For a three-way, completely randomized design with factors A, B, and C and $n_{ijk} > 0$ observations per cell the MGLM is

$$
\begin{aligned}
\mathbf{y}_{ijkm} &= \boldsymbol{\mu} + \boldsymbol{\alpha}_i + \boldsymbol{\beta}_j + \boldsymbol{\tau}_k + (\boldsymbol{\alpha\beta})_{ij} + (\boldsymbol{\beta\tau})_{jk} + (\boldsymbol{\alpha\tau})_{ik} + \boldsymbol{\gamma}_{ijk} + \mathbf{e}_{ijkm} \\
&= \boldsymbol{\mu}_{ijk} + \mathbf{e}_{ijkm} \\
\mathbf{e}_{ijkm} &\sim IN_p\left(\mathbf{0}, \Sigma\right)
\end{aligned}
\tag{4.12.1}
$$

for $i = 1, \ldots, a;\ j = 1, \ldots, b;\ k = 1, \ldots, c;$ and $m = 1, \ldots, n_{ijk} > 0$ which allows for unbalanced, nonorthogonal, connected or disconnected designs. Again the individual effects for the LFR model are not estimable. However, if $n_{ijk} > 0$ then the cell means $\boldsymbol{\mu}_{ijk}$ are estimable and estimated by $\mathbf{y}_{ijk.}$, the cell mean.

In the two-way MANOVA/MANCOVA design we were unable to estimate main effects $\boldsymbol{\alpha}_i$ and $\boldsymbol{\beta}_j$; however, tetrads in the interactions $(\boldsymbol{\gamma}_{ij})$ were estimable. Extending this concept to the three-way design, the "three-way" tetrads have the general structure

$$
\boldsymbol{\psi} = \left(\boldsymbol{\mu}_{ijk} - \boldsymbol{\mu}_{i'jk} - \boldsymbol{\mu}_{ij'k} + \boldsymbol{\mu}_{i'j'k}\right) - \left(\boldsymbol{\mu}_{ijk'} - \boldsymbol{\mu}_{i'jk'} - \boldsymbol{\mu}_{ij'k'} + \boldsymbol{\mu}_{i'j'k'}\right) \tag{4.12.2}
$$

which is no more than a difference in two, two-way tetrads (AB) at levels k and k' of factor C. Thus, a three-way interaction may be interpreted as the difference of two, two-way interactions. Replacing the FR parameters $\boldsymbol{\mu}_{ijk}$ in $\boldsymbol{\psi}$ above with the LFR model parameters, the contrast in (4.12.2) becomes a contrast in the parameters $\boldsymbol{\gamma}_{ijk}$. Hence, the three-way interaction hypotheses for the three-way design becomes

$$
H_{ABC} = \left(\boldsymbol{\gamma}_{ijk} - \boldsymbol{\gamma}_{i'jk} - \boldsymbol{\gamma}_{ij'k} + \boldsymbol{\gamma}_{i'j'k}\right) - \left(\boldsymbol{\gamma}_{ijk'} - \boldsymbol{\gamma}_{i'jk'} - \boldsymbol{\gamma}_{ij'k'} + \boldsymbol{\gamma}_{i'j'k'}\right) = \mathbf{0} \tag{4.12.3}
$$

for all triples $\left(i, i', j, j', k, k'\right)$. Again all main effects are confounded by interaction; two-way interactions are also confounded by three-way interactions. If the three-way test of interaction is not significant, the tests of two-way interactions depend on whether the two-way cell means are created as weighted or unweighted marginal means of $\widehat{\boldsymbol{\mu}}_{ijk}$. This design is considered by Timm and Mieczkowski (1997, p. 296).

A common situation for two-factor designs is to have nested rather than crossed factors. These designs are incomplete because if factor B is nested within factor A, every level of B does not appear with every level of factor A. This is a disconnected design. However, letting $\boldsymbol{\beta}_{(i)j} = \boldsymbol{\beta}_j + \boldsymbol{\gamma}_{ij}$ represent the fact that the jth level of factor B is nested within the ith level of factor A, the MLGL model for the two-way nested design is

$$
\begin{aligned}
\mathbf{y}_{ijk} &= \boldsymbol{\mu} + \boldsymbol{\alpha}_i + \boldsymbol{\beta}_{(i)j} + \mathbf{e}_{ijk} \text{ (LFR)} \\
&= \boldsymbol{\mu}_{ij} + \mathbf{e}_{ijk} \text{ (FR)} \\
\mathbf{e}_{ijk} &\sim IN_p\ (\mathbf{0},\ \Sigma)
\end{aligned}
\tag{4.12.4}
$$

for $i = 1,\ 2, \ldots, a;\ j = 1,\ 2, \ldots, b_i;$ and $k = 1,\ 2, \ldots, n_{ij} > 1$.

While one can again apply general theory to obtain estimable functions, it is easily seen that $\boldsymbol{\mu}_{ij} = \boldsymbol{\mu} + \boldsymbol{\alpha}_i + \boldsymbol{\beta}_{(i)j}$ is estimable and estimated by the cell mean, $\widehat{\boldsymbol{\mu}}_{ij} = \mathbf{y}_{ij.}$. Furthermore, linear combinations of estimable functions are estimable. Thus, $\boldsymbol{\psi} = \boldsymbol{\mu}_{ij} - \boldsymbol{\mu}_{ij'} =$

$\boldsymbol{\beta}_{(i)j} - \boldsymbol{\beta}_{(i)j'}$ for $j \neq j'$ is estimable and estimated by $\widehat{\boldsymbol{\psi}} = \mathbf{y}_{ij.} - \mathbf{y}_{ij'.}$. Hence, the hypothesis of no difference in treatment levels B at each level of Factor A is testable. The hypothesis is written as

$$H_{B(A)} : \text{all } \boldsymbol{\beta}_{(i)j} \text{ are equal} \tag{4.12.5}$$

for $i = 1, 2, \ldots, a$. By associating $\boldsymbol{\beta}_{(i)j} \equiv \boldsymbol{\beta}_j + \boldsymbol{\gamma}_{ij}$, the degrees of freedom for the test is $v_{B(A)} = (b-1) + (a-1)(b-1) = a(b-1)$ if there were an equal number of levels of B at each level of A. However, for the design in (4.12.3) we have b_i levels of B at each level of factor A, or a one-way design at each level. Hence, the overall degrees of freedom is obtained by summing over the a one-way designs so that $v_{B(A)} = \sum_{i=1}^{a}(b_i - 1)$.

To construct tests of A, observe that one must be able to estimate $\boldsymbol{\psi} = \boldsymbol{\alpha}_i - \boldsymbol{\alpha}_{i'}$. However, taking simple differences we see that the differences are confounded by the effects $\boldsymbol{\beta}_{(i)j}$. Hence, tests of differences in A are not testable. The estimable functions and their estimates have the general structure

$$\begin{aligned} \boldsymbol{\psi} &= \sum_i \sum_j t_{ij}\left(\boldsymbol{\mu} + \boldsymbol{\alpha}_i + \boldsymbol{\beta}_{(i)j}\right) \\ \widehat{\boldsymbol{\psi}} &= \sum_i \sum_j t_{ij}\mathbf{y}_{ij.} \end{aligned} \tag{4.12.6}$$

so that the parametric function

$$\boldsymbol{\psi} = \boldsymbol{\alpha}_i - \boldsymbol{\alpha}_{i'} + \sum_j t_{ij}\boldsymbol{\beta}_{(i)j} - \sum_j t_{i'j}\boldsymbol{\beta}_{(i')j} \tag{4.12.7}$$

is estimated by

$$\widehat{\boldsymbol{\psi}} = \sum_j \mathbf{y}_{ij.} - \sum_j \mathbf{y}_{i'j.} \tag{4.12.8}$$

if we make the $\sum_j t_{ij} = \sum_j t_{i'j} = 1$. Two sets of weights are often used. If the unequal n_{ij} are the result of the treatment administered, the $t_{ij} = n_{ij}/n_{i+}$. Otherwise, the weights $t_{ij} = 1/b_i$ are used. This leads to weighted and unweighted tests of H_A. For the LFR model, the test of A becomes

$$H_A : \text{all } \boldsymbol{\alpha}_i + \sum_j t_{ij}\boldsymbol{\beta}_{(i)j} \text{ are equal} \tag{4.12.9}$$

which shows the confounding. In terms of the FR model, the tests are

$$\begin{aligned} H_{A^*} &: \text{all } \overline{\boldsymbol{\mu}}_{i.} \text{ are equal} \\ H_A &: \text{all } \boldsymbol{\mu}_{i.} \text{ are equal} \end{aligned} \tag{4.12.10}$$

where $\overline{\boldsymbol{\mu}}_{i.}$ is a weighted marginal mean that depends on the n_{ij} cell frequencies and $\boldsymbol{\mu}_{i.}$ is an unweighted average that depends on the number of nested levels b_i of effect B within each level of A. In SAS, one uses the Type I and Type III options to generate the correct hypothesis test matrices. To verify this, one uses the E option on the MODEL statement to check Type I and Type III estimates. This should always be done when sample sizes are unequal.

For the nested design given in (4.12.4), the $r(\mathbf{X}) = q = \sum_{i=1}^{a} b_i$ so that

$$v_e = \sum_{i=1}^{a}\sum_{j=1}^{b_i}(n_{ij} - 1)$$

and the error matrix is

$$\mathbf{E} = \sum_i \sum_j \sum_k \left(\mathbf{y}_{ijk} - \mathbf{y}_{ij.}\right)\left(\mathbf{y}_{ijk} - \mathbf{y}_{ij.}\right)' \tag{4.12.11}$$

for each of the tests $H_{B(A)}$, H_{A^*}, and H_A.

One can easily extend the two-way nested design to three factors A, B, and C. A design with B nested in A and C nested in B as discussed in Scheffé (1959, p. 186) has a natural multivariate extension

$$\begin{aligned}
\mathbf{y}_{ijkm} &= \boldsymbol{\mu} + \boldsymbol{\alpha}_i + \boldsymbol{\beta}_{(i)j} + \boldsymbol{\tau}_{(ij)k} + \mathbf{e}_{ijkm} \qquad &\text{(LFR)}\\
&= \boldsymbol{\mu}_{ijk} + \mathbf{e}_{ijkm} &\text{(FR)}\\
\mathbf{e}_{ijkm} &\sim IN_p\ (\mathbf{0}, \Sigma)
\end{aligned}$$

where $i = 1, 2, \ldots, a$; $j = 1, 2, \ldots, b_i$; $k = 1, 2, \ldots, n_{ij}$ and $m = 1, 2, \ldots, m_{ijk}$.

Another common variation of a nested design is to have both nested and crossed factors, a partially nested design. For example, B could be nested in A, but C might be crossed with A and B. The MGLM for this design is

$$\begin{aligned}
\mathbf{y}_{ijkm} &= \boldsymbol{\mu} + \boldsymbol{\alpha}_i + \boldsymbol{\beta}_{(i)j} + \boldsymbol{\tau}_k + \boldsymbol{\gamma}_{ik} + \boldsymbol{\delta}_{(i)jk} + \mathbf{e}_{ijkm}\\
\mathbf{e}_{ijkm} &\sim IN_p\ (\mathbf{0}, \Sigma)
\end{aligned} \tag{4.12.12}$$

over some indices $(i,\ j,\ k,\ m)$.

Every univariate design with crossed and nested fixed effects, a combination of both, has an identical multivariate counterpart. These designs and special designs like fractional factorial, crossover designs, balanced incomplete block designs, Latin square designs, Youden squares, and numerous others may be analyzed using PROC GLM. Random and mixed models also have natural extensions to the multivariate case and are discussed in Chapter 6.

4.13 Complex Design Examples

a. *Nested Design (Example 4.13.1)*

In the investigation of the data given in Table 4.9.1, suppose teachers are nested within classes. Also suppose that the third teacher under noncontract classes was unavailable for the study. The design for the analysis would then be a fixed effects nested design represented diagrammatically as follows

TABLE 4.13.1. Multivariate Nested Design

Classes	A_1				A_2					
Teachers	B_1		B_2		B'_1		B'_2		B'_3	
	R	C	R	C	R	C	R	C	R	C
	9	14	11	15	10	21	11	23	8	17
	8	15	12	18	12	22	14	27	7	15
	11	16	10	16	9	19	13	24	10	18
	9	17	9	17	10	21	15	26	8	17
	9	17	9	18	14	23	14	24	7	19

	T_1	T_2	T_3	T_4	T_5
Noncontrast Classes	×	×			
Contrast Classes			×	×	×

where the × denotes collected data. The data are reorganized as in Table 4.13.1 where factor A, classes, has two levels and factor B, teachers, is nested within factor A. The labels R and C denote the variables reading rate and reading comprehension, as before.

Program m4_13_1a.sas contains the PROC GLM code for the analysis of the multivariate fixed effects nested design. The model for the observation vector $\mathbf{y}_{ijk}$ is

$$\begin{aligned} \mathbf{y}_{ijk} &= \boldsymbol{\mu} + \boldsymbol{\alpha}_i + \boldsymbol{\beta}_{(i)j} + \mathbf{e}_{ijk} \\ \mathbf{e}_{ijk} &\sim IN_2(\mathbf{0}, \Sigma) \end{aligned} \tag{4.13.1}$$

where $a = 2$, $b_1 = 2$, $b_2 = 3$, and $n_{ij} = 5$ for the general model (4.12.4). The total number of observations for the analysis is $n = 25$.

While one may test for differences in factor A (classes), this test is confounded by the effects $\boldsymbol{\beta}_{(i)j}$. For our example, H_A is

$$H_A : \boldsymbol{\alpha}_1 + \sum_{j=1}^{b_1} n_{ij}\boldsymbol{\beta}_{(1)j}/n_{1+} = \boldsymbol{\alpha}_2 + \sum_{j=1}^{b_1} n_{2j}\boldsymbol{\beta}_{(2)j}/n_{2+} \tag{4.13.2}$$

where $n_{1+} = 10$ and $n_{2+} = 15$. This is seen clearly in the output from the estimable functions. While many authors discuss the test of (4.12.5) when analyzing nested designs, the tests of interest are the tests for differences in the levels of B within the levels of A. For the design under study, these tests are

$$\begin{aligned} H_{B(A_1)} &: \boldsymbol{\beta}_{(1)1} = \boldsymbol{\beta}_{(1)2} \\ H_{B(A_2)} &: \boldsymbol{\beta}_{(2)1} = \boldsymbol{\beta}_{(2)2} = \boldsymbol{\beta}_{(2)3} \end{aligned} \tag{4.13.3}$$

TABLE 4.13.2. MANOVA for Nested Design

Source	df	SSCP
H_A: Classes	1	$\mathbf{H}_A = \begin{bmatrix} 7.26 & 31.46 \\ 31.46 & 136.33 \end{bmatrix}$
$H_{B(A)}$: Teachers with Classes	3	$\mathbf{H}_B = \begin{bmatrix} 75.70 & 105.30 \\ 105.30 & 147.03 \end{bmatrix}$
$H_{B(A_1)}$	1	$\mathbf{H}_{B(A_1)} = \begin{bmatrix} 2.5 & 2.5 \\ 2.5 & 2.5 \end{bmatrix}$
$H_{B(A_1)}$	2	$\mathbf{H}_{B(A_1)} = \begin{bmatrix} 73.20 & 102.80 \\ 102.80 & 144.53 \end{bmatrix}$
Error	20	$\mathbf{E} = \begin{bmatrix} 42.8 & 20.8 \\ 20.8 & 42.0 \end{bmatrix}$

The tests in (4.13.3) are "planned" comparisons associated with the overall test

$$H_{B(A)} : \text{all } \boldsymbol{\beta}_{(i)j} \text{ are equal} \tag{4.13.4}$$

The MANOVA statement in PROC GLM by default performs tests of H_A and $H_{B(A)}$. To test (4.13.3), one must construct the test using a CONTRAST statement and a MANOVA statement with $\mathbf{M} = \mathbf{I}_2$. Table 4.13.2 summarizes the MANOVA output for the example. Observe that $\mathbf{H}_{B(A)} = \mathbf{H}_{B(A_1)} + \mathbf{H}_{B(A_2)}$ and that the hypothesis degrees of freedom for $\mathbf{H}_{B(A)}$ add to $\mathbf{H}_{B(A)}$. More generally, $v_A = a-1$, $v_{B(A)} = \sum_{i=1}^{a}(b_i - 1)$, $v_{B(A_i)} = b_i - 1$, and $v_e = \sum_i \sum_j (n_{ij} - 1)$.

Solving the characteristic equation $|\mathbf{H} - \lambda\mathbf{E}| = 0$ for each hypotheses in Table 4.13.2 one may test each overall hypothesis. For the nested design, one tests H_A and $H_{B(A)}$ at some level α. The tests of $H_{B(A_i)}$ are tested at α_i where the $\sum_i \alpha_i = \alpha$. For this example, suppose $\alpha = 0.05$, the $\alpha_i = 0.025$. Reviewing the p-values for the overall tests, the test of $H_A \equiv C$ and $H_{B(C)} \equiv T\,(C)$ are clearly significant. The significance of the overall test is due to differences between teachers in contract classes and not noncontract classes. The p-value for $H_{B(A_1)} \equiv T\,(C1)$ and $H_{B(A_1)} \equiv T\,(C2)$ are 0.4851 and 0.0001, respectively.

With the rejections of an overall test, the overall test criterion determines the simultaneous confidence intervals one may construct to determined the differences in parametric functions that led to rejection. Letting $\psi = \mathbf{c}'\mathbf{B}\mathbf{m}$, $\widehat{\psi} = \mathbf{c}'\widehat{\mathbf{B}}\mathbf{m}$, $\widehat{\sigma}^2_{\widehat{\psi}} = (\mathbf{m}'\mathbf{S}\mathbf{m})\,\mathbf{c}'\left(\mathbf{X}'\mathbf{X}\right)^{-}\mathbf{c}$, then we again have that with probability $1-\alpha$, for all ψ,

$$\widehat{\psi} - c_\alpha \widehat{\sigma}_{\widehat{\psi}} \leq \psi \leq \widehat{\psi} + c_\alpha \widehat{\sigma}_{\widehat{\psi}} \tag{4.13.5}$$

where for the largest root criterion

$$c_\alpha^2 = \left(\frac{\theta^\alpha}{1-\theta^\alpha}\right) v_e = \lambda^\alpha v_e$$

For our example,

$$\mathbf{B} = \begin{bmatrix} \boldsymbol{\mu}' \\ \boldsymbol{\alpha}_1' \\ \boldsymbol{\alpha}_2' \\ \boldsymbol{\beta}_{(1)1}' \\ \boldsymbol{\beta}_{(1)2}' \\ \boldsymbol{\beta}_{(2)1}' \\ \boldsymbol{\beta}_{(2)2}' \\ \boldsymbol{\beta}_{(2)3}' \end{bmatrix} = \begin{bmatrix} \mu_{11} & \mu_{12} \\ \alpha_{11} & \alpha_{12} \\ \alpha_{21} & \alpha_{22} \\ \beta_{(1)11} & \beta_{(1)12} \\ \beta_{(1)21} & \beta_{(1)22} \\ \beta_{(2)11} & \beta_{(2)12} \\ \beta_{(2)22} & \beta_{(2)22} \\ \beta_{(2)31} & \beta_{(2)32} \end{bmatrix} \tag{4.13.6}$$

for the LFR model and $\mathbf{B} = [\mu_{ij}]$ for a FR model. Using the SOLUTION and E3 option on the model statement, one clearly sees that contrasts in the $\boldsymbol{\alpha}_i$ are confounded by the effects $\boldsymbol{\beta}_{(i)j}$. One normally only investigates contrasts in the $\boldsymbol{\alpha}_i$ for those tests of $H_{B(A_i)}$ that are nonsignificant. For pairwise comparisons $\boldsymbol{\psi} = \boldsymbol{\beta}_{(i)j} - \boldsymbol{\beta}_{(i)j'}$ for $i = 1, 2, \ldots, a$ and $j \neq j'$ the standard error has the simple form

$$\widehat{\sigma}_{\widehat{\psi}}^2 = (\mathbf{m}'\mathbf{S}\mathbf{m}) \left(\frac{1}{n_{ij}} + \frac{1}{n_{ij'}} \right)$$

To locate significance following the overall tests, we use several approaches. The largest difference appears to be between Teacher 2 and Teacher 3 for both the rate and comprehension variables. Using the largest root criterion, the TRANSREG procedure, and IML code, with $\alpha = 0.025$ the approximate confidence set for reading comprehension is (4.86, 10.34). Locating significance comparison using CONTRAST statement also permits the location of significant comparisons.

Assuming the teacher factor is random and the class factor is fixed leads to a mixed MANOVA model. While we have included in the program the PROC GLM code for the situation, we postpone discussion until Chapter 6.

b. *Latin Square Design (Example 4.13.2)*

For our next example, we consider a multivariate Latin square design. The design is a generalization of a randomized block design that permits double blocking that reduces the mean square within in a design by controlling for two nuisance variables. For example, suppose an investigator is interested in examining a concept learning task for five experimental treatments that may be adversely effected by days of the week and hours of the day. To investigate treatments the following Latin square design may be employed

Hours of Day

	1	2	3	4	5
Monday	T_2	T_5	T_4	T_3	T_1
Tuesday	T_3	T_1	T_2	T_5	T_4
Wednesday	T_4	T_2	T_3	T_1	T_5
Thursday	T_5	T_3	T_1	T_4	T_2
Friday	T_1	T_4	T_5	T_2	T_3

where each treatment condition T_i appears only once in each row and column. The Latin square design requires only d^2 observations where d represents the number of levels per factor. The Latin square design is a balanced incomplete three-way factorial design. An additive three-way factorial design requires d^3 observations.

The multivariate model for an observation vector $\mathbf{y}_{ijk}$ for the design is

$$\begin{aligned} \mathbf{y}_{ijk} &= \boldsymbol{\mu} + \boldsymbol{\alpha}_i + \boldsymbol{\beta}_j + \boldsymbol{\gamma}_k + \mathbf{e}_{ijk} \\ \mathbf{e}_{ijk} &\sim IN_p(\mathbf{0}, \Sigma) \end{aligned} \tag{4.13.7}$$

for $(i, jk) \in D$ where D is a Latin square design. Using a MR model to analyze a Latin square design with d levels, the rank of the design matrix $\mathbf{X}$ is $r(\mathbf{X}) = 3(d-1) + 1 = 3d - 2$ so that $v_e = n - r(\mathbf{X}) = d^2 - 3d + 2 = (d-1)(d-2)$. While the individual effects in (4.13.7) are not estimable, contrasts in the effects are estimable. This is again easily seen when using PROC GLM by using the option E3 on the MODEL statement.

To illustrate the analysis of a Latin square design, we use data from a concept learning study in the investigation of five experimental treatments ($T_1, T_2, \ldots, T_5$) for the two blocking variables day of the week and hours in the days as previously discussed. The dependent variables are the number of treats to criterion used to measure learning (V_1) and number of errors in the test set on one presentation 10 minutes later (V_2) used to measure retention. The hypothetical data are provided in Table 4.13.3 and are in the file Latin.dat. The cell indexes represent the days of the week, hours of the day, and the treatment, respectively. The SAS code for the analysis is given in program m4_13_1b.sas.

In the analysis of the Latin square design, both blocking variables are nonsignificant. Even though they were not effective in reducing variability between blocks, the treatment effect is significant. The **H** and **E** matrices for the test of no treatment differences are

$$\mathbf{H} = \begin{bmatrix} 420.80 & 48.80 \\ 48.80 & 177.04 \end{bmatrix}$$

$$\mathbf{E} = \begin{bmatrix} 146.80 & 118.00 \\ 118.00 & 422.72 \end{bmatrix}$$

TABLE 4.13.3. Multivariate Latin Square

Cell	V_1	V_2	Cell	V_1	V_2
112	8	4	333	4	17
125	18	8	341	8	8
134	5	3	355	14	8
143	8	16	415	11	9
151	6	12	423	4	15
213	1	6	431	14	17
221	6	19	444	1	5
232	5	7	452	7	8
245	18	9	511	9	14
254	9	23	524	9	13
314	5	11	535	16	23
322	4	5	542	3	7
			553	2	10

Solving $|\mathbf{H} - \lambda\mathbf{E}| = 0$, $\lambda_1 = 3.5776$ and $\lambda_2 = 0.4188$. Using the /CANONICAL option, the standardized and structure (correlation) vectors for the test of treatments follow

Standardized Structure

$$\begin{bmatrix} 1.6096 \\ -0.5128 \end{bmatrix} \begin{bmatrix} 0.0019 \\ 0.9513 \end{bmatrix} \quad \begin{bmatrix} 0.8803 \\ -0.001 \end{bmatrix} \begin{bmatrix} 0.0083 \\ 0.1684 \end{bmatrix}$$

indicating that only the first variable is contributing to the differences in treatments. Using Tukey's method to evaluate differences, all pairwise differences are significant for V_2 while only the comparison between T_2 and T_5 is significant for variable V_1.

Reviewing the Q-Q plots and test statistics, the assumption of multivariate normality seems valid.

Exercises 4.13

1. Box (1950) provides data on tire wear for three factors: road surface, filler type, and proportion of filler. Two observations of the wear at 1000, 2000, and 3000 revolutions were collected for all factor combinations. The data for the study is given in Table 4.13.4.

   (a) Analyze the data using a factorial design. What road filler produces the least wear and in what proportion?

   (b) Reanalyze the data assuming filler is nested within road surface.

2. The artificial data set in file three.dat contains data for a nonorthogonal three-factor MANOVA design. The first three variables represent the factor levels A, B, and C; and, the next two data items represent two dependent variables.

TABLE 4.13.4. Box Tire Wear Data

Road Surface	Filler	25% Tire Wear 1	2	3	50% Tire Wear 1	2	3	75% Tire Wear 1	2	3
1	F_1	194	192	141	233	217	171	265	252	207
		208	188	165	241	222	201	261	283	191
	F_2	239	127	90	224	123	79	243	117	100
		187	105	85	243	123	110	226	125	75
2	F_1	155	169	151	198	187	176	235	225	166
		173	152	141	177	196	167	229	270	183
	F_2	137	82	77	229	94	78	155	76	91
		160	82	83	98	89	48	132	105	69

(a) Assuming observation loss is due to treatments, analyze the data using Type I tests.

(b) Assuming that observation loss is not due to treatment, analyze the data using Type III tests.

4.14 Repeated Measurement Designs

In Chapter 3 we discussed the analysis of a two group profile design where the vector of p responses were commensurate. In such designs, interest focused on parallelism of profiles, differences between groups, and differences in the means for the p commensurate variables. A design that is closely related to this design is the repeated measures design. In these designs, a random sample of subjects are randomly assigned to several treatment groups, factor A, and measured repeatedly over p traits, factor B. Factor A is called the between-subjects factor, and factor B is called the within-subjects factor. In this section, we discuss the univariate and multivariate analysis of one-way repeated measurement designs and extended linear hypotheses. Examples are illustrated in Section 4.15. Growth curve analysis of repeated measurements data is discussed in Chapter 5. Doubly multivariate repeated measurement designs in which vectors of observations are observed over time are discussed in Chapter 6.

a. *One-Way Repeated Measures Design*

The data for the one-way repeated measurement design is identical to the setup shown in Table 3.9.4. The vectors

$$\mathbf{y}'_{ij} = \left[y_{ij1}, y_{ij2}, \ldots, y_{ijp}\right] \sim IN_p\left(\boldsymbol{\mu}_i, \Sigma\right) \tag{4.14.1}$$

represent the vectors of p repeated measurements of the jth subject within the ith treatment group ($i = 1, 2, \ldots, a$). Assigning n_i subjects per group, the subscript $j = 1, 2, \ldots, n_i$ represents subjects within groups and $n = \sum_{i=1}^{a} n_i$ is the total number of subjects in the study. Assuming all $\Sigma_i = \Sigma$ for $i = 1, \ldots, a$, we assume homogeneity of the covariance matrices. The multivariate model for the one-way repeated measurement design is identical to the one-way MANOVA design so that

$$\begin{aligned} \mathbf{y}_{ij} &= \boldsymbol{\mu} + \boldsymbol{\alpha}_i + \mathbf{e}_{ij} = \boldsymbol{\mu}_i + \mathbf{e}_{ij} \\ \mathbf{e}_{ij} &\sim IN_p\left(\boldsymbol{\mu}_i, \Sigma\right) \end{aligned} \tag{4.14.2}$$

For the one-way MANOVA design, the primary hypothesis of interest was the test for differences in treatment groups. In other words, the hypothesis tested that all mean vectors $\boldsymbol{\mu}_i$ are equal. For the two-group profile analysis and repeated measures designs, the primary hypothesis is the test of parallelism or whether there is a significant interaction between treatment groups (Factor A) and trials (Factor B). To construct the hypothesis test matrices $\mathbf{C}$ and $\mathbf{M}$ for the test of interaction, the matrix $\mathbf{C}$ used to compare groups in the one-way MANOVA design is combined with the matrix $\mathbf{M}$ used in the two group profile analysis, similar to (3.9.35). With the error matrix $\mathbf{E}$ defined as in the one-way MANOVA and $\mathbf{H} = (\mathbf{C}\widehat{\mathbf{B}}\mathbf{M})'\left(\mathbf{C}\left(\mathbf{X}'\mathbf{X}\right)^{-}\mathbf{C}'\right)^{-1}(\mathbf{C}\widehat{\mathbf{B}}\mathbf{M})$ where $\widehat{\mathbf{B}}$ is identical to $\widehat{\mathbf{B}}$ for the MANOVA model, the test of interaction is constructed. The parameters for the test are

$$\begin{aligned} s &= \min\left(v_h,\ u\right) = \min\left(a-1,\ p-1\right) \\ M &= \left(|v_h - u| - 1\right)/2 = \left(|a-p| - 1\right)/2 \\ N &= \left(v_e - u - 1\right)/2 = \left(n - a - p\right)/2 \end{aligned}$$

since $v_h = r\left(\mathbf{C}\right) = (a-1)$, $u = r(\mathbf{M})$, and $v_e = n - r\left(\mathbf{X}\right) = n - a$.

If the test of interaction is significant in a repeated measures design, the unrestrictive multivariate test of treatment group differences and the unrestrictive multivariate test of equality of the p trial vectors are not usually of interest.

If the test of interaction is not significant, signifying that treatments and trials are not confounded by interaction, the structure of the elements μ_{ij} in $\mathbf{B}$ are additive so that

$$\mu_{ij} = \mu + \alpha_i + \beta_j \qquad i = 1, \ldots, a;\ j = 1, 2, \ldots, p \tag{4.14.3}$$

When this is the case, we may investigate the restrictive tests

$$\begin{aligned} H_\alpha &: \text{all } \alpha_i \text{ are equal} \\ H_\beta &: \text{all } \beta_i \text{ are equal} \end{aligned} \tag{4.14.4}$$

Or, using the parameters μ_{ij}, the tests become

$$\begin{aligned} H_\alpha &: \mu_{1.} = \mu_{2.} = \ldots = \mu_{a.} \\ H_\beta &: \mu_{.1} = \mu_{.2} = \ldots = \mu_{.p} \\ H_{\beta^*} &: \overline{\mu}_{.1} = \overline{\mu}_{.2} = \ldots = \overline{\mu}_{.p} \end{aligned} \tag{4.14.5}$$

where $\mu_{i.} = \sum_{j=1}^{p} \mu_{ij}/p$, $\mu_{.j} = \sum_{i=1}^{a} \mu_{ij}/a$, and $\overline{\mu}_{.j} = \sum_{i=1}^{a} n_i \mu_{ij}/n$.

To test H_α, the matrix $\mathbf{C}$ is identical to the MANOVA test for group differences, and the matrix $\mathbf{M}' = [1/p, 1/p, \ldots, 1/p]$. The test is equivalent to testing the equality of the a independent group means, or a one-way ANOVA analysis for treatment differences.

The tests H_β and $H_{\beta *}$ are extensions of the tests of conditions, H_C and H_C^W, for the two group profile analysis. The matrix $\mathbf{M}$ is selected equal to the matrix $\mathbf{M}$ used in the test of parallelism and the matrices $\mathbf{C}$ are, respectively,

$$\begin{aligned} \mathbf{C}_\beta &= [1/a, 1/a, \ldots, 1/a] \text{ for } H_\beta \\ \mathbf{C}_{\beta *} &= [n_1/n, n_2/n, \ldots, n_a/n] \text{ for } H_{\beta *} \end{aligned} \tag{4.14.6}$$

Then, it is easily verified that the test statistics follow Hotelling's T^2 distribution where

$$\begin{aligned} T_\beta^2 &= a^2 \left(\sum_{i=1}^{a} 1/n_i \right)^{-1} \mathbf{y}_{..}' \mathbf{M} \left(\mathbf{M}'\mathbf{S}\mathbf{M} \right)^{-1} \mathbf{M}'\mathbf{y}_{..} \\ T_{\beta *}^2 &= n \overline{\mathbf{y}}_{..}' \mathbf{M} \left(\mathbf{M}'\mathbf{S}\mathbf{M} \right)^{-1} \mathbf{M}'\overline{\mathbf{y}}_{..} \end{aligned}$$

are distributed as central T^2 with degrees of freedom $(p-1,\ v_e = n-a)$ under the null hypotheses H_β and $H_{\beta *}$, respectively, where $\overline{\mathbf{y}}_{..}$ and $\mathbf{y}_{..}$ are the weighted and unweighted sample means

$$\mathbf{y}_{..} = \sum_i \mathbf{y}_{i.}/a \text{ and } \overline{\mathbf{y}}_{..} = \sum_i n_i \mathbf{y}_{i.}/n$$

Following the rejection of the test of AB, simultaneous confidence intervals for tetrads in the μ_{ij} are easily established using the same test criterion that was used for the overall test. For the tests of H_β and $H_{\beta *}$, Hotelling T^2 distribution is used to establish confidence intervals. For the test of H_α, standard ANOVA methods are available.

To perform a multivariate analysis of a repeated measures design, the matrix Σ for each group must be positive definite so that $p \geq n_i$ for each group. Furthermore, the analysis assumes an unstructured covariance matrix Σ for the repeated measures. When the matrix Σ is homogeneous and has a simplified (Type H) structure, the univariate mixed model analysis of the multiple group repeated measures design is more powerful.

The univariate mixed model for the design assumes that the subjects are random and nested within the fixed factor A, which is crossed with factor B. The design is called a split-plot design where factor A is the whole plot and factor B is the repeated measures or split-plot, Kirk (1995, Chapter 12). The univariate (split-plot) mixed model is

$$y_{ijk} = \mu + \alpha_i + \beta_k + \gamma_{ik} + s_{(i)j} + e_{ijk} \tag{4.14.7}$$

where $s_{(i)j}$ and e_{ijk} are jointly independent, $s_{(i)j} \sim IN\left(0, \sigma_s^2\right)$ and $e_{ijk} \sim IN\left(0, \sigma_e^2\right)$. The parameters α_i, β_j, and γ_{ij} are fixed effects representing factors A, B, and AB. The parameter $s_{(i)j}$ is the random effect of the jth subject nested within the ith group. The structure of the cov $\left(\mathbf{y}_{ij}\right)$ is

$$\begin{aligned} \Sigma &= \sigma_s^2 \mathbf{J}_p + \sigma_e^2 \mathbf{I}_p \\ &= \rho^2 \mathbf{J}_p + (1-\rho)\, \sigma^2 \mathbf{I}_p \end{aligned} \tag{4.14.8}$$

where $\sigma^2 = \sigma_e^2 + \sigma_s^2$ and the intraclass correlation $\rho = \sigma_s^2 / \left(\sigma_s^2 + \sigma_e^2\right)$. The matrix Σ is said to have equal variances and covariances that are equal or uniform, intraclass structure. Thus, while $\text{cov}(\mathbf{y}_{ij}) \neq \sigma^2 \mathbf{I}$, univariate ANOVA procedures remain valid.

More generally, Huynh and Feldt (1970) showed that to construct exact F tests for B and AB using the mixed univariate model the necessary and sufficient condition is that there exists an orthogonal matrix $\mathbf{M}_{p \times (p-1)}$ $\left(\mathbf{M}'\mathbf{M} = \mathbf{I}_{p-1}\right)$ such that

$$\mathbf{M}' \Sigma \mathbf{M} = \sigma^2 \mathbf{I}_{p-1} \tag{4.14.9}$$

so that Σ satisfies the sphericity condition. Matrices which satisfy this structure are called Type H matrices. In the context of repeated measures designs, (4.14.9) is sometimes called the circularity condition. When one can capitalize on the structure of Σ, the univariate F test of mean treatment differences is more powerful than the multivariate test of mean vector differences since the F test is one contrast of all possible contrasts for the multivariate test. The mixed model exact F tests of B are more powerful than the restrictive multivariate tests H_β $\left(H_{\beta^*}\right)$. The univariate mixed model F test of AB is more powerful than the multivariate test of parallelism since these tests have more degrees of freedom v, $v = r\left(\mathbf{M}\right) v_e$ where v_e is the degrees of freedom for the corresponding multivariate tests. As shown by Timm (1980a), one may easily recover the mixed model tests of B and AB from the restricted multivariate test of B and the test of parallelism. This is done automatically by using the REPEATED option in PROC GLM. The preferred procedure for the analysis of the mixed univariate model is to use PROC MIXED.

While the mixed model F tests are most appropriate if Σ has Type H structure, we know that the preliminary tests of covariance structure behave poorly in small samples and are not robust to nonnormality. Furthermore, Boik (1981) showed that the Type I error rate of the mixed model tests of B and AB are greatly inflated when Σ does not have Type H structure. Hence, he concludes that the mixed model tests should be avoided.

An alternative approach to the analysis of the tests of B and AB is to use the Greenhouse and Geisser (1959) or Huynh and Feldt (1970) approximate F tests. These authors propose factors ϵ and $\widetilde{\epsilon}$ to reduce the numerator and denominator degrees of freedom of the mixed model F tests of B and AB to correct for the fact that Σ does not have Type H structure. In a simulation study conducted by Boik (1991), he shows that while the approximate tests are near the Type I nominal level α, they are not as powerful as the exact multivariate tests so he does not recommend their use. The approximate F tests are also used in studies in which p is greater than n since no multivariate test exists in this situation. Keselman and Keselman (1993) review simultaneous test procedures when approximate F tests are used.

An alternative formulation of the analysis of repeated measures data is to use the univariate mixed linear model. Using the FR cell means model, let $\mu_{jk} = \mu + \alpha_j + \beta_k + \gamma_{jk}$. For this representation, we have interchanged the indices i and j. Then, the vector of repeated measures $\mathbf{y}_{ij}' = \left[y_{ij1}, y_{ij2}, \ldots, y_{ijp}\right]$ where $i = 1, 2, \ldots, n_j$ denotes the ith subject nested within the jth group (switched the role of i and j) so that $s_{i(j)}$ is the random component of subject i within group j; $j = 1, 2, \ldots, a$. Then,

$$\mathbf{y}_{ij} = \boldsymbol{\theta}_j + \mathbf{1}_p s_{(i)j} + \mathbf{e}_{ij} \tag{4.14.10}$$

where $\boldsymbol{\theta}_j = [\mu_{j1}, \mu_{j2}, \ldots, \mu_{jp}]$ is a linear model for the vector $\mathbf{y}_{ij}$ of repeated measures with a fixed component and a random component. Letting $i = 1, 2, \ldots, n$ where $n = \sum_j n_j$ and δ_{ij} be an indicator variable such that $\delta_{ij} = 1$ if subject i is from group j and $\delta_{ij} = 0$ otherwise where $\boldsymbol{\delta}'_i = [\delta_{i1}, \delta_{i2}, \ldots, \delta_{ia}]$, (4.4.10) has the univariate mixed linear model structure

$$\mathbf{y}_i = \mathbf{X}_i \boldsymbol{\beta} + \mathbf{Z}_i \mathbf{b}_i + \mathbf{e}_i \tag{4.14.11}$$

where

$$\underset{p\times 1}{\mathbf{y}_i} = \begin{bmatrix} y_{i1} \\ y_{i2} \\ \vdots \\ y_{ip} \end{bmatrix}, \quad \underset{p\times a}{\mathbf{X}_i} = [\mathbf{I}_p \otimes \boldsymbol{\delta}_i], \quad \underset{pa\times 1}{\boldsymbol{\beta}} = \begin{bmatrix} \mu_{11} \\ \mu_{21} \\ \vdots \\ \mu_{ap} \end{bmatrix}$$

$$\mathbf{Z}_i = \mathbf{1}_p \quad \text{and} \quad b_i = s_{i(j)}$$

and $\mathbf{e}'_i = [e_{ij1}, e_{ij2}, \ldots, e_{ijp}]$. For the vector $\mathbf{y}_i$ of repeated measurements, we have as in the univariate ANOVA model that

$$\begin{aligned} E(\mathbf{y}_i) &= \mathbf{X}_i \boldsymbol{\beta} \\ \operatorname{cov}(\mathbf{y}_i) &= \mathbf{Z}_i \operatorname{cov}(\mathbf{b}_i) \mathbf{Z}'_i + \operatorname{cov}(\mathbf{e}_i) \\ &= \mathbf{J}_p \sigma_s^2 + \sigma_e^2 \mathbf{I}_p \end{aligned}$$

which is a special case of the multivariate mixed linear model to be discussed in Chapter 6. In Chapter 6, we will allow more general structures for the cov $(\mathbf{y}_i)$ and missing data.

In repeated measurement designs, one may also include covariates. The covariates may enter the study in two ways: (a) a set of baseline covariates are measured on all subjects or (b) a set of covariates are measured at each time point so that they vary with time. In situation (a), one may analyze the repeated measures data as a MANCOVA design. Again, the univariate mixed linear model may be used if Σ has Type H structure. When the covariates are changing with time, the situation is more complicated since the MANCOVA model does not apply. Instead one may use the univariate mixed ANCOVA model or use the SUR model. Another approach is to use the mixed linear model given in (4.14.11) which permits the introduction of covariates that vary with time. We discuss these approaches in Chapters 5 and 6.

b. *Extended Linear Hypotheses*

When comparing means in MANOVA/MANCOVA designs, one tests hypotheses of the form $H : \mathbf{CBM} = \mathbf{0}$ and obtains simultaneous confidence intervals for bilinear parametric functions $\psi = \mathbf{c}'\mathbf{Bm}$. However, all potential contrasts of the parameters of $\mathbf{B} = [\mu_{ij}]$ may not have the bilinear form. To illustrate, suppose in a repeated measures design that one is interested in the multivariate test of group differences for a design with three groups and three variables so that

$$\mathbf{B} = \begin{bmatrix} \mu_{11} & \mu_{12} & \mu_{13} \\ \mu_{21} & \mu_{22} & \mu_{23} \\ \mu_{31} & \mu_{32} & \mu_{33} \end{bmatrix} \tag{4.14.12}$$

Then for the multivariate test of equal group means

$$H_G: \begin{bmatrix} \mu_{11} \\ \mu_{12} \\ \mu_{13} \end{bmatrix} = \begin{bmatrix} \mu_{21} \\ \mu_{22} \\ \mu_{23} \end{bmatrix} = \begin{bmatrix} \mu_{31} \\ \mu_{32} \\ \mu_{33} \end{bmatrix} \tag{4.14.13}$$

one may select $\mathbf{C} \equiv \mathbf{C}_o$ and $\mathbf{M} \equiv \mathbf{M}_o$ where

$$\mathbf{C}_o = \begin{bmatrix} 1 & -1 & 0 \\ 0 & 1 & -1 \end{bmatrix} \text{ and } \mathbf{M}_o = \mathbf{I}_3 \tag{4.14.14}$$

to test $H_G : \mathbf{C}_o\mathbf{B}\mathbf{M}_o = \mathbf{0}$. Upon rejection of H_G suppose one is interested in comparing the diagonal means with the average of the off diagonal means. Then,

$$\psi = (\mu_{11} + \mu_{22} + \mu_{33}) - \left[(\mu_{12} + \mu_{21}) + (\mu_{13} + \mu_{31}) + (\mu_{23} + \mu_{32})\right]/2 \tag{4.14.15}$$

This contrast may not be expressed in the bilinear form $\boldsymbol{\psi} = \mathbf{c}'\mathbf{B}\mathbf{m}$. However, for a generalized contrast matrix $\mathbf{G}$ defined by Bradu and Gabriel (1974), where the coefficients in each row and column sum to one, the contrast in (4.14.15) has the general form

$$\psi = \operatorname{tr}(\mathbf{G}\mathbf{B}) = \operatorname{tr} \begin{bmatrix} 1 & -.5 & -.5 \\ -.5 & 1 & -.5 \\ -.5 & -.5 & 1 \end{bmatrix} \begin{bmatrix} \mu_{11} & \mu_{12} & \mu_{13} \\ \mu_{21} & \mu_{22} & \mu_{23} \\ \mu_{31} & \mu_{32} & \mu_{33} \end{bmatrix} \tag{4.14.16}$$

Thus, we need to develop a test of the contrast, $H_\psi : \operatorname{tr}(\mathbf{G}\mathbf{B}) = 0$.

Following the multivariate test of equality of vectors across time or conditions

$$H_C: \begin{bmatrix} \mu_{11} \\ \mu_{21} \\ \mu_{31} \end{bmatrix} = \begin{bmatrix} \mu_{12} \\ \mu_{22} \\ \mu_{32} \end{bmatrix} = \begin{bmatrix} \mu_{13} \\ \mu_{23} \\ \mu_{33} \end{bmatrix} \tag{4.14.17}$$

where $\mathbf{C} \equiv \mathbf{C}_o = \mathbf{I}_3$ and $\mathbf{M} \equiv \mathbf{M}_o = \begin{bmatrix} 1 & 0 \\ -1 & 1 \\ 0 & -1 \end{bmatrix}$, suppose upon rejection of H_C that the contrast

$$\psi = (\mu_{11} - \mu_{12}) + (\mu_{22} - \mu_{23}) + (\mu_{31} - \mu_{33}) \tag{4.14.18}$$

is of interest. Again ψ may not be represented in the bilinear form $\psi = \mathbf{c}'\mathbf{Bm}$. However, for the column contrast matrix

$$\mathbf{G} = \begin{bmatrix} 1 & 0 & 1 \\ -1 & 1 & 0 \\ 0 & -1 & -1 \end{bmatrix} \tag{4.14.19}$$

we observe that $\psi = \text{tr}\,(\mathbf{GB})$. Hence, we again need a procedure to test H_ψ : tr $(\mathbf{GB}) = 0$.

Following the test of parallelism

$$\begin{bmatrix} 1 & -1 & 0 \\ 0 & 1 & -1 \end{bmatrix} \begin{bmatrix} \mu_{11} & \mu_{12} & \mu_{13} \\ \mu_{21} & \mu_{22} & \mu_{23} \\ \mu_{31} & \mu_{32} & \mu_{33} \end{bmatrix} \begin{bmatrix} 1 & 0 \\ -1 & 1 \\ 0 & -1 \end{bmatrix} = \mathbf{0} \tag{4.14.20}$$
$$\mathbf{C}_o\,\mathbf{B}\,\mathbf{M}_o = \mathbf{0}$$

suppose we are interested in the significance of the following tetrads

$$\psi = (\mu_{21} + \mu_{12} - \mu_{31} - \mu_{22}) + (\mu_{32} + \mu_{23} - \mu_{13} - \mu_{22}) \tag{4.14.21}$$

Again, ψ may not be expressed as a bilinear form. However, there does exist a generalized contrast matrix

$$\mathbf{G} = \begin{bmatrix} 0 & 1 & -1 \\ 1 & -2 & 1 \\ -1 & 1 & 0 \end{bmatrix}$$

such that $\psi = \text{tr}\,(\mathbf{GB})$. Again, we want to test H_ψ tr $(\mathbf{GB}) = 0$.

In our examples, we have considered contrasts of an overall test H_o : $\mathbf{C}_o\,\mathbf{B}\,\mathbf{M}_o = \mathbf{0}$ where $\psi = \text{tr}\,(\mathbf{GB})$. Situations arise where $\mathbf{G} = \sum_i \gamma_i \mathbf{G}_i$, called intermediate hypotheses since they are defined by a spanning set $\{\mathbf{G}_i\}$. To illustrate, suppose one was interested in the intermediate hypothesis

$$\begin{aligned} \omega_H : \quad & \mu_{11} = \mu_{21} \\ & \mu_{12} = \mu_{22} = \mu_{32} \\ & \mu_{23} = \mu_{33} \end{aligned}$$

To test ω_H, we may select matrices $\mathbf{G}_i$ as follows

$$\mathbf{G}_1 = \begin{bmatrix} 1 & -1 & 0 \\ 0 & 0 & 0 \\ 0 & 0 & 0 \end{bmatrix}, \quad \mathbf{G}_2 = \begin{bmatrix} 0 & 0 & 0 \\ 1 & -1 & 0 \\ 0 & 0 & 0 \end{bmatrix}$$

$$\mathbf{G}_3 = \begin{bmatrix} 0 & 0 & 0 \\ 0 & 1 & -1 \\ 0 & 0 & 0 \end{bmatrix}, \quad \mathbf{G}_4 = \begin{bmatrix} 0 & 0 & 0 \\ 0 & 0 & 0 \\ 0 & 1 & -1 \end{bmatrix} \tag{4.14.22}$$

The intermediate hypothesis ω_H does not have the general linear hypothesis structure, $\mathbf{C}_o\mathbf{B}\mathbf{M}_o = \mathbf{0}$.

Our illustrations have considered a MANOVA or repeated measures design in which each subject is observed over the same trials or conditions. Another popular repeated measures design is a crossover (of change-over) design in which subjects receive different treatments over different time periods. To illustrate the situation, suppose one wanted to investigate two treatments A and B, for two sequences AB and BA, over two periods (time). The parameter matrix for this situation is given in Figure 4.14.1. The design is a 2×2 crossover design where each subject receives treatments during a different time period. The subjects "cross-over" or "change-over"

		Periods (time) 1	2
Sequence	AB	μ_{11} A	μ_{12} B
	BA	μ_{21} B	μ_{22} A

Figure 4.14.1 2×2 Cross-over Design

from one treatment to the other. The FR parameter matrix for this design is

$$\mathbf{B} = \begin{bmatrix} \mu_{11} & \mu_{12} \\ \mu_{21} & \mu_{22} \end{bmatrix} = \begin{bmatrix} \mu_A & \mu_B \\ \mu_B & \mu_A \end{bmatrix} \tag{4.14.23}$$

where index i = sequence and index j = period. The nuisance effects for crossover designs are the sequence, period, and carryover effects. Because a 2×2 crossover design is balanced for sequence and period effects, the main problem with the design is the potential for a differential carryover effect. The response at period two may be the result of the direct effect (μ_A or μ_B) plus the indirect effect (λ_B or λ_A) of the treatment at the prior period. Then, $\mu_B = \mu_A + \lambda_A$ and $\mu_A = \mu_B + \lambda_B$ at period two. The primary test of interest for the 2×2 crossover design is whether $\psi = \mu_A - \mu_B = 0$; however, this test is confounded by λ_A and λ_B since

$$\begin{aligned}\psi &= \left(\frac{\mu_{11} + \mu_{22}}{2}\right) - \left(\frac{\mu_{12} + \mu_{21}}{2}\right)\\ &= \mu_A - \mu_B + (\lambda_A - \lambda_B)/2\end{aligned}$$

This led Grizzle (1965) to recommend testing $H : \lambda_A = \lambda_B$ before testing for treatment effects. However, Senn (1993) shows that the two step process adversely effects the overall Type I familywise error rate. To guard against this problem, a multivariate analysis is proposed. For the parameter matrix in (4.14.23) we suggest testing for no difference in the mean vectors across the two periods

$$H_p : \begin{bmatrix} \mu_{11} \\ \mu_{21} \end{bmatrix} = \begin{bmatrix} \mu_{12} \\ \mu_{22} \end{bmatrix} \tag{4.14.24}$$

using

$$\mathbf{C} \equiv \mathbf{C}_o = \begin{bmatrix} 1 & 0 \\ 0 & 1 \end{bmatrix} \text{ and } \mathbf{M} = \mathbf{M}_o \begin{bmatrix} 1 \\ -1 \end{bmatrix} \tag{4.14.25}$$

Upon rejecting H_p, one may investigate the contrasts

$$\begin{aligned}\psi_1 &: \mu_{11} - \mu_{22} = 0 \quad \text{or} \quad \lambda_A = 0\\ \psi_2 &: \mu_{21} - \mu_{12} = 0 \quad \text{or} \quad \lambda_B = 0\end{aligned} \tag{4.14.26}$$

Failure to reject either $\psi_1 = 0$ or $\psi_2 = 0$, we conclude that the difference is due to treatment. Again, the joint test of $\psi_1 = 0$ and $\psi_2 = 0$ does not have the bilinear form, $\psi_i \neq \mathbf{c}_i'\mathbf{B}\mathbf{m}_i$. Letting $\boldsymbol{\beta} = \text{vec}(\mathbf{B})$, the contrasts ψ_1 and ψ_2 may be written as H_ψ : $\mathbf{C}_\psi \boldsymbol{\beta} = \mathbf{0}$ where H_ψ becomes

$$\mathbf{C}_\psi \boldsymbol{\beta} = \begin{bmatrix} 1 & 0 & 0 & -1 \\ 0 & 1 & -1 & 0 \end{bmatrix} \begin{bmatrix} \mu_{11} \\ \mu_{21} \\ \mu_{12} \\ \mu_{22} \end{bmatrix} = \begin{bmatrix} 0 \\ 0 \\ 0 \\ 0 \end{bmatrix} \tag{4.14.27}$$

Furthermore, because $\mathbf{K} = \mathbf{M}_o' \otimes \mathbf{C}_o$ for the matrices $\mathbf{M}_o$ and $\mathbf{C}_o$ in the test of H_p, ψ_1 and ψ_2 may be combined into the overall test

$$\boldsymbol{\gamma} = \mathbf{C}_* \boldsymbol{\beta} = \begin{bmatrix} 1 & 0 & 0 & -1 \\ 0 & 1 & -1 & 0 \\ 1 & 0 & -1 & 0 \\ 0 & 1 & 0 & -1 \end{bmatrix} \begin{bmatrix} \mu_{11} \\ \mu_{21} \\ \mu_{12} \\ \mu_{22} \end{bmatrix} = \begin{bmatrix} 0 \\ 0 \\ 0 \\ 0 \end{bmatrix} \tag{4.14.28}$$

where the first two rows of $\mathbf{C}_*$ are the contrasts for ψ_1 and ψ_2 and the last two rows of $\mathbf{C}_*$ is the matrix $\mathbf{K}$.

An alternative representation for (4.14.27) is to write the joint test as

$$\begin{aligned}
\psi_1 &= \text{tr}\left[\begin{pmatrix} 1 & 0 \\ 0 & -1 \end{pmatrix}\begin{pmatrix} \mu_{11} & \mu_{12} \\ \mu_{21} & \mu_{22} \end{pmatrix}\right] = 0 \\
\psi_2 &= \text{tr}\left[\begin{pmatrix} 0 & 1 \\ -1 & 0 \end{pmatrix}\begin{pmatrix} \mu_{11} & \mu_{12} \\ \mu_{21} & \mu_{22} \end{pmatrix}\right] = 0 \\
\psi_3 &= \text{tr}\left[\begin{pmatrix} 1 & 0 \\ -1 & 0 \end{pmatrix}\begin{pmatrix} \mu_{11} & \mu_{12} \\ \mu_{21} & \mu_{22} \end{pmatrix}\right] = 0 \\
\psi_4 &= \text{tr}\left[\begin{pmatrix} 0 & 1 \\ 0 & -1 \end{pmatrix}\begin{pmatrix} \mu_{11} & \mu_{12} \\ \mu_{21} & \mu_{22} \end{pmatrix}\right] = 0
\end{aligned} \tag{4.14.29}$$

so that each contrast has the familiar form: $\psi_i = \text{tr}\,(\mathbf{G}_i\mathbf{B}) = 0$. This suggests representing the overall test of no difference in periods and no differential carryover effect as the intersection of the form tests described in (4.14.29). In our discussion of the repeated measures design, we also saw that contrasts of the form $\psi = \text{tr}\,(\mathbf{GB}) = 0$ for some matrix $\mathbf{G}$ arose naturally. These examples suggest an extended class of linear hypotheses. In particular all tests are special cases of the hypothesis

$$\omega_H = \bigcap_{\mathbf{G}\epsilon\mathbf{G}_o} \{\text{tr}\,(\mathbf{GB}) = 0\} \tag{4.14.30}$$

where $\mathbf{G}_o$ is some set of $p \times q$ matrices that may form k linear combinations of the parameter matrix $\mathbf{B}$. The matrix decomposition described by (4.14.29) is called the extended multivariate linear hypotheses by Mudholkar, Davidson and Subbaiah (1974). The family ω_H includes the family of all maximal hypotheses $H_o : \mathbf{C}_o\mathbf{B}\mathbf{M}_o = \mathbf{0}$, all minimal hypotheses of the form $\text{tr}\,(\mathbf{GB}) = 0$ where the $r\,(\mathbf{G}) = 1$ and all intermediate hypotheses where $\mathbf{G}$ is a linear combination of $\mathbf{G}_i \subseteq \mathbf{G}_o$. To test ω_H, they developed an extended T_o^2 and largest root statistic and constructed $100\,(1-\alpha)\,\%$ simultaneous confidence intervals for all contrasts $\psi = \text{tr}\,(\mathbf{GB}) = 0$. To construct a test of ω_H, they used the UI principal. Suppose a test statistic $t_\psi\,(\mathbf{G})$ may be formed for each minimal hypotheses $\omega_M \subseteq \omega_H$. The overall hypothesis ω_H is rejected if

$$t\,(\mathbf{G}) = \sup_{\mathbf{G}\epsilon\mathbf{G}_o} t_\psi\,(\mathbf{G}) \geq c_\psi\,(\alpha) \tag{4.14.31}$$

is significant for some minimal hypothesis where the critical value $c_\psi\,(\alpha)$ is chosen such that the $P\left(t\,(\mathbf{G}) \leq c_\psi\,(\alpha)\,|\omega_H|\right) = 1-\alpha$.

To develop a test of ω_H, Mudholkar, Davidson and Subbaiah (1974) relate $t_\psi\,(\mathbf{G})$ to symmetric gauge functions (sgf) to generate a class of invariant tests discussed in some detail in Timm and Mieczkowski (1997). Here, a more heuristic argument will suffice.

Consider the maximal hypothesis in the family ω_H, $H_o : \mathbf{C}_o\mathbf{B}\mathbf{M}_o = \mathbf{0}$. To test H_o, we let

$$\begin{aligned}
\mathbf{E}_o &= \mathbf{M}_o'\mathbf{Y}'(\mathbf{I}-\mathbf{X}\left(\mathbf{X}'\mathbf{X}\right)^{-1}\mathbf{X}')\mathbf{Y}\mathbf{M}_o \\
\mathbf{W}_o &= \mathbf{C}_o\left(\mathbf{X}'\mathbf{X}\right)^{-1}\mathbf{C}_o' \\
\widehat{\mathbf{B}} &= \left(\mathbf{X}'\mathbf{X}\right)^{-1}\mathbf{X}'\mathbf{Y} \\
\widehat{\mathbf{B}}_o &= \mathbf{C}_o\widehat{\mathbf{B}}\mathbf{M}_o \\
\mathbf{H}_o &= \widehat{\mathbf{B}}_o'\mathbf{W}_o^{-1}\widehat{\mathbf{B}}_o
\end{aligned} \tag{4.14.32}$$

and relate the test of H_o to the roots of $|\mathbf{H}_o - \lambda\mathbf{E}_o| = 0$. We also observe that

$$\begin{aligned}
\psi &= \operatorname{tr}(\mathbf{G}\mathbf{B}) = \operatorname{tr}(\mathbf{M}_o\mathbf{G}_o\mathbf{C}_o\mathbf{B}) = \operatorname{tr}(\mathbf{G}_o\mathbf{C}_o\mathbf{B}\mathbf{M}_o) = \operatorname{tr}(\mathbf{G}_o\mathbf{B}_o) \\
\widehat{\psi} &= \operatorname{tr}(\mathbf{G}\widehat{\mathbf{B}}) = \operatorname{tr}(\mathbf{G}_o\widehat{\mathbf{B}}_o)
\end{aligned} \tag{4.14.33}$$

for some matrix $\mathbf{G}_o$ in the family. Furthermore, for some $\mathbf{G}_o$, $\widehat{\psi}$ is maximal. To maximize $t(\mathbf{G})$ in (4.14.31), observe that the

$$\operatorname{tr}(\mathbf{G}_o\widehat{\mathbf{B}}_o) = \left[\operatorname{tr}(\mathbf{E}_o^{1/2}\mathbf{G}_o\mathbf{W}_o^{1/2})(\mathbf{W}_o^{-1/2}\widehat{\mathbf{B}}_o\mathbf{E}_o^{-1/2})\right] \tag{4.14.34}$$

Also recall that for the matrix norm for a matrix $\mathbf{M}$ is defined as $\|\mathbf{M}\|_p = \left[\sum_i \lambda_i^{p/2}\right]^{1/2}$ where λ_i is a root of $\mathbf{M}'\mathbf{M}$. Thus, to maximize $t(\mathbf{G})$, we may relate the function $\operatorname{tr}(\mathbf{G}_o\widehat{\mathbf{B}}_o)$ to a matrix norm. Letting

$$\begin{aligned}
\mathbf{M}' &= \mathbf{E}_o^{1/2}\mathbf{G}_o\mathbf{W}_o^{1/2} \\
\mathbf{M}'\mathbf{M} &= \mathbf{E}_o^{1/2}\mathbf{G}_o\mathbf{W}_o^{1/2}\mathbf{G}_o'\mathbf{E}_o^{1/2},
\end{aligned}$$

the $\|\mathbf{M}\|_p$ depends on the roots of $\left|\mathbf{H} - \lambda\mathbf{E}_o^{-1}\right| = 0$ for $\mathbf{H} = \mathbf{G}_o\mathbf{W}_o\mathbf{G}_o'$. For $p = 2$, the $\left[\operatorname{tr}\left(\mathbf{G}_o\mathbf{W}_o\mathbf{G}_o'\mathbf{E}_o\right)\right]^{1/2} = \left(\sum_i \lambda_i\right)^{1/2}$ where the roots $\lambda_i = \lambda_i\left(\mathbf{G}_o\mathbf{W}_o\mathbf{G}_o'\mathbf{E}_o\right) = \lambda_i\left(\mathbf{H}\mathbf{E}_o\right)$ are the roots of $\left|\mathbf{H} - \lambda\mathbf{E}_o^{-1}\right| = 0$. Furthermore observe that for $\mathbf{A} = \mathbf{W}_o^{-1/2}\widehat{\mathbf{B}}_o\mathbf{E}_o^{-1/2}$ that $\mathbf{A}'\mathbf{A} = \mathbf{E}_o^{-1}\widehat{\mathbf{B}}_o'\mathbf{W}_o^{-1}\widehat{\mathbf{B}}_o\mathbf{E}_o^{-1/2}$ and that the $\|\mathbf{A}\|_p = (\sum_i \theta_i^{p/2})^{1/p}$. For $p = 2$, the θ_i are roots of $|\mathbf{H}_o - \theta\mathbf{E}_o| = 0$, the maximal hypothesis. To test $H_o : \mathbf{C}_o\mathbf{B}\mathbf{M}_o = \mathbf{0}$, we use $T_o^2 = v_e \operatorname{tr}\left(\mathbf{H}_o\mathbf{E}_o^{-1}\right)$. For $p = 1$, the test of H_o is related to the largest root of $|\mathbf{H}_o - \theta\mathbf{E}_o| = 0$. These manipulations suggest forming a test statistic with $t(\mathbf{G}) = t_\psi(\mathbf{G}) = |\widehat{\psi}|/\widehat{\sigma}_{\widehat{\psi}}$ and to reject $\psi = \operatorname{tr}(\mathbf{G}\mathbf{B}) = 0$ if $t(\mathbf{G})$ exceeds $c_\psi(\alpha) = c_\alpha$ where c_α depends on the root and trace criteria for testing H_o. Letting $s = \min(v_h, u)$, $M = (|v_h - u| - 1)/2$, $N = (v_e - u - 1)/2$ where $v_h = r(\mathbf{C}_o)$ and $u = r(\mathbf{M}_o)$, we would reject $\omega_m : \psi = \operatorname{tr}(\mathbf{G}\mathbf{B}) = 0$ if $|\widehat{\psi}|/\widehat{\sigma}_{\widehat{\psi}} > c_\alpha$ where $\widehat{\sigma}_{\widehat{\psi}} \equiv \sigma_{\text{Trace}} = \left(\sum_i \lambda_i\right)^{1/2}$ and $\widehat{\sigma}_{\widehat{\psi}} \equiv \sigma_{\text{Root}} = \sum_i \lambda_i^{1/2}$ for λ_i that solve the characteristic equation $\left|\mathbf{H} - \lambda\mathbf{E}_o^{-1}\right| = 0$ for $\mathbf{H} = \mathbf{G}_o\mathbf{W}_o\mathbf{G}_o'$ and $\operatorname{tr}(\mathbf{G}_o\widehat{\mathbf{B}}_o) = \operatorname{tr}(\mathbf{G}\widehat{\mathbf{B}})$ for some matrix $\mathbf{G}_o$. Using Theorem 3.5.1, we may construct simultaneous confidence intervals for parametric function $\psi = \operatorname{tr}(\mathbf{G}\mathbf{B})$.

Theorem 4.14.1. Following the overall test of $H_o : \mathbf{C}_o\mathbf{B}\mathbf{M}_o = 0$, approximate $1 - \alpha$ simultaneous confidence sets for all contrasts $\psi = \operatorname{tr}(\mathbf{G}\mathbf{B}) = 0$ using the extended trace or root criterion are as follows

$$\widehat{\psi} - c_\alpha\widehat{\sigma}_{\widehat{\psi}} \quad \leq \psi \leq \quad \widehat{\psi} + c_\alpha\widehat{\sigma}_{\widehat{\psi}}$$

where for the

Root Criterion

$$c_\alpha^2 \approx \tfrac{v_1}{v_2} F^{1-\alpha}(v_1, v_2)$$

$$v_1 = \max(v_h, u) \text{ and } v_2 = v_e - v_1 + v_h$$

$$\widehat{\sigma}_{\widehat{\psi}} \equiv \widehat{\sigma}_{\text{Root}} = \textstyle\sum_i \lambda_i^{1/2}$$

and the

Trace Criterion

$$c_\alpha^2 \approx \tfrac{sv_1}{v_2} F^{1-\alpha}(v_1,\ v_2)$$

$$v_1 = s(2M + s + 1) \text{ and } v_2 = 2(sN + 1)$$

$$\widehat{\sigma}_{\widehat{\psi}} \equiv \widehat{\sigma}_{\text{Trace}} = \left(\textstyle\sum_i \lambda_i\right)^{1/2}$$

The λ_i are the roots of $\left|\mathbf{H} - \lambda_i \mathbf{E}_o^{-1}\right| = 0$, $\mathbf{E}_o$ is the error SSCP matrix for testing $\mathbf{H}_o$, M, N, v_h and u are defined in the test of H_o and $\mathbf{H} = \mathbf{G}_o\mathbf{W}_o\mathbf{G}_o'$ for some $\mathbf{G}_o$, and the $\text{tr}(\mathbf{G}_o\widehat{\mathbf{B}}_o) = \text{tr}(\mathbf{G}\widehat{\mathbf{B}}) = \widehat{\psi}$.

Theorem 4.14.1 applies to the subfamily of maximal hypotheses H_o and to any minimal hypothesis that has the structure $\psi = \text{tr}(\mathbf{GB}) = 0$. However, intermediate extended multivariate linear hypotheses depend on a family of $\mathbf{G}_i$ so that $\mathbf{G} = \sum_{i=1}^{k} \eta_i \mathbf{G}_i$ for some vector $\boldsymbol{\eta}' = [\eta_1, \eta_2, \ldots, \eta_p]$. Thus, we must maximize $\mathbf{G}$ over the $\mathbf{G}_i$ as suggested in (4.14.31) to test intermediate hypotheses. Letting $\boldsymbol{\tau}' = [\tau_i]$ and the estimate be defined as $\widehat{\boldsymbol{\tau}} = [\widehat{\tau}_i]$ where

$$\widehat{\tau}_i = \text{tr}(\mathbf{G}_{oi}'\widehat{\mathbf{B}}_o) = \text{tr}(\mathbf{G}_i'\widehat{\mathbf{B}})$$
$$\mathbf{T} = [t_{ij}] \quad \text{where } t_{ij} = \text{tr}(\mathbf{G}_{oi}\mathbf{W}_o\mathbf{G}_{oj}'\mathbf{E}_o)$$

and $t(\mathbf{G}) = [t_\psi(\mathbf{G})]^2 = (\boldsymbol{\eta}'\boldsymbol{\tau})^2 / \boldsymbol{\eta}'\mathbf{T}\boldsymbol{\eta}$, Theorem 2.6.10 is used to find the supremum over all vectors $\boldsymbol{\eta}$.

Letting $\mathbf{A} \equiv \widehat{\boldsymbol{\tau}}\widehat{\boldsymbol{\tau}}'$ and $\mathbf{B} \equiv \mathbf{T}$, the maximum is the largest root of $|\mathbf{A} - \lambda\mathbf{B}| = 0$ or $\lambda_1 = \lambda_1(\mathbf{AB}^{-1}) = \lambda_1\left(\widehat{\boldsymbol{\tau}}'\widehat{\boldsymbol{\tau}}\mathbf{T}^{-1}\right) = \widehat{\boldsymbol{\tau}}'\mathbf{T}^{-1}\widehat{\boldsymbol{\tau}}$. Hence, an intermediate extended multivariate linear hypothesis $\omega_H : \psi = 0$ is rejected if $t(\mathbf{G}) = \widehat{\boldsymbol{\tau}}'\mathbf{T}^{-1}\widehat{\boldsymbol{\tau}} > c_{\psi(\alpha)}^2$ is the trace or largest root critical value for some maximal hypothesis. For this situation approximate $100(1-\alpha)\%$ simultaneous confidence intervals for $\psi = \mathbf{a}'\boldsymbol{\tau}$ are given by

$$\mathbf{a}'\widehat{\boldsymbol{\tau}} - c_\alpha^2\sqrt{\frac{\mathbf{a}'\mathbf{T}\mathbf{a}}{n}} \leq \mathbf{a}'\boldsymbol{\tau} \leq \mathbf{a}'\widehat{\boldsymbol{\tau}} + c_\alpha^2\sqrt{\frac{\mathbf{a}'\mathbf{T}\mathbf{a}}{n}} \tag{4.14.35}$$

for arbitrary vectors $\mathbf{a}$. The value c_α^2 may be obtained as in Theorem 4.14.1.

We have shown how extended multivariate linear hypotheses may be tested using an extended T_o^2 or largest root statistic. In our discussion of the 2 × 2 crossover design we illustrated an alternative representation of the test of some hypothesis in the family ω_H. In particular, by vectorizing **B**, the general expression for ω_H is

$$\omega_H : \mathbf{C}_* \operatorname{vec}(\mathbf{B}) = \mathbf{C}_*\boldsymbol{\beta} = \mathbf{0} \tag{4.14.36}$$

Letting $\boldsymbol{\gamma} = \mathbf{C}_*\boldsymbol{\beta}$, and assuming a MVN for the vows of **Y**, the distribution of $\widehat{\boldsymbol{\gamma}} \sim N_v[\boldsymbol{\gamma},\ \mathbf{C}_* \left(\mathbf{D}'\Omega^{-1}\mathbf{D}\right)^{-1} \mathbf{C}_*']$ where $v = r(\mathbf{C}_*)$, $\mathbf{D} = \mathbf{I}_p \otimes \mathbf{X}$ and $\Omega = \Sigma \otimes \mathbf{I}_n$. Because Σ is unknown, we must replace it by a consistent estimate that converges in probability to Σ. Two candidates are the ML estimate $\widehat{\Sigma} = \mathbf{E}_o/n$ and the unbiased estimate $\mathbf{S} = \mathbf{E}_o/[n - r(\mathbf{X})]$. Then, as a large sample approximation to the LR test of ω_H we may use Wald's large sample chi-square statistic given in (3.6.12)

$$\mathrm{X}^2 = (\mathbf{C}_*\widehat{\boldsymbol{\beta}})'[\mathbf{C}_*(\mathbf{D}'\widehat{\Omega}^{-1}\mathbf{D})^{-1}\mathbf{C}_*']^{-1}(\mathbf{C}_*\widehat{\boldsymbol{\beta}}) \sim \chi_v^2 \tag{4.14.37}$$

where $v = r(\mathbf{C}_*)$. If an inverse does not exist, we use a g-inverse. For $\mathbf{C}_* = \mathbf{M}_o' \otimes \mathbf{C}_o$, this is a large sample approximation to T_o^2 given in (3.6.28) so that it may also be considered an alternative to the Mudholkar, Davidson and Subbaiah (1974) procedure. While the two procedures are asymptotically equivalent, Wald's statistic may be used to establish approximate $100(1-\alpha)\%$ simultaneous confidence intervals for all contrasts $\mathbf{c}_*'\boldsymbol{\beta} = \psi$. For the Mudholkar, Davidson and Subbaiah (1974) procedure, two situations were dealt with differently, minimal and maximal hypotheses, and intermediate hypotheses.

4.15 Repeated Measurements and Extended Linear Hypotheses Example

a. *Repeated Measures (Example 4.15.1)*

The data used in the example are provided in Timm (1975, p. 454) and are based upon data from Allen L. Edwards. The experiment investigates the influence of three drugs, each at a different dosage levels, on learning. Fifteen subjects are assigned at random to the three drug groups and five subjects are tested with each drug on three different trials. The data for the study are given in Table 4.15.1 and in file Timm 454.dat. It contains response times for the learning tasks. Program m4_15_1.sas is used to analyze the experiment.

The multivariate linear model for the example is $\underset{15\times 3}{\mathbf{Y}} = \underset{15\times 3}{\mathbf{X}}\ \underset{3\times 3}{\mathbf{B}} + \underset{15\times 3}{\mathbf{E}}$ where the parameter matrix **B** has the structure

$$\mathbf{B} = \begin{bmatrix} \mu_{11} & \mu_{12} & \mu_{13} \\ \mu_{21} & \mu_{22} & \mu_{23} \\ \mu_{31} & \mu_{32} & \mu_{33} \end{bmatrix} = \begin{bmatrix} \boldsymbol{\mu}_1' \\ \boldsymbol{\mu}_2' \\ \boldsymbol{\mu}_3' \end{bmatrix} = [\mu_{ij}] \tag{4.15.1}$$

Given multivariate normality and equality of the covariance matrices for the three independent groups, one may test the hypothesis of equal group mean vectors.

$$H_G : \boldsymbol{\mu}_1 = \boldsymbol{\mu}_2 = \boldsymbol{\mu}_3 \tag{4.15.2}$$

This test is identical to the one-way MANOVA hypothesis. Using PROC GLM, this is tested with the MANOVA statement and h = group. Setting $\alpha = 0.05$, the test of H_G is rejected since $F = 3.52$ on (6, 20) degrees of freedom has a p-value of 0.0154 for the test using Wilks' Λ criterion. All multivariate criteria lead one to reject H_G.

To test for parallelism of the regressions curves for the three groups, **C** is identical to the matrix used to test H_G; however, that post matrix **M** is selected such that

$$\mathbf{M} = \begin{bmatrix} 1 & 0 \\ -1 & 1 \\ 0 & -1 \end{bmatrix}$$

Then, the test of parallelism is

$$H_P : \begin{bmatrix} \mu_{11} - \mu_{12} \\ \mu_{12} - \mu_{13} \end{bmatrix} = \begin{bmatrix} \mu_{21} - \mu_{22} \\ \mu_{22} - \mu_{23} \end{bmatrix} = \begin{bmatrix} \mu_{31} - \mu_{32} \\ \mu_{32} - \mu_{33} \end{bmatrix} \tag{4.15.3}$$

In PROC GLM, H_P is tested using

$$m = \begin{pmatrix} 1 & -1 & 0, \\ 0 & 1 & -1 \end{pmatrix}$$

TABLE 4.15.1. Edward's Repeated Measures Data

	Subject	Trials 1	2	3
Drug Group 1	1	2	4	7
	2	2	6	10
	3	3	7	10
	4	7	9	11
	5	6	9	12
Drug Group 2	1	5	6	10
	2	4	5	10
	3	7	8	11
	4	8	9	11
	5	11	12	13
Drug Group 3	1	3	4	7
	2	3	6	9
	3	4	7	9
	4	8	8	10
	5	7	10	10

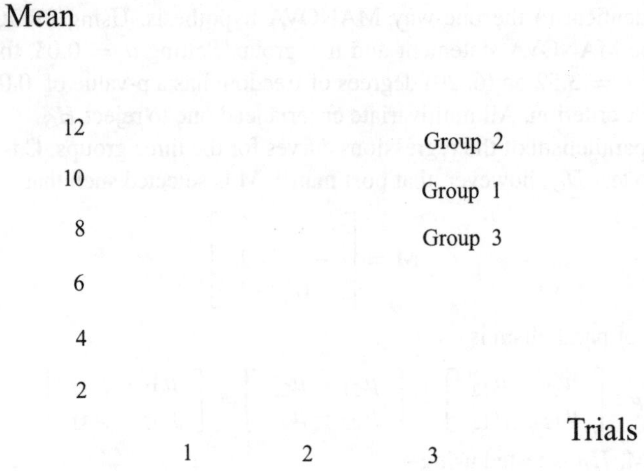

Mean

12
10
8
6
4
2

Group 2
Group 1
Group 3

Trials

1 2 3

FIGURE 4.15.1. Plot of Means Edward's Data

For Wilks' Λ criterion, $\Lambda = 0.4602$ or $F = 2.61$ the p-value for the test is 0.0636. We conclude that there is no interaction between groups and trials. A plot of the means is provided in Figure 4.15.1
While the plots do cross, for $\alpha = 0.05$ the result is not statistically significant.

One may also test that the mean vectors across trials are equal

$$H_C = \begin{bmatrix} \mu_{11} \\ \mu_{21} \\ \mu_{31} \end{bmatrix} = \begin{bmatrix} \mu_{12} \\ \mu_{22} \\ \mu_{32} \end{bmatrix} = \begin{bmatrix} \mu_{13} \\ \mu_{23} \\ \mu_{33} \end{bmatrix} \tag{4.15.4}$$

To test H_C with PROC GLM, one must use a CONTRAST statement in SAS with $\mathbf{C} \equiv \mathbf{I}$ while $\mathbf{M}$ is used to construct differences across trials. For Wilks' Λ criterion, $\Lambda = 0.0527$ with p-value < 0.0001 so that the test is significant.

The three multivariate tests we have just considered have assumed equal, unknown, and unstructured covariance matrices for the three groups under study. The most important test in a repeated measurement design is the test of parallelism or equivalently whether or not there exists a significant interaction between groups and trials. If we can further establish that the circularity condition in (4.14.9) is valid for our study, a uniformly more powerful test of parallelism (group $\times$ trial interaction) exists with the design. To test for sphericity/circularity using PROC GLM, one must use the REPEATED statement with the option PRINTE. This is found in program m14_15_1.sas with the heading "Univariate Tests given Parallelism and Sphericity" We see that Mauchly's test of sphericity for orthogonal components is not rejected (p-value $= 0.1642$). The test of trial$*$group given sphericity has a p-value of 0.0569. This value is less than the corresponding p-value for the multivariate test with p-value 0.0636, but still larger that the nominal $\alpha = 0.05$ level so that we fail to reject the test of parallelism or interaction. Given a nonsignificant interaction, one may average over drug groups to test for differences in trials given parallelism, H_β in (4.14.5). And, one may average over trials to test for differences in drug groups given parallelism,

H_α in (4.14.5). From a MR model perspective, the test of H_α is constructed using

$$\mathbf{C} = \begin{bmatrix} 1 & -1 & 0 \\ 0 & 1 & -1 \end{bmatrix} \quad \text{and} \quad \mathbf{M} = \begin{bmatrix} 1/3 \\ 1/3 \\ 1/3 \end{bmatrix}$$

Using PROC GLM in program m4_15_1.sas this test is labeled "Test of Groups given Parallelism". Testing H_α, $F = 1.13$ with degrees of freedom (2, 12) and p-value of 0.3567, a non-significant result. Observe, that this is merely a contrast following the test of H_G of the one-way MANOVA design. To test H_β from a MR model perspective, one may select

$$\mathbf{C} = \begin{bmatrix} 1, & 1, & 1 \end{bmatrix} \quad \text{and} \quad \mathbf{M} = \begin{bmatrix} 1 & 0 & -1 \\ 0 & 1 & -1 \end{bmatrix}$$

or equivalently, one may select

$$\mathbf{C} = \begin{bmatrix} 1/3, & 1/3, & 1/3 \end{bmatrix} \quad \text{and} \quad \mathbf{M} = \begin{bmatrix} .707107 & -.408258 \\ 0 & .816497 \\ -.707107 & -.408248 \end{bmatrix}$$

so that $\mathbf{M}'\mathbf{M} = \mathbf{I}$. This test is labeled the "Test of Intercept" in SAS. It is associated with the MANOVA statement with $h =$ _all_. For this test,

$$\begin{aligned} \mathbf{H} &= \begin{bmatrix} 163.3334 & 13.4738 \\ 13.4738 & 1.1115 \end{bmatrix} \\ \mathbf{E} &= \begin{bmatrix} 15.000 & 1.7318 \\ 1.7317 & 4.9998 \end{bmatrix} \end{aligned} \tag{4.15.5}$$

For Wilks' Λ criterion, $\Lambda = 0.0839$ with p-value 0.0001. One may think of this test as a contrast following the multivariate test of H_C. Because $s = 1$,

$$\Lambda = \frac{1}{1 + T^2/v_e}$$

this is seen to be a multivariate test of equal correlated means and not a univariate test of independent means as was the case for the test of H_α.

The tests of H_α and H_β were tested given parallelism and assuming an unknown general structure for Σ. However, we saw that $\mathbf{M}'\Sigma\mathbf{M} = \sigma^2\mathbf{I}$. As with the test of parallelism, we may be able to construct a more powerful test of H_α and H_β given sphericity. This is the case. In program m4_15_1.sas we have labeled the tests as "Univariate Tests given Parallelism and Sphericity". They are obtained by using the REPEATED statement. Reviewing the output from PROC GLM one observes that SAS generates tests of Sphericity, a test of no trials effect, a test of no trials*group effect, and a test of group differences. In addition, an ANOVA table is produced for tests of trials and trials*group interaction. Further investigation of the results, we see that the test of trials is our test of H_β, $\Lambda = 0.0839$. The test of no trials*group effect is our test of H_P given in (4.15.3) and $\Lambda = 0.4602$. These are multivariate tests and do not require the sphericity condition to hold.

Next, we have a test for group different with $F = 1.13$. This is the test of H_α. It also does not require sphericity. Finally , we have an ANOVA table for the tests of trials and trials∗groups. These are derived from the multivariate tests H_β and H_P given $\mathbf{M}'\mathbf{M} = \mathbf{I}$ and $\mathbf{M}'\Sigma\mathbf{M} = \sigma^2\mathbf{I}$. To illustrate, recall that the tests of H_β produced **H** and **E** in (4.15.5) with $v_h = 1$ and $v_e = 12$. Suppose we divide the diagonal elements of **H** by $v_h = 1$ and average, then we have $MS_h = 82.222$. Similarly, if we divide the diagonal elements of **E** by $v_e = 12$ and average $MS_e = 0.8333$. These are the mean squares in the ANOVA table labels "trials" and "Error (trials)". To determine the univariate degrees of freedom, the $r(\mathbf{M})$ is used. That is, $r(\mathbf{M})\, v_h = 2(1 = 2$ and $r(\mathbf{M})\, v_e = 2(12) = 24$. This demonstrates that the univariate test of trials is directly determined from the test of H_P given parallelism. One may similarly demonstrate that the univariate tests of interaction given sphericity may be obtained from the multivariate test H_P. Details are provided by Boik (1988). Given that sphericity is satisfied, that univariate tests are uniformly more powerful than the multivariate tests H_β and H_P. When sphericity is not satisfied, some recommend using F adjusted tests where the degrees of freedom of the univariate tests are multiplied by the Geisser-Greenhouse Epsilon, $\widehat{\epsilon}$, or the Huynh-Feldt Epsilon, $\widetilde{\epsilon}$. While this may be appropriate under some conditions of lack of sphericity, it is not necessarily better than the multivariate tests which may be conservative. In general, adjustments are to be avoided, Boik (1991).

Given sphericity, one may also analyze the data in Table 4.15.1 as a univariate split plot design where subjects are random and all other factors are fixed. This is a mixed ANOVA model. For this approach, the data vectors must be reorganized into univariate observations. This is accomplished in the DATA step SPLIT. We have also included in program m4_15_1.sas SAS statements for PROC GLM and PROC MIXED for the univariate analysis. We will discuss the code for this approach in Chapter 6.

Scanning the output, one observes that the univariate ANOVA results for the fixed effects: groups, treat, and trial*group using the random/test results in PROC GLM and the test of fixed effects in PROC MIXED are in agreement with the output generated by PROC GLM using the REPEATED statement. If the univariate assumptions are satisfied, one should use PROC MIXED for testing fixed effects, obtaining confidence intervals for fixed parameters, and for estimating variance components. The output from PROC GLM should only be used to test the significance of variance components and to compute expected mean squares. This is made clear in Chapter 6.

b. Extended Linear Hypotheses (Example 4.15.2)

Returning to our analysis of Edward's data assuming an unknown structure for Edwards Σ, one may find that following a multivariate test of the form $\mathbf{CBM} = \mathbf{0}$ that a contrast involving the elements of **B** may be of interest and does not have a simple bilinear structure. This leads one to investigate extended linear hypotheses. In SAS, one does not have a procedure to analyze extended linear hypotheses. In program m4_15_1.sas we include PROC IML code to test H_G, H_C, H_P, and to obtain simultaneous confidence intervals for a few contrasts in the parameters that do not have a simple bilinear structure. For example, following the test of H_G, we may want to test $\psi = 0$ for ψ defined in (4.14.15). Alternatively, suppose we are interested in testing

$$\begin{aligned} \omega_H: \quad & \mu_{11} = \mu_{21} \\ & \mu_{12} = \mu_{22} = \mu_{32} \\ & \mu_{13} = \mu_{33} \end{aligned} \tag{4.15.6}$$

This has the structure $\psi = \text{Tr}\,(\mathbf{G}_i\mathbf{B})$ for $\mathbf{G}_i$ defined in (4.14.22). Following the test of H_C, suppose we wanted to test $\psi = 0$ for ψ defined in (4.14.18). Finally, suppose following the test of H_P we wanted to investigate ψ defined in (4.14.21).

In program m4_15_1.sas, we first test a maximal hypothesis of the form $H_o : \mathbf{C}_o\mathbf{B}\mathbf{M}_o = \mathbf{0}$ which is identical to a multivariate test. For the test of H_G in (4.14.13), we estimate $\mathbf{B}$ and $\mathbf{E}_o$. Then for ψ in (4.14.15),

$$\widehat{\psi} = \text{tr}(\mathbf{G}\widehat{\mathbf{B}}) = -2.5$$

for $\mathbf{G}$ in (4.14.16). Using Theorem 4.14.1, we solve $\mid \mathbf{H} - \boldsymbol{\lambda}\mathbf{E}_o \mid = 0$ using the EIGVAL function in PROC IML. This yields $\lambda_1 = 6.881$, $\lambda_2 = 2.119$, and $\lambda_3 = .0000158$. Then, for

$$\widehat{\sigma}_{\text{Root}} = \sum_i \lambda_i^{1/2}$$

$$\widehat{\sigma}_{\text{Trace}} = \left(\sum_i \lambda_i\right)^{1/2}$$

the extended root statistics are

$$\mid \widehat{\psi} \mid / \widehat{\sigma}_{\text{Root}} = 0.6123$$

$$\mid \widehat{\psi} \mid / \widehat{\sigma}_{\text{Trace}} = 0.8333$$

Evaluating c_α^2 in Theorem 4.14.1, 0.9891365 and 1.3319921 are, respectively, the critical values for the Root and Trace tests as shown in the output. Because the test statistic does not exceed the critical value, the contrast is not significantly different from zero. Both intervals contain zero

Root	Interval	(−6.5384, 1.5384)
Trace	Interval	(−6.4960, 1.4960)

To test (4.15.6) is more complicated since this involves the maximization of $\mathbf{G}$ over $\mathbf{G}_i$. Letting $\boldsymbol{\tau}' = [\tau_1, \tau_2, \tau_3, \tau_4]$ where $\widehat{\boldsymbol{\tau}}' = [\widehat{\tau}_1, \widehat{\tau}_2, \widehat{\tau}_3, \widehat{\tau}_4]$

$$\widehat{\tau}_i = \text{tr}(\mathbf{G}_i'\widehat{\mathbf{B}})$$

$$T = \left[t_{ij}\right] = \text{tr}(\mathbf{G}_i\mathbf{W}_o\mathbf{G}'_j\mathbf{E}_o)$$

we find that

$$\widehat{\boldsymbol{\tau}}' = [\ -3, \quad -1, \quad 1, \quad 2\ ]$$

$$\mathbf{T} = \begin{bmatrix} 29.6 & 26 & -13 & -.7 \\ 26 & 27.2 & -13.6 & -7.6 \\ -13 & -13.6 & 27.2 & 15.2 \\ -7 & -7.6 & 15.2 & 10.4 \end{bmatrix}$$

and that the test statistic $t(\mathbf{G}) = \widehat{\boldsymbol{\tau}}'\mathbf{T}^{-1}\widehat{\boldsymbol{\tau}} = 2.2081151$. The critical constant for the trace criterion is the square of c_α^2 for the contrast ψ in (4.14.15) or $(1.33199)^2 = 1.774203$. Thus, the test of ω_H in (4.15.6) is significant.

Following the test of parallelism, H_P, the code in program m4_15_1.sas is provided to obtain a confidence interval for ψ given in (4.14.21) and for (4.14.18) follow the test of H_C using both the Root and Trace criteria. The test that

$$\psi = (\mu_{21} + \mu_{12} - \mu_{31} - \mu_{22}) + (\mu_{32} + \mu_{23} - \mu_{13} - \mu_{22}) = 0$$

is nonsignificant since the intervals include zero

$$\text{Root}: (-2.7090, 4.7900)$$
$$\text{Trace}: (-2.7091, 4.7091)$$

while the test that

$$\psi = (\mu_{11} + \mu_{12}) + (\mu_{22} - \mu_{23}) + (\mu_{31} + \mu_{33}) = 0$$

is significant since the intervals do not include zero

$$\text{Root}: (-10.2992, -9.701)$$
$$\text{Trace}: (-10.2942, -9.7050)$$

While extended linear hypotheses allow us to analyze tests that do not have a single bilinear form, a more direct procedure is to use Wald's large sample chi-square statistic given in (4.14.37) or for small samples we may use an F approximation

$$F = X^2 / r(\mathbf{C}_*) \,\dot{\sim}\, F(v, v_e)$$

where $v_e = n - r(\mathbf{X})$. To illustrate, we test (4.15.6). Letting $\boldsymbol{\beta} = \text{vec}(\mathbf{B})$, the matrix $\mathbf{C}_*$ for the test is

$$\mathbf{C}_* = \begin{bmatrix} 1 & 0 & 0 & -1 & 0 & 0 & 0 & 0 & 0 \\ 0 & 1 & 0 & 0 & -1 & 0 & 0 & 0 & 0 \\ 0 & 0 & 0 & 0 & 1 & 0 & 0 & -1 & 0 \\ 0 & 0 & 0 & 0 & 0 & 1 & 0 & 0 & -1 \end{bmatrix}$$

Then for $\mathbf{D} = \mathbf{I}_p \otimes \mathbf{X}$ and $\widehat{\Omega} = \widehat{\Sigma} \otimes \mathbf{I}_n$, $X^2 = 66.67$ and $F = 16.67$. For this example, the chi-square or the more exact F approximation are both significant and consistent with the test using the test statistic $t(\mathbf{G}) = \widehat{\boldsymbol{\tau}}'\mathbf{T}^{-1}\widehat{\boldsymbol{\tau}}$ for the hypothesis.

Again one may use CONTRAST statements to locate significance of tests when using PROC GLM or with IML code, obtain simultaneous confidence sets as illustrated in the previous examples in this chapter. Finally, one may evaluate multivariate normality. For our example, we have no reason to believe the data are not multivariate normal.

Exercises 4.15

1. In a study of learning 15 rats were randomly assigned to three different reinforcement schedules ad then given a maze to run under form experimental conditions.

The sequence in which the four conditions were presented in the experiment was randomized independently for each animal. The dependent variable for the study was the number of seconds taken to run the maze. The data for the study follow.

Reinforcement Schedule	Rat	Conditions 1	2	3	4
1	1	29	20	21	18
	2	24	15	10	8
	3	31	19	10	31
	4	41	11	15	42
	5	30	20	27	53
2	1	25	17	19	17
	2	20	12	8	8
	3	35	16	9	28
	4	35	8	14	40
	5	26	18	18	51
3	1	10	18	16	14
	2	9	10	18	11
	3	7	18	19	12
	4	8	19	20	5
	5	11	20	17	6

(a) Analyze the data using both univariate and multivariate methods and discuss your findings.

(b) Modify program m4_15_1.sas to test that $\psi = (\mu_{11} - \mu_{12}) + (\mu_{22} \mp \mu_{23}) + (\mu_{33} \pm \mu_{34}) = 0$, an extended linear hypothesis.

4.16 Robustness and Power Analysis for MR Models

When testing hypotheses for differences in means using the fixed effects, under normality, the F test is the uniformly most powerful, invariant, unbiased (UMPIU) test, Lehmann (1994). The unbiasedness property ensures that the power of the test for parameters under the alternative is greater than or equal to the size of the test α. The invariance property states that test statistics satisfy special group structure that leave the test statistics invariant. Finally, the UMP condition states that the test maximizes the power uniformly for all parameters in the alternative hypothesis.

In general, UMPIU tests are difficult to find since they depend on the dimension of the subspace occupied by the mean vectors. For example, when analyzing one and two group hypotheses regarding mean vectors, Hotelling's T^2 statistic is UMPIU, Anderson (1984). In this situation, $s = \min(v_h, u) = 1$. When $u = 1$, the means $\boldsymbol{\mu}_1, \boldsymbol{\mu}_2, \ldots, \boldsymbol{\mu}_q$ have a unique order and when $v_h = 1$ the q u-dimensional vectors lie on a line, an $s = 1$ dimensional subspace occupied by the mean vectors so that they may again be uniquely ordered.

For $s > 1$, there is no uniformly most powerful test. The four test statistics are maximally invariant statistics in that they depend on the s eigenvalues of $\mathbf{HE}^{-1}$ (Muirhead, 1982, Theorem 6.1.4). Furthermore, when testing H : $\mathbf{CBM} = \mathbf{0}$ Perlman and Olkin (1980) showed that the tests are unbiased. However, it is in general difficult to determine the power of the test statistics since the distribution function of the roots of $\mathbf{HE}^{-1}$ is complicated.

In the population, the $E(\mathbf{Y}) = \mathbf{XB}$ for the MGLM and under multivariate normality, the noncentrality parameter of the Wishart distribution under the null hypothesis is

$$\Gamma = (\mathbf{CBM})' (\mathbf{C} (\mathbf{X}'\mathbf{X})^{-1} \mathbf{C}')^{-1} (\mathbf{CBM}) (\mathbf{M}'\Sigma\mathbf{M})^{-1} = \Delta\Sigma^{-1} \tag{4.16.1}$$

the population counterpart of $\mathbf{HE}^{-1}$. Thus, the nonzero eigenvalues $\gamma_1, \gamma_2, \ldots, \gamma_s$ of $\Delta\Sigma^{-1}$ represent the s-dimensional subspace spanned by the population vectors $\boldsymbol{\mu}_1, \boldsymbol{\mu}_2, \ldots, \boldsymbol{\mu}_q$ where $s = \min(v_h, u) = r(\Delta)$. An indication of the rank of Δ may be obtained by examining the sample roots of $\mathbf{HE}^{-1}$ by solving $|\mathbf{H}-\lambda\mathbf{E}| = 0$. The population roots $\delta_1 \geq \delta_2 \geq \delta_3 \geq \ldots \geq \delta_s > 0$ lie between two extreme situations: (1) $\delta > 0$ and $\delta_i = 0$ for $i \geq 2$ and (2) $\delta_1 = \delta_2 = \ldots = \delta_s > 0$ where the vectors are equally diffuse in s dimensions. Other configurations correspond to different relations among the roots $\delta_1 > \delta_2 > \ldots > \delta_s$.

When $\delta_1 > 0$ and all other $\delta_i = 0$, Roy's largest root statistic tends to out perform the other test procedures where the relation among the test statistics are as follows $\theta \geq U^{(s)} \geq \Lambda \geq V^{(s)}$. For fixed u and as $v_e \longrightarrow \infty$, the statistics based on Λ, $V^{(s)}$ and $U^{(s)}$ are equivalent since they all depend on an asymptotic noncentral chi-square distribution with degree of freedom $v = v_h\, u$ and noncentrality parameter $\delta = \mathrm{Tr}(\Delta)$. Olson (1974) found that the tests are equivalent, independent of δ if $v_e \geq 10 v_h u$.

For small sample sizes using various approximations to the noncentral distribution of the roots of $\mathbf{HE}^{-1}$, Pillai and Jaysachandran (1967), Lee (1971), Olson (1974), Muller and Peterson (1984), Muller, LaVange, Ramey and Ramey (1992) and Schatzoff (1966) found that when the roots $\delta_1 \geq \delta_2 \geq \ldots \geq \delta_s > 0$ have different configuration that the ordering of power is generally as follows for the four criteria: $V^{(s)} \geq \Lambda \geq U^{(s)} \geq \theta$ with little differences in power for the criteria $V^{(s)}$, Λ and $U^{(s)}$. As a general rule, one should use Roy's largest root test if $\delta_1 \gg \delta_2$, use $V^{(s)}$ if it is known that the δ_i are equal, use $U^{(s)}$ if it is known that the δ_i are very unequal, and with no knowledge use the LRT criterion Λ. Even though $V^{(s)}$ is the locally best invariant unbiased (LBIU) test for testing $\mathbf{CBM} = \mathbf{0}$, it does not always have an advantage in terms of power over the other criteria, except when the δ_i are diffuse. This is not the case for multivariate mixed models, Zhou and Mathew (1993).

While there is no optimal multivariate test for testing $\mathbf{CBM} = \mathbf{0}$, it is of some interest to investigate how sensitive the four test statistics are to violations of normality and homogeniety of covariance matrices. Under homogeniety of the covariance matrices, we know from the multivariate central limit theorem, that all statistics are reasonably robust to nonnormality given large sample sizes; by large, we mean that $v_e/p > 20$. For smaller sample sizes this is not the case.

Robustness studies for departures from MVN that compare the four test criteria have been conducted by Ito (1980), Korin (1972), Mardia (1971), O'Brien, Parente, and Schmitt

(1982), and Olson (1974, 1975) among others, while Davis (1980, 1982) investigated Λ and θ. Olson (1974) also compared the four statistics when the equality of covariance matrices does not hold.

For moderate levels of skewness and kurtosis, O'Brien et al. (1982) concluded that all four statistic are reasonable robust to nonnormality, effecting neither the Type I error rate α, the size of the test, or power. This was also found to be the case for Λ and θ studied by Davis (1980, 1982). In general, increased skewness tends to increase the size of the test, α, while an increase in kurtosis tends to reduce α. Olson (1974) found that outliers tend to reduce α, and ordered the robustness of the tests: $V^{(s)} \geq \Lambda \geq U^{(s)} \geq \theta$. This is the same ordering of the asymptotic power of the tests assuming a MVN. Positive kurtosis tends to reduce the power of all the tests.

When one does not have homogeniety of the covariance matrix, the simulation study conducted by Olson (1974) found that this effects α more than nonnormality. The rates are significantly increased with $V^{(s)}$ being most robust and θ the least, Olson (1975).

The simulation studies tend to show that the statistic most robust to lack of normality and homogeniety of covariance matrices is $V^{(s)}$; however, α increases as the heterogeneity of Σ_i increases. The power is reduced only marginally by outliers, and the Type I error rate is only moderately effected by lack of normality, Olson (1974).

Often a researcher wants to evaluate the power of a multivariate test and determine the sample size for a study. Using noncentral F approximations, Muller and Peterson (1984) showed how to approximate the distributions of $V^{(s)}$ and $U^{(s)}$ using Rao's transformation given in (3.5.3). For the noncentral case, Muller and Peterson (1984) calculated the noncentrality parameter of the F distribution by replacing sample values with population values. To illustrate, recall that for a univariate GLM that the noncentrality parameter for the noncentral F distribution is

$$\begin{aligned} \gamma &= (\mathbf{C}\boldsymbol{\beta})' \left(\mathbf{C}\left(\mathbf{X}'\mathbf{X}\right)^{-1}\mathbf{C}'\right)^{-1} (\mathbf{C}\boldsymbol{\beta})/\sigma^2 \\ &= v_1 \; MS_h^* \,/\, MS_e^* \\ &= v_1 \; F^*(v_1,\; v_2) \end{aligned} \tag{4.16.2}$$

where F^* is the F statistic replaced by population values. Using (3.5.3), we let $v_1 = pv_h$ and

$$\Lambda_* = \frac{|v_e \Sigma|}{|v_e \Sigma + v_h \Delta|} \tag{4.16.3}$$

the population noncentrality $\Gamma = \Delta\Sigma^{-1}$ for $\Delta = (\mathbf{CBM})' \left(\mathbf{C}\left(\mathbf{X}'\mathbf{X}\right)^{-1}\mathbf{C}'\right)^{-1} (\mathbf{CBM})$. Then

$$\gamma = pv_h \left[\frac{1-(\Lambda_*)^{1/d}}{(\Lambda_*)^{1/d}}\right]\left(\frac{fd-2\lambda}{pv_h}\right) \tag{4.16.4}$$

is the approximate noncentrality parameter of the noncentral F distribution for the Λ criterion where f, d and λ are given in (3.5.3). Hence with $\mathbf{X}$, $\mathbf{B}$, $\mathbf{C}$, $\mathbf{M}$ and Σ known since $\Delta = (\mathbf{CBM})' \left(\mathbf{C}\left(\mathbf{X}'\mathbf{X}\right)^{-1}\mathbf{C}'\right)^{-1} (\mathbf{CBM})$ one may calculate γ. Using the SAS function PROBF, the approximate power of the Λ criterion is

$$\text{power} = 1-\beta = 1 - \text{PROBF}\left(F^{1-\alpha},\; v_1,\; v_2,\; \gamma\right) \tag{4.16.5}$$

where $v_1 = pv_h$, $v_2 = fd - 2\lambda$ in (3.5.3), and $F^{1-\alpha}$ is the upper critical value of the F distribution under H_o : $\mathbf{CBU} = \mathbf{0}$. Muller and Peterson (1984) use Rao's approximation to also obtain estimates γ_v and γ_u for $V^{(s)}$ and $U^{(s)}$. Their approach does not use the approximation used for F in Theorem 3.5.1. The population values for the other criteria are defined as

$$V_*^{(s)} = \operatorname{tr}\left[(v_h \Delta)(v_e \Sigma + v_h \Delta)^{-1}\right]$$
$$U_*^{(s)} = \operatorname{tr}\left[(v_h \Delta)(v_e \Sigma)^{-1}\right]$$

Then

$$\gamma_v = v_1[V_*^{(s)}/(s - V_*^{(s)})](v_2/v_1)$$
$$\gamma_u = v_1(U_*^{(s)}/s)(v_3/v_1)$$

where $v_1 = pv_h$, $s = \min(v_h, \ p)$, $v_2 = s(v_e - v_h + s)$ and $v_3 = s(v_e - v_h - 1)$. Muller et al. (1992) extend the procedure to models with repeated measurements using the program power.sas developed by Keith E. Muller. All power calculations while approximate are accurate to approximately two digits of accuracy when compared to other procedures that use asymptotic expansions of the test statistics to evaluate power for the noncentral Wishart distribution. However, the approximation requires the design matrix $\mathbf{X}$ to contain fixed variables. A study similar to that conducted by Gatsonis and Sampson (1989) for a random MR design matrix has yet to be investigated. Because we now have p dependent variables and not one, the guidelines developed by Gatsonis and Sampson are not applicable. One rule may be to increase the sample size by 5p, if the number of independent variables is small (below 10).

4.17 Power Calculations—Power.sas

To calculate the approximate power for a MGLM, one must be able to specify values for the unknown parameter matrix in $\mathbf{B}$ for the linear model $\mathbf{Y} = \mathbf{XB} + \mathbf{E}$ and the unknown covariance matrix Σ. Because interest may involve several hypotheses of the form H : $\mathbf{CBM} - \Theta = \mathbf{0}$ must also be defined. Finally, the estimated value for power depends on the test criterion and the values of α. We use the program power.sas discussed by Muller et al. (1992) to illustrate how one may estimate power for a repeated measurement design. The program is similar to the program MV power illustrated by O'Brien and Muller (1993).

To use the power program, we first have to relate the notation of the program with the matrices in this text. The matrix $\mathbf{U}$ is our matrix $\mathbf{M}$. The default value for $\mathbf{U}$ is the identity matrix $\mathbf{I}_p$. When $\mathbf{U}$ is not equal to $\mathbf{I}$, it must be proportional to an orthonormal matrix. For repeated measures designs, $\mathbf{U}'\mathbf{U} = \mathbf{I}_{p-1}$. The matrix $\mathbf{B}$ is defined as beta, the matrix sigma = Σ and $\Theta \equiv$ THETAO. For our examples, we have taken the matrix Θ as a matrix of zeros. To define the design matrix $\mathbf{X}$, one may input $\mathbf{X}$ as a full rank design matrix or using the cell means model define $\mathbf{X}$ with an essence design matrix (ESSENCEX) and use REPN to denote the number of times each row of the ESSENCEX matrix occurs. For a cell means model, $\mathbf{X}'\mathbf{X} = \text{diag}\,[N_i]$. Finally the matrix $\mathbf{C}$ is defined to represent the overall test or contrast of interest. In program m4_17_1.sas we use the program to calculate the power for the repeated measures design analyzed in Section 4.15 for $\alpha = 0.05$.

To plan the experiment for the three group repeated measures design, suppose the researcher specifies that the parameter matrix $\mathbf{B}$ is

$$\mathbf{B} = \begin{bmatrix} 4 & 7 & 10 \\ 7 & 8 & 11 \\ 5 & 7 & 9 \end{bmatrix}$$

and based on a pilot study that the unknown covariance matrix Σ is

$$\Sigma = \begin{bmatrix} 74 & 65 & 35 \\ 65 & 68 & 38 \\ 35 & 38 & 26 \end{bmatrix}$$

For $n_i = 5$, 10, and 15 subjects in each of the groups, the experimenter wants to evaluate the associated power of several tests for $\alpha = 0.05$.

Assuming a multivariate model, the power for the test of equal group means require

$$\mathbf{C} = \begin{bmatrix} 1 & 0 & -1 \\ 0 & 1 & -1 \end{bmatrix} \quad \text{and} \quad \mathbf{U} = \mathbf{I}_3$$

For the test of equality of vectors across the repeated measurements,

$$\mathbf{C} = \mathbf{I}_3 \quad \text{and} \quad \mathbf{U} = \begin{bmatrix} .707107 & .408248 \\ -.707107 & .408248 \\ 0 & -.816497 \end{bmatrix}$$

The test of parallelism or no interaction requires

$$\mathbf{C} = \begin{bmatrix} 1 & 0 & -1 \\ 0 & 1 & -1 \end{bmatrix} \quad \text{and} \quad \mathbf{U} = \begin{bmatrix} .707107 & .408248 \\ -.707107 & .408248 \\ 0 & -.816497 \end{bmatrix}$$

Given no interaction, the researcher also wanted to evaluate the power of the test for differences in groups given parallelism, the ANOVA test of means, so that

$$\mathbf{C} = \begin{bmatrix} 1 & 0 & -1 \\ 0 & 1 & -1 \end{bmatrix} \quad \text{and} \quad \mathbf{U} = \begin{bmatrix} 1/3 \\ 1/3 \\ 1/3 \end{bmatrix}$$

and finally the test for differences in conditions given parallelism is evaluated using

$$\mathbf{C} = \begin{bmatrix} 1/3 & 1/3 & 1/3 \end{bmatrix} \quad \text{and} \quad \mathbf{U} = \begin{bmatrix} .707107 & .408248 \\ -.707107 & .408248 \\ 0 & -.816497 \end{bmatrix}$$

Recalling that the mixed model univariate analysis of repeated measures design requires the structures of $\mathbf{U}'\Sigma\mathbf{U} = \lambda\mathbf{I}$ for the test of conditions given parallelism and the test

TABLE 4.17.1. Power Calculations—Σ

ANOVA Means	Power	GG_PWR
5	0.061	
10	0.076	
15	0.092	
Conditions/Parallelism		
5	0.731	0.878
10	0.982	0.997
15	0.999	1
Interaction		
5	0.091	0.082
10	0.156	0.132
15	0.229	0.185
Group Mean Vectors		
5	0.114	
10	0.248	
15	0.397	
Condition Mean Vectors		
5	0.480	
10	0.906	
15	0.991	

of interaction, program power.sas also calculates the approximate power for the Geisser-Greenhouse (GG) adjusted test $\widehat{\epsilon}$, and the expected value of $\widehat{\epsilon}$. The program computes the approximate power of the GG and Huynh-Feldt (HF) corrections to the exact mixed model solution whenever the $r\left(\mathbf{U}\right) \geq 2$. The output from the program is given in Table 4.17.1

For a repeated measurement design, the primary hypothesis of interest is the test of interaction, and given no interaction, the test of differences in condition. While 10 subjects per group appears adequate for the test of conditions given parallelism, the test of interaction has low power. For adequate power, one would need over over 60 subjects per treatment group. Charging Σ to

$$\Sigma_1 = \begin{bmatrix} 7.4 & 6.5 & 3.5 \\ 6.5 & 6.8 & 3.8 \\ 3.5 & 3.8 & 2.6 \end{bmatrix}$$

which is closer to the sample estimate $\mathbf{S}$ for the example, power of the multivariate that of interaction with 10 subjects per treatment group becomes 0.927. The power results for all test using Σ_1 are shown in Table 4.17.2.

This illustrates the critical nature Σ plays in estimating power for hypotheses regarding $\mathbf{B}$. Before one develops scenarios regarding $\mathbf{B}$, a reliable estimate of Σ is critical in the design of multivariate experiments. We have also assumed that the independent variables are fixed in our example. In many multivariate regression applications we know that the independent variables are random. While this does not effect hypothesis testing and the estimation of parameters for the MGLM, the effect of treating random independent variables as fixed in power and sample size calculations is in general unknown. A study of multivari-

TABLE 4.17.2. Power Calculations—Σ_1

ANOVA Means	Power	GG_PWR
5	0.0175	
10	0.356	
15	0.504	
Conditions/Parallelism		
5	1	1
10	1	1
15	1	1
Interaction		
5	0.547	0.476
10	0.927	0.454
15	0.993	0.971
Group Mean Vectors		
5	0781	
10	0.998	
15	1	
Condition Mean Vectors		
5	1	1
10	1	1
15	1	1

ate power and sample size that compares exact power and sample size calculations with approximate values obtained assuming fixed independent variables, similar to the study by Gatsonis and Sampson (1989), has not be conduction for the MGLM. We have assumed that the relationship between the random and fixed univariate general linear models extend in a natural way to multivariate general linear models. The guideline of increasing the sample size by five when the number of independent variables is less than ten in multiple linear regression studies may have to be modified to take into account the number of dependent variables in multivariate studies.

Exercises 4.17

1. Use program power.sas to estimate the power calculation in Exercise 3.11, problem 1.
2. Determine the approximate power for the one-way MANOVA design discussed in Section 4.5 using sample estimates for $\mathbf{B}$, Σ and $\alpha = 0.05$ with and without the outlier. Discuss your findings.

4.18 Testing for Mean Differences with Unequal Covariance Matrices

Procedures for testing for differences in mean vectors for two groups were introduced in Chapter 3 and extended in Section 4.6 for independent groups. However, these procedures may not be used for repeated measures designs and more general MR models. In this section, using results due to Nel and van der Merwe (1988) and Nel (1997) methods for more general designs are developed.

Using results due to Johansen (1980), Keselman, Carriere, and Lix (1993) developed an extension of (3.9.23) to analyze multiple group repeated measurement designs which they called a Welch-James (WJ) test. Assuming a one-way repeated measurement design with Σ replaced by Σ_1 in (4.14.1), the test statistic for testing

$$H : \mathbf{C}\boldsymbol{\mu} = \mathbf{0} \tag{4.18.1}$$

where $\boldsymbol{\mu} = (\boldsymbol{\mu}_1, \boldsymbol{\mu}_2, \ldots, \boldsymbol{\mu}_a)'$, $\boldsymbol{\mu}_i = (\boldsymbol{\mu}_{i1}, \boldsymbol{\mu}_{i2}, \ldots, \boldsymbol{\mu}_{ip})'$, and $\mathbf{C}$ is a contrast matrix of dimension $r \times ap$ with rank $r = r(\mathbf{C})$ is

$$X^2_{WJ} = (\mathbf{C}\overline{\mathbf{y}})' (\mathbf{CSC})^{-1} (\mathbf{C}\overline{\mathbf{y}}) \tag{4.18.2}$$

The matrix $\mathbf{S} = \operatorname{diag}[\mathbf{S}_i/n_i, \mathbf{S}_2/n_2, \ldots, \mathbf{S}_a/n_a]$. $\mathbf{S}_i$ is the sample covariance matrix for the ith group, and $\overline{\mathbf{y}} = (\overline{\mathbf{y}}_1, \overline{\mathbf{y}}_2, \ldots, \overline{\mathbf{y}}_a)'$ is a vector of stacked means for the a independent groups. The test statistic X^2_{JW} when divided by c may be approximated by a F distribution, following (3.9.23), with $v_1 = r = R(\mathbf{C})$ and $v_2 = v_1 (v_1 + 2) / 3A$ degrees of freedom. The constants c and A are

$$c = v_1 + 2A - 6A / (v_1 + 2) \tag{4.18.3}$$

$$A = \tfrac{1}{2} \textstyle\sum_{i=1}^{a} \left[\operatorname{tr}\left\{ \mathbf{SC}' (\mathbf{CSC}')^{-1} \mathbf{CQ}_i \right\}^2 + \operatorname{tr}\left\{ \mathbf{SC}' (\mathbf{CSC}')^{-1} \mathbf{CQ}_i \right\}^2 \right] / (n_i - 1)$$

The matrix $\mathbf{Q}_i$ is a block diagonal matrix of dimension $(ap \times ap)$, corresponding to the ith groups, with the ith block equal to a $p \times p$ identity matrix and zero otherwise.

Keselman et al. (1993) show that when repeated measures main effects are tested, the WJ test controls the Type I error rate. For the interaction between groups and conditions, it fails to control the Type I error rate. Tests for between groups were not investigated.

Coombs and Algina (1996) extended the Brown and Forsythe (1974) univariate test to one-way MANOVA designs by replacing univariate means with mean vectors and variances with covariance matrices. Using the MR model $\mathbf{Y} = \mathbf{XB} + \mathbf{E}$ for testing $H : \mathbf{CBM} = \mathbf{0}$ and using results due to Nel and van der Merwe (1986) and Nel (1997), the multivariate Brown-Forsythe (BF) statistic for testing differences in means is obtained by solving $| \mathbf{H} - \boldsymbol{\lambda}\widetilde{\mathbf{E}} | = 0$. Where

$$\begin{aligned} \mathbf{H} &= (\mathbf{C}\widehat{\mathbf{B}}\mathbf{M})' (\mathbf{C} (\mathbf{X}'\mathbf{X})^{-1} \mathbf{C}')^{-1} (\mathbf{C}\widehat{\mathbf{B}}\mathbf{M}) \\ \widetilde{\mathbf{E}} &= \left[\sum_{i=1}^{a} c_i (\mathbf{M}'\mathbf{S}_i\mathbf{M}) \right] \left[\frac{v_e^* (p-1)}{v_h \widehat{f}} \right] \end{aligned} \tag{4.18.4}$$

where $\mathbf{S}_i$ is an sample estimate of Σ_i, $c_i = (N - n_i) / N$, $v_h = r(\mathbf{C})$,

$$v_e^* = \frac{\operatorname{tr}\left[\sum_{i=1}^{a} (c_i \mathbf{M}'\mathbf{S}_i\mathbf{M})\right]^2 + \left[\operatorname{tr} \sum_{i=1}^{a} (c_i \mathbf{M}'\mathbf{S}_i\mathbf{M})\right]^2}{\sum_{i=1}^{a} \{\operatorname{tr}(c_i \mathbf{M}'\mathbf{S}_i\mathbf{M})^2 + [\operatorname{tr}(c_i \mathbf{M}'\mathbf{S}_i\mathbf{M})]^2\} / (n_i - 1)} \tag{4.18.5}$$

where letting NUM denote the numerator of v_e^* the value of $\widehat{f}$ is

$$\begin{aligned}
\widehat{f} &= NUM / \sum_{i=1}^{a} \left[\operatorname{tr} A^2 + (\operatorname{tr} A)^2 \right] + B \\
A &= (1 - 2c_i)^{1/2} \mathbf{M}'\mathbf{S}_i\mathbf{M} \\
B &= \left[\operatorname{tr} \sum_i (c_i \mathbf{M}'\mathbf{S}_i\mathbf{M}) \right]^2 + \operatorname{tr} \left[\sum_i (c_i \mathbf{M}'\mathbf{S}_i\mathbf{M}_i) \right]^2
\end{aligned} \tag{4.18.6}$$

Following Boik (1991), we reject H if for

$$\begin{aligned}
T_o^2 &= v_e^* \operatorname{tr}(\mathbf{H}\widetilde{\mathbf{E}})^{-1} \\
\frac{2\,(sN+1)}{s^2\,(2M+s+1)} T_o^2 / v_e^* &> F^{1-\alpha}\,(v_1, v_2)
\end{aligned} \tag{4.18.7}$$

where $s = \min\left(v_h^*, p\right)$, $N = \left(v_e^* - p - 1\right)/2$, $M = \left(\mid v_h^* - p \mid -1\right)/2$, $v_1 = s(2M + s + 1)$, and $v_2 = s(2N + s + 1)$. Results reported by Algeria (1994) favor the BF test, even though the numerator degrees of freedom were not corrected in his study.

Exercises 4.18

1. For Edward's data in Section 4.15, test the hypotheses: H_G Edwards(4.14.2), H_P (5.15.3) and H_C (4.14.4) using the test statistic X_{WJ}^2 defined in (4.18.2).and statistic T_o^2 defined in (4.18.7). Compare the test results obtained using PROC GLM.

2. Reanalyze the data in Exercise 4.15, Problem 1, using the statistic X_{WJ}^2.

5

Seemingly Unrelated Regression Models

5.1 Introduction

In the MR regression model, the design matrix $\mathbf{X}$ is common to each variable. This limitation does not permit one to associate different design matrices with each dependent variable which in many regression problems may lead to overfitting some variables. To correct this problem, we must be able to model each dependent variable separately within a common, overall model. Using the vec $(\cdot)$ operator on the columns of the data matrix $\mathbf{Y}$, Zellner (1962, 1963) formulated the seemingly unrelated regression (SUR) model as p correlated regression models. Srivastava (1966, 1967) called the design the multiple-design multivariate (MDM) model. Hecker (1987) formulates the model using the vec $(\cdot)$ operator on the rows of $\mathbf{Y}_{n\times p}$ and calls the model the completely general MANOVA (CGMANOVA) model. In this chapter, we review the general theory of the SUR model and show how the model may be used to estimate parameters and test hypothesis in complex design situations including the generalized MANOVA (GMANOVA) model developed by Potthoff and Roy (1964), repeated measurement designs with changing covariates, and mixed MANOVA-GMANOVA designs. In addition, goodness of fit tests, tests for nonadditivity, and sum of profile models are discussed. Finally, the multivariate SUR (MSUR) is reviewed.

5.2 The SUR Model

a. *Estimation and Hypothesis Testing*

To extend the MR model to a SUR model, we represent each dependent variable in the MR model as a univariate model. Following (3.6.29), we represent the mean of **Y** as

$$\begin{aligned} E\left(\mathbf{Y}_{n\times p}\right) &= \left[\mathbf{X}_1\boldsymbol{\beta}_1, \mathbf{X}_2\boldsymbol{\beta}_2, \ldots, \mathbf{X}_p\boldsymbol{\beta}_p\right] \\ &= \left[\mathbf{X}_1,\ \mathbf{X}_2, \ldots, \mathbf{X}_p\right]\begin{bmatrix} \boldsymbol{\beta}_1 & \mathbf{0} & \cdots & \mathbf{0} \\ \mathbf{0} & \boldsymbol{\beta}_2 & \cdots & \mathbf{0} \\ \vdots & \vdots & & \vdots \\ \mathbf{0} & \mathbf{0} & \cdots & \boldsymbol{\beta}_p \end{bmatrix} \\ &= \mathbf{XB} \end{aligned} \tag{5.2.1}$$

where $\mathbf{X} = \left[\mathbf{X}_1, \mathbf{X}_2, \ldots, \mathbf{X}_p\right]$ and $\mathbf{B} = \text{diag}\left[\boldsymbol{\beta}_j\right]$ is a block diagonal matrix. The matrices $\mathbf{X}_j$ are design matrices for each variable for $j = 1, 2, \ldots, p$. Hence, for the jth column of **Y** the $E\left(\mathbf{y}_j\right) = \mathbf{X}_j\boldsymbol{\beta}_j$ and the $\text{cov}\left(\mathbf{y}_j, \mathbf{y}_{j'}\right) = \sigma_{ij}\mathbf{I}_{jj'}$. Letting $\mathbf{y} = \text{vec}\left(\mathbf{Y}\right)$ so that the matrix **Y** is stacked columnwise, we have a GLM representation for the SUR as follows

$$\begin{aligned} E\left(\mathbf{y}\right) &= \mathbf{D}\boldsymbol{\beta} \\ \text{cov}\left(\mathbf{y}\right) &= \Sigma \otimes \mathbf{I} \end{aligned} \tag{5.2.2}$$

$$\underset{np\times\sum_{j=1}^{p} q_j}{\mathbf{D}} = \begin{bmatrix} \mathbf{X}_1 & \mathbf{0} & \cdots & \mathbf{0} \\ \mathbf{0} & \mathbf{X}_2 & \cdots & \mathbf{0} \\ \vdots & \vdots & & \vdots \\ \mathbf{0} & \mathbf{0} & \cdots & \mathbf{X}_p \end{bmatrix} = \bigoplus_{j=1}^{p} \mathbf{X}_j$$

where $\mathbf{X}_j \left(n \times q_j\right)$ has full rank q_j and $q^* = \sum_j q_j$. In formulation (5.2.2), the columns of **Y** are modeled so that we have p correlated regression models. Again, each row of **Y** is assumed to follow a MVN with common covariance structure Σ.

To estimate $\boldsymbol{\beta}$ in (5.2.2), we employ the ML estimate or GLS estimate of $\boldsymbol{\beta}$ given in (3.6.8)

$$\widehat{\boldsymbol{\beta}} = (\mathbf{D}'\left(\Sigma \otimes \mathbf{I}_n\right)^{-1}\mathbf{D})^{-1}\mathbf{D}'\left(\Sigma \otimes \mathbf{I}_n\right)^{-1}\mathbf{y} = (\mathbf{D}'\Omega^{-1}\mathbf{D})^{-1}\mathbf{D}'\Omega^{-1}\mathbf{y} \tag{5.2.3}$$

for $\Omega = \Sigma \otimes \mathbf{I}_n$. Because Σ is unknown, a feasible generalized least squares (FGLS) estimate is obtained by replacing Σ with a consistent estimator. Zellner (1962) proposed replacing Σ by $\widehat{\Sigma} = \mathbf{S} = \left[s_{ij}\right]$ where

$$s_{ij} = \frac{1}{n-q}\mathbf{y}_i'\left(\mathbf{I}_n - \mathbf{X}_i\left(\mathbf{X}_i'\mathbf{X}_i\right)^{-1}\mathbf{X}_i'\right)\left(\mathbf{I}_n - \mathbf{X}_j\left(\mathbf{X}_j'\mathbf{X}_j\right)^{-1}\mathbf{X}_j'\right)\mathbf{y}_j \tag{5.2.4}$$

and $q = r\left(\mathbf{X}_j\right)$, assuming the number of variables is the same for each model. Relaxing this condition, we may let $q = 0$ or let $q = q_{ij}$ where

$$q_{ij} = \text{tr}\left[\left(\mathbf{I}_n - \mathbf{X}_i\left(\mathbf{X}_i'\mathbf{X}_i\right)^{-1}\mathbf{X}_i'\right)\left(\mathbf{I}_n - \mathbf{X}_j\left(\mathbf{X}_j'\mathbf{X}_j\right)^{-1}\mathbf{X}_j'\right)\right] \tag{5.2.5}$$

Substituting $\widehat{\Sigma}$ for Σ in (5.2.3), the FGLS estimator of $\boldsymbol{\beta}$ is $\widehat{\widehat{\boldsymbol{\beta}}}$ defined as

$$\widehat{\widehat{\boldsymbol{\beta}}} = \begin{bmatrix} \widehat{\widehat{\boldsymbol{\beta}}}_1 \\ \widehat{\widehat{\boldsymbol{\beta}}}_2 \\ \vdots \\ \widehat{\widehat{\boldsymbol{\beta}}}_p \end{bmatrix} = \begin{bmatrix} s^{11}\mathbf{X}_1'\mathbf{X}_1 & \cdots & s^{1p}\mathbf{X}_1'\mathbf{X}_p \\ s^{21}\mathbf{X}_2'\mathbf{X}_1 & \cdots & s^{2p}\mathbf{X}_2'\mathbf{X}_p \\ \vdots & & \vdots \\ s^{p1}\mathbf{X}_p'\mathbf{X}_1 & \cdots & s^{pp}\mathbf{X}_p'\mathbf{X}_p \end{bmatrix}^{-1} \begin{bmatrix} \mathbf{X}_1' \sum_{j=1}^{p} s^{1j}\mathbf{y}_j \\ \mathbf{X}_2' \sum_{j=1}^{p} s^{2j}\mathbf{y}_j \\ \vdots \\ \mathbf{X}_p' \sum_{j=1}^{p} s^{pj}\mathbf{y}_j \end{bmatrix} \tag{5.2.6}$$

where $\widehat{\Sigma}^{-1} = \left[s^{ij}\right]$. The asymptotic covariance matrix of $\widehat{\widehat{\boldsymbol{\beta}}}$, is

$$\left[\mathbf{D}'\left(\Sigma^{-1} \otimes \mathbf{I}_n\right)\mathbf{D}\right]^{-1} = \begin{bmatrix} \sigma^{11}\mathbf{X}_1'\mathbf{X}_1 & \cdots & \sigma^{1p}\mathbf{X}_1'\mathbf{X}_p \\ \vdots & & \vdots \\ \sigma^{p1}\mathbf{X}_p'\mathbf{X}_1 & \cdots & \sigma^{pp}\mathbf{X}_p'\mathbf{X}_p \end{bmatrix}^{-1} \tag{5.2.7}$$

To estimate $\boldsymbol{\beta}$, we first estimate Σ with $\widehat{\Sigma}$ such that $\widehat{\Sigma} \xrightarrow{p} \Sigma$, and then obtain the estimate $\widehat{\widehat{\boldsymbol{\beta}}}$, a two-stage process. The ML estimate is obtained by iterating the process. Having calculated $\widehat{\widehat{\boldsymbol{\beta}}}$ in the second step, we can re-estimate $\Omega = \Sigma \otimes \mathbf{I}_n$ as

$$\widehat{\Omega}_2 = (\mathbf{y} - \mathbf{D}\widehat{\widehat{\boldsymbol{\beta}}}_1)(\mathbf{y} - \mathbf{D}\widehat{\widehat{\boldsymbol{\beta}}}_1)'$$

where $\widehat{\widehat{\boldsymbol{\beta}}}_1$ is the two stage estimate. Then a revised estimate of $\boldsymbol{\beta}$ is

$$\widehat{\widehat{\boldsymbol{\beta}}}_2 = (\mathbf{D}'\widehat{\Omega}_2\mathbf{D})^{-1}\mathbf{D}'\widehat{\Omega}_2^{-1}\mathbf{y}$$

Continuing in this manner, the ith iteration is

$$\widehat{\Omega}_i = (\mathbf{y} - \mathbf{D}\widehat{\widehat{\boldsymbol{\beta}}}_{i-1})(\mathbf{y} - \mathbf{D}\widehat{\widehat{\boldsymbol{\beta}}}_{i-1})'$$
$$\widehat{\widehat{\boldsymbol{\beta}}}_i = (\mathbf{D}'\widehat{\Omega}_i\mathbf{D})^{-1}\mathbf{D}'\widehat{\Omega}_i^{-1}\mathbf{y}$$

The process continues until $\widehat{\widehat{\boldsymbol{\beta}}}_i$ converges such that $||\widehat{\widehat{\boldsymbol{\beta}}}_i - \widehat{\widehat{\boldsymbol{\beta}}}_{i-1}||^2 < \epsilon$ for some $\epsilon > 0$. This yields the iterative feasible generalized least squares estimate (IFGLSE). Park (1993) shows that the MLE and IFGLSE of $\boldsymbol{\beta}$ are mathematically equivalent.

To test hypotheses of the form $H : \mathbf{C}_*\boldsymbol{\beta} = \boldsymbol{\xi}$, Wald's statistic defined in (3.6.12) may be used. With $\widehat{\Sigma} = \left[s_{ij}\right]$, $\widehat{\widehat{\boldsymbol{\beta}}}$ the estimate of $\boldsymbol{\beta}$ and $\mathbf{D}$ defined in (5.2.2),

$$W = (\mathbf{C}_*\widehat{\widehat{\boldsymbol{\beta}}} - \boldsymbol{\xi})'\{\mathbf{C}_*[\mathbf{D}'(\widehat{\Sigma} \otimes \mathbf{I}_n)^{-1}\mathbf{D}]^{-1}\mathbf{C}_*'\}^{-1}(\mathbf{C}_*\widehat{\widehat{\boldsymbol{\beta}}} - \boldsymbol{\xi}) \tag{5.2.8}$$

Under H ($\mathbf{C}_*\boldsymbol{\beta} = \boldsymbol{\xi}$), W has an asymptotic chi-square distribution with degrees of freedom $v_h = r\,(\mathbf{C}_*)$. Alternatively, Zellner (1962) proposed the approximate F statistic under H as

$$F^* = (W/v_h)\,/MS_e \tag{5.2.9}$$
$$MS_e = (\mathbf{y} - \mathbf{D}\widehat{\widehat{\boldsymbol{\beta}}})'\widehat{\Omega}^{-1}(\mathbf{y} - \mathbf{D}\widehat{\widehat{\boldsymbol{\beta}}})/v_e$$
$$v_e = n - \sum_j q_j$$

The statistic F^* is calculated in the SAS procedure SYSLIN to test $H : \mathbf{C}_* \boldsymbol{\beta} = \boldsymbol{\xi}$ by using the STEST statement.

In (5.2.1) and (5.2.2), the vectors in $\mathbf{Y}_{n \times p}$ were equated using a columnwise expansion of $\mathbf{Y}$. Alternatively, suppose we consider each row in $\mathbf{Y}$ given (5.2.1). Then we may write (5.2.1) as

$$\underset{p\times 1}{\mathbf{y}_i} = \underset{p\times m}{\mathbf{D}_i} \; \underset{m\times 1}{\boldsymbol{\theta}} + \underset{p\times 1}{\mathbf{e}_i}$$

$$\mathbf{D}_i = \begin{bmatrix} \mathbf{x}'_{i1} & \mathbf{0}' & \cdots & \mathbf{0}' \\ \mathbf{0} & \mathbf{x}'_{i2} & \cdots & \vdots \\ \vdots & \vdots & & \vdots \\ \mathbf{0}' & \mathbf{0}' & \cdots & \mathbf{x}'_{ip} \end{bmatrix}$$

where $\mathbf{x}_{ij}$ is a $q_j \times 1$ vector of variables, $i = 1, 2, \ldots, n$ and $j = 1, 2, \ldots, p$, $q^* = \sum_j q_j$ and $\boldsymbol{\theta}' = \left[\boldsymbol{\beta}'_1, \boldsymbol{\beta}'_2, \ldots, \boldsymbol{\beta}'_p\right]$ is the parameter vector. For this representation of the SUR or MDM model, we roll out $\mathbf{Y}$ row-wise so that

$$\underset{np\times 1}{\mathbf{y}^*} = \underset{np\times m}{\mathbf{A}} \; \underset{m\times 1}{\boldsymbol{\theta}} + \underset{np\times 1}{\mathbf{e}} \tag{5.2.10}$$

where $\mathbf{y}^* = \text{vec}\left(\mathbf{Y}'\right)$, $\mathbf{e} = \left[\mathbf{e}'_1, \mathbf{e}'_2, \ldots, \mathbf{e}'_n\right]'$, $\mathbf{A} = \left[\mathbf{D}'_1, \mathbf{D}'_2, \ldots, \mathbf{D}'_n\right]'$ and the cov $(\mathbf{y}^*) = \mathbf{I} \otimes \Sigma$. We call (5.2.11), the general SUR model. For (5.2.11), we again may obtain a FGLSE of $\boldsymbol{\theta}$

$$\widehat{\widehat{\boldsymbol{\theta}}} = [\mathbf{A}'(\mathbf{I}_n \otimes \widehat{\Sigma})^{-1}\mathbf{A}]^{-1}\mathbf{A}'(\mathbf{I}_n \otimes \widehat{\Sigma})^{-1}\mathbf{y}^* \tag{5.2.11}$$

where $\widehat{\Sigma}$ is again any consistent estimate of Σ. One may employ Zellner's estimate of Σ or even a naive estimate based on only within-subject variation

$$\widehat{\Sigma} = \mathbf{Y}'(\mathbf{I} - \mathbf{1}\left(\mathbf{1}'\mathbf{1}\right)^{-1}\mathbf{1}')\mathbf{Y}/n \tag{5.2.12}$$

To test hypotheses of the form $H : \mathbf{C}\boldsymbol{\theta} = \boldsymbol{\theta}_o$, one may again use Wald's statistic

$$W = (\mathbf{C}\widehat{\widehat{\boldsymbol{\theta}}} - \boldsymbol{\theta}_o)'\{\mathbf{C}[\mathbf{A}'(\mathbf{I}_n \otimes \widehat{\Sigma}^{-1})\mathbf{A}]^{-1}\mathbf{C}'\}(\mathbf{C}\widehat{\widehat{\boldsymbol{\theta}}} - \boldsymbol{\theta}_o) \tag{5.2.13}$$

where $W \mathrel{\dot\sim} \chi^2(v_h)$ and $v_h = r(\mathbf{C})$. Or following Zellner (1962), an F-like statistic as in (5.2.9) may be used.

b. *Prediction*

When developing a regression model, one is also interested in the prediction of future observations. This is also the case for the SUR model, however, it is complicated by the fact that the future observations are dependent on the observed observations. Suppose we are interested in predicting m future observations defined by the matrix $\mathbf{Y}_f$. To predict the matrix of future observations, we let the $mp \times 1$ vector $\mathbf{y}_f = \text{vec}\ (\mathbf{Y}_f)$, and Ω_f represent the $mp \times mp$ covariance matrix of the future observations. Because the vector $\mathbf{y}_f$ is not independent of $\mathbf{y}$, we let $\mathbf{W}_{mp \times np}$ represent the covariance matrix between $\mathbf{y}$ and $\mathbf{y}_f$. Then for the joint vector $\mathbf{z}' = (\mathbf{y}', \mathbf{y}'_f)$, the covariance matrix has the following structure

$$\text{cov}\,(\mathbf{z}) = \text{cov}\begin{bmatrix} \mathbf{y} \\ \mathbf{y}_f \end{bmatrix} = \begin{bmatrix} \Omega & \mathbf{W}' \\ \mathbf{W} & \Omega_f \end{bmatrix} = \mathbf{V} \tag{5.2.14}$$

Letting $\mathbf{D}_f$ represent the design matrix for the future observations $\mathbf{y}_f$, one might be tempted to estimate $\mathbf{y}_f$ by $\mathbf{D}_f\widehat{\boldsymbol{\beta}}$ where $\widehat{\boldsymbol{\beta}}$ is defined in (5.2.3). As shown by Goldberger (1962), this is incorrect since it ignores the covariance matrix $\mathbf{W}$. The best linear unbiased predictor (BLUP) of $\mathbf{y}_f$ is the vector

$$\widehat{\mathbf{y}}_f = \mathbf{D}_f\,\widehat{\boldsymbol{\beta}} + \mathbf{W}\Omega^{-1}(\mathbf{y} - \mathbf{D}\widehat{\boldsymbol{\beta}}) \tag{5.2.15}$$

where $\widehat{\boldsymbol{\beta}}$ is defined in (5.2.3). From Theorem 3.3.2, property 5, the expression in (5.2.16) is the estimated mean $E(\mathbf{y}_f|\mathbf{y})$, under joint normality. Hence, the prediction of $\mathbf{y}_f$ depends on knowing both Ω and $\mathbf{W}$, but not Ω_f. To estimate the vector of future observations, one usually replaces the unknown matrices with consistent estimators based upon the sample data. There are however special situations when the prediction of the future observation vector does not depend on the unknown matrices.

If the covariance between the observation vector $\mathbf{y}$ and the future observation vector $\mathbf{y}_f$ is zero so that the covariance matrix $\mathbf{W} = \mathbf{0}$, then the BLUP of the future observation vector $\mathbf{y}_f$ is the vector $\widehat{\mathbf{y}}_f = \mathbf{D}_f\,\widehat{\boldsymbol{\beta}}$ where $\widehat{\boldsymbol{\beta}}$ is defined in (5.2.3). And, the prediction of the future observation vector only depends on knowing the covariance matrix Ω. While this too is an interesting result, it also is not very useful since $\mathbf{W}$ is usually not equal to zero. If however, $\Omega\mathbf{D} = \mathbf{D}\mathbf{F}$ for some nonsingular conformable matrix $\mathbf{F}$, Zyskind (1967) shows that the ML estimate (GLS estimate) of $\boldsymbol{\beta}$ reduces to the OLS estimate of $\boldsymbol{\beta}$. Since $\widehat{\boldsymbol{\beta}}_{OLS} = (\mathbf{D}'\mathbf{D})^{-1}\mathbf{D}'\mathbf{y}$, the estimate of $\mathbf{y}_f$ does not depend on Ω if $\Omega\mathbf{D} = \mathbf{D}\mathbf{F}$. While this is an interesting result, it is not too useful since $\widehat{\mathbf{y}}_f$ in (5.2.16) still depends on $\mathbf{W}\Omega^{-1}$. However, Elian (2000) shows that the estimate of $\mathbf{y}_f$ may be estimated by

$$\widehat{\mathbf{y}}_f = \mathbf{D}_f\,\widehat{\boldsymbol{\beta}}_{OLS} = \mathbf{D}_f(\mathbf{D}'\mathbf{D})^{-1}\mathbf{D}'\mathbf{y} \tag{5.2.16}$$

if and only if the covariance matrix $\mathbf{W}' = \mathbf{D}\mathbf{G} \neq \mathbf{0}$ for $\mathbf{G} = (\mathbf{D}'\Omega^{-1}\mathbf{D})^{-1}\mathbf{D}'\Omega^{-1}\mathbf{W}'$ and $\Omega\mathbf{D} = \mathbf{D}\mathbf{F}$. Thus, while the matrices $\mathbf{W}$ and Ω are both unknown, the BLUP of $\mathbf{y}_f$ does not depend on knowing either matrix. If we can only establish that the covariance matrix may be represented as $\mathbf{W}' = \mathbf{D}\mathbf{G}$, then the estimate of the future observations still depends on the covariance matrix Ω but not $\mathbf{W}$ since

$$\widehat{\mathbf{y}}_f = \mathbf{D}_f\,\widehat{\boldsymbol{\beta}} = \mathbf{D}_f(\mathbf{D}'\Omega^{-1}\mathbf{D})^{-1}\mathbf{D}'\Omega^{-1}\mathbf{y} \tag{5.2.17}$$

Using Rao's unified theory of least squares, one may extend Elian's result for nonsingular matrices Ω. Following Rao (1973b, Eq. 3.23), the BLUP of the vector $\mathbf{y}_f$ is

$$\widehat{\mathbf{y}}_f = \mathbf{D}_f\,\widehat{\boldsymbol{\beta}}_{GM} + \mathbf{W}\Omega^{-}\Omega\mathbf{Q}(\mathbf{Q}\Omega\mathbf{Q})^{-}\mathbf{Q}\mathbf{y} \tag{5.2.18}$$

where the estimate of parameter vector is the Gauss-Markov estimate given by $\widehat{\boldsymbol{\beta}}_{GM} = (\mathbf{D}'\Omega^{-}\mathbf{D})^{-}\mathbf{D}'\Omega^{-}\mathbf{y}$ and $\mathbf{Q} = \mathbf{I} - \mathbf{D}(\mathbf{D}'\mathbf{D})^{-}\mathbf{D}'$ is the symmetric projection matrix. We can further show that the BLUE of $\mathbf{y}_f$ is

$$\widehat{\mathbf{y}}_f = \mathbf{D}_f \widehat{\boldsymbol{\beta}}_{GM} \tag{5.2.19}$$

if and only if $\mathbf{W}\Omega^{-}\Omega\mathbf{Q} = \mathbf{0}$, a generalization of the condition established by Elian (2000). Finally, if in addition we can establish that $\mathbf{D}'\Omega\mathbf{Q} = \mathbf{0}$, then $\widehat{\boldsymbol{\beta}}_{GM} = \widehat{\boldsymbol{\beta}}_{OLS}$ so that the vector of future observations does not depend on knowing any of the matrices $\mathbf{W}$, Ω, or Ω_f.

5.3 Seeming Unrelated Regression Example

In multivariate regression, the entire set of dependent variables are jointly predicted by a set of independent variables. Every dependent variable is related to every independent variable. In seemingly unrelated regressions, different subsets of independent variables are related to each dependent variable. If for a set of p dependent variables $\mathbf{X}_1 = \mathbf{X}_2 = \ldots = \mathbf{X}_p = \mathbf{X}$, the design matrix $\mathbf{D} = \mathbf{I}_p \otimes \Sigma$ and $\widehat{\boldsymbol{\beta}} = \text{vec}(\widehat{\mathbf{B}})$, the SUR model estimate is identical to the MR model solution. The advantage of the SUR model is that it permits one to relate different independent variables to each dependent variable using the correlations among the errors in different equations to improve upon the estimators. The procedure for performing the analysis of the system of equations in SAS is PROC SYSLIN. To use the procedure, new terminology is introduced. The jointly dependent variables are called endogenous variables and are determined by the model. While these variables can appear also on the right side of an equation, we discuss this situation in Chapter 10, structural equation modeling. Exogenous variables are independent variables that are determined by factors outside the model. They do not depend on any of the endogenous variables in the system of equations.

Estimation of the parameters in SUR models is complicated and the SYSLIN procedure provides many options to the researcher. The primary goal of any option is to obtain efficient estimators, estimators with minimum or no bias and smallest variance. The default method of estimation for SUR models is the ordinary least squares (OLS) method. This method ignores any correlation of the errors across equations. Given a correctly specified model, SUR estimates take into account the intercorrelation of errors across equations. In large samples, SUR estimators will always be at least as efficient as OLS estimators. In PROC SYSLIN, the option SUR replaces Σ in (5.2.3) with an unbiased estimator $\mathbf{S}$. The ITSUR option iterates using improved estimates of $\mathbf{S}$ at each stage. The FIML option estimates Σ and $\boldsymbol{\beta}$ jointly by solving the likelihood equations using instrumental variables, Hausman (1975).

Instrument variables are used to predict endogenous variables at the first-stage regression. For the FIML option all predetermined variables that appear in any system are used as instruments. Because the FIML option involves minimizing the determinant of the covariance matrix associated with residuals of the reduced form of the likelihood for the system of equations, $\widehat{\Sigma} \neq \mathbf{S}$. In PROC SYSLIN, ITSUR and FIML yield different estimates, even though they are mathematically equivalent. One may calculate both and choose the method that yields the smallest variance for the estimates. For poorly specified models or models with small sample sizes, the estimate of Σ may be unreliable. In these situations, OLS estimates may be best. In practice, one usually employs several methods of estimation when the exogenous variables are fixed. For random exogenous variables, the FIML estimate of the model parameters are calculated under joint multivariate normality. The other estimation methods used in the SYSLIN procedure to estimate model parameters use only the conditional distribution $\mathbf{y}|\mathbf{x}$. Implicit in the application of methods that employ the conditional distribution to estimate model parameters is the assumption of weak exogeniety for the variable $\mathbf{x}$, as discussed in multivariate regression (MR). Exogeniety is discussed in Chapter 10.

To test hypotheses using the SYSLIN procedure, the TEST or STEST statements are used. It uses the F-like statistic in (5.2.9) for testing hypotheses. The TEST statement is used within a model while the STEST statement relates variables across models. Thus, by using the SYSLIN procedure and the STEST statement, on may evaluate whether models across dependent or independent populations are equal. Recall that PROC REG only considered a single population. The SYSLIN procedure in SAS allows one to evaluate the equality of regression models developed for several groups, the analysis of heterogeneous data sets for independent or dependent groups.

For Rohwer's data in Table 4.3.1, the MR model was used to fit the five independent variables simultaneously to the three dependent variables. Because some independent variables predict the dependent variables in varying degrees, some variables may be included in the model for one or more dependent variables even though they may have low predictive value. Using the same data, we use the SUR model to fit a separate regression equation to each variable with different subsets of exogenous (independent) variables. Program m5_3_1.sas is used for the analysis.

We begin our analysis with the full and reduced multivariate models using ordinary least squares and the PROC SYSLIN. Comparing the output for the SYSLIN procedure and the procedure REG, we see that the results are identical. The STEST statement is used to test that the coefficients for N and SS are zero across equations. As in Chapter 4, this leads us to the reduced model involving the paired associate learning tasks S, NA, and NA. For this reduced model, tests that these coefficients are zero are significant, however, this is not the case within each model. For the SUR model we use NA (named action) to predict PPVT (Peabody Picture Vocabulary Test); S (still) to predict RPMT (Raven Picture Matrices Test); and both to predict SAT (Student Achievement). We use the estimation procedure FIML, ITSUR and SUR to fit the three correlated regression models to the data. While the three estimation methods give similar results, the FIML methods provide estimates of $\boldsymbol{\beta}$ with the smallest standard errors.

To obtain approximate $100\,(1-\alpha)\,\%$ simultaneous confidence intervals for parametric functions $\psi = \mathbf{a}'\boldsymbol{\beta}$, one estimates $\widehat{\psi}$ and $\widehat{\sigma}_{\widehat{\psi}}$ that is consistent with the estimation method and forms the confidence set

$$\widehat{\psi} - c_\alpha \widehat{\sigma}_{\widehat{\psi}} \leq \psi \leq \widehat{\psi} + c_\alpha \widehat{\sigma}_{\widehat{\psi}}$$

Using the F approximation, as implemented in the SYSLIN procedure,

$$c_\alpha^2 = v_h F^{1-\alpha}\,(v_h, v_e)$$

where v_e is the system degrees of freedom, $v_e = n - \sum_j q_j$.

When fitting SUR models, one may again construct multiple linear regression indices of model fit like the coefficient of determination, McElroy (1977). This is the only index provided by the SYSLIN procedure. While one may construct residual plots using PROC SYSLIN, the extensive features of PROC REG are not available.

In multiple linear regression and multivariate regression, model selection was limited to models nested within an overall model. In SUR models, one may want to compare several models that overlap for some set of variables, but the variables in an alternative model are not a nested subset of an overall model. This leads one to nonnested hypotheses first considered by Cox (1961, 1962). We discuss model specification for nonnested SUR models in Section 5.14.

Exercises 5.3

1. Using only the variables PPVT and SAT, fit the following models to the Rohwer dataset:rohwer.dat.

   (a) Model 1: SAT = na ss and PPVT = ns ss

   (b) Model 2: SAT = n ss and PPVT = s ss

   Which model fits the data best?

2. Using the dataset Rohwer2.dat for the $n = 37$ students in a high-socioeconomic-status area (those records marked with a two for the first variable in the file), fit a SUR model to the data selecting SS, NA and N for predicting SAT, SS, N and NA for predicting PPVT, and NS, SS and NA for predicting RMPT.

5.4 The CGMANOVA Model

To develop a model that may be used to test extended multivariate general linear hypotheses, Hecker (1987) proposed the completely general MANOVA (CGMANOVA) model. Hecker transformed the MR model, $\mathbf{Y} = \mathbf{XB} + \mathbf{E}$, using a row-wise expansion. Letting $\mathbf{y}^* = \text{vec}\left(\mathbf{Y}'\right)$ and $\boldsymbol{\theta} = \text{vec}\left(\mathbf{B}'\right)$, the MR model has the univariate structure

$$\mathbf{y}^* = \begin{bmatrix} \mathbf{y}_1 \\ \mathbf{y}_2 \\ \vdots \\ \mathbf{y}_n \end{bmatrix} = \begin{bmatrix} \mathbf{I}_p \otimes \mathbf{x}_1' \\ \mathbf{I}_p \otimes \mathbf{x}_2' \\ \vdots \\ \mathbf{I}_p \otimes \mathbf{x}_n' \end{bmatrix} \boldsymbol{\theta} + \mathbf{e}^* \tag{5.4.1}$$

where $\mathbf{e}^* = \text{vec}\left(\mathbf{E}'\right)$ and $\text{cov}\left(\mathbf{e}^*\right) = \mathbf{I}_n \otimes \Sigma$.

Letting $\mathbf{D}_i = \mathbf{I}_p \otimes \mathbf{x}_i'$ and associating $\boldsymbol{\theta} = \text{vec}\left(\mathbf{B}'\right)$ with $\boldsymbol{\theta}$ in (5.2.11), we observe that Hecker's CGMANOVA model is no more than a general SUR model. Hence, we may estimate parameters and test hypotheses using (5.2.12) and (5.2.14) with $\widehat{\widehat{\boldsymbol{\theta}}}$ replaced by $\widehat{\widehat{\boldsymbol{\theta}}}$, $\mathbf{A} = \left[\mathbf{D}_1', \mathbf{D}_2', \ldots, \mathbf{D}_n'\right]$ where $\mathbf{D}_i = \mathbf{I}_p \otimes \mathbf{x}_i'$, and $\mathbf{x}_i'$ is a row of the design matrix $\mathbf{X}$. If $\mathbf{X}$ is not of full rank, one may use g-inverses.

Using an independent estimate of Σ, Hecker proposed a LR test statistic for testing hypotheses using the CGMANOVA/SUR model; however, he did not derive its distribution. For large sample sizes, his statistic is equivalent to Wald's statistic.

5.5 CGMANOVA Example

In Section 4.15 we showed how one may test hypotheses for the MGLM that do not have the bilinear form $\mathbf{CBM} = \mathbf{0}$ by using the extended linear model. We reanalyze Edward's data in Table 4.15.1 using theEdwards CGMANOVA/SUR model and the SYSLIN procedure, program m5_5_1.sas.

Recall that the parameter matrix $\mathbf{B}$ has the structure

$$\mathbf{B} = \begin{bmatrix} \mu_{11} & \mu_{12} & \mu_{13} \\ \mu_{21} & \mu_{22} & \mu_{23} \\ \mu_{31} & \mu_{32} & \mu_{33} \end{bmatrix} \tag{5.5.1}$$

To fit a SUR model to $\mathbf{B}$, each dependent variable is related to the independent variables x_1, x_2, and x_3 that represent group membership. To test for group differences the hypothesis is

$$H_G : \begin{bmatrix} \mu_{11} \\ \mu_{12} \\ \mu_{13} \end{bmatrix} = \begin{bmatrix} \mu_{21} \\ \mu_{22} \\ \mu_{23} \end{bmatrix} = \begin{bmatrix} \mu_{31} \\ \mu_{32} \\ \mu_{33} \end{bmatrix} \tag{5.5.2}$$

The test for differences in conditions is

$$H_c : \begin{bmatrix} \mu_{11} - \mu_{13} - \mu_{31} + \mu_{33} & \mu_{12} - \mu_{13} - \mu_{32} + \mu_{33} \\ \mu_{21} - \mu_{23} - \mu_{31} + \mu_{33} & \mu_{22} - \mu_{23} - \mu_{32} + \mu_{33} \end{bmatrix} = \mathbf{0} \tag{5.5.3}$$

These are the standard MANOVA tests. One may also construct tests that do not have bilinear form

$$\begin{aligned} H_1 : \psi = (\mu_{11} + \mu_{22} + \mu_{33}) - \mu_{12} + \mu_{21} \\ - \mu_{13} + \mu_{31} - \mu_{32} + \mu_{33} + \mu_{33} = 0 \end{aligned} \tag{5.5.4}$$

$$\begin{aligned} H_2 : \mu_{11} &= \mu_{21} \\ \mu_{12} &= \mu_{22} + \mu_{32} \\ \mu_{23} &= \mu_{33} \end{aligned} \tag{5.5.5}$$

$$H_3 : \psi = (\mu_{11} - \mu_{12}) + (\mu_{22} - \mu_{23}) + (\mu_{31} - \mu_{33}) = 0 \tag{5.5.6}$$

All of the tests may be tested using the STEST command in the SYSLIN procedure. The results are given in Table 5.5.1.

Comparing the approximate SUR model tests with the exact multivariate likelihood ratio tests, we find that the tests for group and conditions are consistent for both procedures. However, the test of parallel profiles is rejected using the SUR model, and is not rejected

TABLE 5.5.1. SUR Model Tests for Edward's Data

Test	NumDF	DenDF	F-Value	p-Value
Group	6	36	23.89	0.0001
Cond	6	36	4.50	0.0017
G×C	4	36	3.06	0.0219
H1	1	36	101.40	0.0001
H2	4	36	6.62	0.0004
H3	1	36	100.00	0.0001

using the MR model. While the two approaches are asymptotically equivalent, this is not the case for small samples. For the tests of H_1, H_2, H_3, both approaches are based upon approximate testing procedures. While we would again expect the results to be asymptotically equivalent, the two approaches give different results for small samples. For large samples, the advantage of the SUR approach for the analysis of extended linear hypothesis is that the tests may be performed using a SAS procedure and one does not need special IML code.

Exercises 5.5

1. Work problem 1, Exercises 4.15 using the CGMANOVA model.

5.6 The GMANOVA Model

a. *Overview*

In Chapter 4 we analyzed repeated measures data using the mixed ANOVA model and the MANOVA model. For both analyses, we had complete data and observations were gathered at regular intervals for all subjects. For trend analysis, orthogonal polynomials always included the degree equal to the number of time points. In this section, we review the generalized MANOVA, (GMANOVA) model, which allows one to fit polynomial growth curves (of low degree) to within-subjects trends and permits the analysis of incomplete data missing completely at random (MCAR). While the GMANOVA model is more flexible than the MANOVA model, it does not permit one to analyze trends with irregularly spaced intervals for each subject, it does not allows one to analyze individual growth curves, and it does not permit the analysis of data that are missing at random (MAR). In Chapter 6, we will discuss a more general model which does not have these limitations, for these reasons, our discussion of the GMANOVA model is intentionally brief.

The GMANOVA model was introduced by Potthoff and Roy (1964) to analyze mean growth using low degree polynomials. The GMANOVA model is written as

$$\underset{n\times p}{\mathbf{Y}} = \underset{n\times k}{\mathbf{X}} \underset{k\times q}{\mathbf{B}} \underset{q\times p}{\mathbf{P}} + \underset{n\times p}{\mathbf{E}} \tag{5.6.1}$$

where

Y is the data matrix,

X is the known design matrix of fixed variables,

B is a fixed unknown matrix of parameters,

P is the within-subject design matrix or rank $q \leq p$,

E is the random error matrix,

and the $n \geq p + q$ rows of **Y** are assumed to be MVN, independent observation vectors with common unknown covariance matrix Σ.

If $p = q$ and $\mathbf{P} = \mathbf{I}$, (5.6.1) reduces to the MR model or the MANOVA model. Or, if **P** is square and nonsingular, the GMANOVA model also reduces to the MR or MANOVA model since $E(\mathbf{Y}_o) = \mathbf{XB}$ for $\mathbf{Y}_o = \mathbf{YP}^{-1}$. If **P** is an orthogonal matrix, then $\mathbf{P}^{-1} = \mathbf{P}'$. The advantage of the GMANOVA model over the MANOVA model for the analysis of repeated measurement data is that q may be less than p which requires the estimation of fewer parameters and hence may result in a more efficient model.

b. Estimation and Hypothesis Testing

Estimation and inference using the GMANOVA model has been discussed by Rao (1965, 1966, 1967), Khatri (1966), Grizzle and Allen (1969) and Gleser and Olkin (1970). For a general overview see Kariya (1985) who considers the extended or restricted GMANOVA models of which the GMANOVA model is a special case.

Representing the GMANOVA model as a GLM given in (3.6.32), the GLS estimate of **B** is

$$\widehat{\mathbf{B}}_{\text{GLS}} = \left(\mathbf{X}'\mathbf{X}\right)^{-1}\mathbf{X}'\mathbf{Y}\Sigma^{-1}\mathbf{P}'(\mathbf{P}\Sigma^{-1}\mathbf{P}')^{-1} \tag{5.6.2}$$

if we let $\mathbf{A} \equiv \mathbf{X}$ and $\mathbf{Z} \equiv \mathbf{P}$, Potthoff and Roy (1964). Under normality, Khatri (1966) showed that the ML estimates of **B** and Σ are

$$\begin{aligned} \widehat{\mathbf{B}}_{\text{ML}} &= \left(\mathbf{X}'\mathbf{X}\right)^{-1}\mathbf{X}'\mathbf{Y}\mathbf{E}^{-1}\mathbf{P}'(\mathbf{P}\mathbf{E}^{-1}\mathbf{P}')^{-1} \\ \text{n}\,\widehat{\Sigma} &= (\mathbf{Y} - \mathbf{X}\widehat{\mathbf{B}}_{\text{ML}}\mathbf{P})'(\mathbf{Y} - \mathbf{X}\widehat{\mathbf{B}}_{\text{ML}}\mathbf{P}) \\ &= \mathbf{E} + \mathbf{W}'\mathbf{Y}'\mathbf{X}\left(\mathbf{X}'\mathbf{X}\right)^{-1}\mathbf{X}'\mathbf{Y}\mathbf{W} \end{aligned} \tag{5.6.3}$$

where

$$\begin{aligned} \mathbf{E} &= \mathbf{Y}'\left(\mathbf{I}_n - \mathbf{X}\left(\mathbf{X}'\mathbf{X}\right)^{-1}\mathbf{X}'\right)\mathbf{Y} \\ \mathbf{W} &= \mathbf{I}_p - \mathbf{E}^{-1}\mathbf{P}'(\mathbf{P}\mathbf{E}^{-1}\mathbf{P}')^{-1}\mathbf{P} \end{aligned}$$

Comparing (5.6.3) with (5.6.2), we see that the GLS estimate of **B** and the ML estimate have the same form, but they are not equal for the GMANOVA model. This was not the case for the MANOVA model.

Under the MR or MANOVA model $E(\mathbf{Y}) = \mathbf{XB}$, the distribution of $\text{vec}(\widehat{\mathbf{B}})$ is multivariate normal; for the GMANOVA model, this is not the case, Gleser and Olkin (1970). The covariance matrices for $\widehat{\boldsymbol{\beta}}_{\text{ML}} = \text{vec}(\widehat{\mathbf{B}}_{\text{ML}})$ and $\widehat{\boldsymbol{\beta}}_{\text{GLS}} = \text{vec}(\widehat{\mathbf{B}}_{\text{GLS}})$ are

$$\begin{aligned}\text{cov}(\widehat{\boldsymbol{\beta}}_{\text{ML}}) &= \left(\mathbf{P}\Sigma^{-1}\mathbf{P}'\right)^{-1} \otimes \frac{n-k-1}{n-k-(p-q)^{-1}} \left(\mathbf{X}'\mathbf{X}\right)^{-1} \\ \text{cov}(\widehat{\boldsymbol{\beta}}_{\text{GLS}}) &= \left(\mathbf{P}\Sigma^{-1}\mathbf{P}'\right)^{-1} \otimes \left(\mathbf{X}'\mathbf{X}\right)^{-1}\end{aligned} \tag{5.6.4}$$

so that the two are equal only if $p = q$, Rao (1967). Because Σ is unknown and must be estimated, the two estimates are approximately equal for large sample sizes.

To test the hypothesis $H\,(\mathbf{CBM} = \mathbf{0})$ under the GMANOVA model is also complicated. Khatri (1966) derived a likelihood ratio (LR) test using a conditional argument. Rao (1965, 1967) transformed the testing problem to a MANCOVA model and developed a conditional LR test. Gleser and Olkin (1970) showed that the Rao-Khatri test was an unconditional test using a canonical form of the model and a univariate argument. To test $H\,(\mathbf{CBM} = \mathbf{0})$ under the GMANOVA model, the model is reduced to an equivalent conditional MANCOVA model

$$E\left(\mathbf{Y}_o \mid \mathbf{Z}\right) = \mathbf{XB} + \mathbf{Z}\Gamma \tag{5.6.5}$$

where

$\mathbf{Y}_{o(n\times q)}$ is a transformed data matrix

$\mathbf{Z}_{n\times h}$ is a matrix of covariates of rank h

$\Gamma_{h\times q}$ is a matrix of unknown regression coefficients

and $\mathbf{X}$ and $\mathbf{B}$ remain defined as in (5.6.1).

To reduce the GMANOVA model to the MANCOVA model, two matrices $\mathbf{H}_1$ of order $p \times q$ with rank q and $\mathbf{H}_2$ of order $p \times (p-q)$ with rank $p-q$ are constructed such that $\mathbf{H}_{p\times p} = \left[\mathbf{H}_1, \mathbf{H}_2\right]$ and $\mathbf{PH}_1 = \mathbf{I}_q$ and $\mathbf{PH}_2 = \mathbf{0}_{p-q}$. Convenient choices for $\mathbf{H}_1$ and $\mathbf{H}_2$ are the matrices

$$\mathbf{H}_1 = \mathbf{P}'\left(\mathbf{PP}'\right)^{-1} \quad \text{and} \quad \mathbf{H}_2 = \mathbf{I} - \mathbf{PH}_1 \tag{5.6.6}$$

Letting $\mathbf{P}$ be a matrix of orthogonal polynomials so the $\mathbf{P}' = \mathbf{P}^{-1}$, then one may set $\mathbf{H}_1 = \mathbf{P}_1'$ where $\mathbf{P}_1$ is a matrix of normalized orthogonal polynomials representing the constant term and all polynomials up to degree $q-1$ and by letting $\mathbf{H}_2$ represent transpose of the remaining normalized rows of $\mathbf{P}$. For example, if

$$\mathbf{P}_{p\times p} = \left[\begin{array}{rrr} .577350 & .577350 & .577350 \\ -.707107 & 0 & .707107 \\ \hline .408248 & -.816497 & .408248 \end{array}\right]$$

then

$$\mathbf{P} = \begin{bmatrix} \mathbf{P}_{1_{q\times p}} \\ \mathbf{P}_{2_{(p-q)\times p}} \end{bmatrix} \tag{5.6.7}$$

so that $\mathbf{H}_1 = \mathbf{P}_1'$ and $\mathbf{H}_2 = \mathbf{P}_2'$. Setting $\mathbf{Y}_o = \mathbf{YH}_1$ and $\mathbf{Z} = \mathbf{YH}_2$, then the $E(\mathbf{Y}_o) = \mathbf{XBPH}_1 = \mathbf{XBI}_q = \mathbf{XB}$ and the $E(\mathbf{Z}) = E(\mathbf{YH}_2) = \mathbf{XBPH}_2 = \mathbf{XB0} = \mathbf{0}$. Thus, the expected value of $\mathbf{Y}_o$ given $\mathbf{Z}$ is seen to have the form given in (5.6.5), Grizzle and Allen (1969, p. 362). Associating model (5.6.5) with the MANCOVA model in (4.4.27) and using the fact that if $\mathbf{PH}_2 = \mathbf{0}$ results in the identity

$$\mathbf{H}_2\left(\mathbf{H}_2'\mathbf{EH}_2\right)^{-1}\mathbf{H}_2' = \mathbf{E}^{-1} - \mathbf{E}^{-1}\mathbf{P}'\left(\mathbf{PE}^{-1}\mathbf{P}'\right)^{-1}\mathbf{PE}^{-1} \tag{5.6.8}$$

It is easily shown that the BLUE or ML estimate of $\mathbf{B}$ for the MANCOVA model is

$$\widehat{\mathbf{B}} = \left(\mathbf{X}'\mathbf{X}\right)^{-1}\mathbf{X}'\mathbf{YE}^{-1}\mathbf{P}'\left(\mathbf{PE}^{-1}\mathbf{P}'\right)^{-1} \tag{5.6.9}$$

which is the ML estimate of $\mathbf{B}$ given in (5.6.3) for the GMANOVA model. Letting $v_e = n - r\,[\mathbf{X},\ \mathbf{Z}] = n-k-h = n-k-(p-q) = n-k-p+q$ and $\mathbf{S} = \mathbf{E}\,/\,v_e$, an alternative form for $\widehat{\mathbf{B}}$ in (5.6.9) is

$$\widehat{\mathbf{B}} = \left(\mathbf{X}'\mathbf{X}\right)^{-1}\mathbf{X}'\mathbf{YS}^{-1}\mathbf{P}'\left(\mathbf{PS}^{-1}\mathbf{P}'\right)^{-1} \tag{5.6.10}$$

where $E(\mathbf{S}) = \Sigma$ for the conditional model or the GMANOVA model. To test hypotheses regarding $\mathbf{B}$ in the GMANOVA model, we use the Rao-Khatri reduced MANCOVA model and the theory of Chapter 4. To test $H\ (\mathbf{CBM} = \mathbf{0})$, the hypothesis and error sum of squares and cross products (SSCP) matrices have the following form

$$\begin{aligned}\widetilde{\mathbf{E}} &= \mathbf{M}'(\mathbf{PE}^{-1}\mathbf{P}')^{-1}\mathbf{M}\\ \widetilde{\mathbf{H}} &= (\mathbf{C}\widehat{\mathbf{B}}\mathbf{M})'(\mathbf{CR}^{-1}\mathbf{C}')^{-1}(\mathbf{C}\widehat{\mathbf{B}}\mathbf{M})\end{aligned} \tag{5.6.11}$$

where

$$\begin{aligned}\mathbf{R}^{-1} &= \left(\mathbf{X}'\mathbf{X}\right)^{-1} + \left(\mathbf{X}'\mathbf{X}\right)^{-1}\mathbf{X}'\mathbf{Y}\left[\mathbf{E}^{-1} - \mathbf{E}^{-1}\mathbf{P}'\left(\mathbf{PE}^{-1}\mathbf{P}'\right)^{-1}\mathbf{PE}^{-1}\right]\mathbf{Y}'\mathbf{X}\left(\mathbf{X}'\mathbf{X}\right)^{-1}\\ \widehat{\mathbf{B}} &= \left(\mathbf{X}'\mathbf{X}\right)^{-1}\mathbf{X}'\mathbf{YS}^{-1}\mathbf{P}'(\mathbf{PS}^{-1}\mathbf{P}')^{-1}\\ \mathbf{E} &= \mathbf{Y}'[\mathbf{I} - \mathbf{X}\left(\mathbf{X}'\mathbf{X}\right)^{-1}\mathbf{X}']\mathbf{Y}\end{aligned}$$

and $\mathbf{S} = \mathbf{E}/v_e$, $v_e = n-k-p+q$, Timm and Mieczkowski (1997, p. 289). The degrees of freedom for the hypothesis is $v_h = r\,(\mathbf{C})$. With $\widetilde{\mathbf{H}}$ and $\widetilde{\mathbf{E}}$ defined in (5.6.11), the four test statistics based on the roots of $|\ \widetilde{\mathbf{H}} - \lambda\widetilde{\mathbf{E}}\ | = 0$ may be used to test $H : \mathbf{CBM} = \mathbf{0}$ and to establish simultaneous confidence intervals for contrasts $\psi = \mathbf{c}'\mathbf{Bm}$ for the GMANOVA model. Naik (1990) develops prediction intervals for the growth curve model.

There are, as discussed by Rao (1966) and Grizzle and Allen (1969), a few serious problems with the application of the GMANOVA model in practice. First, the estimate of $\mathbf{B}$ is only more efficient if the covariances between $\mathbf{Y}_o$ and $\mathbf{Z}$ are not zero. It is necessary to determine whether all or a best subset of the variables in $\mathbf{Z}$ should be included. Second, while inspecting the correlations between $\mathbf{Y}_o$ and $\mathbf{Z}$ may be helpful, preliminary tests on sample data may be required. Furthermore, the selected set of covariates may vary from sample to sample, while the best subset often depends on the structure of Σ. Third, when the number of treatment groups is greater than one, since the covariates are not determined prior to treatment, it is difficult to determine whether the covariates affect the treatment conditions. In addition, there is no reason to expect that the same set of covariates is "best" for all groups. Finally, because $\widetilde{\mathbf{H}}$ does not necessarily have a noncentral Wishart distribution when H is false, power calculation's for the GMANOVA model are not easily approximated, Fujikoshi (1974).

c. *Test of Fit*

When applying the GMANOVA model in practice, one usually has a large number of repeated measures, p, and we would like to determine for what $q < p$ is the GMANOVA model adequate. When $p = q$, the MANOVA model is appropriate. Grizzle and Allen (1969) developed a goodness-of-fit test for the GMANOVA model which tests the adequacy of modeling a polynomial of degree $q - 1$. Thus, the null hypothesis is that we have a GMANOVA model versus the alternative that the model is MANOVA.

To test the adequacy of the model fit to the data, we must choose the most appropriate set of covariates. Recall that for the GMANOVA model, we have that

$$
\begin{aligned}
E(\mathbf{Y}) &= \mathbf{XBP} = \mathbf{X}\Theta \\
E(\mathbf{Z}) &= \mathbf{XBPH}_2 = \mathbf{X}\Theta\mathbf{H}_2 = \mathbf{0}
\end{aligned}
\tag{5.6.12}
$$

where the $r(\mathbf{X}) = k$ and $\mathbf{X}\Theta\mathbf{H}_2 = \mathbf{0}$ if and only if $\Theta\mathbf{H}_2 = \mathbf{0}$. Hence we may evaluate the fit of the model by testing whether $E(\mathbf{Z}) = \mathbf{0}$ or equivalently that $H : \mathbf{I}_k\Theta\mathbf{H}_2 = \mathbf{0}$ for the model $E(\mathbf{Y}) = \mathbf{X}\Theta$. This is the MANOVA model so that the hypothesis and error matrices are

$$
\begin{aligned}
\mathbf{H} &= (\widehat{\Theta}\mathbf{H}_2)'(\mathbf{X}'\mathbf{X})\widehat{\Theta}\mathbf{H}_2 = \mathbf{P}_2\widehat{\Theta}'(\mathbf{X}'\mathbf{X})\widehat{\Theta}\mathbf{P}_2' \\
\mathbf{E} &= \mathbf{P}_2\mathbf{Y}'(\mathbf{I} - \mathbf{X}(\mathbf{X}'\mathbf{X})^{-1})\mathbf{X}'\mathbf{Y}\mathbf{P}_2'
\end{aligned}
\tag{5.6.13}
$$

where $u = r(\mathbf{P}_2) = p - q$, $v_h = r(\mathbf{C}) = k$, $v_e = n - k$ and $\widehat{\Theta} = (\mathbf{X}'\mathbf{X})^{-1}\mathbf{X}'\mathbf{Y}$. Assuming that a first degree polynomial is inadequate, a second degree polynomial is tested. This sequential testing process continues until an appropriate degree polynomial is found or until it is determined that a MANOVA model is preferred over the GMANOVA model. The GMANOVA model is sometimes called a weighted analysis since we are using the "information" in the covariates as weights. The strategy of continuously testing the model adequacy until one attains nonsignificance may greatly increase the overall Type I error for the study. Hence, it is best to have some knowledge about the degree of the polynomial growth curve.

d. *Subsets of Covariates*

In applying the GMANOVA model, we are assuming that the weighted analysis is appropriate. This may not be the case, especially if $\mathbf{Y}_o$ and $\mathbf{Z}$ are uncorrelated. $\mathbf{Y}_o$ and $\mathbf{Z}$ are independent if and only if $\mathbf{H}_1'\Sigma\mathbf{H}_2 = \mathbf{P}_1\Sigma\mathbf{P}_2' = \mathbf{0}$. This occurs if and only if Σ has Rao's mixed-model covariance structure

$$\Sigma = \mathbf{P}'\Gamma\mathbf{P} + \mathbf{Z}'\Delta\mathbf{Z} \tag{5.6.14}$$

where $\Gamma_{q\times q}$ and $\Delta_{(q-p)\times(q-p)}$ are p.d. matrices with $\mathbf{Z}' \equiv \mathbf{P}_2'$, Kariya (1985, p. 70). Given (5.6.14), the ML of $\mathbf{B}$ is

$$\widehat{\mathbf{B}}_{ML} = \left(\mathbf{X}'\mathbf{X}\right)^{-1}\mathbf{X}'\mathbf{Y}\mathbf{H}_1 = \left(\mathbf{X}'\mathbf{X}\right)^{-1}\mathbf{X}'\mathbf{Y}\mathbf{P}_1' \tag{5.6.15}$$

and is the BLUE, as in the MANOVA model. Note that this is equivalent to setting $\Sigma = \mathbf{I}$ in the weighted case so that the solution for $\widehat{\mathbf{B}}$ in (5.6.15) is sometimes called the unweighted estimate. The unweighted estimate is equal to the weighted estimate if and only if Σ is a member of the class of matrices defined by (5.6.14). Rao's covariance structure occurs naturally in univariate mixed models. A LR test for Rao's covariance structure is equivalent to testing that $\mathbf{Y}_o = \mathbf{Y}\mathbf{P}_1'$ and $\mathbf{Z} = \mathbf{Y}\mathbf{P}_2'$ are independent, Kariya (1985, p. 184). Letting

$$\begin{aligned}\mathbf{V} &= \mathbf{P}\mathbf{Y}'[\mathbf{I} - \mathbf{X}\left(\mathbf{X}'\mathbf{X}\right)^{-1}\mathbf{X}']\mathbf{Y}\mathbf{P}' \\ &= \begin{bmatrix} \mathbf{V}_{11} & \mathbf{V}_{12} \\ \mathbf{V}_{21} & \mathbf{V}_{22} \end{bmatrix}\end{aligned}$$

for $\mathbf{P}$ partitioned as in (5.6.7), we have following the test of independence that the LR test for evaluating Rao's covariance structure is

$$\begin{aligned}\Lambda &= \frac{|\mathbf{V}|}{|\mathbf{V}_{11}|\,|\mathbf{V}_{22}|} = \frac{|\mathbf{V}_{22}|\left|\mathbf{V}_{11} - \mathbf{V}_{12}\mathbf{V}_{22}^{-1}\mathbf{V}_{21}\right|}{|\mathbf{V}_{11}|\,|\mathbf{V}_{22}|} \\ &= \left|\mathbf{I} - \mathbf{V}_{11}^{-1}\mathbf{V}_{12}\mathbf{V}_{22}^{-1}\mathbf{V}_{21}\right|\end{aligned} \tag{5.6.16}$$

and $\Lambda \sim U_{(q,\ q-p,\ n-k)}$. For any LR test, Bartlett (1947) showed that if $\Lambda \sim U_{(p,\ v_h,\ v_e)}$ that

$$X_B^2 = -[v_e - (p - v_h + 1)/2] \ln \Lambda \mathrel{\dot\sim} \chi^2_{(pv_h)} \tag{5.6.17}$$

Hence, we may use the chi-square distribution to test for Rao's covariance structure. For any LR test, the approximation is adequate for most problems and if $p^2 + v_h^2 \le f/3$ where $f = v_e - (p + v_h + 1)/2$ the approximation is accurate to three decimal places, Anderson (1984). Hence

$$-[(n-k) - (p+1)/2] \ln \Lambda \mathrel{\dot\sim} \chi^2_{q(q-p)} \tag{5.6.18}$$

may be used to test that Σ has Rao's covariance structure.

Estimates that use some subset of covariates are called partially weighted estimates, Rao (1965, 1967) and Grizzle and Allen (1969). However, they are not very useful in most GMANOVA applications, Vonesh and Chinchilli (1997, p. 217).

e. *GMANOVA vs SUR*

To represent the GMANOVA model as a SUR model, (5.2.9) is used. We let $\mathbf{y}^* = \text{vec}\left(\mathbf{Y}'\right)$ and $\mathbf{A} = \mathbf{X}_{n\times k} \otimes \mathbf{P}'_{p\times q}$ where the $r(\mathbf{X}) = k$ and the $r(\mathbf{P}) = q$. Letting $\boldsymbol{\theta}' = [\boldsymbol{\beta}'_1, \boldsymbol{\beta}'_2, \ldots, \boldsymbol{\beta}'_k]$ and $\mathbf{e}' = [\mathbf{e}_1, \mathbf{e}_2, \ldots, \mathbf{e}_n]$, the GLSE of $\boldsymbol{\theta}$ by (5.2.12) is

$$
\begin{aligned}
\widehat{\boldsymbol{\theta}} &= \left[\mathbf{A}'\left(\mathbf{I}_n \otimes \Sigma\right)^{-1}\mathbf{A}\right]^{-1}\mathbf{A}'\left(\mathbf{I}_n \otimes \Sigma\right)^{-1}\mathbf{y}^* \\
&= \left[\left(\mathbf{X}' \otimes \mathbf{P}\right)\left(\mathbf{I}_n \otimes \Sigma^{-1}\right)\left(\mathbf{X} \otimes \mathbf{P}'\right)\right]^{-1}\left(\mathbf{X}' \otimes \mathbf{P}\right)\left(\mathbf{I}_n \otimes \Sigma^{-1}\right)\mathbf{y}^* \\
&= \left[\left(\mathbf{X}'\mathbf{X}\right) \otimes \left(\mathbf{P}\Sigma^{-1}\mathbf{P}'\right)\right]^{-1}\left(\mathbf{X}' \otimes \mathbf{P}\Sigma^{-1}\right)\mathbf{y}^* \\
&\left[\left(\mathbf{X}'\mathbf{X}\right)^{-1}\mathbf{X}' \otimes \left(\mathbf{P}\Sigma^{-1}\mathbf{P}'\right)^{-1}\mathbf{P}\Sigma^{-1}\right]\mathbf{y}^*
\end{aligned} \tag{5.6.19}
$$

Replacing Σ by $\mathbf{S}$ where $E(\mathbf{S}) = \Sigma$, we see that the ML estimate for $\mathbf{B}$ in the GMANOVA model is equal to the FGLS estimate of $\mathbf{B}$ for the SUR model. Hence, we may use either model to estimate $\mathbf{B}$. For the GMANOVA model, we were able to obtain a LR test of $\mathbf{CBM} = \mathbf{0}$ using the Rao-Khatri reduction. For more complex models, the distribution of the LR test may be very complicated; in these cases, the Wald statistic for the SUR model provides an alternative, as illustrated later in this Chapter and by Timm (1997).

f. *Missing Data*

In MR models and in particular in GMANOVA studies which involve obtaining repeated measurements over time at some sequence of fixed intervals for all subjects, one often encounters missing data. With missing data, one still needs to estimate parameters and test hypotheses. To solve this problem, numerous approaches have been recommended, Little (1992) and Schafer (2000). In particular, one may use some imputation techniques to estimate the missing values and then obtain estimates and tests using the complete data making adjustments in degrees of freedom of test statistics for estimated data. Because single imputation methods do not take into account the uncertainty in estimating the missing values, the resulting estimated variances of the parameter estimates tend to be biased toward zero, Rubin (1987, p. 13). One may use procedures that estimate parameters and incomplete data iteratively from complete data using Markov chain Monte Carlo multiple imputation methods (including Gibbs sampling, data augmentation, monotone data augmentation, Metropolis algorithm, and sampling-importance resampling), likelihood methods, asymptotic GLS methods and Bayesian methods. These latter methods are usually superior to single step imputation methods; however, they are very computer intensive. A very good discussion of multiple imputation methods for missing multivariate data may be found in Schafer and Olsen (1998). For any of these techniques, one must be concerned about the missing data response pattern. To address these problems, Little and Rubin (1987) characterized three missing data response patterns:

1. The pattern of missing responses is independent of both the observed and missing data.

2. The pattern of missing responses is independent of the missing data and only dependent on the observed data.

3. The pattern of missing responses is dependent on the missing data (the effect that the observed data has on the pattern is ignored).

In these situations, we say that the missing data are either (1) missing completely at random (MCAR), (2) missing at random (MAR), or (3) Neither. In situations (1) and (2), we say that the missing data mechanism is "nonignorable". When situation (3) occurs, there is no solution to the missing data problem. In the former case, the MCAR assumption is stronger than the MAR assumption. Under the MCAR assumption, one may use GLS and non-likelihood methods to obtain estimates and tests. For the weaker MAR assumption, one usually uses likelihood or Bayesian methods assuming normality or multinomial data models. If one uses non-likelihood estimates when data are only MAR, estimates for the standard error of estimated parameters are difficult to obtain, Schafer (2000). A test for whether data are MCAR or MAR has been proposed by Little (1988) and Park and Davis (1993).

A procedure for analyzing missing data for the GMANOVA model which assumes data are MCAR has been proposed by Kleinbaum (1973) which uses FGLS estimates and asymptotic theory to develop tests. The method is discussed and illustrated by Timm and Mieczkowski (1997) who provide SAS code for the method. Likelihood methods which assume data are MAR are discussed in Chapter 6. A multiple imputation procedure, PROC MI, is under development by SAS. Also under development is the PROC MIANALYZE that combines the parameter estimates calculated for each imputation data set (using any standard SAS procedure) for making valid univariate and multivariate inferences for model parameters that take into account the uncertainty for data MAR or MCAR. Both procedures will be available in Release 8e of the SAS System and should be in production with Release 9.0.

Because the GMANOVA model is a regression procedure, one may also employ diagnostic methods to evaluate influential observations, Kish and Chinchilli (1990) and Walker (1993). For a discussion of nonparametric procedures, see Puri and Sen (1985).

5.7 GMANOVA Example

When fitting polynomials to data using the simple multiple linear regression model $\mathbf{y} = \mathbf{X}\boldsymbol{\beta} + \mathbf{e}$, the design matrix is usually a nonorthogonal matrix. For example, for three time points

$$\mathbf{X} = \begin{bmatrix} 1 & 1 & 1 \\ 1 & 2 & 4 \\ 1 & 3 & 9 \end{bmatrix} = \begin{bmatrix} \mathbf{x}_1, & \mathbf{x}_2, & \mathbf{x}_3 \end{bmatrix} \tag{5.7.1}$$

Using the Gram-Schmidt process on the columns of $\mathbf{X}$, then

$$\mathbf{q}_i = \mathbf{x}_i - \sum_{j=1}^{i-1} c_{ij}\mathbf{q}_j$$

where $c_{ij} = \mathbf{x}_i'\mathbf{q}_j/\|\mathbf{q}_j\|^2$, so that

$$\mathbf{Q} = \begin{bmatrix} 1 & -1 & 1 \\ 1 & 0 & 2 \\ 1 & 1 & -1 \end{bmatrix} \quad \text{and} \quad \mathbf{Q}'\mathbf{Q} = \begin{bmatrix} 3 & 0 & 0 \\ 0 & 2 & 0 \\ 0 & 0 & 6 \end{bmatrix} \tag{5.7.2}$$

The matrix $\mathbf{Q}'\mathbf{Q}$ is a diagonal matrix with the squared lengths of each column of $\mathbf{Q}$ as diagonal elements. To normalize $\mathbf{Q}$, the columns of $\mathbf{Q}$ are divided by the square root of the diagonal element of $\mathbf{Q}'\mathbf{Q}$. Then

$$\mathbf{Q}/\|\mathbf{Q}'\mathbf{Q}\| = \mathbf{P} = \begin{bmatrix} 1/\sqrt{3} & -1/\sqrt{2} & 1/\sqrt{6} \\ 1/\sqrt{3} & 0 & -2/\sqrt{6} \\ 1/\sqrt{3} & 1/\sqrt{2} & 1/\sqrt{6} \end{bmatrix} \tag{5.7.3}$$

and $\mathbf{P}'\mathbf{P} = \mathbf{I}$ so that $\mathbf{P}$ is an orthogonal matrix, $\mathbf{P}' = \mathbf{P}^{-1}$.

When fitting a model to $\mathbf{y}$ using $\mathbf{X}$, the model has the form $y = \beta_0 + \beta_1 x + \beta_2 x^2$. Transforming the matrix $\mathbf{X}$ to $\mathbf{Q}$ a matrix $\mathbf{G}^{-1}$ is formed so that $\mathbf{Q} = \mathbf{X}\mathbf{G}^{-1}$ and $\boldsymbol{\xi} = \mathbf{G}\boldsymbol{\beta}$. The transformed model has the form $y = \xi_0 + \xi_1 (x - 2) + \xi_2 \left(3x^2 - 12x + 10\right)$ for $x = 1, 2,$ and 3 as in (5.7.1). Evaluating the linear and quadratic polynomials at $x = 1, 2,$ and 3, the second and third columns of $\mathbf{Q}$ are obtained. The parameter ξ_0 is associated with the constant term. The coefficients in $\mathbf{Q}$ are the familiar constant, linear, and quadratic trend contrasts in models involving the analysis of mean trends in ANOVA. Normalizing the polynomial contrasts, one obtains estimates of the orthonormalized coefficients ξ_0', ξ_1', and ξ_2'. The matrix $\mathbf{P}$ in the GMANOVA model usually employs an orthogonal matrix $\mathbf{P}$ of rank $q < p$ in the product $\mathbf{XBP}$. To obtain the matrix $\mathbf{P}$ using SAS, the function ORPOL is used in the IML procedure. In some experimental situations, data are collected at unequal intervals. Thus, one needs to construct orthogonal polynomial contrasts to analyze trends for unequally spaced data. The function ORPOL also permits one to construct polynomials for the analysis of unequally spaced data.

a. *One Group Design (Example 5.7.1)*

While testing hypotheses for growth curve models is complicated, estimation of model parameters is straightforward. This is because the ML estimate of $\mathbf{B}$ using the GMANOVA model is identical to the SUR model estimator by replacing Σ with the unbiased estimator $\mathbf{S}$. To illustrate the analysis of growth data, we analyze the Elston and Grizzle (1962) ramus bone data given in Table 3.7.3, discussed in Example 3.7.3. These data represent the growth of ramus bone lengths of 20 boys at the ages 8, 8.5, 9, and 9.5. The data are given in ramus.dat and analyzed using program m5_7_1.sas with the outliers.

When fitting a polynomial to growth data, one begins by plotting the means at each time point to visually estimate the lowest degree of the polynomial that appears to capture the mean trend in the data. Using the PROC SUMMARY, the means are calculated, output, and plotted. Because these are four time points, one may fit polynomials that represent constant, linear, quadratic, or cubic trend. The order of the polynomial is 0, 1, 2, or 3. The maximum order is one less than the number of observed time points. From the plot of the means, a linear model appears to adequately model the growth trend for ramus heights.

An informal lack-of-fit test for the adequacy of a linear model may be obtained using PROC GLM and the REPEATED statement. The procedure will sequentially fit polynomial contrasts to the means. Choosing $\alpha = 0.05$ and dividing α by $q - 1$, one fits polynomial contrasts of increasing order to the means. As q increases from $q = 0$ to $q - 1$ where q is the number of time points, one retains polynomials for all significant orders and stops with the highest degree of nonsignificance. Setting $\alpha = 0.05$, $\alpha^* = 0.05/\,(q - 1) = 0.05/3 = 0.0267$. Since the second degree polynomial has a p-value of 0.6750, we conclude that a linear trend is adequate for the ramus data.

Assuming a linear model for our example, the growth curve model has the structure for $E\,(\mathbf{Y}) = \mathbf{XBP}$ as

$$\underset{20\times 4}{E\,(\mathbf{Y})} = \underset{20\times 1}{\mathbf{1}}\ \underset{1\times 2}{[\beta_0, \beta_1]}\ \underset{2\times 4}{\mathbf{P}} \tag{5.7.4}$$

where

$$\mathbf{P}' = \begin{bmatrix} 0.5 & -0.670820 \\ 0.5 & -0.223609 \\ 0.5 & 0.223607 \\ 0.5 & 0.670820 \end{bmatrix} \tag{5.7.5}$$

so that $q < p$.

To estimate $\mathbf{B}$ for our example, we use formula (5.6.20) for the SUR model using PROC IML. Later we estimate $\widehat{\mathbf{B}}$ using (5.6.9) using PROC GLM to illustrate the equivalence of the models. The estimate of $\widehat{\mathbf{B}}$ is

$$\widehat{\mathbf{B}}' = \begin{bmatrix} \beta_0 \\ \beta_1 \end{bmatrix} = \begin{bmatrix} 100.09916 \\ 2.081342 \end{bmatrix} \tag{5.7.6}$$

Alternatively, using the matrix

$$\mathbf{Q}' = \begin{bmatrix} 1 & -1 \\ 1 & -1 \\ 1 & 1 \\ 1 & 3 \end{bmatrix} \tag{5.7.7}$$

the parameter estimate for the regression coefficients is

$$\widehat{\Xi}' = \begin{bmatrix} 50.0496 \\ 0.4654 \end{bmatrix} \tag{5.7.8}$$

To obtain $\widehat{\Xi}$ from $\widehat{\mathbf{B}}$, one divides 100.09916 of $\widehat{\mathbf{B}}'$ by $\sqrt{4} = 2$ and 2.08134 by the $\sqrt{20} = 4.472$, the square root of the diagonal elements of $\mathbf{QQ}'$ in (5.7.7).

To test for model fit using $\mathbf{H}$ and $\mathbf{E}$ defined in (5.6.13), PROC GLM is used with the transformed variables $yt3$ and $yt4$. These variables are created from the 4×4 matrix of orthogonal polynomials in the PROC IML routine. The matrix $\mathbf{H}_1 = \mathbf{P}'$ in (5.7.5), and $\mathbf{H}_2$ is associated with the quadratic and cubic normalized polynomials. The matrix of covariates $\mathbf{Z} = \mathbf{YH}_2$ is defined as the variables $yt3$ and $yt4$. To evaluate fit, we are testing that $E\,(\mathbf{Z}) = \mathbf{0}$. With nonsignificance, the higher order terms are zero and the model is said to fit. For our example, the p-value for the test-of-fit for a linear model is 0.9096 indicating that the 2nd and 3rd degree polynomials do not significantly contribute to variation in the means and are considered nonsignificant. For these data, the formal test of fit is consistent with the ad hoc sequential tests of trend using orthogonal polynomials.

To test hypotheses regarding **B** for the GMANOVA model, we use PROC GLM for the MANCOVA model (5.6.5) with the transformed variables. The MANOVA statement is used to test hypotheses using $\widetilde{\mathbf{E}}$ and $\widetilde{\mathbf{H}}$ given in (5.6.11). One may also use PROG REG, and the MTEST statement or one may use the SYSLIN procedure and the TEST and STEST statements. For the procedures GLM and REG, the tests are exact likelihood ratio tests. For the SYSLIN procedure, the F tests are approximate. For the ramus height data, all procedures indicate that the regression coefficients for the linear model are nonzero. Using the estimated standard errors for the elements of **B** and the MANCOVA model, one may contrast confidence intervals for population parameters $\psi = \mathbf{c}'\mathbf{B}\mathbf{m}$ as illustrated in program m4_5_2.sas.

b. *Two Group Design (Example 5.7.2)*

In the ramus height example, a growth curve model was fit to a single group. In many applications, one obtains growth data for observations in several independent groups. The data usually involve repeated measurements data in which trends are fit to all groups simultaneously and tests involving the coefficient matrix **B** for each group are tested for parallelism or coincidence. To illustrate the process of fitting a GMANOVA model to several independent groups. Edward's data are again used. For this example, we have three independent Edwardsgroups and three time points. Program m5_7_2.sas is used to analyze the data.

As with the one group analysis, we begin by plotting the means for each group. The plots and the informal test of polynomial trend indicate that a linear model is appropriate for modeling the trend in means. The p-value for quadratic trend is 0.1284. The GMANOVA model for Edward's data is Edwards

$$E(\mathbf{Y}) = \mathbf{X}\mathbf{B}\mathbf{P}_1$$

where

$$\mathbf{X} = \begin{bmatrix} \mathbf{1}_5 & \mathbf{0} & \mathbf{0} \\ \mathbf{0} & \mathbf{1}_5 & \mathbf{0} \\ \mathbf{0} & \mathbf{0} & \mathbf{1}_5 \end{bmatrix}, \quad \mathbf{B} = \begin{bmatrix} \beta_{10} & \beta_{11} \\ \beta_{20} & \beta_{21} \\ \beta_{30} & \beta_{31} \end{bmatrix} \tag{5.7.9}$$

To reduce the GMANOVA model to a MANCOVA model, $\mathbf{Y}_o = \mathbf{Y}\mathbf{P}_1'$ where

$$\mathbf{P}_1' = \begin{bmatrix} 0.577350 & -0.707107 \\ 0.577350 & 0 \\ 0.577350 & 0.707107 \end{bmatrix} \tag{5.7.10}$$

and the model

$$E(\mathbf{Y}_o) = \mathbf{X}\mathbf{B} + \mathbf{Z}\Gamma$$

is analyzed with $\mathbf{Z} = \mathbf{Y}\mathbf{P}_2'$

$$\mathbf{P}_2' = \begin{bmatrix} 0.577350 \\ -.816497 \\ 0.408248 \end{bmatrix} \tag{5.7.11}$$

For the MANCOVA model, letting $\mathbf{Q}^* = \mathbf{I} - \mathbf{X}\left(\mathbf{X}'\mathbf{X}\right)^{-1}\mathbf{X}'$, $\widehat{\Gamma} = \left(\mathbf{Z}'\mathbf{Q}^*\mathbf{Z}\right)^{-1}\mathbf{Z}'\mathbf{X}\mathbf{Y}_o$ and

$$\begin{aligned}\widehat{\mathbf{B}} &= \left(\mathbf{X}'\mathbf{X}\right)^{-1}\mathbf{X}'\mathbf{Y}_o - \left(\mathbf{X}'\mathbf{X}\right)^{-1}\mathbf{X}_1\mathbf{Z}\widehat{\Gamma} \\ &= \left(\mathbf{X}'\mathbf{X}\right)^{-1}\mathbf{X}'\mathbf{Y}\mathbf{S}^{-1}\mathbf{P}_1'\left(\mathbf{P}_1\mathbf{S}^{-1}\mathbf{P}_1'\right)^{-1} \\ &= \begin{bmatrix} 12.1244 & 4.2464 \\ 17.6669 & 2.5456 \\ 12.1235 & 2.8228 \end{bmatrix}\end{aligned} \tag{5.7.12}$$

by (5.6.10) and associating $\mathbf{P} \equiv \mathbf{P}_1$. In program m5_7_2.sas, $\widehat{\mathbf{B}}$ is estimated using the procedures GLM and SYSLIN by using the transformed data matrix $\mathbf{Y}_o = \mathbf{Y}\mathbf{P}_1'$.

Before testing hypotheses using the linear growth curve model, one usually performs the test-of-fit or model adequacy. The test that the quadratic term is zero has a p-value of 0.1643. Hence a linear model appears appropriate.

To test hypotheses regaining the parameter matrix $\mathbf{B}$ in the GMANOVA model, the MANCOVA model $E\left(\mathbf{Y}_o\right) = \mathbf{X}\mathbf{B} + \mathbf{Z}\Gamma$ in used. Thus, one may test hypotheses of the form $\mathbf{CBM} = \mathbf{0}$ by selecting matrices $\mathbf{C}$ and $\mathbf{M}$ as illustrated in Section 4.5. For growth curve models, one may be interested in testing that the growth curves are the same for the groups for all parameters. For our example, the test is

$$H_C : \begin{bmatrix} \beta_{10} \\ \beta_{11} \end{bmatrix} = \begin{bmatrix} \beta_{20} \\ \beta_{21} \end{bmatrix} = \begin{bmatrix} \beta_{30} \\ \beta_{31} \end{bmatrix} \tag{5.7.13}$$

The hypothesis test matrix for testing H_C is

$$\mathbf{C} = \begin{bmatrix} 1 & 0 & -1 \\ 0 & 1 & -1 \end{bmatrix}$$

where $\mathbf{M} = \mathbf{I}_2$. Another test of interest may be that the growth curves for the groups are parallel. This test is for the example is

$$H_P : \beta_{11} = \beta_{21} = \beta_{31}$$

For the test of parallelism,

$$\mathbf{C} = \begin{bmatrix} 1 & 0 & -1 \\ 0 & 1 & -1 \end{bmatrix} \mathbf{M} = \begin{bmatrix} 0 & 1 \end{bmatrix}$$

Using the LFR model and Wilks' Λ criterion, $\Lambda = 0.3421$ with p-value 0.0241 for the test of H_C and $\Lambda = 0.6648$ with p-value 0.1059 for the test of H_P. Thus, the profiles are parallel, but not coincident.

Also included in program m5_7_2.sas is the code for evaluating $\mathbf{B}$ using the procedure SYSLIN. The approximate F tests of coincidence and parallelism are also provided. Again, one may establish confidence sets for the coefficients in $\mathbf{B}$.

Exercises 5.7

1. Reanalyze the ramus bone length data with the outliers removes. Discuss you findings.

2. The data in file Danford.dat represent radiation dosage levels for four groups of patients for four treatment levels (column one in the file). The next four columns in a design matrix and the next column is a baseline measurement. Using the next five variables in the dataset which represent posttreatment measurements for five days (columns with the first value of (223, 242, 248, 266, and 274), fit a GMANOVA model to the data and evaluate model fit.

3. Lee (1970) gives data for two dependent variables (time on target in seconds and the number of hits on target and five traits to investigate bilateral transfer of are reminiscence of teaching performance under four treatment conditions: (1) distributed practice on a linear circular-tracking task, (2) distributed practice on a nonlinear hexagonal-tracking task, (3) mass practiced on a linear circular-tracking task, and (4) massed practice on a nonlinear hexagonal-tracking task. Subjects, randomly assigned to each group performed on the given task under each condition with one hand for ten traits and then transferred to the other hand for the same number of traits after a prescribed interval. The two sets of measurements taken for the ten traits were blocked into five block of two trials and averaged, yielding five repeated measures for each dependent variable. The data obtained for groups 1 and 2 are given below. Fit a growth curve to each group. Are they the same? Discuss your findings.

Group 1	Time on Target See				
	1	2	3	4	5
1	13.95	12.00	14.20	14.40	13.00
2	18.13	22.60	19.30	18.25	20.45
3	19.65	21.60	19.70	19.55	21.00
4	20.80	21.15	21.25	21.25	20.00
5	17.80	20.00	20.00	19.80	18.20
6	17.35	20.85	20.85	20.10	20.70
7	16.15	16.70	19.25	16.30	18.35
8	19.10	18.35	22.95	22.70	22.65
9	12.05	15.40	14.75	13.45	11.60
10	8.55	9.00	9.10	10.50	9.55
11	7.15	5.85	6.10	7.05	9.15
12	17.85	17.95	19.05	18.40	16.85
13	14.50	17.70	16.00	17.40	17.10
14	22.30	22.30	21.90	21.65	21.45
15	19.70	19.25	19.85	18.00	17.80
16	13.25	17.40	18.75	18.40	18.80

Group 1	Hits on Target				
	1	2	3	4	5
1	31.50	17.50	36.50	35.50	34.00
2	22.50	12.00	17.50	19.00	16.50
3	18.50	18.00	21.50	18.50	14.501
4	20.50	18.50	17.00	16.50	16.50
5	29.00	21.00	19.00	23.00	21.00
6	22.00	15.50	18.00	18.00	22.50
7	36.00	29.50	22.00	26.00	25.50
8	18.00	9.50	10.50	10.50	14.50
9	28.00	30.50	17.50	31.50	28.00
10	36.00	37.00	36.00	36.00	33.00
11	13.50	32.00	33.00	32.50	36.50
12	23.00	26.00	20.00	21.50	30.00
13	31.00	31.50	33.00	26.00	29.50
14	16.00	14.00	16.00	19.50	18.00
15	32.00	25.50	24.00	30.00	26.50
16	21.50	24.00	22.00	20.50	21.50

5.8 Tests of Nonadditivity

In a two-way MANOVA design with one observation per cell, one may want to evaluate whether a linear model is additive

$$\mathbf{y}_{ij} = \boldsymbol{\mu} + \boldsymbol{\alpha}_i + \boldsymbol{\beta}_j + \mathbf{e}_{ij} \tag{5.8.1}$$

or nonadditive

$$\mathbf{y}_{ij} = \boldsymbol{\mu} + \boldsymbol{\alpha}_i + \boldsymbol{\beta}_j + \boldsymbol{\gamma}_{ij} + \mathbf{e}_{ij} \tag{5.8.2}$$

However, there are no degrees of freedom to test for no interaction. To develop a test of no interaction, McDonald (1972) extended the test of Tukey (1949) following Milliken and Graybill (1971) to the multivariate case.

In extending the test to the multivariate case, McDonald (1972) replaced $\boldsymbol{\gamma}_{ij}$ in (5.8.2) with the same function of $\boldsymbol{\alpha}_i$ and $\boldsymbol{\beta}_j$ a variable at a time. For example, $\boldsymbol{\gamma}_{ij} = \boldsymbol{\alpha}_i \odot \boldsymbol{\beta}_j$. Writing $\boldsymbol{\gamma}_{ij} = \gamma_k(\boldsymbol{\alpha}_i \odot \boldsymbol{\beta}_j)$, one may test $H : \gamma_k = 0$ for $k = 1, 2, \ldots, p$ to determine whether an additive model would be satisfactory for a single variable.

Thus, McDonald (1972) proposed an "expanded" SUR model of the form

$$\begin{aligned} \mathbf{y}_j &= [\mathbf{X}, \mathbf{F}_j] \begin{bmatrix} \boldsymbol{\beta}_j^* \\ \boldsymbol{\gamma}_j \end{bmatrix} + \mathbf{e}_j \\ \operatorname{cov}(\mathbf{y}_j, \mathbf{y}_{j'}) &= \sigma_{jj'} \mathbf{I}_n \end{aligned} \tag{5.8.3}$$

for each variable $j = 1, 2, \ldots, p$ where $\mathbf{F}_j$ are known $(n \times k_j)$ matrix functions of $\mathbf{X}\boldsymbol{\beta}_j^*$. Observe that this is not a linear model since $\mathbf{F}_j$ is a function of $\mathbf{X}\boldsymbol{\beta}_j^*$. However, suppose we assume that $\widehat{\mathbf{F}}_j = \mathbf{X}\widehat{\boldsymbol{\beta}}_j^*$ and that $\widehat{\mathbf{F}}_j$ is known. Then testing $H : \boldsymbol{\gamma}_j = \mathbf{0}$ or no interaction is equivalent to testing for no covariates in the ANCOVA model.

Continuing with this approach, we write (5.8.3) as a multivariate model. Then (5.8.3) becomes

$$
\begin{aligned}
E(\underset{n\times p}{\mathbf{Y}}) &= \underset{n\times q}{\mathbf{X}}\ \underset{q\times p}{\mathbf{B}} + \left[\mathbf{F}_1, \mathbf{F}_2, \ldots, \mathbf{F}_p\right] \begin{bmatrix} \boldsymbol{\gamma}_1 & \mathbf{0} & \cdots & \mathbf{0} \\ \mathbf{0} & \boldsymbol{\gamma}_2 & \cdots & \mathbf{0} \\ \vdots & \vdots & & \vdots \\ \mathbf{0} & \mathbf{0} & \cdots & \boldsymbol{\gamma}_p \end{bmatrix} \\
&= \underset{n\times q}{\mathbf{X}}\ \underset{q\times p}{\mathbf{B}} + \underset{n\times h}{\mathbf{F}}\ \underset{h\times p}{\Gamma}
\end{aligned} \tag{5.8.4}
$$

where $\mathbf{F}_j$ is an $n \times k_j$ matrix of known constants, $h = \sum_j k_j$, $r(\mathbf{X}) = r \le q$ and the rank $[\mathbf{X},\ \mathbf{F}]$ is $r + h < n$ in the vector space spanned by the rows of $\mathbf{XB}$ and $\mathbf{F}$.

By assuming each row of $\mathbf{Y}$ is sampled from a p-variate MVN with covariance matrix Σ, the interaction hypothesis becomes

$$H : \boldsymbol{\gamma}_j = \mathbf{0} \quad \text{for } j = 1, 2, \ldots, p$$

substituting $\widehat{\mathbf{F}} = \mathbf{X}\widehat{\mathbf{B}}$ in model (5.8.4), we have that $\widehat{\Gamma} = (\widehat{\mathbf{F}}'\mathbf{Q}\widehat{\mathbf{F}})^{-1}\mathbf{F}'\mathbf{QY}$ for $\mathbf{Q} = \mathbf{I} - \mathbf{X}\left(\mathbf{X}'\mathbf{X}\right)^{-}\mathbf{X}'$. Thus, testing for nonadditivity is equivalent to testing $H : \Gamma = \mathbf{0}$. From the MANCOVA model, the hypothesis and error matrices for the test of nonadditivity are

$$
\begin{aligned}
\mathbf{H} &= \widehat{\Gamma}'(\widehat{\mathbf{F}}'\mathbf{Q}\widehat{\mathbf{F}})\widehat{\Gamma} \\
&= \mathbf{Y}'\mathbf{Q}\widehat{\mathbf{F}}(\widehat{\mathbf{F}}'\mathbf{Q}\widehat{\mathbf{F}})^{-1}\widehat{\mathbf{F}}'\mathbf{QY} \\
\mathbf{E} &= \mathbf{Y}'\mathbf{QY} - \widehat{\Gamma}'(\widehat{\mathbf{F}}'\mathbf{Q}\widehat{\mathbf{F}})\widehat{\Gamma}
\end{aligned} \tag{5.8.5}
$$

where $v_h = h$ and $v_e = n - r - h$. Hence, to perform the test of nonadditivity, one estimates $\mathbf{F}$ using $\widehat{\mathbf{F}} = \mathbf{X}\widehat{\mathbf{B}}$. Assuming $\widehat{\mathbf{F}}$ is a matrix of known covariates, one tests the hypothesis $H : \Gamma = \mathbf{0}$ using the MANCOVA model. Given the structure of $\mathbf{F}$ and Γ, the matrices involving $\mathbf{F}$ have the structure $\mathbf{F} = \left[\mathbf{F}_1\boldsymbol{\gamma}_1, \ldots, \mathbf{F}_p\boldsymbol{\gamma}_p\right]$ so that $\mathbf{H}$ is partitioned into h matrices $\mathbf{H}_i$ with one degree of freedom. Each $\mathbf{H}_i$ with $\mathbf{E}$ is used to test that $\boldsymbol{\gamma}_i = \mathbf{0}$ for the p variables simultaneously.

To illustrate, consider a two-way design with one observation per cell where $i = 1, 2, \ldots, a$ a levels of factor A and $j = 1, 2, \ldots, b$ levels of factor B. Suppose $\mathbf{F} = [\mathbf{f}_1, \mathbf{f}_2, \ldots, \mathbf{f}_p]$ where $\mathbf{f}_k = \gamma_k\left(\boldsymbol{\alpha}_i \odot \boldsymbol{\beta}_j\right)$ for $k = 1, 2, \ldots, p$. Then $\widehat{\mathbf{f}}_k = \gamma_k\left(\widehat{\boldsymbol{\alpha}}_i \odot \widehat{\boldsymbol{\beta}}_j\right)$ and $v_h \equiv h = p$ and $v_e = n - r - h = ab - (a + b - 1) - p = ab - a - b + 1 - p$ since the $r(\mathbf{X}) = a + b - 1$. To test $H : \Gamma = \mathbf{0}$, $\mathbf{H}$ and $\mathbf{E}$ in (5.8.5) are evaluated. If one has covariates in the two-way design, they may be incorporated into the design matrix $\mathbf{X}$. The rank of $\mathbf{X}$ is increased by the number of independent covariates.

The test of nonadditivity developed by McDonald (1972) and Kshirsagar (1988, 1993) assume multivariate normality. Khattree and Naik (1990) develop a likelihood ratio test that only requires that each observation follows a multivariate elliptical symmetric distribution.

5.9 Testing for Nonadditivity Example

Designs with one observation per cell are common in experimental design. In particular, randomized block designs usually have one observation per cell. For the analysis of such designs, one usually assumes a model with no interaction. Tukey (1949) developed a test for nonadditivity which was generalized by Scheffé (1959, p. 144, prob. 4.19). Milliken and Graybill (1970, 1971) extended the test in a very general manner by considering an extended linear model (ELM) which allows one to analyze certain nonlinear models in the parameters using a linear model. Their tests are special cases of the test developed by McDonald (1972). Setting $p = 1$ in the general case, we have the ELM

$$\begin{aligned} \underset{n\times 1}{\mathbf{y}} &= \underset{n\times q}{\mathbf{X}}\ \underset{q\times 1}{\boldsymbol{\beta}} + \underset{n\times h}{\mathbf{F}_1}\ \underset{h\times 1}{\boldsymbol{\gamma}_1} + \underset{n\times 1}{\mathbf{e}} \\ \mathbf{e} &\sim IN\left(\mathbf{0}, \sigma^2\mathbf{I}\right) \end{aligned} \tag{5.9.1}$$

where $\mathbf{F}_1 = \left[f_{ij}\left(\cdot\right)\right]$ is a matrix of known function of the unknown elements of $\mathbf{X}\boldsymbol{\beta}$, the $r\left(\mathbf{X}\right) = r$, and the $r\left[\mathbf{X}_1\mathbf{F}_1\right] = r + h < n$.

For a simple univariate randomized block design with two blocks and three treatments, the design given in (5.9.1) may be written as

$$\underset{(6\times 1)}{\begin{bmatrix} y_{11} \\ y_{12} \\ y_{21} \\ y_{22} \\ y_{31} \\ y_{32} \end{bmatrix}} = \underset{(6\times 1)}{\begin{bmatrix} 1 & 1 & 0 & 0 & 1 & 0 \\ 1 & 1 & 0 & 0 & 0 & 1 \\ 1 & 0 & 1 & 0 & 1 & 0 \\ 1 & 0 & 1 & 0 & 0 & 1 \\ 1 & 0 & 0 & 1 & 1 & 0 \\ 1 & 0 & 0 & 1 & 0 & 1 \end{bmatrix}} \underset{(6\times 1)}{\begin{bmatrix} \mu \\ \alpha_1 \\ \alpha_2 \\ \alpha_3 \\ \beta_1 \\ \beta_2 \end{bmatrix}} + \underset{(6\times 1)}{\begin{bmatrix} \alpha_1\beta_1 \\ \alpha_1\beta_2 \\ \alpha_2\beta_1 \\ \alpha_2\beta_2 \\ \alpha_3\beta_1 \\ \alpha_3\beta_2 \end{bmatrix}} \underset{1\times 1}{\gamma} + \underset{(6\times 1)}{\begin{bmatrix} e_{11} \\ e_{12} \\ e_{21} \\ e_{22} \\ e_{31} \\ e_{12} \end{bmatrix}} \tag{5.9.2}$$

which represents the model

$$y_{ij} = \mu + \alpha_i + \beta_j + \alpha_i\beta_j + e_{ij}, e_{ij} \sim IN\left(0, \sigma^2\right)$$

Then a test of $H\left(\gamma = 0\right)$ may be used to assess the significance of the nonlinear parameter $\alpha_i\beta_j$. If $\gamma = 0$, the additive model may be satisfactory. For the ELM given in (5.9.1), the test of interaction becomes $H : \boldsymbol{\gamma} = \mathbf{0}$ for h covariates.

If $\mathbf{F}_1$ was known, we see that (5.9.1) is identical to the ANCOVA model. However, $\mathbf{F}_1$ is unknown and is a function of $\mathbf{X}\widehat{\boldsymbol{\beta}}$. Recalling that $\widehat{\boldsymbol{\beta}} = \left(\mathbf{X}'\mathbf{X}\right)^{-}\mathbf{X}'\mathbf{y}$ for any g-inverse $\left(\mathbf{X}'\mathbf{X}\right)^{-}$ and since $\mathbf{F}_1 = [f_{ij}(\mathbf{X}\widehat{\boldsymbol{\beta}}_j)]$, we see $\mathbf{F}_1$ does not depend on $\left(\mathbf{X}'\mathbf{X}\right)^{-}$. This follows from the fact that $\mathbf{X}\widehat{\boldsymbol{\beta}} = \mathbf{X}\left(\mathbf{X}'\mathbf{X}\right)^{-}\mathbf{X}'\mathbf{y} = \mathbf{P}\mathbf{y}$ is unique because $\mathbf{P}$ is a projection matrix. Replacing $\mathbf{F}$ in (5.8.5) with $\widehat{\mathbf{F}}_1$, an F-statistic for testing $H : \boldsymbol{\gamma} = \mathbf{0}$ is

$$F = \frac{H/h}{E/\left(n - r - h\right)} \sim F\left(h, n - r - h\right) \tag{5.9.3}$$

For the univariate randomized block design, this reduces to Tukey's test of additivity for product type interactions. Hence, to perform Tukey's test of additivity, one estimates $\mathbf{F}_1$ and employs the analysis of covariance model to test that the covariate is zero.

For the design given in (5.9.2), we selected a product interaction. Adding constraints to the linear model, $\mathbf{F}_1 = \left[f_{ij}\right]$ has the familiar interaction form $f_{ij} = (y_{i.} - y_{..})\left(y_{.j} - y_{..}\right)$. Or, removing the constants, a more convenient form for f_{ij} is $f_{ij} = y_{i.}y_{.j}$. Again, this is a product "interaction". The general model may be used to test for other functions of the elements of $\mathbf{X}\widehat{\boldsymbol{\beta}}$.

As an example of the univariate procedure, we consider the data provided in Problem 1, Exercise 6.6. The experiment analyzes the number of trials to criterion for a learning experiment for five-letter (F) words and seven-letter (S) words. To test for no interaction assuming all factors in the design are fixed, Tukey's test is performed for each variable individually. For the interaction functions, we set $f_{ij} = y_{i.}y_{.j}$ for each variable. Program m5_9_1.sas is used to perform the tests of nonadditivity using PROC GLM. The data are given in the datafile add.dat.

For each variable, we test that a single covariate is zero. For the five-letter variable F, the covariate is ZF. The test that the coefficient associated with ZF is zero has p-value 0.2777. For the seven letter variable S the p-value for the covariate ZS is 0.7873. Thus, we conclude that the model is additive for each variable individually.

To illustrate the multivariate test of nonadditivity, the data for the multivariate randomized block design given in Problem 2, Exercise 6.6, are used. The experiment involves two factors: four levels of SES and four ethnic groups. The pilot study is used to analyze three dependent variables: Mathematics (Mat), English (Eng) and General Knowledge (GK). The assumed model for the design is $\mathbf{y}_{ij} = \boldsymbol{\mu} + \boldsymbol{\alpha}_i + \boldsymbol{\beta}_j + \mathbf{e}_{ij}$ where all factors are fixed. Letting $\mathbf{F}_k = \gamma_k\left(\boldsymbol{\alpha}_i \odot \boldsymbol{\beta}_j\right)$ $k = 1, 2, 3$ the general model has the structure

$$E(\mathbf{Y}) = \mathbf{XB} + \mathbf{F}\Gamma \tag{5.9.4}$$

for $\mathbf{F} = [\mathbf{F}_1, \mathbf{F}_2, \mathbf{F}_3]$. Letting $\widehat{\mathbf{F}}_k = (\mathbf{y}_{i.} \odot \mathbf{y}_{.j})$, we observe that $v_e = n - r(X) - h = 16-7-3 = 6$. Program m5_9_1.sas is used to perform the multivariate test of nonadditivity using the MANCOVA. The data are given in file add2.dat.

As we saw with the analysis of the MANCOVA in Examples 4.5.2 and 4.9.2, PROC GLM does not provide a joint test that the matrix of covariates are zero. It performs tests for each covariate individually. For this example, $s = \min(v_h, p) = 1$, $M = (\mid p-h \mid -1)/2 = 0.5$ and $N = (v_e - p - 1)/2 = 1$. For the test of nonadditivity, the p-value for the three tests are 0.2175, 0.1299, and 0.6935 for each covariate. These tests indicate that the model is additive. To evaluate the joint test that $\boldsymbol{\gamma}_1 = \boldsymbol{\gamma}_2 = \boldsymbol{\gamma}_3 = 0$, the procedure REG is used. However, the design matrix must be reparameterized to full rank using PROC TRANSREG. For the joint test, $s = 3$, $M = -0.5$, and $N = 1$. The p-values for the overall joint test differ for each of the multivariate criteria. The p-values for Wilks', BLT, and the BNP criteria are: 0.1577, 0.1486, 0.2768. The p-value for Roy's test criterion is 0.0205. These results do not clearly indicate that the model is additive since one of the multivariate criteria rejects the joint test of additivity. However, because the majority of the criteria lead to a nonsignificant result and the p-value for Roy's test is not significant at the 0.01 level, we conclude that the model is additive.

Exercises 5.9

1. Show that H and E in (5.9.3) have the form

$$H = \frac{\left[\sum_i \sum_j y_{ij} (y_{i.} - y_{..}) \left(y_{.j} - y_{..}\right)\right]^2}{\sum_i (y_{i.} - y_{..})^2 \sum_j \left(y_{.j} - y_{..}\right)^2}$$

$$E = \sum_i \sum_j \left(y_{.j} - y_{..} - y_{.j} + y_{..}\right)^2 - H$$

and that $v_e = (a - q)(b - 1) - 1$ where a and b denote the number of blocks and treatments, respectively.

2. Perform the test of multivariate nonadditivity using the data in Exercises 4.9, Problem 1.

3. Perform the test of multivariate nonadditivity using the data in Exercises 6.6, Problem 2.

5.10 Lack of Fit Test

A major disadvantage of the MR model is that the design matrix is common for all variables. Hence, when fitting a polynomial function to a set of dependent variables, many variables may be overfit. Performing the multivariate lack of fit test may lead to underfitting a subset of variables. Levy and Neill (1990) develop multivariate lack of fit tests for the SUR model.

To develop a lack of fit test for the SUR model, we use representation (5.2.1) to test

$$H : E(\mathbf{Y}) = \mathbf{XB} \tag{5.10.1}$$

where $\mathbf{B}$ is created to fit different sets of independent variables to each dependent variable or different order polynomials. Following the multivariate lack of fit test, the pooled error matrix

$$\mathbf{E} = \mathbf{Y}'(\mathbf{I}_n - \mathbf{X}\left(\mathbf{X}'\mathbf{X}\right)^- \mathbf{X}')\mathbf{Y} \tag{5.10.2}$$

is partitioned into two independent components, pure error and lack of fit error written as $\mathbf{E}_{PE}$ and $\mathbf{E}_{LF}$, respectively. The pure error component is an estimate of Σ based upon the within or replicated observations and $\mathbf{E}_{LF}$ is obtained from $\mathbf{E}$ by subtraction, $\mathbf{E}_{LF} = \mathbf{E} - \mathbf{E}_{PE}$.

To compute $\mathbf{E}_{PE}$, suppose there are c different replicated rows in $\mathbf{X}_j$ each occurring n_i times where $i = 1, 2, \ldots, c$ and $1 \leq c \leq n$. Grouping the rows together, let $k = \sum_{i=1}^{c} n_i$ be the first rows of $\mathbf{X}$. Then to calculate $\mathbf{E}_{PE}$, we may use the centering matrix

$$\mathbf{K} = \left[\mathbf{C}_1, \mathbf{C}_2, \ldots, \mathbf{C}_c, \mathbf{0}_{n \times k}\right] \tag{5.10.3}$$

where $\mathbf{C}_i = \mathbf{I}_{n_i} - \mathbf{J}_{n_i}/n_i$ for $i = 1, 2, \ldots, c$ so that

$$\mathbf{E}_{PE} = \mathbf{Y}'\mathbf{K}\mathbf{Y} \equiv \mathbf{Y}'\mathbf{Y} - \mathbf{Y}'\mathbf{V}^{-1}\mathbf{Y} \tag{5.10.4}$$

for $\mathbf{V}^{-1} = \text{diag}\,[n_1, n_2, \ldots, n_c, \mathbf{0}]$ with degrees of freedom in $v_{PE} = k - c \geq p$. The matrix $\mathbf{E}_{LF}$ has degrees of freedom $v_{LF} = n - q^* - v_{PE} \geq p$ where $q^* = r\,(\mathbf{X})$.

To test for lack of fit for the SUR model, we solve

$$\mid \mathbf{E}_{LF} - \lambda \mathbf{E}_{PE} \mid = 0 \tag{5.10.5}$$

and use any of the standard criteria where

$$\begin{aligned} s &= \min\,(v_{\text{LF}}, p) \\ M &= (\mid v_{LF} - p \mid -1)\,/2 \\ N &= (v_{PE} - p - 1)\,/2 \end{aligned}$$

following Khuri (1985). Alternative test statistics have been proposed by Levy and Neill (1990). The simulation study by Levy and Neill (1990) show that the Bartlett-Lawley-Hotelling test statistic is most stable.

5.11 Sum of Profile Designs

Frequently with the analysis of complete repeated measurement data, one wants to analyze treatment effects using low order polynomials with covariates that may be modeled differently or with covariates that may be changing with time. To create a model allowing one to model treatment effects with low order polynomials and baseline covariates, Chinchilli and Elswick (1985) proposed a mixed MANOVA-GMANOVA model of the form

$$E\,(\mathbf{Y} \mid \mathbf{Z}) = \mathbf{XBP} + \mathbf{Z}\Gamma \tag{5.11.1}$$

They were able to obtain ML estimates for the model parameters $\mathbf{B}$, Γ and Σ, and developed likelihood ratio tests for the three hypotheses assuming a MVN distribution

$$\begin{aligned} H_1 &: \mathbf{C}_1\mathbf{B}\mathbf{M}_1 = \mathbf{0} \qquad (5.11.2) \\ H_2 &: \mathbf{C}_2\Gamma = \mathbf{0} \\ H_3 &: H_1 \cap H_2 = \mathbf{0} \end{aligned}$$

They also developed goodness of fit tests for the model comparing it to the GMANOVA and MANOVA models. Timm and Mieczkowski (1997) developed SAS code for the LR and goodness of fit tests for the MANOVA-GMANOVA model.

Patel (1986) considered a model similar to the mixed MANOVA-GMANOVA model that did not allow for polynomial growth, but did permit covariates that change with time. Patel's model has the general structure

$$E\,(\mathbf{Y}) = \mathbf{XB} + \sum_{j=1}^{r} \mathbf{Z}_j \Gamma_j \tag{5.11.3}$$

While Patel was unable to obtain a closed form solution for the ML estimates of Σ, $\mathbf{B}$, and Γ_j, he obtained estimates that converge to ML estimates. He also developed asymptotic LR tests of (5.11.2) under his model.

Verbyla and Venables (1988) extended Patel's model to a sum of GMANOVA profiles model of the form

$$E(\underset{n\times p}{\mathbf{Y}}) = \sum_{i=1}^{r} \mathbf{X}_i\mathbf{B}_i\mathbf{M}_i \tag{5.11.4}$$

called a sum of profiles model. This model has no closed form ML solution for Σ and $\mathbf{B}_i$. Hence, Verbyla and Venables represent (5.11.4) as a SUR model to estimate model parameters. von Rosen (1989) developed necessary and sufficient conditions for the sum of profiles model in (5.12.4) to have a closed form solution. Except for the simple structure considered by Chinchilli and Elswick (1985), his result involves nested design matrices which have limited practical application. In addition, the distribution of the LR statistics is complicated and would have to be approximated using, for example, Bartlett's chi-square approximation.

To estimate parameters and to test hypotheses for the sum of profiles model, we follow the approach recommended by Verbyla and Venables (1988). Given a consistent estimate of Σ, we use Hecker's (1987) CGMANOVA model which we saw is a general SUR model to study models of the form given in (5.11.4). Recalling that for $\mathbf{Y} = \mathbf{ABC}$, $\mathbf{Y}' = \mathbf{C}'\mathbf{B}'\mathbf{A}'$ so that $\text{vec}\left(\mathbf{Y}'\right) = \left(\mathbf{A}\otimes\mathbf{C}'\right)\text{vec}\left(\mathbf{B}'\right)$, we vectorize $\mathbf{Y}$ and $\mathbf{B}$ row-wise so that $\mathbf{y}^* = \text{vec}\left(\mathbf{Y}'\right)$ and $\boldsymbol{\beta}_i^* = \text{vec}\left(\mathbf{B}_i'\right)$, then (5.11.4) may be written as a CGMANOVA/SUR model

$$\begin{aligned} E\left(\mathbf{y}^*\right) &= \left[\left(\mathbf{X}_1\otimes\mathbf{M}_1'\right), \ldots, \left(\mathbf{X}_r\otimes\mathbf{M}_r'\right)\right] \begin{bmatrix} \boldsymbol{\beta}_1^* \\ \boldsymbol{\beta}_2^* \\ \vdots \\ \boldsymbol{\beta}_r^* \end{bmatrix} \\ &= \mathbf{A}\boldsymbol{\theta} \end{aligned} \tag{5.11.5}$$

Using (5.2.12) to estimate $\boldsymbol{\theta}$ with $\widehat{\widehat{\boldsymbol{\theta}}}$ where $\widehat{\Sigma}$ is defined in (5.2.13), tests of the form H : $\mathbf{C}\boldsymbol{\theta} = \boldsymbol{\theta}_o$ are easily tested using the Wald statistic defined in (5.2.14).

The CGMANOVA/SUR model may be used to estimate parameters and to test hypotheses in any multivariate fixed effects design for which LR tests may be difficult to obtain. One merely vectorizes the model, obtains a consistent estimate of Σ and applies Wald's statistic. To illustrate, we consider the multivariate SUR (MSUR) model.

5.12 The Multivariate SUR (MSUR) Model

In (5.2.1), we developed the SUR model as p correlated regression models. Replacing each column vector of $\mathbf{Y}_{n\times p}$ by a matrix $\mathbf{Y}_i$ of order $(n \times p_i)$, the MSUR model has the general form

$$E\left[\mathbf{Y}_1, \mathbf{Y}_2, \ldots, \mathbf{Y}_q\right] = \left[\mathbf{X}_1\Theta_{11}, \mathbf{X}_2\Theta_{22}, \ldots, \mathbf{X}_q\Theta_{qq}\right]$$

$$= \left[\mathbf{X}_1, \mathbf{X}_2, \ldots, \mathbf{X}_q\right] \begin{bmatrix} \Theta_{11} & \mathbf{0} & \cdots & \mathbf{0} \\ \mathbf{0} & \Theta_{22} & \cdots & \mathbf{0} \\ \vdots & \vdots & & \vdots \\ \mathbf{0} & \mathbf{0} & \cdots & \Theta_{qq} \end{bmatrix}$$

$$E\left(\mathbf{Y}\right) = \mathbf{X}\Theta \tag{5.12.1}$$

where $\mathbf{X}_i$ has order $(n \times k_i)$, Θ_{ii} has order $(n \times p_i)$ and the cov $\left(\mathbf{Y}_i, \mathbf{Y}_j\right) = \Sigma_{ij}$ is of order $\left(p_i \times p_j\right)$, $\Sigma = \left[\Sigma_{ij}\right]$ is of order $p \times p$ where $p = \sum_{i=1}^{q} p_i$. For the jth matrix $\mathbf{Y}_j$ in $\mathbf{Y} = \left[\mathbf{Y}_1, \mathbf{Y}_2, \ldots, \mathbf{Y}_q\right]$, we have that $E\left(\mathbf{Y}_j\right) = \mathbf{X}_j\Theta_{jj}$ where the cov $\left(\mathbf{Y}_j\right) = \Sigma_{jj}$. If $p_1 = p_2 = \ldots = p_q = p_o$ (say), then (5.12.1) has the simple form

$$E\begin{bmatrix} \mathbf{Y}_1 \\ \mathbf{Y}_2 \\ \vdots \\ \mathbf{Y}_q \end{bmatrix} = \begin{bmatrix} \mathbf{X}_1 & \mathbf{0} & \cdots & \mathbf{0} \\ \mathbf{0} & \mathbf{X}_2 & \cdots & \mathbf{0} \\ \vdots & \vdots & & \vdots \\ \mathbf{0} & \mathbf{0} & \cdots & \mathbf{X}_q \end{bmatrix} \begin{bmatrix} \Theta_{11} \\ \Theta_{22} \\ \vdots \\ \Theta_{qq} \end{bmatrix}$$

a matrix generalization of (5.2.2).

To represent (5.12.1) as a GLM, we let

$$\mathbf{U} = \left[\mathbf{E}_1, \mathbf{E}_2, \ldots, \mathbf{E}_q\right] \qquad \mathbf{U} = \left[e_{i1}, e_{i2}, \ldots, e_{ip_i}\right]$$
$$\mathbf{Y} = \left[\mathbf{Y}_1, \mathbf{Y}_2, \ldots, \mathbf{Y}_q\right] \qquad \mathbf{Y}_i = \left[y_{i1}, y_{i2}, \ldots, y_{ip_i}\right]$$
$$\Theta = \left[\Theta_{11}, \Theta_{22}, \ldots, \Theta_{qq}\right] \qquad \Theta_{jj} = \left[\Theta_{jj1}, \Theta_{jj2}, \ldots, \Theta_{jjp_j}\right]$$

where

$$\mathbf{e} = \text{vec}\left(\mathbf{U}\right) = \left[\text{vec}\left(\mathbf{U}_1\right), \ldots, \text{vec}\left(\mathbf{U}_q\right)\right]$$
$$\mathbf{y} = \text{vec}\left(\mathbf{Y}\right) = \left[\text{vec}\left(\mathbf{Y}_1\right), \ldots, \text{vec}\left(\mathbf{Y}_q\right)\right]$$
$$\boldsymbol{\theta} = \text{vec}\,\Theta = \left[\text{vec}\left(\Theta_{11}\right), \ldots, \text{vec}\left(\Theta_{qq}\right)\right]$$

so that $\mathbf{y}_{ij}$ is the jth column of $\mathbf{Y}_i$ and θ_{jjt} is the tth column of Θ_{jj}, then (5.12.1) becomes

$$\mathbf{y} = \mathbf{D}\boldsymbol{\theta} + \mathbf{e} \tag{5.12.2}$$
$$\mathbf{D} = \text{diag}\left[\left(\mathbf{I}_{p_1} \otimes \mathbf{X}_1\right), \ldots, \left(\mathbf{I}_{p_q} \otimes \mathbf{X}_q\right)\right]$$

and diag $[\mathbf{A}_i]$ is a block diagonal matrix with $\mathbf{A}_i = \left(\mathbf{I}_{p_i} \otimes \mathbf{X}_i\right)$. The ML estimate of $\boldsymbol{\theta}$ is

$$\widehat{\boldsymbol{\theta}} = \left[\mathbf{D}'\left(\Sigma \otimes \mathbf{I}_n\right)^{-1}\mathbf{D}\right]^{-1} \mathbf{D}'\left(\Sigma \otimes \mathbf{I}_n\right)^{-1}\mathbf{y} \tag{5.12.3}$$

A FGLS estimate of $\boldsymbol{\theta}$ is

$$\widehat{\widehat{\boldsymbol{\theta}}} = \left[\mathbf{D}'(\widehat{\Sigma} \otimes \mathbf{I}_n)^{-1}\mathbf{D}\right]^{-1} \mathbf{D}'\left(\widehat{\Sigma} \otimes \mathbf{I}_n\right)^{-1}\mathbf{y} \tag{5.12.4}$$

where $\widehat{\Sigma} \xrightarrow{p} \Sigma$. Following (5.2.4), we may $\widehat{\Sigma} = [s_{ij}]$ with elements

$$s_{ij} = \frac{1}{n - q_{ij}} \left[\mathbf{Y}_i' (\mathbf{I}_n - \mathbf{X}_i \left(\mathbf{X}_i' \mathbf{X}_i \right)^{-1} \mathbf{X}_i') (\mathbf{I}_n - \mathbf{X}_j (\mathbf{X}_j \mathbf{X}_j')^{-1} \mathbf{X}_j') \mathbf{Y}_j \right] \tag{5.12.5}$$

where $q_{ij} = 0$ or

$$q_{ij} = \text{tr} \left[(\mathbf{I}_n - \mathbf{X}_i \left(\mathbf{X}_i' \mathbf{X}_i \right)^{-1} \mathbf{X}_i') (\mathbf{I}_n - \mathbf{X}_j (\mathbf{X}_j' \mathbf{X}_j)^{-1} \mathbf{X}_j') \right]$$

Since (5.13.2) is a GLM, the Wald statistic

$$W = (\mathbf{C}\widehat{\widehat{\boldsymbol{\theta}}} - \boldsymbol{\theta}_o)' \{ \mathbf{C} [\mathbf{D}' (\widehat{\Sigma} \otimes \mathbf{I}_n)^{-1} \mathbf{D}]^{-1} \mathbf{C}' \}^{-1} (\mathbf{C}\widehat{\widehat{\boldsymbol{\theta}}} - \boldsymbol{\theta}_o) \tag{5.12.6}$$

may be used to test

$$H : \mathbf{C}\boldsymbol{\theta} = \boldsymbol{\theta}_o \tag{5.12.7}$$

where the $r(\mathbf{C}) = v_h$ and $\mathbf{D}$ is defined in (5.12.2).

For the MSUR model, observations were stacked column-wise, for the CGMANOVA model, observations were stated row-wise. The SUR model may be used for either representation, depending on the application. For growth curve models, it is convenient to stack vectors row-wise. This is illustrated with an example in the following section.

5.13 Sum of Profile Example

To illustrate the CGMANOVA/SUR model, data analyzed by Chinchilli and Elswick (1985) and Timm and Mieczkowski (1997, p. 404) from Danford et al. (1960) are utilized. The data set is given in the file Danford.dat. It consists of 45 patients with cancerous lesions who were subjected to whole-body x-radiation. The radiation dosage was at four levels, a control level with six patients and three treatment levels with 14, 15, and 10 patients. A baseline measurement was taken at day 0 (pretreatment), and then measurements were taken daily for ten consecutive days (posttreatment). For our example, only the first five measurements are used. The first variable in the file represents the control and three treatment levels of radiation (1, 2, 3, 4); the next four columns (values 0 or 1) represent a full rank design matrix $\mathbf{X}$ without the covariate; the next column is the baseline covariate; the next columns $(y1 - y10)$ denote the 10 posttreatment measurements; and the last column is set to one so that a common model may be easily fit to the entire group of patients. Program m5_13_1.sas contains the code for the analysis.

Because the data set has a baseline variable, one may not fit a growth curve model to the data since the growth curve model does not permit covariates. A natural model for the data is the MANOVA-GMANOVA model given in (5.11.1). The structure for the model is

$$E(\underset{45\times5}{\mathbf{Y}}) = \underset{45\times4}{\mathbf{X}} \; \underset{4\times q}{\mathbf{B}} \; \underset{q\times5}{\mathbf{P}} + \underset{54\times1}{\mathbf{Z}} \; \underset{1\times5}{\Gamma} \tag{5.13.1}$$

Again, one must decide upon the degree of the polynomial $q \leq p$ in the matrix $\mathbf{P}$. Ignoring the covariate, and plotting the five time points, a linear model appears appropriate. Thus, we consider fitting the model

$$\underset{45\times5}{E(\mathbf{Y})} = \underset{45\times4}{\mathbf{X}} \; \underset{4\times2}{\mathbf{B}} \; \underset{2\times5}{\mathbf{P}} + \underset{54\times1}{\mathbf{Z}} \; \underset{1\times5}{\Gamma} \tag{5.13.2}$$

to the data.

While the MANOVA-GMANOVA model is appropriate for these data and has a closed form solution, the more general sum of profiles model has no simple likelihood ratio test solution since ML estimates of model parameters are difficult to obtain. The more flexible solution is to represent the model as a CGMANOVA/SUR model using (5.11.5). To evaluate model fit using the SUR model approach, one may fit two nested models and use a chi-square or F-like statistic to compare the two models. Tests for comparing two nonnested models are developed in the next section.

To transform the model given in (5.13.2) to a SUR model, we use (5.11.5) with $\mathbf{M}_1' = \mathbf{P}$ and $\mathbf{M}_3' = \mathbf{I}$. For the 45 patients, the repeated measurement row-vectors are stacked columnwise to form the linear model

$$\begin{array}{ccccccc} \mathbf{y} & = & \mathbf{A} & \boldsymbol{\theta} & + & \mathbf{e} \\ \begin{bmatrix} \mathbf{y}_1 \\ \mathbf{y}_2 \\ \vdots \\ \mathbf{y}_{45} \end{bmatrix} & = & \left[\mathbf{X}\otimes\mathbf{P}', \mathbf{Z}\otimes\mathbf{I}_5\right] & \begin{bmatrix} \beta_{11} \\ \beta_{12} \\ \beta_{21} \\ \beta_{22} \\ \beta_{31} \\ \beta_{32} \\ \beta_{41} \\ \beta_{42} \\ \gamma_1 \\ \gamma_2 \\ \gamma_3 \\ \gamma_4 \\ \gamma_5 \end{bmatrix} & + & \begin{bmatrix} \mathbf{e}_1 \\ \mathbf{e}_2 \\ \vdots \\ \mathbf{e}_{45} \end{bmatrix} \end{array} \tag{5.13.3}$$

for the example. For each patient, we have the model

$$\underset{5\times1}{\mathbf{y}_i} = \underset{5\times13}{\mathbf{A}} \; \underset{13\times1}{\boldsymbol{\theta}} + \underset{5\times1}{\mathbf{e}} \qquad i = 1, 2, \ldots, 45 \tag{5.13.4}$$

where $\mathbf{A}_i$ are sets of five rows of the matrix $\mathbf{A} = \left(\mathbf{X}\otimes\mathbf{P}', \mathbf{Z}\otimes\mathbf{I}_5\right)$.

In program m5_13_1.sas we estimate $\boldsymbol{\theta}$ using formula (5.2.12). As a convenient estimate of Σ, we use the naive estimate

$$\widehat{\Sigma} = \mathbf{Y}'(\mathbf{I} - \mathbf{1}\left(\mathbf{1}'\mathbf{1}\right)^{-1}\mathbf{1}')\mathbf{Y}/n$$

based upon the within-patient variation. To test hypotheses, the Wald statistic in (5.2.14) is used.

From the output, the matrices $\mathbf{B}$ and Γ are reconstructed from $\boldsymbol{\theta}$. The estimates for the parameter matrices are

$$\mathbf{B}' = \begin{bmatrix} 148.00 & 107.71 & 141.46 & 132.80 \\ 32.04 & 18.07 & 15.82 & 5.57 \end{bmatrix}$$

$$\Gamma = [\ 0.780, \quad 0.883, \quad 0.921, \quad 0.922, \quad 0.988\ ]$$

The ML estimates for $\mathbf{B}'$ and Γ' obtained by Chinchilli and Elswick (1985) are

$$\mathbf{B}'_{ML} = \begin{bmatrix} 150.89 & 102.80 & 139.87 & 132.19 \\ 32.39 & 17.27 & 15.33 & 5.36 \end{bmatrix}$$

$$\Gamma_M = [\ 0.782 \quad 0.887 \quad 0.925 \quad 0.928 \quad 0.994\ ]$$

as calculated in Timm and Mieczkowski (1977, p. 411). Comparing the two estimators, we have close agreement for the two approaches, even using a naive "plug in" estimate for Σ. We may also use the SYSLIN procedure to fit the model, however, it is rather cumbersome since it involves 45 equations.

To test hypotheses, we use formula (5.2.14). To illustrate the method we evaluate whether the profiles are coincident using (5.13.3). For $\boldsymbol{\theta}$ in (5.13.3), the matrix $\mathbf{C}$ has the form

$$\mathbf{C} = \begin{bmatrix} 1 & 0 & 0 & 0 & 0 & 0 & -1 & 0 & 0 & 0 & 0 & 0 & 0 \\ 0 & 1 & 0 & 0 & 0 & 0 & 0 & -1 & 0 & 0 & 0 & 0 & 0 \\ 0 & 0 & 1 & 0 & 0 & 0 & -1 & 0 & 0 & 0 & 0 & 0 & 0 \\ 0 & 0 & 0 & 1 & 0 & 0 & 0 & -1 & 0 & 0 & 0 & 0 & 0 \\ 0 & 0 & 0 & 0 & 1 & 0 & -1 & 0 & 0 & 0 & 0 & 0 & 0 \\ 0 & 0 & 0 & 0 & 0 & 1 & 0 & -1 & 0 & 0 & 0 & 0 & 0 \end{bmatrix}$$

where the $r(\mathbf{C}) = 6$. The p-value for the test of coincidence is 0.5731 so that we conclude coincident profiles. The exact LR test has a p-value of 0.34.

To test that the covariate is nonzero, we may test $H : \Gamma = \mathbf{0}$. Then, $\mathbf{C} = [\mathbf{0}_5, \mathbf{I}_5]$ with rank 5. Because the p-value for the test is 0.0001, we conclude that the baseline measurement should be included in the model. The exact likelihood ratio test of $H : \Gamma = \mathbf{0}$ involves a matrix inverted Dirichlet distribution which is difficult to evaluate. Using the sum of two chi-square distributions, Timm and Mieczkowski (1997, p. 353) suggest an approximation. Using the approximation for the exact likelihood ratio test, the estimated p-value is 0.0123. Thus, we conclude using either approach to retain the baseline measurement.

Because the SUR model provides the researcher with the analysis of complex linear models when exact likelihood ratio test procedures may be difficult to implement, we strongly recommend its use. Timm (1997) provides additional examples for the approach.

Exercises 5.13

1. The Danford data set contains one outlier, remove the observation and rerun program m5_13_1. Discuss your findings.

2. Using $n = 45$ subjects in the file Danford.data with all 10 consecutive posttreatment measurements, fit a GMANOVA-MANOVA model group using a SUR model.

5.14 Testing Model Specification in SUR Models

When fitting a SUR model to a data set, one may use the Wald statistic or an approximate F statistic to test that some subset of the variables should be excluded from the model. To apply these statistics, the requirement for the procedure is that the alternative model must be nested within the overall model. The approach presumes that the "true" model exists within the family of models being considered. This assumption in some applications may not be true; thus, the true model may not be found. Cox (1961, 1962) viewed the problem of model evaluation as the process of selecting among alternative models instead of restricting ones search within a given family. This leads one to the more general problem of evaluating nonnested hypotheses. A review of the literature in provided by McAleer (1995). Most tests developed for nonnested models are not exact, but depend on asymptotic theory.

In testing hypotheses regarding nested linear models, one considers a linear model

$$\Omega_o : \mathbf{y} = \mathbf{X}\boldsymbol{\beta} + \mathbf{e} \tag{5.14.1}$$

and an alternative model

$$\omega : \mathbf{y} = \mathbf{X}\boldsymbol{\beta}_1 + \mathbf{e} \tag{5.14.2}$$

where under ω, for $\boldsymbol{\beta}' = \left(\boldsymbol{\beta}_1', \boldsymbol{\beta}_2'\right), \boldsymbol{\beta}_2 = \mathbf{0}$. Under H_0, ω is considered the true model and under H_1, Ω_o is the true model. More generally, suppose one wants to evaluate two models

$$\begin{aligned} H_1 &: \mathbf{y} = \mathbf{X}\boldsymbol{\beta} + \mathbf{e}_1 \\ H_2 &: \mathbf{y} = \mathbf{Z}\boldsymbol{\gamma} + \mathbf{e}_2 \end{aligned} \tag{5.14.3}$$

where the intersection of $\mathbf{X}$ and $\mathbf{Z}$ is not null, some variables must be shared by the two models. The models in (5.14.3) are not nested. Cox (1961, p. 106, 1962, p. 406) states that two models are nonnested (or separate) if neither may be obtained from the other by the "imposition of appropriate restrictions, or as a limiting form of a suitable approximation". When one model is nested within another, the likelihood ratio under the null model is zero, and there is only one null model. This is not the case for nonnested models. Each model is considered to be the null model temporarily. While both procedures are concerned with model specification, in the nested situation evaluation is among rivals within a family while nonnested tests evaluate model specification between two competing families. Nonnested tests allow one to determine whether one or both models are misspecified. In the nested model situation, only one model is misspecified.

In considering (5.14.3), we want a test of H_1 vs. H_2 where rejection of H_1 allows us to test the viability of H_2 by testing H_2 vs. H_1. To conclude that specification H_2 is "more correct" than H_1, we must fail to reject H_2 compared to H_1. Or, conversely interchanging H_1 and H_2 we may conclude that H_1 is a "more correct" specification than H_2. If we reject both when comparing H_2 vs. H_1, and H_1 vs. H_2, then neither model provides an adequate specification and both models are considered misspecified. Failure to reject neither, one must postpone judgement regarding model specification.

Assuming $H_1 \equiv H_2$ against H_2 vs. H_1 in (5.14.3) where both models are SUR models so that $\mathbf{e}_1 \equiv \mathbf{e}_0 \sim N(\mathbf{0}, \Omega_0 = \Sigma_0 \otimes \mathbf{I})$ and $\mathbf{e}_2 \equiv \mathbf{e}_1 \sim N(\mathbf{0}, \Omega_1 = \Sigma_1 \otimes \mathbf{I})$, Cox (1961, 1962) developed a modified likelihood ratio test of H_0 vs. H_1. Following Pesaran and Deaton

(1978), Cox's test statistic is

$$T_0 = \frac{n}{2} \ln\left\{\mid \widehat{\Sigma}_1 \mid / \mid \widehat{\Sigma}_{10} \mid\right\} \tag{5.14.4}$$

The covariance matrix $\widehat{\Sigma}_1$ is the MLE of Σ given H_1. To estimate Σ_{10}, one estimates Σ under the difference in the likelihoods ($L_{10} = L_0 - L_1$) which is like minimizing the errors under H_0 (ω) and under H_1 (Ω) and taking the difference as an error estimate. To estimate Σ_{10}, we consider the artificial regression model

$$\widehat{f} = \mathbf{Z}\boldsymbol{\gamma} + \text{error} \tag{5.14.5}$$

where $\widehat{f} = \mathbf{X}\widehat{\boldsymbol{\beta}}$ represents fitted values of $\mathbf{y}$ under H_0. Then

$$\widehat{\Sigma}_{10} = \widehat{\Sigma}_{\text{error}} + \widehat{\Sigma}_0 \tag{5.14.6}$$

The maximum likelihood estimate of $\boldsymbol{\gamma}$ in (5.14.5) is represented as $\widetilde{\boldsymbol{\gamma}}$.

Cox shows that T_0 in (5.14.4) is asymptotically normal with mean zero and variance $\widehat{\sigma}_0^2 = \text{var}(T_0)$ when H_0 is true. The statistic $Z_{01} = T_0/\sigma_0 \mathrel{\dot\sim} N(0, 1)$. The formula for the estimated variance $\widehat{\sigma}_0^2$ given H_0 is

$$\widehat{\sigma}_0^2 = \mathbf{d}_0'\widehat{\Omega}_0^{-1}\mathbf{d}_0 \tag{5.14.7}$$

where

$$\begin{aligned}
\widehat{\Omega}_0 &= \widehat{\Sigma}_0 \otimes \mathbf{I} \\
\mathbf{d}_0 &= [\mathbf{I} - \mathbf{X}(\mathbf{X}'\widehat{\Omega}_0^{-1}\mathbf{X})^{-1}\mathbf{X}'\widehat{\Omega}_0^{-1}]\widehat{\mathbf{h}}_0 \\
\widehat{\mathbf{h}}_0 &= (\widehat{\Sigma}_0\widehat{\Sigma}_{10}^{-1} \otimes \mathbf{I})(\mathbf{X}\widehat{\boldsymbol{\beta}} - \mathbf{Z}\widetilde{\boldsymbol{\gamma}})
\end{aligned} \tag{5.14.8}$$

as developed by Pesaran and Deaton (1978). In a similar manner, we may calculate the statistics T_1, $\widehat{\sigma}_1^2$ and Z_{10} by interchanging the role of $\mathbf{X}$ with $\mathbf{Z}$, testing H_1 vs. H_0.

For multiple linear regression models, Godfrey and Pesaran (1983) recommend replacing ML estimates of the covariance matrices with unbiased estimates when $n < 40$ and the number of variables is small, less than 3. For $n \geq 20$, the statistic maintained a Type I error rate near the nominal level. For SUR models, the same relationships should be maintained for each variable, however, exact guideline need to be established. Replacing ML estimates with unbiased estimates $\mathbf{S}$, the mean of $Z_{01} = T_0/\widehat{\sigma}_0$ is nearest to zero under H_0 leading to fewer Type I errors.

A major disadvantage of Cox's procedure is that the result is asymptotic. For multiple linear regression models, Fisher and McAleer (1981) and McAleer (1983) developed exact tests based on the ELM of Milliken and Graybill (1970). We extend their approach to the SUR model.

To develop a test for evaluating (5.14.3), we let $H_1 \equiv H_0$ and $H_2 \equiv H_1$ and write (5.14.3) as

$$\begin{aligned}
H_0 &: \mathbf{y} = \mathbf{X}\boldsymbol{\beta} + \mathbf{u}_0 & \mathbf{u}_0 &\sim N_{np\times 1}(\mathbf{0}, \Omega_0 = \Sigma_0 \otimes \mathbf{I}) \\
H_1 &: \mathbf{y} = \mathbf{Z}\boldsymbol{\gamma} + \mathbf{u}_1 & \mathbf{u}_1 &\sim N_{np\times 1}(\mathbf{0}, \Omega_1 = \Sigma_1 \otimes \mathbf{I})
\end{aligned}$$

Next, we combine H_0 and H_1 to form the artificial linear model

$$\begin{aligned}\alpha_0\,(\mathbf{y}-\mathbf{X}\boldsymbol{\beta})+\alpha_1\,(\mathbf{y}-\mathbf{Z}\boldsymbol{\gamma})&=\mathbf{u}\\ \mathbf{y}&=\mathbf{X}\boldsymbol{\beta}_*+\alpha_1\mathbf{Z}\boldsymbol{\gamma}+\mathbf{u}\end{aligned} \tag{5.14.9}$$

where $\mathbf{u}=\alpha_0\mathbf{u}_0+\alpha_1\mathbf{u}_1$, $\alpha_0+\alpha_1=1$ and $\boldsymbol{\beta}_*=\alpha_0\boldsymbol{\beta}_0$. Letting $\alpha_0=-\alpha_1/\alpha_0$ and $\mathbf{u}/\alpha_0$, we may write (5.14.9) as an artificial SUR model

$$\mathbf{y}=\mathbf{X}\boldsymbol{\beta}+(\mathbf{y}-\mathbf{Z}\boldsymbol{\gamma})\,\alpha_*+\mathbf{u} \tag{5.14.10}$$

where $\mathbf{u}_*\sim N\left(\mathbf{0},\Omega_*=\Omega_0+\alpha_1^2\Omega_1\right)$. Using (5.10.10), testing $H_0:\alpha_*=0$ is equivalent to testing $H_0:\alpha_1=0$ and each is testing H_0 in (5.14.9). Equation (5.10.10) is similar to the ELM given in (5.9.1) with $\mathbf{y}-\mathbf{Z}\boldsymbol{\gamma}\equiv\mathbf{F}_1$. Thus, (5.10.10) is an extended SUR model. As in the test of nonadditivity, we must estimate $\mathbf{Z}\boldsymbol{\gamma}$ to test $H_0:\alpha_*=0$.

Under H_0, the predicted value of $\mathbf{y}$ is $\mathbf{X}\widehat{\boldsymbol{\beta}}$. For the SUR model,

$$\mathbf{X}\widehat{\boldsymbol{\beta}}=\mathbf{X}(\mathbf{X}'\widehat{\Omega}_0^{-1}\mathbf{X})^{-1}\mathbf{X}'\widehat{\Omega}_0^{-1}\mathbf{y}=\widehat{\mathbf{f}} \tag{5.14.11}$$

say. Replacing $\mathbf{y}$ of H_1 in (5.10.9) with $\widehat{\mathbf{f}}$, the estimate of $\boldsymbol{\gamma}$ is obtained from the artificial SUR model

$$\widehat{\mathbf{f}}=\mathbf{Z}\boldsymbol{\gamma}+\text{error} \tag{5.14.12}$$

Letting Ω_{10} represent the covariance matrix for the SUR model,

$$\widetilde{\boldsymbol{\gamma}}=\left(\mathbf{Z}'\widehat{\Omega}_{10}^{-1}\mathbf{Z}\right)^{-1}\mathbf{Z}'\widehat{\Omega}_{10}^{-1}\widehat{\mathbf{f}}$$

the predicted value of $\widehat{\mathbf{f}}$ defined as $\mathbf{g}_{10}$ (say) is

$$\begin{aligned}\widetilde{\mathbf{g}}_{10}&=\mathbf{Z}\widetilde{\boldsymbol{\gamma}}\\ &=\mathbf{z}(\mathbf{z}'\widehat{\Omega}_{10}^{-1}\mathbf{Z})^{-1}\mathbf{Z}'\widehat{\Omega}_{10}^{-1}\widehat{\mathbf{f}}\end{aligned} \tag{5.14.13}$$

Replacing $\mathbf{y}-\mathbf{Z}\boldsymbol{\gamma}$ in (5.14.10) with $\left(\widehat{f}-\widetilde{\mathbf{g}}_{10}\right)\equiv\widehat{\mathbf{F}}$, the residual of (5.14.12), we obtain the extended SUR model

$$\underset{np\times 1}{\mathbf{y}}=\underset{np\times k}{\mathbf{X}}\;\underset{k\times 1}{\boldsymbol{\beta}}+\underset{np\times 1}{\widehat{\mathbf{F}}}\;\alpha_*+\underset{np\times 1}{\mathbf{u}_*} \tag{5.14.14}$$

Letting $\mathbf{X}_*=\left[\mathbf{X},\widehat{\mathbf{F}}\right]$, model (5.14.13) has the structure of the SUR model

$$\mathbf{y}=\mathbf{X}_*\boldsymbol{\theta}+\mathbf{u}_*$$

where $\boldsymbol{\theta}'=\left(\boldsymbol{\beta}',\alpha_*\right)$ and $\mathbf{u}_*\sim N_{np}\left(\mathbf{0},\Omega_*=\Sigma_*\otimes\mathbf{I}\right)$. To test $H_0:\alpha_*=0$ we may set $\mathbf{C}=\left(\mathbf{0}',1\right)$ and use F^* in (5.2.9) where

$$\begin{aligned}\widehat{\widehat{\theta}}&=\left(\mathbf{X}_*'\widehat{\Omega}_*^{-1}\mathbf{X}_*\right)^{-1}\mathbf{X}_*\widehat{\Omega}_*^{-1}\mathbf{y}\\ \widehat{\Omega}_*&=\widehat{\Sigma}_*\otimes\mathbf{I}\end{aligned}$$

and $\widehat{\Sigma}_0 \xrightarrow{p} \Sigma$. As n becomes large, $\sqrt{F^*} \xrightarrow{d} N(0, 1)$ under H_0 since $\nu_h F^* \xrightarrow{d} \chi^2_{(\nu_h)}$ and $\nu_h = 1$ so that the $\sqrt{F^*}$ is asymptotically equivalent to Cox's test. For known Ω_*, F^* is an exact test. Because F^* is asymptotically equivalent to Rao's score test, it has maximum local power, St. Laurent (1990).

Example 5.14.1 (Nonnested Tests) *To implement Cox's modified LR test procedure and the $\sqrt{F^*}$ statistic developed using the extended SUR model, we again use Rohwer's data in Table 4.3.1. This data was used to illustrate the MR model and the SUR model. When using the multivariate regression procedure or the SUR modeling methods to fit models to data, alternative models are evaluated as subsets of an overall model.*

Because the Raven Picture Matrices Test (RPMT) is not strongly related to the PA learning tasks, we delete this variable from the data set and we consider investigating the relationship of the achievement variables SAT and PPVT to different subsets of paired associate (PA) learning tasks. Under H_0, we relate SAT to N and SS, and PPVT to S and SS. And, under the alternative H_1, SAT is related to NA and SS while PPVT is related to NS and SS. For the SUR models given in (5.14.9), the dependent vector $\mathbf{y}' = (\mathbf{y}_1', \mathbf{y}_2')$ contains the variables $\mathbf{y}_1 = \{\text{SAT}\}$ and $\mathbf{y}_2 = \{\text{PPVT}\}$. Under H_0, $\mathbf{X}$ contains $\mathbf{X}_1 = \{\text{N, SS}\}$ and $\mathbf{X}_2 = \{\text{S, SS}\}$ and under $H_1, \mathbf{Z}$ contains $\mathbf{Z}_1 = \{\text{NA, SS}\}$ and $\mathbf{Z}_2 = \{\text{NS, SS}\}$for the two models. To analyze the two models, program m5_14_1.sas is used.

Program m5_14_1.sas uses the SYSLIN procedure, and IML code to test H_0 vs H_1 and H_1 vs. H_0 for both Cox's test and the $\sqrt{F^*}$ statistic. The program is divided into two sections: Cox's test procedure and the F-like test procedure. The program begins by calculating inputs to T_0 in (5.14.4) for testing H_0 vs H_1 by obtaining unbiased estimates $\mathbf{S}_0$, $\mathbf{S}_1$, and $\mathbf{S}_{10}$ using the SYSLIN procedure with the option ISUR. It also calculates the values of the covariance estimates $\mathbf{S}_0$, $\mathbf{S}_1$, and $\mathbf{S}_{10}$ for testing H_1 vs. H_0 by interchanging $\mathbf{X}$ and $\mathbf{Z}$. Output from this program step is passed to the second IML step using the new Output Delivery System available in Version 8 of SAS to calculate T_0, $\widehat{\sigma}_0^2$ and Z_{01} for testing H_0 vs H_1. The process is repeated to calculate T_0, σ_1^2, Z_{10} for testing H_1 vs H_0. Testing H_0 vs H_1, $Z_{01} = -15.19455$ and for testing H_1 vs H_0, $Z_{10} = -133.0896$. Hence, using Cox's modified LR test procedure we would conclude that both models are misspecified.

Evaluating the same models using the $\sqrt{F^*}$ statistic, we find that $F^* = 22.1945$ and $\sqrt{F^*} = 4.711$ for testing H_0 vs H_1. For testing H_1 vs H_0, $F^* = 12.9697$ and $\sqrt{F^*} = 3.6013$. Using the small sample statistic, we also conclude that both models are misspecified for Rohwer's data.

For this example, the sample size $n = 32$ is small which usually leads to a Type I error rate for Cox's test that is less than the nominal level α. The Type I error for Cox's test is usually less than the nominal level for sample sizes $n \leq 50$. The Type I error rate for statistic $\sqrt{F^*}$ appears to always be near the nominal level of the test for sample sizes larger than 25. For $n \geq 100$, both test procedures always appear to have Type I error rates near the nominal level of the test, Timm and Al-Subaihi (2001). Except for very small sample sizes ($n \leq 25$), the two procedures have approximately the same power. For large sample sizes, $n \geqslant 200$, the two tests are equivalent.

Exercises 5.14

1. Use the file Rohwer2.dat for the $n = 37$ subjects in the high SES area to fit the models proposed for the $n = 32$ low SES students.

5.15 Miscellanea

In applying the MR, SUR, GMANOVA and other multivariate linear models in the analysis of multivariate data, the covariance matrix Σ was unknown and estimated jointly with the parameters associated with the mean, $E(\mathbf{Y})$. The primary focus of the analysis was to estimate the location parameters and Σ, and to test hypotheses regarding the mean. In many applications in the behavioral and social sciences, one often constrains Σ to a specific form. Then, one is interested in estimating both the covariance structure of Σ and the location parameters. Jöreskog (1970, 1973) developed a very general multivariate model for this problem and called it the analysis of covariance structures (ACOVS). His covariance structure model allows one to estimate the covariance structure as well as the location parameters and, in large samples, test hypotheses about the structure of Σ. Jöreskog's model for the specification of $E(\mathbf{Y})$ is identical to the GMANOVA model given in (5.6.1); however, he further assumes that Σ has a general covariance structure of the form

$$\Sigma = \mathbf{A}\left(\Gamma\Phi\Gamma' + \Psi\right)\mathbf{A}' + \Theta \tag{5.15.1}$$

where the matrices $\mathbf{A}_{p\times q} = [\alpha_{ik}]$, $\Gamma_{g\times r} = [\gamma_{km}]$, the symmetric matrix $\Phi_{r\times r} = [\phi_{ms}]$, and the diagonal matrices $\Psi_{g\times g} = \text{diag}[\psi_h]$ and $\Theta_{p\times p} = \text{diag}[\theta_i]$ are parameter matrices. To apply the general model in practice, $\mathbf{Y}$ determines n and p, and $\mathbf{X}$ and $\mathbf{P}$ are determined by the application as in the GMANOVA model. However, the parameters in $\mathbf{B}$, $\mathbf{A}$, Γ, Φ, Ψ, and Θ are permitted to be of three types: *(i)* fixed parameters that have known values; *(ii)* constrained parameters that are unknown but equal to one or more of the other parameters; and *(iii)* free parameters that are unknown, and constrained to be equal to other parameters.

An important problem with Jöreskog's model is the indeterminacy of Σ. That is, if $\mathbf{A}$ is replaced by $\mathbf{AT}_1^{-1}$, Γ by $\mathbf{T}_1\Gamma\mathbf{T}_2^{-1}$, Φ by $\mathbf{T}_2\Phi\mathbf{T}_2'$, Ψ by $\mathbf{T}_1\Psi\mathbf{T}_1'$ with Θ unchanged, the structure of the matrix Σ is unaffected. This is the case for all nonsingular matrices $\mathbf{T}_1$ $(g \times g)$ and $\mathbf{T}_2$ $(r \times r)$ such that $\mathbf{T}_1\Psi\mathbf{T}_1'$ is diagonal. To eliminate the indeterminacy, constraints are imposed on parameters; some values are assigned known values, and others are allowed to be free to vary to ensure that all parameters are uniquely estimable. More will be said about this problem when we address specific applications. For now, we will assume that all indeterminacies have been eliminated so that $\mathbf{T}_1 = \mathbf{T}_2 = \mathbf{I}$ is the only transformation that preserves the specifications regarding the fixed and constrained parameters.

To estimate $\mathbf{B}$, $\mathbf{A}$, Γ, Φ, Ψ, and Θ in the Jöreskog model, Jöreskog minimizes the expression $-2\log L/n$ where L is the likelihood for the MVN model. In particular, letting

$$T(\mathbf{B}) = (\mathbf{Y} - \mathbf{XBP})'(\mathbf{Y} - \mathbf{XBP})/n \tag{5.15.2}$$

the function to be minimized is

$$F_{ML} = -2\log L/n = \log|\Sigma| + \text{tr}\left(\mathbf{T}\Sigma^{-1}\right) \tag{5.15.3}$$

where $\mathbf{T} \equiv T(\mathbf{B})$. The function F is a function of $\mathbf{B}$, $\mathbf{A}$, Γ, Φ, Ψ, and Θ, $\mathbf{T}$ is a function of $\mathbf{B}$ as given in (5.15.2), and Σ is a function of $\mathbf{A}$, Γ, Φ, Ψ, and Θ by (5.15.1). To obtain full information ML estimates for the model parameters in Jöreskog's model is complicated and requires numerical iterative methods. The ML procedure described in Jöreskog (1973) was programed by using a modified Fletcher-Powell algorithm.

To test hypotheses using the model, one lets H_0 be any hypotheses concerning the parametric structure of the general model and H_1 an alternative, usually the GMANOVA model. Letting F_0 be the minimum of F_{ML} under H_o and F_1 the minimum of F_{ML} under H_1, the likelihood ratio test of minus two times the likelihood ratio in terms of F_{ML} becomes

$$X^2 = n(F_0 - F_1) \tag{5.15.4}$$

and has an asymptotic chi-squared distribution under H_0 with degrees of freedom υ equal to the difference in the number of independent parameters estimated under H_1 and H_0. When the alternative is the GMANOVA model, where $\widehat{\Sigma}$ is defined in (5.6.3) and the statistic in (5.14.4) becomes

$$X^2 = n(F_0 - \log|\widehat{\Sigma}| - p) \tag{5.15.5}$$

where $\upsilon^* = kp + p(p+1)/2 - \upsilon_o$ and υ_o is the number of independent parameters estimated under H_o.

We have introduced the Jöreskog's analysis of covariance structures (ACOVS) model to illustrate a very general model which includes the MR, MANOVA, MANCOVA and GMANOVA models as special cases. It may also be used to analyze many applications that arise in classical test theory, the estimation of variance components, path analysis models, complex economic models and linear structural equation models. We will discuss some of these models in the next chapters. A popular special class of the covariance structure model is the Linear Structural Relations (LISREL) model, Jöreskog (1977, 1979) Structural equation modeling (SEM) is discussed in Chapter 10.

6
Multivariate Random and Mixed Models

6.1 Introduction

In our discussion of the MR model the parameter matrix **B** contained only fixed parameters and the analysis of univariate fixed effect designs for orthogonal and nonorthogonal MANOVA and MANCOVA designs extended in a natural manner to the multivariate case. The MANOVA model also provided an alternative for the analysis of repeated measurement designs, if there are no covariates changing with time, and allowed the within-subject covariance matrix to have a general structure. The SUR model was introduced to permit the analysis of repeated measures data with changing covariates. The GMANOVA model was also used to analyze repeated measurements, however, exact tests exist only when there is complete data and the within-subject covariance matrix is unstructured.

In this chapter we discuss a generalization of the GMANOVA model, called the random coefficient regression model, which will permit the analysis of longitudinal repeated measurements data that are MAR. We also discuss the univariate general linear mixed model which allows one to model the within-subject covariance matrix in repeated measurement designs and to analyze multi-response repeated measurements. Next, we discuss a MGLM that allows the parameter matrix to be random or mixed and to contain both random and fixed parameters. For orthogonal designs, univariate models generalize in a natural manner to the multivariate case; this is not the case for nonorthogonal designs which require a multivariate analogue of Satterthwaite's approximation. Finally, we conclude the chapter with a discussion of a general multivariate hierarchical linear model.

6.2 Random Coefficient Regression Models

a. Model Specification

The GMANOVA and SUR models were used to fit polynomials to repeated measurement data to model the behavior of each group rather than individuals. The analysis required complete data for exact tests and if data are missing, it was MCAR. To permit the analysis of repeated measurements data that are gathered at irregular intervals and MAR, we employ the random coefficient regression model considered independently by Rao (1965), Swamy (1971), Lindley and Smith (1972) and Laird and Ware (1982). To specify the model, we use a two-stage hierarchical linear model (TSHLM) where at stage one we model the within-subject observations and at stage two we model the random regression coefficients. The model has the general structure

$$\begin{aligned}\mathbf{y}_i &= \mathbf{Z}_i\boldsymbol{\beta}_i + \mathbf{e}_i \quad i = 1, 2, \ldots, n \\ \boldsymbol{\beta}_i &= \mathbf{Q}_i\boldsymbol{\beta} + \mathbf{a}_i\end{aligned} \tag{6.2.1}$$

The vector $\mathbf{y}_i' = \left[y_{i1}, y_{i2}, \ldots, y_{ir_i}\right]$ is a vector of r_i repeated measurements for the ith subject. The matrix $\mathbf{Z}_i$ is the within-subject design matrix of order $r_i \times h$ of full rank h, and the $r\left(\mathbf{Z}_i\right) = h \leq r_i$. The vector $\boldsymbol{\beta}_i$ $(h \times 1)$ is a vector of regression coefficients for the ith subject and $\mathbf{e}_i \sim IN_{r_i}\left(\mathbf{0}, \sigma^2\mathbf{I}_{r_i}\right)$. The vector $\boldsymbol{\beta}_i$ is being modeled in the second stage as a linear regression equation where $\mathbf{Q}_i$ $(h \times q)$ is the between-subject design matrix and $\boldsymbol{\beta}_{q\times 1}$ is a vector of fixed population parameters. The vectors $\mathbf{a}_i \sim IN_h\left(\mathbf{0}, \Phi\right)$ and are independent of $\mathbf{e}_i$.

Combining the equations in (6.2.1) into a single model, we have that

$$\begin{aligned}\mathbf{y}_i &= \left(\mathbf{Z}_i\mathbf{Q}_i\right)\boldsymbol{\beta} + \mathbf{Z}_i\mathbf{a}_i + \mathbf{e}_i \quad i = 1, 2, \ldots, n \\ &= \mathbf{X}_i\boldsymbol{\beta} + \mathbf{Z}_i\mathbf{a}_i + \mathbf{e}_i \\ E\left(\mathbf{y}_i\right) &= \mathbf{X}_i\boldsymbol{\beta} \\ \operatorname{cov}\left(\mathbf{y}_i\right) &= \mathbf{Z}_i\Phi\mathbf{Z}_i' + \sigma^2\mathbf{I}_{r_i} = \Omega_i\end{aligned} \tag{6.2.2}$$

where the jth element of $\mathbf{y}_i$ has the simple linear structure

$$y_{ij} = \boldsymbol{\beta}'\mathbf{x}_{ij} + \mathbf{a}_i'z_{ij} + e_{ij} \qquad i = 1, 2, \ldots, n \quad j = 1, 2, \ldots, r_i \tag{6.2.3}$$

the $E\left(e_{ij}\right) = 0$, $\operatorname{var}\left(e_{ij}\right) = \sigma^2$ and the $\operatorname{cov}\left(\mathbf{a}_i\right) = \Phi$. Letting $a_i = \mathbf{a}_i'z_{ij}$ where the $\operatorname{var}\left(a_i\right) = \sigma_a^2$ (say), a simple application of (6.2.3) is the one-way mixed random effects model. Replacing $\mathbf{y}_{ij}$ and $\mathbf{e}_{ij}$ by $p\times 1$ random vectors and letting $\boldsymbol{\beta}'\mathbf{x}_{ij} = \mathbf{B}\mathbf{x}_{ij}$ where $\mathbf{B}_{p\times q}$ is a matrix of unknown fixed parameters, we have the multivariate extension of the mixed random effects models. Alternatively, we may write (6.2.2) using a single vector equation. Letting $N = \sum_i r_i$,

$$\mathbf{y}_{N\times 1} = \begin{bmatrix}\mathbf{y}_1 \\ \mathbf{y}_2 \\ \vdots \\ \mathbf{y}_n\end{bmatrix}, \quad \mathbf{X} = \begin{bmatrix}\mathbf{X}_1 \\ \mathbf{X}_2 \\ \vdots \\ \mathbf{X}_n\end{bmatrix}, \quad \mathbf{a} = \begin{bmatrix}\mathbf{a}_1 \\ \mathbf{a}_2 \\ \vdots \\ \mathbf{a}_n\end{bmatrix}$$

$$\mathbf{Z} = \bigoplus_{i=1}^{n} \mathbf{Z}_i = \begin{bmatrix} \mathbf{Z}_1 & \mathbf{0} & \cdots & \mathbf{0} \\ \mathbf{0} & \mathbf{Z}_2 & \cdots & \mathbf{0} \\ \vdots & \vdots & & \vdots \\ \mathbf{0} & \mathbf{0} & \cdots & \mathbf{Z}_n \end{bmatrix} \quad \text{and} \quad \mathbf{e} = \begin{bmatrix} \mathbf{e}_1 \\ \mathbf{e}_2 \\ \vdots \\ \mathbf{e}_n \end{bmatrix}$$

a linear (univariate) model for (6.2.2) becomes

$$\begin{aligned} \mathbf{y} &= \mathbf{X}\boldsymbol{\beta} + \mathbf{Z}\mathbf{a} + \mathbf{e} \\ E(\mathbf{y}) &= \mathbf{X}\boldsymbol{\beta} \\ \operatorname{cov}(\mathbf{y}) &= \mathbf{Z}(\mathbf{I}_n \otimes \Phi)\mathbf{Z}' + \left(\bigoplus_{i=1}^{n} \sigma^2 \mathbf{I}_{r_i}\right) \\ &= \mathbf{Z}(\mathbf{I}_n \otimes \Phi)\mathbf{Z}' + \mathbf{I}_N \otimes \sigma^2 \\ &= \mathbf{Z}(\mathbf{I}_n \otimes \Phi)\mathbf{Z}' + \sigma^2 \mathbf{I}_N \\ &= \Omega \end{aligned} \tag{6.2.4}$$

where $\Omega = \bigoplus_{i=1}^{n} \Omega_i$, $\Omega_i = \mathbf{Z}_i \Phi \mathbf{Z}_i' + \sigma^2 \mathbf{I}_{r_i}$ and $\mathbf{y} \sim N_N(\mathbf{X}\boldsymbol{\beta}, \Omega)$. Model (6.2.4) is linear in the parameters, but is more complicated that the GLM since now it involves the random component vector $\mathbf{a}$. Model (6.2.4) is a mixed linear model with covariance structure Ω.

b. *Estimating the Parameters*

To estimate the fixed parameters σ^2, Φ, and $\boldsymbol{\beta}$ in the random coefficient regression model (6.2.4) is more complicated than estimating $\boldsymbol{\beta}$ and σ^2 in the GLM since we have the additional matrix Φ of unknown covariances. In addition, the prediction of $\mathbf{y}_i$ becomes more complicated. In the GLM, we could obtain confidence intervals for the elements of $E(\mathbf{y}) = \mathbf{X}\boldsymbol{\beta}$ and prediction intervals for $\widehat{\mathbf{y}}$ not in the domain of $\mathbf{X}$, or future observations. In the random coefficient regression model, confidence intervals for the elements of $E(\mathbf{y}_i)$ depend on estimates of $\boldsymbol{\beta}$, Φ, and σ^2. In addition, prediction intervals for the elements of $\widehat{\mathbf{y}}_i$ over the domain of collected data depend on an estimate for the random vectors $\mathbf{a}_i$, called the best linear unbiased predictor (BLUP). To see this, recall that $\mathbf{y}_i = \mathbf{X}_i\boldsymbol{\beta} + \mathbf{Z}_i\mathbf{a}_i + \mathbf{e}_i$ where $E(\mathbf{y}_i) = \mathbf{X}_i\boldsymbol{\beta}$. Hence, a confidence interval for the elements of $E(\mathbf{y}_i)$ depend on $\widehat{\mathbf{y}}_i = \mathbf{X}_i\widehat{\boldsymbol{\beta}}$ and the variability of $\widehat{\mathbf{y}}_i - \mathbf{y}_i$ over the domain $\mathbf{X}_i$. The

$$\operatorname{cov}(\widehat{\mathbf{y}}_i - \mathbf{y}_i) = \mathbf{X}_i \left(\operatorname{cov}\widehat{\boldsymbol{\beta}}\right) \mathbf{X}_i' + \mathbf{Z}_i \Phi \mathbf{Z}_i' + \sigma^2 \mathbf{I}_{r_i} \tag{6.2.5}$$

which depends on $\widehat{\boldsymbol{\beta}}$, Φ and σ^2. However, a prediction interval for $\mathbf{y}_i$ depends on $\widehat{\mathbf{a}}_i$ since $\widehat{\mathbf{y}}_i = \mathbf{X}_i\widehat{\boldsymbol{\beta}} + \mathbf{Z}_i\widehat{\mathbf{a}}_i$. The predictive variance of $\widehat{\mathbf{y}}_i - \mathbf{y}_i$,

$$\operatorname{cov}(\widehat{\mathbf{y}}_i - \mathbf{y}_i) = \mathbf{X}_i \left(\operatorname{cov}\widehat{\boldsymbol{\beta}}\right) \mathbf{X}_i' + \mathbf{Z}_i \operatorname{cov}(\widehat{\mathbf{a}}_i - \mathbf{a}_i) \mathbf{Z}_i' + \sigma^2 \mathbf{I}_{r_i} \tag{6.2.6}$$

depends on the BLUP estimate $\widehat{\mathbf{a}}_i$ and the variance of $\widehat{\mathbf{a}}_i - \mathbf{a}_i$.

To estimate the parameter vector $\boldsymbol{\beta}$ in (6.2.4), we first assume that Φ and σ^2 are known. Then, the ML or GLS estimate of $\boldsymbol{\beta}$ is

$$\begin{aligned}\widehat{\boldsymbol{\beta}} &= \left(\mathbf{X}'\Omega\mathbf{X}\right)^{-1}\mathbf{X}'\Omega^{-1}\mathbf{y}\\ &= \sum_{i=1}^{n}\left(\mathbf{X}_i'\Omega_i^{-1}\mathbf{X}_i\right)^{-1}\left[\sum_{i=1}^{n}\mathbf{X}_i'\Omega_i^{-1}\mathbf{y}_i\right]\end{aligned} \tag{6.2.7}$$

and the

$$\operatorname{cov}(\widehat{\boldsymbol{\beta}}) = \left(\mathbf{X}'\Omega^{-1}\mathbf{X}\right)^{-1} = \left(\sum_{i=1}^{n}\mathbf{X}_i'\Omega_i^{-1}\mathbf{X}_i\right)^{-1} \tag{6.2.8}$$

The distribution of $\widehat{\boldsymbol{\beta}}$ is MVN provided $\mathbf{e}$ is multivariate normal.

To estimate $\boldsymbol{\beta}$, σ^2, and the nonredundent elements of Φ represented by $\boldsymbol{\phi} = [\text{vec}\,(\Phi)]'$ or $\boldsymbol{\theta}_s' = \left(\boldsymbol{\phi}',\ \sigma^2\right)$, called the variance components of the model, requires the use of a computer algorithm since the likelihood equations have no simple closed form solution. Commonly used algorithms to estimate $\boldsymbol{\beta}$ and $\boldsymbol{\theta}$ include the EM (Expectation-Maximization) algorithm described by Laird and Ware (1982) based on the work of Dempster, Laird and Rubin (1977), the Newton-Raphson and Fisher scoring algorithms described by Jennrich and Schluchter (1986), and the ridge-stabilizing, sweep-based, Newton-Raphson algorithm described by Wolfinger, Tobias, and Sall (1994) and used in the SAS procedure PROC MIXED. Due to the EM algorithm's slow convergence for estimating the elements of $\boldsymbol{\theta}$, especially when the parameter $\boldsymbol{\theta}$ is near the boundary of the parameter space, Newton-Raphson-based methods are preferred, Lindstrom and Bates (1988).

The maximum likelihood estimate of $\boldsymbol{\theta}$, $\widehat{\boldsymbol{\theta}}_{ML}$, is biased in small samples and since $\widehat{\boldsymbol{\beta}}$ depends of $\widehat{\boldsymbol{\theta}}$, we represent $\widehat{\boldsymbol{\beta}}$ as $\widehat{\boldsymbol{\beta}}(\widehat{\boldsymbol{\theta}})$. Furthermore, the cov $\widehat{\boldsymbol{\beta}}$ is biased downward because of the variability introduced by working with $\widehat{\Omega}_i$, by using $\widehat{\boldsymbol{\theta}}_{ML}$, instead of Ω_i is not taken into account in the approximation for $(\mathbf{X}_i'\Omega_i^{-1}\mathbf{X}_i)^{-1}$. To reduce the bias in the cov $\widehat{\boldsymbol{\beta}}$, Harville (1977) advocates the use of the restricted (residual) maximum likelihood (REML) estimates for $\boldsymbol{\theta}$, $\widehat{\boldsymbol{\theta}}_{REML}$, which may also lead to a biased estimate, but tends to reduce the downward bias in the cov $\widehat{\boldsymbol{\beta}}$ if $\widehat{\Omega}_i \equiv \widehat{\boldsymbol{\theta}}_{REML}$. REML estimates for the elements of $\boldsymbol{\theta}$ are obtained by maximizing the reduced likelihood equation obtained by minimizing the log-likelihood of the transformed residual contrasts $\boldsymbol{\psi}_i = \left(\mathbf{I} - \mathbf{X}_i\left(\mathbf{X}_i'\mathbf{X}_i\right)^{-}\mathbf{X}_i'\right)\mathbf{y}_i$ of the original data $\mathbf{y}_i$, Searle, Casella and McCulloch (1992) and McCulloch and Searle (2001).

We have indicated how one may obtain a ML estimate for $\boldsymbol{\beta}$ by using $\widehat{\boldsymbol{\beta}}(\widehat{\boldsymbol{\theta}}_{ML})$ and indirectly a REML estimate represented as $\widehat{\boldsymbol{\beta}}(\widehat{\boldsymbol{\theta}}_{REML})$. The estimation of $\mathbf{a}_i$ is more complicated. By an extension of the Gauss-Markov theorem for the general linear mixed model, Harville (1976) showed that the BLUP of $\mathbf{a}_i$ (also called the empirical Bayes estimate of $\mathbf{a}_i$) is

$$\widehat{\mathbf{a}}_i = \Phi\mathbf{Z}_i'\Omega_i^{-1}\left(\mathbf{y}_i - \mathbf{X}_i\widehat{\boldsymbol{\beta}}\,(\boldsymbol{\theta})\right) \tag{6.2.9}$$

where $\Omega_i = \mathbf{Z}_i\Phi\mathbf{Z}_i' + \sigma^2\mathbf{I}$ and $\widehat{\boldsymbol{\beta}}\,(\boldsymbol{\theta})$ is the ML or GLS estimate of $\boldsymbol{\beta}$. The covariance matrix of $\widehat{\mathbf{a}}_i - \mathbf{a}_i$ is

$$\begin{aligned}\text{cov}\left(\widehat{\mathbf{a}}_i - \mathbf{a}_i\right) &= \Phi - \text{cov}\left(\widehat{\mathbf{a}}_i\right) \\ \text{cov}\left(\widehat{\mathbf{a}}_i\right) &= \Phi \mathbf{Z}_i' \mathbf{P}_i \mathbf{Z}_i \Phi \\ \mathbf{P}_i &= \Omega_i^{-1} - \Omega_i^{-1} \mathbf{X}_i \left(\sum_{i=1}^{n} \mathbf{X}_i' \Omega_i^{-1} \mathbf{X}_i \right)^{-1} \mathbf{X}_i^{-1} \Omega_i^{-1}\end{aligned} \tag{6.2.10}$$

To estimate the cov $\left(\widehat{\mathbf{a}}_i - \mathbf{a}_i\right)$, one replaces Ω_i with $\widehat{\boldsymbol{\theta}}_{\text{REML}}$ or $\widehat{\boldsymbol{\theta}}_{\text{ML}}$, Laird and Ware (1982). Using (6.2.10) with $\widehat{\Omega}_i$ substituted for Ω_i, one may evaluate (6.2.5) and (6.2.6) to establish $100\,(1-\alpha)\,\%$ simultaneous confidence intervals for a single element of $\mathbf{y}_i$. For a Bayesian analysis of the random coefficient regression model, one is referred to Smith (1973) and Laird and Ware (1982). Henderson (1963) obtains estimates for $\boldsymbol{\beta}$ and $\mathbf{a}$ by solving the mixed model equations (e.g. Searle, Casella and McCulloch, 1992). We discuss this approach in the next section.

c. *Hypothesis Testing*

To test hypotheses regarding the model parameters of the random coefficient regression model, we consider two situations: (a) observations are taken at regular intervals and contain no missing data so that $r_1 = r_2 = \ldots = r_n = r$ (say) and (b) observations are gathered at irregular intervals so that the r_i are unequal. For situation (a), exact tests and confidence intervals exist for the model while in situation (b), tests are based on large sample theory. In many applications of the random coefficient regression model, the between-subject design matrix $\mathbf{Q}_i$ has the simple structure $\mathbf{Q}_i = \mathbf{I}_h \otimes \mathbf{q}_i'$ where $\mathbf{q}_i'$ is a known $k \times 1$ vector of between subject coefficients usually of the form $\mathbf{q}_i' = (0, 0, 0, 1_k, 0, \ldots, 0)$ with the value one in the kth position. Setting $q = hk$, $\mathbf{Z}_i = \mathbf{Z}$ with $r_i = r$ for all i and letting $\mathbf{Y}'$ represent the data matrix with $\mathbf{y}_i$ as its ith column, we may write $\mathbf{Y}$ as

$$\underset{r\times n}{\mathbf{Y}'} = \underset{r\times h}{\mathbf{Z}'}\ \underset{h\times k}{\mathbf{B}'}\ \underset{k\times n}{\mathbf{Q}'} + \underset{r\times n}{\mathbf{A}'} \tag{6.2.11}$$

where

$$\begin{aligned}\mathbf{Q}' &= [\mathbf{q}_1, \mathbf{q}_2, \ldots, \mathbf{q}_n] \\ \mathbf{A}' &= [\mathbf{a}_1, \mathbf{a}_2, \ldots, \mathbf{a}_n] \\ \mathbf{a}_i &= \mathbf{y}_i - \mathbf{X}_i \boldsymbol{\beta}\end{aligned}$$

Letting $\boldsymbol{\beta} = \text{vec}\left(\mathbf{B}'\right)$in model (6.2.11), model (6.2.2) has the same structure as the GMANOVA model of Chapter 5, equation (5.6.1). The ML estimate of $\mathbf{B}$ assuming $\mathbf{a}_i \sim N(\mathbf{0}, \Sigma)$ where $\Sigma = \mathbf{Z}\Phi\mathbf{Z}' + \sigma^2\mathbf{I}_r$ is identical to the ML estimate of $\boldsymbol{\beta}$ in (6.2.2) so that $\widehat{\boldsymbol{\beta}} = \text{vec}(\widehat{\mathbf{B}}')$ as shown by Laird, Lange and Stram (1987). Furthermore, exact tests exist for testing hypotheses regarding $\mathbf{B}$ of the form $\mathbf{CBM} = \mathbf{0}$ or $\left(\mathbf{M}' \otimes \mathbf{C}\right) \text{vec}\,(\mathbf{B}) = \mathbf{0}$ where $\mathbf{C}$ is a contrast matrix for the GMANOVA model. If the matrices $\mathbf{Z}_i$ are allowed to vary among individuals, an approximate analysis may be accomplished using a SUR model with complete data.

To test hypotheses regarding $\boldsymbol{\beta}$ in (6.2.1) where the r_i are not equal and MAR requires the use of large sample theory. To test hypotheses of the form

$$H : \mathbf{C}\boldsymbol{\beta} = \mathbf{0} \tag{6.2.12}$$

one may again use Wald's statistic

$$\begin{aligned} W &= (\mathbf{C}\widehat{\boldsymbol{\beta}})'[\mathbf{C}(\mathbf{X}'\widehat{\Omega}^{-1}\mathbf{X})^{-}\mathbf{C}']^{-1}(\mathbf{C}\widehat{\boldsymbol{\beta}}) \\ &\sim \chi^2(v_h) \end{aligned} \tag{6.2.13}$$

where $v_h = \text{rank}\,(\mathbf{C})$, using the linear model (6.2.4). Following Theil (1971), one may also use Wald's F statistic approximation

$$F = W/v_h \mathrel{\dot\sim} F\,(v_h,\ v_e) \tag{6.2.14}$$

where $v_e \doteq n - \text{rank}\,[\mathbf{X}, \mathbf{Z}]$, the default procedure of PROC MIXED. Alternatively, the denominator degrees of freedom may be approximated using Satterthwaite's (1946) procedure. His approximation states that if

$$\widehat{\psi} = \mathbf{c}'(\mathbf{X}'\widehat{\Omega}^{-1}\mathbf{X})^{-}\mathbf{c} = \sum_{i=1}^{s} \alpha_i \theta_i \tag{6.2.15}$$

for s variance components, then the degrees of freedom for the contrast $\widehat{\psi}$ is approximated by

$$\widehat{v}_e = \frac{\left(\sum_{i=1}^{s} \alpha_i \widehat{\theta}_i\right)^2}{\left(\sum_{i=1}^{s} \alpha_i \widehat{\theta}_i\right)^2 / v_i} \tag{6.2.16}$$

where α_i are known constants and $\widehat{\theta}_i$ are estimates of the variance components in the vector $\boldsymbol{\theta}_{s\times 1}$ where $s = h\,(h+1)\,/2 + 1$. This is the procedure used in PROC MIXED by employing the option DDFM = SATTERTH on the MODEL statement. For a discussion of Satterthwaite's approximation, one may consult Searle et al. (1992, p. 134) or Verbeke and Molenberghs (1997, p. 279). An elementary introduction is included in Neter, Kutner, Nachtsheim and Wasserman (1996, p. 971) and Dean and Voss (1997). The default option for calculating the denominator degrees of freedom is DDFM = CONTAIN. This method searches for random effects that contain an appropriate fixed effect and a random component and selects the term with the lowest value as the degrees of freedom.

Given the relationship (6.2.11) so that $\left(\mathbf{I}_h \otimes \mathbf{q}_i'\right) = \mathbf{Q}_i$ in (6.2.1), Vonesh and Carter (1987) show that the hypothesis $H : \left(\mathbf{M}' \otimes \mathbf{C}\right) \boldsymbol{\beta} = \mathbf{0}$ or equivalently that $H : \mathbf{CBM} = \mathbf{0}$ may be tested using an F statistic that is an approximation to testing H under the GMANOVA model given in (6.2.11) by using the Bartlett-Pillai-Hotelling trace criterion. That is, by Theorem 3.5.1 assuming multivariate normality, an asymptotically valid test of H is to use

$$\begin{aligned} F &= W/\,(2_s N_o - s + 2)\,(s v_h u)\,(n-k) \\ &\sim F\,(v_h u,\ 2s N_o - s + 2) \end{aligned} \tag{6.2.17}$$

where $v_h = \text{rank}\,(\mathbf{C})$, $u = \text{rank}\,(\mathbf{M})$, $s = \min,\,(v_h,\ u)$ and $N_o = (n-k-u)\,/2$. It is unknown how this approximation compares with Satterthwaite's approximation used in PROC MIXED. However, both procedures are better than Wald's chi-square statistic.

Testing the hypothesis that a variance component in $\boldsymbol{\theta}$ is zero is a more difficult problem, Self and Liang (1987). This is due to the fact that the asymptotic distribution of likelihood ratio statistics under the null hypothesis that the component is zero is no longer chi-square, but a mixture of chi-square distributions. While Stram and Lee (1994) have suggested corrections, PROC MIXED generates Z tests based on asymptotic likelihood properties. The tests are generally unreliable. The procedure also produces confidence limits for variance components using the Satterthwaite procedure where the θ_i in (6.2.15) are replaced by expected mean squares and α_i are known coefficients. To obtain Z tests and approximate standard errors and confidence intervals for components of variance, the options COVTEST and CL are used on the procedure statement.

6.3 Univariate General Linear Mixed Models

a. *Model Specification*

While the random coefficient regression model given in (6.2.1) or more compactly in (6.2.2) permits us to generalize the GMANOVA model in the analysis of growth curves involving MAR repeated measurements data, the structure of the within-subject covariance matrix for $\mathbf{e}_i \sim IN_{r_i}\left(\mathbf{0},\ \sigma^2\mathbf{I}_{r_i}\right)$ is somewhat restrictive. In addition, the second equation of model (6.2.1) requires all elements of each $\boldsymbol{\beta}_i$ to be random. To relax these requirements, we may define each $\boldsymbol{\beta}_i$ as

$$\boldsymbol{\beta}_i = \mathbf{Q}_i\boldsymbol{\beta} + \mathbf{A}_i\mathbf{a}_i \tag{6.3.1}$$

where the matrix $\mathbf{A}_i$ contains 0's and 1's to select the elements in $\boldsymbol{\beta}_i$ that are random and allow the others to be fixed. Furthermore, we assume that the errors $\mathbf{e}_i \sim IN_{r_i}\ (\mathbf{0},\ \Psi_i)$ where Ψ_i is a general $r_i \times r_i$ covariance matrix so that the covariance structure of $\mathbf{y}_i$ is $\Omega_i = \mathbf{Z}_i\Phi\mathbf{Z}_i' + \Psi_i$. Combining (6.3.1) with the first stage equation given in (6.2.1), the univariate mixed effects linear model becomes

$$\begin{aligned}
\mathbf{y}_i &= (\mathbf{Z}_i\mathbf{Q}_i)\,\boldsymbol{\beta}_i + (\mathbf{Z}_i\mathbf{A}_i)\,\mathbf{a}_i + \mathbf{e} \\
&= \mathbf{X}_i\boldsymbol{\beta} + \widetilde{\mathbf{Z}}_i\mathbf{a}_i + \mathbf{e}_i \quad i = 1, 2, \ldots, n \\
E\,(\mathbf{y}_i) &= \mathbf{X}_i\boldsymbol{\beta} \\
\operatorname{cov}(\mathbf{y}_i) &= \widetilde{\mathbf{Z}}_i\Phi\widetilde{\mathbf{Z}}_i' + \Psi_i = \Omega_i
\end{aligned} \tag{6.3.2}$$

Comparing (6.3.2) with (6.2.2), we observe that the linear structures for the two models are very similar since letting $\mathbf{A}_i = \mathbf{I}$ and $\Psi_i = \sigma^2\mathbf{I}_{r_i}$, the random coefficient regression model is a special case of the univariate mixed effects linear model. For notational convenience, we let $\widetilde{\mathbf{Z}}_i \equiv \mathbf{Z}_i$ which is a known design matrix linking $\mathbf{a}_i$ with $\mathbf{y}_i$, Laird and Ware (1982). Thus, the general structure for the univariate linear mixed model is

$$\begin{aligned}
\mathbf{y}_i &= \mathbf{X}_i\boldsymbol{\beta} + \mathbf{Z}_i\mathbf{a}_i + \mathbf{e}_i \quad i = 1, 2, \ldots, n \\
\mathbf{a}_i &\sim IN_h\,(\mathbf{0},\ \Phi) \\
\mathbf{e}_i &\sim IN_{r_i}\,(\mathbf{0},\ \Psi_i)
\end{aligned} \tag{6.3.3}$$

where $\mathbf{X}_i$ of order $r_i \times q$ is a known design matrix, $\boldsymbol{\beta}_{q\times 1}$ is a vector of fixed effect population parameters, $\mathbf{Z}_i$ of order $n \times h$ is a known design matrix, and $\mathbf{a}_i$ and $\mathbf{e}_i$ are independent random vectors. Other names for model (6.3.3) include hierarchical linear models, multilevel models, panel data models, variance component models, or simply mixed effects linear models, Searle et al. (1992), and Kreft and de Leeuw (1998).

Model (6.3.3) is more general than the random coefficient regression model because it provides for alternative intrasubject covariance matrices Ψ_i, thus leading to the estimation of more parameters. Following (6.2.4) with $N = \sum_i r_i$, we may express the model as

$$
\begin{aligned}
\begin{bmatrix} \mathbf{y}_1 \\ \mathbf{y}_2 \\ \vdots \\ \mathbf{y}_n \end{bmatrix} &= \begin{bmatrix} \mathbf{X}_1 \\ \mathbf{X}_2 \\ \vdots \\ \mathbf{X}_n \end{bmatrix} \boldsymbol{\beta} + \\
\underset{N\times 1}{\mathbf{y}} &= \underset{N\times q}{\mathbf{X}} \underset{q\times 1}{\boldsymbol{\beta}} + \\
&\begin{bmatrix} \mathbf{Z}_1 & \mathbf{0} & \cdots & \mathbf{0} \\ \mathbf{0} & \mathbf{Z}_2 & \cdots & \mathbf{0} \\ \vdots & \vdots & & \vdots \\ \mathbf{0} & \mathbf{0} & \cdots & \mathbf{Z}_n \end{bmatrix} \begin{bmatrix} \mathbf{a}_1 \\ \mathbf{a}_2 \\ \vdots \\ \mathbf{a}_n \end{bmatrix} + \begin{bmatrix} \mathbf{e}_1 \\ \mathbf{e}_2 \\ \vdots \\ \mathbf{e}_n \end{bmatrix} \\
&\underset{N\times hn}{\mathbf{Z}} \quad \underset{hn\times 1}{\mathbf{a}} + \underset{N\times 1}{\mathbf{e}}
\end{aligned}
\tag{6.3.4}
$$

where

$$
\begin{aligned}
\operatorname{cov}(\mathbf{a}) &= \mathbf{I}_n \otimes \Phi = \operatorname{Diag}[\Phi, \Phi, \ldots, \Phi] = \mathbf{G} \\
\operatorname{cov}(\mathbf{e}) &= \bigoplus_{i=1}^{n} \Psi_i = \operatorname{Diag}[\Psi_1, \Psi_2, \ldots, \Psi_n] = \Psi
\end{aligned}
\tag{6.3.5}
$$

$\operatorname{cov}(\mathbf{y}) = \Omega = \mathbf{Z}\mathbf{G}\mathbf{Z}' + \Psi = \bigoplus_{i=1}^{n} \Omega_i$ where $\Omega_i = \mathbf{Z}_i \Phi \mathbf{Z}_i' + \Psi_i$. Again, we let $\boldsymbol{\theta}$ represent the nonredundent variances and covariances of Ψ_i, $i = 1, \ldots, n$ and Φ.

One may again obtain ML estimates of the parameters in (6.3.4) and (6.3.5) using the ML procedure which again requires the use of a computer algorithm. To obtain estimates of $\boldsymbol{\beta}$ and $\mathbf{a}$, one may also solve the mixed model normal equations following Henderson (1963)

$$
\begin{bmatrix} \mathbf{X}'\widehat{\Psi}^{-1}\mathbf{X} & \mathbf{X}'\widehat{\Psi}^{-1}\mathbf{Z} \\ \mathbf{Z}'\widehat{\Psi}^{-1}\mathbf{X} & \mathbf{Z}'\widehat{\Psi}^{-1}\mathbf{Z} + \mathbf{G}^{-1} \end{bmatrix} \begin{bmatrix} \widehat{\boldsymbol{\beta}} \\ \widehat{\mathbf{a}} \end{bmatrix} = \begin{bmatrix} \mathbf{X}'\widehat{\Psi}^{-1}\mathbf{y} \\ \mathbf{Z}'\widehat{\Psi}^{-1}\mathbf{y} \end{bmatrix}
\tag{6.3.6}
$$

which upon simplification yields the solution

$$
\begin{aligned}
\widehat{\boldsymbol{\beta}} &= (\mathbf{X}'\widehat{\Omega}^{-1}\mathbf{X})^{-1}\mathbf{X}'\widehat{\Omega}^{-1}\mathbf{y} \\
&= \left(\sum_{i=1}^{n} \mathbf{X}_i'\widehat{\Omega}_i^{-1}\mathbf{X}_i\right)^{-1} \left(\sum_{i=1}^{n} \mathbf{X}_i'\widehat{\Omega}_i^{-1}\mathbf{y}_i\right) \\
\widehat{\mathbf{a}} &= \widehat{\mathbf{G}}\mathbf{Z}'\widehat{\Omega}^{-1}(\mathbf{y} - \mathbf{X}\widehat{\boldsymbol{\beta}})
\end{aligned}
\tag{6.3.7}
$$

so that

$$\widehat{\mathbf{a}}_i = \widehat{\Phi}\mathbf{Z}_i'\widehat{\Omega}_i^{-1}(\mathbf{y}_i - \mathbf{X}_i\widehat{\boldsymbol{\beta}}) \tag{6.3.8}$$

which, except for arbitrary Ψ_i, has the same structure as the random coefficient models. Thus, we may obtain ML or REML estimates of $\boldsymbol{\beta}$ represented as $\widehat{\boldsymbol{\beta}}(\widehat{\boldsymbol{\theta}}_{\text{ML}})$ and $\widehat{\boldsymbol{\beta}}(\widehat{\boldsymbol{\theta}}_{\text{REML}})$, respectively. Or we may obtain the BLUP of $\mathbf{a}$, represented as $\widehat{\mathbf{a}}(\widehat{\boldsymbol{\theta}}_{\text{ML}})$, or the empirical Bayes (EB) estimate represented as $\widehat{\mathbf{a}}(\widehat{\boldsymbol{\theta}}_{\text{REML}})$, depending on the likelihood used to obtain the estimates for the components of variance of the vector $\boldsymbol{\theta}$.

Using (6.3.4), motivation for the BLUP $\widehat{\mathbf{a}}$ follows from the fact that if

$$\begin{aligned} \mathbf{y} &= \mathbf{X}\boldsymbol{\beta} + \mathbf{Z}\mathbf{a} + \mathbf{e} \\ \operatorname{cov}(\mathbf{y}) &= \mathbf{Z}\mathbf{G}\mathbf{Z}' + \Psi = \Omega \\ \operatorname{cov}(\mathbf{y},\ \mathbf{a}) &= \mathbf{Z}\mathbf{G} \quad \text{and} \quad \mathbf{G} = \mathbf{I}_n \otimes \Phi \end{aligned} \tag{6.3.9}$$

$$\begin{pmatrix} \mathbf{a} \\ \mathbf{y} \end{pmatrix} \sim N\left\{ \begin{pmatrix} \mathbf{0} \\ \mathbf{X}\boldsymbol{\beta} \end{pmatrix}, \begin{pmatrix} \mathbf{G}' & \mathbf{G}\mathbf{Z}' \\ \mathbf{Z}\mathbf{G} & \Omega \end{pmatrix} \right\}$$

Then the conditional mean of $\mathbf{a}|\mathbf{y}$ is

$$\begin{aligned} E\,(\mathbf{a} \mid \mathbf{y}) &= \mathbf{G}\mathbf{Z}'\Omega^{-1}\,(\mathbf{y} - \mathbf{X}\boldsymbol{\beta}) \\ \operatorname{cov}(\mathbf{a} \mid \mathbf{y}) &= \mathbf{G} - \mathbf{G}\mathbf{Z}'\left(\mathbf{Z}\mathbf{G}\mathbf{Z}' + \Psi\right)^{-1}\mathbf{Z}\mathbf{G} \end{aligned} \tag{6.3.10}$$

Furthermore, the conditional distribution of $\mathbf{y}$ given $\mathbf{a}$ is

$$(\mathbf{y} \mid \mathbf{a}) \sim N(\mathbf{X}\boldsymbol{\beta} + \mathbf{Z}\mathbf{a}, \Psi) \tag{6.3.11}$$

so that

$$(\mathbf{y}_i \mid \mathbf{a}_i) \sim N(\mathbf{X}_i\boldsymbol{\beta} + \mathbf{Z}_i\mathbf{a}_i,\ \Psi_i)$$

for $i = 1, 2, \ldots, n$.

b. Covariance Structures and Model Fit

A unique feature of the mixed linear model is that one may select various structures for the covariance matrix $\Omega = \bigoplus_{i=1}^{n} \Omega_i$ when estimating the model parameters $\boldsymbol{\beta}$ and $\mathbf{a}$. Because Ω_i has two components, Φ and Ψ_i, one must specify the structure of each when obtaining ML and REML estimates under normality. Using PROC MIXED, the population structure of these matrices are described using the RANDOM and REPEATED statements, respectively. The TYPE = option is used to define Φ and Ψ_i for the random vectors $\mathbf{a}_i$ and $\mathbf{e}_i$, respectively. The population structure of the covariance matrix depends on the application, a partial list for these components is provided in Table 6.3.1. More will be said about their selection in the examples discussed in Section 6.4. References that include numerous examples with discussion of PROC MIXED and the SAS macro %glimmix used to analyze univariate nonlinear non-normal data include Littell et al. (1996), Verbeke and Molenberghs (1977), Vonesh and Chinchilli (1997), and Brown and Prescott (1999). Technical

TABLE 6.3.1. Structured Covariance Matrix

TYPE	STRUCTURE
Unstructured TYPE = UN	$\begin{pmatrix} \sigma_1^2 & \sigma_{12} & \sigma_{13} \\ & \sigma_2^2 & \sigma_{23} \\ & & \sigma_3^2 \end{pmatrix}$
Simple TYPE = SIM	$\begin{pmatrix} \sigma^2 & 0 & 0 \\ & \sigma^2 & 0 \\ & & \sigma^2 \end{pmatrix} = \sigma^2 \mathbf{I}$
Variance Components TYPE = VC	$\begin{pmatrix} \sigma_1^2 & 0 & 0 \\ & \sigma_2^2 & 0 \\ & & \sigma_3^2 \end{pmatrix}$
Compound Symmetry TYPE = CS	$\begin{pmatrix} \sigma_1^2 + \sigma_2^2 & \sigma_2^2 & \sigma_2^2 \\ & \sigma_1^2 + \sigma_2^2 & \sigma_2^2 \\ & & \sigma_1^2 + \sigma_2^2 \end{pmatrix}$ $\sigma^2 (1 - \rho) \mathbf{I}_p + \sigma^2 \rho \mathbf{J}_p$
where	$\sigma^2 = \sigma_1^2 + \sigma_2^2$ and $\rho = \sigma_2^2 / \left(\sigma_1^2 + \sigma_2^2\right)$
Autoregressive (1st order) TYPE = AR(1)	$\begin{pmatrix} \sigma^2 & \rho\sigma^2 & \rho^2\sigma^2 \\ & \sigma^2 & \rho\sigma_2 \\ & & \sigma^2 \end{pmatrix}$
Toeplitz TYPE = TOEP	$\begin{pmatrix} \sigma^2 & \sigma_{12} & \sigma_{13} \\ & \sigma^2 & \sigma_{12} \\ & & \sigma^2 \end{pmatrix}$
Compound Symmetry Heterogeneous[1] TYPE = CSH	$\begin{pmatrix} \sigma_1^2 & \rho\sigma_1\sigma_2 & \rho\sigma_1\sigma_3 \\ & \sigma_2^2 & \rho\sigma_2\sigma_3 \\ & & \sigma_3^2 \end{pmatrix}$

[1]Also available for AR(1) and Toeplitz

documentation of the SAS procedure MIXED is contained in the on-line documentation distributed with Version 8 and in the SAS manuals, SAS (1992, 1997).

When fitting a mixed model, one again needs to evaluate model fit and model assumptions. Assuming for the moment that normality assumptions are valid, one needs to evaluate the structure of the within-structure covariance matrix, evaluate the need for random effects, and determine an appropriate mean structure for the model.

To evaluate the structure of the covariance matrix in fitting a mixed model with the same number of fixed effects, one calculates the value of the REML log-likelihood or the ML log-likelihood values under normality for a given model. Then using the log-likelihood after adjusting for the number of variance components in $\boldsymbol{\theta}$, one may define the information criterion as

$$AIC = \log L(\widehat{\boldsymbol{\theta}}) - v \tag{6.3.12}$$

where v is the number of variance components estimated, Akaike (1974). As noted in Chapter 4, with this definition of AIC a model with the largest information is now considered best. An alternative form of the information criterion was proposed by Schwarz (1978) known as Schwarz's Bayesian information criterion (SBC) and is defined as

$$\text{SBC} = \begin{cases} \log L(\widehat{\boldsymbol{\theta}}_{\text{ML}}) - v \ln N/2 \\ \log L(\widehat{\boldsymbol{\theta}}_{\text{REML}}) - v \ln N^*/2 \end{cases} \tag{6.3.13}$$

where $N = \sum_i r_i$, $N^* = N - q$ and q is the number of elements in $\boldsymbol{\beta}$. These criteria, as well as the criteria

$$\text{HQIC} = \begin{cases} \log L(\widehat{\boldsymbol{\theta}}_{\text{ML}}) - v \ln (\ln N) \\ \log L(\widehat{\boldsymbol{\theta}}_{\text{REML}}) - v \ln (\ln N^*) \end{cases}$$
$$\text{CAIC} = \begin{cases} \log L(\widehat{\boldsymbol{\theta}}_{\text{ML}}) - v \ (\ln N + 1)/2 \\ \log L(\widehat{\boldsymbol{\theta}}_{\text{REML}}) - v \ (\ln N^* + 1)/2 \end{cases} \tag{6.3.14}$$

developed by Hannan and Quinn (1979) and Bozdogan (1987) are calculated in PROC MIXED. When evaluating covariance structures, the model with the largest AIC, SBC, HQIC or CAIC is considered the one with the best fit. These criteria are used to help differentiate among several nested or non-nested models with the same fixed effects and are not to be used as tests of significance.

To compare two nested models with the same fixed effects where the full model has θ_F variance components with v_F degrees of freedom and the reduced model has θ_R components with v_R degrees of freedom, a large sample likelihood ratio tests is used. The criterion is

$$X^2 = 2\left[\ \log L_R - \ \log L_F\right] \dot\sim \chi^2 (v) \tag{6.3.15}$$

where $v = v_R - v_F$, and L_F and L_R are the likelihoods for the reduced and full models, respectively. The PROC MIXED software calculates a Null Model LRT chi-square statistic by default which compares the full model $\mathbf{y} = \mathbf{X}\boldsymbol{\beta} + \mathbf{Za} + \mathbf{e}$ with the reduced model $\mathbf{y} = \mathbf{X}\boldsymbol{\beta} + \mathbf{e}$ to test whether one needs to model the covariance structure for the data. This is an approximate test for structure since this tends to preserve the size of the test at or below the level α while only inflating the standard errors of fixed effects, Altham (1984).

c. *Model Checking*

In fixed effect univariate and multivariate regression models, procedures for the evaluation of model assumptions, outliers and influential observations were discussed. For the mixed model, the situation is complicated by the random effects $\mathbf{a}_i \sim N(\mathbf{0}, \Phi)$ and the fact that $\mathbf{e}_i \sim N(\mathbf{0}, \Psi_i)$ have heterogenous covariance structure Ψ_i.

Recall that the basic assumptions for the mixed model $\mathbf{y}_i = \mathbf{X}_i\boldsymbol{\beta} + \mathbf{Z}_i\mathbf{a}_i + \mathbf{e}_i$ are that

$$\begin{aligned} \mathbf{e}_i &\sim IN_{r_i}(\mathbf{0}, \Psi_i) \\ \mathbf{a}_i &\sim IN_h(\mathbf{0}, \Phi) \\ \mathbf{y}_i &\sim IN_{r_i}\left(\mathbf{X}_i\boldsymbol{\beta}, \Omega_i = \mathbf{Z}_i\Phi\mathbf{Z}_i' + \Psi_i\right) \end{aligned} \tag{6.3.16}$$

In most applications, we assume the error vectors $\mathbf{y}_i$ are mutually independent so that $\Psi_i \equiv \sigma^2 \mathbf{I}_{r_i}$. Then,

$$\Omega_i = \mathbf{Z}_i \Phi \mathbf{Z}_i' + \sigma^2 \mathbf{I}_{r_i} \tag{6.3.17}$$

when $\mathbf{X}_i \equiv \mathbf{Z} \otimes \mathbf{q}_i'$ and $r_i = r$ so that $\Psi_i = \sigma^2 \mathbf{I}_r$ then the mixed model is a GMANOVA model and the procedures discussed by Vonesh and Carter (1987) may be used to evaluate model assumptions. When this is not the case, the situation is more complicated, Lange and Ryan (1989).

For mixed models, as with multiple linear regression models, evaluation of each model fit to a data set needs to be evaluated. This may be accomplished by fitting alternative models and evaluating overall fit using the $\text{tr}(\widehat{\Omega})$ or the $\mid \widehat{\Omega} \mid$ where $\widehat{\Omega} = \bigoplus_{i=1}^{n} \widehat{\Omega}_i$ for various models. The model with the smallest overall generalized variance may be selected or, one may use information criteria.

Evaluating residuals for outliers, normality, and influential observations is complicated by the fact that we have two sources of random variation, $\mathbf{a}_i$ and $\mathbf{e}_i$. In multiple linear regression, one fits the model $\mathbf{y} = \mathbf{X}\boldsymbol{\beta} + \mathbf{e}$ and exams residuals $\widehat{\mathbf{e}} = \mathbf{y} - \mathbf{X}\widehat{\boldsymbol{\beta}}$ to evaluate assumptions regarding $\mathbf{e}$. In the mixed model, the residuals $\widehat{\mathbf{r}} = \mathbf{y} - \mathbf{X}\widehat{\boldsymbol{\beta}} = \mathbf{Z}\mathbf{a} + \mathbf{e}$ model both the random and error components. Furthermore, $\widehat{\mathbf{e}} = \widehat{\mathbf{r}} - \mathbf{Z}\widehat{\mathbf{a}}$ depends on the variability of both $\widehat{\mathbf{r}}$ and $\widehat{\mathbf{a}}$. Thus, in mixed models we have two residuals to evaluate $\widehat{\mathbf{r}}$ and $\widehat{\mathbf{e}}$. This is usually accomplished in two steps. First, we evaluate whether the random effects are normally distributed. Next we investigate the predicted residuals $\widehat{\mathbf{e}} = \mathbf{y} - \mathbf{X}\widehat{\boldsymbol{\beta}} - \mathbf{Z}\widehat{\mathbf{a}}$.

Another complication in the investigation of residuals for the mixed model is that an entire vector of subjects may be an outlier or only one element of the vector may be an outlier. Again, while each component may be normal, this does not imply that the entire vector is multivariate normal.

To investigate mixed model assumptions, we consider the one-way mixed random effects model discussed by Lange and Ryan (1989). The model is

$$y_{ij} = \mu + a_i + e_{ij} \qquad i = 1, 2, \ldots, n; \; j = 1, 2, \ldots, r_i \tag{6.3.18}$$

where $e_{ij} \sim IN\left(0, \; \sigma^2\right)$ and $a_i \sim IN\left(0, \; \sigma_a^2\right)$. The linear model for the ith group is

$$\begin{aligned} \begin{bmatrix} y_{i1} \\ y_{i2} \\ \vdots \\ y_{ir_i} \end{bmatrix} &= \mathbf{1}_{r_i}\mu + \mathbf{1}_{r_i} a_i + \begin{bmatrix} e_{i1} \\ e_{i2} \\ \vdots \\ e_{ir_i} \end{bmatrix} \\ \mathbf{y}_i &= \mathbf{1}_{r_i}\mu + \mathbf{1}_{r_i} a_i + \mathbf{e}_i \end{aligned} \tag{6.3.19}$$

where the parameter μ is fixed.

For each vector $\mathbf{y}_i$, the covariance matrix has compound symmetry structure

$$\Omega_i = \sigma^2 \mathbf{I}_{r_i} + \sigma_a^2 \mathbf{J}_{r_i} \tag{6.3.20}$$

so that the observations within the same group, class or cluster are correlated

$$\begin{aligned} \text{var}\left(y_{ij}\right) &= \sigma^2 + \sigma_a^2 \\ \text{cov}(y_{ij}, \; y_{ij'}) &= \sigma_a^2 \left(j \neq j'\right) \end{aligned} \tag{6.3.21}$$

The correlation of any two observations within the same group is

$$\rho = \sigma_a^2 / \left(\sigma^2 + \sigma_a^2\right)$$

called the intraclass correlation. In addition, the variance ratio is defined as $\omega = \sigma_a^2 / \sigma^2$. Multiplying Ω_i by σ^{-2}, the resulting matrix

$$\mathbf{V}_i = \sigma^{-2}\Omega_i = \mathbf{I}_{r_i} + \omega \mathbf{J}_{r_i} \tag{6.3.22}$$

depends on σ^2 through the variance ratio ω. Also observe that $\mathbf{V}_i^{-1} = \sigma^2\Omega_i^{-1}$.

Because Ω_i in (6.3.20) has compound symmetry structure, its inverse and determinant have simple forms

$$\begin{aligned} \Omega_i^{-1} &= \sigma^{-2}\left[\mathbf{I}_{r_i} - \frac{\sigma_a^2}{r_i\sigma_a^2 + \sigma^2}\mathbf{J}_{r_i}\right] \\ &= \sigma^{-2}\mathbf{I}_{r_i} - \frac{\sigma_a^2}{f_i\sigma^4}\mathbf{J}_{r_i} \\ |\Omega_i| &= f_i\sigma^{2r_i} \end{aligned} \tag{6.3.23}$$

where $f_i = 1 + r_i\omega$. The $|\Omega_i|$ is the generalized variance of $\mathbf{y}_i$. The factor f_i is an inflation factor so that the generalized variance as an index of fit will become large as r_i increases, independent of the variance ratio $\omega = \sigma_a^2 / \sigma^2$.

In working with $\mathbf{V}_i^{-1} = \sigma^{-2}\Omega_i^{-1}$, we have the following useful result found in Longford (1993a, p. 43).

Theorem 6.3.1. For arbitrary vectors $\mathbf{v}$, $\mathbf{v}_1$, and $\mathbf{v}_2$ of order $r_i \times 1$, the quadratic forms $\mathbf{v}'\Omega_i^{-1}\mathbf{1}_{r_i}$ and $\mathbf{v}_1'\Omega_i^{-1}\mathbf{v}_2$ have the structure

$$\begin{aligned} \mathbf{v}'\Omega_i^{-1}\mathbf{1}_{r_i} &= \mathbf{v}'V_i^{-1}\mathbf{1}_{r_i}' / \sigma^2 \\ &= \mathbf{v}'\mathbf{1}_{r_i}\left(1 - r_i\omega f_i^{-1}\right) / \sigma^2 \\ &= \mathbf{v}'\mathbf{1}_{r_i}' / f_i\sigma^2 \\ \mathbf{v}_1'\Omega_i^{-1}\mathbf{v}_2 &= \left(\mathbf{v}_1'\mathbf{v}_2 - \omega f_i^{-1}\mathbf{v}_1'\mathbf{1}_{r_i}\mathbf{v}_2'\mathbf{1}_{r_2}\right) / \sigma^2 \end{aligned}$$

where $f_i = 1 + r_i\omega$.

Continuing with the one-way mixed random effects model, the BLUP of a_i, assuming all parameters are known, using (6.3.8) and Theorem 6.3.1 with the arbitrary vector $\mathbf{v}$ associated with $\mathbf{e}_i = \mathbf{y}_i - \mathbf{X}_i\boldsymbol{\beta}$, is

$$
\begin{aligned}
\widehat{a}_i &= \sigma_a^2 \mathbf{1}'\Omega_i^{-1}\mathbf{e}_i \\
&= \sigma_a^2 \mathbf{1}'\mathbf{V}_i^{-1}\mathbf{e}_i \,/\, \sigma^2 \\
&= \sigma_a^2 \sigma^{-2} f_i^{-1} \mathbf{e}_i' \mathbf{1} r_i \\
&= \sigma_a^2 \sigma^{-2} (1 - r_i\omega)^{-1}\, \mathbf{e}_i' \mathbf{1}_{r_i} \\
&= \frac{\sigma_a^2}{\sigma^2 + r_i\sigma_a^2} \sum_{i=1}^{r_i} (y_{ij} - \mu) \\
&= \left[\frac{\sigma_a^2}{\sigma^2/r_i + \sigma_a^2}\right]\left(\frac{1}{r_i}\right) \sum_{j=1}^{r_i} (y_{ij} - \mu)
\end{aligned}
\tag{6.3.24}
$$

the result given by Lange and Ryan (1989). Given $\widehat{a}_i$ in (6.3.24) and again making use of Theorem 6.3.1, Lange and Ryan estimate the variance of $\widehat{a}_i$ as

$$
\begin{aligned}
\text{var}\,(\widehat{a}_i) &= \Phi \mathbf{Z}_i' \Omega_i^{-1} \Omega_i \Omega_i^{-1} \mathbf{Z}_i \Phi \\
&= \Phi \mathbf{Z}_i' \Omega_i^{-1} Z_i' \Phi \\
&= \sigma_a^2 \left(\mathbf{1}_{r_i}' \Omega_i^{-1} \mathbf{1}_{r_i}\right) \sigma_a^2 \\
&= \sigma_a^4 \,/\, (\sigma_a^2 + \sigma^2 \,/\, r_i)
\end{aligned}
\tag{6.3.25}
$$

Because the parameters are not known, an alternative estimate for the var $(\widehat{a}_i)$ is

$$
\text{var}\,(a_i) - \text{var}\,(\widehat{a}_i) = \sigma_a^2 - \sigma_a^4 \,/\left(\sigma_a^2 + \sigma^2 \,/\, r_i\right)
\tag{6.3.26}
$$

However, notice that the quantity

$$
\begin{aligned}
\sigma_a^2 - \sigma_a^4 / \left(\sigma_a^2 + \sigma^2 / r_i\right) &= \sigma_a^2 - \frac{\sigma_a^2 r_i \omega}{f_i} \\
&= \sigma_a^2 \left[1 - \frac{r_i\omega}{f_i}\right] \\
&= \sigma_a^2 \left[(f_i - r_i\omega) \,/\, f_i\right] \\
&= \sigma_a^2 \,/\, f_i
\end{aligned}
\tag{6.3.27}
$$

This may also be seen by using (6.3.10). That is, under joint multivariate normality of $\mathbf{a}_i$ and $\mathbf{y}_i$, the conditional distribution of a_i in the one-way mixed random effects model given the data is

$$
\begin{aligned}
a_i \mid \mathbf{y}_i &\sim N\left(\sigma_a^2 \mathbf{1}'\Omega_i^{-1}\mathbf{e}_i,\, \sigma_a^2 - \sigma_a^4 \mathbf{1}_{r_i}' \Omega_i^{-1} \mathbf{1}_{r_i}\right) \\
&\sim N\left(\frac{\sigma_a^2 \mathbf{1}_{r_i}' \mathbf{e}_i}{\sigma^2 f_i},\, \frac{\sigma_a^2}{f_i}\right) \\
&\sim N\left(\omega f_i^{-1} \mathbf{1}_{r_i}' \mathbf{e}_i,\, \sigma^2 \omega f_i^{-1}\right)
\end{aligned}
\tag{6.3.28}
$$

Furthermore, given that σ^2 and σ_a^2 are known the conditional distribution of $\mathbf{e}_i$ given the data is

$$\begin{aligned}\mathbf{e}_i \mid \mathbf{y} &\sim \mathbf{N}\left(\sigma^2\Omega_i^{-1}\mathbf{e}_i, \sigma^2 - \sigma^4\Omega_i^{-1}\right)\\ &\sim N\left(\sigma^2\Omega_i^{-1}\mathbf{e}_i, \sigma^4\Omega_i^{-1}\right)\end{aligned} \tag{6.3.29}$$

Forming the standardized variables

$$z_i = \frac{\sigma_a^2 1'\Omega_i^{-1}e_i}{\sqrt{\sigma_a^2 / f_i}} = \frac{\overline{y}_i - \mu}{\sqrt{\sigma^2 / f_i + \sigma_a^2}} \qquad i = 1, 2, \ldots, n \tag{6.3.30}$$

for model (6.3.19) and replacing population parameters with ML estimates, Lange and Ryan (1989) recommend comparing the empirical distribution of z_i with a standard normal to evaluate the distribution of a_i. Replacing σ^2 and Ω_i^{-1} by their ML estimates, one may use chi-square and Beta plots to evaluate the multivariate normality of $\mathbf{e}_i$ using (6.3.29). For appropriate choices of arbitrary vectors $\mathbf{c}$, Lange and Ryan(1989) investigate Q-Q plots of the variables

$$\mathrm{v}_i = \mathbf{c}'\mathbf{e}_i / \sqrt{\mathbf{c}' \operatorname{cov}(\mathbf{e}_i)\, \mathbf{c}} \qquad i = 1, 2, \ldots, n \tag{6.3.31}$$

For the general mixed model under normality, one may derive the conditional distributions of $\mathbf{a}_i$ and $\mathbf{e}_i$ given the data and all known parameters, Longford (1993a). The conditional distributions have the structure

$$\begin{aligned}\mathbf{a}_i \mid \mathbf{y} &\sim N\left(\Phi\mathbf{F}_i\mathbf{Z}_i\mathbf{e}_i,\ \Phi\mathbf{F}_i^{-1}\right)\\ \mathbf{e}_i \mid \mathbf{y} &\sim N\left(\sigma^2\Omega_i^{-1}\mathbf{e}_i,\ \sigma^4\Omega_i^{-1}\right)\\ \mathbf{F}_i &= \mathbf{I}_{r_i} + \sigma^{-2}\mathbf{Z}_i'\mathbf{Z}_i\Phi\\ \Omega_i^{-1} &= \sigma^{-2}\left(\mathbf{I}_{r_i} - \sigma^2\mathbf{Z}_i\Phi\mathbf{F}_i^{-1}Z_i'\right)\\ |\Omega_i| &= \sigma^{2r_i}\ |\mathbf{F}_i|\end{aligned} \tag{6.3.32}$$

given $\mathbf{a}_i \sim IN(\mathbf{0}, \Phi)$ and $\mathbf{e}_i \sim IN\left(\mathbf{0}, \sigma^2\mathbf{I}_{r_i}\right)$. The matrix $\mathbf{F}_i$ is a generalization of the inflation factor. Replacing population parameters by ML estimates, one may examine these distributions for multivariate normality. In addition, influence measures for mixed models may be investigated to locate influential observations on fixed effects. Christensen, Pearson and Johnson (1992) investigate removing an entire vector, a global influence. They do not address the problem of the influence of an element within a vector, called local influence. For additional details and discussion of residual analysis and influence measures in mixed models, one may consult Goldstein (1995) and Verbeke and Molenberghs (1997).

In mixed models, one is usually interested in fitting a model to data and then estimating the fixed parameter $\boldsymbol{\beta}$, testing hypotheses regarding $\boldsymbol{\beta}$, and estimating variance components. In some studies, one may also be interested in the fitted values $\widehat{\mathbf{y}}_i = \mathbf{X}_i\widehat{\boldsymbol{\beta}} + \mathbf{Z}_i\widehat{\mathbf{a}}_i$ and in the prediction of future observations. To predict future observations, one uses the procedure discussed in Chapter 5.2, sub-section (b) since model (6.3.9) is a GLM with arbitrary covariance structure Ω. The effect of departures from normality, covariance structure misspecification, lack of homogeneity of covariance structures Φ, and the identification of influential observations has only begun to be investigated in recent years.

d. Balanced Variance Component Experimental Design Models

In our formulation of the mixed model given by (6.3.4), the vectors $\mathbf{y}_i$ contained an unequal number of elements. In many applications $r_i = r$ for all i as in growth curve models with complete data. Such models are called balanced designs. In experimental design settings, r is the number of replications or observations in a cell of a design and i indexes groups or factors. Starting with the GLM for experimental designs, Hartley and Rao (1967) formulated the univariate mixed linear model used in experimental designs with equal cell frequencies, also called variance component models, as

$$\underset{N\times 1}{\mathbf{y}} = \underset{N\times 1}{\mathbf{X}} \underset{q\times 1}{\boldsymbol{\beta}} + \underset{N\times h}{\mathbf{Z}} \underset{h\times 1}{\mathbf{u}} + \mathbf{e} \tag{6.3.33}$$

where the matrices $\mathbf{X}$ and $\mathbf{Z}$ are known, $\boldsymbol{\beta}$ is a vector of fixed effects, $\mathbf{u}$ is a random vector of random effects and $\mathbf{e}$ is a vector of random errors. Using the Hartley and Rao's formulation, one usually partitions $\mathbf{u}$ into m sub-vectors

$$\mathbf{u}' = \left[\mathbf{u}_1', \mathbf{u}_2', \ldots, \mathbf{u}_m'\right] \tag{6.3.34}$$

where each component $\mathbf{u}_i$ represents, for example, random effects, main effect, interactions and nested factors in the design. Similarly, the matrix $\mathbf{Z}$ is partitioned conformable with $\mathbf{u}$ so that

$$\mathbf{Z} = [\mathbf{Z}_1, \mathbf{Z}_2, \ldots, \mathbf{Z}_m] \tag{6.3.35}$$

where each matrix $\mathbf{Z}_i$ is of order $N \times q_i$. Then, model (6.3.33) becomes

$$\mathbf{y} = \mathbf{X}\boldsymbol{\beta} + \mathbf{Z}\mathbf{u} + \mathbf{e} = \mathbf{X}\boldsymbol{\beta} + \sum_{i=1}^{m} \mathbf{Z}_i\mathbf{u}_i + \mathbf{e} \tag{6.3.36}$$

Or, letting $\mathbf{u}_0 = \mathbf{e}$ and $\mathbf{Z}_0 = \mathbf{I}_N$ (6.3.36) becomes

$$\mathbf{y} = \mathbf{X}\boldsymbol{\beta} + \sum_{i=0}^{m} \mathbf{Z}_i\mathbf{u}_i \tag{6.3.37}$$

For $\widetilde{\mathbf{e}} = \sum_{i=0}^{m} \mathbf{Z}_i\mathbf{u}_i$, a GLM for $\mathbf{y}$ becomes $\mathbf{y} = \mathbf{X}\boldsymbol{\beta} + \widetilde{\mathbf{e}}$.

Assuming $\mathbf{e} \sim N_N\left(\mathbf{0}, \sigma^2\mathbf{I}_N\right)$ and $\mathbf{u}_i \sim N_{q_i}\left(\mathbf{0}, \sigma_i^2\mathbf{I}_{q_i}\right)$ where $\mathbf{e}$ and $\mathbf{u}_i$ are mutually independent, the covariances structure for $\mathbf{y}$ is

$$\begin{aligned} \operatorname{cov}(\mathbf{y}) &= \mathbf{Z}\Phi\mathbf{Z}' + \sigma^2\mathbf{I}_N \\ &= \sum_{i=1}^{m} \sigma_i^2\mathbf{Z}_i\mathbf{Z}_i' + \sigma^2\mathbf{I}_N \\ &= \sum_{i=0}^{m} \sigma_i^2\mathbf{Z}_i\mathbf{Z}_i' = \Omega \end{aligned} \tag{6.3.38}$$

where $\Phi = \operatorname{diag}\left[\sigma_i^2\mathbf{I}_{q_i}\right]$ and $\sigma^2 = \sigma_0^2$. Thus, $\widetilde{\mathbf{e}} \sim N(\mathbf{0}, \Omega)$ for the model $\mathbf{y} = \mathbf{X}\boldsymbol{\beta} + \widetilde{\mathbf{e}}$. Variance component models are special cases of the more general formulation given in (6.3.4). To see this, we again consider the univariate one-way random mixed model.

Example 6.3.1 (Univariate One-way Mixed Model) *For this model, the model equation is*

$$y_{ij} = \mu + a_i + e_{ij} \qquad i = 1, 2, \ldots, a \text{ and } j = 1, 2, \ldots, r$$

where y_{ij} is the jth observation in the ith class or group, μ is a fixed parameter, a_i is a random effect for the ith group, e_{ij} is the random error for observation j within group i. Furthermore, $a_i \sim IN\left(0,\ \sigma_a^2\right)$, $e_{ij} \sim IN\left(0,\ \sigma^2\right)$, and a_i and e_{ij} are mutually independent. A general linear model for the one-way design is

$$\begin{aligned}
\mathbf{y} &= \mathbf{X}\boldsymbol{\beta} + \mathbf{Z}\mathbf{u} + \mathbf{e} \\
&= (\mathbf{I}_a \otimes \mathbf{1}_r)\,\mu + (\mathbf{I}_a \otimes \mathbf{1}_r)\,\mathbf{u} + \mathbf{e} \\
\operatorname{cov}(\mathbf{y}) &= \mathbf{Z}\Phi\mathbf{Z}' + \Psi \\
&= (\mathbf{I}_a \otimes \mathbf{1}_r)\,\sigma_a^2\mathbf{I}_a\,(\mathbf{I}_a \otimes \mathbf{1}_r)'\,\sigma^2\mathbf{I}_{ar} \\
&= \sigma_a^2\,(\mathbf{I}_a \otimes \mathbf{J}_r) + \sigma^2\mathbf{I}_{ar} = \Omega
\end{aligned}$$

which is the structure given in (6.3.38). And, rewriting the cov (**y**) *as*

$$\begin{aligned}
\operatorname{cov}(\mathbf{y}) &= \sigma_a^2\,(\mathbf{I}_a \otimes \mathbf{J}_r) + \sigma^2\,(\mathbf{I}_a \otimes \mathbf{I}_r) \\
&= \mathbf{I}_a \otimes \left(\sigma^2\mathbf{I}_n + \sigma_a^2\mathbf{J}_r\right) \\
&= \mathbf{I}_a \otimes \Omega^* \\
&= \Omega
\end{aligned}$$

where $\Omega_1 = \Omega_2 = \ldots = \Omega_a = \Omega^*$, Ω has block diagonal structure given in (6.3.20) with $r_1 = r_2 = \ldots = r_a = r$. Hence, $\Phi = \sigma_a^2\mathbf{J}_r$ and $\Psi_i = \sigma^2\mathbf{I}_r$ so that the $\operatorname{cov}(\mathbf{y}) = \mathbf{Z}\Phi\mathbf{Z}' + \Psi = \Omega = \bigoplus_{i=1}^{a}\Omega_i$ *where $\Omega_i = \sigma_a^2\mathbf{J}_r + \sigma^2\mathbf{I}_r$ for $i = 1, 2, \ldots, a$.*

e. *Multilevel Hierarchical Models*

A common application of (6.3.3) in the social and behavioral sciences is to model multilevel data, Goldstein (1995). The approach is to represent a student level outcome y_{ij} to a pair of linked models. Given n_j observations within the jth school, the linear model for the level-1 unit with n_j observations is

$$\mathbf{y}_j = \mathbf{X}_j\boldsymbol{\beta}_j + \mathbf{e}_j \tag{6.3.39}$$

The matrices $\mathbf{X}_j$ contain student level variables such as SES, ethnic origin, gender, and IQ. The level-2 unit, school, for $j = 1, 2, \ldots, n$ are considered random units modeled as

$$\boldsymbol{\beta}_j = \mathbf{Z}_j\boldsymbol{\gamma} + \mathbf{u}_j \tag{6.3.40}$$

The matrices $\mathbf{Z}_j$ contain schools level variables such as class size, percentage of minority students, and student-teacher ratio's. Combining (6.3.40) with (6.3.39), we have that

$$\begin{aligned}
\mathbf{y}_j &= \mathbf{X}_j\left(\mathbf{Z}_j\boldsymbol{\gamma} + \mathbf{u}_j\right) + \mathbf{e}_j \\
&= \left(\mathbf{X}_j\mathbf{Z}_j\right)\boldsymbol{\gamma} + \left[\mathbf{X}_j\mathbf{u}_j + \mathbf{e}_j\right] \\
&= \text{fixed part} + \text{random part}
\end{aligned}$$

which has the mixed model structure of (6.3.3).

Example 6.3.2 (Multilevel Linear Model) *To formulate a multilevel model, suppose the relationship among schools is linear*

$$y_{ij} = \alpha_j + \beta_j x_{ij} + e_{ij}$$

where (α_j, β_j) *are unknown parameters and* $e_{ij} \sim N(0, \sigma_e^2)$*. At level-2 of the model, the coefficients* α_j *and* β_j *become random variables with unknown covariance structure. In multilevel notation, one replaces* α_j *with* β_{0j} *and* β_j *with* β_{1j}*, and suppose that the random coefficients are related to a covariate using the linear model*

$$\beta_{0j} = \beta_{00} + \gamma_{01} z_j + u_{0j} \text{ and } \beta_{1j} = \beta_{10} + \gamma_{11} z_j + u_{1j}$$

where u_{0j} *and* u_{1j} *are random,* $u_{ij} \sim N(0, \sigma_{ui}^2)$ *and* z_j *is a covariate at the school level. Combining the results into a single equation, the multilevel model for the outcome variables* y_{ij} *becomes*

$$\begin{aligned} y_{ij} &= (\beta_{00} + \gamma_{01} z_j + u_{0j}) + (\beta_{10} + \gamma_{11} z_j + u_{1j}) x_{ij} + e_{ij} \\ &= [\beta_{00} + \gamma_{01} z_j + \beta_{10} + \gamma_{11} z_j x_{ij}] + [u_{0j} + u_{1j} x_{ij} + e_{ij}] \\ &= \text{fixed part} + \text{random part} \end{aligned}$$

where

$$e_{ij} \sim N\left(0, \sigma_e^2\right)$$

$$\begin{bmatrix} u_{0j} \\ u_{1j} \end{bmatrix} \sim N\left(\begin{bmatrix} 0 \\ 0 \end{bmatrix}_1 \begin{bmatrix} \sigma_{u0}^2 & \sigma_{u01} \\ \sigma_{u10} & \sigma_{u1}^2 \end{bmatrix} \right)$$

so that the errors u_{0j} *and* u_{1j} *are not independent.*

When formulating multilevel models, care must be exercised in the scaling of the variables x_{ij} and coding the variables z_{ij}. For example, x_{ij} may be centered using the grand mean $\overline{x}$ or the school mean $\overline{x}_j$. The parameterization of the model depends on the application. For a discussion of scaling and coding in multilevel modeling, consult Kreft and de Leeuw (1998). Bryk and Raudenbush (1992), Goldstein (1995) and Singer (1998) discuss hierarchical, multilevel models in some detail.

The SAS procedure PROC MIXED uses formulation (6.3.4) to analyze mixed models by defining $\mathbf{Z}$ and specifying the covariance structures of $\mathbf{G}$ and Ψ. When $\mathbf{Z}$ contains indicator variables and $\mathbf{G}$ contains variance components on it diagonal only and $\Psi = \sigma^2 \mathbf{I}_N$ then variance component models result. If $\mathbf{Z} = \mathbf{0}$ and $\Psi = \sigma^2 \mathbf{I}_n$ we obtain the GLM with fixed effects.

f. Prediction

In (6.3.9), the $E(\mathbf{y}_{N \times 1}) = \mathbf{X}\boldsymbol{\beta}$ since $E(\mathbf{a}) = \mathbf{0}$ and the cov $(\mathbf{y}) = \Omega$ where the covariance matrix of the observation vector $\mathbf{y}$ is nonsingular. This is the general Gauss-Markov setup for the observation vector. If the future $m \times 1$ vector $\mathbf{y}_f$ is not independent of $\mathbf{y}$, and

$\mathbf{W}_{m\times N}$ represents the covariance matrix between $\mathbf{y}$ and $\mathbf{y}_f$, then by (5.2.16) the BLUP for the future observation vector $\mathbf{y}_f$ is the vector

$$\widehat{\mathbf{y}}_f = \mathbf{X}\widehat{\boldsymbol{\beta}} + \mathbf{W}\Omega^{-1}(\mathbf{y} - \mathbf{X}\widehat{\boldsymbol{\beta}}) \tag{6.3.41}$$

where $\widehat{\boldsymbol{\beta}}$ is defined in (6.3.7). Thus, the predicted value of the future observation is again seen to be dependent on both $\mathbf{W}$ and Ω.

6.4 Mixed Model Examples

A simple application of the random coefficient (RC) regression model defined in (6.2.2) is to fit a linear model to an independent variable y_{ij} and a covariate x_{ij}. The subscript j represents the jth response on the ith unit. The unit may be a subject, an animal, a patient, a batch, a variety, a treatment, a company, etc. For a linear relation between y_{ij} and x_{ij}, the RC linear regression model is

$$\begin{aligned}
y_{ij} &= \beta_{0i} + \beta_{1i}x_{ij} + e_{ij} \\
\boldsymbol{\beta}_i &= \begin{bmatrix} \beta_{0i} \\ \beta_{1i} \end{bmatrix} \sim \text{IN}\left[\begin{bmatrix} \beta_0 \\ \beta_1 \end{bmatrix}, \Phi\right] \\
\Phi &= \begin{bmatrix} \sigma_0^2 & \sigma_{01} \\ \sigma_{10} & \sigma_1^2 \end{bmatrix} \\
e_{ij} &\sim \text{IN}\left(0, \sigma_e^2\right)
\end{aligned} \tag{6.4.1}$$

for $i = 1, 2, \ldots, n$ and $j = 1, 2, \ldots, r_i$. Letting $a_{0i} = \left(\beta_{0i} - \beta_0\right)$ and $a_{1i} = \left(\beta_{1i} - \beta_1\right)$, the linear RC regression model is written as a mixed linear model

$$\begin{aligned}
y_{ij} &= \beta_0 + a_{0i} + \left(\beta_1 + a_{1i}\right)x_{ij} + e_{ij} \\
&= \beta_0 + \beta_1 x_{ij} + a_{0i} + a_{1i}x_{ij} + e_{ij}
\end{aligned} \tag{6.4.2}$$

Setting $\boldsymbol{\beta}' = \left[\beta_0, \beta_1\right]$,

$$\begin{aligned}
\mathbf{a}_i &= \begin{bmatrix} a_{0i} \\ a_{1i} \end{bmatrix} = \begin{bmatrix} \beta_{0i} - \beta_0 \\ \beta_{1i} - \beta_1 \end{bmatrix} = \boldsymbol{\beta}_i - \boldsymbol{\beta} \\
\mathbf{a}_i &\sim \text{IN}\,(\mathbf{0}, \Phi)
\end{aligned}$$

and letting

$$\mathbf{X}_i = \begin{bmatrix} 1 & x_{i1} \\ 1 & x_{i2} \\ \vdots & \vdots \\ 1 & x_{ir_i} \end{bmatrix}$$

the model in (6.4.2) has the form specified in (6.2.2) with $\mathbf{Z}_i \equiv \mathbf{X}_i$. The covariance structure for the model is

$$\Omega_i = \mathbf{Z}_i' \begin{bmatrix} \sigma_0^2 & \sigma_{01} \\ \sigma_{10} & \sigma_1^2 \end{bmatrix} \mathbf{Z}_i + \begin{bmatrix} \sigma_e^2 & 0 \\ 0 & \sigma_e^2 \end{bmatrix} \tag{6.4.3}$$

so that the vector of variance components $\boldsymbol{\theta}' = \left[\sigma_0^2, \sigma_{01}, \sigma_1^2, \sigma_e^2\right]$.

For this example the pairs of observations $\left(y_{ij}, x_{ij}\right)$ are observed for $i = 1, 2, \ldots, n$ units and $j = 1, 2, \ldots, r_i$ responses. In the RC model, the random regression lines vary about an overall unknown mean, $E\left(y \mid x\right) = \beta_0 + \beta_1 x$. Because $a_{0i} = \beta_{01} - \beta_0$ and $a_{1i} = \beta_{1i} - \beta_1$, the estimates of a_{0i} and a_{1i}, called Best Linear Unbiased Predictors (BLUP), are deviation scores.

In model (6.4.2), y_{ij} was defined with random and fixed intercepts, and slopes. And the variable x_{ij} was a continuous fixed variable. Now assume that x_{ij} is a random classification variable that is nested within the levels of i. Then, letting $a_{0i} = x_{(i)j}$ the statistical model becomes a mixed linear model with general structure given in (6.3.3) or for our example we have that

$$\begin{aligned} y_{ij} &= \beta_0 + \beta_1 x_{ij} + a_{1i} x_{ij} + x_{(i)j} + e_{ij} \\ &= \beta_0 + \beta_1 x_{ij} + a_{1i} x_{ij} + e_{ij}^* \end{aligned} \tag{6.4.4}$$

where $e_{ij}^* = x_{(i)j} + e_{ij}$. Assuming $x_{(i)j}$ and e_{ij} are jointly independent, and that $x_{(i)j} \sim \text{IN}\left(0, \sigma_x^2\right)$ and $e_{(i)j} \sim \text{IN}\left(0, \sigma_e^2\right)$, the covariance structure for e_{ij}^* is

$$e_{ij}^* \sim \text{IN}\left(0, \begin{bmatrix} \sigma_x^2 + \sigma_e^2 & \sigma_x^2 \\ \sigma_x^2 & \sigma_x^2 + \sigma_e^2 \end{bmatrix}\right)$$

where $x_{(i)j}$ is the random effect of the jth level of x within unit i, the $\text{var}(e_{ij}^*) = \sigma_x^2 + \sigma_e^2$, and the $\text{cov}(e_{(ij_1)}, e_{ij_2}) = \sigma_x^2$. The covariance matrix has compound symmetry structure. Letting $\boldsymbol{\beta}' = \left[\beta_0, \beta_1\right]$ and $a_{1i} = \beta_{1i} - \beta_1$ denote a random slope effect such that $a_{1i} \sim \text{IN}(0, \sigma_1^2)$ where $\mathbf{y}_i$ represents the $(r_i \times 1)$ vector of observations for the ith unit $i = 1, 2, \ldots, n$ and

$$\mathbf{X}_i = \begin{bmatrix} 1 & x_{i1} \\ 1 & x_{i2} \\ \vdots & \vdots \\ 1 & x_{ir_i} \end{bmatrix} \qquad \mathbf{Z}_i = \begin{bmatrix} 1 \\ 1 \\ \vdots \\ 1 \end{bmatrix}$$

the mean and covariance structure for $\mathbf{y}_i$ is

$$\begin{aligned} E\left(\mathbf{y}_i\right) &= \mathbf{X}_i \boldsymbol{\beta} \\ \text{cov}\left(\mathbf{y}_i\right) &= \mathbf{Z}_i' \Phi \mathbf{Z}_i + \Psi_i = \Omega_i \end{aligned}$$

where $\Phi = \sigma_1^2$ and Ψ_i has compound symmetry structure

$$\begin{aligned} \Omega_i &= \sigma_1^2 \mathbf{J}_{r_i} + \sigma_x^2 \mathbf{J}_{r_i} + \sigma_e^2 \mathbf{I}_{r_i} \\ &= \sigma_1^2 \mathbf{J}_{r_i} + \Psi_i \end{aligned}$$

The vector of random components $\boldsymbol{\theta}' = [\sigma_1^2, \sigma_x^2, \sigma_e^2]$. Model (6.4.4) may arise if i represents random diets, and j represents individuals nested within diets and one wants to investigate the relationship between weight and age. Fuller and Battese (1973) call this model a one-fold linear model with nested error structure.

In the specification for y_{ij} in (6.4.4), we replaced the random intercept a_{0i} with a nested random variable $x_{(i)j}$. Now suppose we remove $a_{1i}x_{ij}$ from the model, then our revised model is

$$\begin{aligned} y_{ij} &= \beta_0 + \beta_1 x_{ij} + a_{0i} + e_{ij} \\ &= \beta_0 + a_{0i} + \beta_1 x_{ij} + e_{ij} \end{aligned} \tag{6.4.5}$$

where β_0 is a fixed intercept (overall mean), β_1 is the slope for the fixed covariate x_{ij}, and a_{0i} represents the random intercept (treatment) effect. This is a mixed ANCOVA model. Assuming $a_{0i} \equiv \alpha_i$ and $\beta_0 \equiv \mu$, model (6.4.5) is a LFR model. Letting $\mu_i = \beta_0 + \alpha_i$, the model is a FR model. The models are represented as follows

$$\begin{array}{ll} \text{LFR} & y_{ij} = \mu + \alpha_i + \beta_1 x_{ij} + e_{ij} \\ \text{FR} & y_{ij} = \mu_i + \beta_1 x_{ij} + e_{ij} \end{array} \tag{6.4.6}$$

Both models are fixed effect ANCOVA models. To obtain estimates for intercepts and slopes using PROC MIXED, one must use the no intercept (NOINT) option. To obtain a test that all α_i are zero, the LFR model is used. The parameter μ has no meaning in the LFR model. To provide meaning, one often reparameterizes the model as

$$\begin{aligned} y_{ij} &= \mu + \alpha_i + \beta_1 x_{ij} + e_{ij} \\ &= \mu - \beta_1 x_{..} + \alpha_i + \beta_1 (x_{ij} - x_{..}) + e_{ij} \\ &= \mu_{..} + \alpha_i + \beta_1 (x_{ij} - x_{..}) + e_{ij} \end{aligned}$$

where $x_{..}$ is the unweighted mean of the covariate and β_1 is the slope parameter for the common regression equation. The parameter $\mu_{..}$ becomes an overall mean by adding the restriction to the model that the sum of the treatment effects α_i is zero. This is the familiar adjusted means form of the ANCOVA model.

This simple example illustrates the flexibility of the RC regression/mixed linear model. We illustrate the relationships among models (6.4.2), (6.4.4), (6.4.5) and (6.4.6) with a data set using PROC MIXED.

a. *Random Coefficient Regression (Example 6.4.1)*

This example is a modification of a problem discussed in SAS (1997, p. 684) based upon a phamaceutical stability experiment from Obenchain (1990). The study involves the recording of drug potency (expressed as a percentage of the claim on the label), the dependent variables (y_{ij}), for several months of shelf life (0, 1, 3, 6, 9 and 12), the independent variable (x_{ij}). Three batches of product sampled from many products were examined. The potency levels may differ in both initial potency (intercept) and rate of degradation (slope). Since batches are random, variation in each batch may differ from an overall fixed level of initial potency and rate of potency loss as shelf life increases. The data for the study are shown in Table 6.4.1. The program for this example is labeled m6_4_1.sas.

TABLE 6.4.1. Pharmaceutical Stability Data

Batch	Month	Potency Ratio					
1	0	101.2	103.3	103.3	102.1	104.4	102.4
1	1	98.8	99.4	99.7	99.5	.	.
1	3	98.4	99.0	97.3	99.8	.	.
1	6	101.5	100.2	101.7	102.7	.	.
1	9	96.3	97.2	97.2	96.3	.	.
1	12	97.3	97.9	96.8	97.7	97.7	96.7
1	0	102.6	102.7	102.4	102.1	102.9	102.6
2	1	99.1	99.0	99.9	100.6	.	.
2	3	105.7	103.3	103.4	104.0	.	.
2	6	101.3	101.5	100.9	101.4	.	.
2	9	94.1	96.5	97.2	95.6	.	.
2	12	93.1	92.8	95.4	92.2	92.2	93.0
3	0	105.1	103.9	106.1	104.1	103.7	104.6
3	1	102.2	102.0	100.8	99.8	.	.
3	3	101.2	101.8	100.8	102.6	.	.
3	6	101.1	102.0	100.1	100.2	.	.
3	9	100.9	99.5	102.2	100.8	.	.
3	12	97.8	98.3	96.9	98.4	96.9	96.5

The model for the analysis is represented using (6.4.2), where y_{ij} $\equiv$ potency; x_{ij} $\equiv$ months (a continuous variable), β_0 and β_1 represent the fixed slopes and intercepts, and a_{0i} and a_{1i} represent the random slopes and intercepts. There are $n = 3$ batches (units) for level i and an unequal number of observations r_i at each level.

To run PROC MIXED, the method of estimation for the variance components σ_0^2, σ_{01}^2, σ_1^2 is specified using the METHOD= option. The default is METHOD=REML; the options COVTEST, CL and ALPHA = are used to obtain estimates for the variance components, asymptotic standard errors, asymptotic Z tests and approximate $1 - \alpha$ confidence intervals for each element in Φ. The MODEL statement specifies the design matrix $\mathbf{X}$; the intercept is always included by default. The SOLUTION and OUTP= options request that fixed effects be printed with t tests and the fitted values $\widehat{\mathbf{y}} = \mathbf{X}\widehat{\boldsymbol{\beta}} + \mathbf{Z}\widehat{\mathbf{a}}$ be stored in the data set defined by OUTP=. These plots are used to investigate the presence of outliers. The option DDFM = SATTERTH corrects the degrees of freedom for tests of fixed effects for both t tests and F tests. The RANDOM statement defines the structure of the random effects in the model, the covariance structure Φ, and the "subjects" in the data set. The TYPE = UN provides estimates of the random variance components $\sigma_{12} \equiv \sigma_{01} \neq 0$, $\sigma_1^2 \equiv \sigma_0^2$ and $\sigma_2^2 \equiv \sigma_1^2$ as shown in Table 6.3.1. Thus, we have variance component estimates for the slope, intercept and the covariance between the slope and intercept. The SUBJECT = BATCH option specifies that the slopes and intercepts are independent across batches, but correlated within a batch. The REPEATED statement is used to define Ψ_i; with no statement, $\Psi_i = \sigma_e^2 \mathbf{I}_{r_i}$. Using (6.2.9), the option SOLUTION prints the BLUPs for the random effects in the model, $a_{0i} = \beta_{0i} - \beta_0$ and $a_{1i} = \beta_{1i} - \beta_0$. More generally, $\widehat{\mathbf{a}}_i =$

$\widehat{\Phi}\mathbf{Z}_i'\widehat{\Omega}_i^{-1}(\mathbf{y}_i - \mathbf{X}_i\widehat{\boldsymbol{\beta}})$. The standard errors are obtained using (6.2.10).

In fitting model (6.4.2) to the data in Table 6.4.1, we find the estimates for Φ, σ_e^2 and $\boldsymbol{\beta}$ as

$$\widehat{\Phi} = \begin{bmatrix} 0.9768 & -0.1045 \\ -0.1045 & 0.0372 \end{bmatrix} \quad \widehat{\boldsymbol{\beta}} = \begin{bmatrix} 102.70 \\ -0.5259 \end{bmatrix} \tag{6.4.7}$$

$$\widehat{\sigma}_e^2 = 3.2932$$

The components of variance are labeled "UN(i, j)" in the table "Covariance Parameter Estimates"; the estimate of error variance is labeled "Residual". The elements of the fixed parameter $\boldsymbol{\beta}$ are in the table labeled "Solution for Fixed Effects". Also provided for the fixed effects are approximate t tests that the coefficients are zero. The tests are obtained by dividing the parameter estimate by its corresponding estimated (asymptotic) standard error; the degrees of freedom for each test is set to that appearing in the "Test of Fixed Effects" table calculated using (6.2.16). For our example, the test that the slope is zero is only marginally significant using $\alpha = 0.05$, the p-value is 0.0478. Associated with the variance components are Z tests that the components are zero. These asymptotic tests have limited value unless the sample is large; they are nonsignificant for our example. Of interest is the confidence intervals created for each component. These are formed using a first-order Satterthwaite approximation and the chi-square distribution. Because we set $\alpha = 0.10$, an approximate 90% confidence interval for each component is generated. Since the interval for the covariance σ_{01} contains zero, one may consider fitting an alternative model with TYPE = VC for Φ, a diagonal matrix. We will discuss this further later.

Finally, the table of "Solution for Random Effects" provide estimates for $\widehat{\mathbf{a}}_i = \widehat{\boldsymbol{\beta}}_i - \widehat{\boldsymbol{\beta}}_0$ which are deviations of the random coefficients about the mean. For each batch, the deviations are as follows

Batch	Intercept	Slope
1	−1.0010	0.1287
2	0.3934	−0.2060
3	0.6076	−0.0773

(6.4.8)

Because $\alpha = 0.083$ on the random statement, the confidence intervals for the six random effects are approximate simultaneous 95% confidence intervals for the random effects, using the Bonferroni inequality. Thus, the random variation of each batch about $\boldsymbol{\beta}$ does not appear to be significant. The "Test of Fixed Effects" in the output is using (6.2.14) and (6.2.16) to test whether the fixed slope parameter is nonzero. Because we have only one parameter, the F test is identical to the t test for the coefficient with p-value = 0.0478. The analysis suggests no serious variation from batch to batch and that potency degradation is linear.

While PROC MIXED does not directly provide any methods for model checking, one may plot the predicted residuals $\widehat{\mathbf{e}} - \mathbf{y} - \mathbf{X}\widehat{\boldsymbol{\beta}} - \mathbf{Z}\widehat{\mathbf{a}} = \widehat{\mathbf{r}}$ versus the fitted values $\widehat{\mathbf{y}} = \mathbf{X}\widehat{\boldsymbol{\beta}} - \mathbf{Z}\widehat{\mathbf{a}}$

and investigate the plot for outliers. This is accomplished using the PROC PLOT. One may also use PROC UNIVARIATE to evaluate the normality of $\widehat{\mathbf{e}}$ and $\widehat{\mathbf{a}}$; however, $\widehat{\mathbf{e}}$ is confounded with $\widehat{\mathbf{a}}$. The plot of residuals do not indicate the presence of outliers in the data.

In model (6.4.2), the independent variable $x_{ij} \equiv$ MONTH is a continuous variable. To define a new classification variable $x_{(i)j}$ that is qualitative, the MONTHC variable is created in the PROC DATA step. It may be treated as a factor in the linear model. To associate model (6.4.4) with this example, the random parameter $a_{1i} \equiv$ MONTH and the random intercept is removed from the model. This permits a more complex specification for the errors e^*_{ij} (a correlated error structure); for example, compound symmetry. To fit model (6.4.4), the RANDOM statement only includes MONTH; the intercept has been removed. By default TYPE = VC, SUBJECT = BATCH and the covariance parameter is MONTH. The REPEATED statement models Ψ_i. The TYPE = CS specifics the covariance structure to be compound symmetry, Table 6.3.1. The option SUB = MONTHC (BATCH) specifies that the structure pertains to each submatrix of each MONTHC within each BATCH so that Ψ is block diagonal. For this model, we have used both the RANDOM and REPEATED statements in order to structure $\Omega = \bigoplus_{i=1}^n \Omega_i$, TYPE = VC structures Φ and TYPE = CS structures Ψ_i.

Fitting the new model,

$$\widehat{\Psi}_i = \begin{bmatrix} \widehat{\sigma}_e^2 + \widehat{\sigma}_x^2 & \widehat{\sigma}_x^2 \\ \widehat{\sigma}_x^2 & \widehat{\sigma}_e^2 + \widehat{\sigma}_x^2 \end{bmatrix}$$

where $\widehat{\sigma}_e^2 = 0.7967$, $\widehat{\sigma}_x^2 = \widehat{\sigma}^2_{\text{MONTHC}} = 3.7411$ and $\widehat{\Phi} = \widehat{\sigma}_1^2 = \widehat{\sigma}^2_{\text{MONTHC}} = 0.0124$. Thus, while the error variance had decreased, the component of variance due to the nested errors (MONTHC) has increased. The random slope component of variance, σ_1^2, is about the same for the two models. The fixed effects vector for the model with nested, compound symmetry error is

$$\widehat{\boldsymbol{\beta}} = \begin{bmatrix} 102.56 \\ -0.5003 \end{bmatrix} \tag{6.4.9}$$

Comparing these with $\widehat{\boldsymbol{\beta}}$ in (6.4.7), there is little difference in the estimates; however, there are larger differences in the estimated standard errors. The t test for the fixed slope parameter has p-value $= 0.0579$, larger than the $\alpha = 0.05$ nominal level.

Because model (6.4.4) is not nested within model (6.4.2), we may only compare the models using the information criteria. A comparison of AIC (-141.451) and SBC(-145.061) for this model with the nonnested model with unknown covariance structure (-179.164 and -183.978, respectively) favors the model with errors that are nested with compound symmetry structure.

In modeling the pharmaceutical stability data, Φ had unknown structure and $\Psi_i = \sigma_e^2 \mathbf{I}_{r_i}$. For the nested compounded symmetry model, Ψ_i and Φ were both changed. Assuming $\Psi_i = \sigma_e^2 \mathbf{I}_{r_i}$, we now consider setting the covariance between the random intercept and slope to zero. Thus, we return to our first model, changing TYPE = UN to TYPE = VC.

The matrix

$$\Phi = \begin{bmatrix} \sigma_0^2 & 0 \\ 0 & \sigma_1^2 \end{bmatrix}$$

and the linear model is given as in (6.4.2). Now, Φ, σ_0^2 and $\boldsymbol{\beta}$ are

$$\widehat{\Phi} = \begin{bmatrix} 0.8129 & 0 \\ 0 & 0.0316 \end{bmatrix} \quad \widehat{\boldsymbol{\beta}} = \begin{bmatrix} 102.70 \\ -0.5259 \end{bmatrix} \tag{6.4.10}$$

$$\widehat{\sigma}_e^2 = 3.3059$$

While $\widehat{\boldsymbol{\beta}}$ is identical for the two models, the reduction in the estimated standard errors yield a p-value = 0.0398 for the slope parameter.

Comparing the information criteria for this model, AIC $= -178.422$ and SBC $= -182.033$. with the TYPE = UN model, (-179.164, -183.978, respectively) the TYPE = VC model is slightly better. However, because this model is nested within the former model, we may use (6.3.15) to evaluate model fit. Under the reduced model we assume $\sigma_{01} = 0$, and we are testing $H : \sigma_{01} = 0$ vs A. $\sigma_{01} \neq 0$. For TYPE = VC, the reduced model, -2 (Residual Log-likelihood) = 350.8449. For the full model, TYPE = UN, the corresponding value is 350.3281. Comparing the difference of 0.5168 to an asymptotic chi-square distribution with one degree of freedom, we fail to reject H. Thus, there is little evidence to believe that $\sigma_{01} \neq 0$. Observe however, that the model with correlated error and compound symmetry structure fits the data better than the TYPE = VC model.

In fitting model (6.4.2) with the two covariance structures for Φ, TYPE = VC and TYPE = UN, we concluded that the covariance term $\sigma_{01} = 0$ and that the random slopes do not vary significantly from the fixed slope. This motivates one to consider fitting model (6.4.5) to the data; a mixed ANCOVA model, ignoring a nested model structure, for the moment. Model (6.4.5) is identical to model (6.4.6), if we consider α_i to be random. We illustrate the two formulations in program m6_4_1.sas.

For the mixed ANCOVA model, the variance components are $\sigma^2_{\text{INTERCEPT}} = \sigma^2_{\text{BATCH}} = 0.8911$ and $\widehat{\sigma}_e^2 = 3.8412$. The parameter vector of fixed effects is

$$\boldsymbol{\beta} = \begin{bmatrix} 102.70 \\ -0.5259 \end{bmatrix}$$

Now the test of $H : \beta_1 = 0$ is significant with p-value = 0.0001. Again, the random variation in the random intercepts (batches) do not vary significantly about the fixed intercept. Because this model is nested within the model with random slopes and intercepts and TYPE = VC structure for Φ, we may compare the models using the difference in likelihood tests. The difference $358.9971 - 350.8449 = 8.1522$ for $\alpha = 0.05$ is compared to a chi-square distribution with one degree of freedom: $\chi^2_\alpha = 3.84$. Since $8.1522 > 3.84$, we reject the mixed ANCOVA model for these data.

We also include in program m6_4_1.sas the analysis of the univariate fixed effects ANCOVA model using PROC MIXED. The NOINT option permits one to obtain slope and intercept estimates; however, the fixed effect test of BATCH must be ignored. The test of equal BATCH effects is not being tested using the NOINT option. The test of BATCH is testing $\mu_1 = \mu_2 = \mu_3$. The test of parallelism is tested using the fixed effects test of interaction. For our example, the p-value = 0.0009. Hence, the ANCOVA model is not appropriate for these data. One could consider a fixed effects model with unequal slopes; the intraclass covariance model is discussed by Timm and Mieczkowski (1997, p. 67).

When fitting any RC regression model or a mixed linear model to a data set, one should explore the covariance structure of the data by comparing several linear mixed models.

In Example 6.4.1, we showed how the general (univariate) linear mixed model in (6.3.4) may be used to analyze models with random coefficients. The model may also be used to analyze univariate models with random effects, mixed effects, and models with repeated measurements. We next consider an orthogonal/nonorthogonal mixed linear model discussed by Milliken and Johnson (1992, p. 290).

b. *Generalized Randomized Block Design (Example 6.4.2)*

In this example, a company wanted to replace machines (TREAT) to make a new component in one of its factories. Three different machines were analyzed to evaluate the productivity of the machines, six employees (BLOCK) were randomly selected to operate each of the machines on three independent occasions. Thus, treatments and blocks are crossed factors. The dependent variable (SCORE), takes into account the number and quality of the components produced. Because the machines were pre-determined and not sampled from a population, the machine $\equiv$ treatment factor is fixed. Because the employees were sampled from the company at large, persons $\equiv$ blocks are random. In addition, the interaction between treatment and blocks is random. This design is the familiar completely general randomized block (CGRB), nonorthogonal, two-way mixed (Model III) ANOVA design. The statistical mixed linear model for the design is

$$y_{ijk} = \mu + \alpha_i + b_j + (\alpha b)_{ij} + e_{ijk} \tag{6.4.11}$$

for $i = 1, 2, \ldots, a$; $j = 1, 2, \ldots, b$; and $k = 1, 2, \ldots, n_{ij}$. The parameter μ is an overall constant, α_i is the fixed effect of treatment i, b_j is the random effect for block j (to more easily identify the random effects we do not use Greek letters) and $(\alpha b)_{ij}$ is the random interaction between treatments and blocks. The random components are assumed to be normally distributed: $b_j \sim \text{IN}(0, \sigma_b^2)$, $(\alpha b) \sim \text{IN}(0, \sigma_{\alpha b}^2)$ and $e_{ijk} \sim \text{IN}\left(0, \sigma_e^2\right)$.

The data for the model given in (6.4.11) are included in Table 6.4.2. For the completely balanced, orthogonal design, $i = 1, 2, 3$; $j = 1, 2, \ldots, 6$ and $r_{ij} = 3$. For model (6.3.3), the number of replications $r = 3$. To create a nonorthogonal design the number of observations per cell vary; the observations with an asterisk (*) were randomly deleted for a nonorthogonal design and reanalyzed. The SAS code for the example is contained in program m6_4_2.sas.

To analyze model (6.4.11) using the data in Table 6.4.2, two basic approaches have been put forth. The first technique is called the ANOVA or method-of-moments procedure discussed in Searle (1971, Chapter 11). It involves obtaining sums of squares (SS) assuming a fixed model and evaluating the expected mean sum of squares (EMS) under a "set of

TABLE 6.4.2. CGRB Design (Milliken and Johnson, 1992, p. 285)

	Treat 1	Treat 2	Treat 3
	52.0	62.1*	67.5
Block 1	52.8*	62.6*	67.2
	53.1*	64.0	66.9
	52.8	59.7	61.5
Block 2	51.8	60.0	61.7
	53.8*	59.0	62.3
	60.0	68.6	70.8
Block 3	60.2*	65.8	70.6
	58.4*	69.7*	71.0
	51.1	63.2	64.1
Block 4	52.3	62.8	66.2
	50.3*	62.2	64.0
	50.9	64.8	72.1
Block 5	51.8	65.0	72.0
	51.4	64.5*	71.1
	46.4	43.7	62.0
Block 6	44.8	44.2	61.4
	49.2	43.0	60.5

mixed model assumptions". To estimate variance components, one equates the EMS with the observed mean squares; this results in a system of equations that must be solved to obtain method-of-moment estimators (estimates obtained by the ANOVA method) which may be negative. Assuming the y_{ijk} are normally distributed, for most balanced designs exact F tests that a variance component is zero versus the alternative that it is larger than zero may be tested by forming an appropriate ratio of expected mean squares. However, the tests depend on the mixed model specification; in particular, whether or not the mixed model includes or does not include side conditions or constraints on the fixed parameters, Hocking (1985, p. 330), Searle (1971, p. 400) and Voss (1999), and Hinkelmann et al. (2000). To compute EMSs in SAS, one may use PROC GLM. Because PROC GLM uses Hartley's (1967) method of synthesis, which does not impose side conditions on the fixed parameters, the EMSs are calculated assuming a LFR linear model with no restrictions. In some balanced designs, there does not always exist simple ratio's of expected mean squares for testing variance components. Then, one creates quasi-F tests using the Satterthwaite approximation to estimate the denominator degrees of freedom for the test.

A second approach to the analysis of (6.4.11) is to use (6.3.4). For balanced models, the covariance structure has the general form shown in (6.3.38). This approach requires a stronger assumption on the ANOVA model, we must now assume that the y_{ijk} are jointly multivariate normal. Then we may obtain ML and REML estimates of variance components. For balanced designs, REML estimates often agree with method-of-moment estimates. The tests of variance components are more problematic. As indicated earlier, the Z test has limited value unless one uses corrections or has large sample sizes. One may use F tests with appropriate Satterthwaite corrections to the denominator degrees of freedom for balanced designs. For nonorthogonal designs, forming the difference likelihood ratio test as in (6.3.15) is often used. Two models are fit to the data, one with the random components of interest set to zero and the other with the corresponding component nonzero. Then, a large sample chi-square test on one degree of freedom is used to test that a component is zero. Approximate $(1-\alpha)\,\%$ confidence intervals may be constructed for variance components by using a chi-square distribution and the Satterthwaite (1946) approximation procedure. These are generated in PROC MIXED by using the CL option on the PROC MIXED statement. An improved procedure for obtaining confidence intervals for variance components has been proposed by Ting et al. (1990), however, it is not currently supported in SAS. A discussion of the method is provided in Neter et al. (1996, p. 973).

To analyze the data in Table 6.4.2, SAS code for both the analysis of the orthogonal and nonorthogonal designs is provided using both PROC GLM and PROC MIXED. Because of improved algorithms in SAS, the output for this example does not agree with those provided by Milliken and Johnson (1992, Chapter 23). In addition, they did not use PROC MIXED for their analysis. We briefly review the input and output for this example.

When using PROC GLM to analyze a mixed model, the MODEL statement contains both fixed and random effects. The RANDOM statement includes only random effects. And, the /TEST option creates the F statistics. When formulating the RANDOM effect components, we have specified that both the BLOCK and BLOCK * TREAT factors as random. If one would only specify the BLOCK effect as random and not the interaction term, different results would be obtained. The constructed F tests are similar to the analysis of model (6.4.11) with restrictions imposed on the fixed parameters. Recall that the EMS depend on the set of mixed model assumptions. When the interaction term is considered fixed, the test of treatment is confounded with interaction so that no valid test of treatment is available as noted in PROC GLM for this situation. For balanced, orthogonal designs, the SAS output for the EMS and ANOVA tests agree with those discussed by Milliken and Johnson (1992, p. 286). The REML estimates of variance components in PROC MIXED agree with the methods-of-moment estimates. Because of rounding error in the interpolation of the chi-square critical values, the confidence intervals for variance components produced by PROC MIXED differ slightly from those of Milliken and Johnson.

In the PROC MIXED procedure, the random components are not included in the MODEL statement. The DDFM option is used to adjust the F tests for testing fixed effects. Observe that the F test for treatments using PROC GLM and that using PROC MIXED have the same value, $F = 20.58$ with p-value $= 0.0003$. This is because the design is balanced. This F statistic would not be obtained using PROC GLM if one incorrectly considered the interaction term as fixed and not random. Even with the stronger model assumptions, the two procedures are in agreement when exact F tests exist for balanced designs.

The analysis of nonorthogonal mixed linear models is more complicated. Now we have several methods for calculating sums of squares (TYPE I, II, III and IV) and numerous pro-

TABLE 6.4.3. ANOVA Table for Nonorthogonal CGRB Design

Source	(dfh, dfe)	F	EMS	P-value
TREAT	(2, 10.04)	16.5666	$\sigma_e^2 + 2.137\,\sigma_{\alpha b}^2 + Q(\alpha)$	0.0007
BLOCK	(5, 10.02)	5.1659	$\sigma_e^2 + 2.2408\,\sigma_{\alpha b}^2 + 6.7224\,\sigma_b^2$	0.0133
BLOCK*TREAT	(10, 26)	46.3364	$\sigma_e^2 + 2.3162\,\sigma_{\alpha b}^2$	0.0001

posed methods/rules for calculating expected mean squares, Milliken and Johnson (1992, Chapters 10 and 18) and Searle (1971, p. 389). The procedure GLM calculates TYPE I, II, III, or IV sums of squares by specifying E1, E2, E3, or E4 on the MODEL statement. The default is TYPE III sum of squares. For nonorthogonal designs, all cells must be filled when the TYPE III option is used. While PROC GLM develops F tests for both fixed and random components, the tests for random components become extremely unreliable with more incomplete data. Because PROC MIXED only develops tests using TYPE III sums of squares, all fixed effect tests are unweighted.

Comparing the ANOVA table produced by PROC GLM for the unbalanced, nonorthogonal design in Table 6.4.2, with that provided by Milliken and Johnson (1992, p. 290), one observes that the results do not agree. This is due to the fact that their EMSs are calculated from sums of squares obtained by the method of fitting constants or Henderson's Method III procedure. This is equivalent to calculating EMSs using TYPE I sum of squares in SAS. The test of fixed effects are weighted tests that depend on the cell frequencies. The unweighted test of treatment differences obtained employing PROC GLM result in a quasi F test. The output from PROC GLM is shown in Table 6.4.3.

Using PROC MIXED, the test of treatment differences results in an approximate F statistic, $F = 19.97$ with degrees of freedom, $df = (2, 10.1)$. Neither PROC MIXED or PROC GLM generates an exact test for evaluating differences in treatments. Both are approximations to an F-distribution. The F statistic in PROC GLM is a quasi F test, the numerator and denominator MSs are not independent. In PROC MIXED, the fixed effects are weighted least squares estimates and F defined in (6.2.14) with the correction in (6.2.16) approximately follows an F-distribution. It is currently unknown which procedure is best. Also observe the very large difference in the p-value for the Z tests in PROC MIXED compared to the quasi F tests for testing $H : \sigma^2_{\text{BLOCK}} = 0$ and the exact F test for testing $H : \sigma^2_{\text{BLOCK}*\text{TREAT}} = 0$ with p-values 0.0133 and 0.0001, respectively. The p-values for the Z tests in PROC MIXED are 0.1972 and 0.0289, respectively. While both tests for BLOCK differences may be unreliable, the Z tests are useless. The REML estimates for the variance components are $\widehat{\sigma}_b^2 = 22.4551$, $\widehat{\sigma}_{\alpha b}^2 = 14.2340$, and $\widehat{\sigma}_e^2 = 0.8709$ are again close to the estimates reported by Milliken and Johnson using method-of-moment method: $\widehat{\sigma}_b^2 = 21.707$, $\widehat{\sigma}_{\alpha b}^2 = 17.079$, and $\widehat{\sigma}_e^2 = 0.873$.

To obtain approximate 95% simultaneous confidence intervals for differences in unweighted treatment means, the LSMEANS statement is used in PROC MIXED with the TUKEY option to control the Type I error rate at the $\alpha = 0.05$ nominal level.

c. *Repeated Measurements (Example 6.4.3)*

For our next example, we consider a repeated measurements design with unbalanced and missing data, Milliken and Johnson (1992, p. 385). This design was discussed briefly in Chapter 3 with complete data. The study involves an experiment to determine the effects of a drug on the scores obtained by depressed patients on a test designed to measure depression. Two patients were in the placebo group, and three were assigned to the drug group. The mixed ANOVA model for the design is

$$\begin{aligned} y_{ijk} &= \mu + \alpha_i + s_{(i)j} + \beta_j + (\alpha\beta)_{ij} + e_{ijk} \\ i &= 1, 2, \ldots, a;\ j = 1, 2, \ldots, r_i, k = 1, 2, \ldots, r_{ij} \end{aligned} \tag{6.4.12}$$

where $s_{(i)j} \sim \text{IN}\left(0, \sigma_s^2\right)$, $e_{ijk} \sim \text{IN}\left(0, \sigma_e^2\right)$ are jointly independent; a nonorthogonal split-plot design. Patients ≡ subjects are random and nested within treatments, drug groups. For the study, two patients did not return for examination often the second week. The data for the analysis are given in Table 6.4.4.

To analyze the data in Table 6.4.4, program m6_4_3.sas is used with code for both PROC MIXED and PROC GLM. Because two patients did not return for re-examination, a TYPE III analysis is appropriate for these data. The missing observations are not due to treatment. The ANOVA table from PROC GLM is provided in Table 6.4.5. The results are in agreement with those discussed by Milliken and Johnson (1992, p. 394) using TYPE III sums of squares.

To test for differences in drug treatment differences, PROC GLM constructed a quasi F test. The statistic follows an F-distribution only approximately since the numeration and denominator mean squares are no longer independent. Testing for differences between Week 1 and Week 2, observe that the test is confounded by interaction and should thus be avoided. The test of no significant interaction is an exact F test for this design. All tests of fixed effects using PROC MIXED are approximate. We have again selected DDFM = SATTERTH which calculates the degrees of freedom from the data. This option should always be used with unbalanced designs. For balanced, orthogonal designs, PROC MIXED makes available other options, Littell et al. (1996). However, no clear guidelines have been provided by the SAS Institute to recommend one over another except for computational convenience. One option that should never be used is DDFM = RESIDUAL because it tends to generally overestimate the denominator degrees of freedom.

TABLE 6.4.4. Drug Effects Repeated Measures Design

Treatment	Subject	Week 1	Week 2
Placebo	1	24	18
	2	22	—
Drug	1	25	22
	2	23	—
	3	26	24

TABLE 6.4.5. ANOVA Table Repeated Measurements

Source	(dfh, dfe)	F	EMS	P-value
TREAT	(1, 3.10)	6.5409	$\sigma_e^2 + 1.1111\,\sigma_s^2 + Q\,(\alpha, \alpha * \beta)$	0.0806
PATIENT (TREAT)	(3, 1)	11.2222	$\sigma_e^2 + 1.3333\,\sigma_s^2$	0.2152
WEEK	(1, 1)	96.3333	$\sigma_e^2 + \mathrm{Q}(\alpha, \alpha * \beta)$	0.0646
TREAT*WEEK	(1, 1)	16.3333	$\sigma_e^2 + Q\,(\alpha * \beta)$	0.1544
ERROR	1	0.25	σ_e^2	

d. *HLM Model (Example 6.4.4)*

In a one-way mixed ANOVA model, an observation y_{ij} for the ith student in the jth school is represented as

$$
\begin{aligned}
y_{ij} &= \mu + a_j + e_{ij} \\
a_j &\sim \mathrm{IN}\left(0, \sigma_a^2\right) \\
e_{ij} &\sim \mathrm{IN}\left(0, \sigma_e^2\right)
\end{aligned}
\tag{6.4.13}
$$

where a_j and e_{ij} are independent random variables for $i = 1, 2, \ldots, n_j$ and $j = 1, 2, \ldots, J$ Thus,

$$
\operatorname{var} y_{ij} = \sigma_e^2 + \sigma_a^2
$$

$$
\operatorname{cov}\left(y_{i'j}, y_{ij}\right) = \begin{cases} \sigma_a^2 & i \neq i' \\ 0 & \text{otherwise} \end{cases}
$$

Letting

$$
\mathbf{y} = \begin{bmatrix} y_{11} \\ y_{21} \\ y_{12} \\ y_{22} \end{bmatrix} = \begin{bmatrix} \sigma_e^2 + \sigma_a^2 & \sigma_a^2 & 0 & 0 \\ \sigma_a^2 & \sigma_e^2 + \sigma_a^2 & 0 & 0 \\ 0 & 0 & \sigma_e^2 + \sigma_a^2 & \sigma_a^2 \\ 0 & 0 & \sigma_a^2 & \sigma_e^2 + \sigma_a^2 \end{bmatrix}
$$

the covariance matrix for $\mathbf{y}$ has block diagonal structure. The intraclass coefficient between $y_{ij}, y_{i'j}$ is

$$
\rho = \frac{\sigma_a^2}{\sigma_e^2 + \sigma_a^2}
$$

This is a measure of the proportion of the total variability in the y_{ij} that is accounted for by the variability in the a_j. When $\sigma_e^2 = 0$, then $\sigma_y^2 = \sigma_a^2$ and all the variance in y is accounted for by the variation in a_j. There is no random error and y is totally reliable.

One may also formulate y_{ij} in (6.4.13) as a multilevel, hierarchical model where the $level - 1$, student factor is nested within the $level - 2$, school factor. At the first level, we represent the outcome y_{ij} as an intercept for the student's school plus a random error

$$
y_{ij} = \alpha_{0j} + e_{ij} \tag{6.4.14}
$$

For the second level of the model, the school level intercepts one represented as deviations about a constant μ

$$(\alpha_{0j} - \mu) = a_{0j} \sim N\left(0, \sigma_\alpha^2\right) \tag{6.4.15}$$

Substituting (6.4.15) into (6.4.14), the multilevel model is a random effects ANOVA model. Timm and Mieczkowski (1997, Chapter 9) and Singer (1998) consider in some detail the analysis of multilevel, hierarchical models using PROC MIXED.

Exercises 6.4

1. Using the Elston and Grizzle (1962) ramus bone data for a random sample of boys discussed in Example 3.7.3, fit a linear random coefficient model to the growth data. The dependent variable is the ramus length measurement and the independent variable is age.

   (a) Assume the covariance structure for the random intercept and slope is unknown.

   (b) Assume the random coefficients are uncorrelated

   (c) Which model fits the data better?

2. To investigate the effect of drugs on maze learning, two rats were randomly assigned to five drug levels (0, 1, 2, and 3) and the time to run the maze for 8 runs are reported for two rats in the table below. Use program m6_4_1.sas to analyze the data and summarize your findings.

DRUG	RAT	RUNS							
0	1	6	14	12	10	18	10	4	8
	2	8	4	10	18	12	14	7	8
1	1	20	18	26	22	16	28	20	22
	2	22	20	28	16	22	26	18	20
2	1	12	28	16	30	26	24	28	10
	2	10	28	24	26	30	16	28	12
3	1	30	36	36	22	26	38	34	28
	2	38	46	39	47	52	54	49	56

3. An experimenter was interested in investigating two methods of instruction in teaching map reading where students are randomly assigned to three instructors. The covariate measure in this experiment is the score on an achievement test on map reading prior to the training, the criterion measure is the score on a comparable form of the achievement test after training is completed. The data for the study follow.

	Instructor		Instructor 2		Instructor 3	
	X	Y	X	Y	X	Y
Method 1	40	95	30	85	50	90
	35	80	40	100	40	85
	40	95	45	85	40	90
	50	105	40	90	30	80
	45	100	40	90	40	85
Method 2	50	100	50	100	45	95
	30	95	30	90	30	85
	35	95	40	95	25	75
	45	110	45	90	50	105
	30	88	40	95	35	85

(a) Perform both an ANOVA and ANCOVA analysis of the data.

(b) State the mathematical models and associated assumptions. Discuss.

(c) How would the analysis change if instructors were random?

(d) How would the analysis change if instructors are nested within methods and random?

4. In the construction of a projective test, 40 more or less ambiguous pictures of two or more human figures were used. In each picture, the sex of at least one of the figures was only vaguely suggested. In a study of the influence of the introduction of extra cues into the pictures, one set of 40 was retouched so that the vague figure looked slightly more like a woman, in another set each was retouched to make the figure look slightly more like a man. A third set of the original pictures was used as a control. The forty pictures were administered to a group of 18 randomly selected male college students and an independent group of 18 randomly selected female college students. Six members of each group saw the pictures with female cues, six the picture with males cues, and six the original pictures. Each subject was scored according to the number of pictures in which he interpreted the indistinct figure as a female. The results follow.

	Female Cues		Male Cues		No Cues	
Female Subjects	29	36	14	5	22	25
	35	33	8	7	20	30
	28	38	10	10	23	32
Male Subjects	25	35	3	5	18	7
	31	32	8	9	15	11
	26	34	4	6	8	10

Carry out an analysis of variance, assuming (a)both factors are fixed, (b) both factors are random, and (c) that blocks (sex) is random and that the factor cues is fixed.

5. An automobile company was interested in the comparative efficiency of three different sales approaches to one of their products. They selected a random sample of 10 different large cities, and then assigned the various selling approaches of random to three agencies within the same city. The results, in terms of sales volumes over a fixed period for each agency, was as follows.

		Approach		
		A	B	C
City	1	38	27	28
	2	47	45	48
	3	40	24	29
	4	32	23	33
	5	41	34	26
	6	39	23	31
	7	38	29	34
	8	42	30	25
	9	45	30	25
	10	41	27	34

Does there seem to be a significant difference in the three sales approaches? If so, estimate the effect sizes.

6. In a concept formation study, 12 subjects were randomly assigned to three different experimental conditions and then given four trials in the solution of a problem. For each trial, the number of minutes taken to solve the problem was recorded. The results were as follows.

		Trial Number			
Condition	Subject	1	2	3	4
1	1	9	8	8	5
	2	12	11	11	4
	3	15	18	13	10
	4	14	15	12	7
2	5	20	11	12	9
	6	15	15	14	10
	7	12	19	18	7
	8	13	10	15	9
3	9	8	10	6	6
	10	10	5	5	9
	11	8	7	7	8
	12	9	13	12	6

What do these results indicate?

7. Marascuilo (1969) gives data from an experiment designed to investigate the effects of variety on the learning of the mathematical concepts of Boolean set unit and intersection. For the study, students were selected from fifth and seventh grades from four schools representing different socioeconomic areas. The students were then randomly assigned to four experimental conditions that varied in two ways. Subjects in the small variety (S) conditions were given eight problems to solve with each problem repeated six times, while the students in the large-variety (L) conditions solved 48 different problems. Half the students were given familiar geometric forms (G), and the remaining students were given nonsense forms (N) generated from random numbers. The data for the experiment follow.

5th Grade								7th Grade							
School 1				School 2				School 1				School 2			
S		L		S		L		S		L		S		L	
N	G	N	G	N	G	N	G	N	G	N	G	N	G	N	G
18	9	11	20	38	33	35	21	3	25	14	16	19	39	44	41
38	32	6	13	19	22	31	36	10	27	8	6	41	41	39	40
24	6	6	10	44	13	26	34	14	2	7	15	28	36	38	44
17	4	2	0	40	21	27	30	25	21	39	9	40	45	36	46

(a) What are the mathematical model and the statistical assumptions for the design? Test to see that the assumptions are satisfied.

(b) Form an ANOVA table and summarize your results.

(c) Construct appropriate simultaneous confidence intervals for contrasts using $\alpha = 0.05$.

6.5 Mixed Multivariate Models

The extension of univariate ANOVA and ANCOVA models to MANOVA and MANCOVA designs with fixed parameters extended in a natural manner to multivariate designs when an observed random variable was replaced by a vector containing p random variables. Nonorthogonal (unbalanced) designs extended in a natural manner to the multivariate case. The analysis of these designs in SAS were performed using the procedure PROC GLM. And while PROC GLM may be used to analyze the GMANOVA model, the PROC MIXED procedure is preferred since it permits the analysis of treatment effects with time varying covariates, the evaluation of covariance structures and the analysis of repeated measurements with data MAR.

When analyzing overall fixed treatment effects for balanced variance component designs, we saw that we can use PROC MIXED or PROC GLM. While PROC MIXED always constructed the correct F ratio, one could also obtain the appropriate ratio by examination of the Type III expected mean squares in PROC GLM. For unbalanced designs or for the analysis of arbitrary contrasts in the fixed effects, PROC MIXED is used with Satterthwaite's approximation to adjust the denominator degrees of freedom. Because all parameters are considered fixed in PROC GLM, variances components are not estimated so that the estimates of contrast variances are incorrect for mixed models.

The analysis of univariate variance component designs extend in a natural manner to the multivariate case; however, their analyses is more complicated, except for the balanced case. For balanced designs, overall tests of fixed or random effects may be constructed by inspection of the TYPE III expected SSCP matrices using PROC GLM. All the multivariate criteria may be used to test for overall significance, however, Wilks' test is no longer a Likelihood Ratio test since the likelihood must now be maximized subject to the constraint that all variance covariances matrices of random effects are nonnegative definite. The development of multivariate tests of fixed effects are based on the assumption that there exists a random effect whose mean square error matrix has the same expected value as the effect being tested when the null hypothesis is true. When this is not the case, one has to approximate the distribution of the linear combination of mean square error matrices with a central Wishart distribution by equating the first two moments of the distributions, Boik (1988, 1991) and Tan and Gupta (1983). Because $\widehat{\psi} = \mathbf{c}'\widehat{\mathbf{B}}\mathbf{m}$ depends on the random components, the variance of $\widehat{\psi}$ also depends on the random components.

a. *Model Specification*

The general linear multivariate model for balanced multivariate component of variance designs following (6.3.37) has the general structure

$$\mathbf{Y} = \mathbf{XB} + \sum_{j=1}^{m} \mathbf{Z}_j \mathbf{U}_j + \mathbf{E} \tag{6.5.1}$$

where $\mathbf{Y}_{n\times p}$ is the data matrix, $\mathbf{X}_{n\times q}$ is a known design matrix, $\mathbf{B}_{q\times p}$ is a matrix of unknown fixed effects, $\mathbf{Z}_j$ are known $n \times r_j$ matrices of rank r_j, $\mathbf{U}_j$ are random effect matrices of order $r_j \times p$, and $\mathbf{E}_{n\times p}$ is a matrix of random errors. We further assume that the rows $\mathbf{u}'_j$ of $\mathbf{U}_j$ and $\mathbf{e}'_j$ of $\mathbf{E}$ are distributed MVN as follows, $\mathbf{e}'_j \sim IN_p\left(\mathbf{0}, \Sigma_e\right)$ and $\mathbf{u}'_j \sim IN_p\left(\mathbf{0}, \Sigma_j\right)$ for $j = 1, 2, \ldots, m$ so that $E\left(\mathbf{Y}\right) = \mathbf{XB}$. Letting $\Sigma_e \equiv \Sigma_0$, $\mathbf{V}_j = \mathbf{Z}_j\mathbf{Z}'_j$, and $\mathbf{y}^* = \operatorname{vec}\left(\mathbf{Y}'\right)$, the covariance of $\mathbf{y}^*$ is

$$\begin{aligned}
\Omega = \operatorname{cov}\left(\mathbf{y}^*\right) &= \operatorname{cov}\left[\sum_{j=1}^{m} \left(\mathbf{Z}_i \otimes \mathbf{I}\right) \operatorname{vec} \mathbf{u}'_j\right] + \operatorname{cov}\left[\operatorname{vec}\left(\mathbf{E}'\right)\right] \\
&= \sum_{j=1}^{m} \left(\mathbf{Z}_q \otimes \mathbf{I}\right)\left(\mathbf{I} \otimes \Sigma_j\right)\left(\mathbf{Z}_j \otimes \mathbf{I}\right)' + \mathbf{I}_n \otimes \Sigma_o \\
&= \sum_{j=1}^{m} \left(\mathbf{Z}_j\mathbf{Z}'_j \otimes \Sigma_j\right) + \mathbf{I}_n \otimes \Sigma_o \\
&= \sum_{j=0}^{m} \mathbf{V}_j \otimes \Sigma_j
\end{aligned} \tag{6.5.2}$$

by setting $\mathbf{V}_0 \equiv \mathbf{I}_n$ where each Σ_j is nonnegative definite and Ω is positive definite, Mathew (1989).

Because the multivariate component of variance models have an equal number of observations per cell, are balanced, an alternative structure for model (6.5.1) as discussed in Chapter 3, expression (3.6.36) is

$$\begin{aligned} \mathbf{Y} &= \mathbf{XB} + \mathbf{ZU} + \mathbf{E} \\ &= \sum_i \mathbf{K}_i \mathbf{B}_i + \sum_{j=0}^{m} \mathbf{K}_j \mathbf{U}_j \end{aligned} \tag{6.5.3}$$

where the matrices $\mathbf{K}_i$ or $\mathbf{K}_j$ have Kronecker product structure of the form

$$(\mathbf{A}_1 \otimes \mathbf{A}_2 \otimes \ldots \otimes \mathbf{A}_h)$$

where $\mathbf{A}_i$ $(i = 1, 2, \ldots, h)$ and each $\mathbf{A}_i$ is an identity matrix or a vector of $1s$. Expression (6.5.3) is the multivariate extension of the univariate mixed models considered by Searle et al. (1992). The covariance structure for model (6.5.3) is identical to that given by (6.5.2). However, the matrices $\mathbf{V}_j$ have the Kronecker product structure $\mathbf{V}_j = \mathbf{H}_1 \otimes \mathbf{H}_2 \otimes \ldots \otimes \mathbf{H}_h$ where $\mathbf{H}_i$ is an identity matrix $\mathbf{I}$ or a matrix $\mathbf{J}$ so the $\mathbf{V}_j$ commute, Khuri (1982). This structure for Ω was discussed in Chapter 3, expression (3.6.38).

To illustrate (6.5.1), we examine the overparameterized balanced one-way mixed MANOVA model. The model is similar to the fixed effect model considered in Chapter 4 except that the fixed effect parameter vector is replaced by a random vector of treatment effects since treatments are sampled from a large population of treatments. Letting $\mathbf{y}_{ij}$ represent a $p \times 1$ vector of observations, $\boldsymbol{\mu}' = [\mu_1, \mu_2, \ldots, \mu_p]$ denote a vector of constants, and $\mathbf{a}_i' = [a_{i1}, a_{i2}, \ldots, a_{ip}]$ a vector of random effects, the overparameterized multivariate variance components MANOVA model becomes

$$\mathbf{y}_{ij} = \boldsymbol{\mu} + \mathbf{a}_i + \mathbf{e}_{ij} \tag{6.5.4}$$

where $\mathbf{e}_{ij}$ are random $p \times 1$ vectors of errors, the $\mathbf{a}_i$ and $\mathbf{e}_{ij}$ are mutually independent, and $\mathbf{a}_i \sim IN_p(\mathbf{0}, \Sigma_a)$ and $\mathbf{e}_{ij} \sim IN_p(\mathbf{0}, \Sigma_e)$. Letting the data matrix and error matrix have the structure

$$\underset{n \times p}{\mathbf{Y}} = [y_{11}, y_{12}, \ldots, y_{1r}, y_{21}, y_{22}, \ldots, y_{2r}, y_{k1}, y_{k2}, \ldots, y_{kr}]'$$

$$\underset{n \times p}{\mathbf{E}} = [e_{11}, e_{12}, \ldots, e_{1r}, e_{21}, e_{22}, \ldots, e_{2r}, e_{k1}, e_{k2}, \ldots, e_{kr}]'$$

where $n = k_r$, model (6.5.1) has the matrix form

$$\begin{aligned} \mathbf{Y} &= \mathbf{XB} + \mathbf{ZU} + \mathbf{E} \\ &= \underset{kr \times 1}{\mathbf{1}} \underset{1 \times p}{[\mu_1, \mu_2, \ldots, \mu_p]} + \\ & \quad \underset{kr \times k}{\operatorname{diag}[\mathbf{1}_r, \mathbf{1}_r, \ldots, \mathbf{1}_r]} \begin{bmatrix} a_{11} & a_{12} & \cdots & a_{1p} \\ a_{21} & a_{22} & \cdots & a_{2p} \\ \vdots & \vdots & & \vdots \\ a_{k1} & a_{k2} & \cdots & a_{kp} \end{bmatrix} + \mathbf{E} \end{aligned} \tag{6.5.5}$$

Or, letting $\mathbf{a}' = [a_1, a_2, \ldots, a_p]$ the model has the simple Kronecker structure

$$\begin{aligned}\underset{n\times p}{\mathbf{Y}} &= (\mathbf{1}_k \otimes \mathbf{1}_r)\,\boldsymbol{\mu}' + (\mathbf{I}_k \otimes \mathbf{1}_r)\,\mathbf{a} + \mathbf{E} \\ &= \mathbf{K}_1\boldsymbol{\mu}' + \mathbf{K}_2\mathbf{a} + \mathbf{E}\end{aligned} \tag{6.5.6}$$

where $\mathbf{K}_1$ and $\mathbf{K}_2$ have the Kronecker product structure $\mathbf{A}_1 \otimes \mathbf{A}_2$ where $\mathbf{A}_i$ is either the identity matrix or a vector of ones. Furthermore,

$$\begin{aligned}E(\mathbf{Y}) &= (\mathbf{1}_k \otimes \mathbf{1}_r)\,\boldsymbol{\mu}' \\ \operatorname{cov}\left(\operatorname{vec}\mathbf{a}'\right) &= \mathbf{I}_k \otimes \Sigma_a \\ \operatorname{cov}\left(\operatorname{vec}\mathbf{E}'\right) &= \mathbf{I}_n \otimes \Sigma_e\end{aligned} \tag{6.5.7}$$

Using (6.5.6), the structure of Ω for the one-way model is

$$\begin{aligned}\Omega &= \mathbf{K}_2\mathbf{K}_2' \otimes \Sigma_a + \mathbf{I}_n \otimes \Sigma_e \\ &= (\mathbf{I}_k \otimes \mathbf{1}_r)\,(\mathbf{I}_k \otimes \mathbf{1}_r)' \otimes \Sigma_a + \mathbf{I}_n \otimes \Sigma_e \\ &= (\mathbf{I}_k \otimes \mathbf{J}_r) \otimes \Sigma_a + (\mathbf{I}_k \otimes \mathbf{I}_r) \otimes \Sigma_e \\ &= \mathbf{V}_1 \otimes \Sigma_a + \mathbf{V}_0 \otimes \Sigma_0 \\ &= \sum_{j=0}^{1} (\mathbf{V}_j \otimes \Sigma_j)\end{aligned} \tag{6.5.8}$$

letting $\Sigma_1 \equiv \Sigma_a$. Alternatively, observe the Ω has the block diagonal structure $\Omega = \mathbf{I}_k \otimes (\mathbf{J}_r \otimes \Sigma_a + \mathbf{I}_r \otimes \Sigma_e) = \operatorname{diag}[\mathbf{J}_r \otimes \Sigma_a + \mathbf{I}_r \otimes \Sigma_e]$, a generalization of the univariate model.

b. *Hypothesis Testing*

As with the univariate mixed model, the primary hypothesis of interest for the one-way multivariate variance component MANOVA model is whether the covariance matrix Σ_a for the random effects is significantly different from zero.

$$H_1 : \Sigma_a = \mathbf{0} \quad \text{vs} \quad H_A : \Sigma_a \neq \mathbf{0}$$

A test which does not arise in the univariate case is the investigation of the rank of Σ_a, Schott and Saw (1984).

To construct tests of the fixed effects of the form H_o : $\mathbf{CBM} = \mathbf{0}$ or the covariance components $H_j : \Sigma_j = \mathbf{0}$ for $j = 1, 2, \ldots, m$, for MANOVA designs with structure (6.5.3) or (6.5.1), Mathew (1989) using Theorem 3.4.5 establishes conditions for the existence and uniqueness for the SSCP partition of $\mathbf{Y}'\mathbf{Y}$,

$$\mathbf{Y}'\mathbf{Y} = \sum_{i=1}^{t} \mathbf{Y}'\mathbf{P}_i\mathbf{Y} + \sum_{i=1}^{s} \mathbf{Y}'\mathbf{Q}_i\mathbf{Y} \tag{6.5.9}$$

where $\mathbf{P} = \mathbf{X}\left(\mathbf{X}'\mathbf{X}\right)^{-}\mathbf{X}'$ is a projection matrix (symmetric and idempotent), $\mathbf{Q} = \mathbf{I} - \mathbf{P}$, $\mathbf{P} = \sum_{i=1}^{t}\mathbf{P}_i$, $\mathbf{Q} = \sum_{i=1}^{s}\mathbf{Q}_i$ where $\mathbf{P}_i$ and $\mathbf{Q}_i$ are orthogonal projection matrices,

$\mathbf{P}_i\mathbf{P}_j = \mathbf{0}$ and $\mathbf{Q}_i\mathbf{Q}_j = \mathbf{0}$ $(i \neq j)$, that permits one to test hypotheses of the form given by H_o and H_j for multivariate variance components models (6.5.1).

Theorem 6.5.1 *Given a partition of* $\mathbf{Y}'\mathbf{Y}$ *of the form specified by (6.5.9) that satisfies Theorem 3.4.5 for the MANOVA model (6.5.1), the partition is unique if and only if there exists real numbers* $\alpha_{ij} > 0$ *and* $\delta_{ij} > 0$ *such that*

$$\mathbf{V}_j = \sum_{i=1}^{t} \mathbf{P}_i\mathbf{V}_j\mathbf{P}_i + \sum_{i=1}^{s} \mathbf{Q}_i\mathbf{V}_j\mathbf{Q}_i + \sum_{i=1}^{t} \alpha_{ij}\mathbf{P}_i + \sum_{i=1}^{t} \delta_{ij}\mathbf{Q}_i$$

for $i = 1, 2, \ldots, t$ *and* $j = 0, 1, 2, \ldots, m$.

By construction of the matrix SSCP matrix $\mathbf{Y}'\mathbf{Y}$, $\mathbf{P}_i^2 = \mathbf{P}_i$ and $\mathbf{P}_i\mathbf{Q}_i = \mathbf{0}$ for all i; hence, the expression for $\mathbf{V}_j$ in Theorem 6.5.1 is equivalent to the two conditions that (i) $\mathbf{P}_1, \mathbf{P}_2, \ldots, \mathbf{P}_t, \mathbf{V}_0, \mathbf{V}_1, \ldots, \mathbf{V}_k$ commute and (ii) $\alpha_{ij} > 0$ exist such that $\mathbf{P}_i\mathbf{V}_j\mathbf{P}_i = \alpha_{ij}\ \mathbf{P}_i$ for $i = 1, 2, \ldots, t$ and $j = 0, 1, \ldots, m$ given by Mathew (1989).

To apply Theorem 6.5.1 to the one-way design, we partition the vector space into the spaces $\mathbf{1}$, $A|\mathbf{1}$ and $(A|\mathbf{1})^{\perp}$ and write the projection matrices using Kronecker product notation so that $\mathbf{I} = \mathbf{P}_1 + \mathbf{Q}_2 + \mathbf{Q}_3$. Then it is easily verified, as seen in Example 2.6.2, that the general expressions for $\mathbf{P}_i$ and $\mathbf{Q}_i$ are

$$\begin{aligned}
\mathbf{P}_1 &= \mathbf{1}_{kr}\left(\mathbf{1}_{kr}'\mathbf{1}_{kr}\right)^{-1}\mathbf{1}_{kr}' = \frac{1}{kr}\left(\mathbf{1}_r \otimes \mathbf{1}_k\right) = \frac{\mathbf{J}_{kr}}{kr} \\
&= \left(\mathbf{J}_k \otimes \mathbf{J}_r\right)/kr \\
\mathbf{Q}_1 &= \frac{1}{r}\left(\mathbf{I}_k \otimes \mathbf{1}_r\right)\left(\mathbf{I}_k \otimes \mathbf{1}_r\right)' - \mathbf{J}_{rk}/kr \\
&= \frac{1}{r}\left(\mathbf{I}_k \otimes \mathbf{J}_r\right) - \mathbf{J}_{rk}/kr \\
\mathbf{Q}_2 &= \left(\mathbf{I}_k \otimes \mathbf{I}_r\right) - \left(\mathbf{I}_k \otimes \mathbf{J}_r\right)/r
\end{aligned} \tag{6.5.10}$$

where the rank $(\mathbf{P}_1) = 1$, rank $(\mathbf{Q}_1) = k - 1$ and the rank $(\mathbf{Q}_2) = kr - k = k\,(r-1)$. Since $\mathbf{I} = \mathbf{P}_1 + \mathbf{Q}_1 + \mathbf{Q}_2$, the partition for $\mathbf{Y}'\mathbf{Y}$ is

$$\mathbf{Y}'\mathbf{Y} = \mathbf{Y}'\mathbf{P}_1\mathbf{Y} + \mathbf{Y}'\mathbf{Q}_1\mathbf{Y} + \mathbf{Y}'\mathbf{Q}_2\mathbf{Y}$$

where $\mathbf{Q} = \mathbf{Q}_1 + \mathbf{Q}_2$ and $\mathbf{P}_1$, $\mathbf{Q}_1$ and $\mathbf{Q}_2$ are orthogonal projection matrices such that $\mathbf{P}_i^2 = \mathbf{P}_i$, $\mathbf{Q}_i^2 = \mathbf{Q}_i$, $\mathbf{P}_i\mathbf{P}_j = \mathbf{0}\,(i \neq j)$, $\mathbf{Q}_i\mathbf{Q}_j = \mathbf{0}\,(i \neq j)$ and $\mathbf{P}_i\mathbf{Q}_j = \mathbf{0}$. Furthermore, $\mathbf{P}_1$ commutes with $\mathbf{V}_1 = \mathbf{I}_k \otimes \mathbf{J}_r$ and $\mathbf{V}_0 = \mathbf{I}_k \otimes \mathbf{I}_r = \mathbf{I}_n$. Finally, $\mathbf{P}_1\mathbf{V}_j\mathbf{P}_1 = \alpha_{1j}\ \mathbf{P}_1$ where $\alpha_{1j} = r$ and $\alpha_{10} = 1$ as required by Theorem 6.5.1.

In the proof of Theorem 6.5.1, Mathew (1989) shows that $\mathbf{Y}'\mathbf{P}_i\mathbf{Y}$ have independent noncentral Wishart distributions, that $\mathbf{Y}'\mathbf{Q}_i\mathbf{Y}$ have independent central Wishart distributions, and that SSCP matrices are mutually independent. That is that

$$\begin{aligned}
\mathbf{Y}'\mathbf{P}_i\mathbf{Y} &\sim IW_p\left(v_i, \Gamma_i, \Lambda_i\right) \\
\mathbf{Y}'\mathbf{Q}_i\mathbf{Y} &\sim IW_p\left(v_i, \mathbf{0}, \Lambda_i\right)
\end{aligned} \tag{6.5.11}$$

where

$$\Gamma_i = \mathbf{B}'\mathbf{X}'\mathbf{P}_i\mathbf{X}\mathbf{B}\left(\sum_{j=0}^{m}\alpha_{ij}\Sigma_j\right)^{-1}$$

$$\Lambda_i = \sum_{j=0}^{m}\alpha_{ij}\Sigma_j \tag{6.5.12}$$

$$v_i = \text{rank}\,(\mathbf{P}_i) \text{ or the rank}\,(\mathbf{Q}_i)$$

Furthermore, the expected value of the SSCP matrices are

$$\begin{aligned} E\left(\mathbf{Y}'\mathbf{P}_i\mathbf{Y}\right) &= v_i\Lambda_i + \mathbf{B}'\mathbf{X}'\mathbf{P}_i\mathbf{X}\mathbf{B} \qquad i = 1, 2, \ldots, t \\ E\left(\mathbf{Y}'\mathbf{Q}_i\mathbf{Y}\right) &= v_i\Lambda_i \qquad i = 1, 2, \ldots, s \end{aligned} \tag{6.5.13}$$

Details are provided by Khuri et al. (1998).

Evaluating the expectations of the matrix quadratic forms for the one-way model,

$$\begin{aligned} E\left(\mathbf{Y}'\mathbf{Q}_1\mathbf{Y}\right) &= (k-1)\left(r\Sigma_1 + \Sigma_0\right) \\ &= (k-1)\left(r\Sigma_a + \Sigma_e\right) \\ E\left(\mathbf{Y}'\mathbf{Q}_2\mathbf{Y}\right) &= k(r-1)\,\Sigma_0 = k(r-1)\,\Sigma_e \end{aligned} \tag{6.5.14}$$

Using (6.5.14) to solve for Σ_a and Σ_e, an unbiased estimate for the covariance matrices Σ_a and Σ_e are

$$\begin{aligned} \mathbf{S}_a &= \frac{1}{r}\left[\frac{\mathbf{Y}'\mathbf{Q}_1\mathbf{Y}}{k-1} - \frac{\mathbf{Y}'\mathbf{Q}_2\mathbf{Y}}{k(r-1)}\right] \\ &= \frac{1}{r}\left[\mathbf{E}_1/v_1 - \mathbf{E}_2/v_2\right] \\ \mathbf{S}_e &= \mathbf{Y}'\mathbf{Q}_2\mathbf{Y}/k(n-1) \\ &= \mathbf{E}_2/v_2 \\ &= \mathbf{Y}'\left[\mathbf{I} - \mathbf{X}\left(\mathbf{X}'\mathbf{X}\right)^{-}\mathbf{X}'\right]\mathbf{Y}/v_e \end{aligned} \tag{6.5.15}$$

where $v_e = n - r(\mathbf{X}) = k(r-1)$ and $v_i = \text{rank}\,(\mathbf{Q}_i)$.

While the matrices $\mathbf{S}_a$ and $\mathbf{S}_e$ are not necessarily nonnegative definite, the matrix $\mathbf{S}_e$ is *p.d.* provided $n - r(\mathbf{X}) = v_e \geq p$. Making this assumption, Anderson (1985) and Amemiya (1985) independently show how to modify $\mathbf{S}_a$ to create a REML estimate for Σ_a. The procedure is as follows. By solving the characteristic equation $|\mathbf{E}_1 - \lambda\mathbf{E}_2| = 0$ there exists a nonsingular matrix $\mathbf{T}$ such that $\mathbf{T}'\mathbf{E}_1\mathbf{T} = \Lambda$ and $\mathbf{T}'\mathbf{E}_2\mathbf{T} = \mathbf{I}$ so that $\mathbf{E}_1 = \left(\mathbf{T}'\right)^{-1}\Lambda\,(\mathbf{T})^{-1}$ and $\mathbf{E}_2 = \left(\mathbf{T}'\right)^{-1}\mathbf{T}^{-1}$ where Λ contain the eigenvalues $\lambda_1 > \lambda_2 > \ldots > \lambda_p$. Letting $\mathbf{F} = \mathbf{T}^{-1}$, the matrix $\mathbf{E}_1$ may be written as $\mathbf{E}_1 = \mathbf{F}'(\Lambda\,/\,v_1 - \mathbf{I}\,/v_2)\mathbf{F}$. If any λ_i in Λ is negative, its value is set to zero and the new matrix is represented by Δ. Then,

$$\mathbf{S}_{\text{REML}} = \mathbf{F}'\Delta\mathbf{F}/r \tag{6.5.16}$$

is a REML estimate of Σ_a that is positive semidefinite by construction, Mathew, Niyogi and Sinha (1994). Srivastava and Kubokawa (1999) propose improved estimators for jointly estimating both Σ_a and Σ_e. Calvin and Dykstra (1991 a, b) suggest other procedures for estimating covariance matrix components for more complex multivariate components of variances designs requiring computer algorithms.

To test the hypothesis $H_1 : \Sigma_a = \mathbf{0}$, one uses (6.5.14) to create a MANOVA table similar to the fixed effects model using the formula given in (6.5.13) to evaluate the expected values of SSCP matrices. For the one-way MANOVA the hypothesis SSCP matrix for testing $H_o : \boldsymbol{\alpha}_1 = \boldsymbol{\alpha}_2 = \ldots = \boldsymbol{\alpha}_k$ is used to test $H_1 : \Sigma_a = \mathbf{0}$. In general, one locates mean square matrices for testing hypotheses by examining the Type III expected mean square matrices provided in PROC GLM. As in the univariate mixed model, when testing for fixed effects, one may not be able to locate random effects whose mean square error matrix has the same expected value as the effect being tested when the null hypothesis is true. Thus, one may have to combine mean square error matrices to construct an approximate test. Because $\mathbf{Y}'\mathbf{Q}_i\mathbf{Y} \sim \mathbf{W}_p(v_i, \Lambda_i)$ where Λ_i is defined in (6.5.12), we need to approximate the distribution of $\widehat{\mathbf{L}} = \sum_i c_i \left(\mathbf{Y}'\mathbf{Q}_i\mathbf{Y}_i\right)/v_i$ with a central Wishart distribution $W_p(v, \Lambda)$ where $\mathbf{L}$ is some linear combination of mean square error matrices.

To approximate the distribution of $\widehat{\mathbf{L}}$, one equates the first two sample moments of $\widehat{\mathbf{L}}$, its expected value and variance, to those of a $W_p(v, \ \Lambda)$. Recall that by Theorem 2.4.6. if $\mathbf{Q}_i \sim W_p(v_i, \ \Sigma)$ that $E(\mathbf{Q}_i) = v_i \Sigma$ and the cov $(\text{vec}\,\mathbf{Q}_i) = 2v_i \left(\mathbf{I}_{p^2} + \mathbf{K}\right)(\Sigma \otimes \Sigma)$. Tan and Gupta (1983) estimate v and Λ by equating the expected value and determinant (generalized variance) of $\widehat{\mathbf{L}}$ to those of $W_p(v, \ \Lambda)$ which they call a multivariate Satterthwaite approximation. This procedure was employed by Nel and van der Merwe (1986) for the multivariate Behrens-Fisher problem discussed in Chapter 3. Khuri, Mathew and Nel (1994) give necessary and sufficient conditions for $\widehat{\mathbf{L}}$ to have a central Wishart distribution. The approximation is illustrated in Khuri, Mathew and Sinha (1998). An alternative approximation has been suggested by Boik (1988, 1991) who recommends equating the expected value and the trace of $\widehat{\mathbf{L}}$ to those of $W_p(v, \ \Lambda)$ instead of the determinant. In general the two procedures will lead to different estimates of $\widehat{v}$. For tests of random effects, hypothesis and error matrices may need to be constructed. The value of $\widehat{v}$ using the generalized variance criterion is given in Khuri et al. (1998, p. 271).

c. *Evaluating Expected Mean Square*

While we have provided general expressions for evaluating the expected mean squares in mixed models, their evaluation is not straight forward. To facilitate their evaluation, one may use rules. Results may differ depending upon whether restrictions are incorporated into the model and whether interactions of fixed and random effects are considered fixed or random, Searle (1971, pp. 389-404) and Dean and Voss (1999, pp. 627-628), and others discussed in Chapter 6. The SAS procedure PROC GLM does not use rules to evaluate expected mean squares. Instead, it uses the computer syntheses procedure developed by Hartley (1967) to evaluate the expected value of quadratic forms directly for each sum of squares (Type I, II, III and IV) a variable at a time, Milliken and Johnson (1992, Chapters 18 and 22). When one cannot locate appropriate expected mean square matrices to perform tests, PROC GLM does not currently (Version 8) provide any type of multivariate Satterthwaite approximation.

d. *Estimating the Mean*

In many random effect multivariate variance component models, the only fixed parameter is $\boldsymbol{\mu}$. Using the projection matrix $\mathbf{P}_1$, we observe that

$$\widehat{\boldsymbol{\mu}} = \mathbf{P}_1\mathbf{X}\widehat{\mathbf{B}} = \mathbf{P}_1\mathbf{X}\left(\mathbf{X}'\mathbf{X}\right)^{-1}\mathbf{X}'\mathbf{Y} = \mathbf{P}_1\mathbf{P} = \mathbf{P}_1\mathbf{Y} = \overline{\mathbf{y}}.$$

Furthermore, $\mathbf{P}_1\mathbf{V}_j\mathbf{Q} = \mathbf{0}$ so that $\overline{\mathbf{y}}$ and $\mathbf{QY}$ are uncorrelated. Under MVN assumptions, $\overline{\mathbf{y}}$ is BLUE and its components have minimal variance.

For the more general case with fixed effects $\mathbf{P}_1, \mathbf{P}_2, \ldots, \mathbf{P}_t$ and random effects $\mathbf{Q}_i$, the jth column of $\mathbf{Y}$ induces a balanced univariate model of the form

$$\mathbf{y}_j = \mathbf{X}\boldsymbol{\beta}_j + \textstyle\sum_{i=0}^m \mathbf{Z}_i\mathbf{u}_i \qquad j = 1, 2, \ldots, p$$

where the var $\left(\mathbf{y}_j\right) = \sum_{i=0}^m \mathbf{Z}_i\mathbf{Z}_i'\sigma_{ij}^2$. Even though the estimates $\widehat{\boldsymbol{\beta}}_j$ are correlated, for each model the OLS estimate of $\boldsymbol{\beta}_j$ and the GLS estimate of $\boldsymbol{\beta}_j$ are equal for balanced designs with no covariates, Searle et al. (1992, p. 160). In addition, one may obtain BLUP's a variable at a time.

Because $\widehat{\psi} = \mathbf{C}\widehat{\mathbf{B}}\mathbf{M}$ depends on the random effects in the model, the covariance of $\widehat{\psi}$ depends on the estimates of the covariance matrices of the random effects. Letting $\mathbf{A}' = \mathbf{M}'\mathbf{X}\left(\mathbf{X}'\mathbf{X}\right)^{-1}$, the covariance matrix of $\widehat{\psi}$ is

$$\sum_{j=0}^{m} \mathbf{C}\mathbf{V}_j\mathbf{C}' \otimes \mathbf{A}'\Sigma_j\mathbf{A}$$

which depends on $\mathbf{V}_j$ and Σ_j.

e. *Repeated Measurements Model*

A useful application of the multivariate variance components model in the social sciences is in the analysis of the multivariate split-plot mixed model. For this model, the $p \times 1$ observation vector represents the p measures taken on the jth subject in group i, under treatment condition k, where $i = 1, 2, \ldots, g$, $j = 1, 2, \ldots, r$, and $k = 1, 2, \ldots, t$. The mixed model is written as

$$\mathbf{y}_{ijk} = \boldsymbol{\mu} + \boldsymbol{\alpha}_i + \boldsymbol{\beta}_k + (\boldsymbol{\alpha\beta})_{ik} + \mathbf{s}_{(i)j} + \mathbf{e}_{ijk} \tag{6.5.17}$$

Furthermore, we assume that the random components $\mathbf{s}_{(i)j}$ and $\mathbf{e}_{ijk}$ are mutually independent with distributions that are MVN: $s_{(i)j} \sim IN\left(\mathbf{0}\ \Sigma_s\right)$ and $\mathbf{e}_{ijk} \sim IN\left(\mathbf{0},\ \Sigma_e\right)$. Letting $\mathbf{y}_{ij} = \left[\mathbf{y}_{ij1}, \mathbf{y}_{ij2}, \ldots, \mathbf{y}_{ijt}\right]$ and $\mathbf{y}_{ij}^* = \text{vec}\left(\mathbf{y}_{ij}\right)$ denote the $pt \times 1$ vector formed from $p \times t$ matrix $\mathbf{y}_{ij}$, the $\text{cov}(\mathbf{y}_{ij}^*) = \Omega$ for all i and j. The matrix Ω is the $pt \times pt$ covariance matrix for each group. The organization for the design is given in Table 6.5.1.

TABLE 6.5.1. Multivariate Repeated Measurements

Treatment Group	Subject	Conditions (time) 1	2	$\cdots$	t
	1	$\mathbf{y}_{111}$	$\mathbf{y}_{112}$	$\cdots$	$\mathbf{y}_{11t}$
	2	$\mathbf{y}_{121}$	$\mathbf{y}_{122}$	$\cdots$	$\mathbf{y}_{12t}$
1	$\vdots$	$\vdots$	$\vdots$		$\vdots$
	r	$\mathbf{y}_{1r1}$	$\mathbf{y}_{1r2}$	$\cdots$	$\mathbf{y}_{1rt}$
	1	$\mathbf{y}_{211}$	$\mathbf{y}_{212}$	$\cdots$	$\mathbf{y}_{21t}$
	2	$\mathbf{y}_{221}$	$\mathbf{y}_{222}$	$\cdots$	$\mathbf{y}_{22t}$
2	$\vdots$	$\vdots$	$\vdots$		$\vdots$
	r	$\mathbf{y}_{2r1}$	$\mathbf{y}_{2r2}$	$\cdots$	$\mathbf{y}_{2rt}$
	$\vdots$	$\vdots$	$\vdots$		$\vdots$
	1	$\mathbf{y}_{g11}$	$\mathbf{y}_{g12}$	$\cdots$	$\mathbf{y}_{g1t}$
	2	$\mathbf{y}_{g21}$	$\mathbf{y}_{g22}$	$\cdots$	$\mathbf{y}_{g2t}$
g	$\vdots$	$\vdots$	$\vdots$		$\vdots$
	r	$\mathbf{y}_{gr1}$	$\mathbf{y}_{gr2}$	$\cdots$	$\mathbf{y}_{grt}$

Representing the model in matrix form where $\mathbf{Y} = [\mathbf{y}_{11}^*, y_{12}^*, \ldots, y_{gr}^*]$, the mixed model for $\mathbf{Y} = [\mathbf{y}_{11}^*,\ y_{12}^*, \ldots, y_{gr}^*]$, the mixed model for the design is

$$
\begin{aligned}
\underset{n \times pt}{\mathbf{Y}} &= \underset{n\times q}{\mathbf{X}}\ \underset{q\times pt}{\mathbf{B}} + (\mathbf{I}_g \otimes \mathbf{I}_r \otimes \mathbf{1}_t)\mathbf{s}_{(i)j} + \underset{n\times pt}{\mathbf{E}} \\
E\left(\mathbf{Y}\right) &= \mathbf{XB} \\
\operatorname{cov}\left(\operatorname{vec} \mathbf{Y}'\right) &= \left(\mathbf{I}_g \otimes \mathbf{I}_r \otimes \mathbf{J}_t\right) \otimes \Sigma_s + \left(\mathbf{I}_g \otimes \mathbf{I}_r\right) \otimes \Sigma_e \\
&= \mathbf{I}_{gr} \otimes (\mathbf{J}_t \otimes \Sigma_s + \Sigma_e) \\
&= \mathbf{I}_{gr} \otimes \Omega_{pt\times pt}
\end{aligned}
\tag{6.5.18}
$$

so that Ω has compound symmetry structure.

Following the one-way MANOVA mixed model, one may construct unbiased estimates of Σ_s and Σ_e. Evaluating the expected values of the SSCP matrices for treatments, conditions, treatment by conditions interaction, subjects within groups and error, test matrices for the design are easily established. An example is given in the next section using the data in Timm (1980). This model has been considered by Reinsel (1982, 1985), and by Arnold (1979), Mathew (1989) and Alalouf (1980).

Critical to the analysis of the split-plot multivariate mixed model (MMM) analysis is the structure of Ω, Thomas (1983). For our model, $\Omega = \mathbf{I}_t \otimes \Sigma_e + \mathbf{1}_t\mathbf{1}_t' \otimes \Sigma_s$ has compound symmetry structure since

$$\Omega = \begin{bmatrix} \Sigma_1 & \Sigma_2 & \cdots & \Sigma_2 \\ \Sigma_2 & \Sigma_1 & \cdots & \Sigma_2 \\ \vdots & \vdots & & \vdots \\ \Sigma_2 & \Sigma_2 & \cdots & \Sigma_1 \end{bmatrix} \tag{6.5.19}$$

where $\Sigma_1 = \Sigma_e + \Sigma_s$ and $\Sigma_2 = \Sigma_s$. More generally, let $\mathbf{M}$ be a contrast matrix across conditions such that $\mathbf{M}'\mathbf{M} = \mathbf{I}_{t-1}$ and $\mathbf{M}'\mathbf{1}_t = \mathbf{0}$. Letting $u = t - 1$ and

$$\mathbf{V} = \left(\mathbf{M}' \otimes \mathbf{I}_p\right) \Omega \left(\mathbf{M} \otimes \mathbf{I}_p\right) \tag{6.5.20}$$

Boik (1988) and Pavar (1987) showed that a MMM analysis is valid if and only if the matrix $\mathbf{V}$ satisfies the multivariate sphericity condition

$$\mathbf{V} = \mathbf{I}_u \otimes \Sigma \tag{6.5.21}$$

for some *p.d.* matrix $\Sigma_{p\times p}$. Matrices Ω in this class are called Type H matrices, Huynh and Feldt (1970). Clearly, if $\Omega = \mathbf{I}_t \otimes \Sigma_e + \left(\mathbf{1}\mathbf{1}' \otimes \Sigma_s\right)$, then

$$\left(\mathbf{M}' \otimes \mathbf{I}_p\right) \Omega \left(\mathbf{M} \otimes \mathbf{I}_p\right) = \mathbf{I}_u \otimes \Sigma_e$$

so that if Ω has compound symmetry structure, it satisfies the multivariate sphericity condition. In chapter 3 we showed how to test for multivariate sphericity.

Vaish (1994) in his Doctoral dissertation develops a very general characterization of Ω in (6.5.19) which ensures that one may analyze repeated measurements data using the MMM. He shows that Ω must have the general structure.

$$\Omega = (\mathbf{I}_t - t^{-1}\mathbf{J}_t) \otimes \Sigma_e + t^{-1}(\mathbf{1}_t \otimes \mathbf{I}_p)\mathbf{H}' + t^{-1}\mathbf{H}(\mathbf{1}_t' \otimes \mathbf{I}_p) - t^{-1}\mathbf{1}_t\mathbf{1}_t' \otimes \overline{\mathbf{H}} \tag{6.5.22}$$

where $r\left(\mathbf{H}_{p_t \times p}\right) \leq p$, $\overline{\mathbf{H}} = \sum_{i=1}^{t} \mathbf{H}_i/t$ and each $p \times p$ matrix $\mathbf{G}_i$ is formed so that $\mathbf{H} = [\mathbf{H}_1', \mathbf{H}_2', \ldots, \mathbf{H}_t']'$.

6.6 Balanced Mixed Multivariate Models Examples

The SAS procedure GLM may be used to test hypotheses in balanced MANOVA designs. Following the univariate procedure, if an exact F tests exist for the univariate design for testing for fixed effects or random components a variable at a time (the denominator of the F statistic depends on a single mean square error term), then one may construct an exact multivariate test for the design. The test is constructed using the MANOVA statement with the components $H =$ and $E =$ specified, the options defined by h and e are defined using the factors in the design. From the factor names, PROC GLM creates the hypothesis test matrix $\mathbf{H}$ and error matrix $\mathbf{E}$, solving $\mid \mathbf{H} - \lambda\mathbf{E} \mid = 0$ to develop multivariate test statistics. In Version 8.0 of SAS, one may not combine matrices to create "quasi" multivariate tests since no multivariate Satterthwaite approximation for degrees of freedom has been provided. Also, no procedure is currently available to obtain estimates for random covariance matrices.

a. Two-way Mixed MANOVA

Khuri et al. (1998) provide simulated data to analyze a balanced, two-way, random effects, model II MANOVA design

$$\begin{aligned} \mathbf{y}_{ijk} &= \boldsymbol{\mu} + \boldsymbol{\alpha}_i + \boldsymbol{\beta}_j + (\boldsymbol{\alpha\beta})_{ij} + \mathbf{e}_{ijk} \\ i &= 1, 2, \ldots, a; \;\; j = 1, 2, \ldots, b; \;\; k = 1, 2, \ldots, n \end{aligned} \tag{6.6.1}$$

where factors A and B each have four levels, and $n = 3$ bivariate normal observation vectors. The random components $\boldsymbol{\alpha}_i, \boldsymbol{\beta}_j, (\boldsymbol{\alpha\beta})_{ij}$, and $\mathbf{e}_{ijk}$ are independent and follow MVN distributions with zero mean vectors and random covariance matrices $\Sigma_\alpha, \Sigma_\beta, \Sigma_{\alpha\beta}$, and Σ_e. Assuming the parameters $\boldsymbol{\alpha}_i$ are fixed rather than random, model (6.6.1) becomes a mixed, a Model III, MANOVA model. The data and tests of hypotheses for both designs is included in program m6_6_1.sas.

To analyze (6.6.1) assuming $\boldsymbol{\alpha}_i$ are fixed and $\boldsymbol{\beta}_j$ are random, the RANDOM statement includes both factors B and A*B. Reviewing the univariate output, an exact F test for the corresponding univariate model exists for testing H_A : all $\alpha_i = 0$ and $H_B : \sigma^2_B = 0$ if the denominator of the F-statistic contain the interaction mean square. To create the corresponding MANOVA tests: H_A : all $\boldsymbol{\alpha}_i = \mathbf{0}$ and $H_A : \Sigma_\beta = \mathbf{0}$ the MANOVA statement with $h = AB$ and $e = A * B$ is used. For the test of $H_{AB} : \Sigma_{\alpha\beta} = \mathbf{0}$, the default error mean squares matrix is used. The expected mean squares for the Mixed Model III design are shown in Table 6.6.1

All multivariate criteria show that the tests of H_A, H_B, and H_{AB} are rejected. Because contrasts in the vectors $\boldsymbol{\alpha}_i$ involve elements of $\Sigma_{\alpha\beta}$, PROC MIXED is required. Also included in program m6_6_1.sas is the code for the example discussed by Khuri et al. (1998, p. 273).

While one many use the methods-of-moment procedure to estimate the random covariance matrices, they may not be positive definite.

b. Multivariate Split-Plot Design

For this example, model (6.5.17) is analyzed using data from Timm (1980). The data were provided by Dr. Thomas Zullo in the School of Dental Medicine, University of Pittsburgh. Nine subjects were selected from the class of dental students and assigned to two orthopedic treatment groups. Each group was evaluated by studying the effectiveness of three adjustments that represented the position and angle of the mandible over three occasions (conditions). The data for the study are shown in Table 6.6.2 and provided in file mixed.dat.

For model (6.5.17), $g = 2$ (groups), $r = 9$ (subject) and $t = 3$ (conditions). Each observation is a vector of the three measurements on the mandible. To construct exact multivariate tests for differences in occasions and for the interaction between groups and occasions, the matrix $\mathbf{V}$ in (6.5.20) must satisfy the multivariate sphericity (M-sphericity) condition given in (6.5.21). This was tested in Example 3.8.7. More will be said regarding this test and what it means in Section 6.7. For now, we assume that the data satisfy the circularity condition. This being the case, program m6_6_2.sas is used to perform the analysis for the data.

TABLE 6.6.1. Expected Mean Square Matrix

Source	df	EMS
A	3	$\Sigma_e + 3\Sigma_{\alpha\beta} + Q\,(A)$
B	3	$\Sigma_e + 3\Sigma_{\alpha\beta} + 12\Sigma_\beta$
AB	9	$\Sigma_e + 3\Sigma_{\alpha\beta}$
Error	32	

TABLE 6.6.2. Individual Measurements Utilized to Assess the Changes in the Vertical Position and Angle of the Mandible at Three Occasion

	SOr-Me (mm)			ANS-Me (mm)			Pal-MP angle (degrees)		
Group	1	2	3	1	2	3	1	2	3
	117.0	117.5	118.5	59.0	59.0	60.0	10.5	16.5	16.5
	109.0	110.5	111.0	60.0	61.5	61.5	30.5	30.5	30.5
	117.0	120.0	12.5	60.0	61.5	62.0	23.5	23.5	23.5
	122.0	126.0	127.0	67.5	70.5	71.5	33.0	32.0	32.5
T_1	116.0	118.5	119.5	61.5	62.5	63.5	24.5	24.5	24.5
	123.0	126.0	127.0	65.5	61.5	67.5	22.0	22.0	22.0
	130.5	132.0	134.5	68.5	69.5	71.0	33.0	32.5	32.0
	126.5	128.5	130.5	69.0	71.0	73.0	20.0	20.0	20.0
	113.9	116.5	117.9	57.9	59.0	60.5	25.0	25.0	24.5
	128.0	129.0	131.5	67.0	67.5	69.0	24.0	24.0	24.0
	116.5	120.0	121.5	63.5	65.0	66.0	28.5	29.5	29.5
	121.5	125.5	127.5	64.5	67.5	69.0	26.5	27.0	27.0
	109.5	112.0	114.0	54.0	55.5	57.0	18.0	18.5	19.0
T_2	133.0	136.0	137.5	72.0	73.3	75.5	34.5	34.5	34.5
	120.0	124.5	126.0	62.5	65.0	66.0	26.0	26.0	26.0
	129.5	133.5	134.5	65.0	68.0	69.0	18.5	18.5	18.5
	122.0	124.0	125.5	64.5	65.5	66.0	18.5	18.5	18.5
	125.0	127.0	128.0	65.5	66.5	67.0	21.5	21.5	21.6

TABLE 6.6.3. Expected Mean Squares for Model (6.5.17)

Source	df	EMS
GROUP	1	$\Sigma_e + 3\Sigma_{S(G)} + Q\,(G, G * C)$
SUBJ(GROUP)	16	$\Sigma_e + 3\Sigma_{S(G)}$
COND	2	$\Sigma_e + Q(, G * C)$
GROUP*COND	2	$\Sigma_e + Q(G * C)$
ERROR	32	Σ_e

TABLE 6.6.4. MMM Analysis Zullo's Data

HYPOTHESIS	Wilks Λ	df	F	df	p-value
GROUP	0.8839	(3, 1, 16)	0.6132	(3, 14)	0.6176
COND	0.0605	(3, 2, 32)	30.6443	(6, 60)	< 0.0001
GROUP*COND	0.8345	(3, 2, 32)	0.9469	(6, 60)	0.4687

TABLE 6.6.5. Summary of Univariate Output

Variable	F-value	p-value
SOr(Y_1)	76.463/0.472 = 162.1	< 0.001
ANS(Y_2)	31.366/0.852 = 36.82	< 0.001
Pal (Y_3)	0.795/0.773 = 1.03	0.3694

For the univariate analysis of mixed linear models with an equal number of subjects per group and no missing data, the tests for fixed effects using PROC MIXED and PROC GLM are identical. However, PROC GLM may not used to obtain standard errors for contrasts since no provision is made in the procedure for random effects. Because there is no PROC MIXED for the multivariate model, we must use PROC GLM to construct multivariate tests. Again one reviews the univariate expected mean squares to form the appropriate matrices **H** and **E** to test for the fixed effects: Conditions, Groups and Group by Condition interaction. The expected mean squares are provided in Table 6.6.3

From the expected mean squares in Table 6.6.3, we see that the test for differences in conditions is confounded by interaction (G*C). Thus, to interpreted the test of conditions, we must have a nonsignificant interaction. The MANOVA table using Wilks' Λ criterion is given in Table 6.6.4

Because the test of interaction is nonsignificant, we may want to investigate contrasts in conditions. In most mixed models involving both fixed and random effects, PROC GLM does not correctly calculate the correct standard errors for contrasts since it considers all factors fixed. Because the test for conditions and interactions only involve Σ_e, this is not a problem for this design. Because it is usually a problem for more general designs, we recommend using PROC MIXED. That is, using the Bonferroni inequality one divides the overall α by the number of variables. For this problem, $\alpha^* = 0.05/3 = 0.0167$. Then, one may use the LSMEAN statement to obtain approximate confidence intervals as illustrated in program m6_6_2.sas. For pairwise comparisons across conditions for the variable sor_me, the mean differences follow

$$\begin{aligned} -3.35 &\leq C1 - C2 \leq -1.99 \\ -4.74 &\leq C1 - C3 \leq -3.38 \\ -2.07 &\leq C2 - C3 \leq -0.71 \end{aligned}$$

Pairwise comparisons for the variable ANS_ME are also significant. Thus, we conclude that significant differences exist for conditions for two of the three variables and that there is no significant differences between treatment groups.

From the MMM, one may obtain the univariate split-plot F-ratios, one variable at a time by constructing the appropriate error mean squares. The results are provided in Table 6.6.5

Exercises 6.6

1. In a learning experiment with four male and four females, 150 trials were given to each subject in the following manner.

(1)	Morning	With training	Nonsense words
(2)	Afternoon	With training	Letter word
(3)	Afternoon	No training	Nonsense words
(4)	Morning	No training	Letter words

Using the number of trials to criterion for five-letter (F) and seven-letter (S) "words," the data for the randomized block design follows.

Treatment Conditions

Blocks		1		2		3		4	
M	F	120	F	90	F	140	F	70
	S	130	S	100	S	150	S	85
F	F	70	F	30	F	100	F	20
	S	80	S	60	S	110	S	35

Using $\alpha = .05$ for each test, test for differences in Treatment Conditions and for Block differences.

2. In a pilot study designed to investigate the mental abilities of four ethnic groups and four socioeconomic-status (SES) classifications, high SES males (HM), high SES females (HF) low SES males (LM), and low SES females (LF), the following table of data on three different measures, mathematics (MAT), English (ENG), and general knowledge (GK), was obtained.

Ethnic Groups

	I			II		
SES	MAT	ENG	GK	MAT	ENG	GK
HM	80	60	70	85	65	75
HF	85	65	75	89	69	80
LM	89	78	81	91	73	82
LF	92	81	85	100	80	90

	III			IV		
SES	MAT	ENG	GK	MAT	ENG	GK
HM	90	70	80	95	75	85
HF	94	76	85	96	80	90
LM	99	81	90	100	85	97
LF	105	84	91	110	90	101

Carry out a multivariate, analysis-of-variance procedure to investigate differences in ethnic groups (I, II, III, and IV).

3. In an experiment designed to investigate two driver-training programs, students in the eleventh grade in schools I, II, and III were trained with one program and eleventh-grade students in schools IV, V, and VI were trained with another program. After the completion of a 6-week training period, a test measuring knowledge (K) of traffic laws and driving ability (A) was administered to the 138 students in the study.

Program 1						Program 2					
S1(I)		S2(II)		S3(III)		S1(IV)		S2(V)		S3(VI)	
K	A	K	A	K	A	K	A	K	A	K	A
48	66	36	43	82	51	54	79	46	46	21	5
68	16	24	24	79	55	30	79	13	13	51	16
28	22	24	24	82	46	9	36	59	76	52	43
42	21	37	30	65	33	23	79	26	42	53	54
73	10	78	21	33	68	18	66	38	84	11	9
46	13	82	60	79	12	15	82	29	65	34	52
46	17	56	24	33	48	16	89	12	47	40	16
76	15	24	12	75	48	14	82	6	56	43	48
52	11	82	63	33	35	48	83	15	46	11	12
44	64	78	34	67	28	31	65	18	34	39	69
33	14	44	39	67	52	11	74	41	23	32	25
43	25	92	34	67	37	41	88	26	33	33	26
76	60	68	85	33	35	56	67	15	29	45	12
76	18	43	50	33	30	51	93	54	50	27	12
36	49	68	28	67	33	23	83	36	83	49	21
39	75	53	90	67	54	16	76	27	82	44	56
76	16	76	41	75	63	40	69	64	79	64	18
73	11	35	10	83	93	21	37	55	67	27	11
34	12	24	11	94	92	88	82	10	70	79	57
68	63	23	21	79	61	7	32	12	65	47	64
52	69	66	65	89	53	38	75	34	93	17	29
46	22	76	78	94	92	20	77	21	83	39	42
26	18	34	36	93	91	13	70	34	81	21	11

Test for differences between Programs.

4. Using as blocking variables ability tracks and teaching machines, an investigator interested in the evaluation of four teaching units in science employed the following Latin-square design.

		Teaching Machines			
		1	2	3	4
	1	T_2	T_1	T_3	T_4
Ability	2	T_4	T_3	T_1	T_2
Groups	3	T_1	T_4	T_2	T_3
	4	T_3	T_2	T_4	T_1

The treatments T_1, T_2, T_3 and T_4 are four version of measuring astronomical distance in the solar system and beyond the solar system. The dependent variables for the study are subtest scores on one test designed to measure the students ability in determining solar system distances within (W) and beyond (B) the solar system. The data for the study follow.

Cell	W	B	Cell	W	B
112	13	15	311	10	5
121	40	4	324	20	16
133	31	16	332	17	16
144	37	10	343	12	4
214	25	20	413	24	15
233	30	18	422	20	13
231	22	6	434	19	14
242	25	18	441	29	20

Use the mixed MANOVA model for the Latin-square design to test the main-effect hypothesis. What are your conclusions?

6.7 Double Multivariate Model (DMM)

In the last section we reviewed several MMM's and illustrated the analysis of vector valued repeated measurement data that require a restrictive structure for Ω for a valid analysis. We may relax the requirement that Ω be a member of the class of multivariate covariance matrices with Type H structure and now only require the Ω be positive definite. Then, using the data structure given in Table 6.5.1, the multivariate repeated measures may be modeled using the fixed effect MGLM

$$\mathbf{Y} = \mathbf{XB} + \mathbf{E} \tag{6.7.1}$$

where $\mathbf{Y}_{n \times pt}$, and the matrix $\mathbf{Y}$ is organized as $\mathbf{Y} = \left[\mathbf{y}_{11}^*, \mathbf{y}_{12}^*, \ldots, \mathbf{y}_{gr}^*\right]$ so that variables are nested within conditions (time). The matrix $\mathbf{B}$ is an unknown parameter matrix, and $\mathbf{E}$

is a random error matrix such that

$$\mathbf{e}^* = \text{vec}\left(\mathbf{E}'\right) \sim N_{pq}\left(\mathbf{0}\ \mathbf{I}_n \otimes \Omega\right) \tag{6.7.2}$$

so that each row of $\mathbf{Y}$ has covariance structure $\Omega_{pt \times pt}$.

For model (6.7.1), hypotheses have the general structure

$$H : \mathbf{CB}\left(\mathbf{M} \otimes \mathbf{I}_p\right) = \mathbf{0} \tag{6.7.3}$$

where the matrix $\mathbf{C}_{g \times g}$ has rank $g = v_h$, $\mathbf{M}_{t \times (t-1)}$ is a orthogonal contrast matrix such that $\mathbf{M}'\mathbf{M} = I_{t-1}$ and $\mathbf{M}'\mathbf{1} = \mathbf{0}$ so that $\mathbf{A} = \mathbf{M} \otimes \mathbf{I}_p$ has rank $p\,(t-1)$. To test hypotheses, one again forms the hypotheses and error matrices $\mathbf{H}$ and $\mathbf{E}$. For $\mathbf{H}$ in (6.7.3) and model (6.7.1), the matrices $\mathbf{H}$ and $\mathbf{E}$ are constructed as follows

$$\begin{aligned}\mathbf{H} &= \left[\mathbf{C}\widehat{\mathbf{B}}\left(\mathbf{M} \otimes \mathbf{I}_p\right)\right]' \left[\mathbf{C}\left(\mathbf{X}'\mathbf{X}\right)^{-}\mathbf{C}'\right]^{-1} \left[\mathbf{C}\widehat{\mathbf{B}}\left(\mathbf{M} \otimes \mathbf{I}_p\right)\right] \\ \mathbf{E} &= \left(\mathbf{M}' \otimes \mathbf{I}_p\right)\mathbf{Y}'\left[\mathbf{I} - \mathbf{X}\left(\mathbf{X}'\mathbf{X}\right)^{-}\mathbf{X}'\right]\mathbf{Y}\left(\mathbf{M} \otimes \mathbf{I}_p\right)\end{aligned} \tag{6.7.4}$$

where $\widehat{\mathbf{B}} = \left(\mathbf{X}'\mathbf{X}\right)^{-1}\mathbf{X}'\mathbf{Y}$. The matrices $\mathbf{H}$ and $\mathbf{E}$ have independent Wishart distributions with degrees of freedom $v_h = \text{rank}\,\mathbf{C}$ and $v_e = n - r\,(\mathbf{X})$.

A disadvantage of the DMM is that Ω has arbitrary structure. In many applications of repeated measurements, Ω has a simple structure as opposed to a general structure, Crowder and Hand (1990). A very simple structure for Ω is that $\Omega = \mathbf{I}_t \otimes \Sigma_e$ which implies that the measurements observed over time are uncorrelated and that the covariance structure of the p variables is arbitrary and represented by Σ_e. Then, Ω has Type H structure and satisfies the multivariate sphericity (M-sphericity) condition so that an analysis would employ the MMM approach.

When the MMM approach is not valid, but one does not want to assume a general structure, an alternative structure for Ω is to assume that

$$\Omega = \Sigma_t \otimes \Sigma_e \tag{6.7.5}$$

suggested by Boik (1991), Galecki (1994), and Naik and Rao (2001). If Ω satisfies (6.7.5), it does not satisfy the M-sphericity condition since

$$\mathbf{V} = \mathbf{M}'\Sigma_t\mathbf{M} \otimes \Sigma_e = \mathbf{W} \otimes \Sigma_e \neq \mathbf{I}_{(t-1)} \otimes \Sigma_e$$

Hence, a MMM analysis does not produce exact tests. Also observe that $\mathbf{V} = \mathbf{W} \otimes \Sigma_e$ has Kronecker structure. Boik (1991) showed that if $\mathbf{V}$ has Kronecker structure that his ϵ-adjusted test provides a reasonable approximation to a MMM test. And, that the adjustment factor is invariant to choosing a trace or generalized variance criterion when equating the first two moments of the hypothesis and error matrices to obtain a Satterthwaite approximation. The ϵ-adjustment to the MMM tests involves correcting the hypothesis and error degrees of freedom of the Wishart distribution for the tests of conditions and interactions, the within-subject tests that depend on multivariate sphericity. In particular $v_h^* = \widehat{\epsilon} v_h$ and $v_e^* = \widehat{\epsilon} v_e$ where v_e and v_h are the associated degrees of freedom for the MMM tests. To calculate $\widehat{\epsilon}$, we define the generalized trace operator, Thompson (1973).

Definition 6.7.1 *If* $\mathbf{W}$ *is a* $(qp \times qp)$ *matrix and* W_{ij} *are* $p \times p$ *submatrices of* $\mathbf{W}$ *for* $i, j = 1, 2, \ldots, q$, *then the generalized trace operator of* $\mathbf{W}$ *is the* $q \times q$ *matrix*

$$T_q(\mathbf{W}) = \left[\operatorname{tr}(\mathbf{W}_{ij})\right]$$

Interchanging q *and* p, *the matrix* $T_p(W)$ *is a* $p \times p$ *matrix containing submatrices of order* $q \times q$.

Letting $\mathbf{A} = \mathbf{M} \otimes \mathbf{I}_p$ and $\Phi = \mathbf{A}'\widehat{\Omega}\mathbf{A}$ where Φ is partitioned into $p \times p$ submatrices $\mathbf{E}_{ij}$, Boik (1991) purposed a new estimator of ϵ as

$$\widehat{\epsilon} = \left[\operatorname{Tr}(\Phi)\right]^2 / p\,(t-1)\operatorname{tr}(\Phi) \tag{6.7.6}$$

where

$$\Phi = \left\{\mathbf{I}_p \otimes \left[T_u(\Phi)\right]^{-1/2}\right\} \Phi \left\{\mathbf{I}_p \otimes \left[T_u(\Phi)\right]^{-1/2}\right\}$$

and $\widehat{\Omega}$ is an unbiased estimator of Ω. This new estimator may only be used with the Lawley-Hotelling trace criterion, T_o^2. Unlike the ϵ-adjustment of Boik (1988), this new adjustment is invariant under linear transformation of the p dependent variables.

To test the hypothesis that Ω has Kronecker structure versus the alternative that Ω has general structure

$$H : \Omega = \Sigma_t \otimes \Sigma_e \quad \text{vs} \quad A : \Omega = \Sigma \tag{6.7.7}$$

the likelihood ratio criterion is

$$\lambda = |\widehat{\Sigma}|^{n/2} / |\widehat{\Sigma}|^{np/2} |\widehat{\Sigma}_t|^{nt/2} \tag{6.7.8}$$

where $\widehat{\Sigma}_t$ and $\widehat{\Sigma}_e$ are the ML estimates of Ω under H and $\widehat{\Sigma}$ is the estimate of Ω under the alternative. Because the likelihood equations $\widehat{\Sigma}_e$ and $\widehat{\Sigma}_t$ have no closed form, an iterative procedure must be used to obtain the ML estimates. The ML estimate of Σ is $\mathbf{E}/n = \widehat{\Sigma}$ where $\mathbf{E}$ is the error SSCP matrix. The procedure PROC MIXED in SAS may be used to obtain the ML estimates of Σ_e and Σ_t. Alternatively, the algorithm suggested by Boik (1991) may be used. To test (6.7.7), the large sample chi-square statistic is

$$X^2 = -2 \log \lambda \mathrel{\dot\sim} \chi^2(v) \tag{6.7.9}$$

where $v = (p-1)(t-1)\left[(p+1)(t+1)+1\right]/2$.

One may also test the hypothesis that Ω has multivariate sphericity structure versus the alternative that Ω has Kronecker structure

$$\begin{aligned} H &: \Omega = \mathbf{I}_{(t-1)} \otimes \Sigma_e \\ A &: \Omega = \mathbf{W} \otimes \Sigma_e \end{aligned} \tag{6.7.10}$$

where $\mathbf{W} = \mathbf{M}'\Sigma_t\mathbf{M}$ for some matrix $\mathbf{M}$, $\mathbf{M}'\mathbf{M} = \mathbf{I}_{t-1}$ and $\mathbf{M}'\mathbf{1} = \mathbf{0}$. The likelihood ratio statistic is

$$\lambda = |\widehat{\mathbf{W}}|^{np/2} |\widehat{\Sigma}_e|^{n(t-1)/2} / |\widehat{\Omega}_o|^{n(t-1)} \tag{6.7.11}$$

where

$$\widehat{\Omega}_o = \left(\mathbf{M} \otimes \mathbf{I}_p\right)' \widehat{\Omega} \left(\mathbf{M} \otimes \mathbf{I}_p\right)$$
$$= \left(\mathbf{M} \otimes \mathbf{I}_p\right)' \left[\sum_{i=1}^{n} (\mathbf{y}_i - \overline{\mathbf{y}}) (\mathbf{y}_i - \overline{\mathbf{y}})' \right] \left(\mathbf{M} \otimes \mathbf{I}_p\right) \tag{6.7.12}$$

and $\widehat{\mathbf{W}}$ and $\widehat{\Sigma}_e$ are ML estimates of $\mathbf{W}$ and Σ_e under A. The large sample chi-square statistic to test (6.7.10) is

$$X^2 = -2 \log \lambda \,\dot{\sim}\, \chi^2 (v) \tag{6.7.13}$$

where $v = t(t-1)/2$. One may use PROC MIXED to estimate $\widehat{\mathbf{W}}$ and $\widehat{\Sigma}_e$.

In summary, the analysis of multivariate repeated measures data should employ the DMM if Ω has general structure. If Ω satisfies the multivariate sphericity condition so that $\mathbf{V} = (\mathbf{M} \otimes \mathbf{I}) \Omega (\mathbf{M} \otimes \mathbf{I}) = \mathbf{I}_u \otimes \Sigma$, then the MMM should be used. If Ω has Kronecker structure, the ϵ-adjusted developed by Boik (1988, 1991) may be used. As an alternative to the ϵ-adjusted method, one may also analyze multivariate repeated measurements using PROC MIXED. Then, the DMM is represented as a mixed model using (6.3.4) by stacking the g, $pt \times 1$ vectors one upon the other. Because the design is balanced, hypotheses $\mathbf{CB}\left(\mathbf{M} \otimes \mathbf{I}_p\right) = \mathbf{0}$ are represented as $((\mathbf{M} \otimes \mathbf{I}_p)' \otimes \mathbf{C}) \text{ vec } (\mathbf{B}) = \mathbf{0}$ and may be tested using an approximate Sattherwaite F statistic. This is the approach used in PROC MIXED.

Replacing the matrix $\mathbf{I}_n$ in (6.7.2) with a matrix $\mathbf{W}_n$, Young, Seaman and Meaux (1999) develop necessary and sufficient conditions for the structure of $\mathbf{W}_n$ for exact multivariate tests of H given in (6.7.3) for the DMM. They develop exact tests for covariance structure of the form $\mathbf{W}_n \otimes \Omega$, calling such structures independence distribution-preserving (IDP).

6.8 Double Multivariate Model Examples

When analyzing a DMM, one may organize the vector observations for each subject with p variables nested in t occasions (O) as in (6.6.1) so that for each occasion O_i we have

$$\begin{array}{c} O_i \\ \mathrm{v}_1, \mathrm{v}_2, \ldots, \mathrm{v}_p \end{array} \tag{6.8.1}$$

for $i = 1, 2, \ldots, t$. Then, the post matrix for testing (6.7.3) has the form $\mathbf{A} = \mathbf{M} \otimes \mathbf{I}_p$ where the $r(\mathbf{M}) = t - 1 = u$. Alternatively, the multivariate observations may be represented as in Table 6.6.1. Then, the t occasions (conditions) are nested within each of the variables so that

$$\begin{array}{c} \mathrm{v}_i \\ O_1, O_2, \ldots, O_p \end{array} \tag{6.8.2}$$

for $i = 1, 2, \ldots, p$. For this organization, the post matrix $\mathbf{P}$ for testing

$$H : \mathbf{CBP} = \mathbf{0}$$

has the structure $\mathbf{P} = \mathbf{I}_p \otimes \mathbf{M}$. This has a convenient block structures for $\mathbf{P}$; however, the M-sphericity condition has the reverse Kronecker structure, $\mathbf{V} = \Sigma_e \otimes \mathbf{I}_u$. Both representations are used in the literature and we have used both in this text. Details regarding each choice is discussed further by Timm and Mieczkowski (1997, Chapter 6) and will not be duplicated here.

a. Double Multivariate MANOVA (Example 6.8.1)

Using the data in Table 6.6.1, we illustrate the analysis of Dr. Thomas Zullo's data. Because occasions are nested within variables, the post matrix **P** will be used to test hypotheses. Assuming a FR model, the parameter matrix **B** has the structure

$$\mathbf{B} = \begin{bmatrix} \mu_{11}\mu_{12}\mu_{13} & \vdots & \mu_{14}\mu_{15}\mu_{16} & \vdots & \mu_{17}\mu_{18}\mu_{19} \\ & \vdots & & \vdots & \\ \mu_{21}\mu_{22}\mu_{23} & \vdots & \mu_{24}\mu_{25}\mu_{26} & \vdots & \mu_{27}\mu_{28}\mu_{29} \end{bmatrix} \tag{6.8.3}$$

where each block of parameters represents three repeated measurements on each variable.

The first hypothesis test of interest for these data is to test whether the profiles for the two treatment groups are parallel, by testing for an interaction between conditions and groups. The hypothesis is

$$\begin{aligned} \mathbf{H}_{GC} : [\mu_{11} - \mu_{13}, \mu_{12} - \mu_{13}, \ldots, \mu_{17} - \mu_{18}, \mu_{18} - \mu_{19}] \\ = [\mu_{21} - \mu_{23}, \mu_{22} - \mu_{23}, \ldots, \mu_{27} - \mu_{28}, \mu_{28} - \mu_{29}] \end{aligned} \tag{6.8.4}$$

The matrices **C** and $\mathbf{P} = (\mathbf{I}_3 \otimes \mathbf{M})$ of the form $\mathbf{CBP} = \mathbf{0}$ to test H_{GC} are

$$\mathbf{C} = (1, -1), \mathbf{P} = \begin{bmatrix} \mathbf{M} & \mathbf{0} & \mathbf{0} \\ \mathbf{0} & \mathbf{M} & \mathbf{0} \\ \mathbf{0} & \mathbf{0} & \mathbf{M} \end{bmatrix} \quad \text{and} \quad \mathbf{M} = \begin{bmatrix} 1 & 0 \\ 0 & 1 \\ -1 & -1 \end{bmatrix} \tag{6.8.5}$$

To analyze the DMM in SAS, the matrix **P** is defined as $m \equiv \mathbf{P}'$ so that the columns of **P** become the rows of m. This organization of the data does not result in a convenient block structure of **H** and **E**. Instead, one must employ (6.8.1) as suggested by Boik (1988). In either case, if **M** is normalized so that $\mathbf{M}'\mathbf{M} = \mathbf{I}$ and $\mathbf{M}'\mathbf{1} = \mathbf{0}$, the MMM easily obtained from the DMM as illustrated by Timm (1980) and Boik (1988).

Continuing with the DMM analysis, the overall hypothesis for testing differences in Groups, H_{G^*}, is

$$H_{G^*} : \boldsymbol{\mu}_1 = \boldsymbol{\mu}_2 \tag{6.8.6}$$

and the matrices **C** and **P** to test this hypothesis are

$$\mathbf{C}_{G^*} = [1, -1] \text{ and } \mathbf{P} = \mathbf{I}_2 \tag{6.8.7}$$

The hypothesis to test for vector differences in conditions becomes

$$H_{C^*} : \begin{bmatrix} \mu_{11} \\ \mu_{21} \\ \mu_{14} \\ \mu_{24} \\ \mu_{17} \\ \mu_{27} \end{bmatrix} = \begin{bmatrix} \mu_{12} \\ \mu_{22} \\ \mu_{15} \\ \mu_{25} \\ \mu_{18} \\ \mu_{28} \end{bmatrix} = \begin{bmatrix} \mu_{13} \\ \mu_{23} \\ \mu_{16} \\ \mu_{26} \\ \mu_{19} \\ \mu_{29} \end{bmatrix} \tag{6.8.8}$$

The test matrices to test (6.8.8) are

$$\mathbf{C}_{C^*} = \mathbf{I}_2 \text{ and } \mathbf{P} = \begin{bmatrix} 1 & 0 & 0 & 0 & 0 & 0 \\ -1 & 1 & 0 & 0 & 0 & 0 \\ 0 & -1 & 0 & 0 & 0 & 0 \\ 0 & 0 & 1 & 0 & 0 & 0 \\ 0 & 0 & -1 & 1 & 0 & 0 \\ 0 & 0 & 0 & -1 & 0 & 0 \\ 0 & 0 & 0 & -1 & 0 & 0 \\ 0 & 0 & 0 & 0 & 1 & 0 \\ 0 & 0 & 0 & 0 & -1 & 1 \\ 0 & 0 & 0 & 0 & 0 & -1 \end{bmatrix}$$

Tests (6.8.6) and (6.8.8) do not require parallelism of profiles. However, given parallelism, tests for differences between groups and among conditions are written as

$$H_{G^*} \begin{bmatrix} \sum_{j=1}^{3} \mu_{1j}/3 \\ \sum_{j=4}^{6} \mu_{1j}/3 \\ \sum_{j=7}^{9} \mu_{1j}/3 \end{bmatrix} = \begin{bmatrix} \sum_{j=1}^{3} \mu_{2j}/3 \\ \sum_{j=4}^{6} \mu_{2j}/3 \\ \sum_{j=7}^{9} \mu_{2j}/3 \end{bmatrix} \tag{6.8.9}$$

and

$$H_{C^*} \begin{bmatrix} \sum_{i=1}^{2} \mu_{i1}/2 \\ \sum_{i=1}^{2} \mu_{i4}/2 \\ \sum_{i=1}^{2} \mu_{i7}/2 \end{bmatrix} = \begin{bmatrix} \sum_{i=1}^{2} \mu_{i2}/2 \\ \sum_{i=1}^{2} \mu_{i5}/2 \\ \sum_{i=1}^{2} \mu_{i8}/2 \end{bmatrix} = \begin{bmatrix} \sum_{i=1}^{2} \mu_{i3}/2 \\ \sum_{i=1}^{2} \mu_{i6}/2 \\ \sum_{i=1}^{2} \mu_{i9}/2 \end{bmatrix} \tag{6.8.10}$$

respectively. The hypothesis test matrices to text H_G and H_C are

$$\mathbf{C}_G = [1, -1] \quad \text{and} \quad \mathbf{P}' = \begin{bmatrix} 1 & 1 & 1 & 0 & 0 & 0 & 0 & 0 & 0 \\ 0 & 0 & 0 & 1 & 1 & 1 & 0 & 0 & 0 \\ 0 & 0 & 0 & 0 & 0 & 0 & 1 & 1 & 1 \end{bmatrix} \tag{6.8.11}$$

$$\mathbf{C}_G = (1/2, -1/2) \quad \text{and} \quad \mathbf{P}' = \begin{bmatrix} 1 & 0 & 0 & 0 & 0 & 0 \\ 0 & 0 & 0 & 1 & 0 & 0 \\ -1 & 0 & 0 & -1 & 0 & 0 \\ 0 & 1 & 0 & 0 & 0 & 0 \\ 0 & 0 & 0 & 0 & 1 & 0 \\ 0 & -1 & 0 & 0 & -1 & 0 \\ 0 & 0 & 0 & 0 & 0 & 0 \\ 0 & 0 & 1 & 0 & 0 & 1 \\ 0 & 0 & -1 & 0 & 0 & -1 \end{bmatrix} \tag{6.8.12}$$

TABLE 6.8.1. DMM Results, Dr. Zullo's Data

Hypothesis	Wilks' Λ	df	F	df	p-value
GC	0.5830	(6,1,16)	1.3114	(6,11)	0.3292
G*	0.4222	(9,1,16)	1.216	(9,8)	(0.3965)
C*	0.0264	(6.2.16)	9.4595	(12,22)	< 0.0001
G	0.8839	(3,1,16)	0.6132	(3.14)	0.6176
C	0.0338	(6,1,16)	52.4371	(6,11)	< 0.0001

Normalization of the matrix **P** in (6.8.11) and (6.8.12) within each variable allows one to again obtain MMM results from the multivariate tests, given parallelism. This is not the case for the multivariate tests of H_{G^*} and H_{C^*} that do not assume parallelism of profiles. Some standard statistical packages for the analysis of the DMM design, like SPSS, do not test H_{G^*} or H_{C^*} but assume the parallelism condition, and hence are testing H_G and H_C.

The SAS code to perform these tests is provided in program m6_8_1.sas. The results of the tests are given in Table 6.8.1 using only Wilks' Λ criterion.
Because the test of parallelism is not significant, tests of conditions and group differences given parallelism are appropriate. From these tests one may derive MMM tests of H_C and H_{GC} given multivariate sphericity. The MMM tests of group differences is identical to the test of H_G using the DMM and does not depend on the sphericity assumption.

Thomas (1983) derived the likelihood ratio test of multivariate sphericity and Boik (1988) showed that it was the necessary and sufficient condition for the MMM tests to be valid. Boik (1988) also developed ϵ-adjusted multivariate tests of H_C and H_{GC} when multivariate sphericity is not satisfied. Recall that for the simple split-plot design F-ratios are exact if and only if across groups. For the DMM using the data organization (6.8.2), the condition becomes

$$(\mathbf{I}_p \otimes \mathbf{M})' \, \Omega \, (\mathbf{I}_p \otimes \mathbf{M}) = \Sigma_e \otimes \mathbf{I}_u \tag{6.8.13}$$

where the contrast matrix **M** is orthogonal; using organization (6.8.1), the M-sphericity condition becomes

$$(\mathbf{M} \otimes \mathbf{I}_p)' \, \Omega \, (\mathbf{M} \otimes \mathbf{I}_p) = \mathbf{I}_u \otimes \Sigma_e \tag{6.8.14}$$

The LR test statistic for testing (6.8.14) given in (3.8.48) is

$$\lambda = |\mathbf{E}|^{n/2} / \; q^{-1} \left| \sum_{i=1}^{q} \mathbf{E}_{ii} \right|^{nq/2}$$

where $u \equiv q = r\,(\mathbf{M})$ was illustrated in Example 3.8.7. The code is included in program m6_8_1.sas for completeness. Because the M-sphericity test is rejected for these data, the MMM should not be used to test hypotheses.

In many applications of the DMM, the matrix $\widehat{\Omega} = \mathbf{E}/v_e$ is not positive definite. This occurs when the number of subjects in each treatment group is less than the product of the number of occasions times the number of variables. When this occurs, or if the M-sphericity condition is not met, one may use Boik's (1988, 1991) ϵ-adjusted MMM procedure to test H_C and H_{GC}.

Using (6.7.6) with the data organized as in (6.8.1), Boik's (1991) revised estimator of ϵ for these data is $\widehat{\epsilon} = 0.66667$ as calculated in program m6_8_1.sas using PROC IML. Multiplying T_o^2 obtained under multivariate sphericity by $\widehat{\epsilon}$, Boik (1991) shows that the value of the test statistic and the degrees of freedom should be adjusted for testing H_C and H_{GC} (for example, the T_o^2 statistic for testing H_{GC} given M-sphericity 13.7514, is multiplied by $\widehat{\epsilon}$). Boik's new approximation to the T_o^2 statistic results in a p-value for the test of H_{GC} of 0.667. The p-value for the test of H_{GC} assuming incorrectly M-sphericity, is 0.4862 (Example 6.6.2). The p-value for the DMM is 0.3292. Using Boik's (1988) adjustment, $\widehat{\epsilon}^* = 0.73051$ for these data, the p-value for the test of H_{GC} using T_o^2 is 0.482, Timm and Mieczkowski (1997, p. 278). From the simulation results reported by Boik (1991), it appears that the poor performance of the new ϵ-adjusted MMM test is due to the fact that Ω does not have Kronecker structure.

When analyzing designs with multivariate vector observations over occasions, one must ensure that the data are MVN and that the covariance matrices are homogeneous across treatment groups. Given these assumptions, the MMM is most powerful under multivariate sphericity. When multivariate sphericity does not hold, neither the ϵ-adjusted MMM or the DMM is uniformly most powerful. For small sample sizes and Kronecker structure of Ω, Boik's (1991) newly proposed ϵ-adjusted test of the MMM is preferred to the DMM. However, if Kronecker structure can be established, it is better to incorporate the structure into the model, Galecki (1994). Models with certain types of Kronecker structure may be tested using PROC MIXED.

Two important situations in which a Kronecker structure may facilitate the analysis of longitudinal data occur in DMM designs where one has a single within factor (C), and vector observations at each level of factor C. This leads to a condition factor and a second factor that is associated with the variable under study. This also occurs in split-split plot designs where one obtains repeated observations over two within-subject, crossed factors. In both situations, we often assume that the covariance matrix at any level of one within factor is the same for every level of the other within factor. This leads to the analysis of a mixed model in which the within-subject covariance may be considered with differing profiles for each covariance matrix. Because the Kronecker product of covariance matrices are only unique up to a scaler ($\mathbf{A} \otimes \mathbf{B} = \alpha\mathbf{A} \otimes \mathbf{B}/\alpha$ for $\alpha > 0$), PROC MIXED rescales one covariance matrix so that a single diagonal element is set equal to one in the analyses of mixed models with Kronecker structure.

b. Split-Plot Design (Example 6.8.2)

For an example of Kronecker structure, we consider a univariate split-split plot design. The classical univariate mixed model is

$$\begin{aligned} y_{ijkm} &= \mu + \alpha_i + \beta_k + \gamma_m + (\alpha\beta)_{ik} + (\alpha\gamma)_{im} \\ &\quad + (\beta\gamma)_{km} + (\alpha\beta\gamma)_{ikm} + s_{(i)j} \\ &\quad + (\beta s)_{(i)jk} + (\gamma s)_{(i)jm} + e_{(i)jkm} \end{aligned} \tag{6.8.15}$$

$i = 1, 2, \ldots, I$; $j = 1, 2, \ldots, J$; $k = 1, 2, \ldots, K$; and $M = 1, 2, \ldots, M$ where $s_{(i)j} \sim$ IN $(0, \sigma_s^2)$, $\beta s_{(i)jk} \sim$ IN $(0, \sigma^2 \beta s)$, $(\gamma s)_{(i)jm} \sim$ IN $\left(0, \sigma_{\beta s}^2\right)$, $\epsilon_{(i)jkm} \sim$ IN $(0, \sigma_e^2)$, and all random components are jointly independent. For the within-subject structure, we have a fixed effect factorial design, and the random subjects are nested within fixed treatment levels. To fix our ideas, the data in Table 6.8.2 are used, Timm (1980, p. 61).

For the data in Table 6.8.2, $I = 2$, $K = 3$, $M = 3$ and $J = 10$ subjects are nested within the two treatment levels. The necessary and sufficient conditions for exact univariate F tests for this design is that the within subject covariance matrix is homogeneous across groups, and that the three circularity conditions are satisfied

$$\begin{aligned} \mathbf{M}_B'\Omega\mathbf{M}_B &= \lambda\mathbf{I} \\ \mathbf{M}_C'\Omega\mathbf{M}_C &= \lambda\mathbf{I} \\ \mathbf{M}_{BC}'\Omega\mathbf{M}_{BC} &= \lambda\mathbf{I} \end{aligned} \tag{6.8.16}$$

where the matrices $\mathbf{M}_i$ are normalized contrast matrices such that $\mathbf{M}_i'\mathbf{M}_i = \mathbf{I}$, of reduced order.

The assumptions given in (6.8.16) are somewhat restrictive. As an alternative, we analyze the data in Table 6.8.2 assuming that the covariance matrix Ω has Kronecker structure, $\Omega = \Sigma_B \otimes \Sigma_C$, over the within subject factors B and C. Program m6_8_2.sas is used for the analysis.

The (3×3) covariance matrix Σ_B is the unknown covariance matrix for factor B and the (3×3) unknown covariance matrix for factor C is Σ_C. For the data in Table 6.8.2, we consider two models for Ω: (1) both Σ_B and Σ_C have unknown structure and (2) the structure for Σ_B is unknown, but Σ_C has compound symmetry structure. This latter situation occurs frequently with DMM designs when the covariance structure for variables is unknown and the covariance structure for occasions satisfies compound symmetry (CS). Structure (1) and (2) above are represented in PROC MIXED as follows: TYPE=UN@UN and TYPE=UN@CS.

To use PROC MIXED for designs with repeated measurements with Kronecker structure, there must be two distinct within factors. The CLASS variables for the example are: treatment (treat), factor A (a), and factor B (b). The multivariate observations must be organized into a single vector for processing. The MODEL statement models the means for the multivariate observations. When specifying the model statement, one usually includes all combinations of variables defined in the CLASS statement.

We fit two models to the data, one covariance matrix with unknown Kronecker structure and the other that includes a matrix with CS structure. Comparing the AIC information criteria for the models, the model with unknown structure for Σ_B and Σ_C appears to fit better. The AIC information criteria are -544.4 and -582.3, respectively. Because one model is nested in the other, we may also calculate the likelihood ratio difference test statistic, $X^2 = 1150.527 - 1066.800 \doteq 83.73$ on four degrees of freedom to compare the models. Using $\alpha = 0.05$, $\chi^2 = 9.48$ so that the model with CS structure is rejected. The ANOVA table for the fixed effect tests are given in Table 6.8.3

Also provided in the output are ML estimates for Σ_B and Σ_C where Σ_C has been standardized by setting one component to one. Contrasts for differences in treatments and interactions, BC, are also provided in the program.

TABLE 6.8.2. Factorial Structure Data

		B_1			B_2			B_3		
		C_1	C_2	C_3	C_1	C_2	C_3	C_1	C_2	C_3
	S_1	20	21	21	32	42	37	32	32	32
	S_2	67	48	29	43	56	48	39	40	41
	S_3	37	31	25	27	28	30	31	33	34
	S_4	42	49	38	37	36	28	19	27	35
A_1	S_5	57	45	32	27	21	25	30	29	29
	S_6	39	39	38	46	54	43	31	29	28
	S_7	43	32	20	33	46	44	42	37	31
	S_8	35	34	34	39	43	39	35	39	42
	S_9	41	32	23	37	51	39	27	28	30
	S_{10}	39	32	24	30	35	31	26	29	32
	S'_1	47	36	24	31	36	29	21	24	27
	S'_2	53	43	32	40	48	47	46	50	54
	S'_3	38	35	33	38	42	45	48	48	49
	S'_4	60	51	61	54	67	60	53	52	50
A_2	S'_5	37	36	35	40	45	40	34	40	46
	S'_6	59	48	37	45	52	44	36	44	52
	S'_7	67	50	33	47	61	46	31	41	50
	S'_8	43	35	27	32	36	35	33	33	32
	S'_9	64	59	53	58	62	51	40	42	43
	S'_{10}	41	38	34	41	47	42	37	41	46

TABLE 6.8.3. ANOVA for Split-Split Plot Design Ω-Unknown Kronecker Structure

Source	NDF	DDF	Type III F	p-value
TREAT	1	16.6	9.52	0.0069
B	2	14.6	1.58	0.2402
C	2	18.0	34.91	0.0001
BC	4	14.1	43.55	0.0001
TREAT*B	2	14.6	0.09	0.9131
TREAT*C	2	18.0	0.23	0.7957
TREAT*B*C	4	14.1	0.84	0.5204

TABLE 6.8.4. ANOVA for Split-Split Plot Design Ω-Compound Symmetry Structure

Source	NDF	DDF	Type III F	p-value
TREAT	1	18	9.49	0.0064
B	2	36	3.64	0.0364
C	2	36	12.19	< 0.0001
B*C	4	72	34.79	< 0.0001
TREAT*B	2	36	0.11	0.8986
TREAT*C	2	36	0.18	0.8385
TREAT*B*C	4	72	0.96	0.4349

Continuing with the example, the design is next analyzed using PROC MIXED and PROC GLM assuming compound symmetry as defined in (6.8.16). The ANOVA table is shown in Table 6.8.4. To ensure that the degrees of freedom for the F tests in PROC GLM are identical to those generated by PROC MIXED, the option DDFM = CONTAIN is used on the MODEL statement. The option DDFM = SATTERTH may also be used; however, the F tests will no longer agree since this procedure uses the matrices of the random and fixed effects to determine the degrees of freedom of the denominator and not just a "syntactical" name as in the CONTAIN option.

Also included in program m6_8_2.sas is the SAS code for a multivariate analysis of model (6.8.15) where the MR model $\mathbf{Y} = \mathbf{XB} + \mathbf{E}$ assumes an unknown structure for Ω. The contrast matrices $\mathbf{C}$ and $\mathbf{M}$ for testing hypotheses of the form $H : \mathbf{CBM} = \mathbf{0}$ are given in the program. For example, to test interaction hypothesis B*C the matrices $\mathbf{C}$ and $\mathbf{M}$

$$\mathbf{C} = [1, 1] \quad \text{and} \quad \mathbf{M} = \begin{bmatrix} 1 & 0 & 0 & 0 \\ -1 & 1 & 0 & 0 \\ 0 & -1 & 0 & 0 \\ -1 & 0 & 1 & 0 \\ 1 & -1 & -1 & 1 \\ 0 & 1 & 0 & -1 \\ 0 & 0 & -1 & 0 \\ 0 & 0 & 1 & -1 \\ 0 & 0 & 0 & 1 \end{bmatrix}$$

are used. As illustrated in Timm (1980), the univariate F-ratio's may be obtained by averaging appropriately normalized hypothesis and error sums of squares and products matrices. That is, if the post matrix $\mathbf{M}$ for the test of interaction B*C is normalized such that $\mathbf{M}'\mathbf{M} = \mathbf{I}$, then univariate F test for interaction becomes

$$F_{BC} = \frac{r\,(\mathbf{M})\,\mathrm{tr}\,(\mathbf{H})\,/v_{BC}}{r\,(\mathbf{M})\,\mathrm{tr}\,(\mathbf{E})\,/v_e} = \frac{MS_{AB}}{MS_e}$$

TABLE 6.8.5. MANOVA for Split-Split Plot Design Ω-Unknown Structure

Source	NDF	DDF	Type III F	p-value
PARL	8	11	0.3250	0.9392
B*C	4	15	12.8803	< 0.0001
C	2	17	17.6789	< 0.0001
B	2	17	4.0825	0.0356
TREAT	1	18	8.5443	0.0091

The MANOVA test results for the multivariate tests of parallelism (PARL), B*C, C, B and TREAT using F-ratio's are given in Table 6.8.5.

All three models indicate that the treatment effect interactions with B, with C, and with B*C are nonsignificant. The test of PARL in Table 6.8.4 is testing these hypotheses simultaneously. The tests of C and B*C are also consistent for the three approaches. However, the tests of treatment (TREAT) and of B differ for the three approaches. Recall that the mixed model ANOVA procedure assumes a compound symmetry structure for Ω, the MR model assumes no structure, and the mixed model ANOVA with $\Omega = \Sigma_B \otimes \Sigma_C$ assumes a Kronecker structure for Ω. The most appropriate analysis of these data depends on the structure of Ω given multivariate normality and homogeneity of covariance structure across treatments.

The likelihood ratio test given in (6.7.8) that Ω has Kronecker structure versus the alternative that Ω has general structure is tested in program m6_8_2.sas. Since this test is rejected, the most appropriate analysis for the data analysis is to employ a MR model.

Exercises 6.8

1. Lee (1970) gives data for two dependent variables (time on target in seconds and the number of hits on target) and five trials to investigate bilateral transfer of reminiscence of teaching performance under four treatment conditions: (1) distributed practice on a linear circular-tracking task, (2) distributed practice on a nonlinear hexagonal-tracking task, (3) massed practice on a linear circular-tracking task, and (4) massed practice on a nonlinear hexagonal-tracking task. Subjects, randomly assigned to each group, performed on the given task under each condition with one hand for ten traits and then transferred to the other hand for the same number of trials after a prescribed interval. The two sets of measurement taken for the ten trials were blocked into five blocks of two trials and averaged, yielding five repeated measures for each dependent variable. The data obtained for groups 1 and 2 are given below.

Group 1	Time on Target/Sec.				
	1	2	3	4	5
1	13.95	12.00	14.20	14.40	13.00
2	18.15	22.60	19.30	18.25	20.45
3	19.65	21.60	19.70	19.55	21.00
4	30.80	21.35	21.25	21.25	20.90
5	17.80	20.05	20.35	19.80	18.30
6	17.35	20.85	20.95	20.30	20.70
7	16.15	16.70	19.25	16.50	18.55
8	19.10	18.35	22.95	22.70	22.65
9	12.05	15.40	14.75	13.45	11.60
10	8.55	9.00	9.10	10.50	9.55
11	7.35	5.85	6.20	7.05	9.15
12	17.85	17.95	19.05	18.40	16.85
13	14.50	17.70	16.00	17.40	17.10
14	22.30	22.30	21.90	21.65	21.45
15	19.70	19.5	19.85	18.00	17.80
16	13.25	17.40	18.75	18.40	18.80

Group 1	Hits on Target				
	1	2	3	4	5
1	31.50	37.50	36.50	35.50	34.00
2	22.50	12.00	17.50	19.00	16.50
3	18.50	18.00	21.50	18.50	14.50
4	20.50	18.50	17.00	16.50	16.50
5	29.00	21.00	19.00	23.00	21.00
6	22.00	15.50	18.00	18.00	22.50
7	36.00	29.50	22.00	26.00	25.50
8	18.00	9.50	10.50	10.50	14.50
9	28.00	30.50	37.50	31.50	28.00
10	36.00	37.00	36.00	36.00	33.00
11	33.50	32.00	33.00	32.50	36.50
12	23.00	26.00	20.00	21.50	30.00
13	31.00	31.50	33.00	26.00	29.50
14	16.00	14.00	16.00	19.50	18.00
15	32.00	25.50	24.00	30.00	26.50
16	23.50	24.00	22.00	20.50	21.50

Group 2	Time on Target/Sec				
	1	2	3	4	5
1	11.30	13.25	11.90	11.30	9.40
2	6.70	6.50	4.95	4.00	6.65
3	13.70	18.70	16.10	16.20	17.55
4	14.90	15.95	15.40	15.60	15.45
5	10.90	12.10	12.10	13.15	13.35
6	7.55	11.40	12.15	13.00	11.75
7	12.40	14.30	15.80	15.70	15.85
8	12.85	14.45	15.00	14.80	13.35
9	7.50	10.10	12.40	12.40	14.95
10	8.85	9.15	10.70	10.05	9.50
11	12.95	12.25	12.00	12.05	11.35
12	3.35	6.70	6.60	6.70	6.60
13	7.75	8.25	10.40	9.20	10.40
14	14.25	16.20	15.25	17.60	16.25
15	11.40	14.85	17.20	17.15	16.05
16	11.60	13.75	13.25	12.80	10.90

Group 2	Hits on Target				
	1	2	3	4	5
1	49.00	46.50	44.00	45.50	50.00
2	32.50	42.50	46.00	43.00	47.00
3	47.00	49.50	50.50	45.50	48.50
4	42.50	46.50	46.50	44.00	41.50
5	24.00	44.00	43.00	44.50	41.00
6	42.50	42.50	53.50	46.00	54.50
7	48.00	46.00	44.50	42.00	45.00
8	39.00	42.50	47.50	37.50	34.50
9	36.00	43.00	30.00	29.50	39.50
10	40.00	34.50	35.50	35.00	38.00
11	44.50	56.00	53.50	52.50	56.00
12	41.50	36.00	44.50	45.00	43.50
13	33.50	51.50	49.00	43.00	47.00
14	50.50	51.50	51.00	47.50	46.50
15	54.50	54.00	52.00	49.00	49.50
16	43.00	52.50	45.50	47.00	43.00

(a) Arrange the data matrix such that the first five columns represent the measurements on the first validate and the last five columns the second dependent variable, so that the observation matrix is a 32×10 matrix. Letting

$$\mathbf{B} = \begin{bmatrix} \beta_{10} & \beta_{11} & \beta_{12} & \beta_{13} & \beta_{14} & \theta_{10} & \theta_{11} & \theta_{12} & \theta_{13} & \theta_{14} \\ \beta_{20} & \beta_{21} & \beta_{22} & \beta_{23} & \beta_{24} & \theta_{20} & \theta_{21} & \theta_{22} & \theta_{23} & \theta_{24} \end{bmatrix}$$

so that $p_1 = p_2 = q = 5$ and $p_3 = 0$, test the overall hypothesis of parallelism.

$$\mathbf{H}_0 : \begin{bmatrix} \beta_{11} \\ \beta_{12} \\ \beta_{13} \\ \beta_{14} \\ -- \\ \theta_{11} \\ \theta_{12} \\ \theta_{13} \\ \theta_{14} \end{bmatrix} \begin{bmatrix} \beta_{21} \\ \beta_{22} \\ \beta_{23} \\ \beta_{24} \\ -- \\ \theta_{21} \\ \theta_{22} \\ \theta_{23} \\ \theta_{24} \end{bmatrix} \text{ or } . \begin{bmatrix} \boldsymbol{\beta}_1 \\ \boldsymbol{\theta}_1 \end{bmatrix} = \begin{bmatrix} \boldsymbol{\beta}_2 \\ \boldsymbol{\theta}_2 \end{bmatrix}$$

(b) Test for differences in the five repeated measurement occasions and for groups.

(c) Repeat (b) given parallelism.

(d) Analyze the data assuming a Kronecker structure of the type AR(1) over time.

2. SAS (1990a, example 9, p. 988) provides data for two responses, $Y1$ and $Y2$, measured three times for each subject (at pre, post, and follow-up). Each subject reviewed one of three treatments: A, B, or the control C. The data follow.

		$Y1$			$Y2$		
Treat	Subject	PR	PO	FU	PR	PO	FU
A	1	3	13	9	0	0	9
A	2	0	14	10	6	6	3
A	3	4	6	17	8	2	6
A	4	7	7	13	7	6	4
A	5	3	12	11	6	12	6
A	6	10	14	8	13	3	8
B	1	9	11	17	8	11	27
B	2	4	16	13	9	3	26
B	3	8	10	9	12	0	18
B	4	5	9	13	3	0	14
B	5	0	15	11	3	0	25
B	6	4	11	14	4	2	9
C	1	10	12	15	4	3	7
C	2	2	8	12	8	7	20
C	3	4	9	10	2	0	10
C	4	10	8	8	5	8	14
C	5	11	11	11	1	0	11
C	6	1	5	15	8	9	10

(a) Analyze the data using a DMM, a MMM, an a ϵ-adjusted MMM.

(b) Analyze the data assuming a Kronecker structure over the within dimension.

3. Reanalyze Zullo's data set assuming a Kronecker structure for the covariance matrix.

6.9 Multivariate Hierarchical Linear Models

The general univariate hierarchical linear model developed by Laird and Ware (1982) and illustrated using PROC MIXED can be used to analyze univariate components of variance models, random coefficient regression models, repeated measurements observations that contain data MCAR and time varying covariates under numerous covariance structures. In Section 6.5, we extended the univariate components of variance model to multivariate balanced designs and showed how the model may be used to analyze complete multivariate repeated measurements data when the covariance matrix has Type H structure. The DMM in Section 6.7 provided for the analysis of balanced multivariate repeated measures with arbitrary structure. However, the vector observations were not permitted to contain missing vector outcomes or random time-varying covariates. To permit missing observations and time-varying covariates, we can stack the observation vectors one upon another and analyze the data using PROC MIXED. This approach is discussed by Goldstein (1995). Longford (1993a) extends the model by adding measurement error to the model. His model may be fit to growth data using a general SEM. Following Amemiya (1994), we extend the model of Laird and Ware (1982) to the multivariate setting.

Following the univariate model, we introduce the multivariate random regression coefficient model. The model has the two stage structure

$$\begin{aligned} \mathbf{Y}_i &= \mathbf{Z}_i\mathbf{B}_i + \mathbf{E}_i \quad i = 1, 2, \ldots, n \\ \mathbf{B}_i &= \mathbf{Q}_i\mathbf{B} + \mathbf{A}_i \end{aligned} \tag{6.9.1}$$

The matrix $\mathbf{Y}_i$ of order $r_i \times p$ is a matrix of p-variables observed over r_i time points for the ith subject. Hence, the outcomes in $\mathbf{Y}_i$ are response vectors over an unequal number of time points. Here the data are assumed to be only MAR. The matrix $\mathbf{Z}_i$ of order $r_i \times h$ is a known within-subject design matrix with the $r(\mathbf{Z}_i) = h \leq r_i \geq p$. The matrix $\mathbf{B}_i$ $(h \times p)$ is an unknown matrix of regression coefficients for the ith subject and the rows $\mathbf{e}_i'$ of $\mathbf{E}$ are distributed $IN_p(\mathbf{0},\ \Sigma_e)$. In the second stage, the matrix $\mathbf{B}_i$ is being modeled as a multivariate regression model where the rows $\mathbf{a}_i'$ of $\mathbf{A}_i$ are mutually independent of $\mathbf{e}_i'$ and $\mathbf{a}_i' \sim IN_p(\mathbf{0}, \Sigma_a)$. The matrix $\mathbf{B}_{q\times 1}$ is an unknown matrix of regression parameters and the matrix $\mathbf{Q}_i$ $(h \times q)$ is known and fixed.

Combining the equations in (6.9.1) into a single model, we have that

$$\begin{aligned} \mathbf{Y}_i &= (\mathbf{Z}_i\mathbf{Q}_i)\,\mathbf{B} + \mathbf{Z}_i\mathbf{A}_i + \mathbf{E}_i \\ \mathbf{Y}_i &= \mathbf{X}_i\mathbf{B} + \mathbf{Z}_i\mathbf{A}_i + \mathbf{E}_i \end{aligned} \tag{6.9.2}$$

which is a multivariate generalization of (6.3.2). Letting $N = \sum_i r_i$, the multivariate hierarchical linear model may be written as

$$\begin{bmatrix} \mathbf{Y}_1 \\ \mathbf{Y}_2 \\ \vdots \\ \mathbf{Y}_n \end{bmatrix} = \begin{bmatrix} \mathbf{X}_1 \\ \mathbf{X}_2 \\ \vdots \\ \mathbf{X}_n \end{bmatrix} \mathbf{B} + \begin{bmatrix} \mathbf{Z}_1 & \mathbf{0} & \cdots & \mathbf{0} \\ \mathbf{0} & \mathbf{Z}_2 & \cdots & \mathbf{0} \\ \vdots & \vdots & & \vdots \\ \mathbf{0} & \mathbf{0} & \cdots & \mathbf{Z}_n \end{bmatrix} \begin{bmatrix} \mathbf{A}_1 \\ \mathbf{A}_2 \\ \vdots \\ \mathbf{A}_n \end{bmatrix} + \begin{bmatrix} \mathbf{E}_1 \\ \mathbf{E}_2 \\ \vdots \\ \mathbf{E}_n \end{bmatrix}$$

$$\underset{N \times q}{\mathbf{Y}} = \underset{N \times q}{\mathbf{X}} \underset{q \times p}{\mathbf{B}} + \underset{N \times hp}{\mathbf{Z}} \underset{hp \times p}{\mathbf{A}} + \underset{N \times p}{\mathbf{E}} \tag{6.9.3}$$

a multivariate extension of the Laird and Ware (1982) model. The covariance structure for the model is

$$\begin{aligned} \operatorname{cov}(\operatorname{vec} \mathbf{Y}') &= (\mathbf{Z} \otimes \mathbf{I}_p) \operatorname{cov}(\operatorname{vec} \mathbf{A}') + \mathbf{I}_N \otimes \Sigma_e \\ &= (\mathbf{Z} \otimes \mathbf{I}_p)(\mathbf{I}_h \otimes \Sigma_a)(\mathbf{Z} \otimes \mathbf{I}_p)' + \mathbf{I}_N \otimes \Sigma_e \\ &= (\mathbf{Z} \otimes \mathbf{I}_p)\,\Phi\,(\mathbf{Z} \otimes \mathbf{I}_p)' + \mathbf{I}_N \otimes \Sigma_e \\ &= \Omega \end{aligned} \tag{6.9.4}$$

Letting $\Omega = \bigoplus_{i=1}^{n} \Omega_i$, the matrix Ω has block diagonal structure where

$$\Omega_i = (\mathbf{Z}_i \otimes \mathbf{I}_p)\,\Phi\,(\mathbf{Z}_i \otimes \mathbf{I}_p)' + \mathbf{I}_{r_i} \otimes \Sigma_e \tag{6.9.5}$$

and $\Phi_{hp \times hp} = \mathbf{I}_h \otimes \Sigma_a$.

If the variance component matrices Ω_i are known, one may obtain a generalized least square estimator of the fixed effect matrix $\mathbf{B}$ as follows.

$$\begin{aligned} \widehat{\mathbf{B}} &= \left(\mathbf{X}'\Omega^{-1}\mathbf{X}\right)^{-1} \mathbf{X}'\Omega^{-1}\mathbf{Y} \\ &= \left[\sum_{i=1}^{N} \left(\mathbf{X}_i'\Omega_i^{-1}\mathbf{X}'\right)\right]^{-1} \sum_{i=1}^{N} \mathbf{X}_i'\Omega_i^{-1}\mathbf{Y}_i \end{aligned} \tag{6.9.6}$$

Because Ω is unknown, it must be estimated. Again, one may obtain ML and REML estimates of covariance components as shown by Thum (1997) who develops a quasi-Newton algorithm requiring only the first derivatives of the log-likelihood. Currently, no SAS procedure performs the necessary calculations.

Setting $r_1 = r_2 = \ldots = r_n = r$, it is easily shown that the multivariate variance components model is a special case of model (6.9.3). A typical observation has the linear structure

$$\underset{p\times 1}{\mathbf{y}_{ij}} = \underset{p\times q}{\mathbf{B}'} \underset{q\times h}{\mathbf{x}_{ij}} + \underset{p\times h}{\mathbf{A}'_i} \underset{h\times 1}{\mathbf{z}_{ij}} + \mathbf{e}_{ij}$$

$$\begin{aligned} &\text{cov}\left(\mathbf{e}_{ij}\right) = \Sigma \\ &\text{cov}\left[\text{vec}\left(\mathbf{A}'_i\right)\right] = \Phi \end{aligned} \qquad i = 1, 2, \ldots, n,\ j = 1, 2, \ldots, r_i$$

Because the r_i are not equal, exact tests of **B** and variance components do not exist.

6.10 Tests of Means with Unequal Covariance Matrices

For all tests of means involving the MR model, the assumption of normality and equal covariance matrices have been assumed. When this is not the case, adjustments to test statistics have been proposed in Chapters 3 and 4. Because the DMM is a MR model, the adjustments proposed in Chapter 4 are also applicable for this design. The problems reported by Keselman et al. (1993) for repeated measurement designs also apply to multivariate repeated measurement designs.

7

Discriminant and Classification Analysis

7.1 Introduction

Given two or more groups or populations and a set of associated variables one often wants to locate a subset of the variables and associated functions of the subset that leads to maximum separation among the centroids of the groups. The exploratory multivariate procedure of determining variables and a reduced set of functions called discriminants or discriminant functions is called discriminant analysis. Discriminants that are linear functions of the variables are called linear discriminant functions (LDF). The number of functions required to maintain maximum separation for a subset of the original variables is called the rank or dimensionality of the separation. The goals of a discriminant analysis are to construct a set of discriminants that may be used to describe or characterize group separation based upon a reduced set of variables, to analyze the contribution of the original variables to the separation, and to evaluate the degree of separation. Fisher (1936) developed the technique to create a linear discriminant function to establish maximum separation among three species of iris flowers based upon four measurements. In personnel management one may want to discriminate among groups of professionals based upon a skills inventory. In medicine one may want to discriminate among persons who are at high risk or low risk for a specific disease. In a community, the mayor may want to evaluate how far apart several interest groups are on specific issues and to characterize the groups. In industry, one may want to determine when processes are in-control and out-of-control.

A multivariate technique closely associated with discriminant analysis is classification analysis. Classification analysis is concerned with the development of rules for allocating or assigning observations to one or more groups. While one may intuitively expect a good discriminant to also accurately predict group membership for an observation, this may not be the case. A classification rule usually requires more knowledge about the parametric structure of the groups. The goal of classification analysis is to create rules for assigning observations to groups that minimize the total probability of misclassification or the average cost of misclassification. Because linear discriminant functions are often used to develop classification rules, the goals of the two processes tend to overlap and some authors use the term classification analysis instead of discriminant analysis. Because of the close association between the two procedures we treat them together in this chapter. A topic closely associated with classification analysis is cluster analysis. The primary difference between the two is that variables used to develop classification rules are applied to a known number of groups. In cluster analysis variables are used to create associations (similarities) or distances (dissimilarities) to try to determine the unknown number of distinct groups. Cluster analysis is discussed in Chapter 9.

In Section 3.9, a two group linear discriminant function was developed to obtain a contrast in the sample mean vectors that led to maximum separation of the sample group means when comparing the significance of two normal population means $\boldsymbol{\mu}_1$ and $\boldsymbol{\mu}_2$ with a common covariance matrix. We recommended calculating the correlation between the discriminant function and each variable to assist in locating a "meaningful" mean difference, a variable-at-a-time, independent of the other variables in the model when one found $\boldsymbol{\psi} = \boldsymbol{\mu}_1 - \boldsymbol{\mu}_2$ to be significantly different from zero. The linear discriminant function located the exact contrast, up to a constant of proportionality, that led to significance. Correlations of variables with the linear discriminant function were used as a post-hoc data analysis tool. We also suggested a simple linear classification rule that in general tends to minimize the overall probability of misclassification under multivariate normality. In testing $H : \boldsymbol{\mu}_1 = \boldsymbol{\mu}_2$, we did not try to evaluate how each variable contributed to separation, reduce the number of variables to maintain maximum separation, or try to develop a classification rule to minimize misclassification. These problems are considered in this chapter.

In this chapter we discuss the topics of discrimination and classification for two or more populations using continuous multivariate observations and multivariate normality is usually assumed. For a more comprehensive discussion, books by Lachenbruch (1975), Hand (1981), McLachlan (1992), and Huberty (1994) may be consulted.

7.2 Two Group Discrimination and Classification

For a two group discriminant analysis, we assume that we have two independent samples from two multivariate normal populations with common covariance matrix Σ and unknown means $\boldsymbol{\mu}_1$ and $\boldsymbol{\mu}_2$. Thus, $\mathbf{y}_{ij} \sim IN_p(\boldsymbol{\mu}_i, \Sigma)$ where $i = 1, 2$; $j = 1, 2, \ldots, n_i$ and each observation $\mathbf{y}_{ij}$ is a $p \times 1$ vector of random variables. Fisher's two group linear discriminant function L is the linear combination of the variables that provides for the maximum separation between the groups

$$L = \mathbf{a}'\mathbf{y} = \sum_{j=1}^{p} a_j y_j \tag{7.2.1}$$

From (3.9.10), a vector $\mathbf{a} \equiv \mathbf{a}_s$ that provides maximum separation between the discriminant scores $L_{ij} = \mathbf{a}_s' \mathbf{y}_{ij}$ for the two group situation is the vector

$$\mathbf{a}_s = \mathbf{S}^{-1} \left(\overline{\mathbf{y}}_1 - \overline{\mathbf{y}}_2 \right) \tag{7.2.2}$$

where $\overline{\mathbf{y}}_i$ is the sample mean for the observations in group i $(i = 1, 2)$ and $\mathbf{S}$ is the unbiased estimate of Σ. Evaluating the discriminant scores at the group mean vectors $\overline{\mathbf{y}}_i$, the difference in the mean discriminant scores is exactly Mahalanobis' D^2 statistic

$$\begin{aligned} D^2 &= \overline{L}_1 - \overline{L}_2 = \mathbf{a}_s' \overline{\mathbf{y}}_1 - \mathbf{a}_s' \overline{\mathbf{y}}_2 \\ &= \mathbf{a}_s' \left(\overline{\mathbf{y}}_1 - \overline{\mathbf{y}}_2 \right) \\ &= \left(\overline{\mathbf{y}}_1 - \overline{\mathbf{y}}_2 \right)' \mathbf{S}^{-1} \left(\overline{\mathbf{y}}_1 - \overline{\mathbf{y}}_2 \right) \end{aligned} \tag{7.2.3}$$

Hence, if $T^2 = (n_1 n_2 / n_1 + n_2) D^2$ is significant then we have good separation of the sample group centroids for the two populations. It is also easily established that the square of the univariate student t^2 statistic using the mean discriminant scores for two groups with $\mathbf{a} \equiv \mathbf{a}_s$ is equal to T^2 since

$$t^2 = \frac{\left(\overline{L}_1 - \overline{L}_2 \right)^2}{s_L^2 \left(\frac{1}{n_1} + \frac{1}{n_2} \right)} = \left(\frac{n_1 n_2}{n_1 + n_2} \right) \left(\overline{\mathbf{y}}_1 - \overline{\mathbf{y}}_2 \right)' \mathbf{S}^{-1} \left(\overline{\mathbf{y}}_1 - \overline{\mathbf{y}}_2 \right) = T^2 \tag{7.2.4}$$

where s_L^2 is the sample estimate of the common population variance for discriminant scores. Thus, a simple t test on the discriminant scores is equivalent to calculating Hotelling's T^2 statistic for testing for the difference in the mean vectors for two groups.

a. Fisher's Linear Discriminant Function

Letting the discriminant scores $L_{ij} \equiv y$ represent the dependent variable and the dummy independent variables $x_1 = -1$ and $x_2 = 1$ for the two groups, we may fit a regression equation $y = \alpha + \beta \text{x} + e$ to the discriminant scores. Testing $H : \beta = 0$ in the regression problem is equivalent to testing $\rho = 0$ using the t^2 statistic

$$t^2 = \frac{(n-2)\, r^2}{1 - r^2} \tag{7.2.5}$$

Equating t^2 in (7.2.5) with t^2 in (7.2.4), observe that

$$\begin{aligned} \frac{n_1 n_2 D^2}{n_1 + n_2} &= \frac{(n-2)\, r^2}{1 - r^2} \\ D^2 &= \frac{(n_1 - n_2)(n-2)\, r^2}{(n_1 n_2)\left(1 - r^2\right)} \\ r^2 &= \frac{n_1 n_2 D^2}{(n_1 + n_2)(n-2) + n_1 n_2 D^2} \end{aligned} \tag{7.2.6}$$

where $n = n_1 + n_2$. Performing the "pseudoregression" on the "dummy" variables

$$c_1 = \frac{n_2}{n_1 + n_2} \text{ and } c_2 = c_1 - 1 = -\frac{n_1}{n_1 + n_2} \tag{7.2.7}$$

for groups 1 and 2, respectively (Fisher's codes with mean zero), one may regress the original variables on the dummy variables. Then r^2 in (7.2.6) is replaced by R_p^2, the squared multiple correlation or coefficient of determination found by regressing the independent variable c_i on the dependent variables $y_1, y_2, \ldots, y_p$, Fisher (1936). Letting $\mathbf{b}' = [b_1, b_2, \ldots, b_p]$ be the vector of estimated regression coefficients (ignoring the intercept), the vector $\mathbf{b}$ is seen to be proportional to $\mathbf{a}_s$

$$\begin{aligned} \mathbf{b} &= \left[\frac{n_1 n_2}{(n_1 + n_2)(n_1 + n_2 - 2) + n_1 n_2 D_p^2} \right] \mathbf{a}_s \\ \mathbf{a}_s &= \left[\frac{(n_1 + n_2)(n_1 + n_2 - 2)}{n_1 n_2} + D_p^2 \right] \mathbf{b} \end{aligned} \tag{7.2.8}$$

and $R_p^2 = (\bar{\mathbf{y}}_1 - \bar{\mathbf{y}}_2)' \mathbf{b}$, Siotani, Hayakawa and Fujikoshi (1985, Section 9.4). Thus, one may obtain $\mathbf{b}$ and R^2 from $\mathbf{a}$ and D_p^2. Or, by (7.2.6) and (7.2.8), one may find $\mathbf{a}$ and D_p^2 using the dummy variables in (7.2.7) as the dependent variable in a multiple regression analysis.

b. *Testing Discriminant Function Coefficients*

The LDF $L = \mathbf{a}_s'\mathbf{y}$ is an estimate of a population LDF $L = \boldsymbol{\alpha}'\mathbf{y}$ where $\boldsymbol{\alpha} = \Sigma^{-1}(\boldsymbol{\mu}_1 - \boldsymbol{\mu}_2)$. Having found a sample LDF that leads to significant separation between two groups, one may be interested in reducing the number of variables and yet maintain significant discrimination. The strategy differs from the two group MANOVA design where one retains all variables and investigates contrasts for significance.

To test that some subset of the variables in $\mathbf{y}$ do not contribute to maximum separation, the population LDF is partitioned as $\boldsymbol{\alpha}'\mathbf{y} = \boldsymbol{\alpha}_1'\mathbf{y}_1 + \boldsymbol{\alpha}_2'\mathbf{y}_2$ where $\mathbf{y}_1$ is a $(q \times 1)$ subvector of the $(p \times 1)$ vector $\mathbf{y}$. To test that the variables $y_{q+1}, \ldots, y_p$ do not contribute to separation, $H : \boldsymbol{\alpha}_2 = \mathbf{0}$ is tested. For this test, Rao (1970) developed the F statistic

$$\begin{aligned} F &= \frac{v_e - p + 1}{p - q} \frac{\left(\frac{n_1 n_2}{n_1 + n_2}\right)\left(D_p^2 - D_q^2\right)}{v_e + \left(\frac{n_1 n_2}{n_1 + n_2}\right) D_q^2} \\ &= \frac{v_e - p + 1}{p - q} \left(\frac{T_p^2 - T_q^2}{v_e + T_q^2} \right) \\ &= \frac{v_e - p + 1}{p - q} \left(\frac{R_p^2 - R_q^2}{1 - R_p^2} \right) \\ &\sim F_{(p-q,\ v_e - p + 1)} \end{aligned} \tag{7.2.9}$$

where the subscripts p and q on D^2, T^2, and R^2 represent the number of variables in the LDF. A proof of this result is given in Siotani et al. (1985, p. 404).

From the regression approach, equation (7.2.9) is used to examine the extent to which the prediction of the dummy dependent variable c can be improved by including more variables in the model. $1 - R_q^2$ is the amount of variability in c not accounted for by $y_1, y_2, \ldots, y_q$, and $1 - R_p^2$ is a measure of the variability not accounted for by $y_1, y_2, \ldots, y_p$. Thus, the difference

$$\left(1 - R_q^2\right) - \left(1 - R_p^2\right) = R_p^2 - R_q^2$$

is a measure of the reduction in the variability due to the other variables. Or for $q = p - 1$ variables, the difference in the squared multiple correlation coefficients is an estimate of the squared population part (also called semipartial) correlation coefficient for variable p given the $p - 1$ variable with the dependent variable c.

Recall that the proportional reduction of the variation in c remaining after all other variables $y_1, y_2, \ldots, y_q$ have been used (accounted for by $y_{q+1}, \ldots, y_p$) is given by

$$R^2_{c(q+1,\ldots,\,p)\cdot(1,\,2,\ldots,q)} = \frac{R_p^2 - R_q^2}{1 - R_q^2} \tag{7.2.10}$$

and is called the squared partial multiple correlation or the coefficient of partial multiple determination of c with $y_{q+1}, \ldots, y_p$ removing the influence of $y_1, y_2, \ldots, y_q$. Relating (7.2.10) to (7.2.9), we have that

$$F = \frac{R^2_{c(q+1,\ldots,p)\cdot(1,\,2,\ldots,q)} \,/\, (p-q)}{1 - R^2_{c(q+1,\ldots,p)\cdot(1,\ldots,q)} \,/\, (v_e - p - 1)} \tag{7.2.11}$$

as another representation of the F statistic involving the squared partial multiple correlation coefficient.

When $q = p - 1$ in (7.2.9), we are testing $H : \alpha_p = 0$, the significance of the pth variable given that the other variables are included in the LDF. Again using the regression approach, this is the partial F test. From (4.2.51), it is seen to be related to the test of additional information using the Λ criterion. Or, using (7.2.10), the partial F statistic is seen to be related to the square of the partial correlation coefficient

$$r^2_{cp\cdot(1,\,2,\ldots,p-1)} = \left(R_p^2 - R_{p-1}^2\right) / \left(1 - R_{p-1}^2\right) \tag{7.2.12}$$

Thus, the test of the significance of a single variable is related to a partial correlation and not the simple correlation of each variable with the LDF which ignores the other variables in the model, Rencher (1988). The numerator of the ratio is (7.2.12) is the sample estimate of the squared population part (semi-partial) correlation coefficient for variable p which is written as $r^2_{c(p.1,2,\ldots,p-1)}$.

When one deletes variables from the LDF, one must re-calculate the LDF with the variables excluded and the significance of T^2 based on the remaining variables for the two sample case as discussed in Section 3.9. This test is testing the hypothesis $H : \alpha_q = 0$, having removed variables $y_{q+1}, \ldots, y_p$.

Given the relation of the two group LDF to the regression model, one may again use forward, backward and stepwise regression procedures to develop a MANOVA model with some subset of variables $y_1, y_2, \ldots, y_p$ using partial F statistics. In general, stepwise procedures are not recommended since the tests are not independent and usually result in too many variables being selected, Rencher and Larson (1980).

Because a LDF is not unique, one may not test that two LDFs are equal. However, one may want to evaluate whether a LDF for a replicated study is proportional to a standard. A test of $H : LDF_1 \propto LDF_0$ was developed by Siotani et al. (1985, p. 402). Associating T_q^2 (7.2.9) with the standard so that $q = 1$, the test statistic is

$$F = \frac{v_e - p + 1}{p - 1} \left(\frac{T_p^2 - T_1^2}{v_e + T_1^2} \right) \sim F_{(p-1,\, v_e-p+1)} \tag{7.2.13}$$

where T_p^2 is Hotelling's T^2 statistic based on the new set of variables and T_1^2 is the statistic based on the standard.

c. *Classification Rules*

To develop a classification rule for classifying an observation $\mathbf{y}$ into one or the other population in the two group case requires some new notation. First, we let $f_1(\mathbf{y})$ and $f_2(\mathbf{y})$ represent the probability density functions (*pdfs*) associated with the random vector $\mathbf{Y}$ for populations π_1 and π_2, respectively. We let p_1 and p_2 be the prior probabilities that $\mathbf{y}$ is a member of π_1 and π_2, respectively, where $p_1 + p_2 = 1$. And, we let $c_1 = C(2 \mid 1)$ and $c_2 = C(1 \mid 2)$ represent the misclassification cost of assigning an observation from π_2 to π_1, and from π_1 to π_2, respectively. Then, assuming the *pdfs* $f_1(\mathbf{y})$ and $f_2(\mathbf{y})$ are known, the total probability of misclassification (TPM) is equal to p_1 times the probability of assigning an observation to π_2 given that it is from π_1, $P(2 \mid 1)$, plus p_2 times the probability that an observation is classified into π_1 given that it is from π_2, $P(1 \mid 2)$. Hence,

$$TPM = p_1 P(2 \mid 1) + p_2 P(1 \mid 2) \tag{7.2.14}$$

The optimal error rate (OER) is the error rate that minimizes the TPM. Taking costs into account, the average or expected cost of misclassification is defined as

$$ECM = p_1 P(2 \mid 1)\, C(2 \mid 1) + p_2 P(1 \mid 2)\, C(1 \mid 2) \tag{7.2.15}$$

A reasonable classification rule is to make the ECM as small as possible. In practice costs of misclassification are usually unknown.

To assign an observation $\mathbf{y}$ to π_1 or π_2, Fisher (1936) employed his LDF. To apply the rule, he assumed that $\Sigma_1 = \Sigma_2 = \Sigma$ and because he did not assume any *pdf*, Fisher's rule does not require normality. He also assumed that $p_1 = p_2$ and that $C(1 \mid 2) = C(2 \mid 1)$. Using (7.2.3), we see that $D^2 > 0$ so that $\overline{L}_1 - \overline{L}_2 > 0$ and $\overline{L}_1 > \overline{L}_2$. Hence, if

$$L = \mathbf{a}_s'\mathbf{y} = (\overline{\mathbf{y}}_1 - \overline{\mathbf{y}}_2)'\, \mathbf{S}^{-1}\mathbf{y} > \frac{\overline{L}_1 + \overline{L}_2}{2} \tag{7.2.16}$$

we would assign the observation to π_1. However,

$$\frac{\overline{L}_1 + \overline{L}_2}{2} = \frac{1}{2}\left(\overline{\mathbf{y}}_1 - \overline{\mathbf{y}}_2\right)' \mathbf{S}^{-1}\left(\overline{\mathbf{y}}_1 + \overline{\mathbf{y}}_2\right) \tag{7.2.17}$$

and letting

$$\widehat{W} = L - \frac{\left(\overline{L}_1 + \overline{L}_2\right)}{2} = \left(\overline{\mathbf{y}}_1 + \overline{\mathbf{y}}_2\right)' \mathbf{S}^{-1}\left[\mathbf{y} - \frac{1}{2}\left(\overline{\mathbf{y}}_1 + \overline{\mathbf{y}}_2\right)\right] \tag{7.2.18}$$

Fisher's classification rule becomes

$$\begin{aligned} &\text{Assign } \mathbf{y} \text{ to } \pi_1 \text{ if } \widehat{W} > 0 \\ &\text{Assign } \mathbf{y} \text{ to } \pi_2 \text{ if } \widehat{W} < 0 \end{aligned} \tag{7.2.19}$$

the rule given in Section 3.9.

Welch (1939) showed that the optimal classification rule for two groups which minimizes the TPM is to allocate $\mathbf{y}$ to π_1 if

$$p_1 f_1(\mathbf{y}) > p_2 f_2(\mathbf{y}) \tag{7.2.20}$$

and to assign $\mathbf{y}$ to π_2, otherwise. It is interesting to note that (7.2.20) is equivalent to assigning an observation to the population with the largest posterior probability $P(\pi_i \mid \mathbf{y})$. By Bayes' Theorem, the posterior probabilities are

$$\begin{aligned} P(\pi_i \mid \mathbf{y}) &= \frac{P(\pi_i)\, P(\mathbf{y} \mid \pi_i)}{\sum_i P(\pi_1)\, P(\mathbf{y} \mid \pi_i)} \\ &= \frac{p_i f_i(\mathbf{y})}{\sum_i p_i f_i(\mathbf{y})} \end{aligned}$$

Because the denominators of $P(\pi_i \mid \mathbf{y})$ are identical for $i = 1, 2$, one may classify observations using (7.2.20) by equivalently calculating the posterior probabilities.

Assuming π_1 is $N_p(\boldsymbol{\mu}_1, \Sigma)$ and π_2 is $N_p(\boldsymbol{\mu}_2, \Sigma)$, the classification rule using (7.2.20) becomes

$$W = (\boldsymbol{\mu}_1 - \boldsymbol{\mu}_2)' \Sigma^{-1}\left[\mathbf{y} - \frac{1}{2}(\boldsymbol{\mu}_1 + \boldsymbol{\mu}_2)\right] > \log\left(\frac{p_2}{p_1}\right) \tag{7.2.21}$$

Since $\boldsymbol{\mu}_1, \boldsymbol{\mu}_2$ and Σ are usually unknown they must be estimated. Substituting sample estimates for the population parameters, the rule becomes

$$\begin{aligned} &\text{Assign } \mathbf{y} \text{ to } \pi_1 \text{ if } \widehat{W} > \log\left(\frac{p_2}{p_1}\right) \\ &\text{Assign } \mathbf{y} \text{ to } \pi_2 \text{ if } \widehat{W} < \log\left(\frac{p_2}{p_1}\right) \end{aligned} \tag{7.2.22}$$

This rule is also called Anderson's classification rule, Anderson (1951) and Wald (1944). The distribution of $\widehat{W}$ is unknown so that an optimal rule using $\widehat{W}$ has not been developed, Anderson (1984, p. 210).

Welch's optimal rule requires known population parameters. Using $\widehat{W}$, we have substituted sample estimates; hence, Anderson's classification rule and Fisher's rule assuming $p_1 = p_2$ are not optimal, but only asymptotically optimal. The rule minimizes the TPM as the sample sizes become large.

To take costs into account, the rule that minimizes the ECM is to assign $\mathbf{y}$ to π_1 if

$$C(2 \mid 1)\, p_1 f_1(\mathbf{y}) > C(1 \mid 2)\, p_2 f_2(\mathbf{y}) \tag{7.2.23}$$

and to π_2, otherwise. For two normal populations, the rule becomes

$$\begin{aligned} &\text{Assign } \mathbf{y} \text{ to } \pi_1 \text{ if } \widehat{W} > \log\left[\frac{C(1 \mid 2)}{C(2 \mid 1)}\right]\left(\frac{p_2}{p_1}\right) \\ &\text{Assign } y \text{ to } \pi_2 \text{ if } \widehat{W} < \log\left[\frac{C(1 \mid 2)}{C(2 \mid 1)}\right]\left(\frac{p_2}{p_1}\right) \end{aligned} \tag{7.2.24}$$

Anderson (1984, p. 202). Again, if the ratios of the costs and prior probabilities are both one, then rule (7.2.24) reduces to Fisher's classification rule. In summary, Fisher's rule has some optimal properties if $\Sigma_1 = \Sigma_2$, $p_1 = p_2$ and $c_1 = c_2$ provided samples are obtained from normal populations. When this is not the case, Fisher's rule is to be avoided, Krzanowski (1977).

When $\Sigma_1 \neq \Sigma_2$, and samples are from normal populations, a quadratic rule that minimizes the ECM is to allocate $\mathbf{y}$ to π_1 if

$$\begin{aligned} \widehat{Q} = \frac{1}{2}\log\left(\frac{|\mathbf{S}_2|}{|\mathbf{S}_1|}\right) - \frac{1}{2}\left(\overline{\mathbf{y}}_1'\mathbf{S}_1^{-1}\overline{\mathbf{y}}_1 - \mathbf{y}_2' S_2^{-1}\mathbf{y}_2\right) \\ + \left(\overline{\mathbf{y}}_1'\mathbf{S}_1^{-1} - \overline{\mathbf{y}}_2'\mathbf{S}_2^{-1}\right)\mathbf{y} - \frac{1}{2}\mathbf{y}'\left(\mathbf{S}_1^{-1} - \mathbf{S}_2^{-1}\right)\mathbf{y} \\ > \log\left[\frac{C(1 \mid 2)}{C(2 \mid 1)}\left(\frac{p_2}{p_1}\right)\right] \end{aligned} \tag{7.2.25}$$

and to π_2, otherwise, Anderson (1984, p. 234). When $\Sigma_1 \neq \Sigma_2$, one should not use Fisher's LDF.

Classification rules that make assumptions regarding the *pdfs* $f_1(\mathbf{y})$ and $f_2(\mathbf{y})$, such as normality, are called parametric procedures. While Fisher's LDF is nonparametric, it is only asymptotically optimal under normality and requires the covariance matrices to be equal. When this is not the case, a quadratic rule is used under normality. Hence, the quadratic rule is also a parametric procedure.

Welch's optimal rule that minimizes the TPM does not require one to specify the *pdfs*; thus, if one could estimate the empirical form of the density using the data, the optimal rule would be to allocate $\mathbf{y}$ to the population π_i $(i = 1, 2)$ when

$$c_i p_i \widehat{f_i}(\mathbf{y}) \text{ is a maximum} \tag{7.2.26}$$

General procedures used to estimate density functions empirically are called kernel density estimation methods. A review of kernel density estimation may be found in Silverman (1986) and Scott (1992). A study of linear, quadratic, and kernel classification procedures was conducted by Remme, Habbema, and Hermans (1980) for the two group case.

TABLE 7.2.1. Classification/Confusion Table

Actual Population	Predicted π_1	Population π_2	Sample Size
π_1	n_{1C}	n_{1E}	n_1
π_2	n_{2E}	n_{2C}	n_2

They showed that linear and quadratic rules are superior to kernel classification methods when parametric assumptions hold. When the parametric assumptions are violated, kernel density estimation procedures were preferred. One may always use linear, quadratic, and kernel density estimation methods and compare the confusion matrix in Table 7.2.1 when evaluating classification rules.

Another classification method that is a nonparametric data driven procedure is the k-nearest-neighbor classification rule first proposed by Fix and Hodges (1951). Their procedure calculates k Mahalanobis' distances in the neighborhood of an observation $\mathbf{y}$. Evaluating the k distances, k_1 points may be associated with observations in π_1 and k_2 points will be associated with observation in π_2 where $k_1 + k_2 = k$. Taking costs and sample sizes into account, an observation $\mathbf{y}$ is allocated to π_1 if

$$\frac{k_1/n_1}{k_2/n_2} > \left[\frac{C\,(1 \mid 2)}{C\,(2 \mid 1)}\right]\left(\frac{p_2}{p_1}\right) \tag{7.2.27}$$

and to π_2, otherwise. One usually tries several values of k and evaluates its effectiveness to classify observations using Table 7.2.1.

d. *Evaluating Classification Rules*

Given a classification rule, one wants to evaluate how well the rule performs in assigning an observation to the correct population. Given that $f_1\,(\mathbf{y})$ and $f_2\,(\mathbf{y})$ are known (along with their associated population parameters), the TPM expression given in (7.2.14) may be evaluated to obtain the actual error rate (AER). Because the specification of $f_1\,(\mathbf{y})$ and $f_2\,(\mathbf{y})$ is seldom known one generally cannot obtain the AER, but must be satisfied with an estimate. The simplest nonparametric method is to apply the classification rule to the sample and to generate a classification or confusion table as shown in Table 7.2.1. This is called the substitution or resubstitution method.

Then, the observed error rate or apparent error rate (APER) is defined as the ratio of the total number of misclassified observations to the total

$$\text{APER} = \frac{n_{1E} + n_{2E}}{n_1 + n_2} \tag{7.2.28}$$

The apparent correct error rate is 1 – APER. The APER is an estimate of the probability that a classification rule based on a given sample will misclassify a future observation. Unfortunately because the same data are being used to both construct and evaluate the classification rule, the APER tends to underestimate the AER.

To eliminate the bias in the APER, one may split the sample into two parts. The "training" sample and the "validation" sample. Then, the classification rule is created using the training sample and the apparent error rate is determined using the validation sample. This is sometimes called the holdout, resubstitution method. The primary disadvantages of this procedure are that (1) it requires a large sample, and (2) since the classification rule is based upon a subset of the sample, it may be a poor estimate of the population classification function, depending on the split.

An alternative nonparametric approach that seems to work better than the holdout method is Lachenbruch's leave-one-out, cross-validation method developed in his doctoral dissertation and discussed by Lachenbruch and Mickey (1968). It is a one-at-a-time, crossvalidation procedure that goes as follows.

1. Starting with π_1, omit one observation from the sample and develop a classification rule based upon the $n_1 - 1$ and n_2 sample observations.
2. Classify the holdout observation using the rule estimated in step 1.
3. Continue this process until all observations are classified and let $n_{1E}^{(HC)}$ denote the number of misclassified observations in population π_1.
4. Repeat steps 1 to 3 with population π_2 and count the misclassify observation from $\pi_2, n_{2E}^{(HC)}$.

Then, the estimated apparent error rate $\widehat{\text{APAR}}$ is defined as

$$\widehat{\text{APAR}} = \frac{n_{1E}^{(HC)} + n_{2E}^{(HC)}}{n_1 + n_2} \tag{7.2.29}$$

and is nearly an unbiased estimator of the expected average error rate in that $E(\widehat{\text{APAR}}) = E\,(\text{AER}) = E\,(\text{TPM})$. An estimate of the actual error rate based on all possible samples, often called the true error rate. The quantities $\widehat{p}\,(2 \mid 1) = n_{1E}^{(HC)}/n_1$ and $\widehat{p}\,(2/1) = n_{2E}^{(HC)}/n_2$ are estimates of the misclassification probabilities.

The nonparametric one-at-a-time, crossvalidation method is a jackknife-like method proposed to reduce the bias in the AER using the naive substitution method, McLachlan (1992, p. 345). Efron and Gong (1983) develop a jackknife estimator and compared it to the leave-one-out crossvalidation method and bootstrap estimators. For the two-group discrimination problem, Davison and Hall (1992) show that the AERs for the one-at-a-time holdout and bootstrap resampling procedures are essentially the same. Gnaneshanandam and Krzanowski (1990) compare several estimates of the AER and conclude that the one-at-a-time, holdout, crossvalidation method performs as well as resampling procedures, although they are at times more variable. The method is always superior to the substitution method for both normal and nonnormal populations.

While one may use many strategies to construct classification rules two group classification problems, one should always develop a confusion/classification table, Table 7.2.1, to aid in the selection. There is no optional rule. Whatever procedure is selected, one must remember that all procedures are sensitive to outliers in the data.

7.3 Two Group Discriminant Analysis Example

Three procedures are available in SAS for discriminant analysis: CANDISC, DISCRIM and STEPDISC. Assuming multivariate normality and common covariance matrices, the procedures CANDISC and STEPDISC are used to obtain a discriminant function based on all variables or a subset of variables that maximizes separation between centroid means. The procedure DISCRIM calculates discriminant functions employing various criteria in order to classify observations into groups under both normality with equal covariance matrices, normality with unequal covariance matrices, and various nonparametric methods. The METHOD = option on the procedure statement determines the classification rule (criterion). To evaluate the performance of the method selected one must select CROSSVALIDATE to obtain the estimated $\widehat{\text{APAR}}$ rates. The default method in SAS for estimating the estimated apparent error rate is to use the naive resubstitution method.

a. *Egyptian Skull Data (Example 7.3.1)*

For this discriminant analysis example, four measurements on male Egyptian skulls from five different periods categorized into two periods B.C. and A.D. are used to evaluate separation and classification of skull formation. The data were obtained from the DASL Web site, http://lib.stat.com/ Datafiles/EgyptianSkulls. The data are included in Thomson and Randall-Maciver (1905) and may be found in the file skulls.dat. The data consists of 150 cases with four skull measurements

MB	–	Maximum Breadth
BH	–	Basibregmatic Height
BL	–	Basialveslour Length
NH	–	Nasal Height

The approximate year of skull formation is also included. Negative years represent the B.C. period and positive years represent the A.D. period. Year was used to assign the observations to two groups: $1 = B.C.$ and $2 = A.D.$. Program m7_3_1.sas includes the SAS code for the analysis.

As discussed in Example 3.9.1, PROC GLM is used to test $H : \boldsymbol{\mu}_1 = \boldsymbol{\mu}_2$ using the MANOVA statement. The test of equality is rejected with p-value 0.0003. Thus, $\boldsymbol{\mu}_1 \neq \boldsymbol{\mu}_2$. Solving $|\mathbf{H} - \lambda\mathbf{E}| = 0$ using the /CANONICAL option, Table 7.3.1 contains the discriminant structure vectors for the skull data.

The correlation structure and magnitude of the standardized coefficients in Table 7.3.1 indicate that the difference in centroids for the two groups is primarily due to the variables MB, BH, and BL.

The canonical variate output for PROC CANDISC is very similar to that produced by PROC GLM; however, because DISTANCE is included on the procedure statement, it calculates the Mahalanobis distance between mean centroids. Given $\Sigma_1 = \Sigma_2 = \Sigma$, the squared distance is

$$D^2 = \left(\overline{\mathbf{y}}_1 - \overline{\mathbf{y}}_2\right)' \Sigma^{-1} \left(\overline{\mathbf{y}}_1 - \overline{\mathbf{y}}_2\right) = 0.97911$$

TABLE 7.3.1. Discriminant Structure Vectors, $H : \boldsymbol{\mu}_1 = \boldsymbol{\mu}_2$

	Within Structure ρ	Standardized Vector $\mathbf{a}_{wsa}$	Raw Vector $\mathbf{a}_{ws}$
MB	−0.5795	−0.4972	−0.10165601
BH	0.5790	0.4899	0.09917666
BL	0.7210	0.5917	0.11001651
NH	−0.1705	−0.1738	−0.05416834

as given in formula (7.2.3). The canonical correlation, $\widehat{\rho}_c$, is obtained using the relation

$$\widehat{\rho}_c = \left[\widehat{\lambda}_1 / \left(1 + \widehat{\lambda}_1\right)\right]^{1/2} = \left(\frac{0.1587}{1 + 0.1587}\right)^{1/2} = 0.370161$$

where $\widehat{\lambda}_1$ is the root of $| \mathbf{H} - \lambda \mathbf{E} | = \mathbf{0}$, $\widehat{\rho}_c^2 = 0.137019$. Since $\widehat{\rho}_c^2 = \widehat{\theta}_1$ where $\widehat{\theta}_1$ is the root of $| \mathbf{H} - \theta\left(\mathbf{H} + \mathbf{E}\right) | = 0$, $\widehat{\rho}_c^2$ is the proportion of the total sum of squares for the discriminant scores that is due to the differences between groups. Or, it is the amount of variation between groups that is explained by the discriminating variables. For the two group case, $\widehat{\rho}_c^2$ is identical to R_p^2 for the multiple regression of a dichotomous group-membership variable, c_i on the p predictors. More will be said about canonical correlations in Chapter 8. For now, observe that

$$\Lambda = \left(1 + \widehat{\lambda}_1\right)^{-1} = \left(1 + \widehat{\theta}_1\right) = 1 + \widehat{\rho}_c^2$$

so that testing $H : \boldsymbol{\mu}_1 = \boldsymbol{\mu}_2$ is identical to testing $H : \rho_c = 0$. If $\boldsymbol{\mu}_1 \neq \boldsymbol{\mu}_2$, then $\rho_c \neq 0$ so that the canonical correlation ρ_c corresponding to the canonical discriminant function is significantly different from zero.

Using the OUT = option on the CANDISC statement, the discriminant scores for the example are stored in the SAS data set defined by OUT =. To plot the scores, PROC CHART is used. This generates a histogram of the scores for each group, providing a visual graph of separation. PROC TTEST is included to test that the means of the two groups using discriminant scores are equal. For our example, the test statistic squared is

$$t^2 = (4.8475)^2 = 23.499 = T^2 = n_1 n_2 D^2 / \left(n_1 + n_2\right)$$

where $n_1 = 120$, $n_2 = 30$, and $D^2 = 0.97911$. This verifies formula (7.2.4).

To test whether or not a variable in the discriminant function significantly contributes to separation, given the other variables are included in the function, the procedure STEPDISC is used with METHOD = BACKWARD. Setting $q = p - 1$, we are testing that $\alpha_p = 0$. The F statistic for removal for the variable NH is $F = 0.568$ with p-value 0.4521 so that NH does not appear to be contributing to the separation of the centroids and may be considered for exclusion from the discriminant function. If excluded, the squared distance between group centroids is marginally reduced from 0.97911 to 0.95120.

To invoke Fisher's discriminant function for classification of observations into the two groups, PROC DISCRIM is used with all four variables. The option METHOD = NORMAL generates Fisher's classification rule using (7.2.19) with $p_1 = p_2 = 0.5$ and equal

TABLE 7.3.2. Discriminant Functions

Variable	Group 1	Group 2
CONSTANT	−928.878	−921.869
MB	5.999	6.100
BH	4.803	4.705
BL	3.191	3.082
NH	2.121	2.173

TABLE 7.3.3. Skull Data Classification/Confusion Table

Actual Population	Predicted 1	Predicted 2	Sample Size
1	85	35	120
2	14	16	30

costs of misclassification, $C(1 \mid 2) = C(2 \mid 1)$. If $\Sigma_1 \neq \Sigma_2$, a quadratic rule is formulated as defined in (7.2.25). To test $H : \Sigma_1 = \Sigma_2$, the likelihood ratio test is performed using Box's M test given in (3.8.5). To perform the test, the option POOL = TEST is required and pooling occurs only if the test is not significant at the level 0.1. Because this test is not robust to nonnormality, it may reject the hypothesis of equal covariance matrices because of nonnormal data, however, Fisher's rule does not require normality. For nonnormal data and $\Sigma_1 = \Sigma_2$, the option POOL = YES may be used. For the skull data, the p-value for the chi-square test is 0.4024 so that a pooled covariance matrix is used to construct the criterion $\widehat{W}$ in (7.2.22). The program calculates $\widehat{W}_1$ and $\widehat{W}_2$ for each group as

$$\text{constant} = -\frac{1}{2}\overline{\mathbf{x}}_j'\mathbf{S}^{-1}\overline{\mathbf{x}}_j$$

$$\text{coefficient vector} = \mathbf{S}^{-1}\overline{\mathbf{x}}_j$$

so that $\widehat{W}$ is obtained by subtraction. The individual discriminant function coefficients are given in Table 7.3.2.

To evaluate Fisher's rule for the skull data, we employ the CROSSVALIDATE-option. Included in SAS is the Classification/Confusion Table 7.2.1 where $\widehat{\text{APAR}}$ is estimated using a smoothed modification of (7.2.29). Table 7.3.3 is the Classification/Confusion table for the skull data.

Using Table 7.3.3, the estimated error rate is $\widehat{\text{APAR}} = (14 + 15)/150 = 0.3267$. The SAS procedure reports a smoothed error rate estimate resulting in an estimator with smaller variance, Glick (1978). The smoothed estimate is 37.92%. For three variables, the smoothed error rate is 34.58%. To identify the misclassified observations, the option CROSSLISTERR is used. Since one may expect on average 1/3 of the skulls to be misclassified, Fisher's LDF rule does not appear to adequately succeed in classifying the skulls into the two groups. The example illustrates that significant separation does not ensure good classification.

b. *Brain Size (Example 7.3.2)*

In this example we are interested in the development of a discriminant function to distinguish between males and females using brain size data and other variables. The analysis is based upon data from Willerman et al. (1991) collected on 20 male and 20 female right-handed Anglo psychology students at a large southwestern university. The sample of 40 students are from a larger population of students with SAT scores larger than 1350 or lower than 940. The subjects took four subtests of the Wechster Adult Intelligence Scale-Revised test. The scores recorded were Full Scale IQ (FSIQ), Verbal IQ (VIQ) and Performance IQ (PIQ). The researchers also obtained 19 Magnetic Resonance Imaging (MRI) scans and recorded the total pixel count of the scans. The data stored in the file brainsize.dat was obtained from the DASL (http://lib.stat.cmu.edu/DASL/Datafiles/Brainsize) and are given in Table 7.3.4. The researchers withheld the weights of two students and the height of one subject for reasons of confidentiality.

Program m7_3_2.sas contains the SAS code to develop both a linear and quadratic discriminant function for these data. Following the steps of the previous example, we perform a MANOVA analysis on the data using PROC GLM and PROC CANDISC. Rejecting $H : \mu_1 = \mu_2$, the squared Mahalanobis distance between the means is $D^2 = 7.15884$. The discriminant function structure is given in Table 7.3.5.

Because the square of the canonical correlation $\widehat{\rho}_c^2 = 0.653248$, about 65.3% of the variation between groups is explained by the discriminant function. In Example 7.3.1, the value was only 13.77%. However, the variable PIQ does not appear to contribute to the separation in the mean centroids. This is confirmed using PROC STEPDISC with METHOD = BACKWARD; the procedure recommends dropping the variable PIQ.

For these data, $\Sigma_1 = \Sigma_2$ so that we may construct Fisher's linear classification rule. However, we also illustrate the construction of a quadratic rule. For the quadratic function, METHOD = NORMAL and POOL = NO. The smoothed value of $\widehat{\text{APAR}}$ for the linear rule is 12.5% while the quadratic rule yields $\widehat{\text{APAR}} = 20.07\%$. Thus, if $\Sigma_1 = \Sigma_2$, the linear rule yields less errors and has greater efficiency than the quadratic rule. Also observe how the resubstitution method under estimates the error rate. Its value is 5%.

Having obtained a reasonable discriminant rule for a set of data, one may use $\widehat{W}$ to classify future observations into groups.

Exercises 7.3

1. Examples 7.3.1 and 7.3.2 were each demonstrated using the option CROSSVALIDATE. One may also assign observations using posterior probabilities, option POSTERR. Run both examples using this option and discuss the results.

2. Investigate Examples 7.3.1 and 7.3.2 using a nonparametric discriminant method available in PROC DISCRIM.

TABLE 7.3.4. Willeran et al. (1991) Brain Size Data

Gender	FSIQ	VIQ	PIQ	Weight	Height	MRI
Male	140	150	124	166*	72.5	100121
Male	139	123	150	143	73.3	1038437
Male	133	129	128	172	68.8	965353
Male	89	93	84	134	66.3	904858
Male	133	114	147	172	68.8	955466
Male	141	150	128	151	70.0	1079549
Male	135	129	124	155	69.1	924059
Male	100	96	102	178	73.5	945088
Male	80	77	86	180	70.0	889083
Male	83	83	86	166*	71.4*	892420
Male	97	107	84	186	76.5	905940
Male	139	145	128	132	68.0	955003
Male	141	145	131	171	72.0	935494
Male	103	96	110	187	77.0	1062462
Male	144	145	137	191	67.0	949589
Male	103	96	110	192	75.5	997925
Male	90	96	86	181	69.0	679987
Male	140	150	124	144	70.5	949395
Male	81	90	74	148	74.0	930016
Male	89	91	89	179	75.5	935863
Female	133	132	124	118	64.5	816932
Female	137	132	134	147	65.0	951545
Female	99	90	110	146	69.0	928799
Female	138	136	131	138	64.5	991305
Female	92	90	98	175	66.0	854258
Female	132	129	124	118	64.5	833868
Female	140	120	147	155	70.5	856472
Female	96	100	90	146	66.0	878897
Female	83	71	96	135	68.0	865363
Female	132	132	120	127	68.5	852244
Female	101	112	84	136	66.3	808020
Female	135	129	134	122	62.0	790619
Female	91	86	102	114	63.0	831772
Female	85	90	84	140	68.0	798612
Female	77	83	72	106	63.0	793549
Female	130	126	124	159	66.5	866662
Female	133	126	132	127	62.5	857782
Female	83	90	81	143	66.5	834344
Female	133	129	128	153	66.5	948066
Female	88	86	94	139	64.5	893983

*Estimated Value

TABLE 7.3.5. Discriminant Structure Vectors, $H : \boldsymbol{\mu}_1 = \boldsymbol{\mu}_2$

	Within Structure ρ	Standardized Vector $\mathbf{a}_{wsa}$	Raw Vector $\mathbf{a}_{ws}$
FSIQ	0.0476	−1.8430	−0.07652898
VIQ	0.0913	1.6274	0.06891079
PIQ	0.0189	0.3593	0.01598925
Weight	0.6067	0.4676	0.02023707
Height	0.7618	0.8036	0.20244671
MRI	0.6164	0.5935	0.00000821

7.4 Multiple Group Discrimination and Classification

To develop linear discriminant functions in the multiple group case, we again assume that we have samples of random p-vectors from normal populations with common covariance matrices Σ. That $\mathbf{y}_{ij} \sim IN_p(\boldsymbol{\mu}_i, \Sigma)$ $i = 1, 2, \ldots, k$ and $j = 1, 2, \ldots, n_i$ where $n = \sum_{i=1}^{k} n_i$. These assumptions are identical to the one-way MANOVA design discussed in Section 4.4.

a. *Fisher's Linear Discriminant Function*

To generalize Fisher's procedure to k groups, we seek to construct linear combinations of the variables $L = \mathbf{a}'\mathbf{y}$ called discriminants or LDFs that maximize the separation of the k population mean vectors using the sample such that the ratio of between group variation is maximum relative to the within group variation. Letting

$$\mathbf{E} = \sum_{i=1}^{k} \sum_{j=1}^{n_i} \left(\mathbf{y}_{ij} - \mathbf{y}_{i.}\right)\left(\mathbf{y}_{ij} - \mathbf{y}_{i.}\right)' \tag{7.4.1}$$

$$\mathbf{H} = \sum_{i=1}^{k} n_i \left(\mathbf{y}_{i.} - \overline{\mathbf{y}}_{..}\right)\left(\mathbf{y}_{i.} - \overline{\mathbf{y}}_{..}\right)'$$

represent the between-hypothesis and within-error matrices, we have from Theorems 2.6.10 and 2.6.8 that the solution to the eigenequation $|\mathbf{H} - \lambda\mathbf{E}| = 0$ yield eigenvalues $\widehat{\lambda}_m$ and associated vectors $\mathbf{a}_m$ for $m = 1, 2, \ldots, s = \min(k-1, p)$ that maximizes the ratios, $\mathbf{a}_m'\mathbf{H}\mathbf{a}_m / \mathbf{a}_m'\mathbf{E}\mathbf{a}_m$. By Theorem 2.6.8, $\mathbf{a}_m'\mathbf{E}\mathbf{a}_m = \mathbf{0}$ for $\mathbf{a}_m \neq \mathbf{a}_{m'}$ so that vectors are orthogonal in the metric of $\mathbf{E}$ and uncorrelated. The s uncorrelated functions $L_m = \mathbf{a}_m'\mathbf{y}$ are called the linear discriminant functions and have been constructed to provide maximum separation in the means $\boldsymbol{\mu}_i$, based upon the sample. Because the coefficients of the eigenvectors are not unique, following the two group case, the eigenvectors are standardized as

$$\mathbf{a}_{wsa} = (\operatorname{diag} \mathbf{S})^{1/2}\, \mathbf{a}_m \,/\, \sqrt{\mathbf{a}_m'\mathbf{E}\mathbf{a}_m} \tag{7.4.2}$$

for $m = 1, 2, \ldots, s$.

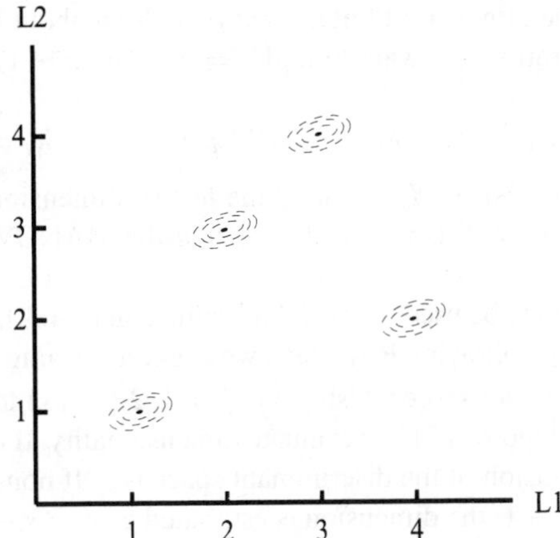

FIGURE 7.4.1. Plot of Discriminant Functions

b. *Testing Discriminant Functions for Significance*

The goals for a multiple group discriminant analysis are similar to the two group case in that we want to examine separation, and locate a subset of the variables that separates the groups as well as the original set. However, there is an additional complication with the multiple group situation. There is no longer one principal discriminant function, but s discriminants.

To test the hypothesis of mean differences in the MANOVA design, we could use Roy's largest root criterion, Wilks' Λ criterion or a trace criterion. Following rejection, we investigated contrasts in the means using all possible subsets of variables to describe the significant result. Significance could occur in any one of s dimensions. To investigate this more fully, if the means are all equal in the population, then the noncentrality matrix Γ for the one-way MANOVA would be zero and the rank of Γ equals zero. If the means $\boldsymbol{\mu}_i$ vary on a line, then one discriminant function could be used to characterized the separation in the k means and the $r(\Gamma) = 1$. Continuing, if the k means lie in some s-dimensional space so that the $r(\Gamma) = s$, it would take s discriminants to achieve significant discrimination.

To see what we mean more clearly, suppose we have four populations and $p = 10$ variables. Evaluating the three population discriminants for the four populations, suppose the discriminant plot shown in Figure 7.4.1 results (L_1 is drawn $\perp$ to L_2 for convenience). For our example, suppose $\lambda_1 > \lambda_2 > \lambda_3 = 0$ so that the $r(\Gamma) = 2$. Then, while L_1 may be adequate to separate the group means, the two functions taken together provide a clearer picture of separation. If $\lambda_1 = \lambda_2 = .5$, then the ratio $\lambda_1 / \sum_{i=1}^{4} \lambda_i$ only accounts for 50% of the discrimination while two eigenvalues account for 100% of the separation. While L_1 clearly separates all groups, L_2 only separates group 3 and 4 from 1 and 2.

To evaluate whether the first discriminant function in Figure 7.4.1 is significant based upon a sample, we might use Roy's largest root test. Assuming that $\widehat{\lambda}_1 \xrightarrow{p} \lambda_1$ and that $L_1 = \widehat{\mathbf{a}}_1 \mathbf{y} \xrightarrow{p} \boldsymbol{\alpha}_1 \mathbf{y}$, then the significance suggests that the population value $\lambda_1 > 0$ and that the means are significantly separated in at least one dimension. Thus the rank of Γ is at least one. However, the largest root criterion says nothing about the higher dimensions. To ensure maximal separation, we want to find $s = \min(p, \ k-1)$. Thus, the hypothesis of interest is

$$H_d : \lambda_1 \geq \ldots \geq \lambda_d > 0 \text{ and } \lambda_{d+1} = \ldots = \lambda_s = 0 \tag{7.4.3}$$

where $d = \text{rank}(\Gamma)$. The test of H_d is called the test of dimensionality. The hypotheses $H_o : \lambda_1 = \lambda_2 = \ldots = \lambda_s = 0$ is identical to testing the MANOVA hypothesis of equal population means.

While Roy's criterion may be used to test H_o as defined in (7.4.3.), rejection only ensures that $d \geq 1$. To determine d following Roy's test, we suggest forming $\binom{k}{2}$ contrasts involving the k vectors of s discriminant scores. Using $\widehat{\lambda}_j$ $(j = 1, 2, \ldots, s)$ to order the scores, one sequentially applies the stepdown FIT to evaluate dimensionality. If all s stepdown tests are significant, then the dimension of the discriminant space is s. If nonsignificance is attained for some $j = 2, 3, \ldots, s-1$, the dimension is estimated as $d = s - j$.

While Roy's test may not be used to determine dimensionality, one may use any of the other criteria to evaluate dimensionality. Following Fujikoshi (1977), we define

$$\begin{aligned}
\Lambda_d &= \prod_{i=d+1}^{s} (1 + \widehat{\lambda}_i)^{-1} = \prod_{i=d+1}^{s} (1 - \widehat{\theta}_i) \\
U_d &= \sum_{i=d+1}^{s} \widehat{\lambda}_i = \sum_{i=d+1}^{s} \widehat{\theta}_i / (1 - \widehat{\theta}_i) \\
V_d &= \sum_{i=d+1}^{s} \widehat{\lambda}_i / (1 + \widehat{\lambda}_i) = \sum_{i=d+1}^{s} \widehat{\theta}_i
\end{aligned} \tag{7.4.4}$$

for $d = 0, 1, 2, \ldots, s-1$, to test H_d. Then provided H_d is true, Fujikoshi (1977) shows that the criteria

$$\begin{aligned}
X_B^2 &= -\left[v_e - \frac{1}{2}(p - v_h + 1) \right] \ln \Lambda_d \\
X_L^2 &= -\left[v_e - d - \frac{1}{2}(p - v_h + 1) + \sum_{i=1}^{d} \widehat{\theta_i^{-1}} \right] \ln \Lambda_d \\
X_{(BLH)}^2 &= -\left[v_e - (p - v_h + 1) + \sum_{i=1}^{d} \widehat{\theta_i^{-1}} \right] \ln U_d \\
X_{(BNP)}^2 &= -\left[v_e - 2d + \sum_{i=1}^{d} \widehat{\theta_i^{-1}} \right] \ln V_d
\end{aligned} \tag{7.4.5}$$

are asymptotically distributed as chi-squared distributions with $v = (v_h - d)(p - d)$ degrees of freedom where $v_e = n - k$ and $v_h = k - 1$. The statistics X_B^2 and X_L^2 were first proposed by Bartlett (1947) and Lawley (1959). Fujikoshi (1977) verified Lawley's correction factor and developed the chi-squared approximations corresponding to the multivariate trace criteria due to Bartlett-Lawley-Hotelling and Bartlett-Nanda-Pillai. The dimension d for the discriminant space is the value at which d becomes nonsignificant. If $d = s$ is significant, then all dimensions are needed for discrimination. Again, these tests are sequential. The size of the test α_i is selected to control the test at some nominal level α. Because the test criteria are not equivalent, the value of $d \le s$ may not be the same using the different criteria.

c. *Variable Selection*

Implicit in our discussion of estimating the dimensionality of the discriminant space is that all p variables contribute to discrimination in some unknown dimension. This may not be the case, thus one may want to reduce the number of variables to improve the precision of the model. For the multiple group case, variable selection becomes more complicated since both p and s are both unknown.

A preliminary investigation of the standardized discriminant coefficients may lead one to delete some variables in the analysis, namely those with small absolute values in all significant dimensions. Deleting the variables and performing a reanalysis may lead to significance with fewer variables and a value of s that is less than or equal to the estimated value obtained using all variables. One may also use the average size of the absolute value of the coefficients to "order" the variables in an ad hoc manner to assess a variable's contribution to separation performing a sequential analysis by deleting variables one-at-a-time.

Alternatively, we may use some type of stepwise, forward elimination, or backward elimination procedure to determine whether a subset of the dependent variables maintain significant mean differences. To develop a test that some subset of the dependent variables do not contribute to overall separation one partitions the dependent variables into two subsets $\mathbf{y}' = [\mathbf{y}_1', \mathbf{y}_2']$. Then one evaluates the contribution $\mathbf{y}_2$ makes to the separation of the group means given that variables $\mathbf{y}_1$ are included in the MANOVA model, a conditional or stepwise MANOVA. This is similar to testing that $\mathbf{B}_2 = \mathbf{0}$ in the MR model considered in Section 4.2, except that one is evaluating the contribution of the dependent set of variables $\mathbf{y}_2$ given that the subset of dependent variables $\mathbf{y}_1$ are included in the MANOVA model. Letting $\mathbf{y}_1$ contain $p - q$ variables and $\mathbf{y}_2$ contain q variables, one partitions $\mathbf{E}$ and $\mathbf{H}$ as

$$\mathbf{E} = \begin{bmatrix} \mathbf{E}_{11} & \mathbf{E}_{12} \\ \mathbf{E}_{21} & \mathbf{E}_{22} \end{bmatrix}, \quad \mathbf{H} = \begin{bmatrix} \mathbf{H}_{11} & \mathbf{H}_{12} \\ \mathbf{H}_{21} & \mathbf{H}_{22} \end{bmatrix} \tag{7.4.6}$$

where $\mathbf{T} = \mathbf{H} + \mathbf{E}$. Using Wilks' Λ criterion, Rao (1973a, p. 551) shows that one may factor Wilks' full model criterion as

$$\begin{aligned} \Lambda_F &= \frac{|\mathbf{E}|}{|\mathbf{E} + \mathbf{H}|} = \frac{|\mathbf{E}_{11}|\left|\mathbf{E}_{22} - \mathbf{E}_{21}\mathbf{E}_{11}^{-1}\mathbf{E}_{12}\right|}{|\mathbf{T}_{11}|\left|\mathbf{T}_{22} - \mathbf{T}_{21}\mathbf{T}_{11}^{-1}\mathbf{T}_{12}\right|} \\ &= \Lambda_R \Lambda_{F|R} \end{aligned} \tag{7.4.7}$$

Rao's test that $\mathbf{y}_2$ given $\mathbf{y}_1$ provides no additional information in discrimination is tested using

$$\Lambda_{F|R} \sim U_{p-q, v_h, v_e} \tag{7.4.8}$$

where $v_h = k - 1$ and $v_e = n - k - q$. To evaluate the contribution of a single variable,

$q = p - 1$ so that the partial Λ statistic in (7.4.7), may be related to the partial F statistic

$$F = \left(\frac{n-k-p+1}{k-1}\right)\frac{1-\Lambda_{F\,|R}}{\Lambda_{F\,|R}} \tag{7.4.8}$$

which has an F distribution with $v_h^* = k - 1$ and $v_e^* = n - k - p + 1$,the partial F statistic in (4.2.51). Given some order of importance for the variables $y_1, y_2, \ldots, y_p$ the F statistic in (7.4.8) may be used to evaluate the redundance of y_p given $y_1, y_2, \ldots, y_{p-1}$. This process may be performed in a stepdown manner decreasing the error degrees of freedom at each step and stopping the process when significance is attained at some step. Because the process is sequential, the Type I error rate should be controlled at some α level where $1 - \alpha = \prod_{i=1}^{p} (1 - \alpha_i)$.

While one may also use (7.4.8) to develop a stepwise selection procedure as suggested by Hawkins (1976) in multiple group discriminant analysis problems, as we found in multiple and multivariate regression they must be used with caution since critical variables may be deleted from the study. A review of selection methods using other criteria with an analysis of alternative techniques is included in McKay and Campbell (1982 a,b), Krishnaiah (1982) and Huberty (1994). McLachlan (1992, Chapter 12) provides a nice overview.

d. *Classification Rules*

To develop a classification rule for an observation $\mathbf{y}$ for the multiple group case involves $\pi_1, \pi_2, \ldots, \pi_k$ populations and $f_i(\mathbf{y})$ for $(i = 1, 2, \ldots k)$ probability density functions. Furthermore, we have $p_1, p_2, \ldots, p_k$ prior probabilities that an observation is from population π_i where the $\sum_i p_i = 1$. The cost of assigning an observation to population π_j when it belongs to π_i is represented as $C(j \mid i)$ for $i,\ j = 1, 2, \ldots, k$. Finally, we let $P(j \mid i)$ denote the probability of classifying an observation into π_j given that it should be in π_i. Then, the $P(i \mid i) = 1 - \sum_{j=1}^{k} P(j \mid i)$ for $i \neq j$. With this notation for the k group case, the total probability of misclassification (TPM) and the expected cost of misclassification become

$$TPM = \sum_{i=1}^{k} p_i \left[\sum_{\substack{j=1 \\ j\neq i}}^{k} P(j \mid i)\right] \tag{7.4.9}$$

$$ECM = \sum_{i=1}^{k} p_i \left[\sum_{\substack{j=1 \\ j\neq i}}^{k} C(j \mid i)\, P(j \mid i)\right] \tag{7.4.10}$$

Again, assuming known probability density functions, the optimal rule for classifying an observation $\mathbf{y}$ into one of k populations that minimizes the TPM is to allocate $\mathbf{y}$ to π_i if

$$p_i\ f_i\ (\mathbf{y}) > p_j\ f_j\ (\mathbf{y}) \qquad \text{for all } j = 1, 2, \ldots, k \tag{7.4.11}$$

so that $p_i\ f_i(\mathbf{y})$ is maximum. If all p_i are equal, the rule is called the maximum likelihood rule since $f_i\ (\mathbf{y})$ is a likelihood for an observation $\mathbf{y}$. Using Bayes' theorem, it is also seen to be equivalent to assigning observations based upon a maximum posterior probability. Assuming $\pi_i \equiv N_p\left(\boldsymbol{\mu}_i,\ \Sigma\right)$ and letting

$$D_i^2\left(\mathbf{y}\right) = \left(\mathbf{y} - \overline{\mathbf{y}}_i\right)' \mathbf{S}^{-1} \left(\mathbf{y} - \overline{\mathbf{y}}_i\right) \tag{7.4.12}$$

The rule in (7.4.11) becomes: Assign $\mathbf{y}$ to π_i for which

$$L_i\left(\mathbf{y}\right) = -D_i^2\left(\mathbf{y}\right)/2 + \log p_i \tag{7.4.13}$$

is a maximum. This is easily established by evaluating the log $p_i f_i(\mathbf{y})$, ignoring constant terms, and estimating $\widehat{\boldsymbol{\mu}}_i$ with $\overline{\mathbf{y}}_i$ and Σ by the unbiased common estimate $\mathbf{S}$. Equivalently, one may assign an observation to π_i to the population with the maximal posterior probability

$$P\left(\pi_i \mid \mathbf{y}\right) = \frac{p_i e^{-D_i^2(\mathbf{y})/2}}{\sum_{j=1}^{k} p_j e^{-D_j^2(\mathbf{y})/2}} \tag{7.4.14}$$

Because the population parameters are being estimated, rules (7.4.13) and (7.4.14) are only asymptotically optimal.

Given that all the populations are multivariate normal, with a common covariance matrix Σ, and equal prior probabilities p_i, classification based upon a maximum value of $-D_i^2(\mathbf{y})$ or that $\mathbf{y}$ is closest to $\overline{\mathbf{y}}_i$ is equivalent to classifying an observation based upon Fisher's s discriminant functions using the eigenvectors of $|\mathbf{H} - \lambda\mathbf{E}| = 0$. This is discussed by Kshirsagar and Arseven (1975), Green (1979) and Johnson and Wichern (1998, Section 11.7).

When the covariance matrices are not equal, one may again develop quadratic classification rules for normal populations. Now the rules depend on $\mathbf{S}_i$ and $|\mathbf{S}_i|$. Letting

$$\widetilde{D}_i^2\left(\mathbf{y}\right) = (\mathbf{y} - \overline{\mathbf{y}}_i)'\mathbf{S}_i^{-1}(\mathbf{y} - \overline{\mathbf{y}}_i) \tag{7.4.15}$$

the posterior probabilities becomes

$$P\left(\pi_i \mid \mathbf{y}\right) = \frac{p_i \left|\mathbf{S}_i\right|^{-1/2} e^{-\widetilde{D}_i^2(\mathbf{y})/2}}{\sum_{j=1}^{k} p_j \left|\mathbf{S}_j\right|^{-1/2} e^{-\widetilde{D}_i^2(\mathbf{y})/2}} \tag{7.4.16}$$

Given any classification rule for multiple populations, one must again evaluate the AER. One again constructs a confusion table for the k groups using the one-at-a-time, crossvalidation procedure to obtain an APER. While the APER is usually used to evaluate a classification rule, it may also be used in variable selection, Gnaneshanandam and Krzanowski (1989).

e. *Logistic Discrimination and Other Topics*

We have been primarily concerned with discrimination and classification assuming a multivariate normal model for the variables in each group. The method may also be applied with repeated measurement data and with adjustments for covariates, McLachlan (1992). However, one often finds that the variables in a study are not continuous, but categorical and continuous. If the group membership variable is categorical, then logistic discrimination may be performed using logistic regression, Anderson (1982) and Bull and Donner (1987). Logistic regression in SAS may be performed using the procedure LOGISTIC. For a discussion of procedures for logistic regression in SAS, one may consult Stokes, Davis, and Koch (2000). Numerous examples discriminant analysis are also provided by Khattree and Naik (2000).

If a random p-vector $\mathbf{Y}$ follows a multivariate normal distribution and a random vector $\mathbf{X}$ follows a multinomial distribution with k categories, then the joint distribution vector $\mathbf{Y}$ and $\mathbf{X}$ is called a p-variate normal mixture distribution. Classifying objects using p-variate normal mixture distributions is called discriminant analysis with partially classified data or Discrimix analysis, Flury (1997).

While linear discriminant functions used for classification are reasonably robust to nonnormality, this is not the case for discrimination. Both discrimination and classification are adversely affected by having heavy tailed distributions and outliers, Campbell (1978, 1982). For a discussion of nonparametric discrimination and classification procedures consult Koffler and Penfield (1979, 1982) and McLachlan (1992). Newer methods of classification and discrimination with unknown distributions include the use of neural networks while classification and regression tress (CART) are used for classification. For an introduction of these newer procedures consult Stern (1996) and Breiman, Friedman, Olshen and Stone (1984).

7.5 Multiple Group Discriminant Analysis Example

To illustrate multiple group discriminant analysis method, data from Lubischew (1962) are used. The data were obtained from the DASL Web site http://lib.stat.cmu.edu/DASL/Datafiles/FleaBeetles and is provided in the file fleabeetles.dat. The data consists of the maximum width of the aedeagus in the forepart (in microns), and the front angle of the aedeagus (in units of 7.5°) measurements on three species of male flea beetles Chactocnema:Coninnna (Con), Heikertingeri (Hei), and Heptapotamica (Hep). The goal of the study is to construct a classification rule to distinguish among the three species. Program m7_5_1.sas contains the SAS code for the analysis.

To begin our analysis, PROC GLM is used to evaluate separation by testing $H : \boldsymbol{\mu}_1 = \boldsymbol{\mu}_2 = \boldsymbol{\mu}_3$. This test assumes all $\Sigma_i = \Sigma$ and multivariate normality. The test is not required for the establishment of a classification rule; however, it provides an indication of separation. Evaluation of separation is also provided using the PROC CANDISC.

The sample means for the three species and an estimate of $\mathbf{S}$ for the flea beetles data follow.

$$\overset{\text{Con}}{\overline{\mathbf{x}}_1 = \begin{bmatrix} 146.19048 \\ 14.09524 \end{bmatrix}} \qquad \overset{\text{Hei}}{\overline{\mathbf{x}}_2 = \begin{bmatrix} 124.64516 \\ 14.29032 \end{bmatrix}}$$

$$\overset{\text{Hep}}{\overline{\mathbf{x}}_3 = \begin{bmatrix} 138.27273 \\ 10.09091 \end{bmatrix}} \qquad \mathbf{S} = \begin{bmatrix} 23.02392262 & \\ -0.55961773 & 1.01429299 \end{bmatrix}$$

TABLE 7.5.1. Discriminant Structure Vectors, $H : \boldsymbol{\mu}_1 = \boldsymbol{\mu}_2 = \boldsymbol{\mu}_3$

	Within Structure	Standardized Vector	Raw Vector
WIDTH	0.7802	1.5258	0.147406745
ANGLE	−0.7116	−1.3394	−0.62527665

TABLE 7.5.2. Squared Mahalanobis Distances Flea Beetles $H : \boldsymbol{\mu}_1 = \boldsymbol{\mu}_2 = \boldsymbol{\mu}_3$

From Species	Con	Hei	Hep
Con	0		(Sym)
Hei	20.26956	0	
Hep	20.32363	23.01849	0

Testing the hypothesis $H : \boldsymbol{\mu}_1 = \boldsymbol{\mu}_2 = \boldsymbol{\mu}_3$ using either PROC GLM, PROC CANDISC or PROC DISCRIM, the overall test of equal means is rejected using all multivariate criteria. Because $s = \min(p, k-1) = (2, 2) = 2$, there are two eigenvalues for the problem. Solving $| \mathbf{H} - \lambda\mathbf{E} | = \mathbf{0}$, the estimated eigenvalues are $\widehat{\lambda}_1 = 4.2929$ and $\widehat{\lambda}_2 = 2.9937$. The GLM procedure for the MANOVA analysis uses Wilks' Λ criterion with X_B^2 defined in (7.4.5) is used to test (7.4.3), sequentially. For our example, both roots in the population appear to be nonzero. Because squared canonical correlation, ρ_i^2, are related to the roots of $| \mathbf{H} - \theta(\mathbf{H} + \mathbf{E}) | = 0$ in that $\rho_i^2 = \theta_i$, the test of dimensionality is labeled a test of canonical correlations using the /CANONICAL option. This will be investigated further in Chapter 8. The structure of the discriminant structure for the test of equal centroids is given in Table 7.5.1.

The entries in Table 7.5.1 indicate that both variables appear to contribute to separation equally. From PROC CANDISC, the squared Mahalanobis distances between species are calculated and shown in Table 7.5.2.

Comparing the pairwise mean differences, all pairs appear to be widely separated. A plot of the data using the two discriminant functions is provided in Figure 7.5.1, obtained using the PROC PLOT procedure. Also provided in the output are one-dimensional histograms used to display marginal separation. The plots indicates clear separation of the three species: $C = con$, $H = Hei$ and $P = Hep$.

The lines D_{ij} have been added to the plot and will be discussed shortly.

To assess the importance of each variable to separation, PROC DISCRIM is used with the option METHOD = BACKWARD. To evaluate the contribution of a single variable, the partial F statistics defined in (7.4.8) are calculated for each variable (F = 124 and F = 119; df =2, 70) are significant (p-value < 0.0001), neither variable should be removed from the analysis.

The final step in our analysis is to create a classification rule to assign the beetles to each group. For this we use PROC DISCRIM. Confirming that $\Sigma_1 = \Sigma_2 = \Sigma$, we use the option POOL = TEST. The p-value for the hypothesis 0.7719 indicates that we may use Fisher's LDF with $\Sigma_1 = \Sigma_2$ for classification. The three subgroups are also seen to

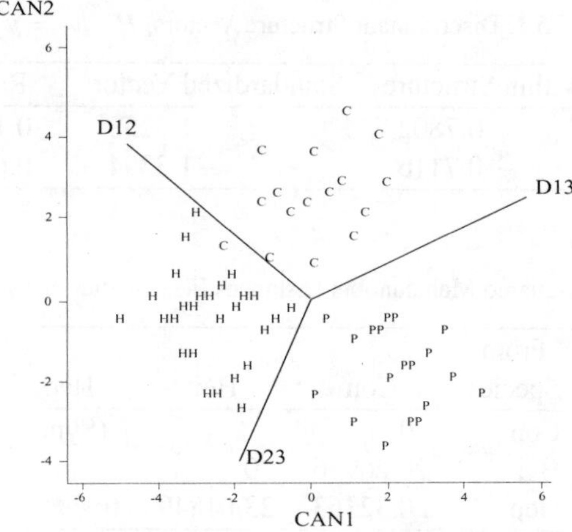

FIGURE 7.5.1. Plot of Flea Beetles Data in the Discriminant Space

TABLE 7.5.3. Fisher's LDFs for Flea Beetles

	Con	Hei	Hep
CONSTANT	−619.746	−487.284	−505.619
WIDTH	6.778	5.834	6.332
ANGLE	17.636	17.308	13.442

be multivariate normal. Setting $p_1 = p_2 = p_3 = 1/3$ with $D_i(\mathbf{y})$ defined in (7.4.12), the LDF rule is created using (7.4.13). That is, an observation $\mathbf{y}$ is assigned to population π_i for which

$$L_i(\mathbf{y}) = -\frac{1}{2}(\overline{\mathbf{y}} - \overline{\mathbf{y}}_i)' \mathbf{S}^{-1} (\mathbf{y} - \overline{\mathbf{y}}_i) + \log p_i$$

is maximal. Ignoring the term $\ln(p_i)$ and since $\mathbf{y}'\mathbf{S}^{-1}\mathbf{y}/2$ is the same for each $L_i(\mathbf{y})$, SAS calculates

$$\begin{aligned} L_i^*(\mathbf{y}) &= \overline{\mathbf{y}}_i'\mathbf{S}^{-1}\mathbf{y} - \overline{\mathbf{y}}_i'\mathbf{S}^{-1}\overline{\mathbf{y}}_i/2 \\ &= \overline{\mathbf{y}}_i'\mathbf{S}^{-1}(\mathbf{y} - \overline{\mathbf{y}}_i/2) \\ &= \text{coefficient vector} - \text{constant} \end{aligned}$$

The linear functions $L_i^*(\mathbf{y})$ are given in Table 7.5.3

The function $L_i^*(\mathbf{y})$ are used to partition the sample space into three regions by forming the differences

$$D_{ij} = L_i^*(\mathbf{y}) - L_j^*(\mathbf{y})$$

for $i \neq j$.

Creating the differences D_{ij}, observe that if $L_1^*(\mathbf{y}) > L_2^*(y)$ and $L_1^*(\mathbf{y}) > L_3^*(\mathbf{y})$ then $D_{12}(\mathbf{y}) > 0$ and $D_{13}(\mathbf{y}) > 0$ and an observation would be assigned to group 1. By the same logic, if $D_{12}(\mathbf{y}) < 0$ and $D_{23}(\mathbf{y}) < 0$, an observation is assigned to group 2. If $D_{13}(\mathbf{y}) < 0$

TABLE 7.5.4. Classification/Confusion Matrix for Species

	Con	Hei	Hep	
Con	20	1	0	21
Hei	0	31	0	31
Hep	0	0	22	22

and $D_{23}(\mathbf{y}) < 0$, an observation is assigned to group 3. For our example, Group 1 $\equiv$ Con $\equiv$ (C), Group 2 $\equiv$ Hei $\equiv$ (H), Group 3 $\equiv$ Hep $\equiv$ (P). Relabeling the axis in Figure 7.5.1, the functions D_{ij} represent three lines in the plane that separate the three groups where CAN 1 $\equiv$ WIDTH and CAN 2 $\equiv$ ANGLE. The lines have been super-imposed onto Figure 7.5.1. From Figure 7.5.1, we observe that only one observation is misclassified. This is shown in the Classification/Confusion table generated by SAS, Table 7.5.4.

The estimated apparent error rate for the LDF rule is 0.0159. In this example, we have excellent classification and separation.

Exercises 7.5

1. Use the Egyptian skull data for the B.C. years: -4000, -3300, -1850, -200 and the A.D. year 150 to evaluate group separation. And, obtain a classification rule to assign skulls to groups. Discuss your findings.

2. Using the option POOL = NO on the PROC DISCRIM statement, obtain a quadratic classification rule for the Flea Beetles data. What do you observe?

3. Develop a discriminant function for the pottery data described in Exercises 4.5, Problem 4.

4. In his seminal paper on discriminant analysis, Fisher (1936) developed his LDF to analyze three species of flowers: (1) *Iris Setosa, (2) Iris Vericolor, and (3) Iris Virginica*. For the 50 specimens of the three species the variables: sepal length (SL), sepal width (SW), petal length (PL) and petal width (PW)are measured in mm. The data are provided in the data set iris.dat where the variables are ordered: SL, SW, PL, PL and Species (1, 2, 3). Obtain two linear and quadratic discriminant functions for the data, and evaluate which of the two functions classify the species in two dimensions best.

8

Principal Component, Canonical Correlation, and Exploratory Factor Analysis

8.1 Introduction

In the previous chapter on discriminant analysis we generated linear combinations of variables called discriminant functions. Applying the discriminant functions to multivariate data obtained from several independent groups resulted in discriminant scores in a smaller dimensional discriminant space. The uncorrelated scores were used to evaluate group differences (separation) and to classify (allocate) observations into groups. In this chapter we investigate three statistical procedures that commonly involve a single group. These procedures were designed to reduce the dimensionality of the data space in order to discover, visualize, and interpret dependences among sets of variables, or to help stabilize the measurements for additional statistical analysis such as regression analysis (Chapter 4) or cluster analysis (Chapter 9).

8.2 Principal Component Analysis

Principal component analysis (PCA) was first introduced by Karl Pearson in the early 1900's. Formal treatment of the method is due to Hotelling (1933) and Rao (1964). In PCA a set of p correlated variables is transformed to a smaller set of uncorrelated hypothetical constructs called principal components (PCs). The PCs are used to discover and interpret the dependences that exist among the variables, and to examine relationships that may exist among individuals. The PCs may be used to stabilize estimates, evaluate multivariate normality, and to detect outliers. While we review the basic theory in this chapter, an extensive discussion of the topic is provided by Jolliffe (1986), Jackson (1991), and Basilevsky (1994).

a. Population Model for PCA

Given an observation vector $\mathbf{Y}'_{1\times p} = [Y_1, Y_2, \ldots, Y_p]$ with mean $\boldsymbol{\mu}$ and covariance matrix Σ of full rank p, the goal of PCA is to create a new set of variables called principal components (PCs) or principal variates. The principal components are linear combinations of the variables of the vector $\mathbf{Y}$ that are uncorrelated such that the variance of the jth component is maximal.

The first principal component of the vector $\mathbf{Y}$ is the linear combination

$$Z_1 = \mathbf{p}'_1\mathbf{Y} \tag{8.2.1}$$

such that the variance of Z_1 is maximal. To determine the first linear combination of $\mathbf{Y}$, a vector $\mathbf{p}$ is sought such that

$$\text{var}(Z_1) = \text{var}\left(\mathbf{p}'_1\mathbf{Y}\right) = \mathbf{p}'_1\Sigma\mathbf{p}_1 \tag{8.2.2}$$

is maximal, subject to the constraint that $\mathbf{p}'_1\mathbf{p}_1 = 1$. The condition that $\mathbf{p}'_1\mathbf{p}_1 = 1$ is imposed to ensure the uniqueness (except for sign) of the principal component. From Theorem 2.6.10, the vector that maximizes (8.2.2), subject to the constraint that $\mathbf{p}'_1\mathbf{p}_1 = 1$, is the characteristic vector associated with the largest root of the eigenequation

$$|\Sigma - \lambda\mathbf{I}| = 0 \tag{8.2.3}$$

The largest variance of Z_1, is the largest root λ_1 of (8.2.3).

To determine the second principal component, the linear combination

$$Z_2 = \mathbf{p}'_2\mathbf{Y} \tag{8.2.4}$$

is constructed such that it is uncorrelated with Z_1 and has maximal variance. For Z_2 to be uncorrelated with Z_1, the covariance between Z_2 and Z_1 must be zero. However, $\Sigma\mathbf{p}_1 = \mathbf{p}_1\lambda_1$ so that the

$$\text{cov}(Z_2,\ Z_1) = \mathbf{p}'_2\Sigma\mathbf{p}_1 = \mathbf{p}'_2\mathbf{p}_1\lambda_1 = 0 \tag{8.2.5}$$

implies that $\mathbf{p}'_2\mathbf{p}_1 = 0$. Furthermore, if $\mathbf{p}_2$ is the second eigenvector of (8.2.3), then $\Sigma\mathbf{p}_2 = \mathbf{p}_2\lambda_2$ and the

$$\text{var}(Z_2) = \mathbf{p}'_2\Sigma\mathbf{p}_2 = \lambda_2 \tag{8.2.6}$$

where $\lambda_1 \geq \lambda_2$. More generally, by Theorem 2.6.5 (Spectral Decomposition Theorem) there exists an orthogonal matrix $\mathbf{P}$ $\left(\mathbf{P}'\mathbf{P} = \mathbf{I}\right)$ such that

$$\mathbf{P}'\Sigma\mathbf{P} = \Lambda = \text{diag}\,[\lambda_i] \tag{8.2.7}$$

where $\lambda_1 \geq \lambda_2 \geq \ldots \geq \lambda_p \geq 0$. Setting

$$\mathbf{Z} = \mathbf{P}'\mathbf{Y} \tag{8.2.8}$$

the $E(\mathbf{Z}) = \mathbf{P}'\boldsymbol{\mu}$ and the $\text{cov}(\mathbf{Z}) = \Lambda$ where the jth element Z_j of $\mathbf{Z}$ is the jth principal component of $\mathbf{Y}$. Thus, the raw score form of the jth component is

$$Z_j = \mathbf{p}'_j\mathbf{Y} = \sum_{i=1}^{p} p_{ij}Y_i \tag{8.2.9}$$

where $\mathbf{p}_j$ is the jth column of $\mathbf{P}$ so that $\mathbf{p}'_j$ is the jth row of $\mathbf{P}'$. In general the mean of Z_j is not zero. Letting $\mathbf{Y}_d = \mathbf{Y} - \boldsymbol{\mu}$, the components $C_j = \mathbf{p}'_j \mathbf{Y}_d = \mathbf{p}'_j (\mathbf{Y} - \boldsymbol{\mu})$ have mean zero and variance λ_j. To standardize the components with regard to both location and scale, the standardized components

$$Z_j^* = \mathbf{p}'_j (\mathbf{Y} - \boldsymbol{\mu}) / \sqrt{\lambda_j} \qquad j = 1, 2, \ldots, p \tag{8.2.10}$$

are constructed so that the $E(Z_j^*) = 0$ and the $\text{var}(Z_j^*) = 1$.

The construction of principal components do not require that the variables in $\mathbf{Y}$ have a multivariate normal distribution. However, assuming that $\mathbf{Y} \sim N_p(\mathbf{0}, \Sigma)$, recall that the constant density ellipsoids associated with $\mathbf{Y}$ have the form $\mathbf{Y}'\Sigma^{-1}\mathbf{Y} = Q > 0$ centered at the mean $\boldsymbol{\mu} = \mathbf{0}$. If $\boldsymbol{\mu} \neq \mathbf{0}$, one may use deviation scores $\mathbf{Y}_d$. Under the transformation $\mathbf{Y} = \mathbf{PZ}$, the principal components have the simplified ellipsoid structure $\mathbf{Z}'\Lambda^{-1}\mathbf{Z} = Q$ where Λ is a diagonal matrix with elements $1/\lambda_j$ and λ_j is an eigenvalue of Σ. Letting θ_{ij} define the direction cosine of the ith old axis with the jth new axis, the matrix $\mathbf{P}$ has the general form

$$\mathbf{P} = \left[\cos\theta_{ij}\right] = \left[\mathbf{p}_1, \mathbf{p}_2, \ldots, \mathbf{p}_p\right]$$

so that the transformation $\mathbf{Z} = \mathbf{P}'\mathbf{Y}$ represents a rigid rotation of the old axes into the new principal axes. If λ_1 is the largest eigenvalue of Σ, then the major axis of the ellipsoid lies in the direction of $\mathbf{p}_1$. For the bivariate case illustrated in Figure 3.3.1, the axes x_1 and x_2 represent the principal components. For the population covariance matrix Σ^{-1} given in Figure 3.3.1 with $\rho = 1/2$, the matrix $\mathbf{P}$ formed from the column eigenvectors of $|\Sigma - \lambda\mathbf{I}| = 0$ is the matrix

$$\mathbf{P} = \left[\cos\theta_{ij}\right] = \begin{bmatrix} \cos 45^\circ & \cos 135^\circ \\ \cos 45^\circ & \cos 45^\circ \end{bmatrix} = \begin{bmatrix} 1/\sqrt{2} & -1/\sqrt{2} \\ 1/\sqrt{2} & 1/\sqrt{2} \end{bmatrix}$$

using direction cosines. The matrix $\mathbf{P}'$ rotates the old axes into the new axes so that the component scores in the new coordinate system are uncorrelated and have maximum variance. Thus, normal PCA is a statistical procedure for transforming a MVN distribution into a set of independent univariate normal distributions, since under normality zero correlation implies independence.

Subtracting the mean $\boldsymbol{\mu}$ from the vector $\mathbf{Y}$, the p PCs are represented as follows

$$\begin{aligned} C_1 &= \mathbf{p}'_1(\mathbf{Y} - \boldsymbol{\mu}) = p_{11}(Y_1 - \mu_1) + \ldots + p_{p1}\left(Y_p - \mu_p\right) \\ C_2 &= \mathbf{p}'_2(\mathbf{Y} - \boldsymbol{\mu}) = p_{12}(Y_1 - \mu_1) + \ldots + p_{p2}\left(Y_p - \mu_p\right) \\ &\vdots \\ C_p &= \mathbf{p}'_p(\mathbf{Y} - \boldsymbol{\mu}) = p_{1p}(Y_1 - \mu_1) + \ldots + p_{pp}\left(Y_p - \mu_p\right) \end{aligned} \tag{8.2.11}$$

where the rank of Σ is p. To obtain standardized components, one divides C_j by $\sqrt{\lambda_j}$ for $j = 1, 2, \ldots, p$. The components are summarized in Table 8.2.1.

Using vector notation, the relationship between $\mathbf{Y}$ and the PCs are

$$\begin{aligned} \mathbf{Z} &= \mathbf{P}'\mathbf{Y} \\ \mathbf{C} &= \mathbf{P}'(\mathbf{Y} - \boldsymbol{\mu}) \\ \mathbf{Z}^* &= \Lambda^{-1/2}\mathbf{P}'(\mathbf{Y} - \boldsymbol{\mu}) = \Lambda^{-1/2}\mathbf{C} \end{aligned} \tag{8.2.12}$$

TABLE 8.2.1. Principal Component Loadings

Variables	Components C_1	C_2	$\dots$	C_p
Y_1	p_{11}	p_{12}	$\dots$	p_{1p}
Y_2	p_{21}	p_{22}	$\dots$	p_{2p}
$\vdots$	$\vdots$	$\vdots$		$\vdots$
Y_p	p_{p1}	p_{p2}	$\dots$	p_{pp}
Eigenvectors	$\mathbf{p}_1$	$\mathbf{p}_2$	$\dots$	$\mathbf{p}_p$
Eigenvalues	λ_1	λ_2	$\dots$	λ_p

TABLE 8.2.2. Principal Component Covariance Loadings (Pattern Matrix)

Variables	Components C_1	C_2	$\dots$	C_p
Y_1	$p_{11}\sqrt{\lambda_1} = q_{11}$	$p_{12}\sqrt{\lambda_2} = q_{12}$	$\dots$	$p_{1p}\sqrt{\lambda_p} = q_{1p}$
Y_2	$p_{21}\sqrt{\lambda_1} = q_{21}$	$p_{22}\sqrt{\lambda_2} = q_{22}$	$\dots$	$p_{2p}\sqrt{\lambda_p} = q_{2p}$
$\vdots$	$\vdots$	$\vdots$		$\vdots$
Y_p	$p_{p1}\sqrt{\lambda_1} = q_{p1}$	$p_{p2}\sqrt{\lambda_2} = q_{p2}$	$\dots$	$p_{pp}\sqrt{\lambda_p} = q_{pp}$

Alternatively, one may also construct the data vector from the PCs

$$\begin{aligned}\mathbf{Y} &= \mathbf{PZ}\\ \mathbf{Y} &= \boldsymbol{\mu} + \mathbf{PC} \qquad (8.2.13)\\ \mathbf{Y} &= \boldsymbol{\mu} + \mathbf{P}\Lambda^{1/2}\mathbf{Z}^* = \boldsymbol{\mu} + \mathbf{QZ}^*\end{aligned}$$

The matrix $\mathbf{Q}$ in (8.2.13) is called the covariance loading matrix since the covariance between $\mathbf{Y}$ and Z_j^*, where $Z_j^* = C_j/\sqrt{\lambda_j}$, is

$$\text{cov}\left(\mathbf{Y},\ Z_j^*\right) = \text{cov}\left(\mathbf{Y}, \mathbf{p}_j'\mathbf{Y}/\sqrt{\lambda_j}\right) = \Sigma\mathbf{p}_j = \lambda_j\mathbf{p}_j/\sqrt{\lambda_j}$$

so that the covariance between variable i and the jth standardized PC is

$$\text{cov}\left(Y_i,\ Z_j^*\right) = p_{ij}\sqrt{\lambda_{ij}} = q_{ij} \qquad (8.2.14)$$

The covariance loadings are shown in Table 8.2.2.

Selecting only a subset of the components $C_1, C_2, \dots, C_k$ where $k \ll p$, observe that variable Y_i may be estimated as

$$Y_i \approx \mu_i + \sum_{j=1}^{k} q_{ij}Z_j^* = \widetilde{Y}_i$$

so that

$$\widetilde{\mathbf{Y}} = \boldsymbol{\mu} + \mathbf{Q}\mathbf{Z}^* \tag{8.2.15}$$

where $\mathbf{Q} = [q_{ij}]$. Letting

$$Q = (\mathbf{Y} - \boldsymbol{\mu} - \widetilde{\mathbf{Y}})'(\mathbf{Y} - \boldsymbol{\mu} - \widetilde{\mathbf{Y}})$$

and assuming $\mathbf{Y} \sim N_p(\boldsymbol{\mu}, \Sigma)$, Jackson and Mudholkar (1979) show that

$$(Q/\theta_1)^{h_o} \mathrel{\dot\sim} N\left[1 + \frac{\theta_2 h_o (h_o - 1)}{\theta_1^2}, \frac{2\theta_2 h_o^2}{\theta_1^2}\right] \tag{8.2.16}$$

is approximately univariate normal where

$$\begin{aligned} \theta_m &= \textstyle\sum_{j=k+1}^{p} \lambda_j^m \qquad m = 1, 2, 3 \\ h_o &= 1 - \left(2\theta_1\theta_3 / 3\theta_2^2\right) \end{aligned}$$

A significant value of Q using sample estimates for $\boldsymbol{\mu}$ and Σ suggests that $\mathbf{Y}$ may be an outlier. Rao (1964) suggests investigating the distances

$$\begin{aligned} D_i^2 &= (\mathbf{Y} - \widetilde{\mathbf{Y}})'(\mathbf{Y} - \widetilde{\mathbf{Y}}) \\ &= \sum_{j=k+1}^{p} \left(q_{ij} Z_j^*\right)^2 \mathrel{\dot\sim} \chi^2 (p - k) \end{aligned} \tag{8.2.17}$$

An informal plot of D_i^2 $(i = 1, 2, \ldots, n)$ may be used to detect outliers. Hawkins (1974) suggests a weighted form of (8.2.17) by dividing each term D_i^2 by $\widehat{\lambda}_j$ to improve convergence to a chi-square distribution. If an outlier is detected, one would exclude it from the estimation of Σ as discussed in chapter 3. Gnanadesikan (1977) reviews other methods for the detection of outliers in PCA.

b. *Number of Components and Component Structure*

Defining the total (univariate) variance as the trace of Σ and recalling that the sum of the roots of $|\Sigma - \lambda \mathbf{I}| = 0$ is equal to the tr(Σ), observe that

$$\begin{aligned} \operatorname{tr}(\Sigma) &= \sigma_{11} + \sigma_{22} + \ldots + \sigma_{pp} \\ &= \lambda_1 + \lambda_2 + \ldots + \lambda_p \end{aligned}$$

is maximized by the PCs. Because the generalized variance of p variables is the $|\Sigma| = \prod_{i=1}^{p} \lambda_i$, we also see that the geometric mean

$$\overline{\lambda}_g = |\Sigma|^{1/p} = \left(\prod_{i=1}^{p} \lambda_i\right)^{1/p} \tag{8.2.18}$$

is maximized by the PCs.

TABLE 8.2.3. Principal Components Correlation Structure

Variables	Components C_1	C_2	$\cdots$	C_k
Y_1	$p_{11}\sqrt{\lambda_1}\,/\,\sigma_1$	$p_{12}\sqrt{\lambda_2}\,/\,\sigma_1$	$\cdots$	$p_{1k}\sqrt{\lambda_k}\,/\,\sigma_1$
Y_2	$p_{21}\sqrt{\lambda_1}\,/\,\sigma_2$	$p_{22}\sqrt{\lambda_2}\,/\,\sigma_2$	$\cdots$	$p_{2k}\sqrt{\lambda_k}\,/\,\sigma_2$
$\vdots$	$\vdots$	$\vdots$	$\cdots$	$\vdots$
Y_p	$p_{p1}\sqrt{\lambda_1}\,/\,\sigma_p$	$p_{p1}\sqrt{\lambda_2}\,/\,\sigma_p$	$\cdots$	$p_{pk}\sqrt{\lambda_k}\,/\,\sigma_p$

Since the tr $(\Sigma)-\sum_{j=1}^{k}\lambda_j = \sum_{j=k+1}^{p}\lambda_j$, the proportion of the total univariate variance accounted for by k PCs is

$$\rho_k^2 = \frac{\sum_{j=1}^{k}\lambda_j}{\sum_{j=1}^{p}\lambda_j} = \frac{\sum_{j=1}^{k}\lambda_j}{\text{tr}\,(\Sigma)} \tag{8.2.19}$$

which may be used as a criterion for selecting a subset of k components from p. One usually wants to account for 70% – 80% of the total univariate variance with a few PCs.

One may use either the geometric mean or arithmetic mean of the eigenvalues as a criterion for retaining $k \ll p$ components. Later in this chapter we discuss how these criteria and formal statistical tests are used to evaluate the number of components one should retain.

Having selected a set of k components where $k \ll p$, we saw that the PCs may be used to estimate the original data using the covariance loadings in Table 8.2.2. Alternatively, one may also want to evaluate the contribution of the ith variable to the jth component. The covariance loadings in Table 8.2.2 may not be used as they depend on $\sigma_{ii} = \sigma_i^2$. Large values for covariance loadings may only reflect differences in population variances. To remove this dependence, one divides the covariance loadings by $\sigma_i = \sqrt{\sigma_{ii}}$. This is no more than calculating the correlation between Y_i and Z_j since the

$$\rho_{Y_i,\,Z_j} = \frac{\text{cov}\left(Y_i,\ Z_j\right)}{\sqrt{\sigma_{ii}}\sqrt{\lambda_j}} = \frac{p_{ij}\sqrt{\lambda_j}}{\sigma_i} \tag{8.2.20}$$

One can create a matrix of these correlations, the structure matrix, shown in Table 8.2.3.

Observe that if all $\sigma_i = 1$ in Table 8.2.3, meaning all variables are standardized, then the loading coefficients are correlations.

The column entries in Table 8.2.3 represent the contribution of variable Y_i on the jth component ignoring the other variables in the linear combination of variables. This same result was encountered in discriminant analysis. Where instead of evaluating correlations, the variables were standardized which was equivalent to adjusting the "raw" eigenvectors by multiplying each coefficient by σ_i. This may not be done in PCA since PCs as shown in Table 8.2.1 are not invariant to linear transformations of the original variables. Rescaling may result in an entirely different set of components; the least important component or variable may become the most important under rescaling. Because the population correlation matrix is a rescaling of the data to standardized variables, the components of standardized variables using a population correlation matrix $\mathbf{P}_\rho$ may lead to very different results. Letting δ_j be the eigenvalues of the equation $|\mathbf{P}_\rho - \delta\mathbf{I}| = 0$ where $\mathbf{P}_\rho$ is a rescaling of Σ, its eigenvalues and eigenvectors may be very different from those obtained using Σ. This is because under a rescaling $\mathbf{Y}^* = \mathbf{DY}$ where $\mathbf{D}$ is a nonsingular diagonal matrix, the roots and vectors of $|\mathbf{D}\Sigma\mathbf{D} - \delta\mathbf{I}| = |\mathbf{P}_\rho - \delta\mathbf{I}| = 0$ are not a simple rescaling of the roots and vectors of $|\Sigma - \lambda\mathbf{I}| = 0$. Thus, the principle component weights are not scale free, Lawley and Maxwell (1971, p. 18). The scaling of the original variables should precede any PCA. Choosing $\mathbf{P}_\rho$ over Σ may not always be the correct decision since it may destroy natural commensurate variability, as illustrated by Naik and Khattree (1996).

An advantage to using $\mathbf{P}_\rho$ instead of Σ is that $\mathbf{P}_\rho$ is scale free since a change in scale of the variables does not effect the elements ρ_{ij} of $\mathbf{P}_\rho$. When analyzing a population correlation matrix, the $\text{tr}\left(\mathbf{P}_\rho\right) = p$ so that $\delta_j \,/\, p$ is a measure of the importance of the jth component. And since the $\sum_{j=1}^{p} \delta_j = p$, the criterion $\bar{\delta} > 1$ is often used in selecting components using $\mathbf{P}_\rho$. Furthermore, since $\sigma_i^2 = 1$ for standardized variables the matrix of covariance loadings (pattern matrix) is identical to the correlation structure matrix when using a population correlation matrix. By dividing the components based on $\mathbf{P}_\rho$ by the $\sqrt{\delta_j}$, both the principal components and the variables are standardized, have mean zero and variance one.

Yet another index that one may calculate when creating PCs is the prediction of the p components on Y_i also called square of the population multiple correlation between Y_i and the p components of $\mathbf{Z}' = \left[Z_1, Z_2, \ldots, Z_p\right]$. Using the definition of multiple correlation, the square of the multiple correlation between Y_i and the vector $\mathbf{Z}' = \left[Z_1, Z_2, \ldots, Z_p\right]$ is

$$\rho^2_{Y_i(Z_1,\, Z_2, \ldots, Z_p)} = \frac{\boldsymbol{\sigma}'_{Y_i\mathbf{Z}} \quad \Sigma^{-1}_{\mathbf{ZZ}} \quad \boldsymbol{\sigma}_{Y_i\mathbf{Z}}}{\sigma_i^2} \tag{8.2.21}$$

Using the fact that $\Sigma_{\mathbf{ZZ}} = \text{diag}\left[\lambda_j\right]$ and (8.2.14), (8.2.21) becomes

$$\rho^2_{Y_i\mathbf{Z}} = \sum_{j=1}^{p} q_{ij}^2 \,/\, \sigma_i^2 = \sum_{j=1}^{p} \lambda_j p_{ij}^2 \,/\, \sigma_i^2 = \sum_{j=1}^{p} \rho^2_{Y_i Z_j} \tag{8.2.22}$$

Thus, for $k \ll p$ one may use the entries in Table 8.2.2 to evaluate how well a subset of components predict Y_i. The residual variance is $\sigma_i^2 - \sum_{j=1}^{k} q_{ij}^2$.

Example 8.2.1 *To illustrate the calculation of principal components, we consider the* 3×3 *covariance matrix*

$$\Sigma = \begin{bmatrix} 26.64 & & \text{(Sym)} \\ 8.25 & 9.85 & \\ 18.29 & 8.27 & 22.08 \end{bmatrix}$$

with corresponding population correlation matrix

$$\mathbf{P}_\rho = \begin{bmatrix} 1.00000 & & \text{(Sym)} \\ 0.52596 & 1.00000 & \\ 0.75413 & 0.56077 & 1.00000 \end{bmatrix}$$

Solving the eigenequation $|\ \Sigma - \lambda \mathbf{I}\ | = 0$, *the roots and vectors of* Σ *are*

$$\Lambda = \text{diag}\left[\ 46.6182, \quad 6.5103, \quad 5.4414\ \right]$$

$$\mathbf{P} = \begin{bmatrix} 0.71056 & -0.61841 & 0.33567 \\ 0.30705 & 0.70175 & 0.64286 \\ 0.63311 & 0.35372 & -0.68852 \end{bmatrix}$$

the quantities in Table 8.2.1. Using 8.2.20, the correlation between variable Y_i *and each component* Z_j, *as shown in Table 8.2.3, are obtained. The structure/pattern matrix is*

$$\mathbf{F}_{\Sigma} = \begin{bmatrix} 0.93996 & -0.30571 & 0.15171 \\ 0.66799 & 0.57051 & 0.47781 \\ 0.91994 & 0.19207 & -0.34180 \end{bmatrix}$$

Using (8.2.9), the proportion of total univariate variance accounted for by each component follows

Proportion	0.7959	0.1112	0.0927
Cumulative	0.7959	0.9071	1.0000

For the population correlation matrix, the equation $|\ \mathbf{P}_\rho - \delta \mathbf{I}\ | = 0$ *is solved. For our example, the results follow.*

$$\Delta = \text{diag}\left[\ 2.2331, \quad 0.5227, \quad 0.2442\ \right]$$

$$\mathbf{P} = \begin{bmatrix} 0.59543 & -0.42647 & 0.68087 \\ 0.52896 & 0.84597 & 0.06730 \\ 0.60470 & -0.32008 & -0.72931 \end{bmatrix}$$

$$\mathbf{F}_\rho = \begin{bmatrix} 0.88979 & -0.30832 & 0.33647 \\ 0.79046 & 0.61161 & 0.03326 \\ 0.90364 & -0.23141 & -0.36040 \end{bmatrix}$$

Proportion	0.7444	0.1742	0.0814
Cumulative	0.7444	0.9186	1.0000

Comparing the two results, we see that even though the matrices $\mathbf{F}_\Sigma$ and $\mathbf{F}_\rho$ *are similar they are not the same. In particular, the signs on the second component are clearly not the same. One must decide before any application of principal component analysis which matrix to analyze. A principal component analysis of a covariance matrix gives more weight to variables with larger variances. A principal component analysis of a covariance matrix is equivalent to an analysis of a weighted correlation matrix, where the weight for each variable is equal to its sample variance. In general, variables with larger weights tend to have larger loadings on the first component and smaller residual correlations than variables with smaller weights. Conversely, a principal component analysis of a correlation matrix is equivalent to an unweighted analysis of a covariance matrix. For this example, two components account for most of the variance. For other problems, the number of important components retained when analyzing the two matrices may also be very different. The choice of which to analyze depends on the goals of the application. Program m8_2_1.sas was used to perform the calculations in Example 8.2.1 employing both the PROC FACTOR and the PROC PRINCOMP. In principal component analysis, the variance explained by a component is equal to the eigenvalue of either the covariance or the correlation matrix used in the analysis. In PROC FACTOR, observe that the variance explained by each factor is labeled weighted or unweighted when one uses the covariance matrix. The usual method for computing the variance accounted for by a factor is to take the sum of squares of the corresponding column of the factor pattern matrix output by PROC FACTOR, resulting in an unweighted variance. If the square of each loading is multiplied by the variance of the variable (the weight) before the sum is taken, the result is the weighted variance explained by a component, the eigenvalue of the covariance matrix. The unweighted variance accounted for by each factor when one is analyzing the covariance matrix is equivalent to the analysis one obtains by analyzing the correlation matrix This is confirmed in our example by comparing the eigenvalues obtained in the analysis of the correlation matrix with the unweighted variance explained by each factor when we analyzed the covariance matrix. The values are 2.233, 0.523, 0.244 and 2.176, 0.456, 0.368, respectively.*

c. *Principal Components with Covariates*

In many applications, one may have a set of variables $\mathbf{Y}' = [Y_1, Y_2, \ldots, Y_p]$ and a set of covariates, $\mathbf{X}' = [X, X, \ldots, X_q]$. For this situation we would again like to find components $Z_j = \mathbf{p}_j'\mathbf{Y}$ such that the joint set of variables Z_j and $\mathbf{X}$ accounts for the variation in $\mathbf{Y}$. We now consider the joint covariance matrix

$$\Sigma = \begin{bmatrix} \Sigma_{yy} & \Sigma_{yx} \\ \Sigma_{xy} & \Sigma_{xx} \end{bmatrix} \tag{8.2.23}$$

Regressing $\mathbf{Y}$ on $\mathbf{X}$, the matrix

$$\begin{aligned} \Sigma_{y.x} &= \Sigma_{yy} - \Sigma_{yx}\Sigma_{xx}^{-1}\Sigma_{xy} \\ &= [\sigma_{ij.\,p+1,\ldots,p+q}] \end{aligned} \tag{8.2.24}$$

is the residual matrix of partial variances and covariances. Since Σ_{yy} is the covariance matrix for $\mathbf{Y}$, the matrix

$$\Sigma_{yx}\Sigma_{xx}^{-1}\Sigma_{xy} = [\alpha_{ij}] \tag{8.2.25}$$

is the dispersion in **Y** explained by **X**. Thus, we are led to consider the eigenequation

$$\left|\Sigma_{y.x} - \theta \mathbf{I}\right| = 0 \tag{8.2.26}$$

given a set of covariates **X** and a set of variables **Y**. Using equation (8.2.26), we find its eigenvalues $\theta_1, \theta_2, \ldots, \theta_p$ and eigenvectors $\mathbf{p}_1, \mathbf{p}_2, \ldots, \mathbf{p}_p$. The quantities

$$Z_1 = \mathbf{p}_1'\mathbf{Y},\ Z_2 = \mathbf{p}_2'\mathbf{Y}, \ldots, Z_p = \mathbf{p}_p'\mathbf{Y} \tag{8.2.27}$$

are defined as the PCs of **Y** given **X** or partial principal components (PPCs). Now, letting

$$\mathbf{P} = \left[\mathbf{p}_1, \mathbf{p}_2, \ldots, \mathbf{p}_p\right] \text{and } \Theta = \text{diag}\left[\theta_i\right]$$

we have that

$$\begin{aligned}
\mathbf{P}'\Sigma_{y\,x}\mathbf{P} &= \Theta \\
\Sigma_{yy} - \Sigma_{y\,x}\Sigma_{xx}^{-1}\Sigma_{x\,y} &= \mathbf{P}\Theta\mathbf{P}' \\
\Sigma_{yy} &= \Sigma_{yx}\Sigma_{xx}^{-1}\Sigma_{xy} + \mathbf{P}\Theta\mathbf{P}'
\end{aligned} \tag{8.2.28}$$

Equation (8.2.28) suggests that the variation in **Y** may be partitioned into variation due to **X** and that due to the PPCs. Now, the total univariate variance of **Y** may be written as

$$\begin{aligned}
\text{tr}\left(\Sigma_{yy}\right) &= \text{tr}\left(\Sigma_{yx}\Sigma_{xx}^{-1}\Sigma_{xy}\right) + \text{tr}\left(\mathbf{P}\Theta\mathbf{P}'\right) \\
\sigma_1^2 + \sigma_2^2 + \cdots + \sigma_p^2 &= \alpha_{11} + \alpha_{12} + \cdots + \alpha_{pp} + \left(\theta_1 + \theta_2 + \cdots + \theta_p\right)
\end{aligned} \tag{8.2.29}$$

To obtain a value of $k \ll p$, we may use the ratio

$$\rho_k^2 = \frac{\alpha_{11} + \alpha_{22} + \cdots + \alpha_{pp} + \theta_1 + \theta_2 + \cdots + \theta_k}{\sigma_1^2 + \sigma_2^2 + \cdots + \sigma_p^2} \tag{8.2.30}$$

following (8.2.19). Again we would like the ratio to account for 70% – 80% of the total variance.

Because we are using the covariance matrix $\Sigma_{y.x}$ to create components and the cov(**Y**, $\mathbf{p}_j'$ (**Y.X**)) $= \mathbf{p}_j'\Sigma_{1.2} = \theta_j\mathbf{p}_j$, we define the standardized partial principal components as

$$Z_j^* = \mathbf{P}_j'\left[\left(\mathbf{Y} - \boldsymbol{\mu}_y\right) - \Sigma_{y\,x}\Sigma_{xx}^{-1}\left(\mathbf{X} - \boldsymbol{\mu}_x\right)\right]/\sqrt{\theta_i} \tag{8.2.31}$$

Finally, letting

$$\mathbf{q}_j = \mathbf{p}_j\sqrt{\theta_j} \quad \text{and} \quad \mathbf{Q} = [\mathbf{q}_1, \mathbf{q}_2, \ldots, \mathbf{q}_k] \tag{8.2.32}$$

as we did when we had no covariates, Table 8.2.4 may be constructed to summarize the relationships between the PPCs $Z_1^*, Z_2^*, \ldots, Z_k^*$ and the variables $Y_1, Y_2, \ldots, Y_p$.

Table 8.2.4 summarizes the relationship between the partial principal components and the variables. The correlations may be used to interpret the partial principal components; the coefficients α_{ij} and β_{ij} represent the variance accounted for by X and the partial principal components, respectively. The square of the multiple correlations $(\alpha_{ii} + \beta_{ii})/\sigma_i$ permits

TABLE 8.2.4. Partial Principal Components

Variable	Z_1^*,	Z_2^*	, . . . ,	Z_k^*	X	Variance Explained by Components
Y_1	$\frac{q_{11}}{\sigma_1}$	$\frac{q_{12}}{\sigma_1}$	$\cdots$	$\frac{q_{1k}}{\sigma_1}$	α_{11}	$q_{11}^2 + \cdots + q_{1k}^2 = \beta_{11}$
Y_2	$\frac{q_{21}}{\sigma_2}$	$\frac{q_{22}}{\sigma_2}$	$\cdots$	$\frac{q_{2k}}{\sigma_2}$	α_{22}	$q_{21}^2 + \cdots + q_{2k}^2 = \beta_{22}$
$\vdots$	$\vdots$	$\vdots$		$\vdots$	$\vdots$	$\vdots$ $\vdots$
Y_p	$\frac{q_{p1}}{\sigma_p}$	$\frac{q_{p2}}{\sigma_p}$	$\cdots$	$\frac{q_{pk}}{\sigma_p}$	α_{pp}	$q_{p1}^2 + \cdots + q_{pk}^2 = \beta_{pp}$

one to evaluate the fit by replacing the original variable Y_i with a subset of partial principal components. The reconstructed observation vector $\mathbf{Y}$ for $\mathbf{Q}$ with $k \ll p$ columns is

$$\widetilde{\mathbf{Y}} = \boldsymbol{\mu}_y + \Sigma_{y\,\mathrm{x}}\Sigma_{\mathrm{xx}}^{-1}\Sigma_{\mathrm{x}\,y}\,(\mathbf{X} - \boldsymbol{\mu}_{\mathrm{x}}) + \mathbf{Q}\mathbf{Z}^* \tag{8.2.33}$$

where

$$\mathbf{Z}^* = \Theta^{-1/2}\mathbf{P}'\left[(\mathbf{Y} - \boldsymbol{\mu}_y) - \Sigma_{y\,\mathrm{x}}\Sigma_{\mathrm{xx}}^{-1}\Sigma_{\mathrm{x}\,y}\,(\mathbf{X} - \boldsymbol{\mu}_{\mathrm{x}})\right] \tag{8.2.34}$$

which reflects both the variation due to $\mathbf{X}$ and the partial principal components.

As in the situation with principal components with no covariates, the PPCs are not invariant to linear transformations of $\mathbf{Y}$. However, to standardize variables and components, the population matrix of partial correlations may be used to obtain eigenvectors and eigenvalues.

d. *Sample PCA*

In the preceding section we discussed the theory of PCA as it applied to an infinite population of measures. In practice, a random sample of n individuals are obtained on p variables. The data for a PCA consists of an $(n \times p)$ data matrix $\mathbf{Y}$ and an $(n \times q)$ data matrix $\mathbf{X}$ of q covariates. To perform a PCA, one employs the unbiased estimator $\mathbf{S}$ for Σ or the sample correlation matrix $\mathbf{R}$. Selecting between $\mathbf{S}$ and $\mathbf{R}$ depends on whether the measurements are commensurate. If the scales of measurements are commensurate, one should analyze $\mathbf{S}$, otherwise $\mathbf{R}$ is used. Never use $\mathbf{R}$ if the scales are commensurate since by forcing all variables to have equal sample variance one may not be able to locate those components that maximize the sample dispersion.

Replacing Σ with $\mathbf{S}$, one solves $|\mathbf{S} - \lambda\mathbf{I}| = 0$ to obtain eigenvalues and eigenvectors usually represented as $\widehat{\lambda}_j$ and $\widehat{\mathbf{p}}_j$. Thus, the sample eigenvectors become $\widehat{\mathbf{P}}$ and the sample eigenvalues become $\widehat{\Lambda} = \mathrm{diag}[\widehat{\lambda}_j]$. Replacing population values with sample estimates, one may construct Tables 8.2.1, 8.2.2, and 8.2.3. Because $\mathbf{Z}^*$ is a matrix, the formula for the standardized principal components for all n individuals is

$$\mathbf{Z}_{n\times p}^* = \mathbf{Y}_d\widehat{\mathbf{P}}\widehat{\Lambda}^{-1/2} = \mathbf{Y}_d\widehat{\mathbf{Q}} \tag{8.2.35}$$

where $\mathbf{Y}_d$ is the matrix of deviation scores after subtracting the sample means and $\widehat{\mathbf{Q}}$ are the sample covariance loading given in Table 8.3.2. Alternatively the data matrix $\mathbf{Y}_d$ becomes

$$\begin{aligned}\mathbf{Y}_d &= \mathbf{Z}^* \left(\widehat{\mathbf{P}}\widehat{\Lambda}^{1/2}\right)^{-1} \\ &= \mathbf{Z}^* \widehat{\Lambda}^{-1/2}\widehat{\mathbf{P}}' \\ &= \mathbf{Z}^*\widehat{\mathbf{Q}}'\end{aligned} \tag{8.2.36}$$

Letting $\mathbf{D}^{-1/2} = \left[\text{diag}\,\mathbf{S}\right]^{-1/2}$, the sample correlation matrix is used in the analysis so that the equation $|\mathbf{R} - \delta\mathbf{I}| = 0$ is solved. Again we let $\widehat{\mathbf{P}} = [\widehat{\mathbf{p}}_1, \widehat{\mathbf{p}}_2, \ldots, \widehat{\mathbf{p}}_k]$ be the sample eigenvectors of $\mathbf{R}$. Then,

$$\mathbf{Z}^* = \mathbf{Z}_d\widehat{\mathbf{P}}\widehat{\Delta}^{-1/2} = \mathbf{Z}_d\widehat{\mathbf{Q}} \tag{8.2.37}$$

Here, $\widehat{\mathbf{Q}}$ is a matrix of correlations. To obtain the variables from the PCs, the equation becomes

$$\begin{aligned}\mathbf{Z}_d &= \mathbf{Z}^* \left(\widehat{\mathbf{P}}\widehat{\Delta}^{-1/2}\right)^{-1} \\ &= \mathbf{Z}^* \widehat{\Delta}^{1/2}\widehat{\mathbf{P}}' \\ &= \mathbf{Z}^*\widehat{\mathbf{Q}}'\end{aligned} \tag{8.2.38}$$

To evaluate how many components to retain when using either $\mathbf{S}$ or $\mathbf{R}$ is complicated since $\widehat{\lambda}_j$ and $\widehat{\delta}_j$ are only estimates of λ_j and δ_j. Assuming that Σ and $\mathbf{P}_\rho$ are nonsingular, the dimensionality of the component space is k if $\lambda_{k+1} = \lambda_{k+2} = \ldots = \lambda_p = 0$ or it is m if $\delta_{m+1} = \delta_{m+2} = \ldots = \delta_p = 0$. In general, $k \leq m$ and is equal to m only if $\sigma_1^2 = \sigma_2^2 = \ldots = \sigma_p^2 = 1$. Before developing formal tests of dimensionality, we consider some ad hoc methods.

Two popular rules of thumb are to use either the geometric mean or the arithmetic mean of the sample eigenvalues. That is to retain a component if

$$\begin{aligned}&(a)\ \widehat{\lambda}_j > \overline{\lambda}_g = \left(\prod_{i=1}^{p}\widehat{\lambda}_i\right)^{1/p} \quad \text{or} \quad \widehat{\lambda}_j > \overline{\lambda} = \sum_{i=1}^{p}\widehat{\lambda}_i/p \\ &(b)\ \widehat{\delta}_j > \overline{\delta}_g = \left(\prod_{i=1}^{p}\widehat{\delta}_i\right)^{1/p} \quad \text{or} \quad \widehat{\delta}_j > \overline{\delta} = \sum_{j=1}^{p}\widehat{\delta}_j/p = 1\end{aligned} \tag{8.2.39}$$

when analyzing $\mathbf{S}$ or $\mathbf{R}$, respectively. Using the simple average, we are saying that one should retain a component whose estimated variance is larger than the average variance of all the original variables. The geometric mean adjusts for outliers in the sample estimates and with no outliers is less than or equal to the mean. There is no sound statistical basis for these rules. For example, when analyzing $\mathbf{R}$ retaining components when $\widehat{\delta}_j > 1$ for m components may cumulatively account for more variation than the original variables, an absurd result.

A more parsimonious rule would be to retain a number that accounts for 70% – 80% of the total univariate variance. To help evaluate the percentage, one may construct a scree plot. The plot was suggested by Cattell (1966). The plot is a graph that compares the estimated eigenvalue versus the number of components k ($k = 1, 2, \ldots, p$). The graph usually has a sharp elbow where both the smaller roots and the larger roots lie on a line. Fitting

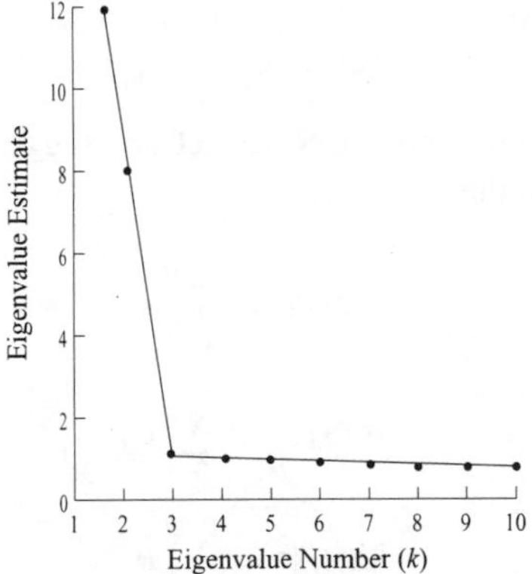

FIGURE 8.2.1. Ideal Scree Plot

a spline regression model to the sample roots, one is estimating the location of the knot. A graph of an ideal scree plot is shown in Figure 8.2.1. The graph reflects a sharp decline in estimated roots where the smaller (lower) and larger (upper) roots may be estimated by two lines. Bentler and Yuan (1996) have developed a maximum likelihood test that the lower roots lie on a line. Jolliffe (1986) and Basilevsky (1994) discuss other methods using resampling techniques to evaluate how many components to retain.

When covariates are included in a PCA, the covariance matrix for a sample has the general structure

$$\mathbf{S} = \begin{bmatrix} \mathbf{S}_{yy} & \mathbf{S}_{y\,\mathrm{x}} \\ \mathbf{S}_{\mathrm{x}\,y} & \mathbf{S}_{\mathrm{xx}} \end{bmatrix} \tag{8.2.40}$$

where the sample residual covariance matrix is

$$\mathbf{S}_{y.x} = \mathbf{S}_{yy} - \mathbf{S}_{yx}\mathbf{S}_{xx}^{-1}\mathbf{S}_{xy} \tag{8.2.41}$$

Letting

$$\mathbf{S}_{yx}\mathbf{S}_{xx}^{-1}\mathbf{S}_{xy} = [a_{ij}] \tag{8.2.42}$$

and $\widehat{\theta}_i$ the eigenvalues $| \ \mathbf{S}_{y.x} - \theta\mathbf{I} \ | = 0$, the standardized partial principal components have the form

$$\mathbf{Z}^* = \left(\mathbf{Y}_d - \mathbf{X}_d\mathbf{S}_{xx}^{-1}\mathbf{S}_{xy}\right)\widehat{\mathbf{P}}\widehat{\Theta}^{-1/2} \tag{8.2.43}$$

where $\mathbf{Y}_d$ and $\mathbf{X}_d$ are deviation matrices. To evaluate the number of roots to retain, one may investigate the sample percentages of the ratio given in (8.2.30). Because $\mathrm{tr}\left(\mathbf{S}_{yx}\mathbf{S}_{xx}^{-1}\mathbf{S}_{xy}\right)$ and $\sum_{i=1}^{p} s_{ii}$ is constant, one may also construct scree plots for $\widehat{\theta}_i$. Letting $\mathbf{D}^{-1/2} = \left[\mathrm{diag}\,\mathbf{S}_{y.x}\right]^{-1/2}$, one may analyze $\mathbf{R}_{y.x}$ using standardized variables for both $\mathbf{Y}$ and $\mathbf{X}$.

McCabe (1984) uses the idea of partial principal components to delete variables from a PCA. Partitioning a set of dependent variables as $\mathbf{Y}' = \left[\mathbf{Y}_1', \mathbf{Y}_2'\right]$ where $\mathbf{Y}_1$ contains q variables and $\mathbf{Y}_2$ contains $p - q$ variables, he investigates the eigenvalues of the residual covariance matrix

$$\mathbf{S}_{2.1} = \mathbf{S}_{22} - \mathbf{S}_{21}\mathbf{S}_{11}^{-1}\mathbf{S}_{12} \tag{8.2.44}$$

where

$$|\mathbf{S}| = |\mathbf{S}_{11}|\,|\mathbf{S}_{2.1}| \tag{8.2.45}$$

Letting $\widehat{\theta}_1, \widehat{\theta}_2, \ldots, \widehat{\theta}_{p-q}$ be the roots of $\mathbf{S}_{2.1}$, McCabe suggests evaluating $2^p - 1$ choices for q. Given q, we have that the

$$|\mathbf{S}_{2.1}| = \prod_{j=1}^{p-q} \widehat{\theta}_j$$

$$\operatorname{tr} |\mathbf{S}_{2.1}| = \sum_{j=1}^{p-q} \widehat{\theta}_j$$

$$\|\mathbf{S}_{2.1}\|^2 = \sum_{j=1}^{p-q} \widehat{\theta}_j^2$$

Depending on the criterion selected, McCabe shows that one should discard the subset of variables in $\mathbf{Y}_2$ for which the selected criterion is minimal. To visualize the criteria one can construct a plot. Jolliffe (1972, 1973) suggests using PCs to select subsets of variables for prediction.

e. Plotting Components

Having completed a component analysis using $\mathbf{S}$, it is often informative to plot the correlations between the components and the variables in a two dimensional component space. This allows one a visual evaluation of determining how variables cluster.

If one is analyzing $\mathbf{R}$, both PC scores and variables are normalized to one. Hence, one may create a two-dimensional joint plot of loadings and scores. Such a plot is termed a biplot by Gabriel (1971). Friendly (1991) has developed a SAS MACRO to generate biplots.

f. Additional Comments

We have assumed that PC analysis only applies to one group, Krzanowski (1982) suggests procedures for comparing PCs across several groups while Keramidas, Devlin and Gnanadesikan (1987) develop Q-Q plots for comparing components. Hastie and Stuetzle (1989) extend principal components to principal curves. Principal component analysis using discrete data is called correspondence analysis. Greenacre (1984), van Rijckevarsel and de Leeuw (1988), and Blasius and Greenacre (1998) discuss this topic. Principal component analysis with constraints and covariates for both subjects and variables are discussed by Takane and Shibayama (1991) and Takane, Kiers and de Leeuw (1995).

g. Outlier Detection

Principal component analysis is sensitive to the presence of outliers. An extreme outlier may generate a single component. We conclude this section with an example to show how principal component analysis may be used to detect outliers in a $n \times p$ data matrix. If one cannot remove the outlier from the data matrix, one may use a robust estimate of the covariance matrix or the correlation matrix. In Section 8.3, we show how principal component analysis is used to explore data structures defined by a covariance or correlation matrix.

Example 8.2.2 (Outlier Detection) *As cheese matures, a chemical process takes place to give it flavor. The data set cheese.dat from Moore and McCabe (1993, p. 792) contains measurements of acetic acid (Acetic), hydrogen sulfide (H2S) and lactic (Lactic) acid and an index of taste for a sample of mature cheddar cheese. The taste variable was obtained based upon several tasters. The variables Acetic and H2S are the natural logarithm of Acetic and H2S concentrations, respectively. We use this data to determine whether there are outliers in the* $n = 30$ *observations using the method of principal components. The data set and SAS code for this example are given in program m8_2_2.sas.*

To determine outliers for these data, expression (8.2.17) is evaluated for each observation. Using the SAS procedure PRINCOMP, one PC accounts for 99% of the total variance. Hence, $p - k = 4 - 1 = 3$ *is the dimension of the residual space. By using the STD option on the PRINCOMP statement, the unit length eigenvectors are divided by the square root of the eigenvalues to create component scores* Z_j^**. Hawkin's* D_i^2 *may be calculated by using the SAS function USS. The distances are output and compared to a chi-square critical value using* $\alpha = 0.05$ *with degrees of freedom equal one,* $\chi^2_{1-\alpha}(3) = 7.81473$*. Comparing each* D_i^2 *with the critical value, no vector observation appears to be an outlier.*

Exercises 8.2

1. For Example 8.2.1, use (8.2.22) to evaluate how well $k = 2$ components predict each variable. These multiple correlation coefficients are called the final communality estimates in PROC FACTOR. They are the sum of squares of each row of the factor pattern matrix in SAS. Run example m8_2_1.sas with NFACT = 2.

2. For the population correlation matrix

$$\mathbf{P}_\rho = \begin{bmatrix} 1.00 & & & & \\ .086 & 1.000 & & & \\ -.031 & .187 & 1.000 & & \\ -.034 & .242 & .197 & 1.000 & \\ .085 & .129 & .080 & .327 & 1.000 \end{bmatrix}$$

   based upon 151 observations, construct Table 8.2.3 for two eigenvalues larger than 1.

   (a) What is the correlation between the second component and the first variable?

   (b) What proportion of the total variance is accounted for by the first and second components?

   (c) Express the component as a linear combination of the five variables such that the mean of the component scores is 0 and the variance 1.

3. Use the Principal Component Analysis method as outlined in Example 8.2.2 to identify outliers in the Ramus data set (Ramus.dat) examined in Example 3.7.3.

4. For the multivariate regression data in Table 4.3.1, Chapter 4, how many partial principal components are needed to account for 70%-90% of the univariate variance of $\mathbf{Y}$ given $\mathbf{X} = \{N, S, NS, NA, SS\}$. In the SAS procedures FACTOR or PRINCOMP, the partial covariance or correlation matrix is analyzed using the PARTIAL command.

5. In Example 8.2.2, we computed D_i^2 using Hawkin's method to test for outliers using the covariance matrix and $\alpha = 0.05$. What happens if you use $\mathbf{R}$? Discuss your findings.

6. For the data in Example 8.2.2, calculate $\left(Q_i/\widehat{\theta}_1\right)^{ho}$ using (8.2.16) to evaluate whether any vector observation is an outlier.

8.3 Principal Component Analysis Examples

As an exploratory data analysis technique, principal component analysis may be used to detect outliers, to uncover data structures that account for a large percentage of the total variance, and to create new hypothetical constructs that may be employed to predict or classify observations into groups. In the social sciences, PCA is often employed in the analyses of tests or questionnaires. Often a test battery or a questionnaire is administered to individuals to characterize and evaluate their behaviors, attitudes or feelings. PCA is used to uncover complex dimensions of the instrument based on the total variance accounted for by the constructs. Insight into the data set is achieved by analyzing either a sample covariance or correlation matrix; when the scales of subtests are not commensurate, the correlation matrix is used.

a. *Test Battery (Example 8.3.1)*

Shin (1971) collected data on intelligence, creativity and achievement for 116 students in the eleventh grade in suburban Pittsburgh. The Otis Quick Scoring Mental Ability Test, Guilford's Divergent Productivity Battery, and Kropp and Stoker's Lisbon Earthquake Achievement Test were used to gather one IQ score, six creativity measures, and six achievement measures for each subject. In addition to the IQ variable (1), the variables included in the creativity test were ideational fluency (2), spontaneous flexibility (3), associational fluency (4), expressional fluency (5), originality (6), and elaboration (7). The achievement measures were knowledge (8), comprehension (9), application (10), analysis (11), synthesis (12), and evaluation (13). The correlation matrix for the study is presented in Table 8.3.1.

Program m8_3_1.sas was used to analyze the correlation matrix in Table 8.3.1 using the SAS procedure FACTOR. The commands METHOD = PRIN and PRIORS = ONE are required for a PCA. The keyword SCREE requests a scree plot. Coefficients in the pattern matrix whose absolute value is larger than 0.40 are marked with an asterisk (*).

TABLE 8.3.1. Matrix of Intercorrelations Among IQ, Creativity, and Achievement Variables

Tests													
1	1.00												(Sym)
2	.16	1.00											
3	.32	.71	1.00										
4	.24	.12	.12	1.00									
5	.43	.34	.45	.43	1.00								
6	.30	.27	.33	.24	.33	1.00							
7	.43	.21	.11	.42	.46	.32	1.00						
8	.67	.13	.27	.21	.39	.27	.38	1.00					
9	.63	.18	.24	.15	.36	.33	.26	.62	1.00				
10	.57	.08	.14	.09	.25	.13	.23	.44	.66	1.00			
11	.59	.10	.16	.09	.25	.12	.28	.58	.66	.64	1.00		
12	.45	.13	.23	.42	.50	.41	.47	.46	.47	.37	.53	1.00	
13	.24	.08	.15	.36	.28	.21	.26	.30	.24	.19	.29	.58	1.00

The output in Table 8.3.2 is formatted using the Output Delivery System (ODS) in SAS and PROC FORMAT. The coefficients are rounded and multiplied by 100 (Version 8 no longer supports the ROUND and FLAG options available in Version 6.12).

Examination of the results in Table 8.3.2 indicates that the first component is a general measure of mental ability where only one measure, ideational fluency, has a correlation considerably lower than 50. The second component is bipolar, comparing low levels of creativity with low levels of achievement. Note that the correlations of both synthesis and evaluation with component two are low. This second component could be termed verbal creativity. The third component compares higher levels of creativity and achievement with lower levels of creativity and achievement; thus, component three is a higher-level cognitive component consisting of high levels of creativity and achievement.

Since the total variance acquired by the three components is only 64.4% using the components for data reduction is not too meaningful. The analysis does lead to a better understanding of creativity, achievement, and intelligence variables in an experimental setting in that, perhaps creativity may be considered a learning aptitude that might affect the relationship between achievement and general intelligence.

PROC FACTOR retained only those components with sample eigenvalues larger than 1. Note that the correspondences scree plot in Figure 8.3.1 has an elbow between the third and fourth roots or possibly between the fourth and fifth roots. A researcher should not be bound by a strict rule for determining the number of components, but allow flexibility in PCA. In this example, with a fourth component, over 70% of the total variance is accounted for in the sample and a natural interpretation can be given the fourth component.

b. *Semantic Differential Ratings (Example 8.3.2)*

Di Vesta and Walls (1970) studied mean semantic differential ratings given by fifth-grade children for 487 words. The semantic differential ratings were obtained on the follow-

TABLE 8.3.2. Summary of Principal-Component Analysis Using 13 × 13 Correlation Matrix

	Tests	Components 1	2	3
1	(IQ)	79∗	−22	14
2	(Ideational fluency)	36	62∗	52∗
3	(Spontaneous flexibility)	47∗	56∗	54∗
4	(Associational fluency)	45∗	34	−55∗
5	(Expressional fluency)	66∗	38	−07
6	(Originality)	50∗	36	−02
7	(Elaboration)	59∗	18	−34
8	(Knowledge)	75∗	−23	08
9	(Comprehension)	76∗	−34	22
10	(Application)	64∗	−48∗	21
11	(Analysis)	71∗	−48∗	13
12	(Synthesis)	76∗	04	−36
13	(Evaluation)	51∗	09	−46∗
	Eigenvalues	5.11	1.81	1.45
	Percentage of total variance	39.3	13.9	11.2
	Cumulative percentage of total variance	39.3	53.2	64.4

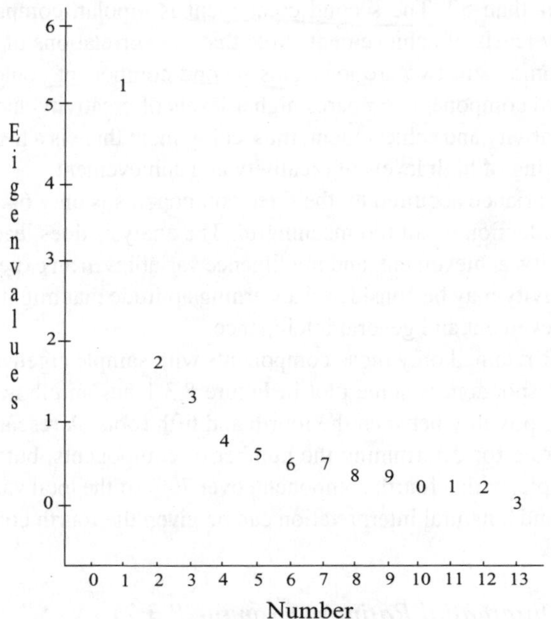

FIGURE 8.3.1. Scree Plot of Eigenvalues Shin Data

TABLE 8.3.3. Intercorrelations of Ratings Among the Semantic Differential Scale

Scale								
1	1.00							(Sym)
2	.95	1.00						
3	.96	.98	1.00					
4	.68	.70	.68	1.00				
5	.33	.35	.31	.52	1.00			
6	.60	.63	.61	.79	.61	1.00		
7	.21	.19	.19	.43	.31	.42	1.00	
8	.30	.31	.31	.57	.29	.57	.68	1.00

TABLE 8.3.4. Summary of Principal-Component Analysis Using 8 × 8 Correlation Matrix

		Components		
	Variables	1	2	3
1	(Friendly/Unfriendly)	87*	−42*	−13
2	(Good/Bad)	88*	−42*	−10
3	(Nice/Awful)	87*	−44*	−15
4	(Brave/Not Brave)	89*	10	07
5	(Big/Little)	58*	26	72*
6	(Strong/Week)	86*	19	22
7	(Moving/Still)	49*	70*	−29
8	(Fast/Slow)	61*	61*	−32
	Eigenvalues	4.76	1.53	.81
	Percentage of total variance	59.6	19.1	10.1
	Cumulative percentage of total variance	59.6	78.7	88.8

ing eight scales: friendly/unfriendly (1), good/bad (2), nice/awful (3), brave/not brave (4), big/little (5), strong/weak (6), moving/still (7), and fast/slow (8). Table 8.3.3 shows the intercorrelations among mean semantic differential ratings for one list of 292 words.

Using program m8_3_2.sas, the correlation matrix in Table 8.3.3 is analyzed. To extract three components, the statement NFACT = 3 is needed. The output is provided in Table 8.3.4. The first component in Table 8.3.4 represents an overall response-set component of the subjects to the list of words. The second component is bipolar, comparing evaluative behavior (friendly/unfriendly, good/bad, and nice/awful) with activity judgements (moving/still and fast/slow). The third component is dominated by fourth variable (big/little), with a weight of 72*, and may be termed a size component. Again, coefficient larger than 0.40 in absolute value are output with an asterik (*).

Often components are rotated using an orthogonal matrix **T** of weights, even though the maximum variance criterion is destroyed by such transformations. This is done because being able to interpret the "rotated" components is more important than preserving maximum variance. In SAS, rotation is accomplished using the statement ROTATE = VAXIMAX. For

TABLE 8.3.5. Covariance Matrix of Ratings on Semantic Differential Scales

Scale								
1	1.44							(Sym)
2	1.58	1.93						
3	1.55	1.83	1.82					
4	.61	.73	.69	.56				
5	.23	.28	.24	.23	.34			
6	.56	.68	.64	.46	.28	.61		
7	.19	.20	.17	.24	.13	.25	.56	
8	.21	.25	.24	.24	.10	.25	.29	.33

TABLE 8.3.6. Summary of Principal-Component Analysis Using 8 × 8 Covariance Matrix

		Components	
	Variables	1	2
1	(Friendly/Unfriendly)	97*	-13
2	(Good/Bad)	98*	−12
3	(Nice/Awful)	98*	−16
4	(Brave/Not Brave)	78*	42*
5	(Big/Little)	41*	47*
6	(Strong/Weak)	72*	51*
7	(Moving/Still)	28	79*
8	(Fast/Slow)	41*	70*
	Eigenvalues	5.76	.95
	Percentage of total variance	75.6	12.5
	Cumulative percentage of total variance	75.6	88.4

this example, rotation does not facilitate interpretation. Rotation is discussed in more detail in Section 8.9.

We also analyzed the sample covariance matrix for the data presented in Table 8.3.5. The output is provided in Table 8.3.6. Comparing the component analysis using **S**, with the analysis using **R**, we note that used only two components are required to account for the same proportion of variance in the sample using **R**. The first component is again a general response-set component, but in this case it is dominated by evaluation behavior. The second component is an activity component.

Using the first two eigenvectors of $|\mathbf{S} - \boldsymbol{\lambda}\mathbf{I}| = 0$, the first two components are represented by

$$
\begin{aligned}
C_1 &= .484\,(y_1 - \overline{y}_1) + .568\,(y_2 - \overline{y}_2) + .550\,(y_3 - \overline{y}_3) + .243\,(y_4 - \overline{y}_4) \\
&\quad + .100\,(y_5 - \overline{y}_5) + .232\,(y_6 - \overline{y}_6) + .087\,(y_7 - \overline{y}_7) + .097\,(y_8 - \overline{y}_8) \\
C_2 &= -.156\,(y_1 - \overline{y}_1) - .178\,(y_2 - \overline{y}_2) - .218\,(y_3 - \overline{y}_3) + .323\,(y_4 - \overline{y}_4) \\
&\quad + .284\,(y_5 - \overline{y}_5) + .409\,(y_6 - \overline{y}_6) + .610\,(y_7 - \overline{y}_7) + .415\,(y_8 - \overline{y}_8)
\end{aligned}
$$

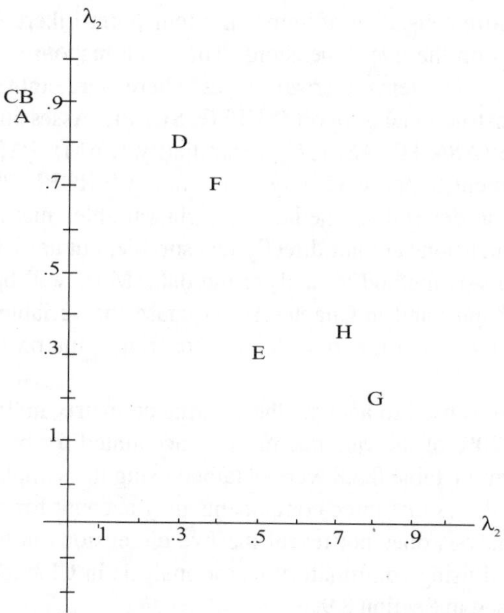

FIGURE 8.3.2. Plot of First Two Components Using **S**

for the raw data matrix given by Di Vesta and Walls.

After a PCA, it is often helpful to plot the structure matrix in the component space. This allows evaluation of those variables that tend to be associated with one another. For example, using the matrix **S** in the Di Vesta and Walls study, Figure 8.3.2 shows such a plot. Inspection of the plot indicates that variables 1, 2, and 3, here represented as A, B, and C, form a cluster while the rest of the variables do not form a distinct cluster. In PROC FACTOR, the plot is obtained by using the NPLOT statement.

c. *Performance Assessment Program (Example 8.3.3)*

This example examines the impact of a Performance Assessment Program (PAP). A questionnaire was designed to evaluate the positive and negative impact that the Program had on a teachers' instructional and assessment methods. The goal of the PAP questionnaire, administered to 265 teachers in elementary school in grades 3, 5, and 8, was to "measure" the following five constructs, Hansen (1999).

1. Teachers' support for the PAP.
2. Teachers' emphasis on outcomes/change in instruction and assessment.
3. Teachers' familiarity with PAP.
4. PAP's impact on instruction/assessment.
5. PAP's impact on professional development.

The PAP questionnaire consisting of items on a four-point Likert Scale, was developed to provide information on the five dimensions. The mean responses of the teachers were calculated from responses to items to create scales. There were eight variables for the five scales: General and Instructional Support (SUPPG, SUPPI), Assessment and Instructional Emphasis and change (ASMTE, ASMTC), Familiarity (FAM), PAP Impact (PAP), and Professional Development Support Activity and Amount (PROF1, PROF2). The primary goal of the study was to determine whether the eight variables measured the five dimensions. Because the dimensions are not directly measurable, but are latent traits, one might employ the factor analysis method to analyze the data. More will be said about this approach later in this chapter and in Chapter 10. Because the variables were created from mean responses over items on a Likert scale, the covariance matrix in Table 8.3.7 is used in the PCA.

Program m8_3_3.sas is used to analyze the covariance matrix in Table 8.3.7. Using the criterion that at least 70% of the variance must be accounted for by the components, the three components given in Table 8.3.8 were obtained using the sample covariance matrix.

The pattern matrix shows that three components may account for over 70% of the variance; however, the analysis does not reveal the five dimensions in the PAP study. These data will be reanalyzed using confirmatory factor analysis in Chapter 10 and exploratory factor analysis discussed in Section 8.9.

Instead of analyzing $\mathbf{S}$, suppose we chose to analyze $\mathbf{R}$ for these data. The output is provided in Table 8.3.9.

Note that both solutions account for about 70% of the variance, the factor patterns for the two matrices are very different and neither recover the five dimensions in the questionnaire.

Exercises 8.3

1. Using a random sample of 502 twelfth-grade students from the Project Talent survey (supplied by William W. Cooley at the University of Pittsburgh), data were collected on 11 tests: (1) general information test, part 1, (2) general information test, part II, (3) English, (4) reading comprehension, (5) creativity, (6) mechanical reasoning, (7) abstract reasoning, (8) mathematics, (9) sociability inventory, (10) physical science interest inventory, and (11) office work interest inventory. Tests (1) through (5) were verbal ability tests, tests (6) through (8) were nonverbal ability tests, and tests (9) through (11) were interest measures (for a description of the variables, see Cooley and Lohnes, 1971). The correlation matrix for the 11 variables and 502 subjects is given in Table 8.3.10. Use PCA to reduce the number of variables in Table 8.3.9 and interpret your findings.

2. For the protein consumption data in the file protein.dat, described in Example 9.4.1, use the covariance matrix and correlation matrix for the food groups to reduce the consumption variables to a few components. Can you name the components and determine whether the variables tend to cluster by food group for both the covariance matrix and correlation matrix? Discuss you findings.

TABLE 8.3.7. PAP Covariance Matrix

Variable								
SUPPG	0.39600							(Sym)
SUPPI	0.24000	0.44100						
ASMTE	0.03310	0.04383	0.21400					
ASMTC	0.03983	0.05242	0.12200	0.24400				
FAM	0.09002	0.08361	0.08825	0.08792	0.360			
PAP	0.12700	0.15700	0.11300	0.11900	0.208	0.365		
PROF1	0.06587	0.09230	0.09137	0.09284	0.200	0.197	0.480	
PROF2	0.10800	0.14000	0.11000	0.06940	0.175	0.184	0.199	0.634

TABLE 8.3.8. Component Using **S** in PAP Study

	Components		
Variables	1	2	3
Supp 1	52∗	69∗	01
Supp 2	58∗	68∗	−02
Asmt 1	50∗	−21	17
Asmt 2	45∗	−14	34
Fam	70∗	−22	28
Impt	78∗	−02	28
Prof 1	68∗	−35	30
Prof 2	72∗	−23	−65∗
Eigenvalues	1.3041	0.5193	0.4034
% Variance	41.6	16.6	12.9
Cumulative	41.6	58.2	71.1

TABLE 8.3.9. PAP Components Using **R** in PAP Study

	Components		
Variables	1	2	3
Supp 1	51∗	69∗	21
Supp 2	55∗	66∗	20
Asmt 1	61∗	−43∗	41∗
Asmt 2	58∗	−40∗	54∗
Fam	72∗	−13	−34
Impt	81∗	−01	−06
Prof 1	65∗	−18	−42∗
Prof 2	60∗	−02	−35
Eigenvalues	3.2362	1.3151	0.9570
% Variance	40.1	16.4	12.0
Cumulative	40.5	56.9	68.9

TABLE 8.3.10. Project Talent Correlation Matrix

Scale											
1	1.000										(Sym)
2	.861	1.000									
3	.492	.550	1.000								
4	.698	.765	.613	1.000							
5	.644	.621	.418	.595	1.000						
6	.661	.519	.160	.413	.522	1.000					
7	.487	.469	.456	.530	.433	.451	1.000				
8	.761	.649	.566	.641	.556	.547	.517	1.000			
9	−.011	.062	.083	.021	.001	−.075	.007	.030	1000		
10	.573	.397	.094	.275	.340	.531	.202	.500	.055	1.000	
11	−.349	−.234	.109	−.087	−.119	−.364	−.079	−.191	.084	−.246	1.000

8.4 Statistical Tests in Principal Component Analysis

Distributional assumptions are not required when using principle component analysis as an exploratory data analysis tool to improve ones understanding of a set of p variables. However, to use PCA as a quasi-confirmatory procedure, tests of hypotheses and confidence intervals for population roots and vectors may be of interest. To perform statistical tests, we assume that the n p-vectors are sampled from a MVN distribution, $\mathbf{Y}_i \sim N_p(\boldsymbol{\mu}, \ \Sigma)$.

a. *Tests Using the Covariance Matrix*

Using a result due to Anderson (1963) and Girshick (1939), one may construct confidence intervals for the eigenvalues of Σ. They showed that as $n \longrightarrow \infty$

$$\frac{\lambda_j - \widehat{\lambda}_j}{\lambda_j \sqrt{2/(n-1)}} \overset{d}{\longrightarrow} IN(0, 1) \text{ for } j = 1, 2, \ldots, p \tag{8.4.1}$$

where the sample variance of $\widehat{\lambda}_j \approx \lambda_j \sqrt{2/(n-1)}$. Solving (8.4.1) for λ_j, an approximate $100(1-\alpha)\%$ confidence interval for λ_j is

$$\frac{\widehat{\lambda}_j}{1 + Z^{1-\alpha/2}\sqrt{2/(n-1)}} \leq \lambda_j \leq \frac{\widehat{\lambda}_j}{1 - Z^{1-\alpha/2}\sqrt{2/(n-1)}} \tag{8.4.2}$$

where $Z^{1-\alpha/2}$ is the upper $\alpha/2$ critical value for a standard normal variate provided in Table I in Appendix A. To control the overall error rate at some nominal level α, one may use the Bonferroni or Šidák inequalities to adjust the α-level for each root to obtain confidence sets for $k \ll p$ roots. For example, one may use $Z^{1-\alpha^*/2}$ where $\alpha^* = \alpha/k$ and $\alpha = 0.05$.

If the multiplicity of λ_j is m in the population, Anderson (1963) showed that an approximate confidence interval for λ_j is

$$\frac{\overline{\lambda}}{1 + Z^{1-\alpha/2}\sqrt{2/(n-1)\,m}} \leq \lambda_j \leq \frac{\overline{\lambda}}{1 - Z^{1-\alpha/2}\sqrt{2/(n-1)\,m}} \tag{8.4.3}$$

where $\overline{\lambda} = \sum_{j=1}^{m} \widehat{\lambda}_j / m$, the sample mean of the estimates of the roots with population multiplicity m.

Using (8.4.3), one may construct a confidence interval for the population average $\mu_\lambda = \sum_{j=1}^{p} \lambda_j / p$ estimated by $\overline{\lambda}$. Clearly, as $n \longrightarrow \infty$

$$\frac{\overline{\lambda} - \mu_\lambda}{\mu_\lambda \sqrt{2/p\,(n-1)}} \xrightarrow{d} N\,(0,\ 1) \tag{8.4.4}$$

so that a 100 $(1 - \alpha)$ % confidence interval for the "average" root criterion is

$$\frac{\overline{\lambda}}{1 + Z^{\alpha/2}\sqrt{2/p\,(n-1)}} \leq \mu_\lambda \leq \frac{\overline{\lambda}}{1 - Z^{\alpha/2}\sqrt{2/p\,(n-1)}} \tag{8.4.5}$$

Example 8.4.1 *For the Di Vesta and Walls' data, an approximate confidence interval for the first eigenvalue is*

$$\frac{5.774}{1 + 1.96\sqrt{2/191}} \leq \lambda_1 \leq \frac{5.774}{1 - 1.96\sqrt{2/291}}$$
$$4.97 \leq \lambda_1 \leq 6.89$$

Using (8.4.5), a 95% confidence interval for μ_λ *is*

$$\frac{31.449}{1 + 1.96\sqrt{2/8\,(191)}} \leq \mu_\lambda \leq \frac{31.449}{1 - 1.96\sqrt{2/8\,(191)}}$$
$$28.06 \leq \mu_\lambda \leq 35.76$$

Using (8.2.39) to retain components, one may want to choose $\overline{\lambda}$ *in the interval for* μ_λ.

If some of the elements of $\mathbf{p}_j$ are near zero in the population, one may want to test that the population covariance loading vector has a specified value $\mathbf{p}_{oj}$. To test the hypothesis

$$H : \mathbf{p}_j = \mathbf{p}_{oj} \tag{8.4.6}$$

for the jth distinct root, when all the roots are distinct $(\lambda_1 > \lambda_2 > \ldots > \lambda_p)$, we may again use a result due to Anderson (1963). He shows that $\sqrt{n-1}\,(\widehat{\mathbf{p}}_j - \mathbf{p}_{oj})$ has a limiting MVN distribution with mean zero and covariance matrix

$$\sum_{\substack{j=1 \\ j \neq j'}}^{p} \left[\frac{\lambda_{j'} \lambda_j}{(\lambda_{j'} - \lambda_j)^2} \right] \mathbf{p}_j \mathbf{p}_{j'}' \tag{8.4.7}$$

Anderson showed that

$$X^2 = (n-1)\left(\widehat{\lambda}_j \mathbf{p}'_{oj}\mathbf{S}^{-1}\mathbf{p}_{oj} + \widehat{\lambda}_j^{-1}\mathbf{p}'_{oj}\mathbf{S}\mathbf{p}_{oj} - 2\right) \tag{8.4.8}$$

converges asymptotically to a χ^2 distribution, when H is true, with $v = p - 1$ degrees of freedom. Thus, the null hypothesis in (8.4.6) is rejected at some level α if $X^2 > \chi^2_{1-\alpha}(p-1)$. Again, one must adjust the size of the test if one is interested in more than one eigenvector.

Testing that the eigenvectors in (8.4.6) have a specified value is only valid if the roots λ_j are distinct. Anderson (1963) developed a test to determine whether the roots are distinct. Before addressing this problem, recall that if a covariance matrix $\Sigma = \sigma^2\mathbf{I}$ then $\sigma_1^2 = \sigma_2^2 = \ldots = \sigma_p^2 = \sigma^2$ or equivalently $\lambda_1 = \lambda_2 = \ldots = \lambda_p = \lambda$ so that the rank of Σ or its dimensionality is one. This is the test of sphericity (3.8.27) discussed in Chapter 3. Replacing $|\mathbf{S}| = \prod_i \widehat{\lambda}_i$ and $\operatorname{tr}(\mathbf{S})/p = \overline{\lambda}$, the test criterion becomes, using Bartlett's correction,

$$\begin{aligned} X^2 &= \left[(n-1) - \left(2p^2+p+2\right)/6p\right]\left[p \log \overline{\lambda} - \log \textstyle\prod_{j=1}^p \widehat{\lambda}_j\right] \\ &\;\dot\sim\; \chi^2\left[(p-1)(p+2)/2\right] \end{aligned} \tag{8.4.9}$$

Rejecting the sphericity hypothesis, some subset of the σ_i^2 may remain equal or equivalently some subset of m adjacent roots may be equal. Anderson (1963) considered the problem of testing the equality of any subset of adjacent roots being equal when he developed the test of (8.4.6). This includes the test that all roots are equal or that only the smallest $m = p - k$ roots are equal. Modifying Anderson's procedure following Bartlett, the hypothesis

$$H : \lambda_{k+1} = \lambda_{k+2} = \ldots = \lambda_{k+m} \tag{8.4.10}$$

that any m roots of Σ are equal may be tested using

$$X^2 = \left[(n-k-1) - \left(2m^2+m+2\right)/6m\right]\left(m \log \overline{\lambda} - \sum_{j=k+1}^{k+m} \log \widehat{\lambda}_j\right) \tag{8.4.11}$$

where

$$\overline{\lambda} = \sum_{j=k+1}^{k+m} \widehat{\lambda}_j / m \tag{8.4.12}$$

When (8.4.10) is true, X^2 converges to a chi-square distribution with

$$v = (m-1)(m+2)/2$$

degrees of freedom. When $k = 0$, the test reduces to Bartlett's (1954) test of sphericity. If $k + m = p$ or $m = p - k$, (8.4.11) reduces to the criterion proposed by Bartlett (1950, 1954) and Lawley (1956) for testing the equality of the last $p - k$ roots

$$H : \lambda_{k+1} = \lambda_{k+2} = \ldots = \lambda_p \tag{8.4.13}$$

Using Lawley's (1956) correction factor, the test of (8.4.13) becomes

$$X^2 = \left\{ (n-k-1) - \left(2m^2+2m+2\right)/6m + \sum_{j=1}^{k} \left(\overline{\lambda}/\lambda_j - \overline{\lambda}\right)^2 \right\} \left[m \log \overline{\lambda} - \sum_{j=k+1}^{p} \log \widehat{\lambda}_j \right] \tag{8.4.14}$$

where $m = p - k$ which is similar to the factor suggested by Lawley in testing for dimensionality in discriminant analyses. James (1969) confirmed Lawley's result, but claims it may be conservative.

Given that one ignores $p - k = m$ components, one may want to evaluate whether the proportion of total univariate variance as defined in (8.2.19) is larger than ρ_o

$$H : \rho_k^2 \geq \rho_o \tag{8.4.15}$$

for $k = 1, 2, \ldots, p$ and $0 < \rho_o < 1$. Letting

$$\begin{aligned} R_k^2 &= \sum_{j=1}^{k} \widehat{\lambda}_j / \operatorname{tr}(\mathbf{S}) \\ \sigma^2 &= 2\left[\rho_o^2 \sum_{j=1}^{k} \lambda_j^2 + \left(1-\rho_o\right)^2 \sum_{j=k+1}^{p} \lambda_j^2\right] / \operatorname{tr}(\Sigma) \end{aligned} \tag{8.4.16}$$

Fujikoshi (1980) shows that

$$\sqrt{n}\left(R_k^2 - \rho_o\right) \xrightarrow{d} N\left(0,\ \sigma^2\right) \tag{8.4.17}$$

Substituting sample estimates for population parameter an approximate test of (8.4.15) results.

Example 8.4.2 *Employing Dr. Di Vesta and Walls' data, the sample roots of* **S** *are*

$$\widehat{\lambda}_1 = 5.7735 \quad \widehat{\lambda}_4 = 0.1869 \quad \widehat{\lambda}_7 = 0.0803$$

$$\widehat{\lambda}_2 = 0.9481 \quad \widehat{\lambda}_5 = 0.1167 \quad \widehat{\lambda}_8 = 0.0314$$

$$\widehat{\lambda}_3 = 0.3564 \quad \widehat{\lambda}_6 = 0.0967$$

To test the hypothesis $H_o : \lambda_2 = \lambda_3$ versus $H_1 : \lambda_2 \neq \lambda_3$, $k = 1$ and $m = 2$. Hence, formula (8.4.11) becomes

$$\begin{aligned} X^2 &= \left[(292-1-1) - \frac{2(2)^2+2+2}{6(2)}\right](2\log 0.6523 - \log 0.9481 - \log 0.3564) \\ &= 289\,(-.8547 + .05330 + 1.0317) \\ &= 66.56 \end{aligned}$$

Since $v = (m-1)(m+2)/2 = 2$, the critical value for the test is $\chi^2_{1-\alpha}(2) = 5.991$ for $\alpha = 0.05$. Thus, $\lambda_2 \neq \lambda_3$ so that the directions of λ_2 and λ_3 are distinct.

Replacing $(n-1)$ with $N = n - q - 1$ where q is the number of fixed covariates, tests of the roots and vectors of $\left|\Sigma_{y.x} - \theta \mathbf{I}\right| = 0$ given $\mathbf{X}$, follow directly.

b. *Tests Using a Correlation Matrix*

General tests regarding the population correlation matrix $\mathbf{P}_\rho$ and the roots and vectors of $\left|\mathbf{P}_\rho - \delta \mathbf{I}\right| = 0$ are complex. In Chapter 3 we developed a test of independence which we showed was equivalent to testing the null hypothesis that the population correlation matrix equals $\mathbf{I}$, $H : \mathbf{P}_\rho = \mathbf{I}$. We also showed how one may test that $\Sigma = \sigma^2 \left[(1-\rho)\,\mathbf{I} + \rho \mathbf{J}\right]$, has equal variances and equal covariances, i.e. that Σ has compound symmetry structure. To test the equivalent hypotheses using a correlation matrix

$$\begin{aligned} H &: \mathbf{P}_\rho = (1-\rho)\,\mathbf{I} + \rho \mathbf{J} \\ H &: \rho_{ij} = \rho \text{ for } i \neq j \\ H &: \delta_2 = \delta_3 = \ldots = \delta_p \end{aligned} \tag{8.4.18}$$

is more complicated. Bartlett (1950, 1951, 1954), Anderson (1963), Lawley (1963) and Aitken, Nelson and Reinfort (1968) suggested tests of (8.4.18). Gleser (1968) showed that only Lawley's test is asymptotically independent of ρ. For $p \leq 6$ Lawley's statistic converges to a chi-square distribution for n as small as 25 (Aitken, et al., 1968).

Lawley's statistic to test H defined in (8.4.18) is

$$X^2 = \frac{n-1}{\lambda^2} \left[\sum_{i<j}\sum \left(r_{ij} - \overline{r}\right)^2 - v \sum_k \left(\overline{r}_k - \overline{r}\right)^2 \right] \tag{8.4.19}$$

where

$$\begin{aligned} \overline{r}_k &= \frac{\sum_{i<k}^p r_{ik}}{p-1}, \quad \overline{r} = \frac{2\sum_{i<j} r_{ij}}{p\,(p-1)} \\ v &= (p-1)^2 \left(1 - \lambda^2\right) / \left[p - (p-2)\,\lambda^2\right] \end{aligned} \tag{8.4.20}$$

and $\lambda = 1 - \rho$ is substituted into (8.4.20) for ρ given in (8.4.18). Under H, $X^2 \xrightarrow{d} \chi^2(v)$ where $v = (p+1)(p-2)/2$.

The test statistic in (8.4.19) is used to test that the dimensionality of $\mathbf{P}_\rho$ is one. To test the hypotheses that the smallest $m = p - k$ roots are equal

$$H : \delta_{k+1} = \delta_{k+2} = \ldots = \delta_p \tag{8.4.21}$$

is more complicated. Lawley (1956) suggested a very approximate procedure provided $\delta_1, \delta_2, \ldots, \delta_k$ are large and $\overline{\delta} = \sum_{j=k+1}^p \delta_j / m$ is small, a condition that is difficult to verify. Lawley's test statistic is

$$X_k^2 = \left\{(n-k-1) - \left(2m^2 + 2m + 2\right)/6m + \sum_{j=1}^k \left(\overline{\delta}/\widehat{\delta}_j - \overline{\delta}\right)^2\right\}\left[m \log \overline{\delta} - \sum_{j=k+1}^p \log \widehat{\delta}_j\right] \tag{8.4.22}$$

Under H with $m = p - k$ and $\delta_k \gg \overline{\delta}$, $X_k^2 \xrightarrow{d} \chi^2(v)$ with $v = (m-1)(m+2)/2$ provided the ratio $n/p \approx 10$, $n > 100$ and $k \leq 4$. When this is not the case, an improved approximation developed by Schott (1988) may be used. Schott approximates the distribution of X_k^2 by $c\chi_d^2$ where c and d are estimated by using the first two moments of X_k^2. An alternative may be to obtain the adjustment using the bootstrap procedure suggested by Rocke (1989). Schott (1991) has also developed a test that an eigenvector of a population correlation matrix $\mathbf{P}_\rho$ has a specified value.

Exercises 8.4

1. For the covariance matrix given in Example 8.2.1, do the following.
   (a) Find an appropriate 95% confidence interval for the first eigenvalue. How would you use this to test the hypothesis that the first eigenvalue has a specified variance?
   (b) Test the hypothesis $H_0 : \lambda_2 = \lambda_3$ that the last two eigenvalues are equal.
   (c) Using $\alpha = .05$, test the hypothesis $H_0 : \Sigma = \Sigma_0$ that Σ has a specified value, if Σ_0 is defined by

$$\Sigma_0 = \begin{bmatrix} 40.43 & & \\ 9.38 & 10.40 & \\ 30.33 & 7.14 & 30.46 \end{bmatrix}$$

2. For the sample correlation matrix given by

$$\mathbf{R} = \begin{bmatrix} 1.000 & & \\ .586 & 1.000 & \\ .347 & .611 & 1.000 \end{bmatrix}$$

constructed from $N = 101$ observations, do each of the following.

   (a) Test the hypothesis $H_0 : \rho_{ij} = \rho$, for an $i \neq j$, using $\alpha = .05$.
   (b) If the test in part a is accepted, test the equality of the last two roots, $H_0 : \delta_2 = \delta_3$.
   (c) The matrix $\mathbf{R}$ appears to have simplex form

$$\begin{bmatrix} 1 & & \\ \rho & 1 & \\ \rho^2 & \rho & 1 \end{bmatrix}$$

with $\rho = .6$. Using $\alpha = .05$, test the hypothesis that

$$\mathbf{P} = \begin{bmatrix} 1.00 & & \\ .60 & 1.00 & \\ .36 & .60 & 1.00 \end{bmatrix}.$$

3. For a sample correlation matrix of rank 10 computed from 500 observations, the following sample eigenvalues were obtained: $\widehat{\delta}_i = 2.969, 1.485, 1.230, .825, .799, .670, .569, .569, .555, .480, .439$. Test the hypothesis that the last six roots are equal, $H : \delta_5 = \delta_6 = \ldots = \delta_{10}$.

8.5 Regression on Principal Components

In a MR model, the elements of $\widehat{\mathbf{B}} = \left[\widehat{\boldsymbol{\beta}}_1, \ldots, \widehat{\boldsymbol{\beta}}_p\right]$ become numerically unstable when $\mathbf{X}'\mathbf{X}$ is singular leading to large standard errors of estimate; this is termed multicollinearity. To reduce the effect of multicollinearity, one may remove variables, locate outliers, or perform some type of modified regression procedure like ridge regression, Hoerl and Kennard (1970 a, b). Another alternative is to replace the independent variables with a fewer number of uncorrelated PCs. The advantage of this method over non standard regression alternatives is that classical tests of significance may be performed on the reparameterized model.

When analyzing growth curves, $\widehat{\mathbf{B}}$ and the cov(vec $\widehat{\mathbf{B}}$) depend on $\mathbf{S}^{-1}$ so that as $\mathbf{S}^{-1}$ becomes unstable for highly correlated dependent variables, the standard errors of elements of $\widehat{\mathbf{B}}$ may become unstable even though $(\mathbf{X}'\mathbf{X})^{-1}$ is well behaved. To correct this situation, Rao (1948) suggested replacing the original variables with PCs.

While the MR model does not take into account the correlation that exists among the dependent variables, a more serious problem occurs when the number of variables $p > n$ and the number of elements in $\mathbf{B}$ is large. To correct this situation, one may fit a reduced-rank MR model to $\mathbf{Y}$, Reinsel and Velu (1998). Alternatively, one may reduce the number of variables by replacing $\mathbf{Y}$ with PCs, Dempster (1969, p. 241).

Given a MR model $\mathbf{Y} = \mathbf{XB} + \mathbf{E}$, recall that the $\text{cov}(\text{vec}\,\widehat{\mathbf{B}}) = \Sigma \otimes \left(\mathbf{X}'\mathbf{X}\right)^{-1}$ so that the elements of $\widehat{\mathbf{B}}$ will have large standard errors under multicollinearity. Reparameterizing the MR model, suppose we replace $\mathbf{X}$ with the PCs $\mathbf{Z} = \mathbf{X}\widehat{\mathbf{P}}$ where $\widehat{\mathbf{P}}$ diagonalizes $\mathbf{X}'\mathbf{X}$, $\widehat{\mathbf{P}}' \left(\mathbf{X}'\mathbf{X}\right) \mathbf{P} = \Lambda$. Then, writing the MR model in terms of PCs, the PC regression model becomes

$$
\begin{aligned}
\mathbf{Y} &= \mathbf{ZP}'\mathbf{B} + \mathbf{E} \\
&= \mathbf{ZB}^* + \mathbf{E}
\end{aligned}
\tag{8.5.1}
$$

Then,

$$
\begin{aligned}
\widehat{\mathbf{B}}^* &= \left(\mathbf{Z}'\mathbf{Z}\right)^{-1}\mathbf{Z}'\mathbf{Y} = \widehat{\Lambda}^{-1}\mathbf{Z}'\mathbf{Y} \\
\text{cov}(\text{vec}\,\widehat{\mathbf{B}}^*) &= \Sigma \otimes \left(\mathbf{Z}'\mathbf{Z}\right)^{-1} = \Sigma \otimes \Lambda^{-1}
\end{aligned}
\tag{8.5.2}
$$

so that using PCs as independent variables, instability is immediately seen when one or more of the eigenvalues λ_i are small. To correct this situation, one may replace $\mathbf{X}$ by a few components. Because the components were obtained independent of $\mathbf{Y}$, the first k components may not be "best" subset. Thus, one may want to utilize some stepwise procedure to obtain a subset of components that maximizes prediction and reduces multicollinearity, Dempster (1963) and Jackson (1991).

a. GMANOVA Model

In formulating the GMANOVA model

$$\underset{n\times p}{\mathbf{Y}} = \underset{n\times k}{\mathbf{X}} \ \underset{k\times q}{\mathbf{B}} \ \underset{q\times p}{\mathbf{Q}} + \underset{n\times p}{\mathbf{E}} \tag{8.5.3}$$

the ML estimate of $\mathbf{B}$ under normality is

$$\widehat{\mathbf{B}} = \left(\mathbf{X}'\mathbf{X}\right)^{-1} \mathbf{X}'\mathbf{Y}\mathbf{S}^{-1}\mathbf{Q}' \left(\mathbf{Q}\mathbf{S}^{-1}\mathbf{Q}'\right)^{-1} \tag{8.5.4}$$

and the covariance matrix of $\widehat{\mathbf{B}}$ is

$$\text{cov}(\text{vec}\,\widehat{\mathbf{B}}) = \frac{n-k-1}{n-k-1-(p-q)} \left(\mathbf{X}'\mathbf{X}\right)^{-1} \otimes \left(\mathbf{Q}\mathbf{S}^{-1}\mathbf{Q}'\right)^{-1} \tag{8.5.5}$$

as shown by Rao (1967) and Grizzle and Allen (1969). Hence, if $\mathbf{S}^{-1}$ is unstable due to highly correlated variables and $(\mathbf{X}'\mathbf{X})$ is ill-conditioned, the standard errors of $\widehat{\mathbf{B}}$ may be large. To correct this situation, let $\widehat{\mathbf{P}}$ be the eigenvectors of $\mathbf{Y}'\mathbf{Y}$ and since $\mathbf{S}$ is a rescaling of $\mathbf{Y}'\mathbf{Y}$

$$(n-k-1)\,\mathbf{S} = \mathbf{Y}'(\mathbf{I} - \mathbf{X}\left(\mathbf{X}'\mathbf{X}\right)^{-1}\mathbf{X}')\mathbf{Y} \tag{8.5.6}$$

the eigenvectors of $\mathbf{S}$ and $\mathbf{Y}'\mathbf{Y}$ are the same. Letting $\widehat{\Lambda}$ be the eigenvalues of $\mathbf{S}$ and $\mathbf{Z} = \mathbf{Y}\widehat{\mathbf{P}}$, one may reparameterized (8.5.3) to the principal component model

$$\begin{aligned} \mathbf{Y}\widehat{\mathbf{P}} &= \mathbf{X}\mathbf{B}\mathbf{Q}\widehat{\mathbf{P}} + \mathbf{E}\widehat{\mathbf{P}} \\ \mathbf{Z} &= \mathbf{X}\mathbf{B}\mathbf{Q}\widehat{\mathbf{P}}\mathbf{E}_o \end{aligned} \tag{8.5.7}$$

by postmultipling (8.5.3) by $\widehat{\mathbf{P}}$ and setting $\mathbf{E}_o = \mathbf{E}\widehat{\mathbf{P}}$. Then, since $\widehat{\mathbf{P}}$ orthogonalizes $\mathbf{S}$, the cov $\mathbf{Z} = \mathbf{P}\Sigma\mathbf{P}'$ and $\mathbf{S}^{-1} = \widehat{\mathbf{P}}\Lambda^{-1}\widehat{\mathbf{P}}'$. To stabilize $\mathbf{Y}$, one may replace the dependent variables by a subset of PCs.

One may also apply this procedure to the MR model by replacing $\mathbf{Y}$ with a set of uncorrelated components. Then, one may fit a regression model to each variable, one at a time using different design matrices as an alternative to the SUR model. Another option is to regress the components (one at a time) on the set of $\mathbf{X}$ variables simultaneously. Or, one may fit all component C_i and $X_1, X_2, \ldots, X_k$ simultaneously, a variation of latent root regression, Gunst, Webster and Mason (1976).

b. The PCA Model

Most authors do not associate a model with PCA. If $\mathbf{Y}$ is a random vector and $\mathbf{Z} = \mathbf{P}'\mathbf{Y}$ is a PC, we have the identity that

$$\mathbf{Y} = \mathbf{P}\mathbf{Z} = \mathbf{P}\mathbf{P}'\mathbf{Y} \tag{8.5.8}$$

Replacing $\mathbf{P}$ with $k \ll p$ eigenvectors, observe that

$$\begin{aligned} \mathbf{Y} &= \mathbf{P}_k\mathbf{P}_k'\mathbf{Y} + \mathbf{e} \\ \mathbf{Y} &= \mathbf{A}\mathbf{B}\mathbf{Y} + \mathbf{e} \end{aligned} \tag{8.5.9}$$

where $\mathbf{A} = \mathbf{P}_k$, $\mathbf{B} = \mathbf{P}'_k$ and the rank of $\mathbf{C} = \mathbf{AB}$ is $k \ll p$. This is the reduced-rank multivariate regression model. Letting $\widetilde{\mathbf{Y}} = \mathbf{ABY}$, then $\mathbf{Y} = \widetilde{\mathbf{Y}} + \mathbf{e}$ so that the analysis of residuals in PCA is a reduced-rank MR model, Reinsel and Velu (1998, p. 36).

8.6 Multivariate Regression on Principal Components Example

To illustrate multivariate regression on principal components, data from Smith, Gnanadesikan and Hughes (1962) are utilized. The data are biochemical measurements on urine samples of men. The variables measured are

$y_1 = $ pH,	$y_8 = $ chloride (mg/ml),
$y_2 = $ modified createnine coefficient,	$y_9 = $ bacon (μ g/ml),
$y_3 = $ pigment createnine,	$y_{10} = $ choline (μg/ml),
$y_4 = $ phosphate (mg/ml),	$y_{11} = $ copper (μg/ml),
$y_5 = $ calcium (mg/ml),	$x_1 = $ volume (ml),
$y_6 = $ phosphours (mg/ml),	$x_2 = $ (specific gravity $-$ 1) $\times 10^3$
$y_7 = $ createnine (mg/ml),	$x_3 = $ weight

for $n = 45$ patients. The data were used in Chapter 4 to illustrate the analysis of a one-way MANCOVA design with two covariates and are provided in the file SGH.dat. The weight variable was used to classify the patients into four groups. To illustrate multivariate regression on principal components, we hypothesize a reverse regression model, $\mathbf{X} = \mathbf{YB} + \mathbf{E}$, using the eleven dependent variables to predict the three independent variables. This design leads to an ill-conditioned $\left(\mathbf{Y}'\mathbf{Y}\right)^{-1}$ matrix. In particular, the variance inflation factor of 13.98 on variable y_6 is evidenced in the output of program m8_6_1.sas. Instead of modifying the set of predictor variables, we perform a PCA on the covariance matrix of the prediction variables. We have chosen to retain the first four components that account for 97% of the variance. The SCORE parameter on the PROC FACTOR statement informs the SAS procedure to save the coefficients in the work SAS data set using OUTSTAT. The SAS procedure SCORE is used to calculate component scores which are output to the SAS data set WORK.SCORE and printed. Finally, we regress $x_1, x_2,$ and x_3 on the four new orthogonal principal components. Alternatively, we may use the program MulSubSel to select the best subset of principal components for the MR model as illustrated in Section 4.3. While we have lost some predictive power by using the four components, the model is better behaved. We have a model that is no longer ill-conditioned with reasonable predictive capability. Wilks' Λ criterion for regression on the original variable had a value of $\Lambda = 0.01086$, so that $\eta^2 = 1 - \Lambda = 0.99$. For regression on the PCs, Wilks' $\Lambda = 0.09966$ and $\eta^2 = 0.90$, a small reduction in η^2.

We have chosen to illustrate the development of a regression model on PCs using PROC FACTOR. One may also use the PROC PRINCOMP for the analysis.

Exercises 8.6

1. Use variables y_3, y_4, y_6 y_7 and y_{10} and x_1 and x_2, to develop a multivariate principal components regression model.

2. Reversing the roles of y_1 and x_1 in Problem 1, develop a model to predict $\mathbf{Y}$ given $\mathbf{X}$.

8.7 Canonical Correlation Analysis

PCA is used to investigate one set of variables with or without covariates. The original variables are replaced by a set of variates called principal components. The components are created to account for maximal variation among the original variables.

A generalization of PCA, developed by Hotelling (1936), is canonical correlation analysis (CCA). The method was developed to investigate relationships between two sets of variables with one or more sets of covariates.

In CCA one is interested in investigating relationships between two sets of variables $\mathbf{Y}' = [Y_1, Y_2, \ldots, Y_p]$ and $\mathbf{X}' = [X_1, X_2, \ldots, X_q]$. In addition one may also have sets of covariates $\mathbf{Z}' = [Z_1, Z_2, \ldots, Z_r]$ and $\mathbf{W}' = [W_1, W_2, \ldots, W_s]$. More will be said about the inclusion of covariates later in this section. The goal of CCA as developed by Hotelling (1936) is to construct two new sets of canonical variates $U = \boldsymbol{\alpha}'\mathbf{Y}$ and $V = \boldsymbol{\beta}'\mathbf{X}$ that are linear combinations of the original variables such that the simple correlation between U and V is maximal, subject to the restriction that each canonical variate U and V has unit variance (to ensure uniqueness, except for sign) and is uncorrelated with other constructed variates within the set.

Canonical correlation analysis may be used to determine whether two sets of variables are independent assuming multivariate normality. As an exploratory tool, it is used as a data reduction method. Given a large number of variables, one may want to locate a few canonical variates in each set to study and reconstruct the intercorrelations among the variables. For example, given a set of ability variables and a set of personality variables, the object of the study may be to determine what sort of personality traits may be associated with various ability domains.

In this section we will present a general overview of the basic theory of CCA, consider some extensions when covariates are included, develop tests of hypotheses, and discuss the interpretation of canonical variates using redundancy analysis. Extensions of CCA to several groups is discussed by Gnanadesikan (1997) while Gittens (1985) provides a general overview of CCA with several applications. The topic is also treated by Basilevsky (1994). CCA with linear constraints for continuous and discrete data is discussed by Yanai and Takane (1992) and Böckenholt and Böckenholt (1990).

a. Population Model for CCA

Given two sets of variables $\mathbf{Y}' = [Y_1, Y, \ldots, Y_p]$ and $\mathbf{X}' = [X_1, X_2, \ldots, X_q]$ with associated covariance, matrices Σ_{yy} and Σ_{xx} where $s = \min(p, q)$, one goal of CCA is to identify canonical variates $U = \boldsymbol{\alpha}'\mathbf{Y}$ and $V = \boldsymbol{\beta}'\mathbf{X}$ such that the correlation between U and V is maximal. Letting the population covariance matrix for the joint vector $(Y_1, Y_2, \ldots, Y_p, X_1, X_2, \ldots, X_q) = (\mathbf{Y}', \mathbf{X}')$ be defined as

$$\Sigma = \begin{bmatrix} \Sigma_{yy} & \Sigma_{yx} \\ \Sigma_{xy} & \Sigma_{xx} \end{bmatrix} \tag{8.7.1}$$

the correlation between U and V is

$$\rho_{UV} = \frac{\boldsymbol{\alpha}'\Sigma_{yx}\boldsymbol{\beta}}{\sqrt{(\boldsymbol{\alpha}'\Sigma_{yy}\boldsymbol{\alpha})\,(\boldsymbol{\beta}'\Sigma_{xx}\boldsymbol{\beta})}} \tag{8.7.2}$$

Because ρ_{UV} involves the canonical variates U and V, it is called a canonical correlation. Thus, the goal of CCA is to find canonical variates $U_i = \boldsymbol{\alpha}_i'\mathbf{Y}$ and $V_i = \boldsymbol{\beta}_i'\mathbf{X}$ for $i = 1, 2, \ldots, s$ such that the set of variates U_i are uncorrelated and the set of variates V_i are uncorrelated, each having unit variance

$$\text{cov}\,(U_i,\ U_j) = \text{cov}\,(V_i,\ V_j) = \begin{cases} 1 & i = j \\ 0 & i \neq j \end{cases}$$

Furthermore, the covariance or correlation between U_i and V_i is ρ_i, for $i = 1, 2, \ldots, s$ and 0, otherwise

$$\begin{aligned} \text{cov}\,(U_i,\ V_i) &= \text{cov}\,(U_i,\ V_i) = \rho_i \quad i = 1, 2, \ldots, s \\ \text{cov}\,(U_i,\ V_i) &= \text{cov}\,(U_i,\ V_j) = 0 \quad i \neq j \end{aligned}$$

Thus, if U_1, U_2 $(p = 2)$ and V_1, V_2 and V_3 $(q = 3)$ are canonical variates, the correlation matrix for $\mathbf{U}' = [U_1, U_2]$ and $\mathbf{V}' = [V_1, V_2, V_3]$ has the form

$$\begin{array}{c} \\ U_1 \\ U_2 \\ \\ V_1 \\ V_2 \\ V_3 \end{array} \begin{array}{c} \begin{matrix} U_1 & U_2 & & V_1 & V_2 & V_3 \end{matrix} \\ \left[\begin{array}{cc:ccc} 1 & 0 & \rho_1 & 0 & 0 \\ 0 & 1 & 0 & \rho_2 & 0 \\ \hdashline \rho_1 & 0 & 1 & 0 & 0 \\ 0 & \rho_2 & 0 & 1 & 0 \\ 0 & 0 & 0 & 0 & 1 \end{array}\right] \end{array} = \left[\begin{array}{c:c} \mathbf{I}_p & \Lambda \\ \hdashline \Lambda' & \mathbf{I}_q \end{array}\right]$$

To maximize (8.7.2), we assume that both Σ_{yy} and Σ_{xx} are positive definite and let $\Sigma_{yy}^{1/2}$ and $\Sigma_{xx}^{1/2}$ be the square root matrices of Σ_{yy} and Σ_{xx}, respectively. Without loss of generality we assume $p \leq q$ (if this is not the case, the role of $\mathbf{X}$ and $\mathbf{Y}$ are interchanged) and let $\widetilde{\alpha} = \Sigma_{yy}^{-1/2}\alpha$ and $\widetilde{\beta} = \Sigma_{xx}^{-1/2}\beta$. Then from (8.7.2) we must find α and β to maximize the

$$\underset{\alpha,\beta}{\text{cov}}\left(\alpha'\mathbf{Y}, \beta'\mathbf{X}\right)^2 = \frac{(\widetilde{\alpha}'\Sigma_{yy}^{-1/2}\Sigma_{yx}\Sigma_{xx}^{-1/2}\widetilde{\beta})^2}{(\widetilde{\alpha}'\widetilde{\alpha})\,(\widetilde{\beta}'\widetilde{\beta})}$$

Now, let $\phi = \Sigma_{yy}^{-1/2}\Sigma_{yx}\Sigma_{xx}^{-1/2}$ so that

$$\phi\phi' = \Sigma_{yy}^{-1/2}\Sigma_{yx}\Sigma_{xx}^{-1}\Sigma_{xy}\Sigma_{yy}^{-1/2} \quad \text{and} \quad \phi'\phi = \Sigma_{xx}^{-1/2}\Sigma_{xy}\Sigma_{yy}\Sigma_{yx}\Sigma_{xx}^{-1/2}.$$

Then by the Cauchy-Schwarz inequality (Theorem 2.3.4), $(\widetilde{\alpha}'\phi\widetilde{\beta})^2 \leq \left(\widetilde{\alpha}'\phi\phi'\alpha\right)(\widehat{\beta}'\beta)$ and since the roots of $\phi\phi'$ are the same as $\phi'\phi$, the maximum of (8.7.2) may be obtained by solving the eigenequations $\left|\phi\phi' - \rho^2\mathbf{I}\right| = 0$ or $\left|\phi'\phi - \rho^2\mathbf{I}\right| = 0$. In terms of the covariance matrices, we may solve

$$\begin{aligned} \left|\Sigma_{yx}\Sigma_{xx}^{-1}\Sigma_{xy} - \rho^2\Sigma_{yy}\right| &= 0 \\ \left|\Sigma_{xy}\Sigma_{yy}^{-1}\Sigma_{yx} - \rho^2\Sigma_{xx}\right| &= 0 \end{aligned} \tag{8.7.3}$$

for $s = \min(p, q) = p < q$ roots ρ_i^2. Using Theorem 2.6.7, let $\left[\mathbf{a}_1, \mathbf{a}_2, \ldots, \mathbf{a}_p\right] = \mathbf{A}$ be the eigenvectors of the first equation and $\left[\mathbf{b}_1, \mathbf{b}_2, \ldots, \mathbf{b}_q\right] = \mathbf{B}$ be the eigenvectors of the second equation. Then $\mathbf{A}'\Sigma_{yy}\mathbf{A} = \mathbf{I}_p$ and $\mathbf{A}'\Sigma_{yx}\Sigma_{xx}^{-1}\Sigma_{xy}\mathbf{A} = \Lambda$; and $\mathbf{B}'\Sigma_{xx}\mathbf{B} = \mathbf{I}_q$ and $\mathbf{B}'\Sigma_{xy}\Sigma_{yy}^{-1}\Sigma_{yx}\mathbf{B} = \Lambda$. The square root of the roots ρ_i^2 are the canonical correlations between the orthogonal canonical variates $U_i = \alpha_i'\mathbf{Y}$ and the orthogonal variates $V_i = \beta_i'\mathbf{X}$ for $i = 1, 2, \ldots, p$. They have been constructed so that the canonical correlations $\rho_1 \geq \rho_2 \geq \ldots \geq \rho_p$ are maximal. The vector of canonical variates are defined as $\mathbf{U} = \mathbf{A}'\mathbf{Y}$ and $\mathbf{V} = \mathbf{B}'\mathbf{X}$. To avoid solving both equations in (8.7.3), the eigenvectors are related as follows

$$\begin{aligned} \alpha_i &= \Sigma_{yy}^{-1}\Sigma_{yx}\beta_i/\rho_i \\ \beta_i &= \Sigma_{xx}^{-1}\Sigma_{xy}\alpha_i/\rho_i \end{aligned} \tag{8.7.4}$$

for $i = 1, 2, \ldots, p$. The vectors α_i and β_i may be standardized to one, $\alpha_i'\Sigma_{yy}\alpha_i = \beta_i'\Sigma_{xx}\beta_i = 1$, so that the variance of U_i and V_i are unity for $i = 1, 2, \ldots, p$.

In applying PCA, a researcher had to establish the scale of measurement for the analysis. This is not the case for CCA since the canonical correlations are invariant to changes in location and scale; the CCA procedure is scale free. Thus, one may replace the covariance matrices in (8.7.3) with population correlation matrices. The eigenvectors of the two are related by the simple rescaling

$$\begin{aligned} \xi_i &= \left(\text{diag}\,\Sigma_{yy}\right)^{1/2}\alpha_i \\ \delta_i &= \left(\text{diag}\,\Sigma_{xx}\right)^{1/2}\beta_i \end{aligned} \tag{8.7.5}$$

where ξ_i and δ_i are the eigenvectors of

$$\begin{aligned} \left|\mathbf{P}_{yy}^{-1/2}\mathbf{P}_{yx}\mathbf{P}_{xx}^{-1}\mathbf{P}_{xy}\mathbf{P}_{yy}^{-1/2} - \rho^2\mathbf{I}\right| &= 0 \\ \left|\mathbf{P}_{xx}^{-1/2}\mathbf{P}_{xy}\mathbf{P}_{yy}^{-1}\mathbf{P}_{yx}\mathbf{P}_{xx}^{-1/2} - \rho^2\mathbf{I}\right| &= 0 \end{aligned} \tag{8.7.6}$$

Letting $\mathbf{C}$ be the eigenvectors of the first equation in (8.7.6) and $\mathbf{D}$ be the eigenvectors of the second, then

$$\begin{aligned} \mathbf{C}'\mathbf{P}_{yy}\mathbf{C} &= \mathbf{I}_p \\ \mathbf{D}'\mathbf{P}_{xx}\mathbf{D} &= \mathbf{I}_q \\ \mathbf{C}'\mathbf{P}_{yx}\mathbf{D} &= \Lambda \end{aligned} \tag{8.7.7}$$

Since $\mathbf{D} = \mathbf{P}_{xx}^{-1}\mathbf{P}_{xy}\mathbf{C}$, the population intercorrelation matrix has the factorization

$$\begin{aligned} \mathbf{P}_{yx} &= \mathbf{P}_{yy}\mathbf{C}\Lambda\mathbf{D}'\mathbf{P}_{xx} \\ &= \sum_{i=1}^{p} \rho_i \mathbf{L}_i \mathbf{M}_i' \end{aligned} \tag{8.7.8}$$

where $\mathbf{L}_i$ and $\mathbf{M}_i$ are column vectors of $\mathbf{P}_{yy}\ \mathbf{C}$ and $\mathbf{P}_{xx}\ \mathbf{D}$, respectively. Thus, by selecting $k \ll s = \min(p, q) = p$ rescaled variates or factors, the correlations explained by k factors are

$$\mathbf{P}_{yx}(k) = \sum_{i=1}^{k} \rho_i \mathbf{L}_i \mathbf{M}_i' \tag{8.7.9}$$

and the residual matrix is

$$\mathbf{P}_{yx} - \mathbf{P}_{yx}(k) = \sum_{i=k+1}^{p} \rho_{i+1} \mathbf{L}_{i+1} \mathbf{M}_{i+1}' \tag{8.7.10}$$

The residual matrix is zero if the rank of $\mathbf{P}_{yx}$ is k.

These observations suggest that one may employ CCA to study and recover the relationship between $\mathbf{Y}$ and $\mathbf{X}$ by studying the structure and rank of $\mathbf{P}_{yx}$, the matrix of intercorrelations. To formalize the relationship, a linear model is formulated for $\mathbf{Y}$ and $\mathbf{X}$

$$\begin{aligned} \underset{p\times 1}{\mathbf{Y}} &= \boldsymbol{\mu}_y + \underset{p\times k}{\Lambda_y}\ \underset{k\times 1}{\mathbf{f}} + \underset{p\times 1}{\mathbf{e}_y} \\ \underset{q\times 1}{\mathbf{X}} &= \boldsymbol{\mu}_x + \underset{q\times k}{\Lambda_x}\ \underset{k\times 1}{\mathbf{f}} + \underset{q\times 1}{\mathbf{e}_x} \end{aligned} \tag{8.7.11}$$

where $\mathbf{f}$ is a vector of unobserved hypothetical factors, $\mathbf{e}_x$ and $\mathbf{e}_y$ are random error vectors, and Λ_y and Λ_x are matrices of regression weights of rank $k \leq p$. Furthermore, we assume that the elements of $\mathbf{f}$ are uncorrelated and have unit variance; and that $\mathbf{e}_y$, $\mathbf{e}_x$ and $\mathbf{f}$ are mutually uncorrelated with zero means

$$\begin{aligned} E(\mathbf{e}_x) = E(\mathbf{e}_y) &= E(\mathbf{f}) = \mathbf{0} \\ \text{cov}(\mathbf{f}) &= \mathbf{I} \\ \text{cov}(\mathbf{f}, \mathbf{e}_y) = \text{cov}(\mathbf{f}, \mathbf{e}_x) &= \text{cov}(\mathbf{e}_x, \mathbf{e}_y) = \mathbf{0} \end{aligned} \tag{8.7.12}$$

Then, standardizing $\mathbf{Y}$ and $\mathbf{X}$ to variables $\mathbf{Z}_y$ and $\mathbf{Z}_x$, model (8.7.11) and our assumptions imply that the population correlation matrices have the structure

$$\begin{aligned} \mathbf{P}_{yy} &= \Lambda_y \Lambda_y' + \Sigma_1 \\ \mathbf{P}_{xx} &= \Lambda_x \Lambda_x' + \Sigma_2 \\ \mathbf{P}_{yx} &= \Lambda_y \Lambda_x' \quad \text{and} \quad \mathbf{P}_{xy} = \Lambda_x \Lambda_y' \end{aligned} \tag{8.7.13}$$

where the cov $(\mathbf{e}_y) = \Sigma_1$ and the cov $(\mathbf{e}_x) = \Sigma_2$. Furthermore, by Theorem 2.5.2, the rank of $\mathbf{P}_{yx} = r\left(\Lambda_y\Lambda_x'\right) \leq k$ where k is the number of factors common to $\mathbf{Y}$ and $\mathbf{X}$. We may further assume that the rank of $\Lambda_y\Lambda_x'$ is k. For if it is less than k model (8.7.11) may be reformulated to have the number of common factors equal to the rank of $\mathbf{P}_{yx}$.

In expression (8.7.8), the matrices $\mathbf{L} = \mathbf{P}_{yy}\mathbf{C}$ and $\mathbf{M} = \mathbf{P}_{xx}\mathbf{D}$ represent the correlations between the p variables and the canonical variates since

$$\begin{aligned}
\rho_{\mathbf{YU}} = \text{cov}\,(\mathbf{Y},\ \mathbf{U}) &= \left(\text{diag}\,\Sigma_{yy}\right)^{-1/2}\Sigma_{yy}\mathbf{A}\\
&= \left(\text{diag}\,\Sigma_{yy}\right)^{-1/2}\Sigma_{yy}\left(\text{diag}\,\Sigma_{yy}\right)^{-1/2}\mathbf{C}\\
&= \mathbf{P}_{yy}\mathbf{C}\\
\rho_{\mathbf{XV}} = \text{cov}\,(\mathbf{X},\ \mathbf{V}) &= (\text{diag}\,\Sigma_{xx})^{-1/2}\Sigma_{xx}\mathbf{B}\\
&= (\text{diag}\,\Sigma_{xx})^{-1/2}\Sigma_{yy}\,(\text{diag}\,\Sigma_{xx})^{-1/2}\mathbf{D}\\
&= \mathbf{P}_{xx}\mathbf{D}
\end{aligned} \tag{8.7.14}$$

Thus, the vectors $\mathbf{L}_1, \mathbf{L}_2, \ldots, \mathbf{L}_k$ and $\mathbf{M}_1, \mathbf{M}_2, \ldots, \mathbf{M}_k$ fully recover $\mathbf{P}_{yx}$ for some $k \leq p$ where $k = r\left(\mathbf{P}_{yx}\right)$ so that $\rho_{k+1} = \ldots = \rho_p = 0$.

It is often the case that the first few canonical correlations $\rho_1, \rho_1, \rho_2, \ldots, \rho_k$ are large while the other $\rho_{k+1}, \rho_{k+2}, \ldots, \rho_p$ are small, then most of the intercorrelation is explained by the first k components or factors. An index of how much the correlation structure is explained by the first k canonical variates is

$$\rho_k^2 = \sum_{i=1}^{k}\rho_i^2 \Big/ \sum_{i=1}^{p}\rho_i^2 = \sum_{i=1}^{k}\rho_i^2 \Big/ \text{tr}(\mathbf{P}_{yy}^{-1/2}\mathbf{P}_{yx}\mathbf{P}_{xx}^{-1}\mathbf{P}_{xy}\mathbf{P}_{yy}^{-1/2}) \tag{8.7.15}$$

If the $r\left(\mathbf{P}_{yx}\right) = k$, then $\rho_k^2 = 1$. Each ρ_i^2 is a measure of the proportion of variance of U_i explained by V_i.

With the analysis of standardized variables, let $U_i = \mathbf{c}_i'\mathbf{Z}_y$ and $V_i = \mathbf{d}_i'\mathbf{Z}_x$ represent the canonical variates. The standardized coefficients (loadings) are used to evaluate the contribution or influence of each variable to a canonical variate. Converting the coefficients to correlations (structure loadings) using (8.7.14), the contribution of each variable, ignoring the other variables, to the canonical variate may be evaluated. These are conveniently represented in Table 8.7.1 for $k \leq s = \min\,(p,\ q) = p$ canonical variates.

R^2 in each row is equal to 1 if $k = p < q$ and is constructed from the inner product of the row vectors in the structure matrix of correlations. Letting $\mathbf{s}\,(U_1),\ \mathbf{s}\,(U_2), \ldots, \mathbf{s}\,(U_k)$ represent the column vectors in Table 8.7.1 for domain $\mathbf{Y}$ and $\mathbf{s}\,(V_1),\ \mathbf{s}\,(V_2), \ldots, \mathbf{s}\,(V_k)$ the corresponding columns for domain $\mathbf{X}$, the inner products of the column vectors for each domain divided by the number of variables in each domain are

$$\begin{aligned}
U_j^* &= \frac{\boldsymbol{\xi}_i'\mathbf{R}_{yy}\mathbf{R}_{yy}\boldsymbol{\xi}_i}{p} = \frac{[\mathbf{s}\,(U_i)]'\,[\mathbf{s}\,(U_i)]}{p} = \sum_{i=1}^{p}\rho_{Y_iU_j}^2 \Big/ p\\
V_j^* &= \frac{\boldsymbol{\delta}_i'\mathbf{R}_{xx}\mathbf{R}_{xx}\boldsymbol{\delta}_i}{q} = \frac{[\mathbf{s}\,(V_i)]'\,[\mathbf{s}\,(V_i)]}{q} = \sum_{i=1}^{q}\rho_{X_iV_j}^2 \Big/ q
\end{aligned} \tag{8.7.16}$$

for $j = 1, 2, \ldots, k$. Because the variables Y_i and X_i have been standardized, the total variance in domain $\mathbf{Y}$ and domain $\mathbf{X}$ are p and q, respectively. Thus, the quantities in

TABLE 8.7.1. Canonical Correlation Analysis

Domain Y

Variable	Loadings				Structure				R^2
	U_1	U_2	$\cdots$	U_k	U_1	U_2	$\cdots$	U_k	
Z_{y_1}	c_{11}	c_{12}	$\cdots$	c_{1k}	$\rho_{Y_1U_1}$	$\rho_{Y_1U_2}$	$\cdots$	$\rho_{Y_1U_k}$	$\sum_{i=1}^k \rho^2_{Y_1U_i}$
Z_{y_2}	c_{21}	c_{22}	$\cdots$	c_{2k}	$\rho_{Y_2U_1}$	$\rho_{Y_2U_2}$	$\cdots$	$\rho_{Y_2U_k}$	$\sum_{i=1}^k \rho^2_{Y_2U_i}$
$\vdots$	$\vdots$	$\vdots$		$\vdots$	$\vdots$	$\vdots$		$\vdots$	$\vdots$
Z_{y_p}	c_{p1}	c_{p2}	$\cdots$	c_{pk}	$\rho_{Y_pU_1}$	$\rho_{Y_pU_2}$	$\cdots$	$\rho_{Y_pU_k}$	$\sum_{i=1}^k \rho^2_{Y_pU_i}$

Domain X

Variable	Loadings				Structure				R^2
	V_1	V_2	$\cdots$	V_k	V_1	V_2	$\cdots$	V_k	
Z_{x_1}	d_{11}	d_{12}	$\cdots$	d_{1k}	$\rho_{X_1V_1}$	$\rho_{X_1V_2}$	$\cdots$	$\rho_{X_1V_k}$	$\sum_{i=1}^k \rho^2_{X_1V_i}$
Z_{x_2}	d_{21}	d_{22}	$\cdots$	d_{2k}	$\rho_{X_2V_1}$	$\rho_{X_2V_2}$	$\cdots$	$\rho_{X_2V_k}$	$\sum_{i=1}^k \rho^2_{X_2V_i}$
$\vdots$	$\vdots$	$\vdots$		$\vdots$	$\vdots$	$\vdots$		$\vdots$	$\vdots$
Z_{x_q}	d_{q1}	d_{q2}	$\cdots$	d_{qk}	$\rho_{X_qV_1}$	$\rho_{X_qV_2}$	$\cdots$	$\rho_{X_qV_k}$	$\sum_{i=1}^k \rho^2_{X_pV_i}$

(8.7.16) represent the proportion of the total variance in each domain accounted for by each canonical variate or, the average variance in a domain that is accounted for by a canonical variate. The proportion of variance of U_i (V_i) accounted for by V_i (U_i) is ρ_i^2, the square of the canonical correlation or the shared variance.

b. *Sample CCA*

To apply CCA in practice, a sample of n $(p+q)$-vectors are collected on two sets of variables $\mathbf{y}' = [y_1, y_2, \ldots, y_p]$ and $\mathbf{x}' = [x_1, x_2, \ldots, x_q]$ with population mean $\boldsymbol{\mu}' = [\boldsymbol{\mu}_x, \boldsymbol{\mu}_y]$ and covariance matrix Σ. The unbiased estimate for the covariance matrix is the sample covariance matrix

$$\mathbf{S} = \begin{bmatrix} \mathbf{S}_{yy} & \mathbf{S}_{yx} \\ \mathbf{S}_{xy} & \mathbf{S}_{xx} \end{bmatrix} \tag{8.7.17}$$

The corresponding sample correlation matrix is

$$\begin{aligned} \mathbf{R} &= [\text{diag}\,(\mathbf{S}_{yy}, \mathbf{S}_{xx})]^{-1/2}\,\mathbf{S}\,[\text{diag}\,(\mathbf{S}_{yy}, \mathbf{S}_{xx})]^{-1/2} \\ &= \begin{bmatrix} \mathbf{R}_{yy} & \mathbf{R}_{yx} \\ \mathbf{R}_{xy} & \mathbf{R}_{xx} \end{bmatrix} \end{aligned} \tag{8.7.18}$$

Then, solving either

$$\begin{aligned} \left|\mathbf{S}_{yx}\mathbf{S}_{xx}^{-1}\mathbf{S}_{xy} - \rho^2\mathbf{S}_{yy}\right| &= 0 \\ \left|\mathbf{R}_{yx}\mathbf{R}_{xx}^{-1}\mathbf{R}_{xy} - \rho^2\mathbf{R}_{yy}\right| &= 0 \end{aligned} \tag{8.7.19}$$

where $\mathbf{a}_i$ and $\mathbf{b}_i$ are the eigenvectors using $\mathbf{S}$ and $\mathbf{c}_i$ and $\mathbf{d}_i$ are the eigenvectors using $\mathbf{R}$, the relationship between the eigenvectors in the sample are as follows

$$\begin{aligned} r_i\mathbf{a}_i &= \mathbf{S}_{yy}^{-1}\mathbf{S}_{yx}\mathbf{b}_i \\ r_i\mathbf{b}_i &= \mathbf{S}_{xx}^{-1}\mathbf{S}_{xy}\mathbf{a}_i \\ \mathbf{c}_i &= \left(\operatorname{diag}\mathbf{S}_{yy}\right)^{1/2}\mathbf{a}_i \\ \mathbf{d}_i &= \left(\operatorname{diag}\mathbf{S}_{xx}\right)^{1/2}\mathbf{b}_i \end{aligned} \tag{8.7.20}$$

where r_i are the estimates of ρ_i, the population canonical correlations. The sample canonical variates are defined as either raw variates or standardized variates

$$\begin{aligned} &\text{Raw} && U_i = \mathbf{a}_i'\mathbf{y} \\ & && V_i = \mathbf{b}_i'\mathbf{x} \\ &\text{Standardized} && U_i = \mathbf{c}_i'\mathbf{z}_y \\ & && V_i = \mathbf{d}_i'\mathbf{z}_x \end{aligned} \tag{8.7.21}$$

where $\mathbf{z}_y$ and $\mathbf{z}_x$ denote standardized variables.

c. *Tests of Significance*

Assuming $\mathbf{X}_i \sim N_q\left(\boldsymbol{\mu}_x, \Sigma_{xx}\right)$ and $\mathbf{Y}_i \sim N_p\left(\boldsymbol{\mu}_y, \Sigma_{yy}\right)$, we may be interested in testing hypotheses regarding the canonical correlations ρ_i or the square of ρ_i, ρ_i^2. If $\rho_1 \geq \rho_2 \geq \ldots \geq \rho_s \geq 0$, we may be interested in testing that all $\rho_i = 0$ where $s = \min(p, q)$

$$H_o : \rho_1 = \rho_2 = \ldots = \rho_s = 0 \tag{8.7.22}$$

This is the test of independence since if $\Sigma_{yx} = \mathbf{0}$ then $\Sigma_{yy}^{-1/2}\Sigma_{yx}\Sigma_{xx}^{-1}\Sigma_{xy}\ \Sigma_{yy}^{-1/2} = \mathbf{0}$ and all $\rho_i = 0$, or, equivalently that the rank of Σ_{yx} is zero. The likelihood ratio test of independence developed is Chapter 3 based upon Wilks' Λ statistic is

$$\begin{aligned} \Lambda_o &= \frac{|\mathbf{S}|}{\left|\mathbf{S}_{yy}\right|\left|\mathbf{S}_{xx}\right|} = \frac{|\mathbf{R}|}{\left|\mathbf{R}_{yy}\right|\left|\mathbf{R}_{xx}\right|} \\ &= \frac{\left|\mathbf{S}_{xx}\right|\left|\mathbf{S}_{yy} - \mathbf{S}_{yx}\mathbf{S}_{xx}^{-1}\mathbf{S}_{xy}\right|}{\left|\mathbf{S}_{yy}\right|\left|\mathbf{S}_{xx}\right|} \\ &= \frac{\left|\mathbf{S}_{yy} - \mathbf{S}_{yx}\mathbf{S}_{xx}^{-1}\mathbf{S}_{xy}\right|}{\left|\mathbf{S}_{yy}\right|} \end{aligned} \tag{8.7.23}$$

and is identical to the Λ test statistic for testing for no linear relationship between **Y** and **X**, that $\mathbf{B}_1 = \mathbf{0}$ in the MR model. From (4.2.14), we may relate Λ_o to the sample canonical correlations

$$\Lambda_o = \prod_{i=1}^{s} \left(1 - r_i^2\right) \tag{8.7.24}$$

where r_i^2 are the sample eigenvalues of

$$|\mathbf{S}_{yx}\mathbf{S}_{xx}^{-1}\mathbf{S}_{xy} - r_i^2\mathbf{S}_{yy}| = 0 \tag{8.7.25}$$

the square of the sample canonical correlations. Thus, Λ_o may be used to test the hypothesis of independence, that $\mathbf{B}_1 = \mathbf{0}$ in the MR model, or that the rank of Σ_{XY} is zero.

More importantly, we may want to test

$$H_k : \rho_1 \neq 0,\ \rho_2 \neq 0, \ldots, \rho_k \neq 0,\ \rho_{k+1} = \ldots = \rho_s = 0 \tag{8.7.26}$$

or that the rank of Σ_{yx} is k so that we should retain k factors in a CCA. Bartlett (1947) showed that if H_k is true for $k = 0, 1, 2, \ldots, s = \min(p, q)$ so that

$$X_k^2 = -\left\{(n-1) - (p+q+1)/2\right\} \log \Lambda_k \tag{8.7.27}$$

converges asymptotically to a chi-square distribution with $v = (p-k)(q-k)$ degrees of freedom, where

$$\Lambda_k = \prod_{i=k+1}^{s} \left(1 - r_i^2\right) \tag{8.7.28}$$

Following Fujikoshi (1977) one may also test H_k using the other overall test criteria

$$\begin{aligned}
X_L^2 &= -\left[(n-k-1) - \frac{1}{2}(p+q+1) + \sum_{i=1}^{k}\left(1/r_i^2\right)\right] \log \Lambda_k \\
X_{BLH}^2 &= \left[(n-p-q-2) + \sum_{i=1}^{k}\left(1/r_i^2\right)\right] \sum_{j=k+1}^{s} r_i^2/\left(1-r_i^2\right) \\
X_{BNP}^2 &= \left[n - 2k + \sum_{i=1}^{k}\left(1/r_i^2\right)\right] \sum_{j=k+1}^{s} r_i^2
\end{aligned} \tag{8.7.29}$$

which are also distributed as $\chi^2(v)$ with degree of freedom $v = (p-k)(q-k)$. SAS uses the F distribution to approximate (8.7.27) ignoring Lawley's (1959) correction factor.

Because r_i^2 is not an unbiased estimate of ρ_i^2, a better estimate of ρ_i^2 having a bias of $O\left(1/n^2\right)$ was developed by Lawley (1959)

$$\widehat{r}_i = r_i \left\{1 - \left(\frac{1}{n-1}\right)\sum_{\substack{j=1 \\ j\neq i}}^{k}\left(\frac{r_j^2}{r_i^2 - r_j^2}\right) - \frac{s-k}{(n-1)\,r_i^2}\right\}^{1/2} \tag{8.7.30}$$

assuming $\rho_{k+1} = \rho_{k+2} = \ldots = \rho_s = 0$. If the adjustment is negative, it is undefined. This is called the adjusted canonical correlation in the SAS procedure CANCORR.

d. Association and Redundancy

In our discussion of the MR model, a natural extension, although biased, of the multiple correlation coefficient squared, R^2, was given by

$$\eta^2_{yx} = 1 - \Lambda \tag{8.7.31}$$

where Λ was Wilks' Λ criteria used to test for no linear relationships between **Y** and **X** or that $\Sigma_{yx} = \mathbf{0}$, the test of independence under normality. From (8.7.23),

$$\Lambda = \frac{|\mathbf{E}|}{|\mathbf{E}+\mathbf{H}|} = \prod_{i=1}^{s}\left(1 - r_i^2\right) \tag{8.7.32}$$

where $s = \min(p, q)$ in CCA. Hotelling (1936) referred to Λ as the vector alienation coefficient (VAC). However, Rozeboom (1965) called $\eta_{yx} = \sqrt{1-\Lambda}$ the vector correlation coefficient (VCC). In the CCA model, observe that

$$\begin{aligned}\eta^2_{yx} &= 1 - \left|\mathbf{I} - \mathbf{S}_{yy}^{-1/2}\mathbf{S}_{yx}\mathbf{S}_{xx}^{-1}\mathbf{S}_{xy}\mathbf{S}_{yy}^{-1/2}\right| \\ &= 1 - \prod_{i=1}^{s}\left(1 - r_i^2\right) \\ &= \sum_{i=1}^{s} r_i^2 - \sum_{i<j} r_i^2 r_j^2 + \sum_{i<j<k} r_i^2 r_j^2 r_k^2 + \ldots + (-1)^s\, r_1^2 r_2^2 r_s^2\end{aligned} \tag{8.7.33}$$

so that η_{yx} is a generalized measure of association between domain **Y** and domain **X**. From (8.7.33), observe that $\eta^2_{yx} \geq r_1^2$. When $s = 1$ or the number of variables in either of the domains **Y** or **X** is one, $\eta^2_{yx} = r_1^2 = R^2$. Or, the VCC reduces to the multiple correlation coefficient. If **X** and **Y** are uncorrelated, $\eta^2_{yx} \longrightarrow 0$. But, if **X** and **Y** are dependent, $\eta^2_{yx} \longrightarrow 1$ independent of the values of $\rho_2, \rho_3, \ldots, \rho_s$ if $\rho_1 = 1$. Thus r_1 is a better measure of association than η_{yx}. Using (8.7.30), one may reduce the bias in r_1^2.

Instead of η^2_{yx} or r_1^2, Yanai (1974) suggested the generalized coefficient of determination (GCD) defined as

$$\begin{aligned}\widetilde{\eta}^2_{yx} &= \operatorname{tr}(\mathbf{S}_{yy}^{-1/2}\mathbf{S}_{yx}\mathbf{S}_{xx}^{-1}\mathbf{S}_{xy}\mathbf{S}_{yy}^{-1/2})/s \\ &= \sum_{i=1}^{s} r_i^2/s\end{aligned} \tag{8.7.34}$$

where $s = \min(p, q)$. The statistic $0 \leq \widetilde{\eta}^2_{yx} \leq 1$ reduces to r^2 in the case of simple linear regression and R^2 in multiple linear regression. Furthermore,

$$r_s^2 \leq \widetilde{\eta}^2_{yx} \leq r_1^2 \leq \eta^2_{yx} \tag{8.7.35}$$

Takeuchi, Yanai and Mukherjee (1982, p. 251). Thus, as an overall measure of association between two sets of variables one may use $\widetilde{\eta}^2_{yx}$ or r_1^2, Cramer and Nicewander (1979). One may of course replace the sample covariance matrices with sample correlation matrices in the above. Using the δ-method and assuming multivariate normality, Ogasawara (1998) has obtained the asymptotic standard errors for several measures of multivariate association.

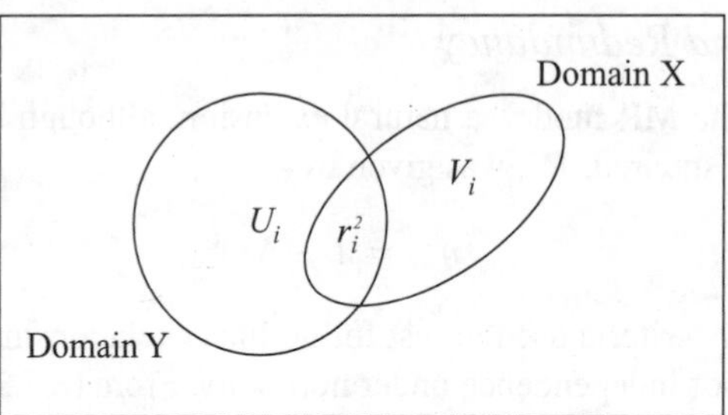

FIGURE 8.7.1. Venn Diagram of Total Variance

An overall measure of association does not help in determining which of the original variables are most influential in the construction of the canonical variates. For this one uses the absolute values of the size of the coefficients in the eigenvectors $\mathbf{c}_i$ and $\mathbf{d}_i$ using standardized variables. One may also investigate the correlations of the canonical variates with each variable as given in (8.7.14). These correlation represent the contribution of a single variable to the canonical construct, ignoring the other variables in the set.

Because we have two sets of variables and two sets of canonical variates the notion of "influence" is more complex since we have within and between domain influence. To study the relationship within a domain, the entries in Table 8.7.1 are replaced by sample estimates. When $k = s = \min(p, q) = p < q$, the R^2 entries in each row sum to 1 so that each canonical variate accounts for a portion of the total sample in their variance in each variable $y_1, y_2, \ldots, y_p$ or $x_1, x_2, \ldots, x_q$ as you move down the structure matrix row by row. Using (8.6.16) with sample values, one may estimate the proportion of the variance in the domain $\mathbf{Y}$ accounted for by each canonical variate $U_1, U_2, \ldots, U_s$. Using domain $\mathbf{X}$, one may estimate the contribution of $V_1, V_2, \ldots, V_s$. Furthermore, $r_1^2, r_2^2, \ldots, r_s^2$ is a measure of variability in U_i accounted for by V_i or conversely. Thus, r_i^2 is a measure of shared variance.

Representing the domains as a Venn diagram, Figure 8.7.1, the total variance of each canonical variate in each domain may be partitioned into three parts: (1) the average unique variance in $\mathbf{Y}$ due to U_i; (2) the shared variance measured by r_i^2; and (3) the average unique variance in $\mathbf{X}$ due to V_i.

Using Figure 8.7.1, Stewart and Love (1968) constructed a redundancy index (RI) to estimate the proportion of the variance in domain $\mathbf{Y}$ that is attributed to $U_i = \mathbf{a}_i'\mathbf{Y}$ and $V_i = \mathbf{b}_i'\mathbf{X}$ by multiplying the average variance in domain $\mathbf{Y}$ by the shared variance as measured by r_i^2. The redundancy of $\mathbf{Y}$ given $\mathbf{X}$ as measured by $U_i \mid V_i$ and r_i^2 is defined as

$$\begin{aligned} RI_{U_i}(\mathbf{Y} \mid V_i) &= \left\{ \frac{[\mathbf{s}(U_i)]'[\mathbf{s}(U_i)]}{p} \right\} r_i^2 \quad i = 1, 2, \ldots, p \\ &= \left(\frac{\mathbf{c}_i'\mathbf{R}_{yy}\mathbf{R}_{yy}\mathbf{c}_i}{p} \right) r_i^2 \end{aligned} \tag{8.7.36}$$

Conversely, the redundancy of $\mathbf{X}$ given $\mathbf{Y}$ as measured by $V_i \mid U_i$ and r_i^2 is

$$RI_{V_i}(\mathbf{X} \mid U_i) = \left\{ \frac{[\mathbf{s}(V_i)]' [\mathbf{s}(V_i)]}{q} \right\} r_i^2 \quad i = 1, 2, \ldots, q$$
$$= \left(\frac{\mathbf{d}_i' \mathbf{R}_{xx} \mathbf{R}_{xx} \mathbf{d}_i}{q} \right) r_i^2 \tag{8.7.37}$$

Because the correlation between Y_i and U_i is not equal to the correlation between X_i and V_i, and because $p \neq q$, the redundancy indexes are asymmetric in that $RI_{U_i}(\mathbf{Y} \mid V_i) \neq RI_{V_i}(\mathbf{X} \mid U_i)$.

Summing the redundancy indexes over all canonical variates within each domain, the total redundancy for $\mathbf{Y}$ given $V_1, V_2, \ldots, V_s$ is

$$R(\mathbf{Y} \mid \mathbf{V}) = \text{tr}\left(\mathbf{S}_{yx}\mathbf{S}_{xx}^{-1}\mathbf{S}_{xy}\right) / \text{tr}\left(\mathbf{S}_{yy}\right)$$
$$= \text{tr}\left(\mathbf{R}_{yx}\mathbf{R}_{xx}^{-1}\mathbf{R}_{xy}\right) / p \tag{8.7.38}$$
$$= \sum_{i=1}^{p} R^2_{Y_i.\mathbf{X}} / p$$

while the redundancy index for $\mathbf{X}$ given $U_1, U_2, \ldots, U_s$ is

$$R(\mathbf{X} \mid \mathbf{U}) = \text{tr}\left(\mathbf{S}_{xy}\mathbf{S}_{yy}^{-1}\mathbf{S}_{yx}\right) / \text{tr}\left(\mathbf{S}_{xx}\right)$$
$$= \text{tr}\left(\mathbf{R}_{xy}\mathbf{R}_{yy}^{-1}\mathbf{R}_{yx}\right) / q \tag{8.7.39}$$
$$= \sum_{i=1}^{q} R^2_{\mathbf{X}_i.\mathbf{Y}} / q$$

Since $R(\mathbf{Y} \mid \mathbf{V}) \neq R(\mathbf{X} \mid \mathbf{U})$, the index allows one to evaluate which domain is more predictable. Because the total redundancy of domain $\mathbf{Y}$ (or $\mathbf{X}$) is equal to the average of the squared multiple correlations $R^2_{Y_i.\mathbf{X}}$ (or $R^2_{X_i.\mathbf{Y}}$) it is not a very good measure of overall association between $\mathbf{Y}$ and $\mathbf{X}$, Cramer and Nicewander (1979). Ogasawara (1998) recommends that one average the redundancy indexes given in (8.7.38) and (8.7.39) to obtain a naive estimator of overall association.

e. *Partial, Part and Bipartial Canonical Correlation*

CCA was originally developed to study the relation between two sets of variables $\mathbf{Y}$ and $\mathbf{X}$. However, one often finds that a set of covariates $\mathbf{Z}$ influences both domains $\mathbf{Y}$ and $\mathbf{X}$. Then one wants to investigate the partial canonical correlation between $\mathbf{Y}$ and $\mathbf{X}$ after removing the linear influence of $\mathbf{Z}$, Roy (1957, p. 40) and Rao (1969). Assuming

$$\mathbf{U} = \begin{bmatrix} \mathbf{Y} \\ \mathbf{X} \\ \mathbf{Z} \end{bmatrix} \sim N_{p+q+r}\left\{ \begin{bmatrix} \mu_1 \\ \mu_2 \\ \mu_3 \end{bmatrix}, \Sigma = \left[\Sigma_{ij}\right] \right\} \tag{8.7.40}$$

the covariance matrix of **Y** and **X** given **Z** is

$$\Sigma_{.3} = \begin{bmatrix} \Sigma_{11.3} & \Sigma_{12.3} \\ \Sigma_{21.3} & \Sigma_{22.3} \end{bmatrix} = \begin{bmatrix} \Sigma_{11} - \Sigma_{13}\Sigma_{33}^{-1}\Sigma_{31} & \Sigma_{12} - \Sigma_{13}\Sigma_{33}^{-1}\Sigma_{32} \\ \Sigma_{21} - \Sigma_{23}\Sigma_{33}^{-1}\Sigma_{31} & \Sigma_{22} - \Sigma_{23}\Sigma_{33}^{-1}\Sigma_{32} \end{bmatrix} \tag{8.7.41}$$

so that a test of partial independence of the two sets of variables **Y** and **X**, after partialing out **Z** from both **Y** and **X** becomes

$$H : \Sigma_{12.3} = \mathbf{0} \tag{8.7.42}$$

To test (8.7.42), one may apply the procedure used to test for independence in (8.7.22) with Σ_{yy}, $\Sigma_{yx}\Sigma_{xy}$, Σ_{xx} replaced by $\Sigma_{11.3}$, $\Sigma_{12.3}$, $\Sigma_{21.3}$ and $\Sigma_{22.3}$, respectively. Either of the determinantal equations

$$\begin{aligned} \left|\Sigma_{12.3}\Sigma_{22.3}^{-1}\Sigma_{21.3} - \rho_{.3}^2\Sigma_{11.3}\right| &= 0 \\ \left|\Sigma_{21.3}\Sigma_{11.3}^{-1}\Sigma_{12.3} - \rho_{.3}^2\Sigma_{22.3}\right| &= 0 \end{aligned} \tag{8.7.43}$$

may be evaluated in the sample with $\Sigma_{ij.3}$ replaced with $\mathbf{S}_{ij.3}$ or $\mathbf{R}_{ij.3}$. The sample roots $r_{i.3}^2$ are called the squares of the partial canonical correlations. The positive square roots $r_{i.3}$ represent the maximal correlation between the partial canonical variates $U_i^* = \mathbf{a}_i'\mathbf{e}_y$ and $V_j^* = \mathbf{b}_i'\mathbf{e}_x$ where $\mathbf{e}_y$ and $\mathbf{e}_x$ represent the residual vectors obtained after regressing **Y** on **Z** and **X** on **Z**. Using the "partial" roots of (8.7.43) to test (8.7.42), the Λ criterion is

$$\Lambda_o' = \prod_{i=1}^{s}\left(1 - r_{i.3}^2\right) = \frac{\left|\mathbf{S}_{11.3} - \mathbf{S}_{12.3}\mathbf{S}_{22.3}^{-1}\mathbf{S}_{21.3}\right|}{|\mathbf{S}_{11.3}|} \tag{8.7.44}$$

where $\Lambda_o' \sim U(p, q, v_e)$ and $v_e = n - r - q - 1$ and r is the number of elements in the partialed out set **Z**. Following Fujikoshi (1977) with

$$\Lambda_k' = \prod_{i=k+1}^{s}\left(1 - r_{i.3}^2\right) \tag{8.7.45}$$

one may test

$$H_k : \rho_{1.3} \neq 0, \rho_{2.3} \neq 0, \ldots, \rho_{k.3} \neq 0, \rho_{(k+1).3} = \cdots = \rho_{s.3} = 0 \tag{8.7.46}$$

where $s = \min(p, q)$. For Lawley's criterion,

$$X_L^2 = -\left[(n - r - k - 1) - \frac{1}{2}(p + q + 1) + \prod_{i=1}^{k} 1/r_{i.3}^2\right] \log \Lambda_k' \tag{8.7.47}$$

Under (8.7.46), X_L^2 is approximately distributed as $\chi^2(v)$ with $v = (p - k)(q - k)$ degrees of freedom for $k = 0, 1, 2, \ldots, s$.

Modifying (8.7.41) to

$$\Sigma_{1(2.3)} = \begin{bmatrix} \Sigma_{11} & \Sigma_{12.3} \\ \Sigma_{21.3} & \Sigma_{22.3} \end{bmatrix} \tag{8.7.48}$$

part canonical correlations and variates may be examined with the variation in the **Z** domain removed from the **X** domain, but not the **Y** domain. This is accomplished by considering the eigenequations

$$\begin{aligned} \left|\Sigma_{12.3}\Sigma_{22.3}^{-1}\Sigma_{21.3} - \rho_{i(2.3)}^2\Sigma_{11}\right| &= 0 \\ \left|\Sigma_{21.3}\Sigma_{11}^{-1}\Sigma_{12.3} - \rho_{i(2.3)}^2\Sigma_{22.3}\right| &= 0 \end{aligned} \tag{8.7.49}$$

This is a generalization of a part correlation in univariate analysis, Timm and Carlson (1976). Our discussion regarding overall multivariate association, redundancy and the interpretation of canonical variates follow immediately for part and partial canonical correlation analysis. To test for part independence, the partial canonical correlations in (8.7.46) are replaced by part canonical correlations.

Timm and Carlson (1976) also discuss the concept of bipartial canonical correlation analysis as an extension of bipartial correlation. Given two univariate variables Y and X, suppose one knows that the association is biased by two sets of covariates that differentially effect Y and X. That is, variable Z influences Y and W influences X. Then forming the residuals after regressing Y on Z and X on W the correlation between the residuals is

$$r_{(1.4)(2.3)} = \frac{r_{12} - r_{14}\,r_{42} - r_{13}\,r_{32} + r_{14}\,r_{43}\,r_{32}}{\sqrt{1-r_{14}^2}\,\sqrt{1-r_{23}^2}} \tag{8.7.50}$$

where $1 \equiv Y,\ 2 \equiv X,\ 3 \equiv Z$ and $4 \equiv W$.

To extend bipartial correlation to multivariate data domains, we assume that $\mathbf{U}' = [\mathbf{Y}', \mathbf{X}', \mathbf{W}', \mathbf{Z}'] \sim N_{p+q+r+t}[\mathbf{0}, \Sigma_{ij}]$ where $i, j = 1, 2, 3, 4$ and construct the sample covariance matrix based upon n observations as follows

$$\mathbf{S}_{(1.4)(2.3)} = \begin{bmatrix} \mathbf{S}_{11} - \mathbf{S}_{14}\mathbf{S}_{44}^{-1}\mathbf{S}_{42} & \mathbf{S}_* \\ \mathbf{S}_*' & \mathbf{S}_{22} - \mathbf{S}_{23}\mathbf{S}_{33}^{-1}\mathbf{S}_{32} \end{bmatrix} \tag{8.7.51}$$

where

$$\mathbf{S}_* = \mathbf{S}_{12} - \mathbf{S}_{14}\mathbf{S}_{44}^{-1}\mathbf{S}_{42} - \mathbf{S}_{13}\mathbf{S}_{33}^{-1}\mathbf{S}_{32} + \mathbf{S}_{14}\mathbf{S}_{44}^{-1}\mathbf{S}_{43}\mathbf{S}_{33}^{-1}\mathbf{S}_{32}$$

Again, $\mathbf{S}_{(1.4)(2.4)}$ has the block structure

$$\mathbf{S}_{(1.4)(2.3)} = \begin{bmatrix} \mathbf{A}_{11} & \mathbf{A}_{12} \\ \mathbf{A}_{21} & \mathbf{A}_{22} \end{bmatrix} \tag{8.7.52}$$

Associating the partitioned matrices in (8.7.51) with those in (8.7.52) one may solve the eigenequation

$$\left|\mathbf{A}_{21}\mathbf{A}_{11}^{-1}\mathbf{A}_{12} - \rho^2\mathbf{A}_{22}\right| = 0 \tag{8.7.53}$$

for the eigenvectors $\mathbf{a}_i$, where

$$\mathbf{b}_i = \mathbf{A}_{22}^{-1}\mathbf{A}_{21}\mathbf{a}_i/\rho_{i(1.4)(2.3)} \tag{8.7.54}$$

are the bipartial canonical variates. The bipartial sample canonical correlations $r_{i(1.4)(2.3)}$ obtained using the sample data are estimates of the population bipartial canonical correlations $\rho_{i(1.4)(2.3)}$ for $i = 1, 2, \ldots, s = \min(p, q)$.

To test the hypotheses of bipartial independence

$$H : \Sigma_{12} - \Sigma_{13}\Sigma_{33}^{-1}\Sigma_{32} - \Sigma_{14}\Sigma_{44}^{-1}\Sigma_{42} + \Sigma_{13}\Sigma_{33}^{-1}\Sigma_{34}\Sigma_{44}^{-1}\Sigma_{42} = \mathbf{0} \tag{8.7.55}$$

or equivalently

$$\begin{aligned} H_k : \quad & \rho_{1(1.4)(2.3)} \neq 0, \ldots, \rho_{k(1.4)(2.3)} \neq 0, \\ & \rho_{(k+1)(1.4)(2.3)} = 0 = \ldots = \rho_{s(1.4)(2.3)} = 0 \end{aligned} \tag{8.7.56}$$

Wilks' Λ statistic

$$\Lambda_k^* = \prod_{i=k+1}^{s}\left(1 - r_{i(1.4)(2.3)}^2\right) \tag{8.7.57}$$

is formed. For Lawley's criterion

$$X_L^2 = -\{[n - \max(r,\ t) - k - 1] - \frac{1}{2}(p + q + 1) + \sum_{i=1}^{k} 1/r_{i(1.4)(2.3)}^2\} \log \Lambda_k^* \tag{8.7.58}$$

where $X_L^2 \mathrel{\dot\sim} \chi^2(p-k)(q-k)$ under H_k in equation (8.7.56).

f. Predictive Validity in Multivariate Regression using CCA

For the multivariate regression model in Chapter 4, we investigated predictive precision using the expected mean square error of prediction. However, we indicated that a measure of linear predictive precision may be defined as the square of the zero-order Pearson correlation between a set of actual observations and the estimates obtained from the calibration sample. For the multivariate regression model, this leads one to the CCA model. Before discussing the multivariate model, recall that for one dependent variable that the population coefficient of determination is identical to the square of the population canonical correlation coefficient. Browne (1975a) developed a formula to estimate the predictive precision of a multiple linear regression model with k random independent variables. Letting ρ_c^2 represent the square of the zero-order Pearson correlation between an observation Y and the predicted value $\widehat{Y}$ based on the linear parametric function $\mathbf{X}\widehat{\boldsymbol{\beta}}$ in the population, Browne (1975a) obtained a formula to estimate of predicted precision without the need for a validation sample under multivariate normality. His estimate for prediction precision follows

$$R_c^2 = \frac{(n-k-3)(R_a^2)^2 + R_a^2}{(n-2k-2)R_a^2 + k} \tag{8.7.59}$$

where R_a^2 is the adjusted estimate of the population coefficient of determination ρ^2. Monte Carlo studies performed by Browne (1975a) show that the estimate results in very small bias. Cotter and Raju (1982) confirmed Browne's result using cross validation procedures. For an estimate of the square of the cross validation correlation coefficient or predictive precision, Browne assumed multivariate normality and random independent variables. Nicholson (1960) proposed the estimate

$$R_c^2 = 1 - \left(\frac{n+k+1}{n}\right)(1 - R_a^2) \tag{8.7.60}$$

of ρ_c^2 assuming fixed independent variables which Uhl and Eisenberg (1970) found to accurately estimate the population parameter ρ_c^2. In multiple linear regression, the adjusted estimate of the coefficient of determination R_a^2 defined in (4.2.36) provides a reasonable estimate of the population coefficient of determination whether the independent variables are considered fixed or random.

To provide a reasonable estimate of predictive precision using the CCA model, we must estimate the square of the correlation between a the observation matrix $\mathbf{Y}_{m \times p}$ and $\widehat{\mathbf{Y}}_{m \times p} = \mathbf{X}\widehat{\mathbf{B}}$ where the $p \times q$ parameter matrix is $\widehat{\mathbf{B}}$ based upon the set of k variables in the calibration sample. These canonical correlations depend on a validation sample. Expression (8.7.30) is used to estimate the p population canonical correlations within the calibration sample. For the multivariate regression model with random $\mathbf{X}$, there are $s = \min(p, q) = p$ (assuming $p \leq q$) estimates of predictive precision, one for each of the p dependent variables. An estimate of predictive precision using the CCA model that does not depend on a validation sample has not yet been developed. However, if one has a validation sample, by (8.7.35), one might use $\widetilde{\eta}^2_{y\widehat{y}}$ as an estimate of predictive precision. It provides a natural estimate of predictive precision where the matrix $\mathbf{Y}_{m \times p}$ is a validation sample and the matrix $\widehat{\mathbf{Y}}_{m \times p}$ is an estimate using the parameter matrix of the calibration sample with $m \doteq n$. One may of course propose other estimates that involve other scalar functions, for example, the Euclidean norm squared. Following Nicholson (1960) with η_a^2 defined in (4.2.39), a reasonable overall estimate of predictive precision for fixed independent variates is the estimate

$$\eta_c^2 = 1 - \left(\frac{n+k+1}{n}\right)(1 - \eta_a^2) \tag{8.7.61}$$

As an extension of Browne's result to the multivariate model, one might consider the estimate

$$\eta_c^2 = \frac{(n-k-3)(\eta_a^2)^2 + \eta_a^2}{(n-2k-2)\eta_a^2 + k} \tag{8.7.62}$$

however, its use in practice requires verification. Ogasawara (1998) provides an overview of multivariate measures of association.

g. *Variable Selection and Generalized Constrained CCA*

To find the best subset of variables in $\mathbf{Y}' = [Y_1, Y_2, \ldots, Y_p]$ and $\mathbf{X}' = [X_1, X_2, \ldots, X_q]$ is complicated since it depends on the measure of multivariate association one employes.

Using (8.7.34), Takeuchi et al. (1982, p. 253) develop a forward selection procedure for entering pairs of variables into a CCA. The pairs of variables are entered to maximize the sum of the squared canonical correlations at each step. With the forward order established, variables are sequentially deleted from least important to most important to maintain a subset of significant canonical variates in order to maintain a significant proportion of explained correlation structure. Any estimate of predictive precision must be determined after one has established a model with reasonable fit.

All of the CCA examples discussed in this section are special cases of the generalized constrained canonical correlation analysis (GCCANO) procedure recently developed by Yanai and Takane (1999). For discrete data, their model may be used for correspondence analysis, part and bi-partial canonical correlation analysis, and for the development of constrained MANOVA and GMANOVA models.

8.8 Canonical Correlation Analysis Examples

a. *Rohwer CCA (Example 8.8.1)*

Professor William D. Rohwer at the University of California at Berkeley selected $n = 37$ kindergarten students in a low-socioeconomic-status area to investigate how well two paired associate (PA) learning tasks using action words versus nonaction words are related to three student achievement tests. The set of variables $\mathbf{Y}' = [NA, SS]$ are the number of items correct out of 20 (on two exposures) of the PA learning tasks using action and still prompt words. The set of variables $\mathbf{X}' = [SAT, PEA, RAV]$ are the scores on a student achievement test (SAT), the Peabody Picture Vocabulary Test (PEA) and the Ravin Progressive Matrices Test (RAV). The sample correlation matrix for the two sets of variables follows.

$$\mathbf{R} = \begin{bmatrix} \mathbf{R}_{yy} & \mathbf{R}_{yx} \\ \mathbf{R}_{xy} & \mathbf{R}_{xx} \end{bmatrix} = \left[\begin{array}{cc|ccc} 1.0000 & & & & \text{(Sym)} \\ .7951 & 1.0000 & & & \\ \hline .2617 & .3341 & 1.0000 & & \\ .6720 & .5876 & .3703 & 1.0000 & \\ .3390 & .3404 & .2114 & .3548 & 1.0000 \end{array}\right] \tag{8.8.1}$$

To analyze the sample correlation matrix in (8.8.1), program m8_8_1.sas is used with the SAS procedure CANCORR. For this example, $p = 2$ and $q = 3$. The correlations of the two sets are represented as 'VAR' variables and as 'WITH' variables in the SAS procedure. Solving (8.7.19), the sample canonical correlations are

$$\begin{aligned} r_1 &= 0.6889 & r_2 &= 0.1936 \\ r_1^2 &= 0.4746 & r_2^2 &= 0.0375 \end{aligned} \tag{8.8.2}$$

The adjusted canonical correlations using (8.7.30) are $\widehat{r}_1 = 0.6612$ and $\widehat{r}_2 = 0.1333$. Using (8.7.15) with sample estimates, we see that one canonical variate in each set accounts for 92.3% of the correlation structure. Using Theorem 2.6.8, the $|\mathbf{R}_{xy}\mathbf{R}_{yy}^{-1}\mathbf{R}_{yx}\mathbf{R}_{xx}^{-1}| = \prod_i r_i^2/\left(1 - r_i^2\right)$. The CANCORR procedure calculates $\widehat{\lambda}_i = r_i^2/\left(1 - r_i^2\right)$ and the ratios of $\widehat{\lambda}_i / \sum_i \widehat{\lambda}_i$ as the proportion of the correlation structure explained by each canonical variate.

Using (8.7.24), and assuming joint normality, the likelihood ratio criterion for testing independence, $H : \Sigma_{yx} = \mathbf{0}$, is

$$\Lambda_o = \prod_{i=1}^{2} \left(1 - r_i^2\right) = 0.5057$$

Relating Λ_o to an F distribution, the p-value for the test of independence is 0.0010. For $\alpha = 0.05$, the test of independence is rejected. Using expression (4.2.39), a measure of association for the multivariate model is estimated by $\eta_a^2 = 0.4437$, or approximately 44.4% of the variance in the Y domain is accounted for by the X domain variables. Using (8.7.61) to estimate the overall cross validation precision of the model, we have that the shrucken estimate is approximately $\eta_c^2 = 0.3356$ using two canonical variates.

To evaluate if we can reduce the number of canonical variates, (8.7.29). The SAS procedure relates X_L^2 to an approximate F distribution. Instead of performing a significance test, one might retain the number of canonical variates that explain 70% $-$ 80% of the correlation structure. For our example, we need only one canonical variate. Using standardized variables,

$$\begin{aligned} U_1 &= 0.7753\,(NA) + 0.2661\,(SS) \\ V_1 &= 0.0520\,(SAT) + 0.8991\,(PEA) + 0.1830\,(RAV) \end{aligned} \tag{8.8.3}$$

The loadings indicate that NA is most important to U_1, while PEA is most important to V_1. Using (8.7.14), one may measure the contribution of each variable to each canonical variate, ignoring the other variables. For our example,

$$\begin{aligned} \boldsymbol{\rho}_{\mathbf{YU}} &= \begin{bmatrix} 0.987 \\ 0.883 \end{bmatrix} \quad \begin{matrix} (NA) \\ (SS) \end{matrix} \\ \boldsymbol{\rho}_{\mathbf{XV}} &= \begin{bmatrix} 0.424 \\ 0.983 \\ 0.513 \end{bmatrix} \quad \begin{matrix} (SAT) \\ (PEA) \\ (RAV) \end{matrix} \end{aligned} \tag{8.8.4}$$

Thus, individually (NA) and (SS) are equally important to U_1 and (PEA) is twice as important to V_1 than either SAT or RAV. The correlations measure the individual effects while the loadings measure the simultaneous effects of the variables on the canonical variates. Averaging the squares of the coefficients in (8.8.4), we may estimate the proportion of the variance in each domain accounted for by each canonical variate. For our example,

$$\begin{aligned} U_1^* &= \frac{(.987)^2 + (.883)^2}{2} = 0.8764 \\ V_1^* &= \frac{(.424)^2 + (.983)^2 + (.513)^2}{3} = 0.4698 \end{aligned}$$

Thus, 88% of the variance in the **Y** set is accounted for by U_1 and only 47% of the **X** set is accounted for by V_1.

Evaluating (8.7.36) and (8.7.37), we may calculate the proportion of the variance the opposite canonical variate accounts for in a given set. That is,

$$RI_{U_1}(Y \mid V_1) = (0.8764)\, r_1^2 = 0.4159$$
$$RI_{V_1}(X \mid U_1) = (0.4698)\, r_1^2 = 0.2230$$

The proportion of the variance in the set $\mathbf{X} = \{SAT, PEA, RAV\}$ accounted for by V_1 is 0.2230 and the proportion of the variance in the set $\mathbf{Y} = \{NA, SS\}$ accounted for by U_1, is 0.4159.

In summary, given the two sets of variables

$$\mathbf{Y} = \{NA, SS\} \text{ and } \mathbf{X} = \{SAT, PEA, RAV\}$$

where

$$U_1 = 0.7753\,(NA) + 0.2662\,(SS)$$

and

$$V_1 = 0.0520\,(SAT) + 0.8991\,(PEA) + 0.1830\,(RAV)$$

it appears that the proportion of variance "in common" to the two canonical variates is about 47% since $r_1^2 = .4746$. However, 88% of the variance in the set **Y** is accounted for by U_1, and only 42% of the variance in **Y** is accounted for by the canonical variance V_1. Similarly, 47% of the variance in the set **X** is accounted for by V_1, but only 23% of the variance in **X** is accounted for by the canonical variate U_1. Averaging the redundancy indices, the overall association between the two data sets is $(0.4159 + 0.2230)/2 = 0.3194$.

b. Partial and Part CCA (Example 8.8.2)

In Exercises 8.3, Problem 1, Project Talent data was analyzed using PCA. The eleven variables may be partitioned into sets of variables as follows.

$$\mathbf{Y} = \{G1, G2\}$$
$$\mathbf{X} = \{VE, VR, VC\}$$
$$\mathbf{Z} = \{NC, NA, NM\}$$
$$\mathbf{W} = \{IS, IP, IO\}$$

The set of variables in domain **Y** (1, 2) are general information tests, the set of variables in domain **X** (3, 4, 5) are verbal ability tests domain **Z** (6, 7, 8) contains nonverbal ability tests, and finally, domain **W** (9, 10, 11) contains interest measures.

For this data set, we illustrate a partial canonical correlation analysis using domains **Y** and **X**, removing the joints effects of set $\mathbf{Q} = [\mathbf{Z}, \mathbf{W}]$ from both domains **Y** and **X**. Thus, (8.7.41) is analyzed with Σ_{ij} replaced by $\mathbf{R}_{ij}$. The matrix of partial correlations is

$$\mathbf{R}_{.3} = \begin{bmatrix} \mathbf{R}_{11.3} & \mathbf{R}_{12.3} \\ \mathbf{R}_{21.3} & \mathbf{R}_{22.3} \end{bmatrix} = \left[\begin{array}{rr|rrr} 1.0000 & & & & (\text{Sym}) \\ 0.6365 & 1.000 & & & \\ \hline -0.0157 & 0.0920 & 1.0000 & & \\ 0.3081 & 0.1301 & 0.1064 & 1.0000 & \\ -0.3055 & -0.1985 & 0.0486 & -0.0265 & 1.0000 \end{array}\right]$$

Alternatively, replacing $\mathbf{R}_{11.3}$ with $\mathbf{R}_{11}$, the matrix of part correlations may be analyzed. Then, $\mathbf{Q}$ is removed from the $\mathbf{X}$ set but not the $\mathbf{Y}$ set. In the SAS procedure, one would input $\mathbf{R}_{1(2.3)}$ and perform a canonical analysis on the modified matrix. The SAS code for both the partial and part canonical correlation analysis is provided in Program m8_2_2.sas. The output is similar in format to Example 8.2.1.

In summary, given the three sets of variables

$$\mathbf{Y} = \{G1, G2\}, \ \mathbf{X} = \{VE, VR, VC\} \text{ and}$$
$$\mathbf{Q} = \{NC, NA, NM, IS, IP, IO\}$$

the significant partial canonical variates are

$$U_1 = 1.1296(G1) - 0.2281(G2)$$
$$V_1 = -0.1342\,(VE) + 0.7278\,(VR) - 0.6625(VC)$$

sharing only 19% of the common variance since $r^2_{1.3} = 0.1897$. The correlation structure for each canonical variate follows.

$$\boldsymbol{\rho}_{\mathbf{YU}} = \begin{bmatrix} 0.9844 \\ 0.4909 \end{bmatrix} \quad \begin{matrix} (G1) \\ (G2) \end{matrix}$$

$$\boldsymbol{\rho}_{\mathbf{XV}} = \begin{bmatrix} -0.0890 \\ 0.7311 \\ -0.0883 \end{bmatrix} \quad \begin{matrix} (VE) \\ (VR) \\ (VC) \end{matrix}$$

Thus, $G1$ is most important to U_1 while VR contributes most to V_1. From the redundancy analysis, 60% of the variance in the $\mathbf{Y}$ set (after removing $\mathbf{Q}$) is account for by U_1 and only 11% is accounted for by V_1. Only 34% of the variance in the $\mathbf{X}$ set (after removing Q) is accounted for by V_1, and only 6% is accounted for by U_1.

Reviewing the output for the part canonical analysis, we see that it is better to remove the influence of the set $\mathbf{Q}$ from $\mathbf{X}$, but not $\mathbf{Y}$. The proportion of shared variance increases to 23% while the variance explained the part canonical variates are almost the same as the partial canonical variates.

To perform a bipartial canonical analysis using the Project Talent data, one would analyze the matrix in (8.7.51) with $\mathbf{S}_{ij}$ replaced by $\mathbf{R}_{ij}$. The resulting matrix may be input into PROC CANCORR.

Exercises 8.8

1. Shin (1971) collected data on intelligence using six creativity measures and six achievement measures for 116 subjects in the eleventh grade in suburban Pittsburgh. The correlation matrix for the study is given in Table 8.3.1 and an explanation of the variables is summarized in Example 8.3.1. Inspection of Table 8.3.1 indicates that four out of the six creativity measures (tests 4, 5, 6, and 7) are highly correlated with synthesis and evaluation (tests 12 and 13), as is IQ (test 1). What else do you notice?

(a) Use canonical correlation analysis to investigate the relationship between the six achievement variables and IQ, and the six creativity variables.

(b) Use canonical correlation analysis to investigate the relationship between the six achievement variables, and IQ and the six creativity variables.

(c) By partialing out the IQ variable from both sets of variables, use partial canonical correlation analysis to analyze the data and compare your results with parts a and b.

(d) Would a part canonical correlation analysis of the data be meaningful? If so, why? If not why not?

(e) Summarize your findings. Include in your discussion an analysis of predictive precision.

2. For the partition of Project Talent variables given in Example 8.8.2, analyze the bi-partial correlation matrix $\mathbf{R}_{(\mathbf{Y}.\mathbf{Z})(\mathbf{X}.\mathbf{W})}$ using PROC CANCORR and summarize your findings.

8.9 Exploratory Factor Analysis

Exploratory factor analysis (EFA) is a causal modeling technique that attempts to "explain" correlations among a set of observed (manifest) variables through the linear combination of a few unknown number of latent (unobserved) random factors. The procedure was originated by the psychologist Charles Spearman in the early 1900's to model human intelligence. Spearman developed the technique to try to understand the causal relationship between the latent human trait intelligence and test scores (grades) obtained in several disciplines. Spearman believed that students' test scores are intercorrelated and that the intercorrelations could be completely explained by a single common latent general intelligence factor g, and that when this factor was removed, the test scores would be uncorrelated. Spearman's model of intelligence assumes that performance on a test (the observed data) is caused by a general unobserved common factor labeled intelligence and a specific or unique factor due to each discipline (test type). Spearman's (1904) single factor model was later generalized by Thurstone (1931, 1947) to multiple factors.

Because both PCA and EFA usually begin with an analysis of the variation of a set of variables as characterized by the correlation or covariance matrix and because both may be used to characterize the variation by a few hypothetical constructs they are often confused. In EFA one may explain all the covariances or correlations with a few common factors that are unobservable or latent. In PCA one needs all components to account for all the covariances or correlations. While a few factors may account for all the intercorrelations (covariances), the same number of factors will not explain as much of the total variance as the same number of principal components. Thus, PCA is concerned with explaining the variance in the variables while EFA is concerned with explaining the covariances. In EFA, the correlation or covariance matrix is partitioned into two parts; that due to the common factors and that due to the unique factors. Any correlations (covariances) not explained by the common factors are associated with the mutually uncorrelated unique (residual) factors. In PCA there is no residual variance, all variance is explained by the components. Another important difference between the two procedures lies in the direction of analysis. In EFA one sets forth a causal model and uses the data to validate the model. This has led to some controversy among statisticians since the power to reject model fit depends on the number of unknown factors and sample size. In PCA there is no causal model since all components are linear combinations of observed data, observed with no error.

The literature associated with EFA is extensive with much of the early work interesting, but obsolete. Early reviews of EFA are included in Harmon (1976) and Mulaik (1972). More recent developments are included in the books by Jöreskog and Sörbom (1979), McDonald (1985), Bollen (1989) and Basilevsky (1994). Statistical treatment of the method is included in Anderson and Rubin (1956) and Lawley and Maxwell (1971). In this section we discuss the exploratory factor analysis model where the structure of the model or underlying theory is not known or specified a priori. The data are used to discover the structure of the model. When a precise theory is set forth for the factor model, the structure of the model is hypothesized a priori and the factor structure is confirmed. This type of factor analysis is called confirmatory factor analysis (CFA) and is discussed in Chapter 10.

a. Population Model for EFA

In EFA we begin with a random observation vector $\mathbf{Y}'_{1\times p} = \left[Y_1, Y_2, \ldots, Y_p\right]$ with mean $\boldsymbol{\mu}$ and covariance matrix Σ. Each vector of p-variables is assumed to have the linear structure

$$\underset{p\times 1}{\mathbf{Y}} - \underset{p\times 1}{\boldsymbol{\mu}} = \underset{p\times k}{\Lambda}\ \underset{k\times 1}{\mathbf{f}} + \underset{p\times 1}{\mathbf{e}} \tag{8.9.1}$$

where $\Lambda = \left[\lambda_{ij}\right]$ is a matrix of regression weights or "loadings" with rank $k \ll p$, $\mathbf{f}$ is a vector of random, unobserved latent factors and $\mathbf{e}$ is a vector of random errors. It is further assumed that $E(\mathbf{Y}) = \boldsymbol{\mu}$, $E(\mathbf{f}) = E(\mathbf{e}) = \mathbf{0}$, $\text{cov}(\mathbf{f}) = \mathbf{I}$, $\text{cov}(\mathbf{Y}) = \Sigma$, $\text{cov}(\mathbf{e}) = \Psi$ is a diagonal matrix with elements $\psi_i > 0$, and that the $\text{cov}(\mathbf{f}, \mathbf{e}) = \mathbf{0}$. These assumptions imply that the covariance matrix Σ for each observation $\mathbf{Y}$ has the structure

$$\begin{aligned}\Sigma &= \text{cov}(\mathbf{Y}) = \text{cov}(\Lambda\, \mathbf{f} + \mathbf{e}) \\ &= \Lambda\, \text{cov}(\mathbf{f})\, \Lambda' + \text{cov}(\mathbf{e}) \\ &= \Lambda\Lambda' + \Psi\end{aligned} \tag{8.9.2}$$

where the covariance matrix Ψ is diagonal, $\Psi = \text{diag}\left[\psi_1, \psi_2, \ldots, \psi_p\right]$. Thus, all the covariances σ_{ij} for $i \neq j$ in Σ depend only on the λ_{ij} through the product $\Lambda\Lambda'$, while the variances σ_{ii} depend on λ_{ij}^2 and ψ_i.

One may also think of (8.9.1) as a linear regression model where

$$\begin{aligned}E(\mathbf{Y} \mid \mathbf{f}) &= \boldsymbol{\mu} + \Lambda\mathbf{f} \\ \text{cov}(\mathbf{Y} \mid \mathbf{f}) &= \Psi = \text{diag}\left[\psi_1, \psi_2, \ldots, \psi_p\right]\end{aligned} \tag{8.9.3}$$

Because the covariance matrix of $\mathbf{Y}$ given $\mathbf{f}$ is diagonal, $\mathbf{f}$ accounts for all the intercorrelations, and hence all linear relationships among the elements of $\mathbf{Y}$. Once the factors $\mathbf{f}$ are partialed out there remains no correlation among the elements of $\mathbf{Y}$. Furthermore, given (8.9.3), the covariance matrix

$$\begin{aligned}\Sigma &= E\left(\mathbf{Y}-\boldsymbol{\mu}\right)\left(\mathbf{Y}-\boldsymbol{\mu}\right)' = E\left(\Lambda\mathbf{f}+\mathbf{e}\right)\left(\Lambda\mathbf{f}+\mathbf{e}\right)' \\ &= \Lambda E\left(\mathbf{f}\mathbf{f}'\right)\Lambda' + \left(\mathbf{e}\mathbf{e}'\right) \\ &= \Lambda\Lambda' + \Psi\end{aligned} \tag{8.9.4}$$

so that the cov $(\mathbf{Y} \mid \mathbf{f}) = \Sigma - \Lambda\Lambda' = \Psi$. Letting $\Psi = \text{diag}\left(\Sigma - \Lambda\Lambda'\right)$, the partial correlation matrix becomes $\Psi^{-1/2}\left(\Sigma - \Lambda\Lambda'\right)\Psi^{-1/2}$.

From (8.9.1) or (8.9.3), we observe that the EFA model makes no distributional assumptions regarding $\mathbf{Y}$. The model parameters are the mean vector $\boldsymbol{\mu}$ and the parameter matrices Λ and Ψ. Later we will make distribution assumptions regarding $\mathbf{Y}$ to test hypotheses of model fit and to obtain maximum likelihood estimates of the model parameters.

Expanding (8.9.1) into a system of linear equations, the model equations became

$$\begin{array}{ccccccccc} Y_1 - \mu_1 & = & \lambda_{11} f_1 & + & \lambda_{12} f_2 & + \ldots + & \lambda_{1k} f_k & + & e_1 \\ \vdots & & \vdots & & \vdots & & \vdots & & \vdots \\ Y_i - \mu_i & = & \lambda_{i1} f_1 & + & \lambda_{i2} f_2 & + \ldots + & \lambda_{ik} f_k & + & e_i \\ \vdots & & \vdots & & \vdots & & \vdots & & \vdots \\ Y_p - \mu_p & = & \lambda_{p1} f_1 & + & \lambda_{p2} f_2 & + \ldots + & \lambda_{pk} f_k & + & e_p \end{array} \tag{8.9.5}$$

where the cov $\left(e_i,\ e_j\right) = 0$ for $i \neq j$, and the λ_{ij} are regression coefficients. For psychologists, the observations Y_i are usually test scores, the λ_{ij} are termed factor loadings, and the residuals e_i are called unique or specific factors. Thus, each test score or observation about the mean is represented as two parts, a common part and a unique part

$$Y_i - \mu_i = c_i + e_i \tag{8.9.6}$$

EFA is used to investigate the unobserved common parts

$$c_i = \lambda_{i1} f_1 + \lambda_{i2} f_2 + \ldots + \lambda_{ik} f_k \tag{8.9.7}$$

whereas PCA is used to study the observed observations Y_i. The unique factor may be regarded as a sum of an error of measurement and a specific factor where the specific factor is associated with a particular test; however, since these do not contribute to any correlation or covariance among test scores, they are omitted in our discussion (see Harmon, 1976).

Corresponding to the structure of Σ given in (8.9.2), Σ is also partitioned into two parts, the common part and the unique part. The variance of a random observation Y_i is

$$\begin{aligned}\sigma_{ii} \equiv \sigma_i^2 &= \lambda_{i1}^2 + \lambda_{i2}^2 + \ldots + \lambda_{ik}^2 + \psi_i \\ \text{var}\left(Y_i\right) &= \text{var}\left(c_i\right) + \text{var}\left(e_i\right) \\ &= h_i^2 + \psi_i\end{aligned} \tag{8.9.8}$$

The variance of the common part of Y_i is represented by h_i^2 and is called the common variance or communality of the response and the var $(e_i) = \psi_i$, the ith diagonal element of Ψ, is termed the unique, or specific variance or the uniqueness of Y_i. The uniqueness is that part of the total variance not accounted for by the common factors, while the communality is that portion of the variance attributed to the common factors.

The covariance between Y_i and Y_j, for $i \neq j$, is

$$\sigma_{ij} = \lambda_{i1}\lambda_{j1} + \lambda_{i2}\lambda_{j2} + \ldots + \lambda_{ik}\lambda_{jk} \tag{8.9.9}$$

where λ_{ij} is the covariance between Y_i and the jth common factor f_j. Using matrix notation, the

$$\begin{aligned} \text{cov}\,(\mathbf{Y},\, \mathbf{f}) &= \text{cov}\,(\Lambda\mathbf{f} + \mathbf{e}, \mathbf{f}) \\ &= \Lambda\,\text{cov}\,(\mathbf{f}, \mathbf{f})\,\Lambda' + \text{cov}\,(\mathbf{e}, \mathbf{f}) \\ &= \Lambda\Lambda' \end{aligned} \tag{8.9.10}$$

since the cov $(\mathbf{e}, \mathbf{f}) = \mathbf{0}$. If standardized variables are analyzed, Σ is a correlation matrix so that (8.9.2) becomes

$$\mathbf{P}_\rho = \Lambda\Lambda' + \Psi \tag{8.9.11}$$

and the loadings λ_{ij} become correlations. Then (8.9.9) becomes

$$\rho_{ij} = \lambda_{i1}\lambda_{j1} + \ldots + \lambda_{ik}\lambda_{jk} \tag{8.9.12}$$

where the diagonal elements of $\mathbf{P}_\rho$ are 1 so that (8.9.8) is written as

$$\begin{aligned} \psi_i &= 1 - h_i^2 \\ &= 1 - \left(\lambda_{i1}^2 + \lambda_{i2}^2 + \ldots + \lambda_{ik}^2\right) \end{aligned} \tag{8.9.13}$$

The quantities $h_i^2 = \sum_{j=1}^k \lambda_{ik}^2$, the inner product of the ith row of Λ with itself, are communalities. They represent the common variance of a variable accounted for by a factor solution.

In our discussion of the EFA model we have assumed that the common factors are orthogonal or uncorrelated as in PCA since the cov $(\mathbf{f}) = \mathbf{I}$. A more general model permits the common factors to be correlated so that the cov $(\mathbf{f}) = \Phi$ where $\Phi \neq \mathbf{I}$. Then Σ has the structure

$$\Sigma = \Lambda\Phi\Lambda' + \Psi \tag{8.9.14}$$

and the resulting common factors are said to be oblique.

A basic property of the EFA model with the structure given by (8.9.14) or (8.9.2) is that the model is scale free under affine transformations of the form $\mathbf{Z} = \mathbf{AY} + \mathbf{b}$ where $\mathbf{A}$ is nonsingular. To see this, observe that the

$$\begin{aligned} \text{cov}\,(\mathbf{Z}) &= \mathbf{A}\,\text{cov}\,(\mathbf{Y})\,\mathbf{A}' \\ &= \mathbf{A}\left(\Lambda\Phi\Lambda'\right)\mathbf{A}' + \mathbf{A}\Psi\mathbf{A}' \\ &= (\mathbf{A}\Lambda)\,\Phi\,(\mathbf{A}\Lambda)' + \mathbf{A}\Psi\mathbf{A}' \\ &= \Lambda_*\Phi\Lambda_*' + \Psi_* \\ &= \Sigma_* \end{aligned}$$

Thus, an affine transformation of the original variables is also applied to Λ and Ψ. This scale free property of the EFA model allows one to analyze either Σ or $\mathbf{P}_\rho$. Even though the model is scale free, some early procedures used to estimate model parameters are not scale free and should be avoided, Harris (1964).

While the EFA model is scale free, the model does not impose sufficient restrictions on the parameters Λ, Φ and Ψ to ensure that Σ may be found to satisfy either (8.9.14) or (8.9.2). The three major problems with the structure are (1) the value of k is unknown, (2) given k, the parameter matrices Λ, Φ, and Ψ are not unique, and (3) even if Ψ is uniquely determined for some k, the factorization of Σ is not unique. These are often called the problem of minimal rank, the problem of equivalent structures, and the transformation problem.

Given a $p \times p$ covariance matrix Σ, a necessary and sufficient condition that there exist k common factors is that there is a matrix Ψ such that the matrix $\Sigma - \Psi$ is $p.s.d.$ and of rank k (see, for example, Anderson and Rubin, 1956, and Jöreskog, 1963). However, Reiersøl (1950, Th 3.3) has shown that if there is a structure for Σ, it is not unique. To determine Ψ uniquely, he infers that the minimum value of k must be attained, although no proof is provided. If $\Sigma - \Psi$ is $p.s.d.$ and of minimal rank k, by the spectral decomposition theorem there exists a matrix Λ such that $\Sigma - \Psi = \Lambda\Lambda'$ so that a structure exists with $\Phi = \mathbf{I}$. Takeuchi, Yanai, and Mukherjee (1982, p. 298) show that Ψ is determined uniquely if

1. $\Psi = \theta \mathbf{I}_p$

2. $\Psi = \theta \left(\text{diag}\, \Sigma^{-1}\right)^{-1}$

Jöreskog (1969a) considered situation (2) and called it image factor analysis.

Given that we have a structure, given k and Ψ, we still have the factor transformation problem. That is, even if a structure exists for some k and the some Ψ, there will be infinitely many others. To see this, let $\mathbf{T}_{k \times k}$ be any nonsingular matrix such that

$$\begin{aligned} \Lambda_* &= \Lambda \mathbf{T}^{-1} \\ \Phi_* &= \mathbf{T} \Phi \mathbf{T}' \end{aligned} \tag{8.9.15}$$

Then,

$$\begin{aligned} \Lambda_* \Phi_* \Lambda_*' + \Psi &= \Lambda \mathbf{T}^{-1} \mathbf{T} \Phi \mathbf{T}' \left(\mathbf{T}'\right)^{-1} \Lambda' + \Psi \\ &= \Lambda \Phi \Lambda' + \Psi \end{aligned}$$

and model (8.9.1) with structure (8.9.14) becomes

$$\mathbf{Y} - \boldsymbol{\mu} = \Lambda_* \mathbf{f}_* + \mathbf{e} \tag{8.9.16}$$

for

$$\begin{aligned} \mathbf{f}_* &= \mathbf{T}\mathbf{f} \\ \Lambda_* \mathbf{f}_* &= \Lambda \mathbf{f} \end{aligned} \tag{8.9.17}$$

This illustrates a fundamental indeterminacy in the EFA model. By observing **Y**, we cannot distinguish between the two models. Thus, without further restrictions on the model, Λ and Φ are not identified. A common restriction is to set $\Phi = \mathbf{I}$. Then the factors are orthogonal and **T** is restricted to an orthogonal matrix. This reduces some of the indeterminacy in the model and is the primary reason for introducing the model with structure Σ given in (8.9.2).

Given the transformation problem in EFA, one usually rotates the k common factors to meaningful constructs. Thurstone (1947, p. 335) described simple structure as follows. The factor matrix Λ should have the following properties.

1. Each row of Λ should have at least one zero.

2. Each column of Λ should have at least k zeros.

3. For all pairs of columns in Λ, there should be several rows in which one loading is zero and one is nonzero; and only a small number of rows with two nonzero elements.

4. If $k \geq 4$, several pairs of columns of Λ should have two zero loadings.

These conditions are not mathematically precise. Reiersøl (1950, Th 9.2) gives precise conditions for the identification of Λ given Ψ. For most practical applications, "simple structure" is discovered using graphical and analytic methods. We will discuss some of the methods available in SAS later when we discuss factor rotation procedures.

Aside from the transformation problem, and the problem of determining k, given a Ψ, there is still the problem of estimating Λ, evaluating model fit, and estimating the factor scores, Anderson and Rubin (1956). To estimate the model parameters, we let $\Phi = \mathbf{I}$ to reduce some indeterminacy in the model. Then

$$\underset{p \times p}{\Sigma} = \underset{p \times k}{\Lambda} \underset{k \times p}{\Lambda'} + \underset{p \times p}{\Psi} \tag{8.9.18}$$

where Σ is the covariance matrix of **Y**; Ψ is the a diagonal, nonsingular covariance matrix of **e**; and $\Lambda\Lambda'$ is the covariance matrix of the common parts of the observation **Y** of rank k such that $\Sigma - \Psi$ is $p.s.d.$ and of minimal rank k. The matrix $\Lambda\Lambda'$ has the same off-diagonal elements as Σ. Because Ψ is nonsingular, each $\psi_i > 0$. When ψ_i is not greater than zero, we have what is known as the Heywood case. In such situations, the EFA model is not appropriate. The goal of EFA is to obtain a few factors that adequately reproduces Σ or $\mathbf{P}_\rho$ if one uses standardized variables.

In (8.9.18) there are $p(p+1)/2$ elements in Σ. They are to be represented in terms of the $p\,(k+1)$ unknown parameters in Λ and Ψ. Because of the transformation problem, Λ may be made to satisfy $k\,(k-1)\,/2$ independent conditions, the number of unique elements in the orthogonal matrix **T**. Thus, the effective number of unknown parameters is not $p + pk$, but $f = (p + pk) - k\,(k-1)\,/2$. Now the degrees of freedom v for the EFA model is given by the number of equations implied by (8.9.18) or the distinct elements in Σ minus the number of free parameters f. Hence $v = (p-k)^2 - (p+k)\,/2$. For $v > 0$, k must satisfy the condition that $p + k < (p-k)^2$, Anderson and Rubin (1956). The $k\,(k-1)\,/2$ conditions imposed on Λ are chosen for mathematical convenience. Two sets of restrictions are that $\Lambda'\Lambda$ is diagonal, also called the principal component characterization of the EFA model. Or, to require $\Lambda'\Psi^{-1}\Lambda$ to be diagonal, the canonical correlation characterization of the EFA model, Rao (1955).

For an estimator to be scale free, suppose we consider a simple rescaling of the data so that $\mathbf{Y}^* = \mathbf{DY}$ where $\mathbf{D}$ is nonsingular. Then $\Lambda^* = \mathbf{D}\Lambda$ and $\Psi^* = \mathbf{D}\Psi\mathbf{D}$ depend on the conditions imposed on Λ and Ψ, Anderson and Rubin (1956). For example if $\Lambda'\Lambda$ is required to be diagonal, then $\Lambda^{*\prime}\Lambda^* = \Lambda'\mathbf{D}^2\Lambda$ is not diagonal under the PC characterization. Alternatively under the canonical correlation characterization, suppose we require $\Lambda'\Psi^{-1}\Lambda$ to be diagonal. Then for $\Lambda^* = \mathbf{D}\Lambda$, the matrix $\Lambda^{*\prime}\Psi^{*-1}\Lambda^*$ is diagonal. This shows that the canonical correlation characterization is scale free as indicated by Swaminathan and Algina (1978).

b. *Estimating Model Parameters*

To estimate the model parameters of the EFA model with structure $\Sigma = \Lambda\Lambda' + \Psi$, one obtains a random sample of n p-vectors to form an $n \times p$ data matrix $\mathbf{Y}_{n\times p}$. Using $\mathbf{Y}$, one constructs the unbiased estimate of Σ defined by $\mathbf{S} = \mathbf{Y}'(\mathbf{I} - \mathbf{1}\left(\mathbf{1}'\mathbf{1}\right)^{-1}\mathbf{1}')\mathbf{Y}/(n-1)$. Or, the sample correlation matrix $\mathbf{R}$. In general, estimation procedures are divided into two broad classes: scale free methods and non-scale free methods. For scale free methods, one may analyze $\mathbf{S}$ or $\mathbf{R}$. While most software packages provide both, non-scale free methods give different results depending on whether $\mathbf{S}$ or $\mathbf{R}$ is analyzed.

In general, non-scale free methods are based on the least squares principle which minimizes the sum of the squares of the elements of $\mathbf{S} - \Lambda\Lambda' - \Psi = \mathbf{S} - \Sigma$. That is the fit function is

$$\begin{aligned} F_{ULS}(\Lambda, \Psi) &= \operatorname{tr}\left[(\mathbf{S}-\Sigma)'(\mathbf{S}-\Sigma)\right] \\ &= \operatorname{tr}(\mathbf{S}-\Sigma)^2 \\ &= \|\mathbf{S}-\Sigma\|^2 \end{aligned} \tag{8.9.19}$$

is minimized for a known value of k where $\Sigma = \Lambda\Lambda' + \Psi$. Methods known as principal component factoring (PCF), principal factor analysis (PFA), iterative principal factor analysis (IPFA), and the minus method fall into this class, Harmon (1976), Mulaik (1972), and Jöreskog (1977). Minimization of (8.9.19) leads to the normal equations

$$\begin{aligned} (\mathbf{S} - \widehat{\Lambda}\widehat{\Lambda}' - \widehat{\Psi})\widehat{\Lambda} &= \mathbf{0} \\ \operatorname{diag}(\mathbf{S} - \Lambda\widehat{\Lambda}' - \widehat{\Psi}) &= \mathbf{0} \end{aligned} \tag{8.9.20}$$

These equations may not be solved directly. Instead, one selects $\widehat{\Psi}_o = \widehat{\Psi}$ and solves the equations

$$\begin{aligned} (\mathbf{S} - \widehat{\Psi}_o)\widehat{\Lambda} &= \widehat{\Lambda}(\widehat{\Lambda}'\widehat{\Lambda}) \\ \widehat{\Psi}_o &= \operatorname{diag}(\mathbf{S} - \Lambda\widehat{\Lambda}') \end{aligned} \tag{8.9.21}$$

in a single step or iteratively. Using the PC characterization that $\widehat{\Lambda}'\widehat{\Lambda}$ is diagonal and setting $\mathbf{S}_r = \mathbf{S} - \widehat{\Psi}_o$, by PCA there exists an orthogonal matrix $\widehat{\mathbf{P}}$ such that

$$\mathbf{S}_r = \widehat{\mathbf{P}}\widehat{\Delta}\widehat{\mathbf{P}}' = (\widehat{\mathbf{P}}\widehat{\Delta}^{1/2})(\widehat{\mathbf{P}}\widehat{\Delta}^{1/2})' = \widehat{\Lambda}\widehat{\Lambda}' \tag{8.9.22}$$

where $\widehat{\Delta}$ is a diagonal matrix containing the roots of $|\mathbf{S}_r - \delta\mathbf{I}| = 0$. By construction,

$$\begin{aligned}
\mathbf{S}_r\widehat{\mathbf{P}} &= \widehat{\mathbf{P}}\widehat{\Delta} \\
\mathbf{S}_r\widehat{\mathbf{P}}\widehat{\Delta}^{1/2} &= \widehat{\mathbf{P}}\widehat{\Delta}^{1/2}(\Delta^{1/2}\mathbf{P}\mathbf{P}\Delta^{1/2}) \\
\mathbf{S}_r\widehat{\Lambda} &= \widehat{\Lambda}(\widehat{\Lambda}'\widehat{\Lambda})
\end{aligned}$$

where $\widehat{\Lambda}'\widehat{\Lambda} = \widehat{\Delta}$ is diagonal. Thus, with an estimate of Ψ for a given k, one has that

$$\widehat{\Lambda} = \mathbf{P}\Delta^{1/2} \text{ and } \Lambda\widehat{\Lambda}' = \mathbf{P}\Delta\mathbf{P}' \tag{8.9.23}$$

The equation $|\mathbf{S}_r - \delta\mathbf{I}| = 0$ corresponds to a PCA of the common part of the test scores, $\mathbf{c} = \Lambda'\mathbf{f}$.

To use the ULS procedure, two assumptions are made: (1) a reasonable estimate of Ψ is available, and (2) k is known. Given $\mathbf{S}$ and treating it as Σ, ignoring sample variation, the number of common factors k is equal to the rank of $\mathbf{S} - \Psi$, if Ψ is known. Since Ψ is not usually known, the rank of $\mathbf{S} - \Psi$ is affected by the elements used in Ψ or by the values of the communalities. Furthermore, the estimates of the loadings are influenced by the number of common factors, which are only determinable if k is known. The common factor structure changes as k changes. But k is not obtained until factoring is complete, and even if k is known, it is not sufficient for Ψ to be uniquely determined. This unfortunate state of affairs has led many researches to avoid the EFA model or to limit its exploration using only the PCF method.

For PCF method, one sets $\widehat{\Psi}_o = \mathbf{I}$ and analyzes $\mathbf{S}$ or $\mathbf{R}$. Then for some number of common factors k, $\mathbf{S} - \widehat{\Lambda}\widehat{\Lambda}' = \widehat{\Psi}$ where $\widehat{\psi}_i = s_{ii} - \sum_{j=1}^{k}\widehat{\lambda}_{ij}^2 = s_{ii} - \widehat{h}_i^2$ and $\widehat{h}_i^2$ is an estimate of the communalities, the variance of the common part of Y_i. To estimate k, one studies the sample eigenvalues of $\mathbf{S} - \widehat{\Psi}$ and analyzes the residual matrix $\mathbf{S} - \widehat{\Lambda}\widehat{\Lambda}' - \widehat{\Psi}$ for various values of k. The $||\mathbf{S} - \widehat{\Lambda}\widehat{\Lambda}' - \widehat{\Psi}||^2 \leq \sum_{i=k+1}^{p}\widehat{\lambda}_i^2$ and the root mean square residual is $||\mathbf{S} - \widehat{\Lambda}\widehat{\Lambda}' - \widehat{\Psi}|| / [p\,(p-1)\,/\,2]^{1/2}$.

Both PFA and IPFA use an "ad hoc" method to estimate the matrix of unique variances Ψ by $\widehat{\Psi}_o$. If one is analyzing $\mathbf{S}$, an estimate the matrix is

$$\widehat{\Psi}_o = (\operatorname{diag}\mathbf{S}^{-1})^{-1} \tag{8.9.24}$$

The rationale for this estimate is that if $\mathbf{S}^{-1}$ is a good estimate of Σ^{-1}, then by the EFA model, for an infinite number of tests,

$$(\Sigma - \widehat{\Psi}_o) = \Lambda\Lambda'$$

exactly and k is the rank of $\Sigma - \widehat{\Psi}_o$. The rank of $\mathbf{S} - \widehat{\Psi}_o$ with $\widehat{\Psi}_o$ estimated by (8.9.24) is known as Guttman's strongest lower bound for the rank of Λ, Mulaik (1972, p. 141). Letting s^{ii} represent the ith diagonal element of $\mathbf{S}^{-1}$, the communality on the diagonal of the matrix $\mathbf{S} - \widehat{\Psi}_o$ has the general form $\widehat{h}_i^2 = s_{ii} - (s^{ii})^{-1}$. When the correlation matrix is analyzed, one may select

$$\widehat{\Psi}_o = \left(\operatorname{diag}\mathbf{R}^{-1}\right)^{-1} \tag{8.9.25}$$

Representing the ith diagonal element of $\mathbf{R}^{-1}$ as r^{ii}, the ith communality has the form

$$\widehat{h}_i^2 = 1 - (r^{ii})^{-1} \tag{8.9.26}$$

and is seen to be equal to the squared multiple correlation (SMC) between Y_i and the remaining $p-1$ variables, R_i^2. The reduced correlation matrix $\mathbf{R} - \widehat{\Psi}_o$ has estimated communalities $\widehat{h}_i^2 = 1 - \widehat{\psi}_i = R_i^2$ or the squared multiple correlations as its diagonal elements. When using a covariance matrix, the diagonal elements are rescaled squared multiple correlations $\widehat{h}_i^2 = s_{ii} - \widehat{\psi}_i = s_{ii} R_i^2$. In IPFA one modifies $\widehat{\Psi}_o$ at each stage after finding $\widehat{\Lambda}$ until $\widehat{\Lambda}_i$ and $\widehat{\Psi}_{i+1}$ agree with $\widehat{\Lambda}_{i-1}$ and $\widehat{\Psi}_i$ to a predetermined number of significant figures. Since the element in the matrix $\widehat{\Lambda}$ are used to estimate the communalities, the communalities are changed at each state of the iterative process. In general, one should avoid using these "ad hoc" methods of analysis since for moderately correlated variables the procedures lead to inconsistent estimates, Basilevsky (1994, p. 365).

One of the first scale free methods for EFA was developed by Lawley (1940) using the maximum likelihood method which assumes that $\mathbf{Y}_i \sim N_p\left(\boldsymbol{\mu},\ \Sigma = \Lambda\Lambda' + \Psi\right)$. Then, the likelihood normal equations are

$$\begin{aligned} \mathbf{S}\widehat{\Psi}^{-1}\widehat{\Lambda} &= \widehat{\Lambda}(\mathbf{I} + \widehat{\Lambda}'\widehat{\Psi}^{-1}\widehat{\Lambda}) \\ \widehat{\Psi} &= \text{diag}(\mathbf{S} - \widehat{\Lambda}\Lambda') \end{aligned} \tag{8.9.27}$$

Again, these equations may not be solved directly; instead an iterative procedure is used. For a review of the method, the reader should consult Lawley and Maxwell (1971), Jöreskog (1967, 1977) and Jöreskog and Lawley (1968). Briefly, to solve (8.9.27) we premultiple the first equation in (8.9.27) by $\Psi^{-1/2}$. This results in the equation

$$(\widehat{\Psi}^{-1/2}\mathbf{S}\widehat{\Psi}^{-1/2})(\Psi^{-1/2}\Lambda) = (\widehat{\Psi}^{-1/2}\Lambda)(\mathbf{I} + \widehat{\Lambda}'\widehat{\Psi}^{-1}\widehat{\Lambda}) \tag{8.9.28}$$

Now using the condition that $\widehat{\Lambda}'\widehat{\Psi}^{-1}\widehat{\Lambda}$ is diagonal because of the transformation problem, the canonical correlation characterization makes the model identifiable, and given k and $\widehat{\Psi}_o$, (8.9.28) implies that $\mathbf{Q} = \Psi_o^{-1/2}\widehat{\Lambda}$ contains the eigenvectors of the $p \times p$ weighted matrix $\widehat{\Psi}_o^{-1/2}\mathbf{S}\widehat{\Psi}_o^{-1/2}$ and that the diagonal matrix $\Delta = \left(\mathbf{I} + \widehat{\Lambda}'\widehat{\Psi}_o^{-1}\widehat{\Lambda}\right)$ contains k nonnegative roots. However, the matrix product $\mathbf{Q}'\mathbf{Q} = \widehat{\Lambda}'\widehat{\Psi}_o^{-1}\widehat{\Lambda} = \Delta - \mathbf{I} \neq \mathbf{I}$. To ensure that $\mathbf{Q}'\mathbf{Q} = \mathbf{I}$, we set $\mathbf{P} = \mathbf{Q}\,(\Delta - \mathbf{I})^{-1/2}$ so that $\mathbf{P}'\mathbf{P} = \mathbf{I}$. Then, $\mathbf{Q}$ becomes

$$\mathbf{Q} = \widehat{\Psi}_o^{-1/2}\Lambda = \mathbf{P}\,(\Delta - \mathbf{I})^{1/2} \tag{8.9.29}$$

or $\mathbf{Q}'\widehat{\Psi}_o^{-1/2}\mathbf{S}\widehat{\Psi}_o^{-1/2}\mathbf{Q} = \Delta$ as required. Hence a solution for the loading matrix from (8.9.29) is

$$\widehat{\Lambda} = \widehat{\Psi}_o^{-1/2}\mathbf{P}\,(\Delta - \mathbf{I})^{1/2} \tag{8.9.30}$$

By forming the matrix product $\widehat{\Lambda}\widehat{\Lambda}'$ and using the equation $\widehat{\Psi} = \text{diag}(\mathbf{S} - \Lambda\widehat{\Lambda})$ for a calculated $\widehat{\Lambda}$, an iterative procedure is established to determine $\widehat{\Psi}$ and $\widehat{\Lambda}$, simultaneously.

A major advantage of the ML method is that it is scale free. That is, if $\widehat{\Lambda}$ is a matrix of loadings for analyzing $\mathbf{S}$, then $\mathbf{D}\widehat{\Lambda}$ are the loadings for analyzing $\mathbf{R}$. A disadvantage of the method is that the matrix $\mathbf{S}$ or $\mathbf{R}$ must be nonsingular. Furthermore, it may result in estimates of ψ_i that are zero or negative, the Heywood or ultra-Heywood situation. When this occurs, this is a strong indication that the EFA model is not an appropriate model for the data.

Another scale free procedure, canonical factor analysis, was developed by Rao (1955). Rather than maximizing the total variance of the variables, as in PCA, Rao utilized CCA in determining k sample canonical variates between $\mathbf{Y}$ and $\mathbf{f}$. Under the CFA model given in (8.7.11), we assume that

$$\begin{bmatrix} \mathbf{Y} \\ \mathbf{f} \end{bmatrix} \sim \left\{ \begin{bmatrix} \boldsymbol{\mu}_y \\ \mathbf{0} \end{bmatrix}, \begin{bmatrix} \Sigma & \Lambda \\ \Lambda' & \mathbf{I} \end{bmatrix} \right\} \tag{8.9.31}$$

and find linear combinations $U_i = \mathbf{a}_i'\mathbf{Y}$ and $V_j = \mathbf{b}_j'\mathbf{f}$ with maximum correlation using (8.6.19). For a sample, we must solve

$$\left|\widehat{\Lambda}\widehat{\Lambda}' - \theta \mathbf{S}\right| = 0 \tag{8.9.32}$$

were $\widehat{\theta}_j$ is the square of the canonical correlation between U_i and V_j. Letting $\widehat{\Lambda}\widehat{\Lambda}' = \widehat{\Sigma} - \widehat{\Psi}$, (8.9.32) becomes

$$\mid \mathbf{S} - \delta\widehat{\Psi} \mid = 0 \tag{8.9.33}$$

for $\widehat{\delta}_j = 1/\left(1 - \widehat{\theta}_j\right)$. Thus letting Δ denote the roots of $\widehat{\Psi}^{-1/2}\mathbf{S}\widehat{\Psi}^{-1/2}$ and $\mathbf{Q}$ the matrix of eigenvectors, CFA is mathematically equivalent to the ML method. That is, the CFA loadings are $\widehat{\Lambda} = \widehat{\Psi}^{-1/2}\mathbf{Q} = \widehat{\Psi}_o^{-1/2}\mathbf{P}\,(\Delta - \mathbf{I})^{1/2}$ and $\mathbf{Q}'\widehat{\Psi}_o^{-1/2}\mathbf{S}\widehat{\Psi}_o^{-1/2}\mathbf{Q} = \Delta$. This demonstrates that Rao's CFA solution is equivalent to the ML solution, without assuming normality. It is also equivalent to Howe's (1955) maximum determinant method that maximizes the determinant of the partial correlation matrix, $\left|\Psi^{-1/2}\left(\Sigma - \Lambda\Lambda'\right)\Psi^{-1/2}\right|$, using model (8.9.3), Morrison (1990).

Instead of maximizing the determinant of the partial correlation matrix

$$\Psi^{-1/2}\left(\Sigma - \Lambda\Lambda'\right)\Psi^{-1/2},$$

suppose we assume Ψ is fixed. Then, for fixed Ψ suppose we minimize the weighted least squares function

$$F\left(\Lambda, \Psi\right) = \operatorname{tr}\left[\Psi_o^{-1}\left(\mathbf{S} - \Sigma\right)\right]^2 \tag{8.9.34}$$

where Ψ_o^{-1} is a "weight" matrix and $\Sigma = \Lambda\Lambda' + \Psi$. Minimizing (8.9.34) for fixed Ψ, one can show that the normal equation is

$$\Psi^{-1}\left(\mathbf{S} - \Lambda\Lambda' - \Psi\right)\Psi^{-1}\Lambda = \mathbf{0} \tag{8.9.35}$$

Multiplying (8.9.35) by Ψ, observe that

$$\mathbf{S}\Psi\widehat{\Lambda} = \widehat{\Lambda}(\mathbf{I} + \widehat{\Lambda}'\widehat{\psi}^{-1}\widehat{\Lambda}) \tag{8.9.36}$$

which is identical to the first equation of the ML solution. Replacing Ψ_o^{-1} with the random weight matrix $\mathbf{S}^{-1}$ which is an estimate of Σ^{-1} where Σ is the true population covariance matrix of $\mathbf{Y}$ and following Jöreskog and Goldberger (1972), (8.9.34) becomes the generalized least squares fit criterion

$$F_{GLS}(\Lambda, \ \Psi) = \text{tr}\left[\mathbf{S}^{-1}(\mathbf{S} - \Sigma)\right]^2 = \text{tr}(\mathbf{I} - \mathbf{S}^{-1}\Sigma)^2 \tag{8.9.37}$$

Minimizing (8.9.37) for fixed Ψ, the equation for the loadings Λ agree with (8.9.36), Jöreskog and Goldberger (1972). Adding the constraint that $\widehat{\Psi}_o = \text{diag}\left(\mathbf{S} - \Lambda\Lambda'\right)$, one can force the GLS estimates, the ML estimates, and the weighted estimates using (8.9.34) to yield identical results, Lee and Jennrich (1979). Thus, for large or small samples and provided $\mathbf{S}$ (or $\mathbf{R}$) is nonsingular ML estimates may be used to estimate Λ and Ψ, given k, for the EFA model. Furthermore, even though ML estimates do not exist for fixed factors (Anderson and Rubin, 1956, p. 129) assuming $\Sigma = \Lambda\Lambda' + \Psi_o$ where Ψ_o is known and used as a "weight" matrix one is led to fixed model estimates which can be made to be "ML like" estimates. For these reasons, we recommend using the ML procedure for the EFA model. The procedure always leads to efficient and consistent estimates of model parameters.

c. *Determining Model Fit*

Under the EFA model, we assume that Σ has the structure $\Sigma = \Lambda\Lambda' + \Psi$ for a given value of k. The alternative is that Σ has any structure Σ. Lawley (1940) derived the likelihood ratio test of

$$H : \Sigma = \Lambda\Lambda' + \Psi \tag{8.9.38}$$

assuming $\mathbf{Y}_i \sim N_p(\boldsymbol{\mu}, \ \Sigma)$. The likelihood ratio statistic for testing H derived by Lawley is an

$$X_k^2 = n\left[\log|\widehat{\Sigma}| - \log|\mathbf{S}| + \text{tr}(\mathbf{S}\widehat{\Sigma}^{-1}) - p\right]$$

If H is true, as $n \longrightarrow \infty$ then X_k^2 converges to a chi-square distribution with $v = [(p-k)^2 - (p+k)]/2$. Observing that the diag $\widehat{\Sigma} = \text{diag}\,\mathbf{S}$ so that the tr $\left(\mathbf{S}^{-1}\widehat{\Sigma}\right) = p$ (Anderson and Rubin, 1956, eq. 8.3) and using Bartlett's (1954) correction, an alternate form of Lawley's statistic is

$$X_k^2 = [n - (2p + 4k + 11)\ /6] \ \log\left(|\widehat{\Sigma}|/|\mathbf{S}|\right) \tag{8.9.39}$$

However, under H the rank of $\Lambda\Lambda'$ is k so that the last $p - k$ roots of $\Lambda\Lambda'$ must be zero. Or, equivalently the roots of $\Psi^{-1/2}\Lambda\Lambda'\Psi^{-1/2}$ must be zero. But under H, $\Sigma = \Lambda\Lambda' + \Psi$ or $\Psi^{-1/2}\Sigma\Psi^{-1/2} = \Psi^{-1/2}\Lambda\Lambda'\Psi^{-1/2} + \mathbf{I}$. By the first ML normal equation, if δ_i is a root of the weighted covariance matrix $\widehat{\Psi}^{-1/2}\mathbf{S}\widehat{\Psi}^{-1/2}$, then $\delta_i = 1 + \gamma_i$ where γ_i is the ith of $\Psi^{-1/2}\Lambda\Lambda'\Psi^{-1/2}$ or the ith diagonal element of $\Gamma = \Lambda'\Psi^{-1}\Lambda$ which is diagonal. Thus, if $\gamma_i = 0$ then $\delta_i = 1$. Hence, testing H is equivalent to testing that the last $p - r$ roots δ_i of $\widehat{\Psi}^{-1/2}\mathbf{S}\widehat{\Psi}^{-1/2}$ differ from one. An alternative form of the test statistic for H for various values of k is

$$X_k^2 = -[n - (2p + 4k + 11)\ /6] \sum_{i=k+1}^{p} \log\left(\widehat{\delta}_i\right) \tag{8.9.40}$$

as given by Anderson and Rubin (1956, p. 136). The test of fit is used to determine the number of factors when using the ML method. However, it tends to over estimate the number of factors. One may also use (8.9.40) when analyzing $\mathbf{P}_\rho$ provided n is large. A test developed by Schott (1988) may be modified for the EFA model for small samples.

Using a formal test of fit may lead one to retain too many factors. One should also construct scree plots of the sample eigenvalues of $\mathbf{S} - \widehat{\Psi}$ and study the elements of $\mathbf{S} - \widehat{\Lambda}\widehat{\Lambda}' - \widehat{\Psi}$, for consistently small values. For various values of k, the root mean square residual (RMS) index

$$\text{RMS} = \sqrt{\frac{\text{tr}(\mathbf{S} - \widehat{\Lambda}\widehat{\Lambda}' - \widehat{\Psi})^2}{p\,(p-1)\,/2}} \tag{8.9.41}$$

is investigated. For strong data, the formal test and ad hoc methods should lead to a consistent value for k. To evaluate whether one has strong data, the variables should not be independent. Thus, $\mathbf{P}_\rho \neq \mathbf{I}$ which may be tested using the procedure discussed in Chapter 3.

An ad hoc procedure for determining "strong" data, available in SAS, is to use Kaiser and Rice's (1974) measure of sampling adequacy (MSA). Their index is designed to evaluate how close $\mathbf{R}^{-1}$ is to a diagonal matrix. It is defined as

$$\text{MSA} = \frac{\sum_{i<j} r_{ij}^2}{\sum_{i<j} r_{ij}^2 + \sum_{i<j} q_{ij}^2} \tag{8.9.42}$$

where $\mathbf{R} = \left[r_{ij}\right]$ and $\mathbf{Q} = \mathbf{D}\mathbf{R}^{-1}\mathbf{D} = \left[q_{ij}\right]$, $\mathbf{D} = [\left(\text{diag}\,\mathbf{R}^{-1}\right)^{1/2}]^{-1}$. As $\mathbf{R} \longrightarrow \mathbf{I}$, the MSA index approaches 0. While Kaiser and Rice (1974) recommend that the MSA be greater than 0.80, the following table developed by them is also provided

MSA	Data Strength
≥ 0.9	Marvelous
0.8	Meritorious
0.7	Middling
0.6	Mediocre
0.5	Miserable
< 0.5	Unacceptable

for evaluating the strength of one's data for analysis by the EFA model.

d. Factor Rotation

In PCA and CCA we did not recommend rotation since these models are uniquely identified and any rotation destroys the criterion of maximum variances in PCA or maximum correlation in CCA. Thus, rotation of components or canonical variates is usually ill-advised. In EFA this is not the case; by design one obtains a solution that is not unique and one rotates for ease of interpretation.

Having determined k and Ψ for the EFA model with orthogonal factors $\mathbf{f}$, one usually tries to discover a loading matrix Λ_* using an orthogonal transformation matrix $\mathbf{P}$ such that $\Lambda_* = \widehat{\Lambda}\mathbf{P}$ where $\mathbf{P}\mathbf{P}' = \mathbf{I}$. Such transformations are orthogonal. Geometrically, the loadings in the ith row of $\widehat{\Lambda}$ are the coordinates of Y_i in the loading space. Using the transformation $\mathbf{P}$, one is trying to find new coordinates that generate a simple structure for the loading matrix, following Thurstone. Recall that the loading matrix represents the covariance between factors and variables, and given high and low values of the loadings on each variable one is trying to investigate patterns to interpret the factors. For this reason, the matrix of loadings is called the pattern matrix. When a correlation matrix is analyzed, the pattern matrix yields correlations between factors and variables. This matrix is called the structure matrix. To interpret factors, one analyzes the pattern matrix.

One of the most popular and effective methods for orthogonal transformations is the varimax method developed by Kaiser (1958). The goal of the varimax method is to obtain factors that have high loadings for a subset of variables on only one factor and zeros for the other factors. This generally leads to easy interpretation since variables are not confounded by factors. To accomplish this, Kaiser's varimax criterion makes the sum of the variances of the loadings in each column of the matrix $\widehat{\Lambda}_*$ large, subject to the constraint that the communality of each variable is unchanged; or letting $\widehat{\Lambda}_* = [\widehat{\lambda}_{ij}]$

$$
\begin{aligned}
V_j &= \sum_{i=1}^{p} (\widehat{\lambda}^2_{ij} - \widehat{\lambda}^2_{.j})^2 / p \\
&= \left[p \sum_{i=1}^{p} \widehat{\lambda}^4_{ij} - \left(\sum_{i=1}^{p} \widehat{\lambda}^2_{ij} \right)^2 \right] / p^2
\end{aligned}
$$

where V_j is the variance of the communality of the variables within factor j, and $\widehat{\lambda}^2_{.j} = \sum_{i=1}^{p} \widehat{\lambda}^2_{ij} / p$ is the average of the squared loadings for factor j. Summing V_j over all factors

$$
V_r = \sum_{j=1}^{k} V_j = \sum_{j=1}^{k} \left[\frac{1}{p} \sum_{i=1}^{p} \widehat{\lambda}^4_{ij} - \frac{1}{p^2} \left(\sum_{i=1}^{p} \widehat{\lambda}^2_{ij} \right)^2 \right] \tag{8.9.43}
$$

yields the raw varimax criterion, Magnus and Neudecker (1999, p. 374).

Since each variable is weighted equally in (8.9.43), the raw varimax criterion is overly influenced by variables with large communalities. To adjust for this, the factors are normalized to unit length, by dividing each loading in a row of the pattern matrix by the communality for the variable, and then returned to their original length; this is Kaiser's normal varimax criterion. Kaiser (1958) developed the matrix $\mathbf{P}$ of direction cosines using an iterative procedure that is applied to all $\binom{k}{2}$ pairs of factors to maximize the normal varimax criterion.

Orthogonal transformations using other criteria have been suggested to facilitate the interpretation of orthogonal factors, Mulaik (1972). For example, the quarimax criterion maximizes the variance across factors (rows of Λ) instead of variables (columns of Λ). This procedure usually results in a single factor with high loadings and other factors with differential loadings across variables. The equamax criterion is an orthogonal transformation that maximizes the weighted sum of row-wise and column-wise variances. While there are numerous other proposed criteria, the varimax procedure usually leads to a satisfactory pattern matrix.

Because orthogonal transformations of Λ using $\mathbf{P}$ to develop Λ_* are rigid transformations of axes that are perpendicular or orthogonal to each other,one may not find a "simple structure" pattern matrix. Then, an alternative to orthogonal transformations are oblique transformations. For this situation, the axes of the factor space do not remain orthogonal.

Using a nonsingular transformation matrix $\mathbf{T}$ such that $\mathbf{f}_* = \mathbf{Tf}$ and $\Lambda_* = \Lambda\mathbf{T}^{-1}$ the $\text{cov}(\mathbf{f}_*) = \mathbf{TT}' \neq \mathbf{I}$ so that the factors are correlated. Oblique transformation "rotate" factor axes differentially. Furthermore, since angles and distances are not preserved, communalities are changed. The variance of the common part of Y_i now has the form

$$\text{var}(c_i) = \sum_{j=1}^{k} \lambda_{ij}^2 + \sum_{j=1}^{k-1} \lambda_{ij}\lambda_{ij+1}\rho\left(f_j, f_{j+1}\right)$$

so that it depends on the correlations among factors. One still uses the pattern matrix to interpret oblique factors, even though they are correlated. In the analysis of $\mathbf{R}$, the pattern matrix is not the same as the structure matrix.

Because oblique transformations allow axes to move differentially, they are sometimes easier to interpret than orthogonal factors. However, they imply a structure for Σ of the form given by (8.9.14), a correlated factor model, and not (8.9.2), an uncorrelated factor model. For an oblique transformation, one may also have to investigate the structure matrix and the intercorrelations among the factors. For further information, the reader should consult Harmon (1976), Mulaik (1972) and McDonald (1985). SAS offers only a few oblique transformations: the orthoblique, oblique procrustes and oblique promax. Each of these start with an orthogonal transformation and through modification result in oblique factors. Currently, oblique rotation that minimize covariance of squared loadings among the columns such as oblimin, a modification of varimax, is not available in the SAS procedure FACTOR, Hakstain and Abel (1974).

e. *Estimating Factor Scores*

Principal components and canonical variates are computed from the observed data matrix $\mathbf{Y}$. However, given the orthogonal EFA model with structure Σ defined in (8.9.2) the factor scores must be estimated. Because the conditional model given in (8.9.3) where $\mathbf{f}$ is considered fixed and the random model given in (8.9.1) where $\mathbf{f}$ is considered random lead to the same structure for $\Sigma = \Lambda\Lambda' + \Psi$, one may estimate factor scores using either the model.

To estimate factor scores using least squares theory, we use formulation (8.9.3) and consider $\mathbf{f}$ fixed. Then, assuming Λ, Ψ and k are known and equal to their estimates, the weighted least squares estimates of $E(\mathbf{Y}|\mathbf{f})$ is

$$\widehat{\mathbf{f}}_i = (\widehat{\Lambda}'\widehat{\Psi}^{-1}\widehat{\Lambda})^{-1}\widehat{\Lambda}'\widehat{\Psi}^{-1}(\mathbf{Y}_i - \overline{\mathbf{y}}) \tag{8.9.44}$$

for $i = 1, 2, \ldots, k$. The matrix of factor scores becomes

$$\widehat{\mathbf{F}} = \mathbf{Y}\widehat{\Psi}^{-1}\widehat{\Lambda}(\widehat{\Lambda}'\widehat{\Psi}^{-1}\widehat{\Lambda})^{-1} \tag{8.9.45}$$

This estimate was proposed by Bartlett (1938).

Although the estimates given in (8.9.45) are BLUE, the procedure assumes that the factors are fixed rather than random variables. Furthermore, $\widehat{\mathbf{f}}$ is an estimate of $E(\mathbf{Y} \mid \mathbf{f})$ and not of $\mathbf{f}$. Finally, they may be inconsistent with ML estimates.

An alternative to the fixed factor approach was proposed by Thompson (1934) known as the regression estimator. Assuming the joint distribution of $\left[\mathbf{Y}', \mathbf{f}'\right]$ is as shown in (8.9.31), Thompson proposed predicting $\mathbf{f}$ using the regression of $\mathbf{f}$ on $\mathbf{Y}$. Given that $\mathbf{F}$ is known and replacing $\mathbf{Y}_{n\times p}$ with deviation scores, $\mathbf{Y}_d$, the MR model relating $\mathbf{Y}_d$ and $\mathbf{F}$ is

$$\underset{n\times k}{\mathbf{F}} = \underset{n\times p}{\mathbf{Y}_d}\ \underset{p\times k}{\mathbf{B}} + \underset{n\times k}{\mathbf{E}} \tag{8.9.46}$$

where $\widehat{\mathbf{B}} = \left(\mathbf{Y}_d'\mathbf{Y}_d\right)^{-1}\mathbf{Y}_d'\mathbf{F}$. Then, the predicted value of $\mathbf{F}$ is

$$\begin{aligned} \widehat{\mathbf{F}} &= \mathbf{Y}_d\widehat{\mathbf{B}} = \mathbf{Y}_d\left(\mathbf{Y}_d'\mathbf{Y}_d\right)^{-1}\mathbf{Y}_d'\mathbf{F} \\ &= \mathbf{Y}_d\mathbf{S}^{-1}\widehat{\Lambda} \end{aligned} \tag{8.9.47}$$

the matrix of estimated factor scores. Assuming $\left[\mathbf{Y}', \mathbf{f}'\right]$ are jointly multivariate normal, the least squares estimate of $\widehat{\mathbf{F}}$ given in (8.9.47) is the ML estimation of the factor scores. Thus, the estimator is consistent with finding ML estimates of the parameters of the EFA model. The matrix $\widehat{\mathbf{F}}$ as given in (8.9.47) are calculated in SAS. $\widehat{\mathbf{F}}$ is a consistent estimator and is not unbiased. Furthermore, $\mathbf{S}$ must be nonsingular and the factor scores are not unique because of the factor inderterminacy problem. $\widehat{\mathbf{F}}$ is usually evaluated for $\widehat{\Lambda} = \widehat{\Lambda}_*$, the rotated loading matrix. For standardized variables, (8.9.47) becomes

$$\widehat{\mathbf{F}} = \mathbf{Z}\mathbf{R}^{-1}\widehat{\Lambda} \tag{8.9.48}$$

For additional remarks on the estimation of factor scores see Harmon (1976), Mulaik (1972), and McDonald (1985).

f. Additional Comments

Topics related to the EFA model include image factor analysis and alpha factor analysis. Following Jöreskog (1969), image factor analysis in SAS is done by performing a PCA on an image covariance matrix. The communality (image) of a variable is the predicted value of the variable obtained by regressing the variable on the remaining variables. There is no indeterminacy due to estimating the communality since it now has a precise meaning. In SAS, squared multiple correlations are inserted on the diagonal of the correlation matrix and the off-diagonals of the matrix are adjusted so that the matrix is $p.s.d.$ Hence, $\mathbf{R}$ must be nonsingular. Alpha factor analysis is a scale invariant technique; however, instead of scaling in the metric of the unique part of the observations, it uses the common part. In alpha factor analysis the data are the population, and the variables are considered a sample from the population of variables. The analysis seeks to find whether factors for a sample of variables hold for a population of variables by constructing common factors that have maximum correlation with factors in the population of variables, Kaiser and Derflinger (1990).

In our presentation, we apply the EFA model to a single population, extensions to several populations are considered by Jöreskog (1971).

8.10 Exploratory Factor Analysis Examples

a. *Performance Assessment Program (PAP—Example 8.10.1)*

For our first example, we re-analyze the correlation matrix **R** associated with the PAP covariance matrix given in Table 8.3.7, Hansen (1999). In PCA we analyzed the total variance of each variable. In factor analysis, the total variance is partitioned into a common variance part $\left(h_i^2\right)$ and a unique variance $\left(\psi_i\right)$. Removing the common variance from **R**, factor analysis is used to analyze the reduced correlation matrix or the common variance in each variable. Thus, the term "common factor" analysis. An initial estimate of the communality when analyzing **R** is $\widehat{h}_i^2 = 1 - \left(r^{ii}\right)^{-1} = 1 - R_i^2$ where R_i^2 is the squared multiple correlation (SMS) between Y_i and the remaining $p - 1$ variables. The statement PRIORS = SMC is used in the SAS procedure FACTOR to obtain the initial SMC estimates. Because the ML method for estimating Ψ and Λ is scale free, the SAS procedure analyzes **R** and not **S**. The ML method is selected using option METHOD = ML.

The first and most difficult problem in EFA is to decide on the number of factors. We have not really addressed this problem since we have assumed the value is known and that $\widehat{\Psi}$ is a perfect estimate of Ψ. Then, the rank of $\mathbf{R} - \Psi$ (the reduced correlation matrix) is k and 100% of the common variance it accounted for by $k < p$ factors. The problem of selecting k is known as the communality problem. Since $\widehat{\Psi}$ is not an exact estimate of Ψ, some of the roots of $\mathbf{R} - \widehat{\Psi}$ will be negative. For guidance in selecting k, Guttman (1954) recommended the number of sample eigenvalues of $\mathbf{R} - \widehat{\Psi}$ that are non-negative, k_G, known as Guttman's strongest lower bound. One should also investigate the scree plots of the eigenvalues of the reduced correlation matrix, perform tests of fit, and investigate the RMS index given in (8.9.41) for several values of k. There is no exact solution to the communality problem. Another criterion is the intepretability criterion. Do the rotated factors make sense?

For the PAP reduced correlation matrix, $k_G = 4$; however, the proportion of common variance accounted for by the fourth latent factor is only 0.07%. The scree plots of the eigenvalues of the reduced correlation matrix have a sharp elbow between the third and fourth sample roots; and the MSA value is 0.781, near the "meritorious" level. Using the rule of parsimony, we investigate both a 2-factor and 3-factor fit. The SAS code is provided in program m8_10_1.sas.

Reviewing the output, the RMS index for the two models are .0198 for the 3-factor solution and .057 for the 2-factor solution. In general, 3-factor fit is better having smaller off-diagonal residuals. In addition, Lawley's statistic of fit accepts a 3-factor solution (p-value = 0.3135) while a 2-factor solution is rejected (p-value < 0.0001). Finally, information criteria due to Akaike (1987), [AIC] and Schwarz (1978) [SIC] support selecting a 3-factor model over a 2-factor model. The criteria use a penalized likelihood function that takes into account both the goodness-of-fit (likelihood) of a model and the number of parameters estimated such that a model with smallest information value is best. For these criteria, the model with a "smaller" value is better. Since $AIC_3 = -5.603$ is less than $AIC_2 = 28$ and $SIC_3 = -30.66 < SIC_2 = 18.29$, we would select the 3-factor solution over the 2-factor solution. Finally, the rotated factor pattern matrix solution recovers more of the five dimensions of the study. The varimax rotated factors are provided in Table 8.10.1.

TABLE 8.10.1. PAP Factors

Variables	Rotated Factors ($\widehat{\lambda}_{ij}$) 1	2	3	Estimated Communality ($\widehat{h}_i^2$)
SUPPG	20	60∗	05	0.40
SUPPI	14	91∗	09	0.85
ASMTE	24	05	75∗	0.62
ASMTC	23	08	63∗	0.46
FAM	75*	11	18	0.60
PAP	64*	30	33	0.61
PROF1	59*	11	19	0.40
PROF2	43*	21	20	0.26
Weighted Eigenvalues	8.2992	3.5427	1.1336	
Proportion	0.6396	0.2730	0.0874	
Cumulative %	64.0	91.3	100%	

Comparing the PCA solution in Table 8.3.9 with the EFA solution in Table 8.10.1, we are able to recover the structure in **R** exactly using 3 factors. However, we did not recover the five dimensions in the questionnaire

b. Di Vesta and Walls (Example 8.10.2)

In program m8_10_2.sas, we analyze the correlation matrix in Table 8.3.4 for the eight semantic differential scale of Di Vesta and Walls (1970). For these data, $k_G = 4$; however, one root dominates the reduced correlation matrix. Even though one root dominates, we fit two factors since the scree plot has a sharp elbow between the second and third eigenvalues. While Kaiser's measure of sampling adequacy may be classified as "meritorious" since $MSA = 0.83$ and 2 factors yield small residuals, the test of fit for a two-factor solution is rejected. Fitting three factors results in a Heywood case, a single factor is trying to account for more than 100% of the total observed common variance. This indicates that these data are not appropriate for a factor analysis.

c. Shin (Example 8.10.3)

These data, like the Di Vesta and Walls example, do not seem to be appropriate for a factor analysis. One again finds that the Heywood condition arises for the three factor solution. In program m8_10_3.sas, we also analyze these data using the non-scale free unweighted least square (ULS) criterion given in (8.9.19). Selecting three factors, the unrotated solution is very similar to the PCA solution. The rotated matrix seems to provide a clearer picture of a three factor solution.

TABLE 8.10.2. Correlation Matrix of 10 Audiovisual Variables

Variable										
v_1	1.00									(Sym)
v_2	.59	1.00								
v_3	.30	.34	1.00							
v_4	.16	.24	.62	1.00						
v_5	−.02	−.13	.28	.37	1.00					
v_6	.00	−.05	.42	.51	.90	1.00				
v_7	.39	.61	.70	.59	.05	.20	1.00			
v_8	.17	.29	.57	.88	.30	.46	.60	1.00		
v_9	−.04	−.14	.28	.33	.93	.86	.04	.28	1.00	
v_{10}	−.04	−.08	.42	.50	.87	.94	.17	.45	.90	1.00

We saw that PCA is not invariant to changes in scales and that no optimal criterion has been developed to determine the number of components. In general non-scale free methods of factor analysis do not improve upon the situation. Indeed, the problem is worse since factor scores relate to latent traits. While scale free EFA methods are widely used in many disciplines for data reduction and the interpretation of latent constructs as an exploratory data analysis tool, we feel that EFA has little utility since factors obtained in practice are usually dependent on the mathematical restriction imposed to attain uniqueness. And, while rotation for interpretation may be critical to an analysis it should not be treated as an end in itself. As an exploratory data analysis tool, factor analysis seeks to uncover "structure" in data. In most disciplines, knowledge regarding the important variables and number of factors is frequently known. One may even know something about the factor pattern. Thus, data analysis is no longer exploratory but confirmatory. The Confirmatory Factor Analysis (CFA) model is discussed in Chapter 10.

Exercises 8.10

1. Bolton (1971) measured 159 deaf individuals on ten skills, four were based on communication skills and six were based upon reception skills. Use the ML factor analysis technique to uncover the structure and discuss your findings. The variables are Reception Skills: v_1 – Unaided Hearing, v_2 – Aided Hearing, v_3 – Speech Reading, v_4 – Reading, v_5 – Manual Signs, v_6 – Fingerspelling, and Communication Skills: v_7 – Oral Speech, v_8 – Writing, v_9 – Manual Signs, v_{10} – Fingerspelling. The correlation matrix is provided in Table 8.10.2.

2. Stankov (1979) analyzed the correlation matrix of 13 audiovisual variables shown in Table 8.10.3 using image analysis and orthoblique rotations. Re-analyze the data using ML factor analysis and PCA with varimax and promax rotations. Compare and interpret your findings.

TABLE 8.10.3. Correlation Matrix of 13 Audiovisual Variables (excluding diagonal)

Variable	2	3	4	5	6	7	8	9	10	11	12	13
1	0.69	0.54	0.37	0.46	0.39	0.37	0.40	0.53	0.37	0.62	0.52	0.29
2		0.31	0.28	0.31	0.21	0.23	0.20	0.32	0.36	0.49	0.29	0.18
3			0.43	0.33	0.42	0.31	0.43	0.46	0.38	0.39	0.39	0.26
4				0.30	0.50	0.52	0.41	0.41	0.37	0.37	0.31	0.27
5					0.49	0.33	0.25	0.28	0.33	0.28	0.36	0.26
6						0.60	0.50	0.36	0.36	0.48	0.37	0.19
7							0.45	0.33	0.22	0.37	0.37	0.29
8								0.31	0.36	0.33	0.28	0.11
9									0.55	0.50	0.28	0.25
10										0.38	0.30	0.21
11											0.40	0.27
12												0.44

Variables

1	Verbal communication	8	Relation perception
2	Experimental evaluation	9	Spatial Scanning
3	Induction (visual)	10	Flexibility of closure
4	Auditory Induction	11	Perceptual speed
5	Memory span	12	Making
6	Temporal tracking	13	Tempo
7	Sound pattern recognition		

9

Cluster Analysis and Multidimensional Scaling

9.1 Introduction

Discriminant analysis is used to evaluate group separation and to develop rules for assigning observations to groups. Cluster analysis is concerned with group identification. The goal of cluster analysis is to partition a set of observations into a distinct number of unknown groups or clusters in such a manner that all observations within a group are similar, while observations in different groups are not similar. If data are represented as an $n \times p$ matrix $\mathbf{Y} = [y_{ij}]$ where

$$\underset{n\times p}{\mathbf{Y}} = \begin{bmatrix} \mathbf{y}_1' \\ \mathbf{y}_i' \\ \vdots \\ \mathbf{y}_n' \end{bmatrix}$$

the goal of cluster analysis is to develop a classification scheme that will partition the rows of $\mathbf{Y}$ into k distinct groups (clusters). The rows of the matrix usually represent items or objects. To uncover the groupings in the data, a measure of nearness, also called a proximity measure needs to be defined. Two natural measures of nearness are the degree of distance or "dissimilarity" and the degree of association or "similarity" between groups. The choice of the proximity measure depends on the subject matter, scale of measurement (nominal, ordinal, interval, ratio), and type of variables (continuous, categorical) being analyzed. In many applications of cluster analysis, one begins with a proximity matrix rather than a data matrix. Given the proximity matrix of order $(n \times n)$ say, the entries may represent dissimilarities $[d_{rs}]$ or similarities $[s_{rs}]$ between the rth and s^{th} objects. Cluster analysis is a tool for classifying objects into groups and is not concerned with the geometric representation of the objects in a low-dimensional space. To explore the dimensionality of the space, one may use multidimensional scaling.

Multidimensional scaling (MDS), like PCA, is a data reduction technique that begins with an $(n \times n)$ proximity matrix of dissimilarities δ_{rs} based upon $(1 \times p)$ vectors and tries to find a set of constructs of lower-dimension k based upon dissimilarities $d_{rs} \approx \delta_{rs}$ for all objects under study. In most applications, the data matrix $\mathbf{Y}_{n \times p}$ is not given. Multidimensional scaling is an exploratory data analysis technique designed to discover the dimensionality of the space of hypothetical constructs called principal coordinates based upon a proximity matrix of perceived dissimilarities (distances). The process of uncovering the geometric representation of the principal coordinates is sometimes referred to as the ordination process, the construction of a low-dimensional plot of the objects under study.

In this chapter we provide an overview of commonly used hierarchical and nonhierarchical clustering methods, and some criteria commonly used to determine the number of clusters. Comprehensive discussions of cluster analysis are included in books on the topic by Anderberg (1973), Hartigan (1975) and Everitt (1993). Metric and nonmetric multidimensional scaling methods are also reviewed. These topics are discussed in more detail by Torgerson (1958), Kruskal and Wish (1978), Davidson (1983), Young (1987), Cox and Cox (1994), and Everitt and Rabe-Hesketh (1997). Both topics are also discussed by Mardia, Kent and Bibby (1979), Seber (1984), and Jobson (1992).

9.2 Proximity Measures

Proximity measures are used to represent the nearness of two objects. If a proximity measure represents similarity, the value of the measure increases as two objects become more similar. Alternatively, if the proximity measure represents dissimilarity the value of the measure decreases in value as two objects become more alike. Letting $\mathbf{y}_r$ and $\mathbf{y}_s$ represent two objects in a p-variate space, an example of a dissimilarity measure is the Euclidean distance between $\mathbf{y}_r$ and $\mathbf{y}_s$. As a measure of similarity, one may use the proportion of the elements in the two vectors that match. More formally, one needs to establish a set of mathematical axioms to create dissimilarity and similarity measures.

a. *Dissimilarity Measures*

Given two objects $\mathbf{y}_r$ and $\mathbf{y}_s$ in a p-dimensional space, a dissimilarity measure satisfies the following conditions.

$$
\begin{aligned}
&1.\quad d_{rs} \geq 0 \text{ for all objects } \mathbf{y}_r \text{ and } \mathbf{y}_s \\
&2.\quad d_{rs} = 0 \text{ if and only if } \mathbf{y}_r = \mathbf{y}_s \qquad (9.2.1)\\
&3.\quad d_{rs} = d_{sr}
\end{aligned}
$$

Condition (3) implies that the measure is symmetric so that the dissimilarity measure that compares $\mathbf{y}_r$ (object r) with $\mathbf{y}_s$ (object s) is the same as the comparison for object s versus object r. Condition (2) requires the measure to be zero whenever object r equals object s, the objects are identical only if $d_{rs} = 0$ and under no other situation. Finally, (1) implies that the measure is never negative. A dissimilarity measure that satisfies conditions (1) to (3) is said to be a semi-metric measure.

For continuous (interval, ratio scale) variables, the most common dissimilarity measure is the Euclidean distance between two objects. Given an $(n \times p)$ matrix $\mathbf{Y}$ with $(1 \times p)$ row vectors $\mathbf{y}_i'$, the square of the Euclidean distance between two rows $\mathbf{y}_r$ and $\mathbf{y}_s$ is defined as

$$d_{rs}^2 = (\mathbf{y}_r - \mathbf{y}_s)' (\mathbf{y}_r - \mathbf{y}_s) = \|\mathbf{y}_r - \mathbf{y}_s\|^2 \tag{9.2.2}$$

The $(n \times n)$ data matrix $\mathbf{D} = [d_{rs}]$ is called the Euclidean distance matrix. Because a change in the units of measurement may cause a variable to dominate the ranking of distances, the Euclidean distance matrix is most effective for variables that are commensurate.

When variables are not commensurate, one may weight the squared differences by $s_j^2 = \sum_{i=1}^{n} (y_{ij} - \overline{y}_{.j})^2 / (n-1)$, $j = 1, 2, ..., p$ where s_j^2 and $\overline{y}_{.j}$ represent estimates of the mean and variance of variable j.

$$d_{rs}^2 = (\mathbf{y}_r - \mathbf{y}_s)' (\text{diag } \mathbf{S})^{-1} (\mathbf{y}_r - \mathbf{y}_s) \tag{9.2.3}$$

This process eliminates the dependence of the analysis on the units of measurement. However, it often causes the within-cluster distances to become larger than the between-cluster differences thus tending to mask clusters, Hartigan (1975, p. 59). This problem also occurs when using the square of Mahalanobis distances as a proximity measure in the metric of the covariance matrix $\mathbf{S}$ defined as follows.

$$d_{rs}^2 = (\mathbf{y}_r - \mathbf{y}_s)' \mathbf{S}^{-1} (\mathbf{y}_r - \mathbf{y}_s) \tag{9.2.4}$$

Mahalanobis distances, whether or not the variables are commensurate, tend to mask clusters even more since correlations tend to be reduced even further, Hartigan (1975, p. 63). There is no satisfactory solution to the inconsistent unit of measurement problem.

Because Euclidean distance is a special case of the Minkowski metric (L_p-norm), the dissimilarity measures may be represented as

$$d_{rs} = \left(\sum_{j=1}^{p} \left| y_{rj} - y_{sj} \right|^{\lambda} \right)^{1/\lambda} \tag{9.2.5}$$

Varying λ in (9.2.5) changes the weight assigned to larger and smaller distances. For $\lambda = 1$, we have city-block distances (L_1-norm) and for $\lambda = 2$, we have Euclidean distances. City-block distances tend to have low sensitivity to outliers.

In defining dissimilarity measures, we did not include the triangular inequality condition that $d_{rs} \le d_{rq} + d_{qs}$ for all points r, s, and q. With this condition, the dissimilarity measures form a metric. While all Minkowski distances form a metric, the requirement is a sufficient condition but not a necessary condition for a proximity measure. For example, we may cluster using the matrix of squared Euclidean distances which do not form a metric or use the norm $d_{rs} = \|\mathbf{y}_r - \mathbf{y}_s\|$, which does form a metric. Replacing the triangular inequality with the condition that $d_{rs} \le \max\left(d_{rq}, d_{qs}\right)$ for all points r, s, and q,

FIGURE 9.2.1. 2 × 2 Contingency Table, Binary Variables

		Row s		Total
		1	0	
Row r	1	a	b	$a+b$
	0	c	d	$c+d$
Total		$a+c$	$b+d$	$p=a+b+c+d$

the semi-metric becomes an ultrametric. It is called an ultrametric since this condition is stronger than the triangular inequality requirement for a metric.

To reduce the size of squared distances (or distances), they are often divided by the number of variables p. Because all dissimilarities are divided by the same number p, this does not effect the clustering results. In addition, the squared Euclidean distances are often mean centered by subtracting from each variable its corresponding mean. Then $\mathbf{Y}_d = [y_{ij} - \overline{y}_{.j}]$ is analyzed. Variable centering does not effect the Euclidean distances since removing the means from the p variables does not change the Euclidean distance between any two objects.

Two other dissimilarity measures that have been proposed when all y_{ij} are positive are the following.

$$\text{Canberra Metric} \qquad d_{rs} = \sum_{j=1}^{p} \left\{ \frac{\mid y_{rj} - y_{sj} \mid}{(y_{rj} + y_{sj})} \right\} \tag{9.2.6}$$

$$\text{Czekanowski Coefficient} \qquad d_{rs} = \frac{\sum_{j=1}^{p} \mid y_{rj} - y_{sj} \mid}{\sum_{j=1}^{p} (y_{rj} + y_{sj})} \tag{9.2.7}$$

Both of these measures are modified L_1-norms and are used when data are skewed and/or contain outliers. A scaled L_1-norm that may be used for data that includes both positive and negative values is Gower's metric

$$\text{Gower Metric} \qquad d_{rs} = \sum_{j=1}^{p} \left| y_{rj} - y_{sj} \right| / R_j \tag{9.2.8}$$

where R_j represents the range of variable j.

We have developed some dissimilarity measures assuming the data are continuous with levels of measurement that are at least interval. For categorical data that have nominal or ordinal scales of measurement, the situation becomes more complex, Anderberg (1973, p. 122). For a simple case, suppose each row $\mathbf{y}_i'$ of $\mathbf{Y}$ contains only binary data. Then, the squared Euclidean distances only provide a count of mismatches, $(1-0)$ or $(0-1)$, and treat the matches, $(1-1)$ and $(0-0)$, equally. When variables are coded either 0 or 1 to indicate the absence or presence of a characteristic, a (2×2) table as shown in Table 9.2.1 may be created to evaluate dissimilarity and similarity measures.

In Figure 9.2.1, the frequencies b and c represent mismatches while frequencies a and c represent matches. Thus the squared Euclidean distance divided by p becomes

$$\sum_{j=1}^{p} (y_{rj} - y_{sj})^2 / p = (b+c) / p = d_{rs}^2 / p$$

for p binary variables. Using any Minkowski metric the value is the same for all $\lambda \geq 1$. Note also that $(b+c)/p = 1 - (a+d)/p$. The quantity $(a+d)/p$ is a measure of similarity in that it reflects the proportion of matches between two $(1 \times p)$ binary vectors. We will say more about similarity measures later. For binary variables, the Canberra metric is identical to the city block metric (L_1-norm) and Czekanowski's coefficient becomes a metric.

In evaluating Czekanowski's metric for binary variables, we have the following expression for the dissimilarity measure

$$d_{rs} = \frac{\sum_{j=1}^{p} \left| y_{rj} - y_{sj} \right|}{\sum_{j=1}^{p} \left(y_{rj} + y_{sj} \right)} = \frac{b+c}{(a+b)+(a+c)} = 1 - \frac{2a}{2a+b+c} \tag{9.2.9}$$

The quantity $2a/(2a+b+c)$ is called Czekanowski's coefficient. It is a measure of similarity in which double weight is given to $(1-1)$ matches, and $(0-0)$ matches are excluded from both the numerator and denominator, thus making them irrelevant in the calculation. We now discuss proximity matrices made up of similarity measures.

b. *Similarity Measures*

Given two objects $\mathbf{y}_r$ and $\mathbf{y}_s$ in a p-dimensional space, a similarity measure satisfies the following conditions.

$$\begin{aligned} &1.\quad 0 \leq s_{rs} \leq 1 \text{ for all objects } \mathbf{y}_r \text{ and } \mathbf{y}_s \\ &2.\quad s_{rs} = 1 \text{ if and only if } \mathbf{y}_r = \mathbf{y}_s \\ &3.\quad s_{rs} = s_{sr} \end{aligned} \tag{9.2.10}$$

Condition (3) again implies the measure is symmetric while conditions (1) and (2) ensure that its always positive and identically one only if objects r and s are identical.

Given a similarity measure that satisfies (9.2.10), one may always create a dissimilarity measure using the relationship that $d_{rs} = 1 - s_{rs}$ or some other decreasing function, however, the new measure may not form a metric. Conversely, given a dissimilarity measure d_{rs} one may want to construct a similarity measure as $s_{rs} = 1/(1+d_{rs})$. Because d_{rs} is unbounded, $0 < s_{rs} \leq 1$ so that s_{rs} never attains the value of zero. Thus, one may not create s_{rs} from d_{rs}. While one may use either similarity measures or dissimilarity measures to cluster the rows of a data matrix, the CLUSTER procedure in SAS only uses dissimilarity measures. In this section, we therefore emphasize converting similarity measures to dissimilarity measures.

A common measure of similarity suggested by some authors is to use the Pearson product moment correlation between objects $\mathbf{y}_r$ and $\mathbf{y}_s$ $r, s = 1, 2, \ldots, n$ defined as

$$q_{rs} = \frac{\sum_{j=1}^{p} \left(y_{rj} - \overline{y}_{r.} \right) \left(y_{sj} - \overline{y}_{s.} \right)}{\left[\sum_{j=1}^{p} \left(y_{rj} - \overline{y}_{r.} \right)^2 \sum_{j=1}^{p} \left(y_{sj} - \overline{y}_{s.} \right)^2 \right]} \tag{9.2.11}$$

where $\overline{y}_{r.} = \sum_j y_{rj}/p$ and $\overline{y}_{s.} = \sum_j y_{sj}/p$. The symbol q_{rs} is being used since we are calculating correlations using the rows of the data matrix. However, because $-1 \leq q_{rs} \leq 1$, it does not satisfy condition (1) in (9.2.10). To correct this situation, one may instead use the quantities $|q_{rs}|$ or $1 - q_{rs}^2$. It has also been suggested to add one to q_{rs} and divide by 2, $q_{rs}^* = (q_{rs} + 1)/2$. Or, equivalently the quantity $1 - q_{rs}$. This still presents a problem since $q_{rs}^* = 1$ does not imply that $\mathbf{y}_r = \mathbf{y}_s$ only that the elements in each vector are perfectly linearly related.

If the matrix $\mathbf{Y}$ is standardized so that $\widetilde{y}_{ij} = (y_{rj} - \overline{y}_{i.})/s_i$ where $s_i = \sum_{j=1}^{p} \widetilde{y}_{sj}^2/(p-1)$ for $i = 1, 2, ..., n$ objects, then one may relate q_{rs} to squared Euclidean distances as follows

$$\begin{aligned} d_{rs}^2 &= \sum_{j=1}^{p} \left(\widetilde{y}_{rj} - \widetilde{y}_{sj}\right)^2 \\ &= \sum_{j=1}^{p} \widetilde{y}_{rj}^2 + \sum_{j=1}^{p} \widetilde{y}_{sj}^2 - 2 \sum_{j=1}^{p} \widetilde{y}_{rj} \widetilde{y}_{sj} \\ &= 2\left(1 - q_{rs}\right) \end{aligned} \tag{9.2.12}$$

so that $d_{rs} = \sqrt{2(1 - q_{rs})}$ becomes a metric. One may then use the quantities d_{rs}, $1 - q_{rs}$ or d_{rs}^2 to cluster rows. As noted by Anderberg (1973, p. 114), standardization makes little sense when the p variables are not commensurate since the measure of similarity q_{rs} has little meaning for different scales of measurement.

Another measure of similarity for rows in $\mathbf{Y}$ is the cosine of the angle θ between two vectors $\mathbf{y}_r$ and $\mathbf{y}_s$ for $r, s = 1, 2, \ldots, n$ defined as

$$\cos\theta = c_{rs} = \mathbf{y}_r'\mathbf{y}_s / \|\mathbf{y}_r\| \, \|\mathbf{y}_s\| \tag{9.2.13}$$

Because $-1 \leq c_{rs} \leq 1$, condition (1) of similarity measures is not satisfied. If one normalizes the elements in each row of $\mathbf{Y}$ so that the $\|\widetilde{\mathbf{y}}_r\|^2 = \|\widetilde{\mathbf{y}}_s\|^2 = 1$, then by the law of cosines

$$\|\widetilde{\mathbf{y}}_r - \widetilde{\mathbf{y}}_s\|^2 = \|\widetilde{\mathbf{y}}_r\|^2 + \|\widetilde{\mathbf{y}}_s\|^2 - 2\,\|\widetilde{\mathbf{y}}_r\|^2\,\|\widetilde{\mathbf{y}}_s\|\cos\theta$$

and the squared distances become

$$d_{rs}^2 = 2\left(1 - c_{rs}\right) \tag{9.2.14}$$

so that one may again employ a dissimilarity measure to cluster rows in $\mathbf{Y}$ with normalization, Anderberg (1973, p. 114).

Similarity measures for binary variables are important in cluster analysis. To construct similarity measures for binary data we again consider the entries in the 2×2 table given in Table 9.2.1. We consider how to weight matches and mismatches because a $(1 - 1)$ match may be more important than a $(0 - 0)$ match since the former implies the presence while the later implies the absence of an attribute. One may even want to discard $(0 - 0)$ matches completely. To provide for the differential weighting of matches and mismatches and the treatment of $(0 - 0)$ matches, Anderberg (1973, p. 89) using Table 9.2.1 summarizes several matching counting schemes as shown in Table 9.2.1

In Table 9.2.1, recall that a and b represent matched pair frequencies while cells c and d represent the frequencies for mismatched pairs when comparing two observations $\mathbf{y}_r$ and $\mathbf{y}_s$.

TABLE 9.2.1. Matching Schemes

Similarity Measure	Weighting of Matches and Mismatches	Coefficient Name	Dissimilarity= (1-Similarity)
1. $\frac{a+d}{p}$	Equal Weight to Matched Pairs With (0 – 0) Matches	Simple Matching	Metric
2. $\frac{2(a+d)}{2(a+d)+b+c}$	Double Weight to Matched Pairs With (0 – 0) Matches	Double Matching	Metric
3. $\frac{a+d}{a+d+2(b+c)}$	Double Weight to Unmatched Pairs With (0 – 0) Matches	Rogers-Tanimoto	Metric
4. $\frac{a}{p}$	Equal Weight to Matches With No(0 – 0) Matches in Numerator	Russell-Rao	Semi-metric
5. $\frac{a}{a+b+c}$	Equal Weight to Matches With No (0 – 0) Matches in Denominator	Jaccard	Metric
6. $\frac{2a}{2a+b+c}$	Double Weight to (1 – 1) Matches With No (0 – 0) Matches in Numerator or Denominator	Czekanowski-Sørensen-Dice	Metric
7. $\frac{a}{a+2(b+c)}$	Double Weight to Unmatched Pairs With No (0 – 0) Matches in Numerator or Denominator	Unnamed	Metric
8. $\frac{a}{a+c}$	Ratio of Matches to Mismatches No (0 – 0) Matches Included	Kulezynski	Semi-metric
9. $\frac{a+d}{b+c}$	Ratio of Matches to Mismatches With (0 – 0) Matches Included	Unnamed	Semi-metric

The coefficients in Table 9.2.1 represent similarity proximity measures, s_{rs}, using various weighting schemes with the inclusion and/or exclusion of $(0-0)$ matches from the numeration and/or denominator. The column titled "Dissimilarity" is defined as $d_{rs} = 1 - s_{rs}$, where s_{rs} is the similarity measure. We indicate in the column whether or not the corresponding dissimilarity measure d_{rs} forms a metric or a semi-metric.

Subsets of similarity (dissimilarity) proximity measures in Table 9.2.1 are monotonically increasing (decreasing). For some agglomerative hierarchical clustering procedures such as the single link and complete link methods, monotonicity is important since the clustering method is invariant to monotonic increasing (decreasing) relationships, Johnson (1967). Anderberg (1973, p. 90) shows that the subgroup of metric proximity measures (1, 2 and 3) and measures (5, 6 and 7) are monotonically related. Clustering algorithms that are monotonically related and yield the same clustering tree diagram (to be discussed later in this chapter) under monotonic transformations of the data. The single link and complete link clustering algorithms under monotonic transformations of proximities will yield the same clustering tree diagram.

In our presentation of dissimilarity and similarity measures, we have discussed measures employed for clustering rows of **Y** containing continuous interval or ratio level data or binary data that commonly occur in multivariate data analysis. Anderberg (1973) develops measures for nominal and ordinal data, and for observation vectors that contain both continuous and discrete data.

c. *Clustering Variables*

Proximity measures for clustering rows of $\mathbf{Y}_{n\times p}$ may also be used to cluster rows of $\mathbf{Y}'_{p\times n}$, or variables. Replacing p with n in Table 9.2.1, all similarity (dissimilarity) measures in Table 9.3.1 may be used to cluster variables. When clustering variables, one is likely to standardize the variables and use some measure of association for clustering. When working with variables, standardization is usually a natural step and a measure of association a_{ij} is used to cluster variables. To convert a_{ij} to a dissimilarity measure, the function $d_{ij} = \sqrt{1 - a_{ij}}$ is often used.

Agresti (1981) provides an overview of several measures of association for nominal and ordinal variables. For binary data, we may use Table 9.2.1 to calculate the Pearson product moment correlation coefficient r_{ij} and the cosine of the angle between variables i and j as follows, Anderberg (1973, pp. 84-85)

$$r_{ij} = \frac{ad - bc}{[(a+b)(c+d)(a+c)(b+d)]^{1/2}} \tag{9.2.15}$$

$$c_{ij} = \left[\left(\frac{a}{a+b}\right)\left(\frac{a}{a+c}\right)\right]^{1/2} \tag{9.2.16}$$

Then, from (9.2.15) and (9.2.16) dissimilarity measures that may be used to cluster variables are $d_{ij}^2 = (1 - r_{ij})$, $d_{ij} = \sqrt{1 - r_{ij}}$, $d_{ij}^2 = (1 - c_{ij})$, or $d_{ij} = \sqrt{1 - c_{ij}}$. Formula (9.2.15) is the familiar phi coefficient. The coefficient c_{ij} is also called the Ochiai coefficient. These measures work most effectively when all $r_{ij} \geq 0$ or all $c_{ij} \geq 0$. In general the entries in Table 9.2.1 are preferred.

9.3 Cluster Analysis

To initiate a cluster analysis one constructs a proximity matrix. The proximity matrix represents the strength of the relationship between pairs of rows in $\mathbf{Y}'_{p\times n}$ or the data matrix $\mathbf{Y}_{n\times p}$. Algorithms designed to perform cluster analysis are usually divided into two broad classes called hierarchical and nonhierarchical clustering methods. Generally speaking, hierarchical methods generate a sequence of cluster solutions beginning with clusters containing a single object and combines objects until all objects form a single cluster; such methods are called agglomerative hierarchical methods. Other hierarchical methods begin with a single cluster and split objects successively to form clusters with single objects; these methods are called diversive hierarchical methods. In both the agglomerative and diversive processes, a tree diagram, or dendogram, is created as a map of the process. The agglomerative hierarchical procedures fall into three broad categories: Linkage, Centroid, and Error Variance methods. Among these procedures, only linkage algorithms may be used to cluster either objects (items) or variables. The other two methods can be used to cluster only objects. Nonhierarchical methods may only be used to cluster items. In this section we review several commonly used agglomerative hierarchical methods: the single linkage (nearest neighbor), complete linkage (farthest neighbor), average linkage (average distance), centroid, and Ward's method. The nonhierarchical k-means method and some procedures for determining the number of clusters are also reviewed.

a. Agglomerative Hierarchical Clustering Methods

Agglomerative hierarchical clustering methods use the elements of a proximity matrix to generate a tree diagram or dendogram as shown in Figure 9.3.1. To begin the process, we start with $n = 5$ clusters which are the branches of the tree. Combining item 1 with 2 reduces the number of clusters by one, from 5 to 4. Joining items 3 and 4, results in 3 clusters. Next, joining item 5 with the cluster (3, 4), results in 2 clusters. Finally, all items are combined to form a single cluster, the root of the tree. As we move from left to right, groups of items are successively combined to form clusters. Once an item or group of items are combined they are never separated. Figure 9.3.1 represents the agglomerative process as we move from left to right. The label at the top of the diagram denotes the number of clusters at each step of the process. Figure 9.3.1 also represents a divisive hierarchical process which begins with the root (a single cluster) and moves from right to left to create the branches (items). In SAS, the procedure TREE is used to construct dendograms.

More generally, given a proximity matrix $\mathbf{D}_{n\times n} = [d_{rs}]$, the steps for the agglomerative hierarchical clustering algorithm are as follows.

1. Begin with n clusters, each containing only a single object.
2. Search the dissimilarity matrix $\mathbf{D}$ for the most similar pair. Let the pair chosen be associated with element d_{rs} so that object r and s are selected.
3. Combine objects r and s into a new cluster (rs) employing some criterion and reduce the number of clusters by 1 by deleting the row and column for objects r and s. Calculate the dissimilarities between the cluster (rs) and all remaining clusters, using the criterion, and add the row and column to the new dissimilarity matrix.
4. Repeat steps 2 and 3, $(n - 1)$ times until all objects form a single cluster. At each step, identify the merged clusters and the value of the dissimilarity at which the clusters are merged.

By changing the criterion in Step 3 above, we obtain several agglomerative hierarchical clustering methods. While we indicated that divisive methods may also be used to cluster

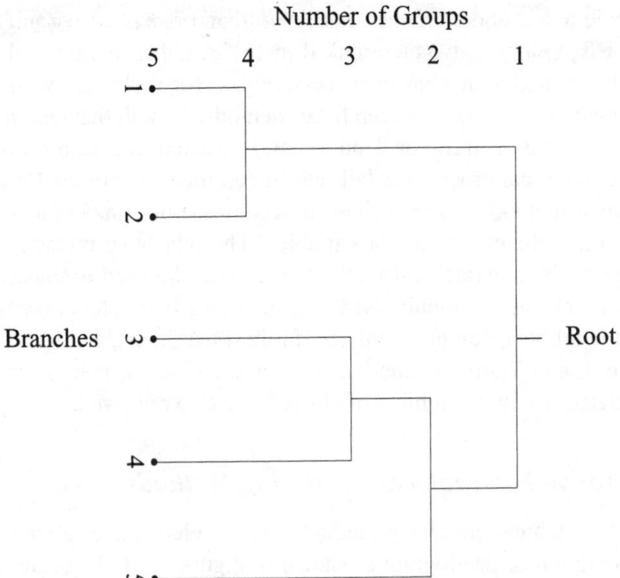

FIGURE 9.3.1. Dendogram for Hierarchical Cluster

objects, the procedure requires a significant number of calculations. For example, to locate the first split in Figure 9.3.1 would require investigating $15 = 2^{n-1} - 1$ partitions for the $n = 5$ objects using some distance criterion. In general divisive methods use large computer resources and splitting criteria are sometimes difficult to establish. A review of divisive techniques is included in Seber (1984). Hartigan (1975) and O'Muircheartaigh and Payne (1977) illustrate the automatic interactive detective (AID) method which splits objects based upon prediction. The classification and regression tree (CART) methodology is closely related to divisive clustering methods, Breiman et al. (1984). The VARCLUS procedure in SAS attempts to divide a set of variables into clusters using a correlation or covariance matrix. We now discuss some agglomerative hierarchical clustering methods.

(a1) Single Link (Nearest-Neighbor) Method

To implement the single link method, one combines objects in clusters using the minimum dissimilarity between clusters. Letting r represent any element in cluster R, $r \in R$, and s be any element in cluster S, $r \in S$, from the clusters in Step 3 of the agglomerative clustering algorithm, distances between R and S are calculated using the rule

$$d_{(R)(S)} = \min \{d_{rs} \mid r \in R \text{ and } s \in S\} \tag{9.3.1}$$

At each step of the process, a dendogram is created representing the ordered distances where objects are joined.

Example 9.3.1 *To illustrate rule (9.3.1), consider the hypothetical dissimilarity matrix*

$$\mathbf{D} = [d_{rs}] = \begin{array}{c} \\ 1 \\ 2 \\ 3 \\ 4 \\ 5 \end{array} \begin{array}{c} \begin{array}{ccccc} 1 & 2 & 3 & 4 & 5 \end{array} \\ \begin{bmatrix} 0 & \boxed{2} & 4 & 7 & 9 \\ 2 & 0 & 8 & 9 & 8 \\ 4 & 8 & 0 & 3 & 7 \\ 7 & 9 & 3 & 0 & 5 \\ 9 & 8 & 7 & 5 & 0 \end{bmatrix} \end{array}$$

Scanning the matrix **D**, *the most similar objects are represented by* $d_{rs} = d_{12} = 2$ *so that objects* 1 and 2 *are jointed to form the cluster* (12). *Following Step 3, we must calculate the minimum values using criterion (9.3.1) as follows.*

$$\begin{aligned} d_{(12)(3)} &= \min\{d_{13}, d_{23}\} = \min\{4, 8\} = 4 \\ d_{(12)(4)} &= \min\{d_{14}, d_{24}\} = \min\{7, 9\} = 7 \\ d_{(12)(5)} &= \min\{d_{15}, d_{25}\} = \min\{9, 8\} = 8 \end{aligned}$$

Deleting the rows in **D** *corresponding to objects* 1 and 2, *and adding the row and column for the cluster* (12), *the new dissimilarity matrix becomes*

$$\mathbf{D}_1 = \begin{array}{c} \\ (1\,2) \\ 3 \\ 4 \\ 8 \end{array} \begin{array}{c} \begin{array}{cccc} (1\,2) & 3 & 4 & 5 \end{array} \\ \begin{bmatrix} 0 & 4 & 7 & 8 \\ 4 & 0 & \boxed{3} & 7 \\ 7 & 3 & 0 & 5 \\ 8 & 7 & 5 & 0 \end{bmatrix} \end{array}$$

The most similar object in $\mathbf{D}_1$ *is* $d_{34} = 3$, *so that we combine element* 3 and 4 *to form the cluster* (34). *Again, we calculate the values*

$$\begin{aligned} d_{(34)(12)} &= \min\left\{d_{(3)(12)}, d_{4(12)}\right\} = \min\{4, 7\} = 4 \\ d_{(34)(5)} &= \min\{d_{35}, d_{45}\} = \min\{7, 5\} = 5 \end{aligned}$$

so that the new dissimilarity matrix is

$$\mathbf{D}_2 = \begin{array}{c} \\ (12) \\ (34) \\ (8) \end{array} \begin{array}{c} \begin{array}{ccc} (12) & (34) & (5) \end{array} \\ \begin{bmatrix} 0 & \boxed{4} & 8 \\ 4 & 0 & 5 \\ 8 & 5 & 0 \end{bmatrix} \end{array}$$

The most similar value in $\mathbf{D}_2$ *is the value* 4 *at which* (34) *is combined with* (12). *Then*

$$d_{((12)(34))5} = \min\left\{d_{(12)(5)}, d_{(34)5}\right\} = \{8, 5\} = 5$$

so that object 5 *is combined with clusters* (12) and (34) *to form the single cluster (1 2 3 4 5) at the value* 5. *The dendogram for nearest dissimilarities follows.*
The groupings and ordered dissimilarities at which clusters are joined are included in the tree diagram at the left margin.

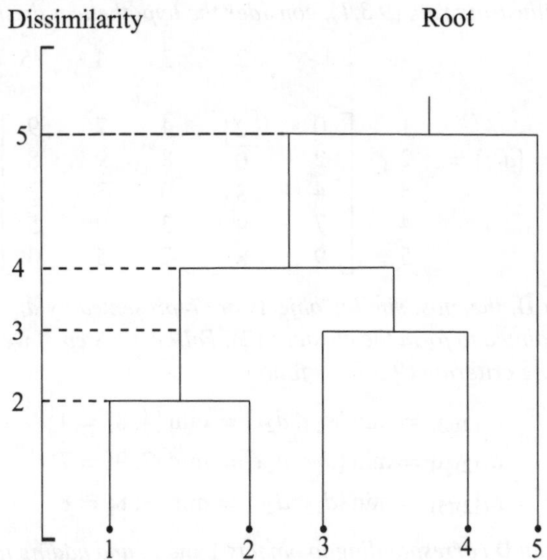

FIGURE 9.3.2. Dendogram for Single Link Example

(a2) Complete Link (Farthest-Neighbor) Method

In the single link method, dissimilarities were replaced using minimum values. For the complete link procedure, maximum values are calculated instead. Letting $r \in R$ and $s \in S$, where R and S are two clusters, distances between clusters R and S are calculated using the rule

$$d_{(R)(S)} = \max\{d_{rs} \mid r \in R \text{ and } s \in S\} \tag{9.3.2}$$

An example with illustrate the procedure.

Example 9.3.2 *To illustrate rule (9.3.2.), we consider the same dissimilarity matrix discussed in Example 9.3.1.*

$$\mathbf{D} = [d_{rs}] = \begin{matrix} \\ 1 \\ 2 \\ 3 \\ 4 \\ 5 \end{matrix} \begin{matrix} \begin{matrix} 1 & 2 & 3 & 4 & 5 \end{matrix} \\ \begin{bmatrix} 0 & \boxed{2} & 4 & 7 & 9 \\ 2 & 0 & 8 & 9 & 8 \\ 4 & 8 & 0 & 3 & 7 \\ 7 & 9 & 3 & 0 & 5 \\ 9 & 8 & 7 & 5 & 0 \end{bmatrix} \end{matrix}$$

From **D** *we again see that $d_{rs} = 2$ represents the most similar objects. Using (9.3.2.), we replace minimum values with maximum values*

$$d_{(12)(3)} = \max\{d_{13}, d_{23}\} = d_{23} = 8$$
$$d_{(12)(4)} = \max\{d_{14}, d_{24}\} = d_{24} = 9$$
$$d_{(12)(5)} = \max\{d_{15}, d_{25}\} = d_{15} = 9$$

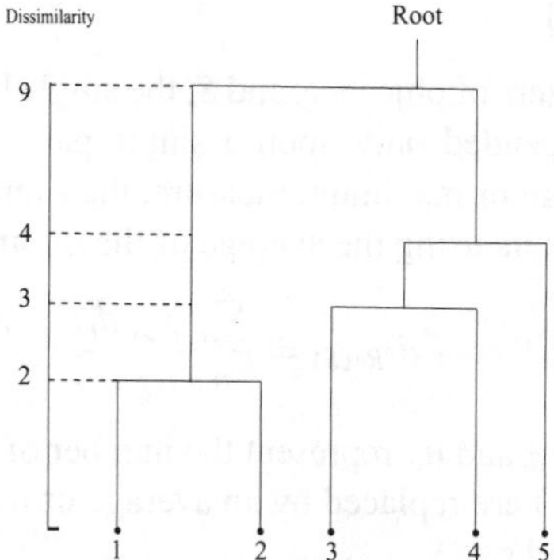

FIGURE 9.3.3. Dendogram for Complete Link Example

so that the new dissimilarity matrix is

$$\mathbf{D}_1 = \begin{array}{c} (12) \\ 3 \\ 4 \\ 5 \end{array} \overset{\begin{array}{cccc} (12) & 3 & 4 & 5 \end{array}}{\begin{bmatrix} 0 & 8 & 9 & 9 \\ 8 & 0 & \boxed{3} & 7 \\ 9 & 3 & 0 & 5 \\ 9 & 7 & 5 & 0 \end{bmatrix}}$$

The most similar object in $\mathbf{D}_1$ *is* $d_{34} = 3$ *so that* 3 *is combined with* 4. *Using the maximum rule,*

$$d_{(34)(12)} = \max\left\{d_{(3)(12)}, d_{4(12)}\right\} = d_{4(12)} = 9$$
$$d_{(34)(5)} = \max\{d_{35}, d_{45}\} = d_{35} = 7$$

so that the modified dissimilarities matrix is

$$\mathbf{D}_2 = \begin{array}{c} (12) \\ (34) \\ (5) \end{array} \overset{\begin{array}{ccc} (12) & (34) & (5) \end{array}}{\begin{bmatrix} 0 & 9 & 9 \\ 9 & 0 & 7 \\ 9 & 7 & 0 \end{bmatrix}}$$

In $\mathbf{D}_2$, (34) *is combined with* 5. *And finally, all elements are combined at the value* 9. *The dendogram for the process follows*

Comparing Figure 9.3.3. with Figure 9.3.2, observe that they do not yield the same result. Objects 3 and 4 *are joined to object 5 by the complete link process while in the single link process they were joined to objects* 1 and 2. *While the two methods do not result in the same dendogram, the tree diagrams are not changed by monotonic transformations of the proximity measures.*

(a3) Average Link Method

When comparing two clusters of objects R and S, the single link and complete link methods of combining clusters depended only upon a single pair of objects within each cluster. Instead of using a minimum or maximum measure, the average link method calculates the distance between two clusters using the average of the dissimilarities in each cluster

$$d_{(R)(S)} = \frac{\sum_r \sum_s d_{rs}}{n_R n_S} \tag{9.3.3}$$

where $r \in R$, $s \in S$, and n_R and n_S represent the number of objects in each cluster. Hence, the dissimilarities in Step 3 are replaced by an average of $n_R n_S$ dissimilarities between all pairs of elements $r \in R$ and $s \in S$.

(a4) Centroid Method

In the average link method, the distance between two clusters is defined as an average of dissimilarity measures. Alternatively, suppose cluster R contains n_R elements and cluster S contains n_S elements. Then, the centroids for the two item clusters are

$$\bar{\mathbf{y}}_r = \frac{\sum_r \mathbf{y}_r}{n_R} = \begin{bmatrix} \bar{y}_{r1} \\ \bar{y}_{r2} \\ \vdots \\ \bar{y}_{rp} \end{bmatrix} \quad \text{and} \quad \bar{\mathbf{y}}_s = \frac{\sum_s \mathbf{y}_s}{n_S} = \begin{bmatrix} \bar{y}_{s1} \\ \bar{y}_{s2} \\ \vdots \\ \bar{y}_{sp} \end{bmatrix} \tag{9.3.4}$$

and the square of the Euclidean distance between the two clusters is $d_{rs}^2 = \left\|\bar{\mathbf{y}}_r - \bar{\mathbf{y}}_s\right\|^2$. For the centroid agglomerative process, one begins with any dissimilarity matrix $\mathbf{D}$ (in SAS, the distances are squared unless one uses the NOSQUARE option). Then, the two most similar clusters are combined using the weighted average of the two clusters. Letting T represent the new cluster, the centroid of T is

$$\bar{\mathbf{y}}_t = \left(n_R \bar{\mathbf{y}}_r + n_S \bar{\mathbf{y}}_s\right) / \left(n_R + n_S\right) \tag{9.3.5}$$

The centroid method is called the median method if an unweighted average of the centroids is used, $\bar{\mathbf{y}}_t = \left(\bar{\mathbf{y}}_r + \bar{\mathbf{y}}_s\right)/2$. The median method is preferred when $n_R >> n_S$ or $n_S \gg n_R$.

Letting the dissimilarity matrix $\mathbf{D} = \left[d_{rs}^2\right]$ where $d_{rs}^2 = \|\mathbf{y}_r - \mathbf{y}_s\|^2$, suppose the elements $r \in R$ and $s \in S$ are combined into a cluster T where

$$\bar{\mathbf{y}}_t = \left(n_R \bar{\mathbf{y}}_s + n_S \bar{\mathbf{y}}_s\right) / \left(n_R + n_S\right).$$

Then to calculate the square of the Euclidean distance between cluster T and the centroid $\bar{\mathbf{y}}_u$ of a third cluster U, the following formula may be used

$$d_{tu}^2 = \left(\frac{n_R}{n_R + n_S}\right) d_{rs}^2 + \left(\frac{n_S}{n_R + n_S}\right) d_{ru}^2 - \left(\frac{n_R n_S}{n_R + n_S}\right) d_{rs}^2 \tag{9.3.6}$$

This is a special case of a general algorithm for updating proximity measures for the single link, complete link, average link, centroid and median methods developed by Williams and Lance (1977).

(a5) Ward's (Incremental Sum of Squares) Method

Given n objects with p variables, the sum of squares within clusters where each object forms its own group is zero. For all objects in a single group, the sum of squares within clusters the sum of squares error is equal to the total sum of squares

$$SSE = \sum_{i=1}^{n} (\mathbf{y}_i - \overline{\mathbf{y}})' (\mathbf{y}_i - \overline{\mathbf{y}}) = \sum_{i=1}^{n} \| \mathbf{y}_i - \overline{\mathbf{y}} \|^2 = T \tag{9.3.7}$$

Thus, the sum of squares within clusters is between zero and SSE. Ward's method for forming clusters joins objects based upon minimizing the minimal increment in the within or error sum of squares. At each step of the process, $n(n-1)/2$ pairs of clusters are formed and the two objects that increase the sum of squares for error least are joined. The process is continued until all objects are joined. The dendogram is constructed based upon the minimum increase in the sum of squares for error.

To see how the process works, let

$$SSE_r = \sum_r \left\| \mathbf{y}_r - \overline{\mathbf{y}}_r \right\|^2, r \in R$$

$$SSE_s = \sum_s \left\| \mathbf{y}_s - \overline{\mathbf{y}}_s \right\|^2, s \in S$$

for clusters R and S. Combining cluster R and S to form cluster T, the error sum of squares for cluster T is

$$SSE_t = \sum_t \left\| \mathbf{y}_t - \overline{\mathbf{y}}_t \right\|^2, t \in T$$

where $\overline{\mathbf{y}}_t = \left(n_R \overline{\mathbf{y}}_r + n_S \overline{\mathbf{y}}_s\right) / (n_R + n_S)$. Then, the incremental increase in joining R and S to form cluster T is $SSE_t - (SSE_r + SSE_s)$. Or, letting SSE_t be the total sum of squares and $SSE_r + SSE_s$ the within cluster sum of squares, the incremental increase in the error sum of squares is no more than the between cluster sum of squares. The incremental between cluster sum of squares (IBCSS) is

$$\begin{aligned}(IBCSS)_{(R)(S)} &= n_R \left\| \mathbf{y}_r - \overline{\mathbf{y}}_r \right\|^2 + n_S \left\| \mathbf{y}_s - \overline{\mathbf{y}}_s \right\|^2 \\ &= \left(\frac{n_R n_S}{n_R + n_S} \right) \left\| \overline{\mathbf{y}}_r - \overline{\mathbf{y}}_s \right\|^2\end{aligned} \tag{9.3.8}$$

For clusters with one object, (9.3.8) becomes $d_{rs}^2/2$. Hence, starting with a dissimilarity matrix $\mathbf{D} = \left[d_{rs}^2\right]$ where d_{rs}^2 is the square of the Euclidean distances (the default in SAS for Ward's method) between objects r and s, the two most similar objects are combined and the new incremental sum of squares proximity matrix has elements $p_{rs} = d_{rs}^2/2$. Combining objects r and s to form a new cluster with mean $\overline{y}_t$ using (9.3.5), the incremental increase in the error sum of squares may be calculated using the formula developed by Williams and Lance (1977) as

$$p_{tu} = \left[(n_U + n_R)\, p_{ru} + (n_U + n_S)\, p_{su} - n_u p_{rs}\right] / \left(n_u + n_R + n_S\right) \tag{9.3.9}$$

b. Nonhierarchical Clustering Methods

In hierarchical clustering the number of clusters is unknown. The process is initiated with a dissimilarity proximity matrix and once an object is assigned to a cluster it is never reallocated. Linkage methods may be used to cluster either items or variables. In nonhierarchical clustering, methods are only applied to cluster items. The process is initiated using the raw data matrix **Y** and not a dissimilarity matrix, **D**. One has to know a priori the number of clusters k which are either cluster centroids or seeds, and observations are reassigned using some criterion with reallocation terminating based upon some stopping rule.

Nonhierarchical clustering methods usually follow the following steps.

1. Select k p-dimensional centroids or seeds (clusters).
2. Assign each observation to the nearest centroid using some L_p-norm, usually the Euclidean distance.
3. Reassign each observation to one of the k clusters based upon some criterion.
4. Stop if there is no reallocation of observations or if reassignment meets some convergence criterion; otherwise, return to Step 2.

In implementing a nonhierarchical clustering method one may vary the method for selecting the k clusters, employ different L_p-norms, and vary the criterion for reallocating observations to clusters to attain cluster stability.

To initiate a nonhierarchical cluster method, one first selects k centroids or seeds. The k seeds may be the first k observations, the first k observations at some defined level of separation, k random seeds, k initial seeds that may be replaced based upon some replacement algorithm, and other variations. Once the seeds are selected, each of the observations are evaluated for assignment or reassignment based upon some convergence criterion. While numerous criteria use multivariate statistics that involve the determinant and trace of the within-cluster and between-cluster variability, MacQueen's k-means algorithm is used in the SAS procedure FASTCLUS with modification using nearest centroid sorting, Anderberg (1973). The basic steps are as follows.

1. Select k seeds.
2. Assign each of the $n - k$ observations to the nearest seed and recalculate the cluster centroid (mean, median, or other depending on the L_p-norm).
3. Repeat Step 2 until all observations are assigned or until changes in cluster centroids become small (no reassignments are made in cluster membership).

In Step 2 of the k-means process, the seed may or may not be updated. Two tests may be made for seed replacement. An observation may replace one of a pair of seeds if the distance between the seeds is less than the distance between an observation and the nearest seed. The former seed becomes an observation in the recalculation of the centroid. If an observation fails this test, one may invoke a second test. The observation replaces the nearest seed if the smallest Euclidean distance from the observation to all seeds other than the nearest one is greater than the shortest distance from the nearest seed to all seeds. For one pass of the data, all observations are associated with k clusters. This process is repeated until all changes in cluster seeds become small based upon a convergence criterion.

When clustering items, hierarchical and nonhierarchical clustering methods may be combined to facilitate the identification of clusters. As a first step, one may use a hierarchical procedure to identify the seeds and number of clusters, these may be input into the nonhierarchical procedure to refine the cluster solution.

c. *Number of Clusters*

Given a cluster analysis solution, one needs to evaluate various indices to determine the number of clusters k. For some value of k, one wants to determine whether the clusters are sufficiently separated so as to illustrate minimal overlap. Thus, evaluating a cluster solution for k clusters is similar to trying to determine the dimension of the space in a PCA or the number of factors in an EFA. In those situations, recall that one constructed informal SCREE plots and under MVN, performed some test of dimensionality.

This is also the situation in cluster analysis. At each step of the clustering process a cluster is formed by joining two observations, by joining an observation and a previous cluster, or by joining two previously formed clusters. At each step of the joining process the number of clusters decrease from $k = n$ to $k = 1$. The "distance" between two clusters joined beginning with n clusters and ending with one is the Euclidean distance or dissimilarity measure between clusters R and S. For example, for the single, complete, and average link methods it is the minimum, maximum or average Euclidean distance, respectively. For Ward's method, it is the between cluster sum of squares as given in (9.3.8). As the number of clusters decreases from n to 1, the value for the "distance" measure should increase since it should be largest when two dissimilar clusters are joined. A shape elbow in the plot of distances versus the number of clusters may be an indication of the number of clusters. If one could construct an index of separation or a "pseudo" test statistic and plot its values as clusters are joined, a large change in the index when plotted against the number of clusters may help to locate a range of values or even one value for the number of clusters. There is no exact procedure for determining the number of clusters since even with random data one may find spurious clusters.

In regression analysis the coefficient of determination, R^2, is a measure of the total variance in the dependent variable accounted for by the independent variables. In an ANOVA study involving the analysis of trends, R^2 or more correctly $\widehat{\eta}^2$, is defined as the ratio of between sum of squares to the total sum of squares and is a measure of the total variation in the dependent variable accounted for by the group means. Thus, in cluster analysis we may construct an index of R^2 as the number of clusters change. For n clusters, the total sum of squares is $T = \sum_{i=1}^{n} \|\mathbf{y}_i - \overline{\mathbf{y}}\|^2$ and the within cluster sum of squares for cluster k (C_k) is $SSE_k = \sum_i \left\|\mathbf{y}_i - \overline{\mathbf{y}}_k\right\|^2$. Then, R^2 for k clusters is defined as

$$R_k^2 = \frac{T - \sum_k SSE_k}{T} \tag{9.3.10}$$

For n clusters, each $SSE_k = 0$ so that $R^2 = 1$. As the number of clusters decrease from

n to 1 they should become more widely separated. A large decrease in R_k^2 would represent a distinct join. Alternatively, joining clusters R and S one might compute the incremental change in R^2 or

$$SR^2 = R_k^2 - R_{k-1}^2 \tag{9.3.11}$$

called the semipartial R^2 index. The statistic SR^2 compares the ratio of $SSE_t - (SSE_r + SSE_s)$ where clusters C_R and C_S are jointed to form C_T to the total sum of squares $T = \sum_{i=1}^{n} \|\mathbf{y}_i - \bar{\mathbf{y}}\|^2$. The larger the increment, the larger the "loss of homogeneity" or the more separated the clusters.

The goal of cluster analysis is to find a small number of homogeneous clusters. For $n = 1$ cluster, the pooled within variance for all variables is the average of the variances for each variable so that $s^2 = \sum_{i=1}^{n} \|\mathbf{y}_i - \bar{\mathbf{y}}\|^2 / p\,(n-1)$ and the root mean square standard deviation is s = RMSSTD. For any new cluster C_k with n_k observations,

$$[\text{RMSSTD}]^2 = \sum_{i=1}^{n_k} \left\|\mathbf{y}_i - \bar{\mathbf{y}}_k\right\|^2 / p\,(n_k - 1) \tag{9.3.12}$$

is the pooled variance of all variables forming a cluster at a given step. Large values of the pooled variance suggest that the clusters are not homogeneous. Thus, a shape decrease to near zero for some $k < n$ suggest the formation of homogeneous clusters.

Under multivariate normality and independence of the n p-vectors for $\Sigma = \sigma^2\mathbf{I}$, one may test that the centroids of k clusters show significant separation using an ANOVA F statistic. One may also evaluate whether two means are significantly separated at any level of the clustering hierarchy as one proceeds through a cluster analysis using a t statistic. Because independence and MVN are rarely satisfied, the statistics are called pseudo F and pseudo t^2 statistics. The pseudo F statistic is defined as

$$F_k^* = \frac{\left(T - \sum_k SSE_k\right) / (k-1)}{\sum_k SSE_k / (n-k)} \tag{9.3.13}$$

If F_k^* decreases with k, one may not use the statistic to estimate k. However, if F_k^* decreases with k and attains a maximum, the value of k at the maximum or immediately prior to the point may be a candidate for the value of k. The pseudo t^2 statistic is defined as

$$\text{pseudo}\, t^2 = \frac{[SSE_t - (SSE_r + SSE_s)]\,(n_R + n_S - 2)}{SSE_r + SSE_s} \tag{9.3.14}$$

for joining cluster C_R with C_S each having n_R and n_S elements. Again, one may plot the pseudo values versus the number of clusters. If the values are irregular at each join as the number of clusters decrease, then it is not a good index for evaluating the number of clusters. However, if the plot looks like a hockey stick the value of $k + 1$ that caused the slope to change is a candidate for the number of clusters.

A number of statistics are generated by cluster analysis programs that may be plotted to heuristically evaluate how many clusters are generated by the clustering process. Some indices generated by the procedure CLUSTER in SAS include

1. Pseudo F and t^2 statistics (PSEUDO).
2. The root mean square standard deviation (RMSSTD).
3. R^2 and semipartial R^2 (RSQUARE).
4. Centroid Distances (NONORM).

d. *Additional Comments*

Cluster analysis is an exploratory data analysis methodology. It tries to discover how objects may or may not be combined. The analysis depends on the amount of random noise in the data, the existence of outliers in the data, the variables selected for the analysis, the proximity measure used, the spatial properties of the data, and the clustering method employed. Because there is no optimal method, we briefly review some properties of the clustering methods studied by Milligan (1980, 1981) and Milligan and Cooper (1985, 1986).

A major advantage of hierarchical clustering methods over nonhierarchical methods is that one does not have to know or guess the number of clusters. Thus, hierarchical method are often called exploratory while nonhierarchical are often called confirmatory. We view the methods as complementary whenever coordinate (interval level) data are used.

All of the hierarchical methods depend on the proximity measure used in an analysis. Independent of the proximity measure, all hierarchical methods are effected by chaining. Objects tend to be assigned to an existing cluster rather than initiating a new cluster. This is especially the case for the single link method which is also very sensitive to errors of measurement, but somewhat robust to outliers. The complete link and Ward's method tend to find compact clusters of nearly equal size with the clustering solution adversely affected by outliers. Comparing the single, complete, and average link methods, the centroid and Ward methods, Milligan (1980) found the average link hierarchical clustering method to be the preferred method. The study by Milligan and Cooper (1985) found the pseudo F index to be the most helpful in identifying the number of clusters.

A major problem with a cluster analysis solution is the validity of the solution. To assess the validity of a solution one may use internal criteria, external criteria and replication or cross-validation methods. A discussion of these issues are presented by Milligan (1981) and Milligan and Cooper (1986).

9.4 Cluster Analysis Examples

Cluster analysis is used to categorize objects such as patients, voters, products, institutions, countries, and cities among others into homogeneous groups based upon a vector of p variables. The clustering of objects into homogeneous groups depends on the scale of the variables, the algorithm used for clustering, and the criterion used to estimate the number of clusters. In general, variables with very large variances relative to others and outliers have an adverse effect on cluster analysis methods. They tend to dominate the proximity measure.

The CLUSTER procedure in SAS performs a hierarchical cluster analysis using either the data matrix $\mathbf{Y}_{n\times p}$ or a matrix $\mathbf{D} = [d_{rs}]$ of dissimilarity measures. The matrix $\mathbf{Y}$ is used to calculate Euclidean distances, $d_{rs} = \|\mathbf{y}_r - \mathbf{y}_s\|$ where $\mathbf{y}_r$ and $\mathbf{y}_s$ are rows of $\mathbf{Y}$. Distances are used by the SINGLE and COMPLETE link methods. The AVERAGE, CENTROID and WARD methods use d_{rs}^2. To replace d_{rs}^2, with distances d_{rs}, one may use the NOSQUARE option. The NONORM option prevents distances from being divided by the average of all the Euclidean distances for all observations. To evaluate the number of clusters, one may always plot the criterion used to join clusters versus the number of clusters. In SAS, the variable is stored in the SAS data set defined by OUTTREE = with the name _HEIGHT_. When _HEIGHT_ is plotted against the number of clusters, _NCL_, one is looking for a sharp elbow to try to estimate the number of clusters. Using the PSEUDO option, one may also plot the pseudo F (_PSF_) and t^2 (_PST2_) statistics versus _NCL_. One usually looks for large values (peaks) to try to locate the number of clusters for most clustering methods. When t^2 is maximum at _NCL_, the number of cluster near _NCL_ + 1 is a possible choice. Because the single link algorithm tends to truncate the tails of distributions, the pseudo F and t statistics should not be used For all clustering methods, one may plot R^2, semi-partial R^2, root mean square standard deviations and other criteria available in the OUTTREE = data set versus the number of clusters when using continuous data. The plots are illustrated in the examples.

a. *Protein Consumption (Example 9.4.1)*

For our first example, data on protein consumption in twenty-five European countries in nine food groups are analyzed. The data were obtained from The Data and Story Library (DASL) Found on the Web site http://lib.stat.cmu.edu/DASL/Data files/Protein.htlm. The data are measurements of protein consumption in nine food groups (Redmeat, Whitemeat, Eggs, Milk, Fish, Cereals, Starch, Nuts, and Fruits and Vegetables). The data are shown in Table 9.4.1 and are available in the data file protein.dat. The list of twenty-five countries represented by the numbers 1 to 25 in Table 9.4.1 follow.

1. Albania
2. Austria
3. Belgium
4. Bulgaria
5. Czechoslovakia
6. Denmark
7. EGermany
8. Finland
9. France
10. Greece
11. Hungry
12. Ireland
13. Italy
14. Netherlands
15. Norway
16. Poland
17. Portugal
18. Romania
19. Spain
20. Sweden
21. Switzerland
22. UK
23. USSR
24. WGerman
25. Yugoslavia

To initiate a hierarchical clustering method in SAS, the METHOD = option is used on the PROC CLUSTER statement along with other options. Most are names of indices that may be plotted against the number of clusters to help to determine the number of clusters. The SIMPLE option provides descriptive statistics for the variables when using a data matrix $\mathbf{Y}$ as input.

TABLE 9.4.1. Protein Consumption in Europe

Country Number	RMEAT	WMEAT	EGGS	MILK	FISH	CERL	STARCH	NUTS	FR_VEG
1	10.1	1.4	0.5	8.9	0.2	42.3	0.6	5.5	1.7
2	8.9	14.0	4.3	19.9	2.1	28.0	3.6	1.3	4.3
3	13.5	9.3	4.1	17.5	4.5	26.6	5.7	2.1	4.0
4	7.8	6.0	1.6	8.3	1.2	56.7	1.1	3.7	4.2
5	9.7	11.4	2.8	12.5	2.0	34.3	5.0	1.1	4.0
6	10.6	10.8	3.7	25.0	9.9	21.9	4.8	0.7	2.4
7	8.4	11.6	3.7	11.1	5.4	24.6	6.5	0.8	3.6
8	9.5	4.9	2.7	33.7	5.8	26.3	5.1	1.0	1.4
9	18.0	9.9	3.3	19.5	5.7	28.1	4.8	2.4	6.5
10	10.2	3.0	2.8	17.6	5.9	41.7	2.2	7.8	6.5
11	5.3	12.4	2.9	9.7	0.3	40.1	4.0	5.4	4.2
12	13.9	10.0	4.7	25.8	2.2	24.0	6.2	1.6	2.9
13	9.0	5.1	2.9	13.7	3.4	36.8	2.1	4.3	6.7
14	9.5	13.6	3.6	23.4	2.5	22.4	4.2	1.8	3.7
15	9.4	4.7	2.7	23.3	9.7	23.0	4.6	1.6	2.7
16	6.9	10.2	2.7	19.3	3.0	36.1	5.9	2.0	6.6
17	6.2	3.7	1.1	4.9	14.2	27.0	5.9	4.7	7.9
18	6.2	6.3	1.5	11.1	1.0	49.6	3.1	5.3	2.8
19	7.1	3.4	3.1	8.6	7.0	29.2	5.7	5.9	7.2
20	9.9	7.8	3.5	24.7	7.5	19.5	3.7	1.4	2.0
21	13.1	10.1	3.1	23.8	2.3	25.6	2.8	2.4	4.9
22	17.4	5.7	4.7	20.6	4.3	24.3	4.7	3.4	3.3
23	9.3	4.6	2.1	16.6	3.0	43.6	6.4	3.4	2.9
24	11.4	12.5	4.1	18.8	3.4	18.6	5.2	1.5	3.8
25	4.4	5.0	1.2	9.5	0.6	55.9	3.0	5.7	3.2

When all variables are measured on the same scale, one may compare the standard deviations to evaluate whether or not some variables may dominate the clustering procedure. When variables are measured on different scales, one may estimate the coefficient of variation ($CV = s/\overline{x}$) for each variable. When variables display large variations, one may need to employ standardized variables. This is accomplished by using the STD option on the PROC CLUSTER statement in SAS. The bimodality index for each variable is calculated as

$$BM = \left(m_3^2 + 1\right) / \left[m_4 + 3\,(n-1)^2 / (n-2)\,(n-3)\right]$$

where m_3 and m_4 represent sample skewness and kurtosis, respectively. For a uniform distribution, bimodality is $\approx .55$. Thus, if a sample value is larger than .55, the marginal distribution may be bimodal or multimodal, an indication of variable clustering. Variables with large values may dominate and hence bias the clustering process. The EIGEN option causes the eigenvalues of the covariance matrix (correlation matrix for standardized variables) to be calculated.

Program m9_4_1.sas is used to analyze the protein consumption data. Reviewing the raw data statistics, observe that the variance of the cereal group is very much larger than the other variables and that the bimodality values for all variables appear reasonable. Because of the large variance in the cereal group, we use standardized variables for our analysis, and compare five clustering methods: CENTROID, WARD, AVERAGE, COMPLETE and SINGLE. The eigenvalues of the correlation matrix for the standardized variables indicate variation in three or four dimensions, accounting for 75% to 85% of the variance. To evaluate the number of clusters for each method, plots of the selected criteria are made from variables stored in the SAS data set defined by OUTTREE =. Plots are created using PROC GPLOT. The procedure TREE is used to create a dendogram plot, and after sorting, PROC PRINT outputs cluster/group membership.

For each method, we generate plots for several indices to try to determine an optimal cluster value. In general, shape peaks of pseudo F and t^2 values at cluster joins (peak +1) are candidates for the number of clusters. When reviewing root mean square deviations between two clusters joined at a given step, larger values imply lack of homogeneity so that smaller values are better. A large drop in R^2 and/or in the semi-partial R^2 value may indicate distinct clusters. Finally, when plotting a distance measure versus the number of clusters, one should investigate dramatic changes "elbows" for the number of clusters. Reviewing the plots of indices and tables for several criteria calculated in SAS, Table 9.4.2 displays choices for the number of clusters using various criteria. Reviewing the entries in Table 9.4.2, it appears that we should consider a solution with 4, 5 or 6 clusters. The results for NCL = 6 are given in Table 9.4.3. Reviewing the entries, the four clustering methods only agree for countries assigned to groups 1 and 2, countries in the Balkans (1) and Western Europe (2). There is disagreement for the countries in the other four groups. However, the complete link method locates "natural" clusters: Group 3 - Scandinavia, Group 4 - Eastern Europe, Group 5 - Mediterranean and Group 6 - Iberian. Thus, the complete link method provides a reasonable solution for the protein consumption data.

For our example, we see that the single link method fails to classify the countries into groups. Baker and Huber (1975) show that the single link method has limited value because of its sensitivity to data errors and chaining. They and others recommend the complete link method over the single link method. However, when the single link method works it usually works very well, especially in the field of numerical taxonomy, Jardine and Sibson (1971, p. 151). When data are strongly clustered, the two methods complete and single link usually yield identical results. The poor performance of the other methods (centroid, average, and ward) may be due to the presence of isolated points or outliers. The methods tend to cluster outliers. When performing an exploratory cluster analysis, one must consider several distance measures and clustering algorithms to try to obtain a parsimonious solution if it exists at all.

b. *Nonhierarchical Method (Example 9.4.2)*

For our second illustration, we reanalyze the Protein Consumption data using a nonhierarchical method. It is usually a good practice to follow a hierarchical clustering procedure with a nonhierarchical method to refine or validate the solution obtained. The methods are complementary to each other. For a nonhierarchical cluster analysis the procedure FASTCLUS is used. The SAS code for our example is included in program m9_4_2.sas. To use the procedure, one must specify either the number of clusters, MAXCLUSTER, or the

TABLE 9.4.2. Protein Data Cluster Choices Criteria

	F	t^2	R^2	SPR^2	RMS	Height
Centroid	3, 4	4	3	2	4	5
Average	3	4	3	6	4	4
Ward	3, 4	4	4	2	6	2, 5
Complete	6	4, 7	5	5	6	4
Single	N.A.	5	5	4	7	3

TABLE 9.4.3. Protein Consumption—Comparison of Hierarchical Clustering Methods

Country	CENTROID	AVERAGE	WARD	COMPLETE	SINGLE
Albania*	1	1	1	1	1
Austria*	2	2	2	2	2
Belgium*	2	2	2	2	2
Bulgaria*	1	1	1	1	1
Czechoslovakia	2	2	4	4	2
Denmark	2	2	3	3	2
EGermany	2	2	4	4	2
Finland	2	2	3	3	2
France*	2	2	2	2	2
Greece	3	3	5	5	3
Hungry	5	6	4	4	1
Ireland*	2	2	2	2	2
Italy	3	3	5	5	3
Netherlands*	2	2	2	2	2
Norway	2	2	3	3	2
Poland	2	2	4	4	2
Portugal	6	4	6	6	6
Romania*	1	1	1	1	1
Spain	3	4	6	6	5
Sweden	2	2	3	3	2
Switzerland*	2	2	2	2	2
UK*	2	2	2	2	2
USSR	4	5	4	4	4
WGermany*	2	2	2	2	2
Yugoslavia*	1	1	1	1	1

RADIUS = parameter. The RADIUS option is used to specify the minimum Euclidean distance between observations selected as initial seeds. The REPLACE option controls seed replacement after initial seed selection using rules for seed replacement. For k clusters, one needs k seeds. If the radius option is not met, an observation may not be selected as a seed. The radius values must be small enough to ensure k seeds. The default value is zero. SAS reports the initial seeds and the minimum distance between initial seeds and new seeds for each seed replacement. When the process does not converge in the desired number of iterations (MAXITER = 20), since each iteration results in a reallocation of all observations, one may have to increase the number of iterations, modify the number of clusters, or increase the default converge criterion, the default is CONVERGE = 0.02. Because we are using standardized variables, the STANDARD procedure is used to transform the data matrix to standardized variables.

To analyze the protein consumption data, we selected a four, five and six cluster solution allowing SAS to select the seeds. For the four group solution, the seeds were Albania, Ireland, Portugal and Denmark. For five clusters, the seeds Albania, Austria, Greece, Portugal, and Denmark were selected. For the six group solution, the countries Albania, Austria, Netherland, Portugal, Spain, and Denmark were selected. Table 9.4.4 summarizes the geographic regions for the seeds.

While the four and five group solutions selected seeds in each geographic region, this was not the case for $k = 6$ clusters. To control seed selection, one may use the SEED = option with REPLACE = NONE or to refine a solution, select REPLACE = FULL. To try to validate the complete links hierarchical cluster solution, we ran FASTCLUS with fixed seeds stored in the file seed.dat with one entry in each geographic region: Albania, Denmark, Ireland, Italy, Portugal and the USSR. The assignments by cluster are displayed in Table 9.4.5. The entries marked (CL) reflect the complete link assignments of the countries to six groups. To visualize the clusters, we used the procedure CANDISC, discussed in Chapter 7.

For the four and five cluster solutions, one observes distinct separation among clusters with Portugal as a distinct outlier. While the Scandinavia and Balkans groupings remained distinct for these two solutions, the other regions provide mixed results. Comparing the six cluster nonhierarchical solution with the complete link solution with either random or fixed seeds, we were unable to validate the complete link result using the nonhierarchical procedure even with seeds from the hierarchical solution. The countries France and Spain were not easily separated using fixed seeds. Because the initial partition in any nonhierarchical method is critical to a successful result, the random seed 6 cluster result was further from the complete link solution. Disagreements are noted with an asterisk (*) in Table 9.4.5. This example illustrates the wide variability that can occur in clustering methods when performing exploratory clustering methodologies. The clustering method, proximity measure selected and whether or not to use standardization is critical to the analysis. Any one may have an adverse effect on the clustering solution.

c. *Teacher Perception (Example 9.4.3)*

In this example we cluster variables. The data are reported by Napoir (1972). The study involves administering a questionnaire to primary school teachers in which views on the

TABLE 9.4.4. Geographic Regions for Random Seeds

Four	Five	Six
Balkans	Balkans	Balkans
Western Europe	Western Europe	Western Europe (2)
Iberian	Mediterranean	Iberian (2)
Scandinavia	Iberian	Scandinavia
	Scandinavia	

TABLE 9.4.5. Protein Consumption—Comparison of Nonhierarchical Clustering Methods

Country (CL)	Four	Five	Six (Random)	Six (Fixed)
Albania (1)	1	1	1	1
Austria (2)	2	2	2	2
Belgium (2)	2	2	3*	2
Bulgaria (1)	1	1	1	1
Czechoslovakia (4)	2	2	2*	4
Denmark (3)	3	3	3	3
EGermany (4)	2	2	4	4
Finland (3)	3	3	4*	3
France (2)	2	4	4*	5*
Greece (5)	2	4	5	5
Hungry (4)	1	1	1	1
Ireland (2)	2	2	2	2
Italy (5)	2	4	2*	5
Netherlands (2)	2	2	2	2
Norway (3)	3	3	3	3
Poland (4)	2	2	2*	4
Portugal (6)	4	5	6	6
Romania (1)	1	1	1	1
Spain (6)	3	4	5*	5*
Sweden (3)	3	3	3	3
Switzerland (2)	2	2	2	2
UK (2)	2	2	4*	2
USSR (4)	2	2	2*	4
WGermany (2)	2	2	2	2
Yugoslavia (1)	1	1	1	1

TABLE 9.4.6. Item Clusters for Perception Data

Other Teachers	Parents	Principals	Pupils
10	07	03	14
11	08	04	16
09	06	02	15
12	05	01	13

need for change in the behavior on attitudes of people with whom they associate (principals, parents, other teachers, and pupils) were recorded. The need for change was indicated on a four-point Likert scale. To measure the degree of association between pairs of items, Goodman and Kruskal's (1963) gamma was calculated. The measure of association is calculated like Kendall's tau, except that tied pairs are excluded from the count of total pairs,

$$\gamma = \frac{C - D}{C + D}$$

where C represents the number of concordant pairs and D represents the number of discordant pairs. Like the correlation coefficient, $-1 \leq \gamma \leq 1$. However only monotonicity is required between two variables for the $|\ \gamma\ |= 1$. Thus, it does not require a linear association. To cluster items in this example, program m9_4_3.sas is used where the similarity matrix of gammas is converted to dissimilarity measures using the relation $d_{ij}^2 = 2\left(1 - \gamma_{ij}\right)$, where γ_{ij} is the Goodwin-Kruskal gamma reported by Napoir (1972). The 16×16 similarity data matrix is provided in program m9_4_3.sas.

To perform a cluster analysis on items, the similarity measures are input using the TYPE = CORR dataset. Using the DATA step, similarities are converted to dissimilarities, $d_{ij}^2 = 2\left(1 - \gamma_{ij}\right)$, as illustrated in the code in program m9_4_3.sas. Having converted the similarity matrix to a dissimilarity matrix, the remaining statements follow those provided in Example 9.4.1. We again illustrate the four hierarchical methods: CENTROID, WARD, AVERAGE, COMPLETE, and SINGLE.

Reviewing the output, all clustering methods clustered the items into four groups based upon the dissimilarity measures. The sixteen items dealing with teachers' perceptions of need for change were associated with four groups of people: other teachers, parents, principals, and pupils. The item clusters are shown in Table 9.4.6.

While all clustering methods were able to recover the item associations, it does not provide a spatial configuration for all pairwise relations among the items. Because these data involve judgements, the measures are not exact and (metric) factor analysis is not appropriate for evaluating dimensionality. Even if it were, we would find a four factor solution. Using the technique of multidimensional scaling, we will show how the items may be represented in a two or three dimensional space.

To cluster the items in our example using PROC CLUSTER, the matrix of similarities are converted to a dissimilarity matrix. To analyze the similarity matrix directly in SAS, one may use the PROC VARCLUS to obtain clusters. This procedure also resulted in four clusters. Program m9_4_3a.sas contains the SAS code for the analysis.

d. Cedar Project (Example 9.4.4)

For this example, symptom data on alcohol, drug and tobacco use were collected on children (ages 16-20), mothers and fathers in the Pittsburgh area obtained the Center for Education and Drug Abuse Research (CEDAR). The alcohol, drug, and tobacco symptoms are coded as present (one) or absent (zero) for six levels of alcohol use, seven levels of drug use, and six levels of tobacco use. Some subjects in the sample have a substance use disorder (SUD = 1) while others do not have a disorder (SUD = 0). The object of the study was to determine whether the subjects can be clustered into a few distinct groups. For this data, the symptom match (0 − 0) is given no weight so the Jaccard Coefficient is used for the analysis. The Jaccard Coefficient of dissimilarity was calculated using the SAS macro %DISTANCE. The distance macro uses the routines xmacro.sas and distnew.sas distributed by SAS with Version 8. The code for the example is included in program m9_4_4.sas. To use the distance macro, we set METHOD=DJACCARD to calculate the Jaccard dissimilarity measure and use Ward's hierarchical clustering method to determine the number of clusters. Reviewing the output from the hierarchical solution, we observe three clear clusters. However, they are difficult to interpret using the symptom data. To more easily interpret the hierarchical dendogram, we used the nonhierarchical FASTCLUS and CANDISC procedures. Reviewing the nonhierarchical solution, we observe three distinct clusters of individuals. The canonical variates for the nonhierarchical solution provide a clear two-dimensional plot of the three clusters. The vertical axis appears to represent the Alcohol-Drug (+,−) dimension, while the horizontal axis appears to represent the Drug-Tobacco (−, +) dimension. While there is no clear clusters for SUD and non-SUD participants, or for children, mothers, and fathers we do observe three distinct clusters based upon the binary symptom data.

Exercises 9.4

1. Without standardizing the Protein Consumption Data, perform a hierarchical and nonhierarchical solution for the data matrix. Discuss your findings.

2. Using the data in Example 9.4.4, perform a cluster analysis using the single link, complete link,centroid method, and Ward's method with Euclidean distances and use the %DISTANCE macro with Czekanowski's Coefficient (METHOD=DICE) with Ward's Method to determine the number of clusters in the data set. How do these cluster solutions compare with the solution given in Example 9.4.4. Summarize your finding.

3. Analyze the data in Example 9.4.4 for each group: children, mothers, and fathers. Are the clusters the same?

9.5 Multidimensional Scaling

Nonhierarchical clustering methods use the data matrix $\mathbf{Y}_{n\times p}$ and a dissimilarity matrix $\mathbf{D}_{n\times n}$ to cluster items. Alternatively, hierarchical clustering techniques use only a dissimilarity matrix. Multidimensional scaling (MDS) techniques seek to find a low dimensional coordinate system to represent n objects using only a proximity matrix usually without observing the n $(p \times 1)$ observation vectors where the dimension of the space is $k \ll p$. Given perceptions or judgements regarding n objects, one tries to construct a low dimensional space to represent the judgements.

In classical, metric MDS problems, we assume there exists a data matrix $\mathbf{Y}_{n \times p}$ such that the distances between any two objects r and s is $\delta_{rs} = \|\mathbf{y}_r - \mathbf{y}_s\|$. Because $\mathbf{Y}$ is not usually observed, we try to find a set of observations $\mathbf{Z} = [\mathbf{z}_1, \mathbf{z}_2, \ldots, \mathbf{z}_n]'$ where $\mathbf{z}_i$ is in a k-dimensional space such that the distances $d_{rs} = \|\mathbf{z}_r - \mathbf{z}_s\|$ where $d_{rs} \approx \delta_{rs}$. Thus, a model for the distances may take the form

$$d_{rs} = \alpha + \beta \delta_{rs} + e_{rs} \text{ for all } r, s \tag{9.5.1}$$

where e_{rs} is random measurement error. Generating a scatter plot of the pairs (d_{rs}, δ_{rs}), one may obtain a set of estimated distances d_{rs} using the relationship

$$\widehat{d}_{rs} = \widehat{\alpha} + \widehat{\beta} \delta_{rs} \tag{9.5.2}$$

The fitted values $\widehat{d}_{rs}$ are not distances, but simply numbers fitted to distances and referred to as disparities. Ordering the proximities δ_{rs} so that

$$\delta_{r_1 s_1} < \delta_{r_2 s_2} < \ldots < \delta_{r_m s_m} \tag{9.5.3}$$

where $m = n(n-1)/2$ and $r_i < s_i$ are distinct integers, then by (9.5.2) if

$$\delta_{rs} < \delta_{u\text{v}} \Longrightarrow \widehat{d}_{rs} \leq \widehat{d}_{u\text{v}} \quad \text{for all } r < s, u < \text{v} \tag{9.5.4}$$

This is called the weak monotonicity constraint.

In most applications of multidimensional scaling, exact proximity measures δ_{rs} are not available; however, one may be able to order the proximities as in (9.5.3) where the ordered quantities are perceived judgments regarding n objects. In addition, we may not be able to specify the relationship between δ_{rs} and d_{rs} as in (9.5.1), but only that

$$d_{rs} = f(\delta_{rs}) + e_{rs} \quad \text{for all } r, s \tag{9.5.5}$$

where the function f is an unknown monotonic increasing function with property (9.5.4). Thus, only the rank order of the distances d_{rs} is used to construct δ_{rs}. Because only the rank order or ordinal relationship of the distances d_{rs} is used in finding the configuration to reconstruct δ_{rs}, the scaling process is termed nonmetric. This has nothing to do with the properties of the space which remains a metric space.

a. *Classical Metric Scaling*

Given n objects and a dissimilarity matrix $\Delta = [\delta_{rs}]$, the matrix Δ is said to be Euclidean if there exists a matrix $\mathbf{Y}_{n \times p}$ such that

$$\delta_{rs}^2 = (\mathbf{y}_r - \mathbf{y}_s)' (\mathbf{y}_r - \mathbf{y}_s) \quad r, s = 1, 2, \ldots, n \tag{9.5.6}$$

The Euclidean matrix Δ has zeros on its main diagonal and is not positive semidefinite. However, a positive definite matrix $\mathbf{B}$ can be constructed from the elements δ_{rs} in Δ using the fundamental theorem of MDS which follows, Seber (1984, p. 236).

Theorem 9.5.1 *Given the contrast matrix*

$$\mathbf{B} = [b_{rs}] = \left(\mathbf{I}_n - n^{-1}\mathbf{1}_n\mathbf{1}_n'\right)\mathbf{A}\left(\mathbf{I}_n - n^{-1}\mathbf{1}_n\mathbf{1}_n'\right)$$

where $\Delta^2 = \left[\delta_{rs}^2\right]$ *and* $\mathbf{A} = -\frac{1}{2}\Delta^2$, *the matrix* Δ *is Euclidean if and only if* $\mathbf{B}$ *is positive semidefinite (p.s.d.).*

Expanding $\mathbf{B}$ in Theorem 9.5.1, an element of $\mathbf{B}$ has the "corrected" ANOVA structure

$$\begin{aligned} b_{rs} &= -\frac{1}{2}\left[\delta_{rs}^2 - \overline{\delta}_{r.}^2 - \overline{\delta}_{.s}^2 + \overline{\delta}_{..}^2\right] \\ &= a_{rs} - \overline{a}_{r.} - \overline{a}_{.s} + \overline{a}_{..} \end{aligned} \tag{9.5.7}$$

where $\overline{\delta}_{r.}^2$, $\overline{\delta}_{.s}^2$, and $\overline{\delta}_{..}^2$ are the row, column and overall averages of the two-way matrix Δ^2 so that b_{rs} is calculated directly from the distances δ_{rs}.

Given that the matrix of distances Δ is Euclidean, there exists a configuration $\mathbf{Y}_{n\times p} = [\mathbf{y}_1, \mathbf{y}_2, \ldots, \mathbf{y}_n]'$ such that $-2a_{rs} = \delta_{rs}^2 = \|\mathbf{y}_r - \mathbf{y}_s\|^2$. By Theorem 9.5.1, and using (9.5.7), one can show that an element of $\mathbf{B}$ has the form

$$b_{rs} = (\mathbf{y}_r - \overline{\mathbf{y}})'(\mathbf{y}_s - \overline{\mathbf{y}}) \tag{9.5.8}$$

so that $\mathbf{B} = \mathbf{Y}_d\mathbf{Y}_d'$ where $\mathbf{Y}_d' = [\mathbf{y}_1 - \overline{\mathbf{y}}, \ldots, \mathbf{y}_n - \overline{\mathbf{y}}]$. Because $\overline{\mathbf{y}}$ is the mean of $\mathbf{y}$, the matrix $\mathbf{Y}_d$ is the mean centered matrix.

Conversely, using Theorem 9.5.1 if $\mathbf{B}$ is p.s.d. and of rank p, by the Spectral Decomposition Theorem there is a matrix $\mathbf{P}$ such that $\mathbf{P}'\mathbf{B}\mathbf{P} = \Lambda$ where $\widehat{\lambda}_1 \geq \widehat{\lambda}_2 \geq \ldots \geq \widehat{\lambda}_p$ are the eigenvalues of $\mathbf{B}$ and $\mathbf{P} = [\mathbf{p}_1, \mathbf{p}_2, \ldots, \mathbf{p}_n]$ are the associated eigenvectors. Hence, $\mathbf{B} = \mathbf{P}\Lambda\mathbf{P}' = \mathbf{Z}\mathbf{Z}'$ where $\mathbf{Z} = \mathbf{P}\Lambda^{1/2}$ is the metric scaling solution with $\mathbf{B} = [b_{rs}] = \mathbf{z}_r'\mathbf{z}_s$. Expressing $\mathbf{B}$ in terms of $\mathbf{A}$, observe that

$$\begin{aligned} d_{rs}^2 &= \|\mathbf{z}_r - \mathbf{z}_s\|^2 \\ &= \mathbf{z}_r'\mathbf{z}_r - 2\mathbf{z}_r'\mathbf{z}_s + \mathbf{z}_s'\mathbf{z}_s \\ &= b_{rr} - 2b_{rs} + b_{ss} \\ &= a_{rr} - 2a_{rs} + a_{ss} \\ &= -2a_{rs} \\ &= \delta_{rs}^2 \end{aligned} \tag{9.5.9}$$

by relating b_{rs} to a_{rs} using (9.5.7), and observing that $a_{rr} = -\delta_{rr}^2/2 = 0$ and $a_{ss} = 0$. Hence, we can recover Δ^2 exactly from $\mathbf{Z}$ where the $\|\mathbf{z}_i\|^2 = \lambda_i \|\mathbf{p}_i\| = \lambda_i$. In practice, $k \ll p$ so that $d_{rs} \approx \delta_{rs}$. The vectors $\mathbf{z}_1, \mathbf{z}_2, \ldots, \mathbf{z}_n$ are called the principal coordinates of $\mathbf{Z}$ in k dimensions. Because rotations and translations of the coordinates do not change interpoint distances, the solution is not unique and may be, for example, translated to the origin and rotated for interpretation. To standardize the principal coordinates, one uses the matrix $\mathbf{Z}^* = \mathbf{Z}\Lambda^{-1/2}$ so that $\|\mathbf{z}_i^*\| = 1$.

Recall in cluster analysis that if s_{rs} is a similarity measure and the profiles are mean-centered using the columns of $\mathbf{Y}_{n\times p}$, then $\delta_{rs}^2 = 2(1 - s_{rs})$. Letting $a_{rs} = s_{rs}$ in Theorem 9.5.1, by (9.5.9) the squared distances

$$\begin{aligned} d_{rs}^2 &= \|\mathbf{z}_r - \mathbf{z}_s\|^2 \\ &= a_{rr} - 2a_{ss} + a_{ss} \\ &= 2(1 - s_{rs}) \\ &= \delta_{rs}^2 \end{aligned}$$

so that one may obtain a metric scaling using $\mathbf{B} = [b_{rs}]$ where b_{rs} is defined in (9.5.7) for $a_{rs} = -\delta_{rs}^2/2$ or $a_{rs} = s_{rs}$. The process of finding the principal coordinates of a proximity matrix using either similarity of dissimilarity measures resulting in a low-dimensional plot is called an ordination of the data. In PCA and EFA, the data matrix $\mathbf{Y}$ is provided. In classical, metric MDS one begins with the matrix $\mathbf{B}$ which is a function of dissimilarities or similarities.

To evaluate the goodness-of-fit of a classical, metric MDS solution of dimension k, we consider two criteria proposed by Mardia et al. (1979, p. 406). Letting $\widehat{\lambda}_i$ be an eigenvalue of $\mathbf{B}$ and $\widehat{\mathbf{B}} = \sum_{i=1}^k \lambda_i \mathbf{z}_i \mathbf{z}_i'$ for eigenvectors $\mathbf{z}_i$, they suggest the two fit criteria

$$\phi^2 = \sum_{r,s} \left(\delta_{rs}^2 - d_{rs}^2\right) = 2n \sum_{i=k+1}^{n} \widehat{\lambda}_i \tag{9.5.10}$$

$$\psi^2 = \sum_{r,s} (b_{rs} - \widehat{b}_{rs})^2 = ||\mathbf{B} - \widehat{\mathbf{B}}||^2 = \sum_{i=k+1}^{n} \widehat{\lambda}_i^2$$

The proportion of the distance matrix explained by a k-dimensional solution is

$$\alpha_{1,k} = \sum_{i=1}^{k} \widehat{\lambda}_i \Big/ \sum_{i=1}^{n} \widehat{\lambda}_i \tag{9.5.11}$$

$$\alpha_{2,k} = \sum_{i=1}^{k} \widehat{\lambda}_i^2 \Big/ \sum_{i=1}^{n} \widehat{\lambda}_i^2$$

In most applications, $k = 2$ provides an adequate fit.

In classical, metric scaling the data matrix $\mathbf{Z}_k$ may be viewed as an orthogonal projection of $\mathbf{Y}$ into a subspace of dimension k by estimating measured proximities with distances. Classical metric scaling (principal ordinate analysis) is equivalent to plotting objects using principal components.

b. *Nonmetric Scaling*

In nonmetric scaling, the matrix Δ instead of being made up of measured proximities is usually based upon perceptions or judgments about a set of objects. Thus, one usually knows that one object is perceived to be better than another, but one cannot say by how much. In these situations one may only order the proximity measures as in (9.5.3) such that f is a monotonic function as in (9.5.5) and the constraint (9.5.4) is satisfied.

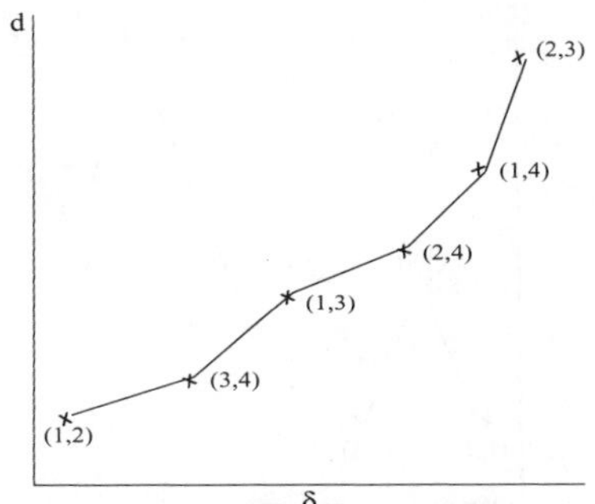

FIGURE 9.5.1. Scatter Plot of Distance Versus Dissimilarities, Given the Monotonicity Constraint

To understand the concepts behind nonmetric scaling, it is helpful to consider a simple example. Consider a situation in which one is given $n = 4$ new cars and asked to rank order $\binom{n}{2} = 6$ dissimilarity measures δ_{rs} with the result that

$$\delta_{12} < \delta_{34} < \delta_{13} < \delta_{24} < \delta_{14} < \delta_{23} \tag{9.5.12}$$

That is cars 1 and 2 are judged to be least dissimilar (most similar) while cars 2 and 4 are most dissimilar (least similar). Using the monotonicity condition given in (9.5.4), suppose that observed Euclidean distance between the points follow the same sequence so that

$$d_{12} \leq d_{34} \leq d_{13} \leq d_{24} \leq d_{14} \leq d_{23} \tag{9.5.13}$$

This means that the order relationship for interpoint distances is identical to the observed dissimilarities. Plotting the pairs (d_{rs}, δ_{rs}) for the above points results in Figure 9.5.1. Observe the natural monotonic chain among all pairs of objects, which is also the case for metric scaling.

Switching d_{34} and d_{13}, and also d_{14} and d_{23}, the revised order becomes

$$d_{12} \leq d_{13} \leq d_{34} \leq d_{24} \leq d_{23} \leq d_{14} \tag{9.5.14}$$

so that the monotonicity constraint is violated. Plotting the pairs when monotonicity is violated, Figure 9.5.2 results.

The scatter plot in Figure 9.5.2 is now not a smooth monotone chain, but more like a "sawtooth" chain. To generate a sequence that satisfies the monotonicity constraint, we may create a set of fitted values $\widehat{d}_{rs}$ such that

$$\widehat{d}_{12} \leq \widehat{d}_{34} \leq \widehat{d}_{13} \leq \widehat{d}_{24} \leq \widehat{d}_{14} \leq \widehat{d}_{23} \tag{9.5.15}$$

Setting $\widehat{d}_{34} = \widehat{d}_{13} = (d_{13} + d_{14})/2$, and $\widehat{d}_{14} = \widehat{d}_{23} = (d_{23} + d_{14})/2$, the new fitted values satisfy the monotonicity condition shown as a dashed (bolder) line in Figure 9.5.2. While the $\widehat{d}_{ij}$ satisfy the monotonicity constraint, they may no longer be distances and are sometimes called pseudodistances.

For a given value of k, one may not be able to find a configuration of points whose pairwise distances (in any L_p-metric) satisfy the monotonicity constraint given in (9.5.4).

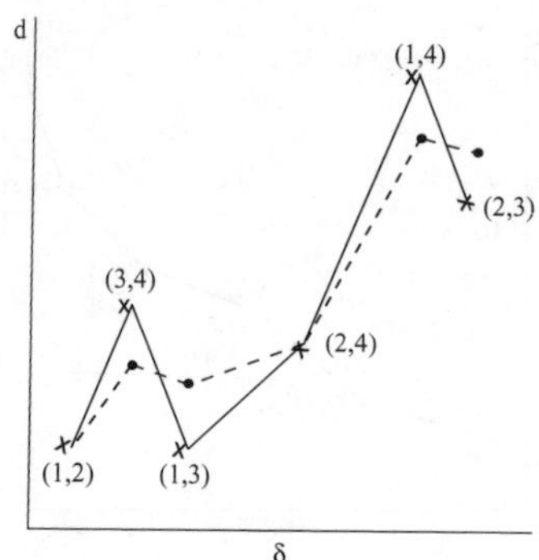

FIGURE 9.5.2. Scatter Plot of Distance Versus Dissimilarities, When the Monotonicity Constraint Is Violated

To evaluate departures from monotonicity Kruskal (1964 a,b) following the metric scaling method recommended the index called the Standardized Residual Sum of Square (STRESS) defined as

$$\text{STRESS} = \left\{ \frac{\sum\sum_{r<s} \left(d_{rs}^2 - \widehat{d}_{rs}^2\right)^2}{\sum\sum_{r<s} d_{rs}^2} \right\}^{1/2} \tag{9.5.16}$$

which is between 0 and 1. Given a set of distances d_{rs}, one selects $\widehat{d}_{rs}$ so that STRESS is minimized subject to the constraint that the pseudodistances $\widehat{d}_{rs}$ are monotonic nondecreasing with the observed δ_{rs} as specified in (9.5.4). The scaling constant in (9.5.4) is the sum of the squared distances $d_{rs.}^2$ To minimize STRESS, Kruskal (1964b, 1977) used a least squares monotone regression procedure and a steepest decent algorithm. Kruskal (1964a) provides the following guidelines for his STRESS index of "goodness" of monotonicity as follows.

Minimum STRESS × 100%	Goodness of Fit
20	Poor
10	Fair
5	Good
2.5	Excellent
0	Perfect

In general, one likes to obtain a value of STRESS ≤ 0.1 for some k.

Observe that Kruskal's STRESS index is similar to a normalized value of ψ in metric scaling. Using normalized deviations of ϕ, Takane, Young and de Leeuw (1977) suggested the index

$$\text{SSTRESS} = \left\{ \frac{\sum\sum_{r<s} \left(d_{rs}^2 - \widehat{d}_{rs}^2\right)^2}{\sum\sum_{r<s} \widehat{d}_{rs}^4} \right\}^{1/2} \tag{9.5.17}$$

Because SSTRESS uses squares of distances instead of distances, it tries harder to fit larger dissimilarities ignoring smaller deviations. Observe that the scaling constant in SSTRESS uses the estimated disparities instead of the observed distances. Both indexes are between 0 and 1. Plotting STRESS or SSTRESS versus k, one may empirically estimate the dimension of the solution at the "elbow" of the hockey stick like plot. The large decline should appear near $k = 2$ or 3.

Algorithms for nonmetric multidimensional scaling are interative and begin with an initial configuration by selecting various values of k beginning with $k = 1$. The initial configuration is usually a metric solution.

The MDS procedure in SAS uses the STRESS criterion defined in (9.5.16) to evaluate goodness of fit (called badness of fit in SAS). The STRESS criterion is obtained using FORMULA $= 1$. The other option, FORMULA $= 2$, in the MDS procedure normalizes (9.5.16) using the mean deviations $\sum\sum_{r<s} \left(d_{rs} - \overline{d}_{rs}\right)^2$ which allows one to relate STRESS to $\sqrt{1 - R^2}$ where R is the multiple correlation coefficient. The algorithm used in SAS to fit a nonmetric model to proximity measures combines monotone regression with a nonlinear least squares regression. The ALSCAL procedure in SAS minimizes SSTRESS using the Newton-Raphson method described in Schiffman, Reynolds and Young (1981).

Most of the initial models for metric and nonmetric scaling are exploratory and descriptive. Assuming the observed elements of $\Delta = \left[\delta_{ij}\right]$ follow some distribution, maximum likelihood methods may be used to scale items allowing one to obtain simultaneous confidences intervals for the elements of Δ. Ramsay (1982) develops maximum likelihood procedures for metric scaling while Brady (1985) discusses the nonmetric situation. De Leeuw and Meulman (1986) use resampling methods to obtain confidence intervals.

c. *Additional Comments*

The MDS methods we have reviewed use only one proximity matrix. If one asked N subjects to rate n objects, we would have a three-way model so that $\Delta = [\delta_{rst}]$. Models that fit distances d_{rst} to δ_{rst} are called individual differences multidimensional scaling (INDSCAL) models, Carroll and Chang (1970). The SAS procedure ALSCAL and MDS perform nonmetric three-way multidimensional scaling. Kruskal and Wish (1978) show that INDSCAL models are in general better than nonmetric MDS models fit to an average dissimilarities matrix since one may weight individuals differentially. De Leeuw and Heiser (1980) show how to fit INDSCAL models with restrictions.

Multidimensional scaling models that fit both objects and variables (items) are called unfolding models, Heiser (1981) and Nishisato (1994, 1996). For these models, individuals provide a rank or preference of n objects so that the matrix $\Delta_{N\times n}$ is rectangular. Meulman (1992) considers a nonlinear data distance model that is closely related to MDS.

We have not discussed correspondence analysis in this book since we have been primarily concerned with continuous variables. Correspondence analysis was introduced by Fisher to analyze frequency data organized in an $I \times J$ two-way table of unscaled frequencies and by Guttman to analyze indicator matrices. Like multidimensional scaling, the goal of correspondence analysis is to represent a contingency table formed from categorical data in a low dimensional space. Carroll, Kumbasar and Romney (1997) show how a variant of correspondence analysis when applied to exact or transformed Euclidean distances is asymptotically equivalent to classical multidimensional scaling. For a discussion of correspondence analysis, the book by Greenacre (1984), Gifi (1990) and Benzécri (1992) may be consulted. Correspondence analysis is available in SAS using the procedure CORRESP. Data analysis techniques that explore dimensionality also involve visualization of the data. Methods for visualizing multivariate data may be found in the non-traditional books by Blasius and Greenacre (1998) and Jambu (1991).

9.6 Multidimensional Scaling Examples

In cluster analysis one begins with distances between objects in order to cluster the objects into groups. In MDS, one tries to discover the dimension of the space that contributes to the differences among the objects based upon imprecise judgements or associations among the objects. Kruskal and Wish (1978) put it very nicely using as an example distance measures among cities on a map. Given a map, one may easily construct distances among cities; however, MDS involves creating a map given only "approximate" distances among cities where the measurements are made in a low dimensional space of unknown dimension. The hope is to be able to display the objects with a spatial orientation that has meaning. MDS employs proximity measures to uncover unknown or hidden structure regarding a set of objects using imprecise measurements.

To perform multidimensional scaling in SAS, we use the MDS procedure. The procedure may be used to fit two- and three-way, metric and nonmetric multidimensional scaling models. The three-way model is the INDSCAL model. Special unfolding models may also be analyzed. Most simple applications of MDS models begin with a dissimilarity matrix (the default in SAS). To convert a similarity matrix to a dissimilarity matrix, the option SIMILAR is used. For example, if the proximity matrix is a correlation matrix the option SIMILAR calculates $\delta_{ij} = 1 - r_{ij}$.

In classical metric scaling, one is given a $n \times n$ Euclidean matrix $\Delta = \left[\delta_{ij}\right]$ of dissimilarities and the goal is to find a matrix $\mathbf{D} = \left[d_{ij}\right]$ that is close to Δ. This is accomplished by using the matrix $\mathbf{B}$ in (9.5.7). The matrix $\mathbf{Z}$ of eigenvectors of $\mathbf{B}$ represent the coordinates of the n points on the ith axis of the Euclidean space. This matrix $\mathbf{Z}$ provides a set of "principal" coordinates resulting in the estimated squared distances d_{ij}^2 so the Δ is approximated by $\mathbf{D}$. That is, $\phi^2 = \sum_{r,s}\left(\delta_{rs}^2 - d_{rs}^2\right)$ is minimized. Implicit in classical metric scaling is the MDS model that $d_{rs} = \delta_{rs} + e_{rs}$. More generally, one may fit a model of the form $d_{rs} = \alpha + \beta\delta_{rs} + e_{rs}$, or $d_{rs} = f\left(\delta_{rs} + e_{rs}\right)$ where f is some arbitrary monotone function. In MDS, the LEVEL = option is used to specify the MDS model using levels of measurement. For metric scaling, LEVEL = ABSOLUTE and the proximity matrix is either a dissimilarity matrix or a similarity matrix. For nonmetric scaling, the option LEVEL = ORDINAL is used. The FIT option allows one to fit distances, squared distances, the logarithm of distances, and the nth power of distances. The FORMULA option governs the badness-of-fit criterion. For FIT = DISTANCE, FORMULA = 1, and LEVEL = ORDINAL, Kruskal's stress formula in (9.5.16) is used for the nonmetric MDS model. More will be said about various options in MDS as we proceed through examples.

TABLE 9.6.1. Road Mileages for Cities

	CH	HO	NY	SF
Chicago	0			(Sym)
Houston	1212	0		
New York	813	1667	0	
San Francisco	2156	1950	2947	0

a. *Classical Metric Scaling (Example 9.6.1)*

The most fundamental application of the MDS model is to reconstruct a map from a matrix $\Delta = \left[\delta_{ij}\right]$ of distances. For our example, we use road miles between four US cities, similar to the flying mile distance example discussed by Kruskal and Wish (1978, p. 8). The road miles are given in Table 9.6.1 among the four U.S. cities: Chicago, Houston, San Francisco and New York. Program m9_6_1.sas performs the analysis.

To analyze the mileages inTable 9.6.1 using PROC MDS, we specify the MDS model as $d_{rs} = \delta_{rs} + e_{rs}$ by using the option LEVEL = ABSOLUTE. The estimated coordinates of the four cities are output to the data set defined in OUT =. Using the DIMENSION = 2 option, we request a two-dimensional solution. To plot the coordinates for the solution, the PROC PLOT statement is used. To save the fitted distances d_{rs}, one must create a data set using option OUTRES =. The FITDATA vector contains the distances d_{rs} and the FITDIST vector contains the δ_{rs} (or transformed δ_{rs} using the FIT option). To evaluate the monotonic relationship between the observed and fitted distances, the PLOT statement is again used.

For this problem, Kruskal's stress fit index, called the badness–of-fit criterion is 0.004817 indicating that the data fit is almost perfect since STRESS $\times 100\% = 0.4$. The coordinates for the four cities are represented in the output labeled configuration. This is obtained by using the option PFINAL. To plot the coordinates, the PROC PLOT DATA = OUT statement is used where _TYPE_'CONFIG'. The Configuration coordinates for the four cities are

$$\mathbf{Z} = \begin{bmatrix} 465.70 & -298.67 \\ -51.28 & 799.08 \\ 1262.82 & -244.23 \\ -1677.23 & -276.19 \end{bmatrix} \begin{matrix} \text{Chicago} \\ \text{Houston} \\ \text{New York} \\ \text{San Francisco} \end{matrix}$$

The default PLOT statement may not scale the axes correctly so that the distance between "tick" marks are approximately equal for both axes. To modify the scaling of tick marks, the VTOH = option is used to adjust the ratio of the vertical distance between tick marks to the horizontal distance. This often requires some experimentation. In addition, the units on each axis must be set using the /HAXIS = VAXIS = option. Finally, the orientation of the plotted objects may not lead to a direct interpretation of physical directional relationships. The MDS plot for the four cities is shown in Figure 9.6.1. Observe that while the "normal"

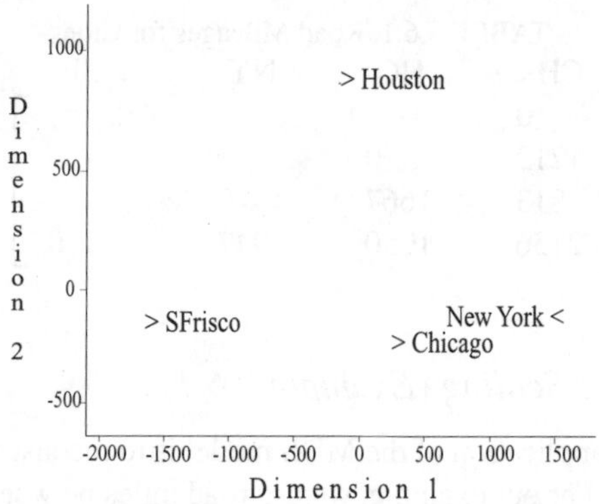

FIGURE 9.6.1. MDS Configuration Plot of Four U.S. Cities

east-west orientation is displayed, this is not the case for the "normal" north-south orientation. The desired orientation occurs if one changes the signs on the second dimension. For interpretation of plots, one may relocate the origin and rotate the axes. This may be accomplished using SAS/INSIGHT and the matrix **Z** in the data set defined by OUT=.

Having obtained a parsimonious configuration, one next inspects the residuals stored in the data set defined in OUTRES =. The PLOT statement FITDIST * RESIDUAL generates a residual plot of the fitted distances versus the residuals. A plot of the residual matrix of quasi-distance residuals resulting in patterns suggests increasing the dimensionality of a solution. Creating a SCREE plot of Stress versus the number of dimensions may also be used to evaluate dimensionality. Often a sharp elbow at some low level of dimensionality may suggest as feasible solution. Finally, one may plot the coordinates (d_{ij}, δ_{ij}); the distances in the MDS space and the original data distances (proximities). This is accomplished using the PLOT statement with FITDIST * FITDATA. A clear monotonic relationship should be evident showing that small dissimilarities correspond to small distances, and large dissimilarities to large distances. For similarities, the plot should represent a monotonic decreasing relationship. If this does not occur, the solution may be incorrect.

b. *Teacher Perception (Example 9.6.2)*

For our next example, we reanalyze the similarity matrix of Napoir (1972) examined in Example 9.4.3. Having found that the items cluster into four groups, we use multidimensional scaling to represent the clusters in a low dimensional space. For the MDS application, the Goodman-Kruskal γ_{ij} similarity measures are used directly. However, one must now specify SIMILAR on the MDS statement. Then, $\delta_{ij} = 1 - \gamma_{ij}$ is the dissimilarity measure analyzed. We investigate solutions in several dimensions and also use the EFA (metric) model discussed in Chapter 8 to explore dimensionality.

Setting DIMENSION = 2, the badness-of-fit criterion value of 0.15 or $100 \times .15 = 15$ indicates a fair to poor fit for the data. Plotting the two-dimensional configuration and joining items using the cluster analysis hierarchy, a compact geometric representation for the four item clusters is evident, Figure 9.6.2. The figure represents wider separation between

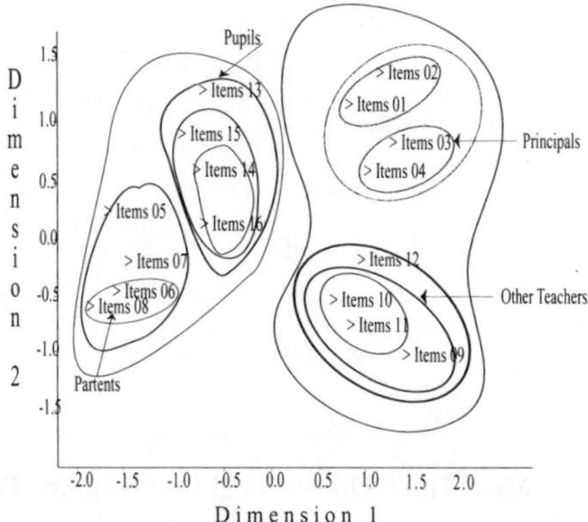

FIGURE 9.6.2. MDS Two-Dimensional Configuration Perception Data

the clusters other teachers and principals than between the parent and pupil clusters.

Figure 9.6.2 enhances the geometric structure of the clusters. However, from the residual plot a distinct cluster of the residuals is evident suggesting that we increase the dimensionality. For a three-dimensional solution, STRESS = 0.08, a slight improvement resulting in a fit results, between good and fair. The matrix **Z** of coordinates for the configuration follows.

$$\mathbf{Z} = \begin{bmatrix} 0.90 & 1.31 & 0.29 \\ 1.35 & 1.56 & -0.15 \\ 1.36 & 1.23 & 0.21 \\ 1.11 & 0.79 & 0.41 \\ -1.25 & 0.35 & 1.15 \\ -1.49 & -0.33 & 0.69 \\ -1.29 & 0.11 & 0.69 \\ -1.48 & -0.22 & 1.30 \\ 1.42 & -1.34 & 0.11 \\ 0.87 & -1.31 & 0.37 \\ 1.12 & -1.25 & 0.57 \\ 1.32 & -1.06 & -0.83 \\ -0.57 & -0.01 & -1.69 \\ -.97 & 0.02 & -0.95 \\ -1.29 & 0.45 & -1.21 \\ -1.10 & -0.32 & -0.96 \end{bmatrix}$$

The coordinates in **Z** are not directly, immediately interpretable. The first two vectors locate the items in the (x, y)-plane and the third vector provides the coordinates of the vertical, perpendicular z-axis. While one may construct a two-dimensional plot, putting the third coordinates within parentheses, the geometric picture may be difficult to interpret. For a clear spatial representation of the clusters, a three-dimensional plot is needed. To create the plot using SAS, one may use SAS/INSIGHT. One may interactively rotate the axes to generate the three-dimensional plot of the item clusters for the perception data, Figure 9.6.3. The three-dimensional plot enhances the separation among clusters and helps to illustrate

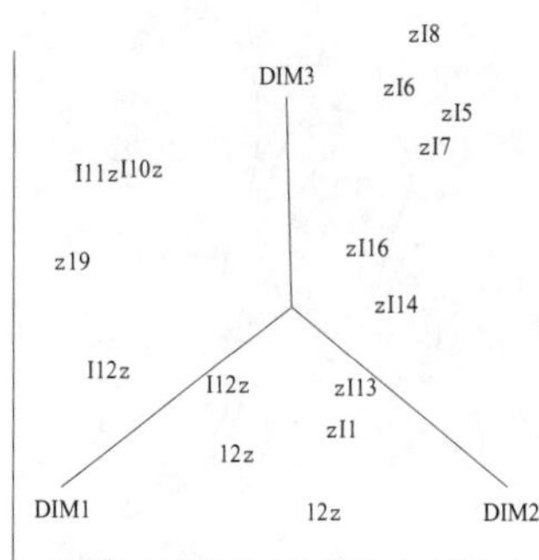

FIGURE 9.6.3. MDS Three-Dimensional Configuration Perception Data

a wider separation between the Parent and Pupils clusters. This was not as evident in the two-dimensional plot.

While a fit for $k = 4$ dimensions reduces the fit value of stress to 0.04, the configuration does not enhance cluster separation or interpretation. Thus, we conclude that a three-dimensional fit adequately represents the structure of the gamma matrix for the perception data.

Embedding a cluster analysis in a low dimensional MDS space often results in a special representation of the objects that reflect the structural aspects of the (dis)similarity matrix using Euclidean distances between items. Recall that the solution for the distance between any two objects has the form

$$\| \mathbf{y}_r - \mathbf{y}_s \|^2 = \| \mathbf{y}_r \|^2 + \| \mathbf{y}_s \|^2 - 2 \| \mathbf{y}_r \| \| \mathbf{y}_s \| \cos \theta_{rs}$$
$$d_{rs}^2 = d_r^2 + d_s^2 + 2 d_r d_s \cos \theta_{rs}$$

For unit vectors $\mathbf{y}_r$ and $\mathbf{y}_s$, the squared distances are

$$d_{rs}^2 = 2\,(1 - \cos \theta_{rs}) = 2\,(1 - a_{rs})$$

where a_{rs} is an association (similarity) measure. In factor analysis, we are not fitting latent factors to d_{rs}^2, but to the associations a_{rs} which are the cosines of angles between objects (items in our example) and not distances. The latent factors, the position vectors in the vector space, are used to simultaneously approximate the pairwise associations, using a linear model for the data. Treating the association measure γ_{ij} as a correlation coefficient we may fit a (metric) EFA model to the matrix. This results in four latent factors with a pattern matrix of dimension four to reconstruct the clusters. Using a MDS model, we were able to recover the clusters in two-dimensions with limited distortion while a three-dimensional solution provided a clear Euclidean spatial representation of the clusters. The EFA model required four dimensions. A metric EFA solution is provided in Table 9.6.2.

In most applications, the metric factor analysis model used to recover dimensionality or structure in a correlation matrix usually results in a higher level of dimensionality. And, no measure of separation is available. For MDS models, one often finds a low dimensional compact solution without severe distortion that is interpreted using a visual process of rotating coordinates. Because γ_{ij} is not a Pearson correlation coefficient, one should analyze

TABLE 9.6.2. Metric EFA Solution for Gamma Matrix

	FACTOR1	FACTOR2	FACTOR3	FACTOR4
I1	74*	25	20	15
I2	84*	1	9	20
I3	82*	12	23	14
I4	73*	22	32	15
I5	21	63*	11	27
I6	13	62*	15	33
I7	15	82*	16	24
I8	6	81*	14	20
I9	27	11	62*	13
I10	16	17	87*	20
I11	15	14	86*	16
I12	30	15	42*	22
I13	21	19	14	62*
I14	16	29	19	79*
I15	15	27	12	64*
I16	12	27	25	70*

the Gamma matrix using a nonmetric factor analysis model that is able to recover rank-order patterns in the matrix, Lingoes and Guttman (1967). However, SAS does not provide this data analysis procedure. The MDS model and the nonmetric EFA model are both able to recover nonlinear patterns in a data matrix with the MDS model fit usually obtained in a lower dimension since it uses distances to recover rank-order patterns instead of cosines of angles between vectors.

c. *Nation (Example 9.6.3)*

This example uses the similarity matrix of perceptions of students in 1968 who rated the overall similarity between twelve nations on a scale from 1 for "very different" to 9 for "very similar", Kruskal and Wish (1978, p. 31). The mean ratings are shown in Table 9.6.2. The higher the mean rating, the more similar the nations.

Using the mean similarity measures, a one, two, and three dimensional MDS model is fit to the data, program m9_6_3.sas. Because the input matrix contains similarity measures, we must again use the option SIMILAR on the PROC MDS statement. However, because the matrix of similarities are not associations, the transformed δ_{rs} are created using the expression $\max(s_{rs}) - s_{rs} = \delta_{rs}$ where the max $s_{rs} = 6.67$ (most similar) for Russia and Yugoslavia. The least similar were China and the Congo where $s_{rs} = 2.39$, the most dissimilar pair. The LEVEL = ORDINAL option performs a monotone transformation on the δ_{rs} and d_{rs} are fit to the transformed proximities.

Fitting four models of dimensions 1, 2, 3, and 4 to the data, the STRESS criteria are 0.39, 0.19, 0.11, and 0.06, respectively. While the Stress criterion is lowest for a four dimensional model, it may not represent the most stable model. To ensure model stability,

TABLE 9.6.3. Mean Similarity Ratings for Twelve Nations

	1	2	3	4	5	6	7	8	9	10	11	12
1	0											(Sym)
2	4.83	0										
3	5.28	4.56	0									
4	3.44	5.00	5.17	0								
5	4.72	4.00	4.11	4.78	0							
6	4.50	4.83	4.00	5.83	3.44	0						
7	3.83	3.33	3.61	4.67	4.00	4.11	0					
8	3.50	3.39	2.94	3.83	4.22	4.50	4.83	0				
9	2.39	4.00	5.50	4.39	3.67	4.11	3.00	4.17	0			
10	3.06	3.39	5.44	4.39	5.06	4.50	4.17	4.61	5.72	0		
11	5.39	2.39	3.17	3.33	5.94	4.28	5.94	6.06	2.56	5.00	0	
12	3.17	3.50	5.11	4.28	4.72	4.00	4.44	4.28	5.06	6.67	3.56	0

Country

1. Brazil	4. Egypt	7. Israel	10. Russia
2. Congo	5. France	8. Japan	11. U.S.A.
3. Cuba	6. India	9. China	12. Yugoslavia

Kruskal and Wish (1978, p. 34) purpose that the number of stimuli minus one should be four times larger than the dimensionality. Letting I represent the number of stimuli and D the dimension, $I \geq 4D + 1$. Thus, model stability usually occurs with no more than $D = 3$ dimensions on average with only $I = 12$ stimuli. Reviewing the residual plots for the two-dimensional model, large residuals indicate an over estimate of distances between nations that are geographically close. Conversely, underestimates of distances between nations that are far apart geographically are negative. For our two-dimensional model residuals larger than 0.90 occur for (India, France), (Israel, France) and (Japan, Israel). While a large negative residual (−0.86) occurs for the pair (Cuba, Brazil). Also, the plot of residuals versus the data does not appear to be random, and reflects a clear pattern.

For a three-dimensional model, the residuals are reduced significantly with no evident pattern. Thus, we conclude that a reasonably stable model occurs in $D = 3$ dimensions. A three-dimensions plot of the result is provided in Figure 9.6.4. The plot was obtained using SAS/INSIGHT.

The three clusters of countries appear to represent the 1968 attitudes of students in that era: Pro-communism countries China, Cuba, Yugoslavia, and Cuba); Under-Developed countries (Egypt, India, Congo); and Pro-Western countries (France, Japan, USA, Brazil and Israel). However, because of the small number of stimuli the model may not be stable.

Exercises 9.6

1. Using the approximate distances between the four Europeans cities: Barcelona, Spain; Berlin, Germany; London, England; and Paris, France construct a map for the cities

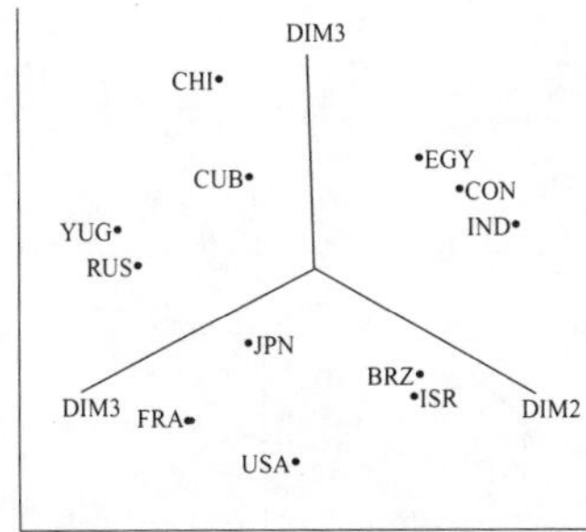

FIGURE 9.6.4. MDS Three-Dimensional Solution - Nations Data

by plotting a pattern of the estimated coordinate. The distance matrix

$$\Delta = \begin{bmatrix} 0 & & & \\ 1550 & 0 & & \\ 1200 & 1000 & 0 & \\ 900 & 950 & 350 & 0 \end{bmatrix}$$

2. The correlation matrix for the Performance Assessment data is

$$R = \begin{bmatrix} 1 & & & & & & & \\ .57 & 1 & & & & & & \\ .11 & .14 & 1 & & & & & \\ .13 & .16 & .52 & 1 & & & & \\ .24 & .21 & .32 & .30 & 1 & & & \\ .33 & .39 & .40 & .40 & .57 & 1 & & \\ .15 & .20 & .29 & .27 & .48 & .47 & 1 & \\ .22 & .26 & .30 & .18 & .37 & .38 & .36 & 1 \end{bmatrix} \begin{matrix} \text{SUPPG} \\ \text{SUPPI} \\ \text{ASMTE} \\ \text{ASMTC} \\ \text{FAM} \\ \text{PAP} \\ \text{PROF1} \\ \text{PROF2} \end{matrix}$$

(a) Cluster the variables and embed the solution in a two-dimensional space.

(b) Using SAS/INSIGHT, obtain a three-dimensional MDS model for the data and interpret your findings.

10
Structural Equation Models

10.1 Introduction

In most of the multivariate linear models analyzed to this point, we formulated relationships between one observed independent set of variables and a set of random dependent variables where the covariance structure of the dependent set involved an unknown and unstructured covariance matrix Σ. Two exceptions to this paradigm included mixed models where Σ was formulated to have structure which contained components of variance and the exploratory factor analysis (EFA) model where Σ depended upon the unknown regression (pattern) coefficients which related unobserved (latent) factors to observed dependent variables.

In this chapter, we study linear relationships between random dependent (endogenous) variables and random independent (exogenous) variables where either can be directly observed (manifest variables) or unobserved hypothetical variables (latent variables). The covariance structure among the observed random variables is used to study the linear structural relations among all the model variables. In the social and behavioral sciences, such models are called "causal" models and involve the analysis of the covariance matrix for the manifest variables derived from a linear structural model. This terminology is unfortunate since most models do not establish causality, but only establish an empirical linear association among the latent and manifest variables under study, Freedman (1987) and Dawid (2000).

Following an overview of path diagrams, basic notation, and the general approach, the topics discussed in this chapter include confirmatory factor analysis (CFA), structural models with manifest variables also called path analysis (PA) or simultaneous equation models, and structural equation models with manifest and latent variables.

The treatment given structural equation modeling (SEM) in this chapter is far from complete. Textbooks by Everitt (1984), McDonald (1985), Bollen (1989), Mueller (1996) and Kaplan (2000) should be consulted for comprehensive coverage. For a review of other texts on the subject, one should consult Steiger (2001), and for a historical overview of the topic, see Bentler (1986). Bentler and Lee (1983) investigate linear models with constraints, and Heise (1986) and Arminger (1998) investigate nonlinear models. For additional advances in SEM, one may also consult the books edited by Bollen and Long (1993), and Marcoulides and Schumacker (1996). Bayesian methods for testing SEM are discussed by Scheines, Hoijtink and Boomsma (1999). Pearl (2000) using graphical models and the logic of intervention clarifies some of the issues, confusion and controversy in the meaning and usage of SEM's. Finally, the analysis of simultaneous equation models of vector observations that vary over individuals and time called panel analysis models is discussed by Hsiao (1986).

10.2 Path Diagrams, Basic Notation, and the General Approach

In multivariate regression models, the two sets of variables in the formulation of the model are random observed dependent variables and fixed (usually non-random) observed independent variables or covariates. While the general theory of the linear model is valid when the independent variables are random, the basic assumption for the model is that the independent variables are distributed independently of error. This assumption, while simple, has significant implications for the analysis of regression models. The assumption allows one to think of a regression model as a structural model which may imply causation. For the model to be structural, we must argue that the unspecified causes (random variables) on the dependent variables, usually included in the error term, are uncorrelated with the random independent variables in the model. This ensures that for models that are misspecified, that the estimated regression coefficients are always consistent whether or not independent variables are included or excluded from the design matrix. This is why randomization is so important in linear models. The implicit assumption of randomization is that experimental manipulation of independent variables affect only the variables in the experimental design and has no effect on the residual causes on the dependent variables. This is not the case for nonexperimental designs, settings in which most "causal models" occur.

In path analysis, or more generally in structural equation modeling (SEM), all variables are defined as random and new terminology is used. The first distinction made among variables in the model is between observed and unobserved random variables. Observed variables are called manifest variables and are directly observed. Latent variables (unobserved variables, factors, true scores) are hypothetical constructs that are not directly measured or observed. In path diagrams, circles or ovals, represent latent variables while squares or rectangles represent observed variables. In addition to manifest and latent variables, the variables in a SEM are characterized as endogenous, exogenous, and as disturbances or errors. Endogenous variables (as in the SUR models using the SYSLIN procedure) are determined within the model; they are caused by other variables within the model. In path diagrams, such variables have a single-headed arrow pointing at them. Exogenous explanatory variables (as in SUR models using the SYSLIN procedure) are treated as predetermined givens and are only influenced by variables outside the model. They usually never have single-headed arrows pointing at them, but instead, they are usually connected by curved lines signifying unanalyzed association (covariation). Random errors or disturbances represent omitted causes on the endogenous variables and are usually taken to be the independent of exogenous variables, but they may be correlated. Merely indicating that a variable is exogenous does not make it exogenous. The concept of exogeniety is important in SEM and regression models with random independent variables since conditional inference (and estimation) is only valid with weak exogeneity. While the assumption of multivariate normality is a necessary condition for weak exogeneity it is not a sufficient condition. Exogeniety is discussed in Section 10.10. An introduction to exogeneity may also be found in the text by Davidson and MacKinnon (1993).

A final distinction made in SEM is whether the model is recursive or nonrecursive. Recursive models are models in which causation is unidirectional. There is no backward causation, causal loops, or bidirectional paths in the model. In nonrecursive models, causation may flow in both directions. This is represented as two separate single-headed straight arrows between the variables. In general, nonrecursive models are more difficult to analyze, Berry (1984). Table 10.2.1 represent some common symbols and relationships among variables in a SEM.

In Table 10.2.1 we have annotated the boxes and circles with letters and Greek symbols to represent the components of the path diagrams using the popular Keesling (1972)-Wiley (1973)-Jöreskeg (1973, 1977) LISREL (linear structural relations) notation which is widely used in many texts. Assuming all variables are standardized to have mean zero (deviation scores), next we define the general LISREL model without an intercept.

The structural model for latent variables is

$$\underset{m\times 1}{\boldsymbol{\eta}_i} = \underset{m\times m}{\mathbf{B}} \underset{m\times 1}{\boldsymbol{\eta}_i} + \underset{m\times n}{\Gamma} \underset{n\times 1}{\boldsymbol{\xi}_i} + \underset{m\times 1}{\boldsymbol{\zeta}_i} \tag{10.2.1}$$

where $\boldsymbol{\eta}_i$, $\boldsymbol{\xi}_i$, and $\boldsymbol{\zeta}_i$ are random vectors of latent endogenous variables, latent exogenous variables, and latent errors for $i = 1, 2, \ldots, N$ observations where $\boldsymbol{\xi}_i \sim N_n(\mathbf{0}, \Phi)$, $\boldsymbol{\zeta}_i \sim N_m(\mathbf{0}, \Psi)$, $\boldsymbol{\zeta}_i$ and $\boldsymbol{\xi}_i$ are independent (uncorrelated), $(\mathbf{I} - \mathbf{B})$ is nonsingular, $\mathbf{B}$ is a matrix with zeros on the diagonal, and the observations are independent. Dropping the subscript, the model has the reduced form

$$\boldsymbol{\eta} = (\mathbf{I} - \mathbf{B})^{-1} \Gamma\boldsymbol{\xi} + (\mathbf{I} - \mathbf{B})^{-1} \boldsymbol{\zeta} = \Pi\boldsymbol{\xi} + \mathbf{e} \tag{10.2.2}$$

where the covariance matrix for the model is

$$\begin{aligned} \Sigma &= \begin{bmatrix} \Sigma_{\eta\eta} & \Sigma_{\eta\xi} \\ \Sigma_{\xi\eta} & \Sigma_{\xi\xi} \end{bmatrix} = \begin{bmatrix} E(\boldsymbol{\eta}\boldsymbol{\eta}') & E(\boldsymbol{\eta}\boldsymbol{\xi}') \\ E(\boldsymbol{\xi}\boldsymbol{\eta}') & E(\boldsymbol{\xi}\boldsymbol{\xi}') \end{bmatrix} \\ &= \begin{bmatrix} (\mathbf{I} - \mathbf{B})^{-1} (\Gamma\Phi\Gamma' + \Psi) (\mathbf{I} - \mathbf{B})^{-1'} & (\mathbf{I} - \mathbf{B})^{-1} \Gamma\Phi \\ \Gamma' (\mathbf{I} - \mathbf{B})^{-1'} \Phi & \Phi \end{bmatrix} \end{aligned} \tag{10.2.3}$$

TABLE 10.2.1. SEM Symbols.

Symbols	Explanation
x y	Square or rectangular boxes represent manifest variables.
ξ η	Circles or ellipses represent latent variables.
ξ η ζ	Latent variable causing another latent variable with latent error.
x_1 x_2 y ε	Two associated manifest variables causing another manifest variable with measurement error.
y_1 ε_1 y_2 ε_2	Nonrecursive relationship betrween manifest variables with correlated errors.
ξ x_1 x_2 x_3 δ_1 δ_2 δ_3	Latent variable causing manifest variables with independent measurement errors.

Relating the unobserved latent constructs to observed variables, the measurement model is

$$\begin{aligned} \underset{(p\times 1)}{\mathbf{y}_i} &= \underset{(p\times m)}{\Lambda_y} \underset{(m\times 1)}{\boldsymbol{\eta}_i} + \underset{(p\times 1)}{\boldsymbol{\epsilon}_i} \\ \underset{(q\times 1)}{\mathbf{x}_i} &= \underset{(q\times n)}{\Lambda_x} \underset{(n\times 1)}{\boldsymbol{\xi}_i} + \underset{(q\times 1)}{\boldsymbol{\delta}_i} \end{aligned} \tag{10.2.4}$$

where $\mathbf{y}_i$ and $\mathbf{x}_i$ are vectors of observed indications of the latent endogenous vectors $\boldsymbol{\eta}_i$ and the latent exogenous vectors $\boldsymbol{\xi}_i$. The vectors $\boldsymbol{\epsilon}_i$ and $\boldsymbol{\delta}_i$ are vectors of measurement errors, and Λ_y and Λ_x are regression coefficients relating $\mathbf{y}$ to $\boldsymbol{\eta}$, and $\mathbf{x}$ to $\boldsymbol{\xi}$, respectively. Finally, we assume that $\boldsymbol{\epsilon}_i \sim N_p(\mathbf{0}, \Theta_\epsilon)$, $\boldsymbol{\delta}_i \sim N_q(\mathbf{0}, \Theta_\delta)$, and that $\boldsymbol{\epsilon}_i, \boldsymbol{\delta}_i, \boldsymbol{\eta}_i, \boldsymbol{\xi}_i$ and $\boldsymbol{\xi}_i$ are mutually independent. Thus, the joint SEM is

$$\begin{aligned} \boldsymbol{\eta} &= \mathbf{B}\boldsymbol{\eta} + \Gamma\boldsymbol{\xi} + \boldsymbol{\zeta} \\ \mathbf{y} &= \Lambda_y\boldsymbol{\eta} + \boldsymbol{\epsilon} \\ \mathbf{x} &= \Lambda_x\boldsymbol{\xi} + \boldsymbol{\delta} \end{aligned} \tag{10.2.5}$$

where the matrices $\mathbf{B}$, Γ, Λ_y, and Λ_x are matrices of direct effects. The covariance matrix of $\mathbf{y}$ is

$$\begin{aligned}\Sigma_{yy} &= \Lambda_y \Sigma_{\eta\eta} \Lambda_y' + \Theta_\epsilon \\ &= \Lambda_y \left[(\mathbf{I} - \mathbf{B})^{-1} \left(\Gamma \Phi \Gamma' + \Psi \right) (\mathbf{I} - \mathbf{B})^{-1\prime} \right] \Lambda_y' + \Theta_\epsilon\end{aligned}$$

The covariance matrix of $\mathbf{x}$ is

$$\Sigma_{xx} = \Lambda_x \Phi \Lambda_x' + \Theta_\delta$$

and the covariance matrix Σ_{xy} is

$$\Sigma_{xy} = \Lambda_x \Sigma_{\xi\eta} \Lambda_y' = \Lambda_x \Phi \Gamma' (\mathbf{I} - \mathbf{B})^{-1\prime} \Lambda_y'$$

Hence, the covariance matrix for the SEM given by (10.2.5) becomes

$$\Sigma(\boldsymbol{\theta}) = \begin{bmatrix} \Lambda_y \left[(\mathbf{I} - \mathbf{B})^{-1} \left(\Gamma \Phi \Gamma' + \Psi \right) (\mathbf{I} - \mathbf{B})^{-1\prime} \right] \Lambda_y' + \Theta_\epsilon & \Lambda_y (\mathbf{I} - \mathbf{B})^{-1} \Gamma \Phi \Lambda_x' \\ \Lambda_x \Phi \Gamma' (\mathbf{I} - \mathbf{B})^{-1\prime} \Lambda_y' & \Lambda_x \Phi \Lambda_x' + \Theta_\delta \end{bmatrix} \tag{10.2.6}$$

where $\boldsymbol{\theta}$ represents the unknown vector of model parameters. Letting $\mathbf{x} \equiv \boldsymbol{\xi}$ so that $\mathbf{x}$ is observed, the matrix Σ_{yy} in (10.2.6) has the ACOVS structure discussed in Chapter 5.

The population covariance matrix $\Sigma(\boldsymbol{\theta})$ in (10.2.6) is a function of all the structural parameters so that the covariance structure is also called a structural model. Letting $\boldsymbol{\theta}$ represent the independent free and distinct (nonredundent) constrained parameters in the parameter matrices $\mathbf{B}$, Φ, Λ_y, Λ_x, Φ, Ψ, Θ_ϵ and Θ_δ, the vector $\boldsymbol{\theta}$ contains the structural parameters to be estimated. The issue of estimability is called the (global) identification problem. A parameter $\boldsymbol{\theta}$ in a SEM is identified if it can be estimated and underidentified (or unidentified) otherwise. If $\boldsymbol{\theta}$ is uniquely estimable, the model is said to be exactly (or just) identified or saturated. Overidentified models yield a family of solutions.

Identification is critical to the analysis of covariance structures. Bollen (1989, p. 89) defines global model identification as follows.

Definition 10.2.1 *If an unknown parameter in $\boldsymbol{\theta}$ can be written as a function of one or more elements of Σ, then the parameters in $\boldsymbol{\theta}$ are identified. If all unknown parameters in $\boldsymbol{\theta}$ are identified, then the model is identified.*

Bollen also gives a definition for model uniqueness, local identification, by considering two vectors $\boldsymbol{\theta}_1$ and $\boldsymbol{\theta}_2$ of an unknown $\boldsymbol{\theta}$.

Definition 10.2.2 *A parameter $\boldsymbol{\theta}$ is locally identified or uniquely defined at a point $\boldsymbol{\theta}_1$ if, in the neighborhood of $\boldsymbol{\theta}_1$ there is no vector $\boldsymbol{\theta}_2$ for which $\Sigma(\boldsymbol{\theta}_1) = \Sigma(\boldsymbol{\theta}_2)$ unless $\boldsymbol{\theta}_1 = \boldsymbol{\theta}_2$.*

This definition of local identification can only be used to detect an unidentified model and is not a sufficient condition for determining whether a model is identified. Clearly, if a pair of vector $\boldsymbol{\theta}_1$ and $\boldsymbol{\theta}_2$ exist such $\Sigma(\boldsymbol{\theta}_1) = \Sigma(\boldsymbol{\theta}_2)$ and $\boldsymbol{\theta}_1 \neq \boldsymbol{\theta}_2$ then the parameter $\boldsymbol{\theta}$ is not identified. Bollen and Jöreskog (1985) show that model uniqueness does not imply global identification and provide several examples.

To solve for the model parameters in a SEM, the number of equations must be larger than or equal to the number of unknowns. The total number of equations as given by the distinct elements in $\Sigma(\boldsymbol{\theta})$ is $\upsilon = (p+q)(p+q+1)/2$. If t is the number of parameters to be estimated, a necessary condition for model identification is that

$$\upsilon^* = \upsilon - t \geq 0 \tag{10.2.7}$$

Provided $\upsilon^* > 0$ and the model is correct and identified, the hypothesis

$$H : \Sigma = \Sigma(\boldsymbol{\theta}) \tag{10.2.8}$$

is said to be testable. This hypothesis tests whether an overidentified SEM is consistent with the observed data and is called an overidentification or goodness-of-fit test.

To analyze a SEM, one begins by specifying the model using a path diagram that relates the latent and observed variables specified using (10.2.5) with covariance structure given in (10.2.6). Following model specification, one next determines whether a model is identified. This is usually a very difficult task since except for some simple structural models, general sufficient conditions for model identification have not been developed. Details regarding model identification for specific models are provided later in this chapter. Given an identified model, one next must estimate the model parameters in $\Sigma(\boldsymbol{\theta})$ for the SEM. To estimate the parameters in $\Sigma(\boldsymbol{\theta})$ for any SEM, one obtains a sample estimate $\mathbf{S}$ of $\Sigma(\boldsymbol{\theta})$ and chooses a scalar error-in-fit function continuous function $F(\mathbf{S}, \Sigma(\boldsymbol{\theta})) \geq 0$. Minimizing the fit function at $\boldsymbol{\theta} = \widehat{\boldsymbol{\theta}}$, the value of the function at $\Sigma(\widehat{\boldsymbol{\theta}}) = \widehat{\Sigma}$ represented as $F(\mathbf{S}, \widehat{\Sigma})$ is a measure of closeness of fit of $\mathbf{S}$ by $\widehat{\Sigma}$. For $\mathbf{S} = \widehat{\Sigma}$, the fit function is defined to be zero so that $\mathbf{S} - \widehat{\Sigma}$ should be approximately $\mathbf{0}$. Two general fit functions used in SEM are a variant of the log likelihood under multivariate normality of the manifest variables, the ML fit function, and several variants of matrix norms that compare $\mathbf{S}$ with weighted estimates of $\Sigma(\widehat{\boldsymbol{\theta}}) = \widehat{\Sigma}$, the weighted least squares (WLS) fit function. The two functions are defined as

$$F_{ML} = \log|\Sigma(\boldsymbol{\theta})| + \operatorname{tr}\left(\mathbf{S}\Sigma(\boldsymbol{\theta})^{-1}\right) - \log(\mathbf{S}) - (p+q) \tag{10.2.9}$$

$$F_{WLS} = \frac{1}{2}\operatorname{tr}\left[\mathbf{W}^{-1}[\mathbf{S} - \Sigma(\boldsymbol{\theta})]\right]^2 = \frac{1}{2}\left\|\mathbf{W}^{-1}[\mathbf{S} - \Sigma(\boldsymbol{\theta})]\right\|^2$$

When $\mathbf{W}^{-1} = \mathbf{S}^{-1}$, weighted least squares is called the generalized least squares (GLS) fit function. If $\mathbf{W}^{-1} = \mathbf{I}$ it becomes the unweighted least squares (ULS) fit function and if $\mathbf{W}^{-1} = \widehat{\Omega}^{-1}$ where $\widehat{\Omega}$ is the asymptotic covariance matrix of the elements of $\mathbf{S}$ discussed in Chapter 3, $\operatorname{cov}(s_{gh}, s_{ij}) \xrightarrow{p} \sigma_{gj}\sigma_{hj} + \sigma_{gj}\sigma_{hi}/N$, the fit function is called the asymptotically distribution free (ADF) fit function, as long as $\mathbf{W}$ is positive definite and the plim $\mathbf{S} = \Sigma$, Browne (1984). Because the estimate for the model parameter $\boldsymbol{\theta}$ using the F_{ML} error-in-fit function is identical to the full information maximum likelihood estimate obtained assuming joint multivariate normality, the estimate $\widehat{\boldsymbol{\theta}}$ is always a consistent estimate of $\boldsymbol{\theta}$, given that the model holds in the population and provided $\boldsymbol{\theta}$ is globally identified. This is not necessarily the case for the other fit functions. The other fit functions also require the vector $\mathbf{x}$ to be weakly exogeneous for the parameters of interest.

Minimization of the functions given in (10.2.9) is complex since it involves a constrained nonlinear system of equations. Jöreskog (1969b, 1973) showed how to minimize F_{ML} using the Davidon-Fletcher-Powell method which only involves first-order derivatives. More

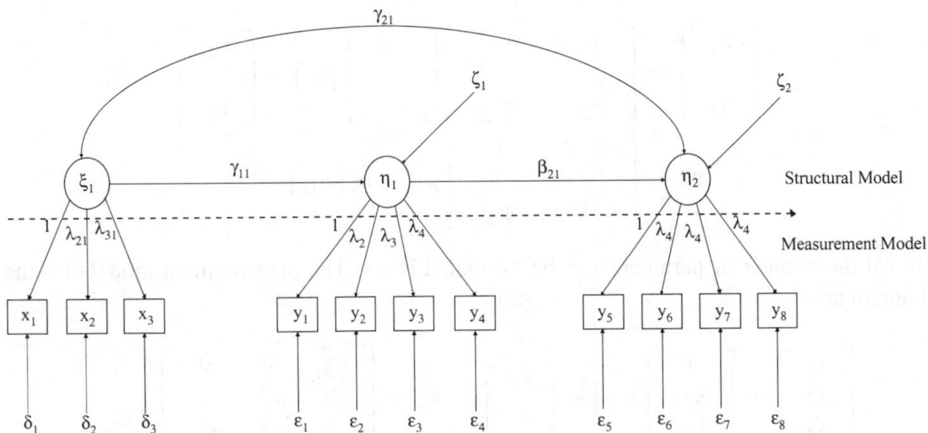

FIGURE 10.2.1. Path Analysis Diagram

recently, Newton-Raphson and Gauss-Newton algorithms that involve second-order derivatives with Levenberg-Marquadt adjustments are used, Thisted (1988).

Given that a model is identified and an error-in-fit function $F(\mathbf{S}, \Sigma(\widehat{\boldsymbol{\theta}}))$ is selected such that $\widehat{\boldsymbol{\theta}}$ is a consistent estimator of $\boldsymbol{\theta}$, the next step in the evaluation of a SEM is the assessment of overall fit. The criteria used to evaluate model fit include chi-square tests, fit indices, root mean square residuals, and others, all of which depend upon the error-in-fit function selected, distribution assumptions of the manifest variables, sample size, whether the fit function selected is scale invariant, and whether the estimation method is scale free. There is no uniformly "best" criterion for evaluating model fit or whether some subset of the model parameters fit better than others. More will be said about this in later sections. Using the fitted model, one relates the model to a theory, compares several models and/or revises the model based upon substantive area expertise. Most research in SEM has been concerned with model fit, the more complex issue of model predictive validity has received little attention. This is because the latent scores in the calibration sample are unobserved. This should change with the release of LISREL 8.30 which allows one to estimate the latent variables in the calibration sample with estimated latent variable scores, Jöreskog et al. (2000, p. 168). The latent variable scores based upon the calibration sample may be compared with the observed scores in the validation sample to evaluate predictive validity of the model. Finally, critical to SEM is the assumption of multivariate normality and the use of an appropriate sample size. One should consult MacCallum, Browne, and Sugawara (1996) and Kaplan (2000) for additional detail on power analysis is SEM.

Figure 10.2.1 illustrates a simple recursive path diagram for three latent variables.

In the path diagram, we have known parameters (values set to one), constrained parameters (the paths from η_1 and η_2 to the effect indicators y_i are equal), and unknown free parameters. The relationships are formulated based upon the knowledge of the substantive area under study.

Assuming the elements in η and ξ are uncorrelated, the latent variables model for Figure 10.2.1 is

$$\begin{bmatrix} \eta_1 \\ \eta_2 \end{bmatrix} = \begin{bmatrix} 0 & 0 \\ \beta_{21} & 0 \end{bmatrix} + \begin{bmatrix} \gamma_{11} \\ \gamma_{21} \end{bmatrix} [\xi_1] + \begin{bmatrix} \zeta_1 \\ \zeta_2 \end{bmatrix}$$

$$\Psi = \begin{bmatrix} \psi_{11} & 0 \\ 0 & \psi_{22} \end{bmatrix}, \quad \Phi = [\phi_{11}]$$

so that the number of parameters to be estimated is six. The measurement models for the diagram are

$$\begin{bmatrix} x_1 \\ x_2 \\ x_3 \end{bmatrix} = \begin{bmatrix} 1 \\ \lambda_{21} \\ \lambda_{31} \end{bmatrix} [\xi_1] + \begin{bmatrix} \delta_1 \\ \delta_2 \end{bmatrix}, \quad \theta_\delta = \begin{bmatrix} \sigma^2_{\delta_1} & 0 & 0 \\ 0 & \sigma^2_{\delta_2} & 0 \\ 0 & 0 & \sigma^2_{\delta_3} \end{bmatrix}$$

$$\begin{bmatrix} y_1 \\ y_2 \\ y_3 \\ y_4 \\ y_5 \\ y_6 \\ y_7 \\ y_8 \end{bmatrix} = \begin{bmatrix} 1 & 0 \\ \lambda_2 & 0 \\ \lambda_3 & 0 \\ \lambda_4 & 0 \\ 0 & 1 \\ 0 & \lambda_2 \\ 0 & \lambda_3 \\ 0 & \lambda_4 \end{bmatrix} \begin{bmatrix} \eta_1 \\ \eta_2 \end{bmatrix} + \begin{bmatrix} \epsilon_1 \\ \epsilon_2 \\ \epsilon_3 \\ \epsilon_4 \\ \epsilon_5 \\ \epsilon_6 \\ \epsilon_7 \\ \epsilon_8 \end{bmatrix}, \quad \Theta_\epsilon = \text{diag}\left[\sigma^2_{\epsilon_i}\right]$$

so that the number of parameters to be estimated is 19. Thus, the total number of free parameters in the model is $t = 6 + 19 = 25$. And, since $t < (p+q)(p+q+1)/2 = 66$, this model is at least not underidentified. Because there is no general necessary condition that allows one to evaluate whether a SEM is identified, we treat the topic with specific applications of the model in the separate sections of this chapter.

The LISREL model defines a SEM by representing relationships among the latent variables (structural model) and the manifest variables (measurement model). An alternative representation requiring only three matrices to define a SEM is McArdle's reticular action model (RAM), McArdle (1980) and McArdle and McDonald (1984). In the RAM, the $p+q=g$ manifest variables and the $m+n=h$ latent variables are organized into a single random vector $\mathbf{v}_{t\times 1}$ where $t = g + h$. The linear model for the system is

$$\underset{t\times 1}{\mathbf{v}} = \underset{t\times t}{\mathbf{A}} \; \underset{t\times t}{\mathbf{v}} + \underset{t\times 1}{\mathbf{u}} \tag{10.2.10}$$

$$\mathbf{u} \sim N_t(\mathbf{0}, \mathbf{P})$$

where $\mathbf{I} - \mathbf{A}$ is nonsingular, and $\mathbf{u}$ and $\mathbf{v}$ are independent. We assume that $E(\mathbf{v}) = \mathbf{0}$ so that all variables are corrected to have mean zero so that moment matrices become covariance matrices.

From (10.2.10), the vector $\mathbf{v}$ may be written as

$$\mathbf{v} = (\mathbf{I} - \mathbf{A})^{-1}\,\mathbf{u} \tag{10.2.11}$$

and partitioning $\mathbf{v}$ so that $\mathbf{v}' = [\mathbf{g}', \mathbf{h}']$ and defining a selection matrix $\mathbf{J} = [\mathbf{I}, \mathbf{0}]$, the manifest variables are related to $\mathbf{v}$ by

$$\mathbf{g} = \mathbf{J}\mathbf{v} = \mathbf{J}\,(\mathbf{I} - \mathbf{A})^{-1}\,\mathbf{u} \tag{10.2.12}$$

The covariance structure for the manifest variables is

$$\Sigma\,(\boldsymbol{\theta}) = \mathbf{J}\,(\mathbf{I} - \mathbf{A})^{-1}\,\mathbf{P}\,(\mathbf{I} - \mathbf{A})^{-1'}\,\mathbf{J}' \tag{10.2.13}$$

To express a LISREL model as a RAM model, we let $\mathbf{v}' = [\mathbf{y}', \mathbf{x}', \boldsymbol{\eta}', \boldsymbol{\xi}']$ so that

$$\begin{bmatrix} \mathbf{y} \\ \mathbf{x} \\ \boldsymbol{\eta} \\ \boldsymbol{\xi} \end{bmatrix} = \begin{bmatrix} \mathbf{0} & \mathbf{0} & \Lambda_y & \mathbf{0} \\ \mathbf{0} & \mathbf{0} & \mathbf{0} & \Lambda_x \\ \mathbf{0} & \mathbf{0} & \mathbf{B} & \Gamma \\ \mathbf{0} & \mathbf{0} & \mathbf{0} & \mathbf{0} \end{bmatrix} \begin{bmatrix} \mathbf{y} \\ \mathbf{x} \\ \boldsymbol{\eta} \\ \boldsymbol{\xi} \end{bmatrix} + \begin{bmatrix} \boldsymbol{\epsilon} \\ \boldsymbol{\delta} \\ \boldsymbol{\zeta} \\ \boldsymbol{\xi} \end{bmatrix} \tag{10.2.14}$$

Letting $\mathbf{J} = \begin{bmatrix} \mathbf{I} & \mathbf{0} & \mathbf{0} & \mathbf{0} \\ \mathbf{0} & \mathbf{I} & \mathbf{0} & \mathbf{0} \end{bmatrix}$ and $\mathbf{g} = \mathbf{J}\mathbf{v}$, the covariance matrix for $\Sigma\,(\boldsymbol{\theta})$ is

$$\Sigma\,(\boldsymbol{\theta}) = \mathbf{J}\,(\mathbf{I} - \mathbf{A})^{-1}\,\mathbf{P}\,(\mathbf{I} - \mathbf{A})^{-1'}\,\mathbf{J}' \tag{10.2.15}$$

where

$$\mathbf{I} - \mathbf{A} = \begin{bmatrix} \mathbf{I} & \mathbf{0} & -\Lambda_y & \mathbf{0} \\ \mathbf{0} & \mathbf{I} & \mathbf{0} & -\Lambda_x \\ \mathbf{0} & \mathbf{0} & \mathbf{I} - \mathbf{B} & \Gamma \\ \mathbf{0} & \mathbf{0} & \mathbf{0} & \mathbf{I} \end{bmatrix} \tag{10.2.16}$$

$$\mathbf{P} = \begin{bmatrix} \boldsymbol{\theta}_\epsilon & & & \\ & \boldsymbol{\theta}_\delta & & \\ & & \Psi & \\ & & & \Phi \end{bmatrix}$$

That the covariance structure for $\Sigma\,(\boldsymbol{\theta})$ in (10.2.15) is identical to $\Sigma\,(\boldsymbol{\theta})$ in (10.2.6) follows by showing that the inverse of $\mathbf{I} - \mathbf{A}$ is

$$(\mathbf{I} - \mathbf{A})^{-1} = \begin{bmatrix} \mathbf{I} & \mathbf{0} & \Lambda_y\,(\mathbf{I} - \mathbf{B})^{-1} & \Lambda_y\,(\mathbf{I} - \mathbf{B})\,\Gamma \\ \mathbf{0} & \mathbf{I} & \mathbf{0} & \Gamma \\ \mathbf{0} & \mathbf{0} & (\mathbf{I} - \mathbf{B})^{-1} & (\mathbf{I} - \mathbf{B})\,\Gamma \\ \mathbf{0} & \mathbf{0} & \mathbf{0} & \mathbf{I} \end{bmatrix} \tag{10.2.17}$$

Substituting (10.2.17) into (10.2.15), the result follows.

Conversely, the RAM model may be written as a LISREL model by setting $\mathbf{B} = \mathbf{A}$, $\Gamma = \mathbf{I}$, $\Phi = \mathbf{P}$, and equating all other matrices to zero. A special case of the RAM model was proposed by Bentler and Weeks (1980). The Bentler-Weeks model states that

$$\boldsymbol{\eta}^* = \mathbf{B}\boldsymbol{\eta}^* + \Gamma\boldsymbol{\xi}^* \tag{10.2.18}$$

where the variables in $\boldsymbol{\eta}^*$ and $\boldsymbol{\xi}^*$ may be manifest or latent, $\mathbf{I} - \mathbf{B}$ is nonsingular and Φ is the covariance matrix of $\boldsymbol{\xi}^*$. For a selection matrix $\mathbf{J}$, the covariance structure for model (10.2.18) is

$$\Sigma(\boldsymbol{\theta}) = \mathbf{J}(\mathbf{I} - \mathbf{A})^{-1}\mathbf{P}(\mathbf{I} - \mathbf{A})^{-1'}\mathbf{J}' \tag{10.2.19}$$

with

$$\mathbf{A} = \begin{bmatrix} \mathbf{B} & \mathbf{0} \\ \mathbf{0} & \mathbf{0} \end{bmatrix} \quad \text{and} \quad \mathbf{P} = \begin{bmatrix} \Gamma\Phi\Gamma' & \Gamma\Phi \\ \Phi\Gamma' & \Phi \end{bmatrix}$$

so that (10.2.18) is a RAM.

In both the formulation of the LISREL and RAM models, the covariance matrix for the manifest variables was a patterned covariance matrix. This motivated McDonald (1980) to propose his Covariance Structure Analysis (COSAN) model. McDonald proposed that each of the g^2 elements of a covariance matrix $\Sigma_{g \times g}$ for a $(g \times 1)$ random vector be expressed as a function of t parameters $\theta_1, \theta_2, \ldots, \theta_t$. In McDonald's notation, his COSAN model is expressed as

$$\Sigma = \left(\prod_{j=1}^{m} \mathbf{F}_j\right) \mathbf{P} \left(\prod_{j=1}^{m} \mathbf{F}_j\right)' \tag{10.2.20}$$

where $\mathbf{F}_j$ is of order $(g_{j-1} \times g_j)$, and $\mathbf{P}$ is symmetric of order n_m. Each element of $\mathbf{F}_j$, or (in some applications) of $\mathbf{F}_j^{-1}$, and each element of $\mathbf{P}$ (or of $\mathbf{P}^{-1}$), is a function of t fundamental parameters $\boldsymbol{\theta}' = [\theta_1, \theta_2, \ldots, \theta_t]$. McDonald (1980) shows that his COSAN model includes as a special case the seemingly more general linear COSAN model analyzed in the CALIS procedure in SAS proposed by Bentler (1976). That is

$$\Sigma = \sum_{j=0}^{m} \left(\prod_{i=1}^{j} \mathbf{B}_i\right) \mathbf{L}_j \left(\prod_{i=1}^{j} \mathbf{B}_i\right) \tag{10.2.21}$$

where the matrices $\mathbf{B}_i$ are defined like $\mathbf{F}_j$ in (10.2.20) and $\mathbf{L}_j$ are defined like $\mathbf{P}$. To show that model (10.2.21) can be obtained from (10.2.20), we set

$$\mathbf{P} = \begin{bmatrix} \mathbf{L}_m & & & \\ & \mathbf{L}_{m-1} & & \\ & & \ddots & \\ & & & \mathbf{L}_0 \end{bmatrix}$$

and select patterned matrices $\mathbf{F}_j$ with structure

$$\mathbf{F}_j = \begin{bmatrix} \mathbf{B}_j & \vdots & \mathbf{I} & \vdots & \\ \cdots & \cdots & \cdots & \cdots & \cdots \\ & \vdots & & \vdots & \\ & \vdots & & \vdots & \end{bmatrix}$$

where the identity matrices $\mathbf{I}$ are constructed to have appropriate orders.

We have shown that the LISREL model is equivalent to the RAM. Using the matrix identity for the inverse of $(\mathbf{I} - \mathbf{A})^{-1}$

$$(\mathbf{I} - \mathbf{A})^{-1} = \sum_{k=0}^{\infty} \mathbf{A}^k = (\mathbf{I} - \mathbf{A}^{k+1})(\mathbf{I} - \mathbf{A})^{-1} \tag{10.2.22}$$

found in Harville (1997, p. 429), the inverse exists if the series converges as $k \longrightarrow \infty$. Then, one observes that (10.2.13) has a structure similar to (10.2.19) and because (10.2.20) is a special case of (10.2.18), one may find a RAM model that is equivalent to a COSAN model. Conversely, one may find a COSAN model that yields the RAM structure as illustrated by McDonald (1985, p. 154). Hence, for a single population the three model representations LISREL, RAM and COSAN are equivalent. The representation selected is usually one dictated by the problem, user preference, and output generated by a software program for the model. While the CALIS procedure in SAS permits one to fit all three representations, the LISREL model documented in Jöreskog and Sörbom (1993) and the RAM model documented by Bentler (1993) and Bentler and Wu (1995) are preferred by many researchers analyzing SEM since they may be used to analyze LISREL models with mean structures (intercepts):

$$\begin{aligned} \boldsymbol{\eta} &= \boldsymbol{\alpha} + \mathbf{B}\boldsymbol{\eta} + \boldsymbol{\Gamma}\boldsymbol{\xi} + \boldsymbol{\zeta} \\ \mathbf{y} &= \boldsymbol{\lambda}_y + \boldsymbol{\Lambda}_y\boldsymbol{\eta} + \boldsymbol{\epsilon} \\ \mathbf{x} &= \boldsymbol{\lambda}_x + \boldsymbol{\Lambda}_x\boldsymbol{\xi} + \boldsymbol{\delta} \end{aligned} \tag{10.2.23}$$

multiple groups models with constraints across groups, and missing data. While the CALIS program may also be used to analyze (10.2.23), it requires equal sample sizes and no missing data. A detail discussion of the use of PROC CALIS may be found in Hatcher (1994). We have tried to employ his notation, called the Bentler-Weeks notation, in our examples in this chapter. A new program for SEM analysis, distributed with SPSS^{TM}, is Amos, short for Analysis of MOment Structures, Arbuckle and Wothke (1999). It has an easy-to-use graphical interface, allows for missing data, and incorporates newer bootstrap methodologies for model evaluation.

10.3 Confirmatory Factor Analysis

In exploratory factor analysis (EFA) discussed in Chapter 8, an observed vector of variables was related to an unobserved (latent) factor using a linear model with uncorrelated

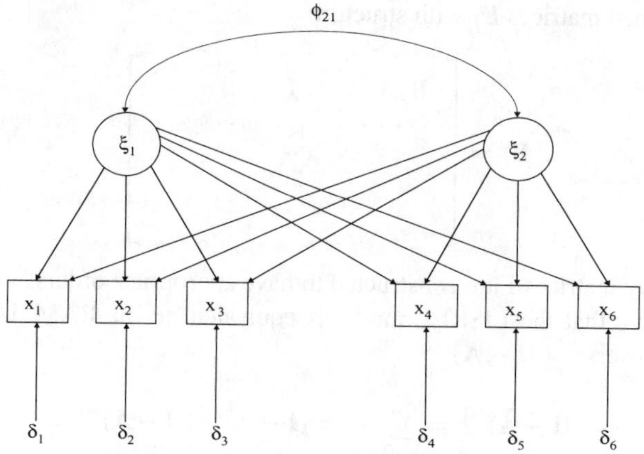

FIGURE 10.3.1. Two Factor EFA Path Diagram

errors. The goal of the analysis was to find a small number of latent factors that account for the observed covariance (correlations) among the p observed variables. Having found a small number of latent factors, all factors were simultaneously transformed to try to determine a small number of "simple structure" orthogonal or oblique factors. If necessary, factor scores could be estimated. When performing an exploratory analysis, the researcher had no prior knowledge of the number of common factors, the pattern of the regression coefficients that related latent factors to variables, or whether the factors were correlated or orthogonal. In addition, the model imposed the restriction that the covariance matrix for errors was diagonal.

Centering the observed data to have mean zero, replacing the observed, manifest vector $\mathbf{y}_{p\times 1}$ with the vector $\mathbf{x}_{q\times 1}$ and the unobserved, latent factor $\mathbf{f}$ with $\boldsymbol{\xi}$, the EFA model using LISREL, SEM notation has the general form

$$\begin{aligned} \underset{q\times 1}{\mathbf{x}} &= \underset{q\times n}{\Lambda_x}\ \underset{n\times 1}{\boldsymbol{\xi}} + \underset{q\times 1}{\boldsymbol{\delta}} \\ \Sigma_{xx} &= \Lambda_x \Phi \Lambda_x' + \Theta_\delta \end{aligned} \tag{10.3.1}$$

where Θ_δ is assumed diagonal. Given a two factor solution with three variables, the EFA model path diagram is shown in Figure 10.3.1.

Using (10.3.1), the model for the diagram is

$$\underset{\mathbf{x}}{\begin{bmatrix} x_1 \\ x_2 \\ x_3 \\ x_4 \\ x_5 \\ x_6 \end{bmatrix}} = \underset{\Lambda_x}{\begin{bmatrix} \lambda_{11} & \lambda_{12} \\ \lambda_{21} & \lambda_{22} \\ \lambda_{31} & \lambda_{32} \\ \lambda_{41} & \lambda_{42} \\ \lambda_{51} & \lambda_{52} \\ \lambda_{61} & \lambda_{62} \end{bmatrix}} \underset{\boldsymbol{\xi}}{\begin{bmatrix} \xi_1 \\ \xi_2 \end{bmatrix}} + \underset{\boldsymbol{\delta}}{\begin{bmatrix} \delta_1 \\ \delta_2 \\ \delta_3 \\ \delta_4 \\ \delta_5 \\ \delta_6 \end{bmatrix}}$$

where

$$\Phi = \begin{bmatrix} \phi_{11} & \phi_{12} \\ \phi_{21} & \phi_{22} \end{bmatrix} \quad \text{and} \quad \Theta_\delta = \text{diag}\left[\sigma^2_{\delta_i}\right]$$

In Figure 10.3.1, observe that each latent factor is related to every variable so that no constraints are imposed on the λ_{ij}. To estimate Λ_x and Θ_δ, recall that the diagonal matrix of unique variances was estimated for various "guessed" values for the minimum number of factors (in the new notation n) and to estimate Λ_x, the residual matrix $\Sigma - \Theta_\delta = \Lambda_x \Lambda_x'$ was solved iteratively, given an initial estimate of Θ_δ. With the number of factors n known, $\widehat{\Theta}_\delta$ and $\widehat{\Lambda}_x$ estimated, the solution was "rotated" to find the $n(n+1)/2$ elements in Φ This multi-step process is used due to the indeterminacy of the structural model.

In confirmatory factor analysis (CFA), we also use model (10.3.1). The major difference is that we specify the number of latent factors n before we begin our analysis; we allow Θ_δ to be a symmetric, unspecified covariance matrix so as to allow for correlated errors of measurement; and we restrict the elements of Λ_x to be associated with specific variables to hypothesize an a priori "pattern" matrix of regression coefficients where the covariance matrix Φ is unspecified. This allows one to estimate all of the qn parameters in Λ_x, the $n(n+1)/2$ parameters in Φ, and the $q(q+1)/2$ parameters in Θ_δ simultaneously, provided the model is identified or overidentified. From (10.2.7), a necessary but not sufficient condition for model identification is that

$$t = qn + n(n+1)/2 + q(q+1)/2 \leq q(q+1)/2 \tag{10.3.2}$$

so that the CFA is usually not identified. For identification or overidentification to occur, one must place restrictions on the parameters in the matrix $\Sigma_{xx} \equiv \Sigma(\boldsymbol{\theta})$.

From equation (10.3.1), implicit in the relationship of ξ_i and δ_i is that δ_i appears only once for each x_i. Because δ_i is not observed, in order to give the latent variable ξ_i a scale, the latent variables must have a metric. This is accomplished in one of two ways. The reference variable solution sets at least one indicator variable, λ_{ij}, in each column of Λ_x to 1, the indicator solution, or the variance of each latent variable is set to 1, the standardized solution. In either case, on average one expects that a unit change in ξ_i implies a unit change in x_i. One may also set some λ_{ij} to zero, constrain some of the λ_{ij} to be equal, and set other parameters to zero or to known values to reduce the number of unknown parameters.

To illustrate the problem of identification, we consider the simple one factor design with $q = 2$ variables and independent measurement errors. The SEM for the design is

$$\begin{bmatrix} x_1 \\ x_2 \end{bmatrix} = \begin{bmatrix} \lambda_{11} \\ \lambda_{21} \end{bmatrix} [\xi_1] + \begin{bmatrix} \delta_1 \\ \delta_2 \end{bmatrix}$$

$$\Sigma = \begin{bmatrix} \sigma_1^2 & \sigma_{12} \\ \sigma_{21} & \sigma_2^2 \end{bmatrix}$$

$$\Sigma(\boldsymbol{\theta}) = \begin{bmatrix} \lambda_{11} \\ \lambda_{21} \end{bmatrix} [\phi_{11}] \begin{bmatrix} \lambda_{11} & \lambda_{12} \end{bmatrix} + \begin{bmatrix} \sigma_{\delta_1}^2 & 0 \\ 0 & \sigma_{\delta_2}^2 \end{bmatrix}$$

$$= \begin{bmatrix} \lambda_{11}^2 \phi_{11} + \sigma_{\delta_1}^2 & \lambda_{11}\phi_{11}\lambda_{12} \\ \lambda_{21}\lambda_{11}\phi_{11} & \lambda_{21}^2 \phi_{11} + \sigma_{\delta_2}^2 \end{bmatrix}$$

so that $\Sigma(\boldsymbol{\theta})$ contains 5 unknown parameters, $\boldsymbol{\theta}' = [\lambda_{11}, \lambda_{21}, \phi_{11}, \sigma^2_{\delta_1}, \sigma^2_{\delta_2}]$, and estimating Σ with $\mathbf{S}$, we have only 3 known values. Hence, we have 3 equations and 5 unknowns so that the model is not identified. The equations have the following form

$$\begin{aligned} s_1^2 &= \lambda_{11}^2\phi_{11} + \sigma^2_{\delta_1} \\ s_{12} &= \lambda_{21}\lambda_{11}\phi_{11} \\ s_2^2 &= \lambda_{21}^2\phi_{11} + \sigma^2_{\delta_2} \end{aligned}$$

To make the model identified, we may set, for example, $\lambda_{21} = \lambda_{11} = 1$ or we may set $\phi_{11} = 1$ and $\sigma^2_{\delta_1} = \sigma^2_{\delta_2} = 1$. For either situation, the system of equations may be solved uniquely for the unknown parameters. Setting one $\lambda_{ij} = 1$ and the other to zero does not make the single factor model identified unless $\phi_{11} = 1$. The problem is that there are too few indication variables for the factor. Increasing the number of indicators from two to three and setting one $\lambda_{ij} = 1$ makes the one factor model identified. Bollen (1989, p. 247) calls this the three-indicator sufficient rule. Extending the reference variable sufficient rule to more than one factor, Λ_x is structured so that each column has at least one $\lambda_{ij} = 1$ and any other row has one and only one nonzero element; at least three indicators are included per factor; and Θ_δ is assumed diagonal with Φ unspecified. For the standardized solution, Φ has ones on its diagonal and none of the elements in Λ_x are set to one.

We next consider a two factor model in which at least two measures have been obtained for the correlated latent exogenous variables ξ_1 and ξ_2 with independent measurement errors. The SEM is

$$\begin{bmatrix} x_1 \\ x_2 \\ x_3 \\ x_4 \end{bmatrix} = \begin{bmatrix} \lambda_{11} & \lambda_{12} \\ \lambda_{21} & \lambda_{22} \\ \lambda_{31} & \lambda_{32} \\ \lambda_{41} & \lambda_{42} \end{bmatrix} \begin{bmatrix} \xi_1 \\ \xi_2 \end{bmatrix} \begin{bmatrix} \delta_1 \\ \delta_2 \\ \delta_3 \\ \delta_4 \end{bmatrix}$$

where

$$\Sigma(\boldsymbol{\theta}) = \Lambda_x \begin{bmatrix} \phi_{11} & \phi_{12} \\ \phi_{21} & \phi_{22} \end{bmatrix} \Lambda_x' + \operatorname{diag}\left[\sigma^2_{\delta_i}\right]$$

and

$$\Sigma = \left[\sigma_{ij}\right] = \begin{bmatrix} \sigma_1^2 & & & (\text{sym}) \\ \sigma_{21} & \sigma_2^2 & & \\ \sigma_{31} & \sigma_{32} & \sigma_3^2 & \\ \sigma_{41} & \sigma_{42} & \sigma_{43} & \sigma_4^2 \end{bmatrix}$$

The number of unknown parameters in Λ_x, Φ and Θ_δ is $15 = t$. The number of known elements in $\mathbf{S}$, an estimate of Σ, is $q(q+1)/2 = 10$. Since, $t > q(q+1)/2$, the model is not identified. To make the model identified, we begin by setting some $\lambda_{ij} = 0$ and some $\lambda_{ij} = 1$ so that each row of Λ_x has one and only one unknown element. For example,

$$\Lambda_x = \begin{bmatrix} 1 & 0 \\ \lambda_{21} & 0 \\ 0 & 1 \\ 0 & \lambda_{42} \end{bmatrix}$$

This scales the latent variables and provides a reasonable factor pattern matrix, as suggested by Wiley (1973, p. 73). It reduces the number of unknown elements t from 15 to 9 so that $t \leq 10$ making the model overidentified. To see whether the model is identified, we must compare the elements of $\Sigma(\boldsymbol{\theta})$ with $\mathbf{S}$ where for the model

$$\Sigma(\boldsymbol{\theta}) = \begin{bmatrix} \phi_{11} + \sigma^2_{\delta_1} & & & \text{(sym)} \\ \lambda_{21}\phi_{11} & \lambda^2_{21}\phi_{11} + \sigma^2_{\delta_2} & & \\ \phi_{21} & \lambda_{21}\phi_{21} & \phi_{22} + \sigma^2_{\delta_3} & \\ \lambda_{42}\phi_{21} & \lambda_{21}\lambda_{42}\phi_{21} & \lambda_{42}\phi_{22} & \lambda^2_{42}\phi_{22} + \sigma^2_{\delta_4} \end{bmatrix}$$

and

$$\mathbf{S} = \begin{bmatrix} s_1^2 & & & \text{(sym)} \\ s_{21} & s_2^2 & & \\ s_{31} & s_{32} & s_3^2 & \\ s_{41} & s_{42} & s_{43} & s_4^2 \end{bmatrix}$$

Equating elements in $\mathbf{S}$ with $\Sigma(\boldsymbol{\theta})$,

$$s_{31} = \phi_{21}$$

$$\lambda_{21}\phi_{21} = s_{32} \implies \lambda_{21} = s_{32}/s_{31}$$

$$s_{41} = \lambda_{42}\phi_{21} \implies \lambda_{42} = s_{41}/s_{31}$$

$$s_{21} = \lambda_{21}\phi_{11} \implies \phi_{11} = s_{31}s_{21}/s_{32}$$

$$s_{43} = \lambda_{42}\phi_{22} \implies \phi_{22} = s_{43}s_{31}/s_{41}$$

and since

$$\begin{aligned} \sigma^2_{\delta_1} &= s_1^2 - \phi_{11} = s_1^2 - s_{31}s_{21}/s_{32} \\ \sigma^2_{\delta_2} &= s_2^2 - \lambda^2_{21}\phi_{11} = s_2^2 - (s_{32}/s_{21})^2 s_{31} \\ \sigma^2_{\delta_3} &= s_3^2 - \phi_{22} = s_3^2 - s_{43}s_{31}/s_{41} \\ \sigma^2_{\delta_4} &= s_4^2 - \lambda^2_{42}\phi_{22} = s_4^2 - (s_{41}/s_{31})(s_{43}s_{31}/s_{41}) \end{aligned}$$

all elements are identified for the two factor model. Alternatively, suppose we can rescale by setting $\phi_{11} = \phi_{22} = 1$, then

$$\Lambda_x = \begin{bmatrix} \lambda_{11} & 0 \\ \lambda_{21} & 0 \\ 0 & \lambda_{32} \\ 0 & \lambda_{42} \end{bmatrix}$$

The model remains identified, but with a different scaling of the latent variables.

Rearranging the elements in $\Sigma(\boldsymbol{\theta})$ using the reference variable solution and letting

$$\Lambda = \begin{bmatrix} \lambda_{21} & 0 \\ 0 & \lambda_{42} \end{bmatrix}, \quad \Delta_1 = \begin{bmatrix} \sigma^2_{\delta_1} & 0 \\ 0 & \sigma^2_{\delta_3} \end{bmatrix} \quad \text{and} \quad \Delta_2 = \begin{bmatrix} \sigma^2_{\delta_2} & 0 \\ 0 & \sigma^2_{\delta_4} \end{bmatrix}$$

Wiley (1973) observed that $\Sigma(\boldsymbol{\theta})$ has the general structure

$$\Sigma(\boldsymbol{\theta}) = \begin{bmatrix} \Phi + \Delta_1 & \Phi\Lambda' \\ \Lambda\Phi & \Lambda\Phi\Lambda' + \Delta_2 \end{bmatrix}$$

Since $\Phi = f(\mathbf{S})$, Φ is identified provided each factor contains at least two indicator variables. Also, since Φ is nonsingular, $\Lambda = (\Lambda\Phi)\Phi^{-1}$ so that Λ is identified and, $\Delta_i = f(\boldsymbol{\phi}, \Lambda)$ so that $\Sigma(\boldsymbol{\theta})$ is identified. These observations substantiate Bollen's (1989, p. 247) two indicator reference variable sufficient Rule 1 for identifying an underidentified CFA model. The rule states that if all $\phi_{ij} \neq 0$ and Θ_δ is diagonal, that there are at least two indicators per factor, and that each row of Λ_x has only one nonzero element while each column has one $\lambda_{ij} = 1$ for each $n > 1$ factors. For the standardized solution, the λ_{ij} set to one are allowed to vary; but, each $\phi_{ii} = 1$. Bollen (1989, p. 247) extends Rule 1 to a Rule 2 that permits $\phi_{ij} \neq 0$ for at least one pair of factors i and j, $i \neq j$, and assuming all other Rule 1 conditions.

Given that a CFA model is overidentified, rules that establish sufficient conditions for model identification when Θ_δ is not diagonal, have not been established. Because of the difficulty associated with global model identification, most researchers depend on tests of local identification using Definition 2.10.2. Local identification, given a model is overidentified, ensures identification in a neighborhood of $\boldsymbol{\theta}_1$, a specific value of $\boldsymbol{\theta}$. Local identification is a necessary but not, in general, a sufficient condition for global identification unless $\sigma[\boldsymbol{\theta}] = \text{vec}[\Sigma(\boldsymbol{\theta})]$ is a convex (concave) function over a closed, bounded convex (concave) parameter space. This condition is not frequently met in practice. To evaluate local identification, most computer programs use the two methods proposed by Wiley (1973, pp. 81-82) which evaluates the rank of the score matrix (it must have rank $r = t$, if $r < t$ then $t - r$ constraints must be added to the SEM) or the inverse of the Fisher information matrix for $\boldsymbol{\theta}$ (it must be nonsingular). The identification problem is a serious and complex issue for any SEM since an estimate $\widehat{\boldsymbol{\theta}}$ of $\boldsymbol{\theta}$ which may be a consistent estimate of $\boldsymbol{\theta}$ for an overidentified SEM is no longer consistent without global identification.

Given a CFA model or any SEM is identified, one next selects a fit function $F(\mathbf{S}, \Sigma(\boldsymbol{\theta}))$, minimizes the function to obtain $\widehat{\boldsymbol{\theta}}$, and employs tests or indices of fit to evaluate the error-in-fit.

The EFA model was scale free under a transformation of the data vector, $\mathbf{x}^* = \mathbf{D}\mathbf{x}$, if (1) the parameters in Λ^* and Θ_δ^* are easily recovered using the scaling matrix $\mathbf{D}$ and (2) the model restrictions imposed on the model parameters (model identification) are also preserved under the rescaling, Swaminathan and Algina (1978). Because the EFA model is scale free using a scaling matrix $\mathbf{D}$, condition (2) is equivalent to the characterization of the model. The GLS and ML estimators are scale free (the canonical correlation characterization of the EFA model) while the ULS estimators are, in general, not scale free (the PC characterization of the model). Defining a fit function $F(\mathbf{S}, \Sigma(\boldsymbol{\theta})) \geq 0$ and equal to zero if and only if $\mathbf{S} = \widehat{\Sigma}$, Swaminathan and Algina (1978) define a fit function to be scale invariant under the nonsingular transformation $\mathbf{D}$ if $F(\mathbf{S}, \Sigma(\boldsymbol{\theta})) = F(\mathbf{DSD}, \mathbf{D}\Sigma(\boldsymbol{\theta})\mathbf{D}) \geq 0$. This implies that the value of the error-in-fit function is uncharged for the analysis of Σ or $\mathbf{P}_\rho$. Using this definition, the fit function is scale invariant. If plim $\mathbf{W}^{-1} = \Sigma^{-1}$, then F_{WLS} is scale invariant so that F_{GLS} and F_{ADF} are scale invariant. Also, if plim diag $\mathbf{W}^{-1} =$ diag Σ^{-1}, the F_{DWLS} fit function is scale invariant. The unweighted least squares fit function F_{ULS} is not scale invariant. However, for any SEM, scale invariance is neither a necessary or sufficient condition for scale freeness. Thus, ML estimators may at times not be scale free while ULS estimators may be scale free under special conditions. The exact conditions for scale freeness are difficult to establish. A necessary condition is that the diag $\Sigma(\widehat{\boldsymbol{\theta}}) =$ diag $\mathbf{S}$ for any consistent estimator $\widehat{\boldsymbol{\theta}}$. Because estimates obtained using the fit functions F_{ML}, F_{GLS} and F_{ADF} are asymptotically efficient when analyzing Σ they are preferred to estimates obtained using F_{DWLS} or F_{ULS}. Also, tests for fit tend to be asymptotically chi-square when analyzing Σ, while tests involving estimates using F_{DWLS} or F_{ULS} have to be adjusted, Browne (1974, 1984).

Assuming the manifest variables are jointly normal and $N > 100$, one may test the goodness-of-fit hypothesis

$$H : \Sigma = \Sigma(\boldsymbol{\theta}) \tag{10.3.3}$$

versus the alternative that Σ is unstructured using a chi-square statistic. If $F(\mathbf{S}, \Sigma(\boldsymbol{\theta}))$is scale invariant, the statistic

$$X^2 = (N-1)\, F(\mathbf{S}, \Sigma(\widehat{\boldsymbol{\theta}})) \tag{10.3.4}$$

converges asymptotically to a chi-square distribution under H, with degrees of freedom $v^* = [(p+q)(p+q+1)/2] - t$. For the CFA model, $v^* = [q(q+1)/2] - t$. The test statistic of model fit is an asymptotic result that is adversely effected by lack of multivariate normality and in particular kurtosis, Browne (1982,1984). When one does not have multivariate normality, one should investigate transformations of the data or a robust estimate of the covariance matrix. The test in (10.3.3) is valid only if the diag $\Sigma(\widehat{\boldsymbol{\theta}}) =$ diag $\mathbf{S}$ when analyzing $\mathbf{R}$, Krane and McDonald (1978). In CFA, $\mathbf{S}$ and not $\mathbf{R}$ should be analyzed unless the units of measurement have no meaning so that $\mathbf{P}_\rho \equiv \Sigma$.

Because the assumption for the chi-square fit test are usually not valid, because more complex models almost always fit better, and since large sample sizes lead to erroneous results; it has limited value, Kaplan (1990). Thus, one usually converts it to an index for distributions with kurtosis equal zero. The converted index is

$$Z = \frac{\sqrt[3]{\frac{X^2}{v^*}} - \left(1 - \frac{2}{9v^*}\right)}{\sqrt{\frac{2}{9v^*}}} \tag{10.3.5}$$

When $|Z| \leq 5$, one usually has a "good fit."

Because most tests of fit are not appropriate for SEMs, there have been over thirty indices of fit proposed in the literature, March, Balla and McDonald (1988). The CALIS procedure calculates over twenty-five indices. For a detail discussion of each, the reader should consult Bollen (1989), Mueller (1996) and Kaplan (2000). The goodness-of-fit index (GFI) and adjusted goodness-of-fit index (AGFI) behave like R^2 and adjusted R_a^2 in multiple regression, Tanaka and Huba (1984). The root mean square residual is also calculated. Bentler's (1990) comparative fit indices and several parsimonious fit indices are calculated, Williams and Holohan (1994). Finally, statistics that measure the variation in the sample covariance accounted for by the model are calculated for all equations, the structural model, and the measurement model.

When evaluating the overall fit of a complex structural equation model, the model often does not fit the data. Most of the procedures compare **S** and $\widehat{\Sigma}$. The goal is to obtain normalized residuals that are less than two. When one finds that a model does not fit the data, it is generally better to apply the principle of parsimony and remove rather than add paths and covariances to the model. Adding parameters more often than not tends to capitalize on chance characteristics. To facilitate the evaluation of alternative nested models one may use likelihood ratio (LR), Lagrange Multiplier (LM), and Wald (W) tests. These tests are only valid under multivariate normality using a scale invariant fit function with large sample sizes ($N > 100$), and for analyzing the structure of Σ and not $\mathbf{P}_\rho$. The likelihood ratio tests fit two models, an unstructued model ($\widehat{\boldsymbol{\theta}}_u$) and a restricted model ($\widehat{\boldsymbol{\theta}}_r$). Comparing the difference in chi-square statistics, one may determine whether or not the restrictions are significant.

The structure of the LR test is

$$X^2 = (N - 1)(F_r - F_u) \tag{10.3.6}$$

where F_r is the scale invariant fit function evaluated at $\widehat{\boldsymbol{\theta}}_r$ and F_u is the same function evaluated at $\widehat{\boldsymbol{\theta}}_u$. Under the restrictions, the statistic $X^2 \,\dot{\sim}\, \chi^2(df)$ with degrees of freedom $df = v_r^* - v_u^*$, the difference in degrees of freedom for the restricted model minus the degrees of freedom for the unrestricted model. Restrictions usually remove coefficients from the model or set free parameters to zero.

The likelihood ratio test procedure requires calculating $\boldsymbol{\theta}$ under the unrestricted model ($\widehat{\boldsymbol{\theta}}_u$) and the restricted model ($\widehat{\boldsymbol{\theta}}_r$). An alternative to the LR test is Rao's score test (Rao, 1947) also called the Lagrange Multiple Test (Silvey, 1959) which uses only $\widehat{\boldsymbol{\theta}}_r$ to evaluate model fit. One can also use Wald's (1943) test statistic which only uses $\widehat{\boldsymbol{\theta}}_u$. For the classical linear regression model with known variance σ^2, the three tests are identical, Buse (1982). This is not the case in general, Davidson and MacKinnon (1993); however, they are asymptotically equivalent. Yuan and Bentler (1999) recommend approximating the distribution of X^2 in (10.3.6) using Hotelling's T^2 statistic.

The Wald (W) test evaluates whether restrictions $r(\boldsymbol{\theta}) = \mathbf{0}$ can be imposed on the model. With restrictions on the model, $r(\widehat{\boldsymbol{\theta}}_r) = \mathbf{0}$ by construction and if the restrictions are valid, then $r(\widehat{\boldsymbol{\theta}}_u) = \mathbf{0}$. The W statistic estimates the change in chi-square that would result from removing a path or covariance from a model. The Lagrange Multiplier (LM) test is evaluating whether restrictions can be removed from the model. The test estimates the reduction in chi-square that would result by allowing a parameter to be estimated. The test is evaluating whether a new path or covariance should be added to the model, MacCallum (1986).

Even with a "good" fit, the path diagram may be incorrect. A different model may represent the same "good" fit. Indeed, for complex models path directions may even be reversed without effecting the fit. Lee and Hershberger (1990) provide a very good discussion and develop some rules for examining a path diagrams to determine if paths can be reversed. To locate internal specification error, one uses modification indices, expected parameter change statistics and cross-validation indexes, Saris, Satorra and Sörbam (1987), and Browne and Cudeck (1989).

10.4 Confirmatory Factor Analysis Examples

a. *Performance Assessment 3 - Factor Model (Example 10.4.1)*

Using a three-factor EFA model and the ML estimation method, we found in Example 8.10.1 that the model was able to recover the sample covariance matrix; however, all five dimensions of the questionnaire instrument were not identified by the solution. Two factors recovered two of the hypothesized dimensions while the third latent factor was confounded by several variables. The development of the EFA model required the specification of only the number of factors and an initial estimate of the unique error variances. We now assume one knows the number of factors and the structure of Λ. Then, using the SEM symbols in Figure 10.1.1, we formulate a three-factor CFA model for the PAP covariance matrix $\Sigma(\boldsymbol{\theta}) = \Lambda_x \Phi \Lambda_x' + \Theta_\delta$ as shown in Figure 10.4.1.

Associated with the SEM in Figure 10.4.1 is the CFA model that follows.

$$\begin{bmatrix} x_1 \\ x_2 \\ x_3 \\ x_4 \\ x_5 \\ x_6 \\ x_7 \\ x_8 \end{bmatrix} = \begin{bmatrix} 1 & 0 & 0 \\ \lambda_{21} & 0 & 0 \\ 0 & 1 & 0 \\ 0 & \lambda_{42} & 0 \\ 0 & 0 & 1 \\ 0 & 0 & \lambda_{63} \\ 0 & 0 & \lambda_{73} \\ 0 & 0 & \lambda_{83} \end{bmatrix} \begin{bmatrix} \xi_1 \\ \xi_2 \\ \xi_3 \end{bmatrix} + \begin{bmatrix} \delta_1 \\ \delta_2 \\ \delta_3 \\ \delta_4 \\ \delta_5 \\ \delta_6 \\ \delta_7 \\ \delta_8 \end{bmatrix} \tag{10.4.1}$$

$$\underset{q\times 1}{\mathbf{x}} = \underset{q\times n}{\Lambda_x} \; \underset{n\times 1}{\xi} + \underset{q\times 1}{\delta}$$

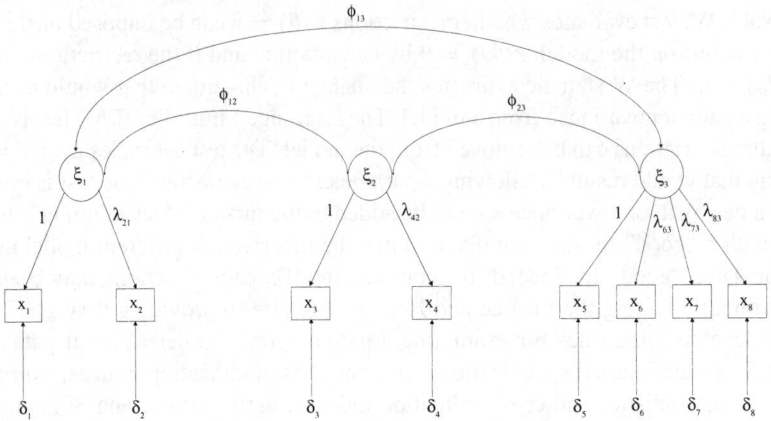

FIGURE 10.4.1. 3-Factor PAP Model

The covariance structure is

$$\Sigma(\boldsymbol{\theta}) = \Lambda_x \Phi \Lambda_x' + \Theta_\delta \tag{10.4.2}$$
$$\Phi = [\phi_{ij}]$$
$$\text{diag}\, \Theta_\delta = [\sigma^2_{\delta_i}]$$

The structure given in (10.4.2) is similar to the structure obtained using the EFA and an oblique rotation. Depending on Φ, the factors may be orthogonal/oblique which supports the general structure given in (10.4.2).

To fit the model given in (10.4.1), we use the procedure CALIS. While the SAS procedure may determine that a model is underidentified, there are no guarantees. Thus, it is best to demonstrate that any proposed CFA model is overidentified. For the model in (10.4.1), $q = 8$ so that by (10.2.7), $v = (8)(9)/2 = 36$. And, counting the number of unknown parameters in Φ, Λ_x and Θ_δ, $t = 6+8+5 = 19$, so that $v - t = 17 \geq 0$. Hence, our model meets the necessary condition for model identification. And because the model meets the two indicator sufficient condition that all $\phi_i \neq 0$, Θ_δ is diagonal, and Λ_x is scaled with one nonzero element per row, the sufficient condition for identification is met. Thus, the model is globally identified.

To analyze the structure given in Figure 10.4.1, the LINEQS statement is used to specify the model equations. The STD statement defines the variances in Θ_δ to be estimated. And, the COV statement is used to identify the elements of Φ to be estimated. Because we are analyzing a covariance matrix the option COV must be included on the PROC CALIS statement. If one does not include COV, the CALIS procedure default is to analyze a correlation matrix. Please be careful. The SAS code for this example is provided in program m10_4_1.sas. While there are many options for the PROC CALIS statement, we have kept this example simple. The option COV is needed to analyze a covariance matrix, EDF is the required degrees of freedom to obtain tests of fit, and the keyword RESIDUAL is needed to output residuals $[r_{ij}] = \mathbf{S} - \widehat{\Sigma}$ for the SEM.

TABLE 10.4.1. 3-Factor PAP Standardized Model

Equations							R^2
x_1	=	0.7148	ξ_1	+	0.6993	δ_1	.51
x_2	=	0.8034	ξ_1	+	0.5954	δ_2	.65
x_3	=	0.7604	ξ_2	+	0.6495	δ_3	.58
x_4	=	0.7021	ξ_2	+	0.7120	δ_4	.49
x_5	=	0.7016	ξ_3	+	.7126	δ_5	.49
x_6	=	0.8151	ξ_3	+	.5796	δ_6	.66
x_7	=	0.6077	ξ_3	+	.7942	δ_7	.37
x_8	=	0.5132	ξ_3	+	.8583	δ_8	.26

When fitting a CFA model, the first task is to evaluate model fit. For large samples and assuming a MVN distribution, one first examines the chi-square statistic. Using (10.3.4), $X^2 = 23.88$ with p-value $= 0.1228$. This statistic indicates a good fit. In many CFA problems the statistic will usually be significant even if the model fits. Thus, it should not be the only index investigated. For the Z test, $\mid Z \mid = 1.16 < 5$.

Examining the root mean square residuals,

$$RMR = \sqrt{2\sum_{i=1}^{q}\sum_{j=1}^{i}\left(s_{ij} - \widehat{\sigma}_{ij}\right)^2 / q\,(q+1)} \tag{10.4.3}$$

for the model, $RMR = 0.0139$ and the mean absolute value of the unstandardized residuals $r_{ij} = s_{ij} - \widehat{\sigma}_{ij}$ is very small (0.009). While our example shows these values to be small, the magnitude of these indices are affected by the differing variable scales. Hence, one examines normalized residuals

$$r_{ij}^{*} = \left(s_{ij} - \widehat{\sigma}_{ij}\right) / \left[\left(\widehat{\sigma}_{ij}\widehat{\sigma}_{jj} + \widehat{\sigma}_{ij}^{2}\right) / N\right]^{1/2} \tag{10.4.4}$$

where N = sample size. Generally, one expects each absolute value of the normalized residuals to be less than 2.00. Also, Bentler's (1990) comparative fit index and Bentler and Bonett's (1980) indices are over 0.9 indicating a reasonable model fit. For a comprehensive evaluation of fit indices, consult Mulaik et al. (1989) and Bollen (1989).

Finally, we investigate the ML estimates for Λ_x, Φ, and Θ_δ. First, observe that no asymptotic standard error is really small (< 0.0005), that all asymptotic t values are large, and that all standardized loadings are larger than 0.60. We conclude that a 3-factor model for the data appears reasonable. The standardized loadings are summarized in Table 10.4.1

For a CFA model, the total variance is equal to the common variance plus the unique (error) variance. And the square of the multiple correlation between a manifest (observed) variable and the latent factor, defined as

$$R_{x_i}^2 = 1 - \frac{\operatorname{var}\widehat{\delta}_i}{\widehat{\sigma}_{ii}} \tag{10.4.5}$$

is a reliability estimate for the observed variable, Bollen (1989, p. 221). Reliability is a measure of how consistently the questionnaire instrument measures what it was designed to measure. Using the standardized model, $R_{x_i}^2 = (\lambda_{ij}^*)^2$, the square of the standardized loadings. The larger the direct standardized effect (excluding errors δ_i), the higher the ξ_i's reliability. From this perspective, the 3-factor model has low reliabilities ranging from 0.26 to 0.65.

To illustrate a poorly fit model, we include in program m 10_4_1.sas the code for fitting a 2-factor model. For this model, the chi-square test of fit is rejected, the Z test value is larger than 5 and one has large normalized residuals.

b. *Performance Assessment 5-Factor Model (Example 10.4.2)*

While the 3-factor CFA model provided a reasonable fit to the PAP data, we were unable to recover the five designed dimensions of the questionnaire instrument. Hence, we next fit a 5-factor CFA to the covariance matrix. For a 5-factor model, (10.4.1) becomes

$$\begin{bmatrix} x_1 \\ x_2 \\ x_3 \\ x_4 \\ x_5 \\ x_6 \\ x_7 \\ x_8 \end{bmatrix} = \begin{bmatrix} 1 & 0 & 0 & 0 & 0 \\ \lambda_{21} & 0 & 0 & 0 & 0 \\ 0 & 1 & 0 & 0 & 0 \\ 0 & \lambda_{42} & 0 & 0 & 0 \\ 0 & 0 & 1 & 0 & 0 \\ 0 & 0 & 0 & 1 & 0 \\ 0 & 0 & 0 & 0 & 1 \\ 0 & 0 & 0 & 0 & \lambda_{85} \end{bmatrix} \begin{bmatrix} \xi_1 \\ \xi_2 \\ \xi_3 \\ \xi_4 \\ \xi_5 \end{bmatrix} + \begin{bmatrix} \delta_1 \\ \delta_2 \\ \delta_3 \\ \delta_4 \\ \delta_5 \\ \delta_6 \\ \delta_7 \\ \delta_8 \end{bmatrix} \tag{10.4.6}$$

$$\underset{q\times 1}{\mathbf{x}} = \underset{q\times n}{\Lambda_x} \; \underset{n\times 1}{\boldsymbol{\xi}} + \underset{q\times 1}{\boldsymbol{\delta}}$$

with covariance structure

$$\begin{aligned} \Sigma(\boldsymbol{\theta}) &= \Lambda_x \Phi \Lambda_x' + \Theta_\delta \\ \Phi &= [\phi_{ij}]\, \phi_{ij} \neq 0 \\ \operatorname{diag} \Theta_\delta &= \left[\sigma_{\delta_1}^2, \sigma_{\delta_2}^2, \sigma_{\delta_3}^2, \sigma_{\delta_4}^2, 0, 0, \sigma_{\delta_7}^2, \sigma_{\delta_8}^2\right] \end{aligned} \tag{10.4.7}$$

Because the observed variables (x_5, x_6) are the same as the latent variables (ξ_3, ξ_4), the error variances are set to zero, $\sigma_{\delta_5}^2 = \sigma_{\delta_6}^2 = 0$. Now the number of model parameters is 24: $15 - \phi_{ij}$, $5 - \sigma_{\delta_i}^2$, and $3 - \lambda_{ij}$. And, $v - t = 36 - 24 = 12 \geq 0$ so that the model meets the necessary condition of model identification. Because all $\phi_{ij} \neq 0$, each row of Λ_x has one nonzero element, and the elements of Θ_δ are set to zero when a factor has less than two indicators, the model is identified. Program m10_4_2.sas is used to analyze the new model.

Reviewing the chi-square fit statistic, the p-value for testing $H : \Sigma = \Sigma(\boldsymbol{\theta})$ is not rejected. The p-value of the test is 0.4528, higher than the 3-factor model. Using the Likelihood Ratio difference test to compare the two models, we consider the 3-factor model to be a nested, restricted model of the unrestricted 5-factor model. Subtracting the chi-square values and the associated degrees of freedom, $X^2 = (23.87 - 11.91) = 11.96$,

TABLE 10.4.2. 5-Factor PAP Standardized Model

Equations							R^2
x_1	=	0.6935	ξ_1	+	0.7205	δ_1	.48
x_2	=	0.8282	ξ_1	+	0.5605	δ_2	.69
x_3	=	0.7562	ξ_2	+	0.6543	δ_3	.57
x_4	=	0.7060	ξ_2	+	0.7082	δ_4	.50
x_5	=	1.0000	ξ_3	+	0.0000	δ_5	1.00
x_6	=	1.0000	ξ_4	+	0.0000	δ_6	1.00
x_7	=	0.6626	ξ_5	+	0.7490	δ_7	.44
x_8	=	0.5444	ξ_5	+	0.8388	δ_8	.30

$df = 17 - 12 = 6$. Since $11.96 > 11.07$, the critical chi-square value for $\alpha = 0.05$, we reject the 3-factor model. To further evaluate whether the 5-factor model is better than the 3-factor model we may compare the information criteria developed by Akaike (1974, 1987), Schwartz (1968) and Bozdogan (1987). As with mixed linear models, the recommendation is to choose a model with largest information. Comparing the criteria for the two models (5-factor vs 3-factor), there is no clear winner. However, the Z test is smaller and most of the fit indices are higher (≈ 1.000) for the 5-factor model. Finally, the average of the normalized residuals is small ($0.22 < 0.25$). Reviewing the loading matrix Λ_x, standard errors and t tests, the 5-factor model fits better than the 3-factor model. The standardized loadings are given in Table 10.4.2.

Comparing the entries in Table 10.4.2 and Table 10.4.1, we see that except for variable $x_1 =$ suppg, all reliability estimates increased and the error variances decreased. Thus, we conclude that the questionnaire instrument appears to recover the five latent domains.

Using the MODIFICATION option on the PROC CALIS statement, the SAS procedure generates Lagrange Multiplier and Wald modification indices, and Stepwise Multivariate Wald tests for fitted models. These tests are discussed by Bollen (1989), and MacCallum, Roznowski and Necowitz (1992). The Wald test is used to estimate the change in the chi-square model statistic that would result by equating a parameter to zero. The effect of eliminating a specific path or setting a "free" covariance parameter to zero. The tests often agree with the t tests calculated for parameters where nonsignificance may result in setting the parameter to zero. In general, t tests and Lagrange Multiplier tests should be used with caution since the action taken may not generalize to other samples. The Lagrange Multiplier and Wald indices are computed for three matrices _GAMMA_, _BETA_, and _PHI_. For the CFA model, the phi matrix contain latent factors and residual error terms and the matrix gamma has as rows indicator variables and as columns factors. The Lagrange Multiplier/Wald Indices estimate the reduction in the chi-square statistic that would result by allowing a parameter to be estimated. That is, the degree to which the model would improve by adding a new parameter (loading/covariance) to the model. The rank order of the ten largest Lagrange multipliers appear at the end of the matrix. For our example, the largest reduction occurs by relating δ_8 and δ_3 or x_7 and f_1. Clearly, we want δ_8 and δ_3 to be uncorrelated; however, the relation x_7 and f_1 suggests that factor 1 may influence variable 7.

As a general practice, when fitting SEM to data, it is generally better to obtain a good model by removing paths to improve or maintain fit than to add paths to a model.

Exercises 10.4

1. Perform CFA on the correlation matrix given in Table 3.10.1 and interpret your results. How do the results compare with the factor structure obtained using EFA.

2. Analyze the correlation matrix in Table 3.10.2 using CFA. Discuss your findings.

3. Analyze the correlation matrix in Table 3.10.2 assuming a correlated factor structure using the CFA model.

10.5 Path Analysis

The CFA model contains one manifest variable and one latent variable. We now consider models commonly found in economics which have only observed, manifest variables. These models are called path analysis (PA) models or simultaneous equation models. Using LISREL notation, the basic model has the structure

$$\underset{p\times 1}{\mathbf{y}} = \underset{p\times p}{\mathbf{B}} \underset{p\times 1}{\mathbf{y}} + \underset{p\times q}{\Gamma} \underset{q\times 1}{\mathbf{x}} + \underset{p\times 1}{\boldsymbol{\zeta}} \tag{10.5.1}$$

where $\boldsymbol{\eta} = \mathbf{y}$ and $\boldsymbol{\xi} = \mathbf{x}$. The matrices $\mathbf{B}$ and Γ contain the direct effect (path) coefficients and from (10.2.3), the covariance structure for the model is

$$\Sigma(\boldsymbol{\theta}) = \begin{bmatrix} (\mathbf{I}-\mathbf{B})^{-1}\left(\Gamma\Phi\Gamma' + \Psi\right)(\mathbf{I}-\mathbf{B})^{-1\prime} & (\mathbf{I}-\mathbf{B})^{-1}\Gamma\Phi \\ \Phi\Gamma'(\mathbf{I}-\mathbf{B})^{-1\prime} & \Phi \end{bmatrix} \tag{10.5.2}$$

For PA models, it is common to have both recursive and nonrecursive models. Recall that a model is recursive if relationships among the variables in $\mathbf{y}$ are one directional while nonrecursive models contain bi-directional relationships and feedback loops. For a recursive model, the matrix $\mathbf{B}$ is always a lower triangular matrix and Ψ is diagonal, otherwise the path analysis model is said to be nonrecursive. Often the vector $\mathbf{y}$ is partitioned into subsets which generates a block structure for $\mathbf{B}$ and Γ. For example, for two subsets of variables the structure of $\mathbf{B}$ and Ψ is as follows.

$$\mathbf{B} = \begin{bmatrix} \mathbf{B}_{11} & \mathbf{0} \\ \mathbf{B}_{21} & \mathbf{B}_{22} \end{bmatrix} \quad \text{and} \quad \Psi = \begin{bmatrix} \Psi_{11} & \mathbf{0} \\ \mathbf{0} & \Psi_{22} \end{bmatrix} \tag{10.5.3}$$

Models with this structure are said to be block-recursive.

When working with path models, one may again convert the structure to reduced form. This expresses the endogenous variables in terms of the random exogenous variables plus

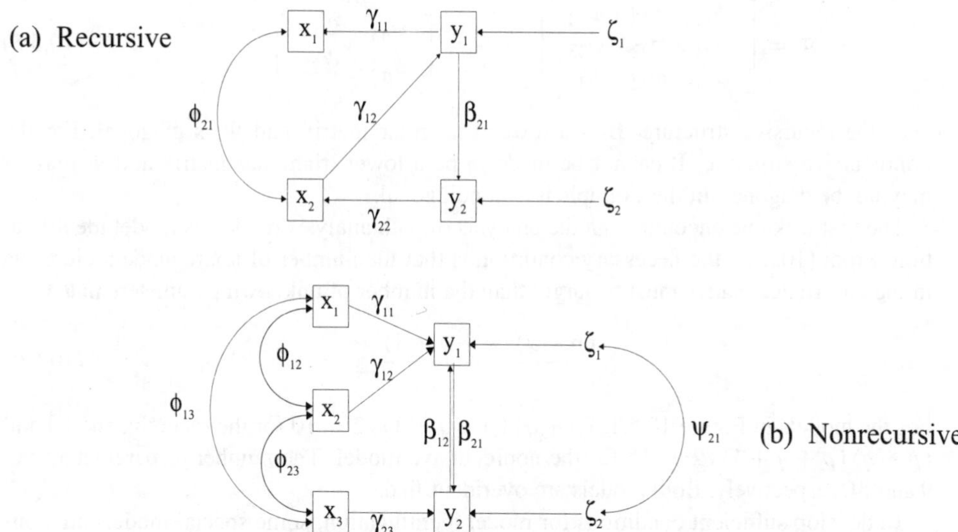

FIGURE 10.5.1. Recursive and Nonrecursive Models

random error, a general nonlinear model. Solving for $\mathbf{y}$ in (10.5.1), the reduced form for the model is

$$\begin{aligned} \mathbf{y} &= (\mathbf{I} - \mathbf{B})^{-1}\, \Gamma \mathbf{x} + (\mathbf{I} - \mathbf{B})^{-1}\, \boldsymbol{\zeta} \\ &= \Pi \mathbf{x} + \mathbf{e} \end{aligned} \tag{10.5.4}$$

The matrix Π contains the reduced form coefficients and the vector $\mathbf{e}$ the reduced form random errors. Model (10.5.4) is the classical conditional regression model.

In Figure 10.5.1, we provided two examples of path diagram. Example (a) is a recursive model and example (b) is a nonrecursive model. For the recursive model, observe that, ζ_1 and ζ_2 are not correlated, which is not the case for the nonrecursive model.

The structural equations for the two models follow.

(a) Recursive Model

$$\begin{bmatrix} y_1 \\ y_2 \end{bmatrix} = \begin{bmatrix} 0 & 0 \\ \beta_{21} & 0 \end{bmatrix} \begin{bmatrix} y_1 \\ y_2 \end{bmatrix} + \begin{bmatrix} \gamma_{11} & \gamma_{12} \\ 0 & \gamma_{22} \end{bmatrix} \begin{bmatrix} x_1 \\ x_2 \end{bmatrix} + \begin{bmatrix} \zeta_1 \\ \zeta_2 \end{bmatrix}$$

$$\Phi = \begin{bmatrix} \phi_{11} & \phi_{12} \\ \phi_{21} & \phi_{22} \end{bmatrix}, \quad \Psi = \begin{bmatrix} \psi_{11} & 0 \\ 0 & \psi_{22} \end{bmatrix}$$

(b) Nonrecursive Model

$$\begin{bmatrix} y_1 \\ y_2 \end{bmatrix} = \begin{bmatrix} 0 & \beta_{12} \\ \beta_{21} & 0 \end{bmatrix} \begin{bmatrix} y_1 \\ y_2 \end{bmatrix} + \begin{bmatrix} \gamma_{11} & \gamma_{12} & 0 \\ 0 & 0 & \gamma_{23} \end{bmatrix} \begin{bmatrix} x_1 \\ x_2 \\ x_3 \end{bmatrix} + \begin{bmatrix} \zeta_1 \\ \zeta_2 \\ \zeta_3 \end{bmatrix}$$

$$\Phi = \begin{bmatrix} \phi_{11} & \phi_{12} & \phi_{13} \\ \phi_{21} & \phi_{22} & \phi_{23} \\ \phi_{31} & \phi_{32} & \phi_{33} \end{bmatrix}, \quad \Psi = \begin{bmatrix} \psi_{11} & \psi_{12} \\ \psi_{21} & \psi_{22} \end{bmatrix} \tag{10.5.5}$$

For the recursive structure, $\mathbf{B}$ is a lower triangular matrix and Ψ is diagonal. For the nonrecursive structure, $\mathbf{B}$ cannot be made to be a lower triangular matrix and Ψ may or may not be diagonal. In the example it is nondiagonal.

The first task one encounters in the analysis of path analysis models is model identification. From (10.2.7), the necessary condition is that the number of nonredundent elements in the covariance matrix must be larger than the number of unknown parameters in $\boldsymbol{\theta}$

$$\frac{(p+q)(p+q+1)}{2} \geq t \tag{10.5.6}$$

For the models in Figure 10.5.1, $(p+q)(p+q+1)/2 = 10$ for the recursive model and $(p+q)(p+q+1)/2 = 15$ for the nonrecursive model. The number of parameters t is 9 and 14, respectively. Both models are overidentified.

To develop sufficient conditions for model identification, some special models are considered. From the reduced form of the structural model if $\mathbf{B} = \mathbf{0}$, then path analysis model is a multivariate regression model. And, Γ, Φ, and Ψ are easily expressed in terms of the observed covariance matrix $\mathbf{S}$. For

$$\mathbf{S} = \begin{bmatrix} \mathbf{S}_{yy} & \mathbf{S}_{yx} \\ \mathbf{S}_{xy} & \mathbf{S}_{xx} \end{bmatrix} \tag{10.5.7}$$

the parameters have the identified structure

$$\Phi = \mathbf{S}_{xx}, \Gamma' = \mathbf{S}_{xx}^{-1}\mathbf{S}_{xy} \quad \text{and} \quad \Psi = \mathbf{S}_{yy} - \mathbf{S}_{yx}\mathbf{S}_{xx}^{-1}\mathbf{S}_{xy} \tag{10.5.8}$$

This rule is not too helpful in practice since most models involve structures with $\mathbf{B} \neq \mathbf{0}$.

A more common situation is to have a recursive model where $\mathbf{B}$ is lower triangular and Ψ is diagonal. The recursive condition is sufficient for model identification. To see this, one merely has to observe that $\mathbf{I} - \mathbf{B}$ is a unit triangular matrix for some ordering of the elements in $\mathbf{y}$, that the inverse of $(\mathbf{I} - \mathbf{B})$ is also triangular, and that both $\mathbf{y}$ and $\mathbf{x}$ are each uncorrelated with $\mathbf{e}$. Then, equating the elements of $\Sigma(\boldsymbol{\theta})$ with $\mathbf{S}$ we have that

$$\begin{aligned} \mathbf{S}_{yy} &= (\mathbf{I} - \mathbf{B})^{-1}\left(\Gamma\Phi\Gamma' + \Psi\right)(\mathbf{I} - \mathbf{B})^{-1\prime} \\ \mathbf{S}_{xy} &= \Phi\Gamma'(\mathbf{I} - \mathbf{B})^{-1\prime} \\ \mathbf{S}_{xx} &= \Phi \end{aligned} \tag{10.5.9}$$

Because Φ is identified, we may determine the elements in Γ, Ψ, and $\mathbf{B}$ using each equation in the simultaneous system one at a time, successively, to solve for all parameters. Because $(\mathbf{I} - \mathbf{B})^{-1}$ is lower triangular, the first equation only depends on the first row of Γ. Thus, one may find $\boldsymbol{\gamma}_1'$, the first row of Γ, and then ψ_{11}. From the second equation, one obtains the second row of Γ, the second row of $\mathbf{B}$ and then ψ_{22}. Continuing, in this manner, all parameters are identified.

In developing sufficient conditions for path models, we placed restrictions on the model parameters, defining $\mathbf{B}$ to be zero or specified the pattern of the coefficients in the matrix $\mathbf{B}$ as lower triangular and Ψ as diagonal for recursive models. For nonrecursive models, we continue with this approach. First, because Φ is always identified we need not concern ourself with this parameter. However, assume Ψ is unconstrained, is free to vary, then necessary and sufficient conditions for identification of nonrecursive models involve a rank condition on the partitioned matrix $\mathbf{G}_{p\times(p+q)} = [\mathbf{I} - \mathbf{B}, \; -\Gamma]$. The rank rule is as follows. Form the matrix $\mathbf{G}$ for $i = 1, 2, \ldots, p$ by deleting all columns of $\mathbf{G}$ that do not have zeros in the row. Then, the necessary and sufficient condition for the nonrecursive model to be identified is that the

$$\text{rank}\,(\mathbf{G}_i) = p - 1 \qquad i = 1, 2, \ldots, p \tag{10.5.10}$$

For a proof of this result, see Greene (2000, p. 667).

To illustrate the use of (10.5.10), consider the nonrecursive example illustrated in Figure 10.5.1. The matrix

$$\mathbf{G} = \begin{bmatrix} 1 & -\beta_{12} & \vdots & \gamma_{11} & \gamma_{12} & 0 \\ & & \vdots & & & \\ -\beta_{21} & 1 & \vdots & 0 & 0 & \gamma_{23} \end{bmatrix} \tag{10.5.11}$$

and the

$$\begin{aligned} \text{rank}\,(\mathbf{G}_1) &= \text{rank}\left[\gamma_{23}\right] = 1 \\ \text{rank}\,(\mathbf{G}_2) &= \text{rank}\left[\gamma_{23}, \gamma_{12}\right] = 1 \end{aligned}$$

so that the nonrecursive model is identified.

Although the identification rule developed for nonrecursive models may be applied in many situations, it does not allow Ψ to be constrained. Bekker and Pollock (1986) have developed additional rules for this case. For very complex nonrecursive models, one must use the general computational methods discussed in CFA. Again, they usually only ensure local identification. One must again have global identification to obtain consistent estimators of $\boldsymbol{\theta}$.

Having established model identification, one once again uses the fit functions defined in (10.2.9) to estimate model parameters. In CFA only one matrix Λ_x of direct effects was estimated. In path analysis, two matrices are involved, $\mathbf{B} = \left[\beta_{ij}\right]$ and $\Gamma = \left[\gamma_{ij}\right]$. These contain the direct effects of y_i on y_j and x_i on x_j, respectively. However, one also has indirect effects and total effects. An indirect effect of y_i or y_j or x_i on y_i is separated by intervening, mediating variables and is represented as products of direct effects. The total effect is the sum of all direct and indirect effects. For the identified recursive system in Figure 10.5.1, the indirect effect of x_1 on y_2 is $\gamma_{11}\beta_{21}$, the direct effect of x_1 on y_1 is γ_{11} so that the total effect is $\gamma_{11}+\gamma_{11}\beta_{21}$. Only, straight arrows are used to trace effects (direct, indirect, and total).

To evaluate the effect of $\mathbf{y}$ on $\mathbf{y}$, recall that $\mathbf{B}$ is the matrix of direct effects. Multiplying the direct matrix $\mathbf{B}$ by itself, the matrix $\mathbf{B}^2$ contains the indirect effects with one intervening variable and $\mathbf{B}^3$ represents the indirect effects with two intervening variables. Because the total (**T**) effect is the sum of the indirect (**I**) effect plus the direct (**D**) effect, the total effect of **y** on **y** is

$$
\begin{aligned}
\mathbf{T_{yy}} &= \mathbf{D_{yy}} + \mathbf{I_{yy}} \\
&= \mathbf{B} + \sum_{k=2}^{\infty} \mathbf{B}^k
\end{aligned}
\tag{10.5.12}
$$

For a recursive model, **B** is a lower triangular matrix with zeros on the diagonal. Since **B** is lower triangular with zeros on the diagonal, it is nilpotent for some k in that $\mathbf{B}^{k+1} = \mathbf{0}$ so that the infinite series in (10.5.12) terminates for recursive models. Then, the total effect of **y** on **y** is

$$
\begin{aligned}
\mathbf{T_{yy}} &= \mathbf{B} + \mathbf{B}^2 + \ldots + \mathbf{B}^k \\
&= \mathbf{B}^0 + \mathbf{B}^1 + \ldots + \mathbf{B}^k - \mathbf{I} \\
&= (\mathbf{I} - \mathbf{B})^{-1} - \mathbf{I}
\end{aligned}
\tag{10.5.13}
$$

using (10.2.21) with $\mathbf{B}^0 \equiv \mathbf{I}$. And, the indirect effect of **y** on **y** is

$$
\begin{aligned}
\mathbf{I_{yy}} &= \mathbf{T_{yy}} = \mathbf{B} \\
&= (\mathbf{I} - \mathbf{B})^{-1} - \mathbf{I} - \mathbf{B}
\end{aligned}
\tag{10.5.14}
$$

To evaluate the effect of **x** on **y**, again assuming a recursive model, we proceed in a similar manner by expressing the total effect as the sum of direct plus indirect effects as

$$
\begin{aligned}
\mathbf{T_{xy}} &= \mathbf{D_{xx}} + \mathbf{I_{xy}} \\
&= \Gamma + \Gamma\mathbf{B} + \Gamma\mathbf{B}^2 + \\
&= \left[\mathbf{I} + \mathbf{B} + \mathbf{B}^2 + \ldots +\right] \Gamma \\
&= (\mathbf{I} - \mathbf{B})^{-1}\,\Gamma
\end{aligned}
\tag{10.5.15}
$$

which is identical to the coefficient matrix for the reduced form of the model. Hence, the indirect effect matrix of **x** on **y** is

$$
\begin{aligned}
\mathbf{I_{xy}} &= \mathbf{T_{xy}} - \Gamma \\
&= (\mathbf{I} - \mathbf{B})^{-1} - \Gamma
\end{aligned}
\tag{10.5.16}
$$

The coefficients for direct, indirect and total effects are summarized in Table 10.5.1

For some nonrecursive models, the indirect and total effects may not exist, Bollen (1989, p. 381). A sufficient condition for convergence of the infinite series for nonrecursive models is that the largest eigenvalue of $\mathbf{BB}'$ is less than one, or $||\mathbf{B}|| < 1$. By the singular value decomposition theorem, a necessary and sufficient condition for the inverse $(\mathbf{I} - \mathbf{B})^{-1}$ to exist is that the largest eigenvalue of **B** is less than one. While this ensures that a solution to the cross-sectional model exits, it does not mean that the system is stable or that it is in equilibrium. To evaluate equilibrium, one must consider a dynamic model.

TABLE 10.5.1. Path Analysis—Direct, Indirect and Total Effects

Component	$\mathbf{x} \longrightarrow \mathbf{y}$	$\mathbf{y} \longrightarrow \mathbf{y}$
Direct	Γ	$\mathbf{B}$
Indirect	$(\mathbf{I}-\mathbf{B})^{-1}\Gamma - \Gamma$	$(\mathbf{I}-\mathbf{B})^{-1} - \mathbf{I} - \mathbf{B}$
Total	$(\mathbf{I}-\mathbf{B})^{-1}\Gamma$	$(\mathbf{I}-\mathbf{B})^{-1} - \mathbf{I}$

The system of equations defined by (10.5.1) may be viewed as a cross-sectional analysis of a dynamic process and as such we are assuming that the system of equations is in equilibrium. To evaluate equilibrium, we consider a simple dynamic simultaneous equation model. A system is considered dynamic if it depends on one or more lagged endogenous variables. A simple dynamic simultaneous equation model has the structure

$$\mathbf{y}_t = \mathbf{B}\mathbf{y}_t + \Gamma\mathbf{x} + \Theta\mathbf{y}_{t-1} + \boldsymbol{\zeta} \tag{10.5.17}$$

where the lagged variable is $\mathbf{y}_{t-1}$ and Θ is a matrix of parameters. We have used the subscript t to denote that the observations vary with time. In more complicated dynamic models, the exogenous variables $\mathbf{x}$ and disturbance terms $\boldsymbol{\zeta}$ may also depend upon time. Letting only the errors in (10.5.17) depend on time and not the exogenous variables, the model is equivalent to a vector autoregressive AR(1) model, Reinsel (1993). The parameter matrix Θ may have diagonal elements that are nonzero. The canonical form for the dynamic system is as follows

$$\mathbf{y}_t = \Pi\mathbf{x} + \Delta\mathbf{y}_{t-1} + \mathbf{e} \tag{10.5.18}$$

where the matrix $\Pi = (\mathbf{I}-\mathbf{B})^{-1}\Gamma$, the matrix $\Delta = (\mathbf{I}-\mathbf{B})^{-1}\Theta$, and vector of errors is $\mathbf{e} = (\mathbf{I}-\mathbf{B})^{-1}\boldsymbol{\zeta}$. By successive substitution into (10.5.18) and using (10.2.22), the dynamic model may be written as follows

$$\begin{aligned}\mathbf{y}_t &= \Delta^t\mathbf{y}_0 + \sum_{s=0}^{t-1}[\Delta^s\Pi\mathbf{x}] + \sum_{s=0}^{t-1}\Delta^s\mathbf{e} \qquad (10.5.19)\\ &= \Delta^t\mathbf{y}_0 + (\mathbf{I}-\Delta^t)(\mathbf{I}-\mathbf{B})^{-1}\Gamma\mathbf{x} + (\mathbf{I}-\Delta^t)(\mathbf{I}-\mathbf{B})^{-1}\boldsymbol{\zeta}\end{aligned}$$

The model indicates that the endogenous observation vector is determined by only the initial vector $\mathbf{y}_0$ since the exogenous variables, and the disturbances do not vary with time. The coefficient matrices included in the bracketed sum are called dynamic multipliers. From the equation (10.5.19), we see that if $\Delta^t \rightarrow \mathbf{0}$ as $t \rightarrow \infty$ and if the matrix $(\mathbf{I}-\mathbf{B})^{-1}$ exists, the solution to the dynamic model is the reduced form solution of the cross-sectional path analysis, simultaneous equation, model given in (10.5.4) and the system is said to be in equilibrium or stable. Since the model considered in (10.5.17) is a vector autoregressive AR(1) process, equilibrium will be attained provided the eigenvalues of Θ are less than one, the stationarity condition in multivariate time series analysis, Reinsel (1993, p. 28). A Monte Carlo study of the behavior of the eigenvalues of $\mathbf{B}$ and Θ was conducted by Kaplan et al. (2000). Their study confirms that evaluation of the eigenvalues of the matrix $\mathbf{B}$ does not provide an accurate assessment of the existence of an equilibrium for a cross-sectional structural equation model. Because the eigenvalues of the matrix Θ are not available in cross-sectional models, one cannot directly evaluate the stability of the process.

Selecting a scale invariant fit function and assuming $\mathbf{x}$ is weakly exogenous for $\mathbf{B}$, Γ, and Ψ, one may estimate the asymptotic standard errors for the parameters. Also of interest may be the standard errors of indirect effects. These, unfortunately are not provided in CALIS, however, by establishing hierarchical models one may evaluate the contribution of variables to a model. The fit statistics reviewed for the CFA model may also be used to evaluate the fit of a path analysis model.

10.6 Path Analysis Examples

In CFA a SEM was used to specify relationships between latent, unobserved variables or factors and observed, manifest variables. In path analysis (PA) all variables are observed, manifest variables. And, a distinction is made between endogenous and exogenous variables. The endogenous (dependent) variables ($\mathbf{y}$) are predicted from variables within the model and the exogenous (independent) variables ($\mathbf{x}$) are only influenced by external variables that are outside the model. In recursive models, the paths are uni-directional while in nonrecursive models they are bi-directional.

a. *Community Structure and Industrial Conflict (Example 10.6.1)*

For our first example, data found in Fox (1984, p. 267), a subset of the variables from a study on community, structure and conflict conducted by Lincoln (1979) are utilized. Lincoln obtained data on the industrial strike activity and characteristics of the surrounding 78 communities. The observational data consisted of four exongenous variables (x_1, x_2, x_3, x_4) and three endogenous variables (y_1, y_2, y_3).

$x_1 - UCON$,	index of metropolitan union concentration (the higher the index, the higher the correlation of union workers employed in a small number of large labor organizations).
$x_2 - EMPCON$,	index of employment concentration in metropolitan areas (the higher the index, the more concentrated the employment).
$x_3 - SIZE$,	logarithm of total employed in the metropolitan area.
$x_4 - UNIZ$,	extent of unionization in the metropolitan area.
$y_1 - STRIKES$,	frequency of recent work stoppages.
$y_2 - STRIKERS$,	number of workers involved.
$y_3 - MANDAYS$,	number of person days lost.

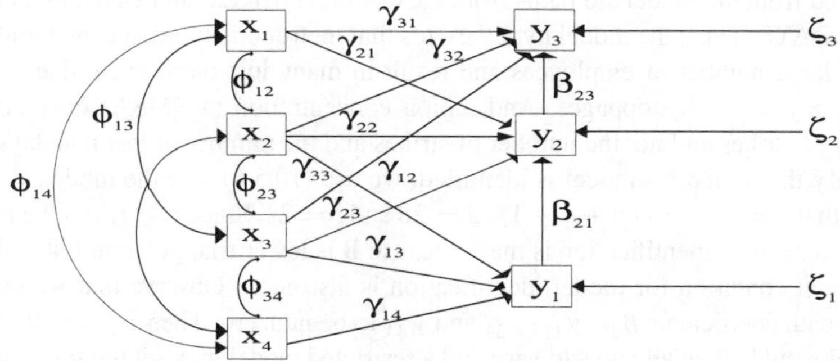

FIGURE 10.6.1. Lincoln's Strike Activity Model in SMSAs

Lincoln's proposed recursive model of strike activity in SMSAs is shown in Figure 10.6.1. The SEM equation for Figure 10.6.1 are

$$\begin{bmatrix} y_1 \\ y_2 \\ y_3 \end{bmatrix} = \begin{bmatrix} 0 & 0 & 0 \\ \beta_{21} & 0 & 0 \\ 0 & \beta_{32} & 0 \end{bmatrix} \begin{bmatrix} y_1 \\ y_2 \\ y_3 \end{bmatrix} +$$

$$\underset{p\times 1}{\mathbf{y}} = \underset{p\times p}{\mathbf{B}} \quad \underset{p\times 1}{\mathbf{y}} +$$

$$\begin{bmatrix} 0 & \gamma_{12} & \gamma_{13} & \gamma_{14} \\ \gamma_{21} & \gamma_{22} & \gamma_{23} & 0 \\ \gamma_{31} & \gamma_{32} & \gamma_{33} & 0 \end{bmatrix} + \begin{bmatrix} x_1 \\ x_2 \\ x_3 \\ x_4 \end{bmatrix} + \begin{bmatrix} \zeta_1 \\ \zeta_2 \\ \zeta_3 \\ \zeta_4 \end{bmatrix}$$

$$\underset{p\times q}{\Gamma} \qquad \underset{q\times 1}{\mathbf{x}} \qquad \underset{p\times 1}{\zeta}$$

and

$$\underset{q\times q}{\Phi} = \begin{bmatrix} \phi_{11} & \phi_{12} & \phi_{13} & \phi_{14} \\ & \phi_{22} & \phi_{23} & \phi_{24} \\ & & \phi_{33} & \phi_{34} \\ & & & \phi_{14} \end{bmatrix} \quad \underset{p\times q}{\Psi} = \operatorname{diag}\left[\psi_{11}, \psi_{22}, \psi_{33}\right]$$

where the general structure for $\Sigma(\boldsymbol{\theta})$ is given in (10.5.2).

The model hypothesizes that UNIZ affects the incidence of strikes while only UCON determines their size and duration. The variables EMPCON and SIZE affect all three conflict variables (STRIKES, STRIKERS, and MANDAYS). MANDAYS is mediated by STRIKERS where STRIKERS is an antecedent to STRIKES. Finally, one would expect all path coefficients to be positive and significant in the model.

Excluded from the model are paths from UCON to STRIKES, and UNIZ to STRIKERS and MANDAYS. Thus, the model hypothesizes that metropolitan union concentration may involve a large number of employees and result in many lost days, but it does not affect the frequency of work stoppages. And, union concentration in SMSAs only effects the frequency of strikes and not the number of strikes and the number of lost mandays.

To verify that Lincoln's model is identified, we use (10.5.6). For the model, $p = 3$ and $q = 4$ so that $v = (p + q)(p + q + 1)/2 = 28$ and $t = 24$. Since $v \geq t$, and the necessary condition for model identification is met. Because $\mathbf{B}$ is lower triangular and Ψ is diagonal, the sufficient condition for model identification is also met. Observe that we could also allow the path coefficients $\beta_{31}, \gamma_{11}, \gamma_{24}$ and γ_{34} to be nonzero. Then $v - t = 0$. Hence, a full model would allow all paths to vary and a restricted model may set the four parameters to zero. When $v = 0$, $\mathbf{S} = \widehat{\Sigma}$ for an exactly identified model. The covariance matrix Σ for Lincoln's strike activity data follows.

$$S = \begin{bmatrix} 0.007744 & & & & & & (\text{Sym}) \\ 0.000635 & 0.000400 & & & & & \\ 0.052401 & 0.005077 & 1.065024 & & & & \\ 0.006624 & 0.001471 & 0.066069 & 0.037636 & & & \\ 0.054564 & 0.012024 & 0.823108 & 0.137249 & 1.809025 & & \\ 0.084675 & 0.015990 & 1.131609 & 0.171958 & 2025220 & 2.496400 & \\ 0.103616 & 0.019572 & 1.325756 & 0.184820 & 1.969703 & 2.567911 & 2.989441 \end{bmatrix}$$

Rather than fitting Lincoln's proposed model, we first fit a model with all elements in $\Gamma = [\gamma_{ij}]$ specified. The model is

$$\begin{aligned} \begin{bmatrix} y_1 \\ y_2 \\ y_3 \end{bmatrix} &= \begin{bmatrix} 0 & 0 & 0 \\ \beta_{21} & 0 & 0 \\ 0 & \beta_{32} & 0 \end{bmatrix} \begin{bmatrix} y_1 \\ y_2 \\ y_3 \end{bmatrix} + \\ \mathbf{y} &= \quad \mathbf{B} \quad \mathbf{y} + \end{aligned}$$

$$\underset{\Gamma}{\begin{bmatrix} \gamma_{11} & \gamma_{12} & \gamma_{13} & \gamma_{14} \\ \gamma_{21} & \gamma_{22} & \gamma_{23} & \gamma_{24} \\ \gamma_{31} & \gamma_{32} & \gamma_{33} & \gamma_{34} \end{bmatrix}} \underset{\mathbf{x}}{\begin{bmatrix} x_1 \\ x_2 \\ x_3 \\ x_4 \end{bmatrix}} + \underset{\zeta}{\begin{bmatrix} \zeta_1 \\ \zeta_2 \\ \zeta_3 \end{bmatrix}} \tag{10.6.1}$$

$$\Phi = [\phi_{ij}] \, \Psi = \text{diag}[\psi_i]$$

For this model, the exogenous x_i are associated with each of the endogenous variables y_i. To analyze the model with all γ_{ij} free to vary using PROC CALIS, the variables are relabeled to specify the LINEQS statements in SAS following the notation of Bentler and Weeks (1980). For details see Hatcher (1994). All variables are relabeled vi where pvivj represents a path coefficient. The letter "c" represents model errors covariances in the COV statement and the letter "e" represents model errors. The CALIS model for the SEM in (10.6.2) is shown in Figure 10.6.2. All variables x_i [v 4, v 5, v 6, and v 7] are related to y_i [v 1, v 2, and v 3]. From Figure 10.6.2, one may construct the LINEQS, STD, and COV statements for the "Full Gamma Matrix" model, program m10_6_1.sas.

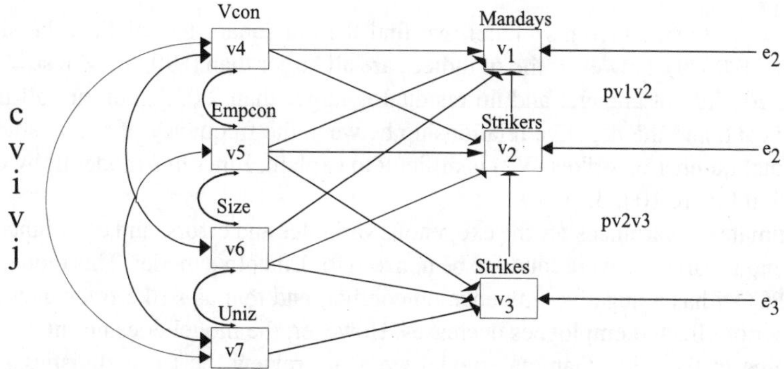

FIGURE 10.6.2. CALIS Model for Eq. (10.6.2).

The steps one employs to evaluate a PA model are similar to the process employed in CFA. One would like to have nonsignificant chi-square and Z tests of fit. Fit indices larger than 0.9, and the $|r_{ij}^*| < 2.0$ for all normalized residuals. Reviewing the fit of our model, the p-value of the chi-square statistic is 0.4803, $Z = 0.0316$, all indices of fit are larger than 0.90 and the $|r_{ij}^*| < 2.00$ with all normalized residuals nearly zero.

Next, the R^2 values of the endogenous variables are reviewed. These measure the proportion of variance accounted for by direct effects of the exogenous variables. The values of R^2 for y_i (v_1, v_2, v_3) are 0.55, 0.40, and 0.57 for each set of x_i (v_4, v_5, v_6, v_7). For the social sciences, these values are reasonable. However, not all the asymptotic t tests that the path coefficients are nonzero are significant, some of the standardized coefficients are moderate in magnitude (< 0.05), and the stepwise multivariate Wald tests indicate that some of the paths may be set to zero.

When trying to fit a SEM in practice, as a general principle, it is usually better to remove paths than to add paths. The stepdown Wald tests estimate the change in the chi-square statistic of model fit that would result by setting a parameter to zero. The Univariate Increment chi-square values represent the change in the chi-square statistic that results when setting a parameter to zero, ordered from lowest to highest. Also provided is the cumulative chi-square effect. Equating a parameter to zero may not improve model fit. It may increase the chi-square statistic so that a model which once fit, may no longer fit. Conversely, if the ratio of the chi-square statistic to its degrees of freedom is reduced, a significant chi-square test of fit may become nonsignificant.

We began with the "Full Gamma" model to see if we can reduce the paths to discover Lincoln's model. This is not the case since one would have to set the paths $PV3V4 = PV1V7 = PV2V7 = 0$. And, while many coefficients are candidates, $PV2V6$, $PV1V4$, $PV1V2$, $PV2V4$, $PV2V5$, $PV3V4$, $PV3V6$, $PV3V5$, only $PV3V4$ is among the entries. The incremental chi-square value of 1.586, indicates a very small change in chi-square. While the "Full Gamma" model provides a "good" fit, it is difficult to interpret the standardized coefficients. In particular, the inverse relationship between union concentration ($V4 \equiv x1$) on the number of recent strikes ($V3 \equiv y3$) and the inverse relationship among the endogenous variables. Setting the coefficients $PV3V4 = PV1V7 =$

$PV2V7 = 0$, to fit Lincoln's model, we find the chi-square test of fit to be significant (p-values = 0.0029). However, the fit indices are all larger than 0.90, the Z test = 2.94, the values of R^2 did not change, and no residual is larger than 2.0. Again, not all the t tests are significant, and the negative relationship between the frequency of recent studies (V3) and the total number of strikes (V2) is difficult to explain. Lincoln's model fit by CALIS is displayed in Figure 10.6.3.

All estimates of variances for the exogenous variables and errors, and covariances among the exogenous variables were found to be nonzero for Lincoln's model. This model suggests that EMPCON has a negative influence on conflict, and that as strike frequency increases the number of effected employees decreases. However, the model does not fit.

Returning to the "Full Gamma" model we again review the t test statistics for values less than 1.96, the magnitude of the standardized path coefficients, and the Wald tests. This may lead one to set $PV1V4 = 0$ in equation one, $PV2V4 = PV2V5 = PV2V6 = 0$ in equation two, and $PV3V4 = PV3V5 = PV3V6 = 0$ equation three. This revised model does fit (chi-square p-value = 0.5690), $|Z| = .1776$, and all fit indices remained larger than 0.90. However, some normalized residuals increased in size, most coefficients increased in size, although not all t tests are larger than 1.96. The R^2 values remained unchanged (0.55, 0.39, 0.55) and the information criteria are smaller than either the "Full Gamma" model or Lincoln's model. Using the CALIS notation, the standardized coefficients are given in Table 10.6.1.

Does the model represented in Table 10.6.1 make sense? Can we interpret our findings and substantiate our results in theory? For this, one investigates the magnitude of the direct and indirect effects, and the signs of the coefficients. The direct effect signs are provided in Figure 10.6.4.

While the revised model in Figure 10.6.4 may provide a reasonable fit, it is very difficult to interpret. This is a frequent problem when analyzing observational data with causal paths not determined by experimentation. Do not make something out of nothing. Obtain another sample, revise your theory and began again.

b. *Nonrecursive Model (Example 10.6.2)*

As a second example of a PA model, an example discussed by Bollen (1989, p. 116) analyzed using LISREL™ and EQS™ is considered. The variables in the model follow.

y_1 — subjective income

y_2 — subjective occupational prestige

y_3 — subjective overall social status

x_1 — income

x_2 — occupational prestige

TABLE 10.6.1. CALIS OUTPUT—Revised Model

$$
\begin{aligned}
V1 &= -0.05 * V2 + 0.04 * V4 - 1.03 * V5 + 1.110 * V6 \\
&\quad + 0.53 * V7 + 0.6698E1 \\
V2 &= -0.39 * V3 + 0.86 * V7 + 0.7821E2 \\
V3 &= 0.74 * V7 + 0.6693E3
\end{aligned}
$$

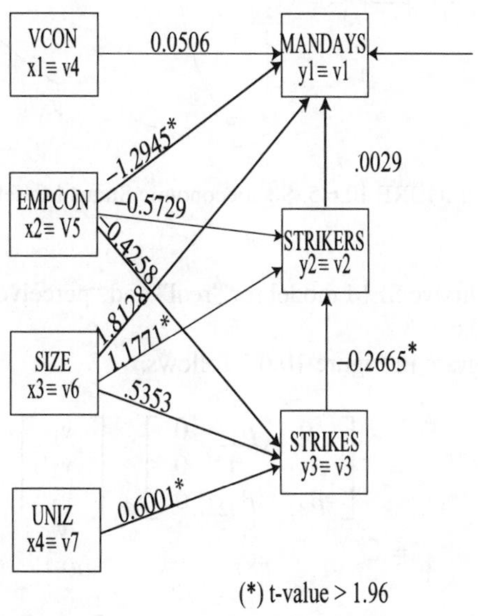

FIGURE 10.6.3. Lincoln's Standardized Strike Activity Model Fit by CALIS.

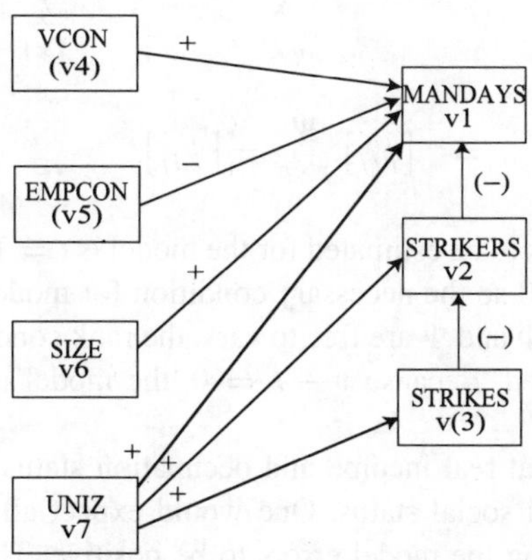

FIGURE 10.6.4. Revised CALIS Model with Signs

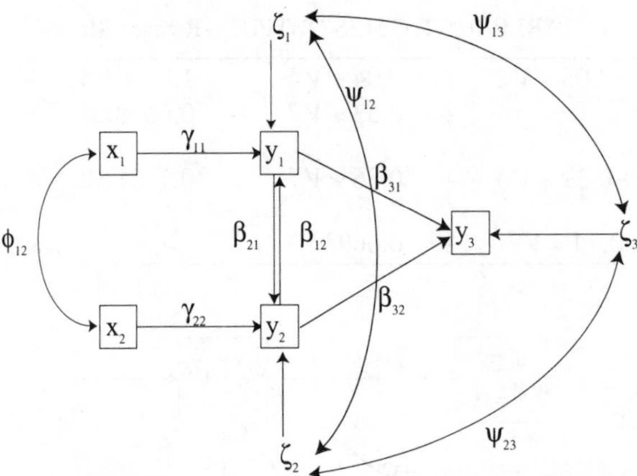

FIGURE 10.6.5. Socioeconomic Status Model

The proposed nonrecursive SEM model for "real" and "perceived" socioeconomic status is provided in Figure 10.6.5.

The SEM for the diagram in Figure 10.6.5 follows.

$$
\begin{aligned}
\underset{3\times 1}{\begin{bmatrix} y_1 \\ y_2 \\ y_3 \end{bmatrix}} &= \begin{bmatrix} 0 & \beta_{12} & 0 \\ \beta_{21} & 0 & 0 \\ \beta_{31} & \beta_{32} & 0 \end{bmatrix} \begin{bmatrix} y_1 \\ y_2 \\ y_3 \end{bmatrix} + \\
\underset{3\times 1}{\mathbf{y}} &= \underset{p\times p}{\mathbf{B}} \quad \underset{p\times 1}{\mathbf{y}} \\
&\begin{bmatrix} \gamma_{11} & 0 \\ 0 & \gamma_{22} \\ 0 & 0 \end{bmatrix} \begin{bmatrix} x_1 \\ x_2 \end{bmatrix} + \begin{bmatrix} \zeta_1 \\ \zeta_2 \\ \zeta_3 \end{bmatrix} \\
&\quad\ \underset{p\times q}{\Gamma} \qquad\quad \underset{q\times 1}{\mathbf{x}} \qquad\ \underset{1\times 1}{\zeta} \\
\underset{q\times q}{\Phi} &= [\phi_{ij}] \ \underset{p\times p}{\Psi} = [\psi_{ij}]
\end{aligned}
\tag{10.6.2}
$$

The number of parameters to be estimated for the model is $t = 15$ and as $v = (p+q)(p+q+1)/2 = 15$, $v - t \geq 0$ so the necessary condition for model identification is satisfied. Since all the elements of Φ and Ψ are free to vary, the rank condition may be used to verify that the model is identified. Because $v - t = 0$, the model is exactly identified so that $\widehat{\Sigma} = \mathbf{S}$.

The model suggests that real income and occupation status (which affect each other), mediate perceived overall social status. One would expect all coefficients to be positive and the correlations among the model errors to be positive. The covariance matrix based on a sample of $N = 432$ white respondents is used for the analysis. The PROC CALIS commands for the analysis are given in program m_10_2.sas. The sample covariance matrix

TABLE 10.6.2. Revised Socioeconomic Status Model

Equation										R²
y_1	=	0.3829	x_1	+	0.9238	ζ_1				0.15
y_2	=	0.3078	y_1	+	0.2293	x_2	+	0.9003	ζ_2	0.19
y_3	=	0.4421	y_1	+	0.5894	y_2	+	0.8753	ζ_3	0.23

for the variables y_1, y_2, y_3, x_1, and x_2 is

$$\mathbf{S} = \begin{bmatrix} 0.499 & & & & (\text{Sym}) \\ 0.166 & 0.410 & & & \\ 0.226 & 0.173 & 0.393 & & \\ 0.564 & 0.259 & 0.382 & 4.831 & \\ 2.366 & 3.840 & 3.082 & 13.656 & 452.711 \end{bmatrix}$$

The results obtained by CALIS are identical to the results published by Bollen (1989, p. 117) using LISREL ™ and EQS™. Because $v - t = 0$, there is no test of model fit. Reviewing the t tests for the fitted model, observe that the coefficients β_{12} $(PV1V2)$, ψ_{21} $(CE1E2)$, ψ_{31} $(CE1E3)$, and ψ_{32} $(CE2E3)$ have asymptotic t values that are less than 1.96 and that the correlations among the model errors are negative, contrary to what one might expect. In addition, the stepwise Wald Test suggests setting $\beta_{12} = 0$.

Rerunning the model setting $\beta_{12} = 0$, the new model is recursive, suggesting that subjective income effects subjective occupational prestige, but not conversely. We continue to allow the errors to be correlated. The new model continues to fit the data and the $\left|r^*_{ij}\right| < 2$. Now, all t tests for the path coefficients are larger than 1.96, and all standardized coefficient are positive. However, the covariances of the model errors remain negative and nonsignificant. The standardized coefficients for the revised model are given in Table 10.6.2.

Finally, we consider setting $\psi_{13} = \psi_{12} = \psi_{23} = 0$. This model does not fit the data as well as the revised model; however, all t tests are significant and the magnitude of the standardized coefficients remained reasonably stable. In addition, the cross validation index for the two models, are not very different, the smaller the index the better the prediction. We would recommend replicating the study with another data set before deciding upon a model for socioeconomic status.

Exercises 10.6

1. Delete from Lincoln's Strike Activity Model variables X_4 $(UNIZ)$ and X_2 $(EMPCON)$.

   (a) Draw the revised path analysis diagram.

   (b) Fit the model to the resulting covariance matrix.

   (c) Discuss your findings.

10.7 Structural Equations with Manifest and Latent Variables

We have reviewed two special applications of the SEM, the CFA model and the path analysis (PA) model. In this section we treat the general model which contains both a structural model and a measurement model. The structural model is identical to the path analysis model except that manifest variables are replaced with latent variables. For the measurement submodel, a CFA model is formulated for both manifest variables **y** and **x**. Recall that the general specification of the model is

$$\begin{aligned} \underset{m\times 1}{\boldsymbol{\eta}} &= \underset{m\times m}{\mathbf{B}}\ \underset{m\times 1}{\boldsymbol{\eta}} + \underset{m\times n}{\Gamma}\ \underset{n\times 1}{\boldsymbol{\xi}} + \underset{m\times 1}{\boldsymbol{\zeta}} \\ \underset{p\times 1}{\mathbf{y}} &= \underset{p\times m}{\Lambda_y}\ \underset{m\times 1}{\boldsymbol{\eta}} + \underset{p\times 1}{\boldsymbol{\epsilon}} \\ \underset{q\times 1}{\mathbf{x}} &= \underset{q\times n}{\Lambda_x}\ \underset{n\times 1}{\boldsymbol{\xi}} + \underset{q\times 1}{\boldsymbol{\delta}} \end{aligned} \tag{10.7.1}$$

where $\Sigma(\boldsymbol{\theta})$ is defined in (10.2.6). The parameters for the model are $\mathbf{B}$, Γ, Λ_y, Λ_x, Φ, Ψ, Θ_ϵ, and Θ_δ.

To determine whether the model is identified, we must establish a sufficient condition for identification since the necessary condition was given in (10.2.7). For this, the model is separated into the measurement and structural models. Letting $\mathbf{x}^* = [\mathbf{y}', \mathbf{x}']'$, the measurement model has the CFA structure

$$\mathbf{x}^* = \begin{bmatrix} \Lambda_y & \mathbf{0} \\ \mathbf{0} & \Lambda_x \end{bmatrix} \begin{bmatrix} \xi_1 \\ \xi_2 \end{bmatrix} + \begin{bmatrix} \boldsymbol{\epsilon} \\ \boldsymbol{\delta} \end{bmatrix} \tag{10.7.2}$$

with parameters Φ, Λ_y, and Λ_x. Using the CFA rules one may establish the identification of this submodel.

Next, one examines the structural model with $\boldsymbol{\xi} = \mathbf{x}$ and $\boldsymbol{\eta} = \mathbf{y}$ to determine whether $\mathbf{B}$, Γ and Ψ are identified using path analysis rules. If each submodel results in an identified model, then the entire model is identified. When the two step rule fails or the specified model does not meet the CFA and path analysis rules, one must use the rank of the score matrix and Fisher's information matrix to evaluate global identification empirically using local identification with several starting values.

Having found that a SEM is identified, one again chooses a fit function to estimate $\boldsymbol{\theta}$, and uses indices to evaluate fit. Replacing **x** with $\boldsymbol{\xi}$ and **y** with $\boldsymbol{\eta}$ in Table 10.5.1, the matrix expressions may be used to determine the direct, indirect and total effects for the latent variables.

To determine the total and indirect effects of $\boldsymbol{\xi}$ on **y**, observe that the direct effects of $\boldsymbol{\xi}$ on **y** are **0**. Hence, the indirect effects of $\boldsymbol{\xi}$ on **y** are equal to the total effects so that

$$\mathbf{T}_{\xi\mathbf{y}} = \mathbf{I}_{\xi\mathbf{y}} = \Lambda_y (\mathbf{I} - \mathbf{B})^{-1} \Gamma \tag{10.7.3}$$

Similarly, the total effects of $\boldsymbol{\eta}$ on $\mathbf{y}$ is

$$\begin{aligned}\mathbf{T}_{\eta\mathbf{y}} &= \Lambda_y + \Lambda_y\mathbf{B} + \Lambda_y\mathbf{B}^2 + \ldots + \Lambda_y\mathbf{B}^k \\ &= \Lambda_y\left[\mathbf{B}^0 + \mathbf{B} + \mathbf{B}^2 + \ldots + \mathbf{B}^k\right] \\ &= \Lambda_y\left(\mathbf{I} - \mathbf{B}\right)^{-1}\end{aligned} \tag{10.7.4}$$

and the indirect effects of $\boldsymbol{\eta}$ on $\mathbf{y}$ is

$$\mathbf{I}_{\eta\mathbf{y}} = \Lambda_y\left(\mathbf{I} - \mathbf{B}\right)^{-1} - \Lambda_y \tag{10.7.5}$$

for recursive models. For nonrecursive models, the effects are only defined under the conditions specified for path analysis models.

10.8 Structural Equations with Manifest and Latent Variables Example

For our next example in this chapter, we consider a longitudinal study of the stability of alienation over time and its relationship to social economic status discussed by Jöreskog (1977) based upon a study by Wheaton, Muthén, Alevin and Summers (1977). Data on 932 individuals were collected in rural Illinois for three years: 1969, 1967, and 1971. The latent variables under study were alienation and social economic status (SES). The indicator variables of alienation consisted of two subscale scores of anomia and powerlessness obtained by Wheaton et al. (1977). For the SES factor, the indicator variables were years of schooling (EDUC) and Duncan's SES index (SEI) administered in 1966. Jöreskog (1977) proposed two models for analysis of alienation and SES, Models A and B. Neither model corresponded to Wheaton et al.'s final model. The models are represented in Figure 10.8.1 for data collected in 1967 and 1971.

The SEM for both models A and B are

$$\begin{bmatrix}\eta_1\\ \eta_2\end{bmatrix} = \begin{bmatrix}0 & 0\\ \beta_{21} & 0\end{bmatrix}\begin{bmatrix}\eta_1\\ \eta_2\end{bmatrix} + \begin{bmatrix}\gamma_{11}\\ \gamma_{21}\end{bmatrix}\xi + \begin{bmatrix}\zeta_1\\ \zeta_2\end{bmatrix}$$

$$\underset{m\times 1}{\boldsymbol{\eta}} = \underset{m\times m}{\mathbf{B}}\;\underset{m\times 1}{\boldsymbol{\eta}} + \underset{m\times n}{\Gamma}\;\underset{n\times 1}{\boldsymbol{\xi}} + \underset{m\times 1}{\boldsymbol{\zeta}}$$

$$\begin{bmatrix}y_1\\ y_2\\ y_3\\ y_4\end{bmatrix} = \begin{bmatrix}1 & 0\\ \lambda_{21} & 0\\ 0 & 1\\ 0 & \lambda_{42}\end{bmatrix}\begin{bmatrix}\eta_1\\ \eta_2\end{bmatrix} + \begin{bmatrix}\epsilon_1\\ \epsilon_2\\ \epsilon_3\\ \epsilon_4\end{bmatrix} \tag{10.8.1}$$

$$\underset{p\times 1}{\mathbf{y}} = \underset{p\times m}{\Lambda_y}\;\underset{m\times 1}{\boldsymbol{\eta}} + \underset{p\times 1}{\boldsymbol{\epsilon}}$$

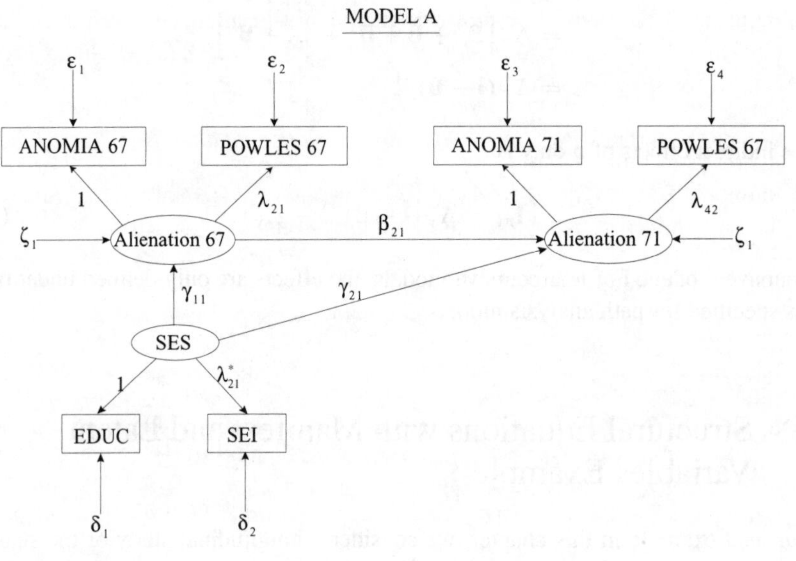

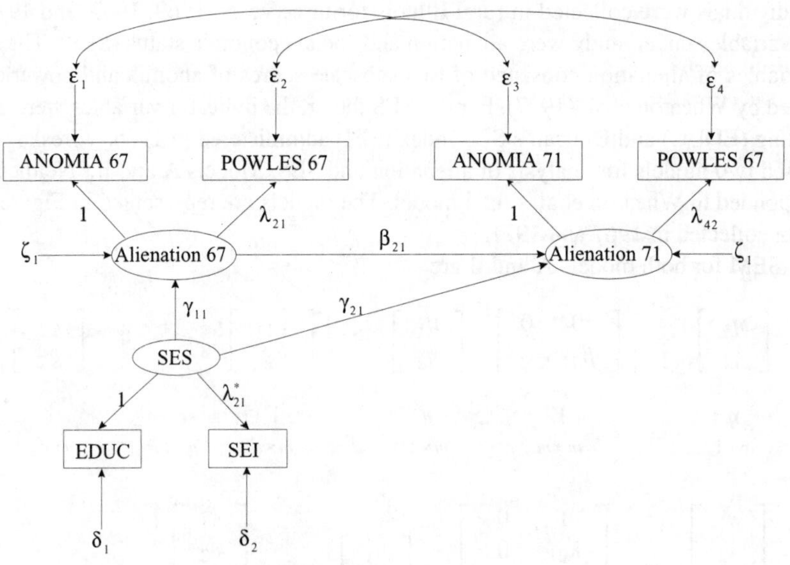

FIGURE 10.8.1. Models for Alienation Stability

$$\begin{bmatrix} x_1 \\ x_2 \end{bmatrix} = \begin{bmatrix} 1 \\ \lambda_{21}^* \end{bmatrix} \xi + \begin{bmatrix} \delta_1 \\ \delta_2 \end{bmatrix}$$

$$\underset{q\times 1}{\mathbf{x}} = \underset{p\times n}{\Lambda_x} \underset{n\times 1}{\boldsymbol{\xi}} + \underset{q\times 1}{\boldsymbol{\delta}}$$

The covariance structures for the models differ. For Model A, $\Phi = \left[\phi_{11}\right] = \text{var}\,\xi_1$, $\Psi = \text{diag}\left[\psi_{11}, \psi_{22}\right]$ where $\psi_{ii} = \text{var}\left(\zeta_i\right)$, and $\Theta_\epsilon = \text{diag}\left[\theta_{ii}\right]$, $\Theta_\delta = \text{diag}\left[\delta_i\right]$ For model B,

$$\Theta_\delta = \begin{bmatrix} \theta_{11} & & & (Sym) \\ 0 & \theta_{22} & & \\ \theta_{31} & 0 & \theta_{33} & \\ 0 & \theta_{42} & 0 & \theta_{44} \end{bmatrix}$$

since the ϵ_i are correlated. The structure of Φ, Ψ, and Θ_ϵ for Model B is identical to the specification given for Model A. Scales for η_1, η_2 and ξ_1 have been set to y_1, y_3, and x_1, respectively. For the model, the observed variables are

$$\begin{aligned} y_1 &= ANOMIA67 & x_1 &= EDUC \\ y_2 &= POWLES67 & x_2 &= SEI \\ y_3 &= ANOMIA71 & & \\ y_4 &= POWLES71 & & \end{aligned}$$

while the latent factors are

$$\begin{aligned} \eta_1 &= \text{Alienation 67} \\ \eta_2 &= \text{Alienation 71} \\ \xi_1 &= SES \end{aligned}$$

The CALIS program to analyze both models is contained in program m10_8_1.sas using the covariance matrix

$$\mathbf{S} = \begin{bmatrix} 11.839 & & & & & (\text{Sym}) \\ 6.947 & 9.364 & & & & \\ 6.819 & 5.091 & 12.532 & & & \\ 4.783 & 5.028 & 7.495 & 9.986 & & \\ -3.834 & -3.889 & -3.841 & -3.625 & 9.60 & \\ -21.899 & -18.831 & -21.748 & -18.775 & 35.522 & 450.283 \end{bmatrix}$$

for the variables y_1, y_2, y_3, y_4, x_1 and x_2.

To determine whether Model A and Model B are identified, the number of observed variances and covariance, $v = (p+q)(p+q+1)/2$ must be greater than or equal to the number of unknown parameters, t_A and t_B, say. For Model A, $t_A = 15$ and for Model B, $t_B = 17$. Since $v = 21$, the necessary condition for model identification is met. Next, one may equate $\Sigma(\boldsymbol{\theta})$ to $\mathbf{S}$ and see if one may estimate all parameters using the observed covariances or one may determine sufficiency in two steps. That is, determine whether the CFA model is identified; and, treating the latent model as a PA model, determine whether it is identified. Using Bollen's Rule 1, the CFA model for Model A is identified. Using Rule 2, the CFA for Model B is identified. Treating the latent model as a PA model in step two, the model is recursive because $\mathbf{B}$ is lower triangular. Both Models are hence identified. The CALIS statements for the analysis of Models A and B, and an alternative Model C are provided in program m10_8_1.sas.

Reviewing the output for Model A, the chi-square statistic of fit, $X^2 = 71.4715$ with $df = 6$, has a p-value < 0.0001 indicating that perhaps the model does not fit. The Z test is also significant. Even though the fit indices are reasonable, the information measures appear large. However, no normalized residuals appear large, although the distribution is skewed. All asymptotic t tests for all covariances and path coefficients are larger than 1.96.

Using the MODIFICATION option, no stepwise Wald tests were generated so that one is not provided with any guidance for deleting paths to "improve" fit. However, the Lagrange Multiplier/Wald Indices in the _PHI_ matrix suggests that one consider adding the covariance parameter θ_{31} to the model. The value of the chi-square statistic is expected to decrease by 63.6999. This path is justified since it represents the covariance between ANOMIA 67 and ANOMIA 71 collected over time on the same subjects. This type of covariance is frequently present in repeated measures data. One might also consider the covariance between POWLESS 67 and POWLESS 71. The expected reduction in chi-square for θ_{42} is 37.2661. This change ensures some model symmetry, and it makes sense. Alas, this is Model B. Note that θ_{12} and θ_{34} are not included, but that θ_{14} and θ_{23} are among the 10 largest entries.

The chi-square statistic for Model B $\left(X^2 = 4.7391 \text{ with } df = 4\right)$ has a p-value of 0.3135. The new model is not rejected and all information criteria for Model B are smaller than Model A. The normalized residuals are well behaved and all asymptotic t tests are significant. However, the stepwise Wald Test suggests setting $\theta_{24} = 0$. This is Model C. Model C is nested within Model B.

The chi-square statistic for Model C $\left(X^2 = 6.3437 \text{ with } df = 5\right)$ has a p-value $= 0.2742$. The information statistics suggest that perhaps Model C is better than A. However, the average of the normalized residuals is marginally larger (0.1722 vs 0.111) for B vs C. A major problem with Model C is that it does not provide the symmetry in the correlated errors which may be more difficult to explain. Finally, if we assume Model B is correct, and because Model C is nested within Model B, we may use the chi-square difference test in (10.4.6) to compare Model B to Model C. The chi-square statistic is $(6.3437 - 4.7391) = 1.60$ with $(5 - 4) = 1$ degrees of freedom. Thus, we fail to reject Model B. We thus, conclude that Model B is a reasonable model for the stability of Alienation.

Exercises 10.8

1. In Figure 10.8.1 (Model A), suppose we remove the background variable SES from the model. What can you say about the estimate for β_{21}?

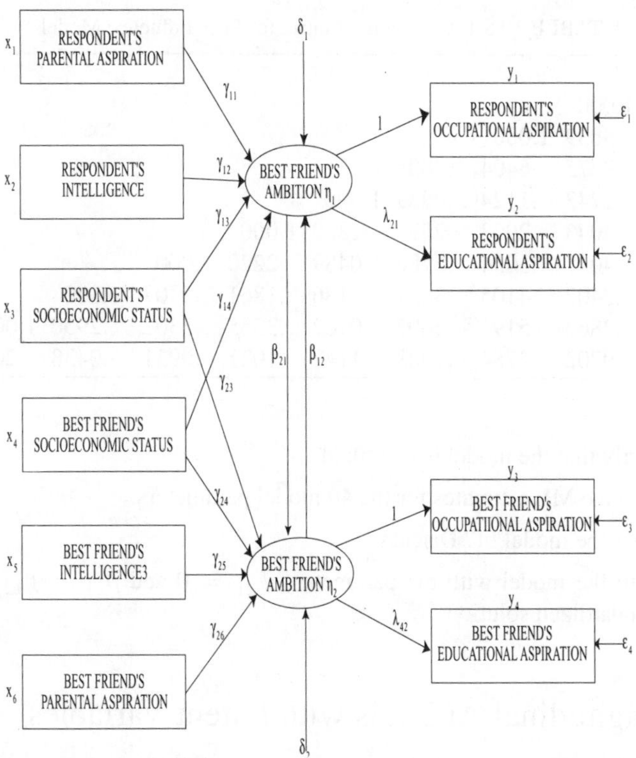

FIGURE 10.8.2. Duncan-Haller-Portes Peer-Influence Model

2. Refit model B in Figure 10.8.1 assuming all errors ϵ_i are correlated. Discuss your findings.

3. Duncan, Haller and Portes (1968) obtained data on $n = 329$ high school students paired with their best friends to analyze peer influence on occupational choice. They assumed that the latent variables η_1 = Respondent's Ambition and η_2 = Best Friend's Ambition were nonrecursive factors related to observed indicators of choice: y_1 = Respondent's occupational aspiration score, y_2 = Respondent's educational aspiration score, y_3 = Best friend's occupational aspiration score, and y_4 = Best friend's educational aspiration. Furthermore, that assumed that the observed variables parental aspiration, intelligence, and socioeconomic status directly influenced the latent ambition factors. That is $\xi_i \equiv x_i$ so that $\Lambda_x = \mathbf{I}$ and $\boldsymbol{\delta} = \mathbf{0}$. The path diagram for their SEM is given in Figure 10.8.2 and the correlations among the peer-influence is given in Table 10.8.1

   (a) Briefly discuss the model.

   (b) Use SEM notation to express the model diagram in Figure 10.6.5 in matrix notation.

TABLE 10.8.1. Correlation Matrix for Peer-Influence Model

y_1	1.000									(Sym)
y_2	.6247	1.000								
y_3	.3269	.3639	1.000							
y_4	.4210	.3275	.6404	1.000						
x_1	.2137	.2742	.1124	.0839	1.000					
x_2	.4105	.4043	.2903	.2599	.1839	1.000				
x_3	.3240	.4047	.3054	.2786	.0489	.2220	1.000			
x_4	.2930	.2407	.4105	.3607	.0186	.1861	.2707	1.000		
x_5	.2995	.2863	.5191	.5007	.0782	.3355	.2302	.2950	1.000	
x_6	.0760	.0702	.2784	.1988	.1147	.1021	.0931	−.0438	.2087	1.000

(c) Verify that the model is identified.

(d) Find the ML estimates for the 40 model parameters.

(e) Does the model fit? Discuss.

(f) Refit the model with the parameters $\psi_{21} = 0$ and $\beta_{21} = \beta_{12}$. Interpret the standardized solution.

10.9 Longitudinal Analysis with Latent Variables

In Chapter 6 we illustrated how one may use the mixed (univariate) linear model to analyze mixed ANOVA designs and to fit growth curves. For $i = 1, 2, \ldots, n$ units, the general model is

$$\mathbf{y}_i = \mathbf{X}_i \boldsymbol{\beta} + \mathbf{Z}_i \mathbf{a}_i + \mathbf{e}_i \tag{10.9.1}$$

To use (10.9.1) to model growth, suppose the relationship between y and x is linear, then for $i = 1, 2, \ldots, n$ and $t = 1, 2, \ldots, T$ we may have that

$$y_{it} = \beta_{0i} + \beta_{1i} x_{it} + e_{it}$$

The variables x_{it} are observed time points for example age or grade levels. Adding a time varying covariate for each subject, we may have that

$$y_{it} = \beta_{01} + \beta_{1i} x_{it} + \theta_i w_{it} + e_{it} \tag{10.9.2}$$

The parameter θ_i varies with individuals. This is the within-subject or level-1 model. For the between-subject or level-2 model, the random coefficients are modeled. For example, we might say that

$$\begin{aligned} \beta_{0i} &= \alpha_{00} + \gamma_{01} z_i + u_{0i} \\ \beta_{1i} &= \alpha_{10} + \gamma_{11} z_i + u_{1i} \\ \theta_i &= \alpha + \gamma z_i + u_i \end{aligned} \tag{10.9.3}$$

Combining (10.9.2) and (10.9.3), observe that

$$\begin{aligned} y_{it} &= \left[\left(\alpha_{00} + \gamma_{01} z_i\right) + \left(\alpha_{10} + \gamma_{11} z_i\right) x_{it} + \left(\alpha + \gamma z_i\right) w_{it}\right] \\ &\quad + \left[u_{0i} + u_{1i} x_{it} + u_i z_i + e_{it}\right] \\ &= \left[\text{fixed part}\right] + \left[\text{random part}\right] \end{aligned}$$

which is a mixed linear model. This model allows covariates to vary with time and individuals, and not all individuals need to be observed at the same time points. Letting y_{ijk} be the observation for the ith individual, at time t and variable k where $\delta_i = 1$ if an observation is obtained on y_k and $\delta_i = 0$, otherwise, then a multivariate multilevel model for change as proposed by Goldstein (1995) becomes

$$\begin{aligned} y_{it} &= \sum_k \delta_k \{\left[\left(\alpha_{00} + \gamma_{01} z_i\right) + \left(\alpha_{10} + \gamma_{11} z_i\right) x_{it} + \left(\alpha + \gamma z_i\right) w_{it}\right] \\ &\quad + \left[u_{0i} + u_{1i} x_{it} + u_i z_i + e_{it}\right]\} \\ &= \sum_k \{\left[\text{fixed part}\right] + \left[\text{random part}\right]\} \end{aligned}$$

To fit the multilevel model into a SEM framework, one first sets $\mathbf{X}_i = \mathbf{X}$ so that each individual is observed at the same time points (not necessarily equally spaced). This was the case for the GMANOVA model. The design is balanced for all individuals. Thus, we have that

$$y_{it} = \beta_{0i} + \beta_{1i} x_t + e_{it}$$

However, not all x_t need to be fixed. For example, x_t may have the values 0 and 1 and the values of x_3 and x_4 may be parameters to be estimated. This permits one to model nonlinear growth. In multilevel modeling, the parameter θ varied across individuals. A multilevel model that permits measurement error has been proposed by Longford (1993), however, SAS does not currently have procedures for the analysis of multilevel models with measurement error. However, one may fit the multilevel model within the SEM framework.

To fit the multilevel model into the SEM framework, the parameter θ is permitted to vary only with time and not individuals. In multilevel modeling, the parameter θ could vary with both individuals and time because z_{it} is always considered fixed. The measurement model becomes

$$y_{it} = \beta_{01} + \beta_{1i} x_t + \theta_t w_{it} + \epsilon_{it} \tag{10.9.4}$$

The equations given by (10.9.3) become the structural part of the SEM

$$\begin{aligned} \beta_{0i} &= \alpha_{00} + \gamma_{01} z_i + u_{0i} \\ \beta_{1i} &= \alpha_{10} + \gamma_{11} z_i + u_{1i} \end{aligned} \tag{10.9.5}$$

In the SEM setup, θ depends on time and not the ith individual. It is a fixed parameter used to evaluate the contribution of the time varying covariate z_{it} on the outcome variable, beyond the growth factors β_{0i} and β_{1i}. Taking expectation of the growth factors, the estimated growth means are

$$\begin{aligned} \widehat{E}\left(\beta_{01}\right) &= \widehat{\alpha}_{00} + \widehat{\gamma}_{01} \overline{z} \\ \widehat{E}\left(\beta_{1i}\right) &= \widehat{\alpha}_{10} + \widehat{\gamma}_{11} \overline{z} \end{aligned}$$

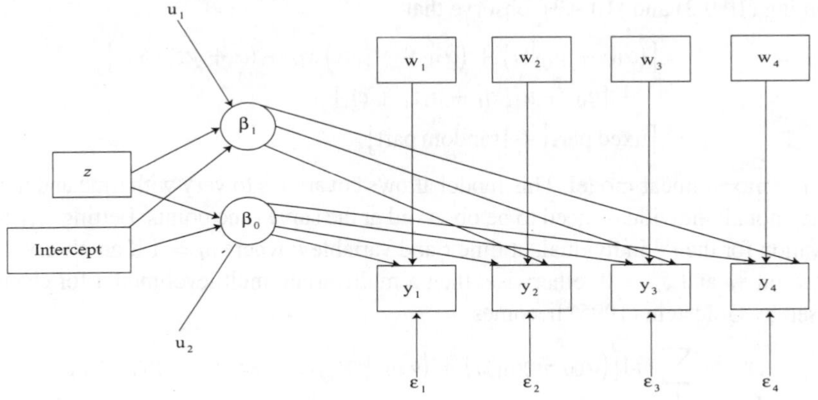

FIGURE 10.9.1. Growth with Latent Variables.

so that the estimated mean outcome is

$$\widehat{E}\left(y_{it}\right)=\widehat{E}\left(\beta_{01}\right)+\widehat{E}\left(\beta_{1i}\right)x_t+\widehat{\theta}_t w_{it}$$

where

$$\widehat{y}_{it}=\widehat{\beta}_{0i}+\widehat{\beta}_{1i}x_{it}+\widehat{\theta}_t w_{it}$$

are predicted outcomes. The parameters β_{0i} represent the initial status of the outcome when time and the covariate is set to zero, and β_{1i} represents the growth rate of the outcome for a unit increase in time. Averaging over subjects, $\widehat{\alpha}_{00}$ is the intercept of the average growth curve and $\widehat{\alpha}_{10}$ is the average growth rate, given no covariates. Covariates are added to account for the variation in the growth factors β_{0i} and β_{1i}. Because one does not observe β_{00} and β_{1i}, the parameters are considered latent factors where $\widehat{\beta}_{0i}$ and $\widehat{\beta}_{1i}$ are latent factor scores. To represent (10.9.4) and (10.9.5) as a SEM, see Figure 10.9.1 with $t=4$ time points. The disturbances u_i may be correlated within level-1 and the errors ϵ_i may be correlated within level-2. In addition, β_0 may be related to β_0 and β_1. Then for example, the second equation in (10.9.5) may be replaced with

$$\beta_{1i}=\alpha+\delta\beta_{0i}+\gamma z_i+u_i$$

Comparing Figure 10.9.1 with Figure 10.2.1, we may associate β_0 and β_1 with η_1 and η_2 then the matrix structural model becomes

$$\boldsymbol{\eta}_i=\boldsymbol{\alpha}+\mathbf{B}\boldsymbol{\eta}_i+\Gamma\mathbf{z}_i+\boldsymbol{\zeta}_i \tag{10.9.6}$$

by replacing $\boldsymbol{\xi}_i$ with the observed covariates $\mathbf{z}_i$.

The matrix form of the measurement model is

$$\mathbf{y}_i=\boldsymbol{\lambda}_y+\Lambda\boldsymbol{\eta}_i+\Theta\mathbf{w}_i+\boldsymbol{\epsilon}_i \tag{10.9.7}$$

where Λ may contain the fixed time point values x_t or for nonlinear growth may contain unknown values.. For example, the values $0, 1, x_3, x_4$ may be used where x_3 and x_4 are parameters to be estimated. The parameter Θ is unknown and must be estimated. Combining

(10.9.6) and (10.9.7) into a single equation, we have that

$$\begin{aligned}\mathbf{y}_i &= \left[\boldsymbol{\lambda}_y + \Lambda\left(\mathbf{I}-\mathbf{B}\right)^{-1}\boldsymbol{\alpha} + \Lambda\left(\mathbf{I}-\mathbf{B}\right)^{-1}\Gamma\mathbf{x}_i\right.\\ &\quad \left. + \Theta\mathbf{x}_i\right] + \left[\Lambda\left(\mathbf{I}-\mathbf{B}\right)^{-1}\boldsymbol{\zeta}_i + \boldsymbol{\epsilon}_i\right] \\ &= \left[\text{fixed part}\right] + \left[\text{random part}\right]\end{aligned} \tag{10.9.8}$$

a complicated mixed model. The time varying covariates w_{it} and the time invariant covariates $\mathbf{z}_i$ are contained in the vector $\mathbf{x}_i$.

To fit Figure 10.9.1 into a SEM, we have that

$$\begin{bmatrix}\beta_0\\ \beta_1\end{bmatrix} = \begin{bmatrix}\alpha_{00}\\ \alpha_{11}\end{bmatrix} + \begin{bmatrix}0 & 0\\ 0 & 0\end{bmatrix}\begin{bmatrix}\beta_0\\ \beta_1\end{bmatrix} + \begin{bmatrix}\gamma_{01}\\ \gamma_{11}\end{bmatrix} z + \begin{bmatrix}\zeta_1\\ \zeta_2\end{bmatrix}$$

$$\begin{bmatrix}y_1\\ y_2\\ y_3\\ y_4\end{bmatrix} = \begin{bmatrix}1 & 0\\ 1 & 1\\ 1 & x_3\\ 1 & x_4\end{bmatrix}\begin{bmatrix}\beta_0\\ \beta_1\end{bmatrix} + \begin{bmatrix}\epsilon_1\\ \epsilon_2\\ \epsilon_3\\ \epsilon_4\end{bmatrix}$$

$$\begin{bmatrix}w_1\\ w_2\\ w_3\\ w_4\end{bmatrix} = \begin{bmatrix}1 & 0 & 0 & 0\\ 0 & 1 & 0 & 0\\ 0 & 0 & 1 & 0\\ 0 & 0 & 0 & 1\end{bmatrix}\begin{bmatrix}\theta_1\\ \theta_2\\ \theta_3\\ \theta_4\end{bmatrix} + \begin{bmatrix}0\\ 0\\ 0\\ 0\end{bmatrix}$$

In the above representation, the common intercept of y_{it} is obtained by restricting the intercepts for each of the outcomes y_t to be equal across time and the mean α_{00} for the latent intercept factor β_{0i} is calculated at zero. Thus, we have that $x_1 = 0$ and $x_2 = 1$. Because x_3 and x_4 may vary, the mean of the slope growth factor (with no covariate) is no longer a constant rate of change over time. It represents a rate of change for a time increment change of one unit. A model is usually first fit with fixed values for x_t with some centering point, are then one considers variable x_t, and finally time invariant covariate may be included in the model.

More generally, using (10.2.2) we may set $\boldsymbol{\lambda}_x = \boldsymbol{\lambda}_y = \mathbf{0}$ and estimate the intercept $\boldsymbol{\alpha}$. Because there is no latent factor $\boldsymbol{\xi}$, the covariance matrix for $\mathbf{y}$ is

$$\text{cov}\left(\mathbf{y}\right) = \Lambda\left(\mathbf{I}-\mathbf{B}\right)^{-1}\Psi\left(\mathbf{I}-\mathbf{B}\right)^{-1\prime}\Lambda' + \Theta_{\epsilon} \tag{10.9.9}$$

where Θ_ϵ is the covariance matrix for the errors and Ψ is the covariance matrix for $\boldsymbol{\zeta}$.

In our SEM example, we used deviation scores and the latent factor η had a mean of zero. Now, the intercept contains a structured mean $\boldsymbol{\alpha}$. The mean of $\mathbf{y}$ has the general structure

$$E\left(\mathbf{y}\right) = \boldsymbol{\lambda}_y + \Lambda\left(\mathbf{I}-\mathbf{B}\right)^{-1}\boldsymbol{\alpha} + \Lambda\left(\mathbf{I}-\mathbf{B}\right)^{-1}\Gamma\mathbf{z} \tag{10.9.10}$$

where $\mathbf{z}$ contains time invariant covariates. Thus, to use the SEM approach to model growth for multilevel models, one must estimate $\boldsymbol{\alpha}$. This is accomplished by adding a dummy variables of all ones to the model, Dunn et al. (1993). Thus, one analyzes the augmented moment matrix

$$\begin{bmatrix} \mathbf{S} + \bar{\mathbf{y}}\bar{\mathbf{y}}' \\ \bar{\mathbf{y}}' & 1 \end{bmatrix}$$

to fit $\boldsymbol{\mu}(\boldsymbol{\theta})$ and $\Sigma(\boldsymbol{\theta})$ to $\bar{\mathbf{y}}$ and $\mathbf{S}$ simultaneously. Here we assume $\bar{\mathbf{y}}$ and $\mathbf{S}$ are asymptotically independent. To analyze the an augmented moment matrix using PROC CALIS, one must use the options UCOV and AUGMENT on the PROC CALIS statement.

The formulation of the multilevel model as a latent variable SEM provides one with a very flexible approach to the analysis of longitudinal data. It may be used to analyze growth processes involving multiple groups and multiple variables, and may include several levels of mediational variables with multiple indicator variables, multiple groups and multiple growth procedure. Illustration of the general approach is discussed by McArdle (1988), Meredith and Tisak (1990), Muthén (1991), Willett and Sayer (1996), and Bijleveld and van der Kamp (1998), Muthén (1997), and Kaplan (2000). MacCallum et al. (1997) illustrate the analysis of multivariate growth curves using the SEM approach and compare the method with a multivariate multilevel modeling approach. The extension of multilevel models for binary responses is considered by Muthén (1984, 1996). The program M plus developed by Muthén and Muthén (1998) performs the necessary calculations for both SEM and multilevel SEM with continuous and categorical data.

10.10 Exogeniety in Structural Equation Models

The analysis of the models in this chapter involve estimating and testing hypotheses regarding model parameters where the observed variables **y** and **x** are both random. However, because we are investigating a system of simultaneous equations a critical assumption in our analysis is the concept of exogeniety. Weak exogeniety allows one to formulate complex relationships about **Y** and **X** defined by a joint probability model, but permits one to estimate model parameters using only the conditional distribution of $\mathbf{Y}|\mathbf{x}$, ignoring the marginal distribution of **X**. This was addressed in Chapter 4 when we considered the multivariate (single equation) regression model and also in Chapter 6 for SUR models.

In path analysis, simultaneous equation modeling, the joint structure of the observed data is represented in (10.5.2). By adding an intercept to (10.5.1), the structural equation model with the endogenous vector **y** and the "exogenous" variable **x** becomes

$$\begin{aligned} \mathbf{y} &= \boldsymbol{\alpha} + \mathbf{B}\mathbf{y} + \Gamma\mathbf{x} + \boldsymbol{\zeta} \qquad (10.10.1) \\ \mathbf{y} &= (\mathbf{I}-\mathbf{B})^{-1}\boldsymbol{\alpha} + (\mathbf{I}-\mathbf{B})^{-1}\Gamma\mathbf{x} + (\mathbf{I}-\mathbf{B})^{-1}\boldsymbol{\zeta} \\ \mathbf{y} &= \Pi_0 + \Pi_1 + \mathbf{e} \end{aligned}$$

a reduced form, vector representation of the multivariate regression (MR) model. The vector Π_0 contain the reduced form model intercepts and the matrix Π_1 contain the reduced slopes. This is equation (10.5.4) with $\Pi = [\Pi_0, \Pi_1]$. The reduced form error vector **e** has the covariance structure, $\text{cov}(\mathbf{e}) = \Sigma_e = (\mathbf{I}-\mathbf{B})^{-1}\Psi(\mathbf{I}-\mathbf{B})'^{-1}$. To estimate model parameters, we may use the likelihood fit function which under multivariate normality of the observed variables leads to full information likelihood estimates for the model parameters. While the other fit functions do not depend on multivariate normality, they do depend on the conditional distribution of $\mathbf{y}|\mathbf{x}$, ignoring the information in $\mathbf{x}$. Ignoring the information in the marginal distribution is only valid if $\mathbf{x}$ is weakly exogenous. The sufficient condition of weak exogeniety permits efficient inference where estimation of the model parameters, tests of hypotheses, and prediction only depend upon the conditional distribution.

Ignoring the latent parameters, the structural model for the observed variables involve the parameter vector $\boldsymbol{\theta} = [\boldsymbol{\mu}_x', \boldsymbol{\mu}_y', \text{vec}(\mathbf{B})', \text{vec}(\Gamma)', \text{vec}(\Psi)', \text{vec}(\Phi)']'$ where an element of the vector $\boldsymbol{\theta}$ is in the parameter space Θ, $\boldsymbol{\theta} \epsilon \Theta$. The factoring of the joint density as $f(\mathbf{Y}, \mathbf{X}; \boldsymbol{\theta}) = f(\mathbf{Y}|\mathbf{X} = \mathbf{x}; \boldsymbol{\lambda}_1) f(\mathbf{X}; \boldsymbol{\lambda}_2)$ is said to operate a sequential cut of the density for joint parameter space $\Theta = (\Lambda_1 \times \Lambda_2)$, a Cartesian product space, if and only if for $\boldsymbol{\lambda} = (\boldsymbol{\lambda}_1', \boldsymbol{\lambda}_2')' = g(\boldsymbol{\theta}) \epsilon (\Lambda_1 \times \Lambda_2) = \Theta$ for $\boldsymbol{\lambda}_1 \epsilon \Lambda_1$ and $\boldsymbol{\lambda}_2 \epsilon \Lambda_2$. Then the parameter vectors $\boldsymbol{\lambda}_1$ and $\boldsymbol{\lambda}_2$ are said to be variation free. That is for any $\boldsymbol{\lambda}_1 \epsilon\ \Lambda_1$, $\boldsymbol{\lambda}_2 \epsilon \Lambda_2$ can attain any value in Λ_2, and conversely, so that the parameters in the vectors $\boldsymbol{\lambda}_1$ and $\boldsymbol{\lambda}_2$ are unconstrained or free to vary. Following Engle, Hendry, and Richard (1983), we have the following definition of weak exogeneity.

Definition 10.10.1 *The variable* $\mathbf{x}$ *is weakly exogeneous for any vector of parameters* $\boldsymbol{\psi}$, *say, if and only if there exists a reparameterization of the elements of* $\boldsymbol{\theta} \in \Theta$ *as* $\boldsymbol{\lambda} = [\boldsymbol{\lambda}_1' \, \boldsymbol{\lambda}_2']' \in (\Lambda_1 \times \Lambda_2)$ *such that (1)* $\boldsymbol{\psi} = f(\boldsymbol{\lambda}_1)$ *alone, and (2) the parameter* $\boldsymbol{\lambda}_1$ *and* $\boldsymbol{\lambda}_2$ *are variation free so that* $\boldsymbol{\lambda} \in (\Lambda_1 \times \Lambda_2)$, *and the joint density has the representation* $f(\mathbf{Y}, \mathbf{X}; \boldsymbol{\theta}) = f(\mathbf{Y}|\mathbf{X} = \mathbf{x}; \boldsymbol{\lambda}_1) f(\mathbf{X}; \boldsymbol{\lambda}_2)$.

If we can assume weak exogeneity for the path analysis, simultaneous equation, model so that $\mathbf{x}$ is weakly exogenous for $\boldsymbol{\lambda}$, then we can restrict our attention to the parameter vector $\boldsymbol{\lambda}_1 = [\text{vec}(\Pi)', \text{vec}(\Sigma_e)']'$ and ignore the vector $\boldsymbol{\lambda}_2 = [\boldsymbol{\mu}_x', \text{vec}(\Phi)']'$. Again, multivariate normality is only a necessary condition for weak exogeneity to hold. As we indicated earlier, the multivariate normal distribution is a member of the class of distributions known as elliptical distributions as is the multivariate t distribution. And, conditional cuts of elliptical distributions are themselves elliptical. However, if the joint distribution of $\mathbf{Y}$ and $\mathbf{X}$ follows an elliptical distribution, the observed variable $\mathbf{x}$ is no longer weakly exogeneous for the parameter vector $\boldsymbol{\lambda}_1$, Bilodeau and Brenner (1999, p. 208). This means that the parameters in the conditional distribution are confounded by the parameters in the marginal distribution. Thus, even if one is only interested in the parameters associated with the conditional distribution, the marginal distribution cannot be ignored. Estimation and inference is no longer valid using only the conditional distribution and one must use an estimation procedure, for example, a full information maximum likelihood method that depends on the joint distribution.

Example 10.10.1 *Assume that the random vector* $\mathbf{Z} = [\mathbf{Y}', \mathbf{X}']'$ *follows a multivariate normal distribution with mean* $\boldsymbol{\mu}_z$ *and partitioned, positive definite, covariance matrix* Ω *defined:*

$$\boldsymbol{\mu}_z = \begin{bmatrix} \boldsymbol{\mu}_y \\ \boldsymbol{\mu}_x \end{bmatrix}, \quad \Omega = \begin{bmatrix} \Sigma_{yy} & \Sigma_{yx} \\ \Sigma_{xy} & \Sigma_{xx} \end{bmatrix}$$

By Theorem 3.3.2, property (5), the parameter vector $\boldsymbol{\theta}$ *of the joint distribution is* $\boldsymbol{\theta}' = [\boldsymbol{\mu}'_y, \boldsymbol{\mu}'_x, vec(\Sigma_{yy})', vec(\Sigma_{yx})', vec(\Sigma_{xx})']$. *Using representation (4.2.65), the parameter vector of the conditional model* $\mathbf{Y}|\mathbf{X} = \mathbf{x}$ *is* $\boldsymbol{\lambda}'_1 = [\boldsymbol{\beta}'_0, vec(\mathbf{B}'_1)', vec(\Sigma)']$ *and the parameter vector for the marginal model is* $\boldsymbol{\lambda}'_2 = [\boldsymbol{\mu}'_x, vec(\Sigma_{xx})']$. *In addition, the factorization of the joint density into the product of the conditional density time the marginal density generates a sequential cut of the joint density since the parameter vectors* $\boldsymbol{\lambda}_1$*and* $\boldsymbol{\lambda}_2$ *are variation free. This is the case since for arbitrary choices of values in* $\boldsymbol{\lambda}_1$*and* $\boldsymbol{\lambda}_2$ *there is a one-to-one correspondence between* $\boldsymbol{\lambda} = [\boldsymbol{\lambda}'_1, \boldsymbol{\lambda}'_2]' \in \Lambda_1 \times \Lambda_2$ *for all admissible values in the parameter space* Θ *since we have that*

$$\boldsymbol{\mu}_z = \begin{bmatrix} \boldsymbol{\beta}_0 + \mathbf{B}'_1\boldsymbol{\mu}_x \\ \boldsymbol{\mu}_x \end{bmatrix}, \quad \Omega = \begin{bmatrix} \Sigma + \mathbf{B}'_1\Sigma\mathbf{B}_1 & \mathbf{B}'_1\Sigma_{yx} \\ \Sigma_{xy}\mathbf{B}_1 & \Sigma_{xx} \end{bmatrix} \epsilon(\Lambda_1 \times \Lambda_2)$$

Thus, $\mathbf{x}$ *is weakly exogenous for the parameter vector* $\boldsymbol{\psi} = f(\boldsymbol{\lambda}_1)$.

Example 10.10.2 *If the random variable* $\mathbf{Z}$ *has an elliptical distribution so that* $\mathbf{Z} \sim E_p(\boldsymbol{\mu}_z, \Theta)$. *Even though the conditional and marginal distributions are both elliptical, the variance of the conditional random variable* $\mathbf{Y}|\mathbf{x}$ *depends on* $\boldsymbol{\mu}_x$ *and* Σ_{xx} *through the product* $g(\mathbf{x})\Sigma$ *where* $g(\mathbf{x})$ *depends on the quadratic form* $(\mathbf{x} - \boldsymbol{\mu}_2)'\Sigma_{xx}^{-1}(\mathbf{x} - \boldsymbol{\mu}_2)'$. *For the a parameter vector* $\boldsymbol{\lambda} = [\boldsymbol{\lambda}'_1, \boldsymbol{\lambda}'_2]' \epsilon \Theta$ *as defined in the multivariate normal example* $\Theta \neq (\Lambda_1 \times \Lambda_2)$ *so that the vectors* $\boldsymbol{\lambda}_1$ *and* $\boldsymbol{\lambda}_2$ *are not variation free. The factoring of the joint density does not generate a sequential cut and* $\mathbf{x}$ *is not weakly exogenous for the parameter* $\boldsymbol{\lambda}_1$.

The notion of exogeniety is often discussed in dynamic simultaneous equation models. For these models, Granger noncausality is also important. A lagged variable $\mathbf{y}_{t-1}$ does not Granger cause $\mathbf{x}_t$ if and only if the probability density function for $f(\mathbf{x}_t|\mathbf{x}_{t-1}, \mathbf{y}_{t-1}) = f(\mathbf{x}_t|\mathbf{x}_{t-1})$. This means that in the conditional distribution, the lagged values $\mathbf{y}_{t-1}$, do not account for variation in the movements in $\mathbf{x}_t$ beyond that already provided by the lagged variable $\mathbf{x}_{t-1}$. Thus, forecasts of the variable $\mathbf{y}_t$ may be made conditional on forecasts of the variable $\mathbf{x}_t$ if $\mathbf{y}_{t-1}$does not Granger cause $\mathbf{x}_t$. However, if one is interested in estimation or inference of parameters in dynamic forecast models, one may estimate the parameters conditionally with respect to $\mathbf{x}_t$ if the variables are weakly exogenous for the parameters in the context of the model in which they are defined. Combining the notions of weak exogeniety and Granger noncausality, we obtain strong exogeniety. If a variable $\mathbf{x}_t$ is weakly exogenous and if $\mathbf{y}_{t-1}$ does not Granger cause $\mathbf{x}_t$, then the variable $\mathbf{x}_t$ is said to be strongly exogeneous. We have only briefly introduced some terms used in forecasting, in estimating parameters in dynamic simultaneous equation models, and in economic policy analysis. For a comprehensive treatment of exogeniety, one should consult the book edited by Ericsson and Iron (1994). Dynamic models are discussed by Hsiao (1986).

Exercises 10.10

1. For the simple dynamic model $y_t = \beta x_t + \epsilon_t$ where $x_t = y_{t-1}\theta + \delta_t$, assume that the errors follow a joint bivariate normal distribution and so that the errors

$\epsilon_t \sim N(0, \sigma_1^2)$, and $\delta_t \sim N(0, \sigma_2^2)$. Show that x is weakly exogenous for β, but that x is not weakly exogenous for the root parameter $\rho = \beta\theta$. The system is dynamically stable if $|\rho| < 1$ and unstable if $|\rho| > 1$. The system oscillates without dampening if $\rho = 1$. This shows that in order to evaluate stability of the simultaneous equation system, you must use the joint bivariate distribution. Or, one cannot evaluate equilibrium by only investigating the conditional distribution since it depends on the parameter θ.

Appendix A
Tables

- TABLE I. Upper Percentage Points of the Standard Normal Distribution, $Z^{1-\alpha}$

  $Z^{1-\alpha}$ denotes the upper $1-\alpha$ critical value for the standard normal distribution. If $X \sim N(0, 1)$and $\alpha = 0.05$, the critical value $Z^{1-\alpha}$ such that the $P\left(X > Z^{1-\alpha}\right) = 0.05$ is read as $Z^{0.95} = 1.645$ from the table.

  Source: Abridged from Table 1. E. S. Pearson and H. O. Hartley (Eds.), *Biometrika tables for statisticians*, Vol 1 (3rd ed.). New York: Cambridge, 1966.

- TABLE II. Upper Percentage Points of the χ^2 Distribution, $\chi^2_{1-\alpha}(v)$

  $\chi^2_{1-\alpha}(v)$ is the upper 1-α critical value for a χ^2 distribution with v degrees of freedom. If $X^2 = \chi^2(v)$ and $\alpha = 0.05$ with $v = 10$, the critical constant such that the $P\left[X^2 > \chi^2_{1-\alpha}(v)\right] = .05$ is $\chi^2_{0.95}(10) = 18.3070$.

  Source: Table 8. E. S. Pearson, and H. O. Hartley (Eds.), *Biometrika tables for statisticians,* Vol 1 (3rd ed.). New York: Cambridge, 1966.

- TABLE III. Upper Percentage Point of Student's Distribution, $t^{1-\alpha}(v)$

  $t^{1-\alpha}(v)$ represent the upper $1-\alpha$ critical value for the t distribution with degree of freedom. If $t \sim t(v)$ and $\alpha = 0.05$ with $v = 10$, the $P\left[t > t^{1-\alpha}(v)\right] = .05$ is $t^{0.95}(10) = 1.812$.

  Source: Table 12, E. S. Pearson and H. O. Hartley (Eds.) *Biometrika tables for statisticians,* Vol 1 (3rd ed.). New York: Cambridge, 1966.

- TABLE IV. Upper Percentage Points of the F Distribution, $F^{1-\alpha}(v_h, v_e)$

  $F^{1-\alpha}(v_h, v_e)$ is the upper $1-\alpha$ critical value of the F distribution with v_h representing the numerator and v_e the determinator degrees of freedom. If $F \sim F^{1-\alpha}(v_h, v_e)$ the critical value for $\alpha = 0.05$, $v_h = 4$, and $v_e = 9$ such that the $P\left[F > F^{1-\alpha}(v_h, v_e)\right] = .05$ is $F^{0..95}(4, 9) = 3.63$. To find the $P\left[F < F^{\alpha}(v_h, v_e)\right]$, the formula $F^{\alpha}(v_h, v_e) = 1/F^{1-\alpha}(v_e, v_h)$ is employed. Since $F^{1-\alpha}(9, 4) = 6.00$, the critical constant is $F^{\alpha}(v_h, v_e) = 1/6 = .167$.

  Source: Table 18, E. S. Pearson, and H. O. Hartley (Eds.), *Biometrika tables for statisticians,* Vol 1 (3rd ed.). New York: Cambridge, 1966.

- TABLE V. Upper Percentage Point for Simultaneous Inference Procedures.

  Upper $\alpha = 0.05$ percentage points for the Scheffé (SCH), Bonferroni-Dunn (BON), Dunn-Šidák (SID) independent t and Šidák multivariate t (STM), Studentized Maximum Modulus distribution, multiple comparison procedures for C comparisons. The critical values c_α in Table V are used to evaluate $\widehat{\psi} \pm c_\alpha \widehat{\sigma}_{\widehat{\psi}}$ for linear parametric functions ψ of the fixed effect linear models where the general structure of for parametric function $\psi = \mathbf{c}'\boldsymbol{\beta}$ for $E(\mathbf{Y}) = \mathbf{X}\boldsymbol{\beta}$. The entries in the table labeled (BON), (SID), and (STM) may also be used to contruct approximate $100(1-\alpha)\%$ simultaneous intervals for C comparisons of the form $\psi_i = \mathbf{c}'\mathbf{B}\mathbf{m}$ for $i = 1, 2, \ldots, C$ comparisons using the MR model, $E(\mathbf{Y}) = \mathbf{XB}$.

  Source: C. Fuchs and A.R. Sampson, Simultaneous confidence intervals for the general linear model, *Biometrics,* 1987, **43**, 457-469.

- TABLE VI. Small Sample Empirical Critical Values for Multivariate Skewness.

  The estimated $1-\alpha$ empirical critical values represented as $\eta_{1-\alpha}$ for $\alpha = 0.10, 0.05$, and 0.01 for and sample sizes n for Mardia's tests of multivariate Skewness.*

- TABLE VII. Small Sample Empirical Critical Values for Multivarite Kurtosis: Lower Values.

  The estimated $1-\alpha$ lower empirical critical values represented as $\eta_{1-\alpha}$ for $\alpha = 0.10$, 0.05, and 0.01 and sample sizes n for Mardia's test of multivariate Kurtosis.*

- TABLE VIII. Small Sample Empirical critical Values for Multivariate Kurtosis: Upper Values.

  The estimated $1-\alpha$ upper empirical critical values represented as $\eta_{1-\alpha}$ for $\alpha = 0.10, 0.05$, and 0.01 and sample sizes n for Mardia's test of multivariate Kurtosis.*

  *Source: J. L. Romeu and A. Ozturk, A comparative study of goodness-of-fit tests for multivariate normality, *Journal of Multivariate Analysis*, 1993, **46**, 309-334.

Table I. Upper Percentage Points of the Standard Normal Distribution $Z^{1-\alpha}$

$1-\alpha$	$Z^{1-\alpha}$	$1-\alpha$	$Z^{1-\alpha}$	$1-\alpha$	$Z^{1-\alpha}$
.50	0.00	.75	0.67	.950	1.645
.51	0.03	.76	0.71	.955	1.695
.52	0.05	.77	0.74	.960	1.751
.53	0.08	.78	0.77	.965	1.812
.54	0.10	.79	0.81	.970	1.881
.55	0.13	.80	0.84	.975	1.960
.56	0.15	.81	0.88	.980	2.054
.57	0.18	.82	0.92	.985	2.170
.58	0.20	.83	0.95	.990	2.326
.59	0.23	.84	0.99	.995	2.576
.60	0.25	.85	1.04	.996	2.652
.61	0.28	.86	1.08	.997	2.748
.62	0.30	.87	1.13	.998	2.878
.63	0.33	.88	1.17	.999	3.090
.64	0.36	.89	1.23		
.65	0.39	.90	1.28	.9995	3.291
.66	0.41	.91	1.34	.9999	3.719
.67	0.44	.92	1.41		
.68	0.47	.93	1.48	.99995	3.891
.69	0.50	.94	1.55	.99999	4.265
.70	0.52				
.71	0.55				
.72	0.58				
.73	0.61				
.74	0.64				

Table II. Upper Percentage Points of the χ^2 Distribution, $\chi^2_{1-\alpha}(v)$

$v \backslash 1-\alpha$	0.750	0.900	0.950	0.975	0.990	0.995	0.999
1	1.32330	2.70554	3.84146	5.02389	6.63490	7.87944	10.828
2	2.77259	4.60517	5.99146	7.37776	9.21034	10.5966	13.816
3	4.10834	6.25139	7.81473	9.34840	11.3449	12.8382	16.266
4	5.38527	7.77944	9.48773	11.1433	13.2767	14.8603	18.467
5	6.62568	9.23636	11.0705	12.8325	15.0863	16.7496	20.515
6	7.84080	10.6446	12.5916	14.4494	16.8119	18.5476	22.458
7	9.03715	12.0170	14.0671	16.0128	18.4753	20.2777	24.322
8	10.2189	13.3616	15.5073	17.5345	20.0902	21.9550	26.125
9	11.3888	14.6837	16.9190	19.0228	21.6660	23.5894	27.877
10	12.5489	15.9872	18.3070	20.4832	23.2093	25.1882	29.588
11	13.7007	17.2750	19.6751	21.9200	24.7250	26.7568	31.264
12	14.8454	18.5493	21.0261	23.3367	26.2170	28.2995	32.909
13	15.9839	19.8119	22.3620	24.7356	27.6882	29.8195	34.528
14	17.1169	21.0641	23.6848	26.1189	29.1412	31.3194	36.123
15	18.2451	22.3071	24.9958	27.4884	30.5779	32.8013	37.697
16	19.3689	23.5418	26.2962	28.8454	31.9999	34.2672	39.252
17	20.4887	24.7690	24.5871	30.1910	33.4087	35.7585	40.790
18	21.6049	25.9894	28.8693	31.5264	34.8053	37.1565	42.312
19	22.7178	27.2036	30.1435	32.8523	36.1909	38.5823	43.820
20	23.8277	28.4120	31.4104	34.1696	37.5662	39.9968	45.315
21	24.9348	29.6151	32.6706	35.4789	38.9322	41.4011	46.797
22	26.0393	30.8133	33.9244	36.7807	40.2894	42.7957	48.268
23	27.1413	32.0069	35.1725	38.0756	41.6384	44.1813	49.728
24	28.2412	33.1962	36.4150	39.3641	42.9798	45.5585	51.179
25	29.3389	34.3816	37.6525	40.6465	44.3141	46.9279	52.618
26	30.4346	35.5632	38.8851	41.9232	45.6417	48.2899	54.052
27	31.5284	36.7412	40.1133	43.1945	46.9629	49.6449	55.476
28	32.6205	37.9159	41.3371	44.4608	48.2782	50.9934	56.892
29	33.7109	39.0875	42.5570	45.7223	49.5879	52.3356	58.301
30	34.7997	40.2560	43.7730	46.9792	50.8922	53.6720	59.703
40	45.6160	51.8051	55.7585	59.3417	63.6907	66.7660	73.402
50	56.3336	63.1671	67.5048	71.4202	76.1539	79.4900	86.661
60	66.9815	74.3970	79.0819	83.2977	88.3794	91.9517	99.607
70	77.5767	85.6270	90.5312	95.0232	100.425	104.215	112.317
80	88.1303	96.5782	101.879	106.629	112.329	116.321	124.839
90	98.6499	107.565	113.145	118.136	124.116	128.299	137.208
100	109.141	118.498	124.342	129.561	135.807	140.169	149.449
X	+0.6745	+1.2816	+1.6449	+1.9600	+2.3263	+2.5758	+3.0902

For $v > 100$, the expression $\chi^2_{1-\alpha} = v\left[1-(2/9v)+X\sqrt{2/9v}\right]^3$ or $\chi^2_{1-\alpha} = (1/2)\left[X+\sqrt{2v-1}\right]^2$ may be used with X defined in the last line of the table as a N(0,1) variable depending on the degree of accuracy desire

Table III. Upper Percentage Points of Student t-distribution, $t^{1-\alpha}(v)$

$v \backslash 1-\alpha$	0.750	0.900	0.950	0.975	0.990	0.995	0.999
1	1.000	3.078	6.314	12.706	31.821	63.657	318.31
2	0.816	1.886	2.920	4.303	6.965	9.925	22.327
3	0.765	1.638	2.353	3.182	4.541	5.841	10.214
4	0.741	1.533	2.132	2.776	3.747	4.604	7.173
5	0.727	1.476	2.015	2.571	3.365	4.032	5.893
6	0.718	1.440	1.943	2.447	3.143	3.707	5.208
7	0.711	1.415	1.895	2.365	2.998	3.499	4.785
8	0.706	1.397	1.860	2.306	2.896	3.355	4.501
9	0.703	1.383	1.833	2.262	2.821	3.250	4.297
10	0.700	1.372	1.812	2.228	2.764	3.169	4.144
11	0.697	1.363	1.796	2.201	2.718	3.106	4.025
12	0.695	1.356	1.782	2.179	2.681	3.055	3.930
13	0.694	1.350	1.771	2.160	2.650	3.012	3.852
14	0.692	1.345	1.761	2.145	2.624	2.977	3.787
15	0.691	1.341	1.753	2.131	2.602	2.947	3.733
16	0.690	1.337	1.746	2.120	2.583	2.921	3.686
17	0.689	1.333	1.740	2.110	2.567	2.898	3.646
18	0.688	1.330	1.734	2.101	2.552	2.878	3.610
19	0.688	1.328	1.729	2.093	2.539	2.861	3.579
20	0.687	1.325	1.725	2.086	2.528	2.845	3.552
21	0.686	1.323	1.721	2.080	2.518	2.831	3.527
22	0.686	1.321	1.717	2.074	2.508	2.819	3.505
23	0.685	1.319	1.714	2.069	2.500	2.807	3.485
24	0.685	1.316	1.711	2.064	2.492	2.797	3.467
25	0.684	1.316	1.708	2.060	2.485	2.787	3.450
26	0.684	1.315	1.706	2.056	2.479	2.779	3.435
27	0.684	1.314	1.703	2.052	2.473	2.771	3.421
28	0.683	1.313	1.701	2.048	2.467	2.763	3.408
29	0.683	1.311	1.699	2.045	2.462	2.756	3.396
30	0.683	1.310	1.697	2.042	2.457	2.750	3.385
40	0.681	1.303	1.684	2.021	2.423	2.704	3.307
60	0.679	1.296	1.671	2.000	2.390	2.660	3.232
120	0.677	1.289	1.658	1.980	2.358	2.167	3.160
∞	0.674	1.282	1.645	1.960	2.326	2.576	3.090

Table IV. Upper Percentage Points of the F-distribution, $F^{1-\alpha}\ (v_h, v_e)$

df		df for (v_h)											
(v_e)	$1-\alpha$	1	2	3	4	5	6	7	8	9	10	11	12
1	.90	39.9	49.5	53.6	55.8	57.2	58.2	58.9	59.4	59.9	60.2	60.5	60.7
	.95	161	200	216	225	230	234	237	239	241	242	243	244
2	.90	8.53	9.00	9.16	9.24	9.29	9.33	9.35	9.37	9.38	9.39	9.40	9.41
	.95	18.5	19.0	19.2	19.2	19.3	19.3	19.4	19.4	19.4	19.4	19.4	19.4
	.99	98.5	99.0	99.2	99.2	99.3	99.3	99.4	99.4	99.4	99.4	99.4	99.4
3	.90	5.54	5.46	5.39	5.34	5.31	5.28	5.27	5.25	5.24	5.23	5.22	5.22
	.95	10.1	9.55	9.28	9.12	9.01	8.94	8.89	8.85	8.81	8.79	8,76	8.74
	.99	34.1	30.8	29.5	28.7	28.2	27.9	27.7	27.5	27.3	27.2	27.1	27.1
4	.90	4.54	4.32	4.19	4.11	4.05	4.01	3.98	3.95	3.94	3.92	3.91	3.90
	.95	7.71	6.94	6.59	6.39	626	616	6.09	6.04	6.00	5.96	5.94	5.91
	.99	21.2	18.0	16.7	16.0	15.5	15.2	15.0	14.8	14.8	14.5	14.4	14.4
5	.90	4.06	3.78	3.62	3,52	3.45	3.40	3.37	3.34	3.32	3.30	3.28	3.27
	.95	6.61	5.79	5.41	5.19	5.05	4.95	4.88	4.82	4.77	4.74	4.71	4.68
	.99	16.3	13.3	12.1	11.4	11.0	10.7	10.5	10.3	10.2	10.1	9.96	9.89
6	.90	3.78	3.46	3.29	3.18	3.11	3.05	3.01	2.98	2.96	2.94	2.92	2.90
	.95	5.99	5.14	4.76	4.53	4.39	4.28	4.21	4.15	4.10	4.06	4.03	4.00
	.99	13.7	10.9	9.78	9.15	8.75	8.47	8.26	8.10	7.98	7/87	7.79	7.72
7	.90	3.59	3.26	3.07	2.96	2.88	2.83	2.78	2.75	2.72	2.70	2.68	2.67
	.95	5.59	4.74	4.35	4.12	3.97	3.87	3.79	3.73	3.68	3.64	3.60	3.57
	.99	12.2	9.55	8.45	7.85	7.46	7.19	6.99	6.84	6.72	6.62	6.54	6.47
8	.90	3.46	3.11	2.92	2.81	2.73	2.67	2.62	2.59	2.56	2.54	2.52	2.50
	.95	5.32	4.46	4.07	3.84	3.69	3.58	3.50	3.44	3.39	3.35	3.31	3.28
	.99	11.3	8.65	7.59	7.01	6.63	6.37	6.18	6.03	5.91	581	5.73	5.67
9	.90	3.36	3.01	2.81	2.69	2.61	2.55	2.51	2.47	2.44	2.42	2.40	2.38
	.95	5.12	4.26	3.86	3.63	3.48	3.37	3.29	3.23	3.18	3.14	3.10	3.07
	.99	10.6	8.02	6.99	6.42	6.06	5.80	5.61	5.47	5.35	5.26	5.18	5.11

Table IV. (continued)

df for (v_h)													df
15	20	24	30	40	50	60	100	120	200	500	∞	$1-\alpha$	(v_e)
61.2	61.7	62.0	62.3	62.5	62.7	62.8	63.0	63.1	63.2	6.33	63.3	.90	1
246	248	249	250	251	252	252	253	253	254	254	254	.95	
9.42	9.44	9.45	9.46	9.47	9.47	9.47	9.48	9.48	9.49	9.49	9.49	.90	2
19.4	19.4	19.5	19.5	19.5	19.5	19.5	19.5	19.5	19.5	19.5	19.5	.95	
99.4	99.4	99.5	99.5	99.5	99.5	99.5	99.5	99.5	99.5	99.5	99.5	.99	
5.20	5.18	5.18	5.17	5.16	5.15	5.15	5.14	5.14	5.14	5.14	5.13	.90	3
8.70	8.66	8.64	8.62	8.59	8.58	8.57	8.55	8.55	8.54	8.53	8.53	.95	
26.9	26.7	26.6	26.5	26.4	26.4	26.3	26.2	26.2	26.2	26.1	26.1	.99	
3.87	3.84	3.83	3.82	3.80	3.80	3.79	3.78	3.78	3.77	3.76	3.76	.90	4
5.86	5.80	5.77	5.75	5.72	5.70	5.69	5.66	5.66	5.65	5.64	5.63	.95	
14.2	14.0	13.9	13.8	13.7	13.7	13.7	13.6	13.6	13.5	13.5	13.5	.99	
3.24	3.21	3.19	3.17	3.16	3.15	3.14	3.13	3.12	3.12	3.11	3.10	.90	5
4.62	4.56	4.53	4.50	4.46	4.44	4.43	4.41	4.40	4.39	4.37	4.36	.95	
9.72	9.55	9.47	9.38	9.29	9.24	9.20	9.13	9.11	9.08	9.04	9.02	.99	
2.87	2.84	2.82	2.80	2.78	2.77	2.76	2.75	2.74	2.73	2.73	2.72	.90	6
3.94	3.87	3.84	3.81	3.77	3.75	3.74	3.71	3.70	3.69	3.68	3.67	.95	
7.56	7.40	7.31	7.23	7.14	7.09	7.06	6.99	6.97	6.93	6.90	6.88	.99	
2.63	2.59	2.58	2.56	2.54	2.52	2.51	2.50	2.49	2.48	2.48	2.47	.90	7
3.51	3.44	3.41	3.83	3.34	3.32	3.30	3.27	3.27	3.25	3.24	3.23	.95	
6.31	6.16	6.07	5.99	5.91	5.86	5.82	5.75	5.74	5.70	5.67	5.65	.99	
2.46	2.42	2.40	2.38	2.36	2.35	2.34	2.32	2.32	2.31	2.30	2.99	.90	8
3.22	3.15	3.12	3.08	3.04	3.02	3.01	2.97	2.97	2.95	2.94	2.93	.95	
5.52	5.36	5.28	5.20	5.12	5.07	5.03	4.96	4.95	4.91	4.88	4.86	.99	
2.34	2.30	2.28	2.25	2.23	2.22	2.21	2.19	2.18	2.17	2.17	2.16	.90	9
3.01	2.94	2.90	2.86	2.83	2.80	2.79	2.76	2.75	2.73	2.72	2.71	.95	
4.96	4.81	4.73	4.65	4.57	4.52	4.48	4.42	4.40	4.36	4.33	4.31	.99	

(continued)

Table IV. (continued)

df (v_e)	$1-\alpha$	df for (v_h) 1	2	3	4	5	6	7	8	9	10	11	12
10	.90	3.29	2.92	2.73	2.61	2.52	2.46	2.41	2.38	2.35	2.32	2.30	2.28
	.95	4.96	4.10	3.71	3.48	3.33	3.22	3.14	3.07	3.02	2.98	2.94	2.91
	.99	10.0	7.56	6.55	5.99	5.64	5.39	5.20	5.06	4.49	4.85	4.77	4.71
11	.90	3.23	2.86	2.66	2.54	2.45	2.39	2.34	2.30	2.27	2.25	2.23	2.21
	.95	4.84	3.98	3.59	3.36	3.20	3.09	3.01	2.95	2.90	2.85	2.82	2.79
	.99	9.65	7.21	6.22	5.67	5.32	5.07	4.89	4.74	4.63	4.54	4.46	4.40
	.75	1.46	1.56	1.56	1.55	1.54	1.53	1.52	1.51	1.51	1.50	1.50	1.49
12	.90	3.18	2.81	2.61	2.48	2.39	2.33	2.28	2.24	2.21	2.19	2.17	2.15
	.95	4.75	3.89	3.49	3.26	3.11	3.00	2.91	2.85	2.80	2.75	2.27	2.69
	.99	9.33	6.93	5.95	5.41	5.06	4.82	4.64	4.50	4.39	4.30	4.22	4.16
13	.90	3.14	2.76	2.56	2.43	2.35	2.28	2.23	2.20	2.16	2.14	2.12	2.10
	.95	4.67	3.81	3.41	3.18	3.03	2.92	2.83	2.77	2.71	2.67	2.63	2.60
	.99	9.07	6.70	5.74	5.21	4.86	4.62	4.44	4.30	4.19	4.10	4.02	3.96
14	.90	3.10	2.73	2.52	2.39	2.31	2.24	2.19	2.15	2.12	2.10	2.08	2.05
	.95	4.60	3.74	3.34	3.11	2.96	2.85	2.76	2.70	2.65	2.60	2.57	2.53
	.99	8.86	6.51	5.56	5.04	4.69	4.46	4.28	4.14	4.03	3.94	3.86	3.80
15	.90	3.07	2.70	2.49	2.36	2.27	2.21	2.16	2.12	2.09	2.06	2.04	2.02
	.95	4.54	3.68	3.29	3.06	2.90	2.79	2.71	2.64	2.59	2.54	2.51	2.48
	.99	8.68	6.36	5.42	4.89	4.56	4.32	4.14	4.00	3.89	3.80	3.73	3.67
16	.90	3.05	2.67	2.46	2.33	2.24	2.18	2.13	2.09	2.06	2.03	2.01	1.99
	.95	4.49	3.63	3.24	3.01	2.85	2.74	2.66	2.59	2.54	2.49	2.46	2.42
	.99	8.53	6.23	5.29	4.77	4.44	4.20	4.03	3.89	3.78	3.69	3.62	3.55
17	.90	3.03	2.64	2.44	2.31	2.22	2.15	2.10	2.06	2.03	2.00	1.98	1.96
	95	4.45	3.59	3.20	2.96	2.81	2.70	2.61	2.55	2.49	2.45	2.41	2.38
	.99	8.40	6.11	5.18	4.67	4.34	4.10	3.93	3.79	3.68	3.59	3.52	3.46
18	.90	3.01	2.62	2.42	2.29	2.20	2.13	2.08	2.04	2.00	1.98	1.96	1.93
	.95	4.41	3.55	3.16	2.93	2.77	2.66	2.58	2.51	2.46	2.41	2.37	2.34
	.99	8.29	6.01	5.09	4.55	4.25	4.01	3.84	3.71	3.60	351	3.43	3.37
19	.90	2.99	2.61	2.40	2.27	2.18	2.11	2.06	2.02	1.98	1.96	.94	1.91
	.95	4.38	3.52	3.13	2.90	2.74	2.63	2.54	2.48	2.42	2.38	234	2.31
	.99	8.18	5.93	5.01	4.50	4.17	3.94	3.77	3.63	3.52	3.43	3.36	3.30
20	.90	2.97	2.59	2.38	2.25	2.16	2.09	2.04	2.00	1.96	1.94	1.92	1.89
	.95	4.35	3.49	3.10	2.87	2.71	2.60	2.51	2.45	2.39	2.35	2.31	2.28
	.99	8.10	5.85	4.94	4.43	4.10	3.87	3.70	3.56	3.46	3.37	3.29	3.23

Table IV. (continued)

df for (v_h)													df
15	20	24	30	40	50	60	100	120	200	500	∞	$1-\alpha$	(v_e)
2.24	2.20	2.18	2.16	2.13	2.12	2.11	2.09	2.08	2.07	2.06	2.06	.90	10
2.85	2.77	2.74	2.70	2.66	2.64	2.62	2.59	2.58	2.56	2.55	2.54	.95	
4.56	4.41	4.33	4.25	4.17	4.12	4.08	4.01	4.00	3.96	3.93	3.91	.99	
2.17	2.12	2.10	2.08	2.05	2.04	2.03	2.00	2.00	1.99	1.98	1.97	.90	11
2.72	2.65	2.61	2.57	2.53	2.51	2.49	2.46	2.45	2.43	2.42	2.40	.95	
4.25	4.10	4.02	3.94	3.86	3.81	3.78	3.71	3.69	3.66	3.62	3.60	.99	
2.10	2.06	2.04	2.01	1.99	1.97	1.96	1.94	1.93	1.92	1.91	1.90	.90	12
2.62	2.54	2.51	2.47	2.43	2.40	2.38	2.35	2.34	2.32	2.31	2.30	.95	
4.01	3.86	3.78	3.70	3.62	3.57	3.54	3.47	3.45	3.41	3.38	3.36	.99	
2.05	2.01	1.98	1.96	1.93	1.92	1.90	1.88	1.88	1.86	1.85	1.85	.90	13
2.53	2.46	2.42	2.38	234	2.31	2.30	2.26	2.25	2.23	2.22	2.21	.95	
3.82	3.66	3.59	3.51	3.43	3.38	3.34	3.27	325	322	319	317	.99	
2.01	1.96	1.94	1.91	1.89	1.87	1.86	1.83	1.83	1.82	1.80	1.80	.90	14
2.46	2.39	2.35	2.31	2.27	2.24	2.22	2.19	2.18	216	2.14	2.13	.95	
3.66	3.51	3.43	3.35	3.27	3.22	3.18	3.11	3.09	3.06	3.03	3.00	.99	
1.97	1.92	1.90	1.87	1.85	1.83	1.82	1.79	1.79	1.77	1.76	1.76	.90	15
2.40	2.33	2.29	2.25	2.20	2.18	2.16	2.12	2.11	2.10	2.08	2.07	.95	
3.52	3.37	3.29	3.21	3.13	3.08	3.05	2.98	2.96	2.92	2.89	2.87	.99	
1.94	1.89	1.87	1.84	1.81	1.79	1.78	1.76	175	1.74	1.73	1.72	.90	16
2.35	2.28	2.24	2.19	2.15	2.12	2.11	2.07	2.06	2.04	2.02	2.01	.95	
3.41	3.26	3.18	3.10	3.02	2.97	2.93	2.86	2.84	2.81	2.78	2.75	.99	
1.91	1.86	1.84	1.81	1.78	1.76	1.75	173	1.72	1.69	1.69	1.69	.90	17
2.31	2.23	2.19	2.15	2.10	2.08	2.06	2.02	2.01	1.99	1.97	1.96	.95	
3.31	3.16	3.08	3.00	2.92	2.87	2.83	2.76	2.75	2.71	2.68	2.65	.99	
1.89	1.84	1.81	1.78	1.75	1.74	1.72	1.70	1.69	1.68	1.67	1.66	.90	18
2.77	2.19	2.15	2.11	2.06	2.04	2.02	1.98	1.97	1.95	1.93	1.92	.95	
3.23	3.08	3.00	2.92	2.84	2.78	2.75	2.68	2.66	2.62	2.59	2.57	.99	
1.86	1.81	1.79	1.76	1.73	1.71	1.70	1.67	167	1.65	1.64	1.63	.90	19
2.23	2.16	2.11	2.07	2.03	2.00	1.98	1.94	1.93	1.91	1.89	1.88	.95	
3.15	3.00	2.92	2.84	2.76	2.71	2.67	2.60	2.58	2.55	2.51	2.49	.99	
1.84	1.79	1.77	1.74	1.71	1.69	1.68	1.65	1.64	1.63	1.62	1.61	.90	20
2.20	2.12	2.08	2.04	1.99	1.97	1.95	1.91	1.90	1.88	1.86	1.84	.95	
3.09	2.94	2.86	2.78	2.69	2.64	2.61	2.54	2.52	2.48	2.44	2.42	.99	

(continued)

Table IV. (continued)

df (v_e)	$1-\alpha$	df for (v_h) 1	2	3	4	5	6	7	8	9	10	11	12
22	.90	2.95	2.56	2.35	2.22	2.13	2.06	2.01	1.97	1.93	1.90	1.88	1.86
	.95	4.30	3.44	3.05	2.82	2.66	2.55	2.46	2.40	2.34	2.30	2.26	22.3
	.99	7.95	5.72	4.82	4.31	3.99	3.76	3.59	3.45	3.35	3.26	3.18	3.12
24	.90	2.93	2.54	2.33	2.19	2.10	2.04	1.98	1.94	1.91	1.88	1.85	1.83
	.95	4.26	3.40	3.01	2.78	2.62	2.51	2.42	2.36	2.30	2.25	2.21	2.18
	.99	7.82	561	4.72	4.22	3.90	3.67	3.50	3.36	3.26	3.17	3.09	3.03
26	.90	2.91	2.52	2.31	2.17	2.08	2.01	1.96	1.92	1.88	1.86	1.84	1.81
	.95	4.23	3.37	2.98	2.74	2.59	2.47	2.39	2.32	2.27	2.22	2.18	2.15
	.99	7.72	5.53	4.64	4.14	3.82	3.59	3.42	3.29	3.18	3.09	3.02	2.96
28	.90	2.89	2.50	2.29	2.16	2.06	2.00	1.94	1.90	1.87	184	1.81	1.79
	.95	4.20	3.34	2.95	2.71	2.56	2.45	2.36	2.29	2.24	2.19	2.15	2.12
	.99	7.64	5.45	4.57	4.07	3.75	3.53	3.36	3.23	3.12	3.03	2.96	2.90
30	.90	2.88	2.49	2.28	2.14	2.05	1.98	1.93	1.88	1.85	1.82	1.79	1.77
	.95	4.17	3.32	2.92	2.69	2.53	2.42	2.33	2.27	2.21	2.16	2.13	2.09
	.99	7.56	5.39	4.51	4.02	3.70	3.47	3.30	3.17	3.07	2.98	2.91	2.84
40	.90	2.84	2.44	2.23	2.09	2.00	1.93	1.87	1.83	1.79	1.76	1.73	1.71
	.95	4.08	3.23	2.84	2.61	2.45	2.34	2.25	2.18	2.12	2.08	2.04	2.00
	.99	7.31	5.18	4.31	3.83	3.51	3.29	3.12	2.99	2.89	2.80	2.73	2.66
60	.90	2.79	2.39	2.18	2.04	1.95	1.87	1.82	1.77	1.74	1.71	1.68	1.66
	.95	4.00	3.15	2.76	2.53	2.37	2.25	2.17	2.10	2.04	1.99	1.95	1.92
	.99	7.08	4.98	4.13	3.65	3.34	3.12	2.95	2.82	2.72	2.63	2.56	2.50
120	.90	2.75	2.35	2.13	1.99	1.90	1.82	1.77	1.72	1.68	1.65	1.62	1.60
	.95	3.92	3.07	2.68	2.45	2.29	2.17	2.09	2.02	1.96	1.91	1.87	1.83
	.99	6.85	4.79	3.95	3.48	3.17	2.96	2.79	2.66	2.56	2.47	2.40	2.34
200	.90	2.73	2.33	2.11	1.97	1.88	1.80	1.75	1.70	1.66	1.63	1.60	1.57
	.95	3.89	3.04	2.65	2.42	2.26	2.14	2.06	1.98	1.93	1.88	1.84	1.80
	.99	6.76	4.71	3.88	3.41	3.11	2.89	2.73	2.60	2.50	2.41	2.34	2.27
∞	.90	2.71	2.30	2.08	1.94	1.85	1.77	1.72	1.67	1.63	1.60	1.57	1.55
	.95	3.84	3.00	2.60	2.37	2.21	2.10	2.01	1.94	1.88	1.83	1.79	1.75
	.99	6.63	4.61	3.78	3.32	3.02	2.80	2.64	2.51	2.41	2.32	2.25	2.18

Table IV. (continued)

df for (v_h)													*df*
15	20	24	30	40	50	60	100	120	200	500		$1-\alpha$	(v_e)
1.81	1.76	1.73	1.70	1.67	1.65	1.64	0.61	1.60	1.59	1.58	1.57	.90	22
2.15	2.07	2.03	1.98	1.94	1.91	1.89	1.85	1.84	1.82	1.80	1.78	.95	
2.98	2.83	2.75	2.67	2.58	2.53	2.50	2.42	2.40	2.36	2.33	2.31	.99	
1.78	1.73	1.70	1.67	1.64	1.62	1.61	1.58	1.57	1.56	1.54	1.53	.90	24
2.11	2.03	1.98	1.94	1.89	1.86	1.84	1.80	1.79	1.77	1.75	1.73	.95	
2.89	2.74	2.66	2.58	2.49	2.44	2.40	2.33	2.31	2.27	2.24	2.21	.99	
1.76	1.71	1.68	1.65	1.61	1.59	1.58	1.55	1.54	1.53	1.51	1.50	.90	26
2.09	1.99	1.95	1.90	1.85	1.82	1.80	1.76	1.75	1.73	1.71	1.69	.95	
2.81	2.66	2.58	2.50	2.42	2.36	2.33	2.25	2.23	2.19	2.16	2.13	.99	
1.74	1.69	1.66	1.63	1.59	1.57	1.56	1.53	1.52	1.50	1.49	1.48	.90	28
2.04	1.69	1.91	1.87	1.82	1.79	1.77	1.73	1.71	1.69	1.67	1.65	.95	
2.75	2.60	2.52	2.44	2.35	2.30	2.26	2.19	2.17	2.13	2.09	2.06	.99	
1.72	1.67	1.64	1.61	1.57	1.55	1.54	1.51	1.50	1.48	1.47	1.46	.90	30
2.01	1.93	1.89	1.84	1.79	1.76	1.74	1.70	1.68	1.66	1.64	1.62	.95	
2.70	2.35	2.47	2.39	2.30	2.55	2.21	2.13	2.11	2.07	2.03	2.01	.99	
1.66	1.61	1.57	1.54	1.51	1.48	1.47	1.43	1.42	1.41	1.39	1.38	.90	40
1.92	1.84	1.79	1.74	1.69	1.66	1.64	1.59	1.58	1.55	1.53	1.51	.95	
2.52	2.37	2.29	2.20	2.11	2.06	2.02	1.94	1.92	1.87	1.83	1.80	.99	
1.60	1.54	1.51	1.48	1.44	1.41	1.40	1.36	1.35	1.33	1.31	1.29	.90	60
1.84	1.75	1.70	1.65	1.59	1.56	1.53	1.48	1.47	1.44	1.41	1.39	.95	
2.35	2.20	2.12	2.03	1.94	1.88	1.84	1.75	1.73	1.68	1.63	1.60	.99	
1.55	1.48	1.45	1.41	1.37	1.34	1.32	1.27	1.26	1.24	1.21	1.19	.90	120
1.75	1.66	1.61	1.55	1.50	1.46	1.43	1.37	1.35	1.32	1.28	1.25	.95	
2.19	2.03	1.95	1.86	1.76	1.70	1.66	1.56	1.53	1.48	1.42	1.38	.99	
1.52	1.46	1.42	1.38	1.34	1.31	1.28	1.24	1.22	1.20	1.17	1.14	.90	200
1.72	1.62	1.57	1.52	1.46	1.41	1.39	1.32	1.29	1.26	1.22	1.19	.95	
2.13	1.97	1.89	1.79	1.69	1.63	1.58	1.48	1.44	1.39	1.33	1.28	.99	
1.49	1.42	1.38	1.34	1.30	1.26	1.24	1.18	1.17	1.13	1.08	1.00	.90	∞
1.67	1.57	1.52	1.46	1.39	1.35	1.32	1.24	1.22	1.17	1.11	1.00	.95	
2.04	1.88	1.79	1.70	1.59	1.52	1.47	1.36	1.32	1.25	1.15	1.00	.99	

Table V. Upper $1-\alpha$ percentage Points for Scheffé, Bonferroni-Dunn, Šidák (independent-t), and STMaxmod (Šidák Multivariate-t) simultantous confidence intervals for ($\alpha = .05$)

		Number of Contrasts (C)									
v_e		2	3	4	5	6	8	10	12	15	20
3	SCH	4.371	5.275	6.039	6.713	7.324	8.412	9.373	10.244	11.426	13.161
	BON	4.177	4.857	5.392	5.841	6.232	6.895	7.453	7.940	8.575	9.465
	SID	4.156	4.826	5.355	5.799	6.185	6.842	7.394	7.876	8.505	9.387
	STM	3.960	4.430	4.764	5.023	5.233	5.562	5.812	6.015	6.259	6.567
4	SCH	3.727	4.447	5.055	5.593	6.081	6.952	7.723	8.423	9.374	10.773
	BON	3.495	3.961	4.315	4.604	4.851	5.261	5.598	5.885	6.254	6.758
	SID	3.481	3.941	4.290	4.577	4.822	5.228	5.562	5.848	6.214	6.714
	STM	3.382	3.745	4.003	4.203	4.366	4.621	4.817	4.975	5.166	5.409
5	SCH	3.402	4.028	4.557	5.025	5.450	6.209	6.881	7.492	8.324	9.548
	BON	3.163	3.534	3.810	4.032	4.219	4.526	4.773	4.983	5.247	5.604
	SID	3.152	3.518	3.791	4.012	4.197	4.501	4.747	4.955	5.219	5.573
	STM	3.091	3.399	3.619	3.789	3.928	4.145	4.312	4.447	4.611	4.819
6	SCH	3.207	3.778	4.248	4.684	5.070	4.760	6.372	6.928	7.686	8.803
	BON	2.969	3.287	3.521	3.707	3.863	4.115	4.317	4.486	4.698	4.981
	SID	2.959	3.274	3.505	3.690	3.845	4.095	4.296	4.464	4.675	4.956
	STM	2.916	3.193	3.389	3.541	3.664	3.858	4.008	4.129	4.275	4.462
7	SCH	3.078	3.611	4.060	4.456	4.816	5.459	6.030	6.549	7.257	8.300
	BON	2.841	3.128	3.335	3.499	3.636	3.855	4.029	4.174	4.355	4.595
	SID	2.832	3.115	3.321	3.484	3.620	3.838	4.011	4.156	4.336	4.574
	STM	2.800	3.056	3.236	3.376	3.489	3.668	3.805	3.916	4.051	4.223
8	SCH	2.986	3.493	3.918	4.294	4.635	5.245	5.785	6.278	6.948	7.938
	BON	2.752	3.016	3.206	3.355	3.479	3.677	3.833	3.962	4.122	4.334
	SID	2.743	3.005	3.193	3.342	3.464	3.661	3.816	3.945	4.105	4.316
	STM	2.718	2.958	3.128	3.258	3.365	3.532	3.600	3.764	3.891	4.052
9	SCH	2.918	3.404	3.812	4.172	4.499	5.083	5.601	6.073	6.715	7.763
	BON	2.685	2.933	3.111	3.250	3.364	3.547	3.690	3.808	3.954	4.146
	SID	2.677	2.923	3.099	3.237	3.351	3.532	3.675	3.793	3.939	4.130
	STM	2.657	2.885	3.046	3.171	3.272	3.430	3.552	3.651	3.770	3.923
10	SCH	2.865	3.335	3.730	4.078	4.394	4.957	5.457	5.912	6.533	7.448
	BON	2.634	2.870	3.038	3.169	3.277	3.448	3.581	3.691	3.827	4.005
	SID	2.626	2.860	3.027	3.157	3.264	3.434	3.568	3.677	3.813	3.989
	STM	2.609	2.829	2.984	3.103	3.199	3.351	3.468	3.562	3.677	3.823
11	SCH	2.822	3.281	3.664	4.002	4.309	4.856	5.342	5.784	6.386	7.275
	BON	2.593	2.820	2.981	3.106	3.208	3.370	3.497	3.600	3.728	3.895
	SID	2.586	2.811	2.970	3.094	3.196	3.358	3.484	3.587	3.715	3.880
	STM	2.571	2.784	2.933	3.048	3.142	3.288	3.400	3.491	3.602	3.743
12	SCH	2.788	3.236	3.611	3.941	4.240	4.774	5.247	5.678	6.265	7.132
	BON	2.560	2.779	2.934	3.055	3.153	3.308	3.428	3.527	3.649	3.807
	SID	2.553	2.770	2.924	3.044	3.141	3.296	3.416	3.515	3.636	3.793
	STM	2.540	2.747	2.892	3.004	3.095	3.236	3.345	3.433	3.541	3.677

Table V (continued)

v_e		2	3	4	5	6	8	10	12	15	20
15	SCH	2.714	3.140	3.496	3.809	4.092	4.596	5.044	5.450	6.004	6.823
	BON	2.490	2.694	2.837	2.947	3.036	3.177	3.286	3.375	3.484	3.624
	SID	2.483	2.685	2.827	2.937	3.026	3.1666	3.275	3.364	3.472	3.612
	STM	2.474	2.669	2.805	2.910	2.994	3.126	3.227	3.309	3.409	3.536
20	SCH	2.643	3.049	3.386	3.682	3.949	4.425	4.845	5.228	5.749	6.518
	BON	2.423	2.613	2.744	2.845	2.927	3.055	3.153	3.233	3.331	3.455
	SID	2.417	2.605	2.736	2.836	2.918	3.045	3.143	3.223	3.320	3.445
	STM	2.411	2.594	2.722	2.819	2.898	3.020	3.114	3.179	3.282	3.399
25	SCH	2.602	2.996	3.322	3.608	3.866	4.324	4.729	5.097	5.598	6.336
	BON	2.385	2.566	2.692	2.787	2.865	2.986	3.078	3.153	3.244	3.361
	SID	2.379	2.558	2.683	2.779	2.856	2.976	3.069	3.144	3.235	3.351
	STM	2.374	2.551	2.673	2.766	2.842	2.959	3.048	3.121	3.208	3.320
30	SCH	2.575	2.961	3.280	3.559	3.811	4.258	4.653	5.010	4.497	6.216
	BON	2.360	2.536	2.657	2.750	2.825	2.941	3.030	3.102	3.189	3.300
	SID	2.354	2.528	2.649	2.742	2.816	2.932	3.021	3.092	3.180	3.291
	STM	2.350	2.522	2.641	2.732	2.805	2.918	3.005	3.075	3.160	3.267
40	SCH	2.542	2.918	3.229	3.500	3.744	4.176	4.558	4.903	5.373	6.064
	BON	2.329	2.499	2.616	2.704	2.776	2.887	2.971	3.039	3.122	3.227
	SID	2.323	2.492	2.608	2.696	2.768	2.878	2.963	3.031	3.113	3.218
	STM	2.321	2.488	2.603	2.690	2.760	2.869	2.952	3.019	3.100	2.203
60	SCH	2.510	2.876	3.178	3.441	3.678	4.096	4.464	4.797	5.248	5.913
	BON	2.299	2.463	2.575	2.660	2.729	2.834	2.915	2.979	3.057	3.156
	SID	2.294	2.456	2.568	2.653	2.721	2.826	2.906	2.971	3.049	3.148
	STM	2.292	2.454	2.564	2.649	2.716	2.821	2.900	2.964	3.041	3.139
∞	-ALL-	2.236	2.388	2.491	2.569	2.631	2.727	2.800	2.858	2.928	3.016

Table VI. Small Sample Empirical Critical Values for Mardia's Skewness Test

Size n	$\rho = 2$ η_{90}	$\rho = 2$ η_{95}	$\rho = 2$ η_{99}	$\rho = 3$ η_{90}	$\rho = 3$ η_{95}	$\rho = 3$ η_{99}	$\rho = 4$ η_{90}	$\rho = 4$ η_{95}	$\rho = 4$ η_{99}
25	5.84	7.37	11.18	12.21	14.26	18.42	21.45	24.63	30.34
50	6.78	8.81	12.24	14.25	16.72	22.53	25.40	28.50	35.81
75	7.04	78.81	12.97	14.86	17.28	23.13	26.28	29.55	36.12
100	7.28	9.17	13.13	17.77	23.31	26.84	30.14	37.50	43.53
125	7.42	9.25	13.25	15.30	17.65	23.85	27.30	30.36	37.48
150	7.47	9.34	13.47	15.47	17.95	23.23	28.66	30.82	37.66
175	7.52	9.31	13.31	15.47	17.96	23.00	27.56	30.67	37.35
200	7.60	9.38	13.43	15.47	17.92	23.28	27.62	30.63	37.68
∞	7.78	9.49	13.28	15.99	18.31	23.21	28.41	31.41	37.57

Size n	$\rho = 5$ η_{90}	$\rho = 5$ η_{95}	$\rho = 5$ η_{99}	$\rho = 6$ η_{90}	$\rho = 6$ η_{95}	$\rho = 6$ η_{98}	$\rho = 8$ η_{90}	$\rho = 8$ η_{95}	$\rho = 8$ η_{99}
25	34.39	38.77	45.63	53.86	57.89	66.15	108.00	113.36	124.22
50	40.98	44.99	54.15	62.09	66.80	76.49	125.02	131.78	146.09
75	43.27	47.52	56.60	65.27	70.01	80.63	130.25	138.00	153.40
100	47.76	57.07	66.35	71.45	82.29	66.35	71.45	139.65	153.73
125	44.42	48.35	56.93	67.60	71.95	82.10	135.10	141.00	155.48
150	44.72	48.86	57.44	67.10	72.58	82.10	135.74	142.34	155.95
175	44.69	48.73	57.09	68.09	72.73	82.08	136.66	143.55	157.78
200	45.08	48.99	57.38	68.20	73.19	83.60	137.09	143.85	156.29
∞	46.06	49.80	57.34	69.92	74.47	83.51	139.56	145.98	157.66

Size n	$\rho = 10$ η_{90}	$\rho = 10$ η_{95}	$\rho = 10$ η_{99}
25	189.94	196.32	209.94
50	219.37	227.97	246.15
75	230.35	238.8	258.68
100	234.00	243.03	261.33
125	237.46	247.69	266.75
150	239.75	248.41	265.31
175	241.47	250.24	267.14
200	241.30	250.20	267.10
∞	246.60	255.19	270.48

Table VII. Small Sample Empirical Critical Values for Mardia's Kurtosis Test Lower Values

Size	$\rho = 2$	$\rho = 2$	$\rho = 2$	$\rho = 3$	$\rho = 3$	$\rho = 3$	$\rho = 4$	$\rho = 4$	$\rho = 4$
n	η_{90}	η_{95}	η_{99}	η_{90}	η_{95}	η_{99}	η_{90}	η_{95}	η_{99}
25	−1.22	−1.33	−1.52	−1.38	−1.49	−1.67	−1.48	−1.61	−1.80
50	−1.35	−1.31	−1.75	−1.49	−1.63	−1.91	−1.58	−1.74	−2.03
75	−1.44	−1.59	−1.91	−1.55	−1.75	−2.05	−1.64	−1.84	−2.17
100	−1.44	−1.62	−1.95	−1.54	−1.75	−2.11	−1.65	−1.86	−2.23
125	−1.50	−1.67	−2.03	−1.57	−1.78	−2.15	−1.65	−1.85	−2.23
150	−1.50	−1.71	−2.12	1.56	−1.75	−2.18	−1.64	−1.86	−2.26
175	−1.49	−1.71	−2.11	−1.59	−1.79	−2.27	−1.69	−1.90	−2.32
200	−1.52	−1.76	−2.14	−1.61	−1.83	−2.21	−1.67	−1.88	−2.33
∞	−1.65	−1.95	−2.58	−1.65	−1.95	−2.58	−1.65	−1.95	−2.58

Size	$\rho = 5$	$\rho = 5$	$\rho = 5$	$\rho = 6$	$\rho = 6$	$\rho = 6$	$\rho = 8$	$\rho = 8$	$\rho = 8$
n	η_{90}	η_{95}	η_{99}	η_{90}	η_{95}	η_{99}	η_{90}	η_{95}	η_{99}
25	−1.61	−1.80	−2.10	−1.69	−1.79	−2.00	−1.87	−1.97	−2.16
50	−1.74	−2.03	−2.09	−1.77	−1.93	−2.20	−1.91	−2.09	−2.38
75	−1.84	−2.17	−2.27	−1.77	−1.97	−2.30	−1.90	−2.11	−2.45
100	−1.86	−2.23	−2.34	−1.78	−1.99	−2.37	−1.90	−2.10	−2.53
125	−1.85	−2.23	−2.35	−1.75	−1.96	−2.34	−1.89	−2.09	−2.53
150	−1.86	−2.26	−2.32	−1.74	−1.97	−2.37	−1.89	−2.13	−2.56
175	−1.90	−2.32	−2.39	−1.77	−2.03	−2.44	−1.86	−2.08	−2.56
200	−1.88	−2.33	−2.38	−1.78	−2.02	−2.43	−1.88	−2.13	−2.59
∞	−1.95	−2.58	−2.68	−1.65	−1.95	−2.58	−1.65	−1.95	−2.58

Size	$\rho = 10$	$\rho = 10$	$\rho = 10$
n	η_{90}	η_{95}	η_{99}
25	−2.04	−2.14	−2.31
50	−2.04	−2.21	−2.49
75	−2.04	−2.23	−2.59
100	−2.01	−2.23	−2.61
125	−2.00	−2.21	−2.62
150	−1.96	−2.22	−2.69
175	−1.98	−2.19	−2.64
200	−1.98	−2.21	−2.67
∞	−1.65	−1.95	−2.58

TableVIII. Small Sample Empirical Critical Values for
Mardia's Kurosis Test Upper Values

Size	$\rho=2$	$\rho=2$	$\rho=2$	$\rho=3$	$\rho=3$	$\rho=3$	$\rho=4$	$\rho=4$	$\rho=4$
n	η_{90}	η_{95}	η_{99}	η_{90}	η_{95}	η_{99}	η_{90}	η_{95}	η_{99}
25	0.87	1.23	2.05	0.63	0.91	1.61	0.49	0.76	1.38
50	1.21	1.60	2.58	1.06	1.45	2.38	0.94	1.28	2.05
75	1.36	1.79	2.80	1.21	1.62	2.49	1.08	1.46	2.27
100	1.43	1.85	2.91	1.35	1.78	2.63	1.25	1.64	2.48
125	1.46	1.90	2.93	1.35	1.74	2.62	1.27	1.72	2.54
150	1.51	1.94	2.81	1.46	1.85	2.74	1.35	1.70	2.59
175	1.55	2.00	2.78	1.46	1.87	2.63	1.38	1.73	2.57
200	1.54	1.95	2.99	1.48	1.89	2.77	1.41	1.75	2.55
∞	1.65	1.95	2.58	1.65	1.95	2.58	1.65	1.95	2.58

Size	$\rho=5$	$\rho=5$	$\rho=5$	$\rho=6$	$\rho=6$	$\rho=6$	$\rho=8$	$\rho=8$	$\rho=8$
n	η_{90}	η_{95}	η_{99}	η_{90}	η_{95}	η_{99}	η_{90}	η_{95}	η_{99}
25	0.26	0.50	1.07	0.09	0.28	0.71	−0.29	−0.08	0.32
50	0.77	1.11	1.85	0.63	0.93	1.57	0.39	0.69	1.27
75	1.03	1.38	2.17	0.89	1.23	1.96	0.73	1.07	1.71
100	1.16	1.57	2.31	1.06	1.41	2.08	0.84	1.16	1.83
125	1.18	1.52	2.27	1.12	1.45	2.18	0.96	1.32	1.98
150	1.28	1.66	2.44	1.18	1.53	2.26	1.02	1.41	2.30
175	1.30	1.71	2.47	1.22	1.60	2.32	1.13	1.44	2.06
200	1.32	1.66	2.44	1.30	1.69	2.38	1.13	1.49	2.12
∞	1.65	1.95	2.58	1.65	1.95	2.58	1.65	1.95	2.58

Size	$\rho=10$	$\rho=10$	$\rho=10$
n	η_{90}	η_{95}	η_{99}
25	−0.65	−0.46	−0.09
50	0.14	0.40	0.96
75	0.48	0.76	1.35
100	0.65	0.96	1.53
125	0.82	1.17	1.84
150	0.90	1.25	1.93
175	0.99	1.40	1.99
200	1.01	1.38	2.06
∞	1.65	1.95	2.58

References

[1] Agrawala, E. (Ed.) (1997). *Machine Recognition of Patterns*. New York: IEEE Press.

[2] Agresti, A. (1981). Measures of nominal-ordinal association. *Journal of the American Statistical Association*, **76**, 524–529.

[3] Aitken, M.A., Nelson, C.W. and Reinfort, K.H. (1968). Tests of Correlation matrices. *Biometrika*, **55**, 327–334.

[4] Akaike, H. (1974). A new look at the statistical model identification. *IEEE Transactions on Automatic Control*, **19**, 716–723.

[5] Akaike, H. (1987). Factor analysis and AIC. *Psychometrika*, **52**, 317–332.

[6] Alalouf, I.S. (1980). A multivariate approach to a mixed linear model. *Journal of the American Statistical Association*, **75**, 194–200.

[7] Alexander, R.A. and Govern, D.M. (1994). Heteroscedasticity in ANOVA. *Journal of Educational Statistics*, **19**, 91–101.

[8] Algina, J. (1994). Some alternative approximate tests for a split-plot design. *Multivariate Behavioral Research*, **29**, 365–384.

[9] Algina, J., Oshima, T.C. and Tang, K.L. (1991). Robustness of Yao's, James', and Johansen's tests under variance-covariance heteroscedasticity and nonnormality. *Journal of Educational Statistics*, **16**, 125–139.

[10] Algina, J. and Tang, K.L. (1988). Type I error ratio for Yao's and James' tests of equality of mean vectors under variance-covariance heteroscedasticity. *Journal of Educational Statistics*, **13**, 281–290.

[11] Altham, P.M.E. (1984). Improving the precision of estimation by fitting a model. *Journal of the Royal Statistical Society, Series B*, **46**, 118–119.

[12] Amemiya, Y. (1985). What should be done when an estimated between-group covariance matrix is not nonnegative definite? *The American Statistician,* **39**, 112–117.

[13] Amemiya, Y. (1994). On multivariate mixed model analysis. In *Multivariate analysis and its applications*, IMS Lecture Notes-Monograph Series, T.W. Anderson, K.T. Fang and I. Olkin (Eds.), **24**, 83–95. Hayward, CA: Institute of Mathematical Statistics.

[14] Anderberg, M.R. (1973). *Cluster analysis for applications*. New York: Academic Press.

[15] Anderson, J.A. (1982). Logistic regression. In *Handbook of statistics, Volumn 2*, P.K. Krishnaiah and L.N. Kanel (Eds.), 169–191. Amsterdam: North -Holland.

[16] Anderson, T.W. (1951). Classification by multivarate analysis. *Psychometrika*, **16**, 631–650.

[17] Anderson, T.W. (1963). Asymptotic theory for principal components. *Annals of Mathematical Statistics*, **34**, 122–148.

[18] Anderson, T.W. (1984). *An introduction to multivariate statistical analysis*, Second Edition. New York: John Wiley.

[19] Anderson, T.W. (1985). Components of variance in MANOVA. In *Multivariate analysis-VI,* P.R. Krishnaiah (Ed.), 1–8. Amsterdam: North-Holland.

[20] Anderson, T.W. and Rubin, H. (1956). Statistical inferences in factor analysis. *Proceedings of the Third Symposium on Mathematical Statistics and Probability*, **V**, 111–150.

[21] Andrews, D.F., Gnanadesikan, R. and Warner, J.L. (1971). Transformations of multivariate data. *Biometrics*, **27**, 825–840.

[22] Angela, D. and Voss, D.T. (1999). *Design and Analysis of Experiments*. New York: Spring-Verlag.

[23] Arbuckle, J.L. and Wothke, W. (1999). *Amos 4.0 User's Guide*. Chicago, IL: Small-Waters Corporation.

[24] Arminger, G. (1998). A Bayesian approach to nonlinear latent variable models using the Gibbs samples and the Metropolis-Hastings algorithm, *Psychometrika*, **63**, 271–300.

[25] Arnold, S.F. (1979). Linear models with exchangeably distributed errors. *Journal of the American Statistical Association*, **74**, 194–199.

[26] Baker, F.B. and Hubert, L.J. (1975). Measuring the power of hierarchical cluster analysis. *Journal of the American Statistical Association*, **70**, 31–38.

[27] Bargman, R.E. (1970). Interpretation and use of a generalized discriminant function. In *Essays in probability and statistics*, R.C. Bose et al. (Eds.), 35–60. Chapel Hill, NC: University of North Carolina.

[28] Barnett, V. and Lewis, T. (1994). *Outliers in statistical data*. New York: John Wiley.

[29] Barrett, B.E. and Gray, J.B. (1994). A computational framework for variable selection in multivariate regression. *Statistics and Computing*, **4**, 203–212.

[30] Barrett, B.E. and Ling, R.F. (1992). General classes of influence measures for multivariate regression. *Journal of the American Statistical Association*, **87**, 184–191.

[31] Bartlett, M.S. (1938). Methods of estimating mental factors. *Nature*, **141**, 609–610.

[32] Bartlett, M.S. (1939). A note on tests of significance in multivariate analysis. *Proceedings of the Cambridge Philosophical Society*, **35**, 180–185.

[33] Bartlett, M.S. (1947). Multivariate analysis. *Journal of the Royal Statistical Society Supplement, Series B*, **9**, 176–197.

[34] Bartlett, M.S. (1950). Tests of significance in factor analysis. *British Journal of Psychology* - Statistics Section, **3**, 77–85.

[35] Bartlett, M.S. (1951). The effect of standardization on a chi-squared approximation in factor analysis. *Biometrika*, **38**, 337–344.

[36] Bartlett, M.S. (1954). A note on the multiplying factors for various χ^2 approximations. *Journal of the Royal Statistical Society, Series B*, **16**, 296–298.

[37] Basilevsky, A. (1994). *Statistical factor analysis and related topics*. New York: John Wiley.

[38] Bedrick, E.J. and Tsai, C.L. (1994). Model selection for multivariate regression in small samples. *Biometrics*, **50**, 226–231.

[39] Bekker, P.A. and Pollock, D.S.G. (1986). Identification of linear stochartic models with covariance restrictions. *Journal of Economics,* **31**, 179–208.

[40] Belsley, D.A., Kuh, E. and Welsch, R.E. (1980). *Regression diagnostics: Identifying influential data and sources of collinearity*. New York: John Wiley.

[41] Bentler, P.M. (1976). Multistructure statistical model applied to factor analysis. *Multivariate Behavioral Research*, **11**, 3–26.

[42] Bentler, P.M. (1986). Structural modeling and psychometrika: An historical perspective on growth and achievements. *Psychometrika*, **51**, 35–51.

[43] Bentler, P.M. (1990). Comparative fit indexes in structural models. *Psychological Bulletin*, **107**, 238–246.

[44] Bentler, P.M. (1993). *EQS structural equations program manual*. Los Angeles, CA: BMDP Statistical Software.

[45] Bentler, P.M. and Bonett, D.G. (1980). Significant tests and goodness of fit in the analysis of covariance structures. *Psychological Bulletin*, **88**, 588–606.

[46] Bentler, P.M. and Lee, S-Y. (1983). Covariance structures under polynomal constraints: Applications to correlation and alpha-type structural models. *Journal of Educational Statistics*, **8**, 207–222.

[47] Bentler, P.M. and Weeks, D.G. (1980). Linear structural equations with latent variables. *Psychometrika*, **45**, 289–308.

[48] Bentler, P.M. and Wu, E.J.C. (1995). *EQS for Windows User's Guide*. Encino, CA: Multivariate Software.

[49] Bentler, P.M. and Yuan, K-H (1996). Test of linear trend in eigenvalues of a covariance matrix with application to data analysis. *British Journal of Mathematical and Statistical Psychology*, **49**, 299–312.

[50] Benzécri, J.P. (1992). Correspondence analysis handbook. New York: Marcel Dekker.

[51] Berndt, E.R. and Savin, N.E. (1977). Conflict among criteria four testing hypotheses in the multivariate linear regression model. *Econometrica*, **45**, 1263–1277.

[52] Berry, W.D. (1984). *Nonrecursive causal models*. Newbury Park, CA: Sage Publications.

[53] Bijleveld, C.C.J.H. and van der Kamp, L.J.Th., (1998). *Longitudinal data analysis: Design, models and methods*. London: Sage *Publications*.

[54] Bilodeau, M. and Brenner, D. (1999). *Theory of multivariate statistics*. New York: Springer-Verlag.

[55] Blasius, J. and Greenacre, M.J. (1998). *Visualization of categorical data*. New York: Academic Press.

[56] Bock, R.D. and Peterson, A.C. (1975). A multivariate correction for attenuation. *Biometrika*, **62**, 673–678.

[57] Böckenholt, U. and Böckenholt, I. (1990). Canonical analysis of contingency tables with linear constraints. *Psychometrika*, **55**, 633–639.

[58] Boik, R.J. (1981). A priori tests in repeated measures designs: Effects of nonsphericity. *Psychometrika*, **46**, 241–255.

[59] Boik, R.J. (1988). The mixed model for multivariate repeated measures: Validity conditions and an approximate test. *Psychometrika*, **53**, 469–486.

[60] Boik, R.J. (1991). Scheffé mixed model for repeated measures a relative efficiency evaluation. *Communications in Statistics - Theory and Methods*, **A20**, 1233–1255.

[61] Boik, R.J. (1998). A local parameterization of orthogonal and semi-orthogonal matrics with applications. *Journal of Multivariate Analysis*, **67**, 244–276.

[62] Bollen, K.A. (1989). *Structural equations with latent variables*. New York: John Wiley.

[63] Bollen, K.A. and Joreskög, K.G. (1985). Uniqueness does not imply identification. *Sociological Methods and Research*, **14**, 155–163.

[64] Bollen, K.A. and Long, J. S. , Eds. (1993). *Testing Structural Equaltion Models*. Newbury Park, CA: Sage Publications.

[65] Bolton, B. (1971). A factor analytical study of communication skills and nonverbal abilities of deaf rehabilitation clients. *Multivariate Behavioral Research*, **6**, 485–501.

[66] Boullion, T.L. and Odell, P.L. (1971). *Generalized inverse matrices*. New York: John Wiley.

[67] Box, G.E.P. (1949). A general distribution theory for a class of likelihood criteria. *Biometrika*, **36**, 317–346.

[68] Box, G.E.P. (1950). Problems in the analysis of growth and wear curves. *Biometrics*, **6**, 362–389.

[69] Box, G.E.P. and Cox, D.R. (1964). An analysis of transformations. *Journal of the Royal Statistical Society*, Series B, **26**, 211–252.

[70] Bozdogan, H. (1987). Model selection and Akaike's information criterion (AIC): The general theory and its analytical extensions. *Psychometrika*, **52**, 345–370.

[71] Bradu, D. and Gabriel, K.R. (1974). Simultaneous statistical inference on interactions in two-way analysis of variance. *Journal of the American Statistical Association*, **69**, 428–436.

[72] Brady, H.E. (1985). Statistical consistency and hypothesis testing for nonmetric multidimensional scaling. *Psychometrika*, **50**, 509–537.

[73] Breiman, L. and Friedman, J.H. (1997). Predicting multivariate responses in multiple linear regression. *Journal of the Royal Statistical Society B*, **59**, 3–54.

[74] Breiman, L., Friedman, J.H., Olshen, R.A., and Stone, C.J. (1984). *Classification and regression trees*. Belmont, CA: Wadsworth.

[75] Breusch, T.S. (1979). Conflict among criteria for testing hypotheses: Extensions and comments. *Econometrica*, **47**, 203–207.

[76] Breusch, T.S. and Pagan, A.R. (1980). The Lagrange multiplier test and its applications to model specification in economics. *Review of Economic Studies*, **47,** 239–254.

[77] Brown, H. and Prescott, R. (1999). Applied *Mixed Models in Medicine.* New York: John Wiley.

[78] Brown, M.B. and Forsythe, A.B. (1974). The small sample behavior of some statistics which test the equality of several means. *Technometrics*, **16**, 129–132.

[79] Browne, M.W. (1974). Generalized least-squares estimators in the analysis of covariance structures. *South African Statistical Journal*, **8**, 1–24.

[80] Browne, M. W. (1975a). Predictive validity of a linear regression equation. *British Journal of Mathematical and Statistical Psychology,* **28**, 79–87.

[81] Browne, M. W. (1975b). Comparison of single sample and cross-validation methods for estimating the mean squared error of prediction in multiple linear regression. *British Journal of Mathematical and Statistical Psychology,* 28, 112–120.

[82] Browne, M.W. (1982). Covariance Structures. In *Topic in applied multivariate analysis,* D.M Hawkins (Ed.), 72–141. Cambridge: Cambridge University Press.

[83] Browne, M.W. (1984). Asymptotic distribution free methods in the analysis of covariance structures. *British Journal of Mathematical and Statistical Psychology*, **37**, 67–83.

[84] Browne, M.W. and Cudeck, R. (1989). Single sample cross-validation indices for covariance structures. *Multivariate Behavioral Research*, **24**, 445–455

[85] Buja, A., Hastie, T., and Tibshirani, R. (1989). Linear smoothers and additive models (with discussion). *Annals of Statistics*, **17**, 453–555.

[86] Bull, S.B. and Donner, A. (1987). The efficiency of multinominal logistic regression compared with multiple group discriminant analysis. *Journal of the American Statistical Association*, **82**, 1118–1122.

[87] Buse, A. (1982). The likelihood ratio, Wald, and Lagrange multiplier tests: An expository note. *The American Statistician*, **36**, 153–157.

[88] Byrk, A.S. and Raudenbush, S.W. (1992). *Hierarchical linear models: Applications and data analysis methods*. Newbury Park, CA: Sage Publications.

[89] Calvin, J.A., and Dykstra, R.L. (1991a). Maximum likelihood estimation of a set of covariance matrices under Loewner order restrictions with applications to balanced multivariate variance components models. *Annals of Statistics*, **19**, 850–869.

[90] Calvin, J.A., and Dykstra, R.L. (1991b). Least squares estimation of covariance matrices in balanced multivariate variance components models. *Journal of American Statistical Association*, **86**, 388–395.

[91] Campbell, N.A. (1978). The influence function as an aid in outlier detection in discriminant analysis. *Applied Statistics*, **27**, 251–258.

[92] Campbell, N.A. (1982). Robust procedures in multivariate analysis II. Robust canonical variate analysis. *Applied Statistics*, **31**, 1–8.

[93] Carmer, S.G. and Swanson, M.R. (1972). An evaluation of ten parwise multiple comparison procedures by Monte Carlo methods. *Journal of the American Statistical Association,* **68**, 66–74.

[94] Carroll, J.D. and Chang, J.J. (1970). Analysis of individual differences in multidimensional scaling via an N-way generalization of 'Eckart-Young' decomposition. *Psychometrika*, **35**, 283–319.

[95] Carroll, J.D. and Green, P. E. (1997). *Mathematical tools for applied multivariate anslysis,* Revised edition. San Diego, CA: Academic Press.

[96] Carroll, J. D., Kumbasar, E. and Romney, A.K. (1997). An equivalence relation between correspondence analysis and classical metric multidimensional scaling for the recovery of Duclidean distances. *British Journal of Mathematical and Statistical Psychology,* **50**, 81–92.

[97] Carter, E.M. and Srivastava, M.S. (1983). Asymptotic non-null distribution for the locally most powerful invariant test for sphericity. *Communications in Statistics - Theory and Methods*, **12**, 1719–1725.

[98] Casella, G. and Berger, R.L. (1990). *Statistical inference*. Belmont, CA: Wadsworth.

[99] Cattell, R.B. (1966). The scree tree for the number of factors. *Multivariate Behavioral Research*, **1**, 245–276.

[100] Chambers, J.M. (1977). *Computational methods for data analysis*. New York: John Wiley.

[101] Chatterjee, S. and Hadi, A.S. (1988). *Sensitivity analysis in linear regression*. New York: John Wiley.

[102] Chinchilli, V.M. and Elswick, R.K. (1985). A mixture of the MANOVA and GMANOVA models. *Communications in Statistics - Theory and Methods*, **14**, 3075–3089.

[103] Christensen, R. (1989). Lack of fit tests based on near or exact replicates. *Annals of Statistics*, **17**, 673–683.

[104] Christensen, R., Pearson, L.M., and Johnson, W. (1992). Case-deletion diagnostics for mixed models. *Technometrics*, **34**, 38–45.

[105] Christensen, W.F. and Rencher, A.C. (1997). A comparison of Type I error rates and power levels for seven solutions to the multivariate Behrens-Fisher problem. *Communications in Statistics - Simulation and Computation*, **26**, 1251–1273.

[106] Coelho, C. A. (1998). The generalized integer gamma distribution - A basis for distributions in multivariate statistics. *Journal of Multivariate Analysis,* **64**, 86–102.

[107] Cook, R.D. and Weisberg, S. (1980). Characterization of an empirical influence function for detecting influential cases in regression. *Technometrics*, **22**, 495–408.

[108] Cooley, W.W. and Lohnes, P.R. (1971). *Multivariate data analysis.* New York: Hohn Wiley.

[109] Coombs, W.T. and Algina, J. (1996). New test statistics for MANOVA/descriptive discriminant analysis. *Educational and Psychological Measurement*, **58**, 382–402.

[110] Cornell, J.E., Young, D.M., Seaman, S.L., and Kirk, R.E. (1992). Power comparisons of eight tests for sphericity in repeated measures designs. *Journal of Educational Statistics*, **17**, 233–249.

[111] Cotter, K. L. and Raju, N. S. (1982). An evaluation of formula-based population squared cross-validity estimation in one sample. *Educational and Psychological Measurement,* **40**, 101–112.

[112] Cox, C.M., Krishnaiah, P.R., Lee, J.C., Reising, J., and Schuurmann, F.J. (1980). A study of finite intersection tests for multiple comparisons of means. In *Multivariate analysis-V,* P.R. Krishnaiah (Ed.), 435–466. New York: North-Holland.

[113] Cox, D.R. (1961). Tests of separate families of hypotheses. In *Proceedings of the Fourth Berkeley Symposium on Mathematical Statistics and Probability, Volume* **1**, 105–123. Berkeley CA: University of California Press.

[114] Cox, D.R. (1962). Further results on tests of separate families of hypotheses. *Journal of the Royal Statistical Society*, **B24**, 406–424.

[115] Cox, T.F. and Cox, M.A.A. (1994). *Multidimensional scaling*. London: Chapman and Hall.

[116] Cramer, E.M. and Nicewander, W.A. (1979). Some symmetric invariant measures of multivariate association. *Psychometrika*, **44**, 43–54.

[117] Crowder, M.J. and Hand, D.J. (1990). *Analysis of repeated measurements*. New York: Chapman and Hall.

[118] Danford, M.B., Hughes, N.N., and McNee, R.C. (1960). On the analysis of repeated-measurements experiments. *Biometrics*, **16**, 547–565.

[119] Davidian, M. and Giltinan, D.M. (1995). *Nonlinear models for repeated measurement data*. New York: Chapman and Hall.

[120] Davidson, L.M. (1983). *Multidimensional scaling*. New York: John Wiley.

[121] Davidson, R. and MacKinnon, J.G. (1993). *Estimation and inference in economics*. New York: Oxford University Press.

[122] Davies, P.L. (1987). Asymptotic behavior of S-estimates of multivariate location and dispersion matrices. *Annals of Statistics*, **15**, 1269–1292.

[123] Davis, A.W. (1980). On the effects of moderate multivariate nonnormality on Wilks' likelihood ratio criterion. *Biometrika*, **67**, 419–427.

[124] Davis, A.W. (1982). On the effects of moderate multivariate nonnormality on Roy's largest root test. *Journal of the American Statistical Association*, **77**, 896–900.

[125] Davison, A.C. and Hall, P. (1992). On the bias and variability of bootstrap and cross-validation estimates of error rate in discrimination problems. *Biometrika*, **79**, 279–284.

[126] Dawid, A.P. (2000). Causal inference without counterfactuals (with discussion). *Journal of the American Statistical Association*, **96**, 407–424.

[127] Dean, A. and Voss, D.T. (1997). *Design and analysis of experiments*. New York: Springer-Verlag.

[128] de Leeuw, J. and Heiser, W. (1980). Multidimensional scaling with restrictions on the configuration. In *Multivariate analysis-V*, P.K. Krishnaiah (Ed.), 501–522. New York: North Holland.

[129] de Leeuw, J. and Meulman, J. (1986). A special jack-knife for multidimensional scaling. *Journal of Classification*, **3**, 97–112.

[130] Dempster, A.P. (1963). Stepwise multivariate analysis of variance based upon principal variables. *Biometrics*, **19**, 478–490.

[131] Dempster, A.P. (1969). *Elements of continuous multivariate analysis*. Reading, MA: Addison-Wesley.

[132] Dempster, A.P., Laird, N.M., and Rubin, D.B. (1977). Maximum likelihood from incomplete data via the EM algorithm. *Journal of the Royal Statistical Society, Series B*, **39**, 1–22.

[133] Devlin, S.J., Gnanadesikan, R. and Kettenring, J.R. (1975). Robust estimation and outlier detection with correlation coefficients. *Biometrika*, **35**, 182–190.

[134] Devlin, S.J., Gnanadesikan, R. and Kettenring, J.R. (1981). Robust estimation of dispersion matrices and principal components. *Journal of the American Statistical Association*, **76**, 354–362.

[135] Dhrymes, P.H. (2000). *Mathematics for Econometrics,* Third edition. New York: Springer-Verlag.

[136] Di Vesta, F.J. and Walls, R.T. (1970). Factor analysis of the semantic attributes of 487 words and some relationships to the conceptual behavior of fifth-grade children. *Journal of Educational Psychology*, **61**, Pt.2, December.

[137] Dufour, J.M. and Dagenais, M.G. (1992). Nonlinear models, rescaling and test invariance. *Journal of Statistical Planning and Inference*, **32**, 111–135.

[138] Duncan, O.D., Haller, H.O., and Portes, A. (1968). Peer influences on aspirations: A reinterpretation. *American Journal of Sociology*, **74**, 119–137.

[139] Dunn, G., Everitt, B.S. and Pickles, A. (1993). *Modeling covariances and latent variables using EQS*. London: Chapman and Hall.

[140] Eaton, M.L. (1985). The Gauss-Markov theorem in multivariate analysis. In *Multivariate Analysis VI,* P.R. Krishnaiah (Ed.), 177–201. Amsterdam: Elsevier Science Publisher B.V.

[141] Efron, B. and Gong, G. (1983). A leisurely look at the bootstrap, the jackknife and cross-validation. *The American Statistician*, **37**, 36–48.

[142] Elian, S.N. (2000) Simple forms of the best linear unbiased predictor in the general linear regression model. *The American Statisticisian,* **54**, 25–28.

[143] Elston, R.C. and Grizzle, J.E. (1962). Estimation of time response curves and their confidence bands. *Biometrics*, **18**, 148–159.

[144] Engle, R.F. (1984), Wald, likelihood ratio and Lagrance multiplier tests in economics. In *Handbook of Economics, Volumn II*, Z.Griliches and M.D. Intrilligator (Eds.), 775–826. Amsterdam: North Holland.

[145] Engle, R.F. , Hendry, D.F., and Richard, J.-F. (1983). Exogeneity. *Econometrica,* **51**, 277–304.

[146] Ericsson, N.R. and Irons, J.S. (Eds.) (1994). *Testing Exogeneity: An introduction.* Oxford: Oxford University Press.

[147] Everitt, B.S. (1984). *An introduction to latent variable methods*. London: Chapman and Hall.

[148] Everitt, B.S. (1993). *Cluster analysis*, Third Edition. London: Edward Arnold.

[149] Everitt, B.S. and Rabe-Hesketh, S. (1997). *The analysis of proximity data*. New York: John Wiley.

[150] Ferguson, T.S. *A course in large sample theory.* New York: Chapman and Hall.

[151] Fisher, G.R. and McAleer, M. (1981). Alternative procedures and associated tests of significance for non-nested hypotheses. *Journal of Economics*, **16**, 103–119.

[152] Fisher, R.A. (1936). The use of multiple measurements in taxonomic problems. *Annals of Eugenics*, **7**, 179–188.

[153] Fix, E. and Hodges, J.L., Jr. (1951). *Discriminatory analysis-nonparametric discrimination: Consistency properties*. Report No. 4, Project No. 21-49-004, School of Aviation Medicine, Randolph Air Force Base, Texas. (Reprinted as pp. 261–279 in Agrawala, 1977.)

[154] Flury, B. (1997). *A first course in multivariate statistics*. New York: Springer-Verlag.

[155] Fox, J. (1984). *Linear statistical models and related methods*. New York: John Wiley.

[156] Freedman, D.A. (1987). As others see us: A case study in path analysis (with discussion). *Journal of Educational Statistics*, **12**, 101–129.

[157] Friedman, J.H. (1991). Multivariate adaptive regression systems. *Annals of Statistics*, **19**, 1–141.

[158] Friendly, M. (1991). *SAS system for statistical graphics*, First Edition. Cary, NC: SAS Institute Inc.

[159] Fuchs, C. and Sampson, A.R. (1987). Simultaneous confidence intervals for the general linear model. *Biometrics*, **43**, 457–469.

[160] Fujikoshi, Y. (1974). Asymptotic expansions of the non-null distributions of three statistics in GMANOVA. *Annals of the Institute of Statistical Mathematics*, **26**, 289–297.

[161] Fujikoshi, Y. (1977). Asymptotic expansions of the distributions of the latent roots in MANOVA and the canonical correlations. *Journal of Multivariate Analysis*, **7**, 386–396.

[162] Fujikoshi, Y. (1980). Asymptotic expansions for the distributions of the sample roots under nonnormality. *Biometrika*, **67**, 45–51.

[163] Fujikoshi, Y. (1993). Two-way ANOVA models with unbalanced data. *Discrete Mathematics*, **115**, 315–334.

[164] Fuller, W.A. (1987). *Measurement error models*. New York: John Wiley.

[165] Fuller, W.A. and Battese, G.E. (1973). Transformations for estimation of linear models with nested-error structure. *Journal of the American Statical Association*, **68**, 626–632.

[166] Gabriel, K.R. (1968). Simultaneous test procedures. *Biometrika*, **55**, 489–504.

[167] Gabriel, K.R. (1971). The biplot graphic display of matrices with applications to principal component analysis. *Biometrika*, **58**, 453–467.

[168] Galecki, A.T. (1994). General class of covariance structures for two or more repeated factors in longitudinal data analysis. *Communications in Statistics - Theory and Methods*, **23**, 3105–3119.

[169] Gatsonis, C. and Sampson, A.R. (1989). Multiple Correlation: Exact Power and Sample Size Calculations. *Psychological Bulletin,* **106**, 516–524.

[170] Geisser, S. and Greenhouse, S.W. (1958). An extension of Box's results in the use of the F distribution in multivariate analysis. *Annals of Mathematical Statistics*, **29**, 885–891.

[171] Gifi, A. (1990). *Nonlinear multivariate analysis*. New York: John Wiley.

[172] Girshick, M.A. (1939). On the sampling theory of roots of determinantal equations. *Annals of Mathematical Statistics*, **10**, 203–224.

[173] Gittens, R. (1985). *Canonical analysis: A review with applications in ecology*. New York: Springer-Verlag.

[174] Gleser, L.J. (1968). Testing a set of correlation coefficients for equality. *Biometrika*, **34**, 513–517.

[175] Gleser, L.J. and Olkin, I. (1970). Linear models in multivariate analysis. In *Essays in probability and statistics*, R.C. Bose, et al. (Eds.), 267–292. Chapel Hill, NC: University of North Carolina.

[176] Glick, N. (1978). Additive estimators for pobabilities of correct classification. *Pattern Recognition*, **10**, 211–222

[177] Gnanadesikan, R. (1997). *Methods for statistical data analysis of multivariate observations*. New York: John Wiley.

[178] Gnanadesikan, R. and Kettenring, J.R. (1972). Robust estimates, residuals, and outlier detection with multiresponse data. *Biometrics*, **28**, 81–124.

[179] Gnaneshanandam, S. and Krzanowski, W.J. (1989). On selecting variables and assessing their performance using linear discriminant analysis. *Austrailian Journal of Statistics*, **31**, 433–447.

[180] Gnaneshanandam, S. and Krzanowski, W.J. (1990). Error-rate estimation in two-group discriminant analysis using the linear discriminant function. *Journal of Statistical Computation and Simulation*, **36**, 157–175.

[181] Godfrey, L.G. and Pesaran, M.H. (1983). Tests of non-nested regression models: Small sample adjustments and Monte Carlo evidence. *Journal of Econometrics*, **21**, 133–154.

[182] Goldberger, A.S. (1962). Best linear unbiased prediction in the generalized linear regression model. *Journal of the American Statistical Association,* **57**, 369–375.

[183] Goldberger, A.S. (1991). *A course in econometrics.* Cambridge, MA: Harvard University Press.

[184] Goldstein, H. (1995). *Multilevel statistical models*, Second Edition. New York: Halstead Press.

[185] Goodman, L.A. and Kruskal, W.H. (1963). Measures of association for cross classifications III: Approximte sample theory. *Journal of the American Statistical Association,* **58**, 310–364.

[186] Goodnight, J.H. (1978). *Tests of hypotheses in fixed-effects linear models*, SAS Technical Report R-101. Cary, NC: SAS Institute Inc.

[187] Graybill, F.A. (1983). *Matrices with applications in statistics*, Second Edition. Belmont, CA: Wadsworth.

[188] Green, B.F. (1979). The two kinds of linear discriminant functions and their relationship. *Journal of Educational Statistics*, **4**, 247–263.

[189] Green, B.F., and Tukey, J. (1960). Complex analysis of variance: General problems. *Psychometrika*, **25**, 127–152.

[190] Greenacre, M.J. (1984). *Theory and applications of correspondence analysis*. London: Academic Press.

[191] Greene, W.H. (2000). *Econometric analysis*, Fourth Edition. New York: Prentice Hall

[192] Greene, W.H. (1993). *Econometric analysis*, Second Edition. New York: Macmillan.

[193] Greenhouse, S.W. and Geisser, S. (1959). On methods in the analysis of profile data. *Psychometrika*, **24**, 95–112.

[194] Grizzle, J.E. (1965). The two-period change-over design and its use in clinical trials. *Biometrics*, **21**, 467–480.

[195] Grizzle, J.E., and Allen, D.M. (1969). Analysis of growth and dose response curves. *Biometrics*, **25**, 357–381.

[196] Gunst, R.F., Webster, J.T. and Mason, R.L. (1976). A comparison of least squares and latent root regression estimators. *Technometrics*, **18**, 74–83.

[197] Guttman, L. (1954). Some necessary conditions for common-factor analysis. *Psychometrika*, **19**, 149–161.

[198] Haerdle, W. (1990). *Applied nonparametric regression*. Cambridge: Cambridge University Press.

[199] Hakstain, A.R. and Abel, R.A. (1974). A further comparison of oblique factor transformation methods. *Psychometrika*, **39**, 429–444.

[200] Hand, D.J. (1981). *Discriminant analysis*. New York: John Wiley.

[201] Hannan, E.J. and Quinn, B.G. (1979). The determination of the order of an autoregression. *Journal of the Royal Statistical Society, Series B*, **41**, 190–195.

[202] Hansen, M.A. (1999). Confirmatory factor analysis of the Maryland State performance assessment program teacher questionnaire for the social studies subject area. Unpublished Master's paper. University of Pittsburgh.

[203] Harmon, H.H. (1976). *Modern factor analysis*, 2nd edition. Chicago, IL: The University of Chicago Press.

[204] Harris, C. (1964). Some recent development in factor analysis. *Educational and Psychological Measurement*, **24**, 193–206.

[205] Harris, P. (1984). An alternative test for multisample sphericity. *Psychometrika*, **49**, 273–275.

[206] Hartigan, J.A. (1975). *Clustering algorithms*. New York: John Wiley.

[207] Hartley, H.O. (1967). Expectations, variances and covariances of ANOVA mean squares by 'synthesis'. *Biometrics*, **23**, 105–114. Corrigenda, **23**, 853.

[208] Hartley, H.O. and Rao, J.N.K. (1967). Maximum likelihood estimation for the mixed analysis of variance models. *Biometrika*, **54**, 93–108.

[209] Harville, D.A. (1976). Extension of the Gauss-Markov theorem to include the estimation of random effects. *Annals of Statistics*, **4**, 384–395.

[210] Harville, D.A. (1977). Maximum likelihood approaches to variance component estimation and to related problems. *Journal of the American Statistical Association*, **72**, 320–340.

[211] Harville, D.A. (1997). *Matrix algebra from a statistician's perspective*. New York: Springer-Verlag.

[212] Hastie, T. and Stuetzle, W. (1989). Principal curves. *Journal of the American Statistical Association*, **84**, 502–516.

[213] Hatcher, L. (1994). *A step-by-step approach to using the SAS system for factor analysis and structural equation modeling*. Cary, NC: SAS Institute Inc.

[214] Hausman, J.A. (1975). An instrumental variable approach to full information estimators for linear and certain nonlinear economic models. *Econometrica*, **43**, 727–738.

[215] Hawkins, D.M. (1974). The detection of errors in multivariate data using principal components. *Journal of the American Statistical Association*, **69**, 340–344.

[216] Hawkins, D.M. (1976). The subset problem in multivariate analysis of variance. *Journal of the Royal Statistical Society, Series B*, **38**, 132–139.

[217] Hayter, A.T. and Tsui, K.L. (1994). Identification in multivariate quality control problems. *Journal of Quality Technology*, **26**, 197–208.

[218] Hecker, H. (1987). A generalization of the GMANOVA-model. *Biometrical Journal*, **29**, 763–790.

[219] Heise, D.R. (1986). Estimating nonlinear models corrected for measurement error. *Sociological Methods and Research*, **14**, 447–472.

[220] Heiser, W.J. (1981). *Unfolding analysis of proximity data*. Luden: Department of Datatheory, University of Leiden.

[221] Henderson, C.R. (1963). Selection index and expected genetic advance. In *Statistical genetics and plant breeding*, W.D. Hanson and H.F. Robinson (Eds.), 141–163. National Academy of Sciences and National Research Council Publication No. *982*, Washington D.C.

[222] Henderson, H.V. and Searle, S.R. (1981). The vec-permutation matrix, the vec-operator and Kronecker products: A review. *Linear and Multilinear Algebra*, **9**, 271–288.

[223] Hines, W.G.S. (1996). Pragmatics of pooling in ANOVA tables. *The American Statistician*, **50**, 127–139 .

[224] Hinkelmann, K., Wolfinger, R.D., and Stroup, W.W. (2000). Letters to the editor, on Voss (1999), with reply. *The American Statistician*, **51**, 228–230.

[225] Hochberg, Y. and Tamhane, A.C. (1987). *Multiple comparison procedures*. New York: John Wiley.

[226] Hocking, R.R. (1985). *The analysis of linear models*. Belmont, CA: Wadsworth.

[227] Hoerl, A.E. and Kennard, R.W. (1970a). Ridge regression: Biased estimation for nonorthogonal problems. *Technometrics*, **12**, 55–67.

[228] Hoerl, A.E. and Kennard, R.W. (1970b). Ridge regression. Applications to nonorthogonal problems. *Technometrics*, **12**, 69–82.

[229] Hossain, A. and Naik, D.N. (1989). Detection of influential observations in multivariate regression. *Journal of Applied Statistics*, **16**, 25–37.

[230] Hotelling, H. (1931). The generalization of Student's ratio. *Annals of Mathematical Statistics*, **2**, 360–378.

[231] Hotelling, H. (1933). Analysis of a complex of statistical variables into principal components. *Journal of Educational Psychology*, **24**, 417–441, 498–520.

[232] Hotelling, H. (1936). Relations between two sets of variates. *Biometrika*, **28**, 321–377.

[233] Howe, W.G. (1955). *Some contributions to factor analysis*. Report No. ORNL-1919. Oak Ridge, Tenn: Oak Ridge National Laboratory.

[234] Hsiao, C. (1986). *Analysis of panal data.* Cambridge. Econometric Society Monographs No. 11. Cambridge University Press.

[235] Huber, P.J. (1970). Studentized robust estimates. In *Nonparametric techniques in statistical inference*, M.L. Puri (Ed.), 453–463. London: Cambridge University Press.

[236] Huber, P.J. (1981). *Robust statistics*. New York: John Wiley.

[237] Huberty, C.J. (1994). *Applied discriminant analysis*. New York: John Wiley.

[238] Hummel, T.J., and Sligo, J. (1971). Empirical comparison of univariate and multivariate analysis of variance procedures. *Psychological Bulletin*, **76**, 49–57.

[239] Huynh, H. and Feldt, L.S. (1970). Conditions under which mean square ratios in repeated measurements designs have exact F-distribution. *Journal of the American Statistical Association*, **65**, 1582–1589.

[240] Huynh, H. and Feldt, L.S. (1976). Estimation of the Box correction for degrees of freedom from sample data in randomized block and split-plot designs. *Journal of Educational Statistics*, **1**, 69–82.

[241] Ito, P.K. (1980). Robustness of ANOVA and MANOVA procedures. In *Handbook of statistics, Volumn 1*, P.R. Krishnaiah (Ed.), 199–236. New York: North Holland.

[242] Jackson, J.E. (1991). *A user's guide to principal components*. New York: Wiley-Interscience.

[243] Jackson, J.E. and Mudholkar, G.S. (1979). Control procedures for residuals associated with principal component analysis. *Technometrics*, **21**, 341–349

[244] Jambu, M. (1991). *Exploratory and multivariate data analysis.* New York: Academic Press.

[245] James, A.T. (1969). Tests of equality of latent roots of the covariance matrix. In *Multivariate analysis-II*, P.K. Krishnaiah (Ed.), 205–218. New York: Academic Press.

[246] James, G.S. (1954). Tests of linear hypotheses in univariate and multivariate analysis when the ratio of the population variance are unknown. *Biometrika*, **41**, 19–43.

[247] Janky, D. G. (2000). Sometimes pooling for analysis of variance hypothesis tests: A review and study of a split-plot model. *The American Statisitician,* **54,** 269–279.

[248] Jardine, N. and Sibson, R. (1971). *Mathematical taxonomy*. New York: John Wiley

[249] Jennrich, R.I. and Schluchter, M.D. (1986). Unbalanced repeated-measures models with structured covariance matrices. *Biometrics*, **42**, 805–820.

[250] Jobson, J.D. (1991). *Applied multivariate data analysis, Volumn I: Regression and experimental design*. New York: Springer-Verlag.

[251] Jobson, J.D. (1992). *Applied multivariate data analysis, Volumn II: Categorical and multivariate methods*. New York: Springer-Verlag.

[252] Johansen, S. (1980). The Welch-James approximation to the distribution of the residual sum of squares in a weighted linear regression. *Biometrika*, **67**, 85–92.

[253] John, S. (1971). Some optimal multivariate tests. *Biometrika*, **58**, 123–127.

[254] Johnson, R.A. and Wichern, D.W. (1998). *Applied multivariate statistical analysis*. New Jersey: Prentice Hall.

[255] Johnson, S.C. (1967). Hierarchical clustering schemes. *Psychometrika*, **32**, 241–254.

[256] Jolliffe, I.T. (1972). Discarding variables in principal component analysis, I. Artificial data. *Applied Statistics*, **21**, 160–173.

[257] Jolliffe, I.T. (1973). Discarding variables in principal component analysis, II. Real data. *Applied Statistics*, **22**, 21–31.

[258] Jolliffe, I.T. (1986). *Principal components analysis*. New York: Springer-Verlag.

[259] Jöreskog, K.G. (1963). *Statistical estimation in factor analysis*. Stockholm: Almguist and Wiksell.

[260] Jöreskog, K.G. (1967). Some contributions to maximum likelihood factor analysis. *Psychometrika*, **32**, 443–482.

[261] Jöreskog, K.G. (1969a). Efficient estimation in image factor analysis. *Psychometrika*, **34**, 51–75.

[262] Jöreskog, K.G. (1969b). A general approach to confirmatory maximum likelihood factor analysis. *Psychometrika*, **39**, 183–202.

[263] Jöreskog, K.G. (1970). A general method for analysis of covariance structures. *Biometrika*, **57**, 239–251.

[264] Jöreskog, K.G. (1971). Simultaneous factor analysis in several population. *Psychometrika*, **59**, 409–426.

[265] Jöreskog, K.G. (1973a). A general method for estimating a linear structural equation system. In *Structural equation models in the social sciences*, A.S. Goldberger and O.D. Duncan (Eds.), 85–112. New York: Academic Press.

[266] Jöreskog, K.G. (1973b). Analysis of Covariance Structures. In *Multivariate analysis-III*, P.R. Krishnaiah (Ed.), 263–285. New York: Academic Press.

[267] Jöreskog, K.G. (1977). Structural equation models in the social sciences: specification, estimation and testing. In *Applications in statistics*, P.R. Krishnaiah (Ed.), 265–287. New York: North-Holland.

[268] Jöreskog, K.G. (1979). Analyzing psychological data by structural analysis of covariance matrices. In *Advances of factor analysis and structural equation models,* K.G. Jöreskog and D. Sörbom (Eds.), 45–100. Cambridge, MA: Abt Associates Inc.

[269] Jöreskog, K.G. and Goldberger, A.G. (1972). Factor analysis by generalized least squares. *Psychometrika,* **37**, 243–260.

[270] Jöreskog, K.G. and Lawley, D.N. (1968). New methods in maximum likelihood factor analysis. *British Journal of Mathematical and Statistical Psychology,* **21**, 85–96.

[271] Jöreskog, K.G. and Sörbom, D. (1979). *Advances in factor analysis and structural equation models*. Cambridge, MA: Abt Associates Inc.

[272] Jöreskog, K.G. and Sörbom, D. (1993). *LISREL* **8**: *User's reference guide*. Chicago: Scientific Software International.

[273] Jöreskog, K.G., Sörbom, D., du Toit, S. and du Toit, M. (2000). *LISREL* **8**: *New statistical features.* (Second printing with revisions). Chicago: Scientific Software Internationall.

[274] Kaiser, H. F. (1958). The varimax criterion for analytic rotation in factor analysis. *Psychometrika*, **23**, 187–200.

[275] Kaiser, H.F. and Derflinger, G. (1990). Same contrasts between maximum likelihood factor analysis and alpha factor analysis. *Applied Psychological Measurement*, **14**, 19–32.

[276] Kaiser, H.F. and Rice, J. (1974). Little jiffy, mark IV. *Educational and Psychological Measurement*, **34**, 111–117.

[277] Kaplan, D. (1990). Evaluating and modifying covariance structure models: A review and recommendation. *Multivariate Behavioral Research,* **25**, 139–155.

[278] Kaplan, D. (2000) *Structural Equation Modeling Foundations and Extensions.* Thousand Oaks, CA: Sage Publications.

[279] Kaplan, D., Harik, P., and Hotchkiss, L. (2000). Cross-sectional estimation of dynamic structural equation models in disequilibrium. In *Structual equation modeling present and future: A Fetschrift in honor of Karl G. Joreskog, 315–339.* Lincolnwood, IL: Scientific Software International.

[280] Kariya, T. (1985). *Testing in the multivariate linear model*. Japan: Kinokuniya.

[281] Kariya, T., Fujikoshi, Y. and Krishnaiah, P.R. (1984). Tests of independence of two multivariate regression equations with different design matrices. *Journal of Multivariate Analysis*, **15**, 383–407.

[282] Keesling, J.W. (1972). *Maximum likelihood approaches to causal analysis*. Ph.D. Dissertation. Department of Education: University of Chicago.

[283] Keramidas, E.M., Devlin, S.J., and Gnanadesikan, R. (1987). A graphical procedure for comparing principal components of several covariance matrices. *Communications in Statistics - Simulation and Computation*, **16**, 161–191.

[284] Keselman, H.J., Carriere, M.C. and Lix, L.M. (1993). Testing repeated measures when covariance matrices are heterogeneous. *Journal of Educational Statistics*, **18**, 305–319

[285] Keselman, H.J. and Keselman, J.C. (1993). Analysis of repeated measurements. In *Applied analysis of variance in behavioral science*, L.K. Edwards (Ed.), 105–145. New York: Marcel Dekker.

[286] Keselman, H.J. and Lix, L.M. (1997). Analyzing multivariate repeated measures designs when covariance matrices are heterogeneous. *British Journal of Mathematical and Statistical Psychology*, **50**, 319–338.

[287] Khatri, C.G. (1966). A note on a MANOVA model applied to problems in growth curves. *Annals of the Institute of Statistical Mathematics*, **18**, 75–86.

[288] Khatri, C.G. and Patel, H.I. (1992). Analysis of multicenter trial using a multivariate approach to a mixed linear model. *Communications in Statistics - Theory and Methods*, **21**, 21–39.

[289] Khattree, R. and Naik, D.N (1990) A note on testing for the interaction in a MANOVA model. *Biometrical Journal,* **32**, 713–716.

[290] Khattree, R. and Naik, D.N (1995). *Applied multivariate statistics with SAS software*. Cary, NC: SAS Institute Inc.

[291] Khattree, R. and Naik, D.N (2000). *Multivariate data reduction and discrimination with SAS software*. Cary, NC: SAS Institute Inc.

[292] Khuri, A.I. (1982). Direct products. A powerful tool for the analysis of balanced data. *Communications in Statistics - Theory and Methods*, **11**, 2903–2920.

[293] Khuri, A.I. (1985). A test of lack of fit in a linear multiresponse model. *Technometrics*, **27**, 213–218.

[294] Khuri, A.I. (1993). *Advanced calculus with applications in statistics*. New York: John Wiley.

[295] Khuri, A.I., Mathew, T. and Nel, D.G. (1994). A test to determine closeness of multivariate Satterthwaite's approximation. *Journal of Multivariate Analysis*, **51**, 201–209.

[296] Khuri, A.I., Mathew, T. and Sinha, B.K. (1998). *Statistical tests for mixed linear models*. New York: John Wiley.

[297] Kim, S. (1992). A practical solution to the multivariate Behrens-Fisher problem. *Biometrika*, **79**, 171–176.

[298] Kirk, R.E. (1995). *Experimental design procedure for the behavioral sciences*, Third Edition. Belmont, CA: Brooks/Cole.

[299] Kish, C.W. and Chinchilli, V.M. (1990). Diagnostics for identifying influential cases in GMANOVA. *Communications in Statistics - Theory and Methods*, **19**, 2683–2704.

[300] Kleinbaum, D.G. (1973). A generalization of the growth curve model which allows missing data. *Journal of Multivariate Analysis,* **3**, 117–124.

[301] Koffler, S.L. and Penfield, D.A. (1979). Nonparametric discrimination procedures for non-normal distributions. *Journal of Statistical Computation and Simulation*, **8**, 281–299.

[302] Koffler, S.L. and Penfield, D.A. (1982). Nonparametric classification based upon inverse normal scores and rank transformations. *Journal of Statistical Computation and Simulation*, **15**, 51–68.

[303] Korin, B.P. (1972). Some comments on the homoscedasticity criterion M and the multivariate analysis of T^2, W and R. *Biometrika*, **59**, 215–216.

[304] Korin, B.P. (1968). On the distribution of a statistic used for listing a covariance matrix. *Biometrika*, **55**, 171–178.

[305] Krane, W.R. and McDonald, R.P. (1978). Scale invariance and the factor analysis of correlation matrices. *British Journal of Mathematical and Statiistical Psychology,* **31**, 218–228.

[306] Kreft, I. and de Leeuw, J. (1998). *Introducing multilevel modeling.* Newbury Park, CA: Sage Publications.

[307] Kres, H. (1983). *Statistical tables for multivariate analysis*. New York: Springer-Verlag.

[308] Krishnaiah, P.R. (1965a). Multiple comparison tests in multiresponse experiments. *Sankhyã Series A*, **27**, 65–72.

[309] Krishnaiah, P.R. (1965b). On the simultaneous ANOVA and MANOVA tests. *Annals of the Institute of Statistical Mathematics*, **17**, 35–53.

[310] Krishnaiah, P.R. (1969). Simultaneous test procedures under general MANOVA models. In *Multivariate analysis-II*, P.R. Krishnaiah (Ed.), 121–143. New York: Academic Press.

[311] Krishnaiah, P.R. (1979). Some developments on simultaneous test procedures. In *Developments in Statistics, Volumn 2*, P.R. Krishnaiah (Ed.), 157–201. New York: North Holland.

[312] Krishnaiah, P.R. (1982). Selection of variables in discriminations. In *Handbook of statistics, Volumn 2*, P.R. Krishnaiah and L.N. Kanal (Eds.), 883–892. Amsterdam: North-Holland.

[313] Krishnaiah, P.R. and Lee, J.C. (1976). On covariance structures. *Sankhyã*, **38**, 357–371.

[314] Krishnaiah, P.R. and Lee, J.C. (1980). Likelihood ratio tests for mean vectors and covariance matrices. In *Handbook of statistics, Volumn 1*, P.R. Krishnaiah (Ed.), 513–570. New York: North Holland.

[315] Kruskal, J.B. (1964a). Multidimensional scaling by optimizing goodness of fit to a nonmetric hypothesis. *Psychometrika*, **29**, 1–27.

[316] Kruskal, J.B. (1964b). Nonmetric multidimensional scaling: A numerical method. *Psychometrika*, **29**, 115–129.

[317] Kruskal, J.B. (1977). Multidimensional scaling and other methods for discovering structure. In *Statistical methods for digital computers, Volumn III*, K. Enslein, A. Ralston and H.S. Wilf (Eds.), 296–339. New York: John Wiley.

[318] Kruskal, J.B. and Wish, M. (1978). *Multidimensional scaling*. Beverly Hills, CA: Sage Publications.

[319] Krzanowski, W.J. (1977). The performance of Fisher's linear discriminant function under non-optimal conditions. *Technometrics*, **19**, 191–200.

[320] Krzanowski, W.J. (1982). Between-groups comparison of principal components - Some sampling results. *Journal of Statistical Computation and Simulation*, **15**, 141–154.

[321] Kshirsagar, A.M. (1988). A note on multivariate linear models with non-additivity. *Australian Journal of Statistics*, **30**, 1–7.

[322] Kshirsagar, A.M. (1993). Tukey's nonadditivity test. In *Applied analysis of variance in behavioral science*, L.K. Edwards (Ed.), 421–435. New York: Marcel Dekker.

[323] Kshirsagar, A.M. and Arseven, E. (1975). A note on the equivalency of two discriminant functions. *The American Statistician*, **29**, 38–39.

[324] Lachenbruch, P.A. (1975). *Discriminant analysis*. New York: Hafner Press.

[325] Lachenbruch, P.A. and Mickey, M.A. (1968). Estimation of error rates in discriminant analysis. *Technometrics*, **10**, 1–10.

[326] Laird, N.M. and Ware, J.H. (1982). Random-effects models for longitudinal data. *Biometrics*, **38**, 963–974.

[327] Laird, N.M., Lange, N. and Stram, D. (1987). Maximum likelihood computations with repeated measures: Application of the EM algorithm. *Journal of the American Statistical Association*, **82**, 97–105.

[328] Lange, N. and Ryan, L. (1989). Assessing normality in random effects models. *Annals of Statistics*, **17**, 624–643.

[329] Lawley, D.N. (1938). A generalization of Fisher's z-test. *Biometrika*, **30**, 180–187.

[330] Lawley, D.N. (1940). The estimation of factor loadings by the method of maximum likelihood. *Proceedings of the Royal Statistical Society of Edenburg, Section A*, **60**, 64–82.

[331] Lawley, D.N. (1956). Tests of significance of the latent roots of covariance and correlation matrices. *Biometrika*, **43**, 128–136.

[332] Lawley, D.N. (1959). Tests of significance in canonical analysis. *Biometrika*, **46**, 59–66.

[333] Lawley, D.N. (1963). On testing a set of correlation coefficients for equality. *Annals of Mathematical Statistics*, **34**, 149–151.

[334] Lawley, D.N. and Maxwell, A.E. (1971). *Factor analysis as a statistical model*. London: Butterworth.

[335] Layard, M.W. (1974). A Monte Carlo comparison of tests for equality of covariance matrices. *Biometrika*, **61**, 461–465.

[336] Lee, J.C., Chang, T.C., and Krishnaiah, P.R. (1977). Approximations to the distribution of the likelihood ratio statistics for listing certain structures on the covariance matrices of real multivariate normal populations. In *Multivariate analysis - IV*, P.R. Krishnaiah (Ed.), 105–118. New York: North-Holland.

[337] Lee, S-Y. and Hershberger, S. (1990). A simple rule for generating equivalent models in covariance structure modeling, *Multivariate Behavioral Research,* **25**, 313–334.

[338] Lee, S-Y. (1971). Asymptotic formulae for the distribution of a multivariate test statistic: Power comparisons of certain tests. *Biometrika*, **58**, 647–651.

[339] Lee, S-Y. and Jennrich, R.I. (1979). A study of algorithums for covariance structure analysis with specific comparisons using factor analysis. *Psychometrika*, **44**, 99–113.

[340] Lee Y.H.K. (1970). Multivariate analysis of variance for analyzing trends in repeated observations. Unpublished doctoral dissertation, University of California, Berkley.

[341] Lehmann, E. L. (1994). *Testing statistical hypotheses,* Second Edition. New York: Chapman and Hall.

[342] Levy, M.S. and Neill, J.W. (1990). Testing for lack of fit in linear multiresponse models based on exact or near replicates. *Communications in Statistics - Theory and Methods*, **19**, 1987–2002.

[343] Lincoln, J. R. (1978). Community structure and industrial conflict: An analysis of strike activity in SMSAs. *American Sociological Review*, **43**, 199–220.

[344] Lindley, D.V. and Smith, A.F.M. (1972). Bayes estimates for the linear model. *Journal of the Royal Statistical Society*, **34**, 1–42.

[345] Lindsay, R. M. and Ehrenberg, A. C. C. (1993). The design of replicatioed studies. *The American Statistician,* **47**, 217–228.

[346] Lindstrom, M.J. and Bates, D.M. (1988). Newton-Raphson and EM algorithms for linear mixed-effects models for repeated-measures data. *Journal of the American Statistical Association,* **83**, 1014–1022.

[347] Lingoes, J.C. and Guttman, L. (1967). Nonmetric factor analysis: A rank reducing alternative to linear factor analysis. *Multivariate Behavioral Research*, **2**, 405–505.

[348] Littell, R.C., Freund, R.J., and Spector, P.C. (1991). *SAS System for linear models*, Third Edition. Cary, NC: SAS Institute Inc.

[349] Littell, R.C., Milliken, G.A., Stroup, W.W. and Wolfinger, R.D. (1996). *SAS system for mixed models*. Cary, NC: SAS Institute Inc.

[350] Little, R.J.A. (1988). A test of missing completely at random for multivariate data with missing values. *Journal of the American Statistical Association*, **83**, 1198–1202.

[351] Little, R.J.A. (1992). Regression with missing X's: A review. *Journal of the American Statistical Association*, **87**, 1227–1237.

[352] Little, R.J.A. and Rubin, D.B. (1987). *Statistical analysis with messy data*. New York: John Wiley.

[353] Longford, N.T. (1993a). *Random coefficient models*. New York: Oxford University Press.

[354] Longford, N. T. (1993b). Regression analysis of multilevel data with measurement error. *British Journal of Mathematical and Statistical Psychology,* **46**, 301–311.

[355] Looney, S.W. (1995). How to use tests for univariate normality to assess multivariate normality. *The American Statistician*, **49**, 64–69.

[356] Looney, S.W. and Gulledge, T.R.,Jr. (1985). Use of the correlation coefficient with normal probability plots. *The American Statistician*, **39**, 75–79.

[357] Lopuaä, H.P. and Rousseeuw, P.J. (1991). Breakdown points of affine estimators of multivariate location and covariance matrices. *Annals of Statistics*, **19**, 229–248.

[358] Lubischew, A.A. (1962). On the use of discriminant functions in taxonomy. *Biometrics*, **18**, 455–477.

[359] MacCallum, R.C., Browne, M.W., and Sugawara, H.M. (1996). Power analysis and determination of sample size for covariance structure modeling. *Psychological Methods*, **1**, 130–149.

[360] MacCallum, R.C. (1986). Specification searches in covariance structure modeling. *Psychological Bulletin*, **100**, 107–120.

[361] MacCallum, R.C., Kim, C., Malarkey, W.B. and Kiecolt-Glaser, J.K. (1997). Studying multivariate change using multilevel models and latent curve models. *Mutivariate Behavioral Research,* **32**, 21–253.

[362] MacCallum, R.C., Roznowski, M. and Necowitz, L.B. (1992). Model modifications covariance structure analysis: The problem of capitalization on chance. *Psychological Bulletin*, 111, 490–504.

[363] MacQueen, J.B. (1967). Some methods for classification and analysis of multivariate observations. In *Proceedings of the 5th Berkeley Symposium in Mathematical Statistics and Probability*, 281–297. Berkeley, CA: University of California Press.

[364] Magnus, J.R. and Neudecker, H. (1979). The commutation matrix: Some properties and applications. *Annals of Statistics*, **7**, 381–394.

[365] Magnus, J.R. and Neudecker, H. (1980). The elimination matrix: Some lemmas and applications. *SIAM Journal of Algebra and Discrete Mathematics*, **1**, 422–449.

[366] Magnus, J.R. and Neudecker, H. (1999). *Matrix differential calculus with applications in statistics and econometrics*. New York: John Wiley.

[367] Mallows, C.L. (1973). Some comments on C_p. *Technometrics*, **15**, 661–675.

[368] Marascuilo, L. (1969). Lecture notes in statistics. University of California, Berkeley. See also Marascuilo and Amster (1966). The effect of variety in children's concept learning. *California Journal of Educational Research*, Vol XVII. No. 3.

[369] March, H.W., Balla, J.R. and McDonald, R.P. (1988). Goodness-of-fit and indices in confirmatory factor analysis. *Psychological Bulletin*, **103**, 391–410.

[370] Marcoulides, G.A. and Schumacker, R.E. (1996). *Advanced structural equation modeling: Issues and techniques*. Mahwah, NJ: Lawrence Erlbaum.

[371] Mardia, K.V. (1970). Measures of multivariate skewness and kurtosis with applications. *Biometrika*, **50**, 519–530.

[372] Mardia, K.V. (1971). The effect of nonnormality on some multivariate tests of robustness to nonnormality in the linear model. *Biometrika*, **58**, 105–121.

[373] Mardia, K.V. (1980). Tests for univariate and multivariate normality. In *Handbook of Statistic, Volumn 1: Analyis of Variance,* P.R. Krishnaiah (Ed.), 279–320 New York: North-Holland.

[374] Mardia, K.V., Kent, J.T. and Bibby, J.M. (1979). *Multivariate analysis*. New York: Academic Press.

[375] Maronna, R.A. (1976). Robust M-estimators of multivariate location and scatter. *Annuals of Statistics*, **4,** 51–67.

[376] Mathew, T. (1989). MANOVA in the multivariate components of variance model. *Journal of Multivariate Analysis,* **29**, 30–38.

[377] Mathew, T., Niyogi, A., and Sinha, B.K. (1994). Improved nonnegative estimation of variance components in balanced multivariate mixed models. *Journal of Multivariate Analysis,* **51**, 83–101.

[378] Mauchly, J.W. (1940). Significance test of sphericity of n-variate normal populations. *Annals of Mathematical Statistics*, **11**, 204–209.

[379] McAleer, M. (1983). Exact tests of a model against nonnested alternatives. *Biometrika*, **70**, 285–288.

[380] McAleer, M. (1995). The significance of testing empirical non-nested models. *Journal of Econometrics*, **67**, 149–171.

[381] McArdle, J.J. (1986). Dynamic but structual equation modeling of reported measures data. In J.R. Nesselroade and R.B. Cattell (Eds.), *Handbook of multivariate experimental psychology*, Vol 2, pp. 561–614. New York: Plenum.

[382] McArdle, J.J. and McDonald, R.P. (1984). Some algebraic properties of the reticular action model for moment structures. *British Journal of Mathematical and Statistical Psychology*, **37**, 234–251.

[383] McCabe, G.P. (1984). Principal variables. *Technometrics,* **26**, 137–141.

[384] McCullagh, P. and Nelder, J.A. (1989). *Generalized linear models*, Second Edition. New York: Chapman and Hall.

[385] McCulloch, C.E. (1982). Symmetric matrix derivatives with applications. *Journal of the American Statistical Association*, **77**, 679–682.

[386] McCulloch, C.E. and Searle, S.R. (2001). *Generalized, linear, and mixed models.* New York: John Wiley.

[387] McDonald, R.P. (1972). A multivariate extension of Tukey's one degree of freedom test for non-additivity. *Journal of American Statistical Association*, **67**, 674–675.

[388] McDonald, R.P. (1980). A simple comprehensive model for the analysis of covariance structures: Some remarks on applications. *British Journal of Mathematical and Statistical Psychology*, **33**, 161–183.

[389] McDonald, R.P. (1985). *Factor analysis and related models*. Hillsdale, NJ: Lawrence Erbaum.

[390] McElroy, M.B. (1977). Goodness of fit for seemingly unrelated regressions: Glahn's $R^2_{y.x}$ and Hooper's $\bar{r}^2$. *Journal of Econometrics*, **6**, 481–487.

[391] McKay, R.J. and Campbell, N.A. (1982a). Variable selection techniques in discriminant analysis I. Description. *British Journal of Mathematics Statistical Psychology*, **35**, 1–29.

[392] McKay, R.J. and Campbell, N.A. (1982b). Variable selection techniques in discriminant analysis II. Allocation. *British Journal of Mathematics Statistical Psychology*, **35**, 30–41.

[393] McLachlan, G.J. (1992). *Discriminant analysis and statistical pattern recognition.* New York: Wiley-Interscience.

[394] McQuarrie, A.D.R. and Tsai, C.L. (1998). *Regression and time series model selection.* New Jersey: World Scientific Publishing Company.

[395] Mendoza, J.L. (1980). A significance test for multisample sphericity. *Psychometrika*, **45**, 495–498.

[396] Meredith, W. and Tisak, J. (1990). Latent curve analysis. *Psychometrika*, **55**, 107–122.

[397] Meulman, J.J. (1992). The integration of multidimensional scaling and multivariate analysis with optional transformations. *Psychometrika,* **54**, 539–565.

[398] Milligan, G.W. (1980). An examination of the effect of six types of error perturbation of fifteen clustering algorithms. *Psychometrika*, **45**, 325–342.

[399] Milligan, G.W. (1981). A Monte Carlo study of thirty interval criterion measures for cluster analysis. *Psychometrika*, **45**, 325–342.

[400] Milligan, G.W. and Cooper, M.C. (1985). An examination of procedures for determining the number of clusters in a data set. *Psychometrika*, **50**, 159–179.

[401] Milligan, G.W. and Cooper, M.C. (1986). A study of the comparability of external criteria for hierarchical cluster analysis. *Multivariate Behavioral Research*, **21**, 441–458.

[402] Milliken, G.A. and Graybill, F.A. (1970). Extensions of the general linear hypothesis model. *Journal of the American Statistical Association*, **65**, 797–807.

[403] Milliken, G.A. and Graybill, F.A. (1971). Tests for interaction in the two-way model with missing data. *Biometrics*, **27**, 1079–1083.

[404] Milliken, G.A. and Johnson, D.E. (1992). *Analysis of messy data, Volumn. 1: Designed experiments.* New York: Chapman and Hall.

[405] Mittelhammer, R.C., Judge, G.C. and Miller, D.J. (2000). *Econometric Foundations.* New York: Cambridge University Press.

[406] Moore, D.S. and McCabe, G. P. (1993). *Introduction to the practice of statistics*, Second Edition. New York: W.H. Freeman.

[407] Morrison, D.F. (1990). *Multivariate statistical methods,* Third Edition. New York: McGraw-Hill.

[408] Mudholkar, G.S., Davidson, M.L., and Subbaiah, P. (1974). Extended linear hypotheses and simultaneous tests in multivariate analysis of variance. *Biometrika*, **61**, 467–478.

[409] Mudholkar, G.S. and Srivastava, D.K. (1996). Trimmed $\tilde{T}^2$: A robust analog of Hotelling's T^2, *Personel Communication*. Department of Statistics, University of Rutgers..

[410] Mudholkar, G.S. and Srivastava, D.K. (1997). Robust analogs of Hotelling's two-sample T^2, *Personel Communication*. Department of Statistic, University of Rutgers..

[411] Mudholkar, G.S. and Subbaiah, P. (1980a). MANOVA multiple comparisons associated with finite intersection tests. In *Multivariate analysis-V,* P.R. Krishnaiah (Ed.), 467–482. New York: North-Holland.

[412] Mudholkar, G.S. and Subbaiah, P. (1980b). A review of step-down procedures for multivariate analysis of variance. In *Multivariate statistical analysis,* R.P. Gupta (Ed.), 161–179. New York: North-Holland.

[413] Mueller, R.O. (1996). *Basic principles of structural equation modeling*. New York: Springer-Verlag.

[414] Muirhead, R.J. (1982). *Aspects of multivariate statistical theory.* New York: John Wiley.

[415] Mulaik, S.A. (1972). The foundations of factor analysis. New York: McGraw-Hill.

[416] Mulaik, S.A., James, L.R., van Alstine, J. Bennett, N., Lind, S. and Stilwell, C.D. (1989). Evaluation of goodness-of-fit indices for structural equation models. *Psychological Bulletin*, **105**, 430–445.

[417] Muller, K.E., LaVange, L.M., Ramey, S.L. and Ramey, C.T. (1992). Power calculations for general linear multivariate models including repeated measures applications. *Journal of the American Statistical Association*, **87**, 1209–1226.

[418] Muller, K.E. and Peterson, B.L. (1984). Practical methods for computing power in testing the multivariate general linear hypothesis. *Computational Statistics and Data Analysis*, **2**, 143–158.

[419] Muthén, B. (1984). A general structural equation model with dichotomous, ordered catigorical and continuous latent variable indicators. *Psychometrika*, **49**, 115–132.

[420] Muthén, B. (1991). Analysis of longitudinal data using latent variable models with varying parameters. In L. Collins and J. Horn (Eds.), *Best methods for the analysis of change. Recent advances, unanswered questions, future directions*, pp. 1–17. Washington D.C.: American Psychological Association.

[421] Muthén, B. (1996). Growth modeling with binary responses. In *Categorical variables in developmental research: Methods and analysis*, A.V. Eye and C. Clogg (Eds.), 37–54. San Diego, CA: Academic Press.

[422] Muthén, B. (1997). Latent variable modeling with longitudinal and multilevel data. In A. Raftery (Ed.), *Sociological Methodology*, pp. 453–487. Boston: Blackwell Publishers.

[423] Muthén, B. and Muthén, L. (1998). *M plus: The comprehensive modeling program for applied researchers*. Los Angeles, CA: Statistical Modeling Innovations.

[424] Myers L. and Dunlap, W.P. (2000). Problems of unequal sample size and unequal variance-covariance matrices with multivariate analyis of variance: A solution following Alexander's Procedure. Paper presented at the Joint Statitical Meetings 2000 of the American Statistical Association in Indianapolis, IN.

[425] Nagarsenker, B.N. (1975). Percentage points of Wilks' L_{vc} criterion. *Communications in Statistics - Theory and Methods*, **4**, 629–641.

[426] Nagarsenker, B.N. and Pillai, K.C.S. (1973a). Distribution of the likelihood ratio criterion for testing a hypothesis specifying a covariance matrix. *Biometrika*, **60**, 359–364.

[427] Nagarsenker, B.N. and Pillai, K.C.S. (1973b). The distribution of the sphericity criterion. *Journal of Multivariate Analysis*, **3**, 226–235.

[428] Naik, D.N. (1990). Prediction intervals for growth curves. *Journal of Applied Statistics,* **17**, 245–254.

[429] Naik, D.N. and Khattree, R. (1996). Revisiting olympic track records: Some practical consideration in principal component analysis. *The American Statistician*, **50**, 140–150.

[430] Naik, D.N. and Rao, S.S. (2001). Analysis of multivariate repeated measures data with a Kronecker product structured covariance matrix. *Journal of Applied Statistics,* **28**, 91–105.

[431] Nanda, D.N. (1950). Distribution of the sum of roots of a determinantal equation under a certain condition. *Annals of Mathematical Statistics*, **21**, 432–439.

[432] Napoir, D. (1972). Nonmetric multidimensional techniques for summated ratings. In *Multidimentional scaling, Volume I Theory*, R.N. Shepard, A.K., Rowney, and S.B. Nerlove (Eds.), 157–178. New York: Seminar Press.

[433] Nath, R. and Pavur, R.J. (1985). A new statistic in the one-way multivariate analysis of variance. *Computational Statistics and Data Analysis*, **22**, 297–315.

[434] Nel, D.G. (1997). Tests for equality of parameter matrices in two multivariate linear models. *Journal of Multivariate Analysis*, **61**, 29–37.

[435] Nel, D.G. and van der Merwe, C.A. (1986). A solution to the multivariate Behrens-Fisher problem. *Communications in Statistics - Theory and Methods*, **15**, 3719–3735.

[436] Nelder, J.A. and Wedderburn, R.W.M. (1972). Generalized linear models. *Journal of the Royal Statistical Society, Series A*, **135**, 370–384.

[437] Nelson, P.R. (1993). Additional theory and methods uses for the analysis of means and extended tables of critical values. *Technometrics*, **35**, 61–71.

[438] Neter, J., Kutner, M.H., Nachtsheim, C.J. and Wasserman, W. (1996). *Applied linear statistical models, regression, analysis of variance and experimental designs*. Fourth Edition. Homewood, IL: Richard D. Irwin, Inc.

[439] Nicholson, G. E. (1960). Prediction in future samples. In *Contributions to Probability and Statistics,* I. Olkin et al. (Eds.), 322–330. Stanford, CA: Stanford University Press.

[440] Nishisato, S. (1994). *Elements of dual scaling: An introduction to practical data analysis*. Hillsdale, NJ: Lawrence Erbaum.

[441] Nishisato, S. (1996). Gleaning in the field of dual scaling. *Psychometrika*, **61**, 559–600.

[442] Obenchain, R.L. (1990). STABLSIM.EXE, Version 9010, unpublished C code. Indianapolis, IN:Eli Lilly and Company.

[443] O'Brien, R.G. and Muller, K.E. (1993). Unified power analysis for t-tests through multivariate hypotheses. In *Applied analysis of variance in behavioral science*, L.K. Edwards (Ed.), 297–344. New York: Marcel Dekker.

[444] O'Brien, P.N., Parente, F.J., and Schmitt, C.J. (1982). A Monte Carlo study on the robustness of four MANOVA criterion test. *Journal of Statistics and Computer Simulation*, **15**, 183–192.

[445] Ogasawara, H. (1998). Standard errors for matrix correlations. *Multivariate Behavioral Research,* **34**, 103–122.

[446] Olson, C.L. (1974). Comparative robustness of six tests in multivariate analysis of variance. *Journal of the American Statistical Association*, **69**, 894–908.

[447] Olson, C.L. (1975). A Monte Carlo investigation of the robustness of multivariate analysis of variance. Psychological Bulletin , 86, 1350-1352.

[448] O'Muircheartaigh, C.A. and Payne, C. (Eds.) (1977). *Exploring data structures.* New York: Wiley.

[449] Park, T. (1993). Equivalence of maximum likelihood estimation and iterative two-stage estimation for seemingly unrelated regression. *Communications in Statistics - Theory and Methods*, **22**, 2285–2296.

[450] Park, T. and Davis, S.S. (1993). A test of the missing data mechanism for repeated categorical data. *Biometrics*, **58**, 545–554.

[451] Patel, H.I. (1983). Use of baseline measurements in the two-period crossover design. *Communications in Statistics - Theory and Methods*, **12**, 2693–2712.

[452] Patel, H.I. (1986). Analysis of repeated measurement designs with changing covariates in clinical trials. *Biometrika*, **73**, 707–715.

[453] Pavur, R.J. (1987). Distribution of multivariate quadratic forms under certain covariance structures. *Canadian Journal of Statistics*, **15**, 169–176.

[454] Pearl, J. (2000). *Causality - Models, Reasoning, and Inference.* Cambridge, UK: Cambridge University Press.

[455] Pearson, E.E. and Hartley, H.O. (Eds.) (1966). *Biometrika tables for statisticians,* Vol 1 (3rd ed.). New York: Cambridge.

[456] Perlman, M.E. and Olkin, I. (1980). Unbiasedness of invariant tests for MANOVA and other multivariate problems. *Annals of Statistics*, **8**, 1326–1341.

[457] Pesaran, M.H. and Deaton, A.S. (1978). Testing non-nested nonlinear regression models. *Econometrica*, **46**, 677–694.

[458] Pillai, K.C.S. (1954). *On some distribution problems in multivariate analysis.* Mimeograph Series No. 88, Institute of Statistics, University of North Carolina, Chapel Hill.

[459] Pillai, K.C.S. (1955). Some new test criteria in multivariate analysis. *Annals of Mathematical Statistics*, **26**, 117–121.

[460] Pillai, K.C.S. (1956). On the distribution of the largest or the smallest root of a matrix in multivariate analysis. *Biometrika*, **43**, 122–127.

[461] Pillai, K.C.S. and Jaysachandran, K. (1967). Power comparisons of tests of two multivariate hypotheses based on four criteria. *Biometrika*, **54**, 195–210.

[462] Postovsky, V.A. (1970). Effects of delay in oral practice at the beginning of second language training. Unpublished doctoral dissertation, University of California, Berkeley.

[463] Potthoff, R.F. and Roy, S.N. (1964). A generalized multivariate analysis of variance model useful especially for growth curve problems. *Biometrika*, **51**, 313–326.

[464] Puri, M.L. and Sen, P.K. (1985). *Nonparametric methods in general linear models.* New York: John Wiley.

[465] Raju, N.S., Bilgic, R., Edwards, J.E., and Fleer, P.R. (1997). Methodology review: Estimation of population validity and cross validity, and the use of equal weights in prediction. Applied Psychological Measurement, **21**, 291–306.

[466] Ramsay, J.O. (1982). Some statistical approaches to multidimensional scaling data (with discussion). *Journal of the Royal Statistical Society, Series A*, **145**, 285–31.

[467] Rao, B. (1969). Partial canonical correlations. *Trabajos de estudestica y de investigacion operative*, **20**, 211–219

[468] Rao, C.R. (1947). Large sample tests of statistical hypotheses concerning several parameters with applications to problems of estimation. *Proceedings of the Cambridge Philosophical Society*, **44**, 50–57.

[469] Rao, C.R. (1948). The utilization of multiple measurement in problems of biological classification (with discussion). *Journal of the Royal Statistical Society*, **B10**, 159–203.

[470] Rao, C.R. (1951) An asymptotic expansion of the distribution of Wilks' criterion. *Bulletin of the International Statistics Institute,33,* 177–180.

[471] Rao, C.R. (1955). Estimation and tests of significance in factor analysis. *Psychometrika*, **20**, 93–111.

[472] Rao, C.R. (1964). The use and interpretation of principal component analysis in applied research. *Sankhyã*, **A26**, 329–358.

[473] Rao, C.R. (1965). The theory of least squares when the parameters are stochastic and its applications to the analysis of growth curves. *Biometrika*, **52**, 447–458.

[474] Rao, C.R. (1966). Covariance adjustment and related problems in multivariate analysis. In *Multivariate Analysis*, P.R. Krishnaiah (Ed.), 87–103. New York: Academic Press.

[475] Rao, C.R. (1967). Least squares theory using an estimated dispersion matrix and its application to measurement of signals. *Proceedings of the Fifth Berkeley Symposium on Mathematical Statistics and Probability, Volumn 1*, 335–371. Berkeley, CA: University of California Press.

[476] Rao, C.R. (1970). Inference on discriminant coeffecients. In *Essays in probability and statistics*, Bose et al. (Eds.), 587–602. Chapel Hill, NC: University of North Carolina.

[477] Rao, C.R. (1973a). *Linear statistical inference and its applications*, Second Edition. New York: John Wiley.

[478] Rao, C.R. (1973b). Representations of best linear unbiased estimators in the Gauss-Markoff model with a singular dispersion matrix. *Journal of Multivariate Analysis,* **3**, 276–292.

[479] Rao, C.R. and Mitra, S.K. (1971). *Generalized inverse of matrices and its applications*. New York: John Wiley.

[480] Rao, C.R. and Yanai, N. (1979). General definition and decomposition of projectors and same application to statistical problems. *Journal of Statistical Planning and Inference*, **3**, 1–17.

[481] Reiersøl, O. (1950). On the identification of parameters in Thurstone's multiple factor analysis. *Psychometrika*, **15**, 121–159.

[482] Reinsel, G.C. (1982). Multivariate repeated-measurements or growth curve models with multivariate random effects covariance structure. *Journal of the American Statistical Association*, **77**, 190–195.

[483] Reinsel, G.C. (1984). Estimation and prediction in a multivariate random effects generalized linear model. *Journal of the American Statistical Association*, **79**, 406–414.

[484] Reinsel, G.C. (1985). Mean squared error properties of empirical Bayes estimators in multivariate random effect generalized linear models. *Journal of the American Statistical Association*, **80**, 642–650.

[485] Reinsel, G.C. (1993). *Elements of multivariate time series analysis*. New York: Springer-Verlag.

[486] Reinsel, G.C. and Velu, R.P. (1998). *Multivariate reduced-rank regression-theory and applications*. New York: Springer-Verlag.

[487] Remme, J., Habbema, J.D.F., and Hermans, J. (1980). A simulation comparison of linear, quadratic, and kernel discrimination. *Journal of Statistical Computing and Simulation*, **11**, 87–106.

[488] Rencher, A.C. (1988). On the use of correlations to interpret canonical functions. *Biometrika*, **75**, 363–365.

[489] Rencher, A.C. (1995). *Methods of multivariate analysis*. New York: John Wiley.

[490] Rencher, A.C. (1998). *Multivariate statistical inference and applications*. New York: John Wiley.

[491] Rencher, A.C. and Larson, S.F. (1980). Bias in Wilks' Λ in stepwise discriminant analysis, *Technometrics*, **22**, 349–356.

[492] Rencher, A.C. and Scott, D.T. (1990). Assessing the contribution of individual variables following rejection of a multivariate hypothesis. *Communications in Statistics -Simulation and Computation*, **19**, 535–553.

[493] Rocke, D.M. (1989). Bootstrap Bartlett adjustment in seemingly unrelated regression. *Journal of the American Statistical Association*, **84**, 598–601.

[494] Romeu, J.L. and Ozturk, A. (1993). A comparative study of goodness-of-fit tests for multivariate normality. *Journal of Multivariate Analysis*, **46**, 309–334.

[495] Rouanet, H. and Lépine, D. (1970). Comparison between treatements in a repeated-measurements design: ANOVA and multivariate methods. *British Journal of Mathematical and Statistical Psychology*, **23**, 147–163.

[496] Rousseeuw, P.J. and Leroy, A.M. (1987). *Robust regression and outlier detection.* New York: John Wiley.

[497] Roy, S.N. (1953). On a heuristic method of test construction and its use in multivariate analysis. *Annals of Mathematical Statistics*, **24**, 220–238.

[498] Roy, S.N. (1957). *Some aspects of multivariate analysis*. New York: Wiley.

[499] Roy, S.N. (1958). Step-down procedure in multivariate analysis. *Annals of Mathematical Statistics*, **29**, 1177–1187.

[500] Roy, S.N. and Bose, R.C. (1953). Simultaneous confidence interval estimation. *Annals of Mathematical Statistics*, **24**, 513–536.

[501] Royston, J.P. (1982). An Extension of Shapiro and Wilk's W test for normality to large samples. *Applied Statistics*, **31**, 115–124.

[502] Royston, J.P. (1992). Approximating the Shapiro-Wilk W test for non-normality. *Journal of Multivariate Analysis Computing*, **2**, 117–119.

[503] Rozeboom, W.W. (1965). Linear correlations between sets of variables. *Psychometrika*, **30**, 57–71.

[504] Rubin, D.B. (1987). *Multiple imputation for nonresponse in surveys*. New York: John Wiley.

[505] St. Laurent, R.T. (1990). The equivalence of the Milliken-Graybill procedure and the score stest. *The American Statistician*, **44**, 36–39.

[506] Sampson, A.R. (1974). A tale of two regressions. *Journal of the American Statistical Association,* **69**, 682–689.

[507] Saris, W.E., Satorra, A. and Sörbam, D. (1987). The detection and correction of specification errors in structural equation models. In *Sociological methodology*, C.C. Clogg (Ed.), 105–129. San Francisco: Jossey-Bass.

[508] SAS Institute Inc. (1990). *SAS procedure guide, Version 6*, Third Edition. Cary, NC: SAS Institute Inc.

[509] SAS Institute Inc. (1992). *SAS Technical Report P-229, SAS/STAT Software: Changes and enhancements, release 6.07*. Cary, NC: SAS Institute Inc.

[510] SAS Institute Inc. (1993). *SAS/INSIGHT user's guide*, Version 6, Second Edition. Cary, N.C.: SAS Institute Inc.

[511] SAS Institute Inc. (1997). *SAS/STAT*. Software: Changes and Enhancements through Release 6.12. Cary, N.C.: SAS Institute, Inc.

[512] Satterthwaite, F.E. (1946). An approximate distribution of estimates of variance components. *Biometrics Bulletin*, **2**, 110–114.

[513] Schafer, J.L. (2000). *Analysis of incomplete multivariate data*. New York: Chapman and Hall.

[514] Schafer, J. L. and Olsen, M.K. (1998). Multiple imputation for multivariate missing=data problems: A data analyst's perspective, *Multivariate Behavioral Research*.**33**, 545–571.

[515] Schatzoff, M. (1966). Sensitivity comparisons among tests of the general linear hypothesis. *Journal of the American Statistic Association*, **61**, 415–435.

[516] Scheffé, H. (1959). *The analysis of variance*. New York: John Wiley.

[517] Scheines, R., Hoijtink, H., and Boomsma, A. (1999). Bayesian estimation and testing of structural equation models. *Psychometrika*, **64**, 37–52.

[518] Schiffman, S.S., Reynolds, M.L., and Young, F.W. (1981). *Introduction to multidimensional scaling: Theory methods and applications*. New York: Academic Press.

[519] Schmidhammer, J.L. (1982). On the selection of variables under regression models using Krishnaiah's finite intersectionn tests. In *Handbook of Statistics,* P.R. Krishnaiah (Ed.), 821–833. Amsterdam: North-Holland.

[520] Schott, J.R. (1988). Testing the equality of the smallest latent roots of a correlation matrix. *Biometrika*, **75**, 794–796.

[521] Schott, J.R. (1991). A test of a specific component of a correlation matrix. *Journal of the American Statistical Association*, **86**, 747–751.

[522] Schott, J.R. (1997). *Matrix analysis for statistics*. New York: John Wiley.

[523] Schott, J.R. and Saw, J.G. (1984). A multivariate one-way classification model with random effects. *Journal of Multivariate Analysis*, **15**, 1–12.

[524] Schwartz, C.J. (1993). The mixed-model ANOVA. The truth, the computer packages, the books-Part I: Balanced Data. *The American Statistician*, **47**, 48–59.

[525] Schwarz, G. (1978). Estimating the dimension of a model. *Annals of Statistics*, **6**, 461–464.

[526] Scott, D.W. (1992). *Multivariate density estimation theory, practice, and visualization*. New York: John Wiley.

[527] Searle, S.R. (1971). *Linear models*. New York: John Wiley.

[528] Searle, S.R. (1982). *Matrix algebra useful for statistics*. New York: John Wiley.

[529] Searle, S.R. (1987). *Linear models for unbalanced data*. New York: John Wiley.

[530] Searle, S.R. (1993). Unbalanced data and cell means models. In *Applied analysis of variance in behavioral science*, L.K. Edwards (Ed.), 375–420. New York: Marcel Dekker.

[531] Searle, S.R., Casella, G. and McCulloch, C.E. (1992). *Variance components*. New York: John Wiley.

[532] Searle, S.R. and Yerex, R.P. (1987). ACO_2: SAS GLM Annotated computer output for analysis of unbalanced data, 2nd Edition. Biometric Unit publication BU-949-M. Cornell University, Ithaca, N.Y.

[533] Seber, G.A.F. (1984). *Multivariate observations*. New York: John Wiley.

[534] Self, S.G. and Liang, K.Y. (1987). Asymptotic properties of maximum likelihood estimators and likelihood ratio tests under nonstandard conditions. *Journal of the American Statistical Association*, **82**, 605–610.

[535] Senn, S. (1993). *Cross-over trials in clinical research*. Chichester, England: John Wiley.

[536] Shapiro, S.S. and Wilk, M.B. (1965). An analysis of variance test for normality. *Biometrika*, **52**, 591–611.

[537] Shapiro, S.S. and Wilk, M.B. and Chen, H. (1968). A comparative study of various tests for normality. *Journal of the American Statistical Association*, **63**, 1343–1372.

[538] Shin, S.H. (1971). Creativity, intelligence and achievement: A study of the relationship between creativity and intelligence, and their effects upon achievement. Unpublished doctoral dissertation, University of Pittsburgh.

[539] Silverman, B.W. (1986). *Density estimation for statistics and data analysis*. New York: Chapman and Hall.

[540] Silvey, S.D. (1959). The Lagrange multiplier test. *Annals of Mathematical Statistics*, **30**, 389–407.

[541] Singer, J.D. (1998). Using SAS PROC MIXED to fit multilevel models, hierarchical models and individual growth models. *Journal of Educational and Behavioral Statistics*, **23**, 323–355.

[542] Singh, A. (1993). Omnibus robust procedures for assessment of multivariate normality and detection of multivariate outliers. In *Multivariate analysis: Future directions*, C.R. Rao (Ed.), 443–482. New York: Elsevier.

[543] Siotani, M., Hayakawa, T. and Fujikoshi, Y. (1985). *Modern multivariate statistical analysis: A graduate course and handbook*. Columbus, OH: American Sciences Press.

[544] Small, N.J.H. (1978). Plotting squared radii. *Biometrika*, **65**, 657–658.

[545] Smith, A.F.M. (1973). A general Bayesian linear model. *Journal of the Royal Statistical Society, Series B,* **35**, 67–75.

[546] Smith, H., Gnanaesikan, R., and Hughes, J.B. (1962) Multivariate analysis of variance (MANOVA). *Biometrics,* **18**, 22–41.

[547] Sparks, R.S., Coutsourides, D. and Troskie, L. (1983). The multivariate C_p. *Communications in Statistics - Theory and Methods*, **12**, 1775–1793.

[548] Sparks, R.S., Zucchini, W. and Coutsourides, D. (1985). On variable selection in multivariable regression. *Communications in Statistics - Theory and Methods*, **14** 1569–1587.

[549] Spearman, C. (1904). General intelligence objectively determined and measured. *American Journal of Psychology*, **15**, 201–293.

[550] Srivastava, J.N. (1966). Some generalizations of multivariate analysis of variances. In *Multivariate Analysis*, P.R. Krishnaiah (Ed.), 129–148. New York: Academic Press.

[551] Srivastava, J.N. (1967). On the extensions of the Gauss-Markoff theorem to complex multivariate linear models. *Annals of the Institute of Statistical Mathematics*, **19**, 417–437.

[552] Srivastava, M.S. (1997). Generalized multivariate analysis of variance models. Technical Report No. 9718, Department of Statistics,University of Toronto. Toronto, Ontario, Canada M5S-3G3.

[553] Srivastava, M.S. and Carter, E.M. (1983). *Applied multivariate statistics*. New York: North-Holland.

[554] Srivastava, M.S. and Khatri, C.G. (1979). *An introduction to multivariate statistics*. New York: North Holland.

[555] Srivastava, M.S. and Kubokawa, T. (1999). Improved nonnegative estimation of multivariate components of variance. *The Annals of Statistics*, **27**, 2008–2032.

[556] Srivastava, M.S. and von Rosen, D. (1998). Outliers in multivariate regression models. *Journal of Multivariate Analysis*, **65**, 195–208.

[557] Stankov, L. (1979). Hierarchical factoring based on image analysis and orthoblique rotations. *Journal of Multivariate Behavioral Research*, **19**, 339–353.

[558] Steiger, J.H. (2001). Driving fast in reverse. *Journal of the American Statistical Association,* **96**, 331–338.

[559] Stein, C. (1960). Multiple regression. In *Contributions to Probability and Statistics,* I. Olkin et al. (Eds.), *424–443.* Stanford, CA: Stanford University Press.

[560] Stern, H.S. (1996). Neural networks in applied statistics. *Technometrics*, **38**, 217–225.

[561] Stewart, D.K. and Love, W.A. (1968). A general canonical correlation index. *Psychological Bulletin*, **70**, 160–163.

[562] Stokes, M. E., Davis, C.S., and Koch, G.G. (2000). *Categorical data analysis using the SAS system,* Second Edition. Cary, NC: SAS Institute Inc.

[563] Styan, G.P.H. (1973). Hadamard products and multivariate statistical analysis. *Linear Algebra and Its Applications*, **6**, 217–240.

[564] Stram, D.O. and Lee, J.W. (1994). Variance components testing in the longitudinal mixed effects mode. *Biometrics*, **50**, 1171–1177.

[565] Sugiura, N. (1972). Locally best invariant test for sphericity and the limiting distribution. *Annals of Mathematical Statistics*, **43**, 1312–1316.

[566] Sugiura, N. (1978). Further analysis of the data by Akaike's information criterion and the finite corrections. *Communications in Statistic-Theory and Methods*, **7**, 13–26.

[567] Swaminathan, H. and Algina, J. (1978). Scale freeness in factor analysis. *Psychometrika*, **43**, 581–583.

[568] Swamy, P. (1971). *Statistical inference in random coefficient regression models*. New York: Springer-Verlag.

[569] Takane, Y., Kiers, H. and de Leeuw, J. (1995). Component analysis with different sets of constraints on different dimensions. *Psychometrika*, **60**, 259–280.

[570] Takane, Y. and Shibayama, T. (1991). Principal component analysis with external information on both subjects and variables. *Psychometrika*, **56**, 97–120.

[571] Takane, Y., Young, F.W., de Leeuw, J. (1977). Non-metric individual differences multidimensional scaling alternating least squares with optimal scaling features. *Psychometrika*, **42**, 7–67.

[572] Takeuchi, K., Yanai, H. and Mukherjee, B.N. (1982). *The foundations of multivariate analysis*. New Delhi: Wiley Eastern Limited.

[573] Tan, W.Y. and Gupta, R.P. (1983). On approximating a linear combination of central Wishart matrices with positive coefficients. *Communications in Statistics - Theory and Methods,* **12**, 2589–2600.

[574] Tanaka, J.S. and Huba, G.J. (1985). A fit indix for covariance structure models under arbitrary GLS estimation. *British Journal of Mathematical and Statistical Psychology*, **38**, 197–201.

[575] Theil, H. (1971). *Principles of econometrics*. New York: Wiley.

[576] Thisted, R.A. (1988). *Elements of statistical computing*. New York: Chapman and Hall.

[577] Thomas, D.R. (1983). Univariate repeated measures techniques applied to multivariate data. *Psychometrika*, **48**, 451–464.

[578] Thompson, G.H. (1934). Hotelling's method modified to give Spearman's *g*. *Journal of Educational Psychology*, **25**, 366–374.

[579] Thompson, R. (1973). The estimation of variance components with an application when records are subject to calling. *Biometrics*, **29**, 527–550.

[580] Thomson, A. and Randall-Maciver, R. (1905). Ancient races of the Thebaid. Oxford: Oxford University Press.

[581] Thum, Y.M. (1997). Hierarchical linear models for multivariate outcomes. *Journal of Educational and Behavioral Statistics*, **22**, 77–108.

[582] Thurstone, L.L. (1931). Multiple factor analysis. *Psychological Review*, **38**, 406–427.

[583] Thurstone, L.L. (1947). *Multiple factor analysis*. Chicago: The University of Chicago Press.

[584] Timm, N.H. (1975). *Multivariate analysis with applications in education and psychology*. Belmont, CA: Brooks/Cole.

[585] Timm, N.H. (1980). Multivariate analysis of repeated measurements. In *Handbook of Statistics, Volumn 1: Analyis of Variance*, P.R. Krishnaiah (Ed.), 41–87. New York: North-Holland.

[586] Timm, N.H. (1993). *Multivariate analysis with applications in education and psychology*. Oberlin, OH: The Digital Printshop.

[587] Timm, N.H. (1995). Simultaneous inference using finite intersection tests: A better mousetrap. *Multivariate Behavioral Research*, **30**, 461–511.

[588] Timm, N.H. (1996). Multivariate quality control using finite intersection tests. *Journal of Quality Technology*, **28**, 233–243.

[589] Timm, N.H. (1997). The CGMANOVA model. *Communication in Statistics - Theory and Methods*, **26**, 1083–1098.

[590] Timm, N.H. (1999). Testing multivariate effect sizes in multiple-endpoint studies. *Multivariate Behavioral Research*, **34**, 457–465

[591] Timm, N.H. and Al-Subaihi, A. (2001). Testing model specification in seemingly unrelated regression models. *Communication in Statistics - Theory and Methods,* **30**, 579–590..

[592] Timm, N.H. and Carlson, J.E. (1975). Analysis of variance through full rank models. *Multivariate Behavioral Research Monograph*, Number 75–1, pp. 120.

[593] Timm, N.H. and Carlson, J.E. (1976). Part and bipartial canonical correlations analysis. *Psychometrika*, **41**, 159–176.

[594] Timm, N.H. and Mieczkowski, T.A. (1997). *Univariate and multivariate general linear models: Theory and applications using SAS software*. Cary, NC: SAS Institute Inc.

[595] Ting, N., Burdick, R.K., Graybill, F.A., Jeyaratnam, S., and Lu, T.F.C. (1990). Confidence intervals on linear combinations of variance components that are unrestricted in sign. *Journal of Statlistical Computing and Simulation*, **35**, 135–143.

[596] Torgerson, W.S. (1958). *Theory and methods of scaling*. New York: John Wiley.

[597] Tubb, A., Parker, A.J., and Nickless, G. (1980). The analysis of Romano-British pottery by atomic absorption spectrophotometry. *Archaeometry*, **22**, 153–171.

[598] Tukey, J.W. (1949). One degree of freedom test for nonadditivity. *Biometrics*, **5**, 232–242.

[599] Uhl, N. and Eisenberg, T. (1970). Predicting shrinkage in the multiple correlation coefficient. *Educational and Psychological Measurement,* **30**, 487–489.

[600] Vaish, A.K. (1994). Invariance properties of statistical tests for dependent observations. Doctoral Dissertation, Old Dominion University.

[601] van Rijckevarsel, J.L.A. and de Leeuw, J. (1988). *Component and correspondence analysis*. Chichester: John Wiley.

[602] Velilla, S. and Barrio, J.A. (1994). A discriminant rule under transformation, *Technometrics*, **36**, 348–353.

[603] Venables, W.N. and Ripley, B.D. (1994). *Modern applied statistics with S-plus*. New York: Springer-Verlag.

[604] Verbeke, G. and Molenberghs, G. (Eds.). (1997). *Linear mixed models in practice: A SAS-oriented approach*. New York: Springer-Verlag.

[605] Verbyla, A.P. and Venables, W.N. (1988). An extension of the growth curve model. *Biometrika*, **75**, 129–138.

[606] Vonesh, E.F. and Carter, R.L. (1987). Efficient inference for a random coefficient growth curve model with unbalanced data. *Biometrics*, **43**, 617–628.

[607] Vonesh, E.F. and Chinchilli, V.M. (1997). *Linear and nonlinear models for the analysis of repeated measurements*. New York: Marcel Dekker.

[608] von Rosen, D. (1989). Maximum likelihood estimators in multivariate linear normal models. *Journal of Multivariate Analysis*, **31**, 187–200.

[609] von Rosen, D. (1991). The growth curve model: A review. *Communications in Statistics - Theory and Methods*, **20**, 2791–2822.

[610] von Rosen, D. (1993). Uniqueness conditions for maximum likelihood estimators in a multivariate linear model. *Journal of Statistical Planning and Inference*, **36**, 269–276.

[611] Voss, D.T. (1999). Resolving the mixed models controversy. *The American Statistician*, **53**, 352–356.

[612] Wald, A. (1943). Tests of statistical hypothesis concerning several parameters when the number of observation is large. *Transactions of the American Mathematical Society*, **54**, 426–482.

[613] Wald, A. (1944). On a statistical problem arising in the classification of an individual into two groups. *Annals of Mathematical Statistics*, **15**, 145–163.

[614] Walker, J. (1993). Diagnostics for the GMANOVA Model. *Personal Communication*, Department of Economics and Social Statistics, Cornell University.

[615] Wall, F. J. (1968) *The generalized variance ration of the U-statistic.* Albuquerque, NM: The Dikewood Corporation.

[616] Wedderburn, R.W.M. (1974). Quasilikelihood functions, generalized linear models and the Gauss-Newton method. *Biometrika*, **61**, 439–447.

[617] Weeks, D.L. and Williams, D.R. (1964). A note on the determination of connectedness in an N-way cross classification. *Technometrics*, **6**, 319–324.

[618] Welch, B.L. (1939). Note on discriminant functions. *Biometrika*, **31**, 218–220.

[619] Weller, S.C. and Romney, A.K. (1990). *Metric scaling: Correspondence analysis.* Newbury Park, CA: Sage Publications.

[620] Westfall, P.H. and Young, S.S. (1993). *Resampling-based multiple testing*. New York: John Wiley.

[621] Wheaton, B. Muthén, B., Alevin, D.F., and Summers, G.F. (1977). Assessing Reliability and Stability in Panel Models. In *Sociological Methodology 1977*, D.R. Heise (Ed.), 84–136. San Francisco: Jossey Bass.

[622] Wiley, D.E. (1973). The identification problem for structural equation models with unmeasured variables. In *Structural equation models in the social sciences*, A.S. Goldberger and O.D. Duncan (Eds.), 69–83. New York: Academic Press.

[623] Wilk, M.B. and Gnanadesikan, R. (1968). Probability plotting methods for the analysis of data. *Biometrika*, **55**, 1–17.

[624] Wilkinson, J.H. (1965). *The algebraic eigenvalues problem*. New York: Oxford Press.

[625] Wilks, S.S. (1932). Certain generalizations in the analysis of variance. *Biometrika*, **24**, 471–494.

[626] Willerman, L., Schultz, R., Rutledge, J.N., and Bigler, E. (1991). In Vivo Brain Size and Intelligence, *Intelligence*, **15**, 223–228.

[627] Willett, J.B. and Sayer, A.G. (1996). Cross-domain analysis of change over time: Combining growth modeling and covariance structural analysis. In G.A. Marcoulides and R.E. Schumacher (Eds.), *Advanced structural equation modeling: Issues and techniques*, pp. 125–158. Mahwah, NJ: Lawrence Erlbaum.

[628] Williams, L.J. and Holohan, P.J. (1994). Parsimony-based fit indices for multiple-dicator models: Do they work? *Structural Equation Modeling*, **1**, 161–189.

[629] Williams, W.T. and Lance, G.N. (1977). Hierarchical clustering methods. In *Statistical methods for digital computers, Volume 3,* K. Ensleim, A. Ralston, and H.S. Wilf (Eds.), 269–295. New York: John Wiley.

[630] Wolfinger, R.D., Tobias, R.D., and Sall, J. (1994). Computing Gaussian likelihoods and their derivatives for general linear mixed models. *SIAM Journal on Scientific Computing*, **15**, 1294–1310.

[631] Wong, O.S., Masaro, J. and Wong, T. (1991). Multivariate version of Cochran's theorem. *Journal of Multivariate Analysis*, **39**, 154–174.

[632] Yanai, H. (1973). Unification of various techniques of multivariate analysis by means of a generalized coefficient of determination (G.C.D.). *The Japanese Journal of Behaviormetrics*, **1**, 45–54 (in Japanese).

[633] Yanai, H. and Takane, Y. (1992). Canonical correlation analysis with linear constraints. *Linear Algebra and Its Applications*, **176**, 75–89.

[634] Yanai, H. and Takane, Y. (1999). Generalized constrained canonical correlation analysis. *Multivariate Behavioral Research* (to appear).

[635] Yao, Y. (1965). An approximate degrees of freedom solution to the multivariate Behrens-Fisher problem. *Biometrika*, **52**, 139–147.

[636] Young, D.M., Seaman, J.W., Jr. and Meaux, L.M. (1999). Independence distribution preserving covariance structures for the multivariate linear model. *Journal of Multivariate Analysis*, **68**, 165–175.

[637] Young, F.W. (1987). *Multidimensional scaling: History, theory and applications*, R.M. Hamer (Ed.), 3–158. Hillsdale, NJ: Lawrence Erlbaum.

[638] Yuan, K-H. and Bentler, P.M. (1999). F tests for mean and covariance structure analysis. *Journal of Educational and Behavioral Statistics*, **24**, 225–243.

[639] Zellner, A. (1962). An efficient method of estimating unrelated regressions and tests for aggregation bias. *Journal of American Statistical Association*, **57**, 348–368.

[640] Zellner, A. (1963). Estimators for seemingly unrelated regression equations: Some exact finite sample results. *Journal of the American Statistical Association*, **58**, 977–992.

[641] Zhou, L. and Mathew, T. (1993). Hypotheses tests for variance components in some multivariate mixed models. *Journal of Statistical Planning and Inference*, **37**, 215–227.

[642] Zyskind, G. (1967). On canonical forms, non-negative covariance matrices and best and simple least squares linear estimators in linear models. *Annals of Mathematical Statistics*, **38**, 1092–1109.

Author Index

Subject Index

Springer Texts in Statistics *(continued from page ii)*

McPherson: Applying and Interpreting Statistics: A Comprehensive Guide, Second Edition
Mueller: Basic Principles of Structural Equation Modeling: An Introduction to LISREL and EQS
Nguyen and Rogers: Fundamentals of Mathematical Statistics: Volume I: Probability for Statistics
Nguyen and Rogers: Fundamentals of Mathematical Statistics: Volume II: Statistical Inference
Noether: Introduction to Statistics: The Nonparametric Way
Nolan and Speed: Stat Labs: Mathematical Statistics Through Applications
Peters: Counting for Something: Statistical Principles and Personalities
Pfeiffer: Probability for Applications
Pitman: Probability
Rawlings, Pantula and Dickey: Applied Regression Analysis
Robert: The Bayesian Choice: From Decision-Theoretic Foundations to Computational Implementation, Second Edition
Robert and Casella: Monte Carlo Statistical Methods
Rose and Smith: Mathematical Statistics with *Mathematica*
Santner and Duffy: The Statistical Analysis of Discrete Data
Saville and Wood: Statistical Methods: The Geometric Approach
Sen and Srivastava: Regression Analysis: Theory, Methods, and Applications
Shao: Mathematical Statistics
Shorack: Probability for Statisticians
Shumway and Stoffer: Time Series Analysis and Its Applications
Terrell: Mathematical Statistics: A Unified Introduction
Timm: Applied Multivariate Analysis
Whittle: Probability via Expectation, Fourth Edition
Zacks: Introduction to Reliability Analysis: Probability Models and Statistical Methods